Applications in Examples and Exercises

BIOLOGICAL

CHEMICAL

CHEMICAL ENGINEERING

CIVIL ENGINEERING

COMMUNICATIONS, COMPUTERS, AND NETWORKS

ELECTRONICS

ENERGY

ENVIRONMENTAL

MATERIALS

MECHANICAL

(Text continued at the back of book.)

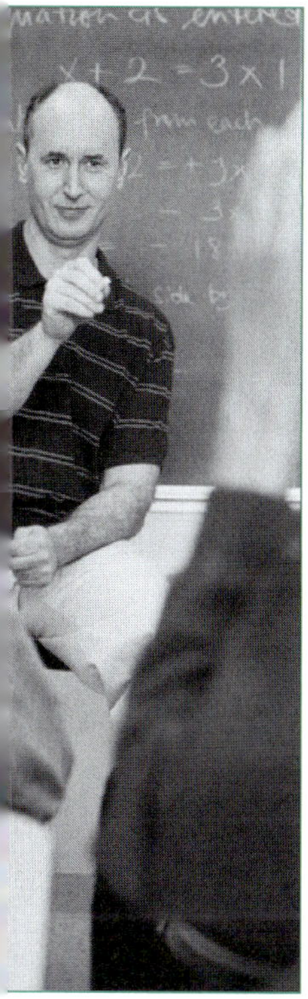

TO THE INSTRUCTOR

WileyPLUS is built around the activities you perform

Prepare & Present

Create outstanding class presentations using a wealth of resources, such as PowerPoint™ slides, image galleries, interactive simulations, and more. Plus you can easily upload any materials you have created into your course, and combine them with the resources Wiley provides you with.

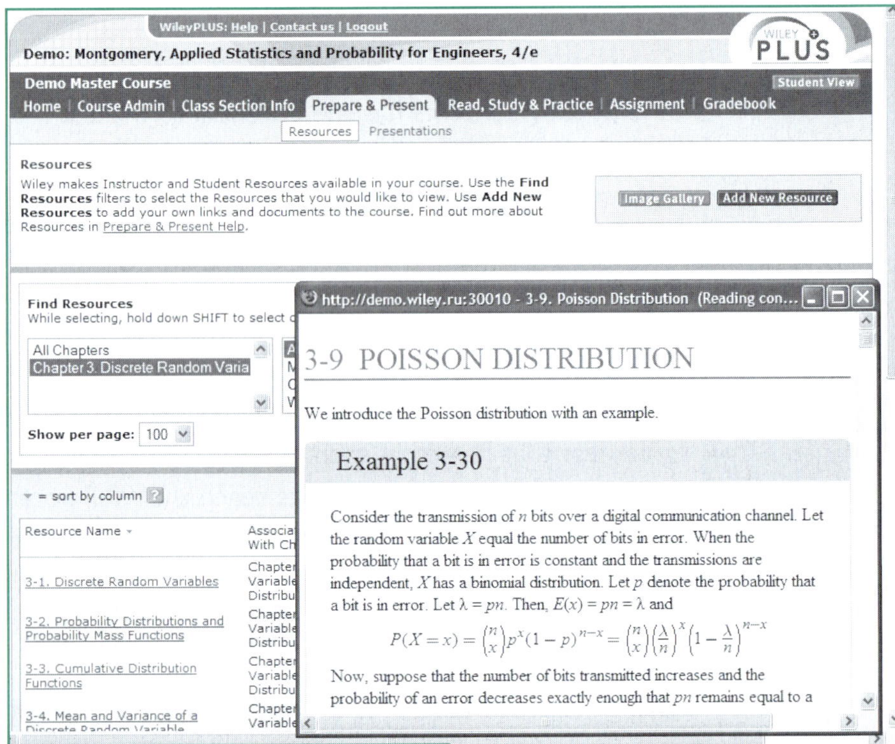

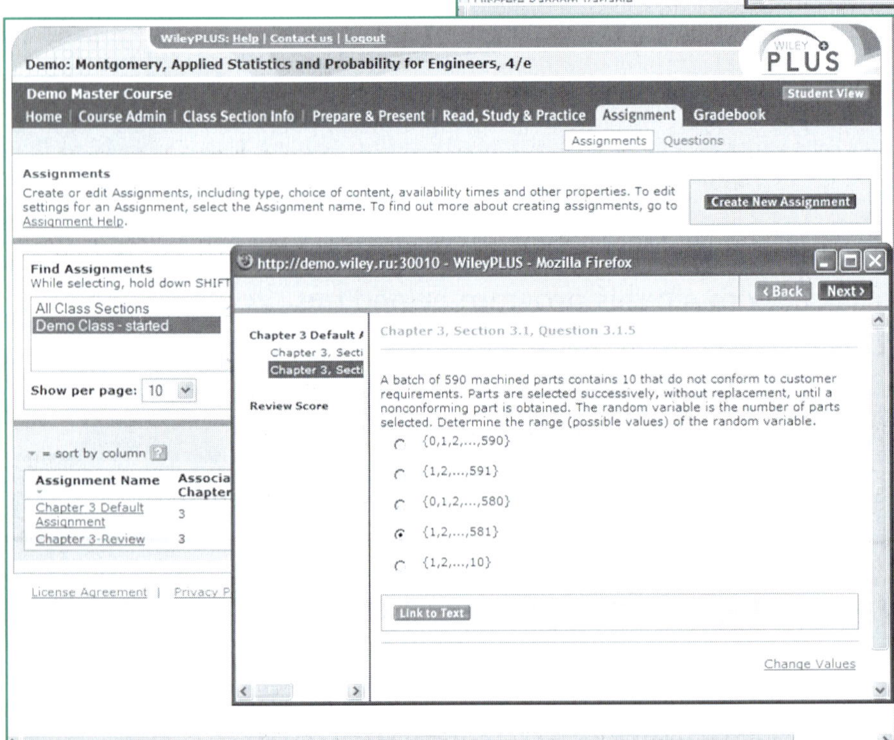

Create Assignments

Automate the assigning and grading of homework or quizzes by using the provided question banks, or by writing your own. Student results will be automatically graded and recorded in your gradebook. *WileyPLUS* also links homework problems to relevant sections of the online text, hints, or solutions—context-sensitive help where students need it most!

* Based on a spring 2005 survey of 972 student users of *WileyPLUS*

n your class each day. With Wiley**PLUS** you can:

Track Student Progress

Keep track of your students' progress via an instructor's gradebook, which allows you to analyze individual and overall class results. This gives you an accurate and realistic assessment of your students' progress and level of understanding.

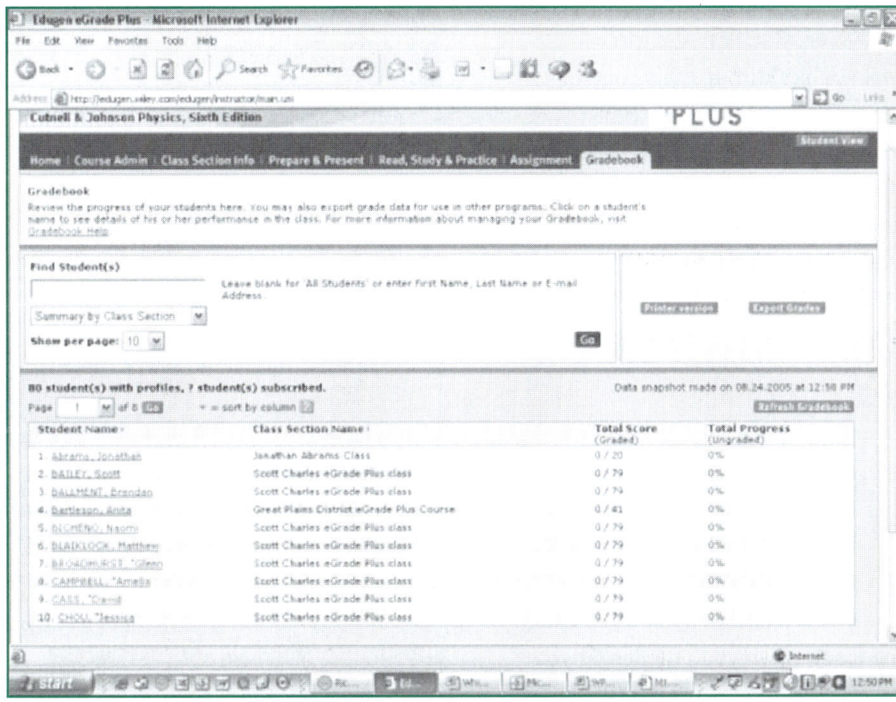

Now Available with WebCT and Blackboard!

Now you can seamlessly integrate all of the rich content and resources available with *WileyPLUS* with the power and convenience of your WebCT or BlackBoard course. You and your students get the best of both worlds with single sign-on, an integrated gradebook, list of assignments and roster, and more. If your campus is using another course management system, contact your local Wiley Representative.

"I studied more for this class than I would have without *WileyPLUS*."

Melissa Lawler, *Western Washington Univ.*

For more information on what *WileyPLUS* can do to help your students reach their potential, please visit

www.wiley.com/college/wileyplus

76% of students surveyed said it made them better prepared for tests. *

TO THE STUDENT

You have the potential to make a difference!

Will you be the first person to land on Mars? Will you invent a car that runs on water? But, first and foremost, will you get through this course?

WileyPLUS is a powerful online system packed with features to help you make the most of your potential, and get the best grade you can!

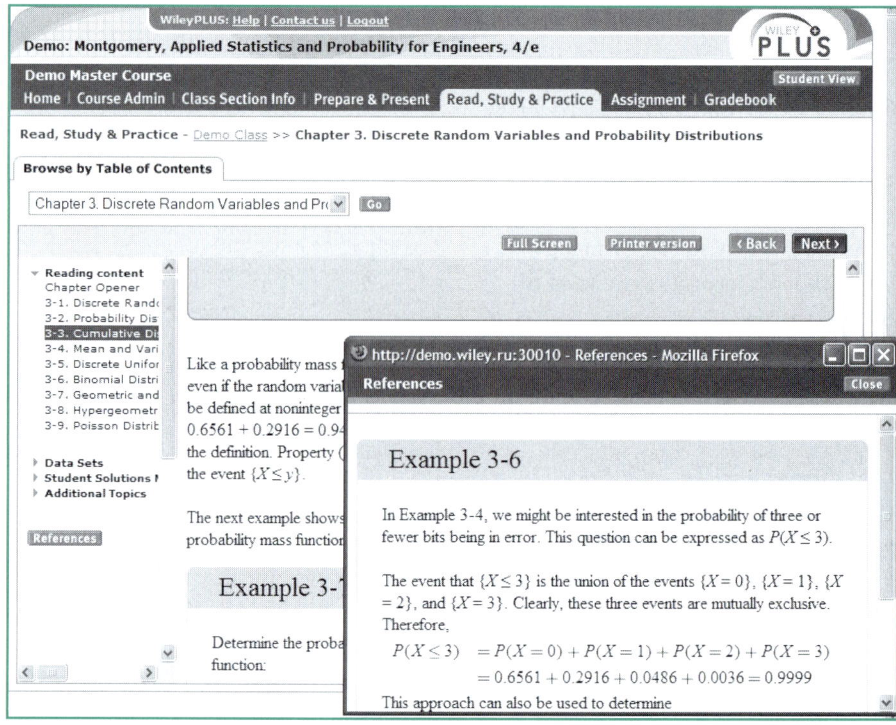

With Wiley**PLUS** you get:

A complete online version of your text and other study resources

Study more effectively and get instant feedback when you practice on your own. Resources like self-assessment quizzes, tutorials, and animations bring the subject matter to life, and help you master the material.

Problem-solving help, instant grading, and feedback on your homework and quizzes

You can keep all of your assigned work in one location, making it easy for you to stay on task. Plus, many homework problems contain direct links to the relevant portion of your text to help you deal with problem-solving obstacles at the moment they come up.

The ability to track your progress and grades throughout the term.

A personal gradebook allows you to monitor your results from past assignments at any time. You'll always know exactly where you stand.

If your instructor uses *WileyPLUS*, you will receive a URL for your class. If not, your instructor can get more information about *WileyPLUS* by visiting www.wiley.com/college/wileyplus

"It has been a great help, and I believe it has helped me to achieve a better grade."
Michael Morris, *Columbia Basin College*

69% of students surveyed said it helped them get a better grade.*

WILEY PLUS

Applied Statistics and Probability for Engineers

Fourth Edition

Douglas C. Montgomery
Arizona State University

George C. Runger
Arizona State University

John Wiley & Sons, Inc.

Cover image from Norm Christiansen

ISBN-13: 978-0-471-79473-8
ISBN-10: 0-471-79473-2

Printed in Asia

10 9 8 7 6 5 4 3 2 1

Other Wiley Books by These Authors

Website: www.wiley.com/college/montgomery

Engineering Statistics, Third Edition
By Montgomery, Runger and Hubele
Introduction to engineering statistics, with topical coverage appropriate for a one-semester course. A modest mathematical level, and an applied approach.

Applied Statistics and Probability for Engineers, Fourth Edition
By Montgomery and Runger
Introduction to engineering statistics, with topical coverage appropriate for either a one- or two-semester course. An applied approach to solving real-world engineering problems.

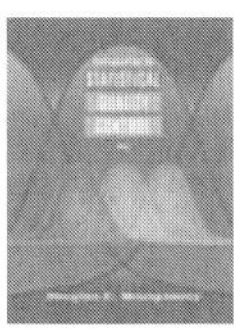

Introduction to Statistical Quality Control, Sixth Edition
By Douglas C. Montgomery
For a first course in statistical quality control. A comprehensive treatment of statistical methodology for quality control and improvement. Some aspects of quality management are also included, such as the six-sigma approach.

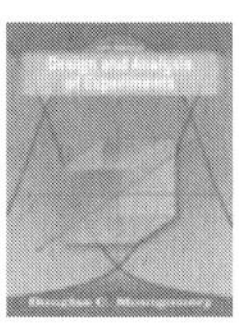

Design and Analysis of Experiments, Fifth Edition
By Douglas C. Montgomery
An introduction to design and analysis of experiments, with the modest prerequisite of a first course in statistical methods. For senior and graduate students, and practitioners, to design and analyze experiments for improving the quality and efficiency of working systems.

Introduction to Linear Regression Analysis, Third Edition
By Montgomery, Peck, and Vining
A comprehensive and thoroughly up-to-date look at regression analysis, still the most widely used technique in statistics today.

Response Surface Methodology: Process and Product Optimization Using Designed Experiments, Second Edition
By Myers and Montgomery Website: www.wiley.com/college/myers
The exploration and optimization of response surfaces, for graduate courses in experimental design, and for applied statisticians, engineers, and chemical and physical scientists.

Generalized Linear Models: With Applications in Engineering and the Sciences
By Myers, Montgomery, and Vining Website: www.wiley.com/college/myers
An introductory text or reference on Generalized Linear Models (GLMs). The range of theoretical topics and applications appeals both to students and practicing professionals.

Preface

INTENDED AUDIENCE

This is an introductory textbook for a first course in applied statistics and probability for undergraduate students in engineering and the physical or chemical sciences. These individuals play a significant role in designing and developing new products and manufacturing systems and processes, and they also improve existing systems. Statistical methods are an important tool in these activities because they provide the engineer with both descriptive and analytical methods for dealing with the variability in observed data. Although many of the methods we present are fundamental to statistical analysis in other disciplines, such as business and management, the life sciences, and the social sciences, we have elected to focus on an engineering-oriented audience. We believe that this approach will best serve students in engineering and the chemical/physical sciences and will allow them to concentrate on the many applications of statistics in these disciplines. We have worked hard to ensure that our examples and exercises are engineering- and science-based, and in almost all cases we have used examples of real data—either taken from a published source or based on our consulting experiences.

We believe that engineers in all disciplines should take at least one course in statistics. Unfortunately, because of other requirements, most engineers will only take one statistics course. This book can be used for a single course, although we have provided enough material for two courses in the hope that more students will see the important applications of statistics in their everyday work and elect a second course. We believe that this book will also serve as a useful reference.

We have retained the relatively modest mathematical level of the first three editions. We have found that engineering students who have completed one or two semesters of calculus should have no difficulty reading almost all of the text. It is our intent to give the reader an understanding of the methodology and how to apply it, not the mathematical theory. We have made many enhancements in this edition, including reorganizing and rewriting major portions of the book and adding a number of new exercises.

ORGANIZATION OF THE BOOK

Perhaps the most common criticism of engineering statistics texts is that they are too long. Both instructors and students complain that it is impossible to cover all of the topics in the book in one or even two terms. For authors, this is a serious issue because there is great variety in both the content and level of these courses, and the decisions about what material to delete without limiting the value of the text are not easy. After struggling with these issues, we decided to divide the text into two components; a set of core topics, many of which are most likely to be covered in an engineering statistics course, and a set of supplementary topics, or topics that will be useful for some but not all courses. The core topics are in the printed book, and the supplementary topics are available on the textbook Web site. Decisions about which topics to include in print and which to include on the Web site were made based on the results of a recent survey of instructors.

Chapter 1 is an introduction to the field of statistics and how engineers use statistical methodology as part of the engineering problem-solving process. This chapter also introduces the reader to some engineering applications of statistics, including building empirical models, designing engineering experiments, and monitoring manufacturing processes. These topics are discussed in more depth in subsequent chapters.

Chapters 2, 3, 4, and 5 cover the basic concepts of probability, discrete and continuous random variables, probability distributions, expected values, joint probability distributions, and independence. We have given a reasonably complete treatment of these topics but have avoided many of the mathematical or more theoretical details.

Chapter 6 begins the treatment of statistical methods with random sampling; data summary and description techniques, including stem-and-leaf plots, histograms, box plots, and probability plotting; and several types of time series plots. Chapter 7 discusses sampling distributions, the central limit theorem, and point estimation of parameters. This chapter also introduces some of the important properties of estimators, the method of maximum likelihood, the method of moments, and Bayesian estimation.

Chapter 8 discusses interval estimation for a single sample. Topics included are confidence intervals for means, variances or standard deviations, and proportions and prediction and tolerance intervals. Chapter 9 discusses hypothesis tests for a single sample. Chapter 10 presents tests and confidence intervals for two samples. This material has been extensively rewritten and reorganized. There is detailed information and examples of methods for determining appropriate sample sizes. We want the student to become familiar with how these techniques are used to solve real-world engineering problems and to get some understanding of the concepts behind them. We give a logical, heuristic development of the procedures, rather than a formal mathematical one.

Chapters 11 and 12 present simple and multiple linear regression. We use matrix algebra throughout the multiple regression material (Chapter 12) because it is the only easy way to understand the concepts presented. Scalar arithmetic presentations of multiple regression are awkward at best, and we have found that undergraduate engineers are exposed to enough matrix algebra to understand the presentation of this material.

Chapters 13 and 14 deal with single- and multifactor experiments, respectively. The notions of randomization, blocking, factorial designs, interactions, graphical data analysis, and fractional factorials are emphasized. Chapter 15 gives a brief introduction to the methods and applications of nonparametric statistics, and Chapter 16 introduces statistical quality control, emphasizing the control chart and the fundamentals of statistical process control.

FEATURED IN THIS BOOK

Definitions, Key Concepts, and Equations

Throughout the text, definitions and key concepts and equations are highlighted by a box to emphasize their importance.

4-4 MEAN AND VARIANCE OF A CONTINUOUS RANDOM VARIABLE

The mean and variance can also be defined for a continuous random variable. Integration replaces summation in the discrete definitions. If a probability density function is viewed as a loading on a beam as in Fig. 4-1, the mean is the balance point.

Mean and Variance

Suppose X is a continuous random variable with probability density function $f(x)$. The **mean** or **expected value** of X, denoted as μ or $E(X)$, is

$$\mu = E(X) = \int_{-\infty}^{\infty} x f(x)\, dx \qquad (4\text{-}4)$$

The **variance** of X, denoted as $V(X)$ or σ^2, is

$$\sigma^2 = V(X) = \int_{-\infty}^{\infty} (x-\mu)^2 f(x)\, dx = \int_{-\infty}^{\infty} x^2 f(x)\, dx - \mu^2$$

The **standard deviation** of X is $\sigma = \sqrt{\sigma^2}$.

he equivalence of the two formulas for variance can be derived as one, as was done for dis ete random variables.

r the copper current measurement in Example 4-1, the mean of X is

$$E(X) = \int_0^{20} x f(x)\, dx = 0.05x^2/2 \Big|_0^{20} = 10$$

e variance of X is

$$V(X) = \int_0^{20} (x-10)^2 f(x)\, dx = 0.05(x-10)^3/3 \Big|_0^{20} = 33.33$$

e expected value of a function $h(X)$ of a continuous random variable is also defined in a raightforward manner.

Expected Value of a Function of a Continuous Random Variable

If X is a continuous random variable with probability density function $f(x)$,

$$E[h(X)] = \int_{-\infty}^{\infty} h(x) f(x)\, dx \qquad (4\text{-}5)$$

Learning Objectives

Learning Objectives at the start of each chapter guide the students in what they are intended to take away from this chapter and serve as a study reference.

LEARNING OBJECTIVES

After careful study of this chapter you should be able to do the following:

1. Determine probabilities from probability density functions
2. Determine probabilities from cumulative distribution functions and cumulative distribution functions from probability density functions, and the reverse
3. Calculate means and variances for continuous random variables
4. Understand the assumptions for each of the continuous probability distributions presented
5. Select an appropriate continuous probability distribution to calculate probabilities in specific applications
6. Calculate probabilities, determine means and variances for each of the continuous probability distributions presented
7. Standardize normal random variables
8. Use the table for the cumulative distribution function of a standard normal distribution to calculate probabilities
9. Approximate probabilities for some binomial and Poisson distributions

Eight-Step Procedure for Hypothesis Testing

The text introduces a sequence of eight steps in applying hypothesis-testing methodology and explicitly exhibits this procedure in examples.

9-1.6 General Procedure for Hypothesis Tests

This chapter develops hypothesis-testing procedures for many practical problems. Use of the following sequence of steps in applying hypothesis-testing methodology is recommended.

1. From the problem context, identify the parameter of interest.
2. State the null hypothesis, H_0.
3. Specify an appropriate alternative hypothesis, H_1.
4. Choose a significance level α.
5. Determine an appropriate test statistic.
6. State the rejection region for the statistic. This step is not necessary if the P-value approach is used.
7. Compute any necessary sample quantities, substitute these into the equation for the test statistic, and compute that value.
8. Decide whether or not H_0 should be rejected and report that in the problem context.

Steps 1–4 should be completed prior to examination of the sample data. This sequence of steps will be illustrated in subsequent sections.

Figures

Numerous figures throughout the text illustrate statistical concepts in multiple formats.

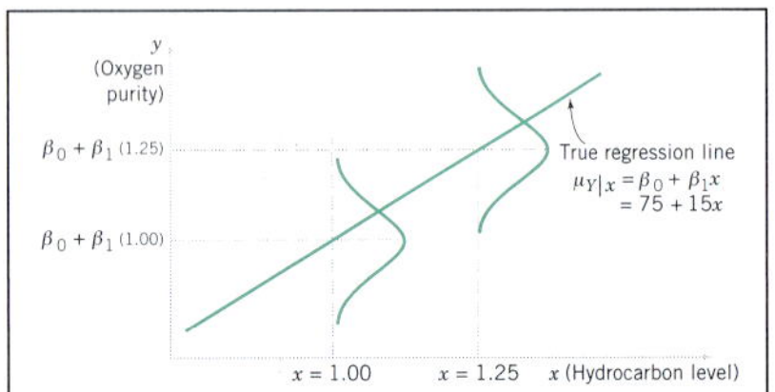

Figure 11-2 The distribution of Y for a given value of x for the oxygen purity-hydrocarbon data.

Minitab Output

Throughout the book, we have presented output from Minitab as typical examples of what can be done with modern statistical software.

Character Stem-and-Leaf Display

Stem-and-leaf of Strength

N = 80 Leaf Unit = 1.0

1	7	6
2	8	7
3	9	7
5	10	1 5
8	11	0 5 8
11	12	0 1 3
17	13	1 3 3 4 5 5
25	14	1 2 3 5 6 8 9 9
37	15	0 0 1 3 4 4 6 7 8 8 8 8
(10)	16	0 0 0 3 3 5 7 7 8 9
33	17	0 1 1 2 4 4 5 6 6 8
23	18	0 0 1 1 3 4 6
16	19	0 3 4 6 9 9
10	20	0 1 7 8
6	21	8
5	22	1 8 9
2	23	7
1	24	5

Figure 6-6 A stem-and-leaf diagram from Minitab.

Example Problems

A set of example problems provides the student with detailed solutions and comments for interesting, real-world situations.

Table 11-1 Oxygen and Hydrocarbon Levels

Observation Number	Hydrocarbon Level x(%)	Purity y(%)
1	0.99	90.01
2	1.02	89.05
3	1.15	91.43
4	1.29	93.74
5	1.46	96.73
6	1.36	94.45
7	0.87	87.59
8	1.23	91.77
9	1.55	99.42
10	1.40	93.65
11	1.19	93.54
12	1.15	92.52
13	0.98	90.56
14	1.01	89.54
15	1.11	89.85
16	1.20	90.39
17	1.26	93.25
18	1.32	93.41
19	1.43	94.98
20	0.95	87.33

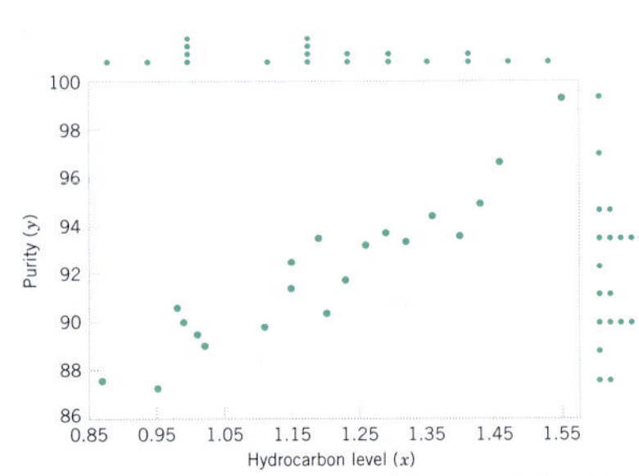

Figure 11-1 Scatter diagram of oxygen purity versus hydrocarbon level from Table 11-1.

Exercises

Each chapter has an extensive collection of exercises, including **end-of-section exercises** that emphasize the material in that section, **supplemental exercises** at the end of the chapter that cover the scope of chapter topics and require the student to make a decision about the approach they will use to solve the problem, and **mind-expanding exercises** that often require the student to extend the text material somewhat or to apply it in a novel situation.

Answers are provided to most odd-numbered exercises in Appendix C in the text, and the WileyPLUS course management systems includes for students complete detailed solutions to selected exercises.

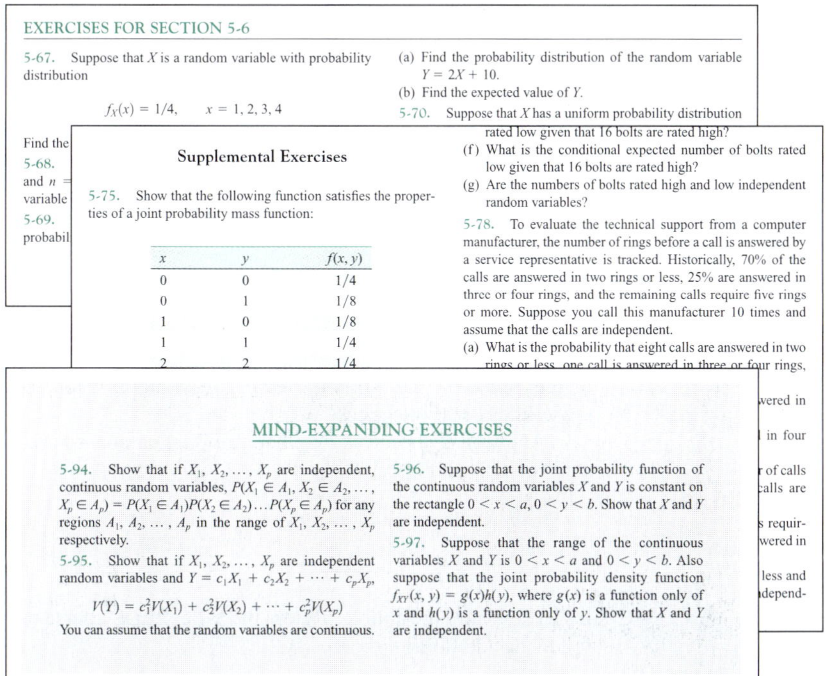

EXERCISES FOR SECTION 5-6

5-67. Suppose that X is a random variable with probability distribution

$$f_X(x) = 1/4, \quad x = 1, 2, 3, 4$$

(a) Find the probability distribution of the random variable $Y = 2X + 10$.
(b) Find the expected value of Y.

5-70. Suppose that X has a uniform probability distribution

Supplemental Exercises

5-75. Show that the following function satisfies the properties of a joint probability mass function:

x	y	$f(x, y)$
0	0	1/4
0	1	1/8
1	0	1/8
1	1	1/4

(f) What is the conditional expected number of bolts rated low given that 16 bolts are rated high?
(g) Are the numbers of bolts rated high and low independent random variables?

5-78. To evaluate the technical support from a computer manufacturer, the number of rings before a call is answered by a service representative is tracked. Historically, 70% of the calls are answered in two rings or less, 25% are answered in three or four rings, and the remaining calls require five rings or more. Suppose you call this manufacturer 10 times and assume that the calls are independent.
(a) What is the probability that eight calls are answered in two

MIND-EXPANDING EXERCISES

5-94. Show that if $X_1, X_2, \ldots, X_p$ are independent, continuous random variables, $P(X_1 \in A_1, X_2 \in A_2, \ldots, X_p \in A_p) = P(X_1 \in A_1)P(X_2 \in A_2)\ldots P(X_p \in A_p)$ for any regions $A_1, A_2, \ldots, A_p$ in the range of $X_1, X_2, \ldots, X_p$ respectively.

5-95. Show that if $X_1, X_2, \ldots, X_p$ are independent random variables and $Y = c_1X_1 + c_2X_2 + \cdots + c_pX_p$,

$$V(Y) = c_1^2V(X_1) + c_2^2V(X_2) + \cdots + c_p^2V(X_p)$$

You can assume that the random variables are continuous.

5-96. Suppose that the joint probability function of the continuous random variables X and Y is constant on the rectangle $0 < x < a$, $0 < y < b$. Show that X and Y are independent.

5-97. Suppose that the range of the continuous variables X and Y is $0 < x < a$ and $0 < y < b$. Also suppose that the joint probability density function $f_{XY}(x, y) = g(x)h(y)$, where $g(x)$ is a function only of x and $h(y)$ is a function only of y. Show that X and Y are independent.

Important Terms and Concepts

At the end of each chapter is a list of important terms and concepts for an easy self-check and study tool.

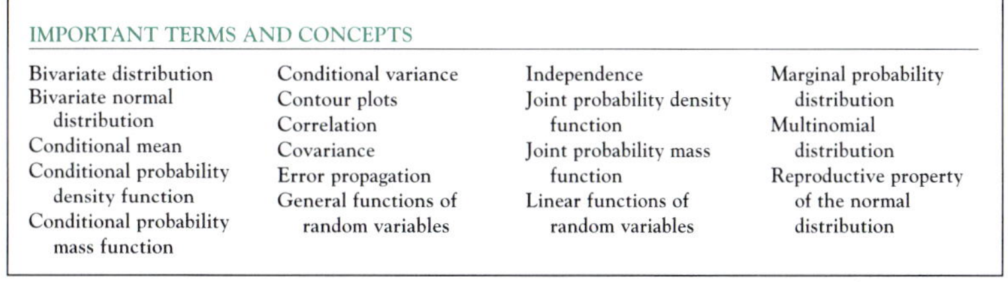

IMPORTANT TERMS AND CONCEPTS

Bivariate distribution	Conditional variance	Independence	Marginal probability distribution
Bivariate normal distribution	Contour plots	Joint probability density function	Multinomial distribution
Conditional mean	Correlation	Joint probability mass function	Reproductive property of the normal distribution
Conditional probability density function	Covariance	Linear functions of random variables	
Conditional probability mass function	Error propagation		
	General functions of random variables		

STUDENT RESOURCES

- **Data Sets** Data Sets for all examples and exercises in the text. Visit the student section of the book Web site at www.wiley.com/college/montgomery to access these materials.
- **Student Solutions Manual** Detailed solutions for selected problems in the book. The Student Solutions Manual may be purchased from the Web site at www.wiley.com/college/montgomery.
- **Section on Logistic Regression**
- **From the Publisher—Statistics Solutions by JustAsk!** Students may also wish to enhance their understanding by using Statistics Solutions, an innovative student-oriented Web site that provides the problem-solving approach and step-by-step solution to marked problems from the text. Included in the Web site are interactive visualizations of statistical theory that are difficult to easily convey on the printed page. The interactive visualizations also act as distribution calculators. Users rave about the simplicity and usefulness of the Web site. Students can access Statistics Solutions on www.wiley.com/college/montgomery.

INSTRUCTOR RESOURCES

The following resources are available only to instructors who adopt the text:

- **Solutions Manual** All solutions to the exercises in the text.
- **Data Sets** Data sets for all examples and exercises in the text.

- **Image Gallery of Text Figures**
- **PowerPoint Lecture Slides**
- **Section on Logistic Regression**

These instructor-only resources are password-protected. Visit the instructor section of the book Web site at www.wiley.com/college/montgomery to register for a password to access these materials.

MINITAB

A Student Version of Minitab is available as an option to purchase in a set with this text. Student versions of software often do not have all the functionality that full versions do. Consequently, student versions may not support all the concepts presented in this text. If you would like to adopt for your course the set of this text with the Student Version of Minitab, please contact your local Wiley representative at www.wiley.com/college/rep.

Alternatively, students may find information about how to purchase the professional version of the software at www.minitab.com.

WileyPLUS

Instructors have the option to adopt the WileyPLUS course management tool along with this text for the following additional resources:

- Your own customizable course Web site, populated with the textbook resources.
- All of the resources that are on the Student and Instructor Companion Sites will also be in the student and instructor sections of your WileyPLUS Web site, so everything will be in one location.
- The entire text in a searchable and linked HTML format. For example, the Important Terms and Concepts at the end of each chapter in the online text link directly to where those concepts are explained in the chapter, for an easy-to-use study tool.
- Online reading and homework assignments, to help you track your students' progress.

For more information about WileyPLUS, visit www.wiley.com/college/wileyplus.

COURSE SYLLABUS SUGGESTIONS

This is a very flexible textbook because instructors' ideas about what should be in a first course on statistics for engineers vary widely, as do the abilities of different groups of students. Therefore, we hesitate to give too much advice but will explain how we use the book.

We believe that a first course in statistics for engineers should be primarily an applied statistics course, not a probability course. In our one-semester course we cover all of Chapter 1 (in one or two lectures); overview the material on probability, putting most of the emphasis on the normal distribution (six to eight lectures); discuss most of Chapters 6 through 10 on confidence intervals and tests (twelve to fourteen lectures); introduce regression models in Chapter 11 (four lectures); give an introduction to the design of experiments from Chapters 13 and 14 (six lectures); and present the basic concepts of statistical process control, including the Shewhart control chart from Chapter 16 (four lectures). This leaves about three to four periods for exams and review. Let us emphasize that the purpose of this course is to introduce engineers to how statistics can be used to solve real-world engineering problems, not to weed out the less mathematically gifted students. This course is not the "baby math-stat" course that is all too often given to engineers.

If a second semester is available, it is possible to cover the entire book, including much of the supplemental material, if appropriate for the audience. It would also be possible to assign and work many of the homework problems in class to reinforce the understanding of the concepts. Obviously, multiple regression and more design of experiments would be major topics in a second course.

USING THE COMPUTER

In practice, engineers use computers to apply statistical methods to solve problems. Therefore, we strongly recommend that the computer be integrated into the class. Throughout the book we have presented output from Minitab as typical examples of what can be done with modern statistical software. In teaching, we have used other software packages, including Statgraphics, JMP, and Statistica. We did not clutter up the book with examples from many different packages because how the instructor integrates the software into the class is ultimately more important than which package is used. All text data is available in electronic form on the textbook Web site. In some chapters, there are problems that we feel should be worked using computer software. We have marked these problems with a special icon in the margin.

In our own classrooms, we use the computer in almost every lecture and demonstrate how the technique is implemented in software as soon as it is discussed in the lecture. Student versions of many statistical software packages are available at low cost, and students can either purchase their own copy or use the products available on the PC local area networks. We have found that this greatly improves the pace of the course and student understanding of the material.

Users should be aware that final answers may differ slightly due to different numerical precision and rounding protocols among softwares.

ACKNOWLEDGMENTS

We would like to express our grateful appreciation to the many organizations and individuals who have contributed to this book. Many instructors who used the first three editions provided excellent suggestions that we have tried to incorporate in this revision. We also thank the following instructors for their insightful reviews and suggestions.

Michael G. Akritas, Penn State University
Barb Barnet, University of Wisconsin-Platteville
Monica Brown, University of Missouri-St. Louis
Max Buot, Carnegie Melon University
Vikram Cariapa, Marquette University
Steven Guffey, Western Virginia University
Kosgallana Gunawardena, University of Wisconsin Oshkosh
Trevor Hale, Ohio University
William Hooper, Clarkson University
Wooseung Jang, University of Missouri-Columbia
Lawrence Kazmierczak, Stevens Institute of Technology
Charles McAllister, Louisiana State University
Mahour Parast, South Dakota School of Mines and Technology
Minnie Patel, San Jose State University
Arunkumar Pennathur, University of Texas at El Paso
Steven Rein, California Polytechnic State University
Manual D. Rossetti, University of Arkansas
Ioulia Rytikova, George Mason University
Ahlam Shalaby, Howard University
Bruce Schmeiser, Purdue University
Jery Stedinger, Cornell University
Jyh-Hone Wang, University of Rhode Island

We would also like to thank the following for their assistance in checking the accuracy and completeness of the exercises and the solutions to exercises.

Abdelaziz Berrado, Arizona State University
Patrick Egbunonu, Queens University
James C. Ford, Ford Consulting Associates
Jing Hu, Arizona State University
Fang Li, Arizona State University
Nuttha Lurponglukana, Arizona State University
Sarah Streett
Yolande Tra, Rochester Institute of Technology

We are also indebted to Dr. Smiley Cheng for permission to adapt many of the statistical tables from his excellent book (with Dr. James Fu), *Statistical Tables for Classroom and Exam Room*. John Wiley and Sons, Prentice Hall, the Institute of Mathematical Statistics, and the editors of Biometrics allowed us to use copyrighted material, for which we are grateful. Thanks are also due to Dr. Lora Zimmer, Dr. Connie Borror, and Dr. Alejandro Heredia-Langner for their outstanding work on the solutions to exercises.

Douglas C. Montgomery

George C. Runger

Contents

1 Statistics in Engineering

CHAPTER OUTLINE

LEARNING OBJECTIVES

After careful study of this chapter you should be able to do the following:

1. Identify the role that statistics can play in the engineering problem-solving process
2. Discuss how variability affects the data collected and used for making engineering decisions
3. Explain the difference between enumerative and analytical studies
4. Discuss the different methods that engineers use to collect data
5. Identify the advantages that designed experiments have in comparison to other methods of collecting engineering data
6. Explain the differences between mechanistic models and empirical models
7. Discuss how probability and probability models are used in engineering and science

1-1 THE ENGINEERING METHOD AND STATISTICAL THINKING

An engineer is someone who solves problems of interest to society by the efficient application of scientific principles. Engineers accomplish this by either refining an existing product or process or by designing a new product or process that meets customers' needs. The **engineering,** or **scientific, method** is the approach to formulating and solving these problems. The steps in the engineering method are as follows:

1. Develop a clear and concise description of the problem.
2. Identify, at least tentatively, the important factors that affect this problem or that may play a role in its solution.

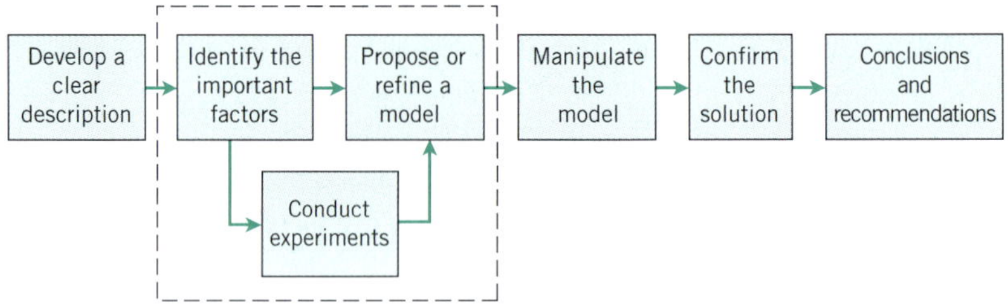

Figure 1-1 The engineering method.

3. Propose a model for the problem, using scientific or engineering knowledge of the phenomenon being studied. State any limitations or assumptions of the model.
4. Conduct appropriate experiments and collect data to test or validate the tentative model or conclusions made in steps 2 and 3.
5. Refine the model on the basis of the observed data.
6. Manipulate the model to assist in developing a solution to the problem.
7. Conduct an appropriate experiment to confirm that the proposed solution to the problem is both effective and efficient.
8. Draw conclusions or make recommendations based on the problem solution.

The steps in the engineering method are shown in Fig. 1-1. Notice that the engineering method features a strong interplay between the problem, the factors that may influence its solution, a model of the phenomenon, and experimentation to verify the adequacy of the model and the proposed solution to the problem. Steps 2–4 in Fig. 1-1 are enclosed in a box, indicating that several cycles or iterations of these steps may be required to obtain the final solution. Consequently, engineers must know how to efficiently plan experiments, collect data, analyze and interpret the data, and understand how the observed data are related to the model they have proposed for the problem under study.

The field of **statistics** deals with the collection, presentation, analysis, and use of data to make decisions, solve problems, and design products and processes. Because many aspects of engineering practice involve working with data, obviously some knowledge of statistics is important to any engineer. Specifically, statistical techniques can be a powerful aid in designing new products and systems, improving existing designs, and designing, developing, and improving production processes.

Statistical methods are used to help us describe and understand **variability.** By variability, we mean that successive observations of a system or phenomenon do not produce exactly the same result. We all encounter variability in our everyday lives, and **statistical thinking** can give us a useful way to incorporate this variability into our decision-making processes. For example, consider the gasoline mileage performance of your car. Do you always get exactly the same mileage performance on every tank of fuel? Of course not—in fact, sometimes the mileage performance varies considerably. This observed variability in gasoline mileage depends on many factors, such as the type of driving that has occurred most recently (city versus highway), the changes in condition of the vehicle over time (which could include factors such as tire inflation, engine compression, or valve wear), the brand and/or octane number of the gasoline used, or possibly even the weather conditions that have been recently experienced. These factors represent potential **sources of variability** in the system. Statistics gives us a framework for describing this variability and for learning about which potential sources of variability are the most important or which have the greatest impact on the gasoline mileage performance.

We also encounter variability in dealing with engineering problems. For example, suppose that an engineer is designing a nylon connector to be used in an automotive engine application. The engineer is considering establishing the design specification on wall thickness at 3/32 inch but is somewhat uncertain about the effect of this decision on the connector pull-off force. If the pull-off force is too low, the connector may fail when it is installed in an engine. Eight prototype units are produced and their pull-off forces measured, resulting in the following data (in pounds): 12.6, 12.9, 13.4, 12.3, 13.6, 13.5, 12.6, 13.1. As we anticipated, not all of the prototypes have the same pull-off force. We say that there is variability in the pull-off force measurements. Because the pull-off force measurements exhibit variability, we consider the pull-off force to be a **random variable.** A convenient way to think of a random variable, say X, that represents a measurement, is by using the model

$$X = \mu + \epsilon \tag{1-1}$$

where μ is a constant and ϵ is a random disturbance. The constant remains the same with every measurement, but small changes in the environment, test equipment, differences in the individual parts themselves, and so forth change the value of ϵ. If there were no disturbances, ϵ would always equal zero and X would always be equal to the constant μ. However, this never happens in the real world, so the actual measurements X exhibit variability. We often need to describe, quantify and ultimately reduce variability.

Figure 1-2 presents a **dot diagram** of these data. The dot diagram is a very useful plot for displaying a small body of data—say, up to about 20 observations. This plot allows us to see easily two features of the data; the **location,** or the middle, and the **scatter** or **variability.** When the number of observations is small, it is usually difficult to identify any specific patterns in the variability, although the dot diagram is a convenient way to see any unusual data features.

The need for statistical thinking arises often in the solution of engineering problems. Consider the engineer designing the connector. From testing the prototypes, he knows that the average pull-off force is 13.0 pounds. However, he thinks that this may be too low for the intended application, so he decides to consider an alternative design with a greater wall thickness, 1/8 inch. Eight prototypes of this design are built, and the observed pull-off force measurements are 12.9, 13.7, 12.8, 13.9, 14.2, 13.2, 13.5, and 13.1. The average is 13.4. Results for both samples are plotted as dot diagrams in Fig. 1-3. This display gives the impression that increasing the wall thickness has led to an increase in pull-off force. However, there are some obvious questions to ask. For instance, how do we know that another sample of prototypes will not give different results? Is a sample of eight prototypes adequate to give reliable results? If we use the test results obtained so far to conclude that increasing the wall thickness increases the strength, what risks are associated with this decision? For example, is it possible that the apparent increase in pull-off force observed in the thicker prototypes is only due to the inherent variability in the system and that increasing the thickness of the part (and its cost) really has no effect on the pull-off force?

Often, physical laws (such as Ohm's law and the ideal gas law) are applied to help design products and processes. We are familiar with this reasoning from general laws to specific cases. But it is also important to reason from a specific set of measurements to more general cases to answer the previous questions. This reasoning is from a **sample** (such as the eight connectors) to a **population** (such as the connectors that will be sold to customers). The

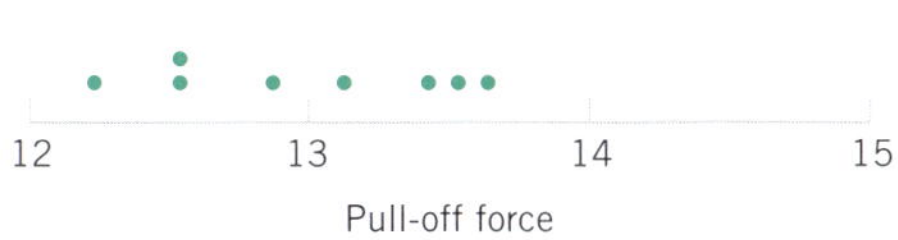

Figure 1-2 Dot diagram of the pull-off force data when wall thickness is 3/32 inch.

Figure 1-3 Dot diagram of pull-off force for two wall thicknesses.

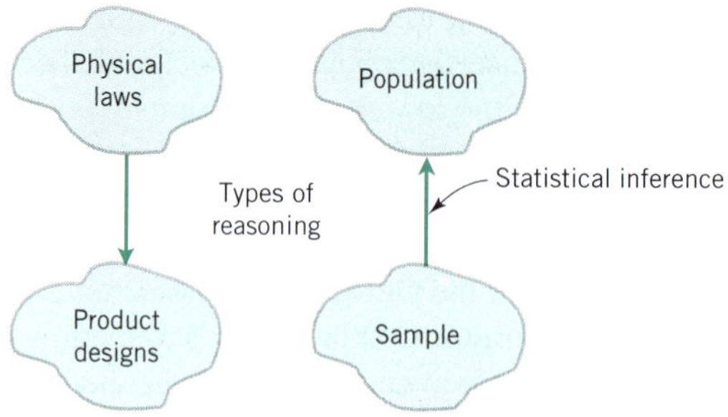

Figure 1-4 Statistical inference is one type of reasoning.

reasoning is referred to as **statistical inference.** See Fig. 1-4. Historically, measurements were obtained from a sample of people and generalized to a population, and the terminology has remained. Clearly, reasoning based on measurements from some objects to measurements on all objects can result in errors (called sampling errors). However, if the sample is selected properly, these risks can be quantified and an appropriate sample size can be determined.

1-2 COLLECTING ENGINEERING DATA

1-2.1 Basic Principles

In the previous section, we illustrated some simple methods for summarizing data. In the engineering environment, the data is almost always a sample that has been selected from some population. Three basic methods of collecting data are

- A **retrospective** study using historical data
- An **observational** study
- A **designed experiment**

An effective data collection procedure can greatly simplify the analysis and lead to improved understanding of the population or process that is being studied. We now consider some examples of these data collection methods.

1-2.2 Retrospective Study

Montgomery, Peck, and Vining (2001) describe an acetone-butyl alcohol distillation column for which concentration of acetone in the distillate or output product stream is an important variable. Factors that may affect the distillate are the reboil temperature, the condensate temperature, and the reflux rate. Production personnel obtain and archive the following records:

- The concentration of acetone in an hourly test sample of output product
- The reboil temperature log, which is a plot of the reboil temperature over time
- The condenser temperature controller log
- The nominal reflux rate each hour

The reflux rate should be held constant for this process. Consequently, production personnel change this very infrequently.

A retrospective study would use either all or a sample of the historical process data archived over some period of time. The study objective might be to discover the relationships among the two temperatures and the reflux rate on the acetone concentration in the output product stream. However, this type of study presents some problems:

1. We may not be able to see the relationship between the reflux rate and acetone concentration, because the reflux rate didn't change much over the historical period.
2. The archived data on the two temperatures (which are recorded almost continuously) do not correspond perfectly to the acetone concentration measurements (which are made hourly). It may not be obvious how to construct an approximate correspondence.
3. Production maintains the two temperatures as closely as possible to desired targets or set points. Because the temperatures change so little, it may be difficult to assess their real impact on acetone concentration.
4. Within the narrow ranges that they do vary, the condensate temperature tends to increase with the reboil temperature. Consequently, the effects of these two process variables on acetone concentration may be difficult to separate.

As you can see, a retrospective study may involve a lot of **data,** but that data may contain relatively little useful **information** about the problem. Furthermore, some of the relevant data may be missing, there may be transcription or recording errors resulting in **outliers** (or unusual values), or data on other important factors may not have been collected and archived. In the distillation column, for example, the specific concentrations of butyl alcohol and acetone in the input feed stream are a very important factor, but they are not archived because the concentrations are too hard to obtain on a routine basis. As a result of these types of issues, statistical analysis of historical data sometimes identifies interesting phenomena, but solid and reliable explanations of these phenomena are often difficult to obtain.

1-2.3 Observational Study

In an observational study, the engineer observes the process or population, disturbing it as little as possible, and records the quantities of interest. Because these studies are usually conducted for a relatively short time period, sometimes variables that are not routinely measured can be included. In the distillation column, the engineer would design a form to record the two temperatures and the reflux rate when acetone concentration measurements are made. It may even be possible to measure the input feed stream concentrations so that the impact of this factor could be studied. Generally, an observational study tends to solve problems 1 and 2 above and goes a long way toward obtaining accurate and reliable data. However, observational studies may not help resolve problems 3 and 4.

1-2.4 Designed Experiments

In a designed experiment the engineer makes *deliberate* or *purposeful changes* in the controllable variables of the system or process, observes the resulting system output data, and then makes an inference or decision about which variables are responsible for the observed changes in output performance. The nylon connector example in Section 1-1 illustrates a designed experiment; that is, a deliberate change was made in the wall thickness of the connector with the objective of discovering whether or not a greater pull-off force could be obtained. Experiments designed with basic principles such as **randomization** are needed to establish **cause and effect** relationships.

Much of what we know in the engineering and physical-chemical sciences is developed through testing or experimentation. Often engineers work in problem areas in which no scientific or engineering theory is directly or completely applicable, so experimentation and observation of the resulting data constitute the only way that the problem can be solved. Even when there is a good underlying scientific theory that we may rely on to explain the phenomena of interest, it is almost always necessary to conduct tests or experiments to confirm that the theory is indeed operative in the situation or environment in which it is being applied. Statistical thinking and statistical methods play an important role in planning, conducting, and analyzing the data from engineering experiments. Designed experiments play a very important role in engineering design and development and in the improvement of manufacturing processes.

For example, consider the problem involving the choice of wall thickness for the nylon connector. This is a simple illustration of a designed experiment. The engineer chose two wall thicknesses for the connector and performed a series of tests to obtain pull-off force measurements at each wall thickness. In this simple **comparative experiment,** the engineer is interested in determining if there is any difference between the 3/32- and 1/8-inch designs. An approach that could be used in analyzing the data from this experiment is to compare the mean pull-off force for the 3/32-inch design to the mean pull-off force for the 1/8-inch design using statistical **hypothesis testing,** which is discussed in detail in Chapters 9 and 10. Generally, a **hypothesis** is a statement about some aspect of the system in which we are interested. For example, the engineer might want to know if the mean pull-off force of a 3/32-inch design exceeds the typical maximum load expected to be encountered in this application, say 12.75 pounds. Thus, we would be interested in testing the hypothesis that the mean strength exceeds 12.75 pounds. This is called a **single-sample hypothesis testing problem.** Chapter 9 presents techniques for this type of problem. Alternatively, the engineer might be interested in testing the hypothesis that increasing the wall thickness from 3/32- to 1/8-inch results in an increase in mean pull-off force. It is an example of a **two-sample hypothesis testing problem.** Two-sample hypothesis testing problems are discussed in Chapter 10.

Designed experiments are a very powerful approach to studying complex systems, such as the distillation column. This process has three factors, the two temperatures and the reflux rate, and we want to investigate the effect of these three factors on output acetone concentration. A good experimental design for this problem must ensure that we can separate the effects of all three factors on the acetone concentration. The specified values of the three factors used in the experiment are called **factor levels.** Typically, we use a small number of levels for each factor, such as two or three. For the distillation column problem, suppose we use a "high," and "low," level (denoted +1 and −1, respectively) for each of the factors. We thus would use two levels for each of the three factors. A very reasonable experiment design strategy uses every possible combination of the factor levels to form a basic experiment with eight different settings for the process. This type of experiment is called a **factorial experiment.** Table 1-1 presents this experimental design.

Table 1-1 The Designed Experiment (Factorial Design) for the Distillation Column

Reboil Temp.	Condensate Temp.	Reflux Rate
−1	−1	−1
+1	−1	−1
−1	+1	−1
+1	+1	−1
−1	−1	+1
+1	−1	+1
−1	+1	+1
+1	+1	+1

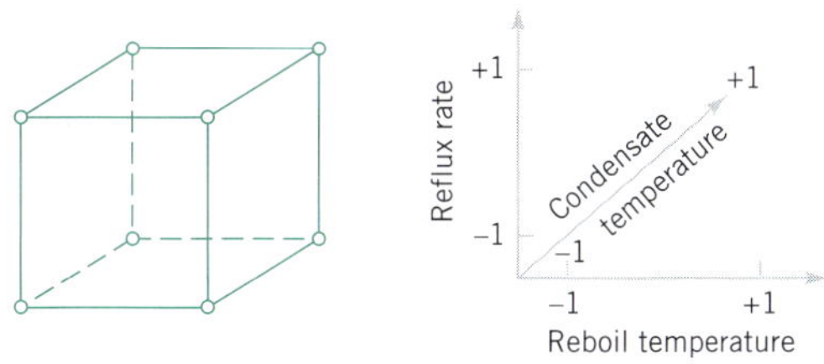

Figure 1-5 The factorial design for the distillation column.

Figure 1-5 illustrates that this design forms a cube in terms of these high and low levels. With each setting of the process conditions, we allow the column to reach equilibrium, take a sample of the product stream, and determine the acetone concentration. We then can draw specific inferences about the effect of these factors. Such an approach allows us to proactively study a population or process.

An important advantage of factorial experiments is that they allow one to detect an **interaction** between factors. Consider only the two temperature factors in the distillation experiment. Suppose that the response concentration is poor when the reboil temperature is *low,* regardless of the condensate temperature. That is, the condensate temperature has no effect when the reboil temperature is *low*. However, when the reboil temperature is *high,* a *high* condensate temperature generates a good response, while a *low* condensate temperature generates a poor response. That is, the condensate temperature changes the response when the reboil temperature is *high*. The effect of condensate temperature depends on the setting of the reboil temperature and these two factors would interact in this case. If the four combinations of *high* and *low* reboil and condensate temperatures were not tested, such an interaction would not be detected.

We can easily extend the factorial strategy to more factors. Suppose that the engineer wants to consider a fourth factor, type of distillation column. There are two types: the standard one and a newer design. Figure 1-6 illustrates how all four factors, reboil temperature, condensate temperature, reflux rate, and column design, could be investigated in a factorial design. Since all four factors are still at two levels, the experimental design can still be represented geometrically as a cube (actually, it's a *hypercube*). Notice that as in any factorial design, all possible combinations of the four factors are tested. The experiment requires 16 trials.

Generally, if there are k factors and they each have two levels, a factorial experimental design will require 2^k runs. For example, with $k = 4$, the 2^4 design in Fig. 1-6 requires 16 tests. Clearly, as the number of factors increases, the number of trials required in a factorial experiment increases rapidly; for instance, eight factors each at two levels would require 256 trials. This quickly becomes unfeasible from the viewpoint of time and other resources. Fortunately, when there are four to five or more factors, it is usually unnecessary to test all possible combinations of factor levels. A **fractional factorial experiment** is a variation of the basic

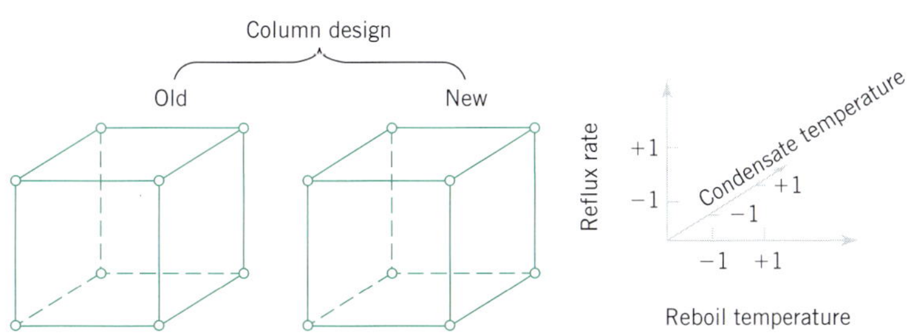

Figure 1-6 A four-factorial experiment for the distillation column.

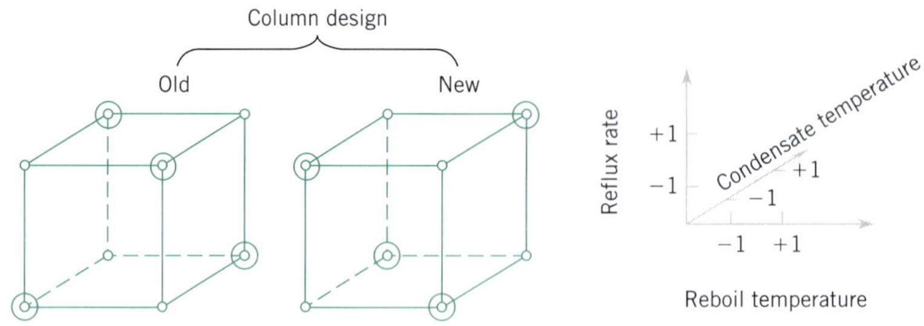

Figure 1-7 A fractional factorial experiment for the connector wall thickness problem.

factorial arrangement in which only a subset of the factor combinations are actually tested. Figure 1-7 shows a fractional factorial experimental design for the four-factor version of the distillation experiment. The circled test combinations in this figure are the only test combinations that need to be run. This experimental design requires only 8 runs instead of the original 16; consequently it would be called a **one-half fraction.** This is an excellent experimental design in which to study all four factors. It will provide good information about the individual effects of the four factors and some information about how these factors interact.

Factorial and fractional factorial experiments are used extensively by engineers and scientists in industrial research and development, where new technology, products, and processes are designed and developed and where existing products and processes are improved. Since so much engineering work involves testing and experimentation, it is essential that all engineers understand the basic principles of planning efficient and effective experiments. We discuss these principles in Chapter 13. Chapter 14 concentrates on the factorial and fractional factorials that we have introduced here.

1-2.5 Observing Processes Over Time

Often data are collected over time. In this case, it is usually very helpful to plot the data versus time in a **time series plot.** Phenomena that might affect the system or process often become more visible in a time-oriented plot and the concept of stability can be better judged.

Figure 1-8 is a dot diagram of acetone concentration readings taken hourly from the distillation column described in Section 1-2.2. The large variation displayed on the dot diagram indicates a lot of variability in the concentration, but the chart does not help explain the reason for the variation. The time series plot is shown in Figure 1-9. A shift in the process mean level is visible in the plot and an estimate of the time of the shift can be obtained.

W. Edwards **Deming,** a very influential industrial statistician, stressed that it is important to understand the nature of variability in processes and systems over time. He conducted an experiment in which he attempted to drop marbles as close as possible to a target on a table. He used a funnel mounted on a ring stand and the marbles were dropped into the funnel. See Fig. 1-10. The funnel was aligned as closely as possible with the center of the target. He then used two different strategies to operate the process. (1) He never moved the funnel. He just

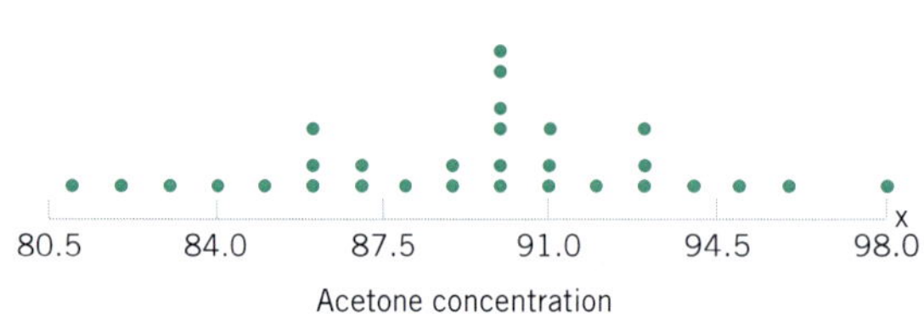

Figure 1-8 The dot diagram illustrates variation but does not identify the problem.

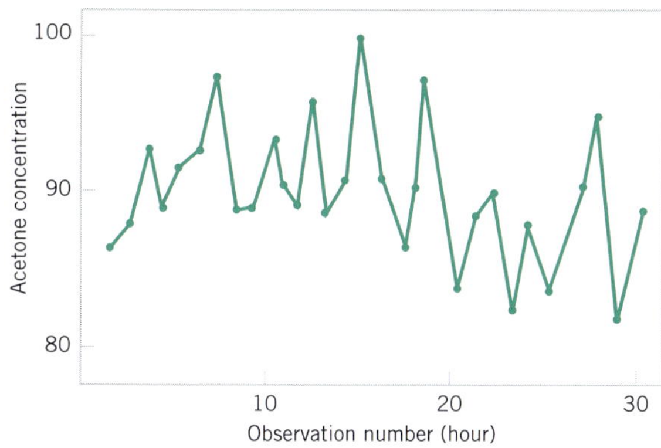

Figure 1-9 A time series plot of concentration provides more information than the dot diagram.

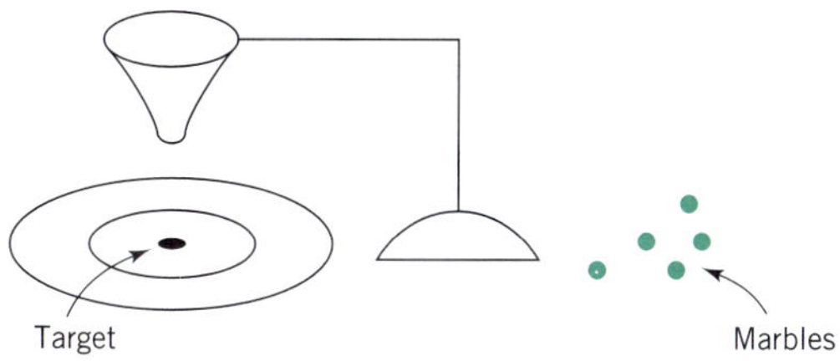

Figure 1-10 Deming's funnel experiment.

dropped one marble after another and recorded the distance from the target. (2) He dropped the first marble and recorded its location relative to the target. He then moved the funnel an equal and opposite distance in an attempt to compensate for the error. He continued to make this type of adjustment after each marble was dropped.

After both strategies were completed, he noticed that the variability of the distance from the target for strategy 2 was approximately 2 times larger than for strategy 1. The adjustments to the funnel increased the deviations from the target. The explanation is that the error (the deviation of the marble's position from the target) for one marble provides no information about the error that will occur for the next marble. Consequently, adjustments to the funnel do not decrease future errors. Instead, they tend to move the funnel farther from the target.

This interesting experiment points out that adjustments to a process based on random disturbances can actually *increase* the variation of the process. This is referred to as **overcontrol** or **tampering.** Adjustments should be applied only to compensate for a nonrandom shift in the process—then they can help. A computer simulation can be used to demonstrate the lessons of the funnel experiment. Figure 1-11 displays a time plot of 100 measurements (denoted as y) from a process in which only random disturbances are present. The target value for the process

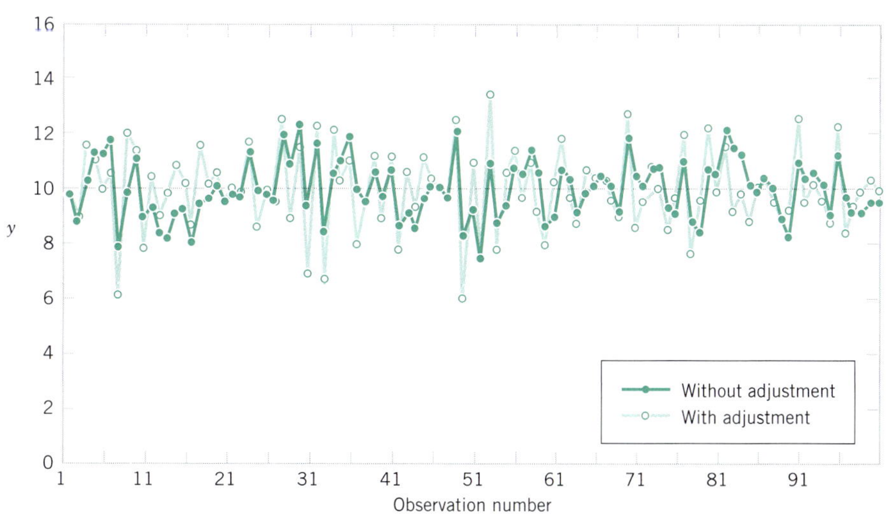

Figure 1-11 Adjustments applied to random disturbances overcontrol the process and increase the deviations from the target.

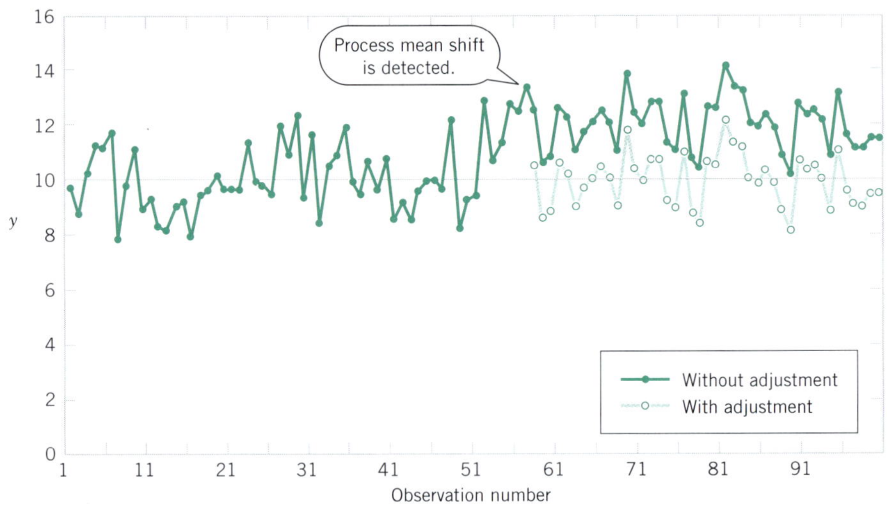

Figure 1-12 Process mean shift is detected at observation number 57, and one adjustment (a decrease of two units) reduces the deviations from target.

is 10 units. The figure displays the data with and without adjustments that are applied to the process mean in an attempt to produce data closer to target. Each adjustment is equal and opposite to the deviation of the previous measurement from target. For example, when the measurement is 11 (one unit above target), the mean is reduced by one unit before the next measurement is generated. The overcontrol has increased the deviations from the target.

Figure 1-12 displays the data without adjustment from Fig. 1-11, except that the measurements after observation number 50 are increased by two units to simulate the effect of a shift in the mean of the process. When there is a true shift in the mean of a process, an adjustment can be useful. Figure 1-12 also displays the data obtained when one adjustment (a decrease of two units) is applied to the mean after the shift is detected (at observation number 57). Note that this adjustment decreases the deviations from target.

The question of when to apply adjustments (and by what amounts) begins with an understanding of the types of variation that affect a process. A **control chart** is an invaluable way to examine the variability in time-oriented data. Figure 1-13 presents a control chart for the concentration data from Fig. 1-9. The **center line** on the control chart is just the average of the concentration measurements for the first 20 samples ($\bar{x} = 91.5$ g/l) when the process is stable. The **upper control limit** and the **lower control limit** are a pair of statistically derived

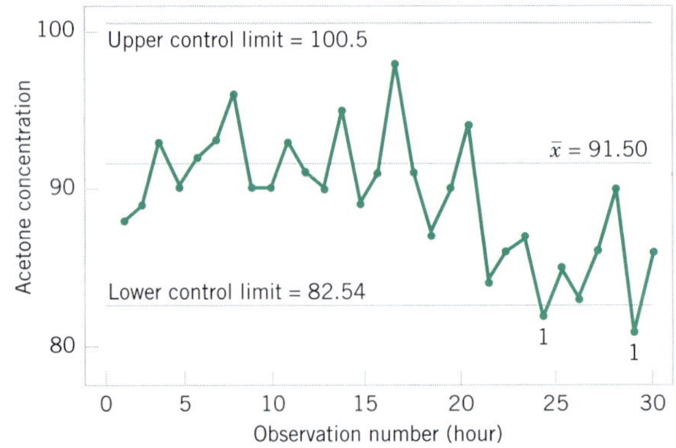

Figure 1-13 A control chart for the chemical process concentration data.

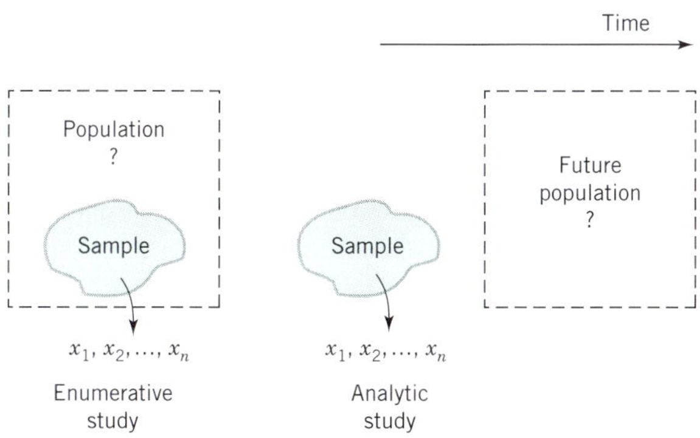

Figure 1-14 Enumerative versus analytic study.

limits that reflect the inherent or natural variability in the process. These limits are located three standard deviations of the concentration values above and below the center line. If the process is operating as it should, without any external sources of variability present in the system, the concentration measurements should fluctuate randomly around the center line, and almost all of them should fall between the control limits.

In the control chart of Fig. 1-13, the visual frame of reference provided by the center line and the control limits indicates that some upset or disturbance has affected the process around sample 20 because all of the following observations are below the center line and two of them actually fall below the lower control limit. This is a very strong signal that corrective action is required in this process. If we can find and eliminate the underlying cause of this upset, we can improve process performance considerably.

Furthermore, W. Edwards Deming pointed out that data from a process are used for different types of conclusions. Sometimes we collect data from a process to evaluate current production. For example, we might sample and measure resistivity on three semiconductor wafers selected from a lot and use this to evaluate the lot. This is called an **enumerative** study. However, in many cases we use data from current production to evaluate future production. We apply conclusions to a conceptual, future population. Deming called this an **analytic** study. Clearly this requires an assumption of a **stable** process and Deming emphasized that control charts were needed to justify this assumption. See Fig. 1-14 as an illustration.

Control charts are a very important application of statistics for monitoring, controlling, and improving a process. The branch of statistics that makes use of control charts is called **statistical process control,** or **SPC.** We will discuss SPC and control charts in Chapter 16.

1-3 MECHANISTIC AND EMPIRICAL MODELS

Models play an important role in the analysis of nearly all engineering problems. Much of the formal education of engineers involves learning about the models relevant to specific fields and the techniques for applying these models in problem formulation and solution. As a simple example, suppose we are measuring the flow of current in a thin copper wire. Our model for this phenomenon might be Ohm's law:

$$\text{Current} = \text{voltage/resistance}$$

or

$$I = E/R \tag{1-2}$$

We call this type of model a **mechanistic model** because it is built from our underlying knowledge of the basic physical mechanism that relates these variables. However, if we performed this measurement process more than once, perhaps at different times, or even on different days, the observed current could differ slightly because of small changes or variations in factors that are not completely controlled, such as changes in ambient temperature, fluctuations in performance of the gauge, small impurities present at different locations in the wire, and drifts in the voltage source. Consequently, a more realistic model of the observed current might be

$$I = E/R + \epsilon \tag{1-3}$$

where ϵ is a term added to the model to account for the fact that the observed values of current flow do not perfectly conform to the mechanistic model. We can think of ϵ as a term that includes the effects of all of the unmodeled sources of variability that affect this system.

Sometimes engineers work with problems for which there is no simple or well-understood mechanistic model that explains the phenomenon. For instance, suppose we are interested in the number average molecular weight (M_n) of a polymer. Now we know that M_n is related to the viscosity of the material (V), and it also depends on the amount of catalyst (C) and the temperature (T) in the polymerization reactor when the material is manufactured. The relationship between M_n and these variables is

$$M_n = f(V, C, T) \tag{1-4}$$

say, where the *form* of the function f is unknown. Perhaps a working model could be developed from a first-order Taylor series expansion, which would produce a model of the form

$$M_n = \beta_0 + \beta_1 V + \beta_2 C + \beta_3 T \tag{1-5}$$

where the β's are unknown parameters. Now just as in Ohm's law, this model will not exactly describe the phenomenon, so we should account for the other sources of variability that may affect the molecular weight by adding another term to the model; therefore

$$M_n = \beta_0 + \beta_1 V + \beta_2 C + \beta_3 T + \epsilon \tag{1-6}$$

is the model that we will use to relate molecular weight to the other three variables. This type of model is called an **empirical model;** that is, it uses our engineering and scientific knowledge of the phenomenon, but it is not directly developed from our theoretical or first-principles understanding of the underlying mechanism.

To illustrate these ideas with a specific example, consider the data in Table 1-2. This table contains data on three variables that were collected in an observational study in a semiconductor manufacturing plant. In this plant, the finished semiconductor is wire bonded to a frame. The variables reported are pull strength (a measure of the amount of force required to break the bond), the wire length, and the height of the die. We would like to find a model relating pull strength to wire length and die height. Unfortunately, there is no physical mechanism that we can easily apply here, so it doesn't seem likely that a mechanistic modeling approach will be successful.

Figure 1-15 presents a three-dimensional plot of all 25 observations on pull strength, wire length, and die height. From examination of this plot, we see that pull strength increases as

Table 1-2 Wire Bond Pull Strength Data

Observation Number	Pull Strength y	Wire Length x_1	Die Height x_2
1	9.95	2	50
2	24.45	8	110
3	31.75	11	120
4	35.00	10	550
5	25.02	8	295
6	16.86	4	200
7	14.38	2	375
8	9.60	2	52
9	24.35	9	100
10	27.50	8	300
11	17.08	4	412
12	37.00	11	400
13	41.95	12	500
14	11.66	2	360
15	21.65	4	205
16	17.89	4	400
17	69.00	20	600
18	10.30	1	585
19	34.93	10	540
20	46.59	15	250
21	44.88	15	290
22	54.12	16	510
23	56.63	17	590
24	22.13	6	100
25	21.15	5	400

both wire length and die height increase. Furthermore, it seems reasonable to think that a model such as

$$\text{Pull strength} = \beta_0 + \beta_1(\text{wire length}) + \beta_2(\text{die height}) + \epsilon$$

would be appropriate as an empirical model for this relationship. In general, this type of empirical model is called a **regression model.** In Chapters 11 and 12 we show how to build these models and test their adequacy as approximating functions. We will use a method for

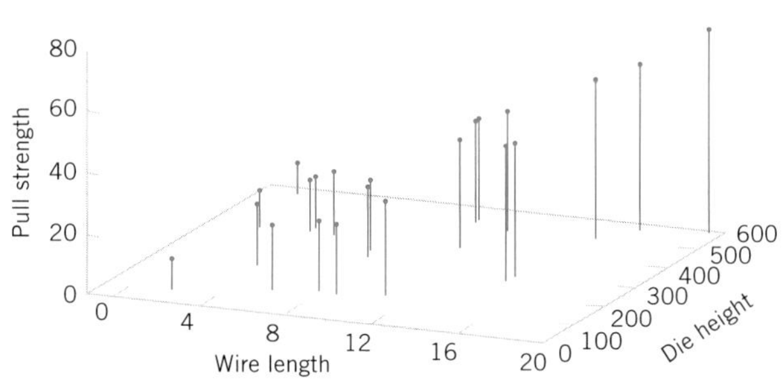

Figure 1-15 Three-dimensional plot of the wire bond pull strength data.

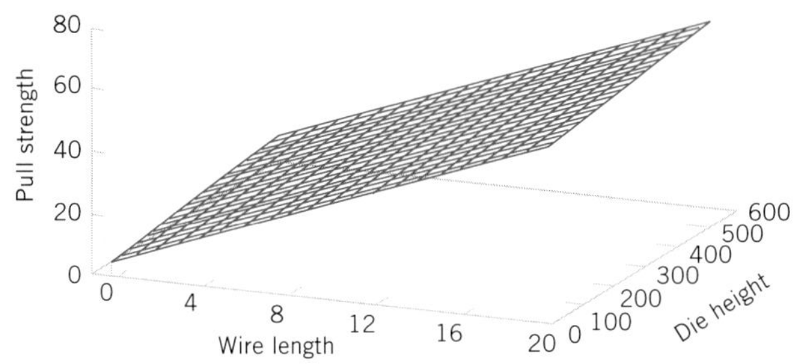

Figure 1-16 Plot of predicted values of pull strength from the empirical model.

estimating the parameters in regression models, called the method of **least squares,** that traces its origins to work by Karl Gauss. Essentially, this method chooses the parameters in the empirical model (the β's) to minimize the sum of the squared distances between each data point and the plane represented by the model equation. Applying this technique to the data in Table 1-2 results in

$$\widehat{\text{Pull strength}} = 2.26 + 2.74(\text{wire length}) + 0.0125(\text{die height}) \tag{1-7}$$

where the "hat," or circumflex, over pull strength indicates that this is an estimated or predicted quantity.

Figure 1-16 is a plot of the predicted values of pull strength versus wire length and die height obtained from Equation 1-7. Notice that the predicted values lie on a plane above the wire length–die height space. From the plot of the data in Fig. 1-15, this model does not appear unreasonable. The empirical model in Equation 1-7 could be used to predict values of pull strength for various combinations of wire length and die height that are of interest. Essentially, the empirical model could be used by an engineer in exactly the same way that a mechanistic model can be used.

1-4 PROBABILITY AND PROBABILITY MODELS

In Section 1-1, it was mentioned that decisions often need to be based on measurements from only a subset of objects selected in a sample. This process of reasoning from a sample of objects to conclusions for a population of objects was referred to as statistical inference. A sample of three wafers selected from a larger production lot of wafers in semiconductor manufacturing was an example mentioned. To make good decisions, an analysis of how well a sample represents a population is clearly necessary. If the lot contains defective wafers, how well will the sample detect this? How can we quantify the criterion to "detect well"? Basically, how can we quantify the risks of decisions based on samples? Furthermore, how should samples be selected to provide good decisions—ones with acceptable risks? **Probability** models help quantify the risks involved in statistical inference, that is, the risks involved in decisions made every day.

More details are useful to describe the role of probability models. Suppose a production lot contains 25 wafers. If all the wafers are defective or all are good, clearly any sample will generate all defective or all good wafers, respectively. However, suppose only one wafer in the lot is defective. Then a sample might or might not detect (include) the wafer. A probability model, along with a method to select the sample, can be used to quantify the risks that the defective wafer is or is not detected. Based on this analysis, the size of the sample might be increased (or decreased). The risk here can be interpreted as follows. Suppose a series of lots, each with exactly one defective wafer, are sampled. The details of the method used to select the sample are postponed until randomness is discussed in the next chapter. Nevertheless, assume that the same size sample (such as three wafers) is selected in the same manner from each lot. The proportion of the lots in which the defective wafer is included in the sample or, more specifically, the limit of this proportion as the number of lots in the series tends to infinity, is interpreted as the probability that the defective wafer is detected.

A probability model is used to calculate this proportion under reasonable assumptions for the manner in which the sample is selected. This is fortunate because we do not want to attempt to sample from an infinite series of lots. Problems of this type are worked in Chapters 2 and 3. More importantly, this probability provides valuable, quantitative information regarding any decision about lot quality based on the sample.

Recall from Section 1-1 that a population might be conceptual, as in an analytic study that applies statistical inference to future production based on the data from current production. When populations are extended in this manner, the role of statistical inference and the associated probability models becomes even more important.

In the previous example, each wafer in the sample was only classified as defective or not. Instead, a continuous measurement might be obtained from each wafer. In Section 1-2.5, concentration measurements were taken at periodic intervals from a production process. Figure 1-8 shows that variability is present in the measurements, and there might be concern that the process has moved from the target setting for concentration. Similar to the defective wafer, one might want to quantify our ability to detect a process change based on the sample data. Control limits were mentioned in Section 1-2.5 as decision rules for whether or not to adjust a process. The probability that a particular process change is detected can be calculated with a probability model for concentration measurements. Models for continuous measurements are developed based on plausible assumptions for the data and a result known as the central limit theorem, and the associated normal distribution is a particularly valuable probability model for statistical inference. Of course, a check of assumptions is important. These types of probability models are discussed in Chapter 4. The objective is still to quantify the risks inherent in the inference made from the sample data.

Throughout Chapters 6 through 15, decisions are based on statistical inference from sample data. Continuous probability models, specifically the normal distribution, are used extensively to quantify the risks in these decisions and to evaluate ways to collect the data and how large a sample should be selected.

IMPORTANT TERMS AND CONCEPTS

Analytic study
Cause and effect
Designed experiment
Empirical model
Engineering method
Enumerative study
Factorial experiment
Fractional factorial experiment
Hypothesis testing
Interaction
Mechanistic model
Observational study
Overcontrol
Population
Probability model
Problem-solving method
Randomization
Retrospective study
Sample
Statistical inference
Statistical process control
Statistical thinking
Tampering
Time series
Variability

2 Introduction to Probability

CHAPTER OUTLINE

LEARNING OBJECTIVES

After careful study of this chapter you should be able to do the following:

1. Understand and describe sample spaces and events for random experiments with graphs, tables, lists, or tree diagrams
2. Interpret probabilities and use probabilities of outcomes to calculate probabilities of events in discrete sample spaces
3. Use permutation and combinations to count the number of outcomes in both an event and the sample space
4. Calculate the probabilities of joint events such as unions and intersections from the probabilities of individual events
5. Interpret and calculate conditional probabilities of events
6. Determine the independence of events and use independence to calculate probabilities
7. Use Bayes' theorem to calculate conditional probabilities
8. Understand random variables

2-1 SAMPLE SPACES AND EVENTS

2-1.1 Random Experiments

If we measure the current in a thin copper wire, we are conducting an experiment. However, in day-to-day repetitions of the measurement the results can differ slightly because of small variations in variables that are not controlled in our experiment, including changes in ambient temperatures, slight variations in gauge and small impurities in the chemical composition of the wire if different locations are selected, and current source drifts. Consequently, this experiment (as well as many we conduct) is said to have a **random** component. In some cases, the random variations, are small enough, relative to our experimental goals, that they can be ignored. However, no matter how carefully our experiment is designed and conducted, the variation is almost always present, and its magnitude can be large enough that the important conclusions from our experiment are not obvious. In these cases, the methods presented in this book for modeling and analyzing experimental results are quite valuable.

Our goal is to understand, quantify, and model the type of variations that we often encounter. When we incorporate the variation into our thinking and analyses, we can make informed judgments from our results that are not invalidated by the variation.

Models and analyses that include variation are not different from models used in other areas of engineering and science. Figure 2-1 displays the important components. A mathematical model (or abstraction) of the physical system is developed. It need not be a perfect abstraction. For example, Newton's laws are not perfect descriptions of our physical universe. Still, they are useful models that can be studied and analyzed to approximately quantify the performance of a wide range of engineered products. Given a mathematical abstraction that is validated with measurements from our system, we can use the model to understand, describe, and quantify important aspects of the physical system and predict the response of the system to inputs.

Throughout this text, we discuss models that allow for variations in the outputs of a system, even though the variables that we control are not purposely changed during our study. Figure 2-2 graphically displays a model that incorporates uncontrollable inputs (noise) that combine with the controllable inputs to produce the output of our system. Because of the uncontrollable inputs, the same settings for the controllable inputs do not result in identical outputs every time the system is measured.

Random Experiment

An experiment that can result in different outcomes, even though it is repeated in the same manner every time, is called a **random experiment.**

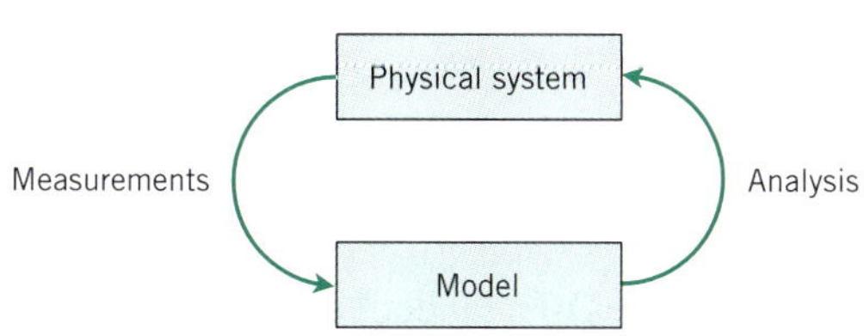

Figure 2-1 Continuous iteration between model and physical system.

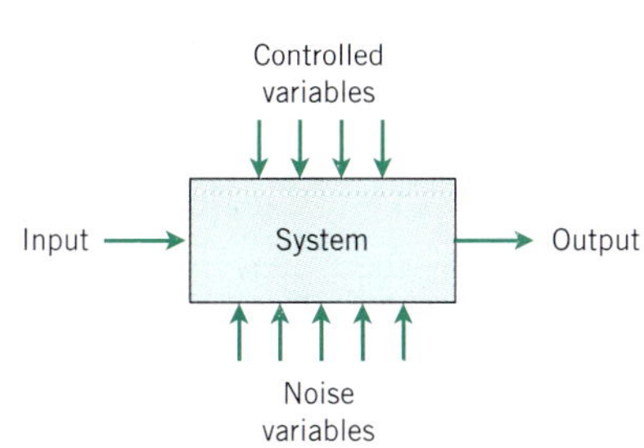

Figure 2-2 Noise variables affect the transformation of inputs to outputs.

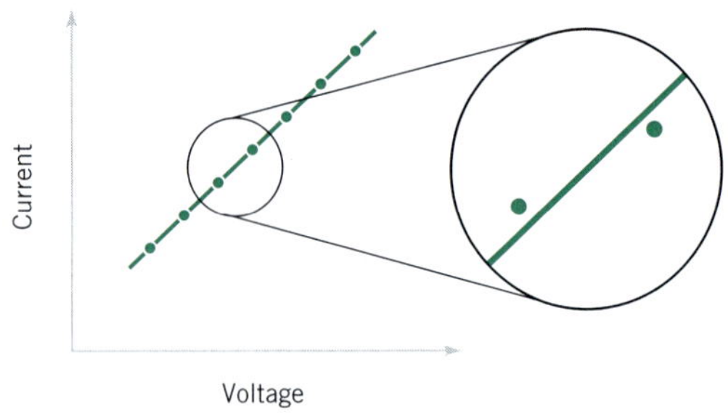

Figure 2-3 A closer examination of the system identifies deviations from the model.

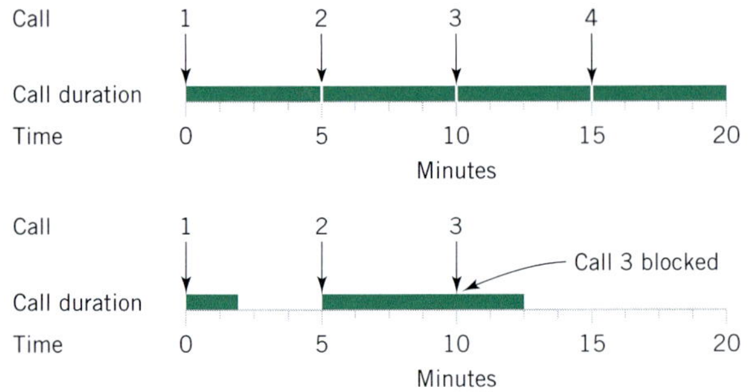

Figure 2-4 Variation causes disruptions in the system.

For the example of measuring current in a copper wire, our model for the system might simply be Ohm's law. Because of uncontrollable inputs, variations in measurements of current are expected. Ohm's law might be a suitable approximation. However, if the variations are large relative to the intended use of the device under study, we might need to extend our model to include the variation. See Fig. 2-3.

As another example, in the design of a communication system, such as a computer or voice communication network, the information capacity available to service individuals using the network is an important design consideration. For voice communication, sufficient external lines need to be purchased from the phone company to meet the requirements of a business. Assuming each line can carry only a single conversation, how many lines should be purchased? If too few lines are purchased, calls can be delayed or lost. The purchase of too many lines increases costs. Increasingly, design and product development is required to meet customer requirements *at a competitive cost.*

In the design of the voice communication system, a model is needed for the number of calls and the duration of calls. Even knowing that on average, calls occur every five minutes and that they last five minutes is not sufficient. If calls arrived precisely at five-minute intervals and lasted for precisely five minutes, one phone line would be sufficient. However, the slightest variation in call number or duration would result in some calls being blocked by others. See Fig. 2-4. A system designed without considering variation will be woefully inadequate for practical use. Our model for the number and duration of calls needs to include variation as an integral component. An analysis of models including variation is important for the design of the phone system.

2-1.2 Sample Spaces

To model and analyze a random experiment, we must understand the set of possible **outcomes** from the experiment. In this introduction to probability, we make use of the basic concepts of sets and operations on sets. It is assumed that the reader is familiar with these topics.

Sample Space

> The set of all possible outcomes of a random experiment is called the **sample space** of the experiment. The sample space is denoted as S.

A sample space is often defined based on the objectives of the analysis.

EXAMPLE 2-1
Molded Plastic Part

Consider an experiment in which you select a molded plastic part, such as a connector, and measure its thickness. The possible values for thickness depend on the resolution of the measuring instrument, and they also depend on upper and lower bounds for thickness. However, it might be convenient to define the sample space as simply the positive real line

$$S = R^+ = \{x \mid x > 0\}$$

because a negative value for thickness cannot occur.

If it is known that all connectors will be between 10 and 11 millimeters thick, the sample space could be

$$S = \{x \mid 10 < x < 11\}$$

If the objective of the analysis is to consider only whether a particular part is low, medium, or high for thickness, the sample space might be taken to be the set of three outcomes:

$$S = \{low, medium, high\}$$

If the objective of the analysis is to consider only whether or not a particular part conforms to the manufacturing specifications, the sample space might be simplified to the set of two outcomes

$$S = \{yes, no\}$$

that indicate whether or not the part conforms.

It is useful to distinguish between two types of sample spaces.

Discrete and Continuous Sample Spaces

A sample space is **discrete** if it consists of a finite or countable infinite set of outcomes. A sample space is **continuous** if it contains an interval (either finite or infinite) of real numbers.

In Example 2-1, the choice $S = R^+$ is an example of a continuous sample space, whereas $S = \{yes, no\}$ is a discrete sample space. As mentioned, the best choice of a sample space depends on the objectives of the study. As specific questions occur later in the book, appropriate sample spaces are discussed.

EXAMPLE 2-2
Manufacturing Specifications

If two connectors are selected and measured, the extension of the positive real line R is to take the sample space to be the positive quadrant of the plane:

$$S = R^+ \times R^+$$

If the objective of the analysis is to consider only whether or not the parts conform to the manufacturing specifications, either part may or may not conform. We abbreviate *yes* and *no* as y and n. If the ordered pair yn indicates that the first connector conforms and the second does not, the sample space can be represented by the four outcomes:

$$S = \{yy, yn, ny, nn\}$$

If we are only interested in the number of conforming parts in the sample, we might summarize the sample space as

$$S = \{0, 1, 2\}$$

As another example, consider an experiment in which the thickness is measured until a connector fails to meet the specifications. The sample space can be represented as

$$S = \{n, yn, yyn, yyyn, yyyyn, \text{and so forth}\}$$

and this is an example of a discrete sample space that is countably infinite.

In random experiments in which items are selected from a batch, we will indicate whether or not a selected item is replaced before the next one is selected. For example, if the batch consists of three items $\{a, b, c\}$ and our experiment is to select two items **without replacement,** the sample space can be represented as

$$S_{\text{without}} = \{ab, ac, ba, bc, ca, cb\}$$

This description of the sample space maintains the order of the items selected so that the outcome *ab* and *ba* are separate elements in the sample space. A sample space with less detail only describes the two items selected $\{\{a, b\}, \{a, c\}, \{b, c\}\}$. This sample space is the possible subsets of two items. Sometimes the ordered outcomes are needed, but in other cases the simpler, unordered sample space is sufficient.

If items are replaced before the next one is selected, the sampling is referred to as **with replacement.** Then the possible ordered outcomes are

$$S_{\text{with}} = \{aa, ab, ac, ba, bb, bc, ca, cb, cc\}$$

The unordered description of the sample space is $\{\{a, a\}, \{a, b\}, \{a, c\}, \{b, b\}, \{b, c\}, \{c, c\}\}$. Sampling without replacement is more common for industrial applications.

Sometimes it is not necessary to specify the exact item selected, but only a property of the item. For example, suppose that there are 5 defective parts and 95 good parts in a batch. To study the quality of the batch, two are selected without replacement. Let *g* denote a good part and *d* denote a defective part. It might be sufficient to describe the sample space (ordered) in terms of quality of each part selected as

$$S = \{gg, gd, dg, dd\}$$

One must be cautious with this description of the sample space because there are many more pairs of items in which both are good than pairs in which both are defective. These differences must be accounted for when probabilities are computed later in this chapter. Still, this summary of the sample space will be convenient when conditional probabilities are used later in this chapter. Also, if there were only one defective part in the batch, there would be fewer possible outcomes

$$S = \{gg, gd, dg\}$$

because *dd* would be impossible. For sampling questions, sometimes the most important part of the solution is an appropriate description of the sample space.

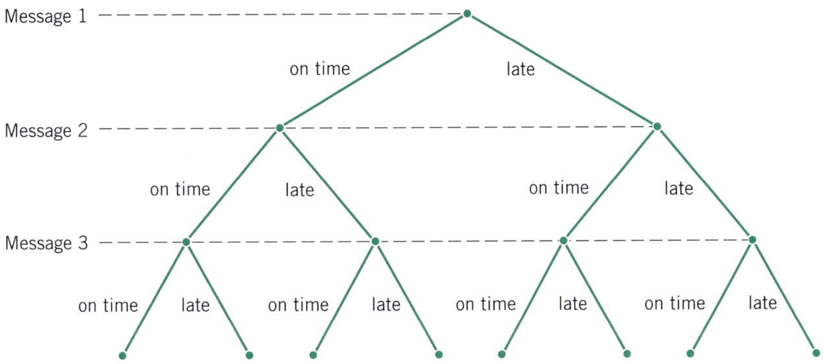

Figure 2-5 Tree diagram for three messages.

Sample spaces can also be described graphically with **tree diagrams.** When a sample space can be constructed in several steps or stages, we can represent each of the n_1 ways of completing the first step as a branch of a tree. Each of the ways of completing the second step can be represented as n_2 branches starting from the ends of the original branches, and so forth.

EXAMPLE 2-3

Each message in a digital communication system is classified as to whether it is received within the time specified by the system design. If three messages are classified, use a tree diagram to represent the sample space of possible outcomes.

Each message can either be received on time or late. The possible results for three messages can be displayed by eight branches in the tree diagram shown in Fig. 2-5.

EXAMPLE 2-4

An automobile manufacturer provides vehicles equipped with selected options. Each vehicle is ordered

With or without an automatic transmission

With or without air-conditioning

With one of three choices of a stereo system

With one of four exterior colors

If the sample space consists of the set of all possible vehicle types, what is the number of outcomes in the sample space? The sample space contains 48 outcomes. The tree diagram for the different types of vehicles is displayed in Fig. 2-6.

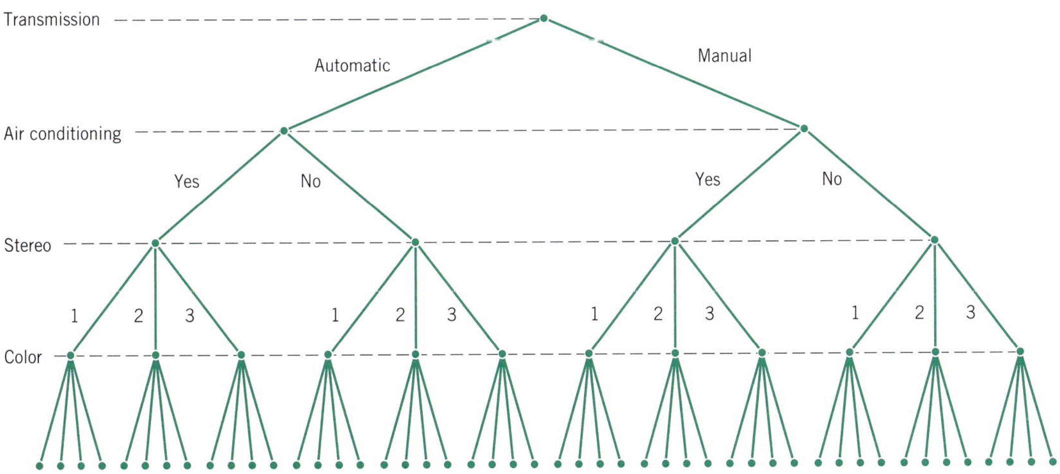

Figure 2-6 Tree diagram for different types of vehicles with 48 outcomes in the sample space.

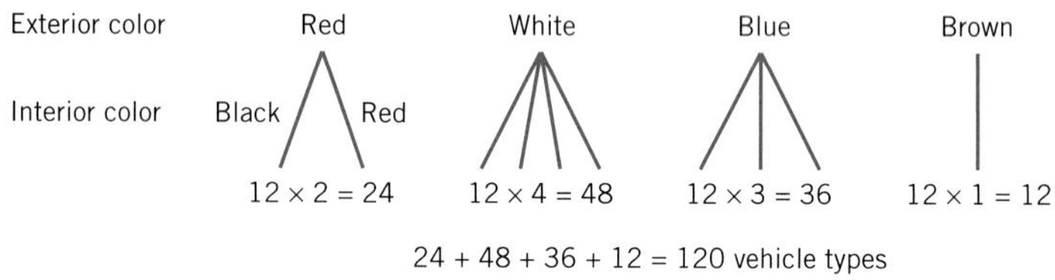

Figure 2-7 Tree diagram for different types of vehicles with interior colors.

EXAMPLE 2-5
Automobile Configurations

Consider an extension of the automobile manufacturer illustration in the previous example in which another vehicle option is the interior color. There are four choices of interior color: red, black, blue, or brown. However,

With a red exterior, only a black or red interior can be chosen.

With a white exterior, any interior color can be chosen.

With a blue exterior, only a black, red, or blue interior can be chosen.

With a brown exterior, only a brown interior can be chosen.

In Fig. 2-6, there are 12 vehicle types with each exterior color, but the number of interior color choices depends on the exterior color. As shown in Fig. 2-7, the tree diagram can be extended to show that there are 120 different vehicle types in the sample space.

2-1.3 Events

Often we are interested in a collection of related outcomes from a random experiment. Related outcomes can be described by subsets of the sample space and set operations can also be applied.

Event

> An **event** is a subset of the sample space of a random experiment.

We can also be interested in describing new events from combinations of existing events. Because events are subsets, we can use basic set operations such as unions, intersections, and complements to form other events of interest. Some of the basic set operations are summarized below in terms of events:

- The **union** of two events is the event that consists of all outcomes that are contained in either of the two events. We denote the union as $E_1 \cup E_2$.
- The **intersection** of two events is the event that consists of all outcomes that are contained in both of the two events. We denote the intersection as $E_1 \cap E_2$.
- The **complement** of an event in a sample space is the set of outcomes in the sample space that are not in the event. We denote the complement of the event E as E'. The notation E^C is also used in other literature to denote the complement.

EXAMPLE 2-6

Consider the sample space $S = \{yy, yn, ny, nn\}$ in Example 2-2. Suppose that the subset of outcomes for which at least one part conforms is denoted as E_1. Then,

$$E_1 = \{yy, yn, ny\}$$

The event in which both parts do not conform, denoted as E_2, contains only the single outcome, $E_2 = \{nn\}$. Other examples of events are $E_3 = \varnothing$, the null set, and $E_4 = S$, the sample space. If $E_5 = \{yn, ny, nn\}$,

$$E_1 \cup E_5 = S \qquad E_1 \cap E_5 = \{yn, ny\} \qquad E_1' = \{nn\}$$

EXAMPLE 2-7

Measurements of the thickness of a plastic connector might be modeled with the sample space $S = R^+$, the set of positive real numbers. Let

$$E_1 = \{x \mid 10 \le x < 12\} \qquad \text{and} \qquad E_2 = \{x \mid 11 < x < 15\}$$

Then,

$$E_1 \cup E_2 = \{x \mid 10 \le x < 15\} \qquad \text{and} \qquad E_1 \cap E_2 = \{x \mid 11 < x < 12\}$$

Also,

$$E_1' = \{x \mid x < 10 \text{ or } 12 \le x\} \qquad \text{and} \qquad E_1' \cap E_2 = \{x \mid 12 \le x < 15\}$$

EXAMPLE 2-8
Polycarbonate Plastic

Samples of polycarbonate plastic are analyzed for scratch and shock resistance. The results from 50 samples are summarized as follows:

		shock resistance	
		high	low
scratch resistance	high	40	4
	low	1	5

Let A denote the event that a sample has high shock resistance, and let B denote the event that a sample has high scratch resistance. Determine the number of samples in $A \cap B$, A', and $A \cup B$.

The event $A \cap B$ consists of the 40 samples for which scratch and shock resistances are high. The event A' consists of the 9 samples in which the shock resistance is low. The event $A \cup B$ consists of the 45 samples in which the shock resistance, scratch resistance, or both are high.

Diagrams are often used to portray relationships between sets, and these diagrams are also used to describe relationships between events. We can use **Venn diagrams** to represent a sample space and events in a sample space. For example, in Fig. 2-8(a) the sample space of the random experiment is represented as the points in the rectangle S. The events A and B are the subsets of points in the indicated regions. Figs. 2-8(b) to 2-8(d) illustrate additional joint events. Figure 2-9 illustrates two events with no common outcomes.

Mutually Exclusive Events

Two events, denoted as E_1 and E_2, such that

$$E_1 \cap E_2 = \varnothing$$

are said to be **mutually exclusive.**

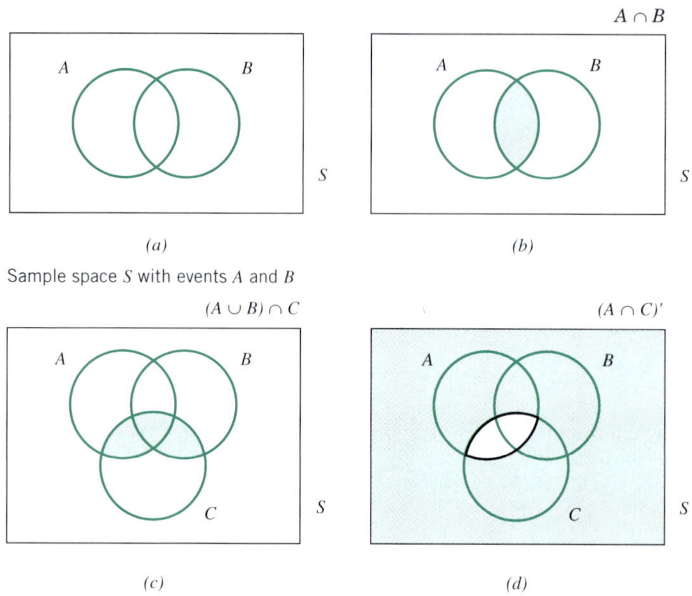

Figure 2-8 Venn diagrams.

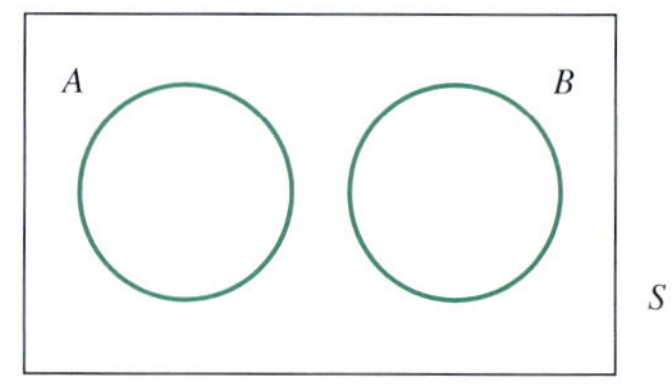

Figure 2-9 Mutually exclusive events.

Additional results involving events are summarized below. The definition of the complement of an event implies that

$$(E')' = E$$

The distributive law for set operations implies that

$$(A \cup B) \cap C = (A \cap C) \cup (B \cap C), \qquad \text{and} \qquad (A \cap B) \cup C = (A \cup C) \cap (B \cup C)$$

DeMorgan's laws imply that

$$(A \cup B)' = A' \cap B' \qquad \text{and} \qquad (A \cap B)' = A' \cup B'$$

Also, remember that

$$A \cap B = B \cap A \qquad \text{and} \qquad A \cup B = B \cup A$$

2-1.4 Counting Techniques

In many of the examples in Chapter 2, it is easy to determine the number of outcomes in each event. In more complicated examples, determining the outcomes that comprise the sample space (or an event) becomes more difficult. Instead, counts of the numbers of outcomes in the sample space and various events are used to analyze the random experiments. These methods are referred to as **counting techniques.** Some simple rules can be used to simplify the calculations.

In Example 2-4, an automobile manufacturer provides vehicles equipped with selected options. Each vehicle is ordered

With or without an automatic transmission

With or without air conditioning

With one of three choices of a stereo system

With one of four exterior colors

The tree diagram in Fig. 2-6 describes the sample space of all possible vehicle types. The size of the sample space equals the number of branches in the last level of the tree and this quantity equals $2 \times 2 \times 3 \times 4 = 48$. This leads to the following useful result.

Multiplication Rule (for counting techniques)

If an operation can be described as a sequence of k steps, and

if the number of ways of completing step 1 is n_1, and

if the number of ways of completing step 2 is n_2 for each way of completing step 1, and

if the number of ways of completing step 3 is n_3 for each way of completing step 2, and so forth,

the total number of ways of completing the operation is

$$n_1 \times n_2 \times \cdots \times n_k$$

EXAMPLE 2-9

In the design of a casing for a gear housing, we can use four different types of fasteners, three different bolt lengths, and three different bolt locations. From the multiplication rule, $4 \times 3 \times 3 = 36$ different designs are possible.

Permutations

Another useful calculation is the number of ordered sequences of the elements of a set. Consider a set of elements, such as $S = \{a, b, c\}$. A **permutation** of the elements is an ordered sequence of the elements. For example, *abc*, *acb*, *bac*, *bca*, *cab*, and *cba* are all of the permutations of the elements of S.

The number of **permutations** of n different elements is $n!$ where

$$n! = n \times (n-1) \times (n-2) \times \cdots \times 2 \times 1 \qquad (2\text{-}1)$$

This result follows from the multiplication rule. A permutation can be constructed by selecting the element to be placed in the first position of the sequence from the n elements, then selecting the element for the second position from the $n - 1$ remaining elements, then selecting the element for the third position from the remaining $n - 2$ elements, and so forth. Permutations such as these are sometimes referred to as linear permutations.

In some situations, we are interested in the number of arrangements of only some of the elements of a set. The following result also follows from the multiplication rule.

Permutations of Subsets

The number of permutations of subsets of r elements selected from a set of n different elements is

$$P_r^n = n \times (n-1) \times (n-2) \times \cdots \times (n-r+1) = \frac{n!}{(n-r)!} \tag{2-2}$$

EXAMPLE 2-10 Printed Circuit Board

A printed circuit board has eight different locations in which a component can be placed. If four different components are to be placed on the board, how many different designs are possible?

Each design consists of selecting a location from the eight locations for the first component, a location from the remaining seven for the second component, a location from the remaining six for the third component, and a location from the remaining five for the fourth component. Therefore,

$$P_4^8 = 8 \times 7 \times 6 \times 5 = \frac{8!}{4!} = 1680 \text{ different designs are possible.}$$

Sometimes we are interested in counting the number of ordered sequences for objects that are not all different. The following result is a useful, general calculation.

Permutations of Similar Objects

The number of permutations of $n = n_1 + n_2 + \cdots + n_r$ objects of which n_1 are of one type, n_2 are of a second type, ..., and n_r are of an rth type is

$$\frac{n!}{n_1!\, n_2!\, n_3! \ldots n_r!} \tag{2-3}$$

EXAMPLE 2-11 Machine Shop Schedule

Consider a machining operation in which a piece of sheet metal needs two identical diameter holes drilled and two identical size notches cut. We denote a drilling operation as d and a notching operation as n. In determining a schedule for a machine shop, we might be interested in the number of different possible sequences of the four operations. The number of possible sequences for two drilling operations and two notching operations is

$$\frac{4!}{2!\,2!} = 6$$

The six sequences are easily summarized: *ddnn*, *dndn*, *dnnd*, *nddn*, *ndnd*, *nndd*.

EXAMPLE 2-12 Bar Codes

A part is labeled by printing with four thick lines, three medium lines, and two thin lines. If each ordering of the nine lines represents a different label, how many different labels can be generated by using this scheme?

From Equation 2-3, the number of possible part labels is

$$\frac{9!}{4!\,3!\,2!} = 1260$$

Combinations

Another counting problem of interest is the number of subsets of r elements that can be selected from a set of n elements. Here, order is not important. These are called **combinations.** Every subset of r elements can be indicated by listing the elements in the set and marking each element with a "*" if it is to be included in the subset. Therefore, each permutation of r *'s and $n - r$ blanks indicate a different subset and the number of these are obtained from Equation 2-3.

For example, if the set is $S = \{a, b, c, d\}$ the subset $\{a, c\}$ can be indicated as

$$\begin{matrix} a & b & c & d \\ * & & * & \end{matrix}$$

Combinations

The number of combinations, subsets of size r that can be selected from a set of n elements, is denoted as $\binom{n}{r}$ or C_r^n and

$$\binom{n}{r} = \frac{n!}{r!(n-r)!} \tag{2-4}$$

EXAMPLE 2-13

A printed circuit board has eight different locations in which a component can be placed. If five identical components are to be placed on the board, how many different designs are possible?

Each design is a subset of the eight locations that are to contain the components. From Equation 2-4, the number of possible designs is

$$\frac{8!}{5!\,3!} = 56$$

The following example uses the multiplication rule in combination with Equation 2-4 to answer a more difficult, but common, question.

EXAMPLE 2-14
Sampling without Replacement

A bin of 50 manufactured parts contains three defective parts and 47 nondefective parts. A sample of six parts is selected from the 50 parts. Selected parts are not replaced. That is, each part can only be selected once and the sample is a subset of the 50 parts. How many different samples are there of size six that contain exactly two defective parts?

A subset containing exactly two defective parts can be formed by first choosing the two defective parts from the three defective parts. Using Equation 2-4, this step can be completed in

$$\binom{3}{2} = \frac{3!}{2!\,1!} = 3 \text{ different ways}$$

Then, the second step is to select the remaining four parts from the 47 acceptable parts in the bin. The second step can be completed in

$$\binom{47}{4} = \frac{47!}{4!\,43!} = 178{,}365 \text{ different ways}$$

Therefore, from the multiplication rule, the number of subsets of size six that contain exactly two defective items is

$$3 \times 178{,}365 = 535{,}095$$

As an additional computation, the total number of different subsets of size six is found to be

$$\binom{50}{6} = \frac{50!}{6!\,44!} = 15{,}890{,}700$$

When probability is discussed in this chapter, the probability of an event is determined as the ratio of the number of outcomes in the event to the number of outcomes in the sample space (for equally likely outcomes). Therefore, the probability that a sample contains exactly two defective parts is

$$\frac{535{,}095}{15{,}890{,}700} = 0.034$$

Note that this example illustrates a common distribution studied in Chapter 3 (hypergeometric distribution).

EXERCISES FOR SECTION 2-1

Provide a reasonable description of the sample space for each of the random experiments in Exercises 2-1 to 2-18. There can be more than one acceptable interpretation of each experiment. Describe any assumptions you make.

2-1. Each of three machined parts is classified as either above or below the target specification for the part.

2-2. Each of four transmitted bits is classified as either in error or not in error.

2-3. In the final inspection of electronic power supplies, three types of nonconformities might occur: functional, minor, or cosmetic. Power supplies that are defective are further classified as to type of nonconformity.

2-4. In the manufacturing of digital recording tape, electronic testing is used to record the number of bits in error in a 350-foot reel.

2-5. In the manufacturing of digital recording tape, each of 24 tracks is classified as containing or not containing one or more bits in error.

2-6. An ammeter that displays three digits is used to measure current in milliamperes.

2-7. A scale that displays two decimal places is used to measure material feeds in a chemical plant in tons.

2-8. The following two questions appear on an employee survey questionnaire. Each answer is chosen from the five-point scale 1 (never), 2, 3, 4, 5 (always).

Is the corporation willing to listen to and fairly evaluate new ideas?

How often are my coworkers important in my overall job performance?

2-9. The concentration of ozone to the nearest part per billion.

2-10. The time until a service transaction is requested of a computer to the nearest millisecond.

2-11. The pH reading of a water sample to the nearest tenth of a unit.

2-12. The voids in a ferrite slab are classified as small, medium, or large. The number of voids in each category is measured by an optical inspection of a sample.

2-13. The time of a chemical reaction is recorded to the nearest millisecond.

2-14. An order for an automobile can specify either an automatic or a standard transmission, either with or without air-conditioning, and any one of the four colors red, blue, black or white. Describe the set of possible orders for this experiment.

2-15. A sampled injection-molded part could have been produced in either one of two presses and in any one of the eight cavities in each press.

2-16. An order for a computer system can specify memory of 4, 8, or 12 gigabytes, and disk storage of 200, 300, or 400 gigabytes. Describe the set of possible orders.

2-17. Calls are repeatedly placed to a busy phone line until a connect is achieved.

2-18. In a magnetic storage device, three attempts are made to read data before an error recovery procedure that repositions the magnetic head is used. The error recovery procedure attempts three repositionings before an "abort" message is sent to the operator. Let

s denote the success of a read operation

f denote the failure of a read operation

F denote the failure of an error recovery procedure

S denote the success of an error recovery procedure

A denote an abort message sent to the operator.

Describe the sample space of this experiment with a tree diagram.

2-19. Three events are shown on the Venn diagram in the following figure:

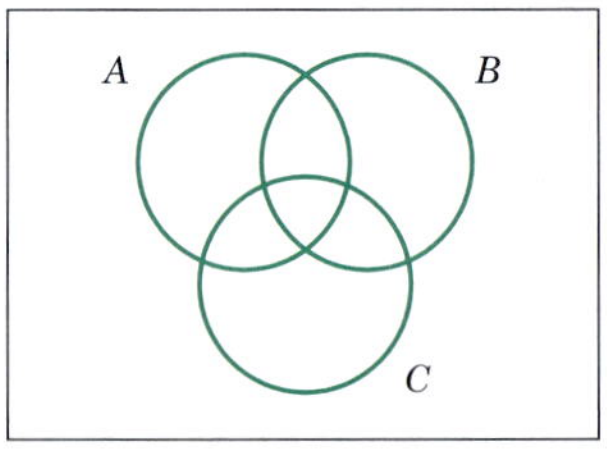

Reproduce the figure and shade the region that corresponds to each of the following events.

(a) A' (b) $A \cap B$ (c) $(A \cap B) \cup C$
(d) $(B \cup C)'$ (e) $(A \cap B)' \cup C$

2-20. Three events are shown on the Venn diagram in the following figure:

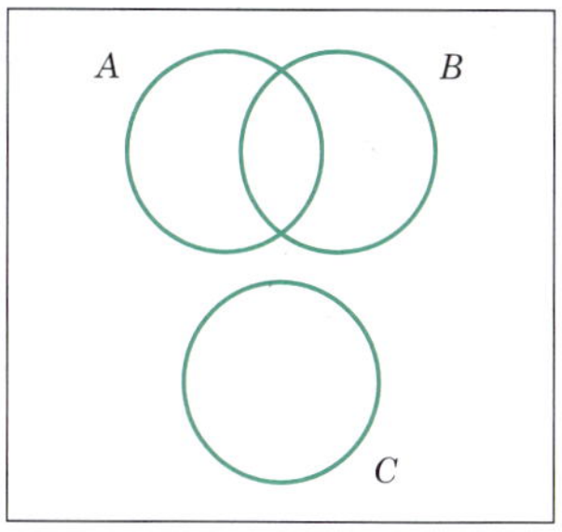

Reproduce the figure and shade the region that corresponds to each of the following events.

(a) A' (b) $(A \cap B) \cup (A \cap B')$
(c) $(A \cap B) \cup C$ (d) $(B \cup C)'$
(e) $(A \cap B)' \cup C$

2-21. A digital scale is used that provides weights to the nearest gram.

(a) What is the sample space for this experiment?

Let A denote the event that a weight exceeds 11 grams, let B denote the event that a weight is less than or equal to 15 grams, and let C denote the event that a weight is greater than or equal to 8 grams and less than 12 grams.

Describe the following events.

(b) $A \cup B$ (c) $A \cap B$
(d) A' (e) $A \cup B \cup C$
(f) $(A \cup C)'$ (g) $A \cap B \cap C$
(h) $B' \cap C$ (i) $A \cup (B \cap C)$

2-22. In an injection-molding operation, the length and width, denoted as X and Y, respectively, of each molded part are evaluated. Let

A denote the event of $48 < X < 52$ centimeters

B denote the event of $9 < Y < 11$ centimeters

C denote the event that a critical length meets customer requirements.

Construct a Venn diagram that includes these events. Shade the areas that represent the following:

(a) A (b) $A \cap B$
(c) $A' \cup B$ (d) $A \cup B$
(e) If these events were mutually exclusive, how successful would this production operation be? Would the process produce parts with $X = 50$ centimeters and $Y = 10$ centimeters?

2-23. Four bits are transmitted over a digital communications channel. Each bit is either distorted or received without distortion. Let A_i denote the event that the ith bit is distorted, $i = 1, \ldots, 4$.

(a) Describe the sample space for this experiment.
(b) Are the A_i's mutually exclusive?

Describe the outcomes in each of the following events:

(c) A_1 (d) A_1'
(e) $A_1 \cap A_2 \cap A_3 \cap A_4$ (f) $(A_1 \cap A_2) \cup (A_3 \cap A_4)$

2-24. In light-dependent photosynthesis, light quality refers to the wavelengths of light that are important. The wavelength of a sample of photosynthetically active radiations (PAR) is measured to the nearest nanometer. The red range is 675–700 nm and the blue range is 450–500 nm. Let A denote the event that PAR occurs in the red range and let B denote the event that PAR occurs in the blue range. Describe the sample space and indicate each of the following events:

(a) A (b) B (c) $A \cap B$ (d) $A \cup B$

2-25. In control replication, cells are replicated over a period of two days. Not until mitosis is completed, can freshly-synthesized DNA be replicated again. Two control mechanisms have been identified—one positive and one negative. Suppose that a replication is observed in three cells. Let A denote the event that all cells are identified as positive and let B denote the event that all cells are negative. Describe the sample space graphically and display each of the following events:

(a) A (b) B
(c) $A \cap B$ (d) $A \cup B$

2-26. Disks of polycarbonate plastic from a supplier are analyzed for scratch and shock resistance. The results from 100 disks are summarized below:

		shock resistance	
		high	low
scratch	high	70	9
resistance	low	16	5

Let A denote the event that a disk has high shock resistance, and let B denote the event that a disk has high scratch resistance. Determine the number of disks in $A \cap B$, A', and $A \cup B$.

2-27. Samples of a cast aluminum part are classified on the basis of surface finish (in microinches) and edge finish. The results of 100 parts are summarized as follows:

		edge finish	
		excellent	good
surface	excellent	80	2
finish	good	10	8

(a) Let A denote the event that a sample has excellent surface finish, and let B denote the event that a sample has excellent edge finish. Determine the number of samples in $A' \cap B$, B', and $A \cup B$.
(b) Assume that each of two samples is to be classified on the basis of surface finish, either excellent or good, edge finish, either excellent or good. Use a tree diagram to represent the possible outcomes of this experiment.

2-28. Samples of emissions from three suppliers are classified for conformance to air-quality specifications. The results from 100 samples are summarized as follows:

		conforms	
		yes	no
supplier	1	22	8
	2	25	5
	3	30	10

Let A denote the event that a sample is from supplier 1, and let B denote the event that a sample conforms to specifications. Determine the number of samples in $A' \cap B$, B', and $A \cup B$.

2-29. The rise time of a reactor is measured in minutes (and fractions of minutes). Let the sample space be positive, real numbers. Define the events A and B as follows:

$$A = \{x \mid x < 72.5\}$$

and

$$B = \{x \mid x > 52.5\}$$

Describe each of the following events:
(a) A' (b) B'
(c) $A \cap B$ (d) $A \cup B$

2-30. A sample of two items is selected without replacement from a batch. Describe the (ordered) sample space for each of the following batches:
(a) The batch contains the items $\{a, b, c, d\}$.
(b) The batch contains the items $\{a, b, c, d, e, f, g\}$.
(c) The batch contains 4 defective items and 20 good items.
(d) The batch contains 1 defective item and 20 good items.

2-31. A sample of two printed circuit boards is selected without replacement from a batch. Describe the (ordered) sample space for each of the following batches:
(a) The batch contains 90 boards that are not defective, 8 boards with minor defects, and 2 boards with major defects.
(b) The batch contains 90 boards that are not defective, 8 boards with minor defects, and 1 board with major defects.

2-32. Counts of the Web pages provided by each of two computer servers in a selected hour of the day are recorded. Let A denote the event that at least 10 pages are provided by server 1 and let B denote the event that at least 20 pages are provided by server 2.
(a) Describe the sample space for the numbers of pages for the two servers graphically.

Show each of the following events on the sample space graph:
(b) A (c) B
(d) $A \cap B$ (e) $A \cup B$

2-33. The rise time of a reactor is measured in minutes (and fractions of minutes). Let the sample space for the rise time of each batch be positive, real numbers. Consider the rise times of *two* batches. Let A denote the event that the rise time of batch 1 is less than 72.5 minutes, and let B denote the event that the rise time of batch 2 is greater than 52.5 minutes.

Describe the sample space for the rise time of two batches graphically and show each of the following events on a two-dimensional plot:
(a) A (b) B'
(c) $A \cap B$ (d) $A \cup B$

2-34. A wireless garage door opener has a code determined by the up or down setting of 12 switches. How many outcomes are in the sample space of possible codes?

2-35. An order for a personal digital assistant can specify any one of five memory sizes, any one of three types of displays, any one of four sizes of a hard disk, and can either include or not include a pen tablet. How many different systems can be ordered?

2-36. In a manufacturing operation, a part is produced by machining, polishing, and painting. If there are three machine tools, four polishing tools, and three painting tools, how many different routings (consisting of machining, followed by polishing, and followed by painting) for a part are possible?

2-37. New designs for a wastewater treatment tank have proposed three possible shapes, four possible sizes, three locations for input valves, and four locations for output valves. How many different product designs are possible?

2-38. A manufacturing process consists of 10 operations that can be completed in any order. How many different production sequences are possible?

2-39. A manufacturing operations consists of 10 operations. However, five machining operations must be completed before any of the remaining five assembly operations can begin. Within each set of five, operations can be completed in any order. How many different production sequences are possible?

2-40. In a sheet metal operation, three notches and four bends are required. If the operations can be done in any order, how many different ways of completing the manufacturing are possible?

2-41. A lot of 140 semiconductor chips is inspected by choosing a sample of five chips. Assume 10 of the chips do not conform to customer requirements.
(a) How many different samples are possible?
(b) How many samples of five contain exactly one nonconforming chip?
(c) How many samples of five contain at least one nonconforming chip?

2-42. In the layout of a printed circuit board for an electronic product, there are 12 different locations that can accommodate chips.
(a) If five different types of chips are to be placed on the board, how many different layouts are possible?

(b) If the five chips that are placed on the board are of the same type, how many different layouts are possible?

2-43. In the laboratory analysis of samples from a chemical process, five samples from the process are analyzed daily. In addition, a control sample is analyzed two times each day to check the calibration of the laboratory instruments.

(a) How many different sequences of process and control samples are possible each day? Assume that the five process samples are considered identical and that the two control samples are considered identical.
(b) How many different sequences of process and control samples are possible if we consider the five process samples to be different and the two control samples to be identical.
(c) For the same situation as part (b), how many sequences are possible if the first test of each day must be a control sample?

2-44. In the design of an electromechanical product, seven different components are to be stacked into a cylindrical casing that holds 12 components in a manner that minimizes the impact of shocks. One end of the casing is designated as the bottom and the other end is the top.

(a) How many different designs are possible?
(b) If the seven components are all identical, how many different designs are possible?
(c) If the seven components consist of three of one type of component and four of another type, how many different designs are possible? (more difficult)

2-45. The design of a communication system considered the following questions:

(a) How many three-digit phone prefixes that are used to represent a particular geographic area (such as an area code) can be created from the digits 0 through 9?
(b) As in part (a), how many three-digit phone prefixes are possible that do not start with 0 or 1, but contain 0 or 1 as the middle digit?
(c) How many three-digit phone prefixes are possible in which no digit appears more than once in each prefix?

2-46. A byte is a sequence of eight bits and each bit is either 0 or 1.

(a) How many different bytes are possible?
(b) If the first bit of a byte is a parity check, that is, the first byte is determined from the other seven bits, how many different bytes are possible?

2-47. In a chemical plant, 24 holding tanks are used for final product storage. Four tanks are selected at random and without replacement. Suppose that six of the tanks contain material in which the viscosity exceeds the customer requirements.

(a) What is the probability that exactly one tank in the sample contains high viscosity material?
(b) What is the probability that at least one tank in the sample contains high viscosity material?
(c) In addition to the six tanks with high viscosity levels, four different tanks contain material with high impurities. What is the probability that exactly one tank in the sample contains high viscosity material and exactly one tank in the sample contains material with high impurities?

2-48. Plastic parts produced by an injection-molding operation are checked for conformance to specifications. Each tool contains 12 cavities in which parts are produced, and these parts fall into a conveyor when the press opens. An inspector chooses 3 parts from among the 12 at random. Two cavities are affected by a temperature malfunction that results in parts that do not conform to specifications.

(a) What is the probability that the inspector finds exactly one nonconforming part?
(b) What is the probability that the inspector finds at least one nonconforming part?

2-49. A bin of 50 parts contains five that are defective. A sample of two is selected at random, without replacement.

(a) Determine the probability that both parts in the sample are defective by computing a conditional probability.
(b) Determine the answer to part (a) by using the subset approach that was described in this section.

2-2 INTERPRETATIONS OF PROBABILITY

2-2.1 Introduction

In this chapter, we introduce probability for **discrete sample spaces**—those with only a finite (or countably infinite) set of outcomes. The restriction to these sample spaces enables us to simplify the concepts and the presentation without excessive mathematics.

Probability is used to quantify the likelihood, or chance, that an outcome of a random experiment will occur. "The chance of rain today is 30%" is a statement that quantifies our feeling about the possibility of rain. The likelihood of an outcome is quantified by assigning a number from the interval [0, 1] to the outcome (or a percentage from 0 to 100%). Higher numbers indicate that the outcome is more likely than lower numbers. A 0 indicates an outcome will not occur. A probability of 1 indicates an outcome will occur with certainty.

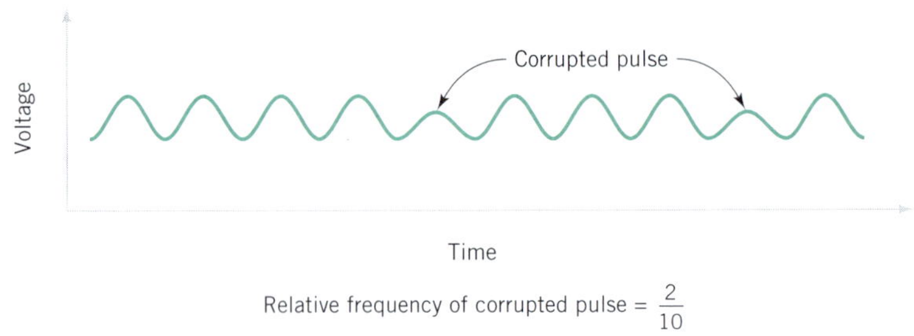

Figure 2-10 Relative frequency of corrupted pulses sent over a communication channel.

The probability of an outcome can be interpreted as our subjective probability, or **degree of belief,** that the outcome will occur. Different individuals will no doubt assign different probabilities to the same outcomes. Another interpretation of probability is based on the conceptual model of repeated replications of the random experiment. The probability of an outcome is interpreted as the limiting value of the proportion of times the outcome occurs in n repetitions of the random experiment as n increases beyond all bounds. For example, if we assign probability 0.2 to the outcome that there is a corrupted pulse in a digital signal, we might interpret this assignment as implying that, if we analyze many pulses, approximately 20% of them will be corrupted. This example provides a **relative frequency** interpretation of probability. The proportion, or relative frequency, of replications of the experiment that result in the outcome is 0.2. Probabilities are chosen so that the sum of the probabilities of all outcomes in an experiment add up to 1. This convention facilitates the relative frequency interpretation of probability. Figure 2-10 illustrates the concept of relative frequency.

Probabilities for a random experiment are often assigned on the basis of a reasonable model of the system under study. One approach is to base probability assignments on the simple concept of equally likely outcomes.

For example, suppose that we will select one laser diode **randomly** from a batch of 100. The sample space is the set of 100 diodes. *Randomly* implies that it is reasonable to assume that each diode in the batch has an equal chance of being selected. Because the sum of the probabilities must equal 1, the probability model for this experiment assigns probability of 0.01 to each of the 100 outcomes. We can interpret the probability by imagining many replications of the experiment. Each time we start with all 100 diodes and select one at random. The probability 0.01 assigned to a particular diode represents the proportion of replicates in which a particular diode is selected. When the model of **equally likely outcomes** is assumed, the probabilities are chosen to be equal.

Equally Likely Outcomes

Whenever a sample space consists of N possible outcomes that are equally likely, the probability of each outcome is $1/N$.

It is frequently necessary to assign probabilities to events that are composed of several outcomes from the sample space. This is straightforward for a discrete sample space.

EXAMPLE 2-15
Laser Diodes

Assume that 30% of the laser diodes in a batch of 100 meet the minimum power requirements of a specific customer. If a laser diode is selected randomly, that is, each laser diode is equally likely to be selected, our intuitive feeling is that the probability of meeting the customer's requirements is 0.30.

Let E denote the subset of 30 diodes that meet the customer's requirements. Because E contains 30 outcomes and each outcome has probability 0.01, we conclude that the probability of E is 0.3. The conclusion matches our intuition. Figure 2-11 illustrates this example.

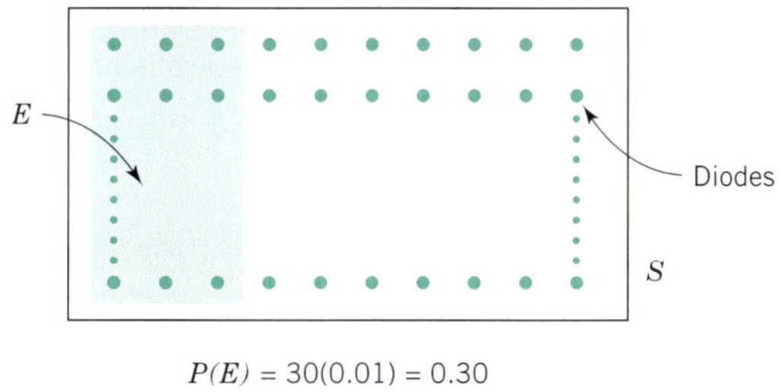

Figure 2-11 Probability of the event E is the sum of the probabilities of the outcomes in E.

For a discrete sample space, the probability of an event can be defined by the reasoning used in the example above.

Probability of an Event

> For a discrete sample space, the *probability of an event E,* denoted as $P(E)$, equals the sum of the probabilities of the outcomes in E.

EXAMPLE 2-16

A random experiment can result in one of the outcomes $\{a, b, c, d\}$ with probabilities 0.1, 0.3, 0.5, and 0.1, respectively. Let A denote the event $\{a, b\}$, B the event $\{b, c, d\}$, and C the event $\{d\}$. Then,

$$P(A) = 0.1 + 0.3 = 0.4$$

$$P(B) = 0.3 + 0.5 + 0.1 = 0.9$$

$$P(C) = 0.1$$

Also, $P(A') = 0.6$, $P(B') = 0.1$, and $P(C') = 0.9$. Furthermore, because $A \cap B = \{b\}$, $P(A \cap B) = 0.3$. Because $A \cup B = \{a, b, c, d\}$, $P(A \cup B) = 0.1 + 0.3 + 0.5 + 0.1 = 1$. Because $A \cap C$ is the null set, $P(A \cap C) = 0$.

EXAMPLE 2-17
Contamination Particles

A visual inspection of a location on wafers from a semiconductor manufacturing process resulted in the following table:

Number of Contamination Particles	Proportion of Wafers
0	0.40
1	0.20
2	0.15
3	0.10
4	0.05
5 or more	0.10

If one wafer is selected randomly from this process and the location is inspected, what is the probability that it contains no particles? If information were available for each wafer, we could define the sample space as the set of all wafers inspected and proceed as in the example with diodes. However, this level of detail is not needed in this case. We can consider the sample space to consist of the six categories that summarize the number of contamination particles on a wafer. Each category has probability equal to the proportion of wafers in the category. The event that there is no contamination particle in the inspected location on the wafer, denoted as E, can be considered to be comprised of the single outcome, namely, $E = \{0\}$. Therefore,

$$P(E) = 0.4$$

What is the probability that a wafer contains three or more particles in the inspected location? Let E denote the event that a wafer contains three or more particles in the inspected location. Then, E consists of the three outcomes {3, 4, 5 or more}. Therefore,

$$P(E) = 0.10 + 0.05 + 0.10 = 0.25$$

Often more than one item is selected from a batch without replacement when production is inspected. In this case, *randomly* selected implies that each possible subset of items is equally likely.

EXAMPLE 2-18 Suppose a batch contains six parts $\{a, b, c, d, e, f\}$ and two parts are selected randomly, without replacement. Suppose that part f is defective, but the others are good. What is the probability that part f appears in the sample?

The sample space consists of all possible (unordered) pairs selected without replacement. From Equation 2-4 or by enumeration there are 15 outcomes. Let E denote the event that part f is in the sample. Then E can be written as $E = \{\{a,f\}, \{b,f\}, \{c,f\}, \{d,f\}, \{e,f\}\}$. Because each outcome is equally likely, $P(E) = 5/15 = 1/3$.

2-2.2 Axioms of Probability

Now that the probability of an event has been defined, we can collect the assumptions that we have made concerning probabilities into a set of **axioms** that the probabilities in any random experiment must satisfy. The axioms ensure that the probabilities assigned in an experiment can be interpreted as relative frequencies and that the assignments are consistent with our intuitive understanding of relationships between relative frequencies. For example, if event A is contained in event B, we should have $P(A) \leq P(B)$. The **axioms do not determine probabilities;** the probabilities are assigned based on our knowledge of the system under study. However, the axioms enable us to easily calculate the probabilities of some events from knowledge of the probabilities of other events.

Axioms of Probability

Probability is a number that is assigned to each member of a collection of events from a random experiment that satisfies the following properties:

If S is the sample space and E is any event in a random experiment,

(1) $P(S) = 1$

(2) $0 \leq P(E) \leq 1$

(3) For two events E_1 and E_2 with $E_1 \cap E_2 = \varnothing$

$$P(E_1 \cup E_2) = P(E_1) + P(E_2)$$

The property that $0 \leq P(E) \leq 1$ is equivalent to the requirement that a relative frequency must be between 0 and 1. The property that $P(S) = 1$ is a consequence of the fact that an outcome from the sample space occurs on every trial of an experiment. Consequently, the relative frequency of S is 1. Property 3 implies that if the events E_1 and E_2 have no outcomes in common, the relative frequency of outcomes in $E_1 \cup E_2$ is the sum of the relative frequencies of the outcomes in E_1 and E_2.

These axioms imply the following results. The derivations are left as exercises at the end of this section. Now,

$$P(\varnothing) = 0$$

and for any event E,

$$P(E') = 1 - P(E)$$

For example, if the probability of the event E is 0.4, our interpretation of relative frequency implies that the probability of E' is 0.6. Furthermore, if the event E_1 is contained in the event E_2,

$$P(E_1) \leq P(E_2)$$

EXERCISES FOR SECTION 2-2

2-50. Each of the possible five outcomes of a random experiment is equally likely. The sample space is $\{a, b, c, d, e\}$. Let A denote the event $\{a, b\}$, and let B denote the event $\{c, d, e\}$. Determine the following:

(a) $P(A)$ (b) $P(B)$
(c) $P(A')$ (d) $P(A \cup B)$
(e) $P(A \cap B)$

2-51. The sample space of a random experiment is $\{a, b, c, d, e\}$ with probabilities 0.1, 0.1, 0.2, 0.4, and 0.2, respectively. Let A denote the event $\{a, b, c\}$, and let B denote the event $\{c, d, e\}$. Determine the following:

(a) $P(A)$ (b) $P(B)$
(c) $P(A')$ (d) $P(A \cup B)$
(e) $P(A \cap B)$

2-52. Orders for a computer are summarized by the optional features that are requested as follows:

	proportion of orders
no optional features	0.3
one optional feature	0.5
more than one optional feature	0.2

(a) What is the probability that an order requests at least one optional feature?
(b) What is the probability that an order does not request more than one optional feature?

2-53. If the last digit of a weight measurement is equally likely to be any of the digits 0 through 9,
(a) What is the probability that the last digit is 0?
(b) What is the probability that the last digit is greater than or equal to 5?

2-54. A part selected for testing is equally likely to have been produced on any one of six cutting tools.
(a) What is the sample space?
(b) What is the probability that the part is from tool 1?
(c) What is the probability that the part is from tool 3 or tool 5?
(d) What is the probability that the part is not from tool 4?

2-55. An injection-molded part is equally likely to be obtained from any one of the eight cavities on a mold.
(a) What is the sample space?
(b) What is the probability a part is from cavity 1 or 2?
(c) What is the probability that a part is neither from cavity 3 nor 4?

2-56. In an acid base titration, a base or acid is gradually added to the other until they have completely neutralized each other. Because acids and bases are usually colorless (as are the water and salt produced in the neutralization reaction), pH is measured to monitor the reaction. Suppose that the equivalence point is reached after approximately 100 mL of a NaOH solution have been added (enough to react with all the acetic acid present) but that replicates are equally likely to indicate from 95 to 104 mL to the nearest mL. Assume that volumes are measured to the nearest mL and describe the sample space.
(a) What is the probability that equivalence is indicated at 100 mL?
(b) What is the probability that equivalence is indicated at less than 100 mL?
(c) What is the probability that equivalence is indicated between 98 and 102 mL (inclusive)?

2-57. In a NiCd battery, a fully charged cell is composed of Nickelic Hydroxide. Nickel is an element that has multiple oxidation states that is usually found in the following states:

nickel charge	proportions found
0	0.17
+2	0.35
+3	0.33
+4	0.15

(a) What is the probability that a cell has at least one of the positive nickel charged options?
(b) What is the probability that a cell is not composed of a positive nickel charge greater than +3?

2-58. A credit card contains 16 digits between 0 and 9. However, only 100 million numbers are valid. If a number is entered randomly, what is the probability that it is a valid number?

2-59. Suppose your vehicle is licensed in a state that issues license plates that consist of three digits (between 0 and 9) followed by three letters (between A and Z). If a license number is selected randomly, what is the probability that yours is the one selected?

2-60. A message can follow different paths through servers on a network. The sender's message can go to one of five servers for the first step, each of them can send to five servers at the second step, each of which can send to four servers at the third step, and then the message goes to the recipient's server.
(a) How many paths are possible?
(b) If all paths are equally likely, what is the probability that a message passes through the first of four servers at the third step?

2-61. Magnesium alkyls are used as homogenous catalysts in the production of linear low density polyethylene (LLDPE), which requires a finer magnesium powder to sustain a reaction. Redox reaction experiments using four different amounts of magnesium powder are performed. Each result may or may not be further reduced in a second step using three different magnesium powder amounts. Each of these results may or may not be further reduced in a third step using three different amounts of magnesium powder.
(a) How many experiments are possible?
(b) If all outcomes are equally likely, what is the probability that the best result is obtained from an experiment that uses all three steps?
(c) Does the result in the previous question change if five or six or seven different amounts are used in the first step? Explain.

2-62. Disks of polycarbonate plastic from a supplier are analyzed for scratch and shock resistance. The results from 100 disks are summarized as follows:

		shock resistance	
		high	low
scratch resistance	high	70	9
	low	16	5

Let A denote the event that a disk has high shock resistance, and let B denote the event that a disk has high scratch resistance. If a disk is selected at random, determine the following probabilities:
(a) $P(A)$ (b) $P(B)$
(c) $P(A')$ (d) $P(A \cap B)$
(e) $P(A \cup B)$ (f) $P(A' \cup B)$

2-63. Samples of emissions from three suppliers are classified for conformance to air-quality specifications. The results from 100 samples are summarized as follows:

		conforms	
		yes	no
supplier	1	22	8
	2	25	5
	3	30	10

Let A denote the event that a sample is from supplier 1, and let B denote the event that a sample conforms to specifications. If a sample is selected at random, determine the following probabilities:
(a) $P(A)$ (b) $P(B)$
(c) $P(A')$ (d) $P(A \cap B)$
(e) $P(A \cup B)$ (f) $P(A' \cup B)$

2-64. An article in the *Journal of Database Management* ["Experimental Study of a Self-Tuning Algorithm for DBMS Buffer Pools," (2005, Vol. 16, pp. 1–20)] provided the workload used in the TPC-C OLTP (Transaction Processing Performance Council's Version C On-Line Transaction Processing) benchmark, which simulates a typical order entry application.

Average Frequencies and Operations in TPC-C

Transaction	Frequency	Selects	Updates	Inserts	Deletes	Non-Unique Selects	Joins
New Order	43	23	11	12	0	0	0
Payment	44	4.2	3	1	0	0.6	0
Order Status	4	11.4	0	0	0	0.6	0
Delivery	5	130	120	0	10	0	0
Stock Level	4	0	0	0	0	0	1

The frequency of each type of transaction (in the second column) can be used as the percentage of each type of transaction. The average number of *selects* operations required for each type of transaction is shown. Let A denote the event of transactions with an average number of *selects* operations of 12 or fewer. Let B denote the event of transactions with an average number of *updates* operations of 12 or fewer. Calculate the following probabilities.

(a) $P(A)$ (b) $P(B)$ (c) $P(A \cap B)$
(d) $P(A \cap B')$ (e) $P(A \cup B)$

2-65. Use the axioms of probability to show the following:
(a) For any event E, $P(E') = 1 - P(E)$.
(b) $P(\emptyset) = 0$
(c) If A is contained in B, then $P(A) \leq P(B)$

2-3 ADDITION RULES

Joint events are generated by applying basic set operations to individual events. Unions of events, such as $A \cup B$; intersections of events, such as $A \cap B$; and complements of events, such as A', are commonly of interest. The probability of a joint event can often be determined from the probabilities of the individual events that comprise it. Basic set operations are also sometimes helpful in determining the probability of a joint event. In this section the focus is on unions of events.

EXAMPLE 2-19
Semiconductor Wafers

Table 2-1 lists the history of 940 wafers in a semiconductor manufacturing process. Suppose one wafer is selected at random. Let H denote the event that the wafer contains high levels of contamination. Then, $P(H) = 358/940$.

Let C denote the event that the wafer is in the center of a sputtering tool. Then, $P(C) = 626/940$. Also, $P(H \cap C)$ is the probability that the wafer is from the center of the sputtering tool and contains high levels of contamination. Therefore,

$$P(H \cap C) = 112/940$$

The event $H \cup C$ is the event that a wafer is from the center of the sputtering tool or contains high levels of contamination (or both). From the table, $P(H \cup C) = 872/940$. An alternative calculation of $P(H \cup C)$ can be obtained as follows. The 112 wafers that comprise the event $H \cap C$ are included once in the calculation of $P(H)$ and again in the calculation of $P(C)$. Therefore, $P(H \cup C)$ can be found to be

$$\begin{aligned} P(H \cup C) &= P(H) + P(C) - P(H \cap C) \\ &= 358/940 + 626/940 - 112/940 = 872/940 \end{aligned}$$

The preceding example illustrates that the probability of A or B is interpreted as $P(A \cup B)$ and that the following general **addition rule** applies.

Table 2-1 Wafers in Semiconductor Manufacturing Classified by Contamination and Location

	Location in Sputtering Tool		
Contamination	Center	Edge	Total
Low	514	68	582
High	112	246	358
Total	626	314	

Probability of a Union

$$P(A \cup B) = P(A) + P(B) - P(A \cap B) \qquad (2\text{-}5)$$

EXAMPLE 2-20
Semiconductor Wafers and Location

The wafers such as those described in Example 2-19 were further classified as either in the "center" or at the "edge" of the sputtering tool that was used in manufacturing, and by the degree of contamination. Table 2-2 shows the proportion of wafers in each category. What is the probability that a wafer was either at the edge or that it contains four or more particles? Let E_1 denote the event that a wafer contains four or more particles, and let E_2 denote the event that a wafer is at the edge.

The requested probability is $P(E_1 \cup E_2)$. Now, $P(E_1) = 0.15$ and $P(E_2) = 0.28$. Also, from the table, $P(E_1 \cap E_2) = 0.04$. Therefore, using Equation 2-1, we find that

$$P(E_1 \cup E_2) = 0.15 + 0.28 - 0.04 = 0.39$$

What is the probability that a wafer contains less than two particles or that it is both at the edge and contains more than four particles? Let E_1 denote the event that a wafer contains less than two particles, and let E_2 denote the event that a wafer is both from the edge and contains more than four particles. The requested probability is $P(E_1 \cup E_2)$. Now, $P(E_1) = 0.60$ and $P(E_2) = 0.03$. Also, E_1 and E_2 are mutually exclusive. Consequently, there are no wafers in the intersection and $P(E_1 \cap E_2) = 0$. Therefore,

$$P(E_1 \cup E_2) = 0.60 + 0.03 = 0.63$$

Recall that two events A and B are said to be mutually exclusive if $A \cap B = \varnothing$. Then, $P(A \cap B) = 0$, and the general result for the probability of $A \cup B$ simplifies to the third axiom of probability.

If A and B are mutually exclusive events,

$$P(A \cup B) = P(A) + P(B) \qquad (2\text{-}6)$$

Table 2-2 Wafers Classified by Contamination and Location

Number of Contamination Particles	Center	Edge	Totals
0	0.30	0.10	0.40
1	0.15	0.05	0.20
2	0.10	0.05	0.15
3	0.06	0.04	0.10
4	0.04	0.01	0.05
5 or more	0.07	0.03	0.10
Totals	0.72	0.28	1.00

Three or More Events

More complicated probabilities, such as $P(A \cup B \cup C)$, can be determined by repeated use of Equation 2-5 and by using some basic set operations. For example,

$$P(A \cup B \cup C) = P[(A \cup B) \cup C] = P(A \cup B) + P(C) - P[(A \cup B) \cap C]$$

Upon expanding $P(A \cup B)$ by Equation 2-5 and using the distributed rule for set operations to simplify $P[(A \cup B) \cap C]$, we obtain

$$\begin{aligned} P(A \cup B \cup C) &= P(A) + P(B) - P(A \cap B) + P(C) - P[(A \cap C) \cup (B \cap C)] \\ &= P(A) + P(B) - P(A \cap B) + P(C) \\ &\qquad - [P(A \cap C) + P(B \cap C) - P(A \cap B \cap C)] \\ &= P(A) + P(B) + P(C) - P(A \cap B) - P(A \cap C) \\ &\qquad - P(B \cap C) + P(A \cap B \cap C) \end{aligned}$$

We have developed a formula for the probability of the union of three events. Formulas can be developed for the probability of the union of any number of events, although the formulas become very complex. As a summary, for the case of three events

$$\begin{aligned} P(A \cup B \cup C) = P(A) + P(B) + P(C) - P(A \cap B) \\ - P(A \cap C) - P(B \cap C) + P(A \cap B \cap C) \end{aligned} \quad (2\text{-}7)$$

Results for three or more events simplify considerably if the events are mutually exclusive. In general, a collection of events, $E_1, E_2, \ldots, E_k$, is said to be mutually exclusive if there is no overlap among any of them.

The Venn diagram for several mutually exclusive events is shown in Fig. 2-12. By generalizing the reasoning for the union of two events, the following result can be obtained:

Mutually Exclusive Events

A collection of events, $E_1, E_2, \ldots, E_k$, is said to be **mutually exclusive** if for all pairs,

$$E_i \cap E_j = \varnothing$$

For a collection of mutually exclusive events,

$$P(E_1 \cup E_2 \cup \ldots \cup E_k) = P(E_1) + P(E_2) + \ldots P(E_k) \quad (2\text{-}8)$$

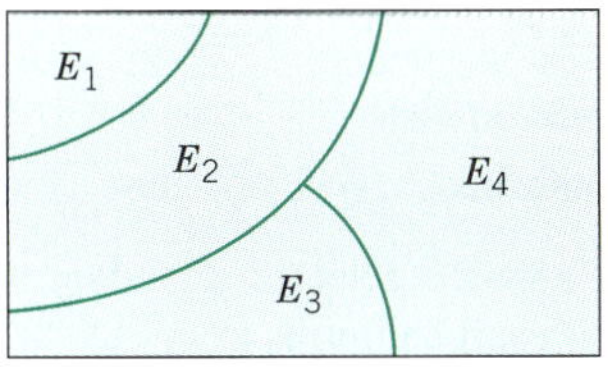

Figure 2-12 Venn diagram of four mutually exclusive events.

EXAMPLE 2-21
pH

A simple example of mutually exclusive events will be used quite frequently. Let X denote the pH of a sample. Consider the event that X is greater than 6.5 but less than or equal to 7.8. This probability is the sum of any collection of mutually exclusive events with union equal to the same range for X. One example is

$$P(6.5 < X \leq 7.8) = P(6.5 < X \leq 7.0) + P(7.0 < X \leq 7.5) + P(7.5 < X \leq 7.8)$$

Another example is

$$P(6.5 < X \leq 7.8) = P(6.5 < X \leq 6.6) + P(6.6 < X \leq 7.1) + P(7.1 < X \leq 7.4) + P(7.4 < X \leq 7.8)$$

The best choice depends on the particular probabilities available.

EXERCISES FOR SECTION 2-3

2-66. If $P(A) = 0.3$, $P(B) = 0.2$, and $P(A \cap B) = 0.1$, determine the following probabilities:

(a) $P(A')$ (b) $P(A \cup B)$ (c) $P(A' \cap B)$
(d) $P(A \cap B')$ (e) $P[(A \cup B)']$ (f) $P(A' \cup B)$

2-67. If A, B, and C are mutually exclusive events with $P(A) = 0.2$, $P(B) = 0.3$, and $P(C) = 0.4$, determine the following probabilities:

(a) $P(A \cup B \cup C)$ (b) $P(A \cap B \cap C)$
(c) $P(A \cap B)$ (d) $P[(A \cup B) \cap C]$
(e) $P(A' \cap B' \cap C')$

2-68. In the article "ACL reconstruction using bone-patellar tendon-bone press-fit fixation: 10-year clinical results" in *Knee Surgery, Sports Traumatology, Arthroscopy* (2005, Vol. 13, pp. 248–255) the following causes for knee injuries were considered:

Activity	Percentage of Knee Injuries
Contact sport	46%
Noncontact sport	44%
Activity of daily living	9%
Riding motorcycle	1%

(a) What is the probability a knee injury resulted from a sport (contact or noncontact)?
(b) What is the probability a knee injury resulted from an activity other than a sport?

2-69. Disks of polycarbonate plastic from a supplier are analyzed for scratch and shock resistance. The results from 100 disks are summarized as follows:

		shock resistance	
		high	low
scratch	high	70	9
resistance	low	16	5

(a) If a disk is selected at random, what is the probability that its scratch resistance is high and its shock resistance is high?
(b) If a disk is selected at random, what is the probability that its scratch resistance is high or its shock resistance is high?
(c) Consider the event that a disk has high scratch resistance and the event that a disk has high shock resistance. Are these two events mutually exclusive?

2-70. Strands of copper wire from a manufacturer are analyzed for strength and conductivity. The results from 100 strands are as follows:

	strength	
	high	low
high conductivity	74	8
low conductivity	15	3

(a) If a strand is randomly selected, what is the probability that its conductivity is high and its strength is high?
(b) If a strand is randomly selected, what is the probability that its conductivity is low or the strength is low?
(c) Consider the event that a strand has low conductivity and the event that the strand has a low strength. Are these two events mutually exclusive?

2-71. The analysis of shafts for a compressor is summarized by conformance to specifications.

		roundness conforms	
		yes	no
surface finish	yes	345	5
conforms	no	12	8

(a) If a shaft is selected at random, what is the probability that the shaft conforms to surface finish requirements?
(b) What is the probability that the selected shaft conforms to surface finish requirements or to roundness requirements?

(c) What is the probability that the selected shaft either conforms to surface finish requirements or does not conform to roundness requirements?
(d) What is the probability that the selected shaft conforms to both surface finish and roundness requirements?

2-72. Cooking oil is produced in two main varieties: mono- and polyunsaturated. Two common sources of cooking oil are corn and canola. The following table shows the number of bottles of these oils at a supermarket:

		type of oil	
		canola	corn
type of unsaturation	mono	7	13
	poly	93	77

(a) If a bottle of oil is selected at random, what is the probability that it belongs to the polyunsaturated category?
(b) What is the probability that the chosen bottle is monounsaturated canola oil?

2-73. A manufacturer of front lights for automobiles tests lamps under a high humidity, high temperature environment using intensity and useful life as the responses of interest. The following table shows the performance of 130 lamps:

		useful life	
		satisfactory	unsatisfactory
intensity	satisfactory	117	3
	unsatisfactory	8	2

(a) Find the probability that a randomly selected lamp will yield unsatisfactory results under any criteria.
(b) The customers for these lamps demand 95% satisfactory results. Can the lamp manufacturer meet this demand?

2-74. A computer system uses passwords that are six characters and each character is one of the 26 letters (a–z) or 10 integers (0–9). Uppercase letters are not used. Let A denote the event that a password begins with a vowel (either a, e, i, o, u) and let B denote the event that a password ends with an even number (either 0, 2, 4, 6, or 8). Suppose a hacker selects a password at random. Determine the following probabilities:
(a) $P(A)$ (b) $P(B)$
(c) $P(A \cap B)$ (d) $P(A \cup B)$

2-4 CONDITIONAL PROBABILITY

Sometimes probabilities need to be reevaluated as additional information becomes available. A useful way to incorporate additional information into a probability model is to assume that the outcome that will be generated is a member of a given event. This event, say A, defines the conditions that the outcome is known to satisfy. Then probabilities can be revised to include this knowledge. The probability of an event B under the knowledge that the outcome will be in event A is denoted as

$$P(B \mid A)$$

and this is called the **conditional probability** of B **given** A.

A digital communication channel has an error rate of one bit per every thousand transmitted. Errors are rare, but when they occur, they tend to occur in bursts that affect many consecutive bits. If a single bit is transmitted, we might model the probability of an error as $1/1000$. However, if the previous bit was in error, because of the bursts, we might believe that the probability that the next bit is in error is greater than $1/1000$.

In a thin film manufacturing process, the proportion of parts that are not acceptable is 2%. However, the process is sensitive to contamination problems that can increase the rate of parts that are not acceptable. If we knew that during a particular shift there were problems with the filters used to control contamination, we would assess the probability of a part being unacceptable as higher than 2%.

In a manufacturing process, 10% of the parts contain visible surface flaws and 25% of the parts with surface flaws are (functionally) defective parts. However, only 5% of parts without surface flaws are defective parts. The probability of a defective part depends on our knowledge of the presence or absence of a surface flaw. Let D denote the event that a part is defective and let F denote the event that a part has a surface flaw. Then, we denote the probability of D given, or assuming, that a part has a surface flaw as $P(D \mid F)$. Because 25% of the parts with surface flaws are

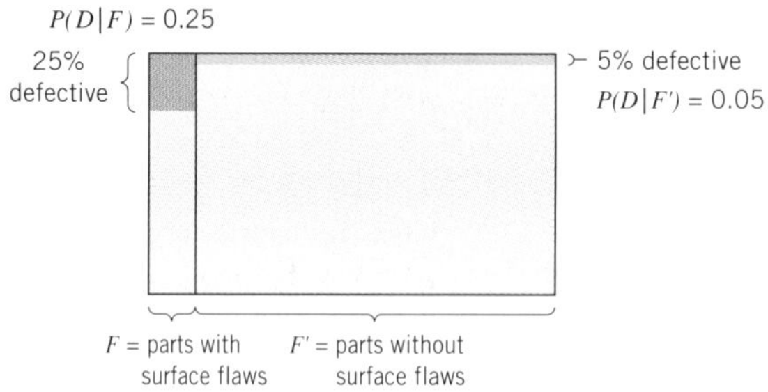

Figure 2-13 Conditional probabilities for parts with surface flaws.

defective, our conclusion can be stated as $P(D|F) = 0.25$. Furthermore, because F' denotes the event that a part does not have a surface flaw and because 5% of the parts without surface flaws are defective, we have that $P(D|F') = 0.05$. These results are shown graphically in Fig. 2-13.

EXAMPLE 2-22 Table 2-3 provides an example of 400 parts classified by surface flaws and as (functionally) defective. For this table the conditional probabilities match those discussed previously in this section. For example, of the parts with surface flaws (40 parts) the number defective is 10. Therefore,

$$P(D|F) = 10/40 = 0.25$$

and of the parts without surface flaws (360 parts) the number defective is 18. Therefore,

$$P(D|F') = 18/360 = 0.05$$

In Example 2-22 conditional probabilities were calculated directly. These probabilities can also be determined from the formal definition of conditional probability.

Conditional Probability

The **conditional probability** of an event B given an event A, denoted as $P(B|A)$, is

$$P(B|A) = P(A \cap B)/P(A) \tag{2-9}$$

for $P(A) > 0$.

This definition can be understood in a special case in which all outcomes of a random experiment are equally likely. If there are n total outcomes,

$$P(A) = (\text{number of outcomes in } A)/n$$

Table 2-3 Parts Classified

		Surface Flaws		
		Yes (event F)	No	Total
Defective	Yes (event D)	10	18	28
	No	30	342	372
	Total	40	360	400

Also,

$$P(A \cap B) = (\text{number of outcomes in } A \cap B)/n$$

Consequently,

$$P(A \cap B)/P(A) = \frac{\text{number of outcomes in } A \cap B}{\text{number of outcomes in } A}$$

Therefore, $P(B|A)$ can be interpreted as the relative frequency of event B among the trials that produce an outcome in event A.

EXAMPLE 2-23
Surface Flaws

Again consider the 400 parts in Table 2-3. From this table

$$P(D|F) = P(D \cap F)/P(F) = \frac{10}{400}\Big/\frac{40}{400} = \frac{10}{40}$$

Note that in this example all four of the following probabilities are different:

$$P(F) = 40/400 \qquad P(F|D) = 10/28$$
$$P(D) = 28/400 \qquad P(D|F) = 10/40$$

Here, $P(D)$ and $P(D|F)$ are probabilities of the same event, but they are computed under two different states of knowledge. Similarly, $P(F)$ and $P(F|D)$ are computed under two different states of knowledge.

The tree diagram in Fig. 2-14 can also be used to display conditional probabilities. The first branch is on surface flaw. Of the 40 parts with surface flaws, 10 are functionally defective and 30 are not. Therefore,

$$P(D|F) = 10/40 \qquad \text{and} \qquad P(D'|F) = 30/40$$

Of the 360 parts without surface flaws, 18 are functionally defective and 342 are not. Therefore,

$$P(D|F') = 18/360 \qquad \text{and} \qquad P(D'|F') = 342/360$$

Random Samples and Conditional Probability

Recall that to select one item randomly from a batch implies that each item is equally likely. If more than one item is selected, *randomly* implies that each element of the sample space is equally likely. For example, when sample spaces were presented earlier in this chapter, sampling with and without replacement were defined and illustrated for the simple case of a batch with three items $\{a, b, c\}$. If two items are selected randomly from this batch without replacement, each of the six outcomes in the ordered sample space

$$S_{\text{without}} = \{ab, ac, ba, bc, ca, cb\}$$

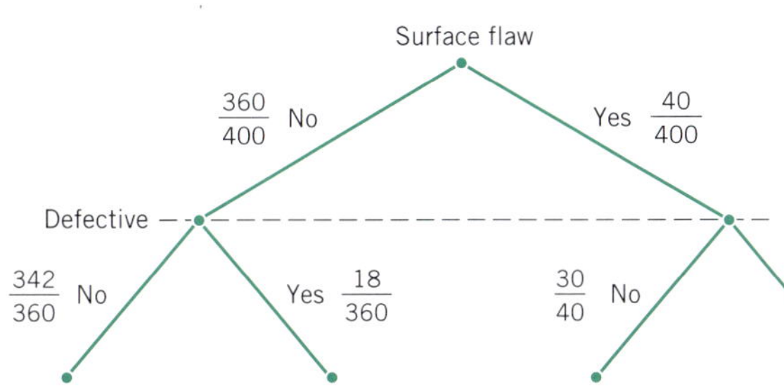

Figure 2-14 Tree diagram for parts classified

has probability 1/6. If the unordered sample space is used, each of the three outcomes in $\{\{a, b\}, \{a, c\}, \{b, c\}\}$ has probability 1/3.

When a sample is selected randomly from a large batch, it is usually easier to avoid enumeration of the sample space and calculate probabilities from conditional probabilities. For example, suppose that a batch contains 10 parts from tool 1 and 40 parts from tool 2. If two parts are selected randomly, without replacement, what is the conditional probability that a part from tool 2 is selected second given that a part from tool 1 is selected first? There are 50 possible parts to select first and 49 to select second. Therefore, the (ordered) sample space has $50 \times 49 = 2450$ outcomes. Let E_1 denote the event that the first part is from tool 1 and E_2 denote the event that the second part is from tool 2. Then

$$P(E_2 \mid E_1) = P(E_1 \cap E_2)/P(E_1)$$

and a count of the number of outcomes in E_1 and the intersection is needed.

Although the answer can be determined from this start, this type of question can be answered more easily with the following result.

Random Samples

> To select randomly implies that at each step of the sample, the items that remain in the batch are equally likely to be selected.

If a part from tool 1 were selected with the first pick, 49 items would remain, 9 from tool 1 and 40 from tool 2, and they would be equally likely to be picked. Therefore, the probability that a part from tool 2 would be selected with the second pick given this first pick is

$$P(E_2 \mid E_1) = 40/49$$

In this manner, other probabilities can also be simplified. For example, let the event E consist of the outcomes with the first selected part from tool 1 and the second part from tool 2. To determine the probability of E, consider each step. The probability that a part from tool 1 is selected with the first pick is $P(E_1) = 10/50$. The conditional probability that a part from tool 2 is selected with the second pick, given that a part from tool 1 is selected first is $P(E_2 \mid E_1) = 40/49$. Therefore,

$$P(E) = P(E_2 \mid E_1)P(E_1) = \frac{40}{49} \cdot \frac{10}{50} = \frac{8}{49}$$

Sometimes a partition of the question into successive picks is an easier method to solve the problem.

EXAMPLE 2-24

A day's production of 850 manufactured parts contains 50 parts that do not meet customer requirements. Two parts are selected randomly without replacement from the batch. What is the probability that the second part is defective given that the first part is defective?

Let A denote the event that the first part selected is defective, and let B denote the event that the second part selected is defective. The probability needed can be expressed as $P(B \mid A)$. If the first part is defective, prior to selecting the second part, the batch contains 849 parts, of which 49 are defective, therefore

$$P(B \mid A) = 49/849$$

EXAMPLE 2-25 Continuing the previous example, if three parts are selected at random, what is the probability that the first two are defective and the third is not defective? This event can be described in shorthand notation as simply $P(ddn)$. We have

$$P(ddn) = \frac{50}{850} \cdot \frac{49}{849} \cdot \frac{800}{848} = 0.0032$$

The third term is obtained as follows. After the first two parts are selected, there are 848 remaining. Of the remaining parts, 800 are not defective. In this example, it is easy to obtain the solution with a conditional probability for each selection.

EXERCISES FOR SECTION 2-4

2-75. Disks of polycarbonate plastic from a supplier are analyzed for scratch and shock resistance. The results from 100 disks are summarized as follows:

		shock resistance	
		high	low
scratch	high	70	9
resistance	low	16	5

Let A denote the event that a disk has high shock resistance, and let B denote the event that a disk has high scratch resistance. Determine the following probabilities:
(a) $P(A)$ (b) $P(B)$
(c) $P(A|B)$ (d) $P(B|A)$

2-76. Samples of skin experiencing desquamation are analyzed for both moisture and melanin content. The results from 100 skin samples are as follows:

		melanin content	
		high	low
moisture	high	13	7
content	low	48	32

Let A denote the event that a sample has low melanin content, and let B denote the event that a sample has high moisture content. Determine the following probabilities:
(a) $P(A)$ (b) $P(B)$
(c) $P(A|B)$ (d) $P(B|A)$

2-77. The analysis of results from a leaf transmutation experiment (turning a leaf into a petal) is summarized by type of transformation completed:

		total textural transformation	
		yes	no
total color	yes	243	26
transformation	no	13	18

(a) If a leaf completes the color transformation, what is the probability that it will complete the textural transformation?
(b) If a leaf does not complete the textural transformation, what is the probability it will complete the color transformation?

2-78. Samples of a cast aluminum part are classified on the basis of surface finish (in microinches) and length measurements. The results of 100 parts are summarized as follows:

		length	
		excellent	good
surface	excellent	80	2
finish	good	10	8

Let A denote the event that a sample has excellent surface finish, and let B denote the event that a sample has excellent length. Determine:
(a) $P(A)$ (b) $P(B)$
(c) $P(A|B)$ (d) $P(B|A)$
(e) If the selected part has excellent surface finish, what is the probability that the length is excellent?
(f) If the selected part has good length, what is the probability that the surface finish is excellent?

2-79. The following table summarizes the analysis of samples of galvanized steel for coating weight and surface roughness:

		coating weight	
		high	low
surface	high	12	16
roughness	low	88	34

(a) If the coating weight of a sample is high, what is the probability that the surface roughness is high?
(b) If the surface roughness of a sample is high, what is the probability that the coating weight is high?
(c) If the surface roughness of a sample is low, what is the probability that the coating weight is low?

2-80. Consider the data on wafer contamination and location in the sputtering tool shown in Table 2-2. Assume that one wafer is selected at random from this set. Let A denote the event that a wafer contains four or more particles, and let B denote the event that a wafer is from the center of the sputtering tool. Determine:
(a) $P(A)$ (b) $P(A|B)$

(c) $P(B)$ (d) $P(B|A)$
(e) $P(A \cap B)$ (f) $P(A \cup B)$

2-81. The following table summarizes the number of deceased beetles under autolysis (the destruction of a cell after its death by the action of its own enzymes) and putrefaction (decomposition of organic matter, especially protein, by microorganisms, resulting in production of foul-smelling matter):

		autolysis	
		high	low
putrefaction	high	14	59
	low	18	9

(a) If the autolysis of a sample is high, what is the probability that the putrefaction is low?
(b) If the putrefaction of a sample is high, what is the probability that the autolysis is high?
(c) If the putrefaction of a sample is low, what is the probability that the autolysis is low?

2-82. A maintenance firm has gathered the following information regarding the failure mechanisms for air conditioning systems:

		evidence of gas leaks	
		yes	no
evidence of electrical failure	yes	55	17
	no	32	3

The units without evidence of gas leaks or electrical failure showed other types of failure. If this is a representative sample of AC failure, find the probability
(a) That failure involves a gas leak
(b) That there is evidence of electrical failure given that there was a gas leak
(c) That there is evidence of a gas leak given that there is evidence of electrical failure

2-83. A lot of 100 semiconductor chips contains 20 that are defective. Two are selected randomly, without replacement, from the lot.
(a) What is the probability that the first one selected is defective?
(b) What is the probability that the second one selected is defective given that the first one was defective?
(c) What is the probability that both are defective?
(d) How does the answer to part (b) change if chips selected were replaced prior to the next selection?

2-84. A batch of 500 containers for frozen orange juice contains 5 that are defective. Two are selected, at random, without replacement from the batch.
(a) What is the probability that the second one selected is defective given that the first one was defective?
(b) What is the probability that both are defective?
(c) What is the probability that both are acceptable?

Three containers are selected, at random, without replacement, from the batch.
(d) What is the probability that the third one selected is defective given that the first and second ones selected were defective?
(e) What is the probability that the third one selected is defective given that the first one selected was defective and the second one selected was okay?
(f) What is the probability that all three are defective?

2-85. A batch of 350 samples of rejuvenated mitochondria contains 8 that are mutated (or defective). Two are selected, at random, without replacement from the batch.
(a) What is the probability that the second one selected is defective given that the first one was defective?
(b) What is the probability that both are defective?
(c) What is the probability that both are acceptable?

2-86. A computer system uses passwords that are exactly seven characters and each character is one of the 26 letters (a–z) or 10 integers (0–9). You maintain a password for this computer system. Let A denote the subset of passwords that begin with a vowel (either a, e, i, o, u) and let B denote the subset of passwords that end with an even number (either 0, 2, 4, 6, or 8).
(a) Suppose a hacker selects a password at random. What is the probability that your password is selected?
(b) Suppose a hacker knows your password is in event A and selects a password at random from this subset. What is the probability that your password is selected?
(c) Suppose a hacker knows your password is in A and B and selects a password at random from this subset. What is the probability that your password is selected?

2-87. If $P(A|B) = 1$, must $A = B$? Draw a Venn diagram to explain your answer.

2-88. Suppose A and B are mutually exclusive events. Construct a Venn diagram that contains the three events A, B, and C such that $P(A|C) = 1$ and $P(B|C) = 0$?

2-5 MULTIPLICATION AND TOTAL PROBABILITY RULES

2-5.1 Multiplication Rule

The probability of the intersection of two events is often needed. The conditional probability definition in Equation 2-9 can be rewritten to provide a formula known as the **multiplication rule** for probabilities.

Multiplication Rule

$$P(A \cap B) = P(B|A)P(A) = P(A|B)P(B) \qquad \text{(2-10)}$$

The last expression in Equation 2-10 is obtained by interchanging A and B.

EXAMPLE 2-26
Machining Stages

The probability that the first stage of a numerically controlled machining operation for high-rpm pistons meets specifications is 0.90. Failures are due to metal variations, fixture alignment, cutting blade condition, vibration, and ambient environmental conditions. Given that the first stage meets specifications, the probability that a second stage of machining meets specifications is 0.95. What is the probability that both stages meet specifications?

Let A and B denote the events that the first and second stages meet specifications, respectively. The probability requested is

$$P(A \cap B) = P(B|A)P(A) = 0.95(0.90) = 0.855$$

Although it is also true that $P(A \cap B) = P(A|B)P(B)$, the information provided in the problem does not match this second formulation.

2-5.2 Total Probability Rule

Sometimes the probability of an event is given under each of several conditions. With enough of these conditional probabilities, the probability of the event can be recovered. For example, suppose that in semiconductor manufacturing the probability is 0.10 that a chip that is subjected to high levels of contamination during manufacturing causes a product failure. The probability is 0.005 that a chip that is not subjected to high contamination levels during manufacturing causes a product failure. In a particular production run, 20% of the chips are subject to high levels of contamination. What is the probability that a product using one of these chips fails?

Clearly, the requested probability depends on whether or not the chip was exposed to high levels of contamination. For any event B, we can write B as the union of the part of B in A and the part of B in A'. That is,

$$B = (A \cap B) \cup (A' \cap B)$$

This result is shown in the Venn diagram in Fig. 2-15. Because A and A' are mutually exclusive, $A \cap B$ and $A' \cap B$ are mutually exclusive. Therefore, from the probability of the union of mutually exclusive events in Equation 2-6 and the Multiplication Rule in Equation 2-10, the following **total probability rule** is obtained.

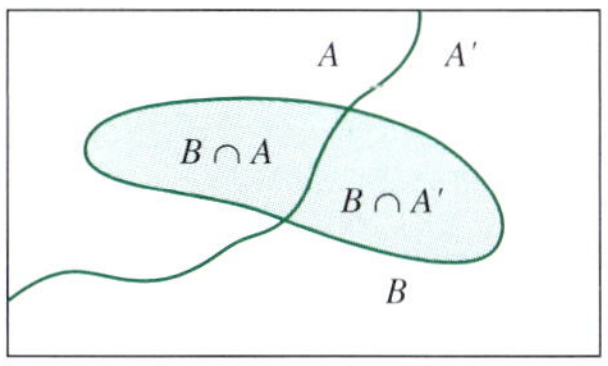

Figure 2-15 Partitioning an event into two mutually exclusive subsets.

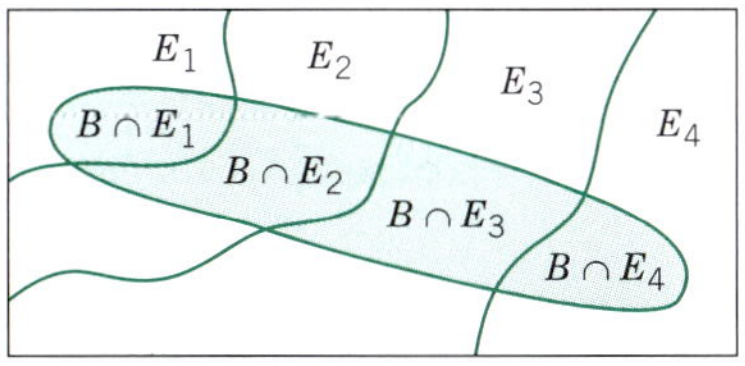

$B = (B \cap E_1) \cup (B \cap E_2) \cup (B \cap E_3) \cup (B \cap E_4)$

Figure 2-16 Partitioning an event into several mutually exclusive subsets.

Total Probability Rule (two events)

For any events A and B,

$$P(B) = P(B \cap A) + P(B \cap A') = P(B|A)P(A) + P(B|A')P(A') \quad (2\text{-}11)$$

EXAMPLE 2-27 Semiconductor Contamination

Consider the contamination discussion at the start of this section. The information is summarized here

Probability of Failure	Level of Contamination	Probability of Level
0.1	High	0.2
0.005	Not High	0.8

Let F denote the event that the product fails, and let H denote the event that the chip is exposed to high levels of contamination. The requested probability is $P(F)$, and the information provided can be represented as

$$P(F|H) = 0.10 \quad \text{and} \quad P(F|H') = 0.005$$
$$P(H) = 0.20 \quad \text{and} \quad P(H') = 0.80$$

From Equation 2-11,

$$P(F) = 0.10(0.20) + 0.005(0.80) = 0.024$$

which can be interpreted as just the weighted average of the two probabilities of failure.

The reasoning used to develop Equation 2-11 can be applied more generally. Because $A \cup A' = S$ we know $(A \cap B) \cup (A' \cap B)$ equals B and because $A \cap A' = \phi$ we know $A \cap B$ and $A' \cap B$ are mutually exclusive. In general, a collection of sets $E_1, E_2, \ldots, E_k$ such that $E_1 \cup E_2 \cup \ldots \cup E_k = S$ is said to be **exhaustive.** A graphical display of partitioning an event B among a collection of mutually exclusive and exhaustive events is shown in Fig. 2-16.

Total Probability Rule (multiple events)

Assume $E_1, E_2, \ldots, E_k$ are k mutually exclusive and exhaustive sets. Then

$$P(B) = P(B \cap E_1) + P(B \cap E_2) + \cdots + P(B \cap E_k)$$
$$= P(B|E_1)P(E_1) + P(B|E_2)P(E_2) + \cdots + P(B|E_k)P(E_k) \quad (2\text{-}12)$$

EXAMPLE 2-28 Semiconductor Failures

Continuing with semiconductor manufacturing, assume the following probabilities for product failure subject to levels of contamination in manufacturing:

Probability of Failure	Level of Contamination
0.10	High
0.01	Medium
0.001	Low

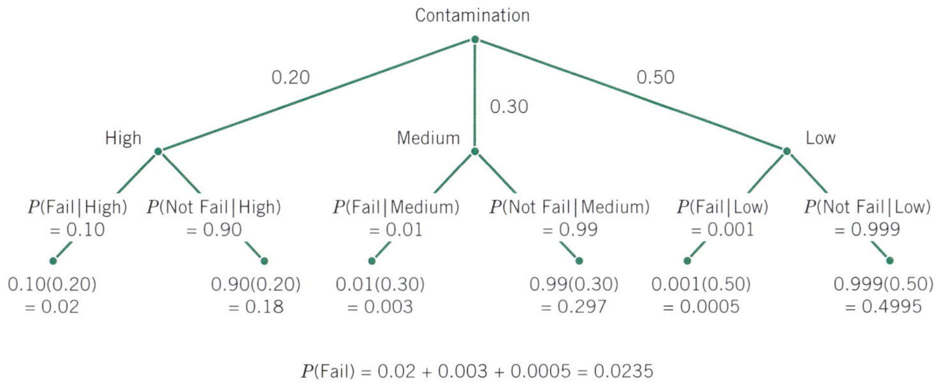

Figure 2-17 Tree diagram for Example 2-28.

In a particular production run, 20% of the chips are subjected to high levels of contamination, 30% to medium levels of contamination, and 50% to low levels of contamination. What is the probability that a product using one of these chips fails? Let

H denote the event that a chip is exposed to high levels of contamination

M denote the event that a chip is exposed to medium levels of contamination

L denote the event that a chip is exposed to low levels of contamination

Then,

$$\begin{aligned} P(F) &= P(F \mid H)P(H) + P(F \mid M)P(M) + P(F \mid L)P(L) \\ &= 0.10(0.20) + 0.01(0.30) + 0.001(0.50) = 0.0235 \end{aligned}$$

The calculations are conveniently organized with the tree diagram in Fig. 2-17.

EXERCISES FOR SECTION 2-5

2-89. Suppose that $P(A \mid B) = 0.4$ and $P(B) = 0.5$. Determine the following:

(a) $P(A \cap B)$

(b) $P(A' \cap B)$

2-90. Suppose that $P(A \mid B) = 0.2$, $P(A \mid B') = 0.3$, and $P(B) = 0.8$. What is $P(A)$?

2-91. The probability is 1% that an electrical connector that is kept dry fails during the warranty period of a portable computer. If the connector is ever wet, the probability of a failure during the warranty period is 5%. If 90% of the connectors are kept dry and 10% are wet, what proportion of connectors fail during the warranty period?

2-92. Suppose 2% of cotton fabric rolls and 3% of nylon fabric rolls contain flaws. Of the rolls used by a manufacturer, 70% are cotton and 30% are nylon. What is the probability that a randomly selected roll used by the manufacturer contains flaws?

2-93. The edge roughness of slit paper products increases as knife blades wear. Only 1% of products slit with new blades have rough edges, 3% of products slit with blades of average sharpness exhibit roughness, and 5% of products slit with worn blades exhibit roughness. If 25% of the blades in manufacturing are new, 60% are of average sharpness, and 15% are worn, what is the proportion of products that exhibit edge roughness?

2-94. In the 2004 presidential election, exit polls from the critical state of Ohio provided the following results:

total	Bush, 2004	Kerry, 2004
no college degree (62%)	50%	50%
college graduate (38%)	53%	46%

What is the probability a randomly selected respondent voted for Bush?

2-95. Computer keyboard failures are due to faulty electrical connects (12%) or mechanical defects (88%). Mechanical defects are related to loose keys (27%) or improper assembly (73%). Electrical connect defects are caused by defective wires (35%), improper connections (13%), or poorly welded wires (52%).

(a) Find the probability that a failure is due to loose keys.

(b) Find the probability that a failure is due to improperly connected or poorly welded wires.

2-96. Heart failures are due to either natural occurrences (87%) or outside factors (13%). Outside factors are related to induced substances (73%) or foreign objects (27%). Natural occurrences are caused by arterial blockage (56%), disease (27%), and infection (e.g., staph infection) (17%).

(a) Determine the probability that a failure is due to induced substance.

(b) Determine the probability that a failure is due to disease or infection.

2-97. A batch of 25 injection-molded parts contains 5 that have suffered excessive shrinkage.

(a) If two parts are selected at random, and without replacement, what is the probability that the second part selected is one with excessive shrinkage?

(b) If three parts are selected at random, and without replacement, what is the probability that the third part selected is one with excessive shrinkage?

2-98. A lot of 100 semiconductor chips contains 20 that are defective.

(a) Two are selected, at random, without replacement, from the lot. Determine the probability that the second chip selected is defective.

(b) Three are selected, at random, without replacement, from the lot. Determine the probability that all are defective.

2-99. An article in the *British Medical Journal* "Comparison of treatment of renal calculi by operative surgery, percutaneous nephrolithotomy, and extracorporeal shock wave lithotripsy," (1986, Vol. 82, pp. 879–892) provided the following discussion of success rates in kidney stone removals. Open surgery had a success rate of 78% (273/350) while a newer method, percutaneous nephrolithotomy (PN), had a success rate of 83% (289/350). This newer method looked better, but the results changed when stone diameter was considered. For stones with diameters less than two centimeters, 93% (81/87) of cases of open surgery were successful compared with only 83% (234/270) of cases of PN. For stones greater than or equal to two centimeters, the success rates were 73% (192/263) and 69% (55/80) for open surgery and PN, respectively. Open surgery is better for both stone sizes, but less successful in total. In 1951, E. H. Simpson pointed out this apparent contradiction (known as **Simpson's Paradox**) but the hazard still persists today. Explain how open surgery can be better for both stone sizes but worse in total.

2-6 INDEPENDENCE

In some cases, the conditional probability of $P(B|A)$ might equal $P(B)$. In this special case, knowledge that the outcome of the experiment is in event A does not affect the probability that the outcome is in event B.

EXAMPLE 2-29
Sampling with Replacement

Suppose a day's production of 850 manufactured parts contains 50 parts that do not meet customer requirements. Suppose two parts are selected from the batch, but the first part is replaced before the second part is selected. What is the probability that the second part is defective (denoted as B) given that the first part is defective (denoted as A)? The probability needed can be expressed as $P(B|A)$.

Because the first part is replaced prior to selecting the second part, the batch still contains 850 parts, of which 50 are defective. Therefore, the probability of B does not depend on whether or not the first part was defective. That is,

$$P(B|A) = 50/850$$

Also, the probability that both parts are defective is

$$P(A \cap B) = P(B|A)P(A) = \left(\frac{50}{850}\right) \cdot \left(\frac{50}{850}\right) = 0.0035$$

EXAMPLE 2-30
Flaws and Functions

The information in Table 2-3 related surface flaws to functionally defective parts. In that case, we determined that $P(D|F) = 10/40 = 0.25$ and $P(D) = 28/400 = 0.07$. Suppose that the situation is different and follows Table 2-4. Then,

$$P(D|F) = 2/40 = 0.05 \quad \text{and} \quad P(D) = 20/400 = 0.05$$

Table 2-4 Parts Classified

		Surface Flaws		
		Yes (event F)	No	Total
Defective	Yes (event D)	2	18	20
	No	38	342	380
	Total	40	360	400

That is, the probability that the part is defective does not depend on whether it has surface flaws. Also,

$$P(F|D) = 2/20 = 0.10 \quad \text{and} \quad P(F) = 40/400 = 0.10$$

so the probability of a surface flaw does not depend on whether the part is defective. Furthermore, the definition of conditional probability implies that

$$P(F \cap D) = P(D|F)P(F)$$

but in the special case of this problem

$$P(F \cap D) = P(D)P(F) = \frac{2}{40} \cdot \frac{2}{20} = \frac{1}{200}$$

The preceding example illustrates the following conclusions. In the special case that $P(B \mid A) = P(B)$, we obtain

$$P(A \cap B) = P(B|A)P(A) = P(B)P(A)$$

and

$$P(A|B) = \frac{P(A \cap B)}{P(B)} = \frac{P(A)P(B)}{P(B)} = P(A)$$

These conclusions lead to an important definition.

Independence (two events)

Two events are **independent** if any one of the following equivalent statements is true:

(1) $P(A|B) = P(A)$

(2) $P(B|A) = P(B)$

(3) $P(A \cap B) = P(A)P(B)$ (2-13)

It is left as a mind-expanding exercise to show that independence implies related results such as

$$P(A' \cap B') = P(A')P(B').$$

The concept of independence is an important relationship between events and is used throughout this text. A mutually exclusive relationship between two events is based only on the outcomes that comprise the events. However, an independence relationship depends on the probability model used for the random experiment. Often, independence is assumed to be part of the random experiment that describes the physical system under study.

EXAMPLE 2-31

A day's production of 850 manufactured parts contains 50 parts that do not meet customer requirements. Two parts are selected at random, without replacement, from the batch. Let A denote the event that the first part is defective, and let B denote the event that the second part is defective.

We suspect that these two events are not independent because knowledge that the first part is defective suggests that it is less likely that the second part selected is defective. Indeed, $P(B|A) = 49/849$. Now, what is $P(B)$? Finding the unconditional $P(B)$ is somewhat difficult because the possible values of the first selection need to be considered:

$$\begin{aligned} P(B) &= P(B|A)P(A) + P(B|A')P(A') \\ &= (49/849)(50/850) + (50/849)(800/850) \\ &= 50/850 \end{aligned}$$

Interestingly, $P(B)$, the unconditional probability that the second part selected is defective, without any knowledge of the first part, is the same as the probability that the first part selected is defective. Yet, our goal is to assess independence. Because $P(B|A)$ does not equal $P(B)$, the two events are not independent, as we suspected.

When considering three or more events, we can extend the definition of independence with the following general result.

Independence (multiple events)

The events $E_1, E_2, \ldots, E_n$ are independent if and only if for any subset of these events $E_{i_1}, E_{i_2}, \ldots, E_{i_k}$,

$$P(E_{i_1} \cap E_{i_2} \cap \cdots \cap E_{i_k}) = P(E_{i_1}) \times P(E_{i_2}) \times \cdots \times P(E_{i_k}) \qquad (2\text{-}14)$$

This definition is typically used to calculate the probability that several events occur assuming that they are independent and the individual event probabilities are known. The knowledge that the events are independent usually comes from a fundamental understanding of the random experiment.

EXAMPLE 2-32
Series Circuit

The following circuit operates only if there is a path of functional devices from left to right. The probability that each device functions is shown on the graph. Assume that devices fail independently. What is the probability that the circuit operates?

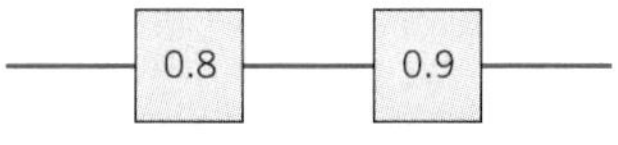

Let L and R denote the events that the left and right devices operate, respectively. There is only a path if both operate. The probability the circuit operates is

$$P(L \text{ and } R) = P(L \cap R) = P(L)P(R) = 0.80(0.90) = 0.72$$

EXAMPLE 2-33

Assume that the probability that a wafer contains a large particle of contamination is 0.01 and that the wafers are independent; that is, the probability that a wafer contains a large particle is not dependent on the characteristics of any of the other wafers. If 15 wafers are analyzed, what is the probability that no large particles are found?

Let E_i denote the event that the ith wafer contains no large particles, $i = 1, 2, \ldots, 15$. Then, $P(E_i) = 0.99$. The probability requested can be represented as $P(E_1 \cap E_2 \cap \cdots \cap E_{15})$. From the independence assumption and Equation 2-14,

$$P(E_1 \cap E_2 \cap \cdots \cap E_{15}) = P(E_1) \times P(E_2) \times \cdots \times P(E_{15}) = 0.99^{15} = 0.86$$

EXAMPLE 2-34
Parallel Circuit

The following circuit operates only if there is a path of functional devices from left to right. The probability that each device functions is shown on the graph. Assume that devices fail independently. What is the probability that the circuit operates?

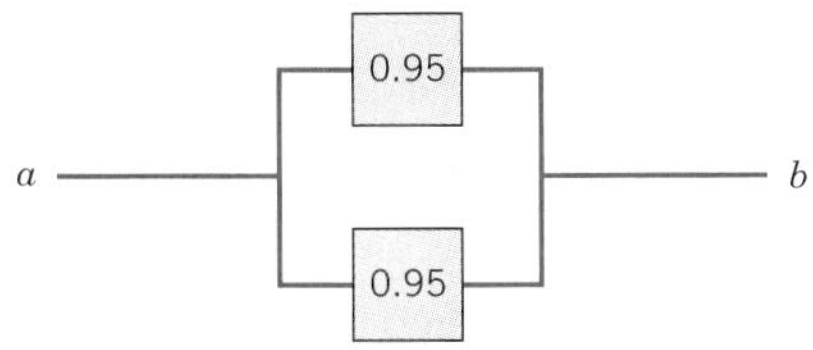

Let T and B denote the events that the top and bottom devices operate, respectively. There is a path if at least one device operates. The probability that the circuit operates is

$$P(T \text{ or } B) = 1 - P[(T \text{ or } B)'] = 1 - P(T' \text{ and } B')$$

A simple formula for the solution can be derived from the complements T' and B'. From the independence assumption,

$$P(T' \text{ and } B') = P(T')P(B') = (1 - 0.95)^2 = 0.05^2$$

so

$$P(T \text{ or } B) = 1 - 0.05^2 = 0.9975$$

EXAMPLE 2-35
Advanced Circuit

The following circuit operates only if there is a path of functional devices from left to right. The probability that each device functions is shown on the graph. Assume that devices fail independently. What is the probability that the circuit operates?

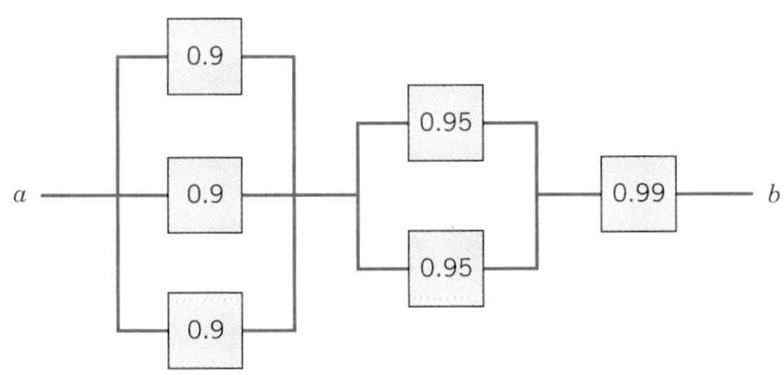

The solution can be obtained from a partition of the graph into three columns. The probability that there is a path of functional devices only through the three units on the left can be determined from the independence in a manner similar to the previous example. It is

$$1 - 0.1^3$$

Similarly, the probability that there is a path of functional devices only through the two units in the middle is

$$1 - 0.05^2$$

The probability that there is a path of functional devices only through the one unit on the right is simply the probability that the device functions, namely, 0.99. Therefore, with the independence assumption used again, the solution is

$$(1 - 0.1^3)(1 - 0.05^2)(0.99) = 0.987$$

EXERCISES FOR SECTION 2-6

2-100. If $P(A|B) = 0.4$, $P(B) = 0.8$, and $P(A) = 0.5$, are the events A and B independent?

2-101. If $P(A|B) = 0.3$, $P(B) = 0.8$, and $P(A) = 0.3$, are the events B and the complement of A independent?

2-102. If $P(A) = 0.2$, $P(B) = 0.2$, and A and B are mutually exclusive, are they independent?

2-103. A batch of 500 containers for frozen orange juice contains 5 that are defective. Two are selected, at random, without replacement, from the batch. Let A and B denote the events that the first and second container selected is defective, respectively.

(a) Are A and B independent events?

(b) If the sampling were done with replacement, would A and B be independent?

2-104. Disks of polycarbonate plastic from a supplier are analyzed for scratch and shock resistance. The results from 100 disks are summarized as follows:

		shock resistance	
		high	low
scratch	high	70	9
resistance	low	16	5

Let A denote the event that a disk has high shock resistance, and let B denote the event that a disk has high scratch resistance. Are events A and B independent?

2-105. Samples of emissions from three suppliers are classified for conformance to air-quality specifications. The results from 100 samples are summarized as follows:

		conforms	
		yes	no
	1	22	8
supplier	2	25	5
	3	30	10

Let A denote the event that a sample is from supplier 1, and let B denote the event that a sample conforms to specifications.

(a) Are events A and B independent?

(b) Determine $P(B|A)$.

2-106. Redundant Array of Inexpensive Discs (RAID) is a technology that uses multiple hard drives to increase the speed of data transfer and provide instant data backup. Suppose that the probability of any hard drive failing in a day is 0.001 and the drive failures are independent.

(a) Suppose you implement a RAID 0 scheme that uses two hard drives each containing a mirror image of the other. What is the probability of data loss? Assume that data loss occurs if both drives fail within the same day.

(b) Suppose you implement a RAID 1 scheme that splits the data over two hard drives. What is the probability of data loss? Assume that data loss occurs if at least one drive fails within the same day.

2-107. The probability that a lab specimen contains high levels of contamination is 0.10. Five samples are checked, and the samples are independent.

(a) What is the probability that none contains high levels of contamination?

(b) What is the probability that exactly one contains high levels of contamination?

(c) What is the probability that at least one contains high levels of contamination?

2-108. In a test of a printed circuit board using a random test pattern, an array of 10 bits is equally likely to be 0 or 1. Assume the bits are independent.

(a) What is the probability that all bits are 1s?

(b) What is the probability that all bits are 0s?

(c) What is the probability that exactly five bits are 1s and five bits are 0s?

2-109. Six tissues are extracted from an ivy plant infested by spider mites. The plant in infested in 20% of its area. A sample is chosen from randomly selected areas on the ivy plant. Assume that the samples are independent.

(a) What is the probability that four successive samples show the signs of infestation?

(b) What is the probability that three out of four successive samples show the signs of infestation?

2-110. A player of a video game is confronted with a series of four opponents and an 80% probability of defeating each opponent. Assume that the results from opponents are independent (and that when the player is defeated by an opponent the game ends).

(a) What is the probability that a player defeats all four opponents in a game?

(b) What is the probability that a player defeats at least two opponents in a game?

(c) If the game is played three times, what is the probability that the player defeats all four opponents at least once?

2-111. In an acid-base titration, a base or acid is gradually added to the other until they have completely neutralized each other. Since acids and bases are usually colorless (as are the water and salt produced in the neutralization reaction), pH is measured to monitor the reaction. Suppose that the equivalence point is reached after approximately 100 mL of a NaOH solution have been added (enough to react with all the acetic acid present) but that replicates are equally likely to indicate from 95 to 104 mL, measured to the nearest mL. Assume that two technicians each conduct titrations independently.

(a) What is the probability that both technicians obtain equivalence at 100 mL?

(b) What is the probability that both technicians obtain equivalence between 98 and 104 mL (inclusive)?

(c) What is the probability that the average volume at equivalence from the technicians is 100 mL?

2-112. A credit card contains 16 digits. It also contains a month and year of expiration. Suppose there are one million users of a credit card with unique card numbers. A hacker randomly selects a 16 digit credit card number.

(a) What is the probability that it belongs to a user?

(b) Suppose a hacker has a 25% chance of correctly guessing the year your card expires and randomly selects one of the 12 months. What is the probability that the hacker correctly selects the month and year of expiration?

2-113. Eight cavities in an injection-molding tool produce plastic connectors that fall into a common stream. A sample is chosen every several minutes. Assume that the samples are independent.

(a) What is the probability that five successive samples were all produced in cavity one of the mold?

(b) What is the probability that five successive samples were all produced in the same cavity of the mold?

(c) What is the probability that four out of five successive samples were produced in cavity one of the mold?

2-114. The following circuit operates if and only if there is a path of functional devices from left to right. The probability that each device functions is as shown. Assume that the probability that a device is functional does not depend on whether or not other devices are functional. What is the probability that the circuit operates?

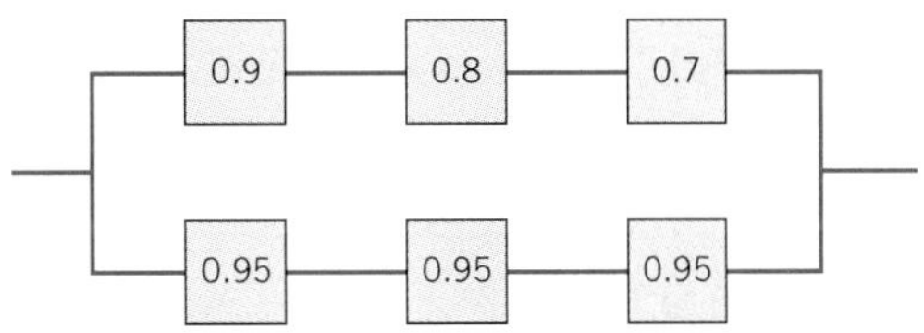

2-115. The following circuit operates if and only if there is a path of functional devices from left to right. The probability each device functions is as shown. Assume that the probability that a device functions does not depend on whether or not other devices are functional. What is the probability that the circuit operates?

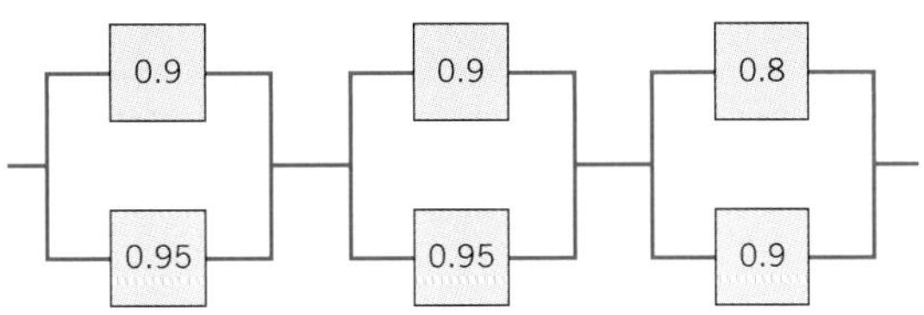

2-7 BAYES' THEOREM

The examples in this chapter indicate that information is often presented in terms of conditional probabilities. These conditional probabilities commonly provide the probability of an event (such as failure) given a condition (such as high or low contamination). But after a random experiment generates an outcome, we are naturally interested in the probability that a condition was present (high contamination) given an outcome (a semiconductor failure). Thomas Bayes addressed this essential question in the 1700s and developed the fundamental result known as **Bayes' theorem.** Do not let the simplicity of the mathematics conceal the importance. There is extensive interest in such probabilities in modern statistical analysis.

From the definition of conditional probability,

$$P(A \cap B) = P(A \mid B)P(B) = P(B \cap A) = P(B \mid A)P(A)$$

Now considering the second and last terms in the expression above, we can write

$$P(A|B) = \frac{P(B|A)P(A)}{P(B)} \quad \text{for} \quad P(B) > 0 \tag{2-15}$$

This is a useful result that enables us to solve for $P(A|B)$ in terms of $P(B|A)$.

EXAMPLE 2-36

Reconsider Example 2-27. The conditional probability a high level of contamination was present when a failure occurred is to be determined. The information from Example 2-27 is summarized here.

Probability of Failure	Level of Contamination	Probability of Level
0.1	High	0.2
0.005	Not High	0.8

The probability of $P(H|F)$ is determined from

$$P(H|F) = \frac{P(F|H)P(H)}{P(F)} = \frac{0.10(0.20)}{0.024} = 0.83$$

The value of $P(F)$ in the denominator of our solution was found from $P(F) = P(F|H)P(H) + P(F|H')P(H')$.

In general, if $P(B)$ in the denominator of Equation 2-15 is written using the Total Probability Rule in Equation 2-12, we obtain the following general result, which is known as **Bayes' Theorem.**

Bayes' Theorem

If $E_1, E_2, \ldots, E_k$ are k mutually exclusive and exhaustive events and B is any event,

$$P(E_1|B) = \frac{P(B|E_1)P(E_1)}{P(B|E_1)P(E_1) + P(B|E_2)P(E_2) + \cdots + P(B|E_k)P(E_k)} \tag{2-16}$$

for $P(B) > 0$

Notice that the numerator always equals one of the terms in the sum in the denominator.

EXAMPLE 2-37
Medical Diagnostic

Because a new medical procedure has been shown to be effective in the early detection of an illness, a medical screening of the population is proposed. The probability that the test correctly identifies someone with the illness as positive is 0.99, and the probability that the test correctly identifies someone without the illness as negative is 0.95. The incidence of the illness in the general population is 0.0001. You take the test, and the result is positive. What is the probability that you have the illness?

Let D denote the event that you have the illness, and let S denote the event that the test signals positive. The probability requested can be denoted as $P(D|S)$. The probability that the test correctly signals someone without the illness as negative is 0.95. Consequently, the probability of a positive test without the illness is

$$P(S|D') = 0.05$$

From Bayes' Theorem,

$$P(D|S) = P(S|D)P(D)/[P(S|D)P(D) + P(S|D')P(D')]$$
$$= 0.99(0.0001)/[0.99(0.0001) + 0.05(1 - 0.0001)]$$
$$= 1/506 = 0.002$$

That is, the probability of you having the illness given a positive result from the test is only 0.002. Surprisingly, even though the test is effective, in the sense that $P(S|D)$ is high and $P(S|D')$ is low, because the incidence of the illness in the general population is low, the chances are quite small that you actually have the disease even if the test is positive.

EXAMPLE 2-38 Bayesian Network

Bayesian networks are used on the Web sites of high-technology manufacturers to allow customers to quickly diagnose problems with products. An oversimplified example is presented here. A printer manufacturer obtained the following probabilities from a database of test results. Printer failures are associated with three types of problems: hardware, software, and other (such as connectors), with probabilities 0.1, 0.6, and 0.3, respectively. The probability of a printer failure given a hardware problem is 0.9, given a software problem is 0.2, and given any other type of problem is 0.5. If a customer enters the manufacturer's Web site to diagnose a printer failure, what is the most likely cause of the problem?

Let the events H, S, and O denote a hardware, software, or other problem, respectively, and let F denote a printer failure. The most likely cause of the problem is the one that corresponds to the largest of $P(H|F)$, $P(S|F)$, and $P(O|F)$. In Bayes' Theorem the denominator is

$$P(F) = P(F|H)P(H) + P(F|S)\,P(S) + P(F|O)P(O) = 0.9(0.1) + 0.2(0.6) + 0.5(0.3) = 0.36$$

Then

$$P(H|F) = P(F|H)P(H)/P(F) = 0.9(0.1)/0.36 = 0.250$$
$$P(S|F) = P(F|S)P(S)/P(F) = 0.2(0.6)/0.36 = 0.333$$
$$P(O|F) = P(F|O)P(O)/P(F) = 0.5(0.3)/0.36 = 0.417$$

Notice that $P(H|F) + P(S|F) + P(O|F) = 1$ because one of the three types of problems is responsible for the failure. Because $P(O|F)$ is largest, the most likely cause of the problem is in the *other* category. A Web site dialog to diagnose the problem quickly should start with a check into that type of problem.

EXERCISES FOR SECTION 2-7

2-116. Suppose that $P(A|B) = 0.7$, $P(A) = 0.5$, and $P(B) = 0.2$. Determine $P(B|A)$.

2-117. Suppose that $P(A|B) = 0.4$, $P(A|B') = 0.2$, $P(B) = 0.8$. Determine $P(B|A)$.

2-118. Software to detect fraud in consumer phone cards tracks the number of metropolitan areas where calls originate each day. It is found that 1% of the legitimate users originate calls from two or more metropolitan areas in a single day. However, 30% of fraudulent users originate calls from two or more metropolitan areas in a single day. The proportion of fraudulent users is 0.01%. If the same user originates calls from two or more metropolitan areas in a single day, what is the probability that the user is fraudulent?

2-119. A new process of more accurately detecting anaerobic respiration in cells is being tested. The new process is important due to its high accuracy, its lack of extensive experimentation, and the fact that it could be used to identify five different categories of organisms: obligate anaerobes, facultative anaerobes, aerotolerant, microaerophiles, and nanaerobes instead of using a single test for each category. The process claims that it can identify obligate anaerobes with 97.8% accuracy, facultative anaerobes with 98.1% accuracy, aerotolerant with 95.9% accuracy, microaerophiles with 96.5% accuracy, and nanaerobes with 99.2% accuracy. If any category is not present, the process does not signal. Samples are prepared for the calibration of the process and 31% of them contain obligate anaerobes, 27% contain facultative anaerobes, 21% contain microaerophiles, 13% contain nanaerobes, and 8% contain aerotolerant. A test sample is selected randomly.

(a) What is the probability that the process will signal?

(b) If the test signals, what is the probability that microaerophiles are present?

2-120. In the 2004 presidential election, exit polls from the critical state of Ohio provided the following results:

	Bush	Kerry
no college degree (62%)	50%	50%
college graduate (38%)	53%	46%

If a randomly selected respondent voted for Bush, what is the probability that the person has a college degree?

2-121. Customers are used to evaluate preliminary product designs. In the past, 95% of highly successful products received good reviews, 60% of moderately successful products received good reviews, and 10% of poor products received good reviews. In addition, 40% of products have been highly successful, 35% have been moderately successful, and 25% have been poor products.

(a) What is the probability that a product attains a good review?

(b) If a new design attains a good review, what is the probability that it will be a highly successful product?

(c) If a product does not attain a good review, what is the probability that it will be a highly successful product?

2-122. An inspector working for a manufacturing company has a 99% chance of correctly identifying defective items and a 0.5% chance of incorrectly classifying a good item as defective. The company has evidence that its line produces 0.9% of nonconforming items.

(a) What is the probability that an item selected for inspection is classified as defective?

(b) If an item selected at random is classified as nondefective, what is the probability that it is indeed good?

2-123. A new analytical method to detect pollutants in water is being tested. This new method of chemical analysis is important because, if adopted, it could be used to detect three different contaminants—organic pollutants, volatile solvents, and chlorinated compounds—instead of having to use a single test for each pollutant. The makers of the test claim that it can detect high levels of organic pollutants with 99.7% accuracy, volatile solvents with 99.95% accuracy, and chlorinated compounds with 89.7% accuracy. If a pollutant is not present, the test does not signal. Samples are prepared for the calibration of the test and 60% of them are contaminated with organic pollutants, 27% with volatile solvents, and 13% with traces of chlorinated compounds. A test sample is selected randomly.

(a) What is the probability that the test will signal?

(b) If the test signals, what is the probability that chlorinated compounds are present?

2-8 RANDOM VARIABLES

We often summarize the outcome from a random experiment by a simple number. In many of the examples of random experiments that we have considered, the sample space has been a description of possible outcomes. In some cases, descriptions of outcomes are sufficient, but in other cases, it is useful to associate a number with each outcome in the sample space. Because the particular outcome of the experiment is not known in advance, the resulting value of our variable is not known in advance. For this reason, the variable that associates a number with the outcome of a random experiment is referred to as a **random variable.**

Random Variable

A **random variable** is a function that assigns a real number to each outcome in the sample space of a random experiment.

Notation is used to distinguish between a random variable and the real number.

A random variable is denoted by an uppercase letter such as X. After an experiment is conducted, the measured value of the random variable is denoted by a lowercase letter such as $x = 70$ milliamperes.

Sometimes a measurement (such as current in a copper wire or length of a machined part) can assume any value in an interval of real numbers (at least theoretically). Then arbitrary precision in the measurement is possible. Of course, in practice, we might round off to the nearest tenth or hundredth of a unit. The random variable that represents this measurement is said to be a **continuous** random variable. The range of the random variable includes all values in an interval of real numbers; that is, the range can be thought of as a continuum.

In other experiments, we might record a count such as the number of transmitted bits that are received in error. Then the measurement is limited to integers. Or we might record that a proportion such as 0.0042 of the 10,000 transmitted bits were received in error. Then the measurement is fractional, but it is still limited to discrete points on the real line. Whenever the measurement is limited to discrete points on the real line, the random variable is said to be a **discrete** random variable.

Discrete and Continuous Random Variables

A **discrete** random variable is a random variable with a finite (or countably infinite) range.
A **continuous** random variable is a random variable with an interval (either finite or infinite) of real numbers for its range.

In some cases, the random variable X is actually discrete but, because the range of possible values is so large, it might be more convenient to analyze X as a continuous random variable. For example, suppose that current measurements are read from a digital instrument that displays the current to the nearest one-hundredth of a milliampere. Because the possible measurements are limited, the random variable is discrete. However, it might be a more convenient, simple approximation to assume that the current measurements are values of a continuous random variable.

Examples of Random Variables

Examples of **continuous** random variables:
electrical current, length, pressure, temperature, time, voltage, weight

Examples of **discrete** random variables:
number of scratches on a surface, proportion of defective parts among 1000 tested, number of transmitted bits received in error

EXERCISES FOR SECTION 2-8

2-124. Decide whether a discrete or continuous random variable is the best model for each of the following variables:

(a) The time until a projectile returns to earth.
(b) The number of times a transistor in a computer memory changes state in one operation.
(c) The volume of gasoline that is lost to evaporation during the filling of a gas tank.
(d) The outside diameter of a machined shaft.
(e) The number of cracks exceeding one-half inch in 10 miles of an interstate highway.
(f) The weight of an injection-molded plastic part.
(g) The number of molecules in a sample of gas.
(h) The concentration of output from a reactor.
(i) The current in an electronic circuit.

Supplemental Exercises

2-125. Samples of laboratory glass are in small, light packaging or heavy, large packaging. Suppose that 2 and 1% of the sample shipped in small and large packages, respectively, break during transit. If 60% of the samples are shipped in large packages and 40% are shipped in small packages, what proportion of samples break during shipment?

2-126. A sample of three calculators is selected from a manufacturing line, and each calculator is classified as either defective or acceptable. Let A, B, and C denote the events that the first, second, and third calculators respectively, are defective.

(a) Describe the sample space for this experiment with a tree diagram.

Use the tree diagram to describe each of the following events:

(b) A (c) B
(d) $A \cap B$ (e) $B \cup C$

2-127. Samples of a cast aluminum part are classified on the basis of surface finish (in microinches) and edge finish. The results of 100 parts are summarized as follows:

		edge finish	
		excellent	good
surface finish	excellent	80	2
	good	10	8

Let A denote the event that a sample has excellent surface finish, and let B denote the event that a sample has excellent edge finish. If a part is selected at random, determine the following probabilities:

(a) $P(A)$ (b) $P(B)$
(c) $P(A')$ (d) $P(A \cap B)$
(e) $P(A \cup B)$ (f) $P(A' \cup B)$

2-128. Shafts are classified in terms of the machine tool that was used for manufacturing the shaft and conformance to surface finish and roundness.

2-129. If A, B, and C are mutually exclusive events, is it possible for $P(A) = 0.3$, $P(B) = 0.4$, and $P(C) = 0.5$? Why or why not?

Tool 1

		roundness conforms	
		yes	no
surface finish conforms	yes	200	1
	no	4	2

Tool 2

		roundness conforms	
		yes	no
surface finish conforms	yes	145	4
	no	8	6

(a) If a shaft is selected at random, what is the probability that the shaft conforms to surface finish requirements or to roundness requirements or is from Tool 1?

(b) If a shaft is selected at random, what is the probability that the shaft conforms to surface finish requirements or does not conform to roundness requirements or is from Tool 2?

(c) If a shaft is selected at random, what is the probability that the shaft conforms to both surface finish and roundness requirements or the shaft is from Tool 2?

(d) If a shaft is selected at random, what is the probability that the shaft conforms to surface finish requirements or the shaft is from Tool 2?

2-130. The analysis of shafts for a compressor is summarized by conformance to specifications:

		roundness conforms	
		yes	no
surface finish conforms	yes	345	5
	no	12	8

(a) If we know that a shaft conforms to roundness requirements, what is the probability that it conforms to surface finish requirements?

(b) If we know that a shaft does not conform to roundness requirements, what is the probability that it conforms to surface finish requirements?

2-131. A researcher receives 100 containers of oxygen. Of those containers, twenty of them have oxygen that is not ionized and the rest are ionized. Two samples are randomly selected, without replacement, from the lot.

(a) What is the probability that the first one selected is not ionized?

(b) What is the probability that the second one selected is not ionized given that the first one was ionized?

(c) What is the probability that both are ionized?

(d) How does the answer in part (b) change if samples selected were replaced prior to the next selection?

2-132. A lot contains 15 castings from a local supplier and 25 castings from a supplier in the next state. Two castings are selected randomly, without replacement, from the lot of 40. Let A be the event that the first casting selected is from the local supplier, and let B denote the event that the second casting is selected from the local supplier. Determine:

(a) $P(A)$ (b) $P(B|A)$
(c) $P(A \cap B)$ (d) $P(A \cup B)$

Suppose three castings are selected at random, without replacement, from the lot of 40. In addition to the definitions of events A and B, let C denote the event that the third casting selected is from the local supplier. Determine:

(e) $P(A \cap B \cap C)$
(f) $P(A \cap B \cap C')$

2-133. In the manufacturing of a chemical adhesive, 3% of all batches have raw materials from two different lots. This occurs when holding tanks are replenished and the remaining portion of a lot is insufficient to fill the tanks.

Only 5% of batches with material from a single lot require reprocessing. However, the viscosity of batches consisting of two or more lots of material is more difficult to control, and 40% of such batches require additional processing to achieve the required viscosity.

Let A denote the event that a batch is formed from two different lots, and let B denote the event that a lot requires additional processing. Determine the following probabilities:

(a) $P(A)$ (b) $P(A')$
(c) $P(B|A)$ (d) $P(B|A')$
(e) $P(A \cap B)$ (f) $P(A \cap B')$
(g) $P(B)$

2-134. Incoming calls to a customer service center are classified as complaints (75% of call) or requests for information (25% of calls). Of the complaints, 40% deal with computer equipment that does not respond and 57% deal with incomplete software installation; and in the remaining 3% of complaints the user has improperly followed the installation instructions. The requests for information are evenly divided on technical questions (50%) and requests to purchase more products (50%).

(a) What is the probability that an incoming call to the customer service center will be from a customer who has not followed installation instructions properly?
(b) Find the probability that an incoming call is a request for purchasing more products.

2-135. A congested computer network has a 0.002 probability of losing a data packet and packet losses are independent events. A lost packet must be resent.

(a) What is the probability that an e-mail message with 100 packets will need any resent?
(b) What is the probability that an e-mail message with 3 packets will need exactly one to be resent?
(c) If 10 e-mail messages are sent, each with 100 packets, what is the probability that at least one message will need some packets to be resent?

2-136. Samples of a cast aluminum part are classified on the basis of surface finish (in microinches) and length measurements. The results of 100 parts are summarized as follows:

		length	
		excellent	good
surface	excellent	80	2
finish	good	10	8

Let A denote the event that a sample has excellent surface finish, and let B denote the event that a sample has excellent length. Are events A and B independent?

2-137. An optical storage device uses an error recovery procedure that requires an immediate satisfactory readback of any written data. If the readback is not successful after three writing operations, that sector of the disk is eliminated as unacceptable for data storage. On an acceptable portion of the disk, the probability of a satisfactory readback is 0.98. Assume the readbacks are independent. What is the probability that an acceptable portion of the disk is eliminated as unacceptable for data storage?

2-138. Semiconductor lasers used in optical storage products require higher power levels for write operations than for read operations. High-power-level operations lower the useful life of the laser.

Lasers in products used for backup of higher speed magnetic disks primarily write, and the probability that the useful life exceeds five years is 0.95. Lasers that are in products that are used for main storage spend approximately an equal amount of time reading and writing, and the probability that the useful life exceeds five years is 0.995. Now, 25% of the products from a manufacturer are used for backup and 75% of the products are used for main storage.

Let A denote the event that a laser's useful life exceeds five years, and let B denote the event that a laser is in a product that is used for backup.

Use a tree diagram to determine the following:

(a) $P(B)$ (b) $P(A|B)$
(c) $P(A|B')$ (d) $P(A \cap B)$
(e) $P(A \cap B')$ (f) $P(A)$
(g) What is the probability that the useful life of a laser exceeds five years?
(h) What is the probability that a laser that failed before five years came from a product used for backup?

2-139. Energy released from three cells breaks the molecular bond and converts ATP (Adenosine Tri-Phosphate) into ADP (adenosine diphosphate). Storage of ATP in muscle cells (even for an athlete) can only sustain maximal muscle power for less than five seconds (a short dash). Three systems are used to replenish ATP: phosphagen system; glycogen-lactic acid system (anaerobic); and aerobic respiration, but the first is only useful for less than 10 seconds, and even the second system provides less than two minutes of ATP. An endurance athlete needs to perform below the anaerobic threshold to sustain energy for extended periods. A sample of 100 individuals is described by the energy system used in exercise at different intensity levels.

	primarily aerobic	
period	yes	no
1	50	7
2	13	30

Let A denote the event that an individual is in period 2, and let B denote the event that the energy is primarily aerobic. Determine the number of individuals in

(a) $A' \cap B$ (b) B' (c) $A \cup B$

2-140. A sample preparation for a chemical measurement is completed correctly by 25% of the lab technicians, completed with a minor error by 70%, and completed with a major error by 5%.

(a) If a technician is selected randomly to complete the preparation, what is the probability it is completed without error?
(b) What is the probability that it is completed with either a minor or a major error?

2-141. In circuit testing of printed circuit boards, each board either fails or does not fail the test. A board that fails the test is then checked further to determine which one of five defect types is the primary failure mode. Represent the sample space for this experiment.

2-142. The data from 200 machined parts are summarized as follows:

	depth of bore	
edge condition	above target	below target
coarse	15	10
moderate	25	20
smooth	50	80

(a) What is the probability that a part selected has a moderate edge condition and a below-target bore depth?
(b) What is the probability that a part selected has a moderate edge condition or a below-target bore depth?
(c) What is the probability that a part selected does not have a moderate edge condition or does not have a below-target bore depth?

2-143. Computers in a shipment of 100 units contain a portable hard drive, CD RW drive, or both according to the following table:

	portable hard drive	
	yes	no
CD RW		
yes	15	80
no	4	1

Let A denote the events that a computer has a portable hard drive and let B denote the event that a computer has a CD RW drive. If one computer is selected randomly, compute
(a) $P(A)$ (b) $P(A \cap B)$ (c) $P(A \cup B)$
(d) $P(A' \cap B)$ (e) $P(A|B)$

2-144. The probability that a customer's order is not shipped on time is 0.05. A particular customer places three orders, and the orders are placed far enough apart in time that they can be considered to be independent events.
(a) What is the probability that all are shipped on time?
(b) What is the probability that exactly one is not shipped on time?
(c) What is the probability that two or more orders are not shipped on time?

2-145. Let E_1, E_2, and E_3 denote the samples that conform to a percentage of solids specification, a molecular weight specification, and a color specification, respectively. A total of 240 samples are classified by the E_1, E_2, and E_3 specifications, where *yes* indicates that the sample conforms.

E_3 yes

		E_2		
		yes	no	Total
E_1	yes	200	1	201
	no	5	4	9
Total		205	5	210

E_3 no

		E_2		
		yes	no	
E_1	yes	20	4	24
	no	6	0	6
Total		26	4	30

(a) Are E_1, E_2, and E_3 mutually exclusive events?
(b) Are E'_1, E'_2, and E'_3 mutually exclusive events?
(c) What is $P(E'_1 \text{ or } E'_2 \text{ or } E'_3)$?
(d) What is the probability that a sample conforms to all three specifications?
(e) What is the probability that a sample conforms to the E_1 or E_3 specification?
(f) What is the probability that a sample conforms to the E_1 or E_2 or E_3 specification?

2-146. Transactions to a computer database are either new items or changes to previous items. The addition of an item can be completed in less than 100 milliseconds 90% of the time, but only 20% of changes to a previous item can be completed in less than this time. If 30% of transactions are changes, what is the probability that a transaction can be completed in less than 100 milliseconds?

2-147. A steel plate contains 20 bolts. Assume that 5 bolts are not torqued to the proper limit. Four bolts are selected at random, without replacement, to be checked for torque.
(a) What is the probability that all four of the selected bolts are torqued to the proper limit?
(b) What is the probability that at least one of the selected bolts is not torqued to the proper limit?

2-148. The following circuit operates if and only if there is a path of functional devices from left to right. Assume devices fail independently and that the probability of *failure* of each device is as shown. What is the probability that the circuit operates?

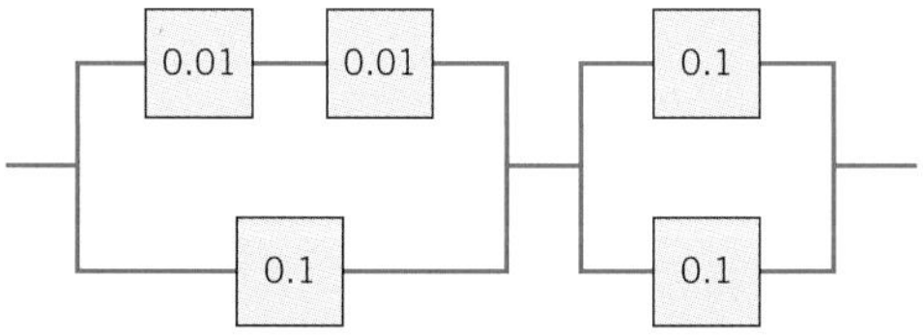

2-149. The probability of getting through by telephone to buy concert tickets is 0.92. For the same event, the probability of accessing the vendor's Web site is 0.95. Assume that these two ways to buy tickets are independent. What is the probability that someone who tries to buy tickets through the Internet and by phone will obtain tickets?

2-150. The British government has stepped up its information campaign regarding foot and mouth disease by mailing brochures to farmers around the country. It is estimated that 99% of Scottish farmers who receive the brochure possess enough information to deal with an outbreak of the disease, but only 90% of those without the brochure can deal with an outbreak. After the first three months of mailing, 95% of the farmers in Scotland received the informative brochure. Compute the probability that a randomly selected farmer will have enough information to deal effectively with an outbreak of the disease.

2-151. In an automated filling operation, the probability of an incorrect fill when the process is operated at a low speed is 0.001. When the process is operated at a high speed, the probability of an incorrect fill is 0.01. Assume that 30% of the containers are filled when the process is operated at a high speed and the remainder are filled when the process is operated at a low speed.

(a) What is the probability of an incorrectly filled container?
(b) If an incorrectly filled container is found, what is the probability that it was filled during the high-speed operation?

2-152. An encryption-decryption system consists of three elements: encode, transmit, and decode. A faulty encode occurs in 0.5% of the messages processed, transmission errors occur in 1% of the messages, and a decode error occurs in 0.1% of the messages. Assume the errors are independent.

(a) What is the probability of a completely defect-free message?
(b) What is the probability of a message that has either an encode or a decode error?

2-153. It is known that two defective copies of a commercial software program were erroneously sent to a shipping lot that has now a total of 75 copies of the program. A sample of copies will be selected from the lot without replacement.

(a) If three copies of the software are inspected, determine the probability that exactly one of the defective copies will be found.
(b) If three copies of the software are inspected, determine the probability that both defective copies will be found.
(c) If 73 copies are inspected, determine the probability that both copies will be found. Hint: Work with the copies that remain in the lot.

2-154. A robotic insertion tool contains 10 primary components. The probability that any component fails during the warranty period is 0.01. Assume that the components fail independently and that the tool fails if any component fails. What is the probability that the tool fails during the warranty period?

2-155. An e-mail message can travel through one of two server routes. The probability of transmission error in each of the servers and the proportion of messages that travel each route are shown in the following table. Assume that the servers are independent.

		probability of error			
	percentage of messages	server 1	server 2	server 3	server 4
route 1	30	0.01	0.015		
route 2	70			0.02	0.003

(a) What is the probability that a message will arrive without error?
(b) If a message arrives in error, what is the probability it was sent through route 1?

2-156. A machine tool is idle 15% of the time. You request immediate use of the tool on five different occasions during the year. Assume that your requests represent independent events.

(a) What is the probability that the tool is idle at the time of all of your requests?
(b) What is the probability that the machine is idle at the time of exactly four of your requests?
(c) What is the probability that the tool is idle at the time of at least three of your requests?

2-157. A lot of 50 spacing washers contains 30 washers that are thicker than the target dimension. Suppose that three washers are selected at random, without replacement, from the lot.

(a) What is the probability that all three washers are thicker than the target?
(b) What is the probability that the third washer selected is thicker than the target if the first two washers selected are thinner than the target?
(c) What is the probability that the third washer selected is thicker than the target?

2-158. Continuation of Exercise 2-157. Washers are selected from the lot at random, without replacement.

(a) What is the minimum number of washers that need to be selected so that the probability that all the washers are thinner than the target is less than 0.10?
(b) What is the minimum number of washers that need to be selected so that the probability that one or more washers are thicker than the target is at least 0.90?

2-159. The following table lists the history of 940 orders for features in an entry-level computer product.

		extra memory	
		no	yes
optional high-	no	514	68
speed processor	yes	112	246

Let A be the event that an order requests the optional high-speed processor, and let B be the event that an order requests extra memory. Determine the following probabilities:

(a) $P(A \cup B)$ (b) $P(A \cap B)$
(c) $P(A' \cup B)$ (d) $P(A' \cap B')$
(e) What is the probability that an order requests an optional high-speed processor given that the order requests extra memory?
(f) What is the probability that an order requests extra memory given that the order requests an optional high-speed processor?

2-160. The alignment between the magnetic tape and head in a magnetic tape storage system affects the performance of the system. Suppose that 10% of the read operations are degraded by skewed alignments, 5% of the read operations are degraded by off-center alignments, and the remaining read operations are properly aligned. The probability of a read error is 0.01 from a skewed alignment, 0.02 from an off-center alignment, and 0.001 from a proper alignment.

(a) What is the probability of a read error?

(b) If a read error occurs, what is the probability that it is due to a skewed alignment?

2-161. The following circuit operates if and only if there is a path of functional devices from left to right. Assume that devices fail independently and that the probability of *failure* of each device is as shown. What is the probability that the circuit does not operate?

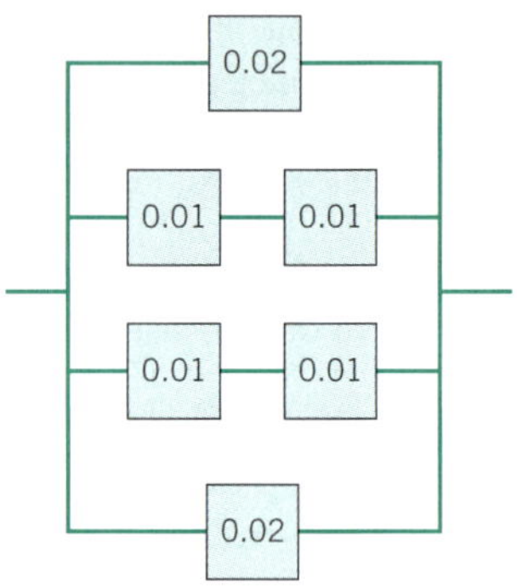

2-162. A company that tracks the use of its web site determined that the more pages a visitor views, the more likely the visitor is to provide contact information. Use the following tables to answer the questions:

Number of pages viewed:	1	2	3	4 or more
Percentage of visitors:	40	30	20	10
Percentage of visitors in each page-view category that provide contact information:	10	10	20	40

(a) What is the probability that a visitor to the web site provides contact information?

(b) If a visitor provides contact information, what is the probability that the visitor viewed four or more pages?

2-163. An article in *Genome Research* "An Assessment of Gene Prediction Accuracy in Large DNA Sequences," (2000, Vol. 10, pp. 1631–1642), considered the accuracy of commercial software to predict nucleotides in gene sequences. The following table shows the number of sequences for which the programs produced predictions and the number of nucleotides correctly predicted (computed globally from the total number of prediction successes and failures on all sequences).

	Number of sequences	Proportion
GenScan	177	0.93
Blastx default	175	0.91
Blastx topcomboN	174	0.97
Blastx 2 stages	175	0.90
GeneWise	177	0.98
Procrustes	177	0.93

Assume the prediction successes and failures are independent between the programs.

(a) What is the probability that all programs predict a nucleotide correctly?

(b) What is the probability that all programs predict a nucleotide incorrectly?

(c) What is the probability that at least one Blastx program predicts a nucleotide correctly?

2-164. A batch contains 36 bacteria cells. Assume that 12 of the cells are not capable of cellular replication. Six cells are selected at random, without replacement, to be checked for replication.

(a) What is the probability that all six cells of the selected cells are able to replicate?

(b) What is the probability that at least one of the selected cells is not capable of replication?

2-165. A computer system uses passwords that are exactly seven characters and each character is one of the 26 letters (a–z) or 10 integers (0–9). Uppercase letters are not used.

(a) How many passwords are possible?

(b) If a password consists of exactly six letters and one number, how many passwords are possible?

(c) If a password consists of five letters followed by two numbers, how many passwords are possible?

2-166. Natural red hair consists of two genes. People with red hair have two dominant genes, two regressive genes, or one dominant and one regressive gene. A group of 1000 people was categorized as follows:

	Gene 2		
Gene 1	Dominant	Regressive	Other
Dominant	5	25	30
Regressive	7	63	35
Other	20	15	800

Let A denote the event that a person has a dominant red-hair gene and let B denote the event that a person has a regressive red hair gene. If a person is selected at random from this group, compute the following.

(a) $P(A)$ (b) $P(A \cap B)$

(c) $P(A \cup B)$ (d) $P(A' \cap B)$ (e) $P(A \mid B)$

(f) Probability that the selected person has red hair

2-167. Two suppliers each supplied 2000 parts and these were evaluated for conformance to specifications. One part type was of greater complexity than the other. The proportion of nonconforming parts of each type are shown in the table.

Supplier	Simple component	Complex assembly	Total
1 Nonconforming	2	10	12
Total	1000	1000	2000
2 Nonconforming	4	6	10
Total	1600	400	2000

One part is selected at random from each supplier. For each supplier, separately calculate the following probabilities:

(a) What is the probability a part conforms to specifications?
(b) What is the probability a part conforms to specifications given it is a complex assembly?
(c) What is the probability a part conforms to specifications given it is a simple component?
(d) Compare your answers for each supplier in part (a) to those in parts (b) and (c) and explain any unusual results.

MIND-EXPANDING EXERCISES

2-168. The alignment between the magnetic tape and head in a magnetic tape storage system affects the performance of the system. Suppose that 10% of the read operations are degraded by skewed alignments, 5% by off-center alignments, 1% by both skewness and off-center, and the remaining read operations are properly aligned. The probability of a read error is 0.01 from a skewed alignment, 0.02 from an off-center alignment, 0.06 from both conditions, and 0.001 from a proper alignment. What is the probability of a read error?

2-169. Suppose that a lot of washers is large enough that it can be assumed that the sampling is done with replacement. Assume that 60% of the washers exceed the target thickness.

(a) What is the minimum number of washers that need to be selected so that the probability that none are thicker than the target is less than 0.10?
(b) What is the minimum number of washers that need to be selected so that the probability that one or more washers are thicker than the target is at least 0.90?

2-170. A biotechnology manufacturing firm can produce diagnostic test kits at a cost of \$20. Each kit for which there is a demand in the week of production can be sold for \$100. However, the half-life of components in the kit requires the kit to be scrapped if it is not sold in the week of production. The cost of scrapping the kit is \$5. The weekly demand is summarized as follows:

	weekly demand			
Number of units	0	50	100	200
Probability of demand	0.05	0.4	0.3	0.25

How many kits should be produced each week to maximize the mean earnings of the firm?

2-171. Assume the following characteristics of the inspection process in Exercise 2-147. If an operator checks a bolt, the probability that an incorrectly torqued bolt is identified is 0.95. If a checked bolt is correctly torqued, the operator's conclusion is always correct. What is the probability that at least one bolt in the sample of four is identified as being incorrectly torqued?

2-172. If the events A and B are independent, show that A' and B' are independent.

2-173. Suppose that a table of part counts is generalized as follows:

		conforms	
		yes	no
supplier	1	ka	kb
	2	a	b

where a, b, and k are positive integers. Let A denote the event that a part is from supplier 1 and let B denote the event that a part conforms to specifications. Show that A and B are independent events.

This exercise illustrates the result that whenever the rows of a table (with r rows and c columns) are proportional, an event defined by a row category and an event defined by a column category are independent.

IMPORTANT TERMS AND CONCEPTS

Addition rule
Axioms of probability
Bayes' theorem
Combination
Conditional probability
Equally likely outcomes
Event
Independence
Multiplication rule
Mutually exclusive events
Outcome
Permutation
Probability
Random experiment
Random variables—discrete and continuous
Sample spaces—discrete and continuous
Simpson's paradox
Total probability rule
Tree Diagram
Venn Diagram
With or without replacement

3 Analysis of Discrete Random Variables

CHAPTER OUTLINE

LEARNING OBJECTIVES

After careful study of this chapter you should be able to do the following:

1. Determine probabilities from probability mass functions and the reverse
2. Determine probabilities from cumulative distribution functions and cumulative distribution functions from probability mass functions, and the reverse
3. Calculate means and variances for discrete random variables
4. Understand the assumptions for each of the discrete probability distributions presented
5. Select an appropriate discrete probability distribution to calculate probabilities in specific applications
6. Calculate probabilities, determine means and variances for each of the discrete probability distributions presented

3-1 DISCRETE RANDOM VARIABLES

Many physical systems can be modeled by the same or similar random experiments and random variables. The distribution of the random variables involved in each of these common systems can be analyzed, and the results of that analysis can be used in different applications and examples. In this chapter, we present the analysis of several random experiments and **discrete random variables** that frequently arise in applications. We often omit a discussion of the underlying sample space of the random experiment and directly describe the distribution of a particular random variable.

EXAMPLE 3-1
Voice Lines

A voice communication system for a business contains 48 external lines. At a particular time, the system is observed, and some of the lines are being used. Let the random variable X denote the number of lines in use. Then, X can assume any of the integer values 0 through 48. When the system is observed, if 10 lines are in use, $x = 10$.

EXAMPLE 3-2

In a semiconductor manufacturing process, two wafers from a lot are tested. Each wafer is classified as *pass* or *fail*. Assume that the probability that a wafer passes the test is 0.8 and that wafers are independent. The sample space for the experiment and associated probabilities are shown in Table 3-1. For example, because of the independence, the probability of the outcome that the first wafer tested passes and the second wafer tested fails, denoted as pf, is

$$P(pf) = 0.8(0.2) = 0.16$$

The random variable X is defined to be equal to the number of wafers that pass. The last column of the table shows the values of X that are assigned to each outcome in the experiment.

EXAMPLE 3-3

Define the random variable X to be the number of contamination particles on a wafer in semiconductor manufacturing. Although wafers possess a number of characteristics, the random variable X summarizes the wafer only in terms of the number of particles.

The possible values of X are integers from zero up to some large value that represents the maximum number of particles that can be found on one of the wafers. If this maximum number is very large, we might simply assume that the range of X is the set of integers from zero to infinity.

Note that more than one random variable can be defined on a sample space. In Example 3-3, we might define the random variable Y to be the number of chips from a wafer that fail the final test.

Table 3-1 Wafer Tests

Outcome			
Wafer 1	Wafer 2	Probability	x
Pass	Pass	0.64	2
Fail	Pass	0.16	1
Pass	Fail	0.16	1
Fail	Fail	0.04	0

EXERCISES FOR SECTION 3-1

For each of the following exercises, determine the range (possible values) of the random variable.

3-1. The random variable is the number of nonconforming solder connections on a printed circuit board with 1000 connections.

3-2. In a voice communication system with 50 lines, the random variable is the number of lines in use at a particular time.

3-3. An electronic scale that displays weights to the nearest pound is used to weigh packages. The display shows only five digits. Any weight greater than the display can indicate is shown as 99999. The random variable is the displayed weight.

3-4. A batch of 500 machined parts contains 10 that do not conform to customer requirements. The random variable is the number of parts in a sample of 5 parts that do not conform to customer requirements.

3-5. A batch of 500 machined parts contains 10 that do not conform to customer requirements. Parts are selected successively, without replacement, until a nonconforming part is obtained. The random variable is the number of parts selected.

3-6. The random variable is the moisture content of a lot of raw material, measured to the nearest percentage point.

3-7. The random variable is the number of surface flaws in a large coil of galvanized steel.

3-8. The random variable is the number of computer clock cycles required to complete a selected arithmetic calculation.

3-9. An order for an automobile can select the base model or add any number of 15 options. The random variable is the number of options selected in an order.

3-10. Wood paneling can be ordered in thicknesses of 1/8, 1/4, or 3/8 inch. The random variable is the total thickness of paneling in two orders.

3-11. A group of 10,000 people are tested for a gene called Ifi202 that has been found to increase the risk for lupus. The random variable is the number of people who carry the gene.

3-12. In an acid-base titration, the milliliters of base that are needed to reach equivalence are measured to the nearest milliliter between 0.1 and 0.15 liters (inclusive).

3-13. The number of mutations in a nucleotide sequence of length 40,000 in a DNA strand after exposure to radiation is measured. Each nucleotide may be mutated.

3-2 PROBABILITY DISTRIBUTIONS AND PROBABILITY MASS FUNCTIONS

Random variables are so important in random experiments that sometimes we essentially ignore the original sample space of the experiment and focus on the probability distribution of the random variable. For example, in Example 3-1, our analysis might focus exclusively on the integers $\{0, 1, \ldots, 48\}$ in the range of X. In Example 3-2, we might summarize the random experiment in terms of the three possible values of X, namely $\{0, 1, 2\}$. In this manner, a random variable can simplify the description and analysis of a random experiment.

The **probability distribution** of a random variable X is a description of the probabilities associated with the possible values of X. For a discrete random variable, the distribution is often specified by just a list of the possible values along with the probability of each. In some cases, it is convenient to express the probability in terms of a formula.

EXAMPLE 3-4
Digital Channel

There is a chance that a bit transmitted through a digital transmission channel is received in error. Let X equal the number of bits in error in the next four bits transmitted. The possible values for X are $\{0, 1, 2, 3, 4\}$. Based on a model for the errors that is presented in the following section, probabilities for these values will be determined. Suppose that the probabilities are

$$P(X = 0) = 0.6561 \qquad P(X = 1) = 0.2916 \qquad P(X = 2) = 0.0486$$

$$P(X = 3) = 0.0036 \qquad P(X = 4) = 0.0001$$

The probability distribution of X is specified by the possible values along with the probability of each. A graphical description of the probability distribution of X is shown in Fig. 3-1.

Suppose a loading on a long, thin beam places mass only at discrete points. See Fig. 3-2. The loading can be described by a function that specifies the mass at each of the discrete

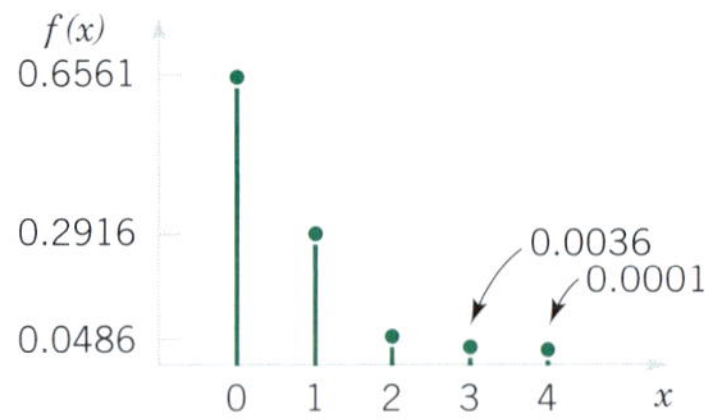

Figure 3-1 Probability distribution for bits in error.

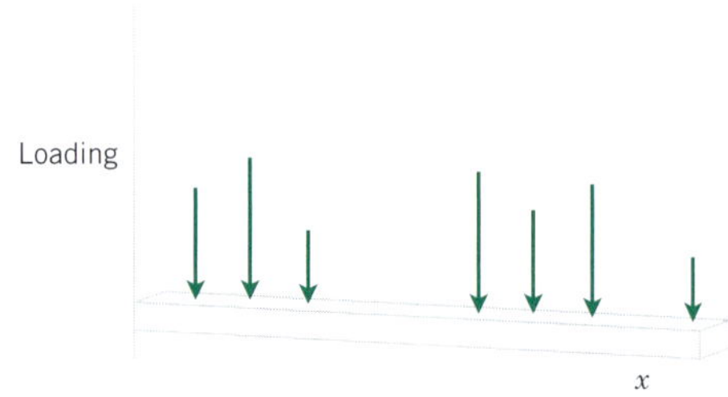

Figure 3-2 Loadings at discrete points on a long, thin beam.

points. Similarly, for a discrete random variable X, its distribution can be described by a function that specifies the probability at each of the possible discrete values for X.

Probability Mass Function

For a discrete random variable X with possible values $x_1, x_2, \ldots, x_n$, a **probability mass function** is a function such that

(1) $f(x_i) \geq 0$

(2) $\sum_{i=1}^{n} f(x_i) = 1$

(3) $f(x_i) = P(X = x_i)$ (3-1)

For example, in Example 3-4, $f(0) = 0.6561, f(1) = 0.2916, f(2) = 0.0486, f(3) = 0.0036$, and $f(4) = 0.0001$. Check that the sum of the probabilities in Example 3-4 is 1.

EXAMPLE 3-5 Wafer Contamination

Let the random variable X denote the number of semiconductor wafers that need to be analyzed in order to detect a large particle of contamination. Assume that the probability that a wafer contains a large particle is 0.01 and that the wafers are independent. Determine the probability distribution of X.

Let p denote a wafer in which a large particle is present, and let a denote a wafer in which it is absent. The sample space of the experiment is infinite, and it can be represented as all possible sequences that start with a string of a's and end with p. That is,

$$s = \{p, ap, aap, aaap, aaaap, aaaaap, \text{and so forth}\}$$

Consider a few special cases. We have $P(X = 1) = P(p) = 0.01$. Also, using the independence assumption

$$P(X = 2) = P(ap) = 0.99(0.01) = 0.0099$$

A general formula is

$$P(X = x) = \underbrace{P(aa \ldots ap)}_{(x-1)a\text{'s}} = 0.99^{x-1}(0.01), \qquad \text{for } x = 1, 2, 3, \ldots$$

Describing the probabilities associated with X in terms of this formula is the simplest method of describing the distribution of X in this example. Clearly $f(x) \geq 0$. The fact that the sum of the probabilities is 1 is left as an exercise. This is an example of a geometric random variable, and details are provided later in this chapter.

EXERCISES FOR SECTION 3-2

3-14. The sample space of a random experiment is $\{a, b, c, d, e, f\}$, and each outcome is equally likely. A random variable is defined as follows:

outcome	a	b	c	d	e	f
x	0	0	1.5	1.5	2	3

Determine the probability mass function of X. Use the probability mass function to determine the following probabilities:

(a) $P(X = 1.5)$ (b) $P(0.5 < X < 2.7)$
(c) $P(X > 3)$ (d) $P(0 \le X < 2)$
(e) $P(X = 0 \text{ or } X = 2)$

For Exercises 3-15 to 3-18 verify that the following functions are probability mass functions, and determine the requested probabilities.

3-15.

x	-2	-1	0	1	2
$f(x)$	1/8	2/8	2/8	2/8	1/8

(a) $P(X \le 2)$ (b) $P(X > -2)$
(c) $P(-1 \le X \le 1)$ (d) $P(X \le -1 \text{ or } X = 2)$

3-16. $f(x) = (8/7)(1/2)^x, \quad x = 1, 2, 3$
(a) $P(X \le 1)$ (b) $P(X > 1)$
(c) $P(2 < X < 6)$ (d) $P(X \le 1 \text{ or } X > 1)$

3-17. $f(x) = \dfrac{2x + 1}{25}, \quad x = 0, 1, 2, 3, 4$
(a) $P(X = 4)$ (b) $P(X \le 1)$
(c) $P(2 \le X < 4)$ (d) $P(X > -10)$

3-18. $f(x) = (3/4)(1/4)^x, \quad x = 0, 1, 2, \ldots$
(a) $P(X = 2)$ (b) $P(X \le 2)$
(c) $P(X > 2)$ (d) $P(X \ge 1)$

3-19. An article in *Knee Surgery, Sports Traumatology, Arthroscopy*, "Arthroscopic meniscal repair with an absorbable screw: results and surgical technique," (2005, Vol. 13, pp. 273–279) cites a success rate more than 90% for meniscal tears with a rim width of less than 3 mm, but only a 67% success rate for tears from 3–6 mm. If you are unlucky enough to suffer a meniscal tear of less than 3 mm on your left knee, and one of width 3–6 mm on your right knee, what is the probability mass function of the number of successful surgeries? Assume the surgeries are independent.

3-20. An optical inspection system is to distinguish among different part types. The probability of a correct classification of any part is 0.98. Suppose that three parts are inspected and that the classifications are independent. Let the random variable X denote the number of parts that are correctly classified. Determine the probability mass function of X.

3-21. In a semiconductor manufacturing process, three wafers from a lot are tested. Each wafer is classified as *pass* or *fail*. Assume that the probability that a wafer passes the test is 0.8 and that wafers are independent. Determine the probability mass function of the number of wafers from a lot that pass the test.

3-22. The space shuttle flight control system called PASS (Primary Avionics Software Set) uses four independent computers working in parallel. At each critical step, the computers "vote" to determine the appropriate step. The probability that a computer will ask for a roll to the left when a roll to the right is appropriate is 0.0001. Let X denote the number of computers that vote for a left roll when a right roll is appropriate. What is the probability mass function of X?

3-23. A disk drive manufacturer estimates that in five years a storage device with 1 terabyte of capacity will sell with probability 0.5, a storage device with 500 gigabytes capacity will sell with a probability 0.3, and a storage device with 100 gigabytes capacity will sell with probability 0.2. The revenue associated with the sales in that year are estimated to be \$50 million, \$25 million, and \$10 million, respectively. Let X be the revenue of storage devices during that year. Determine the probability mass function of X.

3-24. Marketing estimates that a new instrument for the analysis of soil samples will be very successful, moderately successful, or unsuccessful, with probabilities 0.3, 0.6, and 0.1, respectively. The yearly revenue associated with a very successful, moderately successful, or unsuccessful product is \$10 million, \$5 million, and \$1 million, respectively. Let the random variable X denote the yearly revenue of the product. Determine the probability mass function of X.

3-25. The distributor of a machine for cytogenics has developed a new model. The company estimates that when it is introduced into the market, it will be very successful with a probability 0.6, moderately successful with a probability 0.3, and not successful with probability 0.1. The estimated yearly profit associated with the model being very successful is \$15 million and being moderately successful is \$5 million; not successful would result in a loss of \$500,000. Let X be the yearly profit of the new model. Determine the probability mass function of X.

3-26. An assembly consists of two mechanical components. Suppose that the probabilities that the first and second components meet specifications are 0.95 and 0.98. Assume that the components are independent. Determine the probability mass function of the number of components in the assembly that meet specifications.

3-27. An assembly consists of three mechanical components. Suppose that the probabilities that the first, second, and third components meet specifications are 0.95, 0.98, and 0.99. Assume that the components are independent. Determine the probability mass function of the number of components in the assembly that meet specifications.

3-3 CUMULATIVE DISTRIBUTION FUNCTIONS

EXAMPLE 3-6
Digital Channel

In Example 3-4, we might be interested in the probability of three or fewer bits being in error. This question can be expressed as $P(X \leq 3)$.

The event that $\{X \leq 3\}$ is the union of the events $\{X = 0\}$, $\{X = 1\}$, $\{X = 2\}$, and $\{X = 3\}$. Clearly, these three events are mutually exclusive. Therefore,

$$P(X \leq 3) = P(X = 0) + P(X = 1) + P(X = 2) + P(X = 3)$$
$$= 0.6561 + 0.2916 + 0.0486 + 0.0036 = 0.9999$$

This approach can also be used to determine

$$P(X = 3) = P(X \leq 3) - P(X \leq 2) = 0.0036$$

Example 3-6 shows that it is sometimes useful to be able to provide **cumulative probabilities** such as $P(X \leq x)$ and that such probabilities can be used to find the probability mass function of a random variable. Therefore, using cumulative probabilities is an alternate method of describing the probability distribution of a random variable.

In general, for any discrete random variable with possible values $x_1, x_2, \ldots, x_n$, the events $\{X = x_1\}$, $\{X = x_2\}, \ldots, \{X = x_n\}$ are mutually exclusive. Therefore, $P(X \leq x) = \sum_{x_i \leq x} f(x_i)$.

Cumulative Distribution Function

The **cumulative distribution function** of a discrete random variable X, denoted as $F(x)$, is

$$F(x) = P(X \leq x) = \sum_{x_i \leq x} f(x_i)$$

For a discrete random variable X, $F(x)$ satisfies the following properties.

(1) $F(x) = P(X \leq x) = \sum_{x_i \leq x} f(x_i)$

(2) $0 \leq F(x) \leq 1$

(3) If $x \leq y$, then $F(x) \leq F(y)$ (3-2)

Like a probability mass function, a cumulative distribution function provides probabilities. Notice that even if the random variable X can only assume integer values, the cumulative distribution function can be defined at noninteger values. In Example 3-6, $F(1.5) = P(X \leq 1.5) = P\{X = 0\} + P(X = 1) = 0.6561 + 0.2916 = 0.9477$. Properties (1) and (2) of a cumulative distribution function follow from the definition. Property (3) follows from the fact that if $x \leq y$, the event that $\{X \leq x\}$ is contained in the event $\{X \leq y\}$.

The next example shows how the cumulative distribution function can be used to determine the probability mass function of a discrete random variable.

EXAMPLE 3-7
Cumulative Distribution Function

Determine the probability mass function of X from the following cumulative distribution function:

$$F(x) = \begin{cases} 0 & x < -2 \\ 0.2 & -2 \leq x < 0 \\ 0.7 & 0 \leq x < 2 \\ 1 & 2 \leq x \end{cases}$$

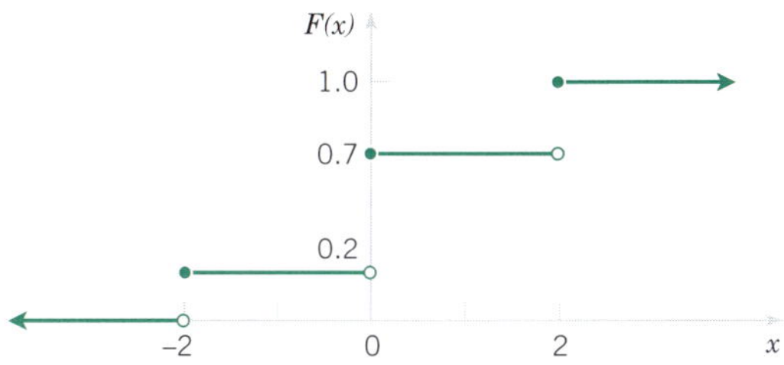

Figure 3-3 Cumulative distribution function for Example 3-7.

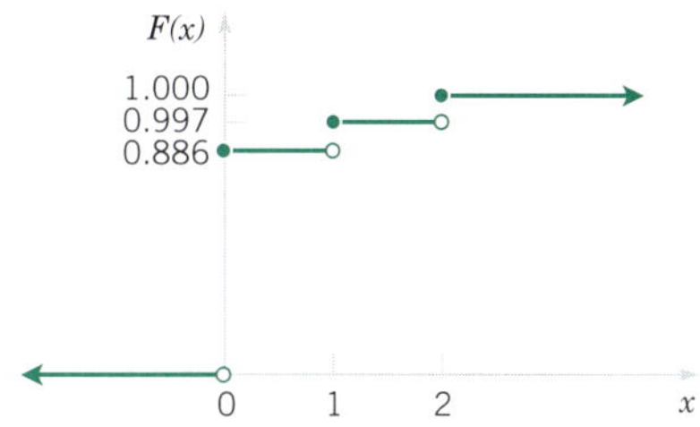

Figure 3-4 Cumulative distribution function for Example 3-8.

Figure 3-3 displays a plot of $F(x)$. From the plot, the only points that receive nonzero probability are -2, 0, and 2. The probability mass function at each point is the change in the cumulative distribution function at the point. Therefore,

$$f(-2) = 0.2 - 0 = 0.2 \qquad f(0) = 0.7 - 0.2 = 0.5 \qquad f(2) = 1.0 - 0.7 = 0.3$$

EXAMPLE 3-8
Sampling without Replacement

Suppose that a day's production of 850 manufactured parts contains 50 parts that do not conform to customer requirements. Two parts are selected at random, without replacement, from the batch. Let the random variable X equal the number of nonconforming parts in the sample. What is the cumulative distribution function of X?

The question can be answered by first finding the probability mass function of X.

$$P(X = 0) = \frac{800}{850} \cdot \frac{799}{849} = 0.886$$

$$P(X = 1) = 2 \cdot \frac{800}{850} \cdot \frac{50}{849} = 0.111$$

$$P(X = 2) = \frac{50}{850} \cdot \frac{49}{849} = 0.003$$

Therefore,

$$F(0) = P(X \le 0) = 0.886$$

$$F(1) = P(X \le 1) = 0.886 + 0.111 = 0.997$$

$$F(2) = P(X \le 2) = 1$$

The cumulative distribution function for this example is graphed in Fig. 3-4. Note that $F(x)$ is defined for all x from $-\infty < x < \infty$ and not only for 0, 1, and 2.

EXERCISES FOR SECTION 3-3

3-28. Determine the cumulative distribution function of the random variable in Exercise 3-14.

3-29. Determine the cumulative distribution function for the random variable in Exercise 3-15; also determine the following probabilities:
(a) $P(X \le 1.25)$ (b) $P(X \le 2.2)$
(c) $P(-1.1 < X \le 1)$ (d) $P(X > 0)$

3-30. Determine the cumulative distribution function for the random variable in Exercise 3-16; also determine the following probabilities:
(a) $P(X < 1.5)$ (b) $P(X \le 3)$
(c) $P(X > 2)$ (d) $P(1 < X \le 2)$

3-31. Determine the cumulative distribution function for the random variable in Exercise 3-21.

3-32. Determine the cumulative distribution function for the random variable in Exercise 3-22.

3-33. Determine the cumulative distribution function for the random variable in Exercise 3-23.

3-34. Determine the cumulative distribution function for the variable in Exercise 3-24.

Verify that the following functions are cumulative distribution functions, and determine the probability mass function and the requested probabilities.

3-35.
$$F(x) = \begin{cases} 0 & x < 1 \\ 0.5 & 1 \le x < 3 \\ 1 & 3 \le x \end{cases}$$

(a) $P(X \le 3)$ (b) $P(X \le 2)$
(c) $P(1 \le X \le 2)$ (d) $P(X > 2)$

3-36. Errors in an experimental transmission channel are found when the transmission is checked by a certifier that detects missing pulses. The number of errors found in an eight-bit byte is a random variable with the following distribution:

$$F(x) = \begin{cases} 0 & x < 1 \\ 0.7 & 1 \le x < 4 \\ 0.9 & 4 \le x < 7 \\ 1 & 7 \le x \end{cases}$$

Determine each of the following probabilities:
(a) $P(X \le 4)$ (b) $P(X > 7)$
(c) $P(X \le 5)$ (d) $P(X > 4)$
(e) $P(X \le 2)$

3-37.
$$F(x) = \begin{cases} 0 & x < -10 \\ 0.25 & -10 \le x < 30 \\ 0.75 & 30 \le x < 50 \\ 1 & 50 \le x \end{cases}$$

(a) $P(X \le 50)$ (b) $P(X \le 40)$
(c) $P(40 \le X \le 60)$ (d) $P(X < 0)$
(e) $P(0 \le X < 10)$ (f) $P(-10 < X < 10)$

3-38. The thickness of wood paneling (in inches) that a customer orders is a random variable with the following cumulative distribution function:

$$F(x) = \begin{cases} 0 & x < 1/8 \\ 0.2 & 1/8 \le x < 1/4 \\ 0.9 & 1/4 \le x < 3/8 \\ 1 & 3/8 \le x \end{cases}$$

Determine the following probabilities:
(a) $P(X \le 1/18)$ (b) $P(X \le 1/4)$
(c) $P(X \le 5/16)$ (d) $P(X > 1/4)$
(e) $P(X \le 1/2)$

3-4 MEAN AND VARIANCE OF A DISCRETE RANDOM VARIABLE

Two numbers are often used to summarize a probability distribution for a random variable X. The mean is a measure of the center or middle of the probability distribution, and the variance is a measure of the dispersion, or variability in the distribution. These two measures do not uniquely identify a probability distribution. That is, two different distributions can have the same mean and variance. Still, these measures are simple, useful summaries of the probability distribution of X.

Mean and Variance

The **mean** or **expected value** of the discrete random variable X, denoted as μ or $E(X)$, is

$$\mu = E(X) = \sum_x x f(x) \tag{3-3}$$

The **variance** of X, denoted as σ^2 or $V(X)$, is

$$\sigma^2 = V(X) = E(X - \mu)^2 = \sum_x (x - \mu)^2 f(x) = \sum_x x^2 f(x) - \mu^2$$

The **standard deviation** of X is $\sigma = \sqrt{\sigma^2}$.

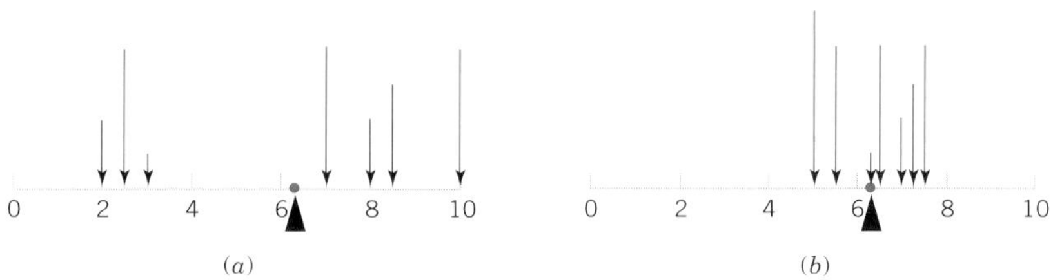

Figure 3-5 A probability distribution can be viewed as a loading with the mean equal to the balance point. Parts (a) and (b) illustrate equal means, but Part (a) illustrates a larger variance.

The mean of a discrete random variable X is a weighted average of the possible values of X, with weights equal to the probabilities. If $f(x)$ is the probability mass function of a loading on a long, thin beam, $E(X)$ is the point at which the beam balances. Consequently, $E(X)$ describes the "center" of the distribution of X in a manner similar to the balance point of a loading. See Fig. 3-5.

The variance of a random variable X is a measure of dispersion or scatter in the possible values for X. The variance of X uses weight $f(x)$ as the multiplier of each possible squared deviation $(x - \mu)^2$. Figure 3-5 illustrates probability distributions with equal means but different variances. Properties of summations and the definition of μ can be used to show the equality of the formulas for variance.

$$V(X) = \sum_x (x - \mu)^2 f(x) = \sum_x x^2 f(x) - 2\mu \sum_x x f(x) + \mu^2 \sum_x f(x)$$

$$= \sum_x x^2 f(x) - 2\mu^2 + \mu^2 = \sum_x x^2 f(x) - \mu^2$$

Either formula for $V(x)$ can be used. Figure 3-6 illustrates that two probability distributions can differ even though they have identical means and variances.

EXAMPLE 3-9
Digital Channel

In Example 3-4, there is a chance that a bit transmitted through a digital transmission channel is received in error. Let X equal the number of bits in error in the next four bits transmitted. The possible values for X are $\{0, 1, 2, 3, 4\}$. Based on a model for the errors that is presented in the following section, probabilities for these values will be determined. Suppose that the probabilities are

$$P(X = 0) = 0.6561 \qquad P(X = 2) = 0.0486 \qquad P(X = 4) = 0.0001$$
$$P(X = 1) = 0.2916 \qquad P(X = 3) = 0.0036$$

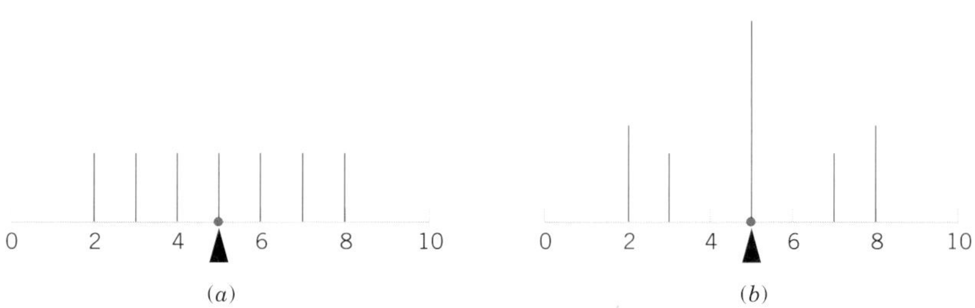

Figure 3-6 The probability distributions illustrated in Parts (a) and (b) differ even though they have equal means and equal variances.

Now

$$\begin{aligned}\mu = E(X) &= 0f(0) + 1f(1) + 2f(2) + 3f(3) + 4f(4)\\ &= 0(0.6561) + 1(0.2916) + 2(0.0486) + 3(0.0036) + 4(0.0001)\\ &= 0.4\end{aligned}$$

Although X never assumes the value 0.4, the weighted average of the possible values is 0.4.

To calculate $V(X)$, a table is convenient.

x	$x - 0.4$	$(x - 0.4)^2$	$f(x)$	$f(x)(x - 0.4)^2$
0	−0.4	0.16	0.6561	0.104976
1	0.6	0.36	0.2916	0.104976
2	1.6	2.56	0.0486	0.124416
3	2.6	6.76	0.0036	0.024336
4	3.6	12.96	0.0001	0.001296

$$V(X) = \sigma^2 = \sum_{i=1}^{5} f(x_i)(x_i - 0.4)^2 = 0.36$$

The alternative formula for variance could also be used to obtain the same result.

EXAMPLE 3-10
Marketing

Two new product designs are to be compared on the basis of revenue potential. Marketing feels that the revenue from design A can be predicted quite accurately to be \$3 million. The revenue potential of design B is more difficult to assess. Marketing concludes that there is a probability of 0.3 that the revenue from design B will be \$7 million, but there is a 0.7 probability that the revenue will be only \$2 million. Which design do you prefer?

Let X denote the revenue from design A. Because there is no uncertainty in the revenue from design A, we can model the distribution of the random variable X as \$3 million with probability 1. Therefore, $E(X) = \$3$ million.

Let Y denote the revenue from design B. The expected value of Y in millions of dollars is

$$E(Y) = \$7(0.3) + \$2(0.7) = \$3.5$$

Because $E(Y)$ exceeds $E(X)$, we might prefer design B. However, the variability of the result from design B is larger. That is,

$$\begin{aligned}\sigma^2 &= (7 - 3.5)^2(0.3) + (2 - 3.5)^2(0.7)\\ &= 5.25 \text{ millions of dollars squared}\end{aligned}$$

Because the units of the variables in this example are millions of dollars, and because the variance of a random variable squares the deviations from the mean, the units of σ^2 are millions of dollars squared. These units make interpretation difficult.

Because the units of standard deviation are the same as the units of the random variable, the standard deviation σ is easier to interpret. In this example, we can summarize our results as "the average deviation of Y from its mean is \$2.29 million."

EXAMPLE 3-11
Messages

The number of messages sent per hour over a computer network has the following distribution:

x = number of messages	10	11	12	13	14	15
$f(x)$	0.08	0.15	0.30	0.20	0.20	0.07

Determine the mean and standard deviation of the number of messages sent per hour.

$$E(X) = 10(0.08) + 11(0.15) + \cdots + 15(0.07) = 12.5$$
$$V(X) = 10^2(0.08) + 11^2(0.15) + \cdots + 15^2(0.07) - 12.5^2 = 1.85$$
$$\sigma = \sqrt{V(X)} = \sqrt{1.85} = 1.36$$

The variance of a random variable X can be considered to be the expected value of a specific function of X, namely, $h(X) = (X - \mu)^2$. In general, the expected value of any function $h(X)$ of a discrete random variable is defined in a similar manner.

Expected Value of a Function of a Discrete Random Variable

If X is a discrete random variable with probability mass function $f(x)$,

$$E[h(X)] = \sum_x h(x)f(x) \qquad (3\text{-}4)$$

EXAMPLE 3-12
Digital Channel

In Example 3-9, X is the number of bits in error in the next four bits transmitted. What is the expected value of the square of the number of bits in error? Now, $h(X) = X^2$. Therefore,

$$E[h(X)] = 0^2 \times 0.6561 + 1^2 \times 0.2916 + 2^2 \times 0.0486 + 3^2 \times 0.0036 + 4^2 \times 0.0001 = 0.52$$

In the previous example, the expected value of X^2 does not equal $E(X)$ squared. However, in the special case that $h(X) = aX + b$ for any constants a and b, $E[h(X)] = aE(X) + b$. This can be shown from the properties of sums in the definition in Equation 3-4.

EXERCISES FOR SECTION 3-4

3-39. If the range of X is the set $\{0, 1, 2, 3, 4\}$ and $P(X = x) = 0.2$ determine the mean and variance of the random variable.

3-40. Determine the mean and variance of the random variable in Exercise 3-14.

3-41. Determine the mean and variance of the random variable in Exercise 3-15.

3-42. Determine the mean and variance of the random variable in Exercise 3-16.

3-43. Determine the mean and variance of the random variable in Exercise 3-17.

3-44. Determine the mean and variance of the random variable in Exercise 3-18.

3-45. Determine the mean and variance of the random variable in Exercise 3-19.

3-46. Determine the mean and variance of the random variable in Exercise 3-20.

3-47. The range of the random variable X is $[0, 1, 2, 3, x]$, where x is unknown. If each value is equally likely and the mean of X is 6, determine x.

3-48. In a NiCd battery, a fully charged cell is composed of Nickelic Hydroxide. Nickel is an element that has multiple oxidation states. Assume the following proportions of the states:

Nickel Charge	Proportions Found
0	0.17
+2	0.35
+3	0.33
+4	0.15

(a) Determine the cumulative distribution function of nickel charge.
(b) Determine the mean and variance of the nickel charge.

3-49. The space shuttle flight control system called PASS (Primary Avionics Software Set) uses four independent computers working in parallel. At each critical step, the computers "vote" to determine the appropriate step. The probability that a computer will ask for a roll to the left when a roll to the right is appropriate is 0.0001. Let X denote the number of computers that vote for a left roll when a right roll is appropriate. What is the mean and variance of X?

3-50. Trees are subjected to different levels of carbon dioxide atmosphere with 6% of the trees in a minimal growth condition at 350 parts per million (ppm), 10% at 450 ppm (slow growth), 47% at 550 ppm (moderate growth), and 37% at 650 ppm (rapid growth). What is the mean and standard deviation of the carbon dioxide atmosphere (in ppm) for these trees in ppm?

3-51. An article in the *Journal of Database Management* "Experimental Study of a Self-Tuning Algorithm for DBMS Buffer Pools," (2005, Vol. 16, pp. 1–20) provided the workload used in the TPC-C OLTP (Transaction Processing Performance Council's Version C On-Line Transaction Processing) benchmark, which simulates a typical order entry application.

The frequency of each type of transaction (in the second column) can be used as the percentage of each type of transaction. The average number of *selects* operations required for each type transaction is shown.

(a) Determine the mean and standard deviation of the number of *selects* operations for a transaction from the distribution of types shown in the table.
(b) Determine the mean and standard deviation of the total number of operations (*selects, updates*, . . . , and *joins*) for a transaction from the distribution of types shown in the table.

Average Frequencies and Operations in TPC-C

Transaction	Frequency	Selects	Updates	Inserts	Deletes	Non-Unique Selects	Joins
New Order	43	23	11	12	0	0	0
Payment	44	4.2	3	1	0	0.6	0
Order Status	4	11.4	0	0	0	0.6	0
Delivery	5	130	120	0	10	0	0
Stock Level	4	0	0	0	0	0	1

3-5 DISCRETE UNIFORM DISTRIBUTION

The simplest discrete random variable is one that assumes only a finite number of possible values, each with equal probability. A random variable X that assumes each of the values $x_1, x_2, \ldots, x_n$, with equal probability $1/n$, is frequently of interest.

Discrete Uniform Distribution

A random variable X has a **discrete uniform distribution** if each of the n values in its range, say, $x_1, x_2, \ldots, x_n$, has equal probability. Then,

$$f(x_i) = 1/n \tag{3-5}$$

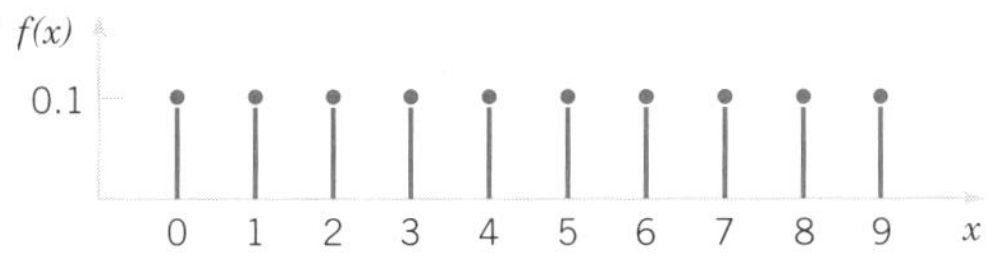

Figure 3-7 Probability mass function for a discrete uniform random variable.

EXAMPLE 3-13

The first digit of a part's serial number is equally likely to be any one of the digits 0 through 9. If one part is selected from a large batch and X is the first digit of the serial number, X has a discrete uniform distribution with probability 0.1 for each value in $R = \{0, 1, 2, \ldots, 9\}$. That is,

$$f(x) = 0.1$$

for each value in R. The probability mass function of X is shown in Fig. 3-7.

Suppose the range of the discrete random variable X is the consecutive integers $a, a + 1, a + 2, \ldots, b$, for $a \le b$. The range of X contains $b - a + 1$ values each with probability $1/(b - a + 1)$. Now,

$$\mu = \sum_{k=a}^{b} k\left(\frac{1}{b - a + 1}\right)$$

The algebraic identity $\sum_{k=a}^{b} k = \frac{b(b+1) - (a-1)a}{2}$ can be used to simplify the result to $\mu = (b + a)/2$. The derivation of the variance is left as an exercise.

Mean and Variance

Suppose X is a discrete uniform random variable on the consecutive integers $a, a + 1, a + 2, \ldots, b$, for $a \le b$. The mean of X is

$$\mu = E(X) = \frac{b + a}{2}$$

The variance of X is

$$\sigma^2 = \frac{(b - a + 1)^2 - 1}{12} \tag{3-6}$$

EXAMPLE 3-14
Number of Voice Lines

As in Example 3-1, let the random variable X denote the number of the 48 voice lines that are in use at a particular time. Assume that X is a discrete uniform random variable with a range of 0 to 48. Then,

$$E(X) = (48 + 0)/2 = 24$$

and

$$\sigma = \{[(48 - 0 + 1)^2 - 1]/12\}^{1/2} = 14.14$$

Equation 3-6 is more useful than it might first appear. If all the values in the range of a random variable X are multiplied by a constant (without changing any probabilities), the mean and standard deviation of X are multiplied by the constant. You are asked to verify this result in an exercise. Because the variance of a random variable is the square of the standard deviation, the variance of X is multiplied by the constant squared. More general results of this type are discussed in Chapter 5.

EXAMPLE 3-15 Proportion of Voice Lines

Let the random variable Y denote the proportion of the 48 voice lines that are in use at a particular time, and X denotes the number of lines that are in use at a particular time. Then, $Y = X/48$. Therefore,

$$E(Y) = E(X)/48 = 0.5$$

and

$$V(Y) = V(X)/48^2 = 0.087$$

EXERCISES FOR SECTION 3-5

3-52. Let the random variable X have a discrete uniform distribution on the integers $0 \le x \le 100$. Determine the mean and variance of X.

3-53. Let the random variable X have a discrete uniform distribution on the integers $1 \le x \le 3$. Determine the mean and variance of X.

3-54. Thickness measurements of a coating process are made to the nearest hundredth of a millimeter. The thickness measurements are uniformly distributed with values 0.15, 0.16, 0.17, 0.18, and 0.19. Determine the mean and variance of the coating thickness for this process.

3-55. Product codes of 2, 3, or 4 letters are equally likely. What is the mean and standard deviation of the number of letters in 100 codes?

3-56. The lengths of plate glass parts are measured to the nearest tenth of a millimeter. The lengths are uniformly distributed, with values at every tenth of a millimeter starting at 590.0 and continuing through 590.9. Determine the mean and variance of lengths.

3-57. Assume that the wavelengths of photosynthetically active radiations (PAR) are uniformly distributed at integer nanometers in the red spectrum from 675 to 700 nm.

(a) What is the mean and variance of the wavelength distribution for this radiation?

(b) If wavelengths are uniformly distributed at integer nanometers from 75 to 100 nanometers, how does the mean and variance of the wavelength distribution compare to the previous part? Explain.

3-58. The probability of an operator entering alphanumeric data incorrectly into a field in a database is equally likely. The random variable X is the number of fields on a data entry form with an error. The data entry form has 28 fields. Is X a discrete uniform random variable? Why or why not.

3-59. Suppose that X has a discrete uniform distribution on the integers 0 through 9. Determine the mean, variance, and standard deviation of the random variable $Y = 5X$ and compare to the corresponding results for X.

3-60. Show that for a discrete uniform random variable X, if each of the values in the range of X is multiplied by the constant c, the effect is to multiply the mean of X by c and the variance of X by c^2. That is, show that $E(cX) = cE(X)$ and $V(cX) = c^2V(X)$.

3-6 BINOMIAL DISTRIBUTION

Consider the following random experiments and random variables:

1. Flip a coin 10 times. Let X = number of heads obtained.
2. A worn machine tool produces 1% defective parts. Let X = number of defective parts in the next 25 parts produced.
3. Each sample of air has a 10% chance of containing a particular rare molecule. Let X = the number of air samples that contain the rare molecule in the next 18 samples analyzed.

4. Of all bits transmitted through a digital transmission channel, 10% are received in error. Let $X =$ the number of bits in error in the next five bits transmitted.
5. A multiple choice test contains 10 questions, each with four choices, and you guess at each question. Let $X =$ the number of questions answered correctly.
6. In the next 20 births at a hospital, let $X =$ the number of female births.
7. Of all patients suffering a particular illness, 35% experience improvement from a particular medication. In the next 100 patients administered the medication, let $X =$ the number of patients who experience improvement.

These examples illustrate that a general probability model that includes these experiments as particular cases would be very useful.

Each of these random experiments can be thought of as consisting of a series of repeated, random trials: 10 flips of the coin in experiment 1, the production of 25 parts in experiment 2, and so forth. The random variable in each case is a count of the number of trials that meet a specified criterion. The outcome from each trial either meets the criterion that X counts or it does not; consequently, each trial can be summarized as resulting in either a success or a failure. For example, in the multiple choice experiment, for each question, only the choice that is correct is considered a success. Choosing any one of the three incorrect choices results in the trial being summarized as a failure.

The terms *success* and *failure* are just labels. We can just as well use A and B or 0 or 1. Unfortunately, the usual labels can sometimes be misleading. In experiment 2, because X counts defective parts, the production of a defective part is called a success.

A trial with only two possible outcomes is used so frequently as a building block of a random experiment that it is called a **Bernoulli trial.** It is usually assumed that the trials that constitute the random experiment are **independent.** This implies that the outcome from one trial has no effect on the outcome to be obtained from any other trial. Furthermore, it is often reasonable to assume that the **probability of a success in each trial is constant.** In the multiple choice experiment, if the test taker has no knowledge of the material and just guesses at each question, we might assume that the probability of a correct answer is 1/4 *for each question.*

EXAMPLE 3-16
Digital Channel

The chance that a bit transmitted through a digital transmission channel is received in error is 0.1. Also, assume that the transmission trials are independent. Let $X =$ the number of bits in error in the next four bits transmitted. Determine $P(X = 2)$.

Let the letter E denote a bit in error, and let the letter O denote that the bit is okay, that is, received without error. We can represent the outcomes of this experiment as a list of four letters that indicate the bits that are in error and those that are okay. For example, the outcome $OEOE$ indicates that the second and fourth bits are in error and the other two bits are okay. The corresponding values for x are

Outcome	x	Outcome	x
$OOOO$	0	$EOOO$	1
$OOOE$	1	$EOOE$	2
$OOEO$	1	$EOEO$	2
$OOEE$	2	$EOEE$	3
$OEOO$	1	$EEOO$	2
$OEOE$	2	$EEOE$	3
$OEEO$	2	$EEEO$	3
$OEEE$	3	$EEEE$	4

The event that $X = 2$ consists of the six outcomes:

$$\{EEOO, EOEO, EOOE, OEEO, OEOE, OOEE\}$$

Using the assumption that the trials are independent, the probability of $\{EEOO\}$ is

$$P(EEOO) = P(E)P(E)P(O)P(O) = (0.1)^2(0.9)^2 = 0.0081$$

Also, any one of the six mutually exclusive outcomes for which $X = 2$ has the same probability of occurring. Therefore,

$$P(X = 2) = 6(0.0081) = 0.0486$$

In general,

$$P(X = x) = (\text{number of outcomes that result in } x \text{ errors}) \text{ times } (0.1)^x(0.9)^{4-x}$$

To complete a general probability formula, only an expression for the number of outcomes that contain x errors is needed. An outcome that contains x errors can be constructed by partitioning the four trials (letters) in the outcome into two groups. One group is of size x and contains the errors, and the other group is of size $n - x$ and consists of the trials that are okay. The number of ways of partitioning four objects into two groups, one of which is of size x, is $\binom{4}{x} = \frac{4!}{x!(4-x)!}$. Therefore, in this example

$$P(X = x) = \binom{4}{x}(0.1)^x(0.9)^{4-x}$$

Notice that $\binom{4}{2} = 4!/[2!\,2!] = 6$, as found above. The probability mass function of X was shown in Example 3-4 and Fig. 3-1.

The previous example motivates the following result.

Binomial Distribution

A random experiment consists of n Bernoulli trials such that

(1) The trials are independent

(2) Each trial results in only two possible outcomes, labeled as "success" and "failure"

(3) The probability of a success in each trial, denoted as p, remains constant

The random variable X that equals the number of trials that result in a success has a **binomial random variable** with parameters $0 < p < 1$ and $n = 1, 2, \ldots$. The probability mass function of X is

$$f(x) = \binom{n}{x} p^x(1-p)^{n-x} \qquad x = 0, 1, \ldots, n \tag{3-7}$$

As in Example 3-16, $\binom{n}{x}$ equals the total number of different sequences of trials that contain x successes and $n - x$ failures. The total number of different sequences that contain x successes and $n - x$ failures times the probability of each sequence equals $P(X = x)$.

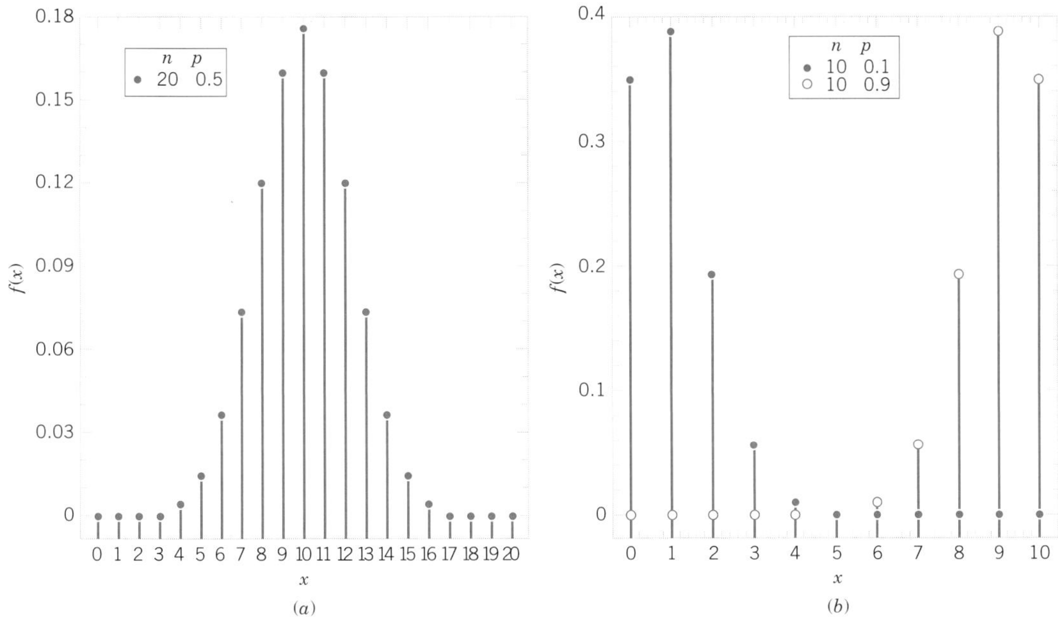

Figure 3-8 Binomial distributions for selected values of n and p.

The probability expression above is a very useful formula that can be applied in a number of examples. The name of the distribution is obtained from the *binomial expansion*. For constants a and b, the binomial expansion is

$$(a + b)^n = \sum_{k=0}^{n} \binom{n}{k} a^k b^{n-k}$$

Let p denote the probability of success on a single trial. Then, by using the binomial expansion with $a = p$ and $b = 1 - p$, we see that the sum of the probabilities for a binomial random variable is 1. Furthermore, because each trial in the experiment is classified into two outcomes, {success, failure}, the distribution is called a "bi"-nomial. A more general distribution, which includes the binomial as a special case, is the multinomial distribution and this is presented in Chapter 5.

Examples of binomial distributions are shown in Fig. 3-8. For a fixed n, the distribution becomes more symmetric as p increases from 0 to 0.5 or decreases from 1 to 0.5. For a fixed p, the distribution becomes more symmetric as n increases.

EXAMPLE 3-17

Several examples using the binomial coefficient $\binom{n}{x}$ follow.

$$\binom{10}{3} = 10!/[3!\,7!] = (10 \cdot 9 \cdot 8)/(3 \cdot 2) = 120$$

$$\binom{15}{10} = 15!/[10!\,5!] = (15 \cdot 14 \cdot 13 \cdot 12 \cdot 11)/(5 \cdot 4 \cdot 3 \cdot 2) = 3003$$

$$\binom{100}{4} = 100!/[4!\,96!] = (100 \cdot 99 \cdot 98 \cdot 97)/(4 \cdot 3 \cdot 2) = 3{,}921{,}225$$

Also recall that $0! = 1$.

EXAMPLE 3-18
Organic Pollution

Each sample of water has a 10% chance of containing a particular organic pollutant. Assume that the samples are independent with regard to the presence of the pollutant. Find the probability that in the next 18 samples, exactly 2 contain the pollutant.

Let X = the number of samples that contain the pollutant in the next 18 samples analyzed. Then X is a binomial random variable with $p = 0.1$ and $n = 18$.
Therefore,

$$P(X = 2) = \binom{18}{2}(0.1)^2(0.9)^{16}$$

Now $\binom{18}{2} = 18!/[2!\ 16!] = 18(17)/2 = 153$. Therefore,

$$P(X = 2) = 153(0.1)^2(0.9)^{16} = 0.284$$

Determine the probability that at least four samples contain the pollutant. The requested probability is

$$P(X \geq 4) = \sum_{x=4}^{18} \binom{18}{x}(0.1)^x(0.9)^{18-x}$$

However, it is easier to use the complementary event,

$$P(X \geq 4) = 1 - P(X < 4) = 1 - \sum_{x=0}^{3} \binom{18}{x}(0.1)^x(0.9)^{18-x}$$

$$= 1 - [0.150 + 0.300 + 0.284 + 0.168] = 0.098$$

Determine the probability that $3 \leq X < 7$. Now

$$P(3 \leq X < 7) = \sum_{x=3}^{6} \binom{18}{x}(0.1)^x(0.9)^{18-x}$$
$$= 0.168 + 0.070 + 0.022 + 0.005$$
$$= 0.265$$

A table of cumulative binomial probabilities is provided in Appendix A and this can simplify some calculations. For example, the binomial distribution in Example 3-16 has $p = 0.1$ and $n = 4$. A probability such as $P(X = 2)$ can be calculated from the table as

$$P(X = 2) = P(X \leq 2) - P(X \leq 1) = 0.9963 - 0.9477 = 0.0486$$

and this agrees with the result obtained previously.

The mean and variance of a binomial random variable can be obtained from an analysis of the independent trials that comprise the binomial experiment. Define new random variables

$$X_i = \begin{cases} 1 \text{ if } i\text{th trial is a success} \\ 0 \text{ otherwise} \end{cases}$$

for $i = 1, 2, \ldots, n$. Then

$$X = X_1 + X_2 + \cdots + X_n$$

Also, it is easy to derive the mean and variance of each X_i as

$$E(X_i) = 1p + 0(1 - p) = p$$

and

$$V(X_i) = (1 - p)^2 p + (0 - p)^2(1 - p) = p(1 - p)$$

Sums of random variables are discussed in Chapter 5 and there the intuitively reasonable result that

$$E(X) = E(X_1) + E(X_2) + \cdots + E(X_n)$$

is derived. Furthermore, for the independent trials of a binomial experiment it is also shown in Chapter 5 that

$$V(X) = V(X_1) + V(X_2) + \cdots + V(X_n)$$

Because $E(X_i) = p$ and $V(X_i) = p(1 - p)$, we obtain the solution $E(X) = np$ and $V(X) = np(1 - p)$.

Mean and Variance

If X is a binomial random variable with parameters p and n,

$$\mu = E(X) = np \qquad \text{and} \qquad \sigma^2 = V(X) = np(1 - p) \tag{3-8}$$

EXAMPLE 3-19

For the number of transmitted bits received in error in Example 3-16, $n = 4$ and $p = 0.1$, so

$$E(X) = 4(0.1) = 0.4 \qquad \text{and} \qquad V(X) = 4(0.1)(0.9) = 0.36$$

and these results match those obtained from a direct calculation in Example 3-9.

EXERCISES FOR SECTION 3-6

3-61. For each scenario described below, state whether or not the binomial distribution is a reasonable model for the random variable and why. State any assumptions you make.

(a) A production process produces thousands of temperature transducers. Let X denote the number of nonconforming transducers in a sample of size 30 selected at random from the process.

(b) From a batch of 50 temperature transducers, a sample of size 30 is selected without replacement. Let X denote the number of nonconforming transducers in the sample.

(c) Four identical electronic components are wired to a controller that can switch from a failed component to one of the remaining spares. Let X denote the number of components that have failed after a specified period of operation.

(d) Let X denote the number of accidents that occur along the federal highways in Arizona during a one-month period.

(e) Let X denote the number of correct answers by a student taking a multiple choice exam in which a student can eliminate some of the choices as being incorrect in some questions and all of the incorrect choices in other questions.

(f) Defects occur randomly over the surface of a semiconductor chip. However, only 80% of defects can be found by testing. A sample of 40 chips with one defect each is tested. Let X denote the number of chips in which the test finds a defect.

(g) Reconsider the situation in part (f). Now, suppose the sample of 40 chips consists of chips with 1 and with 0 defects.

(h) A filling operation attempts to fill detergent packages to the advertised weight. Let X denote the number of detergent packages that are underfilled.
(i) Errors in a digital communication channel occur in bursts that affect several consecutive bits. Let X denote the number of bits in error in a transmission of 100,000 bits.
(j) Let X denote the number of surface flaws in a large coil of galvanized steel.

3-62. Let X be a binomial random variable with $p = 0.2$ and $n = 20$. Use the binomial table in Appendix A to determine the following probabilities.
(a) $P(X \leq 3)$ (b) $P(X > 10)$
(c) $P(X = 6)$ (d) $P(6 \leq X \leq 11)$

3-63. Let X be a binomial random variable with $p = 0.1$ and $n = 10$. Calculate the following probabilities from the binomial probability mass function and also from the binomial table in Appendix A and compare results.
(a) $P(X \leq 2)$ (b) $P(X > 8)$
(c) $P(X = 4)$ (d) $P(5 \leq X \leq 7)$

3-64. The random variable X has a binomial distribution with $n = 10$ and $p = 0.5$. Determine the following probabilities:
(a) $P(X = 5)$ (b) $P(X \leq 2)$
(c) $P(X \geq 9)$ (d) $P(3 \leq X < 5)$

3-65. The random variable X has a binomial distribution with $n = 10$ and $p = 0.01$. Determine the following probabilities.
(a) $P(X = 5)$ (b) $P(X \leq 2)$
(c) $P(X \geq 9)$ (d) $P(3 \leq X < 5)$

3-66. The random variable X has a binomial distribution with $n = 10$ and $p = 0.5$. Sketch the probability mass function of X.
(a) What value of X is most likely?
(b) What value(s) of X is(are) least likely?

3-67. Sketch the probability mass function of a binomial distribution with $n = 10$ and $p = 0.01$ and comment on the shape of the distribution.
(a) What value of X is most likely?
(b) What value of X is least likely?

3-68. Determine the cumulative distribution function of a binomial random variable with $n = 3$ and $p = 1/2$.

3-69. Determine the cumulative distribution function of a binomial random variable with $n = 3$ and $p = 1/4$.

3-70. An electronic product contains 40 integrated circuits. The probability that any integrated circuit is defective is 0.01, and the integrated circuits are independent. The product operates only if there are no defective integrated circuits. What is the probability that the product operates?

3-71. The phone lines to an airline reservation system are occupied 40% of the time. Assume that the events that the lines are occupied on successive calls are independent. Assume that 10 calls are placed to the airline.
(a) What is the probability that for exactly three calls the lines are occupied?
(b) What is the probability that for at least one call the lines are not occupied?
(c) What is the expected number of calls in which the lines are all occupied?

3-72. A multiple choice test contains 25 questions, each with four answers. Assume a student just guesses on each question.
(a) What is the probability that the student answers more than 20 questions correctly?
(b) What is the probability the student answers less than 5 questions correctly?

3-73. A particularly long traffic light on your morning commute is green 20% of the time that you approach it. Assume that each morning represents an independent trial.
(a) Over five mornings, what is the probability that the light is green on exactly one day?
(b) Over 20 mornings, what is the probability that the light is green on exactly four days?
(c) Over 20 mornings, what is the probability that the light is green on more than four days?

3-74. Samples of rejuvenated mitochondria are mutated (defective) in 1% of cases. Suppose 15 samples are studied, and they can be considered to be independent for mutation. Determine the following probabilities. The binomial table in Appendix A can help.
(a) No samples are mutated.
(b) At most one sample is mutated.
(c) More than half the samples are mutated.

3-75. An article in *Information Security Technical Report*, "Malicious software—past, present and future," (2004, Vol. 9, pp. 6–18) provided the following data on the top ten malicious software instances for 2002. The clear leader in the number of registered incidences for the year 2002 was the Internet worm "Klez," and it is still one of the most widespread threats. This virus was first detected on 26 October 2001, and it has held the top spot among malicious software for the longest period in the history of virology.

Place	Name	% Instances
1	I-Worm.Klez	61.22%
2	I-Worm.Lentin	20.52%
3	I-Worm.Tanatos	2.09%
4	I-Worm.BadtransII	1.31%
5	Macro.Word97.Thus	1.19%
6	I-Worm.Hybris	0.60%
7	I-Worm.Bridex	0.32%
8	I-Worm.Magistr	0.30%
9	Win95.CIH	0.27%
10	I-Worm.Sircam	0.24%

The 10 most widespread malicious programs for 2002 *(Source—Kaspersky Labs)*.

Suppose that 20 malicious software instances are reported. Assume that the malicious sources can be assumed to be independent.

(a) What is the probability at least one instance is "Klez"?

(b) What is the probability that three or more instances are "Klez"?

(c) What is the mean and standard deviation of the number of "Klez" instances among the 20 reported?

3-76. Heart failure is due to either natural occurrences (87%) or outside factors (13%). Outside factors are related to induced substances or foreign objects. Natural occurrences are caused by arterial blockage, disease, and infection. Suppose that 20 patients will visit an emergency room with heart failure. Assume that causes of heart failure between individuals are independent.

(a) What is the probability that three individuals have conditions caused by outside factors?

(b) What is the probability that three or more individuals have conditions caused by outside factors?

(c) What is the mean and standard deviation of the number of individuals with conditions caused by outside factors?

3-77. A computer system uses passwords that are exactly six characters and each character is one of the 26 letters (a–z) or 10 integers (0–9). Suppose there are 10000 users on the system with unique passwords. A hacker randomly selects (with replacement) one billion passwords from the potential set in the milliseconds before security software closes the unauthorized access.

(a) What is the distribution of the number of user passwords selected by the hacker?

(b) What is the probability that no user passwords are selected?

(c) What is the mean and variance of the number of user passwords selected?

3-78. **A statistical process control chart example.** Samples of 20 parts from a metal punching process are selected every hour. Typically, 1% of the parts require rework. Let X denote the number of parts in the sample of 20 that require rework. A process problem is suspected if X exceeds its mean by more than three standard deviations.

(a) If the percentage of parts that require rework remains at 1%, what is the probability that X exceeds its mean by more than three standard deviations?

(b) If the rework percentage increases to 4%, what is the probability that X exceeds 1?

(c) If the rework percentage increases to 4%, what is the probability that X exceeds 1 in at least one of the next five hours of samples?

3-79. Because not all airline passengers show up for their reserved seat, an airline sells 125 tickets for a flight that holds only 120 passengers. The probability that a passenger does not show up is 0.10, and the passengers behave independently.

(a) What is the probability that every passenger who shows up can take the flight?

(b) What is the probability that the flight departs with empty seats?

3-80. This exercise illustrates that poor quality can affect schedules and costs. A manufacturing process has 100 customer orders to fill. Each order requires one component part that is purchased from a supplier. However, typically, 2% of the components are identified as defective, and the components can be assumed to be independent.

(a) If the manufacturer stocks 100 components, what is the probability that the 100 orders can be filled without reordering components?

(b) If the manufacturer stocks 102 components, what is the probability that the 100 orders can be filled without reordering components?

(c) If the manufacturer stocks 105 components, what is the probability that the 100 orders can be filled without reordering components?

3-7 GEOMETRIC AND NEGATIVE BINOMIAL DISTRIBUTIONS

3-7.1 Geometric Distribution

Consider a random experiment that is closely related to the one used in the definition of a binomial distribution. Again, assume a series of Bernoulli trials (independent trials with constant probability p of a success on each trial). However, instead of a fixed number of trials, trials are conducted until a success is obtained. Let the random variable X denote the number of trials until the first success. In Example 3-5, successive wafers are analyzed until a large particle is detected. Then, X is the number of wafers analyzed. In the transmission of bits, X might be the number of bits transmitted until an error occurs.

EXAMPLE 3-20
Digital Channel

The probability that a bit transmitted through a digital transmission channel is received in error is 0.1. Assume the transmissions are independent events, and let the random variable X denote the number of bits transmitted *until* the first error.

Then, $P(X = 5)$ is the probability that the first four bits are transmitted correctly and the fifth bit is in error. This event can be denoted as $\{OOOOE\}$, where O denotes an okay bit. Because the trials are independent and the probability of a correct transmission is 0.9,

$$P(X = 5) = P(OOOOE) = 0.9^4 0.1 = 0.066$$

Note that there is some probability that X will equal any integer value. Also, if the first trial is a success, $X = 1$. Therefore, the range of X is $\{1, 2, 3, \ldots\}$, that is, all positive integers.

Geometric Distribution

In a series of Bernoulli trials (independent trials with constant probability p of a success), let the random variable X denote the number of trials until the first success. Then X is a **geometric random variable** with parameter $0 < p < 1$ and

$$f(x) = (1 - p)^{x-1} p \qquad x = 1, 2, \ldots \tag{3-9}$$

Examples of the probability mass functions for geometric random variables are shown in Fig. 3-9. Note that the height of the line at x is $(1 - p)$ times the height of the line at $x - 1$. That is, the probabilities decrease in a geometric progression. The distribution acquires its name from this result.

EXAMPLE 3-21

The probability that a wafer contains a large particle of contamination is 0.01. If it is assumed that the wafers are independent, what is the probability that exactly 125 wafers need to be analyzed before a large particle is detected?

Let X denote the number of samples analyzed until a large particle is detected. Then X is a geometric random variable with $p = 0.01$. The requested probability is

$$P(X = 125) = (0.99)^{124} 0.01 = 0.0029$$

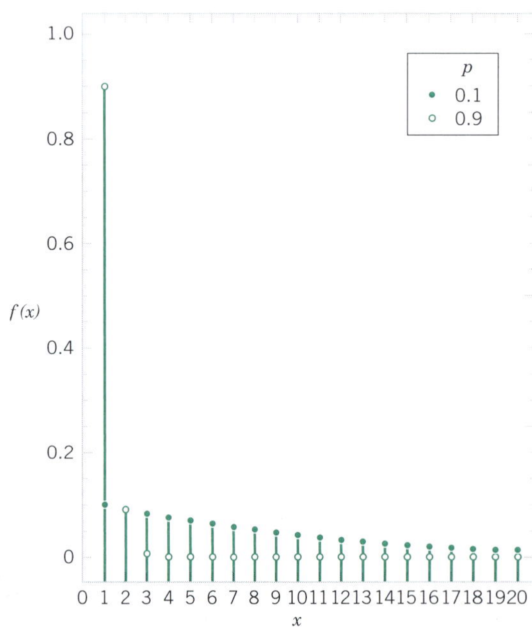

Figure 3-9 Geometric distributions for selected values of the parameter p.

The mean of a geometric random variable is

$$\mu = \sum_{k=1}^{\infty} kp(1-p)^{k-1} = p\sum_{k=1}^{\infty} kq^{k-1}$$

where $q = p - 1$. The right-hand side of the previous equation is recognized to be the partial derivative with respect to q of

$$p\sum_{k=1}^{\infty} q^k = \frac{pq}{1-q}$$

where the last equality is obtained from the known sum of a geometric series. Therefore,

$$\mu = \frac{\partial}{\partial q}\left[\frac{pq}{1-q}\right] = \frac{p}{(1-q)^2} = \frac{p}{p^2} = \frac{1}{p}$$

and the mean is derived. To obtain the variance of a geometric random variable, we can first derive $E(X^2)$ by a similar approach. This can be obtained from partial second derivatives with respect to q. Then the formula $V(X) = E(X^2) - (EX)^2$ is applied. The details are a bit more work and this is left as a mind-expanding exercise.

Mean and Variance

If X is a geometric random variable with parameter p,

$$\mu = E(X) = 1/p \quad \text{and} \quad \sigma^2 = V(X) = (1-p)/p^2 \tag{3-10}$$

EXAMPLE 3-22

Consider the transmission of bits in Example 3-20. Here, $p = 0.1$. The mean number of transmissions until the first error is $1/0.1 = 10$. The standard deviation of the number of transmissions before the first error is

$$\sigma = [(1-0.1)/0.1^2]^{1/2} = 9.49$$

Lack of Memory Property

A geometric random variable has been defined as the number of trials until the first success. However, because the trials are independent, the count of the number of trials until the next success can be started at any trial without changing the probability distribution of the random variable. For example, in the transmission of bits, if 100 bits are transmitted, the probability that the first error, after bit 100, occurs on bit 106 is the probability that the next six outcomes are *OOOOOE*. This probability is $(0.9)^5(0.1) = 0.059$, which is identical to the probability that the initial error occurs on bit 6.

The implication of using a geometric model is that the system presumably will not wear out. The probability of an error remains constant for all transmissions. In this sense, the geometric distribution is said to lack any memory. The **lack of memory property** will be discussed again in the context of an exponential random variable in Chapter 4.

EXAMPLE 3-23
Lack of Memory

In Example 3-20, the probability that a bit is transmitted in error is equal to 0.1. Suppose 50 bits have been transmitted. The mean number of bits until the next error is $1/0.1 = 10$—the same result as the mean number of bits until the first error.

3-7.2 Negative Binomial Distribution

A generalization of a geometric distribution in which the random variable is the number of Bernoulli trials required to obtain r successes results in the **negative binomial distribution.**

EXAMPLE 3-24
Digital Channel

As in Example 3-20, suppose the probability that a bit transmitted through a digital transmission channel is received in error is 0.1. Assume the transmissions are independent events, and let the random variable X denote the number of bits transmitted until the *fourth* error.

Then, X has a negative binomial distribution with $r = 4$. Probabilities involving X can be found as follows. The $P(X = 10)$ is the probability that exactly three errors occur in the first nine trials and then trial 10 results in the fourth error. The probability that exactly three errors occur in the first nine trials is determined from the binomial distribution to be

$$\binom{9}{3}(0.1)^3(0.9)^6$$

Because the trials are independent, the probability that exactly three errors occur in the first 9 trials and trial 10 results in the fourth error is the product of the probabilities of these two events, namely,

$$\binom{9}{3}(0.1)^3(0.9)^6(0.1) = \binom{9}{3}(0.1)^4(0.9)^6$$

The previous result can be generalized as follows.

Negative Binomial Distribution

In a series of Bernoulli trials (independent trials with constant probability p of a success), let the random variable X denote the number of trials until r successes occur. Then X is a **negative binomial random variable** with parameters $0 < p < 1$ and $r = 1, 2, 3, \ldots$, and

$$f(x) = \binom{x-1}{r-1}(1-p)^{x-r}p^r \qquad x = r, r+1, r+2, \ldots. \qquad (3\text{-}11)$$

Because at least r trials are required to obtain r successes, the range of X is from r to ∞. In the special case that $r = 1$, a negative binomial random variable is a geometric random variable. Selected negative binomial distributions are illustrated in Fig. 3-10.

The lack of memory property of a geometric random variable implies the following. Let X denote the total number of trials required to obtain r successes. Let X_1 denote the number of trials required to obtain the first success, let X_2 denote the number of extra trials required to obtain the second success, let X_3 denote the number of extra trials to obtain the third success, and so forth. Then, the total number of trials required to obtain r successes is $X = X_1 + X_2 + \cdots + X_r$. Because of the lack of memory property, each of the random variables $X_1, X_2, \ldots, X_r$ has a geometric distribution with the same value of p. Consequently, a

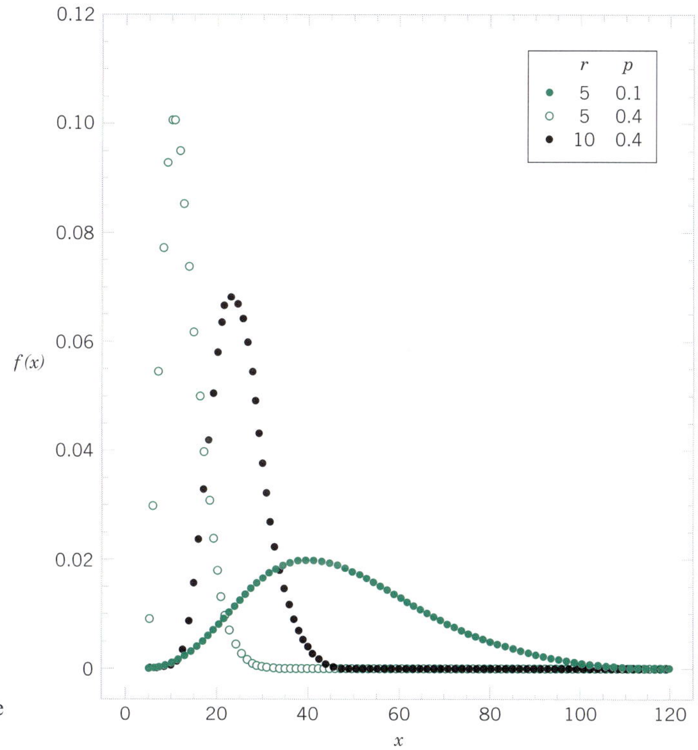

Figure 3-10 Negative binomial distributions for selected values of the parameters r and p.

negative binomial random variable can be interpreted as the sum of r geometric random variables. This concept is illustrated in Fig. 3-11.

Recall that a binomial random variable is a count of the number of successes in n Bernoulli trials. That is, the number of trials is predetermined, and the number of successes is random. A negative binomial random variable is a count of the number of trials required to obtain r successes. That is, the number of successes is predetermined, and the number of trials is random. In this sense, a negative binomial random variable can be considered the opposite, or negative, of a binomial random variable.

The description of a negative binomial random variable as a sum of geometric random variables leads to the following results for the mean and variance. Sums of random variables are studied in Chapter 5.

Mean and Variance

If X is a negative binomial random variable with parameters p and r,

$$\mu = E(X) = r/p \quad \text{and} \quad \sigma^2 = V(X) = r(1-p)/p^2 \qquad (3\text{-}12)$$

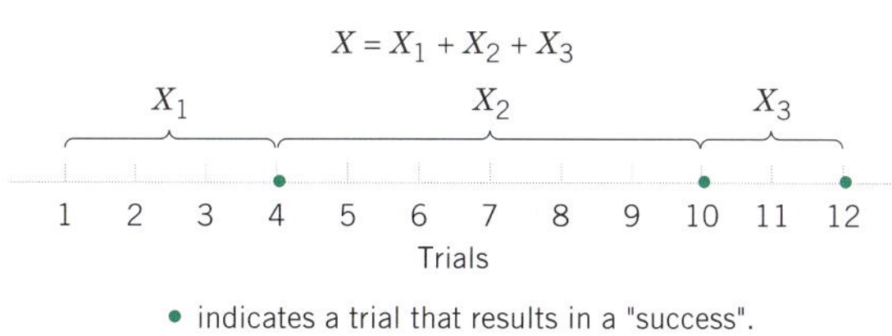

Figure 3-11 Negative binomial random variable represented as a sum of geometric random variables.

EXAMPLE 3-25
Web Servers

A Web site contains three identical computer servers. Only one is used to operate the site, and the other two are spares that can be activated in case the primary system fails. The probability of a failure in the primary computer (or any activated spare system) from a request for service is 0.0005. Assuming that each request represents an independent trial, what is the mean number of requests until failure of all three servers?

Let X denote the number of requests until all three servers fail, and let X_1, X_2, and X_3 denote the number of requests before a failure of the first, second, and third servers used, respectively. Now, $X = X_1 + X_2 + X_3$. Also, the requests are assumed to comprise independent trials with constant probability of failure $p = 0.0005$. Furthermore, a spare server is not affected by the number of requests before it is activated. Therefore, X has a negative binomial distribution with $p = 0.0005$ and $r = 3$. Consequently,

$$E(X) = 3/0.0005 = 6000 \text{ requests}$$

What is the probability that all three servers fail within five requests? The probability is $P(X \leq 5)$ and

$$\begin{aligned} P(X \leq 5) &= P(X = 3) + P(X = 4) + P(X = 5) \\ &= 0.0005^3 + \binom{3}{2} 0.0005^3(0.9995) + \binom{4}{2} 0.0005^3(0.9995)^2 \\ &= 1.25 \times 10^{-10} + 3.75 \times 10^{-10} + 7.49 \times 10^{-10} \\ &= 1.249 \times 10^{-9} \end{aligned}$$

EXERCISES FOR SECTION 3-7

3-81. Suppose the random variable X has a geometric distribution with $p = 0.5$. Determine the following probabilities:
(a) $P(X = 1)$ (b) $P(X = 4)$
(c) $P(X = 8)$ (d) $P(X \leq 2)$
(e) $P(X > 2)$

3-82. Suppose the random variable X has a geometric distribution with a mean of 2.5. Determine the following probabilities:
(a) $P(X = 1)$ (b) $P(X = 4)$
(c) $P(X = 5)$ (d) $P(X \leq 3)$
(e) $P(X > 3)$

3-83. Consider a sequence of independent Bernoulli trials with $p = 0.2$.
(a) What is the expected number of trials to obtain the first success?
(b) After the eighth success occurs, what is the expected number of trials to obtain the ninth success?

3-84. Suppose that X is a negative binomial random variable with $p = 0.2$ and $r = 4$. Determine the following:
(a) $E(X)$ (b) $P(X = 20)$
(c) $P(X = 19)$ (d) $P(X = 21)$
(e) The most likely value for X

3-85. The probability of a successful optical alignment in the assembly of an optical data storage product is 0.8. Assume the trials are independent.
(a) What is the probability that the first successful alignment requires exactly four trials?
(b) What is the probability that the first successful alignment requires at most four trials?
(c) What is the probability that the first successful alignment requires at least four trials?

3-86. In a clinical study, volunteers are tested for a gene that has been found to increase the risk for a disease. The probability that a person carries the gene is 0.1.
(a) What is the probability 4 or more people will have to be tested before 2 with the gene are detected?
(b) How many people are expected to be tested before 2 with the gene are detected?

3-87. Assume that each of your calls to a popular radio station has a probability of 0.02 of connecting, that is, of not obtaining a busy signal. Assume that your calls are independent.
(a) What is the probability that your first call that connects is your tenth call?
(b) What is the probability that it requires more than five calls for you to connect?
(c) What is the mean number of calls needed to connect?

3-88. A player of a video game is confronted with a series of opponents and has an 80% probability of defeating each one. Success with any opponent is independent of previous encounters. The player continues to contest opponents until defeated.
(a) What is the probability mass function of the number of opponents contested in a game?
(b) What is the probability that a player defeats at least two opponents in a game?

(c) What is the expected number of opponents contested in a game?

(d) What is the probability that a player contests four or more opponents in a game?

(e) What is the expected number of game plays until a player contests four or more opponents?

3-89. Heart failure is due to either natural occurrences (87%) or outside factors (13%). Outside factors are related to induced substances or foreign objects. Natural occurrences are caused by arterial blockage, disease, and infection. Assume that causes of heart failure between individuals are independent.

(a) What is the probability that the first patient with heart failure that enters the emergency room has the condition due to outside factors?

(b) What is the probability that third patient with heart failure that enters the emergency room is the first one due to outside factors?

(c) What is the mean number of heart failure patients with the condition due to natural causes that enter the emergency room before the first patient with heart failure from outside factors?

3-90. A computer system uses passwords constructed from the 26 letters (a–z) or 10 integers (0–9). Suppose there are 10000 users on the system with unique passwords. A hacker randomly selects (with replacement) passwords from the potential set.

(a) There are 9900 users with unique six-character passwords on the system and the hacker randomly selects six-character passwords. What is the mean and standard deviation of the number of attempts before the hacker selects a user password?

(b) Suppose there are 100 users with unique three-character passwords on the system and the hacker randomly selects three-character passwords. What is the mean and standard deviation of the number of attempts before the hacker selects a user password?

(c) Comment on the security differences between six- and three-character passwords

3-91. A trading company has eight computers that it uses to trade on the New York Stock Exchange (NYSE). The probability of a computer failing in a day is 0.005, and the computers fail independently. Computers are repaired in the evening and each day is an independent trial.

(a) What is the probability that all eight computers fail in a day?

(b) What is the mean number of days until a specific computer fails?

(c) What is the mean number of days until all eight computers fail in the same day?

3-92. Assume that 20 parts are checked each hour and that X denotes the number of parts in the sample of 20 that require rework. Parts are assumed to be independent with respect to rework.

(a) If the percentage of parts that require rework remains at 1%, what is the probability that hour 10 is the first sample at which X exceeds 1?

(b) If the rework percentage increases to 4%, what is the probability that hour 10 is the first sample at which X exceeds 1?

(c) If the rework percentage increases to 4%, what is the expected number of hours until X exceeds 1?

3-93. A fault-tolerant system that processes transactions for a financial services firm uses three separate computers. If the operating computer fails, one of the two spares can be immediately switched online. After the second computer fails, the last computer can be immediately switched online. Assume that the probability of a failure during any transaction is 10^{-8} and that the transactions can be considered to be independent events.

(a) What is the mean number of transactions before all computers have failed?

(b) What is the variance of the number of transactions before all computers have failed?

3-94. In the process of meiosis, a single parent diploid cell goes through eight different phases. However, only 60% of the processes pass the first six phases and only 40% pass all eight. Assume the results from each phase are independent.

(a) If the probability of a successful pass of each one of the first six phases is constant, what is the probability of a successful pass of a single one of these phases?

(b) If the probability of a successful pass of each one of the last two phases is constant, what is the probability of a successful pass of a single one of these phases?

3-95. Show that the probability density function of a negative binomial random variable equals the probability density function of a geometric random variable when $r = 1$. Show that the formulas for the mean and variance of a negative binomial random variable equal the corresponding results for a geometric random variable when $r = 1$.

3-96. Derive the expression for the variance of a geometric random variable with parameter p. (Formulas for infinite series are required.)

3-8 HYPERGEOMETRIC DISTRIBUTION

In Example 3-8, a day's production of 850 manufactured parts contains 50 parts that do not conform to customer requirements. Two parts are selected at random, without replacement from the day's production. That is, selected units are not replaced before the next selection is made. Let A and B denote the events that the first and second parts are nonconforming, respectively. In Chapter 2, we found $P(B|A) = 49/849$ and $P(A) = 50/850$. Consequently, knowledge that the first part is nonconforming suggests that it is less likely that the second part selected is nonconforming.

This experiment is fundamentally different from the examples based on the binomial distribution. In this experiment, the trials are not independent. Note that, in the unusual case that each unit selected is replaced before the next selection, the trials are independent and there is a constant probability of a nonconforming part on each trial. Then, the number of nonconforming parts in the sample is a binomial random variable.

Let X equal the number of nonconforming parts in the sample. Then

$$
\begin{aligned}
P(X = 0) &= P(\text{both parts conform}) = (800/850)(799/849) = 0.886 \\
P(X = 1) &= P(\text{first part selected conforms and the second part selected does not, or the first part selected does not and the second part selected conforms}) \\
&= (800/850)(50/849) + (50/850)(800/849) = 0.111 \\
P(X = 2) &= P(\text{both parts do not conform}) = (50/850)(49/849) = 0.003
\end{aligned}
$$

As in this example, samples are often selected without replacement. Although probabilities can be determined by the reasoning used in the example above, a general formula for computing probabilities when samples are selected without replacement is quite useful. The counting rules presented in Chapter 2 can be used to justify the formula given below.

Hypergeometric Distribution

A set of N objects contains
K objects classified as successes
$N - K$ objects classified as failures

A sample of size n objects is selected randomly (without replacement) from the N objects, where $K \leq N$ and $n \leq N$.

Let the random variable X denote the number of successes in the sample. Then X is a **hypergeometric random variable** and

$$f(x) = \frac{\binom{K}{x}\binom{N-K}{n-x}}{\binom{N}{n}} \qquad x = \max\{0, n + K - N\} \text{ to } \min\{K, n\} \tag{3-13}$$

The expression $\min\{K, n\}$ is used in the definition of the range of X because the maximum number of successes that can occur in the sample is the smaller of the sample size, n, and the number of successes available, K. Also, if $n + K > N$, at least $n + K - N$ successes must occur in the sample. Selected hypergeometric distributions are illustrated in Fig. 3-12.

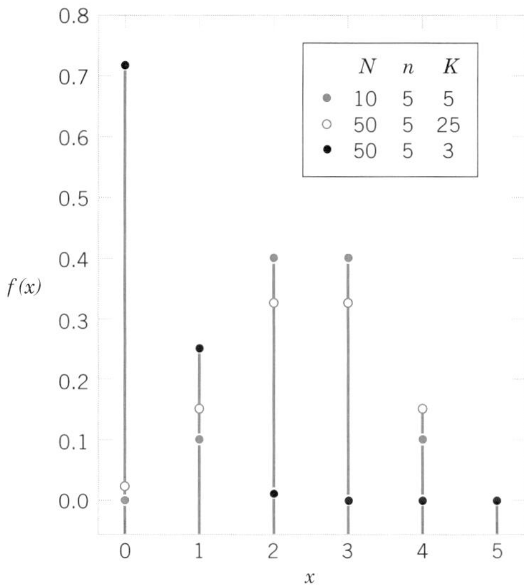

Figure 3-12 Hypergeometric distributions for selected values of parameters N, K, and n.

EXAMPLE 3-26
Sampling without Replacement

The example at the start of this section can be reanalyzed by using the general expression in the definition of a hypergeometric random variable. That is,

$$P(X = 0) = \frac{\binom{50}{0}\binom{800}{2}}{\binom{850}{2}} = \frac{319600}{360825} = 0.886$$

$$P(X = 1) = \frac{\binom{50}{1}\binom{800}{1}}{\binom{850}{2}} = \frac{40000}{360825} = 0.111$$

$$P(X = 2) = \frac{\binom{50}{2}\binom{800}{0}}{\binom{850}{2}} = \frac{1225}{360825} = 0.003$$

EXAMPLE 3-27
Parts from Suppliers

A batch of parts contains 100 parts from a local supplier of tubing and 200 parts from a supplier of tubing in the next state. If four parts are selected randomly and without replacement, what is the probability they are all from the local supplier?

Let X equal the number of parts in the sample from the local supplier. Then, X has a hypergeometric distribution and the requested probability is $P(X = 4)$. Consequently,

$$P(X = 4) = \frac{\binom{100}{4}\binom{200}{0}}{\binom{300}{4}} = 0.0119$$

What is the probability that two or more parts in the sample are from the local supplier?

$$P(X \geq 2) = \frac{\binom{100}{2}\binom{200}{2}}{\binom{300}{4}} + \frac{\binom{100}{3}\binom{200}{1}}{\binom{300}{4}} + \frac{\binom{100}{4}\binom{200}{0}}{\binom{300}{4}}$$

$$= 0.298 + 0.098 + 0.0119 = 0.408$$

What is the probability that at least one part in the sample is from the local supplier?

$$P(X \geq 1) = 1 - P(X = 0) = 1 - \frac{\binom{100}{0}\binom{200}{4}}{\binom{300}{4}} = 0.804$$

The mean and variance of a hypergeometric random variable can be determined from the trials that comprise the experiment. However, the trials are not independent, and so the calculations are more difficult than for a binomial distribution. The results are stated as follows.

Mean and Variance

If X is a hypergeometric random variable with parameters N, K, and n, then

$$\mu = E(X) = np \qquad \text{and} \qquad \sigma^2 = V(X) = np(1 - p)\left(\frac{N - n}{N - 1}\right) \tag{3-14}$$

where $p = K/N$.

Here p is interpreted as the proportion of successes in the set of N objects.

EXAMPLE 3-28

In the previous example, the sample size is 4. The random variable X is the number of parts in the sample from the local supplier. Then, $p = 100/300 = 1/3$. Therefore,

$$E(X) = 4(100/300) = 1.33$$

and

$$V(X) = 4(1/3)(2/3)[(300 - 4)/299] = 0.88$$

For a hypergeometric random variable, $E(X)$ is similar to the mean of a binomial random variable. Also, $V(X)$ differs from the result for a binomial random variable only by the term shown below.

Finite Population Correction Factor

The term in the variance of a hypergeometric random variable

$$\frac{N - n}{N - 1} \tag{3-15}$$

is called the **finite population correction factor.**

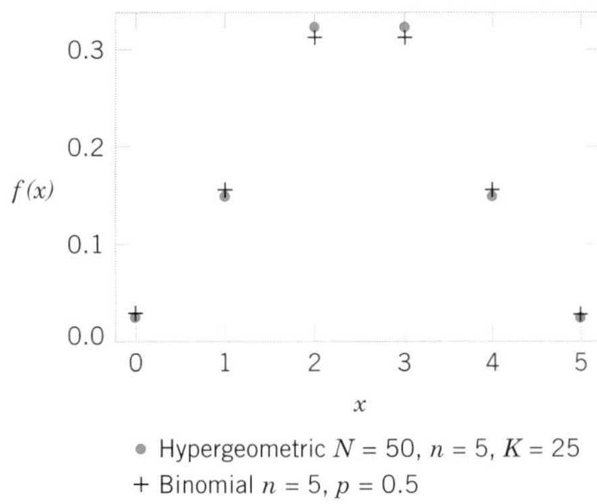

	0	1	2	3	4	5
Hypergeometric probability	0.025	0.149	0.326	0.326	0.149	0.025
Binomial probability	0.031	0.156	0.312	0.312	0.156	0.031

Figure 3-13 Comparison of hypergeometric and binomial distributions.

Sampling with replacement is equivalent to sampling from an infinite set because the proportion of success remains constant for every trial in the experiment. As mentioned previously, if sampling were done with replacement, X would be a binomial random variable and its variance would be $np(1-p)$. Consequently, the finite population correction represents the correction to the binomial variance that results because the sampling is without replacement from the finite set of size N.

If n is small relative to N, the correction is small and the hypergeometric distribution is similar to the binomial. In this case, a binomial distribution can effectively approximate the distribution of the number of units of a specified type in the sample. A case is illustrated in Fig. 3-13.

EXAMPLE 3-29
Customer Sample

A listing of customer accounts at a large corporation contains 1000 customers. Of these, 700 have purchased at least one of the corporation's products in the last three months. To evaluate a new product design, 50 customers are sampled at random from the corporate listing. What is the probability that more than 45 of the sampled customers have purchased from the corporation in the last three months?

The sampling is without replacement. However, because the sample size of 50 is small relative to the number of customer accounts, 1000, the probability of selecting a customer who has purchased from the corporation in the last three months remains approximately constant as the customers are chosen.

For example, let A denote the event that the first customer selected has purchased from the corporation in the last three months, and let B denote the event that the second customer selected has purchased from the corporation in the last three months. Then, $P(A) = 700/1000 = 0.7$ and $P(B|A) = 699/999 = 0.6997$. That is, the trials are approximately independent.

Let X denote the number of customers in the sample who have purchased from the corporation in the last three months. Then, X is a hypergeometric random variable with $N = 1000$, $n = 50$, and $K = 700$. Consequently, $p = K/N = 0.7$. The requested probability is $P(X > 45)$. Because the sample size is small relative to the batch size, the distribution of X can be approximated as binomial with $n = 50$ and $p = 0.7$. Using the binomial approximation to the distribution of X results in

$$P(X > 45) = \sum_{x=46}^{50} \binom{50}{x} 0.7^x (1-0.7)^{50-x} = 0.00017$$

The probability from the hypergeometric distribution is 0.00013, but this requires computer software. The result agrees well with the binomial approximation.

EXERCISES FOR SECTION 3-8

3-97. Suppose X has a hypergeometric distribution with $N = 100$, $n = 4$, and $K = 20$. Determine the following:
(a) $P(X = 1)$
(b) $P(X = 6)$
(c) $P(X = 4)$
(d) Determine the mean and variance of X.

3-98. Suppose X has a hypergeometric distribution with $N = 20$, $n = 4$, and $K = 4$. Determine the following:
(a) $P(X = 1)$
(b) $P(X = 4)$
(c) $P(X \leq 2)$
(d) Determine the mean and variance of X.

3-99. Suppose X has a hypergeometric distribution with $N = 10$, $n = 3$, and $K = 4$. Sketch the probability mass function of X. Determine the cumulative distribution function for X.

3-100. A batch contains 36 bacteria cells and 12 of the cells are not capable of cellular replication. Suppose you examine three bacteria cells selected at random, without replacement.
(a) What is the probability mass function of the number of cells in the sample that can replicate?
(b) What is the mean and variance of the number of cells in the sample that can replicate?
(c) What is the probability that at least one of the selected cells cannot replicate?

3-101. A company employs 800 men under the age of 55. Suppose that 30% carry a marker on the male chromosome that indicates an increased risk for high blood pressure.
(a) If 10 men in the company are tested for the marker in this chromosome, what is the probability that exactly 1 man has the marker?
(b) If 10 men in the company are tested for the marker in this chromosome, what is the probability that more than 1 has the marker?

3-102. Printed circuit cards are placed in a functional test after being populated with semiconductor chips. A lot contains 140 cards, and 20 are selected without replacement for functional testing.
(a) If 20 cards are defective, what is the probability that at least 1 defective card is in the sample?
(b) If 5 cards are defective, what is the probability that at least 1 defective card appears in the sample?

3-103. The analysis of results from a leaf transmutation experiment (turning a leaf into a petal) is summarized by type of transformation completed:

		Total Textural Transformation	
		Yes	No
Total Color Transformation	Yes	243	26
	No	13	18

A naturalist randomly selects three leaves from this set, without replacement. Determine the following probabilities.
(a) Exactly one has undergone both types of transformations.
(b) At least one has undergone both transformations.
(c) Exactly one has undergone one but not both transformations.
(d) At least one has undergone at least one transformation.

3-104. A state runs a lottery in which 6 numbers are randomly selected from 40, without replacement. A player chooses 6 numbers before the state's sample is selected.
(a) What is the probability that the 6 numbers chosen by a player match all 6 numbers in the state's sample?
(b) What is the probability that 5 of the 6 numbers chosen by a player appear in the state's sample?
(c) What is the probability that 4 of the 6 numbers chosen by a player appear in the state's sample?
(d) If a player enters one lottery each week, what is the expected number of weeks until a player matches all 6 numbers in the state's sample?

3-105. Magnetic tape is slit into half-inch widths that are wound into cartridges. A slitter assembly contains 48 blades. Five blades are selected at random and evaluated each day for sharpness. If any dull blade is found, the assembly is replaced with a newly sharpened set of blades.
(a) If 10 of the blades in an assembly are dull, what is the probability that the assembly is replaced the first day it is evaluated?
(b) If 10 of the blades in an assembly are dull, what is the probability that the assembly is not replaced until the third day of evaluation? [*Hint:* Assume the daily decisions are independent, and use the geometric distribution.]
(c) Suppose on the first day of evaluation, two of the blades are dull, on the second day of evaluation six are dull, and on the third day of evaluation, ten are dull. What is the probability that the assembly is not replaced until the third day of evaluation? [*Hint:* Assume the daily decisions are independent. However, the probability of replacement changes every day.]

3-106.
(a) Calculate the finite population corrections for Exercises 3-97 and 3-98. For which exercise should the binomial approximation to the distribution of X be better?
(b) For Exercise 3-97, calculate $P(X = 1)$ and $P(X = 4)$ assuming that X has a binomial distribution and compare these results to results derived from the hypergeometric distribution.
(c) For Exercise 3-98, calculate $P(X = 1)$ and $P(X = 4)$ assuming that X has a binomial distribution and compare these results to the results derived from the hypergeometric distribution.
(d) Use the binomial approximation to the hypergeometric distribution to approximate the probabilities in Exercise 3-102. What is the finite population correction in this exercise?

3-9 POISSON DISTRIBUTION

We introduce the Poisson distribution with an example.

EXAMPLE 3-30

Consider the transmission of n bits over a digital communication channel. Let the random variable X equal the number of bits in error. When the probability that a bit is in error is constant and the transmissions are independent, X has a binomial distribution. Let p denote the probability that a bit is in error. Let $\lambda = pn$. Then, $E(x) = pn = \lambda$ and

$$P(X = x) = \binom{n}{x} p^x (1 - p)^{n-x} = \binom{n}{x} \left(\frac{\lambda}{n}\right)^x \left(1 - \frac{\lambda}{n}\right)^{n-x}$$

Now, suppose that the number of bits transmitted increases and the probability of an error decreases exactly enough that pn remains equal to a constant. That is, n increases and p decreases accordingly, such that $E(X) = \lambda$ remains constant. Then, with some work, it can be shown that

$$\binom{n}{x} \left(\frac{1}{n}\right)^x \rightarrow \frac{1}{x!} \qquad \left(1 - \frac{\lambda}{n}\right)^{-x} \rightarrow 1 \qquad \left(1 - \frac{\lambda}{n}\right)^{n} \rightarrow e^{-\lambda}$$

so that

$$\lim_{n \to \infty} P(X = x) = \frac{e^{-\lambda} \lambda^x}{x!}, \qquad x = 0, 1, 2, \ldots$$

Also, because the number of bits transmitted tends to infinity, the number of errors can equal any nonnegative integer. Therefore, the range of X is the integers from zero to infinity.

The distribution obtained as the limit in the above example is more useful than the derivation above implies. The following example illustrates the broader applicability.

EXAMPLE 3-31
Wire Flaws

Flaws occur at random along the length of a thin copper wire. Let X denote the random variable that counts the number of flaws in a length of L millimeters of wire and suppose that the average number of flaws in L millimeters is λ.

The probability distribution of X can be found by reasoning in a manner similar to the previous example. Partition the length of wire into n subintervals of small length, say, 1 micrometer each. If the subinterval chosen is small enough, the probability that more than one flaw occurs in the subinterval is negligible. Furthermore, we can interpret the assumption that flaws occur at random to imply that every subinterval has the same probability of containing a flaw, say, p. Finally, if we assume that the probability that a subinterval contains a flaw is independent of other subintervals, we can model the distribution of X as approximately a binomial random variable. Because

$$E(X) = \lambda = np$$

we obtain

$$p - \lambda/n$$

That is, the probability that a subinterval contains a flaw is λ/n. With small enough subintervals, n is very large and p is very small. Therefore, the distribution of X is obtained as in the previous example.

Example 3-31 can be generalized to include a broad array of random experiments. The interval that was partitioned was a length of wire. However, the same reasoning can be applied to any interval, including an interval of time, an area, or a volume. For example, counts of (1) particles of contamination in semiconductor manufacturing, (2) flaws in rolls of textiles, (3) calls to a telephone exchange, (4) power outages, and (5) atomic particles emitted from a specimen have all been successfully modeled by the probability mass function in the following definition.

Poisson Distribution

Given an interval of real numbers, assume events occur at random throughout the interval. If the interval can be partitioned into subintervals of small enough length such that

(1) the probability of more than one event in a subinterval is zero,

(2) the probability of one event in a subinterval is the same for all subintervals and proportional to the length of the subinterval, and

(3) the event in each subinterval is independent of other subintervals, the random experiment is called a **Poisson process.**

The random variable X that equals the number of events in the interval is a **Poisson random variable** with parameter $0 < \lambda$, and the probability mass function of X is

$$f(x) = \frac{e^{-\lambda}\lambda^x}{x!} \qquad x = 0, 1, 2, \ldots \tag{3-16}$$

The sum of the probabilities is one because

$$\sum_{k=0}^{\infty} \frac{e^{-\lambda}\lambda^k}{k!} = e^{-\lambda}\sum_{k=0}^{\infty}\frac{\lambda^k}{k!}$$

and the summation on the right-hand side of the previous equation is recognized to be Taylor's expansion of e^x evaluated at λ. Therefore, the summation equals e^{λ} and the right-hand side equals $e^{-\lambda}e^{\lambda} = 1$.

Historically, the term *process* has been used to suggest the observation of a system over time. In our example with the copper wire, we showed that the Poisson distribution could also apply to intervals such as lengths. Figure 3-14 provides graphs of selected Poisson distributions.

It is important to **use consistent units** in the calculation of probabilities, means, and variances involving Poisson random variables. The following example illustrates unit conversions. For example, if the

average number of flaws per millimeter of wire is 3.4, then the

average number of flaws in 10 millimeters of wire is 34, and the

average number of flaws in 100 millimeters of wire is 340.

If a Poisson random variable represents the number of events in some interval, the mean of the random variable must equal the expected number of events in the same length of interval.

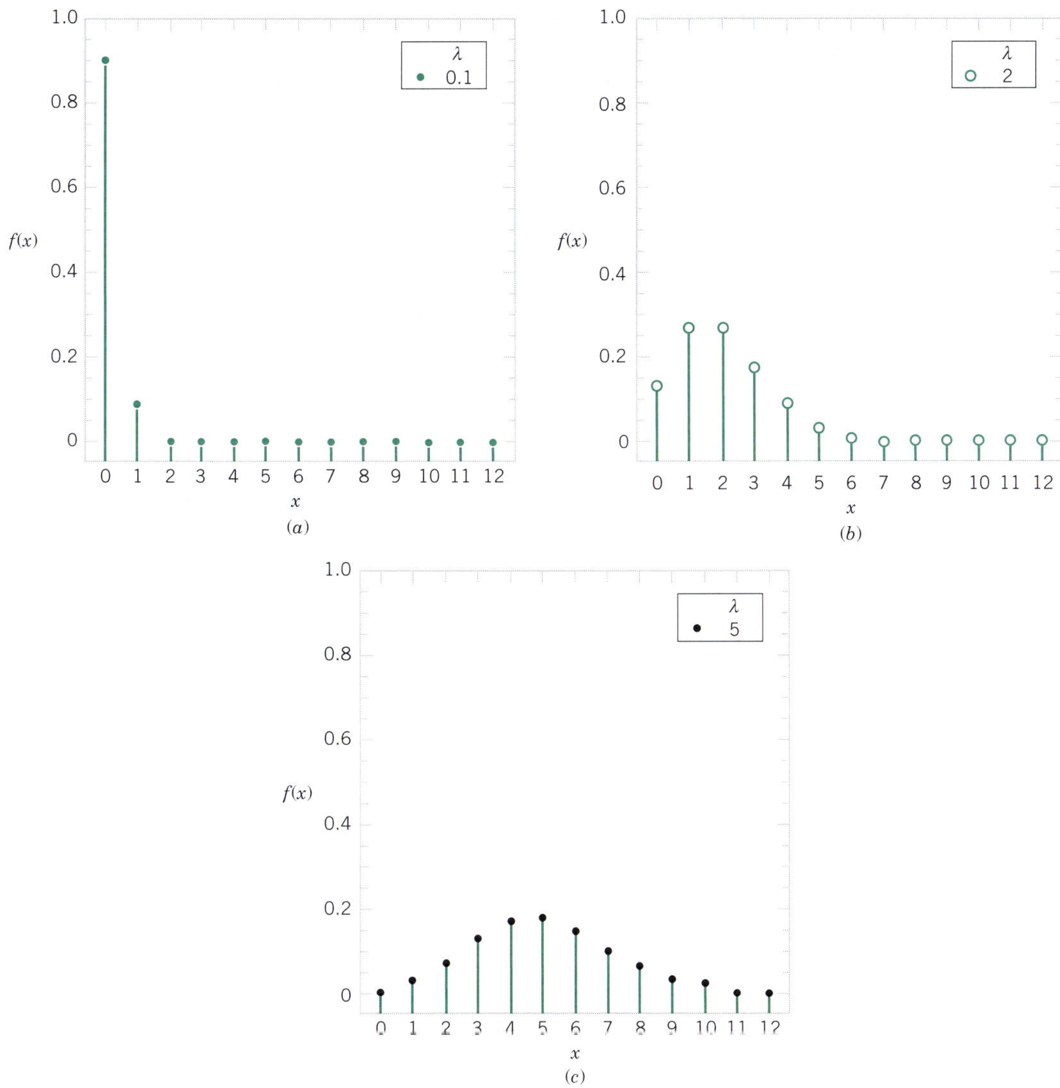

Figure 3-14 Poisson distributions for selected values of the parameters.

EXAMPLE 3-32
Calculations for Wire Flaws

For the case of the thin copper wire, suppose that the number of flaws follows a Poisson distribution with a mean of 2.3 flaws per millimeter. Determine the probability of exactly 2 flaws in 1 millimeter of wire.

Let X denote the number of flaws in 1 millimeter of wire. Then, $E(X) = 2.3$ flaws and

$$P(X = 2) = \frac{e^{-2.3}2.3^2}{2!} = 0.265$$

Determine the probability of 10 flaws in 5 millimeters of wire. Let X denote the number of flaws in 5 millimeters of wire. Then, X has a Poisson distribution with

$$E(X) = 5 \text{ mm} \times 2.3 \text{ flaws/mm} = 11.5 \text{ flaws}$$

Therefore,

$$P(X = 10) = e^{-11.5}\frac{11.5^{10}}{10!} = 0.113$$

Determine the probability of at least 1 flaw in 2 millimeters of wire. Let X denote the number of flaws in 2 millimeters of wire. Then, X has a Poisson distribution with

$$E(X) = 2 \text{ mm} \times 2.3 \text{ flaws/mm} = 4.6 \text{ flaws}$$

Therefore,

$$P(X \geq 1) = 1 - P(X = 0) = 1 - e^{-4.6} = 0.9899$$

EXAMPLE 3-33
CDs

Contamination is a problem in the manufacture of optical storage disks (CDs). The number of particles of contamination that occur on an optical disk has a Poisson distribution, and the average number of particles per centimeter squared of media surface is 0.1. The area of a disk under study is 100 squared centimeters. Find the probability that 12 particles occur in the area of a disk under study.

Let X denote the number of particles in the area of a disk under study. Because the mean number of particles is 0.1 particles per cm^2

$$E(X) = 100 \text{ cm}^2 \times 0.1 \text{ particles/cm}^2 = 10 \text{ particles}$$

Therefore,

$$P(X = 12) = \frac{e^{-10}10^{12}}{12!} = 0.095$$

The probability that zero particles occur in the area of the disk under study is

$$P(X = 0) = e^{-10} = 4.54 \times 10^{-5}$$

Determine the probability that 12 or fewer particles occur in the area of the disk under study. The probability is

$$P(X \leq 12) = P(X = 0) + P(X = 1) + \cdots + P(X = 12) = \sum_{i=0}^{12}\frac{e^{-10}10^i}{i!}$$

Because this sum is tedious to compute, many computer programs calculate cumulative Poisson probabilities. From one such program, $P(X \leq 12) = 0.792$.

The mean of a Poisson random variable is

$$\mu = \sum_{k=1}^{\infty}\frac{ke^{-\lambda}\lambda^k}{k!} = \lambda\sum_{k=1}^{\infty}\frac{e^{-\lambda}\lambda^{k-1}}{(k-1)!}$$

where the summation can start at $k = 1$ because the $k = 0$ term is zero. If a change of variable $j = k - 1$ is used, the summation on the right-hand side of the previous equation is recognized to be the sum of the probabilities of a Poisson random variable and this equals one. Therefore, the previous equation simplifies to $\mu = \lambda$ and the mean is derived.

To obtain the variance of a Poisson random variable we can start with $E(X^2)$ and this equals

$$E(X^2) = \sum_{k=1}^{\infty} \frac{k^2 e^{-\lambda}\lambda^k}{k!} = \lambda \sum_{k=1}^{\infty} \frac{k e^{-\lambda}\lambda^{k-1}}{(k-1)!}$$

Write $k = (k-1) + 1$ to obtain

$$E(X^2) = \lambda \sum_{k=1}^{\infty} \frac{(k-1)e^{-\lambda}\lambda^{k-1}}{(k-1)!} + \lambda \sum_{k=1}^{\infty} \frac{e^{-\lambda}\lambda^{k-1}}{(k-1)!}$$

The summation in the first term on the right-hand side of the previous equation is recognized to be the mean of X and this equals λ so that the first term is λ^2. The summation in the second term on the right-hand side is recognized to be the sum of the probabilities and this equals one. Therefore, the previous equation simplifies to $E(X^2) = \lambda^2 + \lambda$. Because the $V(X) = E(X^2) - (EX)^2$ we have

$$V(X) = \lambda^2 + \lambda - \lambda^2 = \lambda$$

and the variance is derived.

Mean and Variance

If X is a Poisson random variable with parameter λ, then

$$\mu = E(X) = \lambda \quad \text{and} \quad \sigma^2 = V(X) = \lambda \qquad (3\text{-}17)$$

The mean and variance of a Poisson random variable are equal. For example, if particle counts follow a Poisson distribution with a mean of 25 particles per square centimeter, the variance is also 25 and the standard deviation of the counts is 5 per square centimeter. Consequently, information on the variability is very easily obtained. Conversely, if the variance of count data is much greater than the mean of the same data, the Poisson distribution is not a good model for the distribution of the random variable.

EXERCISES FOR SECTION 3-9

3-107. Suppose X has a Poisson distribution with a mean of 4. Determine the following probabilities:
(a) $P(X = 0)$ (b) $P(X \le 2)$
(c) $P(X = 4)$ (d) $P(X = 8)$

3-108. Suppose X has a Poisson distribution with a mean of 0.4. Determine the following probabilities:
(a) $P(X = 0)$ (b) $P(X \le 2)$
(c) $P(X = 4)$ (d) $P(X = 8)$

3-109. Suppose that the number of customers that enter a bank in an hour is a Poisson random variable, and suppose that $P(X = 0) = 0.05$. Determine the mean and variance of X.

3-110. The number of telephone calls that arrive at a phone exchange is often modeled as a Poisson random variable. Assume that on the average there are 10 calls per hour.
(a) What is the probability that there are exactly 5 calls in one hour?
(b) What is the probability that there are 3 or fewer calls in one hour?
(c) What is the probability that there are exactly 15 calls in two hours?
(d) What is the probability that there are exactly 5 calls in 30 minutes?

3-111. Astronomers treat the number of stars in a given volume of space as a Poisson random variable. The density in the Milky Way Galaxy in the vicinity of our solar system is one star per 16 cubic light years.
(a) What is the probability of two or more stars in 16 cubic light years?

(b) How many cubic light years of space must be studied so that the probability of one or more stars exceeds 0.95?

3-112. Data from www.centralhudsonlabs determined the mean number of insect fragments in 225 gram chocolate bars was 14.4, but three brands had insect contamination more than twice the average. See the U.S. Food and Drug Administration–Center for Food Safety and Applied Nutrition for Defect Action Levels for food products. Assume the number of fragments (contaminants) follows a Poisson distribution.

(a) If you consume a 225 gram bar from a brand at the mean contamination level, what is the probability of no insect contaminants?

(b) Suppose you consume a bar that is one-fifth the size tested (45 grams) from a brand at the mean contamination level. What is the probability of no insect contaminants?

(c) If you consume seven 28.35 gram (one ounce) bars this week from a brand at the mean contamination level, what is the probability that you consume one or more insect fragments in more than one bar?

(d) Is the probability of a test result more than twice the mean of 14.4 unusual, or can it be considered typical variation? Explain.

3-113. In 1898 L. J. Bortkiewicz published a book entitled *The Law of Small Numbers*. He used data collected over 20 years to show that the number of soldiers killed by horse kicks each year in each corps in the Prussian cavalry followed a Poisson distribution with a mean of 0.61.

(a) What is the probability of more than one death in a corps in a year?

(b) What is the probability of no deaths in a corps over five years?

3-114. The number of flaws in bolts of cloth in textile manufacturing is assumed to be Poisson distributed with a mean of 0.1 flaw per square meter.

(a) What is the probability that there are two flaws in 1 square meter of cloth?

(b) What is the probability that there is one flaw in 10 square meters of cloth?

(c) What is the probability that there are no flaws in 20 square meters of cloth?

(d) What is the probability that there are at least two flaws in 10 square meters of cloth?

3-115. When a computer disk manufacturer tests a disk, it writes to the disk and then tests it using a certifier. The certifier counts the number of missing pulses or errors. The number of errors on a test area on a disk has a Poisson distribution with $\lambda = 0.2$.

(a) What is the expected number of errors per test area?

(b) What percentage of test areas have two or fewer errors?

3-116. The number of cracks in a section of interstate highway that are significant enough to require repair is assumed to follow a Poisson distribution with a mean of two cracks per mile.

(a) What is the probability that there are no cracks that require repair in 5 miles of highway?

(b) What is the probability that at least one crack requires repair in 1/2 mile of highway?

(c) If the number of cracks is related to the vehicle load on the highway and some sections of the highway have a heavy load of vehicles whereas other sections carry a light load, how do you feel about the assumption of a Poisson distribution for the number of cracks that require repair?

3-117. The number of surface flaws in plastic panels used in the interior of automobiles has a Poisson distribution with a mean of 0.05 flaws per square foot of plastic panel. Assume an automobile interior contains 10 square feet of plastic panel.

(a) What is the probability that there are no surface flaws in an auto's interior?

(b) If 10 cars are sold to a rental company, what is the probability that none of the 10 cars has any surface flaws?

(c) If 10 cars are sold to a rental company, what is the probability that at most one car has any surface flaws?

3-118. The number of failures of a testing instrument from contamination particles on the product is a Poisson random variable with a mean of 0.02 failure per hour.

(a) What is the probability that the instrument does not fail in an 8-hour shift?

(b) What is the probability of at least one failure in a 24-hour day?

Supplemental Exercises

3-119. Let the random variable X be equally likely to assume any of the values $1/8$, $1/4$, or $3/8$. Determine the mean and variance of X.

3-120. Let X denote the number of bits received in error in a digital communication channel, and assume that X is a binomial random variable with $p = 0.001$. If 1000 bits are transmitted, determine the following:

(a) $P(X = 1)$ (b) $P(X \geq 1)$

(c) $P(X \leq 2)$ (d) mean and variance of X

3-121. Batches that consist of 50 coil springs from a production process are checked for conformance to customer requirements. The mean number of nonconforming coil springs in a batch is 5. Assume that the number of nonconforming springs in a batch, denoted as X, is a binomial random variable.

(a) What are n and p?

(b) What is $P(X \leq 2)$?

(c) What is $P(X \geq 49)$?

3-122. An automated egg carton loader has a 1% probability of cracking an egg, and a customer will complain if more than one egg per dozen is cracked. Assume each egg load is an independent event.

(a) What is the distribution of cracked eggs per dozen? Include parameter values.

(b) What is the probability that a carton of a dozen eggs results in a complaint?

(c) What is the mean and standard deviation of the number of cracked eggs in a carton of one dozen?

3-123. A total of 12 cells are replicated. Freshly-synthesized DNA cannot be replicated again until mitosis is completed. Two control mechanisms have been identified—one positive and one negative—that are used with equal probability. Assume that each cell independently uses a control mechanism. Determine the following probabilities.

(a) All cells use a positive control mechanism.
(b) Exactly half the cells use a positive control mechanism.
(c) More than four, but less than seven cells use a positive control mechanism.

3-124. A congested computer network has a 1% chance of losing a data packet and packet losses are independent events. An e-mail message requires 100 packets.

(a) What is the distribution of data packets that must be resent? Include the parameter values.
(b) What is the probability that at least one packet must be resent?
(c) What is the probability that two or more packets must be resent?
(d) What is the mean and standard deviation of the number of packets that must be resent?
(e) If there are 10 messages and each contains 100 packets, what is the probability that at least one message requires two or more packets be resent?

3-125. A particularly long traffic light on your morning commute is green 20% of the time that you approach it. Assume that each morning represents an independent trial.

(a) What is the probability that the first morning that the light is green is the fourth morning that you approach it?
(b) What is the probability that the light is not green for 10 consecutive mornings?

3-126. The probability is 0.6 that a calibration of a transducer in an electronic instrument conforms to specifications for the measurement system. Assume the calibration attempts are independent. What is the probability that at most three calibration attempts are required to meet the specifications for the measurement system?

3-127. An electronic scale in an automated filling operation stops the manufacturing line after three underweight packages are detected. Suppose that the probability of an underweight package is 0.001 and each fill is independent.

(a) What is the mean number of fills before the line is stopped?
(b) What is the standard deviation of the number of fills before the line is stopped?

3-128. The probability that an eagle kills a jackrabbit in a day of hunting is 10%. Assume that results are independent between days.

(a) What is the distribution of the number of days until a successful jackrabbit hunt?
(b) What is the probability that the eagle must wait 5 days for its first successful hunt?
(c) What is the expected number of days until a successful hunt?
(d) If the eagle can survive up to 10 days without food (it requires a successful hunt on the tenth day), what is the probability the eagle is still alive 10 days from now?

3-129. Traffic flow is traditionally modeled as a Poisson distribution. A traffic engineer monitors the traffic flowing through an intersection with an average of 6 cars per minute. To set the timing of a traffic signal the following probabilities are used.

(a) What is the probability of no cars through the intersection within 30 seconds?
(b) What is the probability of three or more cars through the intersection within 30 seconds?
(c) Calculate the minimum number of cars through the intersection so that the probability of this number or fewer cars in 30 seconds is at least 90%.
(d) If the variance of the number of cars through the intersection per minute is 20, is the Poisson distribution appropriate? Explain.

3-130. A shipment of chemicals arrives in 15 totes. Three of the totes are selected at random, without replacement, for an inspection of purity. If two of the totes do not conform to purity requirements, what is the probability that at least one of the nonconforming totes is selected in the sample?

3-131. The probability that your call to a service line is answered in less than 30 seconds is 0.75. Assume that your calls are independent.

(a) If you call 10 times, what is the probability that exactly 9 of your calls are answered within 30 seconds?
(b) If you call 20 times, what is the probability that at least 16 calls are answered in less than 30 seconds?
(c) If you call 20 times, what is the mean number of calls that are answered in less than 30 seconds?

3-132. Continuation of Exercise 3-131.

(a) What is the probability that you must call four times to obtain the first answer in less than 30 seconds?
(b) What is the mean number of calls until you are answered in less than 30 seconds?

3-133. Continuation of Exercise 3-131.

(a) What is the probability that you must call six times in order for two of your calls to be answered in less than 30 seconds?
(b) What is the mean number of calls to obtain two answers in less than 30 seconds?

3-134. The number of messages sent to a computer bulletin board is a Poisson random variable with a mean of 5 messages per hour.

(a) What is the probability that 5 messages are received in 1 hour?
(b) What is the probability that 10 messages are received in 1.5 hours?
(c) What is the probability that less than two messages are received in one-half hour?

3-135. A Web site is operated by four identical computer servers. Only one is used to operate the site; the others are spares that can be activated in case the active server fails. The probability that a request to the Web site generates a failure in the active server is 0.0001. Assume that each request is an independent trial. What is the mean time until failure of all four computers?

3-136. The number of errors in a textbook follows a Poisson distribution with a mean of 0.01 error per page. What is the probability that there are three or less errors in 100 pages?

3-137. The probability that an individual recovers from an illness in a one-week time period without treatment is 0.1. Suppose that 20 independent individuals suffering from this illness are treated with a drug and 4 recover in a one-week time period. If the drug has no effect, what is the probability that 4 or more people recover in a one-week time period?

3-138. Patient response to a generic drug to control pain is scored on a 5-point scale, where a 5 indicates complete relief. Historically the distribution of scores is

1	2	3	4	5
0.05	0.1	0.2	0.25	0.4

Two patients, assumed to be independent, are each scored.
(a) What is the probability mass function of the total score?
(b) What is the probability mass function of the average score?

3-139. In a manufacturing process that laminates several ceramic layers, 1% of the assemblies are defective. Assume that the assemblies are independent.
(a) What is the mean number of assemblies that need to be checked to obtain five defective assemblies?
(b) What is the standard deviation of the number of assemblies that need to be checked to obtain five defective assemblies?

3-140. Continuation of Exercise 3-139. Determine the minimum number of assemblies that need to be checked so that the probability of at least one defective assembly exceeds 0.95.

3-141. Determine the constant c so that the following function is a probability mass function: $f(x) = cx$ for $x = 1, 2, 3, 4$.

3-142. A manufacturer of a consumer electronics product expects 2% of units to fail during the warranty period. A sample of 500 independent units is tracked for warranty performance.
(a) What is the probability that none fails during the warranty period?
(b) What is the expected number of failures during the warranty period?
(c) What is the probability that more than two units fail during the warranty period?

3-143. Messages that arrive at a service center for an information systems manufacturer have been classified on the basis of the number of keywords (used to help route messages) and the type of message, either email or voice. Also, 70% of the messages arrive via email and the rest are voice.

number of keywords	0	1	2	3	4
email	0.1	0.1	0.2	0.4	0.2
voice	0.3	0.4	0.2	0.1	0

Determine the probability mass function of the number of keywords in a message.

3-144. The random variable X has the following probability distribution:

x	2	3	5	8
probability	0.2	0.4	0.3	0.1

Determine the following:
(a) $P(X \leq 3)$ (b) $P(X > 2.5)$
(c) $P(2.7 < X < 5.1)$ (d) $E(X)$
(e) $V(X)$

3-145. Determine the probability mass function for the random variable with the following cumulative distribution function:

$$F(x) = \begin{cases} 0 & x < 2 \\ 0.2 & 2 \leq x < 5.7 \\ 0.5 & 5.7 \leq x < 6.5 \\ 0.8 & 6.5 \leq x < 8.5 \\ 1 & 8.5 \leq x \end{cases}$$

3-146. Each main bearing cap in an engine contains four bolts. The bolts are selected at random, without replacement, from a parts bin that contains 30 bolts from one supplier and 70 bolts from another.
(a) What is the probability that a main bearing cap contains all bolts from the same supplier?
(b) What is the probability that exactly three bolts are from the same supplier?

3-147. Assume the number of errors along a magnetic recording surface is a Poisson random variable with a mean of one error every 10^5 bits. A sector of data consists of 4096 eight-bit bytes.
(a) What is the probability of more than one error in a sector?
(b) What is the mean number of sectors until an error is found?

3-148. An installation technician for a specialized communication system is dispatched to a city only when three or more orders have been placed. Suppose orders follow a Poisson distribution with a mean of 0.25 per week for a city with a population of 100,000 and suppose your city contains a population of 800,000.

(a) What is the probability that a technician is required after a one-week period?

(b) If you are the first one in the city to place an order, what is the probability that you have to wait more than two weeks from the time you place your order until a technician is dispatched?

3-149. From 500 customers, a major appliance manufacturer will randomly select a sample without replacement. The company estimates that 25% of the customers will provide useful data. If this estimate is correct, what is the probability mass function of the number of customers that will provide useful data?

(a) Assume that the company samples 5 customers.

(b) Assume that the company samples 10 customers.

3-150. It is suspected that some of the totes containing chemicals purchased from a supplier exceed the moisture content target. Samples from 30 totes are to be tested for moisture content. Assume that the totes are independent. Determine the proportion of totes from the supplier that must exceed the moisture content target so that the probability is 0.90 that at least one tote in the sample of 30 fails the test.

3-151. Messages arrive to a computer server according to a Poisson distribution with a mean rate of 10 per hour. Determine the length of an interval of time such that the probability that no messages arrive during this interval is 0.90.

3-152. Flaws occur in the interior of plastic used for automobiles according to a Poisson distribution with a mean of 0.02 flaw per panel.

(a) If 50 panels are inspected, what is the probability that there are no flaws?

(b) What is the expected number of panels that need to be inspected before a flaw is found?

(c) If 50 panels are inspected, what is the probability that the number of panels that have one or more flaws is less than or equal to 2?

MIND-EXPANDING EXERCISES

3-153. Derive the convergence results used to obtain a Poisson distribution as the limit of a binomial distribution.

3-154. Show that the function $f(x)$ in Example 3-5 satisfies the properties of a probability mass function by summing the infinite series.

3-155. Derive the formula for the mean and standard deviation of a discrete uniform random variable over the range of integers $a, a + 1, \ldots, b$.

3-156. A company performs inspection on shipments from suppliers in order to defect nonconforming products. Assume a lot contains 1000 items and 1% are nonconforming. What sample size is needed so that the probability of choosing at least one nonconforming item in the sample is at least 0.90? Assume the binomial approximation to the hypergeometric distribution is adequate.

3-157. A company performs inspection on shipments from suppliers in order to detect nonconforming products. The company's policy is to use a sample size that is always 10% of the lot size. Comment on the effectiveness of this policy as a general rule for all sizes of lots.

3-158. Surface flaws in automobile exterior panels follow a Poisson distribution with a mean of 0.1 flaw per panel. If 100 panels are checked, what is the probability that fewer than five panels have any flaws?

3-159. A large bakery can produce rolls in lots of either 0, 1000, 2000, or 3000 per day. The production cost per item is $0.10. The demand varies randomly according to the following distribution:

demand for rolls	0	1000	2000	3000
probability of demand	0.3	0.2	0.3	0.2

Every roll for which there is a demand is sold for $0.30. Every roll for which there is no demand is sold in a secondary market for $0.05. How many rolls should the bakery produce each day to maximize the mean profit?

3-160. A manufacturer stocks components obtained from a supplier. Suppose that 2% of the components are defective and that the defective components occur independently. How many components must the manufacturer have in stock so that the probability that 100 orders can be completed without reordering components is at least 0.95?

IMPORTANT TERMS AND CONCEPTS

Bernoulli trial
Binomial distribution
Cumulative probability distribution function-discrete random variable
Discrete uniform distribution
Expected value of a function of a random variable
Finite population correction factor
Geometric distribution
Hypergeometric distribution
Lack of memory property-discrete random variable
Mean-discrete random variable
Mean-function of a discrete random variable
Negative binomial distribution
Poisson distribution
Poisson process
Probability distribution-discrete random variable
Probability mass function
Standard deviation-discrete random variable
Variance-discrete random variable

4 Analysis of Continuous Random Variables

CHAPTER OUTLINE

LEARNING OBJECTIVES

After careful study of this chapter you should be able to do the following:

1. Determine probabilities from probability density functions
2. Determine probabilities from cumulative distribution functions and cumulative distribution functions from probability density functions, and the reverse
3. Calculate means and variances for continuous random variables
4. Understand the assumptions for each of the continuous probability distributions presented
5. Select an appropriate continuous probability distribution to calculate probabilities in specific applications
6. Calculate probabilities, determine means and variances for each of the continuous probability distributions presented
7. Standardize normal random variables
8. Use the table for the cumulative distribution function of a standard normal distribution to calculate probabilities
9. Approximate probabilities for some binomial and Poisson distributions

4-1 CONTINUOUS RANDOM VARIABLES

Previously, we discussed the measurement of the current in a thin copper wire. We noted that the results might differ slightly in day-to-day replications because of small variations in variables that are not controlled in our experiment—changes in ambient temperatures, small impurities in the chemical composition of the wire, current source drifts, and so forth.

Another example is the selection of one part from a day's production and very accurately measuring a dimensional length. In practice, there can be small variations in the actual measured lengths due to many causes, such as vibrations, temperature fluctuations, operator differences, calibrations, cutting tool wear, bearing wear, and raw material changes. Even the measurement procedure can produce variations in the final results.

In these types of experiments, the measurement of interest—current in a copper wire experiment, length of a machined part—can be represented by a random variable. It is reasonable to model the range of possible values of the random variable by an interval (finite or infinite) of real numbers. For example, for the length of a machined part, our model enables the measurement from the experiment to result in any value within an interval of real numbers. Because the range is any value in an interval, the model provides for any precision in length measurements. However, because the number of possible values of the random variable X is uncountably infinite, X has a distinctly different distribution from the discrete random variables studied previously. The range of X includes all values in an interval of real numbers; that is, the range of X can be thought of as a continuum.

A number of continuous distributions frequently arise in applications. These distributions are described, and example computations of probabilities, means, and variances are provided in the remaining sections of this chapter.

4-2 PROBABILITY DISTRIBUTIONS AND PROBABILITY DENSITY FUNCTIONS

Density functions are commonly used in engineering to describe physical systems. For example, consider the density of a loading on a long, thin beam as shown in Fig. 4-1. For any point x along the beam, the density can be described by a function (in grams/cm). Intervals with large loadings correspond to large values for the function. The total loading between points a and b is determined as the integral of the density function from a to b. This integral is the area under the density function over this interval, and it can be loosely interpreted as the sum of all the loadings over this interval.

Similarly, a **probability density function** $f(x)$ can be used to describe the probability distribution of a **continuous random variable** X. If an interval is likely to contain a value for X, its probability is large and it corresponds to large values for $f(x)$. The probability that X is between a and b is determined as the integral of $f(x)$ from a to b. See Fig. 4-2.

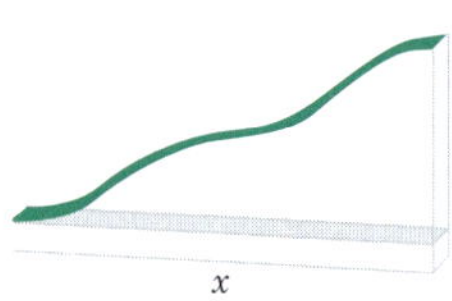

Figure 4-1 Density function of a loading on a long, thin beam.

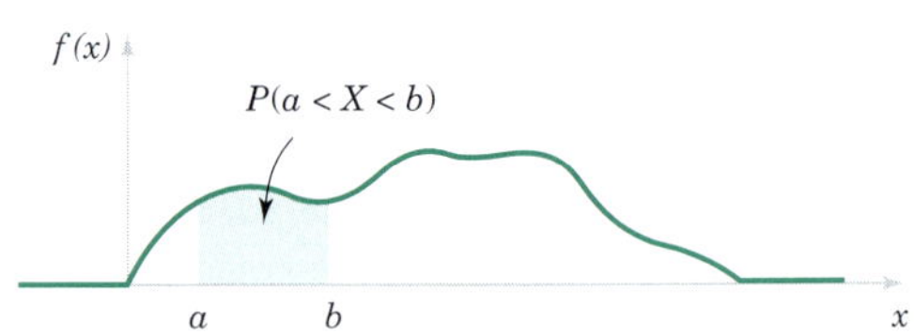

Figure 4-2 Probability determined from the area under $f(x)$.

Probability Density Function

For a continuous random variable X, a **probability density function** is a function such that

(1) $f(x) \geq 0$

(2) $\int_{-\infty}^{\infty} f(x)\, dx = 1$

(3) $P(a \leq X \leq b) = \int_{a}^{b} f(x)\, dx = \text{area under } f(x) \text{ from } a \text{ to } b$

for any a and b (4-1)

A probability density function provides a simple description of the probabilities associated with a random variable. As long as $f(x)$ is nonnegative and $\int_{-\infty}^{\infty} f(x)\, dx = 1$, $0 \leq P(a < X < b) \leq 1$ so that the probabilities are properly restricted. A probability density function is zero for x values that cannot occur and it is assumed to be zero wherever it is not specifically defined.

A **histogram** is an approximation to a probability density function. See Fig. 4-3. For each interval of the histogram, the area of the bar equals the relative frequency (proportion) of the measurements in the interval. The relative frequency is an estimate of the probability that a measurement falls in the interval. Similarly, the area under $f(x)$ over any interval equals the true probability that a measurement falls in the interval.

The important point is that ***f(x)* is used to calculate an area** that represents the probability that X assumes a value in $[a, b]$. For the current measurement example, the probability that X results in [14 mA, 15 mA] is the integral of the probability density function of X over this interval. The probability that X results in [14.5 mA, 14.6 mA] is the integral of the same function, $f(x)$, over the smaller interval. By appropriate choice of the shape of $f(x)$, we can represent the probabilities associated with any continuous random variable X. The shape of $f(x)$ determines how the probability that X assumes a value in [14.5 mA, 14.6 mA] compares to the probability of any other interval of equal or different length.

For the density function of a loading on a long thin beam, because every point has zero width, the loading at any point is zero. Similarly, for a continuous random variable X and *any* value x.

$$P(X = x) = 0$$

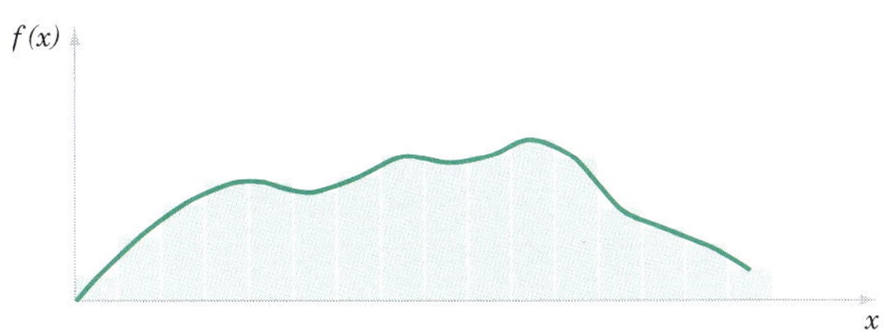

Figure 4-3 Histogram approximates a probability density function.

Based on this result, it might appear that our model of a continuous random variable is useless. However, in practice, when a particular current measurement is observed, such as 14.47 milliamperes, this result can be interpreted as the rounded value of a current measurement that is actually in a range such as $14.465 \le x \le 14.475$. Therefore, the probability that the rounded value 14.47 is observed as the value for X is the probability that X assumes a value in the interval [14.465, 14.475], which is not zero. Similarly, because each point has zero probability, one need not distinguish between inequalities such as $<$ or $\le$ for continuous random variables.

If X is a **continuous random variable,** for any x_1 and x_2,

$$P(x_1 \le X \le x_2) = P(x_1 < X \le x_2) = P(x_1 \le X < x_2) = P(x_1 < X < x_2) \quad (4\text{-}2)$$

EXAMPLE 4-1
Electric Current

Let the continuous random variable X denote the current measured in a thin copper wire in milliamperes. Assume that the range of X is [0, 20 mA], and assume that the probability density function of X is $f(x) = 0.05$ for $0 \le x \le 20$. What is the probability that a current measurement is less than 10 milliamperes?

The probability density function is shown in Fig. 4-4. It is assumed that $f(x) = 0$ wherever it is not specifically defined. The probability requested is indicated by the shaded area in Fig. 4-4.

$$P(X < 10) = \int_0^{10} f(x)\, dx = \int_0^{10} 0.05\, dx = 0.5$$

As another example,

$$P(5 < X < 20) = \int_5^{20} f(x)\, dx = 0.75$$

EXAMPLE 4-2
Hole Diameter

Let the continuous random variable X denote the diameter of a hole drilled in a sheet metal component. The target diameter is 12.5 millimeters. Most random disturbances to the process result in larger diameters. Historical data show that the distribution of X can be modeled by a probability density function $f(x) = 20e^{-20(x-12.5)}$, $x \ge 12.5$.

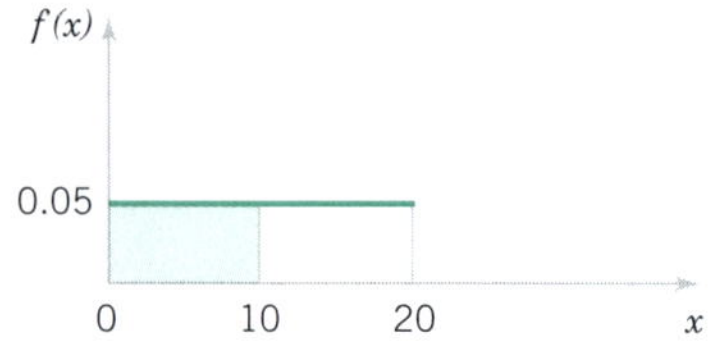

Figure 4-4 Probability density function for Example 4-1.

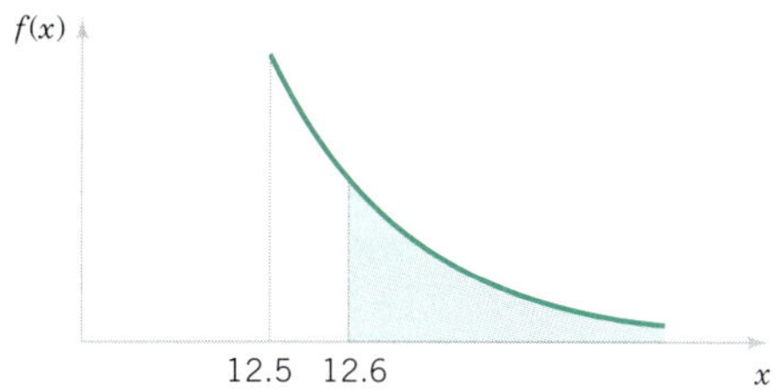

Figure 4-5 Probability density function for Example 4-2.

If a part with a diameter larger than 12.60 millimeters is scrapped, what proportion of parts is scrapped? The density function and the requested probability are shown in Fig. 4-5. A part is scrapped if $X > 12.60$. Now,

$$P(X > 12.60) = \int_{12.6}^{\infty} f(x)\,dx = \int_{12.6}^{\infty} 20e^{-20(x-12.5)}\,dx = -e^{-20(x-12.5)}\Big|_{12.6}^{\infty} = 0.135$$

What proportion of parts is between 12.5 and 12.6 millimeters? Now,

$$P(12.5 < X < 12.6) = \int_{12.5}^{12.6} f(x)\,dx = -e^{-20(x-12.5)}\Big|_{12.5}^{12.6} = 0.865$$

Because the total area under $f(x)$ equals 1, we can also calculate $P(12.5 < X < 12.6) = 1 - P(X > 12.6) = 1 - 0.135 = 0.865$.

EXERCISES FOR SECTION 4-2

4-1. Suppose that $f(x) = e^{-x}$ for $0 < x$. Determine the following probabilities:
(a) $P(1 < X)$ (b) $P(1 < X < 2.5)$
(c) $P(X = 3)$ (d) $P(X < 4)$
(e) $P(3 \leq X)$

4-2. Suppose that $f(x) = e^{-x}$ for $0 < x$.
(a) Determine x such that $P(x < X) = 0.10$.
(b) Determine x such that $P(X \leq x) = 0.10$.

4-3. Suppose that $f(x) = x/8$ for $3 < x < 5$. Determine the following probabilities:
(a) $P(X < 4)$ (b) $P(X > 3.5)$
(c) $P(4 < X < 5)$ (d) $P(X < 4.5)$
(e) $P(X < 3.5 \text{ or } X > 4.5)$

4-4. Suppose that $f(x) = e^{-(x-4)}$ for $4 < x$. Determine the following probabilities:
(a) $P(1 < X)$ (b) $P(2 \leq X < 5)$
(c) $P(5 < X)$ (d) $P(8 < X < 12)$
(e) Determine x such that $P(X < x) = 0.90$.

4-5. Suppose that $f(x) = 1.5x^2$ for $-1 < x < 1$. Determine the following probabilities:
(a) $P(0 < X)$ (b) $P(0.5 < X)$
(c) $P(-0.5 \leq X \leq 0.5)$ (d) $P(X < -2)$
(e) $P(X < 0 \text{ or } X > -0.5)$
(f) Determine x such that $P(x < X) = 0.05$.

4-6. The probability density function of the time to failure of an electronic component in a copier (in hours) is $f(x) = e^{-x/1000}/1000$ for $x > 0$. Determine the probability that
(a) A component lasts more than 3000 hours before failure.
(b) A component fails in the interval from 1000 to 2000 hours.
(c) A component fails before 1000 hours.
(d) Determine the number of hours at which 10% of all components have failed.

4-7. The probability density function of the net weight in pounds of a packaged chemical herbicide is $f(x) = 2.0$ for $49.75 < x < 50.25$ pounds.
(a) Determine the probability that a package weighs more than 50 pounds.
(b) How much chemical is contained in 90% of all packages?

4-8. The probability density function of the length of a hinge for fastening a door is $f(x) = 1.25$ for $74.6 < x < 75.4$ millimeters. Determine the following:
(a) $P(X < 74.8)$
(b) $P(X < 74.8 \text{ or } X > 75.2)$

(c) If the specifications for this process are from 74.7 to 75.3 millimeters, what proportion of hinges meets specifications?

4-9. The probability density function of the length of a metal rod is $f(x) = 2$ for $2.3 < x < 2.8$ meters.

(a) If the specifications for this process are from 2.25 to 2.75 meters, what proportion of the bars fail to meet the specifications?

(b) Assume that the probability density function is $f(x) = 2$ for an interval of length 0.5 meters. Over what value should the density be centered to achieve the greatest proportion of bars within specifications?

4-10. If X is a continuous random variable, argue that $P(x_1 \le X \le x_2) = P(x_1 < X \le x_2) = P(x_1 \le X < x_2) = P(x_1 < X < x_2)$.

4-3 CUMULATIVE DISTRIBUTION FUNCTIONS

An alternative method to describe the distribution of a discrete random variable can also be used for continuous random variables.

Cumulative Distribution Function

The **cumulative distribution function** of a continuous random variable X is

$$F(x) = P(X \le x) = \int_{-\infty}^{x} f(u)\, du \tag{4-3}$$

for $-\infty < x < \infty$.

Extending the definition of $f(x)$ to the entire real line enables us to define the cumulative distribution function for all real numbers. The following example illustrates the definition.

EXAMPLE 4-3 Electric Current

For the copper current measurement in Example 4-1, the cumulative distribution function of the random variable X consists of three expressions. If $x < 0, f(x) = 0$. Therefore,

$$F(x) = 0, \quad \text{for} \quad x < 0$$

and

$$F(x) = \int_{0}^{x} f(u)\, du = 0.05x, \quad \text{for} \quad 0 \le x < 20$$

Finally,

$$F(x) = \int_{0}^{x} f(u)\, du = 1, \quad \text{for} \quad 20 \le x$$

Therefore,

$$F(x) = \begin{cases} 0 & x < 0 \\ 0.05x & 0 \le x < 20 \\ 1 & 20 \le x \end{cases}$$

The plot of $F(x)$ is shown in Fig. 4-6.

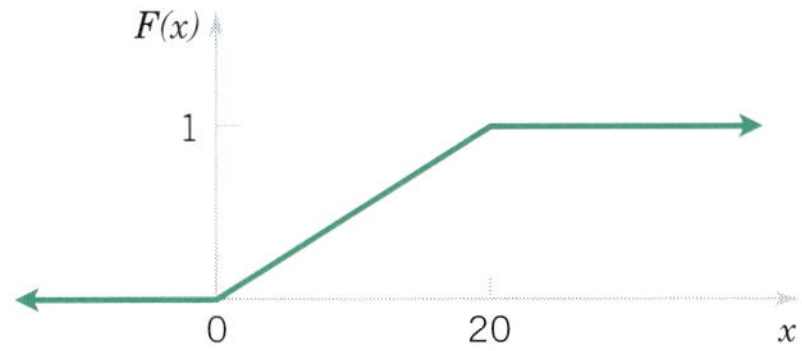

Figure 4-6 Cumulative distribution function for Example 4-3.

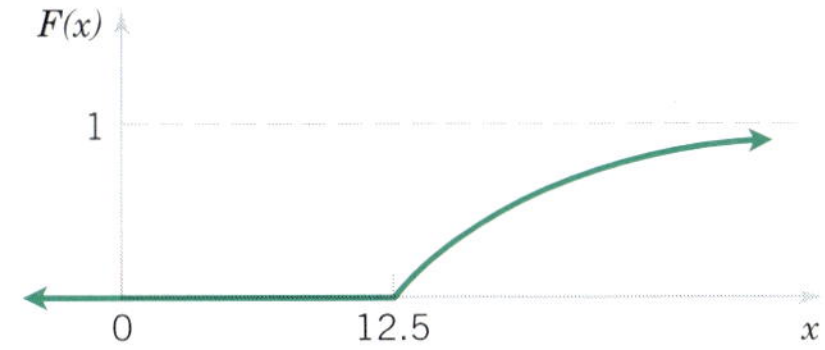

Figure 4-7 Cumulative distribution function for Example 4-4.

Notice that in the definition of $F(x)$ any $<$ can be changed to $\leq$ and vice versa. That is, $F(x)$ can be defined as either $0.05x$ or 0 at the end-point $x = 0$, and $F(x)$ can be defined as either $0.05x$ or 1 at the end-point $x = 20$. In other words, $F(x)$ is a continuous function. For a discrete random variable, $F(x)$ is not a continuous function. Sometimes, a continuous random variable is defined as one that has a continuous cumulative distribution function.

EXAMPLE 4-4
Hole Diameter

For the drilling operation in Example 4-2, $F(x)$ consists of two expressions.

$$F(x) = 0 \quad \text{for} \quad x < 12.5$$

and for $12.5 \leq x$

$$F(x) = \int_{12.5}^{x} 20e^{-20(u-12.5)}\,du = 1 - e^{-20(x-12.5)}$$

Therefore,

$$F(x) = \begin{cases} 0 & x < 12.5 \\ 1 - e^{-20(x-12.5)} & 12.5 \leq x \end{cases}$$

Figure 4-7 displays a graph of $F(x)$.

The probability density function of a continuous random variable can be determined from the cumulative distribution function by differentiating. Recall that the fundamental theorem of calculus states that

$$\frac{d}{dx}\int_{-\infty}^{x} f(u)\,du = f(x)$$

Then, given $F(x)$

$$f(x) = \frac{dF(x)}{dx}$$

as long as the derivative exists.

EXAMPLE 4-5
Reaction Time

The time until a chemical reaction is complete (in milliseconds) is approximated by the cumulative distribution function

$$F(x) = \begin{cases} 0 & x < 0 \\ 1 - e^{-0.01x} & 0 \le x \end{cases}$$

Determine the probability density function of X. What proportion of reactions is complete within 200 milliseconds? Using the result that the probability density function is the derivative of the $F(x)$, we obtain

$$f(x) = \begin{cases} 0 & x < 0 \\ 0.01e^{-0.01x} & 0 \le x \end{cases}$$

The probability that a reaction completes within 200 milliseconds is

$$P(X < 200) = F(200) = 1 - e^{-2} = 0.8647.$$

EXERCISES FOR SECTION 4-3

4-11. Suppose the cumulative distribution function of the random variable X is

$$F(x) = \begin{cases} 0 & x < 0 \\ 0.2x & 0 \le x < 5 \\ 1 & 5 \le x \end{cases}$$

Determine the following:
(a) $P(X < 2.8)$ (b) $P(X > 1.5)$
(c) $P(X < -2)$ (d) $P(X > 6)$

4-12. Suppose the cumulative distribution function of the random variable X is

$$F(x) = \begin{cases} 0 & x < -2 \\ 0.25x + 0.5 & -2 \le x < 2 \\ 1 & 2 \le x \end{cases}$$

Determine the following:
(a) $P(X < 1.8)$ (b) $P(X > -1.5)$
(c) $P(X < -2)$ (d) $P(-1 < X < 1)$

4-13. The gap width is an important property of a magnetic recording head. In coded units, if the width is a continuous random variable over the range from $0 < x < 2$ with $f(x) = 0.5x$, determine the cumulative distribution function of the gap width.

4-14. The probability density function of the time customers arrive at a terminal (in minutes after 8:00 A.M.) is $f(x) = e^{-x/10}/10$ for $0 < x$. Determine the probability that

(a) The first customer arrives by 9:00 A.M.
(b) The first customer arrives between 8:15 A.M. and 8:30 A.M.
(c) Two or more customers arrive before 8:40 A.M. among five that arrive at the terminal. Assume customer arrivals are independent.
(d) Determine the cumulative distribution function and use the cumulative distribution function to determine the probability that the first customer arrives between 8:15 A.M. and 8:30 A.M.

4-15. Determine the cumulative distribution function for the distribution in Exercise 4-1.

4-16. Determine the cumulative distribution function for the distribution in Exercise 4-3.

4-17. Determine the cumulative distribution function for the distribution in Exercise 4-4.

4-18. Determine the cumulative distribution function for the distribution in Exercise 4-6. Use the cumulative distribution function to determine the probability that a component lasts more than 3000 hours before failure.

4-19. Determine the cumulative distribution function for the distribution in Exercise 4-9. Use the cumulative distribution function to determine the probability that a length exceeds 2.7 meters.

Determine the probability density function for each of the following cumulative distribution functions.

4-20. $F(x) = 1 - e^{-2x} \quad x > 0$

4-21.

$$F(x) = \begin{cases} 0 & x < 0 \\ 0.2x & 0 \le x < 4 \\ 0.04x + 0.64 & 4 \le x < 9 \\ 1 & 9 \le x \end{cases}$$

4-22.

$$F(x) = \begin{cases} 0 & x < -2 \\ 0.25x + 0.5 & -2 \le x < 1 \\ 0.5x + 0.25 & 1 \le x < 1.5 \\ 1 & 1.5 \le x \end{cases}$$

4-4 MEAN AND VARIANCE OF A CONTINUOUS RANDOM VARIABLE

The mean and variance can also be defined for a continuous random variable. Integration replaces summation in the discrete definitions. If a probability density function is viewed as a loading on a beam as in Fig. 4-1, the mean is the balance point.

Mean and Variance

Suppose X is a continuous random variable with probability density function $f(x)$. The **mean** or **expected value** of X, denoted as μ or $E(X)$, is

$$\mu = E(X) = \int_{-\infty}^{\infty} x f(x)\, dx \tag{4-4}$$

The **variance** of X, denoted as $V(X)$ or σ^2, is

$$\sigma^2 = V(X) = \int_{-\infty}^{\infty} (x - \mu)^2 f(x)\, dx = \int_{-\infty}^{\infty} x^2 f(x)\, dx - \mu^2$$

The **standard deviation** of X is $\sigma = \sqrt{\sigma^2}$.

The equivalence of the two formulas for variance can be derived as one, as was done for discrete random variables.

EXAMPLE 4-6 Electric Current

For the copper current measurement in Example 4-1, the mean of X is

$$E(X) = \int_{0}^{20} x f(x)\, dx = 0.05x^2/2 \Big|_0^{20} = 10$$

The variance of X is

$$V(X) = \int_{0}^{20} (x - 10)^2 f(x)\, dx = 0.05(x - 10)^3/3 \Big|_0^{20} = 33.33$$

The expected value of a function $h(X)$ of a continuous random variable is also defined in a straightforward manner.

Expected Value of a Function of a Continuous Random Variable

If X is a continuous random variable with probability density function $f(x)$,

$$E[h(X)] = \int_{-\infty}^{\infty} h(x) f(x)\, dx \tag{4-5}$$

EXAMPLE 4-7

In Example 4-1, X is the current measured in milliamperes. What is the expected value of the squared current? Now, $h(X) = X^2$. Therefore,

$$E[h(X)] = \int_{-\infty}^{\infty} x^2 f(x)\,dx = \int_{0}^{20} 0.05x^2\,dx = 0.05\left.\frac{x^3}{3}\right|_0^{20} = 133.33$$

In the previous example, the expected value of X^2 does not equal $E(X)$ squared. However, in the special case that $h(X) = aX + b$ for any constants a and b, $E[h(X)] = aE(X) + b$. This can be shown from the properties of integrals.

EXAMPLE 4-8
Hole Diameter

For the drilling operation in Example 4-2, the mean of X is

$$E(X) = \int_{12.5}^{\infty} xf(x)\,dx = \int_{12.5}^{\infty} x\,20e^{-20(x-12.5)}\,dx$$

Integration by parts can be used to show that

$$E(X) = -xe^{-20(x-12.5)} - \left.\frac{e^{-20(x-12.5)}}{20}\right|_{12.5}^{\infty} = 12.5 + 0.05 = 12.55$$

The variance of X is

$$V(X) = \int_{12.5}^{\infty} (x - 12.55)^2 f(x)\,dx$$

Although more difficult, integration by parts can be used two times to show that $V(X) = 0.0025$.

EXERCISES FOR SECTION 4-4

4-23. Suppose $f(x) = 0.25$ for $0 < x < 4$. Determine the mean and variance of X.

4-24. Suppose $f(x) = 0.125x$ for $0 < x < 4$. Determine the mean and variance of X.

4-25. Suppose $f(x) = 1.5x^2$ for $-1 < x < 1$. Determine the mean and variance of X.

4-26. Suppose that $f(x) = x/8$ for $3 < x < 5$. Determine the mean and variance for x.

4-27. Suppose that contamination particle size (in micrometers) can be modeled as $f(x) = 2x^{-3}$ for $1 < x$. Determine the mean of X.

4-28. Suppose the probability density function of the length of computer cables is $f(x) = 0.1$ from 1200 to 1210 millimeters.
(a) Determine the mean and standard deviation of the cable length.
(b) If the length specifications are $1195 < x < 1205$ millimeters, what proportion of cables are within specifications?

4-29. The thickness of a conductive coating in micrometers has a density function of $600x^{-2}$ for $100\ \mu\text{m} < x < 120\ \mu\text{m}$.
(a) Determine the mean and variance of the coating thickness.
(b) If the coating costs \$0.50 per micrometer of thickness on each part, what is the average cost of the coating per part?

4-30. The probability density function of the weight of packages delivered by a post office is $f(x) = 70/(69x^2)$ for $1 < x < 70$ pounds.
(a) Determine the mean and variance of weight.
(b) If the shipping cost is \$2.50 per pound, what is the average shipping cost of a package?
(c) Determine the probability that the weight of a package exceeds 50 pounds.

4-31. Integration by parts is required. The probability density function for the diameter of a drilled hole in millimeters is $10e^{-10(x-5)}$ for $x > 5$ mm. Although the target diameter is 5 millimeters, vibrations, tool wear, and other nuisances produce diameters larger than 5 millimeters.
(a) Determine the mean and variance of the diameter of the holes.
(b) Determine the probability that a diameter exceeds 5.1 millimeters.

4-5 CONTINUOUS UNIFORM DISTRIBUTION

The simplest continuous distribution is analogous to its discrete counterpart.

Continuous Uniform Distribution

A continuous random variable X with probability density function

$$f(x) = 1/(b - a), \qquad a \leq x \leq b \tag{4-6}$$

is a **continuous uniform random variable.**

The probability density function of a continuous uniform random variable is shown in Fig. 4-8. The mean of the continuous uniform random variable X is

$$E(X) = \int_a^b \frac{x}{b - a}\, dx = \frac{0.5x^2}{b - a}\Bigg|_a^b = \frac{(a + b)}{2}$$

The variance of X is

$$V(X) = \int_a^b \frac{\left(x - \left(\frac{a + b}{2}\right)\right)^2}{b - a}\, dx = \frac{\left(x - \frac{a + b}{2}\right)^3}{3(b - a)}\Bigg|_a^b = \frac{(b - a)^2}{12}$$

These results are summarized as follows.

Mean and Variance

If X is a continuous uniform random variable over $a \leq x \leq b$,

$$\mu = E(X) = \frac{(a + b)}{2} \quad \text{and} \quad \sigma^2 = V(X) = \frac{(b - a)^2}{12} \tag{4-7}$$

EXAMPLE 4-9
Uniform Current

Let the continuous random variable X denote the current measured in a thin copper wire in milliamperes. Assume that the range of X is [0, 20 mA], and assume that the probability density function of X is $f(x) = 0.05, 0 \leq x \leq 20$.

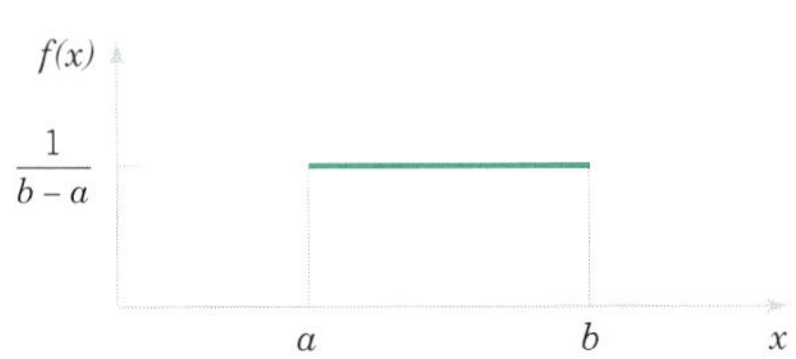

Figure 4-8 Continuous uniform probability density function.

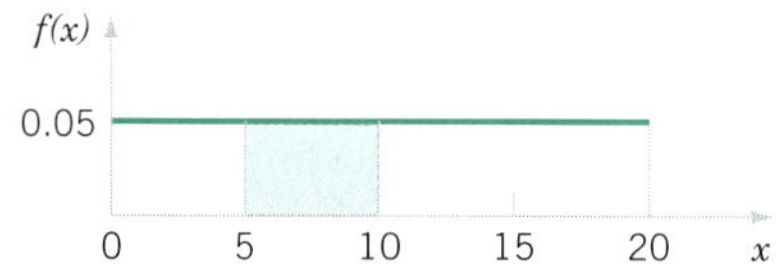

Figure 4-9 Probability for Example 4-9.

What is the probability that a measurement of current is between 5 and 10 milliamperes? The requested probability is shown as the shaded area in Fig. 4-9.

$$P(5 < X < 10) = \int_{5}^{10} f(x)\,dx$$

$$= 5(0.05) = 0.25$$

The mean and variance formulas can be applied with $a = 0$ and $b = 20$. Therefore,

$$E(X) = 10 \text{ mA} \quad \text{and} \quad V(X) = 20^2/12 = 33.33 \text{ mA}^2$$

Consequently, the standard deviation of X is 5.77 mA.

The cumulative distribution function of a continuous uniform random variable is obtained by integration. If $a < x < b$,

$$F(x) = \int_{a}^{x} 1/(b-a)\,du = x/(b-a) - a/(b-a)$$

Therefore, the complete description of the cumulative distribution function of a continuous uniform random variable is

$$F(x) = \begin{cases} 0 & x < a \\ (x-a)/(b-a) & a \le x < b \\ 1 & b \le x \end{cases}$$

An example of $F(x)$ for a continuous uniform random variable is shown in Fig. 4-6.

EXERCISES FOR SECTION 4-5

4-32. Suppose X has a continuous uniform distribution over the interval [1.5, 5.5].
(a) Determine the mean, variance, and standard deviation of X.
(b) What is $P(X < 2.5)$?
(c) Determine the cumulative distribution function.

4-33. Suppose X has a continuous uniform distribution over the interval $[-1, 1]$.
(a) Determine the mean, variance, and standard deviation of X.
(b) Determine the value for x such that $P(-x < X < x) = 0.90$.
(c) Determine the cumulative distribution function.

4-34. The net weight in pounds of a packaged chemical herbicide is uniform for $49.75 < x < 50.25$ pounds.
(a) Determine the mean and variance of the weight of packages.
(b) Determine the cumulative distribution function of the weight of packages.
(c) Determine $P(X < 50.1)$.

4-35. The thickness of a flange on an aircraft component is uniformly distributed between 0.95 and 1.05 millimeters.
(a) Determine the cumulative distribution function of flange thickness.
(b) Determine the proportion of flanges that exceeds 1.02 millimeters.
(c) What thickness is exceeded by 90% of the flanges?
(d) Determine the mean and variance of flange thickness.

4-36. Suppose the time it takes a data collection operator to fill out an electronic form for a database is uniformly between 1.5 and 2.2 minutes.
(a) What is the mean and variance of the time it takes an operator to fill out the form?
(b) What is the probability that it will take less than two minutes to fill out the form?
(c) Determine the cumulative distribution function of the time it takes to fill out the form.

4-37. The thickness of photoresist applied to wafers in semiconductor manufacturing at a particular location on the wafer is uniformly distributed between 0.2050 and 0.2150 micrometers.
(a) Determine the cumulative distribution function of photoresist thickness.
(b) Determine the proportion of wafers that exceeds 0.2125 micrometers in photoresist thickness.
(c) What thickness is exceeded by 10% of the wafers?
(d) Determine the mean and variance of photoresist thickness.

4-38. An adult can lose or gain two pounds of water in the course of a day. Assume that the changes in water weight is uniformly distributed between minus two and plus two pounds in a day. What is the standard deviation of your weight over a day?

4-39. A dolphin show is scheduled to start at 9:00 A.M., 9:30 A.M., and 10:00 A.M. Once the show starts, the gate will be closed. A visitor will arrive at the gate at a time uniformly distributed between 8:30 A.M. and 10:00 A.M. Determine
(a) The cumulative distribution function of the time (in minutes) between arrival and 8:30 A.M.
(b) The mean and variance of the distribution in the previous part.
(c) The probability that a visitor waits less than 10 minutes for a show.
(d) The probability that a visitor waits more than 20 minutes for a show.

4-40. Measurement error that is continuous and uniformly distributed from -3 to $+3$ millivolts is added to the true voltage of a circuit. Then the measurement is rounded to the nearest millivolt so that it becomes discrete. Suppose that the true voltage is 250 millivolts.
(a) What is the probability mass function of the measured voltage?
(b) What is the mean and variance of the measured voltage?

4-6 NORMAL DISTRIBUTION

Undoubtedly, the most widely used model for the distribution of a random variable is a **normal distribution.** Whenever a random experiment is replicated, the random variable that equals the average (or total) result over the replicates tends to have a normal distribution as the number of replicates becomes large. De Moivre presented this fundamental result, known as the **central limit theorem,** in 1733. Unfortunately, his work was lost for some time, and Gauss independently developed a normal distribution nearly 100 years later. Although De Moivre was later credited with the derivation, a normal distribution is also referred to as a **Gaussian distribution.**

When do we average (or total) results? Almost always. For example, an automotive engineer may plan a study to average pull-off force measurements from several connectors. If we assume that each measurement results from a replicate of a random experiment, the normal distribution can be used to make approximate conclusions about this average. These conclusions are the primary topics in the subsequent chapters of this book.

Furthermore, sometimes the central limit theorem is less obvious. For example, assume that the deviation (or error) in the length of a machined part is the sum of a large number of infinitesimal effects, such as temperature and humidity drifts, vibrations, cutting angle variations, cutting tool wear, bearing wear, rotational speed variations, mounting and fixture variations, variations in numerous raw material characteristics, and variation in levels of contamination. If the component errors are independent and equally likely to be positive or negative, the total error can be shown to have an approximate normal distribution. Furthermore, the normal distribution arises in the study of numerous basic physical phenomena. For example, the physicist Maxwell developed a normal distribution from simple assumptions regarding the velocities of molecules.

The theoretical basis of a normal distribution is mentioned to justify the somewhat complex form of the probability density function. Our objective now is to calculate probabilities for a normal random variable. The central limit theorem will be stated more carefully later.

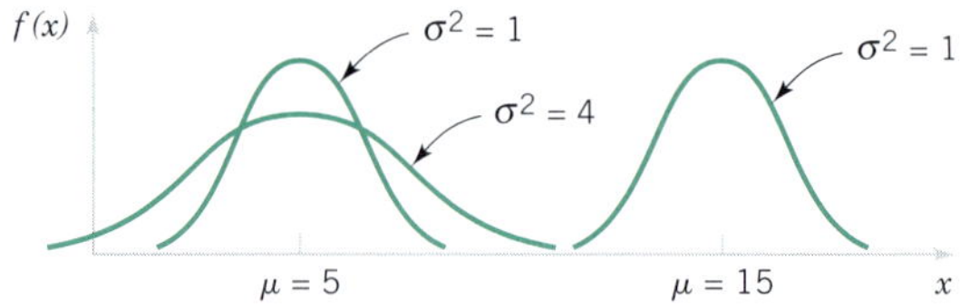

Figure 4-10 Normal probability density functions for selected values of the parameters μ and σ^2.

Random variables with different means and variances can be modeled by normal probability density functions with appropriate choices of the center and width of the curve. The value of $E(X) = \mu$ determines the center of the probability density function and the value of $V(X) = \sigma^2$ determines the width. Figure 4-10 illustrates several normal probability density functions with selected values of μ and σ^2. Each has the characteristic symmetric bell-shaped curve, but the centers and dispersions differ. The following definition provides the formula for normal probability density functions.

Normal Distribution

A random variable X with probability density function

$$f(x) = \frac{1}{\sqrt{2\pi}\sigma} e^{\frac{-(x-\mu)^2}{2\sigma^2}} \qquad -\infty < x < \infty \tag{4-8}$$

is a **normal random variable** with parameters μ, where $-\infty < \mu < \infty$, and $\sigma > 0$. Also,

$$E(X) = \mu \quad \text{and} \quad V(X) = \sigma^2 \tag{4-9}$$

and the notation $N(\mu, \sigma^2)$ is used to denote the distribution. The mean and variance of X are shown to equal μ and σ^2, respectively, at the end of this Section 5-6.

EXAMPLE 4-10

Assume that the current measurements in a strip of wire follow a normal distribution with a mean of 10 milliamperes and a variance of 4 (milliamperes)^2. What is the probability that a measurement exceeds 13 milliamperes?

Let X denote the current in milliamperes. The requested probability can be represented as $P(X > 13)$. This probability is shown as the shaded area under the normal probability density function in Fig. 4-11. Unfortunately, there is no closed-form expression for the integral of a normal probability density function, and probabilities based on the normal distribution are typically found numerically or from a table (that we will later introduce).

Some useful results concerning a normal distribution are summarized below and in Fig. 4-12. For any normal random variable,

$$P(\mu - \sigma < X < \mu + \sigma) = 0.6827$$
$$P(\mu - 2\sigma < X < \mu + 2\sigma) = 0.9545$$
$$P(\mu - 3\sigma < X < \mu + 3\sigma) = 0.9973$$

Also, from the symmetry of $f(x)$, $P(X > \mu) = P(X < \mu) = 0.5$. Because $f(x)$ is positive for all x, this model assigns some probability to each interval of the real line. However, the

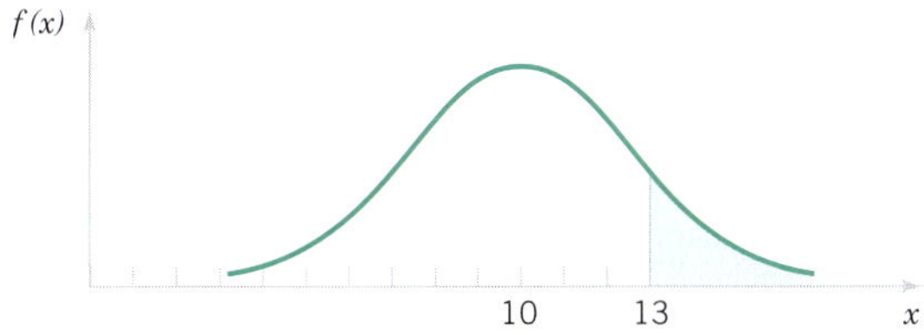

Figure 4-11 Probability that $X > 13$ for a normal random variable with $\mu = 10$ and $\sigma^2 = 4$.

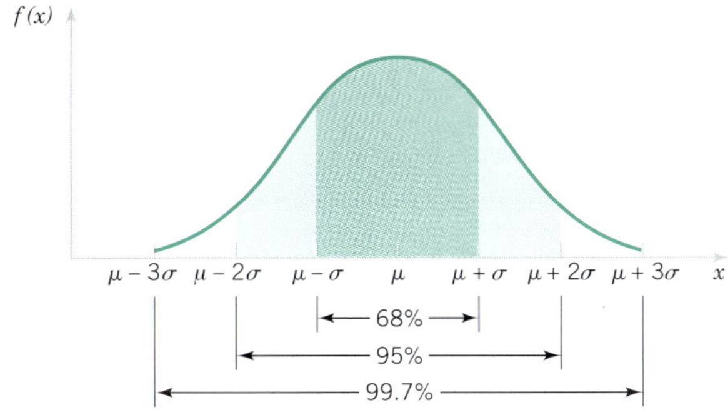

Figure 4-12 Probabilities associated with a normal distribution.

probability density function decreases as x moves farther from μ. Consequently, the probability that a measurement falls far from μ is small, and at some distance from μ the probability of an interval can be approximated as zero.

The area under a normal probability density function beyond 3σ from the mean is quite small. This fact is convenient for quick, rough sketches of a normal probability density function. The sketches help us determine probabilities. Because more than 0.9973 of the probability of a normal distribution is within the interval $(\mu - 3\sigma, \mu + 3\sigma)$, 6σ is often referred to as the **width** of a normal distribution. Advanced integration methods can be used to show that the area under the normal probability density function from $-\infty < x < \infty$ is 1.

Standard Normal Random Variable

A normal random variable with

$$\mu = 0 \quad \text{and} \quad \sigma^2 = 1$$

is called a **standard normal random variable** and is denoted as Z.

The cumulative distribution function of a standard normal random variable is denoted as

$$\Phi(z) = P(Z \le z)$$

Appendix Table III provides cumulative probabilities for a standard normal random variable. Cumulative distribution functions for normal random variables are also widely available in computer packages. They can be used in the same manner as Appendix Table III to obtain probabilities for these random variables. The use of Table III is illustrated by the following example.

EXAMPLE 4-11 Standard Normal Distribution

Assume Z is a standard normal random variable. Appendix Table III provides probabilities of the form $\Phi(z) = P(Z \le z)$. The use of Table II to find $P(Z \le 1.5)$ is illustrated in Fig. 4-13. Read down the z column to the row that equals 1.5. The probability is read from the adjacent column, labeled 0.00, to be 0.93319.

The column headings refer to the hundredth's digit of the value of z in $P(Z \le z)$. For example, $P(Z \le 1.53)$ is found by reading down the z column to the row 1.5 and then selecting the probability from the column labeled 0.03 to be 0.93699.

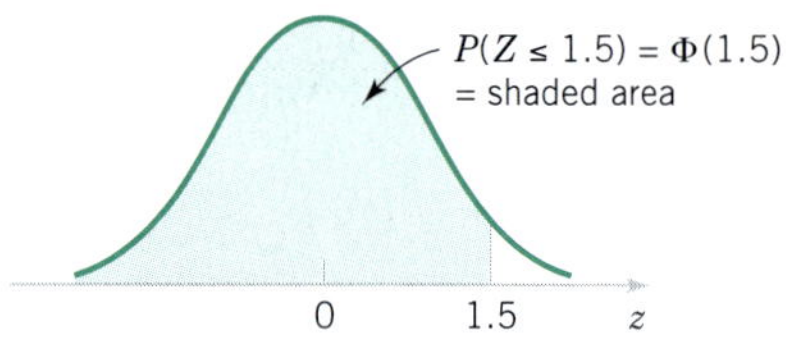

z	0.00	0.01	0.02	0.03
0	0.50000	0.50399	0.50398	0.51197
⋮		⋮		
1.5	0.93319	0.93448	0.93574	0.93699

Figure 4-13 Standard normal probability density function.

Probabilities that are not of the form $P(Z \leq z)$ are found by using the basic rules of probability and the symmetry of the normal distribution along with Appendix Table III. The following examples illustrate the method.

EXAMPLE 4-12

The following calculations are shown pictorially in Fig. 4-14. In practice, a probability is often rounded to one or two significant digits.

(1) $P(Z > 1.26) = 1 - P(Z \leq 1.26) = 1 - 0.89616 = 0.10384$

(2) $P(Z < -0.86) = 0.19490.$

(3) $P(Z > -1.37) = P(Z < 1.37) = 0.91465$

(4) $P(-1.25 < Z < 0.37)$. This probability can be found from the difference of two areas, $P(Z < 0.37) - P(Z < -1.25)$. Now,

$$P(Z < 0.37) = 0.64431 \quad \text{and} \quad P(Z < -1.25) = 0.10565$$

Therefore,

$$P(-1.25 < Z < 0.37) = 0.64431 - 0.10565 = 0.53866$$

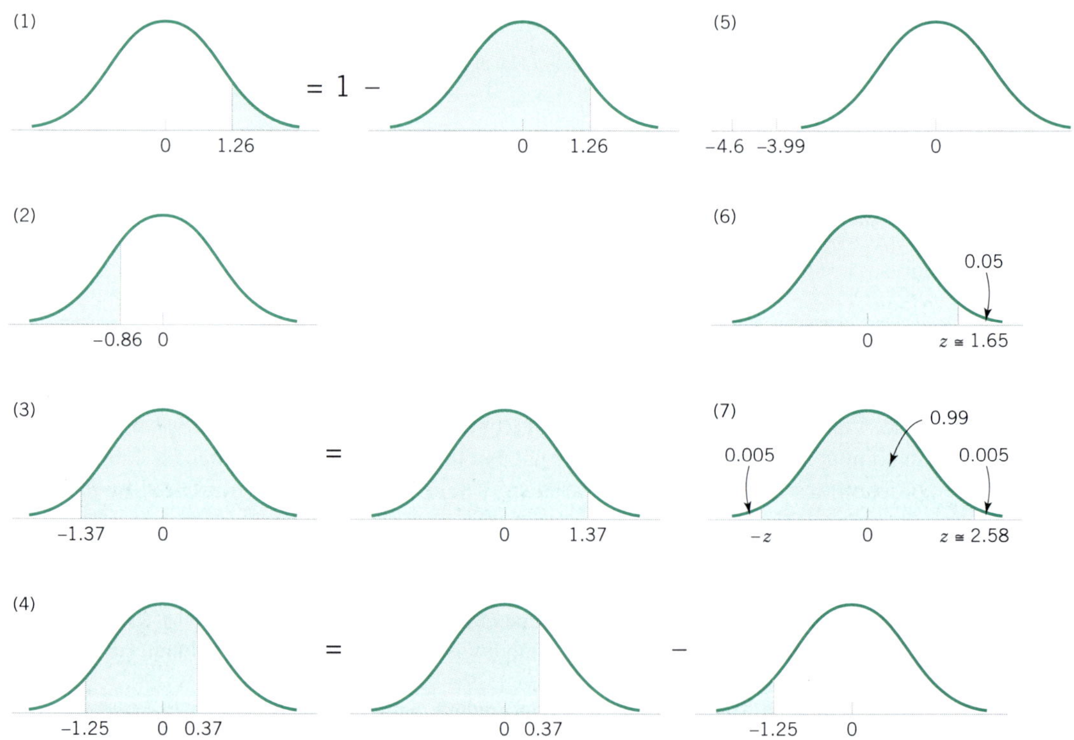

Figure 4-14 Graphical displays for standard normal distributions.

(5) $P(Z \le -4.6)$ cannot be found exactly from Appendix Table II. However, the last entry in the table can be used to find that $P(Z \le -3.99) = 0.00003$. Because $P(Z \le -4.6) < P(Z \le -3.99)$, $P(Z \le -4.6)$is nearly zero.

(6) Find the value z such that $P(Z > z) = 0.05$. This probability expression can be written as $P(Z \le z) = 0.95$. Now, Table II is used in reverse. We search through the probabilities to find the value that corresponds to 0.95. The solution is illustrated in Fig. 4-14. We do not find 0.95 exactly; the nearest value is 0.95053, corresponding to $z = 1.65$.

(7) Find the value of z such that $P(-z < Z < z) = 0.99$. Because of the symmetry of the normal distribution, if the area of the shaded region in Fig. 4-14(7) is to equal 0.99, the area in each tail of the distribution must equal 0.005. Therefore, the value for z corresponds to a probability of 0.995 in Table II. The nearest probability in Table II is 0.99506, when $z = 2.58$.

The preceding examples show how to calculate probabilities for standard normal random variables. To use the same approach for an arbitrary normal random variable would require a separate table for every possible pair of values for μ and σ. Fortunately, all normal probability distributions are related algebraically, and Appendix Table III can be used to find the probabilities associated with an arbitrary normal random variable by first using a simple transformation.

Standardizing a Normal Random Variable

If X is a normal random variable with $E(X) = \mu$ and $V(X) = \sigma^2$, the random variable

$$Z = \frac{X - \mu}{\sigma} \tag{4-10}$$

is a normal random variable with $E(Z) = 0$ and $V(Z) = 1$. That is, Z is a standard normal random variable.

Creating a new random variable by this transformation is referred to as **standardizing.** The random variable Z represents the distance of X from its mean in terms of standard deviations. It is the key step to calculate a probability for an arbitrary normal random variable.

EXAMPLE 4-13 Normally Distributed Current

Suppose the current measurements in a strip of wire are assumed to follow a normal distribution with a mean of 10 milliamperes and a variance of 4 $(\text{milliamperes})^2$. What is the probability that a measurement will exceed 13 milliamperes?

Let X denote the current in milliamperes. The requested probability can be represented as $P(X > 13)$. Let $Z = (X - 10)/2$. The relationship between the several values of X and the transformed values of Z are shown in Fig. 4-15. We note that $X > 13$ corresponds to $Z > 1.5$. Therefore, from Appendix Table II,

$$P(X > 13) = P(Z > 1.5) = 1 - P(Z \le 1.5) = 1 - 0.93319 = 0.06681$$

Rather than using Fig. 4-15, the probability can be found from the inequality $X > 13$. That is,

$$P(X > 13) = P\left(\frac{(X - 10)}{2} > \frac{(13 - 10)}{2}\right) = P(Z > 1.5) = 0.06681$$

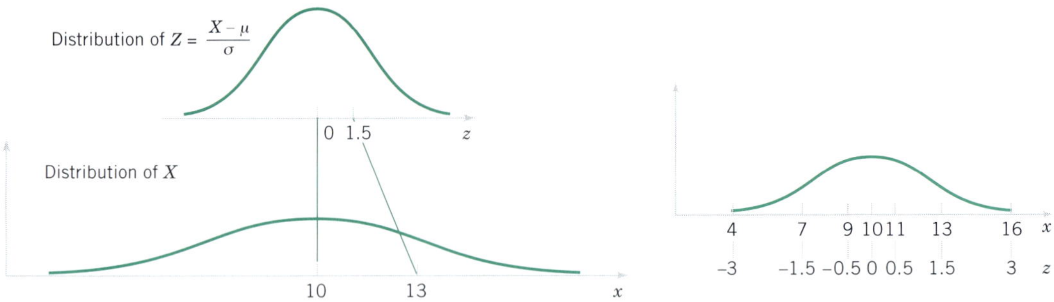

Figure 4-15 Standardizing a normal random variable.

In the preceding example, the value 13 is transformed to 1.5 by standardizing, and 1.5 is often referred to as the **z-value** associated with a probability. The following summarizes the calculation of probabilities derived from normal random variables.

Standardizing to Calculate a Probability

Suppose X is a normal random variable with mean μ and variance σ^2. Then,

$$P(X \le x) = P\left(\frac{X-\mu}{\sigma} \le \frac{x-\mu}{\sigma}\right) = P(Z \le z) \tag{4-11}$$

where Z is a **standard normal random variable,** and $z = \dfrac{(x-\mu)}{\sigma}$ is the **z-value** obtained by **standardizing** X.

The probability is obtained by using Appendix Table II with $z = (x - \mu)/\sigma$.

EXAMPLE 4-14 Normally Distributed Current

Continuing the previous example, what is the probability that a current measurement is between 9 and 11 milliamperes? From Fig. 4-15, or by proceeding algebraically, we have

$$\begin{aligned} P(9 < X < 11) &= P((9-10)/2 < (X-10)/2 < (11-10)/2) \\ &= P(-0.5 < Z < 0.5) = P(Z < 0.5) - P(Z < -0.5) \\ &= 0.69146 - 0.30854 = 0.38292 \end{aligned}$$

Determine the value for which the probability that a current measurement is below this value is 0.98. The requested value is shown graphically in Fig. 4-16. We need the value of x such that $P(X < x) = 0.98$. By standardizing, this probability expression can be written as

$$\begin{aligned} P(X < x) &= P((X-10)/2 < (x-10)/2) \\ &= P(Z < (x-10)/2) \\ &= 0.98 \end{aligned}$$

Appendix Table III is used to find the z-value such that $P(Z < z) = 0.98$. The nearest probability from Table III results in

$$P(Z < 2.05) = 0.97982$$

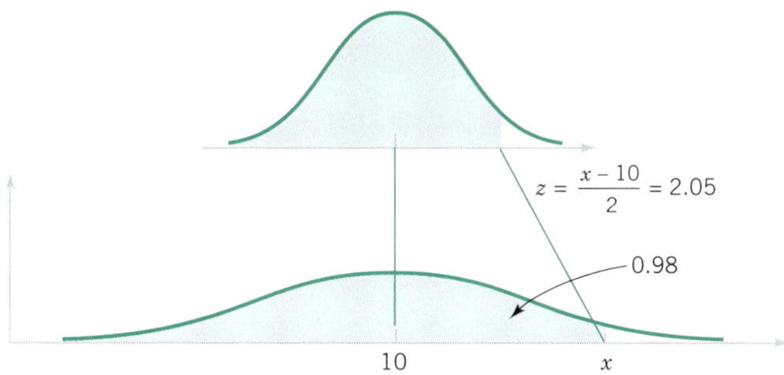

Figure 4-16 Determining the value of x to meet a specified probability.

Therefore, $(x - 10)/2 = 2.05$, and the standardizing transformation is used in reverse to solve for x. The result is

$$x = 2(2.05) + 10 = 14.1 \text{ milliamperes}$$

EXAMPLE 4-15 Signal Detection

Assume that in the detection of a digital signal the background noise follows a normal distribution with a mean of 0 volt and standard deviation of 0.45 volt. The system assumes a digital 1 has been transmitted when the voltage exceeds 0.9. What is the probability of detecting a digital 1 when none was sent?

Let the random variable N denote the voltage of noise. The requested probability is

$$P(N > 0.9) = P\left(\frac{N}{0.45} > \frac{0.9}{0.45}\right) = P(Z > 2) = 1 - 0.97725 = 0.02275$$

This probability can be described as the probability of a false detection.

Determine symmetric bounds about 0 that include 99% of all noise readings. The question requires us to find x such that $P(-x < N < x) = 0.99$. A graph is shown in Fig. 4-17. Now,

$$\begin{aligned} P(-x < N < x) &= P(-x/0.45 < N/0.45 < x/0.45) \\ &= P(-x/0.45 < Z < x/0.45) = 0.99 \end{aligned}$$

From Appendix Table III

$$P(-2.58 < Z < 2.58) = 0.99$$

Therefore,

$$x/0.45 = 2.58$$

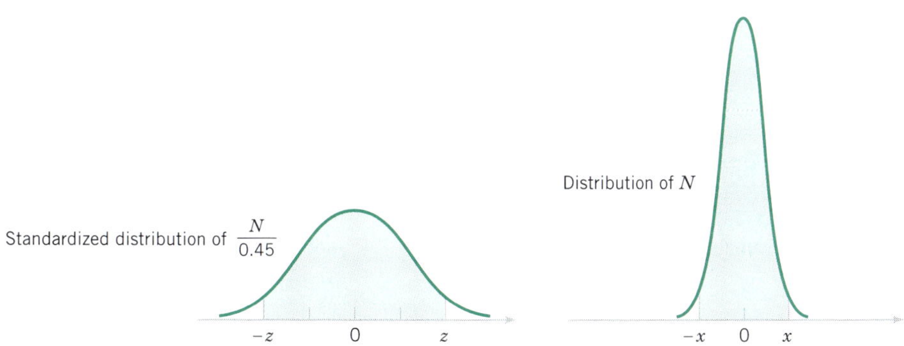

Figure 4-17 Determining the value of x to meet a specified probability.

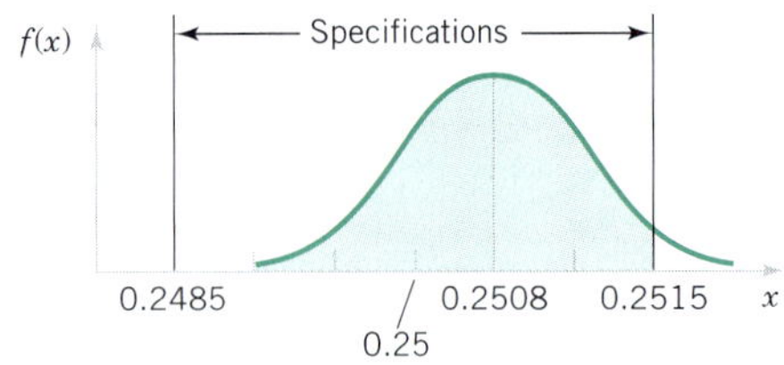

Figure 4-18 Distribution for Example 4-16.

and

$$x = 2.58(0.45) = 1.16$$

Suppose that when a digital 1 signal is transmitted the mean of the noise distribution shifts to 1.8 volts. What is the probability that a digital 1 is not detected? Let the random variable S denote the voltage when a digital 1 is transmitted. Then,

$$P(S < 0.9) = P\left(\frac{S - 1.8}{0.45} < \frac{0.9 - 1.8}{0.45}\right) = P(Z < -2) = 0.02275$$

This probability can be interpreted as the probability of a missed signal.

EXAMPLE 4-16 Shaft Diameter

The diameter of a shaft in an optical storage drive is normally distributed with mean 0.2508 inch and standard deviation 0.0005 inch. The specifications on the shaft are 0.2500 ± 0.0015 inch. What proportion of shafts conforms to specifications?

Let X denote the shaft diameter in inches. The requested probability is shown in Fig. 4-18 and

$$\begin{aligned} P(0.2485 < X < 0.2515) &= P\left(\frac{0.2485 - 0.2508}{0.0005} < Z < \frac{0.2515 - 0.2508}{0.0005}\right) \\ &= P(-4.6 < Z < 1.4) = P(Z < 1.4) - P(Z < -4.6) \\ &= 0.91924 - 0.0000 = 0.91924 \end{aligned}$$

Most of the nonconforming shafts are too large, because the process mean is located very near to the upper specification limit. If the process is centered so that the process mean is equal to the target value of 0.2500,

$$\begin{aligned} P(0.2485 < X < 0.2515) &= P\left(\frac{0.2485 - 0.2500}{0.0005} < Z < \frac{0.2515 - 0.2500}{0.0005}\right) \\ &= P(-3 < Z < 3) \\ &= P(Z < 3) - P(Z < -3) \\ &= 0.99865 - 0.00135 \\ &= 0.9973 \end{aligned}$$

By recentering the process, the yield is increased to approximately 99.73%.

EXERCISES FOR SECTION 4-6

4-41. Use Appendix Table III to determine the following probabilities for the standard normal random variable Z:
(a) $P(Z < 1.32)$ (b) $P(Z < 3.0)$
(c) $P(Z > 1.45)$ (d) $P(Z > -2.15)$
(e) $P(-2.34 < Z < 1.76)$

4-42. Use Appendix Table III to determine the following probabilities for the standard normal random variable Z:
(a) $P(-1 < Z < 1)$ (b) $P(-2 < Z < 2)$
(c) $P(-3 < Z < 3)$ (d) $P(Z > 3)$
(e) $P(0 < Z < 1)$

4-43. Assume Z has a standard normal distribution. Use Appendix Table III to determine the value for z that solves each of the following:
(a) $P(Z < z) = 0.9$ (b) $P(Z < z) = 0.5$
(c) $P(Z > z) = 0.1$ (d) $P(Z > z) = 0.9$
(e) $P(-1.24 < Z < z) = 0.8$

4-44. Assume Z has a standard normal distribution. Use Appendix Table III to determine the value for z that solves each of the following:
(a) $P(-z < Z < z) = 0.95$ (b) $P(-z < Z < z) = 0.99$
(c) $P(-z < Z < z) = 0.68$ (d) $P(-z < Z < z) = 0.9973$

4-45. Assume X is normally distributed with a mean of 10 and a standard deviation of 2. Determine the following:
(a) $P(X < 13)$ (b) $P(X > 9)$
(c) $P(6 < X < 14)$ (d) $P(2 < X < 4)$
(e) $P(-2 < X < 8)$

4-46. Assume X is normally distributed with a mean of 10 and a standard deviation of 2. Determine the value for x that solves each of the following:
(a) $P(X > x) = 0.5$
(b) $P(X > x) = 0.95$
(c) $P(x < X < 10) = 0.2$
(d) $P(-x < X - 10 < x) = 0.95$
(e) $P(-x < X - 10 < x) = 0.99$

4-47. Assume X is normally distributed with a mean of 5 and a standard deviation of 4. Determine the following:
(a) $P(X < 11)$ (b) $P(X > 0)$
(c) $P(3 < X < 7)$ (d) $P(-2 < X < 9)$
(e) $P(2 < X < 8)$

4-48. Assume X is normally distributed with a mean of 5 and a standard deviation of 4. Determine the value for x that solves each of the following:
(a) $P(X > x) = 0.5$ (b) $P(X > x) = 0.95$
(c) $P(x < X < 9) = 0.2$ (d) $P(3 < X < x) = 0.95$
(e) $P(-x < X - 5 < x) = 0.99$

4-49. The compressive strength of samples of cement can be modeled by a normal distribution with a mean of 6000 kilo grams per square centimeter and a standard deviation of 100 kilograms per square centimeter.
(a) What is the probability that a sample's strength is less than 6250 Kg/cm^2?
(b) What is the probability that a sample's strength is between 5800 and 5900 Kg/cm^2?
(c) What strength is exceeded by 95% of the samples?

4-50. The time until recharge for a battery in a laptop computer under common conditions is normally distributed with mean of 260 minutes and a standard deviation of 50 minutes.
(a) What is the probability that a battery lasts more than four hours?
(b) What are the quartiles (the 25% and 75% values) of battery life?
(c) What value of life in minutes is exceeded with 95% probability?

4-51. An article in *Knee Surgery Sports Traumatol Arthrosc,* "Effect of provider volume on resource utilization for surgical procedures," (2005, Vol. 13, pp. 273–279) showed a mean time of 129 minutes and a standard deviation of 14 minutes for ACL reconstruction surgery at high-volume hospitals (with more than 300 such surgeries per year).
(a) What is the probability that your ACL surgery at a high-volume hospital requires a time more than two standard deviations above the mean?
(b) What is the probability that your ACL surgery at a high-volume hospital is completed in less than 100 minutes?
(c) The probability of a completed ACL surgery at a high-volume hospital is equal to 95% at what time?
(d) If your surgery requires 199 minutes, what do you conclude about the volume of such surgeries at your hospital? Explain.

4-52. Cholesterol is a fatty substance that is an important part of the outer lining (membrane) of cells in the body of animals. Its normal range for an adult is 120–240 mg/dl. The Food and Nutrition Institute of the Philippines found that the total cholesterol level for Filipino adults has a mean of 159.2 mg/dl and 84.1% of adults have a cholesterol level below 200 mg/dl (http://www.fnri.dost.gov.ph/facts/p3lipnat.html). Suppose that the total cholesterol level is normally distributed.
(a) Determine the standard deviation of this distribution.
(b) What are the quartiles (the 25% and 75% values) of this distribution?
(c) What is the value of the cholesterol level that exceeds 90% of the population?
(d) An adult is at moderate risk if cholesterol level is more than one but less than two standard deviations above the mean. What percentage of the population is at moderate risk according to this criterion?
(e) An adult is thought to be at high risk if his cholesterol level is more than two standard deviations above the mean. What percentage of the population is at high risk?
(f) An adult has low risk if cholesterol level is one standard deviation or more below the mean. What percentage of the population is at low risk?

4-53. The line width for semiconductor manufacturing is assumed to be normally distributed with a mean of 0.5 micrometer and a standard deviation of 0.05 micrometer.
(a) What is the probability that a line width is greater than 0.62 micrometer?
(b) What is the probability that a line width is between 0.47 and 0.63 micrometer?
(c) The line width of 90% of samples is below what value?

4-54. The fill volume of an automated filling machine used for filling cans of carbonated beverage is normally distributed with a mean of 12.4 fluid ounces and a standard deviation of 0.1 fluid ounce.
(a) What is the probability a fill volume is less than 12 fluid ounces?
(b) If all cans less than 12.1 or greater than 12.6 ounces are scrapped, what proportion of cans is scrapped?

(c) Determine specifications that are symmetric about the mean that include 99% of all cans.

4-55. In the previous exercise, suppose that the mean of the filling operation can be adjusted easily, but the standard deviation remains at 0.1 ounce.
(a) At what value should the mean be set so that 99.9% of all cans exceed 12 ounces?
(b) At what value should the mean be set so that 99.9% of all cans exceed 12 ounces if the standard deviation can be reduced to 0.05 fluid ounce?

4-56. The reaction time of a driver to visual stimulus is normally distributed with a mean of 0.4 seconds and a standard deviation of 0.05 seconds.
(a) What is the probability that a reaction requires more than 0.5 seconds?
(b) What is the probability that a reaction requires between 0.4 and 0.5 seconds?
(c) What is the reaction time that is exceeded 90% of the time?

4-57. The speed of a file transfer from a server on campus to a personal computer at a student's home on a weekday evening is normally distributed with a mean of 60 kilobits per second and a standard deviation of 4 kilobits per second.
(a) What is the probability that the file will transfer at a speed of 70 kilobits per second or more?
(b) What is the probability that the file will transfer at a speed of less than 58 kilobits per second?
(c) If the file is 1 megabyte, what is the average time it will take to transfer the file? (Assume eight bits per byte.)

4-58. The average height of a woman aged 20–74 years is 64 inches in 2002 with an increase of approximately one inch from 1960 (http://usgovinfo.about.com/od/healthcare). Suppose the height of a woman is normally distributed with a standard deviation of 2 inches.
(a) What is the probability a randomly selected woman in this population is between 58 inches and 70 inches?
(b) What are the quartiles of this distribution?
(c) Determine the height that is symmetric about the mean that includes 90% of this population.
(d) What is the probability that five women selected at random from this population all exceed 68 inches?

4-59. In an accelerator center, an experiment needs a 1.41 cm thick aluminum cylinder (http://puhep1.princeton.edu/mumu/target/Solenoid_Coil.pdf). Suppose that the thickness of a cylinder has a normal distribution with a mean 1.41 cm and a standard deviation of 0.01 cm.
(a) What is the probability that a thickness is greater than 1.42 cm?
(b) What thickness is exceeded by 95% of the samples?
(c) If the specifications require that the thickness is between 1.39 cm and 1.43 cm, what proportion of the samples meet specifications?

4-60. The demand for water use in Phoenix in 2003 hit a high of about 442 million gallons per day on June 27, 2003 (http://phoenix.gov/WATER/wtrfacts.html). Water use in the summer is normally distributed with a mean of 310 million gallons per day and a standard deviation of 45 million gallons per day. City reservoirs have a combined storage capacity of nearly 350 million gallons.
(a) What is the probability that a day requires more water than is stored in city reservoirs?
(b) What reservoir capacity is needed so that the probability it is exceeded is 1%?
(c) What amount of water use is exceeded with 95% probability?
(d) Water is provided to approximately 1.4 million people. What is the mean daily consumption per person at which the probability that the demand exceeds the current reservoir capacity is 1%? Assume that the standard deviation of demand remains the same.

4-61. The life of a semiconductor laser at a constant power is normally distributed with a mean of 7000 hours and a standard deviation of 600 hours.
(a) What is the probability that a laser fails before 5000 hours?
(b) What is the life in hours that 95% of the lasers exceed?
(c) If three lasers are used in a product and they are assumed to fail independently, what is the probability that all three are still operating after 7000 hours?

4-62. The diameter of the dot produced by a printer is normally distributed with a mean diameter of 0.002 inch and a standard deviation of 0.0004 inch.
(a) What is the probability that the diameter of a dot exceeds 0.0026 inch?
(b) What is the probability that a diameter is between 0.0014 and 0.0026 inch?
(c) What standard deviation of diameters is needed so that the probability in part (b) is 0.995?

4-63. The weight of a sophisticated running shoe is normally distributed with a mean of 12 ounces and a standard deviation of 0.5 ounce.
(a) What is the probability that a shoe weighs more than 13 ounces?
(b) What must the standard deviation of weight be in order for the company to state that 99.9% of its shoes are less than 13 ounces?
(c) If the standard deviation remains at 0.5 ounce, what must the mean weight be in order for the company to state that 99.9% of its shoes are less than 13 ounces?

4-64. Measurement error that is normally distributed with a mean of zero and a standard deviation of 0.5 gram is added to the true weight of a sample. Then the measurement is rounded to the nearest gram. Suppose that the true weight of a sample is 165.5 grams.
(a) What is the probability that the rounded result is 167 grams?
(b) What is the probability that the rounded result is 167 grams or greater?

4-7 NORMAL APPROXIMATION TO THE BINOMIAL AND POISSON DISTRIBUTIONS

We began our section on the normal distribution with the central limit theorem and the normal distribution as an approximation to a random variable with a large number of trials. Consequently, it should not be a surprise to learn that the normal distribution can be used to approximate binomial probabilities for cases in which n is large. The following example illustrates that for many physical systems the binomial model is appropriate with an extremely large value for n. In these cases, it is difficult to calculate probabilities by using the binomial distribution. Fortunately, the normal approximation is most effective in these cases. An illustration is provided in Fig. 4-19. The area of each bar equals the binomial probability of x. Notice that the area of bars can be approximated by areas under the normal density function.

From Fig. 4-19 it can be seen that a probability such as $P(3 \leq X \leq 7)$ is better approximated by the area under the normal curve from 2.5 to 7.5. This observation provides a method to improve the approximation of binomial probabilities. Because a continuous normal distribution is used to approximate a discrete binomial distribution, the modification is referred to as a **continuity correction.**

EXAMPLE 4-17

In a digital communication channel, assume that the number of bits received in error can be modeled by a binomial random variable, and assume that the probability that a bit is received in error is 1×10^{-5}. If 16 million bits are transmitted, what is the probability that 150 or fewer errors occur?

Let the random variable X denote the number of errors. Then X is a binomial random variable and

$$P(X \leq 150) = \sum_{x=0}^{150} \binom{16{,}000{,}000}{x} (10^{-5})^x (1 - 10^{-5})^{16{,}000{,}000-x}$$

Clearly, the probability in Example 4-17 is difficult to compute. Fortunately, the normal distribution can be used to provide an excellent approximation in this example.

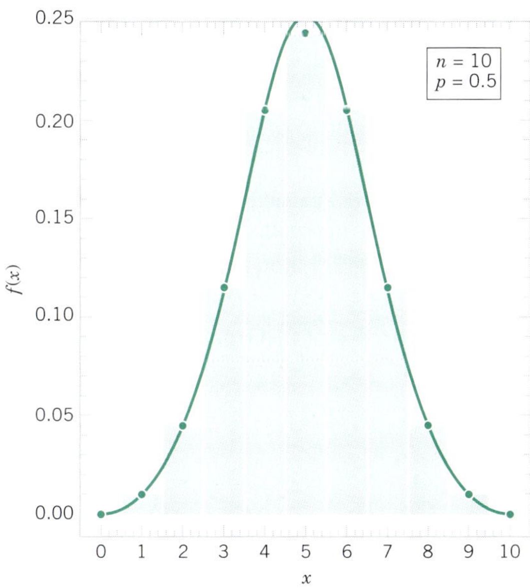

Figure 4-19 Normal approximation to the binomial distribution.

Normal Approximation to the Binomial Distribution

If X is a binomial random variable with parameters n and p

$$Z = \frac{X - np}{\sqrt{np(1-p)}} \tag{4-12}$$

is approximately a standard normal random variable. To approximate a binomial probability with a normal distribution a **continuity correction** is applied as follows

$$P(X \le x) = P(X \le x + 0.5) \cong P\left(Z \le \frac{x + 0.5 - np}{\sqrt{np(1-p)}}\right)$$

and

$$P(x \le X) = P(x - 0.5 \le X) \cong P\left(\frac{x - 0.5 - np}{\sqrt{np(1-p)}} \le Z\right)$$

The approximation is good for $np > 5$ and $n(1-p) > 5$.

Recall that for a binomial variable X, $E(X) = np$ and $V(X) = np(1-p)$. Consequently, the expression in Equation 4-12 is nothing more than the formula for standardizing the random variable X. Probabilities involving X can be approximated by using a standard normal distribution. The approximation is good when n is large relative to p.

A way to remember the approximation is to write the probability in terms of $\le$ or $\ge$ and then add or subtract the 0.5 correction factor to make the probability greater.

EXAMPLE 4-18

The digital communication problem in the previous example is solved as follows:

$$P(X \le 150) = P(X \le 150.5) = P\left(\frac{X - 160}{\sqrt{160(1-10^{-5})}} \le \frac{150.5 - 160}{\sqrt{160(1-10^{-5})}}\right)$$
$$\cong P(Z \le -0.75) = 0.227$$

Because $np = (16 \times 10^6)(1 \times 10^{-5}) = 160$ and $n(1-p)$ is much larger, the approximation is expected to work well in this case.

EXAMPLE 4-19
Normal Approximation to Binomial

Again consider the transmission of bits in Example 4-18. To judge how well the normal approximation works, assume only $n = 50$ bits are to be transmitted and that the probability of an error is $p = 0.1$. The exact probability that 2 or less errors occur is

$$P(X \le 2) = \binom{50}{0} 0.9^{50} + \binom{50}{1} 0.1(0.9^{49}) + \binom{50}{2} 0.1^2(0.9^{48}) = 0.112$$

Based on the normal approximation

$$P(X \le 2) = P\left(\frac{X - 5}{\sqrt{50(0.1)(0.9)}} < \frac{2.5 - 5}{\sqrt{50(0.1)(0.9)}}\right) = P(Z < -1.18) = 0.119$$

Even for a sample as small as 50 bits, the normal approximation is reasonable.

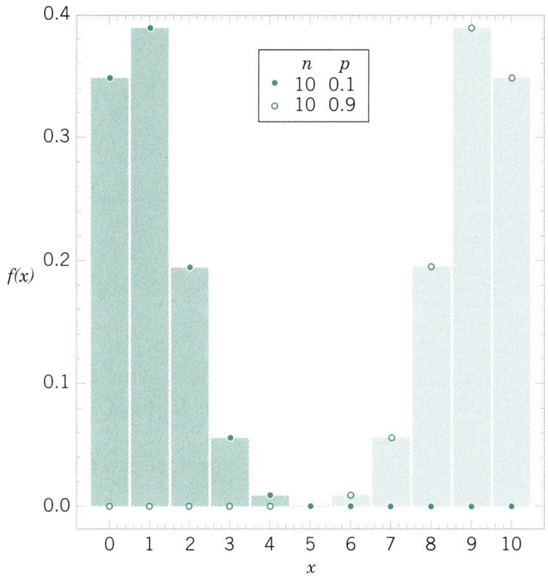

Figure 4-20 Binomial distribution is not symmetrical if p is near 0 or 1.

As another example, $P(8 < X) = P(9 \leq X)$ and this is better approximated as

$$P(9 \leq X) = P(8.5 \leq X) \approx P\left(\frac{8.5 - 5}{2.12} \leq Z\right) = P(1.65 \leq Z) = 0.05$$

We can even approximate $P(X = 5) = P(5 \leq X \leq 5)$ as

$$P(5 \leq X \leq 5) \approx P\left(\frac{5 - 0.5 - 5}{2.12} \leq Z \leq \frac{5 + 0.5 - 5}{2.12}\right) = P(-0.24 \leq Z \leq 0.24) = 0.19$$

and this compares well with the exact answer of 0.1849.

The correction factor is used to improve the approximation. However, if np or $n(1 - p)$ is small, the binomial distribution is quite skewed and the symmetric normal distribution is not a good approximation. Two cases are illustrated in Fig. 4-20.

Recall that the binomial distribution is a satisfactory approximation to the hypergeometric distribution when n, the sample size, is small relative to N, the size of the population from which the sample is selected. A rule of thumb is that the binomial approximation is effective if $n/N < 0.1$. Recall that for a hypergeometric distribution p is defined as $p = K/N$. That is, p is interpreted as the number of successes in the population. Therefore, the normal distribution can provide an effective approximation of hypergeometric probabilities when $n/N < 0.1$, $np > 5$ and $n(1 - p) > 5$. Figure 4-21 provides a summary of these guidelines.

Recall that the Poisson distribution was developed as the limit of a binomial distribution as the number of trials increased to infinity. Consequently, it should not be surprising to find that the normal distribution can also be used to approximate probabilities of a Poisson random variable.

hypergometric distribution	$\approx$ $\frac{n}{N} < 0.1$	binomial distribution	$\approx$ $np > 5$ $n(1-p) > 5$	normal distribution

Figure 4-21 Conditions for approximating hypergeometric and binomial probabilities.

Normal Approximation to the Poisson Distribution

If X is a Poisson random variable with $E(X) = \lambda$ and $V(X) = \lambda$,

$$Z = \frac{X - \lambda}{\sqrt{\lambda}} \tag{4-13}$$

is approximately a standard normal random variable. The approximation is good for

$$\lambda > 5$$

EXAMPLE 4-20 Normal Approximation to Poisson

Assume that the number of asbestos particles in a squared meter of dust on a surface follows a Poisson distribution with a mean of 1000. If a squared meter of dust is analyzed, what is the probability that 950 or fewer particles are found?

This probability can be expressed exactly as

$$P(X \le 950) = \sum_{x=0}^{950} \frac{e^{-1000}\, 1000^x}{x!}$$

The computational difficulty is clear. The probability can be approximated as

$$P(X \le x) = P\left(Z \le \frac{950 - 1000}{\sqrt{1000}}\right) = P(Z \le -1.58) = 0.057$$

EXERCISES FOR SECTION 4-7

4-65. Suppose that X is a binomial random variable with $n = 200$ and $p = 0.4$.

(a) Approximate the probability that X is less than or equal to 70.

(b) Approximate the probability that X is greater than 70 and less than 90.

(c) Approximate the probability $X = 80$.

4-66. Suppose that X is a Poisson random variable with $\lambda = 6$.

(a) Compute the exact probability that X is less than 4.

(b) Approximate the probability that X is less than 4 and compare to the result in part (a).

(c) Approximate the probability that $8 < X < 12$.

4-67. Suppose that X has a Poisson distribution with a mean of 64. Approximate the following probabilities:

(a) $P(X > 72)$

(b) $P(X < 64)$

(c) $P(60 < X \le 68)$

4-68. The manufacturing of semiconductor chips produces 2% defective chips. Assume the chips are independent and that a lot contains 1000 chips.

(a) Approximate the probability that more than 25 chips are defective.

(b) Approximate the probability that between 20 and 30 chips are defective.

4-69. There were 49.7 million people with some type of long-lasting condition or disability living in the United States in 2000. This represented 19.3 percent of the majority of civilians aged five and over (http://factfinder.census.gov). A sample of 1000 persons is selected at random.

(a) Approximate the probability that more than 200 persons in the sample have a disability.

(b) Approximate the probability that between 180 and 300 people in the sample have a disability.

4-70. Phoenix water is provided to approximately 1.4 million people, who are served through more than 362,000 accounts (http://phoenix.gov/WATER/wtrfacts.html). All accounts are metered and billed monthly. The probability that an account has an error in a month is 0.001, and accounts can be assumed to be independent.

(a) What is the mean and standard deviation of the number of account errors each month?

(b) Approximate the probability of fewer than 350 errors in a month.

(c) Approximate a value so that the probability that the number of errors exceeding this value is 0.05.

(d) Approximate the probability of more than 400 errors per month in the next two months. Assume that results between months are independent.

4-71. An electronic office product contains 5000 electronic components. Assume that the probability that each component operates without failure during the useful life of the product is 0.999, and assume that the components fail independently. Approximate the probability that 10 or more of the original 5000 components fail during the useful life of the product.

4-72. A corporate Web site contains errors on 50 of 1000 pages. If 100 pages are sampled randomly, without replacement, approximate the probability that at least 1 of the pages in error are in the sample.

4-73. Suppose that the number of asbestos particles in a sample of 1 squared centimeter of dust is a Poisson random variable with a mean of 1000. What is the probability that 10 squared centimeters of dust contains more than 10,000 particles?

4-74. A high-volume printer produces minor print-quality errors on a test pattern of 1000 pages of text according to a Poisson distribution with a mean of 0.4 per page.

(a) Why are the number of errors on each page independent random variables?
(b) What is the mean number of pages with errors (one or more)?
(c) Approximate the probability that more than 350 pages contain errors (one or more).

4-75. Hits to a high-volume Web site are assumed to follow a Poisson distribution with a mean of 10,000 per day. Approximate each of the following:

(a) The probability of more than 20,000 hits in a day
(b) The probability of less than 9900 hits in a day
(c) The value such that the probability that the number of hits in a day exceed the value is 0.01
(d) Approximate the expected number of days in a year (365 days) that exceed 10,200 hits.
(e) Approximate the probability that over a year (365 days) more than 15 days each have more than 10,200 hits.

4-8 EXPONENTIAL DISTRIBUTION

The discussion of the Poisson distribution defined a random variable to be the number of flaws along a length of copper wire. The distance between flaws is another random variable that is often of interest. Let the random variable X denote the length from any starting point on the wire until a flaw is detected.

As you might expect, the distribution of X can be obtained from knowledge of the distribution of the number of flaws. The key to the relationship is the following concept. The distance to the first flaw exceeds 3 millimeters if and only if there are no flaws within a length of 3 millimeters—simple, but sufficient for an analysis of the distribution of X.

In general, let the random variable N denote the number of flaws in x millimeters of wire. If the mean number of flaws is λ per millimeter, N has a Poisson distribution with mean λx. We assume that the wire is longer than the value of x. Now,

$$P(X > x) = P(N = 0) = \frac{e^{-\lambda x}(\lambda x)^0}{0!} = e^{-\lambda x}$$

Therefore,

$$F(x) = P(X \leq x) = 1 - e^{-\lambda x}, \quad x \geq 0$$

is the cumulative distribution function of X. By differentiating $F(x)$, the probability density function of X is calculated to be

$$f(x) = \lambda e^{-\lambda x}, \quad x \geq 0$$

The derivation of the distribution of X depends only on the assumption that the flaws in the wire follow a **Poisson process.** Also, the starting point for measuring X doesn't matter because the probability of the number of flaws in an interval of a Poisson process depends only on the length of the interval, not on the location. For any Poisson process, the following general result applies.

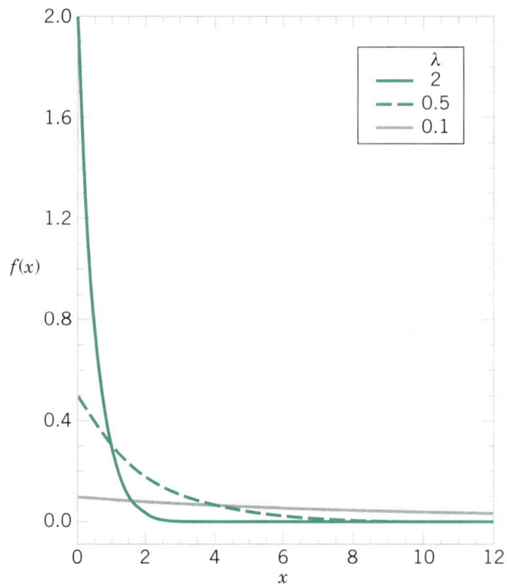

Figure 4-22 Probability density function of exponential random variables for selected values of λ.

Exponential Distribution

> The random variable X that equals the distance between successive events of a Poisson process with mean $\lambda > 0$ is an **exponential random variable** with parameter λ. The probability density function of X is
>
> $$f(x) = \lambda e^{-\lambda x} \quad \text{for} \quad 0 \le x < \infty \tag{4-14}$$

The exponential distribution obtains its name from the exponential function in the probability density function. Plots of the exponential distribution for selected values of λ are shown in Fig. 4-22. For any value of λ, the exponential distribution is quite skewed. The following results are easily obtained and are left as an exercise.

Mean and Variance

> If the random variable X has an exponential distribution with parameter λ,
>
> $$\mu = E(X) = \frac{1}{\lambda} \quad \text{and} \quad \sigma^2 = V(X) = \frac{1}{\lambda^2} \tag{4-15}$$

It is important to **use consistent units** in the calculation of probabilities, means, and variances involving exponential random variables. The following example illustrates unit conversions.

EXAMPLE 4-21 Computer Usage

In a large corporate computer network, user log-ons to the system can be modeled as a Poisson process with a mean of 25 log-ons per hour. What is the probability that there are no log-ons in an interval of 6 minutes?

Let X denote the time in hours from the start of the interval until the first log-on. Then, X has an exponential distribution with $\lambda = 25$ log-ons per hour. We are interested in the probability that X

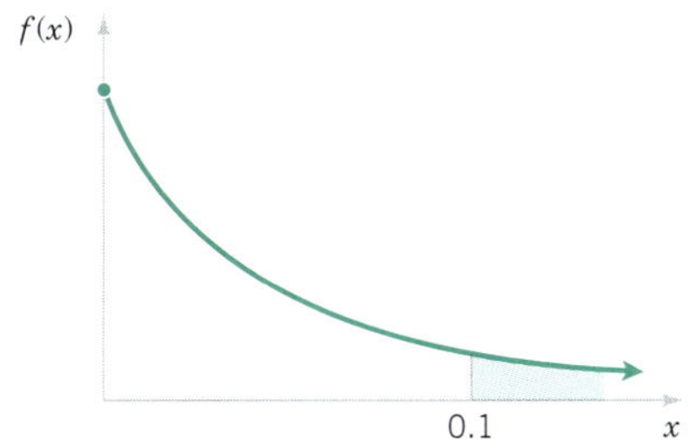

Figure 4-23
Probability for the exponential distribution in Example 4-21.

exceeds 6 minutes. Because λ is given in log-ons per hour, we express all time units in hours. That is, 6 minutes $= 0.1$ hour. The probability requested is shown as the shaded area under the probability density function in Fig. 4-23. Therefore,

$$P(X > 0.1) = \int_{0.1}^{\infty} 25e^{-25x}\, dx = e^{-25(0.1)} = 0.082$$

Also, the cumulative distribution function can be used to obtain the same result as follows:

$$P(X > 0.1) = 1 - F(0.1) = e^{-25(0.1)}$$

An identical answer is obtained by expressing the mean number of log-ons as 0.417 log-ons per minute and computing the probability that the time until the next log-on exceeds 6 minutes. Try it.

What is the probability that the time until the next log-on is between 2 and 3 minutes? Upon converting all units to hours,

$$P(0.033 < X < 0.05) = \int_{0.033}^{0.05} 25e^{-25x}\, dx = -e^{-25x}\Big|_{0.033}^{0.05} = 0.152$$

An alternative solution is

$$P(0.033 < X < 0.05) = F(0.05) - F(0.033) = 0.152$$

Determine the interval of time such that the probability that no log-on occurs in the interval is 0.90. The question asks for the length of time x such that $P(X > x) = 0.90$. Now,

$$P(X > x) = e^{-25x} = 0.90$$

Take the (natural) log of both sides to obtain $-25x = \ln(0.90) = -0.1054$. Therefore,

$$x = 0.00421 \text{ hour} = 0.25 \text{ minute}$$

Furthermore, the mean time until the next log-on is

$$\mu = 1/25 = 0.04 \text{ hour} = 2.4 \text{ minutes}$$

The standard deviation of the time until the next log-on is

$$\sigma = 1/25 \text{ hours} = 2.4 \text{ minutes}$$

In the previous example, the probability that there are no log-ons in a 6-minute interval is 0.082 regardless of the starting time of the interval. A Poisson process assumes that events occur uniformly throughout the interval of observation; that is, there is no clustering of events. If the log-ons are well modeled by a Poisson process, the probability that the first log-on after noon occurs after 12:06 P.M. is the same as the probability that the first log-on after 3:00 P.M. occurs after 3:06 P.M. And if someone logs on at 2:22 P.M., the probability the next log-on occurs after 2:28 P.M. is still 0.082.

Our starting point for observing the system does not matter. However, if there are high-use periods during the day, such as right after 8:00 A.M., followed by a period of low use, a Poisson process is not an appropriate model for log-ons and the distribution is not appropriate for computing probabilities. It might be reasonable to model each of the high- and low-use periods by a separate Poisson process, employing a larger value for λ during the high-use periods and a smaller value otherwise. Then, an exponential distribution with the corresponding value of λ can be used to calculate log-on probabilities for the high- and low-use periods.

Lack of Memory Property

An even more interesting property of an exponential random variable is concerned with conditional probabilities.

EXAMPLE 4-22
Lack of Memory Property

Let X denote the time between detections of a particle with a Geiger counter and assume that X has an exponential distribution with $E(X) = 1.4$ minutes. The probability that we detect a particle within 30 seconds of starting the counter is

$$P(X < 0.5 \text{ minute}) = F(0.5) = 1 - e^{-0.5/1.4} = 0.30$$

In this calculation, all units are converted to minutes. Now, suppose we turn on the Geiger counter and wait 3 minutes without detecting a particle. What is the probability that a particle is detected in the next 30 seconds?

Because we have already been waiting for 3 minutes, we feel that we are "due." That is, the probability of a detection in the next 30 seconds should be greater than 0.3. However, for an exponential distribution, this is not true. The requested probability can be expressed as the conditional probability that $P(X < 3.5 \mid X > 3)$. From the definition of conditional probability,

$$P(X < 3.5 \mid X > 3) = P(3 < X < 3.5)/P(X > 3)$$

where

$$P(3 < X < 3.5) = F(3.5) - F(3) = [1 - e^{-3.5/1.4}] - [1 - e^{-3/1.4}] = 0.035$$

and

$$P(X > 3) = 1 - F(3) = e^{-3/1.4} = 0.117$$

Therefore,

$$P(X < 3.5 \mid X > 3) = 0.035/0.117 = 0.30$$

After waiting for 3 minutes without a detection, the probability of a detection in the next 30 seconds is the same as the probability of a detection in the 30 seconds immediately after starting the counter. The fact that you have waited 3 minutes without a detection does not change the probability of a detection in the next 30 seconds.

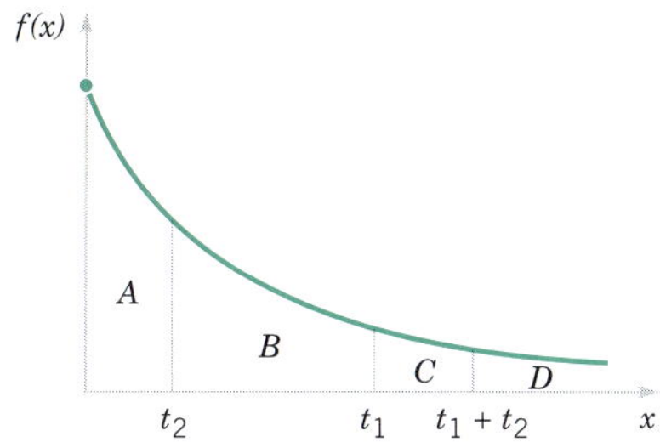

Figure 4-24 Lack of memory property of an exponential distribution.

Example 4-22 illustrates the **lack of memory property** of an exponential random variable and a general statement of the property follows. In fact, the exponential distribution is the only continuous distribution with this property.

Lack of Memory Property

For an exponential random variable X,

$$P(X < t_1 + t_2 \mid X > t_1) = P(X < t_2) \tag{4-16}$$

Figure 4-24 graphically illustrates the lack of memory property. The area of region A divided by the total area under the probability density function $(A + B + C + D = 1)$ equals $P(X < t_2)$. The area of region C divided by the area $C + D$ equals $P(X < t_1 + t_2 \mid X > t_1)$. The lack of memory property implies that the proportion of the total area that is in A equals the proportion of the area in C and D that is in C. The mathematical verification of the lack of memory property is left as a mind-expanding exercise.

The lack of memory property is not that surprising when you consider the development of a Poisson process. In that development, we assumed that an interval could be partitioned into small intervals that were independent. These subintervals are similar to independent Bernoulli trials that comprise a binomial process; knowledge of previous results does not affect the probabilities of events in future subintervals. An exponential random variable is the continuous analog of a geometric random variable, and they share a similar lack of memory property.

The exponential distribution is often used in reliability studies as the model for the time until failure of a device. For example, the lifetime of a semiconductor chip might be modeled as an exponential random variable with a mean of 40,000 hours. The lack of memory property of the exponential distribution implies that the device does not wear out. That is, regardless of how long the device has been operating, the probability of a failure in the next 1000 hours is the same as the probability of a failure in the first 1000 hours of operation. The lifetime L of a device with failures caused by random shocks might be appropriately modeled as an exponential random variable. However, the lifetime L of a device that suffers slow mechanical wear, such as bearing wear, is better modeled by a distribution such that $P(L < t + \Delta t \mid L > t)$ increases with t. Distributions such as the Weibull distribution are often used, in practice, to model the failure time of this type of device. The Weibull distribution is presented in a later section.

EXERCISES FOR SECTION 4-8

4-76. Suppose X has an exponential distribution with $\lambda = 2$. Determine the following:
(a) $P(X \leq 0)$ (b) $P(X \geq 2)$
(c) $P(X \leq 1)$ (d) $P(1 < X < 2)$
(e) Find the value of x such that $P(X < x) = 0.05$.

4-77. Suppose X has an exponential distribution with mean equal to 10. Determine the following:
(a) $P(X > 10)$ (b) $P(X > 20)$
(c) $P(X < 30)$
(d) Find the value of x such that $P(X < x) = 0.95$.

4-78. Suppose X has an exponential distribution with a mean of 10. Determine the following:
(a) $P(X < 5)$
(b) $P(X < 15 \mid X > 10)$
(c) Compare the results in parts (a) and (b) and comment on the role of the lack of memory property.

4-79. Suppose the counts recorded by a Geiger counter follow a Poisson process with an average of two counts per minute.
(a) What is the probability that there are no counts in a 30-second interval?
(b) What is the probability that the first count occurs in less than 10 seconds?
(c) What is the probability that the first count occurs between 1 and 2 minutes after start-up?

4-80. Suppose that the log-ons to a computer network follow a Poisson process with an average of 3 counts per minute.
(a) What is the mean time between counts?
(b) What is the standard deviation of the time between counts?
(c) Determine x such that the probability that at least one count occurs before time x minutes is 0.95.

4-81. The time between calls to a plumbing supply business is exponentially distributed with a mean time between calls of 15 minutes.
(a) What is the probability that there are no calls within a 30-minute interval?
(b) What is the probability that at least one call arrives within a 10-minute interval?
(c) What is the probability that the first call arrives within 5 and 10 minutes after opening?
(d) Determine the length of an interval of time such that the probability of at least one call in the interval is 0.90.

4-82. The life of automobile voltage regulators has an exponential distribution with a mean life of six years. You purchase an automobile that is six years old, with a working voltage regulator, and plan to own it for six years.
(a) What is the probability that the voltage regulator fails during your ownership?
(b) If your regulator fails after you own the automobile three years and it is replaced, what is the mean time until the next failure?

4-83. Suppose that the time to failure (in hours) of fans in a personal computer can be modeled by an exponential distribution with $\lambda = 0.0003$.
(a) What proportion of the fans will last at least 10,000 hours?
(b) What proportion of the fans will last at most 7000 hours?

4-84. The time between the arrival of electronic messages at your computer is exponentially distributed with a mean of two hours.
(a) What is the probability that you do not receive a message during a two-hour period?
(b) If you have not had a message in the last four hours, what is the probability that you do not receive a message in the next two hours?
(c) What is the expected time between your fifth and sixth messages?

4-85. The time between arrivals of taxis at a busy intersection is exponentially distributed with a mean of 10 minutes.
(a) What is the probability that you wait longer than one hour for a taxi?
(b) Suppose you have already been waiting for one hour for a taxi, what is the probability that one arrives within the next 10 minutes?
(c) Determine x such that the probability that you wait more than x minutes is 0.10.
(d) Determine x such that the probability that you wait less than x minutes is 0.90.
(e) Determine x such that the probability that you wait less than x minutes is 0.50.

4-86. The number of stork sightings on a route in South Carolina follows a Poisson process with a mean of 2.3 per year.
(a) What is the mean time between sightings?
(b) What is the probability that there are no sightings within three months (0.25 years)?
(c) What is the probability that the time until the first sighting exceeds six months?
(d) What is the probability of no sighting within three years?

4-87. According to results from the analysis of chocolate bars in Chapter 3, the mean number of insect fragments was 14.4 in 225 grams. Assume the number of fragments follow a Poisson distribution.
(a) What is the mean number of grams of chocolate until a fragment is detected?
(b) What is the probability that there are no fragments in a 28.35 gram (one ounce) chocolate bar?
(c) Suppose you consume seven one-ounce (28.35 grams) bars this week. What is the probability of no insect fragments?

4-88. The distance between major cracks in a highway follows an exponential distribution with a mean of 5 miles.
(a) What is the probability that there are no major cracks in a 10-mile stretch of the highway?
(b) What is the probability that there are two major cracks in a 10-mile stretch of the highway?
(c) What is the standard deviation of the distance between major cracks?
(d) What is the probability that the first major crack occurs between 12 and 15 miles of the start of inspection?
(e) What is the probability that there are no major cracks in two separate 5-mile stretches of the highway?
(f) Given that there are no cracks in the first 5 miles inspected, what is the probability that there are no major cracks in the next 10 miles inspected?

4-89. The lifetime of a mechanical assembly in a vibration test is exponentially distributed with a mean of 400 hours.
(a) What is the probability that an assembly on test fails in less than 100 hours?

(b) What is the probability that an assembly operates for more than 500 hours before failure?
(c) If an assembly has been on test for 400 hours without a failure, what is the probability of a failure in the next 100 hours?
(d) If 10 assemblies are tested, what is the probability that at least one fails in less than 100 hours? Assume that the assemblies fail independently.
(e) If 10 assemblies are tested, what is the probability that all have failed by 800 hours? Assume the assemblies fail independently.

4-90. The time between arrivals of small aircraft at a county airport is exponentially distributed with a mean of one hour.
(a) What is the probability that more than three aircraft arrive within an hour?
(b) If 30 separate one-hour intervals are chosen, what is the probability that no interval contains more than three arrivals?
(c) Determine the length of an interval of time (in hours) such that the probability that no arrivals occur during the interval is 0.10.

4-91. The time between calls to a corporate office is exponentially distributed with a mean of 10 minutes.
(a) What is the probability that there are more than three calls in one-half hour?
(b) What is the probability that there are no calls within one-half hour?
(c) Determine x such that the probability that there are no calls within x hours is 0.01.
(d) What is the probability that there are no calls within a two-hour interval?
(e) If four nonoverlapping one-half hour intervals are selected, what is the probability that none of these intervals contains any call?
(f) Explain the relationship between the results in part (a) and (b).

4-92. Assume that the flaws along a magnetic tape follow a Poisson distribution with a mean of 0.2 flaw per meter. Let X denote the distance between two successive flaws.
(a) What is the mean of X?
(b) What is the probability that there are no flaws in 10 consecutive meters of tape?
(c) Does your answer to part (b) change if the 10 meters are not consecutive?
(d) How many meters of tape need to be inspected so that the probability that at least one flaw is found is 90%?
(e) What is the probability that the first time the distance between two flaws exceeds 8 meters is at the fifth flaw?
(f) What is the mean number of flaws before a distance between two flaws exceeds 8 meters?

4-93. If the random variable X has an exponential distribution with mean θ, determine the following:
(a) $P(X > \theta)$ (b) $P(X > 2\theta)$
(c) $P(X > 3\theta)$
(d) How do the results depend on θ?

4-94. Derive the formula for the mean and variance of an exponential random variable.

4-9 ERLANG AND GAMMA DISTRIBUTIONS

An exponential random variable describes the length until the first count is obtained in a Poisson process. A generalization of the exponential distribution is the length until r counts occur in a Poisson process. Consider the following example.

EXAMPLE 4-23
Processor Failure

The failures of the central processor units of large computer systems are often modeled as a Poisson process. Typically, failures are not caused by components wearing out, but by more random failures of the large number of semiconductor circuits in the units. Assume that the units that fail are immediately repaired, and assume that the mean number of failures per hour is 0.0001. Let X denote the time until four failures occur in a system. Determine the probability that X exceeds 40,000 hours.

Let the random variable N denote the number of failures in 40,000 hours of operation. The time until four failures occur exceeds 40,000 hours if and only if the number of failures in 40,000 hours is three or less. Therefore,

$$P(X > 40{,}000) = P(N \le 3)$$

The assumption that the failures follow a Poisson process implies that N has a Poisson distribution with

$$E(N) = 40{,}000(0.0001) = 4 \text{ failures per 40,000 hours}$$

Therefore,

$$P(X > 40{,}000) = P(N \le 3) = \sum_{k=0}^{3} \frac{e^{-4}4^k}{k!} = 0.433$$

The previous example can be generalized to show that if X is the time until the rth event in a Poisson process then

$$P(X > x) = \sum_{k=0}^{r-1} \frac{e^{-\lambda x}(\lambda x)^k}{k!}$$

Because $P(X > x) = 1 - F(x)$, the probability density function of X equals the negative of the derivative of the right-hand side of the previous equation. After extensive algebraic simplification, the probability density function of X can be shown to equal

$$f(x) = \frac{\lambda^r x^{r-1} e^{-\lambda x}}{(r-1)!}$$

for $x > 0$ and $r = 1, 2, \ldots$. This probability density function defines an **Erlang distribution.** Clearly, an Erlang random variable with $r = 1$ is an exponential random variable.

It is convenient to generalize the Erlang distribution to allow r to assume any nonnegative value. Then the Erlang and some other common distributions become special cases of this generalized distribution. To accomplish this step, the factorial function $(r - 1)!$ has to be generalized to apply to any nonnegative value or r, but the generalized function should still equal $(r - 1)!$ when r is a positive integer.

Gamma Function

The **gamma function** is

$$\Gamma(r) = \int_0^\infty x^{r-1} e^{-x}\, dx, \quad \text{for } r > 0 \tag{4-17}$$

It can be shown that the integral in the definition of $\Gamma(r)$ is finite. Furthermore, by using integration by parts it can be shown that

$$\Gamma(r) = (r-1)\Gamma(r-1)$$

This result is left as an exercise. Therefore, if r is a positive integer (as in the Erlang distribution),

$$\Gamma(r) = (r-1)!$$

Also, $\Gamma(1) = 0! = 1$ and it can be shown that $\Gamma(1/2) = \pi^{1/2}$. The gamma function can be interpreted as a generalization to noninteger values of r of the term $(r - 1)!$ that is used in the Erlang probability density function. Now the Erlang distribution can be generalized.

Gamma Distribution

The random variable X with probability density function

$$f(x) = \frac{\lambda^r x^{r-1} e^{-\lambda x}}{\Gamma(r)}, \quad \text{for } x > 0 \tag{4-18}$$

has a **gamma random variable** with parameters $\lambda > 0$ and $r > 0$. If r is an integer, X has an Erlang distribution.

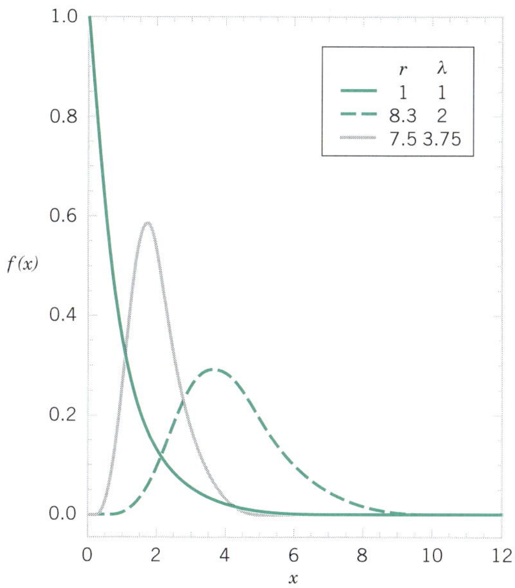

Figure 4-25 Gamma probability density functions for selected values of λ and r.

The parameters λ and r are often called the **scale** and **shape** parameters, respectively. However, one should check the definitions used in software packages. For example, Minitab defines the scale parameter as $1/\lambda$. Sketches of the gamma distribution for several values of λ and r are shown in Fig. 4-25. Many different shapes can be generated from changes to the parameters. Also, the change of variable $u = \lambda x$ and the definition of the gamma function can be used to show that the density integrates to one.

Recall that for an exponential distribution with parameter λ the mean and variance are $1/\lambda$ and $1/\lambda^2$, respectively. An Erlang random variable is the time until the rth event in a Poisson process and the time between events are independent. Therefore, it is plausible that the mean and variance of a gamma random variable multiply the exponential results by r. This motivates the following conclusions. Repeated integration by parts can be used to derive these, but the details are lengthy and omitted.

Mean and Variance

If X is a **gamma random variable** with parameters λ and r,

$$\mu = E(X) = r/\lambda \quad \text{and} \quad \sigma^2 = V(X) = r/\lambda^2 \tag{4-19}$$

EXAMPLE 4-24

The time to prepare a micro-array slide for high-throughput genomics is a Poisson process with a mean of two hours per slide. What is the probability that 10 slides require more than 25 hours to prepare?

Let X denote the time to prepare 10 slides. Because of the assumption of a Poisson process, X has a gamma distribution with $\lambda = 1/2$, $r = 10$, and the requested probability is $P(X > 25)$. The probability can be obtained from software that provides cumulative Poisson probabilities or gamma probabilities.

For the cumulative Poisson probabilities we use the method in Example 4-23 to obtain

$$P(X > 25) = \sum_{k=0}^{9} \frac{e^{-12.5}(12.5)^k}{k!}$$

In Minitab we set the mean = 12.5 and the input = 9 to obtain $P(X > 25) = 0.2014$.

As a check, we use the gamma cumulative probability function in Minitab. Set the shape parameter to 10, the scale parameter to 2 because Minitab uses the inverse of our definition, and the input to 25. The probability computed is $P(X \leq 25) = 0.7986$ and when this is subtracted from one we match with the previous result that $P(X > 25) = 0.2014$.

What are the mean and standard deviation of the time to prepare 10 slides?

The mean time is

$$E(X) = r/\lambda = 10/0.5 = 20$$

The variance of time is

$$V(X) = r/\lambda^2 = 10/0.5^2 = 40$$

so that the standard deviation is $40^{1/2} = 6.32$ hours.

The slides will be completed by what length of time with probability equal to 0.95?

The question asks for x such that

$$P(X \leq x) = 0.95$$

where X is gamma with $\lambda = 0.5$ and $r = 10$. In Minitab, we use the gamma inverse cumulative probability function and set the shape parameter to 10, the scale parameter to 2 because Minitab uses the inverse of our definition, and the probability to 0.95. The solution computed is

$$P(X \leq 31.41) = 0.95$$

Furthermore, the **chi-squared distribution** is a special case of the gamma distribution in which $\lambda = 1/2$ and r equals one of the values $1/2, 1, 3/2, 2, \ldots$. This distribution is used extensively in interval estimation and tests of hypotheses that are discussed in subsequent chapters. The chi-squared distribution is discussed in Chapter 7.

EXERCISES FOR SECTION 4-9

4-95. Use the properties of the gamma function to evaluate the following:
(a) $\Gamma(6)$ (b) $\Gamma(5/2)$
(c) $\Gamma(9/2)$

4-96. Given the probability density function $f(x) = 0.01^3x^2e^{-0.01x}/\Gamma(3)$, determine the mean and variance of the distribution.

4-97. Calls to a telephone system follow a Poisson distribution with a mean of five calls per minute.
(a) What is the name applied to the distribution and parameter values of the time until the tenth call?
(b) What is the mean time until the tenth call?
(c) What is the mean time between the ninth and tenth calls?
(d) What is the probability that exactly four calls occur within one minute?
(e) If 10 separate one-minute intervals are chosen, what is the probability that all intervals contain more than two calls?

4-98. Raw materials are studied for contamination. Suppose that the number of particles of contamination per pound of material is a Poisson random variable with a mean of 0.01 particle per pound.
(a) What is the expected number of pounds of material required to obtain 15 particles of contamination?
(b) What is the standard deviation of the pounds of materials required to obtain 15 particles of contamination?

4-99. The time between failures of a laser in a cytogenics machine is exponentially distributed with a mean of 25,000 hours.
(a) What is the expected time until the second failure?
(b) What is the probability that the time until the third failure exceeds 50,000 hours?

4-100. In a data communication system, several messages that arrive at a node are bundled into a packet before they are transmitted over the network. Assume the messages arrive at the node according to a Poisson process with

$\tau = 30$ messages per minute. Five messages are used to form a packet.

(a) What is the mean time until a packet is formed, that is, until five messages arrived at the node?
(b) What is the standard deviation of the time until a packet is formed?
(c) What is the probability that a packet is formed in less than 10 seconds?
(d) What is the probability that a packet is formed in less than 5 seconds?

4-101. Errors caused by contamination on optical disks occur at the rate of one error every 10^5 bits. Assume the errors follow a Poisson distribution.

(a) What is the mean number of bits until five errors occur?
(b) What is the standard deviation of the number of bits until five errors occur?
(c) The error-correcting code might be ineffective if there are three or more errors within 10^5 bits. What is the probability of this event?

4-102. Calls to the help line of a large computer distributor follow a Poisson distribution with a mean of 20 calls per minute.

(a) What is the mean time until the one-hundredth call?
(b) What is the mean time between call numbers 50 and 80?
(c) What is the probability that three or more calls occur within 15 seconds?

4-103. The time between arrivals of customers at an automatic teller machine is an exponential random variable with a mean of 5 minutes.

(a) What is the probability that more than three customers arrive in 10 minutes?
(b) What is the probability that the time until the fifth customer arrives is less than 15 minutes?

4-104. Use integration by parts to show that $\Gamma(r) = (r-1)\Gamma(r-1)$.

4-105. Show that the gamma density function $f(x, \lambda, r)$ integrates to 1.

4-106. Use the result for the gamma distribution to determine the mean and variance of a chi-square distribution with $r = 7/2$.

4-10 WEIBULL DISTRIBUTION

As mentioned previously, the Weibull distribution is often used to model the time until failure of many different physical systems. The parameters in the distribution provide a great deal of flexibility to model systems in which the number of failures increases with time (bearing wear), decreases with time (some semiconductors), or remains constant with time (failures caused by external shocks to the system).

Weibull Distribution

The random variable X with probability density function

$$f(x) = \frac{\beta}{\delta}\left(\frac{x}{\delta}\right)^{\beta-1} \exp\left[-\left(\frac{x}{\delta}\right)^{\beta}\right], \qquad \text{for } x > 0 \tag{4-20}$$

is a **Weibull random variable** with scale parameter $\delta > 0$ and shape parameter $\beta > 0$.

The flexibility of the Weibull distribution is illustrated by the graphs of selected probability density functions in Fig. 4-26. By inspecting the probability density function, it is seen that when $\beta = 1$, the Weibull distribution is identical to the exponential distribution. Also, the **Raleigh distribution** is a special case when the shape parameter is 2.

The cumulative distribution function is often used to compute probabilities. The following result can be obtained.

Cumulative Distribution Function

If X has a Weibull distribution with parameters δ and β, then the cumulative distribution function of X is

$$F(x) = 1 - e^{-\left(\frac{x}{\delta}\right)^{\beta}} \tag{4-21}$$

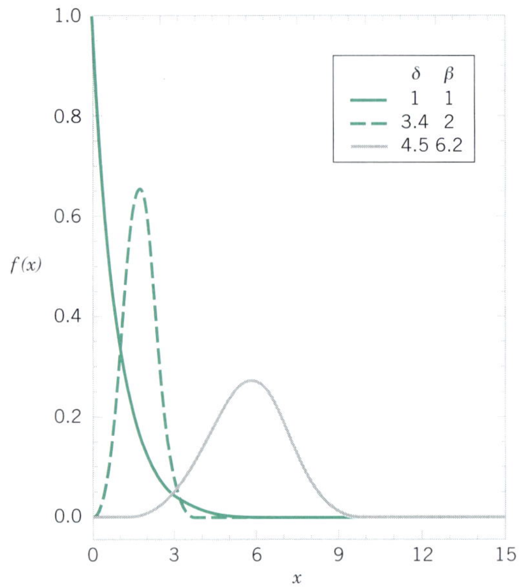

Figure 4-26 Weibull probability density functions for selected values of δ and β.

Also, the following result can be obtained.

Mean and Variance

If X has a Weibull distribution with parameters δ and β,

$$\mu = E(X) = \delta\Gamma\left(1 + \frac{1}{\beta}\right) \quad \text{and} \quad \sigma^2 = V(X) = \delta^2\Gamma\left(1 + \frac{2}{\beta}\right) - \delta^2\left[\Gamma\left(1 + \frac{1}{\beta}\right)\right]^2 \tag{4-21}$$

EXAMPLE 4-25
Bearing Wear

The time to failure (in hours) of a bearing in a mechanical shaft is satisfactorily modeled as a Weibull random variable with $\beta = 1/2$, and $\delta = 5000$ hours. Determine the mean time until failure.

From the expression for the mean,

$$E(X) = 5000\Gamma[1 + (1/0.5)] = 5000\Gamma[3] = 5000 \times 2! = 10{,}000 \text{ hours}$$

Determine the probability that a bearing lasts at least 6000 hours. Now

$$P(x > 6000) = 1 - F(6000) = exp\left[-\left(\frac{6000}{5000}\right)^{1/2}\right] = e^{-1.095} = 0.334$$

Consequently, only 33.4% of all bearings last at least 6000 hours.

EXERCISES FOR SECTION 4-10

4-107. Suppose that X has a Weibull distribution with $\beta = 0.2$ and $\delta = 100$ hours. Determine the mean and variance of X.

4-108. Suppose that X has a Weibull distribution with $\beta = 0.2$ and $\delta = 100$ hours. Determine the following:
(a) $P(X < 10{,}000)$ (b) $P(X > 5000)$

4-109. If X is a Weibull random variable with $\beta = 1$, and $\delta = 1000$, what is another name for the distribution of X and what is the mean of X?

4-110. Assume that the life of a roller bearing follows a Weibull distribution with parameters $\beta = 2$ and $\delta = 10{,}000$ hours.

(a) Determine the probability that a bearing lasts at least 8000 hours.
(b) Determine the mean time until failure of a bearing.
(c) If 10 bearings are in use and failures occur independently, what is the probability that all 10 bearings last at least 8000 hours?

4-111. The life (in hours) of a computer processing unit (CPU) is modeled by a Weibull distribution with parameters $\beta = 3$ and $\delta = 900$ hours.

(a) Determine the mean life of the CPU.
(b) Determine the variance of the life of the CPU.
(c) What is the probability that the CPU fails before 500 hours?

4-112. Assume the life of a packaged magnetic disk exposed to corrosive gases has a Weibull distribution with $\beta = 0.5$ and the mean life is 600 hours.

(a) Determine the probability that a packaged disk lasts at least 500 hours.
(b) Determine the probability that a packaged disk fails before 400 hours.

4-113. The life (in hours) of a magnetic resonance imagining machine (MRI) is modeled by a Weibull distribution with parameters $\beta = 2$ and $\delta = 500$ hours.

(a) Determine the mean life of the MRI.
(b) Determine the variance of the life of the MRI.
(c) What is the probability that the MRI fails before 250 hours?

4-114. An article in the *Journal of the Indian Geophysical Union* titled "Weibull and gamma distributions for wave parameter predictions," (2005, Vol. 9, pp. 55–64) used the Weibull distribution to model ocean wave heights. Assume that the mean wave height at the observation station is 2.5 m and the shape parameter equals 2. Determine the standard deviation of wave height.

4-115. An article in the *Journal of Geophysical Research,* "Spatial and temporal distributions of U.S. of winds and wind power at 80 m derived from measurements, "(2003, vol. 108, pp. 10–1: 10–20) considered wind speed at stations throughout the U.S. A Weibull distribution can be used to model the distribution of wind speeds at a given location. Every location is characterized by a particular shape and scale parameter. For a station at Amarillo, Texas, the mean wind speed at 80 m (the hub height of large wind turbines) in 2000 was 10.3 m/s with a standard deviation of 4.9 m/s. Determine the shape and scale parameters of a Weibull distribution with these properties.

4-11 LOGNORMAL DISTRIBUTION

Variables in a system sometimes follow an exponential relationship as $x = \exp(w)$. If the exponent is a random variable, say $W, X = \exp(W)$ is a random variable and the distribution of X is of interest. An important special case occurs when W has a normal distribution. In that case, the distribution of X is called a **lognormal distribution.** The name follows from the transformation $\ln(X) = W$. That is, the natural logarithm of X is normally distributed.

Probabilities for X are obtained from the transformation to W, but we need to recognize that the range of X is $(0, \infty)$. Suppose that W is normally distributed with mean θ and variance ω^2; then the cumulative distribution function for X is

$$\begin{aligned} F(x) &= P[X \le x] = P[\exp(W) \le x] = P[W \le \ln(x)] \\ &= P\left[Z \le \frac{\ln(x) - \theta}{\omega}\right] = \Phi\left[\frac{\ln(x) - \theta}{\omega}\right] \end{aligned}$$

for $x > 0$, where Z is a standard normal random variable. Therefore, Appendix Table III can be used to determine the probability. Also, $F(x) = 0$, for $x \le 0$.

The probability density function of X can be obtained from the derivative of $F(x)$. This derivative is applied to the last term in the expression for $F(x)$, the integral of the standard normal density function. Furthermore, from the probability density function, the mean and variance of X can be derived. The details are omitted, but a summary of results follows.

Lognormal Distribution

Let W have a normal distribution with mean θ and variance ω^2; then $X = \exp(W)$ is a **lognormal random variable** with probability density function

$$f(x) = \frac{1}{x\omega\sqrt{2\pi}} \exp\left[-\frac{(\ln x - \theta)^2}{2\omega^2}\right] \qquad 0 < x < \infty$$

The mean and variance of X are

$$E(X) = e^{\theta + \omega^2/2} \quad \text{and} \quad V(X) = e^{2\theta + \omega^2}\left(e^{\omega^2} - 1\right) \tag{4-22}$$

The parameters of a lognormal distribution are θ and ω^2, but care is needed to interpret that these are the mean and variance of the normal random variable W. The mean and variance of X are the functions of these parameters shown in (4-22). Figure 4-27 illustrates lognormal distributions for selected values of the parameters.

The lifetime of a product that degrades over time is often modeled by a lognormal random variable. For example, this is a common distribution for the lifetime of a semiconductor laser. A Weibull distribution can also be used in this type of application, and with an appropriate choice for parameters, it can approximate a selected lognormal distribution. However, a lognormal distribution is derived from a simple exponential function of a normal random variable, so it is easy to understand and easy to evaluate probabilities.

EXAMPLE 4-26
Semiconductor Laser

The lifetime of a semiconductor laser has a lognormal distribution with $\theta = 10$ hours and $\omega = 1.5$ hours. What is the probability the lifetime exceeds 10,000 hours?

From the cumulative distribution function for X

$$P(X > 10{,}000) = 1 - P[\exp(W) \le 10{,}000] = 1 - P[W \le \ln(10{,}000)]$$

$$= 1 - \Phi\left(\frac{\ln(10{,}000) - 10}{1.5}\right) = 1 - \Phi(-0.52) = 1 - 0.30 = 0.70$$

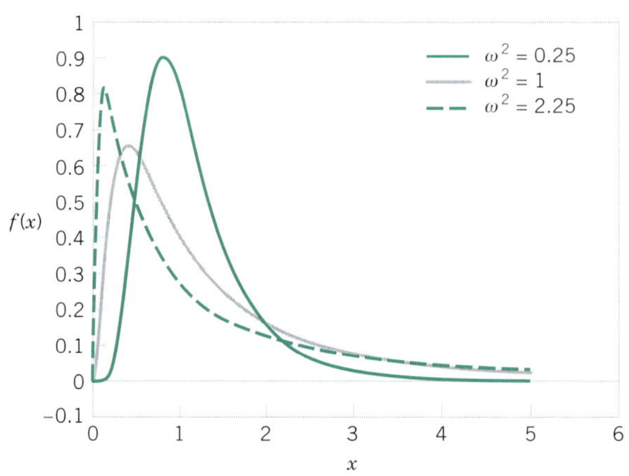

Figure 4-27 Lognormal probability density functions with $\theta = 0$ for selected values of ω^2.

What lifetime is exceeded by 99% of lasers? The question is to determine x such that $P(X > x) = 0.99$. Therefore,

$$P(X > x) = P[\exp(W) > x] = P[W > \ln(x)] = 1 - \Phi\left(\frac{\ln(x) - 10}{1.5}\right) = 0.99$$

From Appendix Table III, $1 - \Phi(z) = 0.99$ when $z = -2.33$. Therefore,

$$\frac{\ln(x) - 10}{1.5} = -2.33 \quad \text{and} \quad x = \exp(6.505) = 668.48 \text{ hours.}$$

Determine the mean and standard deviation of lifetime. Now,

$$E(X) = e^{\theta + \omega^2/2} = \exp(10 + 1.125) = 67{,}846.3$$

$$V(X) = e^{2\theta + \omega^2}(e^{\omega^2} - 1) = \exp(20 + 2.25)[\exp(2.25) - 1] = 39{,}070{,}059{,}886.6$$

so the standard deviation of X is 197,661.5 hours. Notice that the standard deviation of lifetime is large relative to the mean.

EXERCISES FOR SECTION 4-11

4-116. Suppose that X has a lognormal distribution with parameters $\theta = 5$ and $\omega^2 = 9$. Determine the following:
(a) $P(X < 13{,}300)$
(b) The value for x such that $P(X \le x) = 0.95$
(c) The mean and variance of X

4-117. Suppose that X has a lognormal distribution with parameters $\theta = -2$ and $\omega^2 = 9$. Determine the following:
(a) $P(500 < X < 1000)$
(b) The value for x such that $P(X < x) = 0.1$
(c) The mean and variance of X

4-118. Suppose that X has a lognormal distribution with parameters $\theta = 2$ and $\omega^2 = 4$. Determine the following:
(a) $P(X < 500)$
(b) The conditional probability that $X < 1500$ given that $X > 1000$
(c) What does the difference between the probabilities in parts (a) and (b) imply about lifetimes of lognormal random variables?

4-119. The length of time (in seconds) that a user views a page on a Web site before moving to another page is a lognormal random variable with parameters $\theta = 0.5$ and $\omega^2 = 1$.
(a) What is the probability that a page is viewed for more than 10 seconds?
(b) By what length of time have 50% of the users moved to another page?
(c) What is the mean and standard deviation of the time until a user moves from the page?

4-120. Suppose that X has a lognormal distribution and that the mean and variance of X are 100 and 85,000, respectively. Determine the parameters θ and ω^2 of the lognormal distribution. (*Hint:* define $x = \exp(\theta)$ and $y = \exp(\omega^2)$ and write two equations in terms of x and y.)

4-121. The lifetime of a semiconductor laser has a lognormal distribution, and it is known that the mean and standard deviation of lifetime are 10,000 and 20,000, respectively.
(a) Calculate the parameters of the lognormal distribution
(b) Determine the probability that a lifetime exceeds 10,000 hours
(c) Determine the lifetime that is exceeded by 90% of lasers

4-122. An article in *Health and Population: Perspectives and Issues* (2000, Vol. 23, pp. 28–36) used the lognormal distribution to model blood pressure in humans. The mean systolic blood pressure (SBP) in males age 17 was 120.87 mm Hg. If the coefficient of variation (100% × standard deviation/mean) is 9%, what are the parameter values of the lognormal distribution?

4-123. Derive the probability density function of a lognormal random variable from the derivative of the cumulative distribution function.

Supplemental Exercises

4-124. The probability density function of the time it takes a hematology cell counter to complete a test on a blood sample is $f(x) = 0.04$ for $50 < x < 75$ seconds.
(a) What percentage of tests require more than 70 seconds to complete.
(b) What percentage of tests require less than one minute to complete.
(c) Determine the mean and variance of the time to complete a test on a sample.

4-125. The tensile strength of paper is modeled by a normal distribution with a mean of 35 pounds per square inch and a standard deviation of 2 pounds per square inch.
(a) What is the probability that the strength of a sample is less than 40 lb/in^2?

(b) If the specifications require the tensile strength to exceed 30 lb/in^2, what proportion of the samples is scrapped?

4-126. The time it takes a cell to divide (called mitosis) is normally distributed with an average time of one hour and a standard deviation of 5 minutes.

(a) What is the probability that a cell divides in less than 45 minutes?

(b) What is the probability that it takes a cell more than 65 minutes to divide?

(c) By what time have approximately 99% of all cells completed mitosis?

4-127. The length of an injection-molded plastic case that holds magnetic tape is normally distributed with a length of 90.2 millimeters and a standard deviation of 0.1 millimeter.

(a) What is the probability that a part is longer than 90.3 millimeters or shorter than 89.7 millimeters?

(b) What should the process mean be set at to obtain the greatest number of parts between 89.7 and 90.3 millimeters?

(c) If parts that are not between 89.7 and 90.3 millimeters are scrapped, what is the yield for the process mean that you selected in part (b)?

Assume that the process is centered so that the mean is 90 millimeters and the standard deviation is 0.1 millimeter. Suppose that 10 cases are measured, and they are assumed to be independent.

(d) What is the probability that all 10 cases are between 89.7 and 90.3 millimeters?

(e) What is the expected number of the 10 cases that are between 89.7 and 90.3 millimeters?

4-128. The sick-leave time of employees in a firm in a month is normally distributed with a mean of 100 hours and a standard deviation of 20 hours.

(a) What is the probability that the sick-leave time for next month will be between 50 and 80 hours?

(b) How much time should be budgeted for sick leave if the budgeted amount should be exceeded with a probability of only 10%?

4-129. The percentage of people exposed to a bacteria who become ill is 20%. Assume that people are independent. Assume that 1000 people are exposed to the bacteria. Approximate each of the following:

(a) The probability that more than 225 become ill

(b) The probability that between 175 and 225 become ill

(c) The value such that the probability that the number of people that become ill exceeds the value is 0.01

4-130. The time to failure (in hours) for a laser in a cytometry machine is modeled by an exponential distribution with $\lambda = 0.00004$.

(a) What is the probability that the laser will last at least 20,000 hours?

(b) What is the probability that the laser will last at most 30,000 hours?

(c) What is the probability that the laser will last between 20,000 and 30,000 hours?

4-131. When a bus service reduces fares, a particular trip from New York City to Albany, New York, is very popular. A small bus can carry four passengers. The time between calls for tickets is exponentially distributed with a mean of 30 minutes. Assume that each call orders one ticket. What is the probability that the bus is filled in less than 3 hours from the time of the fare reduction?

4-132. The time between process problems in a manufacturing line is exponentially distributed with a mean of 30 days.

(a) What is the expected time until the fourth problem?

(b) What is the probability that the time until the fourth problem exceeds 120 days?

4-133. The life of a recirculating pump follows a Weibull distribution with parameters $\beta = 2$, and $\delta = 700$ hours.

(a) Determine the mean life of a pump.

(b) Determine the variance of the life of a pump.

(c) What is the probability that a pump will last longer than its mean?

4-134. The size of silver particles in a photographic emulsion is known to have a log normal distribution with a mean of 0.001 mm and a standard deviation of 0.002 mm.

(a) Determine the parameter values for the lognormal distribution.

(b) What is the probability of a particle size greater than 0.005 mm?

4-135. Suppose that $f(x) = 0.5x - 1$ for $2 < x < 4$. Determine the following:

(a) $P(X < 2.5)$

(b) $P(X > 3)$

(c) $P(2.5 < X < 3.5)$

(d) Determine the cumulative distribution function of the random variable.

(e) Determine the mean and variance of the random variable.

4-136. The time between calls is exponentially distributed with a mean time between calls of 10 minutes.

(a) What is the probability that the time until the first call is less than 5 minutes?

(b) What is the probability that the time until the first call is between 5 and 15 minutes?

(c) Determine the length of an interval of time such that the probability of at least one call in the interval is 0.90.

(d) If there has not been a call in 10 minutes, what is the probability that the time until the next call is less than 5 minutes?

(e) What is the probability that there are no calls in the intervals from 10:00 to 10:05, from 11:30 to 11:35, and from 2:00 to 2:05?

(f) What is the probability that the time until the third call is greater than 30 minutes?

(g) What is the mean time until the fifth call?

4-137. The CPU of a personal computer has a lifetime that is exponentially distributed with a mean lifetime of six years. You have owned this CPU for three years.

(a) What is the probability that the CPU fails in the next three years?
(b) Assume that your corporation has owned 10 CPUs for three years, and assume that the CPUs fail independently. What is the probability that at least one fails within the next three years?

4-138. Suppose that X has a lognormal distribution with parameters $\theta = 0$ and $\omega^2 = 4$. Determine the following:

(a) $P(10 < X < 50)$
(b) The value for x such that $P(X < x) = 0.05$
(c) The mean and variance of X

4-139. Suppose that X has a lognormal distribution and that the mean and variance of X are 50 and 4000, respectively. Determine the following:

(a) The parameters θ and ω^2 of the lognormal distribution
(b) The probability that X is less than 150

4-140. Asbestos fibers in a dust sample are identified by an electron microscope after sample preparation. Suppose that the number of fibers is a Poisson random variable and the mean number of fibers per squared centimeter of surface dust is 100. A sample of 800 square centimeters of dust is analyzed. Assume a particular grid cell under the microscope represents 1/160,000 of the sample.

(a) What is the probability that at least one fiber is visible in the grid cell?
(b) What is the mean of the number of grid cells that need to be viewed to observe 10 that contain fibers?
(c) What is the standard deviation of the number of grid cells that need to be viewed to observe 10 that contain fibers?

4-141. Without an automated irrigation system, the height of plants two weeks after germination is normally distributed with a mean of 2.5 centimeters and a standard deviation of 0.5 centimeters.

(a) What is the probability that a plant's height is greater than 2.25 centimeters?
(b) What is the probability that a plant's height is between 2.0 and 3.0 centimeters?
(c) What height is exceeded by 90% of the plants?

4-142. Continuation of Exercise 4-141. With an automated irrigation system, a plant grows to a height of 3.5 centimeters two weeks after germination.

(a) What is the probability of obtaining a plant of this height or greater from the distribution of heights in Exercise 4-135.
(b) Do you think the automated irrigation system increases the plant height at two weeks after germination?

4-143. The thickness of a laminated covering for a wood surface is normally distributed with a mean of 5 millimeters and a standard deviation of 0.2 millimeter.

(a) What is the probability that a covering thickness is greater than 5.5 millimeters?
(b) If the specifications require the thickness to be between 4.5 and 5.5 millimeters, what proportion of coverings do not meet specifications?
(c) The covering thickness of 95% of samples is below what value?

4-144. The diameter of the dot produced by a printer is normally distributed with a mean diameter of 0.002 inch. Suppose that the specifications require the dot diameter to be between 0.0014 and 0.0026 inch. If the probability that a dot meets specifications is to be 0.9973, what standard deviation is needed?

4-145. Continuation of Exercise 4-144. Assume that the standard deviation of the size of a dot is 0.0004 inch. If the probability that a dot meets specifications is to be 0.9973, what specifications are needed? Assume that the specifications are to be chosen symmetrically around the mean of 0.002.

4-146. The life of a semiconductor laser at a constant power is normally distributed with a mean of 7000 hours and a standard deviation of 600 hours.

(a) What is the probability that a laser fails before 5,800 hours?
(b) What is the life in hours that 90% of the lasers exceed?
(c) What should the mean life equal in order for 99% of the lasers to exceed 10,000 hours before failure?
(d) A product contains three lasers, and the product fails if any of the lasers fails. Assume the lasers fail independently. What should the mean life equal in order for 99% of the products to exceed 10,000 hours before failure?

4-147. Continuation of Exercise 4-146. Rework parts (a) and (b). Assume that the lifetime is an exponential random variable with the same mean.

4-148. Continuation of Exercise 4-146. Rework parts (a) and (b). Assume that the lifetime is a lognormal random variable with the same mean and standard deviation.

4-149. A square inch of carpeting contains 50 carpet fibers. The probability of a damaged fiber is 0.0001. Assume the damaged fibers occur independently.

(a) Approximate the probability of one or more damaged fibers in 1 square yard of carpeting.
(b) Approximate the probability of four or more damaged fibers in 1 square yard of carpeting.

4-150. An airline makes 200 reservations for a flight that holds 185 passengers. The probability that a passenger arrives for the flight is 0.9 and the passengers are assumed to be independent.

(a) Approximate the probability that all the passengers that arrive can be seated.
(b) Approximate the probability that there are empty seats.
(c) Approximate the number of reservations that the airline should make so that the probability that everyone who arrives can be seated is 0.95. [*Hint*: Successively try values for the number of reservations.]

MIND-EXPANDING EXERCISES

4-151. The steps in this exercise lead to the probability density function of an Erlang random variable X with parameters λ and r, $f(x) = \lambda^r x^{r-1} e^{-\lambda x}/(r-1)!$, $x > 0$, $r = 1, 2, \ldots$.

(a) Use the Poisson distribution to express $P(X > x)$.

(b) Use the result from part (a) to determine the cumulative distribution function of X.

(c) Differentiate the cumulative distribution function in part (b) and simplify to obtain the probability density function of X.

4-152. A bearing assembly contains 10 bearings. The bearing diameters are assumed to be independent and normally distributed with a mean of 1.5 millimeters and a standard deviation of 0.025 millimeter. What is the probability that the maximum diameter bearing in the assembly exceeds 1.6 millimeters?

4-153. Let the random variable X denote a measurement from a manufactured product. Suppose the target value for the measurement is m. For example, X could denote a dimensional length, and the target might be 10 millimeters. The **quality loss** of the process producing the product is defined to be the expected value of $\$k(X - m)^2$, where k is a constant that relates a deviation from target to a loss measured in dollars.

(a) Suppose X is a continuous random variable with $E(X) = m$ and $V(X) = \sigma^2$. What is the quality loss of the process?

(b) Suppose X is a continuous random variable with $E(X) = \mu$ and $V(X) = \sigma^2$. What is the quality loss of the process?

4-154. The lifetime of an electronic amplifier is modeled as an exponential random variable. If 10% of the amplifiers have a mean of 20,000 hours and the remaining amplifiers have a mean of 50,000 hours, what proportion of the amplifiers fail before 60,000 hours?

4-155. Lack of Memory Property. Show that for an exponential random variable X, $P(X < t_1 + t_2 \mid X > t_1) = P(X < t_2)$

4-156. A process is said to be of **six-sigma quality** if the process mean is at least six standard deviations from the nearest specification. Assume a normally distributed measurement.

(a) If a process mean is centered between upper and lower specifications at a distance of six standard deviations from each, what is the probability that a product does not meet specifications? Using the result that 0.000001 equals one part per million, express the answer in parts per million.

(b) Because it is difficult to maintain a process mean centered between the specifications, the probability of a product not meeting specifications is often calculated after assuming the process shifts. If the process mean positioned as in part (a) shifts upward by 1.5 standard deviations, what is the probability that a product does not meet specifications? Express the answer in parts per million.

(c) Rework part (a). Assume that the process mean is at a distance of three standard deviations.

(d) Rework part (b). Assume that the process mean is at a distance of three standard deviations and then shifts upward by 1.5 standard deviations.

(e) Compare the results in parts (b) and (d) and comment.

IMPORTANT TERMS AND CONCEPTS

Chi-squared distribution
Continuity correction
Continuous uniform distribution
Cumulative probability distribution function-continuous random variable
Erlang distribution
Exponential distribution
Gamma distribution
Lack of memory property-continuous random variable
Lognormal distribution
Mean-continuous random variable
Mean-function of a continuous random variable
Normal approximation to binomial and Poisson probabilities
Normal distribution
Probability density function
Probability distribution-continuous random variable
Standard deviation-continuous random variable
Standardizing
Standard normal distribution
Variance-continuous random variable
Weibull distribution

5 Analysis of Joint Probability Distributions

CHAPTER OUTLINE

LEARNING OBJECTIVES

After careful study of this chapter you should be able to do the following:

1. Use joint probability mass functions and joint probability density functions to calculate probabilities
2. Calculate marginal and conditional probability distributions from joint probability distributions
3. Use the multinomial distribution to determine probabilities
4. Interpret and calculate covariances and correlations between random variables
5. Understand properties of a bivariate normal distribution and be able to draw contour plots for the probability density function
6. Calculate means and variances for linear combinations of random variables and calculate probabilities for linear combinations of normally distributed random variables
7. Determine the distribution of a general function of a random variable

In Chapters 3 and 4 we studied probability distributions for a single random variable. However, it is often useful to have more than one random variable defined in a random experiment. For example, in the classification of transmitted and received signals, each signal can be classified as high, medium, or low quality. We might define the random variable X to be the number of high-quality signals received and the random variable Y to be the number of low-quality signals received. In another example, the continuous random variable X can denote the length of one dimension of an injection-molded part, and the continuous random variable Y might denote the length of another dimension. We might be interested in probabilities that can be expressed in terms of both X and Y. For example, if the specifications for X and Y are (2.95 to 3.05) and (7.60 to 7.80) millimeters, respectively, we might be interested in the probability that a part satisfies both specifications; that is, $P(2.95 < X < 3.05$ and $7.60 < Y < 7.80)$.

In general, if X and Y are two random variables, the probability distribution that defines their simultaneous behavior is called a **joint probability distribution.** In this chapter, we investigate some important properties of these joint distributions.

5-1 TWO OR MORE DISCRETE RANDOM VARIABLES

5-1.1 Joint Probability Distributions

For simplicity, we begin by considering random experiments in which only two random variables are studied. In later sections, we generalize the presentation to the joint probability distribution of more than two random variables.

EXAMPLE 5-1
Signal Bars

Calls are made to check the airline schedule at your departure city. You monitor the number of bars of signal strength on your cell phone and the number of times you have to state the name of your departure city before the voice system recognizes the name.

In the first four bits transmitted, let

X denote the number of bars of signal strength on your cell phone

Y denote the number of times you need to state your departure city

By specifying the probability of each of the points in Fig. 5-1, we specify the joint probability distribution of X and Y. Similarly to an individual random variable, we define the range of the random variables (X, Y) to be the set of points (x, y) in two-dimensional space for which the probability that $X = x$ and $Y = y$ is positive.

If X and Y are discrete random variables, the joint probability distribution of X and Y is a description of the set of points (x, y) in the range of (X, Y) along with the probability of each point. The joint probability distribution of two random variables is sometimes referred to as the **bivariate probability distribution** or **bivariate distribution** of the random variables. One way to describe the joint probability distribution of two discrete random variables is

	x = number of bars of signal strength		
y = number of times city name is stated	1	2	3
4	0.15	0.1	0.05
3	0.02	0.1	0.05
2	0.02	0.03	0.2
1	0.01	0.02	0.25

Figure 5-1 Joint probability distribution of X and Y in Example 5-1.

through a joint probability mass function. Also, $P(X = x \text{ and } Y = y)$ is usually written as $P(X = x, Y = y)$.

Joint Probability Mass Function

The **joint probability mass function** of the discrete random variables X and Y, denoted as $f_{XY}(x, y)$, satisfies

(1) $f_{XY}(x, y) \geq 0$

(2) $\sum_x \sum_y f_{XY}(x, y) = 1$

(3) $f_{XY}(x, y) = P(X = x, Y = y)$ (5-1)

Just as the probability mass function of a single random variable X is assumed to be zero at all values outside the range of X, so the joint probability mass function of X and Y is assumed to be zero at values for which a probability is not specified.

5-1.2 Marginal Probability Distributions

If more than one random variable is defined in a random experiment, it is important to distinguish between the joint probability distribution of X and Y and the probability distribution of each variable individually. The individual probability distribution of a random variable is referred to as its **marginal probability distribution.**

In general, the marginal probability distribution of X can be determined from the joint probability distribution of X and other random variables. For example, to determine $P(X = x)$, we sum $P(X = x, Y = y)$ over all points in the range of (X, Y) for which $X = x$. Subscripts on the probability mass functions distinguish between the random variables.

EXAMPLE 5-2
Marginal Distribution

The joint probability distribution of X and Y in Fig. 5-1 can be used to find the marginal probability distribution of X. For example,

$$P(X = 3) = P(X = 3, Y = 1) + P(X = 3, Y = 2) + P(X = 3, Y = 3) + P(X = 3, Y = 4)$$
$$= 0.25 + 0.2 + 0.05 + 0.05 = 0.55$$

The marginal probability distribution for X is found by summing the probabilities in each column, whereas the marginal probability distribution for Y is found by summing the probabilities in each row. The results are shown in Fig. 5-2.

	x = number of bars of signal strength			
y = number of times city name is stated	**1**	**2**	**3**	**Marginal probability distribution of Y**
4	0.15	0.1	0.05	0.3
3	0.02	0.1	0.05	0.17
2	0.02	0.03	0.2	0.25
1	0.01	0.02	0.25	0.28
	0.2	0.25	0.55	
	Marginal probability distribution of X			

Figure 5-2 Marginal probability distributions of X and Y from Fig. 5-1.

Marginal Probability Mass Function

If X and Y are discrete random variables with joint probability mass function $f_{XY}(x, y)$, then the **marginal probability mass functions** of X and Y are

$$f_X(x) = P(X = x) = \sum_y f_{XY}(x, y) \quad \text{and} \quad f_Y(y) = P(Y = y) = \sum_x f_{XY}(x, y) \tag{5-2}$$

where the first sum is over all points in the range of (X, Y) for which $X = x$ and the second sum is over all points in the range of (X, Y) for which $Y = y$

Also, $E(X)$ and $V(X)$ can be obtained by first calculating the marginal probability distribution of X and then determining $E(X)$ and $V(X)$ by the usual method.

In Fig. 5-2, the marginal probability distribution of X is used to obtain the mean as

$$E(X) = 1(0.2) + 2(0.25) + 3(0.55) = 2.35$$

5-1.3 Conditional Probability Distributions

When two random variables are defined in a random experiment, knowledge of one can change the probabilities that we associate with the values of the other. Recall that in Example 5-1, X denotes the number of bars of service and Y denotes the number of times you need to state the name of the departure city. One expects the probability $Y = 1$ to be greater at $X = 3$ bars than at $X = 1$ bar. From the notation for conditional probability in Chapter 2 we can write such conditional probabilities as $P(Y = 1|X = 3)$ and $P(Y = 1|X = 1)$. Consequently, the random variables X and Y are expected to be dependent. Knowledge of the value obtained for X changes the probabilities associated with the values of Y.

Recall that the definition of conditional probability for events A and B is $P(B|A) = P(A \cap B)/P(A)$. This definition can be applied with the event A defined to be $X = x$ and event B defined to be $Y = y$.

EXAMPLE 5-3
Signal Bars

For Example 5-1, X and Y denote the number of bars of signal strength and times you need to state your departure city received, respectively. Then

$$\begin{aligned} P(Y = 1|X = 3) &= P(X = 3, Y = 1)/P(X = 3) \\ &= f_{XY}(3, 1)/f_X(3) = 0.25/0.55 = 0.454 \end{aligned}$$

The probability that $Y = 2$ given that $X = 3$ is

$$\begin{aligned} P(Y = 2|X = 3) &= P(X = 3, Y = 2)/P(X = 3) \\ &= f_{XY}(3, 2)/f_X(3) = 0.2/0.55 = 0.364 \end{aligned}$$

Further work shows that $P(Y = 3|X = 3) = 0.091$ and $P(Y = 4|X = 3) = 0.091$. Note that $P(Y = 1|X = 3) + P(Y = 2|X = 3) + P(Y = 3|X = 3) + P(Y = 4|X = 3) = 1$. This set of probabilities defines the conditional probability distribution of Y given that $X = 3$.

Example 5-3 illustrates that the conditional probabilities that $Y = y$ given that $X = x$ can be thought of as a new probability distribution. The following definition generalizes these ideas.

Conditional Probability Mass Function

Given discrete random variables X and Y with joint probability mass function $f_{XY}(x, y)$ the **conditional probability mass function** of Y given $X = x$ is

$$f_{Y|x}(y) = f_{XY}(x, y)/f_X(x) \qquad \text{for } f_X(x) > 0 \tag{5-3}$$

The function $f_{Y|x}(y)$ is used to find the probabilities of the possible values for Y given that $X = x$. That is, it is the probability mass function for the possible values of Y given that $X = x$.

Because a conditional probability mass function $f_{Y|x}(y)$ is a probability mass function, the following properties are satisfied:

(1) $f_{Y|x}(y) \geq 0$

(2) $\sum_y f_{Y|x}(y) = 1$

(3) $P(Y = y|X = x) = f_{Y|x}(y)$ (5-4)

EXAMPLE 5-4

For the joint probability distribution in Fig. 5-1, $f_{Y|x}(y)$ is found by dividing each $f_{XY}(x, y)$ by $f_X(x)$. Here, $f_X(x)$ is simply the sum of the probabilities in each column of Fig. 5-1. The function $f_{Y|x}(y)$ is shown in Fig. 5-3. In Fig. 5-3, each column sums to one because it is a probability distribution.

Properties of random variables can be extended to a conditional probability distribution of Y given $X = x$. The usual formulas for mean and variance can be applied to a conditional probability mass function.

Conditional Mean and Variance

The **conditional mean** of Y given $X = x$, denoted as $E(Y|x)$ or $\mu_{Y|x}$, is

$$E(Y|x) = \sum_y y f_{Y|x}(y) \tag{5-5}$$

and the **conditional variance** of Y given $X = x$, denoted as $V(Y|x)$ or $\sigma^2_{Y|x}$, is

$$V(Y|x) = \sum_y (y - \mu_{Y|x})^2 f_{Y|x}(y) = \sum_y y^2 f_{Y|x}(y) - \mu^2_{Y|x}$$

	x = number of bars of signal strength		
y = number of times city name is stated	1	2	3
4	0.750	0.400	0.091
3	0.100	0.400	0.091
2	0.100	0.120	0.364
1	0.050	0.080	0.454

Figure 5-3 Conditional probability distributions of Y given $X = x$, $f_{Y|x}(y)$ in Example 5-4.

EXAMPLE 5-5

For the random variables in Example 5-1, the conditional mean of Y given $X = 1$ is obtained from the conditional distribution in Fig. 5-3:

$$E(Y|1) = \mu_{Y|1} = 1(0.05) + 2(0.1) + 3(0.1) + 4(0.75) = 3.55$$

The conditional mean is interpreted as the expected number of times the city name is stated given that one bar of signal is present. The conditional variance of Y given $X = 1$ is

$$V(Y|1) = (1 - 3.55)^2\, 0.05 + (2 - 3.55)^2\, 0.1 + (3 - 3.55)^2\, 0.1 + (4 - 3.55)^2\, 0.75 = 0.748$$

5-1.4 Independence

In some random experiments, knowledge of the values of X does not change any of the probabilities associated with the values for Y.

EXAMPLE 5-6
Independent Random Variables

In a plastic molding operation, each part is classified as to whether it conforms to color and length specifications. Define the random variable X and Y as

$$X = \begin{cases} 1 & \text{if the part conforms to color specifications} \\ 0 & \text{otherwise} \end{cases}$$

$$Y = \begin{cases} 1 & \text{if the part conforms to length specifications} \\ 0 & \text{otherwise} \end{cases}$$

Assume the joint probability distribution of X and Y is defined by $f_{XY}(x, y)$ in Fig. 5-4(a). The marginal probability distributions of X and Y are also shown in Fig. 5-4(a). Note that $f_{XY}(x, y) = f_X(x)f_Y(y)$. The conditional probability mass function $f_{Y|x}(y)$ is shown in Fig. 5-4(b). Notice that for any x, $f_{Y|x}(y) = f_Y(y)$. That is, knowledge of whether or not the part meets color specifications does not change the probability that it meets length specifications.

By analogy with independent events, we define two random variables to be **independent** whenever $f_{XY}(x, y) = f_X(x)\, f_Y(y)$ for all x and y. Notice that independence implies that $f_{XY}(x, y) = f_X(x)f_Y(y)$ for *all* x and y. If we find one pair of x and y in which the equality fails, X and Y are not independent. If two random variables are independent, then

$$f_{Y|x}(y) = \frac{f_{XY}(x, y)}{f_X(x)} = \frac{f_X(x)\, f_Y(y)}{f_X(x)} = f_Y(y)$$

With similar calculations, the following equivalent statements can be shown.

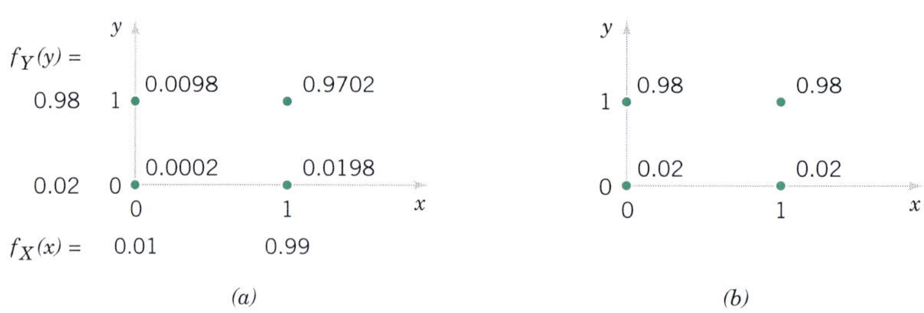

Figure 5-4 (a) Joint and marginal probability distributions of X and Y in Example 5-6. (b) Conditional probability distribution of Y given $X = x$ in Example 5-6.

Independence

For discrete random variables X and Y, if any one of the following properties is true, the others are also true, and X and Y are **independent.**

(1) $f_{XY}(x, y) = f_X(x) f_Y(y)$ for all x and y
(2) $f_{Y|x}(y) = f_Y(y)$ for all x and y with $f_X(x) > 0$
(3) $f_{X|y}(x) = f_X(x)$ for all x and y with $f_Y(y) > 0$
(4) $P(X \in A, Y \in B) = P(X \in A) P(Y \in B)$ for any sets A and B in the range of X and Y, respectively. (5-6)

Rectangular Range for (X, Y)

If the set of points in two-dimensional space that receive positive probability under $f_{XY}(x, y)$ does not form a rectangle, X and Y are not independent because knowledge of X can restrict the range of values of Y that receive positive probability. If the set of points in two-dimensional space that receives positive probability under $f_{XY}(x, y)$ forms a rectangle, independence is possible but not demonstrated. One of the conditions in Equation 5-6 must still be verified.

Rather than verifying independence from a joint probability distribution, knowledge of the random experiment is often used to assume that two random variables are independent. Then, the joint probability mass function of X and Y is computed from the product of the marginal probability mass functions. For example, the time to complete a computer search should be independent of the user's height!

5-1.5 Multiple Discrete Random Variables

EXAMPLE 5-7
Digital Channel

In some cases, more than two random variables are defined in a random experiment, and the concepts presented earlier in the chapter can easily be extended. The notation can be cumbersome and if doubts arise, it is helpful to refer to the equivalent concept for two random variables. Suppose that the quality of each bit received in a digital channel is categorized into one of the four classes, excellent, good, fair, or poor, denoted by E, G, F, and P, respectively. Also, let the random variables X_1, X_2, X_3, and X_4 denote the number of bits that are E, G, F, and P, respectively, in a transmission of 20 bits. In this example, we are interested in the joint probability distribution of four random variables. Because each of the 20 bits is categorized into one of the four classes, only values for x_1, x_2, x_3, and x_4 such that $x_1 + x_2 + x_3 + x_4 = 20$ receive positive probability in the probability distribution.

In general, given discrete random variables $X_1, X_2, X_3, \ldots, X_p$, the joint probability distribution of $X_1, X_2, X_3, \ldots, X_p$ is a description of the set of points $(x_1, x_2, x_3, \ldots, x_p)$ in the range of $X_1, X_2, X_3, \ldots, X_p$, along with the probability of each point. A joint probability mass function is a simple extension of a bivariate probability mass function.

Joint Probability Mass Function

The **joint probability mass function** of $X_1, X_2, \ldots, X_p$ is

$$f_{X_1 X_2 \ldots X_p}(x_1, x_2, \ldots, x_p) = P(X_1 = x_1, X_2 = x_2, \ldots, X_p = x_p) \qquad (5\text{-}7)$$

for all points $(x_1, x_2, \ldots, x_p)$ in the range of $X_1, X_2, \ldots, X_p$.

A marginal probability distribution is a simple extension of the result for two random variables.

Marginal Probability Mass Function

If $X_1, X_2, X_3, \ldots, X_p$ are discrete random variables with joint probability mass function $f_{X_1X_2\ldots X_p}(x_1, x_2, \ldots, x_p)$, the **marginal probability mass function** of any X_i is

$$f_{X_i}(x_i) = P(X_i = x_i) = \sum f_{X_1X_2\ldots X_p}(x_1, x_2, \ldots, x_p) \tag{5-8}$$

where the sum is over the points in the range of $(X_1, X_2, \ldots, X_p)$ for which $X_i = x_i$.

EXAMPLE 5-8

Points that have positive probability in the joint probability distribution of three random variables X_1, X_2, X_3 are shown in Fig. 5-5. The range is the nonnegative integers with $x_1 + x_2 + x_3 = 3$. The marginal probability distribution of X_2 is found as follows.

$$P(X_2 = 0) = f_{X_1X_2X_3}(3, 0, 0) + f_{X_1X_2X_3}(0, 0, 3) + f_{X_1X_2X_3}(1, 0, 2) + f_{X_1X_2X_3}(2, 0, 1)$$
$$P(X_2 = 1) = f_{X_1X_2X_3}(2, 1, 0) + f_{X_1X_2X_3}(0, 1, 2) + f_{X_1X_2X_3}(1, 1, 1)$$
$$P(X_2 = 2) = f_{X_1X_2X_3}(1, 2, 0) + f_{X_1X_2X_3}(0, 2, 1)$$
$$P(X_2 = 3) = f_{X_1X_2X_3}(0, 3, 0)$$

Furthermore, $E(X_i)$ and $V(X_i)$ for $i = 1, 2, \ldots, p$ can be determined from the marginal probability distribution of X_i or from the joint probability distribution of $X_1, X_2, \ldots, X_p$ as follows.

Mean and Variance from Joint Distribution

$$E(X_i) = \sum x_i f_{X_1X_2\ldots X_p}(x_1, x_2, \ldots, x_p)$$

and

$$V(X_i) = \sum (x_i - \mu_{X_i})^2 f_{X_1X_2\ldots X_p}(x_1, x_2, \ldots, x_p) \tag{5-9}$$

where the sum is over all points in the range of $X_1, X_2, \ldots, X_p$.

With several random variables, we might be interested in the probability distribution of some subset of the collection of variables. The probability distribution of $X_1, X_2, \ldots, X_k$, $k < p$ can be obtained from the joint probability distribution of $X_1, X_2, \ldots, X_p$ as follows.

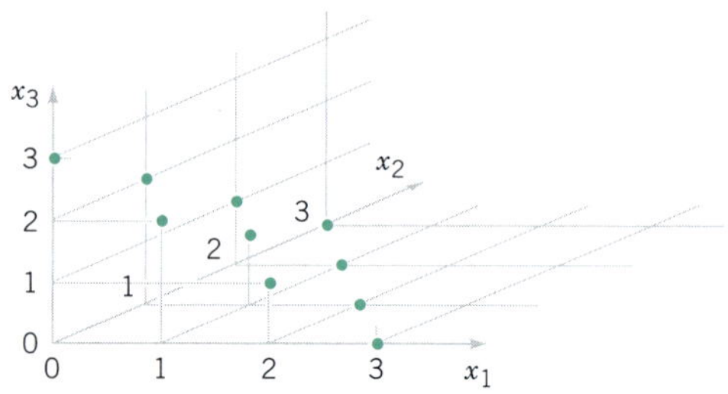

Figure 5-5 Joint probability distribution of X_1, X_2, and X_3.

Distribution of a Subset of Random Variables

If $X_1, X_2, X_3, \ldots, X_p$ are discrete random variables with joint probability mass function $f_{X_1X_2\ldots X_p}(x_1, x_2, \ldots, x_p)$, the **joint probability mass function** of $X_1, X_2, \ldots, X_k$, $k < p$, is

$$f_{X_1X_2\ldots X_k}(x_1, x_2, \ldots, x_k) = P(X_1 = x_1, X_2 = x_2, \ldots, X_k = x_k)$$
$$= \sum P(X_1 = x_1, X_2 = x_2, \ldots, X_k = x_k) \quad (5\text{-}10)$$

where the sum is over all points in the range of $X_1, X_2, \ldots, X_p$ for which $X_1 = x_1$, $X_2 = x_2, \ldots, X_k = x_k$.

That is, $P(X_1 = x_1, X_2 = x_2, \ldots, X_k = x_k)$ is the sum of the probabilities over all points in the range of $X_1, X_2, X_3, \ldots, X_p$ for which $X_1 = x_1$, $X_2 = x_2, \ldots,$ and $X_k = x_k$. An example is presented in the next section. Any k random variables can be used in the definition. The first k simplifies the notation.

Conditional Probability Distributions

Conditional probability distributions can be developed for multiple discrete random variables by an extension of the ideas used for two discrete random variables. For example, the conditional joint probability mass function of X_1, X_2, X_3 given X_4, X_5 is

$$f_{X_1X_2X_3|x_4x_5}(x_1, x_2, x_3) = \frac{f_{X_1X_2X_3X_4X_5}(x_1, x_2, x_3, x_4, x_5)}{f_{X_4X_5}(x_4, x_5)}$$

for $f_{X_4X_5}(x_4, x_5) > 0$. The conditional joint probability mass function of X_1, X_2, X_3 given X_4, X_5 provides the conditional probabilities at all points in the range of X_1, X_2, X_3, X_4, X_5 for which $X_4 = x_4$ and $X_5 = x_5$.

The concept of independence can be extended to multiple discrete random variables.

Independence

Discrete random variables $X_1, X_2, \ldots, X_p$ are **independent** if and only if

$$f_{X_1X_2\ldots X_p}(x_1, x_2, \ldots, x_p) = f_{X_1}(x_1)\, f_{X_2}(x_2) \ldots f_{X_p}(x_p) \quad (5\text{-}11)$$

for *all* $x_1, x_2, \ldots, x_p$.

Similar to the result for bivariate random variables, independence implies that Equation 5-11 holds for *all* $x_1, x_2, \ldots, x_p$. If we find one point for which the equality fails, $X_1, X_2, \ldots, X_p$ are not independent. It can be shown that if $X_1, X_2, \ldots, X_p$ are independent,

$$P(X_1 \in A_1, X_2 \in A_2, \ldots, X_p \in A_p) = P(X_1 \in A_1)P(X_2 \in A_2) \ldots P(X_p \in A_p)$$

for *any* sets $A_1, A_2, \ldots, A_p$.

5-1.6 Multinomial Probability Distribution

A joint probability distribution for multiple discrete random variables that is quite useful is an extension of the binomial. The random experiment that generates the probability distribution consists of a series of independent trials. However, the results from each trial can be categorized into one of k classes.

EXAMPLE 5-9
Digital Channel

We might be interested in a probability such as the following. Of the 20 bits received, what is the probability that 14 are excellent, 3 are good, 2 are fair, and 1 is poor? Assume that the classifications of individual bits are independent events and that the probabilities of E, G, F, and P are 0.6, 0.3, 0.08, and 0.02, respectively. One sequence of 20 bits that produces the specified numbers of bits in each class can be represented as

$$EEEEEEEEEEEEEEGGGFFP$$

Using independence, we find that the probability of this sequence is

$$P(EEEEEEEEEEEEEEGGGFFP) = 0.6^{14}0.3^{3}0.08^{2}0.02^{1} = 2.708 \times 10^{-9}$$

Clearly, all sequences that consist of the same numbers of E's, G's, F's, and P's have the same probability. Consequently, the requested probability can be found by multiplying 2.708×10^{-9} by the number of sequences with 14 E's, three G's, two F's, and one P. The number of sequences is found from Chapter 2 to be

$$\frac{20!}{14!3!2!1!} = 2325600$$

Therefore, the requested probability is

$$P(14E\text{'s, three } G\text{'s, two } F\text{'s, and one } P) = 2325600(2.708 \times 10^{-9}) = 0.0063$$

Example 5-9 leads to the following generalization of a binomial experiment and a binomial distribution.

Multinomial Distribution

Suppose a random experiment consists of a series of n trials. Assume that

(1) The result of each trial is classified into one of k classes.
(2) The probability of a trial generating a result in class 1, class 2, ..., class k is constant over the trials and equal to $p_1, p_2, \ldots, p_k$, respectively.
(3) The trials are independent.

The random variables $X_1, X_2, \ldots, X_k$ that denote the number of trials that result in class 1, class 2, ..., class k, respectively, have a **multinomial distribution** and the joint probability mass function is

$$P(X_1 = x_1, X_2 = x_2, \ldots, X_k = x_k) = \frac{n!}{x_1!x_2!\cdots x_k!} p_1^{x_1} p_2^{x_2} \ldots p_k^{x_k} \tag{5-12}$$

for $x_1 + x_2 + \cdots + x_k = n$ and $p_1 + p_2 + \cdots + p_k = 1$.

The multinomial distribution is considered a multivariable extension of the binomial distribution.

EXAMPLE 5-10
Digital Channel

In Example 5-9, let the random variables X_1, X_2, X_3, and X_4 denote the number of bits that are E, G, F, and P, respectively, in a transmission of 20 bits. The probability that 12 of the bits received are E, 6 are G, 2 are F, and 0 are P is

$$P(X_1 = 12, X_2 = 6, X_3 = 2, X_4 = 0) = \frac{20!}{12!6!2!0!} 0.6^{12}0.3^{6}0.08^{2}0.02^{0} = 0.0358$$

Each trial in a multinomial random experiment can be regarded as either generating or not generating a result in class i, for each $i = 1, 2, \ldots, k$. Because the random variable X_i is the number of trials that result in class i, X_i has a binomial distribution.

Mean and Variance

> If $X_1, X_2, \ldots, X_k$ have a multinomial distribution, the marginal probability distribution of X_i is binomial with
>
> $$E(X_i) = np_i \quad \text{and} \quad V(X_i) = np_i(1 - p_i) \tag{5-13}$$

EXAMPLE 5-11
Marginal Probability Distributions

In Example 5-10, the marginal probability distribution of X_2 is binomial with $n = 20$ and $p = 0.3$. Furthermore, the joint marginal probability distribution of X_2 and X_3 is found as follows. The $P(X_2 = x_2, X_3 = x_3)$ is the probability that exactly x_2 trials result in G and that x_3 result in F. The remaining $n - x_2 - x_3$ trials must result in either E or P. Consequently, we can consider each trial in the experiment to result in one of three classes, $\{G\}$, $\{F\}$, or $\{E, P\}$, with probabilities 0.3, 0.08, and $0.6 + 0.02 = 0.62$, respectively. With these new classes, we can consider the trials to comprise a new multinomial experiment. Therefore,

$$\begin{aligned} f_{X_2X_3}(x_2, x_3) &= P(X_2 = x_2, X_3 = x_3) \\ &= \frac{n!}{x_2!x_3!\,(n - x_2 - x_3)!}(0.3)^{x_2}(0.08)^{x_3}(0.62)^{n-x_2-x_3} \end{aligned}$$

The joint probability distribution of other sets of variables can be found similarly.

EXERCISES FOR SECTION 5-1

5-1. Show that the following function satisfies the properties of a joint probability mass function.

x	y	$f_{XY}(x, y)$
1	1	1/4
1.5	2	1/8
1.5	3	1/4
2.5	4	1/4
3	5	1/8

Determine the following:

(a) $P(X < 2.5, Y < 3)$ (b) $P(X < 2.5)$
(c) $P(Y < 3)$ (d) $P(X > 1.8, Y > 4.7)$
(e) $E(X)$, $E(Y)$, $V(X)$, and $V(Y)$.
(f) Marginal probability distribution of the random variable X.
(g) Conditional probability distribution of Y given that $X = 1.5$.
(h) Conditional probability distribution of X given that $Y - 2$.
(i) $E(Y|X = 1.5)$
(j) Are X and Y independent?

5-2. Determine the value of c that makes the function $f(x, y) = c(x + y)$ a joint probability mass function over the nine points with $x = 1, 2, 3$ and $y = 1, 2, 3$.

Determine the following:

(a) $P(X = 1, Y < 4)$ (b) $P(X = 1)$

(c) $P(Y = 2)$ (d) $P(X < 2, Y < 2)$
(e) $E(X)$, $E(Y)$, $V(X)$, and $V(Y)$.
(f) Marginal probability distribution of the random variable X.
(g) Conditional probability distribution of Y given that $X = 1$.
(h) Conditional probability distribution of X given that $Y = 2$.
(i) $E(Y|X = 1)$
(j) Are X and Y independent?

5-3. Show that the following function satisfies the properties of a joint probability mass function.

x	y	$f_{XY}(x, y)$
−1	−2	1/8
−0.5	−1	1/4
0.5	1	1/2
1	2	1/8

Determine the following:
(a) $P(X < 0.5, Y < 1.5)$ (b) $P(X < 0.5)$
(c) $P(Y < 1.5)$ (d) $P(X > 0.25, Y < 4.5)$
(e) Determine $E(X)$, $E(Y)$, $V(X)$, and $V(Y)$.
(f) Marginal probability distribution of the random variable X.
(g) Conditional probability distribution of Y given that $X = 1$.
(h) Conditional probability distribution of X given that $Y = 1$.
(i) $E(X|y = 1)$
(j) Are X and Y independent?

5-4. Four electronic printers are selected from a large lot of damaged printers. Each printer is inspected and classified as containing either a major or a minor defect. Let the random variables X and Y denote the number of printers with major and minor defects, respectively. Determine the range of the joint probability distribution of X and Y.

5-5. In the transmission of digital information, the probability that a bit has high, moderate, and low distortion is 0.01, 0.10, and 0.95, respectively. Suppose that three bits are transmitted and that the amount of distortion of each bit is assumed to be independent. Let X and Y denote the number of bits with high and moderate distortion out of the three, respectively. Determine
(a) $f_{XY}(x, y)$ (b) $f_X(x)$
(c) $E(X)$ (d) $f_{Y|1}(y)$
(e) $E(Y|X = 1)$ (f) Are X and Y independent?

5-6. A small-business Web site contains 100 pages and 60%, 30%, and 10% of the pages contain low, moderate, and high graphic content, respectively. A sample of four pages is selected without replacement, and X and Y denote the number of pages with moderate and high graphics output in the sample. Determine
(a) $f_{XY}(x, y)$ (b) $f_X(x)$
(c) $E(X)$ (d) $f_{Y|3}(y)$
(e) $E(Y|X = 3)$ (f) $V(Y|X = 3)$
(g) Are X and Y independent?

5-7. A manufacturing company employs two inspecting devices to sample a fraction of their output for quality control purposes. The first inspection monitor is able to accurately detect 99.3% of the defective items it receives, whereas the second is able to do so in 99.7% of the cases. Assume that four defective items are produced and sent out for inspection. Let X and Y denote the number of items that will be identified as defective by inspecting devices 1 and 2, respectively. Assume the devices are independent. Determine
(a) $f_{XY}(x, y)$ (b) $f_X(x)$
(c) $E(X)$ (d) $f_{Y|2}(y)$
(e) $E(Y|X = 2)$ (f) $V(Y|X = 2)$
(g) Are X and Y independent?

5-8. Suppose the random variables X, Y, and Z have the following joint probability distribution

x	y	z	$f(x, y, z)$
1	1	1	0.05
1	1	2	0.10
1	2	1	0.15
1	2	2	0.20
2	1	1	0.20
2	1	2	0.15
2	2	1	0.10
2	2	2	0.05

Determine the following:
(a) $P(X = 2)$ (b) $P(X = 1, Y = 2)$
(c) $P(Z < 1.5)$ (d) $P(X = 1 \text{ or } Z = 2)$
(e) $E(X)$ (f) $P(X = 1|Y = 1)$
(g) $P(X = 1, Y = 1|Z = 2)$ (h) $P(X = 1|Y = 1, Z = 2)$
(i) Conditional probability distribution of X given that $Y = 1$ and $Z = 2$.

5-9. An engineering statistics class has 40 students and 60% are electrical engineering majors, 10% are industrial engineering majors, and 30% are mechanical engineering majors. A sample of four students is selected randomly, without replacement, for a project team. Let X and Y denote the number of industrial engineering and mechanical engineering majors, respectively. Determine the following:
(a) $f_{XY}(x, y)$ (b) $f_X(x)$
(c) $E(X)$ (d) $f_{Y|3}(y)$
(e) $E(Y|X = 3)$ (f) $V(Y|X = 3)$
(g) Are X and Y independent?

5-10. An article in the *Journal of Database Management* "Experimental Study of a Self-Tuning Algorithm for DBMS Buffer Pools," (2005, Vol. 16, pp. 1–20) provided the

Average frequencies and operations in TPC-C

Transaction	Frequency	Selects	Updates	Inserts	Deletes	Non-Unique Selects	Joins
New Order	43	23	11	12	0	0	0
Payment	44	4.2	3	1	0	0.6	0
Order Status	4	11.4	0	0	0	0.6	0
Delivery	5	130	120	0	10	0	0
Stock Level	4	0	0	0	0	0	1

workload used in the TPC-C OLTP (Transaction Processing Performance Council's Version C On-Line Transaction Processing) benchmark, which simulates a typical order entry application. The frequency of each type of transaction (in the second column) can be used as the percentage of each type of transaction. Let X and Y denote the average number of *selects* and *updates* operations, respectively, required for each type transaction. Determine the following:

(a) $P(X < 5)$ (b) $E(X)$

(c) Conditional probability mass function of X given $Y = 0$

(d) $P(X < 6 \mid Y = 0)$ (e) $E(X \mid Y = 0)$

5-11. Based on the number of voids, a ferrite slab is classified as either high, medium, or low. Historically, 5% of the slabs are classified as high, 85% as medium, and 10% as low. A sample of 20 slabs is selected for testing. Let X, Y, and Z denote the number of slabs that are independently classified as high, medium, and low, respectively.

(a) What are the name and the values of the parameters of the joint probability distribution of X, Y, and Z?

(b) What is the range of the joint probability distribution of X, Y, Z?

(c) What is the name and the values of the parameters of the marginal probability distribution of X?

(d) $E(X)$ and $V(X)$.

Determine the following:

(e) $P(X = 1, Y = 17, Z = 3)$

(f) $P(X \leq 1, Y = 17, Z = 3)$

(g) $P(X \leq 1)$ (h) $E(Y)$

(i) $P(X = 2, Z = 3 \mid Y = 17)$ (j) $P(X = 2 \mid Y = 17)$

(k) $E(X \mid Y = 17)$

5-12. An order of 15 printers contains four with a graphics-enhancement feature, five with extra memory, and six with both features. Four printers are selected at random, without replacement, from this set. Let the random variables X, Y, and Z denote the number of printers in the sample with graphics enhancement only, extra memory only, and both, respectively.

(a) Describe the range of the joint probability distribution of X, Y, and Z.

(b) Is the probability distribution of X, Y, and Z multinomial? Why or why not?

(c) Determine the conditional probability distribution of X given that $Y = 2$.

Determine the following:

(d) $P(X = 1, Y = 2, Z = 1)$ (e) $P(X = 1, Y = 1)$

(f) $E(X)$ and $V(X)$

(g) $P(X = 1, Y = 2 \mid Z = 1)$ (h) $P(X = 2 \mid Y = 2)$

(i) Conditional probability distribution of X given that $Y = 0$ and $Z = 3$.

5-13. Four electronic ovens that were dropped during shipment are inspected and classified as containing either a major, a minor, or no defect. In the past, 60% of dropped ovens had a major defect, 30% had a minor defect, and 10% had no defect. Assume that the defects on the four ovens occur independently.

(a) Is the probability distribution of the count of ovens in each category multinomial? Why or why not?

(b) What is the probability that, of the four dropped ovens, two have a major defect and two have a minor defect?

(c) What is the probability that no oven has a defect?

Determine the following:

(d) The joint probability mass function of the number of ovens with a major defect and the number with a minor defect.

(e) The expected number of ovens with a major defect.

(f) The expected number of ovens with a minor defect.

(g) The conditional probability that two ovens have major defects given that two ovens have minor defects.

(h) The conditional probability that three ovens have major defects given that two ovens have minor defects.

(i) The conditional probability distribution of the number of ovens with major defects given that two ovens have minor defects.

(j) The conditional mean of the number of ovens with major defects given that two ovens have minor defects.

5-14. In the transmission of digital information, the probability that a bit has high, moderate, or low distortion is 0.01, 0.04, and 0.95, respectively. Suppose that three bits are transmitted and that the amount of distortion of each bit is assumed to be independent. Let X and Y denote the number of bits with high and moderate distortion out of the three transmitted, respectively. Determine the following:

(a) What is the probability that two bits have high distortion and one has moderate distortion?

(b) What is the probability that all three bits have low distortion?

(c) The probability distribution, mean and variance of X.

(d) The conditional probability distribution, conditional mean and conditional variance of X given that $Y = 2$.

5-15. For the Transaction Processing Performance Council's benchmark in Exercise 5-10. Let X, Y, and Z denote the average number of *selects*, *updates*, and *inserts* operations required for each type of transaction, respectively. Calculate the following:

(a) $f_{XYZ}(x, y, z)$

(b) Conditional probability mass function for X and Y given $Z = 0$

(c) $P(X < 6, Y < 6 | Z = 0)$

(d) $E(X | Y = 0, Z = 0)$

5-16. A marketing company performed a risk analysis for a manufacturer of synthetic fibers and concluded that new competitors present no risk 13% of the time (due mostly to the diversity of fibers manufactured), moderate risk 72% of the time (some overlapping of products), and very high risk (competitor manufactures the exact same products) 15% of the time. It is known that 12 international companies are planning to open new facilities for the manufacture of synthetic fibers within the next three years. Assume the companies are independent. Let X, Y, and Z denote the number of new competitors that will pose no, moderate, and very high risk for the interested company, respectively.

(a) What is the range of the joint probability distribution of X, Y, and Z?

Determine the following:

(b) $P(X = 1, Y = 3, Z = 1)$

(c) $P(Z \leq 2)$

(d) $P(Z = 2 | Y = 1, X = 10)$ (e) $P(Z \leq 1 | X = 10)$

(f) $P(Y \leq 1, Z \leq 1 | X = 10)$ (g) $E(Z | X = 10)$

5-2 TWO CONTINUOUS RANDOM VARIABLES

5-2.1 Joint Probability Distributions

Our presentation of the joint probability distribution of two continuous random variables is similar to our discussion of two discrete random variables. As an example, let the continuous random variable X denote the length of one dimension of an injection-molded part, and let the continuous random variable Y denote the length of another dimension. The sample space of the random experiment consists of points in two dimensions.

We can study each random variable separately. However, because the two random variables are measurements from the same part, small disturbances in the injection-molding process, such as pressure and temperature variations, might be more likely to generate values for X and Y in specific regions of two-dimensional space. For example, a small pressure increase might generate parts such that both X and Y are greater than their respective targets and a small pressure decrease might generate parts such that X and Y are both less than their respective targets. Therefore, based on pressure variations, we expect that the probability of a part with X much greater than its target and Y much less than its target is small. Knowledge of the joint probability distribution of X and Y provides information that is not obvious from the marginal probability distributions.

The joint probability distribution of two continuous random variables X and Y can be specified by providing a method for calculating the probability that X and Y assume a value in any region R of two-dimensional space. Analogous to the probability density function of a single continuous random variable, a **joint probability density function** can be defined over two-dimensional space. The double integral of $f_{XY}(x, y)$ over a region R provides the probability that (X, Y) assumes a value in R. This integral can be interpreted as the volume under the surface $f_{XY}(x, y)$ over the region R.

A joint probability density function for X and Y is shown in Fig. 5-6. The probability that (X, Y) assumes a value in the region R equals the volume of the shaded region in Fig. 5-6. In this manner, a joint probability density function is used to determine probabilities for X and Y.

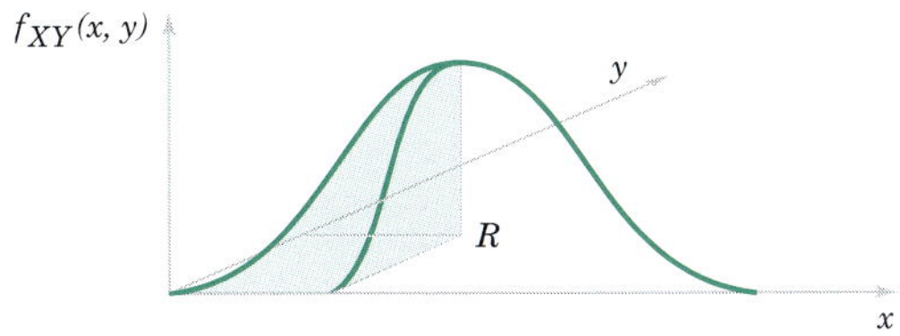

Figure 5-6 Joint probability density function for random variables X and Y. Probability that (X, Y) is in the region R is determined by the volume of $f_{XY}(x, y)$ over the region R.

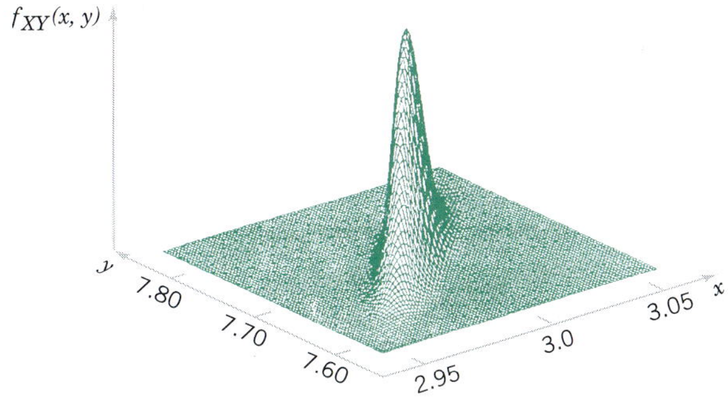

Figure 5-7 Joint probability density function for the lengths of different dimensions of an injection-molded part.

Joint Probability Density Function

A **joint probability density function** for the continuous random variables X and Y, denoted as $f_{XY}(x, y)$, satisfies the following properties:

(1) $f_{XY}(x, y) \geq 0$ for all x, y

(2) $\int_{-\infty}^{\infty} \int_{-\infty}^{\infty} f_{XY}(x, y)\, dx\, dy = 1$

(3) For any region R of two-dimensional space

$$P((X, Y) \in R) = \iint_R f_{XY}(x, y)\, dx\, dy \qquad (5\text{-}14)$$

Typically, $f_{XY}(x, y)$ is defined over all of two-dimensional space by assuming that $f_{XY}(x, y) = 0$ for all points for which $f_{XY}(x, y)$ is not specified.

At the start of this chapter, the lengths of different dimensions of an injection-molded part were presented as an example of two random variables. Each length might be modeled by a normal distribution. However, because the measurements are from the same part, the random variables are typically not independent. A probability distribution for two normal random variables that are not independent is important in many applications and it is presented later in this chapter. If the specifications for X and Y are 2.95 to 3.05 and 7.60 to 7.80 millimeters, respectively, we might be interested in the probability that a part satisfies both specifications; that is, $P(2.95 < X < 3.05, 7.60 < Y < 7.80)$. Suppose that $f_{XY}(x, y)$ is shown in Fig. 5-7. The required probability is the volume of $f_{XY}(x, y)$ within the specifications. Often a probability such as this must be determined from a numerical integration.

EXAMPLE 5-12 Server Access Time

Let the random variable X denote the time until a computer server connects to your machine (in milliseconds), and let Y denote the time until the server authorizes you as a valid user (in milliseconds). Each of these random variables measures the wait from a common starting time and $X < Y$. Assume that the joint probability density function for X and Y is

$$f_{XY}(x, y) = 6 \times 10^{-6} \exp(-0.001x - 0.002y) \qquad \text{for } x < y$$

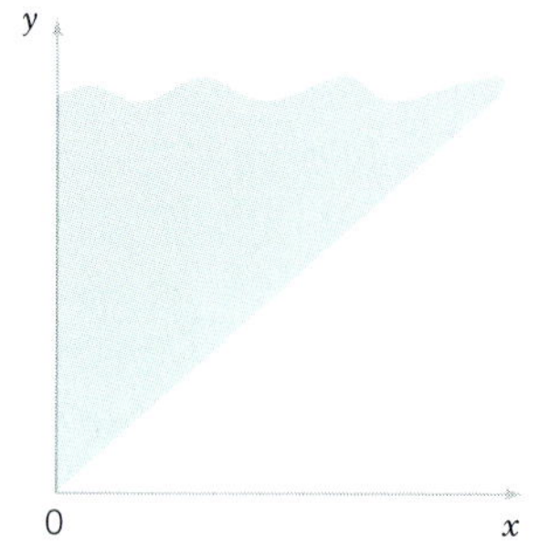

Figure 5-8 The joint probability density function of X and Y is nonzero over the shaded region.

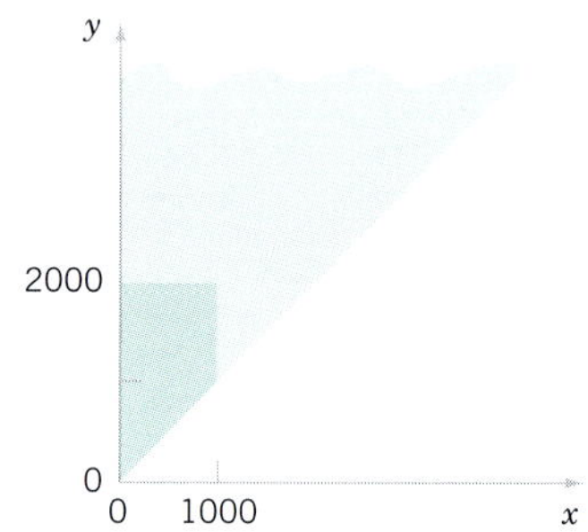

Figure 5-9 Region of integration for the probability that $X < 1000$ and $Y < 2000$ is darkly shaded.

Reasonable assumptions can be used to develop such a distribution, but for now, our focus is only on the joint probability density function.

The region with nonzero probability is shaded in Fig. 5-8. The property that this joint probability density function integrates to 1 can be verified by the integral of $f_{XY}(x, y)$ over this region as follows:

$$\int_{-\infty}^{\infty}\int_{-\infty}^{\infty} f_{XY}(x, y)\, dy\, dx = \int_{0}^{\infty}\left(\int_{x}^{\infty} 6 \times 10^{-6} e^{-0.001x-0.002y}\, dy\right) dx$$

$$= 6 \times 10^{-6}\int_{0}^{\infty}\left(\int_{x}^{\infty} e^{-0.002y}\, dy\right) e^{-0.001x}\, dx = 6 \times 10^{-6}\int_{0}^{\infty}\left(\frac{e^{-0.002x}}{0.002}\right) e^{-0.001x}\, dx$$

$$= 0.003\left(\int_{0}^{\infty} e^{-0.003x}\, dx\right) = 0.003\left(\frac{1}{0.003}\right) = 1$$

The probability that $X < 1000$ and $Y < 2000$ is determined as the integral over the darkly shaded region in Fig. 5-9.

$$P(X \le 1000, Y \le 2000) = \int_{0}^{1000}\int_{x}^{2000} f_{XY}(x, y)\, dy\, dx = 6 \times 10^{-6}\int_{0}^{1000}\left(\int_{x}^{2000} e^{-0.002y}\, dy\right) e^{-0.001x}\, dx$$

$$= 6 \times 10^{-6}\int_{0}^{1000}\left(\frac{e^{-0.002x} - e^{-4}}{0.002}\right) e^{-0.001x}\, dx$$

$$= 0.003\int_{0}^{1000} e^{-0.003x} - e^{-4} e^{-0.001x}\, dx$$

$$= 0.003\left[\left(\frac{1 - e^{-3}}{0.003}\right) - e^{-4}\left(\frac{1 - e^{-1}}{0.001}\right)\right]$$

$$= 0.003(316.738 - 11.578) = 0.915$$

5-2.2 Marginal Probability Distributions

Similar to joint discrete random variables, we can find the marginal probability distributions of X and Y from the joint probability distribution.

Marginal Probability Density Function

If the joint probability density function of continuous random variables X and Y is $f_{XY}(x, y)$, the **marginal probability density functions** of X and Y are

$$f_X(x) = \int_y f_{XY}(x, y)\, dy \quad \text{and} \quad f_Y(y) = \int_x f_{XY}(x, y)\, dx \tag{5-15}$$

where the first integral is over all points in the range of (X, Y) for which $X = x$ and the second integral is over all points in the range of (X, Y) for which $Y = y$

A probability for only one random variable, say, for example, $P(a < X < b)$, can be found from the marginal probability distribution of X or from the integral of the joint probability distribution of X and Y as

$$P(a < X < b) = \int_a^b f_X(x)dx = \int_a^b \left[\int_{-\infty}^{\infty} f(x, y)dy \right] dx = \int_a^b \int_{-\infty}^{\infty} f(x, y)dydx$$

Also, $E(X)$ and $V(X)$ can be obtained by first calculating the marginal probability distribution of X.

EXAMPLE 5-13
Server Access Time

For the random variables that denote times in Example 5-12, calculate the probability that Y exceeds 2000 milliseconds.

This probability is determined as the integral of $f_{XY}(x, y)$ over the darkly shaded region in Fig. 5-10. The region is partitioned into two parts and different limits of integration are determined for each part.

$$P(Y > 2000) = \int_0^{2000} \left(\int_{2000}^{\infty} 6 \times 10^{-6} e^{-0.001x - 0.002y}\, dy \right) dx + \int_{2000}^{\infty} \left(\int_x^{\infty} 6 \times 10^{-6} e^{-0.001x - 0.002y}\, dy \right) dx$$

The first integral is

$$6 \times 10^{-6} \int_0^{2000} \left(\frac{e^{-0.002y}}{-0.002} \Bigg|_{2000}^{\infty} \right) e^{-0.001x}\, dx = \frac{6 \times 10^{-6}}{0.002} e^{-4} \int_0^{2000} e^{-0.001x}\, dx = \frac{6 \times 10^{-6}}{0.002} e^{-4} \left(\frac{1 - e^{-2}}{0.001} \right) = 0.0475$$

The second integral is

$$6 \times 10^{-6} \int_{2000}^{\infty} \left(\frac{e^{-0.002y}}{-0.002} \Bigg|_x^{\infty} \right) e^{-0.001x}\, dx = \frac{6 \times 10^{-6}}{0.002} \int_{2000}^{\infty} e^{-0.003x}\, dx = \frac{6 \times 10^{-6}}{0.002} \left(\frac{e^{-6}}{0.003} \right) = 0.0025$$

Therefore,

$$P(Y > 2000) = 0.0475 + 0.0025 = 0.05.$$

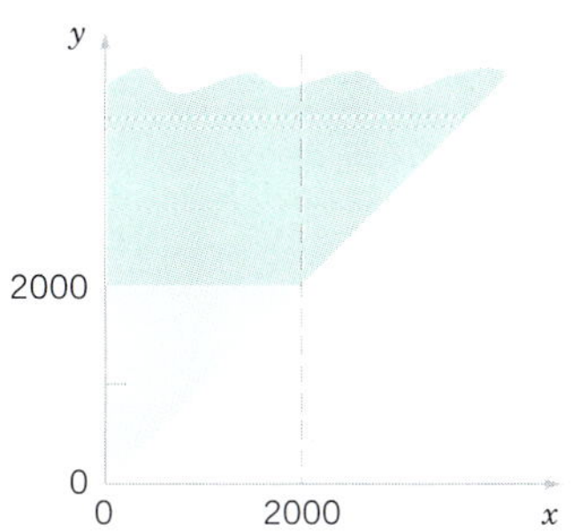

Figure 5-10 Region of integration for the probability that $Y \leq 2000$ is darkly shaded and it is partitioned into two regions with $x < 2000$ and $x > 2000$.

Alternatively, the probability can be calculated from the marginal probability distribution of Y as follows. For $y > 0$

$$f_Y(y) = \int_0^y 6 \times 10^{-6} e^{-0.001x-0.002y}\,dx = 6 \times 10^{-6} e^{-0.002y} \int_0^y e^{-0.001x}\,dx$$

$$= 6 \times 10^{-6} e^{-0.002y} \left(\frac{e^{-0.001x}}{-0.001} \Big|_0^y \right) = 6 \times 10^{-6} e^{-0.002y} \left(\frac{1 - e^{-0.001y}}{0.001} \right)$$

$$= 6 \times 10^{-3} e^{-0.002y} (1 - e^{-0.001y}) \quad \text{for } y > 0$$

We have obtained the marginal probability density function of Y. Now,

$$P(Y > 2000) = 6 \times 10^{-3} \int_{2000}^{\infty} e^{-0.002y}(1 - e^{-0.001y})\,dy$$

$$= 6 \times 10^{-3} \left[\left(\frac{e^{-0.002y}}{-0.002} \Big|_{2000}^{\infty} \right) - \left(\frac{e^{-0.003y}}{-0.003} \Big|_{2000}^{\infty} \right) \right] = 6 \times 10^{-3} \left[\frac{e^{-4}}{0.002} - \frac{e^{-6}}{0.003} \right] = 0.05$$

5-2.3 Conditional Probability Distributions

Analogous to discrete random variables, we can define the conditional probability distribution of Y given $X = x$.

Conditional Probability Density Function

Given continuous random variables X and Y with joint probability density function $f_{XY}(x, y)$, the **conditional probability density function** of Y given $X = x$ is

$$f_{Y|x}(y) = \frac{f_{XY}(x, y)}{f_X(x)} \quad \text{for} \quad f_X(x) > 0 \tag{5-16}$$

The conditional probability density function provides the conditional probabilities for the values of Y given that $X = x$.

Because the conditional probability density function $f_{Y|x}(y)$ is a probability density function for all y in R_x, the following properties are satisfied:

(1) $f_{Y|x}(y) \geq 0$

(2) $\int f_{Y|x}(y)\,dy = 1$

(3) $P(Y \in B | X = x) = \int_B f_{Y|x}(y)\,dy$ for any set B in the range of Y

(5-17)

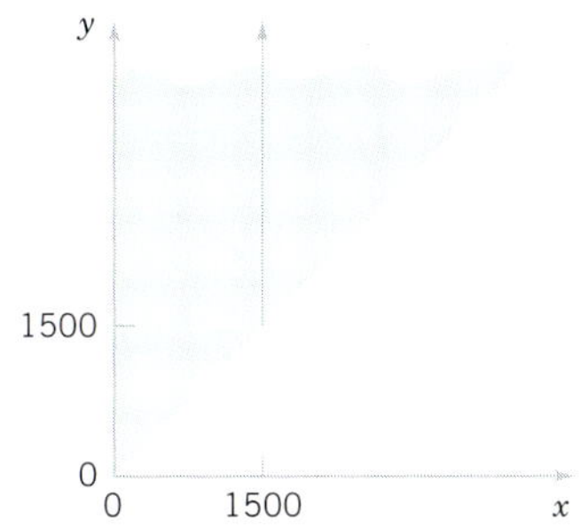

Figure 5-11 The conditional probability density function for Y, given that $x = 1500$, is nonzero over the solid line.

It is important to state the region in which a joint, marginal, or conditional probability density function is not zero. The following example illustrates this.

EXAMPLE 5-14
Conditional Probability

For the random variables that denote times in Example 5-12, determine the conditional probability density function for Y given that $X = x$.

First the marginal density function of x is determined. For $x > 0$

$$f_X(x) = \int_x^{\infty} 6 \times 10^{-6} e^{-0.001x - 0.002y}\, dy = 6 \times 10^{-6} e^{-0.001x} \left(\frac{e^{-0.002y}}{-0.002} \bigg|_x^{\infty} \right)$$

$$= 6 \times 10^{-6} e^{-0.001x} \left(\frac{e^{-0.002x}}{0.002} \right) = 0.003 e^{-0.003x} \quad \text{for} \quad x > 0$$

This is an exponential distribution with $\lambda = 0.003$. Now, for $0 < x$ and $x < y$ the conditional probability density function is

$$f_{Y|x}(y) = f_{XY}(x, y)/f_x(x) = \frac{6 \times 10^{-6} e^{-0.001x - 0.002y}}{0.003 e^{-0.003x}}$$

$$= 0.002 e^{0.002x - 0.002y} \quad \text{for } 0 < x \quad \text{and} \quad x < y$$

The conditional probability density function of Y, given that $x = 1500$, is nonzero on the solid line in Fig. 5-11.

Determine the probability that Y exceeds 2000, given that $x = 1500$. That is, determine $P(Y > 2000 \mid x = 1500)$. The conditional probability density function is integrated as follows:

$$P(Y > 2000 | x = 1500) = \int_{2000}^{\infty} f_{Y|1500}(y)\, dy = \int_{2000}^{\infty} 0.002 e^{0.002(1500) - 0.002y}\, dy$$

$$= 0.002 e^3 \left(\frac{e^{-0.002y}}{-0.002} \bigg|_{2000}^{\infty} \right) = 0.002 e^3 \left(\frac{e^{-4}}{0.002} \right) = 0.368$$

Conditional Mean and Variance

The **conditional mean** of Y given $X = x$, denoted as $E(Y|x)$ or $\mu_{Y|x}$, is

$$E(Y|x) = \int y f_{Y|x}(y)\, dy$$

and the **conditional variance** of Y given $X = x$, denoted as $V(Y|x)$ or $\sigma^2_{Y|x}$, is

$$V(Y|x) = \int (y - \mu_{Y|x})^2 f_{Y|x}(y)\, dy = \int y^2 f_{Y|x}(y)\, dy - \mu^2_{Y|x} \qquad \text{(5-18)}$$

EXAMPLE 5-15
Conditional Mean and Variance

For the random variables that denote times in Example 5-12, determine the conditional mean for Y given that $x = 1500$.

The conditional probability density function for Y was determined in Example 5-14. Because $f_{Y|1500}(y)$ is nonzero for $y > 1500$,

$$E(Y|X = 1500) = \int_{1500}^{\infty} y(0.002e^{0.002(1500)-0.002y})\, dy = 0.002e^{3} \int_{1500}^{\infty} ye^{-0.002y}\, dy$$

Integrate by parts as follows:

$$\int_{1500}^{\infty} ye^{-0.002y}\, dy = y\frac{e^{-0.002y}}{-0.002}\bigg|_{1500}^{\infty} - \int_{1500}^{\infty}\left(\frac{e^{-0.002y}}{-0.002}\right) dy$$

$$= \frac{1500}{0.002}e^{-3} - \left(\frac{e^{-0.002y}}{(-0.002)(-0.002)}\bigg|_{1500}^{\infty}\right)$$

$$= \frac{1500}{0.002}e^{-3} + \frac{e^{-3}}{(0.002)(0.002)} = \frac{e^{-3}}{0.002}(2000)$$

With the constant $0.002e^{3}$ reapplied

$$E(Y|X = 1500) = 2000$$

5-2.4 Independence

The definition of independence for continuous random variables is similar to the definition for discrete random variables. If $f_{XY}(x, y) = f_X(x) f_Y(y)$ for all x and y, X and Y are **independent.** Independence implies that $f_{XY}(x, y) = f_X(x) f_Y(y)$ for *all* x and y. If we find one pair of x and y in which the equality fails, X and Y are not independent.

Independence

For continuous random variables X and Y, if any one of the following properties is true, the others are also true, and X and Y are said to be **independent.**

(1) $f_{XY}(x, y) = f_X(x) f_Y(y)$ for all x and y

(2) $f_{Y|x}(y) = f_Y(y)$ for all x and y with $f_X(x) > 0$

(3) $f_{X|y}(x) = f_X(x)$ for all x and y with $f_Y(y) > 0$

(4) $P(X \in A, Y \in B) = P(X \in A)P(Y \in B)$ for any sets A and B in the range of X and Y, respectively. (5-19)

EXAMPLE 5-16
Dependent Random Variables

For the joint distribution of times in Example 5-12, the

- Marginal distribution of Y was determined in Example 5-13.
- Conditional distribution of Y given $X = x$ was determined in Example 5-14.

Because the marginal and conditional probability densities are not the same for all values of x, property (2) of Equation 5-19 implies that the random variables are not independent. The fact that these variables are not independent can be determined quickly by noticing that the range of (X, Y), shown in Fig. 5-8, is not rectangular. Consequently, knowledge of X changes the interval of values for Y that receives nonzero probability.

EXAMPLE 5-17
Independent Random Variables

Suppose that Example 5-12 is modified so that the joint probability density function of X and Y is $f_{XY}(x, y) = 2 \times 10^{-6} e^{-0.001x - 0.002y}$ for $x \geq 0$ and $y \geq 0$. Show that X and Y are independent and determine $P(X > 1000, Y < 1000)$.

The marginal probability density function of X is

$$f_X(x) = \int_0^\infty 2 \times 10^{-6} e^{-0.001x - 0.002y}\, dy = 0.001\, e^{-0.001x} \quad \text{for } x > 0$$

The marginal probability density function of Y is

$$f_Y(y) = \int_0^\infty 2 \times 10^{-6} e^{-0.001x - 0.002y}\, dx = 0.002\, e^{-0.002y} \quad \text{for } y > 0$$

Therefore, $f_{XY}(x, y) = f_X(x) f_Y(y)$ for all x and y and X and Y are independent.

To determine the probability requested, property (4) of Equation 5-19 and the fact that each random variable has an exponential distribution can be applied.

$$P(X > 1000, Y < 1000) = P(X > 1000)P(Y < 1000) = e^{-1}(1 - e^{-2}) = 0.318$$

Often, based on knowledge of the system under study, random variables are assumed to be independent. Then, probabilities involving both variables can be determined from the marginal probability distributions.

EXAMPLE 5-18
Machined Dimensions

Let the random variables X and Y denote the lengths of two dimensions of a machined part, respectively. Assume that X and Y are independent random variables, and further assume that the distribution of X is normal with mean 10.5 millimeters and variance 0.0025 $(\text{millimeter})^2$ and that the distribution of Y is normal with mean 3.2 millimeters and variance 0.0036 $(\text{millimeter})^2$. Determine the probability that $10.4 < X < 10.6$ and $3.15 < Y < 3.25$.

Because X and Y are independent,

$$P(10.4 < X < 10.6, 3.15 < Y < 3.25) = P(10.4 < X < 10.6)P(3.15 < Y < 3.25)$$

$$= P\left(\frac{10.4 - 10.5}{0.05} < Z < \frac{10.6 - 10.5}{0.05}\right) P\left(\frac{3.15 - 3.2}{0.06} < Z < \frac{3.25 - 3.2}{0.06}\right)$$

$$= P(-2 < Z < 2)P(-0.833 < Z < 0.833) = 0.568$$

where Z denotes a standard normal random variable.

5-2.5 Multiple Continuous Random Variables

As for discrete random variables, in some cases, more than two continuous random variables are defined in a random experiment.

EXAMPLE 5-19
Machined Dimensions

Many dimensions of a machined part are routinely measured during production. Let the random variables, X_1, X_2, X_3, and X_4 denote the lengths of four dimensions of a part. Then, at least four random variables are of interest in this study.

The joint probability distribution of continuous random variables, $X_1, X_2, X_3, \ldots, X_p$ can be specified by providing a method of calculating the probability that $X_1, X_2, X_3, \ldots, X_p$

assume a value in a region R of p-dimensional space. A joint probability density function $f_{X_1 X_2 \ldots X_p}(x_1, x_2, \ldots, x_p)$ is used to determine the probability that $(X_1, X_2, X_3, \ldots, X_p)$ assume a value in a region R by the multiple integral of $f_{X_1 X_2 \ldots X_p}(x_1, x_2, \ldots, x_p)$ over the region R.

Joint Probability Density Function

A **joint probability density function** for the continuous random variables $X_1, X_2, X_3, \ldots, X_p$, denoted as $f_{X_1 X_2 \ldots X_p}(x_1, x_2, \ldots, x_p)$, satisfies the following properties:

(1) $f_{X_1 X_2 \ldots X_p}(x_1, x_2, \ldots, x_p) \geq 0$

(2) $$\int_{-\infty}^{\infty}\int_{-\infty}^{\infty} \cdots \int_{-\infty}^{\infty} f_{X_1 X_2 \ldots X_p}(x_1, x_2, \ldots, x_p)\, dx_1\, dx_2 \ldots dx_p = 1$$

(3) For any region B of p-dimensional space

$$P[(X_1, X_2, \ldots, X_p) \in B] = \int\int_B \cdots \int f_{X_1 X_2 \ldots X_p}(x_1, x_2, \ldots, x_p)\, dx_1\, dx_2 \ldots dx_p \quad (5\text{-}20)$$

Typically, $f_{X_1 X_2 \ldots X_p}(x_1, x_2, \ldots, x_p)$ is defined over all of p-dimensional space by assuming that $f_{X_1 X_2 \ldots X_p}(x_1, x_2, \ldots, x_p) = 0$ for all points for which $f_{X_1 X_2 \ldots X_p}(x_1, x_2, \ldots, x_p)$ is not specified.

EXAMPLE 5-20
Component Lifetimes

In an electronic assembly, let the random variables X_1, X_2, X_3, X_4 denote the lifetimes of four components in hours. Suppose that the joint probability density function of these variables is

$$f_{X_1 X_2 X_3 X_4}(x_1, x_2, x_3, x_4) = 9 \times 10^{-12} e^{-0.001x_1 - 0.002x_2 - 0.0015x_3 - 0.003x_4}$$
$$\text{for } x_1 \geq 0, x_2 \geq 0, x_3 \geq 0, x_4 \geq 0$$

What is the probability that the device operates for more than 1000 hours without any failures? The requested probability is $P(X_1 > 1000, X_2 > 1000, X_3 > 1000, X_4 > 1000)$, which equals the multiple integral of $f_{X_1 X_2 X_3 X_4}(x_1, x_2, x_3, x_4)$ over the region $x_1 > 1000, x_2 > 1000, x_3 > 1000, x_4 > 1000$. The joint probability density function can be written as a product of exponential functions, and each integral is the simple integral of an exponential function. Therefore,

$$P(X_1 > 1000, X_2 > 1000, X_3 > 1000, X_4 > 1000) = e^{-1-2-1.5-3} = 0.00055$$

Suppose that the joint probability density function of several continuous random variables is a constant, say c over a region R (and zero elsewhere). In this special case,

$$\int_{-\infty}^{\infty}\int_{-\infty}^{\infty} \cdots \int_{-\infty}^{\infty} f_{X_1 X_2 \ldots X_p}(x_1, x_2, \ldots, x_p)\, dx_1\, dx_2 \ldots dx_p = c \times (\text{volume of region } R) = 1$$

by property (2) of Equation 5-20. Therefore, $c = 1/\text{volume}(R)$. Furthermore, by property (3) of Equation 5-20

$$P[(X_1, X_2, \ldots, X_p) \in B]$$
$$= \int\int_B \cdots \int f_{X_1X_2\ldots X_p}(x_1, x_2, \ldots, x_p)\, dx_1\, dx_2 \ldots dx_p = c \times \text{volume}\,(B \cap R)$$
$$= \frac{\text{volume}\,(B \cap R)}{\text{volume}\,(R)}$$

When the joint probability density function is constant, the probability that the random variables assume a value in the region B is just the ratio of the volume of the region $B \cap R$ to the volume of the region R for which the probability is positive.

EXAMPLE 5-21
Probability as a Ratio of Volumes

Suppose the joint probability density function of the continuous random variables X and Y is constant over the region $x^2 + y^2 \leq 4$. Determine the probability that $X^2 + Y^2 \leq 1$.

The region that receives positive probability is a circle of radius 2. Therefore, the area of this region is 4π. The area of the region $x^2 + y^2 \leq 1$ is π. Consequently, the requested probability is $\pi/(4\pi) = 1/4$.

Marginal Probability Density Function

If the joint probability density function of continuous random variables $X_1, X_2, \ldots, X_p$ is $f_{X_1X_2\ldots X_p}(x_1, x_2, \ldots, x_p)$, the **marginal probability density function** of X_i is

$$f_{X_i}(x_i) = \int\int_{R_{x_i}} \cdots \int f_{X_1X_2\ldots X_p}(x_1, x_2, \ldots, x_p)\, dx_1\, dx_2 \ldots dx_{i-1}\, dx_{i+1} \ldots dx_p \qquad (5\text{-}21)$$

where the integral is over all points in the range of $X_1, X_2, \ldots, X_p$ for which $X_i = x_i$.

As for two random variables, a probability involving only one random variable, say, for example $P(a < X_i < b)$, can be determined from the marginal probability distribution of X_i or from the joint probability distribution of $X_1, X_2, \ldots, X_p$. That is,

$$P(a < X_i < b) = P(-\infty < X_1 < \infty, \ldots, -\infty < X_{i-1} < \infty, a < X_i < b,$$
$$-\infty < X_{i+1} < \infty, \ldots, -\infty < X_p < \infty)$$

Furthermore, $E(X_i)$ and $V(X_i)$, for $i = 1, 2, \ldots, p,$ can be determined from the marginal probability distribution of X_i or from the joint probability distribution of $X_1, X_2, \ldots, X_p$ as follows.

Mean and Variance from Joint Distribution

$$E(X_i) = \int_{-\infty}^{\infty}\int_{-\infty}^{\infty} \cdots \int_{-\infty}^{\infty} x_i f_{X_1X_2\ldots X_p}(x_1, x_2, \ldots, x_p)\, dx_1\, dx_2 \ldots dx_p$$

and (5-22)

$$V(X_i) = \int_{-\infty}^{\infty}\int_{-\infty}^{\infty} \cdots \int_{-\infty}^{\infty} (x_i - \mu_{X_i})^2 f_{X_1X_2\ldots X_p}(x_1, x_2, \ldots, x_p)\, dx_1\, dx_2 \ldots dx_p$$

The probability distribution of a subset of variables such as $X_1, X_2, \ldots, X_k$, $k < p$, can be obtained from the joint probability distribution of $X_1, X_2, X_3, \ldots, X_p$ as follows.

Distribution of a Subset of Random Variables

If the joint probability density function of continuous random variables $X_1, X_2, \ldots, X_p$ is $f_{X_1 X_2 \ldots X_p}(x_1, x_2, \ldots, x_p)$, the **probability density function** of $X_1, X_2, \ldots, X_k$, $k < p$, is

$$f_{X_1 X_2 \ldots X_k}(x_1, x_2, \ldots, x_k) = \int\int \ldots \int f_{X_1 X_2 \ldots X_p}(x_1, x_2, \ldots, x_p)\, dx_{k+1}\, dx_{k+2} \ldots dx_p \tag{5-23}$$

where the integral is over all points in the range of $X_1, X_2, \ldots, X_p$ for which $X_1 = x_1, X_2 = x_2, \ldots, X_k = x_k$.

Conditional Probability Distribution

Conditional probability distributions can be developed for multiple continuous random variables by an extension of the ideas used for two continuous random variables.

$$f_{X_1 X_2 X_3 | x_4 x_5}(x_1, x_2, x_3) = \frac{f_{X_1 X_2 X_3 X_4 X_5}(x_1, x_2, x_3, x_4, x_5)}{f_{X_4 X_5}(x_4, x_5)}$$

for $f_{X_4 X_5}(x_4, x_5) > 0$.

The concept of independence can be extended to multiple continuous random variables.

Independence

Continuous random variables $X_1, X_2, \ldots, X_p$ are **independent** if and only if

$$f_{X_1 X_2 \ldots X_p}(x_1, x_2 \ldots, x_p) = f_{X_1}(x_1) f_{X_2}(x_2) \ldots f_{X_p}(x_p) \quad \text{for } all\ x_1, x_2, \ldots, x_p \tag{5-24}$$

Similar to the result for only two random variables, independence implies that Equation 5-24 holds for *all* $x_1, x_2, \ldots, x_p$. If we find one point for which the equality fails, $X_1, X_2, \ldots, X_p$ are not independent. It is left as an exercise to show that if $X_1, X_2, \ldots, X_p$ are independent,

$$P(X_1 \in A_1, X_2 \in A_2, \ldots, X_p \in A_p) = P(X_1 \in A_1) P(X_2 \in A_2) \ldots P(X_p \in A_p)$$

for *any* regions $A_1, A_2, \ldots, A_p$ in the range of $X_1, X_2, \ldots, X_p$, respectively.

EXAMPLE 5-22

In Chapter 3, we showed that a negative binomial random variable with parameters p and r can be represented as a sum of r geometric random variables $X_1, X_2, \ldots, X_r$. Each geometric random variable represents the additional trials required to obtain the next success. Because the trials in a binomial experiment are independent, $X_1, X_2, \ldots, X_r$ are independent random variables.

EXAMPLE 5-23 Layer Thickness

Suppose X_1, X_2, and X_3 represent the thickness in micrometers of a substrate, an active layer, and a coating layer of a chemical product. Assume that X_1, X_2, and X_3 are independent and normally distributed with $\mu_1 = 10000$, $\mu_2 = 1000$, $\mu_3 = 80$, $\sigma_1 = 250$, $\sigma_2 = 20$, and $\sigma_3 = 4$, respectively.

The specifications for the thickness of the substrate, active layer, and coating layer are $9200 < x_1 < 10800$, $950 < x_2 < 1050$, and $75 < x_3 < 85$, respectively. What proportion of chemical products meets all thickness specifications? Which one of the three thicknesses has the least probability of meeting specifications?

The requested probability is $P(9200 < X_1 < 10800, 950 < X_2 < 1050, 75 < X_3 < 85$. Because the random variables are independent,

$$P(9200 < X_1 < 10800, 950 < X_2 < 1050, 75 < X_3 < 85)$$
$$= P(9200 < X_1 < 10800)P(950 < X_2 < 1050)P(75 < X_3 < 85)$$

After standardizing, the above equals

$$P(-3.2 < Z < 3.2)P(-2.5 < Z < 2.5)P(-1.25 < Z < 1.25)$$

where Z is a standard normal random variable. From the table of the standard normal distribution, the above equals

$$(0.99862)(0.98758)(0.78870) = 0.7778$$

The thickness of the coating layer has the least probability of meeting specifications. Consequently, a priority should be to reduce variability in this part of the process.

EXERCISES FOR SECTION 5-2

5-17. Determine the value of c such that the function $f(x, y) = cxy$ for $0 < x < 3$ and $0 < y < 3$ satisfies the properties of a joint probability density function.

Determine the following:

(a) $P(X < 2, Y < 3)$ (b) $P(X < 2.5)$
(c) $P(1 < Y < 2.5)$ (d) $P(X > 1.8, 1 < Y < 2.5)$
(e) $E(X)$ (f) $P(X < 0, Y < 4)$
(g) Marginal probability distribution of the random variable X
(h) Conditional probability distribution of Y given that $X = 1.5$
(i) $E(Y|X) = 1.5)$
(j) $P(Y < 2|X = 1.5)$
(k) Conditional probability distribution of X given that $Y = 2$

5-18. Determine the value of c that makes the function $f(x, y) = c(x + y)$ a joint probability density function over the range $0 < x < 3$ and $x < y < x + 2$.

Determine the following:

(a) $P(X < 1, Y < 2)$ (b) $P(1 < X < 2)$
(c) $P(Y > 1)$ (d) $P(X < 2, Y < 2)$
(e) $E(X)$ (f) $V(X)$
(g) Marginal probability distribution of X
(h) Conditional probability distribution of Y given that $X = 1$
(i) $E(Y|X = 1)$
(j) $P(Y > 2|X = 1)$
(k) Conditional probability distribution of X given that $Y = 2$

5-19. Determine the value of c that makes the function $f(x, y) = cxy$ a joint probability density function over the range $0 < x < 3$ and $0 < y < x$.

Determine the following:

(a) $P(X < 1, Y < 2)$ (b) $P(1 < X < 2)$
(c) $P(Y > 1)$ (d) $P(X < 2, Y < 2)$
(e) $E(X)$ (f) $E(Y)$
(g) Marginal probability distribution of X
(h) Conditional probability distribution of Y given $X = 1$
(i) $E(Y|X = 1)$
(j) $P(Y > 2|X = 1)$
(k) Conditional probability distribution of X given $Y = 2$

5-20. Determine the value of c that makes the function $f(x, y) = ce^{-2x-3y}$ a joint probability density function over the range $0 < x$ and $0 < y < x$.

Determine the following:

(a) $P(X < 1, Y < 2)$ (b) $P(1 < X < 2)$
(c) $P(Y > 3)$ (d) $P(X < 2, Y < 2)$
(e) $E(X)$ (f) $E(Y)$
(g) Marginal probability distribution of X
(h) Conditional probability distribution of Y given $X = 1$
(i) $E(Y|X = 1)$
(j) Conditional probability distribution of X given $Y = 2$

5-21. Determine the value of c that makes the function $f(x, y) = ce^{-2x-3y}$ a joint probability density function over the range $0 < x$ and $x < y$.

Determine the following:

(a) $P(X < 1, Y < 2)$ (b) $P(1 < X < 2)$
(c) $P(Y > 3)$ (d) $P(X < 2, Y < 2)$
(e) $E(X)$ (f) $E(Y)$

(g) Marginal probability distribution of X
(h) Conditional probability distribution of Y given $X = 1$
(i) $E(Y|X = 1)$
(j) $P(Y < 2|X = 1)$
(k) Conditional probability distribution of X given $Y = 2$

5-22. The conditional probability distribution of Y given $X = x$ is $f_{Y|x}(y) = xe^{-xy}$ for $y > 0$ and the marginal probability distribution of X is a continuous uniform distribution over 0 to 10.
(a) Graph $f_{Y|X}(y) = xe^{-xy}$ for $y > 0$ for several values of x.
Determine
(b) $P(Y < 2|X = 2)$ (c) $E(Y|X = 2)$
(d) $E(Y|X = x)$ (e) $f_{XY}(x, y)$
(f) $f_Y(y)$

5-23. Two methods of measuring surface smoothness are used to evaluate a paper product. The measurements are recorded as deviations from the nominal surface smoothness in coded units. The joint probability distribution of the two measurements is a uniform distribution over the region $0 < x < 4$, $0 < y$, and $x - 1 < y < x + 1$. That is, $f_{XY}(x, y) = c$ for x and y in the region. Determine the value for c such that $f_{XY}(x, y)$ is a joint probability density function.

Determine the following:
(a) $P(X < 0.5, Y < 0.5)$ (b) $P(X < 0.5)$
(c) $E(X)$ (d) $E(Y)$
(e) Marginal probability distribution of X
(f) Conditional probability distribution of Y given $X = 1$
(g) $E(Y|X = 1)$ (h) $P(Y < 0.5|X = 1)$

5-24. The time between surface finish problems in a galvanizing process is exponentially distributed with a mean of 40 hours. A single plant operates three galvanizing lines that are assumed to operate independently.
(a) What is the probability that none of the lines experiences a surface finish problem in 40 hours of operation?
(b) What is the probability that all three lines experience a surface finish problem between 20 and 40 hours of operation?
(c) Why is the joint probability density function not needed to answer the previous questions?

5-25. A popular clothing manufacturer receives Internet orders via two different routing systems. The time between orders for each routing system in a typical day is known to be exponentially distributed with a mean of 3.2 minutes. Both systems operate independently.
(a) What is the probability that no orders will be received in a 5 minute period? In a 10 minute period?
(b) What is the probability that both systems receive two orders between 10 and 15 minutes after the site is officially open for business?
(c) Why is the joint probability distribution not needed to answer the previous questions?

5-26. The blade and the bearings are important parts of a lathe. The lathe can operate only when both of them work properly. For one kind of lathe, the lifetime of the blade is exponentially distributed with the mean three years; the lifetime of the bearings is also exponentially distributed with the mean four years. Assume each lifetime is independent.
(a) What is the probability that the lathe will operate at least five years?
(b) The lifetime of the lathe exceeds what time with 95% probability?

5-27. Suppose the random variables X, Y, and Z have the joint probability density function $f(x, y, z) = 8xyz$ for $0 < x < 1$, $0 < y < 1$, and $0 < z < 1$. Determine the following:
(a) $P(X < 0.5)$ (b) $P(X < 0.5, Y < 0.5)$
(c) $P(Z < 2)$ (d) $P(X < 0.5 \text{ or } Z < 2)$
(e) $E(X)$
(f) $P(X < 0.5|Y = 0.5)$
(g) $P(X < 0.5, Y < 0.5|Z = 0.8)$
(h) Conditional probability distribution of X given that $Y = 0.5$ and $Z = 0.8$
(i) $P(X < 0.5|Y = 0.5, Z = 0.8)$

5-28. Suppose the random variables X, Y, and Z have the joint probability density function $f_{XYZ}(x, y, z) = c$ over the cylinder $x^2 + y^2 < 4$ and $0 < z < 4$. Determine the constant c so that $f_{XYZ}(x, y, z)$ is a probability density function.

Determine the following:
(a) $P(X^2 + Y^2 < 2)$ (b) $P(Z < 2)$
(c) $E(X)$ (d) $P(X < 1|Y = 1)$
(e) $P(X^2 + Y^2 < 1|Z = 1)$
(f) Conditional probability distribution of Z given that $X = 1$ and $Y = 1$.

5-29. Determine the value of c that makes $f_{XYZ}(x, y, z) = c$ a joint probability density function over the region $x > 0$, $y > 0$, $z > 0$, and $x + y + z < 1$.

Determine the following:
(a) $P(X < 0.5, Y < 0.5, Z < 0.5)$
(b) $P(X < 0.5, Y < 0.5)$
(c) $P(X < 0.5)$
(d) $E(X)$
(e) Marginal distribution of X
(f) Joint distribution of X and Y
(g) Conditional probability distribution of X given that $Y = 0.5$ and $Z = 0.5$
(h) Conditional probability distribution of X given that $Y = 0.5$

5-30. The yield in pounds from a day's production is normally distributed with a mean of 1500 pounds and standard deviation of 100 pounds. Assume that the yields on different days are independent random variables.
(a) What is the probability that the production yield exceeds 1400 pounds on each of five days next week?
(b) What is the probability that the production yield exceeds 1400 pounds on at least four of the five days next week?

5-31. The weights of adobe bricks used for construction are normally distributed with a mean of 3 pounds and a standard deviation of 0.25 pound. Assume that the weights of the bricks are independent and that a random sample of 20 bricks is selected.

(a) What is the probability that all the bricks in the sample exceed 2.75 pounds?

(b) What is the probability that the heaviest brick in the sample exceeds 3.75 pounds?

5-32. A manufacturer of electroluminescent lamps knows that the amount of luminescent ink deposited on one of its products is normally distributed with a mean of 1.2 grams and a standard deviation of 0.03 grams. Any lamp with less than 1.14 grams of luminescent ink will fail to meet customer's specifications. A random sample of 25 lamps is collected and the mass of luminescent ink on each is measured.

(a) What is the probability that at least 1 lamp fails to meet specifications?

(b) What is the probability that 5 lamps or fewer fail to meet specifications?

(c) What is the probability that all lamps conform to specifications?

(d) Why is the joint probability distribution of the 25 lamps not needed to answer the previous questions?

5-3 COVARIANCE AND CORRELATION

When two or more random variables are defined on a probability space, it is useful to describe how they vary together; that is, it is useful to measure the relationship between the variables. A common measure of the relationship between two random variables is the **covariance.** To define the covariance, we need to describe the expected value of a function of two random variables $h(X, Y)$. The definition simply extends that used for a function of a single random variable.

Expected Value of a Function of Two Random Variables

$$E[h(X, Y)] = \begin{cases} \sum\sum h(x, y) f_{XY}(x, y) & X, Y \textit{ discrete} \\ \iint h(x, y) f_{XY}(x, y)\,dx\,dy & X, Y \textit{ continuous} \end{cases} \tag{5-25}$$

That is, $E[h(X, Y)]$ can be thought of as the weighted average of $h(x, y)$ for each point in the range of (X, Y). The value of $E[h(X, Y)]$ represents the average value of $h(X, Y)$ that is expected in a long sequence of repeated trials of the random experiment.

EXAMPLE 5-24
Expected Value of a Function of Two Random Variables

For the joint probability distribution of the two random variables in Fig. 5-12, calculate $E[(X - \mu_X)(Y - \mu_Y)]$.

The result is obtained by multiplying $x - \mu_X$ times $y - \mu_Y$, times $f_{XY}(x, y)$ for each point in the range of (X, Y). First, μ_X and μ_Y are determined from the marginal distributions for X and Y

$$\mu_X = 1 \times 0.3 + 3 \times 0.7 = 2.4$$

and

$$\mu_Y = 1 \times 0.3 + 2 \times 0.4 + 3 \times 0.3 = 2.0$$

Therefore,

$$\begin{aligned} E[(X - \mu_X)(Y - \mu_Y)] = (1 - 2.4)(1 - 2.0) \times 0.1 \\ + (1 - 2.4)(2 - 2.0) \times 0.2 + (3 - 2.4)(1 - 2.0) \times 0.2 \\ + (3 - 2.4)(2 - 2.0) \times 0.2 + (3 - 2.4)(3 - 2.0) \times 0.3 = 0.2 \end{aligned}$$

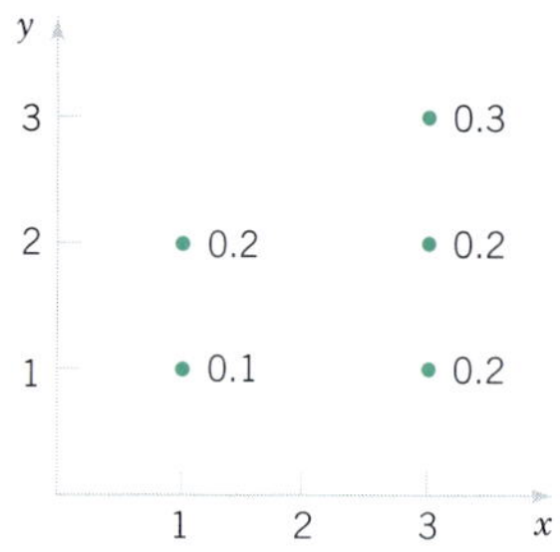

Figure 5-12 Joint distribution of X and Y for Example 5-24.

The covariance is defined for both continuous and discrete random variables by the same formula.

Covariance

The **covariance** between the random variables X and Y, denoted as cov(X, Y) or σ_{XY}, is

$$\sigma_{XY} = E[(X - \mu_X)(Y - \mu_Y)] = E(XY) - \mu_X\mu_Y \tag{5-26}$$

If the points in the joint probability distribution of X and Y that receive positive probability tend to fall along a line of positive (or negative) slope, σ_{XY} is positive (or negative). If the points tend to fall along a line of positive slope, X tends to be greater than μ_X when Y is greater than μ_Y. Therefore, the product of the two terms $x - \mu_X$ and $y - \mu_Y$ tends to be positive. However, if the points tend to fall along a line of negative slope, $x - \mu_X$ tends to be positive when $y - \mu_Y$ is negative, and vice versa. Therefore, the product of $x - \mu_X$ and $y - \mu_Y$ tends to be negative. In this sense, the covariance between X and Y describes the variation between the two random variables. Figure 5-13 shows examples of pairs of random variables with positive, negative, and zero covariance.

Covariance is a measure of **linear relationship** between the random variables. If the relationship between the random variables is nonlinear, the covariance might not be sensitive to the relationship. This is illustrated in Fig. 5-13(d). The only points with nonzero probability are the points on the circle. There is an identifiable relationship between the variables. Still, the covariance is zero.

The equality of the two expressions for covariance in Equation 5-26 is shown for continuous random variables as follows. By writing the expectations as integrals,

$$E[(Y - \mu_Y)(X - \mu_X)] = \int_{-\infty}^{\infty}\int_{-\infty}^{\infty} (x - \mu_X)(y - \mu_Y) f_{XY}(x, y)\, dx\, dy$$

$$= \int_{-\infty}^{\infty}\int_{-\infty}^{\infty} [xy - \mu_X y - x\mu_Y + \mu_X\mu_Y] f_{XY}(x, y)\, dx\, dy$$

Now

$$\int_{-\infty}^{\infty}\int_{-\infty}^{\infty} \mu_X y\, f_{XY}(x, y)\, dx\, dy = \mu_X \left[\int_{-\infty}^{\infty}\int_{-\infty}^{\infty} y f_{XY}(x, y)\, dx\, dy\right] = \mu_X\mu_Y$$

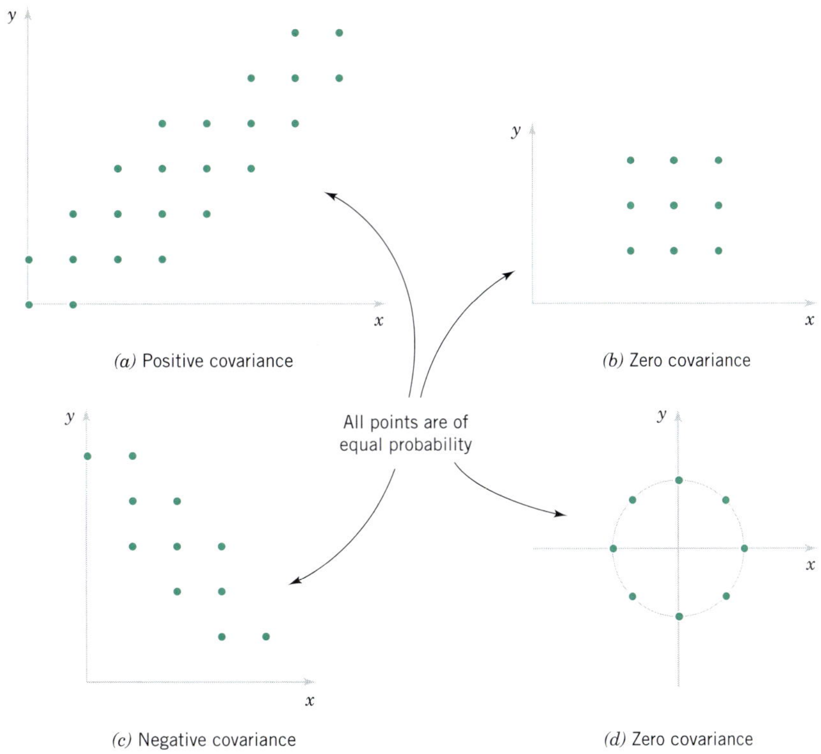

Figure 5-13 Joint probability distributions and the sign of covariance between X and Y.

Therefore,

$$E[(X-\mu_X)(Y-\mu_Y)] = \int_{-\infty}^{\infty}\int_{-\infty}^{\infty} xyf_{XY}(x,y)\,dx\,dy - \mu_X\mu_Y - \mu_X\mu_Y + \mu_X\mu_Y$$

$$= \int_{-\infty}^{\infty}\int_{-\infty}^{\infty} xyf_{XY}(x,y)\,dx\,dy - \mu_X\mu_Y = E(XY) - \mu_X\mu_Y$$

EXAMPLE 5-25

In Example 5-1, the random variables X and Y are the number of signal bars and the number of times you need to state your departure city, respectively. Is the covariance between X and Y positive or negative?

As the signal bars increase, the number of times you need to state your name decreases. Therefore, X and Y have a negative covariance. This can be verified from the joint probability distribution in Fig. 5-1.

There is another measure of the relationship between two random variables that is often easier to interpret than the covariance.

Correlation

The **correlation** between random variables X and Y, denoted as ρ_{XY}, is

$$\rho_{XY} = \frac{\text{cov}(X,Y)}{\sqrt{V(X)V(Y)}} = \frac{\sigma_{XY}}{\sigma_X\sigma_Y} \tag{5-27}$$

Because $\sigma_X > 0$ and $\sigma_Y > 0$, if the covariance between X and Y is positive, negative, or zero, the correlation between X and Y is positive, negative, or zero, respectively. The following result can be shown.

For any two random variables X and Y

$$-1 \leq \rho_{XY} \leq +1 \tag{5-28}$$

The correlation just scales the covariance by the standard deviation of each variable. Consequently, the correlation is a dimensionless quantity that can be used to compare the linear relationships between pairs of variables in different units.

If the points in the joint probability distribution of X and Y that receive positive probability tend to fall along a line of positive (or negative) slope, ρ_{XY} is near $+1$ (or -1). If ρ_{XY} equals $+1$ or -1, it can be shown that the points in the joint probability distribution that receive positive probability fall exactly along a straight line. Two random variables with nonzero correlation are said to be **correlated.** Similar to covariance, the correlation is a measure of the **linear relationship** between random variables.

EXAMPLE 5-26
Covariance

For the discrete random variables X and Y with the joint distribution shown in Fig. 5-14, determine σ_{XY} and ρ_{XY}.

The calculations for $E(XY)$, $E(X)$, and $V(X)$ are as follows.

$$\begin{aligned} E(XY) &= 0 \times 0 \times 0.2 + 1 \times 1 \times 0.1 + 1 \times 2 \times 0.1 + 2 \times 1 \times 0.1 \\ &\quad + 2 \times 2 \times 0.1 + 3 \times 3 \times 0.4 = 4.5 \\ E(X) &= 0 \times 0.2 + 1 \times 0.2 + 2 \times 0.2 + 3 \times 0.4 = 1.8 \\ V(X) &= (0 - 1.8)^2 \times 0.2 + (1 - 1.8)^2 \times 0.2 + (2 - 1.8)^2 \times 0.2 \\ &\quad + (3 - 1.8)^2 \times 0.4 = 1.36 \end{aligned}$$

Because the marginal probability distribution of Y is the same as for X, $E(Y) = 1.8$ and $V(Y) = 1.36$. Consequently,

$$\sigma_{XY} = E(XY) - E(X)E(Y) = 4.5 - (1.8)(1.8) = 1.26$$

Furthermore,

$$\rho_{XY} = \frac{\sigma_{XY}}{\sigma_X \sigma_Y} = \frac{1.26}{(\sqrt{1.36})(\sqrt{1.36})} = 0.926$$

EXAMPLE 5-27
Correlation

Suppose that the random variable X has the following distribution: $P(X = 1) = 0.2$, $P(X = 2) = 0.6$, $P(X = 3) = 0.2$. Let $Y = 2X + 5$. That is, $P(Y = 7) = 0.2$, $P(Y = 9) = 0.6$, $P(Y = 11) = 0.2$. Determine the correlation between X and Y. Refer to Fig. 5-15.

Because X and Y are linearly related, $\rho = 1$. This can be verified by direct calculations: Try it.

For independent random variables, we do not expect any relationship in their joint probability distribution. The following result is left as an exercise.

If X and Y are independent random variables,

$$\sigma_{XY} = \rho_{XY} = 0 \tag{5-29}$$

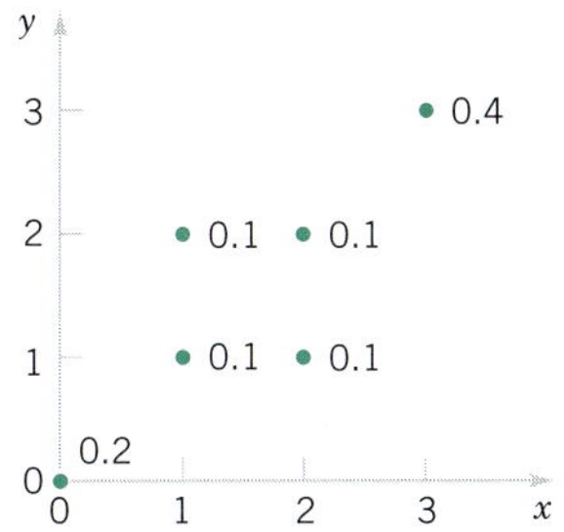

Figure 5-14 Joint distribution for Example 5-26.

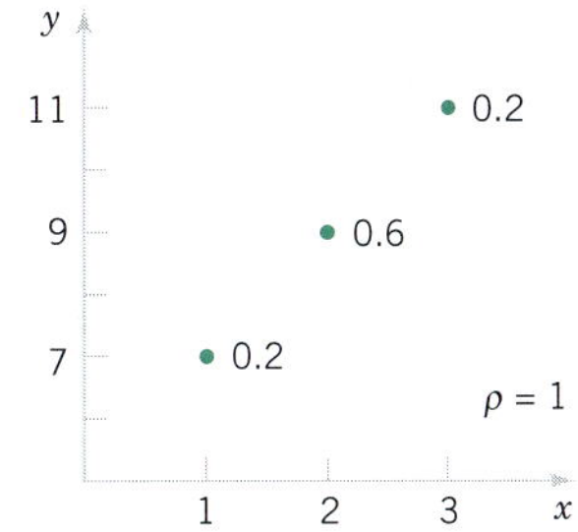

Figure 5-15 Joint distribution for Example 5-27.

EXAMPLE 5-28
Independence Implies Zero Covariance

For the two random variables in Fig. 5-16, show that $\sigma_{XY} = 0$.

The two random variables in this example are continuous random variables. In this case $E(XY)$ is defined as the double integral over the range of (X, Y). That is,

$$E(XY) = \int_0^4 \int_0^2 xy f_{XY}(x, y)\,dx\,dy = \frac{1}{16}\int_0^4 \left[\int_0^2 x^2y^2\,dx\right]dy = \frac{1}{16}\int_0^4 y^2\left[x^3/3\Big|_0^2\right]$$

$$= \frac{1}{16}\int_0^4 y^2[8/3]\,dy = \frac{1}{6}\left[y^3/3\Big|_0^4\right] = \frac{1}{6}[64/3] = 32/9$$

Also,

$$E(X) = \int_0^4 \int_0^2 x f_{XY}(x, y)\,dx\,dy = \frac{1}{16}\int_0^4 y\left[\int_0^2 x^2\,dx\right]dy = \frac{1}{16}\int_0^4 y\left[x^3/3\Big|_0^2\right]dy$$

$$= \frac{1}{16}\left[y^2/2\Big|_0^4\right][8/3] = \frac{1}{6}[16/2] = 4/3$$

$$E(Y) = \int_0^4 \int_0^2 y f_{XY}(x, y)\,dx\,dy = \frac{1}{16}\int_0^4 y^2\left[\int_0^2 x\,dx\right]dy = \frac{1}{16}\int_0^4 y^2\left[x^2/2\Big|_0^2\right]dy$$

$$= \frac{2}{16}\left[y^3/3\Big|_0^4\right] = \frac{1}{8}[64/3] = 8/3$$

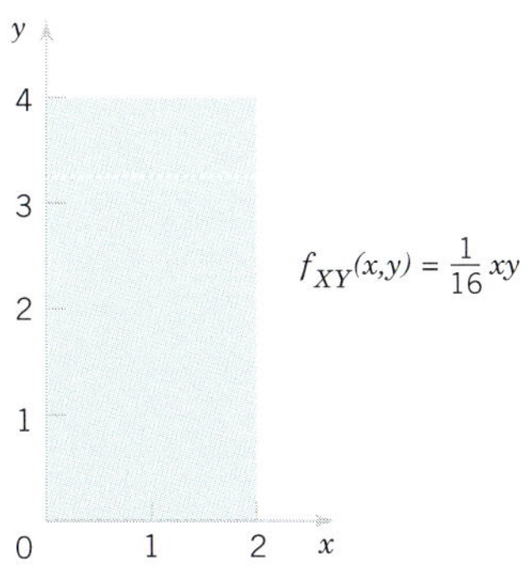

Figure 5-16 Random variables with zero covariance from Example 5-28.

Thus,

$$E(XY) - E(X)E(Y) = 32/9 - (4/3)(8/3) = 0$$

It can be shown that these two random variables are independent. You can check that $f_{XY}(x, y) = f_X(x)f_Y(y)$ for all x and y.

However, if the correlation between two random variables is zero, we *cannot* immediately conclude that the random variables are independent. Figure 5-13(d) provides an example.

EXERCISES FOR SECTION 5-3

5-33. Determine the covariance and correlation for the following joint probability distribution:

x	1	1	2	4
y	3	4	5	6
$f_{XY}(x, y)$	1/8	1/4	1/2	1/8

5-34. Determine the covariance and correlation for the following joint probability distribution:

x	−1	−0.5	0.5	1
y	−2	−1	1	2
$f_{XY}(x, y)$	1/8	1/4	1/2	1/8

5-35. Determine the value for c and the covariance and correlation for the joint probability mass function $f_{XY}(x, y) = c(x + y)$ for $x = 1, 2, 3$ and $y = 1, 2, 3$.

5-36. Determine the covariance and correlation for the joint probability distribution shown in Fig. 5-4(a) and described in Example 5-6.

5-37. Determine the covariance and correlation for X_1 and X_2 in the joint distribution of the multinomial random variables X_1, X_2 and X_3 with $p_1 = p_2 = p_3 = 1/3$ and $n = 3$. What can you conclude about the sign of the correlation between two random variables in a multinomial distribution?

5-38. For the Transaction Processing Performance Council's benchmark in Exercise 5-10, let X, Y, and Z denote the average number of *selects*, *updates*, and *inserts* operations required for each type transaction, respectively. Calculate the following:

(a) Covariance between X and Y

(b) Correlation between X and Y

(c) Covariance between X and Z

(d) Correlation between X and Z

5-39. Determine the value for c and the covariance and correlation for the joint probability density function $f_{XY}(x, y) = cxy$ over the range $0 < x < 3$ and $0 < y < x$.

5-40. Determine the value for c and the covariance and correlation for the joint probability density function $f_{XY}(x, y) = c$ over the range $0 < x < 5$, $0 < y$, and $x - 1 < y < x + 1$.

5-41. Determine the covariance and correlation for the joint probability density function $f_{XY}(x, y) = e^{-x-y}$ over the range $0 < x$ and $0 < y$.

5-42. Determine the covariance and correlation for the joint probability density function $f_{XY}(x, y) = 6 \times 10^{-6} e^{-0.001x-0.002y}$ over the range $0 < x$ and $x < y$ from Example 5-12.

5-43. The joint probability distribution is

x	−1	0	0	1
y	0	−1	1	0
$f_{XY}(x, y)$	1/4	1/4	1/4	1/4

Show that the correlation between X and Y is zero, but X and Y are not independent.

5-44. Suppose X and Y are independent continuous random variables. Show that $\sigma_{XY} = 0$.

5-45. Suppose that the correlation between X and Y is ρ. For constants a, b, c, and d, what is the correlation between the random variables $U = aX + b$ and $V = cY + d$?

5-4 BIVARIATE NORMAL DISTRIBUTION

An extension of a normal distribution to two random variables is an important bivariate probability distribution.

EXAMPLE 5-29 **Bivariate Normal Distribution**

At the start of this chapter, the length of different dimensions of an injection-molded part was presented as an example of two random variables. Each length might be modeled by a normal distribution. However, because the measurements are from the same part, the random variables are typically not independent. A probability distribution for two normal random variables that are not independent is important in many applications. As stated at the start of the chapter, if the specifications for X and Y are 2.95 to 3.05 and 7.60 to 7.80 millimeters, respectively, we might be interested in the probability that a part satisfies both specifications; that is, $P(2.95 < X < 3.05, 7.60 < Y < 7.80)$.

Bivariate Normal Probability Density Function

The probability density function of a **bivariate normal distribution** is

$$f_{XY}(x, y; \sigma_X, \sigma_Y, \mu_X, \mu_Y, \rho) = \frac{1}{2\pi\sigma_X\sigma_Y\sqrt{1-\rho^2}} \exp\left\{\frac{-1}{2(1-\rho^2)}\left[\frac{(x-\mu_X)^2}{\sigma_X^2} - \frac{2\rho(x-\mu_X)(y-\mu_Y)}{\sigma_X\sigma_Y} + \frac{(y-\mu_Y)^2}{\sigma_Y^2}\right]\right\} \quad (5\text{-}30)$$

for $-\infty < x < \infty$ and $-\infty < y < \infty$, with parameters $\sigma_X > 0$, $\sigma_Y > 0$, $-\infty < \mu_X < \infty$, $-\infty < \mu_Y < \infty$, and $-1 < \rho < 1$.

The result that $f_{XY}(x, y; \sigma_X, \sigma_Y, \mu_X, \mu_Y, \rho)$ integrates to 1 is left as an exercise. Also, the bivariate normal probability density function is positive over the entire plane of real numbers.

Two examples of bivariate normal distributions are illustrated in Fig. 5-17 along with corresponding contour plots. Each curve on the contour plots is a set of points for which the probability density function is constant. As seen in the contour plots, the bivariate normal probability density function is constant on ellipses in the (x, y) plane. (We can consider a circle to be a special case of an ellipse.) The center of each ellipse is at the point (μ_X, μ_Y). If $\rho > 0$ $(\rho < 0)$, the major axis of each ellipse has positive (negative) slope, respectively. If $\rho = 0$, the major axis of the ellipse is aligned with either the x or y coordinate axis.

EXAMPLE 5-30

The joint probability density function $f_{XY}(x, y) = \frac{1}{\sqrt{2\pi}} e^{-0.5(x^2+y^2)}$ is a special case of a bivariate normal distribution with $\sigma_X = 1$, $\sigma_Y = 1$, $\mu_X = 0$, $\mu_Y = 0$, and $\rho = 0$. This probability density function is illustrated in Fig. 5-18. Notice that the contour plot consists of concentric circles about the origin.

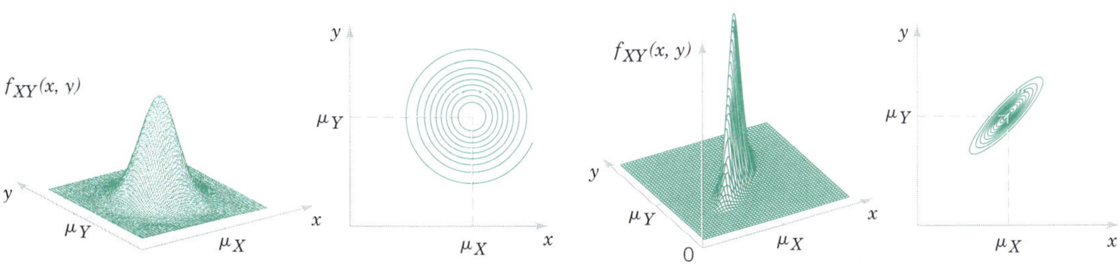

Figure 5-17 Examples of bivariate normal distributions.

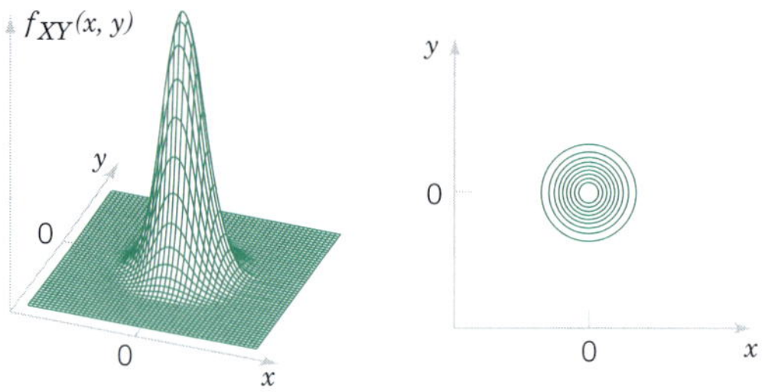

Figure 5-18 Bivariate normal probability density function with $\sigma_X = 1$, $\sigma_Y = 1$, $\rho = 0$, $\mu_X = 0$, and $\mu_Y = 0$.

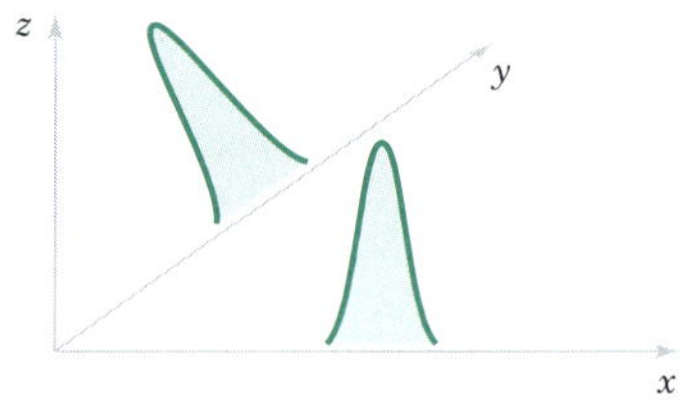

Figure 5-19 Marginal probability density functions of a bivariate normal distribution.

The following results can be shown for a bivariate normal distribution. The details are left as an exercise.

Marginal Distributions of Bivariate Normal Random Variables

If X and Y have a bivariate normal distribution with joint probability density $f_{XY}(x, y; \sigma_X, \sigma_Y, \mu_X, \mu_Y, \rho)$, the **marginal probability distributions** of X and Y are normal with means μ_X and μ_Y and standard deviations σ_X and σ_Y, respectively. (5-31)

Figure 5-19 illustrates that the marginal probability distributions of X and Y are normal.

Conditional Distribution of Bivariate Normal Random Variables

If X and Y have a bivariate normal distribution with joint probability density $f_{XY}(x, y; \sigma_X, \sigma_Y, \mu_X, \mu_Y, \rho)$, the conditional probability distribution of Y given $X = x$ is normal with mean

$$\mu_{Y|x} = \mu_Y - \mu_X \rho \frac{\sigma_Y}{\sigma_X} + \frac{\sigma_Y}{\sigma_X} \rho x$$

and variance

$$\sigma^2_{Y|x} = \sigma^2_Y (1 - \rho^2)$$

Furthermore, as the notation suggests, ρ represents the correlation between X and Y. The following result is left as an exercise.

Correlation of Bivariate Normal Random Variables

If X and Y have a bivariate normal distribution with joint probability density function $f_{XY}(x, y; \sigma_X, \sigma_Y, \mu_X, \mu_Y, \rho)$, the correlation between X and Y is ρ. (5-32)

The contour plots in Fig. 5-17 illustrate that as ρ moves from zero (left graph) to 0.9 (right graph), the ellipses narrow around the major axis. The probability is more concentrated about a line in the (x, y) plane and graphically displays greater correlation between the variables. If $\rho = -1$ or $+1$, all the probability is concentrated on a line in the (x, y) plane. That is, the probability that X and Y assume a value that is not on the line is zero. In this case, the bivariate normal probability density is not defined.

In general, zero correlation does not imply independence. But in the special case that X and Y have a bivariate normal distribution, if $\rho = 0$, X and Y are independent. The details are left as an exercise.

For Bivariate Normal Random Variables Zero Correlation Implies Independence

> If X and Y have a bivariate normal distribution with $\rho = 0$, X and Y are independent. (5-33)

An important use of the bivariate normal distribution is to calculate probabilities involving two correlated normal random variables.

EXAMPLE 5-31
Injection-Molded Part

Suppose that the X and Y dimensions of an injection-molded part have a bivariate normal distribution with $\sigma_X = 0.04$, $\sigma_Y = 0.08$, $\mu_X = 3.00$, $\mu_Y = 7.70$, and $\rho = 0.8$. Then, the probability that a part satisfies both specifications is

$$P(2.95 < X < 3.05, 7.60 < Y < 7.80)$$

This probability can be obtained by integrating $f_{XY}(x, y; \sigma_X, \sigma_Y, \mu_X \mu_Y, \rho)$ over the region $2.95 < x < 3.05$ and $7.60 < y < 7.80$, as shown in Fig. 5-7. Unfortunately, there is often no closed-form solution to probabilities involving bivariate normal distributions. In this case, the integration must be done numerically.

EXERCISES FOR SECTION 5-4

5-46. Let X and Y represent concentration and viscosity of a chemical product. Suppose X and Y have a bivariate normal distribution with $\sigma_X = 4$, $\sigma_Y = 1$, $\mu_X = 2$, and $\mu_Y = 1$. Draw a rough contour plot of the joint probability density function for each of the following values for ρ:
(a) $\rho = 0$ (b) $\rho = 0.8$
(c) $\rho = -0.8$

5-47. Suppose X and Y have a bivariate normal distribution with $\sigma_X = 0.04$, $\sigma_Y = 0.08$, $\mu_X = 3.00$, $\mu_Y = 7.70$, and $\rho = 0$.
Determine the following:
(a) $P(2.95 < X < 3.05)$
(b) $P(7.60 < Y < 7.80)$
(c) $P(2.95 < X < 3.05, 7.60 < Y < 7.80)$

5-48. In an acid-base titration, a base or acid is gradually added to the other until they have completely neutralized each other. Let X and Y denote the milliliters of acid and base needed for equivalence, respectively. Assume X and Y have a bivariate normal distribution with $\sigma_X = 5$ mL, $\sigma_Y = 2$ mL, $\mu_X = 120$ mL, $\mu_Y = 100$ mL, and $\rho = 0.6$.
Determine the following:
(a) Covariance between X and Y
(b) Marginal probability distribution of X
(c) $P(X < 116)$
(d) Conditional probability distribution of X given that $Y = 102$
(e) $P(X < 116 \mid Y = 102)$

5-49. In the manufacture of electroluminescent lamps, several different layers of ink are deposited onto a plastic substrate. The thickness of these layers is critical if specifications regarding the final color and intensity of light are to be met. Let X and Y denote the thickness of two different layers of ink. It is known that X is normally distributed with a mean of 0.1 millimeter and a standard deviation of 0.00031 millimeter, and Y is normally distributed with a mean of 0.23 millimeter and a standard deviation of 0.00017 millimeter. The value of ρ for these variables is equal to zero. Specifications call for a lamp to have a thickness of the ink corresponding to X in the range of 0.099535 to 0.100465 millimeters and Y in the range of 0.22966 to 0.23034 millimeters. What is the probability that a randomly selected lamp will conform to specifications?

5-50. Suppose that X and Y have a bivariate normal distribution with joint probability density function $f_{XY}(x, y; \sigma_X, \sigma_Y, \mu_X, \mu_Y, \rho)$.
(a) Show that the conditional distribution of Y, given that $X = x$ is normal.
(b) Determine $E(Y|X = x)$.
(c) Determine $V(Y|X = x)$.

5-51. If X and Y have a bivariate normal distribution with $\rho = 0$, show that X and Y are independent.

5-52. Show that the probability density function $f_{XY}(x, y; \sigma_X, \sigma_Y, \mu_X, \mu_Y, \rho)$ of a bivariate normal distribution integrates to one. [*Hint:* Complete the square in the exponent and use the fact that the integral of a normal probability density function for a single variable is 1.]

5-53. If X and Y have a bivariate normal distribution with joint probability density $f_{XY}(x, y; \sigma_X, \sigma_Y, \mu_X, \mu_Y, \rho)$, show that the marginal probability distribution of X is normal with mean μ_X and standard deviation σ_X. [*Hint:* Complete the square in the exponent and use the fact that the integral of a normal probability density function for a single variable is 1.]

5-5 LINEAR FUNCTIONS OF RANDOM VARIABLES

A random variable is sometimes defined as a function of one or more random variables. In this section, results for linear functions are highlighted because of their importance in the remainder of the book. For example, if the random variables X_1 and X_2 denote the length and width, respectively, of a manufactured part, $Y = 2X_1 + 2X_2$ is a random variable that represents the perimeter of the part. As another example, recall that the negative binomial random variable was represented as the sum of several geometric random variables.

In this section, we develop results for random variables that are linear combinations of random variables.

Linear Combination

Given random variables $X_1, X_2, \ldots, X_p$ and constants $c_1, c_2, \ldots, c_p$,

$$Y = c_1X_1 + c_2X_2 + \cdots + c_pX_p \tag{5-34}$$

is a **linear combination** of $X_1, X_2, \ldots, X_p$.

Now, $E(Y)$ can be found from the joint probability distribution of $X_1, X_2, \ldots, X_p$ as follows. Assume $X_1, X_2, \ldots, X_p$ are continuous random variables. An analogous calculation can be used for discrete random variables.

$$\begin{aligned}
E(Y) &= \int_{-\infty}^{\infty}\int_{-\infty}^{\infty}\cdots\int_{-\infty}^{\infty} (c_1x_1 + c_2x_2 + \cdots + c_px_p) f_{X_1X_2\ldots X_p}(x_1, x_2, \ldots, x_p)\, dx_1\, dx_2 \ldots dx_p \\
&= c_1 \int_{-\infty}^{\infty}\int_{-\infty}^{\infty}\cdots\int_{-\infty}^{\infty} x_1 f_{X_1X_2\ldots X_p}(x_1, x_2, \ldots, x_p)\, dx_1\, dx_2 \ldots dx_p \\
&\quad + c_2 \int_{-\infty}^{\infty}\int_{-\infty}^{\infty}\cdots\int_{-\infty}^{\infty} x_2 f_{X_1X_2\ldots X_p}(x_1, x_2, \ldots, x_p)\, dx_1\, dx_2 \ldots dx_p + , \ldots, \\
&\quad + c_p \int_{-\infty}^{\infty}\int_{-\infty}^{\infty}\cdots\int_{-\infty}^{\infty} x_p f_{X_1X_2\ldots X_p}(x_1, x_2, \ldots, x_p)\, dx_1\, dx_2 \ldots dx_p
\end{aligned}$$

By using Equation 5-22 for each of the terms in this expression, we obtain the following.

Mean of a Linear Function

If $Y = c_1X_1 + c_2X_2 + \cdots + c_pX_p$,

$$E(Y) = c_1E(X_1) + c_2E(X_2) + \cdots + c_pE(X_p) \tag{5-35}$$

Furthermore, it is left as an exercise to show the following.

Variance of a Linear Function

If $X_1, X_2, \ldots, X_p$ are random variables, and $Y = c_1X_1 + c_2X_2 + \cdots + c_pX_p$, then in general

$$V(Y) = c_1^2V(X_1) + c_2^2V(X_2) + \cdots + c_p^2V(X_p) + 2\sum_{i<j}\sum c_ic_j \operatorname{cov}(X_i, X_j) \quad (5\text{-}36)$$

If $X_1, X_2, \ldots, X_p$ are **independent,**

$$V(Y) = c_1^2V(X_1) + c_2^2V(X_2) + \cdots + c_p^2V(X_p) \quad (5\text{-}37)$$

Note that the result for the variance in Equation 5-37 requires the random variables to be independent. To see why the independence is important, consider the following simple example. Let X_1 denote any random variable and define $X_2 = -X_1$. Clearly, X_1 and X_2 are not independent. In fact, $\rho_{XY} = -1$. Now, $Y = X_1 + X_2$ is 0 with probability 1. Therefore, $V(Y) = 0$, regardless of the variances of X_1 and X_2.

EXAMPLE 5-32 Negative Binomial Distribution

In Chapter 3, we found that if Y is a negative binomial random variable with parameters p and r, $Y = X_1 + X_2 + \cdots + X_r$, where each X_i is a geometric random variable with parameter p and they are independent. Therefore, $E(X_i) = 1/p$ and $V(X_i) = (1 - p)/p^2$. From Equation 5-35, $E(Y) = r/p$ and from Equation 5-37, $V(Y) = r(1 - p)/p^2$.

An approach similar to the one applied in the above example can be used to verify the formulas for the mean and variance of an Erlang random variable in Chapter 4. An important use of equation 5-37 is in **error propagation** and this is presented in the following example.

EXAMPLE 5-33 Error Propagation

A semiconductor product consists of three layers. If the variances in thickness of the first, second, and third layers are 25, 40, and 30 nanometers squared, what is the variance of the thickness of the final product?

Let X_1, X_2, X_3, and X be random variables that denote the thickness of the respective layers, and the final product. Then

$$X = X_1 + X_2 + X_3$$

The variance of X is obtained from equaion 5-37

$$V(X) = V(X_1) + V(X_2) + V(X_3) = 25 + 40 + 30 = 95 \text{ nm}^2$$

Consequently, the standard deviation of thickness of the final product is $95^{1/2} = 9.75$ nm and this shows how the variation in each layer is propagated to the final product.

The particular linear function that represents the average of p random variables, with identical means and variances, is used quite often in subsequent chapters. We highlight the results for this special case.

Mean and Variance of an Average

If $\overline{X} = (X_1 + X_2 + \cdots + X_p)/p$ with $E(X_i) = \mu$ for $i = 1, 2, \ldots, p$

$$E(\overline{X}) = \mu \tag{5-38a}$$

if $X_1, X_2, \ldots, X_p$ are also independent with $V(X_i) = \sigma^2$ for $i = 1, 2, \ldots, p$,

$$V(\overline{X}) = \frac{\sigma^2}{p} \tag{5-38b}$$

The conclusion for $V(\overline{X})$ is obtained as follows. Using Equation 5-37, with $c_i = 1/p$ and $V(X_i) = \sigma^2$, yields

$$V(\overline{X}) = \underbrace{(1/p)^2\sigma^2 + \cdots + (1/p)^2\sigma^2}_{p \text{ terms}} = \sigma^2/p$$

Another useful result concerning linear functions of random variables is a **reproductive property** that holds for independent, normal random variables.

Reproductive Property of the Normal Distribution

If $X_1, X_2, \ldots, X_p$ are independent, normal random variables with $E(X_i) = \mu_i$ and $V(X_i) = \sigma_i^2$, for $i = 1, 2, \ldots, p$,

$$Y = c_1X_1 + c_2X_2 + \cdots + c_pX_p$$

is a normal random variable with

$$E(Y) = c_1\mu_1 + c_2\mu_2 + \cdots + c_p\mu_p$$

and

$$V(Y) = c_1^2\sigma_1^2 + c_2^2\sigma_2^2 + \cdots + c_p^2\sigma_p^2 \tag{5-39}$$

The mean and variance of Y follow from Equations 5-35 and 5-37. The fact that Y has a normal distribution can be obtained from supplemental material on moment-generating functions on the Web site for the book.

EXAMPLE 5-34
Linear Function of Independent Normal Random Variables

Let the random variables X_1 and X_2 denote the length and width, respectively, of a manufactured part. Assume that X_1 is normal with $E(X_1) = 2$ centimeters and standard deviation 0.1 centimeter and that X_2 is normal with $E(X_2) = 5$ centimeters and standard deviation 0.2 centimeter. Also, assume that X_1 and X_2 are independent. Determine the probability that the perimeter exceeds 14.5 centimeters.

Then, $Y = 2X_1 + 2X_2$ is a normal random variable that represents the perimeter of the part. We obtain, $E(Y) = 14$ centimeters and the variance of Y is

$$V(Y) = 4 \times 0.1^2 + 4 \times 0.2^2 = 0.2$$

Now,

$$P(Y > 14.5) = P[(Y - \mu_Y)/\sigma_Y > (14.5 - 14)/\sqrt{0.2}]$$
$$= P(Z > 1.12) = 0.13$$

EXAMPLE 5-35 Beverage Volume

Soft-drink cans are filled by an automated filling machine. The mean fill volume is 12.1 fluid ounces, and the standard deviation is 0.1 fluid ounce. Assume that the fill volumes of the cans are independent, normal random variables. What is the probability that the average volume of 10 cans selected from this process is less than 12 fluid ounces?

Let $X_1, X_2, \ldots, X_{10}$ denote the fill volumes of the 10 cans. The average fill volume (denoted as $\overline{X}$) is a normal random variable with

$$E(\overline{X}) = 12.1 \quad \text{and} \quad V(\overline{X}) = \frac{0.1^2}{10} = 0.001$$

Consequently,

$$P(\overline{X} < 12) = P\left[\frac{\overline{X} - \mu_{\overline{X}}}{\sigma_{\overline{X}}} < \frac{12 - 12.1}{\sqrt{0.001}}\right]$$
$$= P(Z < -3.16) = 0.00079$$

EXERCISES FOR SECTION 5-5

5-54. X and Y are independent, normal random variables with $E(X) = 0$, $V(X) = 4$, $E(Y) = 10$, and $V(Y) = 9$.

Determine the following:

(a) $E(2X + 3Y)$ (b) $V(2X + 3Y)$
(c) $P(2X + 3Y < 30)$ (d) $P(2X + 3Y < 40)$

5-55. X and Y are independent, normal random variables with $E(X) = 2$, $V(X) = 5$, $E(Y) = 6$, and $V(Y) = 8$.

Determine the following:

(a) $E(3X + 2Y)$ (b) $V(3X + 2Y)$
(c) $P(3X + 2Y < 18)$ (d) $P(3X + 2Y < 28)$

5-56. Suppose that the random variable X represents the length of a punched part in centimeters. Let Y be the length of the part in millimeters. If $E(X) = 5$ and $V(X) = 0.25$, what are the mean and variance of Y?

5-57. A plastic casing for a magnetic disk is composed of two halves. The thickness of each half is normally distributed with a mean of 2 millimeters and a standard deviation of 0.1 millimeter and the halves are independent.

(a) Determine the mean and standard deviation of the total thickness of the two halves.
(b) What is the probability that the total thickness exceeds 4.3 millimeters?

5-58. Making handcrafted pottery generally takes two major steps: wheel throwing and firing. The time of wheel throwing and the time of firing are normally distributed random variables with means of 40 min and 60 min and standard deviations of 2 min and 3 min, respectively.

(a) What is the probability that a piece of pottery will be finished within 95 min?
(b) What is the probability that it will take longer than 110 min?

5-59. In the manufacture of electroluminescent lamps, several different layers of ink are deposited onto a plastic substrate. The thickness of these layers is critical if specifications regarding the final color and intensity of light are to be met. Let X and Y denote the thickness of two different layers of ink. It is known that X is normally distributed with a mean of 0.1 millimeter and a standard deviation of 0.00031 millimeter and Y is also normally distributed with a mean of 0.23 millimeter and a standard deviation of 0.00017 millimeter. Assume that these variables are independent.

(a) If a particular lamp is made up of these two inks only, what is the probability that the total ink thickness is less than 0.2337 millimeter?
(b) A lamp with a total ink thickness exceeding 0.2405 millimeters lacks the uniformity of color demanded by the customer. Find the probability that a randomly selected lamp fails to meet customer specifications.

5-60. The width of a casing for a door is normally distributed with a mean of 24 inches and a standard deviation of 1/8 inch. The width of a door is normally distributed with a mean of 23 and 7/8 inches and a standard deviation of 1/16 inch. Assume independence.

(a) Determine the mean and standard deviation of the difference between the width of the casing and the width of the door.

(b) What is the probability that the width of the casing minus the width of the door exceeds 1/4 inch?

(c) What is the probability that the door does not fit in the casing?

5-61. An article in *Knee Surgery Sports Traumatology, Arthroscopy*, "Effect of provider volume on resource utilization for surgical procedures," (2005, Vol. 13, pp. 273–279) showed a mean time of 129 minutes and a standard deviation of 14 minutes for ACL reconstruction surgery for high-volume hospitals (with more than 300 such surgeries per year). If a high-volume hospital needs to schedule 10 surgeries, what is the mean and variance of the total time to complete these surgeries? Assume the times of the surgeries are independent and normally distributed.

5-62. Soft-drink cans are filled by an automated filling machine and the standard deviation is 0.5 fluid ounce. Assume that the fill volumes of the cans are independent, normal random variables.

(a) What is the standard deviation of the average fill volume of 100 cans?

(b) If the mean fill volume is 12.1 ounces, what is the probability that the average fill volume of the 100 cans is below 12 fluid ounces?

(c) What should the mean fill volume equal so that the probability that the average of 100 cans is below 12 fluid ounces is 0.005?

(d) If the mean fill volume is 12.1 fluid ounces, what should the standard deviation of fill volume equal so that the probability that the average of 100 cans is below 12 fluid ounces is 0.005?

(e) Determine the number of cans that need to be measured such that the probability that the average fill volume is less than 12 fluid ounces is 0.01.

5-63. The photoresist thickness in semiconductor manufacturing has a mean of 10 micrometers and a standard deviation of 1 micrometer. Assume that the thickness is normally distributed and that the thicknesses of different wafers are independent.

(a) Determine the probability that the average thickness of 10 wafers is either greater than 11 or less than 9 micrometers.

(b) Determine the number of wafers that needs to be measured such that the probability that the average thickness exceeds 11 micrometers is 0.01.

(c) If the mean thickness is 10 micrometers, what should the standard deviation of thickness equal so that the probability that the average of 10 wafers is either greater than 11 or less than 9 micrometers is 0.001?

5-64. Assume that the weights of individuals are independent and normally distributed with a mean of 160 pounds and a standard deviation of 30 pounds. Suppose that 25 people squeeze into an elevator that is designed to hold 4300 pounds.

(a) What is the probability that the load (total weight) exceeds the design limit?

(b) What design limit is exceeded by 25 occupants with probability 0.0001?

5-65. Weights of parts are normally distributed with variance σ^2. Measurement error is normally distributed with mean zero and variance $0.5\sigma^2$, independent of the part weights, and adds to the part weight. Upper and lower specifications are centered at 3σ about the process mean.

(a) Without measurement error, what is the probability that a part exceeds the specifications?

(b) With measurement error, what is the probability that a part is measured as beyond specifications? Does this imply it is truly beyond specifications?

(c) What is the probability that a part is measured beyond specifications if the true weight of the part is one σ below the upper specification limit?

5-66. A U-shaped component is to be formed from the three parts A, B, and C. The picture is shown in Fig. 5-20. The length of A is normally distributed with a mean of 10 millimeters and a standard deviation of 0.1 millimeter. The thickness of parts B and C is normally distributed with a mean of 2 millimeters and a standard deviation of 0.05 millimeter. Assume all dimensions are independent.

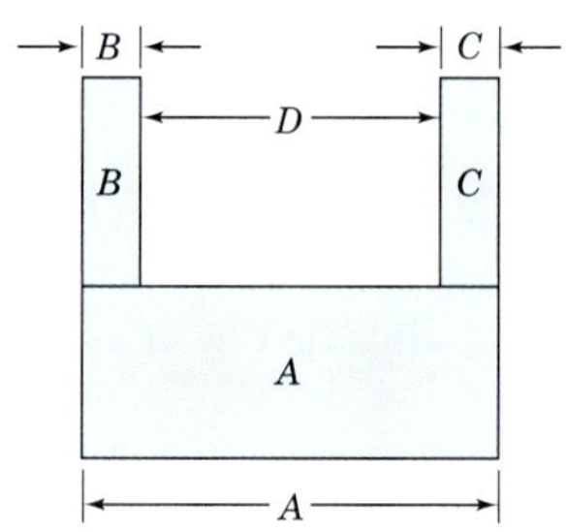

Figure 5-20 Figure for the U-shaped component.

(a) Determine the mean and standard deviation of the length of the gap D.

(b) What is the probability that the gap D is less than 5.9 millimeters?

5-6 GENERAL FUNCTIONS OF RANDOM VARIABLES

In many situations in statistics, it is necessary to derive the probability distribution of a function of one or more random variables. In this section, we present some results that are helpful in solving this problem.

Suppose that X is a discrete random variable with probability distribution $f_X(x)$. Let $Y = h(X)$ be a function of X that defines a one-to-one transformation between the values of X and Y, and we wish to find the probability distribution of Y. By a one-to-one transformation, we mean that each value x is related to one and only one value of $y = h(x)$ and that each value of y is related to one and only one value of x, say, $x = u(y)$, where $u(y)$ is found by solving $y = h(x)$ for x in terms of y.

Now, the random variable Y takes on the value y when X takes on the value $u(y)$. Therefore, the probability distribution of Y is

$$f_Y(y) = P(Y = y) = P[X = u(y)] = f_X[u(y)]$$

We may state this result as follows.

General Function of a Discrete Random Variable

Suppose that X is a **discrete** random variable with probability distribution $f_X(x)$. Let $Y = h(X)$ define a one-to-one transformation between the values of X and Y so that the equation $y = h(x)$ can be solved uniquely for x in terms of y. Let this solution be $x = u(y)$. Then the probability mass function of the random variable Y is

$$f_Y(y) = f_X[u(y)] \tag{5-40}$$

EXAMPLE 5-36
Function of a Discrete Random Variable

Let X be a geometric random variable with probability distribution

$$f_X(x) = p(1 - p)^{x-1}, \qquad x = 1, 2, \ldots$$

Find the probability distribution of $Y = X^2$.

Since $X \geq 0$, the transformation is one to one; that is, $y = x^2$ and $x = \sqrt{y}$. Therefore, Equation 5-40 indicates that the distribution of the random variable Y is

$$f_Y(y) = f(\sqrt{y}) = p(1 - p)^{\sqrt{y}-1}, \qquad y = 1, 4, 9, 16, \ldots$$

We now consider the situation where the random variables are continuous. Let $Y = h(X)$, with X continuous and the transformation is one to one.

General Function of a Continuous Random Variable

Suppose that X is a **continuous** random variable with probability distribution $f_X(x)$. The function $Y = h(X)$ is a one-to-one transformation between the values of Y and X so that the equation $y = h(x)$ can be uniquely solved for x in terms of y. Let this solution be $x = u(y)$. The probability distribution of Y is

$$f_Y(y) = f_X[u(y)]\,|J| \tag{5-41}$$

where $J = u'(y)$ is called the **Jacobian** of the transformation and the absolute value of J is used.

Equation 5-41 is shown as follows. Let the function $y = h(x)$ be an increasing function of x. Now

$$P(Y \le a) = P[X \le u(a)]$$
$$= \int_{-\infty}^{u(a)} f_X(x)\, dx$$

If we change the variable of integration from x to y by using $x = u(y)$, we obtain $dx = u'(y)\, dy$ and then

$$P(Y \le a) = \int_{-\infty}^{a} f_X[u(y)]u'(y)\, dy$$

Since the integral gives the probability that $Y \le a$ for all values of a contained in the feasible set of values for y, $f_X[u(y)]u'(y)$ must be the probability density of Y. Therefore, the probability distribution of Y is

$$f_Y(y) = f_X[u(y)]u'(y) = f_X[u(y)]J$$

If the function $y = h(x)$ is a decreasing function of x, a similar argument holds.

EXAMPLE 5-37
Function of a Continuous Random Variable

Let X be a continuous random variable with probability distribution

$$f_X(x) = \frac{x}{8}, \quad 0 \le x < 4$$

Find the probability distribution of $Y = h(X) = 2X + 4$.

Note that $y = h(x) = 2x + 4$ is an increasing function of x. The inverse solution is $x = u(y) = (y - 4)/2$, and from this we find the Jacobian to be $J = u'(y) = dx/dy = 1/2$. Therefore, from Equation 5-41 the probability distribution of Y is

$$f_Y(y) = \frac{(y-4)/2}{8}\left(\frac{1}{2}\right) = \frac{y-4}{32}, \quad 4 \le y \le 12$$

EXERCISES FOR SECTION 5-6

5-67. Suppose that X is a random variable with probability distribution

$$f_X(x) = 1/4, \quad x = 1, 2, 3, 4$$

Find the probability distribution of $Y = 2X + 1$.

5-68. Let X be a binomial random variable with $p = 0.25$ and $n = 3$. Find the probability distribution of the random variable $Y = X^2$.

5-69. Suppose that X is a continuous random variable with probability distribution

$$f_X(x) = \frac{x}{18}, \quad 0 \le x \le 6$$

(a) Find the probability distribution of the random variable $Y = 2X + 10$.
(b) Find the expected value of Y.

5-70. Suppose that X has a uniform probability distribution

$$f_X(x) = 1, \quad 0 \le x \le 1$$

Show that the probability distribution of the random variable $Y = -2 \ln X$ is chi-squared with two degrees of freedom.

5-71. A random variable X has the following probability distribution:

$$f_X(x) = e^{-x}, \quad x \ge 0$$

(a) Find the probability distribution for $Y = X^2$.
(b) Find the probability distribution for $Y = X^{1/2}$.
(c) Find the probability distribution for $Y = \ln X$.

5-72. The velocity of a particle in a gas is a random variable V with probability distribution

$$f_V(v) = av^2e^{-bv} \qquad v > 0$$

where b is a constant that depends on the temperature of the gas and the mass of the particle.
(a) Find the value of the constant a.
(b) The kinetic energy of the particle is $W = mV^2/2$. Find the probability distribution of W.

5-73. Suppose that X has the probability distribution

$$f_X(x) = 1, \qquad 1 \le x \le 2$$

Find the probability distribution of the random variable $Y = e^X$.

5-74. The random variable X has the probability distribution

$$f_X(x) = \frac{x}{8}, \qquad 0 \le x \le 4$$

Find the probability distribution of $Y = (X - 2)^2$.

Supplemental Exercises

5-75. Show that the following function satisfies the properties of a joint probability mass function:

x	y	$f(x, y)$
0	0	1/4
0	1	1/8
1	0	1/8
1	1	1/4
2	2	1/4

Determine the following:
(a) $P(X < 0.5, Y < 1.5)$ (b) $P(X \le 1)$
(c) $P(X < 1.5)$ (d) $P(X > 0.5, Y < 1.5)$
(e) Determine $E(X)$, $E(Y)$, $V(X)$, and $V(Y)$.
(f) Marginal probability distribution of the random variable X
(g) Conditional probability distribution of Y given that $X = 1$
(h) $E(Y|X = 1)$
(i) Are X and Y independent? Why or why not?
(j) Calculate the correlation between X and Y.

5-76. The percentage of people given an antirheumatoid medication who suffer severe, moderate, or minor side effects are 10, 20, and 70%, respectively. Assume that people react independently and that 20 people are given the medication. Determine the following:
(a) The probability that 2, 4, and 14 people will suffer severe, moderate, or minor side effects, respectively
(b) The probability that no one will suffer severe side effects
(c) The mean and variance of the number of people that will suffer severe side effects
(d) What is the conditional probability distribution of the number of people who suffer severe side effects given that 19 suffer minor side effects?
(e) What is the conditional mean of the number of people who suffer severe side effects given that 19 suffer minor side effects?

5-77. The backoff torque required to remove bolts in a steel plate is rated as high, moderate, or low. Historically, the probability of a high, moderate, or low rating is 0.6, 0.3, or 0.1, respectively. Suppose that 20 bolts are evaluated and that the torque ratings are independent.
(a) What is the probability that 12, 6, and 2 bolts are rated as high, moderate, and low, respectively?
(b) What is the marginal distribution of the number of bolts rated low?
(c) What is the expected number of bolts rated low?
(d) What is the probability that the number of bolts rated low is greater than two?
(e) What is the conditional distribution of the number of bolts rated low given that 16 bolts are rated high?
(f) What is the conditional expected number of bolts rated low given that 16 bolts are rated high?
(g) Are the numbers of bolts rated high and low independent random variables?

5-78. To evaluate the technical support from a computer manufacturer, the number of rings before a call is answered by a service representative is tracked. Historically, 70% of the calls are answered in two rings or less, 25% are answered in three or four rings, and the remaining calls require five rings or more. Suppose you call this manufacturer 10 times and assume that the calls are independent.
(a) What is the probability that eight calls are answered in two rings or less, one call is answered in three or four rings, and one call requires five rings or more?
(b) What is the probability that all 10 calls are answered in four rings or less?
(c) What is the expected number of calls answered in four rings or less?
(d) What is the conditional distribution of the number of calls requiring five rings or more given that eight calls are answered in two rings or less?
(e) What is the conditional expected number of calls requiring five rings or more given that eight calls are answered in two rings or less?
(f) Are the number of calls answered in two rings or less and the number of calls requiring five rings or more independent random variables?

5-79. Determine the value of c such that the function $f(x, y) = cx^2y$ for $0 < x < 3$ and $0 < y < 2$ satisfies the properties of a joint probability density function.

Determine the following:

(a) $P(X < 1, Y < 1)$ (b) $P(X < 2.5)$
(c) $P(1 < Y < 2.5)$ (d) $P(X > 2, 1 < Y < 1.5)$
(e) $E(X)$ (f) $E(Y)$
(g) Marginal probability distribution of the random variable X.
(h) Conditional probability distribution of Y given that $X = 1$.
(i) Conditional probability distribution of X given that $Y = 1$.

5-80. The joint distribution of the continuous random variables X, Y, and Z is constant over the region $x^2 + y^2 \leq 1$, $0 < z < 4$. Determine the following:

(a) $P(X^2 + Y^2 \leq 0.5)$
(b) $P(X^2 + Y^2 \leq 0.5, Z < 2)$
(c) Joint conditional probability density function of X and Y given that $Z = 1$?
(d) Marginal probability density function of X?
(e) Conditional mean of Z given that $X = 0$ and $Y = 0$.
(f) Conditional mean of Z given that $X = x$ and $Y = y$.

5-81. Suppose that X and Y are independent, continuous uniform random variables for $0 < x < 1$ and $0 < y < 1$. Use the joint probability density function to determine the probability that $|X - Y| < 0.5$.

5-82. The lifetimes of six major components in a copier are independent exponential random variables with means of 8000, 10,000, 10,000, 20,000, 20,000, and 25,000 hours, respectively.

(a) What is the probability that the lifetimes of all the components exceed 5000 hours?
(b) What is the probability that at least one component lifetime exceeds 25,000 hours?

5-83. Contamination problems in semiconductor manufacturing can result in a functional defect, a minor defect, or no defect in the final product. Suppose that 20, 50, and 30% of the contamination problems result in functional, minor, and no defects, respectively. Assume that the defects of 10 contamination problems are independent.

(a) What is the probability that the 10 contamination problems result in two functional defects and five minor defects?
(b) What is the distribution of the number of contamination problems that result in no defects?
(c) What is the expected number of contamination problems that result in no defects?

5-84. The weight of adobe bricks for construction is normally distributed with a mean of 3 pounds and a standard deviation of 0.25 pound. Assume that the weights of the bricks are independent and that a random sample of 25 bricks is chosen.

(a) What is the probability that the mean weight of the sample is less than 2.95 pounds?
(b) What value will the mean weight exceed with probability 0.99?

5-85. The length and width of panels used for interior doors (in inches) are denoted as X and Y, respectively. Suppose that X and Y are independent, continuous uniform random variables for $17.75 < x < 18.25$ and $4.75 < y < 5.25$, respectively.

(a) By integrating the joint probability density function over the appropriate region, determine the probability that the area of a panel exceeds 90 square inches.
(b) What is the probability that the perimeter of a panel exceeds 46 inches?

5-86. The weight of a small candy is normally distributed with a mean of 0.1 ounce and a standard deviation of 0.01 ounce. Suppose that 16 candies are placed in a package and that the weights are independent.

(a) What are the mean and variance of package net weight?
(b) What is the probability that the net weight of a package is less than 1.6 ounces?
(c) If 17 candies are placed in each package, what is the probability that the net weight of a package is less than 1.6 ounces?

5-87. The time for an automated system in a warehouse to locate a part is normally distributed with a mean of 45 seconds and a standard deviation of 30 seconds. Suppose that independent requests are made for 10 parts.

(a) What is the probability that the average time to locate 10 parts exceeds 60 seconds?
(b) What is the probability that the total time to locate 10 parts exceeds 600 seconds?

5-88. A mechanical assembly used in an automobile engine contains four major components. The weights of the components are independent and normally distributed with the following means and standard deviations (in ounces):

Component	Mean	Standard Deviation
Left case	4	0.4
Right case	5.5	0.5
Bearing assembly	10	0.2
Bolt assembly	8	0.5

(a) What is the probability that the weight of an assembly exceeds 29.5 ounces?
(b) What is the probability that the mean weight of eight independent assemblies exceeds 29 ounces?

5-89. Suppose X and Y have a bivariate normal distribution with $\sigma_X = 4$, $\sigma_Y = 1$, $\mu_X = 4$, $\mu_Y = 4$, and $\rho = -0.2$. Draw a rough contour plot of the joint probability density function.

5-90. If $$f_{XY}(x, y) = \frac{1}{1.2\pi} \exp\left\{\frac{-1}{0.72}\left[(x-1)^2 - 1.6(x-1)(y-2) + (y-2)^2\right]\right\}$$

determine $E(X)$, $E(Y)$, $V(X)$, $V(Y)$, and ρ by reorganizing the parameters in the joint probability density function.

5-91. The permeability of a membrane used as a moisture barrier in a biological application depends on the thickness of two integrated layers. The layers are normally distributed with means of 0.5 and 1 millimeters, respectively. The standard deviations of layer thickness are 0.1 and 0.2 millimeters, respectively. The correlation between layers is 0.7.

(a) Determine the mean and variance of the total thickness of the two layers.

(b) What is the probability that the total thickness is less than 1 millimeter?

(c) Let X_1 and X_2 denote the thickness of layers 1 and 2, respectively. A measure of performance of the membrane is a function $2X_1 + 3X_2$ of the thickness. Determine the mean and variance of this performance measure.

5-92. The permeability of a membrane used as a moisture barrier in a biological application depends on the thickness of three integrated layers. Layers 1, 2, and 3 are normally distributed with means of 0.5, 1, and 1.5 millimeters, respectively. The standard deviations of layer thickness are 0.1, 0.2, and 0.3, respectively. Also, the correlation between layers 1 and 2 is 0.7, between layers 2 and 3 is 0.5, and between layers 1 and 3 is 0.3.

(a) Determine the mean and variance of the total thickness of the three layers.

(b) What is the probability that the total thickness is less than 1.5 millimeters?

5-93. A small company is to decide what investments to use for cash generated from operations. Each investment has a mean and standard deviation associated with the percentage gain. The first security has a mean percentage gain of 5% with a standard deviation of 2%, and the second security provides the same mean of 5% with a standard deviation of 4%. The securities have a correlation of −0.5, so there is a negative correlation between the percentage returns. If the company invests two million dollars with half in each security, what is the mean and standard deviation of the percentage return? Compare the standard deviation of this strategy to one that invests the two million dollars into the first security only.

MIND-EXPANDING EXERCISES

5-94. Show that if $X_1, X_2, \ldots, X_p$ are independent, continuous random variables, $P(X_1 \in A_1, X_2 \in A_2, \ldots, X_p \in A_p) = P(X_1 \in A_1)P(X_2 \in A_2)\ldots P(X_p \in A_p)$ for any regions $A_1, A_2, \ldots, A_p$ in the range of $X_1, X_2, \ldots, X_p$ respectively.

5-95. Show that if $X_1, X_2, \ldots, X_p$ are independent random variables and $Y = c_1X_1 + c_2X_2 + \cdots + c_pX_p$,

$$V(Y) = c_1^2V(X_1) + c_2^2V(X_2) + \cdots + c_p^2V(X_p)$$

You can assume that the random variables are continuous.

5-96. Suppose that the joint probability function of the continuous random variables X and Y is constant on the rectangle $0 < x < a$, $0 < y < b$. Show that X and Y are independent.

5-97. Suppose that the range of the continuous variables X and Y is $0 < x < a$ and $0 < y < b$. Also suppose that the joint probability density function $f_{XY}(x, y) = g(x)h(y)$, where $g(x)$ is a function only of x and $h(y)$ is a function only of y. Show that X and Y are independent.

IMPORTANT TERMS AND CONCEPTS

Bivariate distribution
Bivariate normal distribution
Conditional mean
Conditional probability density function
Conditional probability mass function
Conditional variance
Contour plots
Correlation
Covariance
Error propagation
General functions of random variables
Independence
Joint probability density function
Joint probability mass function
Linear functions of random variables
Marginal probability distribution
Multinomial distribution
Reproductive property of the normal distribution

6 An Explanation of Random Sampling and Data Description

CHAPTER OUTLINE

LEARNING OBJECTIVES

After careful study of this chapter you should be able to do the following:

1. Compute and interpret the sample mean, sample variance, sample standard deviation, sample median, and sample range
2. Explain the concepts of sample mean, sample variance, population mean, and population variance
3. Construct and interpret visual data displays, including the stem-and-leaf display, the histogram, and the box plot
4. Explain the concept of random sampling
5. Construct and interpret normal probability plots
6. Explain how to use box plots and other data displays to visually compare two or more samples of data
7. Know how to use simple time series plots to visually display the important features of time-oriented data

6-1 NUMERICAL SUMMARIES

Well-constructed data summaries and displays are essential to good statistical thinking, because they can focus the engineer on important features of the data or provide insight about the type of model that should be used in solving the problem. The computer has become an important tool in the presentation and analysis of data. While many statistical techniques require only a hand-held calculator, much time and effort may be required by this approach, and a computer will perform the tasks much more efficiently.

Most statistical analysis is done using a prewritten library of statistical programs. The user enters the data and then selects the types of analysis and output displays that are of interest. Statistical software packages are available for both mainframe machines and personal computers. We will present examples of output from Minitab (one of the most widely-used PC packages), throughout the book. We will not discuss the hands-on use of Minitab for entering and editing data or using commands. This information is found in the software documentation.

We often find it useful to describe data features **numerically.** For example, we can characterize the location or central tendency in the data by the ordinary arithmetic average or mean. Because we almost always think of our data as a sample, we will refer to the arithmetic mean as the **sample mean.**

Sample Mean

If the n observations in a sample are denoted by $x_1, x_2, \ldots, x_n$, the **sample mean** is

$$\bar{x} = \frac{x_1 + x_2 + \cdots + x_n}{n} = \frac{\sum_{i=1}^{n} x_i}{n} \tag{6-1}$$

EXAMPLE 6-1
Sample Mean

Let's consider the eight observations collected from the prototype engine connectors from Chapter 1. The eight observations are $x_1 = 12.6$, $x_2 = 12.9$, $x_3 = 13.4$, $x_4 = 12.3$, $x_5 = 13.6$, $x_6 = 13.5$, $x_7 = 12.6$, and $x_8 = 13.1$. The sample mean is

$$\bar{x} = \frac{x_1 + x_2 + \cdots + x_n}{n} = \frac{\sum_{i=1}^{8} x_i}{8} = \frac{12.6 + 12.9 + \cdots + 13.1}{8}$$

$$= \frac{104}{8} = 13.0 \text{ pounds}$$

A physical interpretation of the sample mean as a measure of location is shown in the dot diagram of the pull-off force data. See Figure 6-1. Notice that the sample mean $\bar{x} = 13.0$ can be thought of as a "balance point." That is, if each observation represents 1 pound of mass placed at the point on the x-axis, a fulcrum located at $\bar{x}$ would exactly balance this system of weights.

The sample mean is the average value of all the observations in the data set. Usually, these data are a **sample** of observations that have been selected from some larger **population** of observations. Here the population might consist of all the connectors that will be manufactured and sold to customers. Recall that this type of population is called a **conceptual** or **hypothetical population,** because it does not physically exist. Sometimes there is an actual physical population, such as a lot of silicon wafers produced in a semiconductor factory.

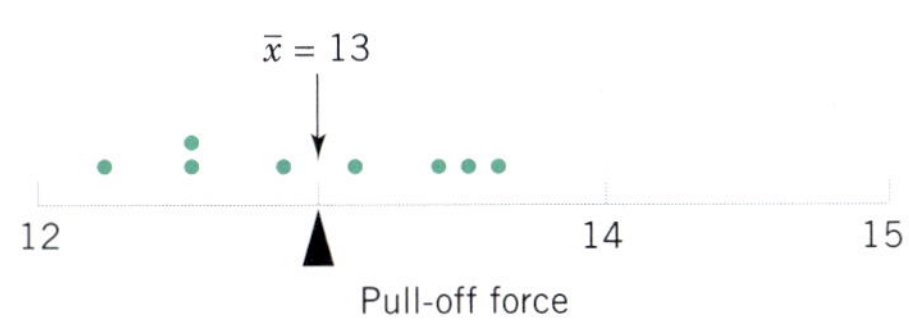

Figure 6-1 The sample mean as a balance point for a system of weights.

In previous chapters we have introduced the mean of a probability distribution, denoted μ. If we think of a probability distribution as a model for the population, one way to think of the mean is as the average of all the measurements in the population. For a finite population with N equally-likely values, the probability mass function is $f(x_i) = 1/N$ and the mean is

$$\mu = \sum_{i=1}^{N} x_i f(x_i) = \frac{\sum_{i=1}^{N} x_i}{N} \tag{6-2}$$

The sample mean, $\bar{x}$, is a reasonable estimate of the population mean, μ. Therefore, the engineer designing the connector using a 3/32-inch wall thickness would conclude, on the basis of the data, that an estimate of the mean pull-off force is 13.0 pounds.

Although the sample mean is useful, it does not convey all of the information about a sample of data. The variability or scatter in the data may be described by the **sample variance** or the **sample standard deviation.**

Sample Variance

If $x_1, x_2, \ldots, x_n$ is a sample of n observations, the **sample variance** is

$$s^2 = \frac{\sum_{i=1}^{n} (x_i - \bar{x})^2}{n - 1} \tag{6-3}$$

The **sample standard deviation,** s, is the positive square root of the sample variance.

The units of measurements for the sample variance are the square of the original units of the variable. Thus, if x is measured in pounds, the units for the sample variance are $(\text{pounds})^2$. The standard deviation has the desirable property of measuring variability in the original units of the variable of interest, x.

How Does the Sample Variance Measure Variability?

To see how the sample variance measures dispersion or variability, refer to Fig. 6-2, which shows the deviations $x_i - \bar{x}$ for the connector pull-off force data. The greater the amount of variability in the pull-off force data, the larger in absolute magnitude some of the deviations $x_i - \bar{x}$ will be. Since the deviations $x_i - \bar{x}$ always sum to zero, we must use a measure of variability that changes the negative deviations to nonnegative quantities. Squaring the deviations is the approach used in the sample variance. Consequently, if s^2 is small, there is relatively little variability in the data, but if s^2 is large, the variability is relatively large.

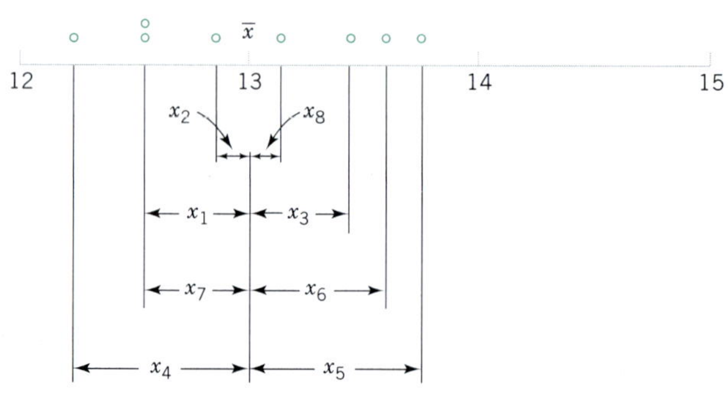

Figure 6-2 How the sample variance measures variability through the deviations $x_i - \bar{x}$.

Table 6-1 Calculation of Terms for the Sample Variance and Sample Standard Deviation

i	x_i	$x_i - \bar{x}$	$(x_i - \bar{x})^2$
1	12.6	−0.4	0.16
2	12.9	−0.1	0.01
3	13.4	0.4	0.16
4	12.3	−0.7	0.49
5	13.6	0.6	0.36
6	13.5	0.5	0.25
7	12.6	−0.4	0.16
8	13.1	0.1	0.01
	104.0	0.0	1.60

EXAMPLE 6-2
Sample Variance

Table 6-1 displays the quantities needed for calculating the sample variance and sample standard deviation for the pull-off force data. These data are plotted in Fig. 6-2. The numerator of s^2 is

$$\sum_{i=1}^{8} (x_i - \bar{x})^2 = 1.60$$

so the sample variance is

$$s^2 = \frac{1.60}{8 - 1} = \frac{1.60}{7} = 0.2286 \text{ (pounds)}^2$$

and the sample standard deviation is

$$s = \sqrt{0.2286} = 0.48 \text{ pounds}$$

Computation of s^2

The computation of s^2 requires calculation of $\bar{x}$, n subtractions, and n squaring and adding operations. If the original observations or the deviations $x_i - \bar{x}$ are not integers, the deviations $x_i - \bar{x}$ may be tedious to work with, and several decimals may have to be carried to ensure numerical accuracy. A more efficient computational formula for the sample variance is obtained as follows:

$$s^2 = \frac{\sum_{i=1}^{n} (x_i - \bar{x})^2}{n - 1} = \frac{\sum_{i=1}^{n} (x_i^2 + \bar{x}^2 - 2\bar{x}x_i)}{n - 1} = \frac{\sum_{i=1}^{n} x_i^2 + n\bar{x}^2 - 2\bar{x}\sum_{i=1}^{n} x_i}{n - 1}$$

and since $\bar{x} = (1/n)\sum_{i=1}^{n} x_i$, this last equation reduces to

$$s^2 = \frac{\sum_{i=1}^{n} x_i^2 - \frac{\left(\sum_{i=1}^{n} x_i\right)^2}{n}}{n - 1} \qquad (6\text{-}4)$$

Note that Equation 6-4 requires squaring each individual x_i, then squaring the sum of the x_i, subtracting $(\sum x_i)^2/n$ from $\sum x_i^2$, and finally dividing by $n - 1$. Sometimes this is called the shortcut method for calculating s^2 (or s).

EXAMPLE 6-3 We will calculate the sample variance and standard deviation using the shortcut method, Equation 6-4. The formula gives

$$s^2 = \frac{\sum_{i=1}^{n} x_i^2 - \frac{\left(\sum_{i=1}^{n} x_i\right)^2}{n}}{n-1} = \frac{1353.6 - \frac{(104)^2}{8}}{7} = \frac{1.60}{7} = 0.2286 \text{ (pounds)}^2$$

and

$$s = \sqrt{0.2286} = 0.48 \text{ pounds}$$

These results agree exactly with those obtained previously.

Analogous to the sample variance s^2, the variability in the population is defined by the **population variance** (σ^2). As in earlier chapters, the positive square root of σ^2, or σ, will denote the **population standard deviation.** When the population is finite and consists of N equally-likely values, we may define the population variance as

$$\sigma^2 = \frac{\sum_{i=1}^{N} (x_i - \mu)^2}{N} \tag{6-5}$$

We observed previously that the sample mean could be used as an estimate of the population mean. Similarly, the sample variance is an estimate of the population variance. In Chapter 7, we will discuss **estimation of parameters** more formally.

Note that the divisor for the sample variance is the sample size minus one ($n - 1$), while for the population variance it is the population size N. If we knew the true value of the population mean μ, we could find the *sample* variance as the average squared deviation of the sample observations about μ. In practice, the value of μ is almost never known, and so the sum of the squared deviations about the sample average $\bar{x}$ must be used instead. However, the observations x_i tend to be closer to their average, $\bar{x}$, than to the population mean, μ. Therefore, to compensate for this we use $n - 1$ as the divisor rather than n. If we used n as the divisor in the sample variance, we would obtain a measure of variability that is, on the average, consistently smaller than the true population variance σ^2.

Another way to think about this is to consider the sample variance s^2 as being based on $n - 1$ **degrees of freedom.** The term *degrees of freedom* results from the fact that the n deviations $x_1 - \bar{x}, x_2 - \bar{x}, \ldots, x_n - \bar{x}$ always sum to zero, and so specifying the values of any $n - 1$ of these quantities automatically determines the remaining one. This was illustrated in Table 6-1. Thus, only $n - 1$ of the n deviations, $x_i - \bar{x}$, are freely determined.

In addition to the sample variance and sample standard deviation, the **sample range,** or the difference between the largest and smallest observations, is a useful measure of variability. The sample range is defined as follows.

Sample Range

If the n observations in a sample are denoted by $x_1, x_2, \ldots, x_n$, the **sample range** is

$$r = \max(x_i) - \min(x_i) \tag{6-6}$$

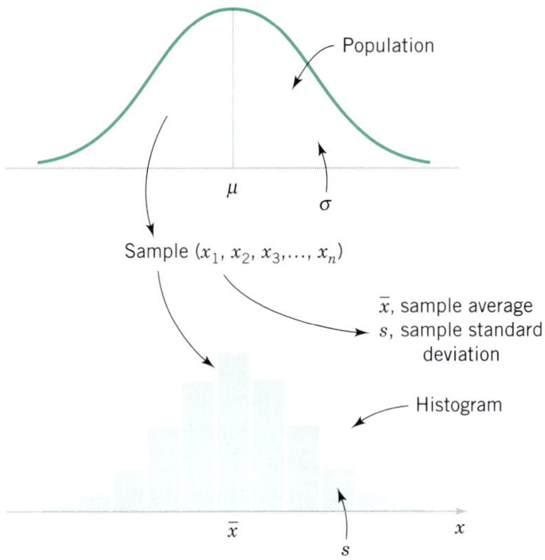

Figure 6-3 Relationship between a population and a sample.

For the pull-off force data, the sample range is $r = 13.6 - 12.3 = 1.3$. Generally, as the variability in sample data increases, the sample range increases.

The sample range is easy to calculate, but it ignores all of the information in the sample data between the largest and smallest values. For example, the two samples 1, 3, 5, 8, and 9 and 1, 5, 5, 5, and 9, both have the same range ($r = 8$). However, the standard deviation of the first sample is $s_1 = 3.35$, while the standard deviation of the second sample is $s_2 = 2.83$. The variability is actually less in the second sample.

Sometimes, when the sample size is small, say $n < 8$ or 10, the information loss associated with the range is not too serious. For example, the range is used widely in statistical quality control where sample sizes of 4 or 5 are fairly common. We will discuss some of these applications in Chapter 16.

In most statistics problems, we work with a sample of observations selected from the population that we are interested in studying. Figure 6-3 illustrates the relationship between the population and the sample.

EXERCISES FOR SECTIONS 6-1

6-1. Eight measurements were made on the inside diameter of forged piston rings used in an automobile engine. The data (in millimeters) are 74.001, 74.003, 74.015, 74.000, 74.005, 74.002, 74.005, and 74.004. Calculate the sample mean and sample standard deviation, construct a dot diagram, and comment on the data.

6-2. In *Applied Life Data Analysis* (Wiley, 1982), Wayne Nelson presents the breakdown time of an insulating fluid between electrodes at 34 kV. The times, in minutes, are as follows: 0.19, 0.78, 0.96, 1.31, 2.78, 3.16, 4.15, 4.67, 4.85, 6.50, 7.35, 8.01, 8.27, 12.06, 31.75, 32.52, 33.91, 36.71, and 72.89. Calculate the sample mean and sample standard deviation.

6-3. The January 1990 issue of *Arizona Trend* contains a supplement describing the 12 "best" golf courses in the state. The yardages (lengths) of these courses are as follows: 6981, 7099, 6930, 6992, 7518, 7100, 6935, 7518, 7013, 6800, 7041, and 6890. Calculate the sample mean and sample standard deviation. Construct a dot diagram of the data.

6-4. An article in the *Journal of Structural Engineering* (Vol. 115, 1989) describes an experiment to test the yield strength of circular tubes with caps welded to the ends. The first yields (in kN) are 96, 96, 102, 102, 102, 104, 104, 108, 126, 126, 128, 128, 140, 156, 160, 160, 164, and 170. Calculate the sample mean and sample standard deviation. Construct a dot diagram of the data.

6-5. An article in *Human Factors* (June 1989) presented data on visual accommodation (a function of eye movement) when recognizing a speckle pattern on a high-resolution CRT screen. The data are as follows: 36.45, 67.90, 38.77, 42.18, 26.72, 50.77, 39.30, and 49.71. Calculate the sample mean and sample standard deviation. Construct a dot diagram of the data.

6-6. The following data are direct solar intensity measurements (watts/m^2) on different days at a location in southern

Spain: 562, 869, 708, 775, 775, 704, 809, 856, 655, 806, 878, 909, 918, 558, 768, 870, 918, 940, 946, 661, 820, 898, 935, 952, 957, 693, 835, 905, 939, 955, 960, 498, 653, 730, and 753. Calculate the sample mean and sample standard deviation. Prepare a dot diagram of this data. Indicate where the sample mean falls on this diagram. Give a practical interpretation of the sample mean.

6-7. The April 22, 1991 issue of *Aviation Week and Space Technology* reports that during Operation Desert Storm, U.S. Air Force F-117A pilots flew 1270 combat sorties for a total of 6905 hours. What is the mean duration of an F-117A mission during this operation? Why is the parameter you have calculated a population mean?

6-8. Preventing fatigue crack propagation in aircraft structures is an important element of aircraft safety. An engineering study to investigate fatigue crack in $n = 9$ cyclically loaded wing boxes reported the following crack lengths (in mm): 2.13, 2.96, 3.02, 1.82, 1.15, 1.37, 2.04, 2.47, 2.60. Calculate the sample mean and sample standard deviation. Prepare a dot diagram of the data.

6-9. An article in the *Journal of Physiology* (2002, Vol. 545, pp. 27–41) "Response of rat muscle to acute resistance exercise defined by transcriptional and translational profiling" studied gene expression as a function of resistance exercise. Expression data (measures of gene activity) from one gene are shown in the following table. One group of rats was exercised for six hours while the other received no exercise. Compute the sample mean and standard deviation of the exercise and no-exercise groups separately. Construct a dot diagram for the exercise and no-exercise groups separately. Comment on any differences between the groups.

6 Hours of Exercise	6 Hours of Exercise	No Exercise	No Exercise
425.313	208.475	485.396	406.921
223.306	286.484	159.471	335.209
388.793	244.242	478.314	
139.262	408.099	245.782	
212.565	157.743	236.212	
324.024	436.37	252.773	

6-10. Exercise 6-5 describes data from an article in *Human Factors* on visual accommodation from an experiment involving a high-resolution CRT screen.

Data from a second experiment using a low-resolution screen were also reported in the article. They are 8.85, 35.80, 26.53, 64.63, 9.00, 15.38, 8.14, and 8.24. Prepare a dot diagram for this second sample and compare it to the one for the first sample. What can you conclude about CRT resolution in this situation?

6-11. The pH of a solution is measured eight times by one operator using the same instrument. She obtains the following data: 7.15, 7.20, 7.18, 7.19, 7.21, 7.20, 7.16, and 7.18. Calculate the sample mean and sample standard deviation. Comment on potential major sources of variability in this experiment.

6-12. An article in the *Journal of Aircraft* (1988) describes the computation of drag coefficients for the NASA 0012 airfoil. Different computational algorithms were used at $M_\infty = 0.7$ with the following results (drag coefficients are in units of drag counts; that is, one count is equivalent to a drag coefficient of 0.0001): 79, 100, 74, 83, 81, 85, 82, 80, and 84. Compute the sample mean, sample variance, and sample standard deviation, and construct a dot diagram.

6-13. The following data are the joint temperatures of the O-rings (°F) for each test firing or actual launch of the space shuttle rocket motor (from *Presidential Commission on the Space Shuttle Challenger Accident*, Vol. 1, pp. 129–131): 84, 49, 61, 40, 83, 67, 45, 66, 70, 69, 80, 58, 68, 60, 67, 72, 73, 70, 57, 63, 70, 78, 52, 67, 53, 67, 75, 61, 70, 81, 76, 79, 75, 76, 58, 31.

(a) Compute the sample mean and sample standard deviation and construct a dot diagram of the temperature data.
(b) Set aside the smallest observation (31°F) and recompute the quantities in part (a). Comment on your findings. How "different" are the other temperatures from this last value?

6-2 STEM-AND-LEAF DIAGRAMS

The dot diagram is a useful data display for small samples, up to (say) about 20 observations. However, when the number of observations is moderately large, other graphical displays may be more useful.

For example, consider the data in Table 6-2. These data are the compressive strengths in pounds per square inch (psi) of 80 specimens of a new aluminum-lithium alloy undergoing evaluation as a possible material for aircraft structural elements. The data were recorded in the order of testing, and in this format they do not convey much information about compressive strength.

Table 6-2 Compressive Strength (in psi) of 80 Aluminum-Lithium Alloy Specimens

105	221	183	186	121	181	180	143
97	154	153	174	120	168	167	141
245	228	174	199	181	158	176	110
163	131	154	115	160	208	158	133
207	180	190	193	194	133	156	123
134	178	76	167	184	135	229	146
218	157	101	171	165	172	158	169
199	151	142	163	145	171	148	158
160	175	149	87	160	237	150	135
196	201	200	176	150	170	118	149

Questions such as "What percent of the specimens fail below 120 psi?" are not easy to answer. Because there are many observations, constructing a dot diagram of these data would be relatively inefficient; more effective displays are available for large data sets.

A **stem-and-leaf diagram** is a good way to obtain an informative visual display of a data set $x_1, x_2, \ldots, x_n$, where each number x_i consists of at least two digits. To construct a stem-and-leaf diagram, use the following steps.

Steps to Construct a Stem-and-Leaf Diagram

(1) Divide each number x_i into two parts: a **stem,** consisting of one or more of the leading digits and a **leaf,** consisting of the remaining digit.

(2) List the stem values in a vertical column.

(3) Record the leaf for each observation beside its stem.

(4) Write the units for stems and leaves on the display.

To illustrate, if the data consist of percent defective information between 0 and 100 on lots of semiconductor wafers, we can divide the value 76 into the stem 7 and the leaf 6. In general, we should choose relatively few stems in comparison with the number of observations. It is usually best to choose between 5 and 20 stems.

EXAMPLE 6-4 Alloy Strength

To illustrate the construction of a stem-and-leaf diagram, consider the alloy compressive strength data in Table 6-2. We will select as stem values the numbers 7, 8, 9, ..., 24. The resulting stem-and-leaf diagram is presented in Fig. 6-4. The last column in the diagram is a frequency count of the number of leaves associated with each stem. Inspection of this display immediately reveals that most of the compressive strengths lie between 110 and 200 psi and that a central value is somewhere between 150 and 160 psi. Furthermore, the strengths are distributed approximately symmetrically about the central value. The stem-and-leaf diagram enables us to determine quickly some important features of the data that were not immediately obvious in the original display in Table 6-2.

In some data sets, it may be desirable to provide more classes or stems. One way to do this would be to modify the original stems as follows: Divide the stem 5 (say) into two new stems, 5L and 5U. The stem 5L has leaves 0, 1, 2, 3, and 4, and stem 5U has leaves 5, 6, 7, 8, and 9. This will double the number of original stems. We could increase the number of original stems by four by defining five new stems: 5z with leaves 0 and 1, 5t (for twos and three) with leaves 2 and 3, 5f (for fours and fives) with leaves 4 and 5, 5s (for six and seven) with leaves 6 and 7, and 5e with leaves 8 and 9.

Stem	Leaf	Frequency
7	6	1
8	7	1
9	7	1
10	5 1	2
11	5 8 0	3
12	1 0 3	3
13	4 1 3 5 3 5	6
14	2 9 5 8 3 1 6 9	8
15	4 7 1 3 4 0 8 8 6 8 0 8	12
16	3 0 7 3 0 5 0 8 7 9	10
17	8 5 4 4 1 6 2 1 0 6	10
18	0 3 6 1 4 1 0	7
19	9 6 0 9 3 4	6
20	7 1 0 8	4
21	8	1
22	1 8 9	3
23	7	1
24	5	1

Stem : Tens and hundreds digits (psi); Leaf: Ones digits (psi)

Figure 6-4 Stem-and-leaf diagram for the compressive strength data in Table 6-2.

EXAMPLE 6-5
Chemical Yield

Figure 6-5 illustrates the stem-and-leaf diagram for 25 observations on batch yields from a chemical process. In Fig. 6-5(a) we have used 6, 7, 8, and 9 as the stems. This results in too few stems, and the stem-and-leaf diagram does not provide much information about the data. In Fig. 6-5(b) we have divided each stem into two parts, resulting in a display that more adequately displays the data. Figure 6-5(c) illustrates a stem-and-leaf display with each stem divided into five parts. There are too many stems in this plot, resulting in a display that does not tell us much about the shape of the data.

Stem	Leaf
6	1 3 4 5 5 6
7	0 1 1 3 5 7 8 8 9
8	1 3 4 4 7 8 8
9	2 3 5

(a)

Stem	Leaf
6L	1 3 4
6U	5 5 6
7L	0 1 1 3
7U	5 7 8 8 9
8L	1 3 4 4
8U	7 8 8
9L	2 3
9U	5

(b)

Stem	Leaf
6z	1
6t	3
6f	4 5 5
6s	6
6e	
7z	0 1 1
7t	3
7f	5
7s	7
7e	8 8 9
8z	1
8t	3
8f	4 4
8s	7
8e	8 8
9z	
9t	2 3
9f	5
9s	
9e	

(c)

Figure 6-5 Stem-and-leaf displays for Example 6-5. Stem: Tens digits. Leaf: Ones digits.

Character Stem-and-Leaf Display

Stem-and-leaf of Strength

N = 80 Leaf Unit = 1.0

1	7	6
2	8	7
3	9	7
5	10	1 5
8	11	0 5 8
11	12	0 1 3
17	13	1 3 3 4 5 5
25	14	1 2 3 5 6 8 9 9
37	15	0 0 1 3 4 4 6 7 8 8 8 8
(10)	16	0 0 0 3 3 5 7 7 8 9
33	17	0 1 1 2 4 4 5 6 6 8
23	18	0 0 1 1 3 4 6
16	19	0 3 4 6 9 9
10	20	0 1 7 8
6	21	8
5	22	1 8 9
2	23	7
1	24	5

Figure 6-6 A stem-and-leaf diagram from Minitab.

Figure 6-6 shows a stem-and-leaf display of the compressive strength data in Table 6-2 produced by Minitab. The software uses the same stems as in Fig. 6-4. Note also that the computer orders the leaves from smallest to largest on each stem. This form of the plot is usually called an **ordered stem-and-leaf diagram.** This is not usually done when the plot is constructed manually because it can be time consuming. The computer adds a column to the left of the stems that provides a count of the observations at and above each stem in the upper half of the display and a count of the observations at and below each stem in the lower half of the display. At the middle stem of 16, the column indicates the number of observations at this stem.

The ordered stem-and-leaf display makes it relatively easy to find data features such as percentiles, quartiles, and the median. The sample **median** is a measure of central tendency that divides the data into two equal parts, half below the median and half above. If the number of observations is even, the median is halfway between the two central values. From Fig. 6-6 we find the 40th and 41st values of strength as 160 and 163, so the median is $(160 + 163)/2 = 161.5$. If the number of observations is odd, the median is the central value. The sample **mode** is the most frequently occurring data value. Figure 6-6 indicates that the mode is 158; this value occurs four times, and no other value occurs as frequently in the sample. If there were more than one value that occurred four times, the data would have multiple modes.

We can also divide data into more than two parts. When an ordered set of data is divided into four equal parts, the division points are called **quartiles.** The *first* or *lower quartile*, q_1, is a value that has approximately 25% of the observations below it and approximately 75% of the observations above. The *second quartile*, q_2, has approximately 50% of the observations below its value. The second quartile is exactly equal to the median. The *third* or *upper quartile*, q_3, has approximately 75% of the observations below its value. As in the case of the median, the quartiles may not be unique. The compressive strength data in Fig. 6-6 contains $n = 80$ observations. Minitab software calculates the first and third quartiles as the $(n + 1)/4$ and $3(n + 1)/4$ ordered observations and interpolates as needed. For example, $(80 + 1)/4 = 20.25$ and $3(80 + 1)/4 = 60.75$. Therefore, Minitab interpolates between the 20th and 21st ordered observation to obtain $q_1 = 143.50$ and between the 60th and 61st observation to obtain $q_3 = 181.00$. In general, the 100*k*th **percentile** is a data value such that

Table 6-3 Summary Statistics for the Compressive Strength Data from Minitab

Variable	N	Mean	Median	StDev	SE Mean
	80	162.66	161.50	33.77	3.78
	Min	Max	Q1	Q3	
	76.00	245.00	143.50	181.00	

approximately $100k\%$ of the observations are at or below this value and approximately $100(1 - k)\%$ of them are above it. Finally, we may use the **interquartile range,** defined as $\text{IQR} = q_3 - q_1$, as a measure of variability. The interquartile range is less sensitive to the extreme values in the sample than is the ordinary sample range.

Many statistics software packages provide data summaries that include these quantities. The output obtained for the compressive strength data in Table 6-2 from Minitab is shown in Table 6-3.

EXERCISES FOR SECTION 6-2

6-14. An article in *Technometrics* (Vol. 19, 1977, p. 425) presents the following data on the motor fuel octane ratings of several blends of gasoline:

88.5	98.8	89.6	92.2	92.7	88.4	87.5	90.9
94.7	88.3	90.4	83.4	87.9	92.6	87.8	89.9
84.3	90.4	91.6	91.0	93.0	93.7	88.3	91.8
90.1	91.2	90.7	88.2	94.4	96.5	89.2	89.7
89.0	90.6	88.6	88.5	90.4	84.3	92.3	92.2
89.8	92.2	88.3	93.3	91.2	93.2	88.9	
91.6	87.7	94.2	87.4	86.7	88.6	89.8	
90.3	91.1	85.3	91.1	94.2	88.7	92.7	
90.0	86.7	90.1	90.5	90.8	92.7	93.3	
91.5	93.4	89.3	100.3	90.1	89.3	86.7	
89.9	96.1	91.1	87.6	91.8	91.0	91.0	

Construct a stem-and-leaf display for these data. Calculate the median and quartiles of these data.

6-15. The following data are the numbers of cycles to failure of aluminum test coupons subjected to repeated alternating stress at 21,000 psi, 18 cycles per second.

1115	865	1015	885	1594	1000	1416	1501
1310	2130	845	1223	2023	1820	1560	1238
1540	1421	1674	375	1315	1940	1055	990
1502	1109	1016	2265	1269	1120	1764	1468
1258	1481	1102	1910	1260	910	1330	1512
1315	1567	1605	1018	1888	1730	1608	1750
1085	1883	706	1452	1782	1102	1535	1642
798	1203	2215	1890	1522	1578	1781	
1020	1270	785	2100	1792	758	1750	

Construct a stem-and-leaf display for these data. Calculate the median and quartiles of these data. Does it appear likely that a coupon will "survive" beyond 2000 cycles? Justify your answer.

6-16. The percentage of cotton in material used to manufacture men's shirts follows. Construct a stem-and-leaf display for the data. Calculate the median and quartiles of these data.

34.2	37.8	33.6	32.6	33.8	35.8	34.7	34.6
33.1	36.6	34.7	33.1	34.2	37.6	33.6	33.6
34.5	35.4	35.0	34.6	33.4	37.3	32.5	34.1
35.6	34.6	35.4	35.9	34.7	34.6	34.1	34.7
36.3	33.8	36.2	34.7	34.6	35.5	35.1	35.7
35.1	37.1	36.8	33.6	35.2	32.8	36.8	36.8
34.7	34.0	35.1	32.9	35.0	32.1	37.9	34.3
33.6	34.1	35.3	33.5	34.9	34.5	36.4	32.7

6-17. The following data represent the yield on 90 consecutive batches of ceramic substrate to which a metal coating has been applied by a vapor-deposition process. Construct a stem-and-leaf display for these data. Calculate the median and quartiles of these data.

94.1	86.1	95.3	84.9	88.8	84.6	94.4	84.1
93.2	90.4	94.1	78.3	86.4	83.6	96.1	83.7
90.6	89.1	97.8	89.6	85.1	85.4	98.0	82.9
91.4	87.3	93.1	90.3	84.0	89.7	85.4	87.3
88.2	84.1	86.4	93.1	93.7	87.6	86.6	86.4
86.1	90.1	87.6	94.6	87.7	85.1	91.7	84.5
95.1	95.2	94.1	96.3	90.6	89.6	87.5	
90.0	86.1	92.1	94.7	89.4	90.0	84.2	
92.4	94.3	96.4	91.1	88.6	90.1	85.1	
87.3	93.2	88.2	92.4	84.1	94.3	90.5	
86.6	86.7	86.4	90.6	82.6	97.3	95.6	
91.2	83.0	85.0	89.1	83.1	96.8	88.3	

6-18. Calculate the sample median, mode, and mean of the data in Exercise 6-14. Explain how these three measures of location describe different features of the data.

6-19. Calculate the sample median, mode, and mean of the data in Exercise 6-15. Explain how these three measures of location describe different features in the data.

6-20. Calculate the sample median, mode, and mean for the data in Exercise 6-16. Explain how these three measures of location describe different features of the data.

6-21. The net energy consumption (in billion of kilowatt hours) for countries in Asia in 2003 is as follows (source: U.S. Department of Energy Web site http://www.eia.doe.gov/emeu). Construct a stem-and-leaf diagram for these data and comment on any important features that you notice. Compute the sample mean, sample standard deviation, and sample median.

	Billion of Kilowatt Hours
Afghanistan	1.04
Australia	200.66
Bangladesh	16.20
Burma	6.88
China	1,671.23
Hong Kong	38.43
India	519.04
Indonesia	101.80
Japan	946.27
Korea, North	17.43
Korea, South	303.33
Laos	3.30
Malaysia	73.63
Mongolia	2.91
Nepal	2.30
New Zealand	37.03
Pakistan	71.54
Philippines	44.48
Singapore	30.89
Sri Lanka	6.80
Taiwan	154.34
Thailand	107.34
Vietnam	36.92
Asia & Oceania	**4,403.92**

6-22. The female students in an undergraduate engineering core course at ASU self-reported their heights to the nearest inch. The data are below. Construct a stem-and-leaf diagram for the height data and comment on any important features that you notice. Calculate the sample mean, sample standard deviation, and the sample median of height.

62 64 66 67 65 68 61 65 67 65 64 63 67
68 64 66 68 69 65 67 62 66 68 67 66 65
69 65 70 65 67 68 65 63 64 67 67

6-23. The shear strengths of 100 spot welds in a titanium alloy follow. Construct a stem-and-leaf diagram for the weld strength data and comment on any important features that you notice. What is the 95th percentile of strength?

5408 5431 5475 5442 5376 5388 5459 5422 5416 5435
5420 5429 5401 5446 5487 5416 5382 5357 5388 5457
5407 5469 5416 5377 5454 5375 5409 5459 5445 5429
5463 5408 5481 5453 5422 5354 5421 5406 5444 5466
5399 5391 5477 5447 5329 5473 5423 5441 5412 5384
5445 5436 5454 5453 5428 5418 5465 5427 5421 5396
5381 5425 5388 5388 5378 5481 5387 5440 5482 5406
5401 5411 5399 5431 5440 5413 5406 5342 5452 5420
5458 5485 5431 5416 5431 5390 5399 5435 5387 5462
5383 5401 5407 5385 5440 5422 5448 5366 5430 5418

6-24. An important quality characteristic of water is the concentration of suspended solid material. Following are 60 measurements on suspended solids from a certain lake. Construct a stem-and-leaf diagram for this data and comment on any important features that you notice. Compute the sample mean, sample standard deviation, and the sample median. What is the 90th percentile of concentration?

42.4 65.7 29.8 58.7 52.1 55.8 57.0 68.7 67.3 67.3
54.3 54.0 73.1 81.3 59.9 56.9 62.2 69.9 66.9 59.0
56.3 43.3 57.4 45.3 80.1 49.7 42.8 42.4 59.6 65.8
61.4 64.0 64.2 72.6 72.5 46.1 53.1 56.1 67.2 70.7
42.6 77.4 54.7 57.1 77.3 39.3 76.4 59.3 51.1 73.8
61.4 73.1 77.3 48.5 89.8 50.7 52.0 59.6 66.1 31.6

6-25. The United States Golf Association tests golf balls to ensure that they conform to the rules of golf. Balls are tested for weight, diameter, roundness, and overall distance. The overall distance test is conducted by hitting balls with a driver swung by a mechanical device nicknamed "Iron Byron" after the legendary great Byron Nelson, whose swing the machine is said to emulate. Following are 100 distances (in yards) achieved by a particular brand of golf ball in the overall distance test. Construct a stem-and-leaf diagram for this data and comment on any important features that you notice. Compute the sample mean, sample standard deviation, and the sample median. What is the 90th percentile of distances?

261.3 259.4 265.7 270.6 274.2 261.4 254.5 283.7
258.1 270.5 255.1 268.9 267.4 253.6 234.3 263.2
254.2 270.7 233.7 263.5 244.5 251.8 259.5 257.5
257.7 272.6 253.7 262.2 252.0 280.3 274.9 233.7
237.9 274.0 264.5 244.8 264.0 268.3 272.1 260.2
255.8 260.7 245.5 279.6 237.8 278.5 273.3 263.7
241.4 260.6 280.3 272.7 261.0 260.0 279.3 252.1
244.3 272.2 248.3 278.7 236.0 271.2 279.8 245.6
241.2 251.1 267.0 273.4 247.7 254.8 272.8 270.5
254.4 232.1 271.5 242.9 273.6 256.1 251.6
256.8 273.0 240.8 276.6 264.5 264.5 226.8
255.3 266.6 250.2 255.8 285.3 255.4 240.5
255.0 273.2 251.4 276.1 277.8 266.8 268.5

6-26. A semiconductor manufacturer produces devices used as central processing units in personal computers. The speed of the device (in megahertz) is important because it determines the price that the manufacturer can charge for the devices. The following table contains measurements on 120 devices. Construct a stem-and-leaf diagram for this data and comment on any important features that you notice. Compute the sample mean, sample standard deviation, and the sample median. What percentage of the devices has a speed exceeding 700 megahertz?

680	669	719	699	670	710	722	663	658	634	720	690
677	669	700	718	690	681	702	696	692	690	694	660
649	675	701	721	683	735	688	763	672	698	659	704
681	679	691	683	705	746	706	649	668	672	690	724
652	720	660	695	701	724	668	698	668	660	680	739
717	727	653	637	660	693	679	682	724	642	704	695
704	652	664	702	661	720	695	670	656	718	660	648
683	723	710	680	684	705	681	748	697	703	660	722
662	644	683	695	678	674	656	667	683	691	680	685
681	715	665	676	665	675	655	659	720	675	697	663

6-27. A group of wine enthusiasts taste-tested a pinot noir wine from Oregon. The evaluation was to grade the wine on a 0 to 100 point scale. The results follow. Construct a stem-and-leaf diagram for this data and comment on any important features that you notice. Compute the sample mean, sample standard deviation, and the sample median. A wine rated above 90 is considered truly exceptional. What proportion of the taste-tasters considered this particular pinot noir truly exceptional?

94	90	92	91	91	86	89	91	91	90
90	93	87	90	91	92	89	86	89	90
88	95	91	88	89	92	87	89	95	92
85	91	85	89	88	84	85	90	90	83

6-28. In their book *Introduction to Linear Regression Analysis* (3rd edition, Wiley, 2001) Montgomery, Peck, and Vining present measurements on $NbOCl_3$ concentration from a tube-flow reactor experiment. The data, in gram − mole per liter $\times 10^{-3}$, are as follows. Construct a stem-and-leaf diagram for this data and comment on any important features that you notice. Compute the sample mean, sample standard deviation, and the sample median.

450	450	473	507	457	452	453	1215	1256
1145	1085	1066	1111	1364	1254	1396	1575	1617
1733	2753	3186	3227	3469	1911	2588	2635	2725

6-29. In Exercise 6-22, we presented height data that was self-reported by female undergraduate engineering students in a core course at ASU. In the same class, the male students self-reported their heights as follows. Construct a comparative stem-and-leaf diagram by listing the stems in the center of the display and then placing the female leaves on the left and the male leaves on the right. Comment on any important features that you notice in this display.

69	67	69	70	65	68	69	70	71	69	66	67	69	75	68	67	68
69	70	71	72	68	69	69	70	71	68	72	69	69	68	69	73	70
73	68	69	71	67	68	65	68	68	69	70	74	71	69	70	69	

6-3 FREQUENCY DISTRIBUTIONS AND HISTOGRAMS

A **frequency distribution** is a more compact summary of data than a stem-and-leaf diagram. To construct a frequency distribution, we must divide the range of the data into intervals, which are usually called **class intervals, cells,** or **bins.** If possible, the bins should be of equal width in order to enhance the visual information in the frequency distribution. Some judgment must be used in selecting the number of bins so that a reasonable display can be developed. The number of bins depends on the number of observations and the amount of scatter or dispersion in the data. A frequency distribution that uses either too few or too many bins will not be informative. We usually find that between 5 and 20 bins is satisfactory in most cases and that the number of bins should increase with n. Choosing the **number of bins** approximately equal to the square root of the number of observations often works well in practice.

A frequency distribution for the comprehensive strength data in Table 6-2 is shown in Table 6-4. Since the data set contains 80 observations, and since $\sqrt{80} \simeq 9$, we suspect that about eight to nine bins will provide a satisfactory frequency distribution. The largest and smallest data values are 245 and 76, respectively, so the bins must cover a range of at least $245 - 76 = 169$ units on the psi scale. If we want the lower limit for the first bin to begin

Table 6-4 Frequency Distribution for the Compressive Strength Data in Table 6-2

Class	$70 \le x < 90$	$90 \le x < 110$	$110 \le x < 130$	$130 \le x < 150$	$150 \le x < 170$	$170 \le x < 190$	$190 \le x < 210$	$210 \le x < 230$	$230 \le x < 250$
Frequency	2	3	6	14	22	17	10	4	2
Relative frequency	0.0250	0.0375	0.0750	0.1750	0.2750	0.2125	0.1250	0.0500	0.0250
Cumulative relative frequency	0.0250	0.0625	0.1375	0.3125	0.5875	0.8000	0.9250	0.9750	1.0000

slightly below the smallest data value and the upper limit for the last bin to be slightly above the largest data value, we might start the frequency distribution at 70 and end it at 250. This is an interval or range of 180 psi units. Nine bins, each of width 20 psi, give a reasonable frequency distribution, so the frequency distribution in Table 6-4 is based on nine bins.

The second row of Table 6-4 contains a **relative frequency distribution.** The relative frequencies are found by dividing the observed frequency in each bin by the total number of observations. The last row of Table 6-4 expresses the relative frequencies on a cumulative basis. Frequency distributions are often easier to interpret than tables of data. For example, from Table 6-4 it is very easy to see that most of the specimens have compressive strengths between 130 and 190 psi and that 97.5 percent of the specimens fail below 230 psi.

The **histogram** is a visual display of the frequency distribution. The stages for constructing a histogram follow.

Constructing a Histogram (Equal Bin Widths)

(1) Label the bin (class interval) boundaries on a horizontal scale.
(2) Mark and label the vertical scale with the frequencies or the relative frequencies.
(3) Above each bin, draw a rectangle where height is equal to the frequency (or relative frequency) corresponding to that bin.

Figure 6-7 is the histogram for the compression strength data. The histogram, like the **stem-and-leaf diagram,** provides a visual impression of the shape of the distribution of the measurements and information about the central tendency and scatter or dispersion in the data. Notice the symmetric, bell-shaped distribution of the strength measurements in Fig. 6-7. This display often gives insight about possible choices of probability distributions to use as a model

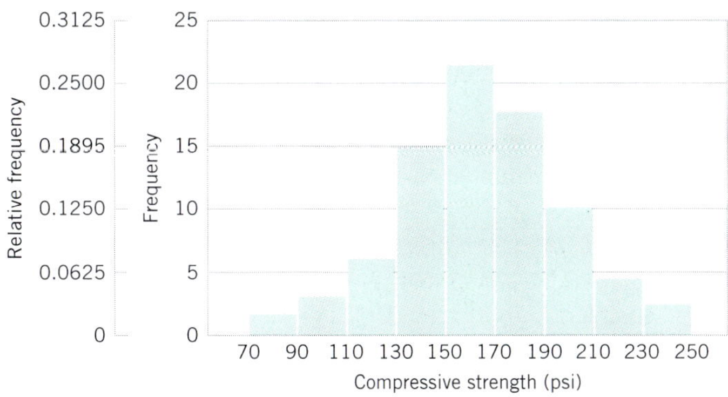

Figure 6-7 Histogram of compressive strength for 80 aluminum-lithium alloy specimens.

for the population. For example, here we would likely conclude that the normal distribution is a reasonable model for the population of compression strength measurements.

Sometimes a histogram with **unequal bin widths** will be employed. For example, if the data have several extreme observations or outliers, using a few equal-width bins will result in nearly all observations falling in just a few of the bins. Using many equal-width bins will result in many bins with zero frequency. A better choice is to use shorter intervals in the region where most of the data falls and a few wide intervals near the extreme observations. When the bins are of unequal width, the rectangle's **area** (not its height) should be proportional to the bin frequency. This implies that the rectangle height should be

$$\text{Rectangle height} = \frac{\text{bin frequency}}{\text{bin width}}$$

In passing from either the original data or stem-and-leaf diagram to a frequency distribution or histogram, we have lost some information because we no longer have the individual observations. However, this information loss is often small compared with the conciseness and ease of interpretation gained in using the frequency distribution and histogram.

Figure 6-8 shows a histogram of the compressive strength data from Minitab with 17 bins. We have noted that histograms may be relatively sensitive to the number of bins and their width. For small data sets, histograms may change dramatically in appearance if the number and/or width of the bins changes. Histograms are more stable for larger data sets, preferably of size 75 to 100 or more. Figure 6-9 shows the Minitab histogram for the compressive strength data with nine bins. This used Minitab's default setting and the result is similar to the original histogram shown in Fig. 6-7. Since the number of observations is moderately large ($n = 80$), the choice of the number of bins is not especially important, and both Figs. 6-8 and 6-9 convey similar information.

Figure 6-10 shows a variation of the histogram available in Minitab, the **cumulative frequency plot.** In this plot, the height of each bar is the total number of observations that are less than or equal to the upper limit of the bin. Cumulative distributions are also useful in data interpretation; for example, we can read directly from Fig. 6-10 that there are approximately 70 observations less than or equal to 200 psi.

When the sample size is large, the histogram can provide a reasonably reliable indicator of the general **shape** of the distribution or population of measurements from which the sample

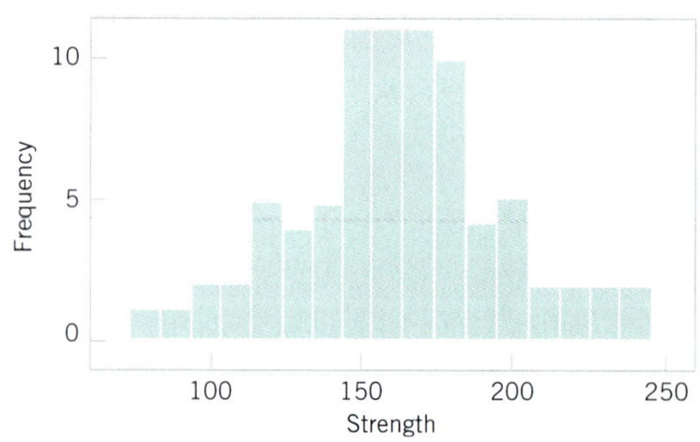

Figure 6-8 A histogram of the compressive strength data from Minitab with 17 bins.

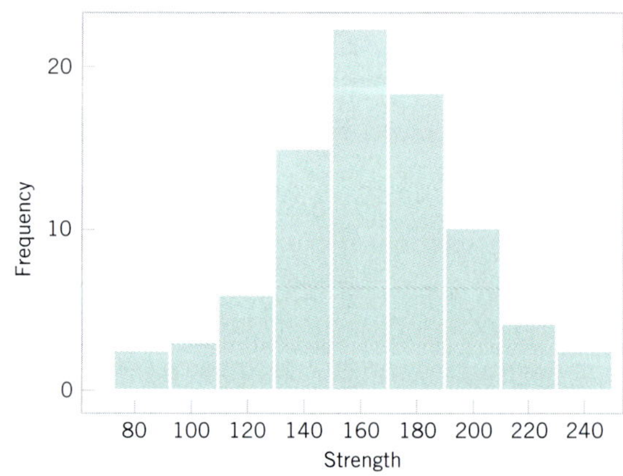

Figure 6-9 A histogram of the compressive strength data from Minitab with nine bins.

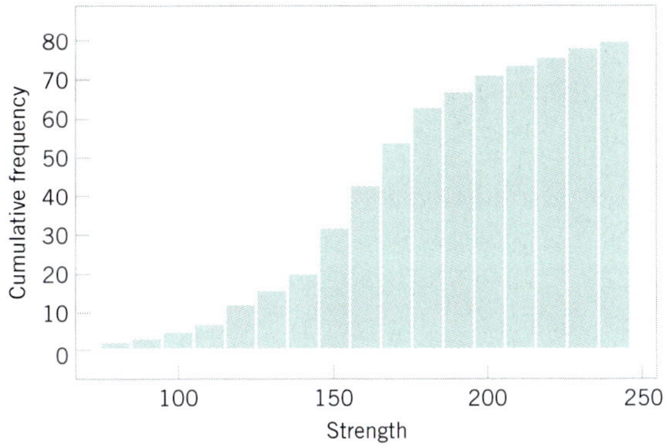

Figure 6-10 A cumulative distribution plot of the compressive strength data from Minitab.

was drawn. Figure 6-11 presents three cases. The median is denoted as $\tilde{x}$. Generally, if the data are symmetric, as in Fig. 6-11(b), the mean and median coincide. If, in addition, the data have only one mode (we say the data are *unimodal*), the mean, median, and mode all coincide. If the data are *skewed* (asymmetric, with a long tail to one side), as in Fig. 6-11(a) and (c), the mean, median, and mode do not coincide. Usually, we find that mode $<$ median $<$ mean if the distribution is skewed to the right, whereas mode $>$ median $>$ mean if the distribution is skewed to the left.

Frequency distributions and histograms can also be used with qualitative or categorical data. In some applications there will be a natural ordering of the categories (such as freshman, sophomore, junior, and senior), whereas in others the order of the categories will be arbitrary (such as male and female). When using categorical data, the bins should have equal width.

EXAMPLE 6-6

Figure 6-12 on the next page presents the production of transport aircraft by the Boeing Company in 1985. Notice that the 737 was the most popular model, followed by the 757, 747, 767, and 707.

A chart of occurrences by category (in which the categories are ordered by the number of occurrences) is sometimes referred to as a **Pareto chart.** An exercise asks you to construct such a chart.

In this section we have concentrated on descriptive methods for the situation in which each observation in a data set is a single number or belongs to one category. In many cases, we work with data in which each observation consists of several measurements. For example, in a gasoline mileage study, each observation might consist of a measurement of miles per gallon, the size of the engine in the vehicle, engine horsepower, vehicle weight, and vehicle length. This is an example of **multivariate data.** In later chapters, we will discuss analyzing this type of data.

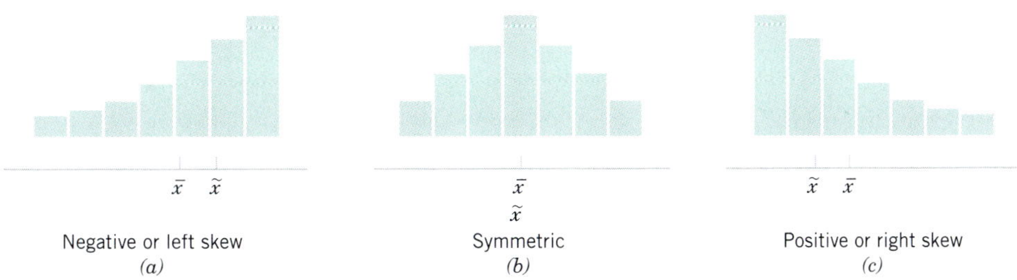

Figure 6-11 Histograms for symmetric and skewed distributions.

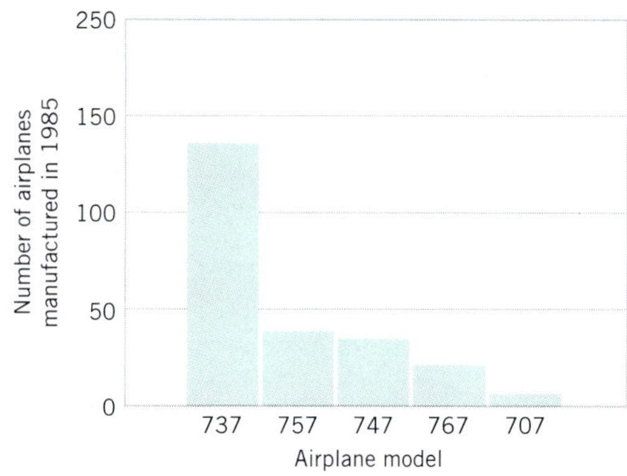

Figure 6-12 Airplane production in 1985. (*Source:* Boeing Company.)

EXERCISES FOR SECTION 6-3

6-30. Construct a frequency distribution and histogram for the motor fuel octane data from Exercise 6-14. Use eight bins.

6-31. Construct a frequency distribution and histogram using the failure data from Exercise 6-15.

6-32. Construct a frequency distribution and histogram for the cotton content data in Exercise 6-16.

6-33. Construct a frequency distribution and histogram for the yield data in Exercise 6-17.

6-34. Construct frequency distributions and histograms with 8 bins and 16 bins for the motor fuel octane data in Exercise 6-14. Compare the histograms. Do both histograms display similar information?

6-35. Construct histograms with 8 and 16 bins for the data in Exercise 6-15. Compare the histograms. Do both histograms display similar information?

6-36. Construct histograms with 8 and 16 bins for the data in Exercise 6-16. Compare the histograms. Do both histograms display similar information?

6-37. Construct a histogram for the energy consumption data in Exercise 6-21.

6-38. Construct a histogram for the female student height data in Exercise 6-22.

6-39. Construct a histogram for the spot weld shear strength data in Exercise 6-23. Comment on the shape of the histogram. Does it convey the same information as the stem-and-leaf display?

6-40. Construct a histogram for the water quality data in Exercise 6-24. Comment on the shape of the histogram. Does it convey the same information as the stem-and-leaf display?

6-41. Construct a histogram for the overall golf distance data in Exercise 6-25. Comment on the shape of the histogram. Does it convey the same information as the stem-and-leaf display?

6-42. Construct a histogram for the semiconductor speed data in Exercise 6-26. Comment on the shape of the histogram. Does it convey the same information as the stem-and-leaf display?

6-43. Construct a histogram for the pinot noir wine rating data in Exercise 6-27. Comment on the shape of the histogram. Does it convey the same information as the stem-and-leaf display?

6-44. **The Pareto Chart.** An important variation of a histogram for categorical data is the Pareto chart. This chart is widely used in quality improvement efforts, and the categories usually represent different types of defects, failure modes, or product/process problems. The categories are ordered so that the category with the largest frequency is on the left, followed by the category with the second largest frequency and so forth. These charts are named after the Italian economist V. Pareto, and they usually exhibit "Pareto's law"; that is, most of the defects can be accounted for by only a few categories. Suppose that the following information on structural defects in automobile doors is obtained: dents, 4; pits, 4; parts assembled out of sequence, 6; parts undertrimmed, 21; missing holes/slots, 8; parts not lubricated, 5; parts out of contour, 30; and parts not deburred, 3. Construct and interpret a Pareto chart.

6-4 BOX PLOTS

The stem-and-leaf display and the histogram provide general visual impressions about a data set, while numerical quantities such as $\bar{x}$ or s provide information about only one feature of the data. The **box plot** is a graphical display that simultaneously describes several important

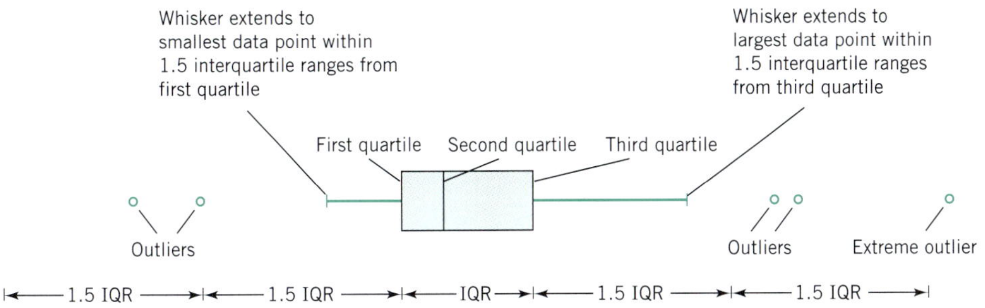

Figure 6-13 Description of a box plot.

features of a data set, such as center, spread, departure from symmetry, and identification of unusual observations or outliers.

A box plot displays the three quartiles, the minimum, and the maximum of the data on a rectangular box, aligned either horizontally or vertically. The box encloses the interquartile range with the left (or lower) edge at the first quartile, q_1, and the right (or upper) edge at the third quartile, q_3. A line is drawn through the box at the second quartile (which is the 50th percentile or the median), $q_2 = \bar{x}$. A line, or **whisker,** extends from each end of the box. The lower whisker is a line from the first quartile to the smallest data point within 1.5 interquartile ranges from the first quartile. The upper whisker is a line from the third quartile to the largest data point within 1.5 interquartile ranges from the third quartile. Data farther from the box than the whiskers are plotted as individual points. A point beyond a whisker, but less than 3 interquartile ranges from the box edge, is called an **outlier.** A point more than 3 interquartile ranges from the box edge is called an **extreme outlier.** See Fig. 6-13. Occasionally, different symbols, such as open and filled circles, are used to identify the two types of outliers. Sometimes box plots are called *box-and-whisker plots*.

Figure 6-14 presents the box plot from Minitab for the alloy compressive strength data shown in Table 6-2. This box plot indicates that the distribution of compressive strengths is fairly symmetric around the central value, because the left and right whiskers and the lengths of the left and right boxes around the median are about the same. There are also two mild outliers at lower strength and one at higher strength. The upper whisker extends to observation 237 because it is the greatest observation below the limit for upper outliers. This limit is $q_3 + 1.5\text{IQR} = 181 + 1.5(181 - 143.5) = 237.25$. The lower whisker extends to observation 97 because it is the smallest observation above the limit for lower outliers. This limit is $q_1 - 1.5\text{IQR} = 143.5 - 1.5(181 - 143.5) = 87.25$.

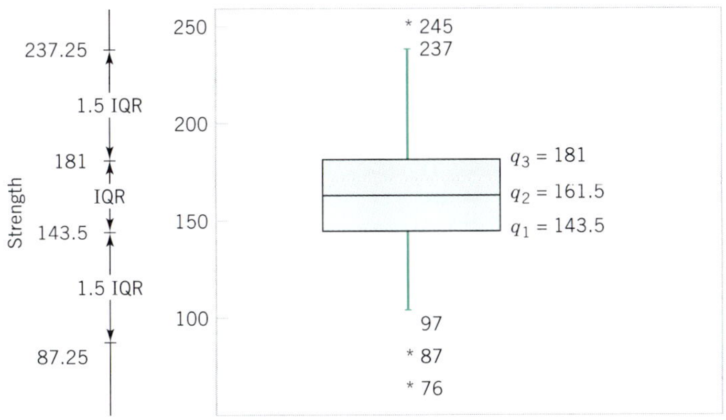

Figure 6-14 Box plot for compressive strength data in Table 6-2.

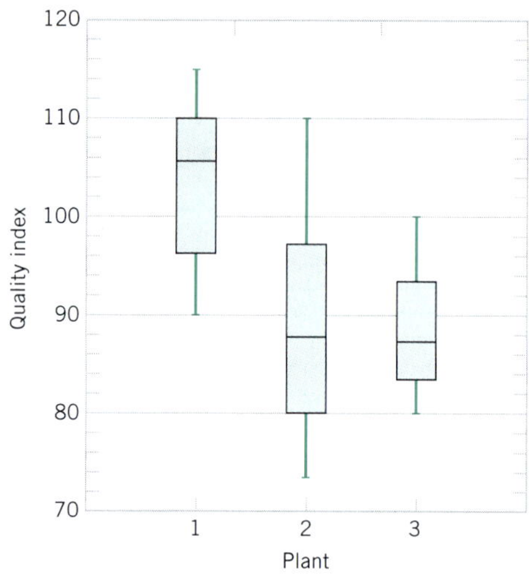

Figure 6-15 Comparative box plots of a quality index at three plants.

Box plots are very useful in graphical comparisons among data sets, because they have high visual impact and are easy to understand. For example, Fig. 6-15 shows the comparative box plots for a manufacturing quality index on semiconductor devices at three manufacturing plants. Inspection of this display reveals that there is too much variability at plant 2 and that plants 2 and 3 need to raise their quality index performance.

EXERCISES FOR SECTION 6-4

6-45. The "cold start ignition time" of an automobile engine is being investigated by a gasoline manufacturer. The following times (in seconds) were obtained for a test vehicle: 1.75, 1.92, 2.62, 2.35, 3.09, 3.15, 2.53, 1.91.

(a) Calculate the sample mean, sample variance, and sample standard deviation.

(b) Construct a box plot of the data.

6-46. An article in the *Transactions of the Institution of Chemical Engineers* (Vol. 34, 1956, pp. 280–293) reported data from an experiment investigating the effect of several process variables on the vapor phase oxidation of naphthalene. A sample of the percentage mole conversion of naphthalene to maleic anhydride follows: 4.2, 4.7, 4.7, 5.0, 3.8, 3.6, 3.0, 5.1, 3.1, 3.8, 4.8, 4.0, 5.2, 4.3, 2.8, 2.0, 2.8, 3.3, 4.8, 5.0.

(a) Calculate the sample mean, sample variance, and sample standard deviation.

(b) Construct a box plot of the data.

6-47. The nine measurements that follow are furnace temperatures recorded on successive batches in a semiconductor manufacturing process (units are °F): 953, 950, 948, 955, 951, 949, 957, 954, 955.

(a) Calculate the sample mean, sample variance, and standard deviation.

(b) Find the median. How much could the largest temperature measurement increase without changing the median value?

(c) Construct a box plot of the data.

6-48. Exercise 6-12 presents drag coefficients for the NASA 0012 airfoil. You were asked to calculate the sample mean, sample variance, and sample standard deviation of those coefficients.

(a) Find the median and the upper and lower quartiles of the drag coefficients.

(b) Construct a box plot of the data.

(c) Set aside the largest observation (100) and rework parts a and b. Comment on your findings.

6-49. Exercise 6-13 presented the joint temperatures of the O-rings (°F) for each test firing or actual launch of the space shuttle rocket motor. In that exercise you were asked to find the sample mean and sample standard deviation of temperature.

(a) Find the median and the upper and lower quartiles of temperature.

(b) Set aside the smallest observation (31°F) and recompute the quantities in part (a). Comment on your findings. How "different" are the other temperatures from this smallest value?

(c) Construct a box plot of the data and comment on the possible presence of outliers.

6-50. Reconsider the motor fuel octane rating data in Exercise 6-14. Construct a box plot of the data and write an interpretation of the plot. How does the box plot compare in interpretive value to the original stem-and-leaf diagram.

6-51. Reconsider the energy consumption data in Exercise 6-21. Construct a box plot of the data and write an interpretation of the plot. How does the box plot compare in interpretive value to the original stem-and-leaf diagram?

6-52. Reconsider the water quality data in Exercise 6-24. Construct a box plot of the concentrations and write an interpretation of the plot. How does the box plot compare in interpretive value to the original stem-and-leaf diagram?

6-53. Reconsider the weld strength data in Exercise 6-23. Construct a box plot of the data and write an interpretation of the plot. How does the box plot compare in interpretive value to the original stem-and-leaf diagram?

6-54. Reconsider the semiconductor speed data in Exercise 6-26. Construct a box plot of the data and write an interpretation of the plot. How does the box plot compare in interpretive value to the original stem-and-leaf diagram?

6-55. Use the data on heights of female and male engineering students from Exercises 6-22 and 6-29 to construct comparative box plots. Write an interpretation of the information that you see in these plots.

6-56. In Exercise 6-45, data was presented on the cold start ignition time of a particular gasoline used in a test vehicle. A second formulation of the gasoline was tested in the same vehicle, with the following times (in seconds): 1.83, 1.99, 3.13, 3.29, 2.65, 2.87, 3.40, 2.46, 1.89, and 3.35. Use this new data along with the cold start times reported in Exercise 6-45 to construct comparative box plots. Write an interpretation of the information that you see in these plots.

6-57. An article in *Nature Genetics* (2003, Vol. 34(1), pp. 85–90) "Treatment-specific changes in gene expression discriminate in vivo drug response in human leukemia cells" studied gene expression as a function of treatments for leukemia. One group received a high dose of the drug while the control group received no treatment. Expression data (measures of gene activity) from one gene are shown in the following table. Construct a box plot for each group of patients. Write an interpretation to compare the information in these plots.

Gene Expression

High Dose	Control	Control	Control
16.1	297.1	820.1	166.5
134.9	491.8	82.5	2258.4
52.7	1332.9	713.9	497.5
14.4	1172	785.6	263.4
124.3	1482.7	114	252.3
99	335.4	31.9	351.4
24.3	528.9	86.3	678.9
16.3	24.1	646.6	3010.2
15.2	545.2	169.9	67.1
47.7	92.9	20.2	318.2
12.9	337.1	280.2	2476.4
72.7	102.3	194.2	181.4
126.7	255.1	408.4	2081.5
46.4	100.5	155.5	424.3
60.3	159.9	864.6	188.1
23.5	168	355.4	563
43.6	95.2	634	149.1
79.4	132.5	2029.9	2122.9
38	442.6	362.1	1295.9
58.2	15.8		
26.5	175.6		
25.1	131.1		

6-5 TIME SEQUENCE PLOTS

The graphical displays that we have considered thus far such as histograms, stem-and-leaf plots, and box plots are very useful visual methods for showing the variability in data. However, we noted in Chapter 1 that time is an important factor that contributes to variability in data, and those graphical methods do not take this into account. A **time series** or **time sequence** is a data set in which the observations are recorded in the order in which they occur. A **time series plot** is a graph in which the vertical axis denotes the observed value of the variable (say x) and the horizontal axis denotes the time (which could be minutes, days, years, etc.) When measurements are plotted as a time series, we often see trends, cycles, or other broad features of the data that could not be seen otherwise.

For example, consider Fig. 6-16(a), which presents a time series plot of the annual sales of a company for the last 10 years. The general impression from this display is that sales show an upward **trend.** There is some variability about this trend, with some years' sales increasing over those of the last year and some years' sales decreasing. Figure 6-16(b) shows the last three years of sales reported by quarter. This plot clearly shows that the annual sales in this

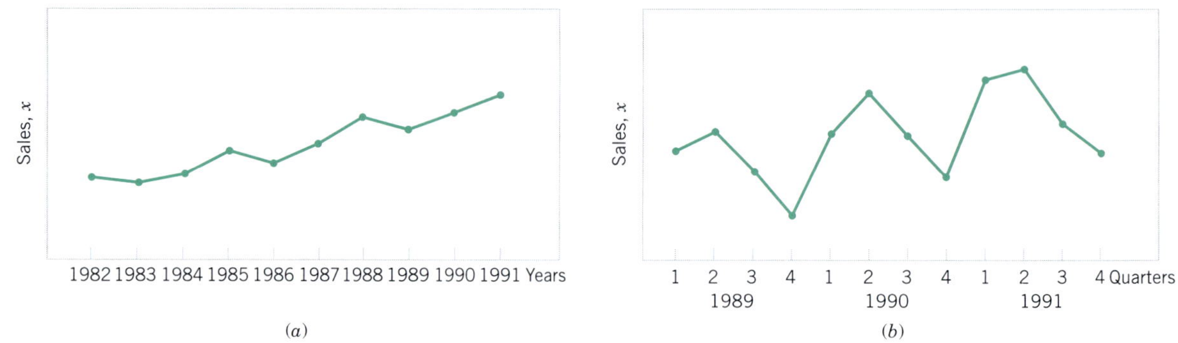

Figure 6-16 Company sales by year (a) and by quarter (b).

business exhibit a **cyclic** variability by quarter, with the first- and second-quarter sales being generally greater than sales during the third and fourth quarters.

Sometimes it can be very helpful to combine a time series plot with some of the other graphical displays that we have considered previously. J. Stuart Hunter (*The American Statistician*, Vol. 42, 1988, p. 54) has suggested combining the stem-and-leaf plot with a time series plot to form a **digidot plot.**

Figure 6-17 shows a digidot plot for the observations on compressive strength from Table 6-2, assuming that these observations are recorded in the order in which they occurred. This plot effectively displays the overall variability in the compressive strength data and simultaneously shows the variability in these measurements over time. The general impression is that compressive strength varies around the mean value of 162.66, and there is no strong obvious pattern in this variability over time.

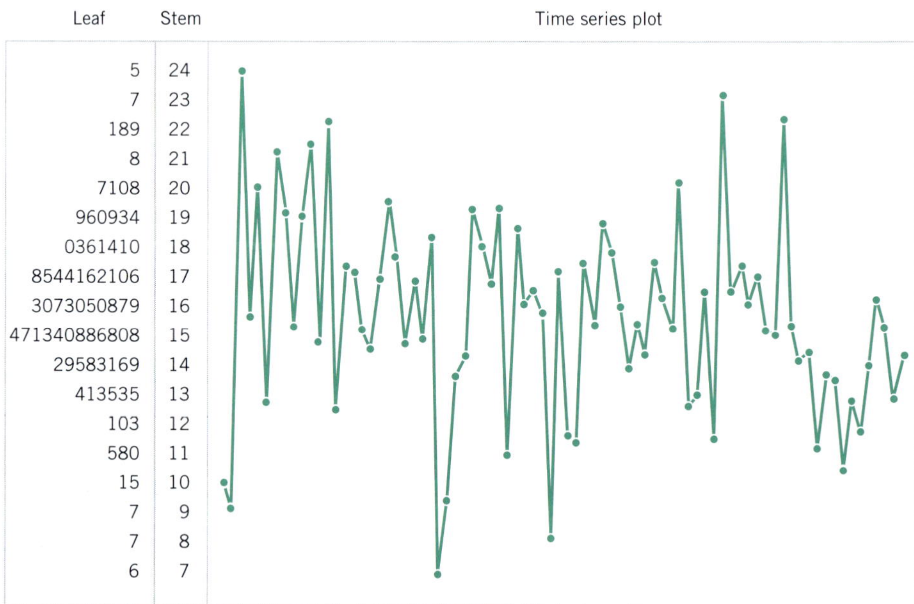

Figure 6-17 A digidot plot of the compressive strength data in Table 6-2.

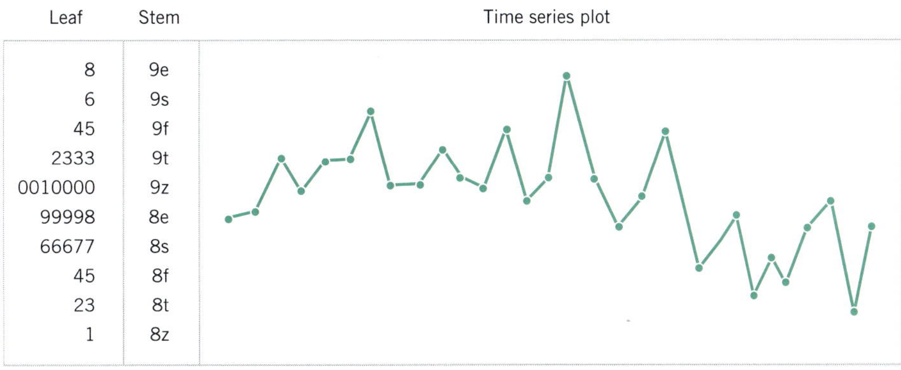

Figure 6-18 A digidot plot of chemical process concentration readings, observed hourly.

The digidot plot in Fig. 6-18 tells a different story. This plot summarizes 30 observations on concentration of the output product from a chemical process, where the observations are recorded at one-hour time intervals. This plot indicates that during the first 20 hours of operation this process produced concentrations generally above 85 grams per liter, but that following sample 20, something may have occurred in the process that results in lower concentrations. If this variability in output product concentration can be reduced, operation of this process can be improved.

EXERCISES FOR SECTION 6-5

6-58. The following data are the viscosity measurements for a chemical product observed hourly (read down, then left to right). Construct and interpret either a digidot plot or a separate stem-and-leaf and time series plot of these data. Specifications on product viscosity are at 48 ± 2. What conclusions can you make about process performance?

47.9	48.6	48.0	48.1	43.0	43.2
47.9	48.8	47.5	48.0	42.9	43.6
48.6	48.1	48.6	48.3	43.6	43.2
48.0	48.3	48.0	43.2	43.3	43.5
48.4	47.2	47.9	43.0	43.0	43.0
48.1	48.9	48.3	43.5	42.8	
48.0	48.6	48.5	43.1	43.1	

6-59. The pull-off force for a connector is measured in a laboratory test. Data for 40 test specimens follow (read down, then left to right). Construct and interpret either a digidot plot or a separate stem-and-leaf and time series plot of the data.

241	203	201	251	236	190
258	195	195	238	245	175
237	249	255	210	209	178
210	220	245	198	212	175
194	194	235	199	185	190
225	245	220	183	187	
248	209	249	213	218	

6-60. In their book *Time Series Analysis, Forecasting, and Control* (Prentice Hall, 1994), G. E. P. Box, G. M. Jenkins, and G. C. Reinsel present chemical process concentration readings made every two hours. Some of these data follow (read down, then left to right).

17.0	16.7	17.1	17.5	17.6
16.6	17.4	17.4	18.1	17.5
16.3	17.2	17.4	17.5	16.5
16.1	17.4	17.5	17.4	17.8
17.1	17.4	17.4	17.4	17.3
16.9	17.0	17.6	17.1	17.3
16.8	17.3	17.4	17.6	17.1
17.4	17.2	17.3	17.7	17.4
17.1	17.4	17.0	17.4	16.9
17.0	16.8	17.8	17.8	17.3

Construct and interpret either a digidot plot or a stem-and-leaf plot of these data.

6-61. The 100 annual Wolfer sunspot numbers from 1770 to 1869 follow. (For an interesting analysis and interpretation of these numbers, see the book by Box, Jenkins, and Reinsel referenced in Exercise 6-60. Their analysis requires some advanced knowledge of statistics and statistical model building.) Read down, then left to right. The 1869 result is 74. Construct and interpret either a digidot plot or a stem-and-leaf and time series plot of these data.

101	31	154	38	83	90
82	7	125	23	132	67
66	20	85	10	131	60
35	92	68	24	118	47
41	10	16	8	62	94
21	8	7	13	98	96
16	2	4	57	124	77
6	0	2	122	96	59
4	1	8	138	66	44
7	5	17	103	64	47
14	12	36	86	54	30
34	14	50	63	39	16
45	35	62	37	21	7
43	46	67	24	7	37
48	41	71	11	4	74
42	30	48	15	23	
28	24	28	40	55	

6-62. In their book *Forecasting and Time Series Analysis*, 2nd edition (McGraw-Hill, 1990), D. C. Montgomery, L. A. Johnson, and J. S. Gardiner analyze the data in Table 6-5, which are the monthly total passenger airline miles flown in the United Kingdom from 1964–1970 (in millions of miles). Comment on any features of the data that are apparent. Construct and interpret either a digidot plot or a separate stem-and-leaf and time series plot of these data.

6-63. The following table shows the number of earthquakes per year of magnitude 7.0 and greater since 1900 (source: Earthquake Data Base System of the U.S. Geological Survey, National Earthquake Information Center, Golden CO).

1900	13	1904	16	1908	18	1912	22
1901	14	1905	26	1909	32	1913	23
1902	8	1906	32	1910	36	1914	22
1903	10	1907	27	1911	24	1915	18

1916	25	1942	27	1968	30	1994	13
1917	21	1943	41	1969	27	1995	20
1918	21	1944	31	1970	29	1996	15
1919	14	1945	27	1971	23	1997	16
1920	8	1946	35	1972	20	1998	12
1921	11	1947	26	1973	16	1999	18
1922	14	1948	28	1974	21	2000	15
1923	23	1949	36	1975	21	2001	16
1924	18	1950	39	1976	25	2002	13
1925	17	1951	21	1977	16	2003	15
1926	19	1952	17	1978	18	2004	15
1927	20	1953	22	1979	15		
1928	22	1954	17	1980	18		
1929	19	1955	19	1981	14		
1930	13	1956	15	1982	10		
1931	26	1957	34	1983	15		
1932	13	1958	10	1984	8		
1933	14	1959	15	1985	15		
1934	22	1960	22	1986	6		
1935	24	1961	18	1987	11		
1936	21	1962	15	1988	8		
1937	22	1963	20	1989	7		
1938	26	1964	15	1990	18		
1939	21	1965	22	1991	16		
1940	23	1966	19	1992	13		
1941	24	1967	16	1993	12		

Construct and interpret either a digidot plot or a separate stem-and-leaf and time series plot of these data.

6-64. The following table shows U.S. petroleum imports, imports as a percentage of total, and Persian Gulf imports as a percentage of all imports by year since 1973 (source: U.S. Department of Energy Web site http://www.eia.doe.gov/). Construct and interpret either a digidot plot or a separate stem-and-leaf and time series plot for each column of data.

Table 6-5 United Kingdom Passenger Airline Miles Flown

Month	1964	1965	1966	1967	1968	1969	1970
Jan.	7.269	8.350	8.186	8.334	8.639	9.491	10.840
Feb.	6.775	7.829	7.444	7.899	8.772	8.919	10.436
Mar.	7.819	8.829	8.484	9.994	10.894	11.607	13.589
Apr.	8.371	9.948	9.864	10.078	10.455	8.852	13.402
May	9.069	10.638	10.252	10.801	11.179	12.537	13.103
June	10.248	11.253	12.282	12.953	10.588	14.759	14.933
July	11.030	11.424	11.637	12.222	10.794	13.667	14.147
Aug.	10.882	11.391	11.577	12.246	12.770	13.731	14.057
Sept.	10.333	10.665	12.417	13.281	13.812	15.110	16.234
Oct.	9.109	9.396	9.637	10.366	10.857	12.185	12.389
Nov.	7.685	7.775	8.094	8.730	9.290	10.645	11.594
Dec.	7.682	7.933	9.280	9.614	10.925	12.161	12.772

Year	Petroleum Imports (thousand barrels per day)	Total Petroleum Imports as Percent of Petroleum Products Supplied (tenth percent)	Petroleum Imports from Persian Gulf as Percent of Total Petroleum Imports (tenth percent)
1973	6256.145	361	135
1974	6112.184	367	170
1975	6055.712	371	192
1976	7312.598	418	251
1977	8807.249	477	278
1978	8363.411	443	265
1979	8456.129	456	244
1980	6909.025	405	219
1981	5995.673	373	203
1982	5113.311	334	136
1983	5051.353	331	87
1984	5436.982	345	93
1985	5067.144	322	61
1986	6223.512	382	146
1987	6677.696	400	161
1988	7402.021	428	208
1989	8060.545	465	230
1990	8017.521	471	245
1991	7626.748	456	241
1992	7887.697	463	225
1993	8620.422	500	206
1994	8996.222	507	192
1995	8834.94	498	178
1996	9478.492	517	169
1997	10161.562	545	172
1998	10708.071	566	199
1999	10852.258	555	227
2000	11459.251	581	217
2001	11871.337	604	232
2002	11530.241	583	196
2003	12264.386	612	203
2004	13145.093	634	189

6-6 PROBABILITY PLOTS

How do we know if a particular probability distribution is a reasonable model for data? Sometimes, this is an important question because many of the statistical techniques presented in subsequent chapters are based on an assumption that the population distribution is of a specific type. Thus, we can think of determining whether data come from a specific probability distribution as **verifying assumptions.** In other cases, the form of the distribution can give insight into the underlying physical mechanism generating the data. For example, in reliability engineering, verifying that time-to-failure data come from an exponential distribution identifies the **failure mechanism** in the sense that the failure rate is constant with respect to time.

Some of the visual displays we have used earlier, such as the histogram, can provide insight about the form of the underlying distribution. However, histograms are usually not really reliable indicators of the distribution form unless the sample size is very large. A **probability plot** is a graphical method for determining whether sample data conform to a hypothesized distribution based on a subjective visual examination of the data. The general procedure is very simple and can be performed quickly. It is also more reliable than the histogram for small to moderate size samples. Probability plotting typically uses special axes that have been scaled for the hypothesized distribution. Software is widely available for the normal, lognormal, Weibull, and various chi-square and gamma distributions. We focus primarily on normal probability plots because many statistical techniques are appropriate only when the population is (at least approximately) normal.

To construct a probability plot, the observations in the sample are first ranked from smallest to largest. That is, the sample $x_1, x_2, \ldots, x_n$ is arranged as $x_{(1)}, x_{(2)}, \ldots, x_{(n)}$, where $x_{(1)}$ is the smallest observation, $x_{(2)}$ is the second smallest observation, and so forth, with $x_{(n)}$ the largest. The ordered observations $x_{(j)}$ are then plotted against their observed cumulative frequency $(j - 0.5)/n$ on the appropriate probability paper. If the hypothesized distribution adequately describes the data, the plotted points will fall approximately along a straight line; if the plotted points deviate significantly from a straight line, the hypothesized model is not appropriate. Usually, the determination of whether or not the data plot as a straight line is subjective. The procedure is illustrated in the following example.

EXAMPLE 6-7
Battery Life

Ten observations on the effective service life in minutes of batteries used in a portable personal computer are as follows: 176, 191, 214, 220, 205, 192, 201, 190, 183, 185. We hypothesize that battery life is adequately modeled by a normal distribution. To use probability plotting to investigate this hypothesis, first arrange the observations in ascending order and calculate their cumulative frequencies $(j - 0.5)/10$ as shown in Table 6-6.

The pairs of values $x_{(j)}$ and $(j - 0.5)/10$ are now plotted on normal probability axes. This plot is shown in Fig. 6-19. Most normal probability plots $100(j - 0.5)/n$ on the left vertical scale and $100[1 - (j - 0.5)/n]$ on the right vertical scale, with the variable value plotted on the horizontal scale. A straight line, chosen subjectively, has been drawn through the plotted points. In drawing the straight line, you should be influenced more by the points near the middle of the plot than by the extreme points. A good rule of thumb is to draw the line approximately between the 25th and 75th percentile points. This is how the line in Fig. 6-19 was determined. In assessing the "closeness" of the points to the straight line, imagine a "fat pencil" lying along the line. If all the points are covered by this imaginary pencil, a

Table 6-6 Calculation for Constructing a Normal Probability Plot

j	$x_{(j)}$	$(j - 0.5)/10$	z_j
1	176	0.05	−1.64
2	183	0.15	−1.04
3	185	0.25	−0.67
4	190	0.35	−0.39
5	191	0.45	−0.13
6	192	0.55	0.13
7	201	0.65	0.39
8	205	0.75	0.67
9	214	0.85	1.04
10	220	0.95	1.64

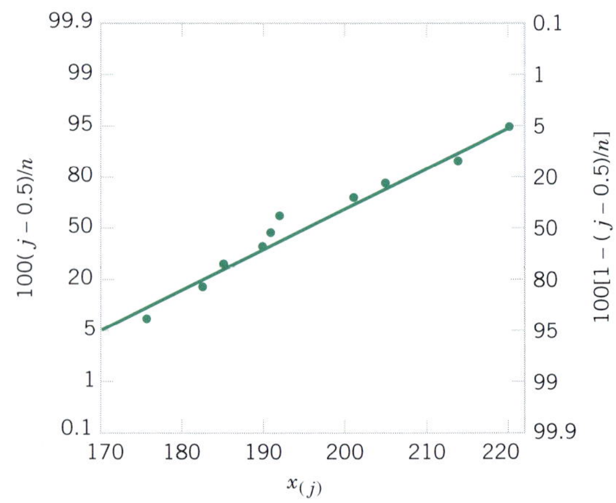

Figure 6-19 Normal probability plot for battery life.

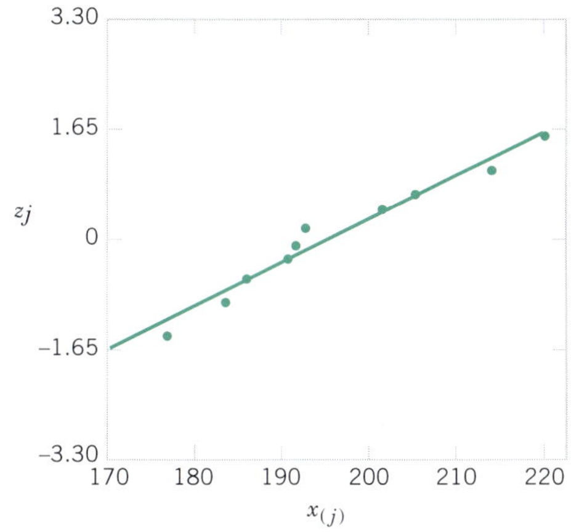

Figure 6-20 Normal probability plot obtained from standardized normal scores.

normal distribution adequately describes the data. Since the points in Fig. 6-19 would pass the "fat pencil" test, we conclude that the normal distribution is an appropriate model.

A **normal probability plot** can also be constructed on ordinary axes by plotting the standardized normal scores z_j against $x_{(j)}$, where the standardized normal scores satisfy

$$\frac{j - 0.5}{n} = P(Z \leq z_j) = \Phi(z_j)$$

For example, if $(j - 0.5)/n = 0.05$, $\Phi(z_j) = 0.05$ implies that $z_j = -1.64$. To illustrate, consider the data from Example 6-4. In the last column of Table 6-6 we show the standardized normal scores. Figure 6-20 presents the plot of z_j versus $x_{(j)}$. This normal probability plot is equivalent to the one in Fig. 6-19.

We have constructed our probability plots with the probability scale (or the z-scale) on the vertical axis. Some computer packages "flip" the axis and put the probability scale on the horizontal axis.

The normal probability plot can be useful in identifying distributions that are symmetric but that have tails that are "heavier" or "lighter" than the normal. They can also be useful in identifying skewed distributions. When a sample is selected from a light-tailed distribution (such as the uniform distribution), the smallest and largest observations will not be as extreme as would be expected in a sample from a normal distribution. Thus if we consider the straight line drawn through the observations at the center of the normal probability plot, observations on the left side will tend to fall below the line, whereas observations on the right side will tend to fall above the line. This will produce an S-shaped normal probability plot such as shown in Fig. 6-21(a). A heavy-tailed distribution will result in data that also produces an S-shaped normal probability plot, but now the observations on the left will be above the straight line and the observations on the right will lie below the line. See Fig. 6-19(b). A positively skewed distribution will tend to produce a pattern such as shown in Fig. 6-19(c), where points on both ends of the plot tend to fall below the line, giving a curved shape to the plot. This occurs because both the smallest and the largest observations from this type of distribution are larger than expected in a sample from a normal distribution.

Even when the underlying population is exactly normal, the sample data will not plot exactly on a straight line. Some judgment and experience are required to evaluate the plot. Generally, if the sample size is $n < 30$, there can be a lot of deviation from linearity in normal

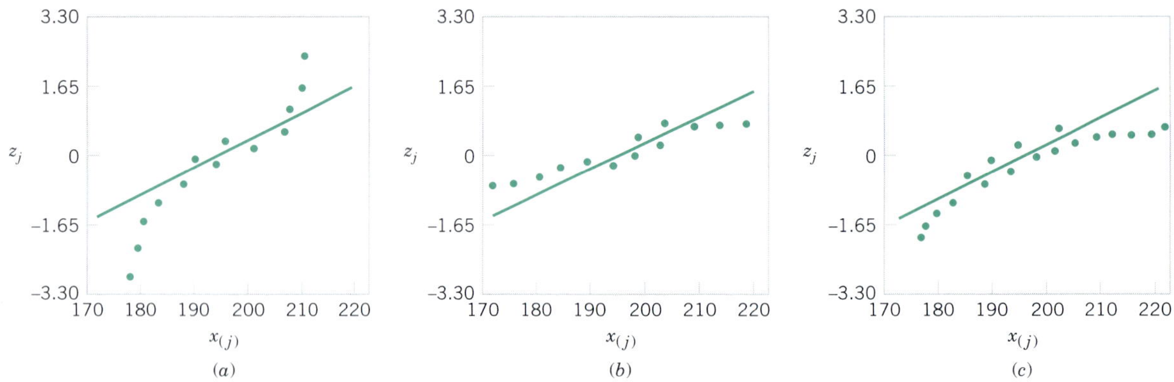

Figure 6-21 Normal probability plots indicating a nonnormal distribution. (a) Light-tailed distribution. (b) Heavy-tailed distribution. (c) A distribution with positive (or right) skew.

plots, so in these cases only a very severe departure from linearity should be interpreted as a strong indication of nonnormality. As n increases, the linear pattern will tend to become stronger, and the normal probability plot will be easier to interpret and more reliable as an indicator of the form of the distribution.

EXERCISES FOR SECTION 6-6

6-65. Construct a normal probability plot of the piston ring diameter data in Exercise 6-1. Does it seem reasonable to assume that piston ring diameter is normally distributed?

6-66. Construct a normal probability plot of the insulating fluid breakdown time data in Exercise 6-2. Does it seem reasonable to assume that breakdown time is normally distributed?

6-67. Construct a normal probability plot of the visual accommodation data in Exercise 6-5. Does it seem reasonable to assume that visual accommodation is normally distributed?

6-68. Construct a normal probability plot of the solar intensity data in Exercise 6-6. Does it seem reasonable to assume that solar intensity is normally distributed?

6-69. Construct a normal probability plot of the O-ring joint temperature data in Exercise 6-13. Does it seem reasonable to assume that O-ring joint temperature is normally distributed? Discuss any interesting features that you see on the plot.

6-70. Construct a normal probability plot of the octane rating data in Exercise 6-14. Does it seem reasonable to assume that octane rating is normally distributed?

6-71. Construct a normal probability plot of the cycles to failure data in Exercise 6-15. Does it seem reasonable to assume that cycles to failure is normally distributed?

6-72. Construct a normal probability plot of the suspended solids concentration data in Exercise 6-24. Does it seem reasonable to assume that the concentration of suspended solids in water from this particular lake is normally distributed?

6-73. Construct two normal probability plots for the height data in Exercises 6-22 and 6-29. Plot the data for female and male students on the same axes. Does height seem to be normally distributed for either group of students? If both populations have the same variance, the two normal probability plots should have identical slopes. What conclusions would you draw about the heights of the two groups of students from visual examination of the normal probability plots?

6-74. It is possible to obtain a "quick and dirty" estimate of the mean of a normal distribution from the fiftieth percentile value on a normal probability plot. Provide an argument why this is so. It is also possible to obtain an estimate of the standard deviation of a normal distribution by subtracting the eighty-fourth percentile value from the fiftieth percentile value. Provide an argument why this is so.

Supplemental Exercises

6-75. The National Oceanic and Atmospheric Administration provided the monthly absolute estimates of global (land and ocean combined) temperature index (degrees C) from 2000–2004. Read January to December from left to right. (source: http://www.ncdc.noaa.gov/oa/climate/research/anomalies/anomalies.html) Construct and interpret either a digidot plot or a separate stem-and-leaf and time series plot of these data.

Year	Global Monthly Temperature											
2000	12.28	12.63	13.22	14.21	15.13	15.82	16.05	16.02	15.29	14.29	13.16	12.47
2001	12.44	12.55	13.35	14.22	15.28	15.99	16.23	16.17	15.44	14.52	13.52	12.61
2002	12.71	12.88	13.49	14.30	15.26	16.02	16.28	16.06	15.44	14.45	13.43	12.56
2003	12.60	12.62	13.30	14.21	15.28	16.02	16.28	16.19	15.54	14.62	13.49	12.77
2004	12.52	12.77	13.41	14.28	15.15	15.92	16.17	16.08	15.48	14.59	13.58	12.71

6-76. The concentration of a solution is measured six times by one operator using the same instrument. She obtains the following data: 63.2, 67.1, 65.8, 64.0, 65.1, and 65.3 (grams per liter).

(a) Calculate the sample mean. Suppose that the desirable value for this solution has been specified to be 65.0 grams per liter. Do you think that the sample mean value computed here is close enough to the target value to accept the solution as conforming to target? Explain your reasoning.

(b) Calculate the sample variance and sample standard deviation.

(c) Suppose that in measuring the concentration, the operator must set up an apparatus and use a reagent material. What do you think the major sources of variability are in this experiment? Why is it desirable to have a small variance of these measurements?

6-77. The table below shows unemployment data for the U.S. that is seasonally adjusted. Construct a time series plot of this data and comment on any features (source: U.S. Bureau of Labor Web site http://data.bls.gov).

6-78. A sample of six resistors yielded the following resistances (ohms): $x_1 = 45, x_2 = 38, x_3 = 47, x_4 = 41, x_5 = 35$, and $x_6 = 43$.

(a) Compute the sample variance and sample standard deviation.

(b) Subtract 35 from each of the original resistance measurements and compute s^2 and s. Compare your results with those obtained in part (a) and explain your findings.

(c) If the resistances were 450, 380, 470, 410, 350, and 430 ohms, could you use the results of previous parts of this problem to find s^2 and s?

6-79. Consider the following two samples:

Sample 1: 10, 9, 8, 7, 8, 6, 10, 6

Sample 2: 10, 6, 10, 6, 8, 10, 8, 6

(a) Calculate the sample range for both samples. Would you conclude that both samples exhibit the same variability? Explain.

(b) Calculate the sample standard deviations for both samples. Do these quantities indicate that both samples have the same variability? Explain.

(c) Write a short statement contrasting the sample range versus the sample standard deviation as a measure of variability.

6-80. An article in *Quality Engineering* (Vol. 4, 1992, pp. 487–495) presents viscosity data from a batch chemical process. A sample of these data follows on p. 226:

Unemployment Percentage

Year	Jan	Feb	Mar	Apr	May	Jun	Jul	Aug	Sep	Oct	Nov	Dec
1995	5.6	5.4	5.4	5.8	5.6	5.6	5.7	5.7	5.6	5.5	5.6	5.6
1996	5.6	5.5	5.5	5.6	5.6	5.3	5.5	5.1	5.2	5.2	5.4	5.4
1997	5.3	5.2	5.2	5.1	4.9	5.0	4.9	4.8	4.9	4.7	4.6	4.7
1998	4.6	4.6	4.7	4.3	4.4	4.5	4.5	4.5	4.6	4.5	4.4	4.4
1999	4.3	4.4	4.2	4.3	4.2	4.3	4.3	4.2	4.2	4.1	4.1	4.0
2000	4.0	4.1	4.0	3.8	4.0	4.0	4.0	4.1	3.9	3.9	3.9	3.9
2001	4.2	4.2	4.3	4.4	4.3	4.5	4.6	4.9	5.0	5.3	5.6	5.7
2002	5.7	5.7	5.7	5.9	5.8	5.8	5.8	5.7	5.7	5.7	5.9	6.0
2003	5.8	5.9	5.8	6.0	6.1	6.3	6.2	6.1	6.1	6.0	5.9	5.7
2004	5.7	5.6	5.7	5.5	5.6	5.6	5.5	5.4	5.4	5.5	5.4	5.4
2005	5.2	5.4	5.2	5.2	5.1	5.0	5.0					

13.3	14.3	14.9	15.2	15.8	14.2	16.0	14.0
14.5	16.1	13.7	15.2	13.7	16.9	14.9	14.4
15.3	13.1	15.2	15.9	15.1	14.9	13.6	13.7
15.3	15.5	14.5	16.5	13.4	15.2	15.3	13.8
14.3	12.6	15.3	14.8	14.1	14.4	14.3	15.6
14.8	14.6	15.6	15.1	14.8	15.2	15.6	14.5
15.2	14.3	15.8	17.0	14.3	14.6	16.1	12.8
14.5	15.4	13.3	14.9	14.3	16.4	13.9	16.1
14.6	15.2	14.1	14.8	16.4	14.2	15.2	16.6
14.1	16.8	15.4	14.0	16.9	15.7	14.4	15.6

(a) Reading down and left to right, draw a time series plot of all the data and comment on any features of the data that are revealed by this plot.
(b) Consider the notion that the first 40 observations were generated from a specific process, whereas the last 40 observations were generated from a different process. Does the plot indicate that the two processes generate similar results?
(c) Compute the sample mean and sample variance of the first 40 observations; then compute these values for the second 40 observations. Do these quantities indicate that both processes yield the same mean level? The same variability? Explain.

6-81. The total net electricity consumption of the U.S. by year from 1980–2002 (in billion kilowatt hours) follows. Net consumption excludes the energy consumed by the generating units. Read left to right.

2094.45, 2147.10, 2086.44, 2150.96, 2285.80, 2323.97, 2368.75, 2457.27, 2578.06, 2754.95, 2826.59, 2988.97, 3068.67, 3157.35, 3246.98, 3294.04, 3425.13, 3494.60, 3604.68, 3553.79, 3659.99

(source: U.S. Department of Energy Web site http://www/eia.doe.gov/emeu/international/electric.html#Consumption Total)

Construct a time series plot of these data. Construct and interpret a stem-and-leaf display of these data.

6-82. Reconsider the data from Exercise 6-80. Prepare comparative box plots for two groups of observations: the first 40 and the last 40. Comment on the information in the box plots.

6-83. The data shown in Table 6-7 are monthly champagne sales in France (1962-1969) in thousands of bottles.
(a) Construct a time series plot of the data and comment on any features of the data that are revealed by this plot.
(b) Speculate on how you would use a graphical procedure to forecast monthly champagne sales for the year 1970.

6-84. The following data are the temperatures of effluent at discharge from a sewage treatment facility on consecutive days:

43	47	51	48	52	50	46	49
45	52	46	51	44	49	46	51
49	45	44	50	48	50	49	50

(a) Calculate the sample mean, sample median, sample variance and sample standard deviation.
(b) Construct a box plot of the data and comment on the information in this display.

6-85. A manufacturer of coil springs is interested in implementing a quality control system to monitor his production process. As part of this quality system, it is decided to record the number of nonconforming coil springs in each production batch of size 50. During 40 days of production, 40 batches of data were collected as follows:

Read data across.

9	12	6	9	7	14	12	4	6	7
8	5	9	7	8	11	3	6	7	7
11	4	4	8	7	5	6	4	5	8
19	19	18	12	11	17	15	17	13	13

(a) Construct a stem-and-leaf plot of the data.
(b) Find the sample average and standard deviation.

Table 6-7 Champagne Sales in France

Month	1962	1963	1964	1965	1966	1967	1968	1969
Jan.	2.851	2.541	3.113	5.375	3.633	4.016	2.639	3.934
Feb.	2.672	2.475	3.006	3.088	4.292	3.957	2.899	3.162
Mar.	2.755	3.031	4.047	3.718	4.154	4.510	3.370	4.286
Apr.	2.721	3.266	3.523	4.514	4.121	4.276	3.740	4.676
May	2.946	3.776	3.937	4.520	4.647	4.968	2.927	5.010
June	3.036	3.230	3.986	4.539	4.753	4.677	3.986	4.874
July	2.282	3.028	3.260	3.663	3.965	3.523	4.217	4.633
Aug.	2.212	1.759	1.573	1.643	1.723	1.821	1.738	1.659
Sept.	2.922	3.595	3.528	4.739	5.048	5.222	5.221	5.591
Oct.	4.301	4.474	5.211	5.428	6.922	6.873	6.424	6.981
Nov.	5.764	6.838	7.614	8.314	9.858	10.803	9.842	9.851
Dec.	7.132	8.357	9.254	10.651	11.331	13.916	13.076	12.670

(c) Construct a time series plot of the data. Is there evidence that there was an increase or decrease in the average number of nonconforming springs made during the 40 days? Explain.

6-86. A communication channel is being monitored by recording the number of errors in a string of 1000 bits. Data for 20 of these strings follow:

Read data across.

3	1	0	1	3	2	4	1	3	1
1	1	2	3	3	2	0	2	0	1

(a) Construct a stem-and-leaf plot of the data.
(b) Find the sample average and standard deviation.
(c) Construct a time series plot of the data. Is there evidence that there was an increase or decrease in the number of errors in a string? Explain.

6-87. Reconsider the golf course yardage data in Exercise 6-3. Construct a box plot of the yardages and write an interpretation of the plot.

6-88. Reconsider the data in Exercise 6-80. Construct normal probability plots for two groups of the data: the first 40 and the last 40 observations. Construct both plots on the same axes. What tentative conclusions can you draw?

6-89. Construct a normal probability plot of the effluent discharge temperature data from Exercise 6-84. Based on the plot, what tentative conclusions can you draw?

6-90. Construct normal probability plots of the cold start ignition time data presented in Exercises 6-45 and 6-56. Construct a separate plot for each gasoline formulation, but arrange the plots on the same axes. What tentative conclusions can you draw?

6-91. Reconsider the golf ball overall distance data in Exercise 6-25. Construct a box plot of the yardage distance and write an interpretation of the plot. How does the box plot compare in interpretive value to the original stem-and-leaf diagram.

6-92. **Transformations.** In some data sets, a transformation by some mathematical function applied to the original data, such as $\sqrt{y}$ or log y, can result in data that are simpler to work with statistically than the original data. To illustrate the effect of a transformation, consider the following data, which represent cycles to failure for a yarn product: 675, 3650, 175, 1150, 290, 2000, 100, 375.

(a) Construct a normal probability plot and comment on the shape of the data distribution.
(b) Transform the data using logarithms; that is, let y^* (new value) = log y (old value). Construct a normal probability plot of the transformed data and comment on the effect of the transformation.

6-93. In 1879, A. A. Michelson made 100 determinations of the velocity of light in air using a modification of a method proposed by the French physicist Foucault. He made the measurements in five trials of 20 measurements each. The observations (in kilometers per second) follow. Each value has 299,000 subtracted from it.

Trial 1

850	900	930	950	980
1000	930	760	1000	960
740	1070	850	980	880
980	650	810	1000	960

Trial 2

960	960	880	850	900
830	810	880	800	760
940	940	800	880	840
790	880	830	790	800

Trial 3

880	880	720	620	970
880	850	840	850	840
880	860	720	860	950
910	870	840	840	840

Trial 4

890	810	800	760	750
910	890	880	840	850
810	820	770	740	760
920	860	720	850	780

Trial 5

890	780	760	790	820
870	810	810	950	810
840	810	810	810	850
870	740	940	800	870

The currently accepted true velocity of light in a vacuum is 299, 792.5 kilometers per second. Stigler (1977, *The Annals of Statistics*) reports that the "true" value for comparison to these measurements is 734.5. Construct comparative box plots of these measurements. Does it seem that all five trials are consistent with respect to the variability of the measurements? Are all five trials centered on the same value? How does each group of trials compare to the true value? Could there have been "startup" effects in the experiment that Michelson performed? Could there have been bias in the measuring instrument?

6-94. In 1789, Henry Cavendish estimated the density of the earth by using a torsion balance. His 29 measurements follow, expressed as a multiple of the density of water.

5.50	5.30	5.47	5.10	5.29	5.65
5.55	5.61	5.75	5.63	5.27	5.44
5.57	5.36	4.88	5.86	5.34	5.39
5.34	5.53	5.29	4.07	5.85	5.46
5.42	5.79	5.62	5.58	5.26	

(a) Calculate the sample mean, sample standard deviation, and median of the Cavendish density data.

(b) Construct a normal probability plot of the data. Comment on the plot. Does there seem to be a "low" outlier in the data?

(c) Would the sample median be a better estimate of the density of the earth than the sample mean? Why?

MIND-EXPANDING EXERCISES

6-95. Consider the airfoil data in Exercise 6-12. Subtract 30 from each value and then multiply the resulting quantities by 10. Now compute s^2 for the new data. How is this quantity related to s^2 for the *original* data? Explain why.

6-96. Consider the quantity $\sum_{i=1}^{n}(x_i - a)^2$. For what value of a is this quantity minimized?

6-97. Using the results of Exercise 6-96, which of the two quantities $\sum_{i=1}^{n}(x_i - \bar{x})^2$ and $\sum_{i=1}^{n}(x_i - \mu)^2$ will be smaller, provided that $\bar{x} \neq \mu$?

6-98. **Coding the Data.** Let $y_i = a + bx_i$, $i = 1, 2, \ldots, n$, where a and b are nonzero constants. Find the relationship between $\bar{x}$ and $\bar{y}$, and between s_x and s_y.

6-99. A sample of temperature measurements in a furnace yielded a sample average (°F) of 835.00 and a sample standard deviation of 10.5. Using the results from Exercise 6-98, what are the sample average and sample standard deviations expressed in °C?

6-100. Consider the sample $x_1, x_2, \ldots, x_n$ with sample mean $\bar{x}$ and sample standard deviation s. Let $z_i = (x_i - \bar{x})/s$, $i = 1, 2, \ldots, n$. What are the values of the sample mean and sample standard deviation of the z_i?

6-101. An experiment to investigate the survival time in hours of an electronic component consists of placing the parts in a test cell and running them for 100 hours under elevated temperature conditions. (This is called an "accelerated" life test.) Eight components were tested with the following resulting failure times:

$$75, 63, 100^+, 36, 51, 45, 80, 90$$

The observation 100^+ indicates that the unit still functioned at 100 hours. Is there any meaningful measure of location that can be calculated for these data? What is its numerical value?

6-102. Suppose that we have a sample $x_1, x_2, \ldots, x_n$ and we have calculated $\bar{x}_n$ and s_n^2 for the sample. Now an $(n + 1)$st observation becomes available. Let $\bar{x}_{n+1}$ and s_{n+1}^2 be the sample mean and sample variance for the sample using all $n + 1$ observations.

(a) Show how $\bar{x}_{n+1}$ can be computed using $\bar{x}_n$ and x_{n+1}.

(b) Show that $ns_{n+1}^2 = (n - 1)s_n^2 + \dfrac{n(x_{n+1} - \bar{x}_n)^2}{n + 1}$

(c) Use the results of parts (a) and (b) to calculate the new sample average and standard deviation for the data of Exercise 6-22, when the new observation is $x_{38} = 64$.

6-103. **Trimmed Mean.** Suppose that the data are arranged in increasing order, T% of the observations are removed from each end and the sample mean of the remaining numbers is calculated. The resulting quantity is called a *trimmed mean.* The trimmed mean generally lies between the sample mean $\bar{x}$ and the sample median $\tilde{x}$. Why?

(a) Calculate the 10% trimmed mean for the yield data in Exercise 6-17.

(b) Calculate the 20% trimmed mean for the yield data in Exercise 6-17 and compare it with the quantity found in part (a).

(c) Compare the values calculated in parts (a) and (b) with the sample mean and median for the yield data. Is there much difference in these quantities? Why?

6-104. **Trimmed Mean.** Suppose that the sample size n is such that the quantity $nT/100$ is not an integer. Develop a procedure for obtaining a trimmed mean in this case.

IMPORTANT TERMS AND CONCEPTS

Box plot	Normal probability plot	Probability plot	Sample variance
Frequency distribution and histogram	Pareto chart	Relative frequency distribution	Stem-and-leaf diagram
Median, quartiles and percentiles	Population mean	Sample mean	Time series plots
Multivariable data	Population standard deviation	Sample standard deviation	
	Population variance		

7 Point Estimation of Parameters and Sampling Distributions

CHAPTER OUTLINE

LEARNING OBJECTIVES

After careful study of this chapter you should be able to do the following:

1. Explain the general concepts of estimating the parameters of a population or a probability distribution
2. Explain the important role of the normal distribution as a sampling distribution
3. Understand the central limit theorem
4. Explain important properties of point estimators, including bias, variance, and mean square error
5. Know how to construct point estimators using the method of moments and the method of maximum likelihood
6. Know how to compute and explain the precision with which a parameter is estimated
7. Know how to construct a point estimator using the Bayesian approach

7-1 INTRODUCTION

The field of statistical inference consists of those methods used to make decisions or to draw conclusions about a **population.** These methods utilize the information contained in a **sample** from the population in drawing conclusions. This chapter begins our study of the statistical methods used for inference and decision making.

Statistical inference may be divided into two major areas: **parameter estimation** and **hypothesis testing.** As an example of a parameter estimation problem, suppose that a structural engineer is analyzing the tensile strength of a component used in an automobile chassis. Since variability in tensile strength is naturally present between the individual components because of differences in raw material batches, manufacturing processes, and measurement procedures (for example), the engineer is interested in estimating the mean tensile strength of the components. In practice, the engineer will use sample data to compute a number that is in some sense a reasonable value (or guess) of the true mean. This number is called a **point estimate.** We will see that it is possible to establish the precision of the estimate.

Now consider a situation in which two different reaction temperatures can be used in a chemical process, say t_1 and t_2. The engineer conjectures that t_1 results in higher yields than does t_2. Statistical hypothesis testing is a framework for solving problems of this type. In this case, the hypothesis would be that the mean yield using temperature t_1 is greater than the mean yield using temperature t_2. Notice that there is no emphasis on estimating yields; instead, the focus is on drawing conclusions about a stated hypothesis.

Suppose that we want to obtain a point estimate of a population parameter. We know that before the data is collected, the observations are considered to be random variables, say $X_1, X_2, \ldots, X_n$. Therefore, any function of the observation, or any **statistic,** is also a random variable. For example, the sample mean $\overline{X}$ and the sample variance S^2 are statistics and they are also random variables.

Since a statistic is a random variable, it has a probability distribution. We call the probability distribution of a statistic a **sampling distribution.** The notion of a sampling distribution is very important and will be discussed and illustrated later in the chapter.

When discussing inference problems, it is convenient to have a general symbol to represent the parameter of interest. We will use the Greek symbol θ (theta) to represent the parameter. The objective of point estimation is to select a single number, based on sample data, that is the most plausible value for θ. A numerical value of a sample statistic will be used as the point estimate.

In general, if X is a random variable with probability distribution $f(x)$, characterized by the unknown parameter θ, and if $X_1, X_2, \ldots, X_n$ is a random sample of size n from X, the statistic $\hat{\Theta} = h(X_1, X_2, \ldots, X_n)$ is called a **point estimator** of θ. Note that $\hat{\Theta}$ is a random variable because it is a function of random variables. After the sample has been selected, $\hat{\Theta}$ takes on a particular numerical value $\hat{\theta}$ called the **point estimate** of θ.

Point Estimator

> A **point estimate** of some population parameter θ is a single numerical value $\hat{\theta}$ of a statistic $\hat{\Theta}$. The statistic $\hat{\Theta}$ is called the **point estimator**.

As an example, suppose that the random variable X is normally distributed with an unknown mean μ. The sample mean is a point estimator of the unknown population mean μ.

That is, $\hat{\mu} = \overline{X}$. After the sample has been selected, the numerical value $\bar{x}$ is the point estimate of μ. Thus, if $x_1 = 25, x_2 = 30, x_3 = 29$, and $x_4 = 31$, the point estimate of μ is

$$\bar{x} = \frac{25 + 30 + 29 + 31}{4} = 28.75$$

Similarly, if the population variance σ^2 is also unknown, a point estimator for σ^2 is the sample variance S^2, and the numerical value $s^2 = 6.9$ calculated from the sample data is called the point estimate of σ^2.

Estimation problems occur frequently in engineering. We often need to estimate

- The mean μ of a single population
- The variance σ^2 (or standard deviation σ) of a single population
- The proportion p of items in a population that belong to a class of interest
- The difference in means of two populations, $\mu_1 - \mu_2$
- The difference in two population proportions, $p_1 - p_2$

Reasonable point estimates of these parameters are as follows:

- For μ, the estimate is $\hat{\mu} = \bar{x}$, the sample mean.
- For σ^2, the estimate is $\hat{\sigma}^2 = s^2$, the sample variance.
- For p, the estimate is $\hat{p} = x/n$, the sample proportion, where x is the number of items in a random sample of size n that belong to the class of interest.
- For $\mu_1 - \mu_2$, the estimate is $\hat{\mu}_1 - \hat{\mu}_2 = \bar{x}_1 - \bar{x}_2$, the difference between the sample means of two independent random samples.
- For $p_1 - p_2$, the estimate is $\hat{p}_1 - \hat{p}_2$, the difference between two sample proportions computed from two independent random samples.

We may have several different choices for the point estimator of a parameter. For example, if we wish to estimate the mean of a population, we might consider the sample mean, the sample median, or perhaps the average of the smallest and largest observations in the sample as point estimators. In order to decide which point estimator of a particular parameter is the best one to use, we need to examine their statistical properties and develop some criteria for comparing estimators.

7-2 SAMPLING DISTRIBUTIONS AND THE CENTRAL LIMIT THEOREM

Statistical inference is concerned with making **decisions** about a population based on the information contained in a random sample from that population. For instance, we may be interested in the mean fill volume of a can of soft drink. The mean fill volume in the population is required to be 300 milliliters. An engineer takes a random sample of 25 cans and computes the sample average fill volume to be $\bar{x} = 298$ milliliters. The engineer will probably decide that the population mean is $\mu = 300$ milliliters, even though the sample mean was 298 milliliters because he or she knows that the sample mean is a reasonable estimate of μ and that a sample mean of 298 milliliters is very likely to occur, even if the true population mean is $\mu = 300$ milliliters. In fact, if the true mean is 300 milliliters, tests of

25 cans made repeatedly, perhaps every five minutes, would produce values of $\bar{x}$ that vary both above and below $\mu = 300$ milliliters.

The link between the probability models in the earlier chapters and the data is made as follows. Each numerical value in the data is the observed value of a random variable. Furthermore, the random variables are usually assumed to be independent and identically distributed. These random variables are known as a random sample.

Random Sample

> The random variables $X_1, X_2, \ldots, X_n$ are a **random sample** of size n if (a) the X_i's are independent random variables, and (b) every X_i has the same probability distribution.

The observed data is also referred to as a random sample, but the use of the same phrase should not cause any confusion.

The primary purpose in taking a random sample is to obtain information about the unknown population parameters. Suppose, for example, that we wish to reach a conclusion about the proportion of people in the United States who prefer a particular brand of soft drink. Let p represent the unknown value of this proportion. It is impractical to question every individual in the population to determine the true value of p. In order to make an inference regarding the true proportion p, a more reasonable procedure would be to select a random sample (of an appropriate size) and use the observed proportion $\hat{p}$ of people in this sample favoring the brand of soft drink.

The sample proportion, $\hat{p}$ is computed by dividing the number of individuals in the sample who prefer the brand of soft drink by the total sample size n. Thus, $\hat{p}$ is a function of the observed values in the random sample. Since many random samples are possible from a population, the value of $\hat{p}$ will vary from sample to sample. That is, $\hat{p}$ is a random variable. Such a random variable is called a **statistic.**

Statistic

> A **statistic** is any function of the observations in a random sample.

We have encountered statistics before. For example, if $X, X_2, \ldots, X_n$ is a random sample of size n, the **sample mean** $\overline{X}$, the **sample variance** S^2, and the **sample standard deviation** S are statistics. Since a statistic is a random variable, it has a probability distribution.

Sampling Distribution

> The probability distribution of a statistic is called a **sampling distribution.**

For example, the probability distribution of $\overline{X}$ is called the **sampling distribution of the mean.** The sampling distribution of a statistic depends on the distribution of the population, the size of the sample, and the method of sample selection. We now present perhaps the most important sampling distribution. Other sampling distributions and their applications will be illustrated extensively in the following two chapters.

Consider determining the sampling distribution of the sample mean $\overline{X}$. Suppose that a random sample of size n is taken from a normal population with mean μ and variance σ^2. Now each observation in this sample, say, $X_1, X_2, \ldots, X_n$, is a normally and independently distributed random variable with mean μ and variance σ^2. Then because linear functions of independent normally distributed random variables are also normally distributed (Chapter 5), we conclude that the sample mean

$$\overline{X} = \frac{X_1 + X_2 + \cdots + X_n}{n}$$

has a normal distribution with mean

$$\mu_{\overline{X}} = \frac{\mu + \mu + \cdots + \mu}{n} = \mu$$

and variance

$$\sigma^2_{\overline{X}} = \frac{\sigma^2 + \sigma^2 + \cdots + \sigma^2}{n^2} = \frac{\sigma^2}{n}$$

If we are sampling from a population that has an unknown probability distribution, the sampling distribution of the sample mean will still be approximately normal with mean μ and variance σ^2/n, if the sample size n is large. This is one of the most useful theorems in statistics, called the **central limit theorem.** The statement is as follows:

Central Limit Theorem

If $X_1, X_2, \ldots, X_n$ is a random sample of size n taken from a population (either finite or infinite) with mean μ and finite variance σ^2, and if $\overline{X}$ is the sample mean, the limiting form of the distribution of

$$Z = \frac{\overline{X} - \mu}{\sigma/\sqrt{n}} \tag{7-1}$$

as $n \rightarrow \infty$, is the standard normal distribution.

The normal approximation for $\overline{X}$ depends on the sample size n. Figure 7-1(a) shows the distribution obtained for throws of a single, six-sided true die. The probabilities are equal (1/6) for all the values obtained, 1, 2, 3, 4, 5, or 6. Figure 7-1(b) shows the distribution of the average score obtained when tossing two dice, and Fig. 7-1(c), 7-1(d), and 7-1(e) show the distributions of average scores obtained when tossing three, five, and ten dice, respectively. Notice that, while the population (one die) is relatively far from normal, the distribution of averages is approximated reasonably well by the normal distribution for sample sizes as small as five. (The dice throw distributions are discrete, however, while the normal is continuous). Although the central limit theorem will work well for small samples ($n = 4, 5$) in most cases, particularly where the population is continuous, unimodal, and symmetric, larger samples will be required in other situations, depending on the shape of the population. In many cases of practical interest, if $n \geq 30$, the normal approximation will be satisfactory regardless of the shape of the population. If $n < 30$, the central limit theorem will work if the distribution of the population is not severely nonnormal.

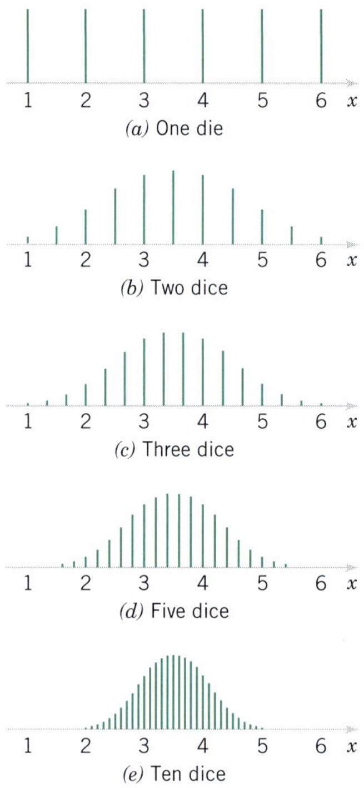

Figure 7-1 Distributions of average scores from throwing dice. [Adapted with permission from Box, Hunter, and Hunter (1978).]

EXAMPLE 7-1
Resistors

An electronics company manufactures resistors that have a mean resistance of 100 ohms and a standard deviation of 10 ohms. The distribution of resistance is normal. Find the probability that a random sample of $n = 25$ resistors will have an average resistance less than 95 ohms.

Note that the sampling distribution of $\overline{X}$ is normal, with mean $\mu_{\overline{X}} = 100$ ohms and a standard deviation of

$$\sigma_{\overline{X}} = \frac{\sigma}{\sqrt{n}} = \frac{10}{\sqrt{25}} = 2$$

Therefore, the desired probability corresponds to the shaded area in Fig. 7-1. Standardizing the point $\overline{X} = 95$ in Fig. 7-2, we find that

$$z = \frac{95 - 100}{2} = -2.5$$

and therefore,

$$P(\overline{X} < 95) = P(Z < -2.5) \\ = 0.0062$$

The following example makes use of the central limit theorem.

EXAMPLE 7-2
Central Limit Theorem

Suppose that a random variable X has a continuous uniform distribution

$$f(x) = \begin{cases} 1/2, & 4 \le x \le 6 \\ 0, & \text{otherwise} \end{cases}$$

Find the distribution of the sample mean of a random sample of size $n = 40$.

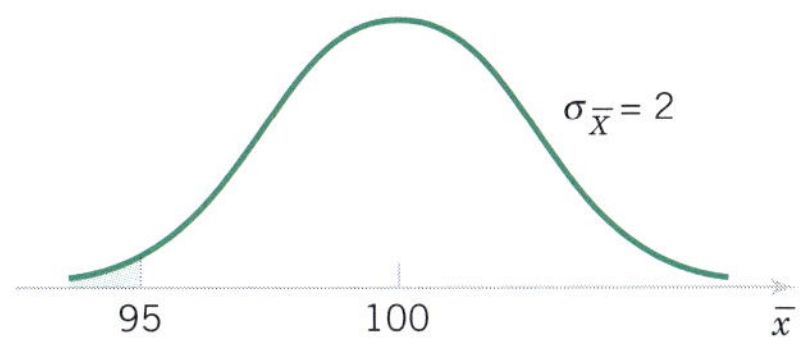

Figure 7-2 Probability for Example 7-1.

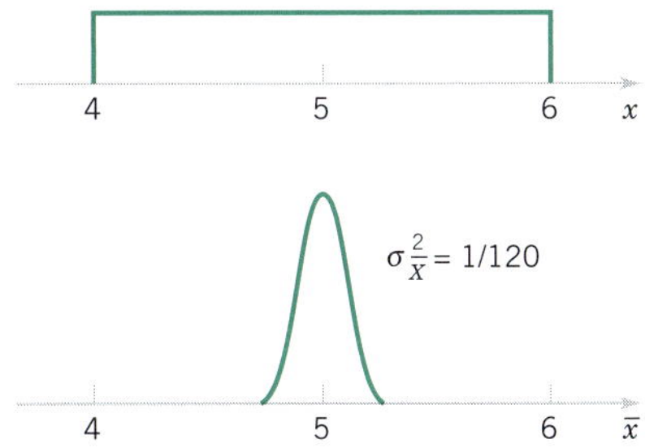

Figure 7-3 The distributions of X and $\overline{X}$ for Example 7-2.

The mean and variance of X are $\mu = 5$ and $\sigma^2 = (6 - 4)^2/12 = 1/3$. The central limit theorem indicates that the distribution of $\overline{X}$ is approximately normal with mean $\mu_{\overline{X}} = 5$ and variance $\sigma^2_{\overline{X}} = \sigma^2/n = 1/[3(40)] = 1/120$. The distributions of X and $\overline{X}$ are shown in Fig. 7-3.

Now consider the case in which we have two independent populations. Let the first population have mean μ_1 and variance σ_1^2 and the second population have mean μ_2 and variance σ_2^2. Suppose that both populations are normally distributed. Then, using the fact that linear combinations of independent normal random variables follow a normal distribution (see Chapter 5), we can say that the sampling distribution of $\overline{X}_1 - \overline{X}_2$ is normal with mean

$$\mu_{\overline{X}_1 - \overline{X}_2} = \mu_{\overline{X}_1} - \mu_{\overline{X}_2} = \mu_1 - \mu_2 \tag{7-2}$$

and variance

$$\sigma^2_{\overline{X}_1 - \overline{X}_2} = \sigma^2_{\overline{X}_1} + \sigma^2_{\overline{X}_2} = \frac{\sigma_1^2}{n_1} + \frac{\sigma_2^2}{n_2} \tag{7-3}$$

If the two populations are not normally distributed and if both sample sizes n_1 and n_2 are greater than 30, we may use the central limit theorem and assume that $\overline{X}_1$ and $\overline{X}_2$ follow approximately independent normal distributions. Therefore, the sampling distribution of $\overline{X}_1 - \overline{X}_2$ is approximately normal with mean and variance given by Equations 7-2 and 7-3, respectively. If either n_1 or n_2 is less than 30, the sampling distribution of $X_1 - X_2$ will still be approximately normal with mean and variance given by Equations 7-2 and 7-3, provided that the population from which the small sample is taken is not dramatically different from the normal. We may summarize this with the following definition.

Approximate Sampling Distribution of a Difference in Sample Means

If we have two independent populations with means μ_1 and μ_2 and variances σ_1^2 and σ_2^2 and if $\overline{X}_1$ and $\overline{X}_2$ are the sample means of two independent random samples of sizes n_1 and n_2 from these populations, then the sampling distribution of

$$Z = \frac{X_1 - X_2 - (\mu_1 - \mu_2)}{\sqrt{\sigma_1^2/n_1 + \sigma_2^2/n_2}} \tag{7-4}$$

is approximately standard normal, if the conditions of the central limit theorem apply. If the two populations are normal, the sampling distribution of Z is exactly standard normal.

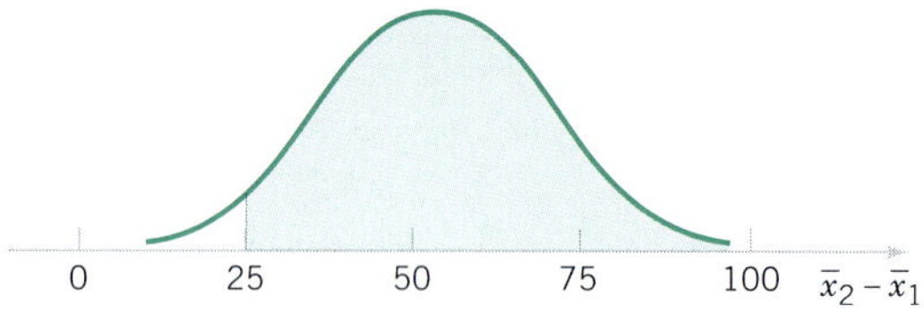

Figure 7-4 The sampling distribution of $\bar{X}_2 - \bar{X}_1$ in Example 7-3.

EXAMPLE 7-3 Aircraft Engine Life

The effective life of a component used in a jet-turbine aircraft engine is a random variable with mean 5000 hours and standard deviation 40 hours. The distribution of effective life is fairly close to a normal distribution. The engine manufacturer introduces an improvement into the manufacturing process for this component that increases the mean life to 5050 hours and decreases the standard deviation to 30 hours. Suppose that a random sample of $n_1 = 16$ components is selected from the "old" process and a random sample of $n_2 = 25$ components is selected from the "improved" process. What is the probability that the difference in the two sample means $\bar{X}_2 - \bar{X}_1$ is at least 25 hours? Assume that the old and improved processes can be regarded as independent populations.

To solve this problem, we first note that the distribution of $\bar{X}_1$ is normal with mean $\mu_1 = 5000$ hours and standard deviation $\sigma_1/\sqrt{n_1} = 40/\sqrt{16} = 10$ hours, and the distribution of $\bar{X}_2$ is normal with mean $\mu_2 = 5050$ hours and standard deviation $\sigma_2/\sqrt{n_2} = 30/\sqrt{25} = 6$ hours. Now the distribution of $\bar{X}_2 - \bar{X}_1$ is normal with mean $\mu_2 - \mu_1 = 5050 - 5000 = 50$ hours and variance $\sigma_2^2/n_2 + \sigma_1^2/n_1 = (6)^2 + (10)^2 = 136$ hours^2. This sampling distribution is shown in Fig. 7-4. The probability that $\bar{X}_2 - \bar{X}_1 \geq 25$ is the shaded portion of the normal distribution in this figure.

Corresponding to the value $\bar{x}_2 - \bar{x}_1 = 25$ in Fig. 7-4, we find that

$$z = \frac{25 - 50}{\sqrt{136}} = -2.14$$

and we find that

$$P(\bar{X}_2 - \bar{X}_1 \geq 25) = P(Z \geq -2.14) = 0.9838$$

EXERCISES FOR SECTION 7-2

7-1. PVC pipe is manufactured with a mean diameter of 1.01 inch and a standard deviation of 0.003 inch. Find the probability that a random sample of $n = 9$ sections of pipe will have a sample mean diameter greater than 1.009 inch and less than 1.012 inch.

7-2. Suppose that samples of size $n = 25$ are selected at random from a normal population with mean 100 and standard deviation 10. What is the probability that the sample mean falls in the interval from $\mu_{\bar{X}} - 1.8\sigma_{\bar{X}}$ to $\mu_{\bar{X}} + 1.0\sigma_{\bar{X}}$?

7-3. A synthetic fiber used in manufacturing carpet has tensile strength that is normally distributed with mean 75.5 psi and standard deviation 3.5 psi. Find the probability that a random sample of $n = 6$ fiber specimens will have sample mean tensile strength that exceeds 75.75 psi.

7-4. Consider the synthetic fiber in the previous exercise. How is the standard deviation of the sample mean changed when the sample size is increased from $n = 6$ to $n = 49$?

7-5. The compressive strength of concrete is normally distributed with $\mu = 2500$ psi and $\sigma = 50$ psi. Find the probability that a random sample of $n = 5$ specimens will have a sample mean diameter that falls in the interval from 2499 psi to 2510 psi.

7-6. Consider the concrete specimens in the previous exercise. What is the standard error of the sample mean?

7-7. A normal population has mean 100 and variance 25. How large must the random sample be if we want the standard error of the sample average to be 1.5?

7-8. Suppose that the random variable X has the continuous uniform distribution

$$f(x) = \begin{cases} 1, & 0 \leq x \leq 1 \\ 0, & \text{otherwise} \end{cases}$$

Suppose that a random sample of $n = 12$ observations is selected from this distribution. What is the approximate probability distribution of $\bar{X} - 6$? Find the mean and variance of this quantity.

7-9. Suppose that X has a discrete uniform distribution

$$f(x) = \begin{cases} 1/3, & x = 1, 2, 3 \\ 0, & \text{otherwise} \end{cases}$$

A random sample of $n = 36$ is selected from this population. Find the probability that the sample mean is greater than 2.1 but less than 2.5, assuming that the sample mean would be measured to the nearest tenth.

7-10. The amount of time that a customer spends waiting at an airport check-in counter is a random variable with mean 8.2 minutes and standard deviation 1.5 minutes. Suppose that a random sample of $n = 49$ customers is observed. Find the probability that the average time waiting in line for these customers is

(a) Less than 10 minutes
(b) Between 5 and 10 minutes
(c) Less than 6 minutes

7-11. A random sample of size $n_1 = 16$ is selected from a normal population with a mean of 75 and a standard deviation of 8. A second random sample of size $n_2 = 9$ is taken from another normal population with mean 70 and standard deviation 12. Let $\overline{X}_1$ and $\overline{X}_2$ be the two sample means. Find

(a) The probability that $\overline{X}_1 - \overline{X}_2$ exceeds 4
(b) The probability that $3.5 \le \overline{X}_1 - \overline{X}_2 \le 5.5$

7-12. A consumer electronics company is comparing the brightness of two different types of picture tubes for use in its television sets. Tube type A has mean brightness of 100 and standard deviation of 16, while tube type B has unknown mean brightness, but the standard deviation is assumed to be identical to that for type A. A random sample of $n = 25$ tubes of each type is selected, and $\overline{X}_B - \overline{X}_A$ is computed. If μ_B equals or exceeds μ_A, the manufacturer would like to adopt type B for use. The observed difference is $\bar{x}_B - \bar{x}_A = 3.5$. What decision would you make, and why?

7-13. The elasticity of a polymer is affected by the concentration of a reactant. When low concentration is used, the true mean elasticity is 55, and when high concentration is used the mean elasticity is 60. The standard deviation of elasticity is 4, regardless of concentration. If two random samples of size 16 are taken, find the probability that $\overline{X}_{\text{high}} - \overline{X}_{\text{low}} \ge 2$.

7-3 GENERAL CONCEPTS OF POINT ESTIMATION

7-3.1 Unbiased Estimators

An estimator should be "close" in some sense to the true value of the unknown parameter. Formally, we say that $\hat{\Theta}$ is an unbiased estimator of θ if the expected value of $\hat{\Theta}$ is equal to θ. This is equivalent to saying that the mean of the probability distribution of $\hat{\Theta}$ (or the mean of the sampling distribution of $\hat{\Theta}$) is equal to θ.

Bias of an Estimator

The point estimator $\hat{\Theta}$ is an **unbiased estimator** for the parameter θ if

$$E(\hat{\Theta}) = \theta \qquad (7\text{-}5)$$

If the estimator is not unbiased, then the difference

$$E(\hat{\Theta}) - \theta \qquad (7\text{-}6)$$

is called the **bias** of the estimator $\hat{\Theta}$.

When an estimator is unbiased, the bias is zero; that is, $E(\hat{\Theta}) - \theta = 0$.

EXAMPLE 7-4
Sample Mean and Variance Are Unbiased

Suppose that X is a random variable with mean μ and variance σ^2. Let $X_1, X_2, \ldots, X_n$ be a random sample of size n from the population represented by X. Show that the sample mean $\overline{X}$ and sample variance S^2 are unbiased estimators of μ and σ^2, respectively.

First consider the sample mean. In Section 5.5 in Chapter 5, we showed that $E(\overline{X}) = \mu$. Therefore, the sample mean $\overline{X}$ is an unbiased estimator of the population mean μ.

Now consider the sample variance. We have

$$E(S^2) = E\left[\frac{\sum_{i=1}^{n}(X_i - \overline{X})^2}{n-1}\right] = \frac{1}{n-1}E\sum_{i=1}^{n}(X_i - \overline{X})^2$$

$$= \frac{1}{n-1}E\sum_{i=1}^{n}(X_i^2 + \overline{X}^2 - 2\overline{X}X_i) = \frac{1}{n-1}E\left(\sum_{i=1}^{n}X_i^2 - n\overline{X}^2\right)$$

$$= \frac{1}{n-1}\left[\sum_{i=1}^{n}E(X_i^2) - nE(\overline{X}^2)\right]$$

The last equality follows the equation for the mean of a linear function in Chapter 5. However, since $E(X_i^2) = \mu^2 + \sigma^2$ and $E(\overline{X}^2) = \mu^2 + \sigma^2/n$, we have

$$E(S^2) = \frac{1}{n-1}\left[\sum_{i=1}^{n}(\mu^2 + \sigma^2) - n(\mu^2 + \sigma^2/n)\right]$$

$$= \frac{1}{n-1}(n\mu^2 + n\sigma^2 - n\mu^2 - \sigma^2) = \sigma^2$$

Therefore, the sample variance S^2 is an unbiased estimator of the population variance σ^2.

Although S^2 is unbiased for σ^2, S is a biased estimator of σ. For large samples, the bias is very small. However, there are good reasons for using S as an estimator of σ in samples from normal distributions, as we will see in the next three chapters when we discuss confidence intervals and hypothesis testing.

Sometimes there are several unbiased estimators of the sample population parameter. For example, suppose we take a random sample of size $n = 10$ from a normal population and obtain the data $x_1 = 12.8$, $x_2 = 9.4$, $x_3 = 8.7$, $x_4 = 11.6$, $x_5 = 13.1$, $x_6 = 9.8$, $x_7 = 14.1$, $x_8 = 8.5$, $x_9 = 12.1$, $x_{10} = 10.3$. Now the sample mean is

$$\overline{x} = \frac{12.8 + 9.4 + 8.7 + 11.6 + 13.1 + 9.8 + 14.1 + 8.5 + 12.1 + 10.3}{10} = 11.04$$

the sample median is

$$\tilde{x} = \frac{10.3 + 11.6}{2} = 10.95$$

and a 10% trimmed mean (obtained by discarding the smallest and largest 10% of the sample before averaging) is

$$\overline{x}_{tr(10)} = \frac{8.7 + 9.4 + 9.8 + 10.3 + 11.6 + 12.1 + 12.8 + 13.1}{8} = 10.98$$

We can show that all of these are unbiased estimates of μ. Since there is not a unique unbiased estimator, we cannot rely on the property of unbiasedness alone to select our estimator. We need a method to select among unbiased estimators. We suggest a method in the following section.

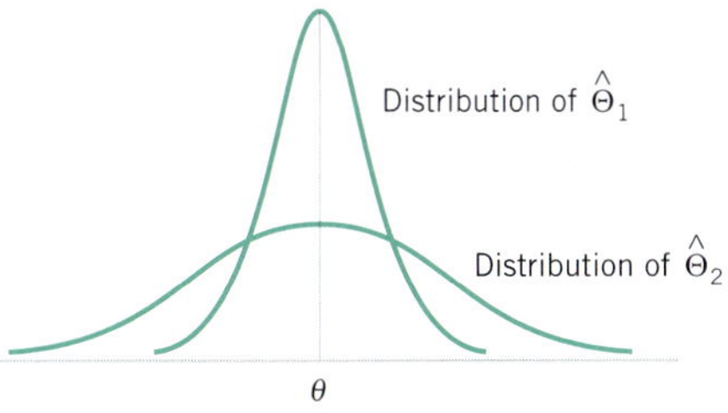

Figure 7-5 The sampling distributions of two unbiased estimators $\hat{\Theta}_1$ and $\hat{\Theta}_2$.

7-3.2 Variance of a Point Estimator

Suppose that $\hat{\Theta}_1$ and $\hat{\Theta}_2$ are unbiased estimators of θ. This indicates that the distribution of each estimator is centered at the true value of θ. However, the variance of these distributions may be different. Figure 7-5 illustrates the situation. Since $\hat{\Theta}_1$ has a smaller variance than $\hat{\Theta}_2$, the estimator $\hat{\Theta}_1$ is more likely to produce an estimate close to the true value θ. A logical principle of estimation, when selecting among several estimators, is to choose the estimator that has minimum variance.

Minimum Variance Unbiased Estimator

If we consider all unbiased estimators of θ, the one with the smallest variance is called the **minimum variance unbiased estimator** (MVUE).

In a sense, the MVUE is most likely among all unbiased estimators to produce an estimate $\hat{\theta}$ that is close to the true value of θ. It has been possible to develop methodology to identify the MVUE in many practical situations. While this methodology is beyond the scope of this book, we give one very important result concerning the normal distribution.

If $X_1, X_2, \ldots, X_n$ is a random sample of size n from a normal distribution with mean μ and variance σ^2, the sample mean $\overline{X}$ is the MVUE for μ.

In situations in which we do not know whether an MVUE exists, we could still use a minimum variance principle to choose among competing estimators. Suppose, for example, we wish to estimate the mean of a population (not necessarily a *normal* population). We have a random sample of n observations $X_1, X_2, \ldots, X_n$ and we wish to compare two possible estimators for μ: the sample mean $\overline{X}$ and a single observation from the sample, say, X_i. Note that both $\overline{X}$ and X_i are unbiased estimators of μ; for the sample mean, we have $V(\overline{X}) = \sigma^2/n$ from Chapter 5 and the variance of any observation is $V(X_i) = \sigma^2$. Since $V(\overline{X}) < V(X_i)$ for sample sizes $n \geq 2$, we would conclude that the sample mean is a better estimator of μ than a single observation X_i.

7-3.3 Standard Error: Reporting a Point Estimate

When the numerical value or point estimate of a parameter is reported, it is usually desirable to give some idea of the precision of estimation. The measure of precision usually employed is the standard error of the estimator that has been used.

Standard Error of an Estimator

> The **standard error** of an estimator $\hat{\Theta}$ is its standard deviation, given by $\sigma_{\hat{\Theta}} = \sqrt{V(\hat{\Theta})}$. If the standard error involves unknown parameters that can be estimated, substitution of those values into $\sigma_{\hat{\Theta}}$ produces an **estimated standard error,** denoted by $\hat{\sigma}_{\hat{\Theta}}$.

Sometimes the estimated standard error is denoted by $s_{\hat{\Theta}}$ or $se(\hat{\Theta})$.

Suppose we are sampling from a normal distribution with mean μ and variance σ^2. Now the distribution of $\overline{X}$ is normal with mean μ and variance σ^2/n, so the standard error of $\overline{X}$ is

$$\sigma_{\overline{X}} = \frac{\sigma}{\sqrt{n}}$$

If we did not know σ but substituted the sample standard deviation S into the above equation, the estimated standard error of $\overline{X}$ would be

$$\hat{\sigma}_{\overline{X}} = \frac{S}{\sqrt{n}}$$

When the estimator follows a normal distribution, as in the above situation, we can be reasonably confident that the true value of the parameter lies within two standard errors of the estimate. Since many point estimators are normally distributed (or approximately so) for large n, this is a very useful result. Even in cases in which the point estimator is not normally distributed, we can state that so long as the estimator is unbiased, the estimate of the parameter will deviate from the true value by as much as four standard errors at most 6 percent of the time. Thus a very conservative statement is that the true value of the parameter differs from the point estimate by at most four standard errors. See Chebyshev's inequality in the supplemental material on the Web site.

EXAMPLE 7-5
Thermal Conductivity

An article in the *Journal of Heat Transfer* (Trans. ASME, Sec. C, 96, 1974, p. 59) described a new method of measuring the thermal conductivity of Armco iron. Using a temperature of 100°F and a power input of 550 watts, the following 10 measurements of thermal conductivity (in Btu/hr-ft-°F) were obtained:

$$41.60, 41.48, 42.34, 41.95, 41.86,$$
$$42.18, 41.72, 42.26, 41.81, 42.04$$

A point estimate of the mean thermal conductivity at 100°F and 550 watts is the sample mean or

$$\overline{x} = 41.924 \text{ Btu/hr-ft-°F}$$

The standard error of the sample mean is $\sigma_{\overline{X}} = \sigma/\sqrt{n}$, and since σ is unknown, we may replace it by the sample standard deviation $s = 0.284$ to obtain the estimated standard error of $\overline{X}$ as

$$\hat{\sigma}_{\overline{X}} = \frac{s}{\sqrt{n}} = \frac{0.284}{\sqrt{10}} = 0.0898$$

Notice that the standard error is about 0.2 percent of the sample mean, implying that we have obtained a relatively precise point estimate of thermal conductivity. If we can assume that thermal conductivity is normally distributed, 2 times the standard error is $2\hat{\sigma}_{\bar{X}} = 2(0.0898) = 0.1796$, and we are highly confident that the true mean thermal conductivity is within the interval 41.924 ± 0.1796, or between 41.744 and 42.104.

7-3.4 Mean Squared Error of an Estimator

Sometimes it is necessary to use a biased estimator. In such cases, the mean squared error of the estimator can be important. The **mean squared error** of an estimator $\hat{\Theta}$ is the expected squared difference between $\hat{\Theta}$ and θ.

Mean Squared Error of an Estimator

The **mean squared error** of an estimator $\hat{\Theta}$ of the parameter θ is defined as

$$\text{MSE}(\hat{\Theta}) = E(\hat{\Theta} - \theta)^2 \tag{7-7}$$

The mean squared error can be rewritten as follows:

$$\begin{aligned}\text{MSE}(\hat{\Theta}) &= E[\hat{\Theta} - E(\hat{\Theta})]^2 + [\theta - E(\hat{\Theta})]^2 \\ &= V(\hat{\Theta}) + (\text{bias})^2\end{aligned}$$

That is, the mean squared error of $\hat{\Theta}$ is equal to the variance of the estimator plus the squared bias. If $\hat{\Theta}$ is an unbiased estimator of θ, the mean squared error of $\hat{\Theta}$ is equal to the variance of $\hat{\Theta}$.

The mean squared error is an important criterion for comparing two estimators. Let $\hat{\Theta}_1$ and $\hat{\Theta}_2$ be two estimators of the parameter θ, and let MSE $(\hat{\Theta}_1)$ and MSE $(\hat{\Theta}_2)$ be the mean squared errors of $\hat{\Theta}_1$ and $\hat{\Theta}_2$. Then the **relative efficiency** of $\hat{\Theta}_2$ to $\hat{\Theta}_1$ is defined as

$$\frac{\text{MSE}(\hat{\Theta}_1)}{\text{MSE}(\hat{\Theta}_2)} \tag{7-8}$$

If this relative efficiency is less than 1, we would conclude that $\hat{\Theta}_1$ is a more efficient estimator of θ than $\hat{\Theta}_2$, in the sense that it has a smaller mean square error.

Sometimes we find that biased estimators are preferable to unbiased estimators because they have smaller mean squared error. That is, we may be able to reduce the variance of the estimator considerably by introducing a relatively small amount of bias. As long as the reduction in variance is greater than the squared bias, an improved estimator from a mean squared error viewpoint will result. For example, Fig. 7-6 shows the probability distribution of a biased estimator $\hat{\Theta}_1$ that has a smaller variance than the unbiased estimator $\hat{\Theta}_2$. An estimate based on $\hat{\Theta}_1$ would more likely be close to the true value of θ than would an estimate based on $\hat{\Theta}_2$. Linear regression analysis (Chapters 11 and 12) is an area in which biased estimators are occasionally used.

An estimator $\hat{\Theta}$ that has a mean squared error that is less than or equal to the mean squared error of any other estimator, for all values of the parameter θ, is called an **optimal** estimator of θ. Optimal estimators rarely exist.

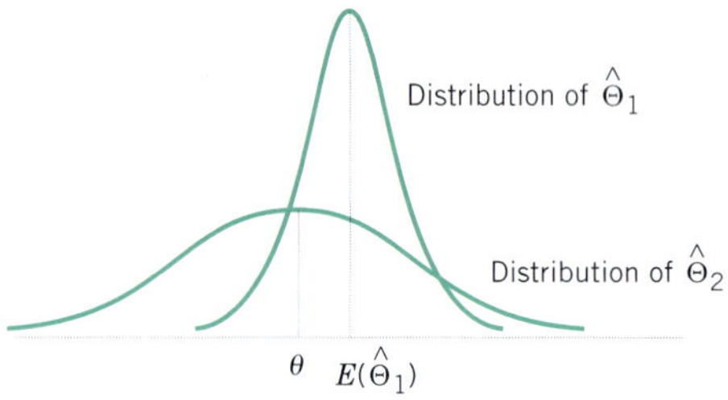

Figure 7-6 A biased estimator $\hat{\Theta}_1$ that has smaller variance than the unbiased estimator $\hat{\Theta}_2$.

EXERCISES FOR SECTION 7-3

7-14. Suppose we have a random sample of size $2n$ from a population denoted by X, and $E(X) = \mu$ and $V(X) = \sigma^2$. Let

$$\overline{X}_1 = \frac{1}{2n}\sum_{i=1}^{2n} X_i \quad \text{and} \quad \overline{X}_2 = \frac{1}{n}\sum_{i=1}^{n} X_i$$

be two estimators of μ. Which is the better estimator of μ? Explain your choice.

7-15. Let $X_1, X_2, \ldots, X_7$ denote a random sample from a population having mean μ and variance σ^2. Consider the following estimators of μ:

$$\hat{\Theta}_1 = \frac{X_1 + X_2 + \cdots + X_7}{7}$$

$$\hat{\Theta}_2 = \frac{2X_1 - X_6 + X_4}{2}$$

(a) Is either estimator unbiased?
(b) Which estimator is best? In what sense is it best? Calculate the relative efficiency of the two estimators.

7-16. Suppose that $\hat{\Theta}_1$ and $\hat{\Theta}_2$ are unbiased estimators of the parameter θ. We know that $V(\hat{\Theta}_1) = 10$ and $V(\hat{\Theta}_2) = 4$. Which estimator is best and in what sense is it best? Calculate the relative efficiency of the two estimators.

7-17. Suppose that $\hat{\Theta}_1$ and $\hat{\Theta}_2$ are estimators of the parameter θ. We know that $E(\hat{\Theta}_1) = \theta$, $E(\hat{\Theta}_2) = \theta/2$, $V(\hat{\Theta}_1) = 10$, $V(\hat{\Theta}_2) = 4$. Which estimator is best? In what sense is it best?

7-18. Suppose that $\hat{\Theta}_1$, $\hat{\Theta}_2$, and $\hat{\Theta}_3$ are estimators of θ. We know that $E(\hat{\Theta}_1) = E(\hat{\Theta}_2) = \theta$, $E(\hat{\Theta}_3) \neq \theta$, $V(\hat{\Theta}_1) = 12$, $V(\hat{\Theta}_2) = 10$, and $E(\hat{\Theta}_3 - \theta)^2 = 6$. Compare these three estimators. Which do you prefer? Why?

7-19. Let three random samples of sizes $n_1 = 20$, $n_2 = 10$, and $n_3 = 8$ be taken from a population with mean μ and variance σ^2. Let S_1^2, S_2^2, and S_3^2 be the sample variances. Show that $S^2 = (20S_1^2 + 10S_2^2 + 8S_3^2)/38$ is an unbiased estimator of σ^2.

7-20. (a) Show that $\sum_{i=1}^{n}(X_i - \overline{X})^2/n$ is a biased estimator of σ^2.
(b) Find the amount of bias in the estimator.
(c) What happens to the bias as the sample size n increases?

7-21. Let $X_1, X_2, \ldots, X_n$ be a random sample of size n from a population with mean μ and variance σ^2.
(a) Show that $\overline{X}^2$ is a biased estimator for μ^2.
(b) Find the amount of bias in this estimator.
(c) What happens to the bias as the sample size n increases?

7-22. Data on pull-off force (pounds) for connectors used in an automobile engine application are as follows: 79.3, 75.1, 78.2, 74.1, 73.9, 75.0, 77.6, 77.3, 73.8, 74.6, 75.5, 74.0, 74.7, 75.9, 72.9, 73.8, 74.2, 78.1, 75.4, 76.3, 75.3, 76.2, 74.9, 78.0, 75.1, 76.8.
(a) Calculate a point estimate of the mean pull-off force of all connectors in the population. State which estimator you used and why.
(b) Calculate a point estimate of the pull-off force value that separates the weakest 50% of the connectors in the population from the strongest 50%.
(c) Calculate point estimates of the population variance and the population standard deviation.
(d) Calculate the standard error of the point estimate found in part (a). Provide an interpretation of the standard error.
(e) Calculate a point estimate of the proportion of all connectors in the population whose pull-off force is less than 73 pounds.

7-23. Data on oxide thickness of semiconductors are as follows: 425, 431, 416, 419, 421, 436, 418, 410, 431, 433, 423, 426, 410, 435, 436, 428, 411, 426, 409, 437, 422, 428, 413, 416.
(a) Calculate a point estimate of the mean oxide thickness for all wafers in the population.
(b) Calculate a point estimate of the standard deviation of oxide thickness for all wafers in the population.
(c) Calculate the standard error of the point estimate from part (a).
(d) Calculate a point estimate of the median oxide thickness for all wafers in the population.
(e) Calculate a point estimate of the proportion of wafers in the population that have oxide thickness greater than 430 angstrom.

7-24. Suppose that X is the number of observed "successes" in a sample of n observations where p is the probability of success on each observation.
(a) Show that $\hat{P} = X/n$ is an unbiased estimator of p.
(b) Show that the standard error of $\hat{P}$ is $\sqrt{p(1 - p)/n}$. How would you estimate the standard error?

7-25. $\overline{X}_1$ and S_1^2 are the sample mean and sample variance from a population with mean μ_1 and variance σ_1^2. Similarly, $\overline{X}_2$ and S_2^2 are the sample mean and sample variance from a second independent population with mean μ_2 and variance σ_2^2. The sample sizes are n_1 and n_2, respectively.

(a) Show that $\overline{X}_1 - \overline{X}_2$ is an unbiased estimator of $\mu_1 - \mu_2$.

(b) Find the standard error of $\overline{X}_1 - \overline{X}_2$. How could you estimate the standard error?

(c) Suppose that both populations have the same variance; that is, $\sigma_1^2 = \sigma_2^2 = \sigma_2$. Show that

$$S_p^2 = \frac{(n_1 - 1)S_1^2 + (n_2 - 1)S_2^2}{n_1 + n_2 - 2}$$

is an unbiased estimator of σ^2.

7-26. Two different plasma etchers in a semiconductor factory have the same mean etch rate μ. However, machine 1 is newer than machine 2 and consequently has smaller variability in etch rate. We know that the variance of etch rate for machine 1 is σ_1^2 and for machine 2 it is $\sigma_2^2 = a\sigma_1^2$. Suppose that we have n_1 independent observations on etch rate from machine 1 and n_2 independent observations on etch rate from machine 2.

(a) Show that $\hat{\mu} = \alpha\, \overline{X}_1 + (1 - \alpha)\, \overline{X}_2$ is an unbiased estimator of μ for any value of α between 0 and 1.

(b) Find the standard error of the point estimate of μ in part (a).

(c) What value of α would minimize the standard error of the point estimate of μ?

(d) Suppose that $a = 4$ and $n_1 = 2n_2$. What value of α would you select to minimize the standard error of the point estimate of μ. How "bad" would it be to arbitrarily choose $\alpha = 0.5$ in this case?

7-27. Of n_1 randomly selected engineering students at ASU, X_1 owned an HP calculator, and of n_2 randomly selected engineering students at Virginia Tech X_2 owned an HP calculator. Let p_1 and p_2 be the probability that randomly selected ASU and Va. Tech engineering students, respectively, own HP calculators.

(a) Show that an unbiased estimate for $p_1 - p_2$ is $(X_1/n_1) - (X_2/n_2)$.

(b) What is the standard error of the point estimate in part (a)?

(c) How would you compute an estimate of the standard error found in part (b)?

(d) Suppose that $n_1 = 200$, $X_1 = 150$, $n_2 = 250$, and $X_2 = 185$. Use the results of part (a) to compute an estimate of $p_1 - p_2$.

(e) Use the results in parts (b) through (d) to compute an estimate of the standard error of the estimate.

7-4 METHODS OF POINT ESTIMATION

The definitions of unbiasedness and other properties of estimators do not provide any guidance about how good estimators can be obtained. In this section, we discuss methods for obtaining point estimators: the method of moments and the method of maximum likelihood. We also discuss Bayesian estimation of parameters. Maximum likelihood estimates are generally preferable to moment estimators because they have better efficiency properties. However, moment estimators are sometimes easier to compute. Both methods can produce unbiased point estimators.

7-4.1 Method of Moments

The general idea behind the method of moments is to equate **population moments,** which are defined in terms of expected values, to the corresponding **sample moments.** The population moments will be functions of the unknown parameters. Then these equations are solved to yield estimators of the unknown parameters.

Moments

Let $X_1, X_2, \ldots, X_n$ be a random sample from the probability distribution $f(x)$, where $f(x)$ can be a discrete probability mass function or a continuous probability density function. The kth **population moment** (or **distribution moment**) is $E(X^k)$, $k = 1, 2, \ldots$. The corresponding kth **sample moment** is $(1/n)\sum_{i=1}^{n} X_i^k$, $k = 1, 2, \ldots$.

To illustrate, the first population moment is $E(X) = \mu$, and the first sample moment is $(1/n)\sum_{i=1}^{n} X_i = \overline{X}$. Thus by equating the population and sample moments, we find that $\hat{\mu} = \overline{X}$. That is, the sample mean is the **moment estimator** of the population mean. In the general case, the population moments will be functions of the unknown parameters of the distribution, say, $\theta_1, \theta_2, \ldots, \theta_m$.

Moment Estimators

> Let $X_1, X_2, \ldots, X_n$ be a random sample from either a probability mass function or probability density function with m unknown parameters $\theta_1, \theta_2, \ldots, \theta_m$. The **moment estimators** $\hat{\Theta}_1, \hat{\Theta}_2, \ldots, \hat{\Theta}_m$ are found by equating the first m population moments to the first m sample moments and solving the resulting equations for the unknown parameters.

EXAMPLE 7-6 Exponential Distribution Moment Estimator

Suppose that $X_1, X_2, \ldots, X_n$ is a random sample from an exponential distribution with parameter λ. Now there is only one parameter to estimate, so we must equate $E(X)$ to $\overline{X}$. For the exponential, $E(X) = 1/\lambda$. Therefore $E(X) = \overline{X}$ results in $1/\lambda = \overline{X}$, so $\hat{\lambda} = 1/\overline{X}$ is the moment estimator of λ.

As an example, suppose that the time to failure of an electronic module used in an automobile engine controller is tested at an elevated temperature to accelerate the failure mechanism. The time to failure is exponentially distributed. Eight units are randomly selected and tested, resulting in the following failure time (in hours): $x_1 = 11.96$, $x_2 = 5.03$, $x_3 = 67.40$, $x_4 = 16.07$, $x_5 = 31.50$, $x_6 = 7.73$, $x_7 = 11.10$, and $x_8 = 22.38$. Because $\bar{x} = 21.65$, the moment estimate of λ is $\hat{\lambda} = 1/\bar{x} = 1/21.65 = 0.0462$.

EXAMPLE 7-7 Normal Distribution Moment Estimator

Suppose that $X_1, X_2, \ldots, X_n$ is a random sample from a normal distribution with parameters μ and σ^2. For the normal distribution $E(X) = \mu$ and $E(X^2) = \mu^2 + \sigma^2$. Equating $E(X)$ to $\overline{X}$ and $E(X^2)$ to $\frac{1}{n}\sum_{i=1}^{n} X_i^2$ gives

$$\mu = \overline{X}, \qquad \mu^2 + \sigma^2 = \frac{1}{n}\sum_{i=1}^{n} X_i^2$$

Solving these equations gives the moment estimators

$$\hat{\mu} = \overline{X}, \qquad \hat{\sigma}^2 = \frac{\sum_{i=1}^{n} X_i^2 - n\left(\frac{1}{n}\sum_{i=1}^{n} X_i\right)^2}{n} = \frac{\sum_{i=1}^{n}(X_i - \overline{X})^2}{n}$$

Notice that the moment estimator of σ^2 is not an unbiased estimator.

EXAMPLE 7-8 Gamma Distribution Moment Estimator

Suppose that $X_1, X_2, \ldots, X_n$ is a random sample from a gamma distribution with parameters r and λ. For the gamma distribution $E(X) = r/\lambda$ and $E(X^2) = r(r+1)/\lambda^2$. The moment estimators are found by solving

$$r/\lambda = \overline{X}, \quad r(r+1)/\lambda^2 = \frac{1}{n}\sum_{i=1}^{n} X_i^2$$

The resulting estimators are

$$\hat{r} = \frac{\overline{X}^2}{(1/n)\sum_{i=1}^{n} X_i^2 - \overline{X}^2} \qquad \hat{\lambda} = \frac{\overline{X}}{(1/n)\sum_{i=1}^{n} X_i^2 - \overline{X}^2}$$

To illustrate, consider the time to failure data introduced following Example 7-6. For this data, $\bar{x} = 21.65$ and $\sum_{i=1}^{8} x_i^2 = 6639.40$, so the moment estimates are

$$\hat{r} = \frac{(21.65)^2}{(1/8)6645.43 - (21.65)^2} = 1.29, \qquad \hat{\lambda} = \frac{21.65}{(1/8)6645.43 - (21.65)^2} = 0.0598$$

When $r = 1$, the gamma reduces to the exponential distribution. Because $\hat{r}$ slightly exceeds unity, it is quite possible that either the gamma or the exponential distribution would provide a reasonable model for the data.

7-4.2 Method of Maximum Likelihood

One of the best methods of obtaining a point estimator of a parameter is the method of maximum likelihood. This technique was developed in the 1920s by a famous British statistician, Sir R. A. Fisher. As the name implies, the estimator will be the value of the parameter that maximizes the **likelihood function.**

Maximum Likelihood Estimator

Suppose that X is a random variable with probability distribution $f(x; \theta)$, where θ is a single unknown parameter. Let $x_1, x_2, \ldots, x_n$ be the observed values in a random sample of size n. Then the **likelihood function** of the sample is

$$L(\theta) = f(x_1; \theta) \cdot f(x_2; \theta) \cdot \cdots \cdot f(x_n; \theta) \tag{7-9}$$

Note that the likelihood function is now a function of only the unknown parameter θ. The **maximum likelihood estimator** (MLE) of θ is the value of θ that maximizes the likelihood function $L(\theta)$.

In the case of a discrete random variable, the interpretation of the likelihood function is clear. The likelihood function of the sample $L(\theta)$ is just the probability

$$P(X_1 = x_1, X_2 = x_2, \ldots, X_n = x_n)$$

That is, $L(\theta)$ is just the probability of obtaining the sample values $x_1, x_2, \ldots, x_n$. Therefore, in the discrete case, the maximum likelihood estimator is an estimator that maximizes the probability of occurrence of the sample values.

EXAMPLE 7-9
Bernoulli Distribution MLE

Let X be a Bernoulli random variable. The probability mass function is

$$f(x; p) = \begin{cases} p^x(1-p)^{1-x}, & x = 0, 1 \\ 0, & \text{otherwise} \end{cases}$$

where p is the parameter to be estimated. The likelihood function of a random sample of size n is

$$\begin{aligned} L(p) &= p^{x_1}(1-p)^{1-x_1} p^{x_2}(1-p)^{1-x_2} \cdots p^{x_n}(1-p)^{1-x_n} \\ &= \prod_{i=1}^{n} p^{x_i}(1-p)^{1-x_i} = p^{\sum_{i=1}^{n} x_i}(1-p)^{n-\sum_{i=1}^{n} x_i} \end{aligned}$$

We observe that if $\hat{p}$ maximizes $L(p)$, $\hat{p}$ also maximizes $\ln L(p)$. Therefore,

$$\ln L(p) = \left(\sum_{i=1}^{n} x_i\right)\ln p + \left(n - \sum_{i=1}^{n} x_i\right)\ln(1-p)$$

Now

$$\frac{d \ln L(p)}{dp} = \frac{\sum_{i=1}^{n} x_i}{p} - \frac{\left(n - \sum_{i=1}^{n} x_i\right)}{1-p}$$

Equating this to zero and solving for p yields $\hat{p} = (1/n)\sum_{i=1}^{n} x_i$. Therefore, the maximum likelihood estimator of p is

$$\hat{P} = \frac{1}{n}\sum_{i=1}^{n} X_i$$

Suppose that this estimator was applied to the following situation: n items are selected at random from a production line, and each item is judged as either defective (in which case we set $x_i = 1$) or nondefective (in which case we set $x_i = 0$). Then $\sum_{i=1}^{n} x_i$ is the number of defective units in the sample, and $\hat{p}$ is the sample proportion defective. The parameter p is the population proportion defective; and it seems intuitively quite reasonable to use $\hat{p}$ as an estimate of p.

Although the interpretation of the likelihood function given above is confined to the discrete random variable case, the method of maximum likelihood can easily be extended to a continuous distribution. We now give two examples of maximum likelihood estimation for continuous distributions.

EXAMPLE 7-10
Normal Distribution MLE

Let X be normally distributed with unknown μ and known variance σ^2. The likelihood function of a random sample of size n, say $X_1, X_2, \ldots, X_n$, is

$$L(\mu) = \prod_{i=1}^{n} \frac{1}{\sigma\sqrt{2\pi}} e^{-(x_i-\mu)^2/(2\sigma^2)} = \frac{1}{(2\pi\sigma^2)^{n/2}} e^{\frac{-1}{2\sigma^2}\sum_{i=1}^{n}(x_i-\mu)^2}$$

Now

$$\ln L(\mu) = -(n/2)\ln(2\pi\sigma^2) - (2\sigma^2)^{-1}\sum_{i=1}^{n}(x_i-\mu)^2$$

and

$$\frac{d \ln L(\mu)}{d\mu} = (\sigma^2)^{-1}\sum_{i=1}^{n}(x_i-\mu)$$

Equating this last result to zero and solving for μ yields

$$\hat{\mu} = \frac{\sum_{i=1}^{n} X_i}{n} = \overline{X}$$

Thus the sample mean is the maximum likelihood estimator of μ. Notice that this is identical to the moment estimator.

EXAMPLE 7-11
Exponential Distribution MLE

Let X be exponentially distributed with parameter λ. The likelihood function of a random sample of size n, say $X_1, X_2, \ldots, X_n$, is

$$L(\lambda) = \prod_{i=1}^{n} \lambda e^{-\lambda x_i} = \lambda^n e^{-\lambda \sum_{i=1}^{n} x_i}$$

The log likelihood is

$$\ln L(\lambda) = n \ln \lambda - \lambda \sum_{i=1}^{n} x_i$$

Now

$$\frac{d \ln L(\lambda)}{d\lambda} = \frac{n}{\lambda} - \sum_{i=1}^{n} x_i$$

and upon equating this last result to zero we obtain

$$\hat{\lambda} = n/\sum_{i=1}^{n} X_i = 1/\overline{X}$$

Thus the maximum likelihood estimator of λ is the reciprocal of the sample mean. Notice that this is the same as the moment estimator.

It is easy to illustrate graphically just how the method of maximum likelihood works. Figure 7-7(a) plots the log of the likelihood function for the exponential parameter from Example 7-11, using the $n = 8$ observations on failure time given following Example 7-6. It is common for the log likelihood function to be negative. We found that the estimate of λ was $\hat{\lambda} = 0.0462$. From Example 7-11, we know that this is a maximum likelihood estimate. Figure 7-7(a) shows clearly that the log likelihood function is maximized at a value of λ that is approximately equal to 0.0462. Notice that the log likelihood function is relatively flat in the region of the maximum. This implies that the parameter is not estimated very precisely. If the parameter were estimated precisely, the log likelihood function would be very peaked at the maximum value. The sample size here is relatively small, and this has led to the imprecision in estimation. This is illustrated in Fig. 7-7(b) where we have plotted the difference in log likelihoods for the maximum value, assuming that the sample sizes were $n = 8$, 20, and 40 but

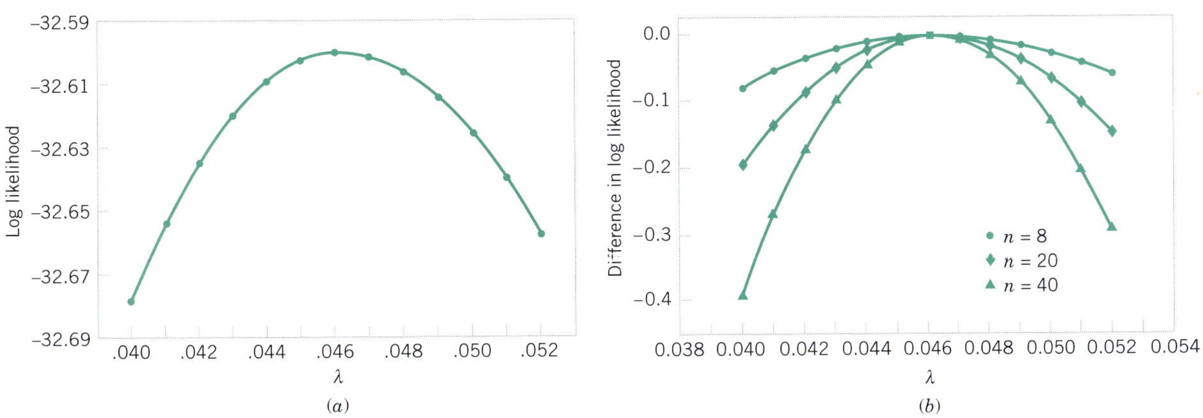

Figure 7-7 Log likelihood for the exponential distribution, using the failure time data. (a) Log likelihood with $n = 8$ (original data). (b) Log likelihood if $n = 8$, 20, and 40.

that the sample average time to failure remained constant at $\bar{x} = 21.65$. Notice how much steeper the log likelihood is for $n = 20$ in comparison to $n = 8$, and for $n = 40$ in comparison to both smaller sample sizes.

The method of maximum likelihood can be used in situations where there are several unknown parameters, say, $\theta_1, \theta_2, \ldots, \theta_k$ to estimate. In such cases, the likelihood function is a function of the k unknown parameters $\theta_1, \theta_2, \ldots, \theta_k$, and the maximum likelihood estimators $\{\hat{\Theta}_i\}$ would be found by equating the k partial derivatives $\partial L(\theta_1, \theta_2, \ldots, \theta_k)/\partial\theta_i$, $i = 1, 2, \ldots, k$ to zero and solving the resulting system of equations.

EXAMPLE 7-12
Normal Distribution MLEs for μ and σ^2

Let X be normally distributed with mean μ and variance σ^2, where both μ and σ^2 are unknown. The likelihood function for a random sample of size n is

$$L(\mu, \sigma^2) = \prod_{i=1}^{n} \frac{1}{\sigma\sqrt{2\pi}} e^{-(x_i-\mu)^2/(2\sigma^2)} = \frac{1}{(2\pi\sigma^2)^{n/2}} e^{\frac{-1}{2\sigma^2}\sum_{i=1}^{n}(x_i-\mu)^2}$$

and

$$\ln L(\mu, \sigma^2) = -\frac{n}{2}\ln(2\pi\sigma^2) - \frac{1}{2\sigma^2}\sum_{i=1}^{n}(x_i - \mu)^2$$

Now

$$\frac{\partial \ln L(\mu, \sigma^2)}{\partial \mu} = \frac{1}{\sigma^2}\sum_{i=1}^{n}(x_i - \mu) = 0$$

$$\frac{\partial \ln L(\mu, \sigma^2)}{\partial(\sigma^2)} = -\frac{n}{2\sigma^2} + \frac{1}{2\sigma^4}\sum_{i=1}^{n}(x_i - \mu)^2 = 0$$

The solutions to the above equations yield the maximum likelihood estimators

$$\hat{\mu} = \bar{X} \qquad \hat{\sigma}^2 = \frac{1}{n}\sum_{i=1}^{n}(X_i - \bar{X})^2$$

Once again, the maximum likelihood estimators are equal to the moment estimators.

Properties of the Maximum Likelihood Estimator

The method of maximum likelihood is often the estimation method that mathematical statisticians prefer, because it produces estimators with good statistical properties. We summarize these properties as follows.

Properties of a Maximum Likelihood Estimator

Under very general and not restrictive conditions, when the sample size n is large and if $\hat{\Theta}$ is the maximum likelihood estimator of the parameter θ,

(1) $\hat{\Theta}$ is an approximately unbiased estimator for θ $[E(\hat{\Theta}) \simeq \theta]$,

(2) the variance of $\hat{\Theta}$ is nearly as small as the variance that could be obtained with any other estimator, and

(3) $\hat{\Theta}$ has an approximate normal distribution.

Properties 1 and 2 essentially state that the maximum likelihood estimator is approximately an MVUE. This is a very desirable result and, coupled with the fact that it is fairly easy to obtain in many situations and has an asymptotic normal distribution (the "asymptotic" means "when n is large"), explains why the maximum likelihood estimation technique is widely used. To use maximum likelihood estimation, remember that the distribution of the population must be either known or assumed.

To illustrate the "large-sample" or asymptotic nature of the above properties, consider the maximum likelihood estimator for σ^2, the variance of the normal distribution, in Example 7-12. It is easy to show that

$$E(\hat{\sigma}^2) = \frac{n-1}{n}\sigma^2$$

The bias is

$$E(\hat{\sigma}^2) - \sigma^2 = \frac{n-1}{n}\sigma^2 - \sigma^2 = \frac{-\sigma^2}{n}$$

Because the bias is negative, $\hat{\sigma}^2$ tends to underestimate the true variance σ^2. Note that the bias approaches zero as n increases. Therefore, $\hat{\sigma}^2$ is an asymptotically unbiased estimator for σ^2.

We now give another very important and useful property of maximum likelihood estimators.

Invariance Property

Let $\hat{\Theta}_1, \hat{\Theta}_2, \ldots, \hat{\Theta}_k$ be the maximum likelihood estimators of the parameters $\theta_1, \theta_2, \ldots, \theta_k$. Then the maximum likelihood estimator of any function $h(\theta_1, \theta_2, \ldots, \theta_k)$ of these parameters is the same function $h(\hat{\Theta}_1, \hat{\Theta}_2, \ldots, \hat{\Theta}_k)$ of the estimators $\hat{\Theta}_1, \hat{\Theta}_2, \ldots, \hat{\Theta}_k$.

EXAMPLE 7-13

In the normal distribution case, the maximum likelihood estimators of μ and σ^2 were $\hat{\mu} = \overline{X}$ and $\hat{\sigma}^2 = \sum_{i=1}^{n}(X_i - \overline{X})^2/n$. To obtain the maximum likelihood estimator of the function $h(\mu, \sigma^2) = \sqrt{\sigma^2} = \sigma$, substitute the estimators $\hat{\mu}$ and $\hat{\sigma}^2$ into the function h, which yields

$$\hat{\sigma} = \sqrt{\hat{\sigma}^2} = \left[\frac{1}{n}\sum_{i=1}^{n}(X_i - \overline{X})^2\right]^{1/2}$$

Thus, the maximum likelihood estimator of the standard deviation σ is *not* the sample standard deviation S.

Complications in Using Maximum Likelihood Estimation

While the method of maximum likelihood is an excellent technique, sometimes complications arise in its use. For example, it is not always easy to maximize the likelihood function because the equation(s) obtained from $dL(\theta)/d\theta = 0$ may be difficult to solve. Furthermore, it may not always be possible to use calculus methods directly to determine the maximum of $L(\theta)$. These points are illustrated in the following two examples.

EXAMPLE 7-14
Uniform Distribution MLE

Let X be uniformly distributed on the interval 0 to a. Since the density function is $f(x) = 1/a$ for $0 \le x \le a$ and zero otherwise, the likelihood function of a random sample of size n is

$$L(a) = \prod_{i=1}^{n}\frac{1}{a} = \frac{1}{a^n}$$

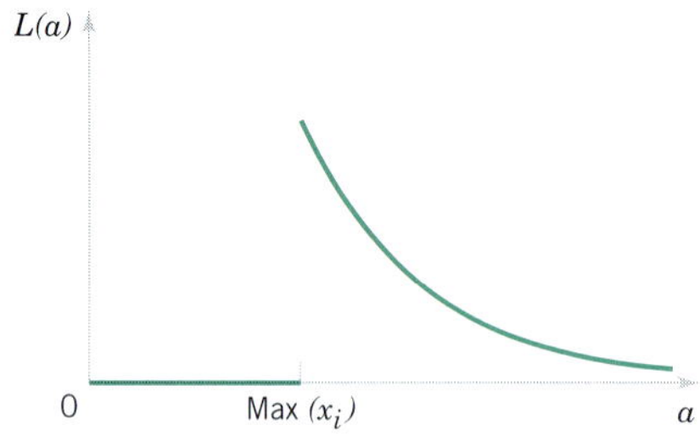

Figure 7-8 The likelihood function for the uniform distribution in Example 7-14.

for

$$0 \le x_1 \le a, 0 \le x_2 \le a, \ldots, 0 \le x_n \le a$$

Note that the slope of this function is not zero anywhere. That is, as long as $\max(x_i) \le a$, the likelihood is $1/a^n$, which is positive, but when $a < \max(x_i)$, the likelihood goes to zero, as illustrated in Fig. 7-8. Therefore, calculus methods cannot be used directly because the maximum value of the likelihood function occurs at a point of discontinuity. However, since $d/da(a^{-n}) = -n/a^{n+1}$ is less than zero for all values of $a > 0$, a^{-n} is a decreasing function of a. This implies that the maximum of the likelihood function $L(a)$ occurs at the lower boundary point. The figure clearly shows that we could maximize $L(a)$ by setting $\hat{a}$ equal to the smallest value that it could logically take on, which is $\max(x_i)$. Clearly, a cannot be smaller than the largest sample observation, so setting $\hat{a}$ equal to the largest sample value is reasonable.

EXAMPLE 7-15
Gamma Distribution MLE

Let $X_1, X_2, \ldots, X_n$ be a random sample from the gamma distribution. The log of the likelihood function is

$$\ln L(r, \lambda) = \ln\left(\prod_{i=1}^{n} \frac{\lambda^r x_i^{r-1} e^{-\lambda x_i}}{\Gamma(r)}\right)$$
$$= nr \ln(\lambda) + (r-1)\sum_{i=1}^{n} \ln(x_i) - n \ln[\Gamma(r)] - \lambda \sum_{i=1}^{n} x_i$$

The derivatives of the log likelihood are

$$\frac{\partial \ln L(r, \lambda)}{\partial r} = n \ln(\lambda) + \sum_{i=1}^{n} \ln(x_i) - n\frac{\Gamma'(r)}{\Gamma(r)}$$
$$\frac{\partial \ln L(r, \lambda)}{\partial \lambda} = \frac{nr}{\lambda} - \sum_{i=1}^{n} x_i$$

When the derivatives are equated to zero, we obtain the equations that must be solved to find the maximum likelihood estimators of r and λ:

$$\hat{\lambda} = \frac{\hat{r}}{\bar{x}}$$
$$n \ln(\hat{\lambda}) + \sum_{i=1}^{n} \ln(x_i) = n\frac{\Gamma'(\hat{r})}{\Gamma(\hat{r})}$$

There is no closed form solution to these equations.

Figure 7-9 shows a graph of the log likelihood for the gamma distribution using the $n = 8$ observations on failure time introduced previously. Figure 7-9(a) shows the **log likelihood surface** as a function of r and λ, and Figure 7-9(b) is a **contour plot.** These plots reveal that the log likelihood is maximized at approximately $\hat{r} = 1.75$ and $\hat{\lambda} = 0.08$. Many statistics computer programs use numerical techniques to solve for the maximum likelihood estimates when no simple solution exists.

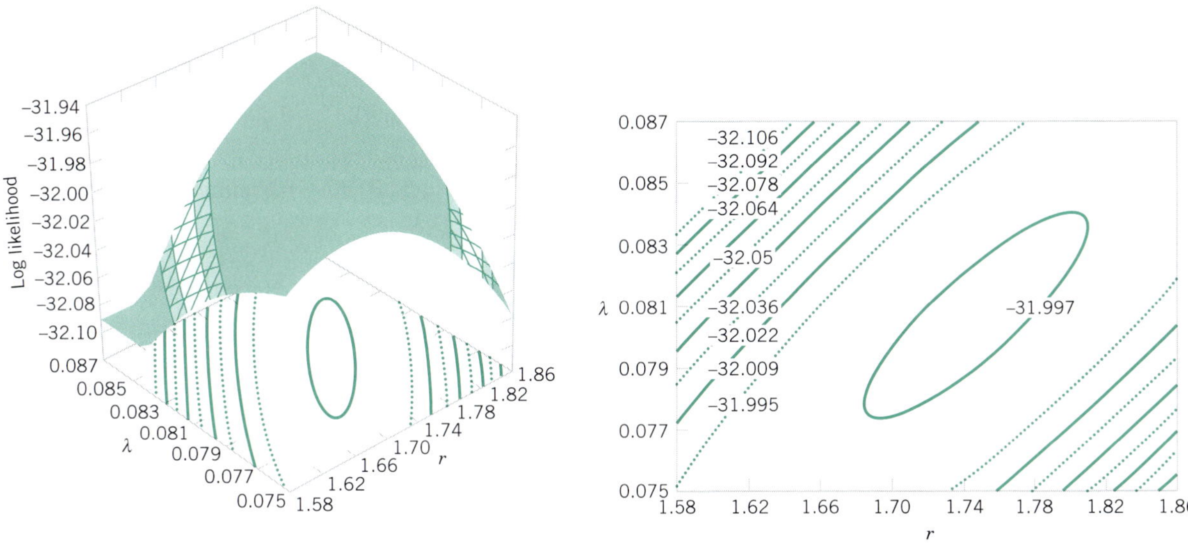

(a) (b)

Figure 7-9 Log likelihood for the gamma distribution using the failure time data. (a) Log likelihood surface. (b) Contour plot.

7-4.3 Bayesian Estimation of Parameters

This book uses methods of statistical inference based on the information in the sample data. In effect, these methods interpret probabilities as relative frequencies. Sometimes we call probabilities that are interpreted in this manner **objective probabilities.** There is another approach to statistical inference, called the **Bayesian** approach, that combines sample information with other information that may be available prior to collecting the sample. In this section we briefly illustrate how this approach may be used in parameter estimation.

Suppose that the random variable X has a probability distribution that is a function of one parameter θ. We will write this probability distribution as $f(x \mid \theta)$. This notation implies that the exact form of the distribution of X is conditional on the value assigned to θ. The classical approach to estimation would consist of taking a random sample of size n from this distribution and then substituting the sample values x_i into the estimator for θ. This estimator could have been developed using the maximum likelihood approach, for example.

Suppose that we have some additional information about θ and that we can summarize that information in the form of a probability distribution for θ, say, $f(\theta)$. This probability distribution is often called the **prior distribution** for θ, and suppose that the mean of the prior is μ_0 and the variance is σ_0^2. This is a very novel concept insofar as the rest of this book is concerned because we are now viewing the parameter θ as a random variable. The probabilities associated with the prior distribution are often called **subjective probabilities,** in that they usually reflect the analyst's degree of belief regarding the true value of θ. The Bayesian approach to estimation uses the prior distribution for θ, $f(\theta)$, and the joint probability distribution of the sample, say $f(x_1, x_2, \ldots, x_n \mid \theta)$, to find a **posterior distribution** for θ, say, $f(\theta \mid x_1, x_2, \ldots, x_n)$. This posterior distribution contains information both from the sample and the prior distribution for θ. In a sense, it expresses our degree of belief regarding the true value of θ after observing the sample data. It is easy conceptually to find the posterior distribution. The joint probability distribution of the sample $X_1, X_2, \ldots, X_n$ and the parameter θ (remember that θ is a random variable) is

$$f(x_1, x_2, \ldots, x_n, \theta) = f(x_1, x_2, \ldots, x_n \mid \theta) f(\theta)$$

and the marginal distribution of $X_1, X_2, \ldots, X_n$ is

$$f(x_1, x_2, \ldots, x_n) = \begin{cases} \sum_{\theta} f(x_1, x_2, \ldots, x_n, \theta), & \theta \text{ discrete} \\ \int_{-\infty}^{\infty} f(x_1, x_2, \ldots, x_n, \theta)\, d\theta, & \theta \text{ continuous} \end{cases}$$

Therefore, the desired distribution is

$$f(\theta \mid x_1, x_2, \ldots, x_n) = \frac{f(x_1, x_2, \ldots, x_n, \theta)}{f(x_1, x_2, \ldots, x_n)}$$

We define the **Bayes estimator** of θ as the value $\tilde{\theta}$ that corresponds to the mean of the posterior distribution $f(\theta \mid x_1, x_2, \ldots, x_n)$.

Sometimes, the mean of the posterior distribution of θ can be determined easily. As a function of θ, $f(\theta \mid x_1, \ldots, x_n)$ is a probability density function and $x_1, \ldots, x_n$ are just constants. Because θ enters into $f(\theta \mid x_1, \ldots, x_n)$ only through $f(x_1, \ldots, x_n, \theta)$ if $f(x_1, \ldots, x_n, \theta)$, as a function of θ is recognized as a well-known probability function, the posterior mean of θ can be deduced from the well-known distribution without integration or even calculation of $f(x_1, \ldots, x_n)$.

EXAMPLE 7-16
Bayes Estimator for the Mean of a Normal Distribution

Let $X_1, X_2, \ldots, X_n$ be a random sample from the normal distribution with mean μ and variance σ^2, where μ is unknown and σ^2 is known. Assume that the prior distribution for μ is normal with mean μ_0 and variance σ_0^2; that is

$$f(\mu) = \frac{1}{\sqrt{2\pi}\sigma_0} e^{-(\mu-\mu_0)^2/(2\sigma_0^2)} = \frac{1}{\sqrt{2\pi\sigma_0^2}} e^{-(\mu^2 - 2\mu_0\mu + \mu_0^2)/(2\sigma_0^2)}$$

The joint probability distribution of the sample is

$$\begin{aligned} f(x_1, x_2, \ldots, x_n \mid \mu) &= \frac{1}{(2\pi\sigma^2)^{n/2}} e^{-(1/2\sigma^2)\sum_{i=1}^{n}(x_i-\mu)^2} \\ &= \frac{1}{(2\pi\sigma^2)^{n/2}} e^{-(1/2\sigma^2)(\Sigma x_i^2 - 2\mu\Sigma x_i + n\mu^2)} \end{aligned}$$

Thus, the joint probability distribution of the sample and μ is

$$\begin{aligned} f(x_1, x_2, \ldots, x_n, \mu) &= \frac{1}{(2\pi\sigma^2)^{n/2}\sqrt{2\pi}\sigma_0} e^{-(1/2)[(1/\sigma_0^2 + n/\sigma^2)\mu^2 - (2\mu_0/\sigma_0^2 + 2\Sigma x_i/\sigma^2)\mu + \Sigma x_i^2/\sigma^2 + \mu_0^2/\sigma_0^2]} \\ &= e^{-(1/2)\left[\left(\frac{1}{\sigma_0^2} + \frac{1}{\sigma^2/n}\right)\mu^2 - 2\left(\frac{\mu_0}{\sigma_0^2} + \frac{\bar{x}}{\sigma^2/n}\right)\mu\right]} h_1(x_1, \ldots, x_n, \sigma^2, \mu_0, \sigma_0^2) \end{aligned}$$

Upon completing the square in the exponent

$$f(x_1, x_2, \ldots, x_n, \mu) = e^{-(1/2)\left(\frac{1}{\sigma_0^2} + \frac{1}{\sigma^2/n}\right)\left[\mu^2 - \left(\frac{(\sigma^2/n)\mu_0}{\sigma_0^2 + \sigma^2/n} + \frac{\bar{x}\sigma_0^2}{\sigma_0^2 + \sigma^2/n}\right)\right]^2} h_2(x_1, \ldots, x_n, \sigma^2, \mu_0, \sigma_0^2)$$

where $h_i(x_1, \ldots, x_n, \sigma^2, \mu_0, \sigma_0^2)$ is a function of the observed values, σ^2, μ_0, and σ_0^2.

Now, because $f(x_1, \ldots, x_n)$ does not depend on μ,

$$f(\mu \mid x_1, \ldots, x_n) = e^{-(1/2)\left(\frac{1}{\sigma_0^2}+\frac{1}{\sigma^2/n}\right)\left[\mu^2-\left(\frac{(\sigma^2/n)\mu_0+\sigma_0^2\bar{x}}{\sigma_0^2+\sigma^2/n}\right)\right]} h_3(x_1, \ldots, x_n, \sigma^2, \mu_0, \sigma_0^2)$$

This is recognized as a normal probability density function with posterior mean

$$\frac{(\sigma^2/n)\mu_0 + \sigma_0^2\bar{x}}{\sigma_0^2 + \sigma^2/n}$$

and posterior variance

$$\left(\frac{1}{\sigma_0^2} + \frac{1}{\sigma^2/n}\right)^{-1} = \frac{\sigma_0^2(\sigma^2/n)}{\sigma_0^2 + \sigma^2/n}$$

Consequently, the Bayes estimate of μ is a weighted average of μ_0 and $\bar{x}$. For purposes of comparison, note that the maximum likelihood estimate of μ is $\hat{\mu} = \bar{x}$.

To illustrate, suppose that we have a sample of size $n = 10$ from a normal distribution with unknown mean μ and variance $\sigma^2 = 4$. Assume that the prior distribution for μ is normal with mean $\mu_0 = 0$ and variance $\sigma_0^2 = 1$. If the sample mean is 0.75, the Bayes estimate of μ is

$$\frac{(4/10)0 + 1(0.75)}{1 + (4/10)} = \frac{0.75}{1.4} = 0.536$$

Note that the maximum likelihood estimate of μ is $\bar{x} = 0.75$.

There is a relationship between the Bayes estimator for a parameter and the maximum likelihood estimator of the same parameter. For large sample sizes, the two are nearly equivalent. In general, the difference between the two estimators is small compared to $1/\sqrt{n}$. In practical problems, a moderate sample size will produce approximately the same estimate by either the Bayes or maximum likelihood method, if the sample results are consistent with the assumed prior information. If the sample results are inconsistent with the prior assumptions, the Bayes estimate may differ considerably from the maximum likelihood estimate. In these circumstances, if the sample results are accepted as being correct, the prior information must be incorrect. The maximum likelihood estimate would then be the better estimate to use.

If the sample results are very different from the prior information, the Bayes estimator will always tend to produce an estimate that is between the maximum likelihood estimate and the prior assumptions. If there is more inconsistency between the prior information and the sample, there will be more difference between the two estimates.

EXERCISES FOR SECTION 7-4

7-28. Let X be a geometric random variable with parameter p. Find the maximum likelihood estimator of p, based on a random sample of size n.

7-29. Consider the Poisson distribution with parameter λ. Find the maximum likelihood estimator of λ, based on a random sample of size n.

7-30. Let X be a random variable with the following probability distribution:

$$f(x) = \begin{cases} (\theta + 1)x^\theta, & 0 \le x \le 1 \\ 0 \quad , & \text{otherwise} \end{cases}$$

Find the maximum likelihood estimator of θ, based on a random sample of size n.

7-31. Consider the shifted exponential distribution

$$f(x) = \lambda e^{-\lambda(x-\theta)}, \qquad x \geq \theta$$

When $\theta = 0$, this density reduces to the usual exponential distribution. When $\theta > 0$, there is only positive probability to the right of θ.

(a) Find the maximum likelihood estimator of λ and θ, based on a random sample of size n.

(b) Describe a practical situation in which one would suspect that the shifted exponential distribution is a plausible model.

7-32. Consider the probability density function

$$f(x) = \frac{1}{\theta^2} x e^{-x/\theta}, \qquad 0 \leq x < \infty, \quad 0 < \theta < \infty$$

Find the maximum likelihood estimator for θ.

7-33. Let $X_1, X_2, \ldots, X_n$ be uniformly distributed on the interval 0 to a. Show that the moment estimator of a is $\hat{a} = 2\overline{X}$. Is this an unbiased estimator? Discuss the reasonableness of this estimator.

7-34. Consider the probability density function

$$f(x) = c(1 + \theta x), \quad -1 \leq x \leq 1$$

(a) Find the value of the constant c.

(b) What is the moment estimator for θ?

(c) Show that $\hat{\theta} = 3\overline{X}$ is an unbiased estimator for θ.

(d) Find the maximum likelihood estimator for θ.

7-35. The Rayleigh distribution has probability density function

$$f(x) = \frac{x}{\theta} e^{-x^2/2\theta}, \qquad x > 0, \qquad 0 < \theta < \infty$$

(a) It can be shown that $E(X^2) = 2\theta$. Use this information to construct an unbiased estimator for θ.

(b) Find the maximum likelihood estimator of θ. Compare your answer to part (a).

(c) Use the invariance property of the maximum likelihood estimator to find the maximum likelihood estimator of the median of the Raleigh distribution.

7-36. Let $X_1, X_2, \ldots, X_n$ be uniformly distributed on the interval 0 to a. Recall that the maximum likelihood estimator of a is $\hat{a} = \max(X_i)$.

(a) Argue intuitively why $\hat{a}$ cannot be an unbiased estimator for a.

(b) Suppose that $E(\hat{a}) = na/(n + 1)$. Is it reasonable that $\hat{a}$ consistently underestimates a? Show that the bias in the estimator approaches zero as n gets large.

(c) Propose an unbiased estimator for a.

(d) Let $Y = \max(X_i)$. Use the fact that $Y \leq y$ if and only if each $X_i \leq y$ to derive the cumulative distribution function of Y. Then show that the probability density function of Y is

$$f(y) = \begin{cases} \dfrac{ny^{n-1}}{a^n}, & 0 \leq y \leq a \\ 0\ , & \text{otherwise} \end{cases}$$

Use this result to show that the maximum likelihood estimator for a is biased.

(e) We have two unbiased estimators for a: the moment estimator $\hat{a}_1 = 2\overline{X}$ and $\hat{a}_2 = [(n + 1)/n] \max(X_i)$, where $\max(X_i)$ is the largest observation in a random sample of size n. It can be shown that $V(\hat{a}_1) = a^2/(3n)$ and that $V(\hat{a}_2) = a^2/[n(n + 2)]$. Show that if $n > 1$, $\hat{a}_2$ is a better estimator than $\hat{a}$. In what sense is it a better estimator of a?

7-37. Consider the Weibull distribution

$$f(x) = \begin{cases} \dfrac{\beta}{\delta}\left(\dfrac{x}{\delta}\right)^{\beta-1} e^{-\left(\frac{x}{\delta}\right)^{\beta}}, & 0 < x \\ 0\ , & \text{otherwise} \end{cases}$$

(a) Find the likelihood function based on a random sample of size n. Find the log likelihood.

(b) Show that the log likelihood is maximized by solving the equations

$$\beta = \left[\frac{\sum_{i=1}^{n} x_i^{\beta} \ln(x_i)}{\sum_{i=1}^{n} x_i^{\beta}} - \frac{\sum_{i=1}^{n} \ln(x_i)}{n}\right]^{-1}$$

$$\delta = \left[\frac{\sum_{i=1}^{n} x_i^{\beta}}{n}\right]^{1/\beta}$$

(c) What complications are involved in solving the two equations in part (b)?

7-38. Reconsider the oxide thickness data in Exercise 7-23 and suppose that it is reasonable to assume that oxide thickness is normally distributed.

(a) Compute the maximum likelihood estimates of μ and σ^2.

(b) Graph the likelihood function in the vicinity of $\hat{\mu}$ and $\hat{\sigma}^2$, the maximum likelihood estimates, and comment on its shape.
(c) Suppose that the sample size was larger ($n = 40$) but the maximum likelihood estimates were numerically equal to the values obtained in part (a). Graph the likelihood function for $n = 40$, compare it to the one from part (b), and comment on the effect of the larger sample size.

7-39. Suppose that X is a normal random variable with unknown mean μ and known variance σ^2. The prior distribution for μ is a normal distribution with mean μ_0 and variance σ_0^2. Show that the Bayes estimator for μ becomes the maximum likelihood estimator when the sample size n is large.

7-40. Suppose that X is a normal random variable with unknown mean μ and known variance σ^2. The prior distribution for μ is a uniform distribution defined over the interval $[a, b]$.
(a) Find the posterior distribution for μ.
(b) Find the Bayes estimator for μ.

7-41. Suppose that X is a Poisson random variable with parameter λ. Let the prior distribution for λ be a gamma distribution with parameters $m + 1$ and $(m + 1)/\lambda_0$.
(a) Find the posterior distribution for λ.
(b) Find the Bayes estimator for λ.

7-42. Suppose that X is a normal random variable with unknown mean and known variance $\sigma^2 = 9$. The prior distribution for μ is normal with $\mu_0 = 4$ and $\sigma_0^2 = 1$. A random sample of $n = 25$ observations is taken, and the sample mean is $\bar{x} = 4.85$.
(a) Find the Bayes estimate of μ.
(b) Compare the Bayes estimate with the maximum likelihood estimate.

7-43. The weight of boxes of candy is a normal random variable with mean μ and variance $1/10$ pound. The prior distribution for μ is normal with mean 5.03 pound, and variance $1/25$ pound. A random sample of 10 boxes gives a sample mean of $\bar{x} = 5.05$ pound.
(a) Find the Bayes estimate of μ.
(b) Compare the Bayes estimate with the maximum likelihood estimate.

7-44. The time between failures of a machine has an exponential distribution with parameter λ. Suppose that the prior distribution for λ is exponential with mean 100 hours. Two machines are observed, and the average time between failures is $\bar{x} = 1125$ hours.
(a) Find the Bayes estimate for λ.
(b) What proportion of the machines do you think will fail before 1000 hours?

Supplemental Exercises

7-45. Transistors have a life that is exponentially distributed with parameter λ. A random sample of n transistors is taken. What is the joint probability density function of the sample?

7-46. Suppose that a random variable is normally distributed with mean μ and variance σ^2, and we draw a random sample of five observations from this distribution. What is the joint probability density function of the sample?

7-47. Suppose that X is uniformly distributed on the interval from 0 to 1. Consider a random sample of size 4 from X. What is the joint probability density function of the sample?

7-48. A procurement specialist has purchased 25 resistors from vendor 1 and 30 resistors from vendor 2. Let $X_{1,1}, X_{1,2}, \ldots, X_{1,25}$ represent the vendor 1 observed resistances, which are assumed to be normally and independently distributed with mean 100 ohms and standard deviation 1.5 ohms. Similarly, let $X_{2,1}, X_{2,2}, \ldots, X_{2,30}$ represent the vendor 2 observed resistances, which are assumed to be normally and independently distributed with mean 105 ohms and standard deviation of 2.0 ohms. What is the sampling distribution of $\bar{X}_1 - \bar{X}_2$? What is the standard error of $\bar{X}_1 - \bar{X}_2$?

7-49. A random sample of 36 observations has been drawn from a normal distribution with mean 50 and standard deviation 12. Find the probability that the sample mean is in the interval $47 \leq \bar{X} \leq 53$. Is the assumption of normality important. Why?

7-50. A random sample of $n = 9$ structural elements is tested for compressive strength. We know that the true mean compressive strength $\mu = 5500$ psi and the standard deviation is $\sigma = 100$ psi. Find the probability that the sample mean compressive strength exceeds 4985 psi.

7-51. A normal population has a known mean 50 and known variance $\sigma^2 = 2$. A random sample of $n = 16$ is selected from this population, and the sample mean is $\bar{x} = 52$. How unusual is this result?

7-52. A random sample of size $n = 16$ is taken from a normal population with $\mu = 40$ and $\sigma^2 = 5$. Find the probability that the sample mean is less than or equal to 37.

7-53. A manufacturer of semiconductor devices takes a random sample of 100 chips and tests them, classifying each chip as defective or nondefective. Let $X_i = 0$ if the chip is nondefective and $X_i = 1$ if the chip is defective. The sample fraction defective is

$$\hat{P} = \frac{X_1 + X_2 + \cdots + X_{100}}{100}$$

What is the sampling distribution of the random variable $\hat{P}$?

7-54. Let X be a random variable with mean μ and variance σ^2. Given two independent random samples of sizes n_1 and n_2, with sample means $\bar{X}_1$ and $\bar{X}_2$, show that

$$\bar{X} = a\bar{X}_1 + (1 - a)\bar{X}_2, \quad 0 < a < 1$$

is an unbiased estimator for μ. If $\overline{X}_1$ and $\overline{X}_2$ are independent, find the value of a that minimizes the standard error of $\overline{X}$.

7-55. A random variable x has probability density function

$$f(x) = \frac{1}{2\theta^3} x^2 e^{-x/\theta}, \qquad 0 < x < \infty, \quad 0 < \theta < \infty$$

Find the maximum likelihood estimator for θ.

7-56. Let $f(x) = \theta x^{\theta - 1}$, $0 < \theta < \infty$, and $0 < x < 1$. Show that $\hat{\Theta} = -n/(\ln \prod_{i=1}^{n} X_i)$ is the maximum likelihood estimator for θ.

7-57. Let $f(x) = (1/\theta)x^{(1-\theta)/\theta}$, $0 < x < 1$, and $0 < \theta < \infty$. Show that $\hat{\Theta} = -(1/n)\sum_{i=1}^{n} \ln(X_i)$ is the maximum likelihood estimator for θ and that $\hat{\Theta}$ is an unbiased estimator for θ.

MIND-EXPANDING EXERCISES

7-58. A lot consists of N transistors, and of these M ($M \le N$) are defective. We randomly select two transistors without replacement from this lot and determine whether they are defective or nondefective. The random variable

$$X_i = \begin{cases} 1, & \text{if the } i\text{th transistor is nondefective} \\ 0, & \text{if the } i\text{th transistor is defective} \end{cases} \qquad i = 1, 2$$

Determine the joint probability function for X_1 and X_2. What are the marginal probability functions for X_1 and X_2? Are X_1 and X_2 independent random variables?

7-59. When the sample standard deviation is based on a random sample of size n from a normal population, it can be shown that S is a biased estimator for σ. Specifically,

$$E(S) = \sigma\sqrt{2/(n-1)}\; \Gamma(n/2)/\Gamma[(n-1)/2]$$

(a) Use this result to obtain an unbiased estimator for σ of the form $c_n S$, when the constant c_n depends on the sample size n.

(b) Find the value of c_n for $n = 10$ and $n = 25$. Generally, how well does S perform as an estimator of σ for large n with respect to bias?

7-60. A collection of n randomly selected parts is measured twice by an operator using a gauge. Let X_i and Y_i denote the measured values for the ith part. Assume that these two random variables are independent and normally distributed and that both have true mean μ_i and variance σ^2.

(a) Show that the maximum likelihood estimator of σ^2 is $\hat{\sigma}^2 = (1/4n)\sum_{i=1}^{n}(X_i - Y_i)^2$.

(b) Show that $\hat{\sigma}^2$ is a biased estimator for σ^2. What happens to the bias as n becomes large?

(c) Find an unbiased estimator for σ^2.

7-61. **Consistent Estimator.** Another way to measure the closeness of an estimator $\hat{\Theta}$ to the parameter θ is in terms of consistency. If $\hat{\Theta}_n$ is an estimator of θ based on a random sample of n observations, $\hat{\Theta}_n$ is consistent for θ if

$$\lim_{n\to\infty} P(|\hat{\Theta}_n - \theta| < \epsilon) = 1$$

Thus, consistency is a large-sample property, describing the limiting behavior of $\hat{\Theta}_n$ as n tends to infinity. It is usually difficult to prove consistency using the above definition, although it can be done from other approaches. To illustrate, show that $\overline{X}$ is a consistent estimator of μ (when $\sigma^2 < \infty$) by using Chebyshev's inequality from the supplemental material on the Web.

7-62. **Order Statistics.** Let $X_1, X_2, \ldots, X_n$ be a random sample of size n from X, a random variable having distribution function $F(x)$. Rank the elements in order of increasing numerical magnitude, resulting in $X_{(1)}, X_{(2)}, \ldots, X_{(n)}$, where $X_{(1)}$ is the smallest sample element ($X_{(1)} = \min\{X_1, X_2, \ldots, X_n\}$) and $X_{(n)}$ is the largest sample element ($X_{(n)} = \max\{X_1, X_2, \ldots, X_n\}$). $X_{(i)}$ is called the ith order statistic. Often the distribution of some of the order statistics is of interest, particularly the minimum and maximum sample values. $X_{(1)}$ and $X_{(n)}$, respectively.

(a) Prove that the cumulative distribution functions of these two order statistics, denoted respectively by $F_{X_{(1)}}(t)$ and $F_{X_{(n)}}(t)$ are

$$F_{X_{(1)}}(t) = 1 - [1 - F(t)]^n$$
$$F_{X_{(n)}}(t) = [F(t)]^n$$

(b) Prove that if X is continuous with probability density function $f(x)$, the probability distributions of $X_{(1)}$ and $X_{(n)}$ are

$$f_{X_{(1)}}(t) = n[1 - F(t)]^{n-1} f(t)$$
$$f_{X_{(n)}}(t) = n[F(t)]^{n-1} f(t)$$

MIND-EXPANDING EXERCISES

(c) Let $X_1, X_2, \ldots, X_n$ be a random sample of a Bernoulli random variable with parameter p. Show that

$$P(X_{(n)} = 1) = 1 - (1 - p)^n$$
$$P(X_{(1)} = 0) = 1 - p^n$$

(d) Let $X_1, X_2, \ldots, X_n$ be a random sample of a normal random variable with mean μ and variance σ^2. Derive the probability density functions of $X_{(1)}$ and $X_{(n)}$.

(e) Let $X_1, X_2, \ldots, X_n$ be a random sample of an exponential random variable of parameter λ. Derive the cumulative distribution functions and probability density functions for $X_{(1)}$ and $X_{(n)}$.

7-63. Let $X_1, X_2, \ldots, X_n$ be a random sample of a continuous random variable with cumulative distribution function $F(x)$. Find

$$E[F(X_{(n)})]$$

and

$$E[F(X_{(1)})]$$

7-64. Let X be a random variable with mean μ and variance σ^2, and let $X_1, X_2, \ldots, X_n$ be a random sample of size n from X. Show that the statistic $V = k\sum_{i=1}^{n-1}(X_{i+1} - X_i)^2$ is an unbiased estimator for σ^2 for an appropriate choice for the constant k. Find this value for k.

7-65. When the population has a normal distribution, the estimator

$$\hat{\sigma} = \text{median}\,(|X_1 - \overline{X}|, |X_2 - \overline{X}|, \cdots, |X_n - X|)/0.6745$$

is sometimes used to estimate the population standard deviation. This estimator is more robust to outliers than the usual sample standard deviation and usually does not differ much from S when there are no unusual observations.

(a) Calculate $\hat{\sigma}$ and S for the data 10, 12, 9, 14, 18, 15, and 16.

(b) Replace the first observation in the sample (10) with 50 and recalculate both S and $\hat{\sigma}$.

7-66. **Censored Data.** A common problem in industry is life testing of components and systems. In this problem, we will assume that lifetime has an exponential distribution with parameter λ, so $\hat{\mu} = 1/\hat{\lambda} = \overline{X}$ is an unbiased estimate of μ. When n components are tested until failure and the data $X_1, X_2, \ldots, X_n$ represent actual lifetimes, we have a complete sample, and $\overline{X}$ is indeed an unbiased estimator of μ. However, in many situations, the components are only left under test until $r < n$ failures have occurred. Let Y_1 be the time of the first failure, Y_2 be the time of the second failure, ..., and Y_r be the time of the last failure. This type of test results in **censored data.** There are $n - r$ units still running when the test is terminated. The total accumulated test time at termination is

$$T_r = \sum_{i=1}^{r} Y_i + (n - r)Y_r$$

(a) Show that $\hat{\mu} = T_r/r$ is an unbiased estimator for μ. [*Hint:* You will need to use the memoryless property of the exponential distribution and the results of Exercise 7-62 for the distribution of the minimum of a sample from an exponential distribution with parameter λ.]

(b) It can be shown that $V(T_r/r) = 1/(\lambda^2 r)$. How does this compare to $V(\overline{X})$ in the uncensored experiment?

IMPORTANT TERMS AND CONCEPTS

Bayes estimator
Bias in parameter estimation
Central limit theorem
Estimator versus estimate
Likelihood function
Maximum likelihood estimator
Mean square error of an estimator
Minimum variance unbiased estimator
Moment estimator
Normal distribution as the sampling distribution of a sample mean
Normal distribution as the sampling distribution of the difference in two sample means
Parameter estimation
Point estimator
Population or distribution moments
Posterior distribution
Prior distribution
Sample moments
Sampling distribution
Standard error and estimated standard error of an estimator
Statistic
Statistical inference
Unbiased estimator

8 Analysis of Statistical Intervals: Single Sample

CHAPTER OUTLINE

LEARNING OBJECTIVES

After careful study of this chapter, you should be able to do the following:

1. Construct confidence intervals on the mean of a normal distribution, using either the normal distribution or the t distribution method
2. Construct confidence intervals on the variance and standard deviation of a normal distribution
3. Construct confidence intervals on a population proportion
4. Use a general method for constructing an approximate confidence interval on a parameter
5. Construct prediction intervals for a future observation

6. Construct a tolerance interval for a normal population
7. Explain the three types of interval estimates: confidence intervals, prediction intervals, and tolerance intervals

8-1 INTRODUCTION

In the previous chapter we illustrated how a parameter can be estimated from sample data. However, it is important to understand how good is the estimate obtained. For example, suppose that we estimate the mean viscosity of a chemical product to be $\hat{\mu} = \bar{x} = 1000$. Now because of sampling variability, it is almost never the case that $\mu = \bar{x}$. The point estimate says nothing about how close $\hat{\mu}$ is to μ. Is the process mean likely to be between 900 and 1100? Or is it likely to be between 990 and 1010? The answer to these questions affects our decisions regarding this process. Bounds that represent an interval of plausible values for a parameter are an example of an interval estimate. Surprisingly, it is easy to determine such intervals in many cases, and the same data that provided the point estimate are typically used.

An interval estimate for a population parameter is called a **confidence interval.** We cannot be certain that the interval contains the true, unknown population parameter—we only use a sample from the full population to compute the point estimate and the interval. However, the confidence interval is constructed so that we have high confidence that it does contain the unknown population parameter. Confidence intervals are widely used in engineering and the sciences.

A **tolerance interval** is another important type of interval estimate. For example, the chemical product viscosity data might be assumed to be normally distributed. We might like to calculate limits that bound 95% of the viscosity values. For a normal distribution, we know that 95% of the distribution is in the interval

$$\mu - 1.96\sigma, \mu + 1.96\sigma \tag{8-1}$$

However, this is not a useful tolerance interval because the parameters μ and σ are unknown. Point estimates such as $\bar{x}$ and s can be used in Equation 8-1 for μ and σ. However, we need to account for the potential error in each point estimate to form a tolerance interval for the distribution. The result is an interval of the form

$$\bar{x} - ks, \bar{x} + ks \tag{8-2}$$

where k is an appropriate constant (that is larger than 1.96 to account for the estimation error). As for a confidence interval, it is not certain that Equation 8-2 bounds 95% of the distribution, but the interval is constructed so that we have high confidence that it does. Tolerance intervals are widely used and, as we will subsequently see, they are easy to calculate for normal distributions.

Confidence and tolerance intervals bound unknown elements of a distribution. In this chapter you will learn to appreciate the value of these intervals. A **prediction interval** provides bounds on one (or more) future observations from the population. For example, a prediction interval could be used to bound a single, new measurement of viscosity—another useful interval. With a large sample size, the prediction interval for normally distributed data tends to the tolerance interval in Equation 8-1, but for more modest sample sizes the prediction and tolerance intervals are different.

Keep the purpose of the three types of interval estimates clear:

- A confidence interval bounds population or distribution parameters (such as the mean viscosity).
- A tolerance interval bounds a selected proportion of a distribution.
- A prediction interval bounds future observations from the population or distribution.

8-2 CONFIDENCE INTERVAL ON THE MEAN OF A NORMAL DISTRIBUTION, VARIANCE KNOWN

The basic ideas of a confidence interval (CI) are most easily understood by initially considering a simple situation. Suppose that we have a normal population with unknown mean μ and known variance σ^2. This is a somewhat unrealistic scenario because typically both the mean and variance are unknown. However, in subsequent sections we will present confidence intervals for more general situations.

8-2.1 Development of the Confidence Interval and Its Basic Properties

Suppose that $X_1, X_2, \ldots, X_n$ is a random sample from a normal distribution with unknown mean μ and known variance σ^2. From the results of Chapter 5 we know that the sample mean $\overline{X}$ is normally distributed with mean μ and variance σ^2/n. We may **standardize** $\overline{X}$ by subtracting the mean and dividing by the standard deviation, which results in the variable

$$Z = \frac{\overline{X} - \mu}{\sigma/\sqrt{n}} \tag{8-3}$$

The random variable Z has a standard normal distribution.

A **confidence interval** estimate for μ is an interval of the form $l \leq \mu \leq u$, where the end-points l and u are computed from the sample data. Because different samples will produce different values of l and u, these end-points are values of random variables L and U, respectively. Suppose that we can determine values of L and U such that the following probability statement is true:

$$P\{L \leq \mu \leq U\} = 1 - \alpha \tag{8-4}$$

where $0 \leq \alpha \leq 1$. There is a probability of $1 - \alpha$ of selecting a sample for which the CI will contain the true value of μ. Once we have selected the sample, so that $X_1 = x_1, X_2 = x_2, \ldots, X_n = x_n$, and computed l and u, the resulting **confidence interval** for μ is

$$l \leq \mu \leq u \tag{8-5}$$

The end-points or bounds l and u are called the **lower-** and **upper-confidence limits,** respectively, and $1 - \alpha$ is called the **confidence coefficient.**

In our problem situation, because $Z = (\overline{X} - \mu)/(\sigma/\sqrt{n})$ has a standard normal distribution, we may write

$$P\left\{-z_{\alpha/2} \leq \frac{\overline{X} - \mu}{\sigma/\sqrt{n}} \leq z_{\alpha/2}\right\} = 1 - \alpha$$

Now manipulate the quantities inside the brackets by (1) multiplying through by $\sigma/\sqrt{n}$, (2) subtracting $\overline{X}$ from each term, and (3) multiplying through by -1. This results in

$$P\left\{\overline{X} - z_{\alpha/2}\frac{\sigma}{\sqrt{n}} \leq \mu \leq \overline{X} + z_{\alpha/2}\frac{\sigma}{\sqrt{n}}\right\} = 1 - \alpha \tag{8-6}$$

From consideration of Equation 8-4, the lower and upper limits of the inequalities in Equation 8-6 are the lower- and upper-confidence limits L and U, respectively. This leads to the following definition.

Confidence Interval on the Mean, Variance Known

If $\bar{x}$ is the sample mean of a random sample of size n from a normal population with known variance σ^2, a $100(1 - \alpha)\%$ CI on μ is given by

$$\bar{x} - z_{\alpha/2}\sigma/\sqrt{n} \leq \mu \leq \bar{x} + z_{\alpha/2}\sigma/\sqrt{n} \tag{8-7}$$

where $z_{\alpha/2}$ is the upper $100\alpha/2$ percentage point of the standard normal distribution.

EXAMPLE 8-1
Metallic Material Transition

ASTM Standard E23 defines standard test methods for notched bar impact testing of metallic materials. The Charpy V-notch (CVN) technique measures impact energy and is often used to determine whether or not a material experiences a ductile-to-brittle transition with decreasing temperature. Ten measurements of impact energy (J) on specimens of A238 steel cut at 60°C are as follows: 64.1, 64.7, 64.5, 64.6, 64.5, 64.3, 64.6, 64.8, 64.2, and 64.3. Assume that impact energy is normally distributed with $\sigma = 1J$. We want to find a 95% CI for μ, the mean impact energy. The required quantities are $z_{\alpha/2} = z_{0.025} = 1.96$, $n = 10$, $\sigma = 1$, and $\bar{x} = 64.46$. The resulting 95% CI is found from Equation 8-7 as follows:

$$\bar{x} - z_{\alpha/2}\frac{\sigma}{\sqrt{n}} \leq \mu \leq \bar{x} + z_{\alpha/2}\frac{\sigma}{\sqrt{n}}$$

$$64.46 - 1.96\frac{1}{\sqrt{10}} \leq \mu \leq 64.46 + 1.96\frac{1}{\sqrt{10}}$$

$$63.84 \leq \mu \leq 65.08$$

That is, based on the sample data, a range of highly plausible values for mean impact energy for A238 steel at 60°C is $63.84J \leq \mu \leq 65.08J$.

Interpreting a Confidence Interval

How does one interpret a confidence interval? In the impact energy estimation problem in Example 8-1 the 95% CI is $63.84 \leq \mu \leq 65.08$, so it is tempting to conclude that μ is within this interval with probability 0.95. However, with a little reflection, it's easy to see that this cannot be correct; the true value of μ is unknown and the statement $63.84 \leq \mu \leq 65.08$ is either correct (true with probability 1) or incorrect (false with probability 1). The correct interpretation lies in the realization that a CI is a *random interval* because in the probability statement defining the end-points of the interval (Equation 8-4), L and U are random variables. Consequently, the correct interpretation of a $100(1 - \alpha)\%$ CI depends on the relative frequency view of probability. Specifically, if an infinite number of random samples are collected and a $100(1 - \alpha)\%$ confidence interval for μ is computed from each sample, $100(1 - \alpha)\%$ of these intervals will contain the true value of μ.

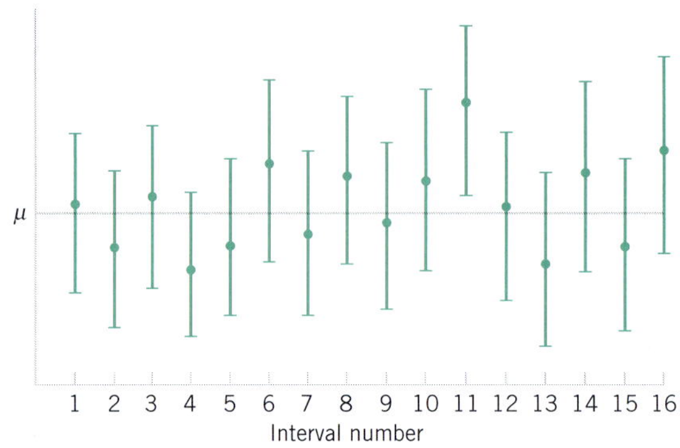

Figure 8-1 Repeated construction of a confidence interval for μ.

The situation is illustrated in Fig. 8-1, which shows several $100(1 - \alpha)\%$ confidence intervals for the mean μ of a normal distribution. The dots at the center of the intervals indicate the point estimate of μ (that is, $\bar{x}$). Notice that one of the intervals fails to contain the true value of μ. If this were a 95% confidence interval, in the long run only 5% of the intervals would fail to contain μ.

Now in practice, we obtain only one random sample and calculate one confidence interval. Since this interval either will or will not contain the true value of μ, it is not reasonable to attach a probability level to this specific event. The appropriate statement is the observed interval $[l, u]$ brackets the true value of μ with **confidence** $100(1 - \alpha)$. This statement has a frequency interpretation; that is, we don't know if the statement is true for this specific sample, but the *method* used to obtain the interval $[l, u]$ yields correct statements $100(1 - \alpha)\%$ of the time.

Confidence Level and Precision of Estimation

Notice in Example 8-1 that our choice of the 95% level of confidence was essentially arbitrary. What would have happened if we had chosen a higher level of confidence, say, 99%? In fact, doesn't it seem reasonable that we would want the higher level of confidence? At $\alpha = 0.01$, we find $z_{\alpha/2} = z_{0.01/2} = z_{0.005} = 2.58$, while for $\alpha = 0.05$, $z_{0.025} = 1.96$. Thus, the **length** of the 95% confidence interval is

$$2(1.96\sigma/\sqrt{n}) = 3.92\sigma/\sqrt{n}$$

whereas the length of the 99% CI is

$$2(2.58\sigma/\sqrt{n}) = 5.16\sigma/\sqrt{n}$$

Thus, the 99% CI is longer than the 95% CI. This is why we have a higher level of confidence in the 99% confidence interval. Generally, for a fixed sample size n and standard deviation σ, the higher the confidence level, the longer the resulting CI.

The length of a confidence interval is a measure of the **precision** of estimation. From the preceeding discussion, we see that precision is inversely related to the confidence level. It is desirable to obtain a confidence interval that is short enough for decision-making purposes and that also has adequate confidence. One way to achieve this is by choosing the sample size n to be large enough to give a CI of specified length or precision with prescribed confidence.

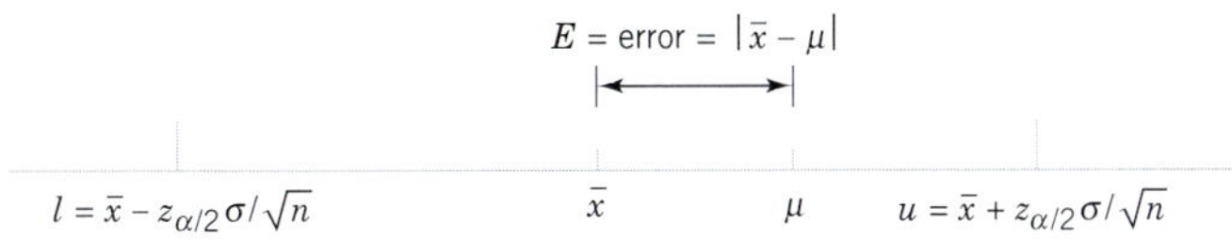

Figure 8-2 Error in estimating μ with $\bar{x}$.

8-2.2 Choice of Sample Size

The precision of the confidence interval in Equation 8-7 is $2z_{\alpha/2}\sigma/\sqrt{n}$. This means that in using $\bar{x}$ to estimate μ, the error $E = |\bar{x} - \mu|$ is less than or equal to $z_{\alpha/2}\sigma/\sqrt{n}$ with confidence $100(1 - \alpha)$. This is shown graphically in Fig. 8-2. In situations where the sample size can be controlled, we can choose n so that we are $100(1 - \alpha)$ percent confident that the error in estimating μ is less than a specified bound on the error E. The appropriate sample size is found by choosing n such that $z_{\alpha/2}\sigma/\sqrt{n} = E$. Solving this equation gives the following formula for n.

Sample Size for Specified Error on the Mean, Variance Known

If $\bar{x}$ is used as an estimate of μ, we can be $100(1 - \alpha)\%$ confident that the error $|\bar{x} - \mu|$ will not exceed a specified amount E when the sample size is

$$n = \left(\frac{z_{\alpha/2}\sigma}{E}\right)^2 \tag{8-8}$$

If the right-hand side of Equation 8-8 is not an integer, it must be rounded up. This will ensure that the level of confidence does not fall below $100(1 - \alpha)\%$. Notice that $2E$ is the length of the resulting confidence interval.

EXAMPLE 8-2
Metallic Material Transition

To illustrate the use of this procedure, consider the CVN test described in Example 8-1, and suppose that we wanted to determine how many specimens must be tested to ensure that the 95% CI on μ for A238 steel cut at 60°C has a length of at most $1.0J$. Since the bound on error in estimation E is one-half of the length of the CI, to determine n we use Equation 8-8 with $E = 0.5$, $\sigma = 1$, and $z_{\alpha/2} = 1.96$. The required sample size is 16

$$n = \left(\frac{z_{\alpha/2}\sigma}{E}\right)^2 = \left[\frac{(1.96)1}{0.5}\right]^2 = 15.37$$

and because n must be an integer, the required sample size is $n = 16$.

Notice the general relationship between sample size, desired length of the confidence interval $2E$, confidence level $100(1 - \alpha)$, and standard deviation σ:

- As the desired length of the interval $2E$ decreases, the required sample size n increases for a fixed value of σ and specified confidence.
- As σ increases, the required sample size n increases for a fixed desired length $2E$ and specified confidence.
- As the level of confidence increases, the required sample size n increases for fixed desired length $2E$ and standard deviation σ.

8-2.3 One-Sided Confidence Bounds

The confidence interval in Equation 8-7 gives both a lower confidence bound and an upper confidence bound for μ. Thus it provides a two-sided CI. It is also possible to obtain one-sided confidence bounds for μ by setting either $l = -\infty$ or $u = \infty$ and replacing $z_{\alpha/2}$ by z_α.

One-Sided Confidence Bounds on the Mean, Variance Known

A $100(1-\alpha)\%$ **upper-confidence bound** for μ is

$$\mu \leq u = \bar{x} + z_\alpha \sigma/\sqrt{n} \tag{8-9}$$

and a $100(1-\alpha)\%$ **lower-confidence bound** for μ is

$$\bar{x} - z_\alpha \sigma/\sqrt{n} = l \leq \mu \tag{8-10}$$

EXAMPLE 8-3 One-Sided Confidence Bound

The same data for impact testing from Example 8-1 is used to construct a lower, one-sided 95% confidence interval for the mean impact energy. Recall that $\bar{x} = 64.46$, $\sigma = 1J$, and $n = 10$. The interval is

$$\bar{x} - z_\alpha \frac{\sigma}{\sqrt{n}} \leq \mu$$

$$64.46 - 1.64 \frac{1}{\sqrt{10}} \leq \mu$$

$$63.94 \leq \mu$$

The lower limit for the two-sided interval in Example 8-1 was 63.84. Because $z_\alpha < z_{\alpha/2}$, the lower limit of a one-sided interval is always greater than the lower limit of a two-sided interval of equal confidence. The one-sided interval does not bound μ from above so that it still achieves 95% confidence with a slightly greater lower limit. If our interest is only in the lower limit for μ, then the one-sided interval is preferred because it provides equal confidence with a greater lower limit. Similarly, a one-sided upper limit is always less than a two-sided upper limit of equal confidence.

8-2.4 General Method to Derive a Confidence Interval

It is easy to give a general method for finding a confidence interval for an unknown parameter θ. Let $X_1, X_2, \ldots, X_n$ be a random sample of n observations. Suppose we can find a statistic $g(X_1, X_2, \ldots, X_n; \theta)$ with the following properties:

1. $g(X_1, X_2, \ldots, X_n; \theta)$ depends on both the sample and θ.
2. The probability distribution of $g(X_1, X_2, \ldots, X_n; \theta)$ does not depend on θ or any other unknown parameter.

In the case considered in this section, the parameter $\theta = \mu$. The random variable $g(X_1, X_2, \ldots, X_n; \mu) = (\bar{X} - \mu)/(\sigma/\sqrt{n})$ and satisfies both conditions above; it depends on the sample and on μ, and it has a standard normal distribution since σ is known. Now one must find constants C_L and C_U so that

$$P[C_L \leq g(X_1, X_2, \ldots, X_n; \theta) \leq C_U] = 1 - \alpha \tag{8-11}$$

Because of property 2, C_L and C_U do not depend on θ. In our example, $C_L = -z_{\alpha/2}$ and $C_U = z_{\alpha/2}$. Finally, you must manipulate the inequalities in the probability statement so that

$$P[L(X_1, X_2, \ldots, X_n) \leq \theta \leq U(X_1, X_2, \ldots, X_n)] = 1 - \alpha \tag{8-12}$$

This gives $L(X_1, X_2, \ldots, X_n)$ and $U(X_1, X_2, \ldots, X_n)$ as the lower and upper confidence limits defining the $100(1 - \alpha)\%$ confidence interval for θ. The quantity $g(X_1, X_2, \ldots, X_n; \theta)$ is often called a "pivotal quantity" because we pivot on this quantity in Equation 8-11 to produce Equation 8-12. In our example, we manipulated the pivotal quantity $(\overline{X} - \mu)/(\sigma/\sqrt{n})$ to obtain $L(X_1, X_2, \ldots, X_n) = \overline{X} - z_{\alpha/2}\sigma/\sqrt{n}$ and $U(X_1, X_2, \ldots, X_n) = \overline{X} + z_{\alpha/2}\sigma/\sqrt{n}$.

8-2.5 Large-Sample Confidence Interval for μ

We have assumed that the population distribution is normal with unknown mean and known standard deviation σ. We now present a **large-sample CI** for μ that does not require these assumptions. Let $X_1, X_2, \ldots, X_n$ be a random sample from a population with unknown mean μ and variance σ^2. Now if the sample size n is large, the central limit theorem implies that $\overline{X}$ has approximately a normal distribution with mean μ and variance σ^2/n. Therefore $Z = (\overline{X} - \mu)/(\sigma/\sqrt{n})$ has approximately a standard normal distribution. This ratio could be used as a pivotal quantity and manipulated as in Section 8-2.1 to produce an approximate CI for μ. However, the standard deviation σ is unknown. It turns out that when n is large, replacing σ by the sample standard deviation S has little effect on the distribution of Z. This leads to the following useful result.

Large-Sample Confidence Interval on the Mean

When n is large, the quantity

$$\frac{\overline{X} - \mu}{S/\sqrt{n}}$$

has an approximate standard normal distribution. Consequently,

$$\bar{x} - z_{\alpha/2}\frac{s}{\sqrt{n}} \leq \mu \leq \bar{x} + z_{\alpha/2}\frac{s}{\sqrt{n}} \tag{8-13}$$

is a **large sample confidence interval** for μ, with confidence level of approximately $100(1 - \alpha)\%$.

Equation 8-13 holds regardless of the shape of the population distribution. Generally n should be at least 40 to use this result reliably. The central limit theorem generally holds for $n \geq 30$, but the larger sample size is recommended here because replacing σ by S in Z results in additional variability.

EXAMPLE 8-4
Mercury Contamination

An article in the 1993 volume of the *Transactions of the American Fisheries Society* reports the results of a study to investigate the mercury contamination in largemouth bass. A sample of fish was selected from 53 Florida lakes and mercury concentration in the muscle tissue was measured (ppm). The mercury concentration values are

1.230	0.490	0.490	1.080	0.590	0.280	0.180	0.100	0.940
1.330	0.190	1.160	0.980	0.340	0.340	0.190	0.210	0.400
0.040	0.830	0.050	0.630	0.340	0.750	0.040	0.860	0.430
0.044	0.810	0.150	0.560	0.840	0.870	0.490	0.520	0.250
1.200	0.710	0.190	0.410	0.500	0.560	1.100	0.650	0.270
0.270	0.500	0.770	0.730	0.340	0.170	0.160	0.270	

The summary statistics from Minitab are displayed below:

Descriptive Statistics: Concentration

Variable	N	Mean	Median	TrMean	StDev	SE Mean
Concentration	53	0.5250	0.4900	0.5094	0.3486	0.0479
Variable	Minimum	Maximum	Q1	Q3		
Concentration	0.0400	1.3300	0.2300	0.7900		

Figure 8-3(a) and (b) presents the histogram and normal probability plot of the mercury concentration data. Both plots indicate that the distribution of mercury concentration is not normal and is positively skewed. We want to find an approximate 95% CI on μ. Because $n > 40$, the assumption of normality is not necessary to use Equation 8-13. The required quantities are $n = 53$, $\bar{x} = 0.5250$, $s = 0.3486$, and $z_{0.025} = 1.96$. The approximate 95% CI on μ is

$$\bar{x} - z_{0.025}\frac{s}{\sqrt{n}} \leq \mu \leq \bar{x} + z_{0.025}\frac{s}{\sqrt{n}}$$

$$0.5250 - 1.96\frac{0.3486}{\sqrt{53}} \leq \mu \leq 0.5250 + 1.96\frac{0.3486}{\sqrt{53}}$$

$$0.4311 \leq \mu \leq 0.6189$$

This interval is fairly wide because there is a lot of variability in the mercury concentration measurements.

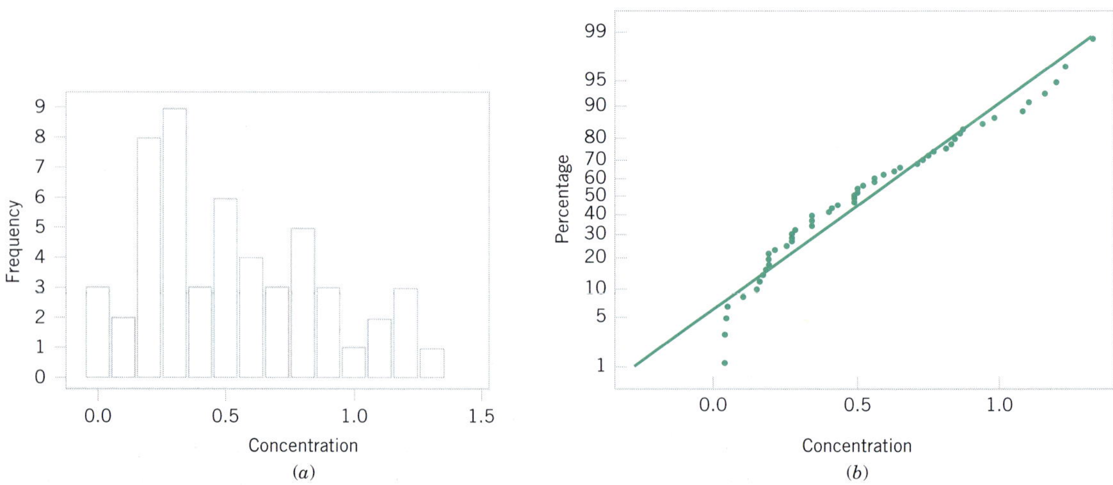

Figure 8-3 Mercury concentration in largemouth bass (a) Histogram. (b) Normal probability plot.

Large-Sample Confidence Interval for a Parameter

The large-sample confidence interval for μ in Equation 8-13 is a special case of a more general result. Suppose that θ is a parameter of a probability distribution and let $\hat{\Theta}$ be an estimator of θ. If $\hat{\Theta}$ (1) has an approximate normal distribution, (2) is approximately unbiased for θ, and (3) has standard deviation $\sigma_{\hat{\Theta}}$ that can be estimated from the sample data, then the quantity $(\hat{\Theta} - \theta)/\sigma_{\hat{\Theta}}$ has an approximate standard normal distribution. Then a large-sample approximate CI for θ is given by

Large-Sample Approximate Confidence Interval

$$\hat{\theta} - z_{\alpha/2}\sigma_{\hat{\Theta}} \leq \theta \leq \hat{\theta} + z_{\alpha/2}\sigma_{\hat{\Theta}} \tag{8-14}$$

Maximum likelihood estimators usually satisfy the three conditions listed above, so Equation 8-14 is often used when $\hat{\Theta}$ is the maximum likelihood estimator of θ. Finally, note that Equation 8-14 can be used even when $\sigma_{\hat{\Theta}}$ is a function of other unknown parameters (or of θ). Essentially, all one does is to use the sample data to compute estimates of the unknown parameters and substitute those estimates into the expression for $\sigma_{\hat{\Theta}}$.

EXERCISES FOR SECTION 8-2

8-1. For a normal population with known variance σ^2, answer the following questions:

(a) What is the confidence level for the interval $\bar{x} - 2.14\sigma/\sqrt{n} \leq \mu \leq \bar{x} + 2.14\sigma/\sqrt{n}$?
(b) What is the confidence level for the interval $\bar{x} - 2.49\sigma/\sqrt{n} \leq \mu \leq \bar{x} + 2.49\sigma/\sqrt{n}$?
(c) What is the confidence level for the interval $\bar{x} - 1.85\sigma/\sqrt{n} \leq \mu \leq \bar{x} + 1.85\sigma/\sqrt{n}$?

8-2. For a normal population with known variance σ^2:

(a) What value of $z_{\alpha/2}$ in Equation 8-7 gives 98% confidence?
(b) What value of $z_{\alpha/2}$ in Equation 8-7 gives 80% confidence?
(c) What value of $z_{\alpha/2}$ in Equation 8-7 gives 75% confidence?

8-3. Consider the one-sided confidence interval expressions for a mean of a normal population.

(a) What value of z_α would result in a 90% CI?
(b) What value of z_α would result in a 95% CI?
(c) What value of z_α would result in a 99% CI?

8-4. A confidence interval estimate is desired for the gain in a circuit on a semiconductor device. Assume that gain is normally distributed with standard deviation $\sigma = 20$.

(a) Find a 95% CI for μ when $n = 10$ and $\bar{x} = 1000$.
(b) Find a 95% CI for μ when $n = 25$ and $\bar{x} = 1000$.
(c) Find a 99% CI for μ when $n = 10$ and $\bar{x} = 1000$.
(d) Find a 99% CI for μ when $n = 25$ and $\bar{x} = 1000$.
(e) How does the length of the CIs computed above change with the changes in sample size and confidence level?

8-5. Consider the gain estimation problem in Exercise 8-4.

(a) How large must n be if the length of the 95% CI is to be 40?
(b) How large must n be if the length of the 99% CI is to be 40?

8-6. Following are two confidence interval estimates of the mean μ of the cycles to failure of an automotive door latch mechanism (the test was conducted at an elevated stress level to accelerate the failure).

$$3124.9 \leq \mu \leq 3215.7 \qquad 3110.5 \leq \mu \leq 3230.1$$

(a) What is the value of the sample mean cycles to failure?
(b) The confidence level for one of these CIs is 95% and the confidence level for the other is 99%. Both CIs are calculated from the same sample data. Which is the 95% CI? Explain why.

8-7. Suppose that $n = 100$ random samples of water from a fresh water lake were taken and the calcium concentration (milligrams per liter) measured. A 95% CI on the mean calcium concentration is $0.49 \leq \mu \leq 0.82$.

(a) Would a 99% CI calculated from the same sample data be longer or shorter?
(b) Consider the following statement: There is a 95% chance that μ is between 0.49 and 0.82. Is this statement correct? Explain your answer.
(c) Consider the following statement: If $n = 100$ random samples of water from the lake were taken and the 95% CI on μ computed, and this process was repeated 1000 times, 950 of the CIs will contain the true value of μ. Is this statement correct? Explain your answer.

8-8. Past experience has indicated that the breaking strength of yarn used in manufacturing drapery material is normally distributed and that $\sigma = 2$ psi. A random sample of nine specimens is tested, and the average breaking strength is found to be 98 psi. Find a 95% two-sided confidence interval on the true mean breaking strength.

8-9. The yield of a chemical process is being studied. From previous experience yield is known to be normally distributed and $\sigma = 3$. The past five days of plant operation have resulted in the following percent yields: 91.6, 88.75, 90.8, 89.95, and 91.3. Find a 95% two-sided confidence interval on the true mean yield.

8-10. The diameter of holes for a cable harness is known to have a normal distribution with $\sigma = 0.01$ inch. A random sample of size 10 yields an average diameter of 1.5045 inch. Find a 99% two-sided confidence interval on the mean hole diameter.

8-11. A manufacturer produces piston rings for an automobile engine. It is known that ring diameter is normally distributed with $\sigma = 0.001$ millimeters. A random sample of 15 rings has a mean diameter of $\bar{x} = 74.036$ millimeters.

(a) Construct a 99% two-sided confidence interval on the mean piston ring diameter.

(b) Construct a 99% lower-confidence bound on the mean piston ring diameter. Compare the lower bound of this confidence interval with the one in part (a).

8-12. The life in hours of a 75-watt light bulb is known to be normally distributed with $\sigma = 25$ hours. A random sample of 20 bulbs has a mean life of $\bar{x} = 1014$ hours.

(a) Construct a 95% two-sided confidence interval on the mean life.

(b) Construct a 95% lower-confidence bound on the mean life. Compare the lower bound of this confidence interval with the one in part (a).

8-13. A civil engineer is analyzing the compressive strength of concrete. Compressive strength is normally distributed with $\sigma^2 = 1000(\text{psi})^2$. A random sample of 12 specimens has a mean compressive strength of $\bar{x} = 3250$ psi.

(a) Construct a 95% two-sided confidence interval on mean compressive strength.

(b) Construct a 99% two-sided confidence interval on mean compressive strength. Compare the width of this confidence interval with the width of the one found in part (a).

8-14. Suppose that in Exercise 8-12 we wanted the error in estimating the mean life from the two-sided confidence interval to be five hours at 95% confidence. What sample size should be used?

8-15. Suppose that in Exercise 8-12 we wanted the total width of the two-sided confidence interval on mean life to be six hours at 95% confidence. What sample size should be used?

8-16. Suppose that in Exercise 8-13 it is desired to estimate the compressive strength with an error that is less than 15 psi at 99% confidence. What sample size is required?

8-17. By how much must the sample size n be increased if the length of the CI on μ in Equation 8-7 is to be halved?

8-18. If the sample size n is doubled, by how much is the length of the CI on μ in Equation 8-7 reduced? What happens to the length of the interval if the sample size is increased by a factor of four?

8-19. An article in the *Journal of Agricultural Science,* "The use of residual maximum likelihood to model grain quality characteristics of wheat with variety, climatic and nitrogen fertilizer effects" (1997, Vol. 128, pp. 135–142) investigated means of wheat grain crude protein content (CP) and Hagberg falling number (HFN) surveyed in the UK. The analysis used a variety of nitrogen fertilizer application (kg N/ha), temperature (°C), and total monthly rainfall (mm). The data shown below describe temperatures for wheat grown at Harper Adams Agricultural College between 1982 and 1993. The temperatures measured in June were obtained as followed:

15.2	14.2	14.0	12.2	14.4	12.5
14.3	14.2	13.5	11.8	15.2	

Assume that the standard deviation is known to be $\sigma = 0.5$.

(a) Construct a 99% two-sided confidence interval on the mean temperature.

(b) Construct a 95% lower-confidence bound on the mean temperature.

(c) Suppose that we wanted to be 95% confident that the error in estimating the mean temperature is less than 2 degrees Celsius. What sample size should be used?

(d) Suppose that we wanted the total width of the two-sided confidence interval on mean temperature to be 1.5 degrees Celsius at 95% confidence. What sample size should be used?

8-3 CONFIDENCE INTERVAL ON THE MEAN OF A NORMAL DISTRIBUTION, VARIANCE UNKNOWN

When we are constructing confidence intervals on the mean μ of a normal population when σ^2 is known, we can use the procedure in Section 8-2.1. This CI is also approximately valid (because of the central limit theorem) regardless of whether or not the underlying population is normal, so long as n is reasonably large ($n \geq 40$, say). As noted in Section 8-2.5, we can even handle the case of unknown variance for the large-sample-size situation. However, when the sample is small and σ^2 is unknown, we must make an assumption about the form of the underlying distribution to obtain a valid CI procedure. A reasonable assumption in many cases is that the underlying distribution is normal.

Many populations encountered in practice are well approximated by the normal distribution, so this assumption will lead to confidence interval procedures of wide applicability. In fact, moderate departure from normality will have little effect on validity. When the assumption is unreasonable, an alternate is to use the nonparametric procedures in Chapter 15 that are valid for any underlying distribution.

Suppose that the population of interest has a normal distribution with unknown mean μ and unknown variance σ^2. Assume that a random sample of size n, say $X_1, X_2, \ldots, X_n$, is available, and let $\overline{X}$ and S^2 be the sample mean and variance, respectively.

We wish to construct a two-sided CI on μ. If the variance σ^2 is known, we know that $Z = (\overline{X} - \mu)/(\sigma/\sqrt{n})$ has a standard normal distribution. When σ^2 is unknown, a logical procedure is to replace σ with the sample standard deviation S. The random variable Z now becomes $T = (\overline{X} - \mu)/(S/\sqrt{n})$. A logical question is what effect does replacing σ by S have on the distribution of the random variable T? If n is large, the answer to this question is "very little," and we can proceed to use the confidence interval based on the normal distribution from Section 8-2.5. However, n is usually small in most engineering problems, and in this situation a different distribution must be employed to construct the CI.

8-3.1 t Distribution

t Distribution

Let $X_1, X_2, \ldots, X_n$ be a random sample from a normal distribution with unknown mean μ and unknown variance σ^2. The random variable

$$T = \frac{\overline{X} - \mu}{S/\sqrt{n}} \tag{8-15}$$

has a t distribution with $n - 1$ degrees of freedom.

The t probability density function is

$$f(x) = \frac{\Gamma[(k+1)/2]}{\sqrt{\pi k}\,\Gamma(k/2)} \cdot \frac{1}{[(x^2/k) + 1]^{(k+1)/2}} \qquad -\infty < x < \infty \tag{8-16}$$

where k is the number of degrees of freedom. The mean and variance of the t distribution are zero and $k/(k - 2)$ (for $k > 2$), respectively.

Several t distributions are shown in Fig. 8-4. The general appearance of the t distribution is similar to the standard normal distribution in that both distributions are symmetric and unimodal, and the maximum ordinate value is reached when the mean $\mu = 0$. However, the t distribution has heavier tails than the normal; that is, it has more probability in the tails than the normal distribution. As the number of degrees of freedom $k \rightarrow \infty$, the limiting form of the t distribution is the standard normal distribution. Generally, the number of degrees of freedom for t are the number of degrees of freedom associated with the estimated standard deviation.

Appendix Table V provides **percentage points** of the t distribution. We will let $t_{\alpha,k}$ be the value of the random variable T with k degrees of freedom above which we find an area

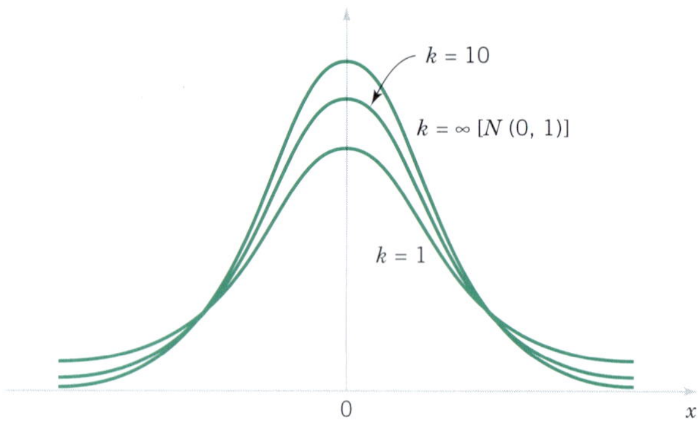

Figure 8-4 Probability density functions of several t distributions.

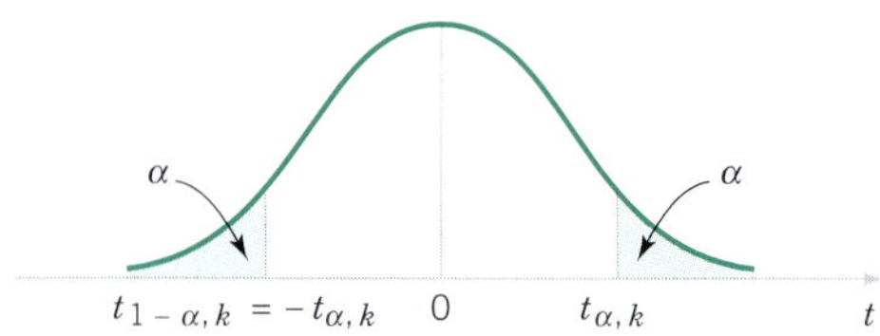

Figure 8-5 Percentage points of the t distribution.

(or probability) α. Thus, $t_{\alpha,k}$ is an upper-tail 100α percentage point of the t distribution with k degrees of freedom. This percentage point is shown in Fig. 8-5. In the Appendix Table V the α values are the column headings, and the degrees of freedom are listed in the left column. To illustrate the use of the table, note that the t-value with 10 degrees of freedom having an area of 0.05 to the right is $t_{0.05,10} = 1.812$. That is,

$$P(T_{10} > t_{0.05,10}) = P(T_{10} > 1.812) = 0.05$$

Since the t distribution is symmetric about zero, we have $t_{1-\alpha,n} = -t_{\alpha,n}$; that is, the t-value having an area of $1 - \alpha$ to the right (and therefore an area of α to the left) is equal to the negative of the t-value that has area α in the right tail of the distribution. Therefore, $t_{0.95,10} = -t_{0.05,10} = -1.812$. Finally, because $t_{\alpha,\infty}$ is the standard normal distribution, the familiar z_α values appear in the last row of Appendix Table V.

8-3.2 t Confidence Interval on μ

It is easy to find a $100(1 - \alpha)$ percent confidence interval on the mean of a normal distribution with unknown variance by proceeding essentially as we did in Section 8-2.1. We know that the distribution of $T = (\overline{X} - \mu)/(S/\sqrt{n})$ is t with $n - 1$ degrees of freedom. Letting $t_{\alpha/2,n-1}$ be the upper $100\alpha/2$ percentage point of the t distribution with $n - 1$ degrees of freedom, we may write:

$$P(-t_{\alpha/2,n-1} \le T \le t_{\alpha/2,n-1}) = 1 - \alpha$$

or

$$P\left(-t_{\alpha/2,n-1} \le \frac{\overline{X} - \mu}{S/\sqrt{n}} \le t_{\alpha/2,n-1}\right) = 1 - \alpha$$

Rearranging this last equation yields

$$P(\overline{X} - t_{\alpha/2,n-1}S/\sqrt{n} \le \mu \le \overline{X} + t_{\alpha/2,n-1}S/\sqrt{n}) = 1 - \alpha \tag{8-17}$$

This leads to the following definition of the $100(1 - \alpha)$ percent two-sided confidence interval on μ.

Confidence Interval on the Mean, Variance Unknown

If $\bar{x}$ and s are the mean and standard deviation of a random sample from a normal distribution with unknown variance σ^2, a **100(1 − α) percent confidence interval on μ** is given by

$$\bar{x} - t_{\alpha/2,n-1}s/\sqrt{n} \leq \mu \leq \bar{x} + t_{\alpha/2,n-1}s/\sqrt{n} \tag{8-18}$$

where $t_{\alpha/2,n-1}$ is the upper $100\alpha/2$ percentage point of the t distribution with $n - 1$ degrees of freedom.

One-sided confidence bounds on the mean of a normal distribution are also of interest and are easy to find. Simply use only the appropriate lower or upper confidence limit from Equation 8-18 and replace $t_{\alpha/2,n-1}$ by $t_{\alpha,n-1}$.

EXAMPLE 8-5 Alloy Adhesion

An article in the journal *Materials Engineering* (1989, Vol. II, No. 4, pp. 275–281) describes the results of tensile adhesion tests on 22 U-700 alloy specimens. The load at specimen failure is as follows (in megapascals):

19.8	10.1	14.9	7.5	15.4	15.4
15.4	18.5	7.9	12.7	11.9	11.4
11.4	14.1	17.6	16.7	15.8	
19.5	8.8	13.6	11.9	11.4	

The sample mean is $\bar{x} = 13.71$, and the sample standard deviation is $s = 3.55$. Figures 8-6 and 8-7 show a box plot and a normal probability plot of the tensile adhesion test data, respectively. These displays provide good support for the assumption that the population is normally distributed. We want to find a 95% CI on μ. Since $n = 22$, we have $n - 1 = 21$ degrees of freedom for t, so $t_{0.025,21} = 2.080$. The resulting CI is

$$\bar{x} - t_{\alpha/2,n-1}s/\sqrt{n} \leq \mu \leq \bar{x} + t_{\alpha/2,n-1}s/\sqrt{n}$$

$$13.71 - 2.080(3.55)/\sqrt{22} \leq \mu \leq 13.71 + 2.080(3.55)/\sqrt{22}$$

$$13.71 - 1.57 \leq \mu \leq 13.71 + 1.57$$

$$12.14 \leq \mu \leq 15.28$$

The CI is fairly wide because there is a lot of variability in the tensile adhesion test measurements.

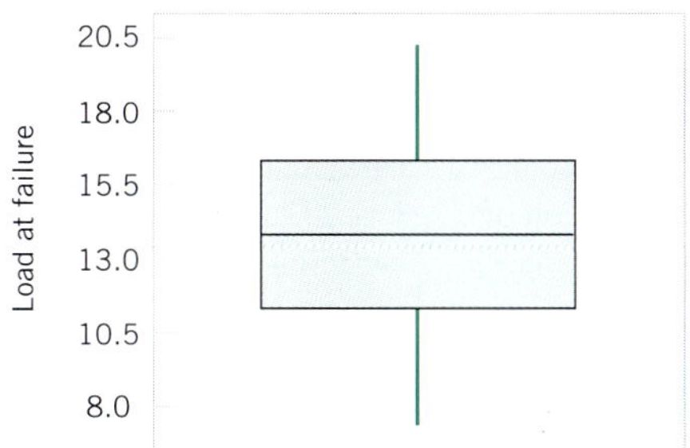

Figure 8-6 Box and whisker plot for the load at failure data in Example 8-5.

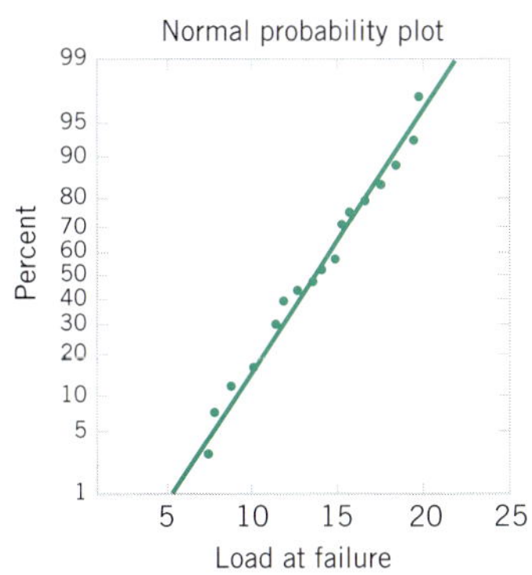

Figure 8-7 Normal probability plot of the load at failure data from Example 8-5.

It is not as easy to select a sample size n to obtain a specified length (or precision of estimation) for this CI as it was in the known-σ case because the length of the interval involves s (which is unknown before the data are collected), n, and $t_{\alpha/2,n-1}$. Note that the t-percentile depends on the sample size n. Consequently, an appropriate n can only be obtained through trial and error. The results of this will, of course, also depend on the reliability of our prior "guess" for σ.

EXERCISES FOR SECTION 8-3

8-20. Find the values of the following percentiles: $t_{0.025,15}$, $t_{0.05,10}$, $t_{0.10,20}$, $t_{0.005,25}$, and $t_{0.001,30}$.

8-21. Determine the t-percentile that is required to construct each of the following two-sided confidence intervals:
(a) Confidence level = 95%, degrees of freedom = 12
(b) Confidence level = 95%, degrees of freedom = 24
(c) Confidence level = 99%, degrees of freedom = 13
(d) Confidence level = 99.9%, degrees of freedom = 15

8-22. Determine the t-percentile that is required to construct each of the following one-sided confidence intervals:
(a) Confidence level = 95%, degrees of freedom = 14
(b) Confidence level = 99%, degrees of freedom = 19
(c) Confidence level = 99.9%, degrees of freedom = 24

8-23. A research engineer for a tire manufacturer is investigating tire life for a new rubber compound and has built 16 tires and tested them to end-of-life in a road test. The sample mean and standard deviation are 60,139.7 and 3645.94 kilometers. Find a 95% confidence interval on mean tire life.

8-24. An Izod impact test was performed on 20 specimens of PVC pipe. The sample mean is $\bar{x} = 1.25$ and the sample standard deviation is $s = 0.25$. Find a 99% lower confidence bound on Izod impact strength.

8-25. A postmix beverage machine is adjusted to release a certain amount of syrup into a chamber where it is mixed with carbonated water. A random sample of 25 beverages was found to have a mean syrup content of $\bar{x} = 1.10$ fluid ounces and a standard deviation of $s = 0.015$ fluid ounces. Find a 95% CI on the mean volume of syrup dispensed.

8-26. An article in *Medicine and Science in Sports and Exercise* "Maximal Leg-Strength Training Improves Cycling Economy in Previously Untrained Men," (2005, Vol. 37 pp. 131–1236) studied cycling performance before and after eight weeks of leg-strength training. Seven previously untrained males performed leg-strength training three days per week for eight weeks (with four sets of five replications at 85% of one repetition maximum). Peak power during incremental cycling increased to a mean of 315 watts with a standard deviation of 16 watts. Construct a 95% confidence interval for the mean peak power after training.

8-27. An article in *Obesity Research* "Impaired pressure natriuresis in obese youths," (2003, Vol. 11, pp. 745–751) described a study in which all meals were provided for 14 lean boys for three days followed by one stress (with a video-game task). The average systolic blood pressure (SBP) during the test was 118.3 mm HG with a standard deviation of 9.9 mm HG. Construct a 99% one-sided upper confidence interval for mean SBP.

8-28. An article in the *Journal of Composite Materials* (December 1989, Vol. 23, p. 1200) describes the effect of delamination on the natural frequency of beams made from composite laminates. Five such delaminated beams were subjected to loads, and the resulting frequencies were as follows (in hertz):

230.66, 233.05, 232.58, 229.48, 232.58

Check the assumption of normality in the population. Calculate a 90% two-sided confidence interval on mean natural frequency.

8-29. The Bureau of Meteorology of the Australian Government provided the mean annual rainfall (in millimeters) in Australia 1983–2002 as follows (http://www.bom.gov.au/climate/change/rain03.txt)

499.2, 555.2, 398.8, 391.9, 453.4, 459.8, 483.7, 417.6, 469.2, 452.4, 499.3, 340.6, 522.8, 469.9, 527.2, 565.5, 584.1, 727.3, 558.6, 338.6

Check the assumption of normality in the population. Construct a 95% confidence interval for the mean annual rainfall.

8-30. The solar energy consumed (intrillion BTU) in the U.S. by year from 1989–2004 (source: U.S. Department of Energy Web site http://www.eia.doe.gov/emeu) is shown in the table below. Read down, then right for year.

55.291	66.458	70.237	65.454
59.718	68.548	69.787	64.391
62.688	69.857	68.793	63.62
63.886	70.833	66.388	63.287

Check the assumption of normality in the population. Construct a 95% confidence interval for the mean solar energy consumed.

8-31. The brightness of a television picture tube can be evaluated by measuring the amount of current required to achieve a particular brightness level. A sample of 10 tubes results in $\bar{x} = 317.2$ and $s = 15.7$. Find (in microamps) a 99% confidence interval on mean current required. State any necessary assumptions about the underlying distribution of the data.

8-32. A particular brand of diet margarine was analyzed to determine the level of polyunsaturated fatty acid (in percentages). A sample of six packages resulted in the following data: 16.8, 17.2, 17.4, 16.9, 16.5, 17.1.
(a) Check the assumption that the level of polyunsaturated fatty acid is normally distributed.

(b) Calculate a 99% confidence interval on the mean μ. Provide a practical interpretation of this interval.

(c) Calculate a 99% lower confidence bound on the mean. Compare this bound with the lower bound of the two-sided confidence interval and discuss why they are different.

8-33. The compressive strength of concrete is being tested by a civil engineer. He tests 12 specimens and obtains the following data.

2216	2237	2249	2204
2225	2301	2281	2263
2318	2255	2275	2295

(a) Check the assumption that compressive strength is normally distributed. Include a graphical display in your answer.

(b) Construct a 95% two-sided confidence interval on the mean strength.

(c) Construct a 95% lower-confidence bound on the mean strength. Compare this bound with the lower bound of the two-sided confidence interval and discuss why they are different.

8-34. A machine produces metal rods used in an automobile suspension system. A random sample of 15 rods is selected, and the diameter is measured. The resulting data (in millimeters) are as follows:

8.24	8.25	8.20	8.23	8.24
8.21	8.26	8.26	8.20	8.25
8.23	8.23	8.19	8.28	8.24

(a) Check the assumption of normality for rod diameter.

(b) Calculate a 95% two-sided confidence interval on mean rod diameter.

(c) Calculate a 95% upper confidence bound on the mean. Compare this bound with the upper bound of the two-sided confidence interval and discuss why they are different.

8-35. An article in *Computers & Electrical Engineering*, "Parallel simulation of cellular neural networks" (1996, Vol. 22, pp. 61–84) considered the speed-up of cellular neural networks (CNN) for a parallel general-purpose computing architecture based on six transputers in different areas. The data follow:

3.775302	3.350679	4.217981	4.030324	4.639692
4.139665	4.395575	4.824257	4.268119	4.584193
4.930027	4.315973	4.600101		

(a) Is there evidence to support the assumption that speed-up of CNN is normally distributed? Include a graphical display in your answer.

(b) Construct a 95% two-sided confidence interval on the mean speed-up.

(c) Construct a 95% lower confidence bound on the mean speed-up.

8-36. The wall thickness of 25 glass 2-liter bottles was measured by a quality-control engineer. The sample mean was $\bar{x} = 4.05$ millimeters, and the sample standard deviation was $s = 0.08$ millimeter. Find a 95% lower confidence bound for mean wall thickness. Interpret the interval you have obtained.

8-37. An article in *Nuclear Engineering International* (February 1988, p. 33) describes several characteristics of fuel rods used in a reactor owned by an electric utility in Norway. Measurements on the percentage of enrichment of 12 rods were reported as follows:

2.94	3.00	2.90	2.75	3.00	2.95
2.90	2.75	2.95	2.82	2.81	3.05

(a) Use a normal probability plot to check the normality assumption.

(b) Find a 99% two-sided confidence interval on the mean percentage of enrichment. Are you comfortable with the statement that the mean percentage of enrichment is 2.95 percent? Why?

8-4 CONFIDENCE INTERVAL ON THE VARIANCE AND STANDARD DEVIATION OF A NORMAL DISTRIBUTION

Sometimes confidence intervals on the population variance or standard deviation are needed. When the population is modeled by a normal distribution, the tests and intervals described in this section are applicable. The following result provides the basis of constructing these confidence intervals.

χ^2 Distribution

Let $X_1, X_2, \ldots, X_n$ be a random sample from a normal distribution with mean μ and variance σ^2, and let S^2 be the sample variance. Then the random variable

$$X^2 = \frac{(n-1)S^2}{\sigma^2} \tag{8-19}$$

has a chi-square (χ^2) distribution with $n - 1$ degrees of freedom.

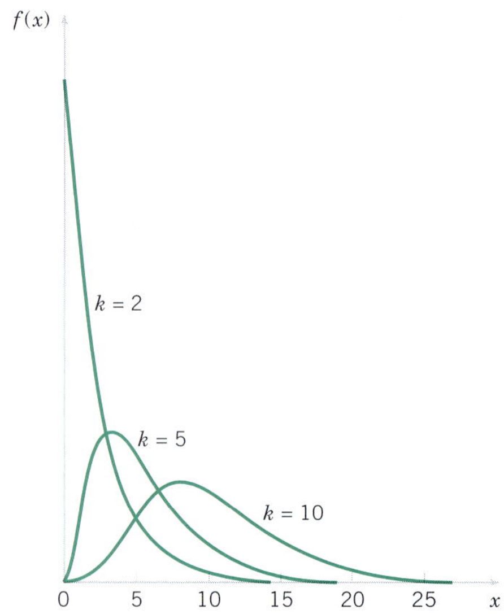

Figure 8-8 Probability density functions of several χ^2 distributions.

The probability density function of a χ^2 random variable is

$$f(x) = \frac{1}{2^{k/2}\Gamma(k/2)} x^{(k/2)-1} e^{-x/2} \qquad x > 0 \tag{8-20}$$

where k is the number of degrees of freedom. The mean and variance of the χ^2 distribution are k and $2k$, respectively. Several chi-square distributions are shown in Fig. 8-8. Note that the chi-square random variable is nonnegative and that the probability distribution is skewed to the right. However, as k increases, the distribution becomes more symmetric. As $k \to \infty$, the limiting form of the chi-square distribution is the normal distribution.

The **percentage points** of the χ^2 distribution are given in Table IV of the Appendix. Define $\chi^2_{\alpha,k}$ as the percentage point or value of the chi-square random variable with k degrees of freedom such that the probability that X^2 exceeds this value is α. That is,

$$P(X^2 > \chi^2_{\alpha,k}) = \int_{\chi^2_{\alpha,k}}^{\infty} f(u)\, du = \alpha$$

This probability is shown as the shaded area in Fig. 8-9(a). To illustrate the use of Table IV, note that the areas α are the column headings and the degrees of freedom k are given in the left column. Therefore, the value with 10 degrees of freedom having an area (probability) of 0.05

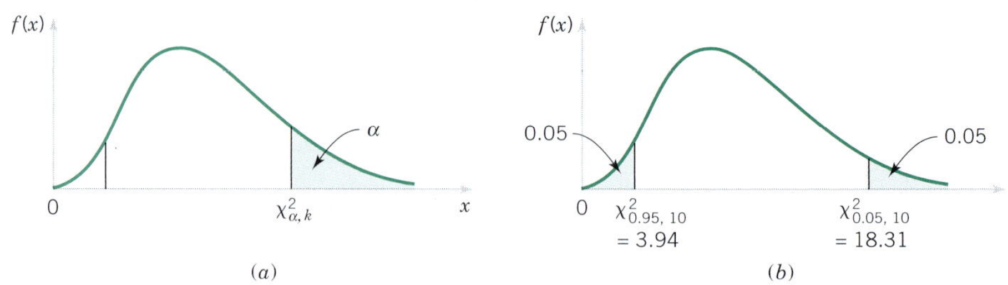

Figure 8-9 Percentage point of the χ^2 distribution. (a) The percentage point $\chi^2_{\alpha,k}$. (b) The upper percentage point $\chi^2_{0.05,10} = 18.31$ and the lower percentage point $\chi^2_{0.95,10} = 3.94$.

to the right is $\chi^2_{0.05,10} = 18.31$. This value is often called an **upper** 5% point of chi-square with 10 degrees of freedom. We may write this as a probability statement as follows:

$$P(X^2 > \chi^2_{0.05,10}) = P(X^2 > 18.31) = 0.05$$

Conversely, a **lower** 5% point of chi-square with 10 degrees of freedom would be $\chi^2_{0.95,10} = 3.94$ (from Appendix A). Both of these percentage points are shown in Figure 8-9(b).

The construction of the $100(1 - \alpha)\%$ CI for σ^2 is straightforward. Because

$$X^2 = \frac{(n-1)S^2}{\sigma^2}$$

is chi-square with $n - 1$ degrees of freedom, we may write

$$P(\chi^2_{1-\alpha/2,n-1} \leq X^2 \leq \chi^2_{\alpha/2,n-1}) = 1 - \alpha$$

so that

$$P\left(\chi^2_{1-\alpha/2,n-1} \leq \frac{(n-1)S^2}{\sigma^2} \leq \chi^2_{\alpha/2,n-1}\right) = 1 - \alpha$$

This last equation can be rearranged as

$$P\left(\frac{(n-1)S^2}{\chi^2_{\alpha/2,n-1}} \leq \sigma^2 \leq \frac{(n-1)S^2}{\chi^2_{1-\alpha/2,n-1}}\right) = 1 - \alpha$$

This leads to the following definition of the confidence interval for σ^2.

Confidence Interval on the Variance

If s^2 is the sample variance from a random sample of n observations from a normal distribution with unknown variance σ^2, then **a $100(1 - \alpha)\%$ confidence interval on σ^2 is**

$$\frac{(n-1)s^2}{\chi^2_{\alpha/2,n-1}} \leq \sigma^2 \leq \frac{(n-1)s^2}{\chi^2_{1-\alpha/2,n-1}} \tag{8-21}$$

where $\chi^2_{\alpha/2,n-1}$ and $\chi^2_{1-\alpha/2,n-1}$ are the upper and lower $100\alpha/2$ percentage points of the chi-square distribution with $n - 1$ degrees of freedom, respectively. A **confidence interval for σ** has lower and upper limits that are the square roots of the corresponding limits in Equation 8-21.

It is also possible to find a $100(1 - \alpha)\%$ lower confidence bound or upper confidence bound on σ^2.

One-Sided Confidence Bounds on the Variance

The $100(1 - \alpha)\%$ lower and upper confidence bounds on σ^2 are

$$\frac{(n-1)s^2}{\chi^2_{\alpha,n-1}} \leq \sigma^2 \quad \text{and} \quad \sigma^2 \leq \frac{(n-1)s^2}{\chi^2_{1-\alpha,n-1}} \tag{8-22}$$

respectively.

EXAMPLE 8-6
Detergent Filling

An automatic filling machine is used to fill bottles with liquid detergent. A random sample of 20 bottles results in a sample variance of fill volume of $s^2 = 0.0153$ (fluid ounces)^2. If the variance of fill volume is too large, an unacceptable proportion of bottles will be under- or overfilled. We will assume that the fill volume is approximately normally distributed. A 95% upper-confidence interval is found from Equation 8-22 as follows:

$$\sigma^2 \leq \frac{(n-1)s^2}{\chi^2_{0.95,19}}$$

or

$$\sigma^2 \leq \frac{(19)0.0153}{10.117} = 0.0287 \text{ (fluid ounce)}^2$$

This last expression may be converted into a confidence interval on the standard deviation σ by taking the square root of both sides, resulting in

$$\sigma \leq 0.17$$

Therefore, at the 95% level of confidence, the data indicate that the process standard deviation could be as large as 0.17 fluid ounce.

EXERCISES FOR SECTION 8-4

8-38. Determine the values of the following percentiles: $\chi^2_{0.05,10}$, $\chi^2_{0.025,15}$, $\chi^2_{0.01,12}$, $\chi^2_{0.95,20}$, $\chi^2_{0.99,18}$, $\chi^2_{0.995,16}$, and $\chi^2_{0.005,25}$.

8-34. Determine the χ^2 percentile that is required to construct each of the following CIs:

(a) Confidence level = 95%, degrees of freedom = 24, one-sided (upper)

(b) Confidence level = 99%, degrees of freedom = 9, one-sided (lower)

(c) Confidence level = 90%, degrees of freedom = 19, two-sided.

8-39. A rivet is to be inserted into a hole. A random sample of $n = 15$ parts is selected, and the hole diameter is measured. The sample standard deviation of the hole diameter measurements is $s = 0.008$ millimeters. Construct a 99% lower confidence bound for σ^2.

8-40. The sugar content of the syrup in canned peaches is normally distributed. A random sample of $n = 10$ cans yields a sample standard deviation of $s = 4.8$ milligrams. Calculate a 95% two-sided confidence interval for σ.

8-41. The percentage of titanium in an alloy used in aerospace castings is measured in 51 randomly selected parts. The sample standard deviation is $s = 0.37$. Construct a 95% two-sided confidence interval for σ.

8-42. An article in *Medicine and Science in Sports and Exercise* "Electrostimulation Training Effects on the Physical Performance of Ice Hockey Players," (2005, Vol. 37, pp. 455–460) considered the use of electromyostimulation (EMS) as a method to train healthy skeletal muscle. EMS sessions consisted of 30 contractions (four-second duration, 85 Hz) and were carried out three times per week for three weeks on 17 ice hockey players. The ten-meter skating performance test showed a standard deviation of 0.09 seconds. Construct a 95% confidence interval of the standard deviation of the skating performance test.

8-43. An article in *Urban Ecosystems*, "Urbanization and warming of Phoenix (Arizona, USA): Impacts, feedbacks and mitigation," (2002, Vol. 6, pp. 183–203) mentions that Phoenix is ideal to study the effects of an urban heat island because it has grown from a population of 300,000 to approximately 3 million over the last 50 years and this is a period with a continuous, detailed climate record. The 50 year averages of the mean annual temperatures at eight sites in Phoenix are shown below. Check the assumption of normality in the population with a probability plot. Construct a 95% confidence interval for the standard deviation over the sites of the mean annual temperatures.

Site	Average Mean Temperature (°C)
Sky Harbor Airport	23.3
Phoenix Greenway	21.7
Phoenix Encanto	21.6
Waddell	21.7
Litchfield	21.3
Laveen	20.7
Maricopa	20.9
Harlquahala	20.1

8-44. An article in *Cancer Research* "Analyses of litter-matched time-to-response data, with modifications for recovery of interlitter information," (1977, Vol. 37, pp. 3863–3868) tested the tumorigenesis of a drug. Rats were randomly selected from litters and given the drug. The times of tumor appearance were recorded as follows:

101, 104, 104, 77, 89, 88, 104, 96, 82, 70, 89, 91, 39, 103, 93, 85, 104, 104, 81, 67, 104, 104, 104, 87, 104, 89, 78, 104, 86, 76, 103, 102, 80, 45, 94, 104, 104, 76, 80, 72, 73

Calculate a 95% confidence interval on the standard deviation of time until a tumor appearance. Check the assumption of normality of the population and comment on the assumptions for the confidence interval.

8-45. An article in *Technometrics* (1999, Vol. 41, pp. 202–211) studied the capability of a gauge by measuring the weight of paper. The data for repeated measurements of one sheet of paper are in the following table. Construct a 95% one-sided, upper confidence interval for the standard deviation of these measurements. Check the assumption of normality of the data and comment on the assumptions for the confidence interval.

Observations				
3.481	3.448	3.485	3.475	3.472
3.477	3.472	3.464	3.472	3.470
3.470	3.470	3.477	3.473	3.474

8-46. An article in the *Australian Journal of Agricultural Research*, "Non-Starch Polysaccharides and Broiler Performance on Diets Containing Soyabean Meal as the Sole Protein Concentrate" (1993, Vol. 44, No. 8, pp. 1483–1499) determined the essential amino acid (Lysine) composition level of soybean meals are as shown below (g/kg):

22.2	24.7	20.9	26.0	27.0
24.8	26.5	23.8	25.6	23.9

(a) Construct a 99% two-sided confidence interval for σ^2.
(b) Calculate a 99% lower confidence bound for σ^2.
(c) Calculate a 90% lower confidence bound for σ.
(d) Compare the intervals that you have computed.

8-5 LARGE-SAMPLE CONFIDENCE INTERVAL FOR A POPULATION PROPORTION

It is often necessary to construct confidence intervals on a population proportion. For example, suppose that a random sample of size n has been taken from a large (possibly infinite) population and that $X(\leq n)$ observations in this sample belong to a class of interest. Then $\hat{P} = X/n$ is a point estimator of the proportion of the population p that belongs to this class. Note that n and p are the parameters of a binomial distribution. Furthermore, from Chapter 4 we know that the sampling distribution of $\hat{P}$ is approximately normal with mean p and variance $p(1-p)/n$, if p is not too close to either 0 or 1 and if n is relatively large. Typically, to apply this approximation we require that np and $n(1-p)$ be greater than or equal to 5. We will make use of the normal approximation in this section.

Normal Approximation for a Binomial Proportion

If n is large, the distribution of

$$Z = \frac{X - np}{\sqrt{np(1-p)}} = \frac{\hat{P} - p}{\sqrt{\dfrac{p(1-p)}{n}}}$$

is approximately standard normal.

To construct the confidence interval on p, note that

$$P(-z_{\alpha/2} \leq Z \leq z_{\alpha/2}) \simeq 1 - \alpha$$

so

$$P\left(-z_{\alpha/2} \le \frac{\hat{P} - p}{\sqrt{\dfrac{p(1-p)}{n}}} \le z_{\alpha/2}\right) \simeq 1 - \alpha$$

This may be rearranged as

$$P\left(\hat{P} - z_{\alpha/2}\sqrt{\frac{p(1-p)}{n}} \le p \le \hat{P} + z_{\alpha/2}\sqrt{\frac{p(1-p)}{n}}\right) \simeq 1 - \alpha \tag{8-23}$$

The quantity $\sqrt{p(1-p)/n}$ in Equation 8-23 is called the *standard error of the point estimator* $\hat{P}$. This was discussed in Chapter 7. Unfortunately, the upper and lower limits of the confidence interval obtained from Equation 8-23 contain the unknown parameter p. However, as suggested at the end of Section 8-2.5, a satisfactory solution is to replace p by $\hat{P}$ in the standard error, which results in

$$P\left(\hat{P} - z_{\alpha/2}\sqrt{\frac{\hat{P}(1-\hat{P})}{n}} \le p \le \hat{P} + z_{\alpha/2}\sqrt{\frac{\hat{P}(1-\hat{P})}{n}}\right) \simeq 1 - \alpha \tag{8-24}$$

This leads to the approximate $100(1 - \alpha)\%$ confidence interval on p.

Approximate Confidence Interval on a Binomial Proportion

If $\hat{p}$ is the proportion of observations in a random sample of size n that belongs to a class of interest, an approximate $100(1 - \alpha)\%$ confidence interval on the proportion p of the population that belongs to this class is

$$\hat{p} - z_{\alpha/2}\sqrt{\frac{\hat{p}(1-\hat{p})}{n}} \le p \le \hat{p} + z_{\alpha/2}\sqrt{\frac{\hat{p}(1-\hat{p})}{n}} \tag{8-25}$$

where $z_{\alpha/2}$ is the upper $\alpha/2$ percentage point of the standard normal distribution.

This procedure depends on the adequacy of the normal approximation to the binomial. To be reasonably conservative, this requires that np and $n(1 - p)$ be greater than or equal to 5. In situations where this approximation is inappropriate, particularly in cases where n is small, other methods must be used. Tables of the binomial distribution could be used to obtain a confidence interval for p. However, we could also use numerical methods based on the binomial probability mass function that are implemented in computer programs.

EXAMPLE 8-7
Crankshaft Bearings

In a random sample of 85 automobile engine crankshaft bearings, 10 have a surface finish that is rougher than the specifications allow. Therefore, a point estimate of the proportion of bearings in the population that exceeds the roughness specification is $\hat{p} = x/n = 10/85 = 0.12$. A 95% two-sided confidence interval for p is computed from Equation 8-25 as

$$\hat{p} - z_{0.025}\sqrt{\frac{\hat{p}(1-\hat{p})}{n}} \le p \le \hat{p} + z_{0.025}\sqrt{\frac{\hat{p}(1-\hat{p})}{n}}$$

or

$$0.12 - 1.96\sqrt{\frac{0.12(0.88)}{85}} \leq p \leq 0.12 + 1.96\sqrt{\frac{0.12(0.88)}{85}}$$

which simplifies to

$$0.05 \leq p \leq 0.19$$

Choice of Sample Size

Since $\hat{P}$ is the point estimator of p, we can define the error in estimating p by $\hat{P}$ as $E = |p - \hat{P}|$. Note that we are approximately $100(1 - \alpha)\%$ confident that this error is less than $z_{\alpha/2}\sqrt{p(1 - p)/n}$. For instance, in Example 8-7, we are 95% confident that the sample proportion $\hat{p} = 0.12$ differs from the true proportion p by an amount not exceeding 0.07.

In situations where the sample size can be selected, we may choose n to be $100(1 - \alpha)\%$ confident that the error is less than some specified value E. If we set $E = z_{\alpha/2}\sqrt{p(1 - p)/n}$ and solve for n, the appropriate sample size is

Sample Size for a Specified Error on a Binomial Proportion

$$n = \left(\frac{z_{\alpha/2}}{E}\right)^2 p(1 - p) \tag{8-26}$$

An estimate of p is required to use Equation 8-26. If an estimate $\hat{p}$ from a previous sample is available, it can be substituted for p in Equation 8-26, or perhaps a subjective estimate can be made. If these alternatives are unsatisfactory, a preliminary sample can be taken, $\hat{p}$ computed, and then Equation 8-26 used to determine how many additional observations are required to estimate p with the desired accuracy. Another approach to choosing n uses the fact that the sample size from Equation 8-26 will always be a maximum for $p = 0.5$ [that is, $p(1 - p) \leq 0.25$ with equality for $p = 0.5$], and this can be used to obtain an upper bound on n. In other words, we are at least $100(1 - \alpha)\%$ confident that the error in estimating p by $\hat{p}$ is less than E if the sample size is

$$n = \left(\frac{z_{\alpha/2}}{E}\right)^2 (0.25) \tag{8-27}$$

EXAMPLE 8-8
Crankshaft Bearings

Consider the situation in Example 8-7. How large a sample is required if we want to be 95% confident that the error in using $\hat{p}$ to estimate p is less than 0.05? Using $\hat{p} = 0.12$ as an initial estimate of p, we find from Equation 8-26 that the required sample size is

$$n = \left(\frac{z_{0.025}}{E}\right)^2 \hat{p}(1 - \hat{p}) = \left(\frac{1.96}{0.05}\right)^2 0.12(0.88) \cong 163$$

If we wanted to be *at least* 95% confident that our estimate $\hat{p}$ of the true proportion p was within 0.05 regardless of the value of p, we would use Equation 8-27 to find the sample size

$$n = \left(\frac{z_{0.025}}{E}\right)^2 (0.25) = \left(\frac{1.96}{0.05}\right)^2 (0.25) \cong 385$$

Notice that if we have information concerning the value of p, either from a preliminary sample or from past experience, we could use a smaller sample while maintaining both the desired precision of estimation and the level of confidence.

One-Sided Confidence Bounds

We may find approximate one-sided confidence bounds on p by a simple modification of Equation 8-25.

Approximate One-Sided Confidence Bounds on a Binomial Proportion

The approximate $100(1 - \alpha)\%$ lower and upper confidence bounds are

$$\hat{p} - z_\alpha \sqrt{\frac{\hat{p}(1-\hat{p})}{n}} \le p \quad \text{and} \quad p \le \hat{p} + z_\alpha \sqrt{\frac{\hat{p}(1-\hat{p})}{n}} \tag{8-28}$$

respectively.

EXERCISES FOR SECTION 8-5

8-47. The fraction of defective integrated circuits produced in a photolithography process is being studied. A random sample of 300 circuits is tested, revealing 13 defectives.

(a) Calculate a 95% two-sided CI on the fraction of defective circuits produced by this particular tool.

(b) Calculate a 95% upper confidence bound on the fraction of defective circuits.

8-48. An article in *Knee Surgery, Sports Traumatology, Arthroscopy* (2005, Vol. 13, pp. 273–279) "Arthroscopic meniscal repair with an absorbable screw: results and surgical technique" showed that only 25 out of 37 tears (67.6%) located between 3 and 6 mm from the meniscus rim were healed.

(a) Calculate a two-sided 95% confidence interval on the proportion of such tears that will heal.

(b) Calculate a 95% lower confidence bound on the proportion of such tears that will heal.

8-49. The 2004 presidential election exit polls from the critical state of Ohio provided the following results. There were 2020 respondents in the exit polls and 768 were college graduates. Of the college graduates, 412 voted for George Bush.

(a) Calculate a 95% confidence interval for the proportion of college graduates in Ohio that voted for George Bush.

(b) Calculate a 95% lower confidence bound for the proportion of college graduates in Ohio that voted for George Bush.

8-50. Of 1000 randomly selected cases of lung cancer, 823 resulted in death within 10 years.

(a) Calculate a 95% two-sided confidence interval on the death rate from lung cancer.

(b) Using the point estimate of p obtained from the preliminary sample, what sample size is needed to be 95% confident that the error in estimating the true value of p is less than 0.03?

(c) How large must the sample be if we wish to be at least 95% confident that the error in estimating p is less than 0.03, regardless of the true value of p?

8-51. An article in *Journal of the American Statistical Association* (1990, Vol. 85, pp. 972–985) measured weight of 30 rats under experiment controls. Suppose that there are 12 underweight rats.

(a) Calculate a 95% two-sided confidence interval on the true proportion of rats that would show underweight from the experiment.

(b) Using the point estimate of p obtained from the preliminary sample, what sample size is needed to be 95% confident that the error in estimating the true value of p is less than 0.02?

(c) How large must the sample be if we wish to be at least 95% confident that the error in estimating p is less than 0.02, regardless of the true value of p?

8-52. A random sample of 50 suspension helmets used by motorcycle riders and automobile race-car drivers was subjected to an impact test, and on 18 of these helmets some damage was observed.

(a) Find a 95% two-sided confidence interval on the true proportion of helmets of this type that would show damage from this test.

(b) Using the point estimate of p obtained from the preliminary sample of 50 helmets, how many helmets must be tested to be 95% confident that the error in estimating the true value of p is less than 0.02?

(c) How large must the sample be if we wish to be at least 95% confident that the error in estimating p is less than 0.02, regardless of the true value of p?

8-53. The Arizona Department of Transportation wishes to survey state residents to determine what proportion of the population would like to increase statewide highway speed limits to 75 mph from 65 mph. How many residents do they need to survey if they want to be at least 99% confident that the sample proportion is within 0.05 of the true proportion?

8-54. A study is to be conducted of the percentage of homeowners who own at least two television sets. How large a sample is required if we wish to be 99% confident that the error in estimating this quantity is less than 0.017?

8-6 GUIDELINES FOR CONSTRUCTING CONFIDENCE INTERVALS

The most difficult step in constructing a confidence interval is often the match of the appropriate calculation to the objective of the study. Common cases are listed in Table 8-1 along with the reference to the section that covers the appropriate calculation for a confidence interval test. Table 8-1 provides a simple road map to help select the appropriate analysis. Two primary comments can help identify the analysis:

1. Determine the parameter (and the distribution of the data) that will be bounded by the confidence interval or tested by the hypothesis.
2. Check if other parameters are known or need to be estimated.

In Chapter 9, we will study a procedure closely related to confidence intervals called hypothesis testing. Table 8-1 can be used for those procedures also. This road map will be extended to more cases in Chapter 10.

8-7 TOLERANCE AND PREDICTION INTERVALS

8-7.1 Prediction Interval for a Future Observation

In some problem situations, we may be interested in predicting a future observation of a variable. This is a different problem than estimating the mean of that variable, so a confidence interval is not appropriate. In this section we show how to obtain a $100(1 - \alpha)\%$ **prediction interval** on a future value of a normal random variable.

Table 8-1 The Roadmap for Constracting Confidence Intervals and Performing Hypothesis Tests, One-Sample Case

Parameter to Be Bounded by the Confidence Interval or Tested with a Hypothesis?	Symbol	Other Parameters?	Confidence Interval Section	Hypothesis Test Section	Comments
Mean of normal distribution	μ	Standard deviation σ known	8-2	9-2	
Mean of arbitrary distribution with large sample size	μ	Sample size large enough that central limit theorem applies and σ is essentially known	8-2.5	9-2.5	Large sample size is often taken to be $n \geq 40$
Mean of normal distribution	μ	Standard deviation σ unknown and estimated	8-3	9-3	
Variance (or standard deviation) of normal distribution	σ^2	Mean μ unknown and estimated	8-4	9-4	
Population Proportion	p	None	8-5	9-5	

Suppose that $X_1, X_2, \ldots, X_n$ is a random sample from a normal population. We wish to predict the value X_{n+1}, a single **future** observation. A point prediction of X_{n+1} is $\overline{X}$, the sample mean. The prediction error is $X_{n+1} - \overline{X}$. The expected value of the prediction error is

$$E(X_{n+1} - \overline{X}) = \mu - \mu = 0$$

and the variance of the prediction error is

$$V(X_{n+1} - \overline{X}) = \sigma^2 + \frac{\sigma^2}{n} = \sigma^2\left(1 + \frac{1}{n}\right)$$

because the future observation, X_{n+1} is independent of the mean of the current sample $\overline{X}$. The prediction error $X_{n+1} - \overline{X}$ is normally distributed. Therefore

$$Z = \frac{X_{n+1} - \overline{X}}{\sigma\sqrt{1 + \frac{1}{n}}}$$

has a standard normal distribution. Replacing σ with S results in

$$T = \frac{X_{n+1} - \overline{X}}{S\sqrt{1 + \frac{1}{n}}}$$

which has a t distribution with $n - 1$ degrees of freedom. Manipulating T as we have done previously in the development of a CI leads to a prediction interval on the future observation X_{n+1}.

Prediction Interval

> **A 100(1 − α)% prediction interval on a single future observation from a normal distribution** is given by
>
> $$\overline{x} - t_{\alpha/2,n-1}\, s\sqrt{1 + \frac{1}{n}} \le X_{n+1} \le \overline{x} + t_{\alpha/2,n-1}\, s\sqrt{1 + \frac{1}{n}} \qquad (8\text{-}29)$$

The prediction interval for X_{n+1} will always be longer than the confidence interval for μ because there is more variability associated with the prediction error than with the error of estimation. This is easy to see because the prediction error is the difference between two random variables ($X_{n+1} - \overline{X}$), and the estimation error in the CI is the difference between one random variable and a constant ($\overline{X} - \mu$). As n gets larger ($n \rightarrow \infty$), the length of the CI decreases to zero, essentially becoming the single value μ, but the length of the prediction interval approaches $2z_{\alpha/2}\sigma$. So as n increases, the uncertainty in estimating μ goes to zero, although there will always be uncertainty about the future value X_{n+1} even when there is no need to estimate any of the distribution parameters.

EXAMPLE 8-9
Alloy Adhesion

Reconsider the tensile adhesion tests on specimens of U-700 alloy described in Example 8-5. The load at failure for $n = 22$ specimens was observed, and we found that $\overline{x} = 13.71$ and $s = 3.55$. The 95% confidence interval on μ was $12.14 \le \mu \le 15.28$. We plan to test a twenty-third specimen.

A 95% prediction interval on the load at failure for this specimen is

$$\bar{x} - t_{\alpha/2,n-1}s\sqrt{1 + \frac{1}{n}} \leq X_{n+1} \leq \bar{x} + t_{\alpha/2,n-1}s\sqrt{1 + \frac{1}{n}}$$

$$13.71 - (2.080)3.55\sqrt{1 + \frac{1}{22}} \leq X_{23} \leq 13.71 + (2.080)3.55\sqrt{1 + \frac{1}{22}}$$

$$6.16 \leq X_{23} \leq 21.26$$

Notice that the prediction interval is considerably longer than the CI.

8-7.2 Tolerance Interval for a Normal Distribution

Consider a population of semiconductor processors. Suppose that the speed of these processors has a normal distribution with mean $\mu = 600$ megahertz and standard deviation $\sigma = 30$ megahertz. Then the interval from $600 - 1.96(30) = 541.2$ to $600 + 1.96(30) = 658.8$ megahertz captures the speed of 95% of the processors in this population because the interval from -1.96 to 1.96 captures 95% of the area under the standard normal curve. The interval from $\mu - z_{\alpha/2}\sigma$ to $\mu + z_{\alpha/2}\sigma$ is called a **tolerance interval.**

If μ and σ are unknown, we can use the data from a random sample of size n to compute $\bar{x}$ and s, and then form the interval $(\bar{x} - 1.96s, \bar{x} + 1.96s)$. However, because of sampling variability in $\bar{x}$ and s, it is likely that this interval will contain less than 95% of the values in the population. The solution to this problem is to replace 1.96 by some value that will make the proportion of the distribution contained in the interval 95% with some level of confidence. Fortunately, it is easy to do this.

Tolerance Interval

A **tolerance interval** for capturing at least γ% of the values in a normal distribution with confidence level $100(1 - \alpha)$% is

$$\bar{x} - ks, \qquad \bar{x} + ks$$

where k is a tolerance interval factor found in Appendix Table XII. Values are given for $\gamma = 90\%$, 95%, and 99% and for 90%, 95%, and 99% confidence.

One-sided tolerance bounds can also be computed. The tolerance factors for these bounds are also given in Appendix Table XII.

EXAMPLE 8-10
Alloy Adhesion

Let's reconsider the tensile adhesion tests originally described in Example 8-5. The load at failure for $n = 22$ specimens was observed, and we found that $\bar{x} = 13.71$ and $s = 3.55$. We want to find a tolerance interval for the load at failure that includes 90% of the values in the population with 95% confidence. From Appendix Table XII the tolerance factor k for $n = 22$, $\gamma = 0.90$, and 95% confidence is $k = 2.264$. The desired tolerance interval is

$$(\bar{x} - ks, \bar{x} + ks) \quad \text{or} \quad [13.71 - (2.264)3.55, 13.71 + (2.264)3.55]$$

which reduces to (5.67, 21.74). We can be 95% confident that at least 90% of the values of load at failure for this particular alloy lie between 5.67 and 21.74 megapascals.

From Appendix Table XII, we note that as $n \to \infty$, the value of k goes to the z-value associated with the desired level of containment for the normal distribution. For example, if we want 90% of the population to fall in the two-sided tolerance interval, k approaches $z_{0.05} = 1.645$ as $n \to \infty$. Note that as $n \to \infty$, a $100(1 - \alpha)\%$ prediction interval on a future value approaches a tolerance interval that contains $100(1 - \alpha)\%$ of the distribution.

EXERCISES FOR SECTION 8-7

8-55. Consider the tire-testing data described in Exercise 8-23. Compute a 95% prediction interval on the life of the next tire of this type tested under conditions that are similar to those employed in the original test. Compare the length of the prediction interval with the length of the 95% CI on the population mean.

8-56. Consider the Izod impact test described in Exercise 8-24. Compute a 99% prediction interval on the impact strength of the next specimen of PVC pipe tested. Compare the length of the prediction interval with the length of the 99% CI on the population mean.

8-57. Consider the syrup dispensing measurements described in Exercise 8-25. Compute a 95% prediction interval on the syrup volume in the next beverage dispensed. Compare the length of the prediction interval with the length of the 95% CI on the population mean.

8-58. Consider the natural frequency of beams described in Exercise 8-28. Compute a 90% prediction interval on the diameter of the natural frequency of the next beam of this type that will be tested. Compare the length of the prediction interval with the length of the 90% CI on the population mean.

8-59. Consider the rainfall in Exercise 8-29. Compute a 95% prediction interval on the rainfall for the next year. Compare the length of the prediction interval with the length of the 95% CI on the population mean.

8-60. Consider the margarine test described in Exercise 8-32. Compute a 99% prediction interval on the polyunsaturated fatty acid in the next package of margarine that is tested. Compare the length of the prediction interval with the length of the 99% CI on the population mean.

8-61. Consider the television tube brightness test described in Exercise 8-31. Compute a 99% prediction interval on the brightness of the next tube tested. Compare the length of the prediction interval with the length of the 99% CI on the population mean.

8-62. Consider the suspension rod diameter measurements described in Exercise 8-34. Compute a 95% prediction interval on the diameter of the next rod tested. Compare the length of the prediction interval with the length of the 95% CI on the population mean.

8-63. Consider the test on the compressive strength of concrete described in Exercise 8-33. Compute a 90% prediction interval on the next specimen of concrete tested.

8-64. Consider the bottle wall thickness measurements described in Exercise 8-36. Compute a 90% prediction interval on the wall thickness of the next bottle tested.

8-65. Consider the fuel rod enrichment data described in Exercise 8-37. Compute a 90% prediction interval on the enrichment of the next rod tested. Compare the length of the prediction interval with the length of the 99% CI on the population mean.

8-66. How would you obtain a one-sided prediction bound on a future observation? Apply this procedure to obtain a 95% one-sided prediction bound on the wall thickness of the next bottle for the situation described in Exercise 8-36.

8-67. Consider the tire-testing data in Exercise 8-23. Compute a 95% tolerance interval on the life of the tires that has confidence level 95%. Compare the length of the tolerance interval with the length of the 95% CI on the population mean. Which interval is shorter? Discuss the difference in interpretation of these two intervals.

8-68. Consider the Izod impact test described in Exercise 8-24. Compute a 99% tolerance interval on the impact strength of PVC pipe that has confidence level 90%. Compare the length of the tolerance interval with the length of the 99% CI on the population mean. Which interval is shorter? Discuss the difference in interpretation of these two intervals.

8-69. Consider the syrup volume data in Exercise 8-25. Compute a 95% tolerance interval on the syrup volume that has confidence level 90%. Compare the length of the tolerance interval with the length of the 95% CI on the population mean.

8-70. Consider the margarine test described in Exercise 8-32. Compute a 99% tolerance interval on the polyunsaturated fatty acid in this particular type of margarine that has confidence level 95%. Compare the length of the tolerance interval with the length of the 99% CI on the population mean. Which interval is shorter? Discuss the difference in interpretation of these two intervals.

8-71. Consider the rainfall data in Exercise 8-29. Compute a 95% tolerance interval that has confidence level 95%. Compare the length of the tolerance interval with the length of the 95% CI on the population mean. Discuss the difference in interpretation of these two intervals.

8-72. Consider the diameter data in Exercise 8-34. Compute a 95% tolerance interval on the diameter of the rods described that has 90% confidence. Compare the length of the tolerance interval with the length of the 95% CI on the population mean. Which interval is shorter? Discuss the difference in interpretation of these two intervals.

8-73. Consider the television tube brightness data in Exercise 8-31. Compute a 99% tolerance interval on the brightness of the television tubes that has confidence level 95%. Compare the length of the tolerance interval with the length of the 99% CI on the population mean. Which interval is shorter? Discuss the difference in interpretation of these two intervals.

8-74. Consider the strength of concrete data in Exercise 8-33. Compute a 90% tolerance interval on the compressive strength of the concrete that has 90% confidence.

8-75. Consider the fuel rod enrichment data described in Exercise 8-37. Compute a 99% tolerance interval on rod enrichment that has confidence level 95%. Compare the length of the tolerance interval with the length of the 95% CI on the population mean.

8-76. Consider the bottle wall thickness measurements described in Exercise 8-36.

(a) Compute a 90% tolerance interval on bottle wall thickness that has confidence level 90%.
(b) Compute a 90% lower tolerance bound on bottle wall thickness that has confidence level 90%. Why would a lower tolerance bound likely be of interest here?

Supplemental Exercises

8-77. Consider the confidence interval for μ with known standard deviation σ:

$$\bar{x} - z_{\alpha_1}\sigma/\sqrt{n} \le \mu \le \bar{x} + z_{\alpha_2}\sigma/\sqrt{n}$$

where $\alpha_1 + \alpha_2 = \alpha$. Let $\alpha = 0.05$ and find the interval for $\alpha_1 = \alpha_2 = \alpha/2 = 0.025$. Now find the interval for the case $\alpha_1 = 0.01$ and $\alpha_2 = 0.04$. Which interval is shorter? Is there any advantage to a "symmetric" confidence interval?

8-78. A normal population has a known mean 50 and unknown variance.

(a) A random sample of $n = 16$ is selected from this population, and the sample results are $\bar{x} = 52$ and $s = 8$. How unusual are these results? That is, what is the probability of observing a sample average as large as 52 (or larger) if the known, underlying mean is actually 50?
(b) A random sample of $n = 30$ is selected from this population, and the sample results are $\bar{x} = 52$ and $s = 8$. How unusual are these results?
(c) A random sample of $n = 100$ is selected from this population, and the sample results are $\bar{x} = 52$ and $s = 8$. How unusual are these results?
(d) Compare your answers to parts (a)–(c) and explain why they are the same or differ.

8-79. A normal population has known mean $\mu = 50$ and variance $\sigma^2 = 5$. What is the approximate probability that the sample variance is greater than or equal to 7.44? less than or equal to 2.56?

(a) For a random sample of $n = 16$.
(b) For a random sample of $n = 30$.
(c) For a random sample of $n = 71$.
(d) Compare your answers to parts (a)–(c) for the approximate probability that the sample variance is greater than or equal to 7.44. Explain why this tail probability is increasing or decreasing with increased sample size.
(e) Compare your answers to parts (a)–(c) for the approximate probability that the sample variance is less than or equal to 2.56. Explain why this tail probability is increasing or decreasing with increased sample size.

8-80. An article in the *Journal of Sports Science* (1987, Vol. 5, pp. 261–271) presents the results of an investigation of the hemoglobin level of Canadian Olympic ice hockey players. The data reported are as follows (in g/dl):

15.3	16.0	14.4	16.2	16.2
14.9	15.7	15.3	14.6	15.7
16.0	15.0	15.7	16.2	14.7
14.8	14.6	15.6	14.5	15.2

(a) Given the following probability plot of the data, what is a logical assumption about the underlying distribution of the data?

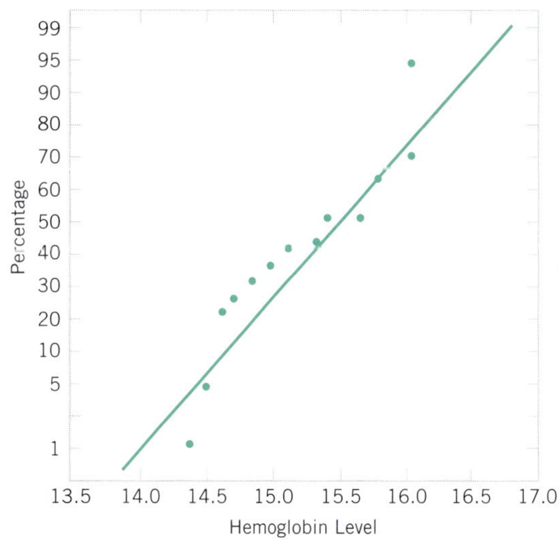

(b) Explain why this check of the distribution underlying the sample data is important if we want to construct a confidence interval on the mean.
(c) Based on this sample data, a 95% confidence interval for the mean is (15.04, 15.62). Is it reasonable to infer that the true mean could be 14.5? Explain your answer.

(d) Explain why this check of the distribution underlying the sample data is important if we want to construct a confidence interval on the variance.

(e) Based on this sample data, a 95% confidence interval for the variance is (0.22, 0.82). Is it reasonable to infer that the true variance could be 0.35? Explain your answer.

(f) Is it reasonable to use these confidence intervals to draw an inference about the mean and variance of hemoglobin levels

(i) of Canadian doctors? Explain your answer.

(ii) of Canadian children ages 6–12? Explain your answer.

8-81. The article "Mix Design for Optimal Strength Development of Fly Ash Concrete" (*Cement and Concrete Research*, 1989, Vol. 19, No. 4, pp. 634–640) investigates the compressive strength of concrete when mixed with fly ash (a mixture of silica, alumina, iron, magnesium oxide, and other ingredients). The compressive strength for nine samples in dry conditions on the twenty-eighth day are as follows (in megapascals):

40.2	30.4	28.9	30.5	22.4
25.8	18.4	14.2	15.3	

(a) Given the following probability plot of the data, what is a logical assumption about the underlying distribution of the data?

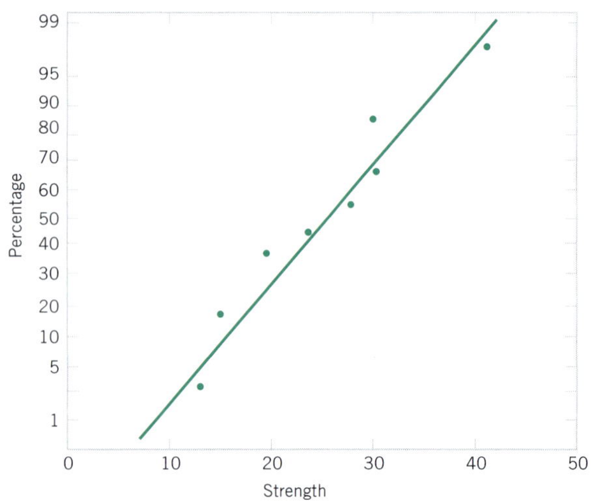

(b) Find a 99% lower one-sided confidence interval on mean compressive strength. Provide a practical interpretation of this interval.

(c) Find a 98% two-sided confidence interval on mean compressive strength. Provide a practical interpretation of this interval and explain why the lower end-point of the interval is or is not the same as in part (b).

(d) Find a 99% upper one-sided confidence interval on the variance of compressive strength. Provide a practical interpretation of this interval.

(e) Find a 98% two-sided confidence interval on the variance of compression strength. Provide a practical interpretation of this interval and explain why the upper end-point of the interval is or is not the same as in part (d).

(f) Suppose that it was discovered that the largest observation 40.2 was misrecorded and should actually be 20.4. Now the sample mean $\bar{x} = 23$ and the sample variance $s^2 = 39.8$. Use these new values and repeat parts (c) and (e). Compare the original computed intervals and the newly computed intervals with the corrected observation value. How does this mistake affect the values of the sample mean, sample variance, and the width of the two-sided confidence intervals?

(g) Suppose, instead, that it was discovered that the largest observation 40.2 is correct, but that the observation 25.8 is incorrect and should actually be 24.8. Now the sample mean $\bar{x} = 25$ and the standard deviation $s = 8.41$. Use these new values and repeat parts (c) and (e). Compare the original computed intervals and the newly computed intervals with the corrected observation value. How does this mistake affect the values of the sample mean, sample variance, and the width of the two-sided confidence intervals?

(h) Use the results from parts (f) and (g) to explain the effect of mistakenly recorded values on sample estimates. Comment on the effect when the mistaken values are near the sample mean and when they are not.

8-82. An operating system for a personal computer has been studied extensively, and it is known that the standard deviation of the response time following a particular command is $\sigma = 8$ milliseconds. A new version of the operating system is installed, and we wish to estimate the mean response time for the new system to ensure that a 95% confidence interval for μ has length at most 5 milliseconds.

(a) If we can assume that response time is normally distributed and that $\sigma = 8$ for the new system, what sample size would you recommend?

(b) Suppose that we are told by the vendor that the standard deviation of the response time of the new system is smaller, say $\sigma = 6$; give the sample size that you recommend and comment on the effect the smaller standard deviation has on this calculation.

8-83. Consider the hemoglobin data in Exercise 8-80. Find the following:

(a) An interval that contains 95% of the hemoglobin values with 90% confidence.

(b) An interval that contains 99% of the hemoglobin values with 90% confidence.

8-84. Consider the compressive strength of concrete data from Exercise 8-81. Find a 95% prediction interval on the next sample that will be tested.

8-85. The maker of a shampoo knows that customers like this product to have a lot of foam. Ten sample bottles of the product are selected at random and the foam heights observed are as follows (in millimeters): 210, 215, 194, 195, 211, 201, 198, 204, 208, and 196.

(a) Is there evidence to support the assumption that foam height is normally distributed?
(b) Find a 95% CI on the mean foam height.
(c) Find a 95% prediction interval on the next bottle of shampoo that will be tested.
(d) Find an interval that contains 95% of the shampoo foam heights with 99% confidence.
(e) Explain the difference in the intervals computed in parts (b), (c), and (d).

8-86. During the 1999 and 2000 baseball seasons, there was much speculation that the unusually large number of home runs that were hit was due at least in part to a livelier ball. One way to test the "liveliness" of a baseball is to launch the ball at a vertical surface with a known velocity V_L and measure the ratio of the outgoing velocity V_O of the ball to V_L. The ratio $R = V_O/V_L$ is called the coefficient of restitution. Following are measurements of the coefficient of restitution for 40 randomly selected baseballs. The balls were thrown from a pitching machine at an oak surface.

0.6248	0.6237	0.6118	0.6159	0.6298	0.6192
0.6520	0.6368	0.6220	0.6151	0.6121	0.6548
0.6226	0.6280	0.6096	0.6300	0.6107	0.6392
0.6230	0.6131	0.6223	0.6297	0.6435	0.5978
0.6351	0.6275	0.6261	0.6262	0.6262	0.6314
0.6128	0.6403	0.6521	0.6049	0.6170	
0.6134	0.6310	0.6065	0.6214	0.6141	

(a) Is there evidence to support the assumption that the coefficient of restitution is normally distributed?
(b) Find a 99% CI on the mean coefficient of restitution.
(c) Find a 99% prediction interval on the coefficient of restitution for the next baseball that will be tested.
(d) Find an interval that will contain 99% of the values of the coefficient of restitution with 95% confidence.
(e) Explain the difference in the three intervals computed in parts (b), (c), and (d).

8-87. Consider the baseball coefficient of restitution data in Exercise 8-86. Suppose that any baseball that has a coefficient of restitution that exceeds 0.635 is considered too lively. Based on the available data, what proportion of the baseballs in the sampled population are too lively? Find a 95% lower confidence bound on this proportion.

8-88. An article in the *ASCE Journal of Energy Engineering* ("Overview of Reservoir Release Improvements at 20 TVA Dams," Vol. 125, April 1999, pp. 1–17) presents data on dissolved oxygen concentrations in streams below 20 dams in the Tennessee Valley Authority system. The observations are (in milligrams per liter): 5.0, 3.4, 3.9, 1.3, 0.2, 0.9, 2.7, 3.7, 3.8, 4.1, 1.0, 1.0, 0.8, 0.4, 3.8, 4.5, 5.3, 6.1, 6.9, and 6.5.

(a) Is there evidence to support the assumption that the dissolved oxygen concentration is normally distributed?
(b) Find a 95% CI on the mean dissolved oxygen concentration.
(c) Find a 95% prediction interval on the dissolved oxygen concentration for the next stream in the system that will be tested.
(d) Find an interval that will contain 95% of the values of the dissolved oxygen concentration with 99% confidence.
(e) Explain the difference in the three intervals computed in parts (b), (c), and (d).

8-89. The tar content in 30 samples of cigar tobacco follows:

1.542	1.585	1.532	1.466	1.499	1.611
1.622	1.466	1.546	1.494	1.548	1.626
1.440	1.608	1.520	1.478	1.542	1.511
1.459	1.533	1.532	1.523	1.397	1.487
1.598	1.498	1.600	1.504	1.545	1.558

(a) Is there evidence to support the assumption that the tar content is normally distributed?
(b) Find a 99% CI on the mean tar content.
(c) Find a 99% prediction interval on the tar content for the next observation that will be taken on this particular type of tobacco.
(d) Find an interval that will contain 99% of the values of the tar content with 95% confidence.
(e) Explain the difference in the three intervals computed in parts (b), (c), and (d).

8-90. A manufacturer of electronic calculators takes a random sample of 1200 calculators and finds that there are eight defective units.

(a) Construct a 95% confidence interval on the population proportion.
(b) Is there evidence to support a claim that the fraction of defective units produced is 1% or less?

8-91. An article in *The Engineer* ("Redesign for Suspect Wiring," June 1990) reported the results of an investigation into wiring errors on commercial transport aircraft that may produce faulty information to the flight crew. Such a wiring error may have been responsible for the crash of a British Midland Airways aircraft in January 1989 by causing the pilot to shut down the wrong engine. Of 1600 randomly selected aircraft, eight were found to have wiring errors that could display incorrect information to the flight crew.

(a) Find a 99% confidence interval on the proportion of aircraft that have such wiring errors.
(b) Suppose we use the information in this example to provide a preliminary estimate of p. How large a sample would be required to produce an estimate of p that we are 99% confident differs from the true value by at most 0.008?
(c) Suppose we did not have a preliminary estimate of p. How large a sample would be required if we wanted to be at least 99% confident that the sample proportion differs

from the true proportion by at most 0.008 regardless of the true value of p?

(d) Comment on the usefulness of preliminary information in computing the needed sample size.

8-92. An article in *Engineering Horizons* (Spring 1990, p. 26) reported that 117 of 484 new engineering graduates were planning to continue studying for an advanced degree. Consider this as a random sample of the 1990 graduating class.

(a) Find a 90% confidence interval on the proportion of such graduates planning to continue their education.
(b) Find a 95% confidence interval on the proportion of such graduates planning to continue their education.
(c) Compare your answers to parts (a) and (b) and explain why they are the same or different.
(d) Could you use either of these confidence intervals to determine whether the proportion is actually 0.25? Explain your answer. *Hint:* Use the normal approximation to the binomial.

8-93. An article in the *Journal of Applied Physiology,* "Humidity does not affect central nervous system oxygen toxicity" (2001, Vol. 91, pp. 1327–1333) reported that central nervous system (CNS) oxygen toxicity can appear in humans on exposure to oxygen pressures >180 kPa. CNS oxygen toxicity can occur as convulsions (similar to epileptic seizures, grand mal) and loss of consciousness without any warning symptoms. CNS oxygen toxicity is a risk encountered in several fields of human activity, such as combat diving with closed-circuit breathing apparatus and diving with mixtures of nitrogen and oxygen (nitrox) or nitrogen, oxygen, and helium (trimix) in sport and professional diving to depths >30 m. The risk of oxygen toxicity is always considered when deep diving is planned. The data shown below demonstrates shortened latencies in a dry atmosphere (<10% humidity) in 11 rats at O_2 of 507 kPa. The data collected is as follows:

22	26	19	27	37	27
14	19	23	18	18	

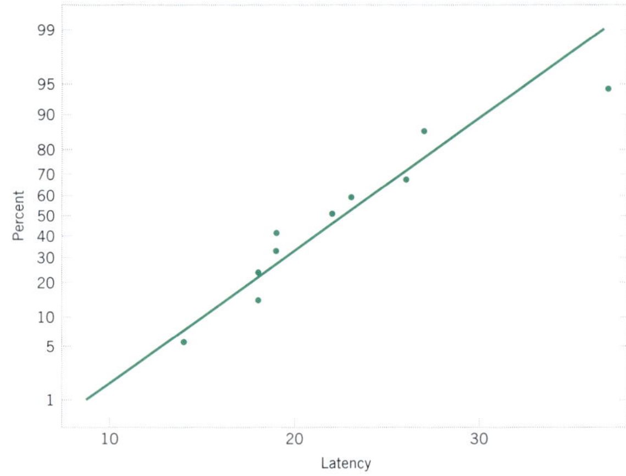

(a) Given the above probability plot of the data, what is a logical assumption about the underlying distribution of the data?
(b) Explain why this check of the distribution underlying the sample data is important if we want to construct a confidence interval on the mean.
(c) Find the 95% confidence interval for the mean.
(d) Explain why this check of the distribution underlying the sample data is important if we want to construct a confidence interval on the variance.
(e) Find the 95% confidence interval for the variance.

8-94. An article in the *Journal of Human Nutrition and Dietetics,* "The Validation of Energy and Protein Intakes by Doubly Labeled Water and 24-Hour Urinary Nitrogen Excretion in Post-Obese Subjects" (1995, Vol. 8, pp. 51–64) showed the energy intake expressed as a basal metabolic rate, BMR (MJ).

5.40	5.67	5.79	6.85	6.92
5.70	6.08	5.48	5.44	5.51

(a) Use a normal probability plot to check the normality assumption.
(b) Find a 99% two-sided confidence interval on the mean BMR.

MIND-EXPANDING EXERCISES

8-95. An electrical component has a time-to-failure (or lifetime) distribution that is exponential with parameter λ, so the mean lifetime is $\mu = 1/\lambda$. Suppose that a sample of n of these components is put on test, and let X_i be the observed lifetime of component i. The test continues only until the rth unit fails, where $r < n$. This results in a **censored** life test. Let X_1 denote the time at which the first failure occurred, X_2 denote the time at which the second failure occurred, and so on. Then the total lifetime that has been accumulated at test termination is

$$T_r = \sum_{i=1}^{r} X_i + (n - r)X_r$$

We have previously shown in Exercise 7-66 that T_r/r is an unbiased estimator for μ.

(a) It can be shown that $2\lambda T_r$ has a chi-square distribution with $2r$ degrees of freedom. Use this fact to develop a $100(1 - \alpha)\%$ confidence interval for mean lifetime $\mu = 1/\lambda$.

(b) Suppose 20 units were put on test, and the test terminated after 10 failures occurred. The failure times (in hours) are 15, 18, 19, 20, 21, 21, 22, 27, 28, 29. Find a 95% confidence interval on mean lifetime.

8-96. Consider a two-sided confidence interval for the mean μ when σ is known;

$$\bar{x} - z_{\alpha_1}\sigma/\sqrt{n} \leq \mu \leq \bar{x} + z_{\alpha_2}\sigma/\sqrt{n}$$

where $\alpha_1 + \alpha_2 = \alpha$. If $\alpha_1 = \alpha_2 = \alpha/2$, we have the usual $100(1 - \alpha)\%$ confidence interval for μ. In the above, when $\alpha_1 \neq \alpha_2$, the interval is not symmetric about μ. The length of the interval is $L = \sigma(z_{\alpha_1} + z_{\alpha_2})/\sqrt{n}$. Prove that the length of the interval L is minimized when $\alpha_1 = \alpha_2 = \alpha/2$. *Hint:* Remember that $\Phi(z_a) = 1 - \alpha$, so $\Phi^{-1}(1 - \alpha) = z_\alpha$, and the relationship between the derivative of a function $y = f(x)$ and the inverse $x = f^{-1}(y)$ is $(d/dy)f^{-1}(y) = 1/[(d/dx)f(x)]$.

8-97. It is possible to construct a **nonparametric tolerance interval** that is based on the extreme values in a random sample of size n from any continuous population. If p is the minimum proportion of the population contained between the smallest and largest sample observations with confidence $1 - \alpha$, it can be shown that

$$np^{n-1} - (n - 1)p^n = \alpha$$

and n is approximately

$$n = \frac{1}{2} + \left(\frac{1 + p}{1 - p}\right)\left(\frac{\chi^2_{\alpha,4}}{4}\right)$$

(a) In order to be 95% confident that at least 90% of the population will be included between the extreme values of the sample, what sample size will be required?

(b) A random sample of 10 transistors gave the following measurements on saturation current (in milliamps): 10.25, 10.41, 10.30, 10.26, 10.19, 10.37, 10.29, 10.34, 10.23, 10.38. Find the limits that contain a proportion p of the saturation current measurements at 95% confidence. What is the proportion p contained by these limits?

8-98. Suppose that $X_1, X_2, \ldots, X_n$ is a random sample from a continuous probability distribution with median $\tilde{\mu}$.

(a) Show that

$$P\{\min(X_i) < \tilde{\mu} < \max(X_i)\} = 1 - \left(\frac{1}{2}\right)^{n-1}$$

Hint: The complement of the event $[\min(X_i) < \tilde{\mu} < \max(X_i)]$ is $[\max(X_i) \leq \tilde{\mu}] \cup [\min(X_i) \leq \tilde{\mu}]$, but $\max(X_i) \leq \tilde{\mu}$ if and only if $X_i \leq \tilde{\mu}$ for all i.]

(b) Write down a $100(1 - \alpha)\%$ confidence interval for the median $\tilde{\mu}$, where

$$\alpha = \left(\frac{1}{2}\right)^{n-1}.$$

8-99. Students in the industrial statistics lab at ASU calculate a lot of confidence intervals on μ. Suppose all these CIs are independent of each other. Consider the next one thousand 95% confidence intervals that will be calculated. How many of these CIs do you expect to capture the true value of μ? What is the probability that between 930 and 970 of these intervals contain the true value of μ?

IMPORTANT TERMS AND CONCEPTS

Chi-squared distribution
Confidence coefficient
Confidence interval
Confidence interval for a population proportion
Confidence interval for the variance of a normal distribution
Confidence intervals for the mean of a normal distribution
Confidence level
Error in estimation
Large sample confidence interval
One-sided confidence bounds
Precision of parameter estimation
Prediction interval
Tolerance interval
Two-sided confidence interval
t distribution

9 Tests of Hypotheses: Single Sample

CHAPTER OUTLINE

LEARNING OBJECTIVES

After careful study of this chapter, you should be able to do the following:

1. Structure engineering decision-making problems as hypothesis tests
2. Test hypotheses on the mean of a normal distribution using either a *Z*-test or a *t*-test procedure
3. Test hypotheses on the variance or standard deviation of a normal distribution

4. Test hypotheses on a population proportion
5. Use the P-value approach for making decisions in hypotheses tests
6. Compute power, type II error probability, and make sample size selection decisions for tests on means, variances, and proportions
7. Explain and use the relationship between confidence intervals and hypothesis tests
8. Use the chi-square goodness of fit test to check distributional assumptions
9. Use contingency table tests

9-1 HYPOTHESIS TESTING

9-1.1 Statistical Hypotheses

In the previous chapter we illustrated how to construct a confidence interval estimate of a parameter from sample data. However, many problems in engineering require that we decide whether to accept or reject a statement about some parameter. The statement is called a **hypothesis,** and the decision-making procedure about the hypothesis is called **hypothesis testing.** This is one of the most useful aspects of statistical inference, since many types of decision-making problems, tests, or experiments in the engineering world can be formulated as hypothesis-testing problems. Furthermore, as we will see, there is a very close connection between hypothesis testing and confidence intervals.

Statistical hypothesis testing and confidence interval estimation of parameters are the fundamental methods used at the data analysis stage of a **comparative experiment,** in which the engineer is interested, for example, in comparing the mean of a population to a specified value. These simple comparative experiments are frequently encountered in practice and provide a good foundation for the more complex experimental design problems that we will discuss in Chapters 13 and 14. In this chapter we discuss comparative experiments involving a single population, and our focus is on testing hypotheses concerning the parameters of the population.

We now give a formal definition of a statistical hypothesis.

Statistical Hypothesis

A **statistical hypothesis** is a statement about the parameters of one or more populations.

Since we use probability distributions to represent populations, a statistical hypothesis may also be thought of as a statement about the probability distribution of a random variable. The hypothesis will usually involve one or more parameters of this distribution.

For example, suppose that we are interested in the burning rate of a solid propellant used to power aircrew escape systems. Now burning rate is a random variable that can be described by a probability distribution. Suppose that our interest focuses on the mean burning rate (a parameter of this distribution). Specifically, we are interested in deciding whether or not the mean burning rate is 50 centimeters per second. We may express this formally as

$$H_0\colon \mu = 50 \text{ centimeters per second}$$
$$H_1\colon \mu \neq 50 \text{ centimeters per second} \tag{9-1}$$

The statement H_0: $\mu = 50$ centimeters per second in Equation 9-1 is called the **null hypothesis,** and the statement H_1: $\mu \neq 50$ centimeters per second is called the **alternative hypothesis.** Since the alternative hypothesis specifies values of μ that could be either greater or less than 50 centimeters per second, it is called a **two-sided alternative hypothesis.** In some situations, we may wish to formulate a **one-sided alternative hypothesis,** as in

$$\begin{aligned} H_0&: \mu = 50 \text{ centimeters per second} \\ H_1&: \mu < 50 \text{ centimeters per second} \end{aligned} \quad \text{or} \quad \begin{aligned} H_0&: \mu = 50 \text{ centimeters per second} \\ H_1&: \mu > 50 \text{ centimeters per second} \end{aligned} \qquad (9\text{-}2)$$

It is important to remember that hypotheses are always statements about the population or distribution under study, not statements about the sample. The value of the population parameter specified in the null hypothesis (50 centimeters per second in the above example) is usually determined in one of three ways. First, it may result from past experience or knowledge of the process, or even from previous tests or experiments. The objective of hypothesis testing then is usually to determine whether the parameter value has changed. Second, this value may be determined from some theory or model regarding the process under study. Here the objective of hypothesis testing is to verify the theory or model. A third situation arises when the value of the population parameter results from external considerations, such as design or engineering specifications, or from contractual obligations. In this situation, the usual objective of hypothesis testing is conformance testing.

A procedure leading to a decision about a particular hypothesis is called a **test of a hypothesis.** Hypothesis-testing procedures rely on using the information in a random sample from the population of interest. If this information is consistent with the hypothesis, we will not reject the hypothesis; however, if this information is inconsistent with the hypothesis, we will conclude that the hypothesis is false. We emphasize that the truth or falsity of a particular hypothesis can never be known with certainty, unless we can examine the entire population. This is usually impossible in most practical situations. Therefore, a hypothesis-testing procedure should be developed with the probability of reaching a wrong conclusion in mind. In our treatment of hypothesis testing, the null hypothesis will always be stated so that it specifies an exact value of the parameter (as in the statement H_0: $\mu = 50$ centimeters per second in Equation 9-1). Testing the hypothesis involves taking a random sample, computing a **test statistic** from the sample data, and then using the test statistic to make a decision about the null hypothesis.

9-1.2 Tests of Statistical Hypotheses

To illustrate the general concepts, consider the propellant burning rate problem introduced earlier. The null hypothesis is that the mean burning rate is 50 centimeters per second, and the alternate is that it is not equal to 50 centimeters per second. That is, we wish to test

$$\begin{aligned} H_0&: \mu = 50 \text{ centimeters per second} \\ H_1&: \mu \neq 50 \text{ centimeters per second} \end{aligned}$$

Suppose that a sample of $n = 10$ specimens is tested and that the sample mean burning rate $\bar{x}$ is observed. The sample mean is an estimate of the true population mean μ. A value of the sample mean $\bar{x}$ that falls close to the hypothesized value of $\mu = 50$ centimeters per second does not conflict with the null hypothesis that the true mean μ is really 50 centimeters per second. On the other hand, a sample mean that is considerably different from 50 centimeters per second is evidence in support of the alternative hypothesis H_1. Thus, the sample mean is the test statistic in this case.

The sample mean can take on many different values. Suppose that if $48.5 \le \bar{x} \le 51.5$, we will not reject the null hypothesis H_0: $\mu = 50$, and if either $\bar{x} < 48.5$ or $\bar{x} > 51.5$, we will reject the null hypothesis in favor of the alternative hypothesis H_1: $\mu \neq 50$. This is illustrated in Fig. 9-1. The values of $\bar{x}$ that are less than 48.5 and greater than 51.5 constitute the **critical region** for the test, while all values that are in the interval $48.5 \le \bar{x} \le 51.5$ form a region for which we will fail to reject the null hypothesis. By convention, this is usually called the **acceptance region.** The boundaries between the critical regions and the acceptance region are called the **critical values.** In our example the critical values are 48.5 and 51.5. It is customary to state conclusions relative to the null hypothesis H_0. Therefore, we reject H_0 in favor of H_1 if the test statistic falls in the critical region and fail to reject H_0 otherwise.

This decision procedure can lead to either of two wrong conclusions. For example, the true mean burning rate of the propellant could be equal to 50 centimeters per second. However, for the randomly selected propellant specimens that are tested, we could observe a value of the test statistic $\bar{x}$ that falls into the critical region. We would then reject the null hypothesis H_0 in favor of the alternate H_1 when, in fact, H_0 is really true. This type of wrong conclusion is called a **type I error.**

Type I Error

Rejecting the null hypothesis H_0 when it is true is defined as a **type I error.**

Now suppose that the true mean burning rate is different from 50 centimeters per second, yet the sample mean $\bar{x}$ falls in the acceptance region. In this case we would fail to reject H_0 when it is false. This type of wrong conclusion is called a **type II error.**

Type II Error

Failing to reject the null hypothesis when it is false is defined as a **type II error.**

Thus, in testing any statistical hypothesis, four different situations determine whether the final decision is correct or in error. These situations are presented in Table 9-1.

Because our decision is based on random variables, probabilities can be associated with the type I and type II errors in Table 9-1. The probability of making a type I error is denoted by the Greek letter α.

Probability of Type I Error

$$\alpha = P(\text{type I error}) = P(\text{reject } H_0 \text{ when } H_0 \text{ is true}) \tag{9-3}$$

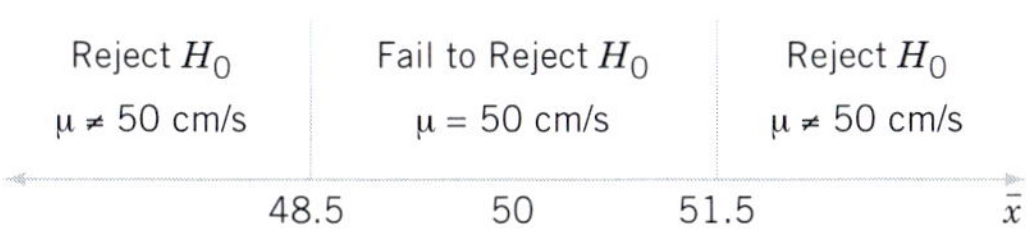

Figure 9-1 Decision criteria for testing H_0: $\mu =$ 50 centimeters per second versus H_1: $\mu \neq 50$ centimeters per second.

Table 9-1 Decisions in Hypothesis Testing

Decision	H_0 Is True	H_0 Is False
Fail to reject H_0	no error	type II error
Reject H_0	type I error	no error

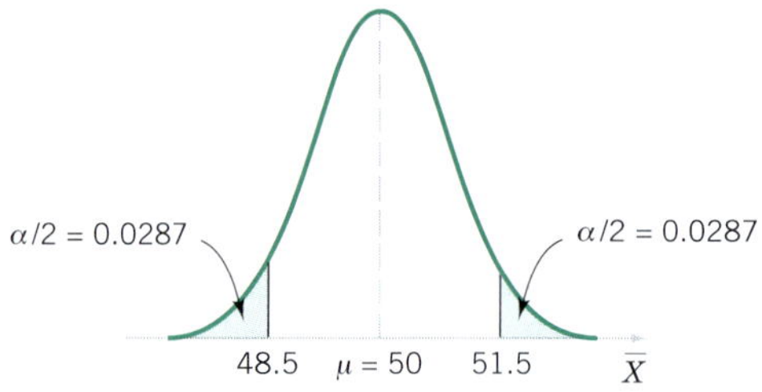

Figure 9-2 The critical region for H_0: $\mu = 50$ versus H_1: $\mu \neq 50$ and $n = 10$.

Sometimes the type I error probability is called the **significance level,** or the **α-error,** or the size of the test. In the propellant burning rate example, a type I error will occur when either $\bar{x} > 51.5$ or $\bar{x} < 48.5$ when the true mean burning rate is $\mu = 50$ centimeters per second. Suppose that the standard deviation of burning rate is $\sigma = 2.5$ centimeters per second and that the burning rate has a distribution for which the conditions of the central limit theorem apply, so the distribution of the sample mean is approximately normal with mean $\mu = 50$ and standard deviation $\sigma/\sqrt{n} = 2.5/\sqrt{10} = 0.79$. The probability of making a type I error (or the significance level of our test) is equal to the sum of the areas that have been shaded in the tails of the normal distribution in Fig. 9-2. We may find this probability as

$$\alpha = P(\bar{X} < 48.5 \text{ when } \mu = 50) + P(\bar{X} > 51.5 \text{ when } \mu = 50)$$

The z-values that correspond to the critical values 48.5 and 51.5 are

$$z_1 = \frac{48.5 - 50}{0.79} = -1.90 \quad \text{and} \quad z_2 = \frac{51.5 - 50}{0.79} = 1.90$$

Therefore

$$\alpha = P(Z < -1.90) + P(Z > 1.90) = 0.0287 + 0.0287 = 0.0574$$

This implies that 5.74% of all random samples would lead to rejection of the hypothesis H_0: $\mu = 50$ centimeters per second when the true mean burning rate is really 50 centimeters per second.

From inspection of Fig. 9-2, notice that we can reduce α by widening the acceptance region. For example, if we make the critical values 48 and 52, the value of α is

$$\begin{aligned}\alpha &= P\left(Z < \frac{48 - 50}{0.79}\right) + P\left(Z > \frac{52 - 50}{0.79}\right) = P(Z < -2.53) + P(Z > 2.53) \\ &= 0.0057 + 0.0057 = 0.0114\end{aligned}$$

We could also reduce α by increasing the sample size. If $n = 16$, $\sigma/\sqrt{n} = 2.5/\sqrt{16} = 0.625$, and using the original critical region from Fig. 9-1, we find

$$z_1 = \frac{48.5 - 50}{0.625} = -2.40 \quad \text{and} \quad z_2 = \frac{51.5 - 50}{0.625} = 2.40$$

Therefore

$$\alpha = P(Z < -2.40) + P(Z > 2.40) = 0.0082 + 0.0082 = 0.0164$$

In evaluating a hypothesis-testing procedure, it is also important to examine the probability of a **type II error,** which we will denote by β. That is,

Probability of Type II Error

$$\beta = P(\text{type II error}) = P(\text{fail to reject } H_0 \text{ when } H_0 \text{ is false}) \tag{9-4}$$

To calculate β (sometimes called the **β-error**), we must have a specific alternative hypothesis; that is, we must have a particular value of μ. For example, suppose that it is important to reject the null hypothesis H_0: $\mu = 50$ whenever the mean burning rate μ is greater than 52 centimeters per second or less than 48 centimeters per second. We could calculate the probability of a type II error β for the values $\mu = 52$ and $\mu = 48$ and use this result to tell us something about how the test procedure would perform. Specifically, how will the test procedure work if we wish to detect, that is, reject H_0, for a mean value of $\mu = 52$ or $\mu = 48$? Because of symmetry, it is necessary only to evaluate one of the two cases—say, find the probability of accepting the null hypothesis H_0: $\mu = 50$ centimeters per second when the true mean is $\mu =$ 52 centimeters per second.

Figure 9-3 will help us calculate the probability of type II error β. The normal distribution on the left in Fig. 9-3 is the distribution of the test statistic $\overline{X}$ when the null hypothesis H_0: $\mu = 50$ is true (this is what is meant by the expression "under H_0: $\mu = 50$"), and the normal distribution on the right is the distribution of $\overline{X}$ when the alternative hypothesis is true and the value of the mean is 52 (or "under H_1: $\mu = 52$"). Now a type II error will be committed if the sample mean $\overline{X}$ falls between 48.5 and 51.5 (the critical region boundaries) when $\mu = 52$. As seen in Fig. 9-3, this is just the probability that $48.5 \le \overline{X} \le 51.5$ when the true mean is $\mu = 52$, or the shaded area under the normal distribution centered at $\mu = 52$. Therefore, referring to Fig. 9-3, we find that

$$\beta = P(48.5 \le \overline{X} \le 51.5 \text{ when } \mu = 52)$$

The z-values corresponding to 48.5 and 51.5 when $\mu = 52$ are

$$z_1 = \frac{48.5 - 52}{0.79} = -4.43 \quad \text{and} \quad z_2 = \frac{51.5 - 52}{0.79} = -0.63$$

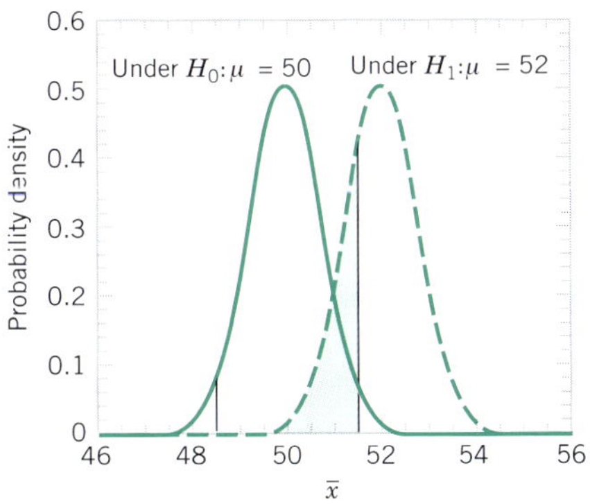

Figure 9-3 The probability of type II error when $\mu = 52$ and $n = 10$.

Therefore

$$\begin{aligned}\beta &= P(-4.43 \le Z \le -0.63) = P(Z \le -0.63) - P(Z \le -4.43)\\ &= 0.2643 - 0.0000 = 0.2643\end{aligned}$$

Thus, if we are testing H_0: $\mu = 50$ against H_1: $\mu \ne 50$ with $n = 10$, and the true value of the mean is $\mu = 52$, the probability that we will fail to reject the false null hypothesis is 0.2643. By symmetry, if the true value of the mean is $\mu = 48$, the value of β will also be 0.2643.

The probability of making a type II error β increases rapidly as the true value of μ approaches the hypothesized value. For example, see Fig. 9-4, where the true value of the mean is $\mu = 50.5$ and the hypothesized value is H_0: $\mu = 50$. The true value of μ is very close to 50, and the value for β is

$$\beta = P(48.5 \le \overline{X} \le 51.5 \text{ when } \mu = 50.5)$$

As shown in Fig. 9-4, the z-values corresponding to 48.5 and 51.5 when $\mu = 50.5$ are

$$z_1 = \frac{48.5 - 50.5}{0.79} = -2.53 \quad \text{and} \quad z_2 = \frac{51.5 - 50.5}{0.79} = 1.27$$

Therefore

$$\begin{aligned}\beta &= P(-2.53 \le Z \le 1.27) = P(Z \le 1.27) - P(Z \le -2.53)\\ &= 0.8980 - 0.0057 = 0.8923\end{aligned}$$

Thus, the type II error probability is much higher for the case where the true mean is 50.5 centimeters per second than for the case where the mean is 52 centimeters per second. Of course, in many practical situations we would not be as concerned with making a type II error if the mean were "close" to the hypothesized value. We would be much more interested in detecting large differences between the true mean and the value specified in the null hypothesis.

The type II error probability also depends on the sample size n. Suppose that the null hypothesis is H_0: $\mu = 50$ centimeters per second and that the true value of the mean is $\mu = 52$. If the sample size is increased from $n = 10$ to $n = 16$, the situation of Fig. 9-5 results.

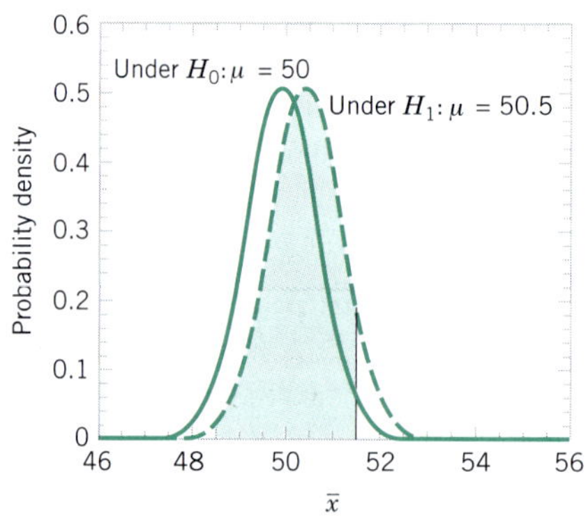

Figure 9-4 The probability of type II error when $\mu = 50.5$ and $n = 10$.

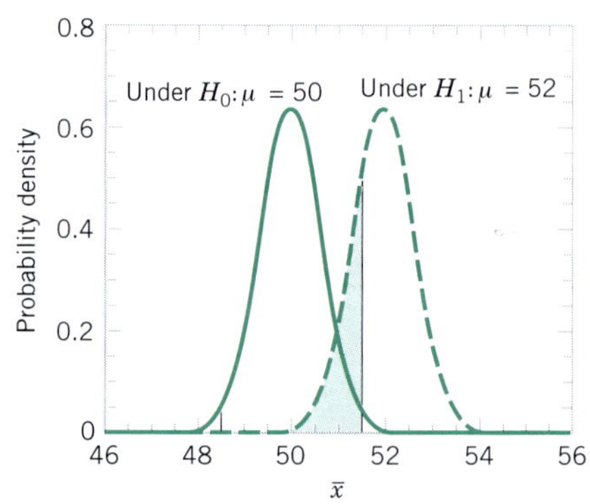

Figure 9-5 The probability of type II error when $\mu = 52$ and $n = 16$.

The normal distribution on the left is the distribution of $\overline{X}$ when the mean $\mu = 50$, and the normal distribution on the right is the distribution of $\overline{X}$ when $\mu = 52$. As shown in Fig. 9-5, the type II error probability is

$$\beta = P(48.5 \leq \overline{X} \leq 51.5 \text{ when } \mu = 52)$$

When $n = 16$, the standard deviation of $\overline{X}$ is $\sigma/\sqrt{n} = 2.5/\sqrt{16} = 0.625$, and the z-values corresponding to 48.5 and 51.5 when $\mu = 52$ are

$$z_1 = \frac{48.5 - 52}{0.625} = -5.60 \quad \text{and} \quad z_2 = \frac{51.5 - 52}{0.625} = -0.80$$

Therefore

$$\begin{aligned}\beta &= P(-5.60 \leq Z \leq -0.80) = P(Z \leq -0.80) - P(Z \leq -5.60)\\ &= 0.2119 - 0.0000 = 0.2119\end{aligned}$$

Recall that when $n = 10$ and $\mu = 52$, we found that $\beta = 0.2643$; therefore, increasing the sample size results in a decrease in the probability of type II error.

The results from this section and a few other similar calculations are summarized in the following table. The critical values are adjusted to maintain equal α for $n = 10$ and $n = 16$. This type of calculation is discussed later in the chapter.

Acceptance Region	Sample Size	α	β at $\mu = 52$	β at $\mu = 50.5$
$48.5 < \bar{x} < 51.5$	10	0.0576	0.2643	0.8923
$48 < \bar{x} < 52$	10	0.0114	0.5000	0.9705
$48.81 < \bar{x} < 51.19$	16	0.0576	0.0966	0.8606
$48.42 < \bar{x} < 51.58$	16	0.0114	0.2515	0.9578

The results in boxes were not calculated in the text but can easily be verified by the reader. This display and the discussion above reveal four important points:

1. The size of the critical region, and consequently the probability of a type I error α, can always be reduced by appropriate selection of the critical values.
2. Type I and type II errors are related. A decrease in the probability of one type of error always results in an increase in the probability of the other, provided that the sample size n does not change.
3. An increase in sample size reduces β, provided that α is held constant.
4. When the null hypothesis is false, β increases as the true value of the parameter approaches the value hypothesized in the null hypothesis. The value of β decreases as the difference between the true mean and the hypothesized value increases.

Generally, the analyst controls the type I error probability α when he or she selects the critical values. Thus, it is usually easy for the analyst to set the type I error probability at (or near) any desired value. Since the analyst can directly control the probability of wrongly rejecting H_0, we always think of rejection of the null hypothesis H_0 as a **strong conclusion.**

On the other hand, the probability of type II error β is not a constant, but depends on the true value of the parameter. It also depends on the sample size that we have selected. Because the type II error probability β is a function of both the sample size and the extent to which the null hypothesis H_0 is false, it is customary to think of the decision to accept H_0 as a **weak conclusion,** unless we know that β is acceptably small. Therefore, rather than saying we "accept H_0", we prefer the terminology "fail to reject H_0". Failing to reject H_0 implies that we have not found sufficient evidence to reject H_0, that is, to make a strong statement. Failing to reject H_0 does not necessarily mean that there is a high probability that H_0 is true. It may simply mean that more data are required to reach a strong conclusion. This can have important implications for the formulation of hypotheses.

An important concept that we will make use of is the **power** of a statistical test.

Power

> The **power** of a statistical test is the probability of rejecting the null hypothesis H_0 when the alternative hypothesis is true.

The power is computed as $1 - \beta$, and **power** can be interpreted as **the probability of correctly rejecting a false null hypothesis.** We often compare statistical tests by comparing their power properties. For example, consider the propellant burning rate problem when we are testing H_0: $\mu = 50$ centimeters per second against H_1: $\mu \neq 50$ centimeters per second. Suppose that the true value of the mean is $\mu = 52$. When $n = 10$, we found that $\beta = 0.2643$, so the power of this test is $1 - \beta = 1 - 0.2643 = 0.7357$ when $\mu = 52$.

Power is a very descriptive and concise measure of the **sensitivity** of a statistical test, where by sensitivity we mean the ability of the test to detect differences. In this case, the sensitivity of the test for detecting the difference between a mean burning rate of 50 centimeters per second and 52 centimeters per second is 0.7357. That is, if the true mean is really 52 centimeters per second, this test will correctly reject H_0: $\mu = 50$ and "detect" this difference 73.57% of the time. If this value of power is judged to be too low, the analyst can increase either α or the sample size n.

9-1.3 One-Sided and Two-Sided Hypotheses

In constructing hypotheses, we will always state the null hypothesis as an equality so that the probability of type I error α can be controlled at a specific value. The alternative hypothesis might be either one-sided or two-sided, depending on the conclusion to be drawn if H_0 is rejected. If the objective is to make a claim involving statements such as greater than, less than, superior to, exceeds, at least, and so forth, a one-sided alternative is appropriate. If no direction is implied by the claim, or if the claim not equal to is to be made, a two-sided alternative should be used.

EXAMPLE 9-1 Propellant Burning Rate

Consider the propellant burning rate problem. Suppose that if the burning rate is less than 50 centimeters per second, we wish to show this with a strong conclusion. The hypotheses should be stated as

$$H_0: \mu = 50 \text{ centimeters per second}$$
$$H_1: \mu < 50 \text{ centimeters per second}$$

Here the critical region lies in the lower tail of the distribution of $\overline{X}$. Since the rejection of H_0 is always a strong conclusion, this statement of the hypotheses will produce the desired outcome if H_0 is rejected. Notice that, although the null hypothesis is stated with an equal sign, it is understood to include any value of μ not specified by the alternative hypothesis. Therefore, failing to reject H_0 does not mean that $\mu = 50$ centimeters per second exactly, but only that we do not have strong evidence in support of H_1.

In some real-world problems where one-sided test procedures are indicated, it is occasionally difficult to choose an appropriate formulation of the alternative hypothesis. For example, suppose that a soft-drink beverage bottler purchases 10-ounce bottles from a glass company. The bottler wants to be sure that the bottles meet the specification on mean internal pressure or bursting strength, which for 10-ounce bottles is a minimum strength of 200 psi. The bottler has decided to formulate the decision procedure for a specific lot of bottles as a hypothesis testing problem. There are two possible formulations for this problem, either

$$H_0: \mu = 200 \text{ psi}$$
$$H_1: \mu > 200 \text{ psi} \tag{9-5}$$

or

$$H_0: \mu = 200 \text{ psi}$$
$$H_1: \mu < 200 \text{ psi} \tag{9-6}$$

Consider the formulation in Equation 9-5. If the null hypothesis is rejected, the bottles will be judged satisfactory; if H_0 is not rejected, the implication is that the bottles do not conform to specifications and should not be used. Because rejecting H_0 is a strong conclusion, this formulation forces the bottle manufacturer to "demonstrate" that the mean bursting strength of the bottles exceeds the specification. Now consider the formulation in Equation 9-6. In this situation, the bottles will be judged satisfactory unless H_0 is rejected. That is, we conclude that the bottles are satisfactory unless there is strong evidence to the contrary.

Which formulation is correct, the one of Equation 9-5 or Equation 9-6? The answer is it depends on the objective of the analysis. For Equation 9-5, there is some probability that H_0 will not be rejected (i.e., we would decide that the bottles are not satisfactory), even though the true mean is slightly greater than 200 psi. This formulation implies that we want the bottle manufacturer to demonstrate that the product meets or exceeds our specifications. Such a formulation could be appropriate if the manufacturer has experienced difficulty in meeting specifications in the past or if product safety considerations force us to hold tightly to the 200 psi specification. On the other hand, for the formulation of Equation 9-6 there is some probability that H_0 will be accepted and the bottles judged satisfactory, even though the true mean is slightly less than 200 psi. We would conclude that the bottles are unsatisfactory only when there is strong evidence that the mean does not exceed 200 psi, that is, when H_0: $\mu = 200$ psi is rejected. This formulation assumes that we are relatively happy with the bottle manufacturer's past performance and that small deviations from the specification of $\mu \geq 200$ psi are not harmful.

In formulating one-sided alternative hypotheses, we should remember that rejecting H_0 is always a strong conclusion. Consequently, we should put the statement about which it is important to make a strong conclusion in the alternative hypothesis. In real-world problems, this will often depend on our point of view and experience with the situation.

9-1.4 P-Values in Hypothesis Tests

One way to report the results of a hypothesis test is to state that the null hypothesis was or was not rejected at a specified α-value or level of significance. For example, in the propellant problem above, we can say that H_0: $\mu = 50$ was rejected at the 0.05 level of significance. This statement of conclusions is often inadequate because it gives the decision maker no idea about whether the computed value of the test statistic was just barely in the rejection region or whether it was very far into this region. Furthermore, stating the results this way imposes the predefined level of significance on other users of the information. This approach may be unsatisfactory because some decision makers might be uncomfortable with the risks implied by $\alpha = 0.05$.

To avoid these difficulties the ***P*-value** approach has been adopted widely in practice. The *P*-value is the probability that the test statistic will take on a value that is at least as extreme as the observed value of the statistic when the null hypothesis H_0 is true. Thus, a *P*-value conveys much information about the weight of evidence against H_0, and so a decision maker can draw a conclusion at *any* specified level of significance. We now give a formal definition of a *P*-value.

***P*-Value**

> The ***P*-value** is the smallest level of significance that would lead to rejection of the null hypothesis H_0 with the given data.

It is customary to call the test statistic (and the data) significant when the null hypothesis H_0 is rejected; therefore, we may think of the *P*-value as the smallest level α at which the data are significant. Once the *P*-value is known, the decision maker can determine how significant the data are without the data analyst formally imposing a preselected level of significance.

Consider the two-sided hypothesis test for burning rate

$$H_0 : \mu = 50 \quad H_1 : \mu \neq 50$$

with $n = 16$ and $\sigma = 2.5$. Suppose that the observed sample mean is $\bar{x} = 51.3$ centimeters per second. Figure 9-6 shows a critical region for this test with critical values at 51.3 and the symmetric value 48.7. The *P*-value of the test is the α associated with this critical region. Any smaller value for α decreases the critical region and the test fails to reject the

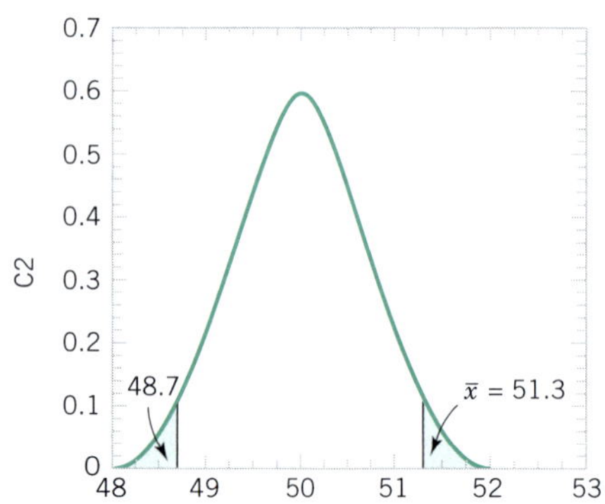

Figure 9-6 *P*-value is area of shaded region when $\bar{x} = 51.3$.

null hypothesis when $\bar{x} = 51.3$. The P-value is easy to compute after the test statistic is observed. In this example

$$\begin{aligned} P\text{-value} &= 1 - P(48.7 < \bar{X} < 51.3) \\ &= 1 - P\left(\frac{48.7 - 50}{2.5/\sqrt{16}} < Z < \frac{51.3 - 50}{2.5/\sqrt{16}}\right) \\ &= 1 - P(-2.08 < Z < 2.08) \\ &= 1 - 0.962 = 0.038 \end{aligned}$$

This null hypothesis would be rejected for any $0.038 < \alpha$, but not rejected for a smaller significance level such as 0.01. P-values are often calculated in this manner. The observed value of the test statistic is used to define a critical region, and the α associated with this critical region is computed.

It is not always easy to compute the exact P-value for a test. However, most modern computer programs for statistical analysis report P-values. We will also show how to approximate the P-value.

9-1.5 Connection between Hypothesis Tests and Confidence Intervals

There is a close relationship between the test of a hypothesis about any parameter, say θ, and the confidence interval for θ. If $[l, u]$ is a $100(1 - \alpha)\%$ confidence interval for the parameter θ, the test of size α of the hypothesis

$$H_0: \theta = \theta_0$$
$$H_1: \theta \neq \theta_0$$

will lead to rejection of H_0 if and only if θ_0 is **not** in the $100(1 - \alpha)\%$ CI $[l, u]$. As an illustration, consider the escape system propellant problem with $\bar{x} = 51.3$, $\sigma = 2.5$, and $n = 16$. The null hypothesis H_0: $\mu = 50$ was rejected, using $\alpha = 0.05$. The 95% two-sided CI on μ can be calculated using Equation 8-7. This CI is $51.3 \pm 1.96(2.5/\sqrt{16})$ and this is $50.075 \leq \mu \leq 52.525$. Because the value $\mu_0 = 50$ is not included in this interval, the null hypothesis H_0: $\mu = 50$ is rejected.

Although hypothesis tests and CIs are equivalent procedures insofar as decision making or **inference** about μ is concerned, each provides somewhat different insights. For instance, the confidence interval provides a range of likely values for μ at a stated confidence level, whereas hypothesis testing is an easy framework for displaying the **risk levels** such as the P-value associated with a specific decision. We will continue to illustrate the connection between the two procedures throughout the text.

9-1.6 General Procedure for Hypothesis Tests

This chapter develops hypothesis-testing procedures for many practical problems. Use of the following sequence of steps in applying hypothesis-testing methodology is recommended.

1. From the problem context, identify the parameter of interest.
2. State the null hypothesis, H_0.
3. Specify an appropriate alternative hypothesis, H_1.

4. Choose a significance level α.
5. Determine an appropriate test statistic.
6. State the rejection region for the statistic. This step is not necessary if the P-value approach is used.
7. Compute any necessary sample quantities, substitute these into the equation for the test statistic, and compute that value.
8. Decide whether or not H_0 should be rejected and report that in the problem context.

Steps 1–4 should be completed prior to examination of the sample data. This sequence of steps will be illustrated in subsequent sections.

In practice, such a formal and (seemingly) rigid procedure is not always necessary. Generally, once the experimenter (or decision maker) has decided on the question of interest and has determined the *design of the experiment* (that is, how the data are to be collected, how the measurements are to be made, and how many observations are required), only three steps are really required:

1. Specify the test statistic to be used (such as Z_0).
2. Specify the location of the critical region (two-tailed, upper-tailed, or lower-tailed).
3. Specify the criteria for rejection (typically, the value of α, or the P-value at which rejection should occur).

These steps are often completed almost simultaneously in solving real-world problems, although we emphasize that it is important to think carefully about each step. That is why we present and use the eight-step process: it seems to reinforce the essentials of the correct approach. While you may not use it every time in solving real problems, it is a helpful framework when you are first learning about hypothesis testing.

Statistical versus Practical Significance

We noted previously that reporting the results of a hypothesis test in terms of a P-value is very useful because it conveys more information than just the simple statement "reject H_0" or "fail to reject H_0". That is, rejection of H_0 at the 0.05 level of significance is much more meaningful if the value of the test statistic is well into the critical region, greatly exceeding the 5% critical value, than if it barely exceeds that value.

Even a very small P-value can be difficult to interpret from a practical viewpoint when we are making decisions because, while a small P-value indicates **statistical significance** in the sense that H_0 should be rejected in favor of H_1, the actual departure from H_0 that has been detected may have little (if any) **practical significance** (engineers like to say "engineering significance"). This is particularly true when the sample size n is large.

For example, consider the propellant burning rate problem of Example 9-1 where we are testing H_0: $\mu = 50$ centimeters per second versus H_1: $\mu \neq 50$ centimeters per second with $\sigma = 2.5$. If we suppose that the mean rate is really 50.5 centimeters per second, this is not a serious departure from H_0: $\mu = 50$ centimeters per second in the sense that if the mean really is 50.5 centimeters per second there is no practical observable effect on the performance of the aircrew escape system. In other words, concluding that $\mu = 50$ centimeters per second when it is really 50.5 centimeters per second is an inexpensive error and has no practical significance. For a reasonably large sample size, a true value of $\mu = 50.5$ will lead to a sample $\bar{x}$ that is close to 50.5 centimeters per second, and we would not want this value of $\bar{x}$ from the sample to result in rejection of H_0. The following display shows the P-value for testing H_0: $\mu = 50$ when we observe $\bar{x} = 50.5$ centimeters per second and the power of the test at $\alpha = 0.05$ when the true mean is 50.5 for various sample sizes n:

Sample Size n	P-value When $\bar{x} = 50.5$	Power (at $\alpha = 0.05$) When True $\mu = 50.5$
10	0.527	0.097
25	0.317	0.170
50	0.157	0.293
100	0.046	0.516
400	6.3×10^{-5}	0.979
1000	2.5×10^{-10}	1.000

The P-value column in this display indicates that for large sample sizes the observed sample value of $\bar{x} = 50.5$ would strongly suggest that H_0: $\mu = 50$ should be rejected, even though the observed sample results imply that from a practical viewpoint the true mean does not differ much at all from the hypothesized value $\mu_0 = 50$. The power column indicates that if we test a hypothesis at a fixed significance level α and even if there is little practical difference between the true mean and the hypothesized value, a large sample size will almost always lead to rejection of H_0. The moral of this demonstration is clear:

> Be careful when interpreting the results from hypothesis testing when the sample size is large, because any small departure from the hypothesized value μ_0 will probably be detected, even when the difference is of little or no practical significance.

EXERCISES FOR SECTION 9-1

9-1. In each of the following situations, state whether it is a correctly stated hypothesis testing problem and why.

(a) H_0: $\mu = 25$, H_1: $\mu \neq 25$

(b) H_0: $\sigma > 10$, H_1: $\sigma = 10$

(c) H_0: $\bar{x} = 50$, H_1: $\bar{x} \neq 50$

(d) H_0: $p = 0.1$, H_1: $p = 0.5$

(e) H_0: $s = 30$, H_1: $s > 30$

9-2. A semiconductor manufacturer collects data from a new tool and conducts a hypothesis test with the null hypothesis that a critical dimension mean width equals 100 nm. The conclusion is to not reject the null hypothesis. Does this result provide strong evidence that the critical dimension mean equals 100 nm? Explain.

9-3. The standard deviation of critical dimension thickness in semiconductor manufacturing is $\sigma = 20$ nm.

(a) State the null and alternative hypotheses used to demonstrate that the standard deviation is reduced.

(b) Assume that the previous test does not reject the null hypothesis. Does this result provide strong evidence that the standard deviation has not been reduced? Explain.

9-4. The mean pull-off force of a connector depends on cure time.

(a) State the null and alternative hypotheses used to demonstrate that the pull-off force is below 25 newtons.

(b) Assume that the previous test does not reject the null hypothesis. Does this result provide strong evidence that the pull-off force is greater than or equal to 25 newtons? Explain.

9-5. A textile fiber manufacturer is investigating a new drapery yarn, which the company claims has a mean thread elongation of 12 kilograms with a standard deviation of 0.5 kilograms. The company wishes to test the hypothesis H_0: $\mu = 12$ against H_1: $\mu < 12$, using a random sample of four specimens.

(a) What is the type I error probability if the critical region is defined as $\bar{x} < 11.5$ kilograms?

(b) Find β for the case where the true mean elongation is 11.25 kilograms.

(c) Find β for the case where the true mean is 11.5 kilograms.

9-6. Repeat Exercise 9-5 using a sample size of $n = 16$ and the same critical region.

9-7. In Exercise 9-5, find the boundary of the critical region if the type I error probability is

(a) $\alpha = 0.01$ and $n = 4$
(b) $\alpha = 0.05$ and $n = 4$
(c) $\alpha = 0.01$ and $n = 16$
(d) $\alpha = 0.05$ and $n = 16$

9-8. In Exercise 9-5, calculate the probability of a type II error if the true mean elongation is 11.5 kilograms and

(a) $\alpha = 0.05$ and $n = 4$
(b) $\alpha = 0.05$ and $n = 16$
(c) Compare the values of β calculated in the previous parts. What conclusion can you draw?

9-9. In Exercise 9-5, calculate the P-value if the observed statistic is

(a) $\bar{x} = 11.25$ (b) $\bar{x} = 11.0$ (c) $\bar{x} = 11.75$

9-10. The heat evolved in calories per gram of a cement mixture is approximately normally distributed. The mean is thought to be 100 and the standard deviation is 2. We wish to test H_0: $\mu = 100$ versus H_1: $\mu \neq 100$ with a sample of $n = 9$ specimens.

(a) If the acceptance region is defined as $98.5 \leq \bar{x} \leq 101.5$, find the type I error probability α.
(b) Find β for the case where the true mean heat evolved is 103.
(c) Find β for the case where the true mean heat evolved is 105. This value of β is smaller than the one found in part (b) above. Why?

9-11. Repeat Exercise 9-10 using a sample size of $n = 5$ and the same acceptance region.

9-12. In Exercise 9-10, find the boundary of the critical region if the type I error probability is

(a) $\alpha = 0.01$ and $n = 9$
(b) $\alpha = 0.05$ and $n = 9$
(c) $\alpha = 0.01$ and $n = 5$
(d) $\alpha = 0.05$ and $n = 5$

9-13. In Exercise 9-10, calculate the probability of a type II error if the true mean heat evolved is 103 and

(a) $\alpha = 0.05$ and $n = 9$
(b) $\alpha = 0.05$ and $n = 5$
(c) Compare the values of β calculated in the previous parts. What conclusion can you draw?

9-14. In Exercise 9-10, calculate the P-value if the observed statistic is

(a) $\bar{x} = 98$ (b) $\bar{x} = 101$ (c) $\bar{x} = 102$

9-15. A consumer products company is formulating a new shampoo and is interested in foam height (in millimeters). Foam height is approximately normally distributed and has a standard deviation of 20 millimeters. The company wishes to test H_0: $\mu = 175$ millimeters versus H_1: $\mu > 175$ millimeters, using the results of $n = 10$ samples.

(a) Find the type I error probability α if the critical region is $\bar{x} > 185$.
(b) What is the probability of type II error if the true mean foam height is 185 millimeters?
(c) Find β for the true mean of 195 millimeters.

9-16. Repeat Exercise 9-15 assuming that the sample size is $n = 16$ and the boundary of the critical region is the same.

9-17. In Exercise 9-15, find the boundary of the critical region if the type I error probability is

(a) $\alpha = 0.01$ and $n = 10$
(b) $\alpha = 0.05$ and $n = 10$
(c) $\alpha = 0.01$ and $n = 16$
(d) $\alpha = 0.05$ and $n = 16$

9-18. In Exercise 9-15, calculate the probability of a type II error if the true mean foam height is 185 millimeters and

(a) $\alpha = 0.05$ and $n = 10$
(b) $\alpha = 0.05$ and $n = 16$
(c) Compare the values of β calculated in the previous parts. What conclusion can you draw?

9-19. In Exercise 9-15, calculate the P-value if the observed statistic is

(a) $\bar{x} = 180$ (b) $\bar{x} = 190$ (c) $\bar{x} = 170$

9-20. A manufacturer is interested in the output voltage of a power supply used in a PC. Output voltage is assumed to be normally distributed, with standard deviation 0.25 Volts, and the manufacturer wishes to test H_0: $\mu = 5$ Volts against H_1: $\mu \neq 5$ Volts, using $n = 8$ units.

(a) The acceptance region is $4.85 \leq \bar{x} \leq 5.15$. Find the value of α.
(b) Find the power of the test for detecting a true mean output voltage of 5.1 Volts.

9-21. Rework Exercise 9-20 when the sample size is 16 and the boundaries of the acceptance region do not change. What impact does the change in sample size have on the results of parts (a) and (b)?

9-22. In Exercise 9-20, find the boundary of the critical region if the type I error probability is

(a) $\alpha = 0.01$ and $n = 8$
(b) $\alpha = 0.05$ and $n = 8$
(c) $\alpha = 0.01$ and $n = 16$
(d) $\alpha = 0.05$ and $n = 16$

9-23. In Exercise 9-20, calculate the P-value if the observed statistic is

(a) $\bar{x} = 5.2$ (b) $\bar{x} = 4.7$ (c) $\bar{x} = 5.1$

9-24. In Exercise 9-20, calculate the probability of a type II error if the true mean output is 5.05 volts and

(a) $\alpha = 0.05$ and $n = 10$
(b) $\alpha = 0.05$ and $n = 16$
(c) Compare the values of β calculated in the previous parts. What conclusion can you draw?

9-25. The proportion of adults living in Tempe, Arizona, who are college graduates is estimated to be $p = 0.4$. To test this hypothesis, a random sample of 15 Tempe adults is selected. If the number of college graduates is between 4 and 8, the hypothesis will be accepted; otherwise, we will conclude that $p \neq 0.4$.

(a) Find the type I error probability for this procedure, assuming that $p = 0.4$.

(b) Find the probability of committing a type II error if the true proportion is really $p = 0.2$.

9-26. The proportion of residents in Phoenix favoring the building of toll roads to complete the freeway system is believed to be $p = 0.3$. If a random sample of 10 residents shows that 1 or fewer favor this proposal, we will conclude that $p < 0.3$.

(a) Find the probability of type I error if the true proportion is $p = 0.3$.

(b) Find the probability of committing a type II error with this procedure if $p = 0.2$.

(c) What is the power of this procedure if the true proportion is $p = 0.2$?

9-27. A random sample of 500 registered voters in Phoenix is asked if they favor the use of oxygenated fuels year-round to reduce air pollution. If more than 400 voters respond positively, we will conclude that at least 60% of the voters favor the use of these fuels.

(a) Find the probability of type I error if exactly 60% of the voters favor the use of these fuels.

(b) What is the type II error probability β if 75% of the voters favor this action?

Hint: use the normal approximation to the binomial.

9-28. If we plot the probability of accepting $H_0{:}\mu = \mu_0$ versus various values of μ and connect the points with a smooth curve, we obtain the **operating characteristic curve** (or the **OC curve**) of the test procedure. These curves are used extensively in industrial applications of hypothesis testing to display the sensitivity and relative performance of the test. When the true mean is really equal to μ_0, the probability of accepting H_0 is $1 - \alpha$.

(a) Construct an OC curve for Exercise 9-15, using values of the true mean μ of 178, 181, 184, 187, 190, 193, 196, and 199.

(b) Convert the OC curve into a plot of the **power function** of the test.

9-2 TESTS ON THE MEAN OF A NORMAL DISTRIBUTION, VARIANCE KNOWN

In this section, we consider hypothesis testing about the mean μ of a single normal population where the variance of the population σ^2 is known. We will assume that a random sample $X_1, X_2, \ldots, X_n$ has been taken from the population. Based on our previous discussion, the sample mean $\overline{X}$ is an **unbiased point estimator** of μ with variance σ^2/n.

9-2.1 Hypothesis Tests on the Mean

Suppose that we wish to test the hypotheses

$$\begin{aligned} H_0&: \mu = \mu_0 \\ H_1&: \mu \neq \mu_0 \end{aligned} \tag{9-7}$$

where μ_0 is a specified constant. We have a random sample $X_1, X_2, \ldots, X_n$ from a normal population. Since $\overline{X}$ has a normal distribution (i.e., the **sampling distribution** of $\overline{X}$ is normal) with mean μ_0 and standard deviation $\sigma/\sqrt{n}$ if the null hypothesis is true, we could construct a critical region based on the computed value of the sample mean $\overline{X}$, as in Section 9-1.2.

It is usually more convenient to *standardize* the sample mean and use a test statistic based on the standard normal distribution. That is, the test procedure for H_0: $\mu = \mu_0$ uses the **test statistic**

Test Statistic

$$Z_0 = \frac{\overline{X} - \mu_0}{\sigma/\sqrt{n}} \tag{9-8}$$

If the null hypothesis H_0: $\mu = \mu_0$ is true, $E(\overline{X}) = \mu_0$, and it follows that the distribution of Z_0 is the standard normal distribution [denoted $N(0, 1)$]. Consequently, if H_0: $\mu = \mu_0$ is true, the probability is $1 - \alpha$ that the test statistic Z_0 falls between $-z_{\alpha/2}$ and $z_{\alpha/2}$, where $z_{\alpha/2}$ is the

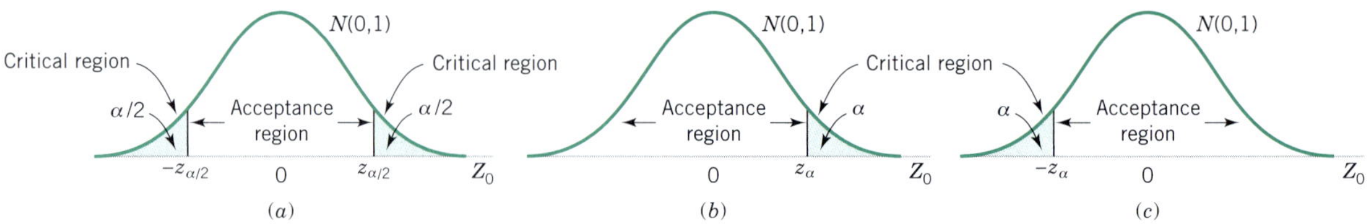

Figure 9-7 The distribution of Z_0 when H_0: $\mu = \mu_0$ is true, with critical region for (a) the two-sided alternative $H_1: \mu \neq \mu_0$, (b) the one-sided alternative $H_1: \mu > \mu_0$, and (c) the one-sided alternative $H_1: \mu < \mu_0$.

$100\alpha/2$ percentage point of the standard normal distribution. The regions associated with $z_{\alpha/2}$ and $-z_{\alpha/2}$ are illustrated in Fig. 9-7(a). Note that the probability is α that the test statistic Z_0 will fall in the region $Z_0 > z_{\alpha/2}$ or $Z_0 < -z_{\alpha/2}$ when H_0: $\mu = \mu_0$ is true. Clearly, a sample producing a value of the test statistic that falls in the tails of the distribution of Z_0 would be unusual if H_0: $\mu = \mu_0$ is true; therefore, it is an indication that H_0 is false. Thus, we should reject H_0 if the observed value of the test statistic z_0 is either

$$z_0 > z_{\alpha/2} \quad \text{or} \quad z_0 < -z_{\alpha/2} \tag{9-9}$$

and we should fail to reject H_0 if

$$-z_{\alpha/2} \leq z_0 \leq z_{\alpha/2} \tag{9-10}$$

The inequalities in Equation 9-10 define the acceptance region for H_0, and the two inequalities in Equation 9-9 define the **critical region** or **rejection region.** The type I error probability for this test procedure is α.

It is easier to understand the critical region and the test procedure, in general, when the test statistic is Z_0 rather than $\overline{X}$. However, the same critical region can always be written in terms of the computed value of the sample mean $\overline{x}$. A procedure identical to the above is as follows:

$$\text{Reject } H_0\text{: } \mu = \mu_0 \text{ if either } \overline{x} > a \text{ or } \overline{x} < b$$

where

$$a = \mu_0 + z_{\alpha/2}\sigma/\sqrt{n} \quad \text{and} \quad b = \mu_0 - z_{\alpha/2}\sigma/\sqrt{n}$$

EXAMPLE 9-2
Propellant Burning Rate

Aircrew escape systems are powered by a solid propellant. The burning rate of this propellant is an important product characteristic. Specifications require that the mean burning rate must be 50 centimeters per second. We know that the standard deviation of burning rate is $\sigma = 2$ centimeters per second. The experimenter decides to specify a type I error probability or significance level of $\alpha = 0.05$ and selects a random sample of $n = 25$ and obtains a sample average burning rate of $\overline{x} = 51.3$ centimeters per second. What conclusions should be drawn?

We may solve this problem by following the eight-step procedure outlined in Section 9-1.6. This results in

1. The parameter of interest is μ, the mean burning rate.
2. H_0: $\mu = 50$ centimeters per second
3. H_1: $\mu \neq 50$ centimeters per second
4. $\alpha = 0.05$

5. The test statistic is

$$z_0 = \frac{\bar{x} - \mu_0}{\sigma/\sqrt{n}}$$

6. Reject H_0 if $z_0 > 1.96$ or if $z_0 < -1.96$. Note that this results from step 4, where we specified $\alpha = 0.05$, and so the boundaries of the critical region are at $z_{0.025} = 1.96$ and $-z_{0.025} = -1.96$.

7. Computations: Since $\bar{x} = 51.3$ and $\sigma = 2$,

$$z_0 = \frac{51.3 - 50}{2/\sqrt{25}} = 3.25$$

8. Conclusion: Since $z_0 = 3.25 > 1.96$, we reject $H_0\!:\mu = 50$ at the 0.05 level of significance. Stated more completely, we conclude that the mean burning rate differs from 50 centimeters per second, based on a sample of 25 measurements. In fact, there is strong evidence that the mean burning rate exceeds 50 centimeters per second.

We may also develop procedures for testing hypotheses on the mean μ where the alternative hypothesis is one-sided. Suppose that we specify the hypotheses as

$$H_0\text{: } \mu = \mu_0 \qquad H_1\text{: } \mu > \mu_0 \tag{9-11}$$

In defining the critical region for this test, we observe that a negative value of the test statistic Z_0 would never lead us to conclude that H_0: $\mu = \mu_0$ is false. Therefore, we would place the critical region in the **upper tail** of the standard normal distribution and reject H_0 if the computed value of z_0 is too large. That is, we would reject H_0 if

$$z_0 > z_\alpha \tag{9-12}$$

as shown in Figure 9-7(b). Similarly, to test

$$H_0\text{: } \mu = \mu_0 \qquad H_1\text{: } \mu < \mu_0 \tag{9-13}$$

we would calculate the test statistic Z_0 and reject H_0 if the value of z_0 is too small. That is, the critical region is in the **lower tail** of the standard normal distribution as shown in Figure 9-7(c), and we reject H_0 if

$$z_0 < -z_\alpha \tag{9-14}$$

Tests on the Mean, Variance Known

Null hypothesis: H_0: $\mu = \mu_0$

Test statistic: $$Z_0 = \frac{\bar{X} - \mu_0}{\sigma/\sqrt{n}}$$

Alternative hypothesis	Rejection criteria
H_1: $\mu \neq \mu_0$	$z_0 > z_{\alpha/2,n-1}$ or $z_0 < -z_{\alpha/2,n-1}$
H_1: $\mu > \mu_0$	$z_0 > z_{\alpha,n-1}$
H_1: $\mu < \mu_0$	$z_0 < -z_{\alpha,n-1}$

P-Values

For the foregoing normal distribution tests it is relatively easy to compute the P-value. If z_0 is the computed value of the test statistic, the P-value is

$$P = \begin{cases} 2[1 - \Phi(|z_0|)] & \text{for a two-tailed test: } H_0\text{: } \mu = \mu_0 \quad H_1\text{: } \mu \neq \mu_0 \\ 1 - \Phi(z_0) & \text{for a upper-tailed test: } H_0\text{: } \mu = \mu_0 \quad H_1\text{: } \mu > \mu_0 \\ \Phi(z_0) & \text{for a lower-tailed test: } H_0\text{: } \mu = \mu_0 \quad H_1\text{: } \mu < \mu_0 \end{cases} \tag{9-15}$$

Here, $\Phi(z)$ is the standard normal cumulative distribution function defined in Chapter 4. Recall that $\Phi(z) = P(Z \leq z)$, where Z is $N(0, 1)$. To illustrate this, consider the propellant problem in Example 9-2. The computed value of the test statistic is $z_0 = 3.25$ and since the alternative hypothesis is two-tailed, the P-value is

$$P\text{-value} = 2[1 - \Phi(3.25)] = 0.0012$$

Thus, H_0: $\mu = 50$ would be rejected at any level of significance $\alpha \geq P\text{-value} = 0.0012$. For example, H_0 would be rejected if $\alpha = 0.01$, but it would not be rejected if $\alpha = 0.001$.

9-2.2 Type II Error and Choice of Sample Size

In testing hypotheses, the analyst directly selects the type I error probability. However, the probability of type II error β depends on the choice of sample size. In this section, we will show how to calculate the probability of type II error β. We will also show how to select the sample size to obtain a specified value of β.

Finding the Probability of Type II Error β

Consider the two-sided hypothesis

$$H_0\text{: } \mu = \mu_0$$
$$H_1\text{: } \mu \neq \mu_0$$

Suppose that the null hypothesis is false and that the true value of the mean is $\mu = \mu_0 + \delta$, say, where $\delta > 0$. The test statistic Z_0 is

$$Z_0 = \frac{\overline{X} - \mu_0}{\sigma/\sqrt{n}} = \frac{\overline{X} - (\mu_0 + \delta)}{\sigma/\sqrt{n}} + \frac{\delta\sqrt{n}}{\sigma}$$

Therefore, the distribution of Z_0 when H_1 is true is

$$Z_0 \sim N\left(\frac{\delta\sqrt{n}}{\sigma}, 1\right) \tag{9-16}$$

The distribution of the test statistic Z_0 under both the null hypothesis H_0 and the alternate hypothesis H_1 is shown in Fig. 9-8. From examining this figure, we note that if H_1 is true, a type II error will be made only if $-z_{\alpha/2} \leq Z_0 \leq z_{\alpha/2}$ where $Z_0 \sim N(\delta\sqrt{n}/\sigma, 1)$. That is, the probability of the type II error β is the probability that Z_0 falls between $-z_{\alpha/2}$ and $z_{\alpha/2}$ *given that H_1 is true*. This probability is shown as the shaded portion of Fig. 9-8. Expressed mathematically, this probability is

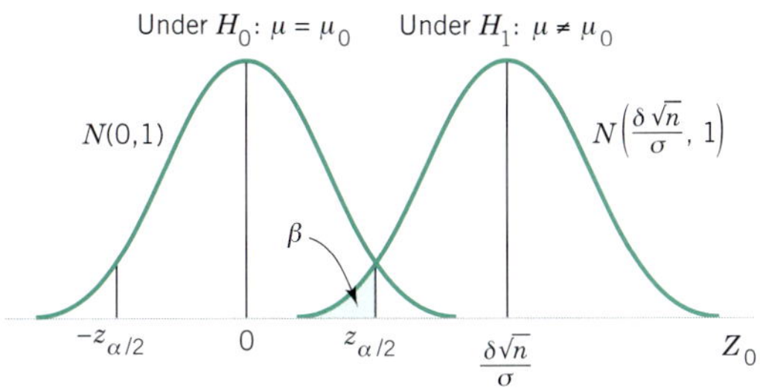

Figure 9-8 The distribution of Z_0 under H_0 and H_1.

Probability of a Type II Error for a Two-Sided Test on the Mean, Variance Known

$$\beta = \Phi\left(z_{\alpha/2} - \frac{\delta\sqrt{n}}{\sigma}\right) - \Phi\left(-z_{\alpha/2} - \frac{\delta\sqrt{n}}{\sigma}\right) \tag{9-17}$$

where $\Phi(z)$ denotes the probability to the left of z in the standard normal distribution. Note that Equation 9-17 was obtained by evaluating the probability that Z_0 falls in the interval $[-z_{\alpha/2}, z_{\alpha/2}]$ when H_1 is true. Furthermore, note that Equation 9-17 also holds if $\delta < 0$, due to the symmetry of the normal distribution. It is also possible to derive an equation similar to Equation 9-17 for a one-sided alternative hypothesis.

Sample Size Formulas

One may easily obtain formulas that determine the appropriate sample size to obtain a particular value of β for a given δ and α. For the two-sided alternative hypothesis, we know from Equation 9-17 that

$$\beta = \Phi\left(z_{\alpha/2} - \frac{\delta\sqrt{n}}{\sigma}\right) - \Phi\left(-z_{\alpha/2} - \frac{\delta\sqrt{n}}{\sigma}\right)$$

or if $\delta > 0$,

$$\beta \simeq \Phi\left(z_{\alpha/2} - \frac{\delta\sqrt{n}}{\sigma}\right) \tag{9-18}$$

since $\Phi(-z_{\alpha/2} - \delta\sqrt{n}/\sigma) \simeq 0$ when δ is positive. Let z_β be the 100β upper percentile of the standard normal distribution. Then, $\beta = \Phi(-z_\beta)$. From Equation 9-18

$$-z_\beta \simeq z_{\alpha/2} - \frac{\delta\sqrt{n}}{\sigma}$$

or

Sample Size for a Two-Sided Test on the Mean, Variance Known

$$n \simeq \frac{(z_{\alpha/2} + z_\beta)^2 \sigma^2}{\delta^2} \quad \text{where} \quad \delta = \mu - \mu_0 \tag{9-19}$$

If n is not an integer, the convention is to round the sample size up to the next integer. This approximation is good when $\Phi(-z_{\alpha/2} - \delta\sqrt{n}/\sigma)$ is small compared to β. For either of the one-sided alternative hypotheses the sample size required to produce a specified type II error with probability β given δ and α is

Sample Size for a One-Sided Test on the Mean, Variance Known

$$n = \frac{(z_\alpha + z_\beta)^2 \sigma^2}{\delta^2} \quad \text{where} \quad \delta = \mu - \mu_0 \qquad \text{(9-20)}$$

EXAMPLE 9-3 Propellant Burning Rate Type II Error

Consider the rocket propellant problem of Example 9-2. Suppose that the true burning rate is 49 centimeters per second. What is β for the two-sided test with $\alpha = 0.05$, $\sigma = 2$, and $n = 25$?

Here $\delta = 1$ and $z_{\alpha/2} = 1.96$. From Equation 9-17

$$\beta = \Phi\left(1.96 - \frac{\sqrt{25}}{\sigma}\right) - \Phi\left(-1.96 - \frac{\sqrt{25}}{\sigma}\right) = \Phi(-0.54) - \Phi(-4.46) = 0.295$$

The probability is about 0.3 that this difference from 50 centimeters per second will not be detected. That is, the probability is about 0.3 that the test will fail to reject the null hypothesis when the true burning rate is 49 centimeters per second.

Suppose that the analyst wishes to design the test so that if the true mean burning rate differs from 50 centimeters per second by as much as 1 centimeter per second, the test will detect this (i.e., reject H_0: $\mu = 50$) with a high probability, say 0.90. Now, we note that $\sigma = 2$, $\delta = 51 - 50 = 1$, $\alpha = 0.05$, and $\beta = 0.10$. Since $z_{\alpha/2} = z_{0.025} = 1.96$ and $z_\beta = z_{0.10} = 1.28$, the sample size required to detect this departure from H_0: $\mu = 50$ is found by Equation 9-19 as

$$n \simeq \frac{(z_{\alpha/2} + z_\beta)^2 \sigma^2}{\delta^2} = \frac{(1.96 + 1.28)^2 2^2}{(1)^2} \simeq 42$$

The approximation is good here, since $\Phi(-z_{\alpha/2} - \delta\sqrt{n}/\sigma) = \Phi(-1.96 - (1)\sqrt{42}/2) = \Phi(-5.20) \simeq 0$, which is small relative to β.

Using Operating Characteristic Curves

When performing sample size or type II error calculations, it is sometimes more convenient to use the **operating characteristic (OC) curves** in Appendix Charts VI*a* and VI*b*. These curves plot β as calculated from Equation 9-17 against a parameter d for various sample sizes n. Curves are provided for both $\alpha = 0.05$ and $\alpha = 0.01$. The parameter d is defined as

$$d = \frac{|\mu - \mu_0|}{\sigma} = \frac{|\delta|}{\sigma} \qquad \text{(9-21)}$$

so one set of operating characteristic curves can be used for all problems regardless of the values of μ_0 and σ. From examining the operating characteristic curves or Equation 9-17 and Fig. 9-8, we note that

1. The further the true value of the mean μ is from μ_0, the smaller the probability of type II error β for a given n and α. That is, we see that for a specified sample size and α, large differences in the mean are easier to detect than small ones.

2. For a given δ and α, the probability of type II error β decreases as n increases. That is, to detect a specified difference δ in the mean, we may make the test more powerful by increasing the sample size.

EXAMPLE 9-4
Propellant Burning Rate Type II Error from OC Curve

Consider the propellant problem in Example 9-2. Suppose that the analyst is concerned about the probability of type II error if the true mean burning rate is $\mu = 51$ centimeters per second. We may use the operating characteristic curves to find β. Note that $\delta = 51 - 50 = 1$, $n = 25$, $\sigma = 2$, and $\alpha = 0.05$. Then using Equation 9-21 gives

$$d = \frac{|\mu - \mu_0|}{\sigma} = \frac{|\delta|}{\sigma} = \frac{1}{2}$$

and from Appendix Chart VII*a*, with $n = 25$, we find that $\beta = 0.30$. That is, if the true mean burning rate is $\mu = 51$ centimeters per second, there is approximately a 30% chance that this will not be detected by the test with $n = 25$.

EXAMPLE 9-5
Propellant Burning Rate Sample Size from OC Curve

Once again, consider the propellant problem in Example 9-2. Suppose that the analyst would like to design the test so that if the true mean burning rate differs from 50 centimeters per second by as much as 1 centimeter per second, the test will detect this (i.e., reject H_0: $\mu = 50$) with a high probability, say, 0.90. This is exactly the same requirement as in Example 9-3, where we used Equation 9-19 to find the required sample size to be $n = 42$. The operating characteristic curves can also be used to find the sample size for this test. Since $d = |\mu - \mu_0|/\sigma = 1/2$, $\alpha = 0.05$, and $\beta = 0.10$, we find from Appendix Chart VII*a* that the required sample size is approximately $n = 40$. This closely agrees with the sample size calculated from Equation 9-19.

In general, the operating characteristic curves involve three parameters: β, d, and n. Given any two of these parameters, the value of the third can be determined. There are two typical applications of these curves:

1. For a given n and d, find β (as illustrated in Example 9-4). This kind of problem is often encountered when the analyst is concerned about the sensitivity of an experiment already performed, or when sample size is restricted by economic or other factors.
2. For a given β and d, find n. This was illustrated in Example 9-5. This kind of problem is usually encountered when the analyst has the opportunity to select the sample size at the outset of the experiment.

Operating characteristic curves are given in Appendix Charts VII*c* and VII*d* for the one-sided alternatives. If the alternative hypothesis is either H_1: $\mu > \mu_0$ or H_1: $\mu < \mu_0$, the abscissa scale on these charts is

$$d = \frac{|\mu - \mu_0|}{\sigma} \tag{9-22}$$

Using the Computer

Many statistics software packages will calculate sample sizes and type II error probabilities. To illustrate, here are some computations from Minitab for the propellant burning rate problem.

Minitab Computations

Power and Sample Size

1-Sample Z Test
Testing mean = null (versus not = null)
Calculating power for mean = null + difference
Alpha = 0.05 Sigma = 2

Difference	Sample Size	Target Power	Actual Power
1	43	0.9000	0.9064

Power and Sample Size

1-Sample Z Test
Testing mean = null (versus not = null)
Calculating power for mean = null + difference
Alpha = 0.05 Sigma = 2

Difference	Sample Size	Target Power	Actual Power
1	28	0.7500	0.7536

Power and Sample Size

1-Sample Z Test
Testing mean = null (versus not = null)
Calculating power for mean = null + difference
Alpha = 0.05 Sigma = 2

Difference	Sample Size	Power
1	25	0.7054

In the first part of the boxed display, we asked Minitab to work Example 9-3, that is, to find the sample size n that would allow detection of a difference from $\mu_0 = 50$ of 1 centimeter per second with power of 0.9 and $\alpha = 0.05$. The answer, $n = 43$, agrees closely with the calculated value from Equation 9-19 in Example 9-3, which was $n = 42$. The difference is due to Minitab using a value of z_β that has more than two decimal places. The second part of the computer output relaxes the power requirement to 0.75. Note that the effect is to reduce the required sample size to $n = 28$. The third part of the output is the solution to Example 9-4, where we wish to determine the type II error probability of (β) or the power $= 1 - \beta$ for the sample size $n = 25$. Note that Minitab computes the power to be 0.7054, which agrees closely with the answer obtained from the OC curve in Example 9-4. Generally, however, the computer calculations will be more accurate than visually reading values from an OC curve.

9-2.3 Large-Sample Test

We have developed the test procedure for the null hypothesis H_0: $\mu = \mu_0$ assuming that the population is normally distributed and that σ^2 is known. In many if not most practical situations σ^2 will be unknown. Furthermore, we may not be certain that the population is well modeled by a normal distribution. In these situations if n is large (say $n > 40$) the sample standard deviation s can be substituted for σ in the test procedures with little effect. Thus, while we have given a test for the mean of a normal distribution with known σ^2, it can be easily converted into a **large-sample test procedure for unknown σ^2** that is valid regardless of the form of the distribution of the population. This large-sample test relies on the central limit theorem just as the large-sample confidence interval on μ that was presented in the previous chapter did. Exact treatment of the case where the population is normal, σ^2 is unknown, and n is small involves use of the t distribution and will be deferred until Section 9-3.

EXERCISES FOR SECTION 9-2

9-29. State the null and alternative hypothesis in each case.

(a) A hypothesis test will be used to potentially provide evidence that the population mean is greater than 10.

(b) A hypothesis test will be used to potentially provide evidence that the population mean is not equal to 7.

(c) A hypothesis test will be used to potentially provide evidence that the population mean is less than 5.

9-30. A hypothesis will be used to test that a population mean equals 7 against the alternative that the population mean does not equal 7 with known variance σ. What are the critical values for the test statistic Z_0 for the following significance levels?

(a) 0.01 (b) 0.05 (c) 0.10

9-31. A hypothesis will be used to test that a population mean equals 10 against the alternative that the population mean is greater than 10 with known variance σ. What is the critical value for the test statistic Z_0 for the following significance levels?

(a) 0.01 (b) 0.05 (c) 0.10

9-32. A hypothesis will be used to test that a population mean equals 5 against the alternative that the population mean is less than 5 with known variance σ. What is the critical value for the test statistic Z_0 for the following significance levels?

(a) 0.01 (b) 0.05 (c) 0.10

9-33. For the hypothesis test H_0: $\mu = 7$ against H_1: $\mu \neq 7$ and variance known, calculate the P-value for each of the following test statistics.

(a) $z_0 = 2.05$ (b) $z_0 = -1.84$ (c) $z_0 = 0.4$

9-34. For the hypothesis test H_0: $\mu = 10$ against H_1: $\mu > 10$ and variance known, calculate the P-value for each of the following test statistics.

(a) $z_0 = 2.05$ (b) $z_0 = -1.84$ (c) $z_0 = 0.4$

9-35. For the hypothesis test H_0: $\mu = 5$ against H_1: $\mu < 5$ and variance known, calculate the P-value for each of the following test statistics.

(a) $z_0 = 2.05$ (b) $z_0 = -1.84$ (c) $z_0 = 0.4$

9-36. The mean water temperature downstream from a power plant cooling tower discharge pipe should be no more than 100°F. Past experience has indicated that the standard deviation of temperature is 2°F. The water temperature is measured on nine randomly chosen days, and the average temperature is found to be 98°F.

(a) Is there evidence that the water temperature is acceptable at $\alpha = 0.05$?

(b) What is the P-value for this test?

(c) What is the probability of accepting the null hypothesis at $\alpha = 0.05$ if the water has a true mean temperature of 104°F?

9-37. A manufacturer produces crankshafts for an automobile engine. The wear of the crankshaft after 100,000 miles (0.0001 inch) is of interest because it is likely to have an impact on warranty claims. A random sample of $n = 15$ shafts is tested and $\bar{x} = 2.78$. It is known that $\sigma = 0.9$ and that wear is normally distributed.

(a) Test H_0: $\mu = 3$ versus H_1: $\mu \neq 3$ using $\alpha = 0.05$.

(b) What is the power of this test if $\mu = 3.25$?

(c) What sample size would be required to detect a true mean of 3.75 if we wanted the power to be at least 0.9?

9-38. A melting point test of $n = 10$ samples of a binder used in manufacturing a rocket propellant resulted in $\bar{x} = 154.2$°F. Assume that melting point is normally distributed with $\sigma = 1.5$°F.

(a) Test H_0: $\mu = 155$ versus H_1: $\mu \neq 155$ using $\alpha = 0.01$.

(b) What is the P-value for this test?

(c) What is the β-error if the true mean is $\mu = 150$?

(d) What value of n would be required if we want $\beta < 0.1$ when $\mu = 150$? Assume that $\alpha = 0.01$.

9-39. The life in hours of a battery is known to be approximately normally distributed, with standard deviation $\sigma = 1.25$ hours. A random sample of 10 batteries has a mean life of $\bar{x} = 40.5$ hours.

(a) Is there evidence to support the claim that battery life exceeds 40 hours? Use $\alpha = 0.05$.

(b) What is the P-value for the test in part (a)?

(c) What is the β-error for the test in part (a) if the true mean life is 42 hours?

(d) What sample size would be required to ensure that β does not exceed 0.10 if the true mean life is 44 hours?

(e) Explain how you could answer the question in part (a) by calculating an appropriate confidence bound on life.

9-40. An engineer who is studying the tensile strength of a steel alloy intended for use in golf club shafts knows that tensile strength is approximately normally distributed with $\sigma = 60$ psi. A random sample of 12 specimens has a mean tensile strength of $\bar{x} = 3450$ psi.

(a) Test the hypothesis that mean strength is 3500 psi. Use $\alpha = 0.01$.

(b) What is the smallest level of significance at which you would be willing to reject the null hypothesis?
(c) What is the β-error for the test in part (a) if the true mean is 3470?
(d) Suppose that we wanted to reject the null hypothesis with probability at least 0.8 if mean strength $\mu = 3500$. What sample size should be used?
(e) Explain how you could answer the question in part (a) with a two-sided confidence interval on mean tensile strength.

9-41. Supercavitation is a propulsion technology for undersea vehicles that can greatly increase their speed. It occurs above approximately 50 meters per second, when pressure drops sufficiently to allow the water to dissociate into water vapor, forming a gas bubble behind the vehicle. When the gas bubble completely encloses the vehicle, supercavitation is said to occur. Eight tests were conducted on a scale model of an undersea vehicle in a towing basin with the average observed speed $\bar{x} = 102.2$ meters per second. Assume that speed is normally distributed with known standard deviation $\sigma = 4$ meters per second.
(a) Test the hypotheses H_0: $\mu = 100$ versus H_1: $\mu < 100$ using $\alpha = 0.05$.
(b) What is the P-value for the test in part (a)?
(c) Compute the power of the test if the true mean speed is as low as 95 meters per second.
(d) What sample size would be required to detect a true mean speed as low as 95 meters per second if we wanted the power of the test to be at least 0.85?
(e) Explain how the question in part (a) could be answered by constructing a one-sided confidence bound on the mean speed.

9-42. A bearing used in an automotive application is suppose to have a nominal inside diameter of 1.5 inches. A random sample of 25 bearings is selected and the average inside diameter of these bearings is 1.4975 inches. Bearing diameter is known to be normally distributed with standard deviation $\sigma = 0.01$ inch.
(a) Test the hypotheses H_0: $\mu = 1.5$ versus H_1: $\mu \neq 1.5$ using $\alpha = 0.01$.
(b) What is the P-value for the test in part (a)?
(c) Compute the power of the test if the true mean diameter is 1.495 inches.
(d) What sample size would be required to detect a true mean diameter as low as 1.495 inches if we wanted the power of the test to be at least 0.9?
(e) Explain how the question in part (a) could be answered by constructing a two-sided confidence interval on the mean diameter.

9-43. Medical researchers have developed a new artificial heart constructed primarily of titanium and plastic. The heart will last and operate almost indefinitely once it is implanted in the patient's body, but the battery pack needs to be recharged about every four hours. A random sample of 50 battery packs is selected and subjected to a life test. The average life of these batteries is 4.05 hours. Assume that battery life is normally distributed with standard deviation $\sigma = 0.2$ hour.
(a) Is there evidence to support the claim that mean battery life exceeds 4 hours? Use $\alpha = 0.05$.
(b) What is the P-value for the test in part (a)?
(c) Compute the power of the test if the true mean battery life is 4.5 hours.
(d) What sample size would be required to detect a true mean battery life of 4.5 hours if we wanted the power of the test to be at least 0.9?
(e) Explain how the question in part (a) could be answered by constructing a one-sided confidence bound on the mean life.

9-3 TESTS ON THE MEAN OF A NORMAL DISTRIBUTION, VARIANCE UNKNOWN

9-3.1 Hypothesis Tests on the Mean

We now consider the case of **hypothesis testing** on the mean of a population with **unknown variance** σ^2. The situation is analogous to Section 8-3, where we considered a **confidence interval** on the mean for the same situation. As in that section, the validity of the test procedure we will describe rests on the assumption that the population distribution is at least approximately normal. The important result upon which the test procedure relies is that if $X_1, X_2, \ldots, X_n$ is a random sample from a normal distribution with mean μ and variance σ^2, the random variable

$$T = \frac{\overline{X} - \mu}{S/\sqrt{n}}$$

has a t distribution with $n - 1$ degrees of freedom. Recall that we used this result in Section 8-3 to devise the t-confidence interval for μ. Now consider testing the hypotheses

$$H_0\colon \mu = \mu_0$$
$$H_1\colon \mu \neq \mu_0$$

We will use the **test statistic**

$$T_0 = \frac{\overline{X} - \mu_0}{S/\sqrt{n}} \tag{9-23}$$

If the null hypothesis is true, T_0 has a t distribution with $n - 1$ degrees of freedom. When we know the distribution of the test statistic when H_0 is true (this is often called the **reference distribution** or the **null distribution**), we can locate the critical region to control the type I error probability at the desired level. In this case we would use the t percentage points $-t_{\alpha/2,n-1}$ and $t_{\alpha/2,n-1}$ as the boundaries of the critical region so that we would reject $H_0\colon \mu = \mu_0$ if

$$t_0 > t_{\alpha/2,n-1} \qquad \text{or if} \quad t_0 < -t_{\alpha/2,n-1}$$

where t_0 is the observed value of the test statistic T_0. The test procedure is very similar to the test on the mean with known variance described in Section 9-2, except that T_0 is used as the test statistic instead of Z_0 and the t_{n-1} distribution is used to define the critical region instead of the standard normal distribution. A summary of the test procedures for both two- and one-sided alternative hypotheses follows:

Tests on the Mean, Variance Unknown

Null hypothesis: $H_0\colon \mu = \mu_0$

Test statistic: $T_0 = \dfrac{\overline{X} - \mu_0}{S/\sqrt{n}}$

Alternative hypothesis	Rejection criteria
$H_1\colon \mu \neq \mu_0$	$t_0 > t_{\alpha/2,n-1}$ or $t_0 < -t_{\alpha/2,n-1}$
$H_1\colon \mu > \mu_0$	$t_0 > t_{\alpha,n-1}$
$H_1\colon \mu < \mu_0$	$t_0 < -t_{\alpha,n-1}$

Figure 9-9 shows the location of the critical region for these situations.

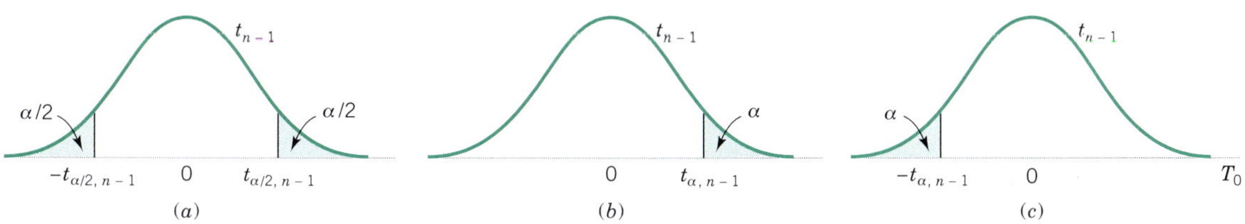

Figure 9-9 The reference distribution for $H_0\colon \mu = \mu_0$ with critical region for (a) $H_1\colon \mu \neq \mu_0$, (b) $H_1\colon \mu > \mu_0$, and (c) $H_1\colon \mu < \mu_0$.

EXAMPLE 9-6
Golf Club Design

The increased availability of light materials with high strength has revolutionized the design and manufacture of golf clubs, particularly drivers. Clubs with hollow heads and very thin faces can result in much longer tee shots, especially for players of modest skills. This is due partly to the "spring-like effect" that the thin face imparts to the ball. Firing a golf ball at the head of the club and measuring the ratio of the outgoing velocity of the ball to the incoming velocity can quantify this spring-like effect. The ratio of velocities is called the coefficient of restitution of the club. An experiment was performed in which 15 drivers produced by a particular club maker were selected at random and their coefficients of restitution measured. In the experiment the golf balls were fired from an air cannon so that the incoming velocity and spin rate of the ball could be precisely controlled. It is of interest to determine if there is evidence (with $\alpha = 0.05$) to support a claim that the mean coefficient of restitution exceeds 0.82. The observations follow:

0.8411	0.8191	0.8182	0.8125	0.8750
0.8580	0.8532	0.8483	0.8276	0.7983
0.8042	0.8730	0.8282	0.8359	0.8660

The sample mean and sample standard deviation are $\bar{x} = 0.83725$ and $s = 0.02456$. The normal probability plot of the data in Fig. 9-10 supports the assumption that the coefficient of restitution is normally distributed. Since the objective of the experimenter is to demonstrate that the mean coefficient of restitution exceeds 0.82, a one-sided alternative hypothesis is appropriate.

The solution using the eight-step procedure for hypothesis testing is as follows:

1. The parameter of interest is the mean coefficient of restitution, μ.
2. H_0: $\mu = 0.82$
3. H_1: $\mu > 0.82$. We want to reject H_0 if the mean coefficient of restitution exceeds 0.82.
4. $\alpha = 0.05$
5. The test statistic is

$$t_0 = \frac{\bar{x} - \mu_0}{s/\sqrt{n}}$$

6. Reject H_0 if $t_0 > t_{0.05,14} = 1.761$

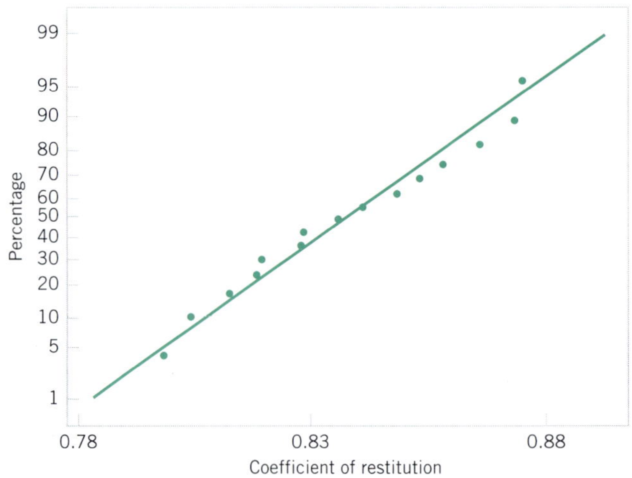

Figure 9-10. Normal probability plot of the coefficient of restitution data from Example 9-6.

7. Computations: Since $\bar{x} = 0.83725$, $s = 0.02456$, $\mu_0 = 0.82$, and $n = 15$, we have

$$t_0 = \frac{0.83725 - 0.82}{0.02456/\sqrt{15}} = 2.72$$

8. Conclusions: Since $t_0 = 2.72 > 1.761$, we reject H_0 and conclude at the 0.05 level of significance that the mean coefficient of restitution exceeds 0.82.

Minitab will conduct the one-sample t-test. The output from this software package is in the following display:

Minitab Computations

One-Sample T: COR

Test of mu = 0.82 vs mu > 0.82

Variable	N	Mean	StDev	SE Mean
COR	15	0.83725	0.02456	0.00634

Variable	95.0% Lower Bound	T	P
COR	0.82608	2.72	0.008

Notice that Minitab computes both the test statistic T_0 and a 95% lower confidence bound for the coefficient of restitution. Because the 95% lower confidence bound exceeds 0.82, we would reject the hypothesis that H_0: $\mu = 0.82$ and conclude that the alternative hypothesis H_1: $\mu > 0.82$ is true. Minitab also calculates a P-value for the test statistic T_0. In the next section we explain how this is done.

9-3.2 *P*-Value for a *t*-Test

The P-value for a t-test is just the smallest level of significance at which the null hypothesis would be rejected. That is, it is the tail area beyond the value of the test statistic t_0 for a one-sided test or twice this area for a two-sided test. Because the t-table in Appendix A contains only 10 critical values for each t distribution, computation of the exact P-value directly from the table is usually impossible. However, it is easy to find upper and lower bounds on the P value from this table.

To illustrate, consider the t-test based on 14 degrees of freedom in Example 9-6. The relevant critical values from Appendix Table IV are as follows:

Critical Value:	0.258	0.692	1.345	1.761	2.145	2.624	2.977	3.326	3.787	4.140
Tail Area:	0.40	0.25	0.10	0.05	0.025	0.01	0.005	0.0025	0.001	0.0005

Notice that $t_0 = 2.72$ in Example 9-6, and that this is between two tabulated values, 2.624 and 2.977. Therefore, the P-value must be between 0.01 and 0.005. These are effectively the upper and lower bounds on the P-value.

Example 9-6 is an upper-tailed test. If the test is lower-tailed, just change the sign of t_0 and proceed as above. Remember that for a two-tailed test the level of significance associated with a particular critical value is twice the corresponding tail area in the column heading. This fact must be taken into account when we compute the bound on the P-value. For example, suppose that $t_0 = 2.72$ for a two-tailed alternate based on 14 degrees of freedom. The value $2.624 < t_0 < 2.977$, so the lower and upper bounds on the P-value would be $0.01 = 2(0.005) < P\text{-value} < 0.02 = 2(0.01)$ for this case.

Finally, most computer programs report P-values along with the computed value of the test statistic. Some hand-held calculators also have this capability. In Example 9-6, Minitab gave the P-value for the value $t_0 = 2.72$ in Example 9-6 as 0.008.

9-3.3 Type II Error and Choice of Sample Size

The type II error probability for tests on the mean of a normal distribution with unknown variance depends on the distribution of the test statistic in Equation 9-23 when the null hypothesis H_0: $\mu = \mu_0$ is false. When the true value of the mean is $\mu = \mu_0 + \delta$, the distribution for T_0 is called the **noncentral *t* distribution** with $n - 1$ degrees of freedom and noncentrality parameter $\delta\sqrt{n}/\sigma$. Note that if $\delta = 0$, the noncentral t distribution reduces to the usual **central *t* distribution.** Therefore, the type II error of the two-sided alternative (for example) would be

$$\begin{aligned}\beta &= P\{-t_{\alpha/2,n-1} \le T_0 \le t_{\alpha/2,n-1} \mid \delta \ne 0\}\\ &= P\{-t_{\alpha/2,n-1} \le T_0' \le t_{\alpha/2,n-1}\}\end{aligned}$$

where T_0' denotes the noncentral t random variable. Finding the type II error probability β for the t-test involves finding the probability contained between two points of the noncentral t distribution. Because the noncentral t-random variable has a messy density function, this integration must be done numerically.

Fortunately, this ugly task has already been done, and the results are summarized in a series of O.C. curves in Appendix Charts VII*e*, VII*f*, VII*g*, and VII*h* that plot β for the t-test against a parameter d for various sample sizes n. Curves are provided for two-sided alternatives on Charts VII*e* and VII*f*. The abscissa scale factor d on these charts is defined as

$$d = \frac{|\mu - \mu_0|}{\sigma} = \frac{|\delta|}{\sigma} \tag{9-24}$$

For the one-sided alternative $\mu > \mu_0$ or $\mu < \mu_0$, we use charts VIG and VIH with

$$d = \frac{|\mu - \mu_0|}{\sigma} = \frac{|\delta|}{\sigma} \tag{9-25}$$

We note that d depends on the unknown parameter σ^2. We can avoid this difficulty in several ways. In some cases, we may use the results of a previous experiment or prior information to make a rough initial estimate of σ^2. If we are interested in evaluating test performance after the data have been collected, we could use the sample variance s^2 to estimate σ^2. If there is no previous experience on which to draw in estimating σ^2, we then define the difference in the mean d that we wish to detect relative to σ. For example, if we wish to detect a small difference in the mean, we might use a value of $d = |\delta|/\sigma \le 1$ (for example), whereas if we are interested in detecting only moderately large differences in the mean, we might select $d = |\delta|/\sigma = 2$ (for example). That is, it is the value of the ratio $|\delta|/\sigma$ that is important in determining sample size, and if it is possible to specify the relative size of the difference in means that we are interested in detecting, then a proper value of d can usually be selected.

EXAMPLE 9-7
Golf Club Design Sample Size

Consider the golf club testing problem from Example 9-6. If the mean coefficient of restitution exceeds 0.82 by as much as 0.02, is the sample size $n = 15$ adequate to ensure that H_0: $\mu = 0.82$ will be rejected with probability at least 0.8?

To solve this problem, we will use the sample standard deviation $s = 0.02456$ to estimate σ. Then $d = |\delta|/\sigma = 0.02/0.02456 = 0.81$. By referring to the operating characteristic curves in Appendix Chart VII*g* (for $\alpha = 0.05$) with $d = 0.81$ and $n = 15$, we find that $\beta = 0.10$, approximately. Thus, the probability of rejecting H_0: $\mu = 0.82$ if the true mean exceeds this by 0.02 is approximately $1 - \beta = 1 - 0.10 = 0.90$, and we conclude that a sample size of $n = 15$ is adequate to provide the desired sensitivity.

Minitab will also perform power and sample size computations for the one-sample *t*-test. Below are several calculations based on the golf club testing problem:

Minitab Computations

Power and Sample Size

1-Sample t Test
Testing mean = null (versus > null)
Calculating power for mean = null + difference
Alpha = 0.05 Sigma = 0.02456

Difference	Sample Size	Power
0.02	15	0.9117

Power and Sample Size

1-Sample t Test
Testing mean = null (versus > null)
Calculating power for mean = null + difference
Alpha = 0.05 Sigma = 0.02456

Difference	Sample Size	Power
0.01	15	0.4425

Power and Sample Size

1-Sample t Test
Testing mean = null (versus > null)
Calculating power for mean = null + difference
Alpha = 0.05 Sigma = 0.02456

Difference	Sample Size	Target Power	Actual Power
0.01	39	0.8000	0.8029

In the first portion of the computer output, Minitab reproduces the solution to Example 9-7, verifying that a sample size of $n = 15$ is adequate to give power of at least 0.8 if the mean coefficient of restitution exceeds 0.82 by at least 0.02. In the middle section of the output, we used Minitab to compute the power to detect a difference between μ and $\mu_0 = 0.82$ of 0.01. Notice that with $n = 15$, the power drops considerably to 0.4425. The final portion of the output is the sample size required for a power of at least 0.8 if the difference between μ and μ_0 of interest is actually 0.01. A much larger n is required to detect this smaller difference.

EXERCISES FOR SECTION 9-3

9-44. A hypothesis will be used to test that a population mean equals 7 against the alternative that the population mean does not equal 7 with unknown variance σ. What are the critical values for the test statistic T_0 for the following significance levels and sample sizes?
(a) $\alpha = 0.01$ and $n = 20$
(b) $\alpha = 0.05$ and $n = 12$
(c) $\alpha = 0.10$ and $n = 15$

9-45. A hypothesis will be used to test that a population mean equals 10 against the alternative that the population mean is greater than 10 with known variance σ. What is the critical value for the test statistic Z_0 for the following significance levels?
(a) $\alpha = 0.01$ and $n = 20$
(b) $\alpha = 0.05$ and $n = 12$
(c) $\alpha = 0.10$ and $n = 15$

9-46. A hypothesis will be used to test that a population mean equals 5 against the alternative that the population mean is less than 5 with known variance σ. What is the critical value for the test statistic Z_0 for the following significance levels?
(a) $\alpha = 0.01$ and $n = 20$
(b) $\alpha = 0.05$ and $n = 12$
(c) $\alpha = 0.10$ and $n = 15$

9-47. For the hypothesis test H_0: $\mu = 7$ against H_1: $\mu \neq 7$ with variance unknown and $n = 20$, approximate the P-value for each of the following test statistics.
(a) $t_0 = 2.05$ (b) $t_0 = -1.84$ (c) $t_0 = 0.4$

9-48. For the hypothesis test H_0: $\mu = 10$ against H_1: $\mu > 10$ with variance unknown and $n = 15$, approximate the P-value for each of the following test statistics.
(a) $t_0 = 2.05$ (b) $t_0 = -1.84$ (c) $t_0 = 0.4$

9-49. For the hypothesis test H_0: $\mu = 5$ against H_1: $\mu < 5$ with variance unknown and $n = 12$, approximate the P-value for each of the following test statistics.
(a) $t_0 = 2.05$ (b) $t_0 = -1.84$ (c) $t_0 = 0.4$

9-50. An article in the *ASCE Journal of Energy Engineering* (1999, Vol. 125, pp. 59–75) describes a study of the thermal inertia properties of autoclaved aerated concrete used as a building material. Five samples of the material were tested in a structure, and the average interior temperature (°C) reported was as follows: 23.01, 22.22, 22.04, 22.62, and 22.59.
(a) Test the hypotheses H_0: $\mu = 22.5$ versus H_1: $\mu \neq 22.5$, using $\alpha = 0.05$. Find the P-value.
(b) Check the assumption that interior temperature is normally distributed.
(c) Compute the power of the test if the true mean interior temperature is as high as 22.75.
(d) What sample size would be required to detect a true mean interior temperature as high as 22.75 if we wanted the power of the test to be at least 0.9?
(e) Explain how the question in part (a) could be answered by constructing a two-sided confidence interval on the mean interior temperature.

9-51. A 1992 article in the *Journal of the American Medical Association* ("A Critical Appraisal of 98.6 Degrees F, the Upper Limit of the Normal Body Temperature, and Other Legacies of Carl Reinhold August Wundrlich") reported body temperature, gender, and heart rate for a number of subjects. The body temperatures for 25 female subjects follow: 97.8, 97.2, 97.4, 97.6, 97.8, 97.9, 98.0, 98.0, 98.0, 98.1, 98.2, 98.3, 98.3, 98.4, 98.4, 98.4, 98.5, 98.6, 98.6, 98.7, 98.8, 98.8, 98.9, 98.9, and 99.0.
(a) Test the hypotheses H_0: $\mu = 98.6$ versus H_1: $\mu \neq 98.6$, using $\alpha = 0.05$. Find the P-value.
(b) Check the assumption that female body temperature is normally distributed.
(c) Compute the power of the test if the true mean female body temperature is as low as 98.0.
(d) What sample size would be required to detect a true mean female body temperature as low as 98.2 if we wanted the power of the test to be at least 0.9?
(e) Explain how the question in part (a) could be answered by constructing a two-sided confidence interval on the mean female body temperature.

9-52. Cloud seeding has been studied for many decades as a weather modification procedure (for an interesting study of this subject, see the article in *Technometrics*, "A Bayesian Analysis of a Multiplicative Treatment Effect in Weather Modification", Vol. 17, pp. 161–166). The rainfall in acre-feet from 20 clouds that were selected at random and seeded with silver nitrate follows: 18.0, 30.7, 19.8, 27.1, 22.3, 18.8, 31.8, 23.4, 21.2, 27.9, 31.9, 27.1, 25.0, 24.7, 26.9, 21.8, 29.2, 34.8, 26.7, and 31.6.
(a) Can you support a claim that mean rainfall from seeded clouds exceeds 25 acre-feet? Use $\alpha = 0.01$. Find the P-value.
(b) Check that rainfall is normally distributed.
(c) Compute the power of the test if the true mean rainfall is 27 acre-feet.
(d) What sample size would be required to detect a true mean rainfall of 27.5 acre-feet if we wanted the power of the test to be at least 0.9?
(e) Explain how the question in part (a) could be answered by constructing a one-sided confidence bound on the mean diameter.

9-53. The sodium content of thirty 300-gram boxes of organic corn flakes was determined. The data (in milligrams) are as follows: 131.15, 130.69, 130.91, 129.54, 129.64, 128.77, 130.72, 128.33, 128.24, 129.65, 130.14, 129.29, 128.71, 129.00, 129.39, 130.42, 129.53, 130.12, 129.78, 130.92, 131.15, 130.69, 130.91, 129.54, 129.64, 128.77, 130.72, 128.33, 128.24, and 129.65.

(a) Can you support a claim that mean sodium content of this brand of cornflakes differs from 130 milligrams? Use $\alpha = 0.05$. Find the P-value.
(b) Check that sodium content is normally distributed.
(c) Compute the power of the test if the true mean sodium content is 130.5 milligrams.
(d) What sample size would be required to detect a true mean sodium content of 130.1 milligrams if we wanted the power of the test to be at least 0.75?
(e) Explain how the question in part (a) could be answered by constructing a two-sided confidence interval on the mean sodium content.

9-54. Consider the baseball coefficient of restitution data first presented in Exercise 8-86.
(a) Does the data support the claim that the mean coefficient of restitution of baseballs exceeds 0.635? Use $\alpha = 0.05$. Find the P-value.
(b) Check the normality assumption.
(c) Compute the power of the test if the true mean coefficient of restitution is as high as 0.64.
(d) What sample size would be required to detect a true mean coefficient of restitution as high as 0.64 if we wanted the power of the test to be at least 0.75?
(e) Explain how the question in part (a) could be answered with a confidence interval.

9-55. Consider the dissolved oxygen concentration at TVA dams first presented in Exercise 8-88.
(a) Test the hypotheses H_0: $\mu = 4$ versus H_1: $\mu \neq 4$. Use $\alpha = 0.01$. Find the P-value.
(b) Check the normality assumption.
(c) Compute the power of the test if the true mean dissolved oxygen concentration is as low as 3.
(d) What sample size would be required to detect a true mean dissolved oxygen concentration as low as 2.5 if we wanted the power of the test to be at least 0.9?
(e) Explain how the question in part (a) could be answered with a confidence interval.

9-56. Reconsider the data from *Medicine and Science in Sports and Exercise* described in Exercise 8-26. The sample size was seven and the sample mean and sample standard deviation were 315 watts and 16 watts, respectively.
(a) Is there evidence that leg strength exceeds 300 watts at significance level 0.05? Find the P-value.
(b) Compute the power of the test if the true strength is 305 watts.
(c) What sample size would be required to detect a true mean of 305 watts if the power of the test should be at least 0.90?
(d) Explain how the question in part (a) could be answered with a confidence interval.

9-57. Reconsider the tire testing experiment described in Exercise 8-23.
(a) The engineer would like to demonstrate that the mean life of this new tire is in excess of 60,000 kilometers. Formulate and test appropriate hypotheses, and draw conclusions using $\alpha = 0.05$.
(b) Suppose that if the mean life is as long as 61,000 kilometers, the engineer would like to detect this difference with probability at least 0.90. Was the sample size $n = 16$ used in part (a) adequate?

9-58. Reconsider the Izod impact test on PVC pipe described in Exercise 8-24. Suppose that you want to use the data from this experiment to support a claim that the mean impact strength exceeds the ASTM standard (one foot-pound per inch). Formulate and test the appropriate hypotheses using $\alpha = 0.05$.

9-59. Reconsider the television tube brightness experiment in Exercise 8-31. Suppose that the design engineer claims that this tube will require at least 300 microamps of current to produce the desired brightness level. Formulate and test an appropriate hypothesis to confirm this claim using $\alpha = 0.05$. Find the P-value for this test. State any necessary assumptions about the underlying distribution of the data.

9-60. Exercise 6-22 gave data on the heights of female engineering students at ASU.
(a) Can you support a claim that mean height of female engineering students at ASU is at least 65 inches? Use $\alpha = 0.05$. Find the P-value.
(b) Check the normality assumption.
(c) Compute the power of the test if the true mean height is 62 inches.
(d) What sample size would be required to detect a true mean height of 64 inches if we wanted the power of the test to be at least 0.8?

9-61. Exercise 6-25 describes testing golf balls for an overall distance standard.
(a) Can you support a claim that mean distance achieved by this particular golf ball exceeds 280 yards? Use $\alpha = 0.05$. Find the P-value.
(b) Check the normality assumption.
(c) Compute the power of the test if the true mean distance is 290 yards.
(d) What sample size would be required to detect a true mean distance of 290 yards if we wanted the power of the test to be at least 0.8?

9-62. Exercise 6-24 presented data on the concentration of suspended solids in lake water.
(a) Test the hypotheses H_0: $\mu = 55$ versus H_1: $\mu \neq 55$, use $\alpha = 0.05$. Find the P-value.
(b) Check the normality assumption.
(c) Compute the power of the test if the true mean concentration is as low as 50.
(d) What sample size would be required to detect a true mean concentration as low as 50 if we wanted the power of the test to be at least 0.9?

9-4 TESTS ON THE VARIANCE AND STANDARD DEVIATION OF A NORMAL DISTRIBUTION

Sometimes hypothesis tests on the population variance or standard deviation are needed. When the population is modeled by a normal distribution, the tests and intervals described in this section are applicable.

9-4.1 Hypothesis Tests on the Variance

Suppose that we wish to test the hypothesis that the variance of a normal population σ^2 equals a specified value, say σ_0^2, or equivalently, that the standard deviation σ is equal to σ_0. Let $X_1, X_2, \ldots, X_n$ be a random sample of n observations from this population. To test

$$H_0\colon \sigma^2 = \sigma_0^2 \qquad H_1\colon \sigma^2 \neq \sigma_0^2 \tag{9-26}$$

we will use the test statistic:

Test Statistic

$$X_0^2 = \frac{(n-1)S^2}{\sigma_0^2} \tag{9-27}$$

If the null hypothesis $H_0\colon \sigma^2 = \sigma_0^2$ is true, the test statistic X_0^2 defined in Equation 9-27 follows the chi-square distribution with $n - 1$ degrees of freedom. This is the reference distribution for this test procedure. Therefore, we calculate χ_0^2, the value of the test statistic X_0^2, and the null hypothesis $H_0\colon \sigma^2 = \sigma_0^2$ would be rejected if

$$\chi_0^2 > \chi^2_{\alpha/2,n-1} \qquad \text{or if} \quad \chi_0^2 < \chi^2_{1-\alpha/2,n-1}$$

where $\chi^2_{\alpha/2,n-1}$ and $\chi^2_{1-\alpha/2,n-1}$ are the upper and lower $100\alpha/2$ percentage points of the chi-square distribution with $n - 1$ degrees of freedom, respectively. Figure 9-11(a) shows the critical region.

The same test statistic is used for one-sided alternative hypotheses. For the one-sided hypothesis

$$H_0\colon \sigma^2 = \sigma_0^2 \qquad H_1\colon \sigma^2 > \sigma_0^2 \tag{9-28}$$

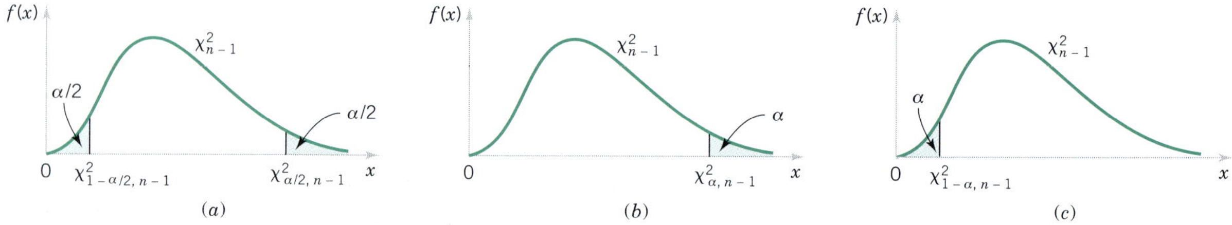

Figure 9-11 Reference distribution for the test of $H_0\colon \sigma^2 = \sigma_0^2$ with critical region values for (a) $H_1\colon \sigma^2 \neq \sigma_0^2$, (b) $H_1\colon \sigma^2 > \sigma_0^2$, and (c) $H_1\colon \sigma^2 < \sigma_0^2$.

we would reject H_0 if $\chi_0^2 > \chi_{\alpha,n-1}^2$, whereas for the other one-sided hypothesis

$$H_0: \sigma^2 = \sigma_0^2 \qquad H_1: \sigma^2 < \sigma_0^2 \tag{9-29}$$

we would reject H_0 if $\chi_0^2 < \chi_{1-\alpha,n-1}^2$. The one-sided critical regions are shown in Figure 9-11(b) and (c).

Tests on the Variance of a Normal Distribution

Null hypothesis: $H_0: \sigma^2 = \sigma_0^2$

Test statistic: $$\chi_0^2 = \frac{(n-1)S^2}{\sigma_0^2}$$

Alternative hypothesis	Rejection criteria
$H_1: \sigma^2 \neq \sigma_0^2$	$\chi_0^2 > \chi_{\alpha/2,n-1}^2$ or $\chi_0^2 < -\chi_{\alpha/2,n-1}$
$H_1: \sigma^2 > \sigma_0^2$	$\chi_0^2 > \chi_{\alpha,n-1}^2$
$H_1: \sigma^2 < \sigma_0^2$	$\chi_0^2 < -\chi_{\alpha,n-1}^2$

EXAMPLE 9-8 Automated Filling

An automated filling machine is used to fill bottles with liquid detergent. A random sample of 20 bottles results in a sample variance of fill volume of $s^2 = 0.0153$ (fluid ounces)^2. If the variance of fill volume exceeds 0.01 (fluid ounces)^2, an unacceptable proportion of bottles will be underfilled or overfilled. Is there evidence in the sample data to suggest that the manufacturer has a problem with underfilled or overfilled bottles? Use $\alpha = 0.05$, and assume that fill volume has a normal distribution.

Using the eight-step procedure results in the following:

1. The parameter of interest is the population variance σ^2.
2. $H_0: \sigma^2 = 0.01$
3. $H_1: \sigma^2 > 0.01$
4. $\alpha = 0.05$
5. The test statistic is

$$\chi_0^2 = \frac{(n-1)s^2}{\sigma_0^2}$$

6. Reject H_0 if $\chi_0^2 > \chi_{0.05,19}^2 = 30.14$.
7. Computations:

$$\chi_0^2 = \frac{19(0.0153)}{0.01} = 29.07$$

8. Conclusions: Since $\chi_0^2 = 29.07 < \chi_{0.05,19}^2 = 30.14$, we conclude that there is no strong evidence that the variance of fill volume exceeds 0.01 (fluid ounces)^2.

Using Appendix Table III, it is easy to place bounds on the P-value of a chi-square test. From inspection of the table, we find that $\chi_{0.10,19}^2 = 27.20$ and $\chi_{0.05,19}^2 = 30.14$. Since $27.20 < 29.07 < 30.14$, we conclude that the P-value for the test in Example 9-8 is in the interval $0.05 < P\text{-value} < 0.10$. The actual P-value from computer software is $P\text{-value} = 0.0649$.

9-4.2 Type II Error and Choice of Sample Size

Operating characteristic curves for the chi-square tests in Section 9-4.1 are provided in Appendix Charts VI*i* through VI*n* for $\alpha = 0.05$ and $\alpha = 0.01$. For the two-sided alternative hypothesis of Equation 9-26, Charts VII*i* and VII*j* plot β against an abscissa parameter

$$\lambda = \frac{\sigma}{\sigma_0} \tag{9-30}$$

for various sample sizes n, where σ denotes the true value of the standard deviation. Charts VI*k* and VI*l* are for the one-sided alternative H_1: $\sigma^2 > \sigma_0^2$, while Charts VII*m* and VII*n* are for the other one-sided alternative H_1: $\sigma^2 < \sigma_0^2$. In using these charts, we think of σ as the value of the standard deviation that we want to detect.

These curves can be used to evaluate the β-error (or power) associated with a particular test. Alternatively, they can be used to **design** a test—that is, to determine what sample size is necessary to detect a particular value of σ that differs from the hypothesized value σ_0.

EXAMPLE 9-9
Automated Filling Sample Size

Consider the bottle-filling problem from Example 9-8. If the variance of the filling process exceeds 0.01 (fluid ounces)^2, too many bottles will be underfilled. Thus, the hypothesized value of the standard deviation is $\sigma_0 = 0.10$. Suppose that if the true standard deviation of the filling process exceeds this value by 25%, we would like to detect this with probability at least 0.8. Is the sample size of $n = 20$ adequate?

To solve this problem, note that we require

$$\lambda = \frac{\sigma}{\sigma_0} = \frac{0.125}{0.10} = 1.25$$

This is the abscissa parameter for Chart VII*k*. From this chart, with $n = 20$ and $\lambda = 1.25$, we find that $\beta \simeq 0.6$. Therefore, there is only about a 40% chance that the null hypothesis will be rejected if the true standard deviation is really as large as $\sigma = 0.125$ fluid ounce.

To reduce the β-error, a larger sample size must be used. From the operating characteristic curve with $\beta = 0.20$ and $\lambda = 1.25$, we find that $n = 75$, approximately. Thus, if we want the test to perform as required above, the sample size must be at least 75 bottles.

EXERCISES FOR SECTION 9-4

9-63. Consider the test of H_0: $\sigma^2 = 7$ against H_1: $\sigma^2 \neq 7$. What are the critical values for the test statistic X_0^2 for the following significance levels and sample sizes?
(a) $\alpha = 0.01$ and $n = 20$
(b) $\alpha = 0.05$ and $n = 12$
(c) $\alpha = 0.10$ and $n = 15$

9-64. Consider the test of H_0: $\sigma^2 = 10$ against H_1: $\sigma^2 > 10$. What are the critical values for the test statistic X_0^2 for the following significance levels and sample sizes?
(a) $\alpha = 0.01$ and $n = 20$
(b) $\alpha = 0.05$ and $n = 12$
(c) $\alpha = 0.10$ and $n = 15$

9-65. Consider the test of H_0: $\sigma^2 = 5$ against H_1: $\sigma^2 < 5$. What are the critical values for the test statistic X_0^2 for the following significance levels and sample sizes?
(a) $\alpha = 0.01$ and $n = 20$
(b) $\alpha = 0.05$ and $n = 12$
(c) $\alpha = 0.10$ and $n = 15$

9-66. Consider the hypothesis test of H_0: $\sigma^2 = 7$ against H_1: $\sigma^2 \neq 7$. Approximate the *P*-value for each of the following test statistics.
(a) $x_0^2 = 25.2$ and $n = 20$
(b) $x_0^2 = 15.2$ and $n = 12$
(c) $x_0^2 = 23.0$ and $n = 15$

9-67. Consider the test of H_0: $\sigma^2 = 5$ against H_1: $\sigma^2 < 5$. Approximate the *P*-value for each of the following test statistics.
(a) $x_0^2 = 25.2$ and $n = 20$
(b) $x_0^2 = 15.2$ and $n = 12$
(c) $x_0^2 = 4.2$ and $n = 15$

9-68. Consider the hypothesis test of H_0: $\sigma^2 = 10$ against H_1: $\sigma^2 > 10$. Approximate the P-value for each of the following test statistics.
(a) $x_0^2 = 25.2$ and $n = 20$
(b) $x_0^2 = 15.2$ and $n = 12$
(c) $x_0^2 = 4.2$ and $n = 15$

9-69. The data from *Medicine and Science in Sports and Exercise* described in Exercise 8-42 considered ice hockey player performance after electrostimulation training. In summary, there were 17 players and the sample standard deviation of performance was 0.09 seconds.
(a) Is there strong evidence to conclude that the standard deviation of performance time exceeds the historical value of 0.75 seconds? Use $\alpha = 0.05$. Find the P-value for this test.
(b) Discuss how part (a) could be answered by constructing a 95% one-sided confidence interval for σ.

9-70. The data from *Technometrics* described in Exercise 8-45 considered the variability in repeated measurements of the weight of a sheet of paper. In summary, the sample standard deviation from 15 measurements was 0.0083 grams.
(a) Does the measurement standard deviation differ from 0.01 grams at $\alpha = 0.05$? Find the P-value for this test.
(b) Discuss how part (a) could be answered by constructing a confidence interval for σ.

9-71. Reconsider the percentage of titanium in an alloy used in aerospace castings from Exercise 8-41. Recall that $s = 0.37$ and $n = 51$.
(a) Test the hypothesis H_0: $\sigma = 0.25$ versus H_1: $\sigma \neq 0.25$ using $\alpha = 0.05$. State any necessary assumptions about the underlying distribution of the data. Find the P-value.
(b) Explain how you could answer the question in part (a) by constructing a 95% two-sided confidence interval for σ.

9-72. Data from an Izod impact test was described in Exercise 8-24. The sample standard deviation was 0.25 and $n = 20$ specimens were tested.
(a) Test the hypothesis that $\sigma = 0.10$ against an alternative specifying that $\sigma \neq 0.10$, using $\alpha = 0.01$, and draw a conclusion. State any necessary assumptions about the underlying distribution of the data.
(b) What is the P-value for this test?
(c) Could the question in part (a) have been answered by constructing a 99% two-sided confidence interval for σ^2?

9-73. Data for tire life was described in Exercise 8-23. The sample standard deviation was 3645.94 kilometers and $n = 16$.
(a) Can you conclude, using $\alpha = 0.05$, that the standard deviation of tire life is less than 4000 kilometers? State any necessary assumptions about the underlying distribution of the data. Find the P-value for this test.
(b) Explain how you could answer the question in part (a) by constructing a 95% one-sided confidence interval for σ.

9-74. If the standard deviation of hole diameter exceeds 0.01 millimeters, there is an unacceptably high probability that the rivet will not fit. Suppose that $n = 15$ and $s = 0.008$ millimeters.
(a) Is there strong evidence to indicate that the standard deviation of hole diameter exceeds 0.01 millimeters? Use $\alpha = 0.01$. State any necessary assumptions about the underlying distribution of the data. Find the P-value for this test.
(b) Suppose that the actual standard deviation of hole diameter exceeds the hypothesized value by 50%. What is the probability that this difference will be detected by the test described in part (a).
(c) If σ is really as large as 0.0125 millimeters, what sample size will be required to defect this with power of at least 0.8?

9-75. Recall the sugar content of the syrup in canned peaches from Exercise 8-40. Suppose that the variance is thought to be $\sigma^2 = 18$ (milligrams)^2. Recall that a random sample of $n = 10$ cans yields a sample standard deviation of $s = 4.8$ milligrams.
(a) Test the hypothesis H_0: $\sigma^2 = 18$ versus H_1: $\sigma^2 \neq 18$ using $\alpha = 0.05$. Find the P-value for this test.
(b) Suppose that the actual standard deviation is twice as large as the hypothesized value. What is the probability that this difference will be detected by the test described in part (a)?
(c) Suppose that the true variance is $\sigma^2 = 40$. How large a sample would be required to detect this difference with probability at least 0.90?

9-5 TESTS ON A POPULATION PROPORTION

It is often necessary to test hypotheses on a population proportion. For example, suppose that a random sample of size n has been taken from a large (possibly infinite) population and that $X(\leq n)$ observations in this sample belong to a class of interest. Then $\hat{P} = X/n$ is a point estimator of the proportion of the population p that belongs to this class. Note that n and p are the parameters of a binomial distribution. Furthermore, from Chapter 7 we know that the sampling distribution of $\hat{P}$ is approximately normal with mean p and variance $p(1 - p)/n$, if p is not too close to either 0 or 1 and if n is relatively large. Typically, to apply this approximation we require that np and $n(1 - p)$ be greater than or equal to 5. We will give a large-sample test that makes use of the normal approximation to the binomial distribution.

9-5.1 Large-Sample Tests on a Proportion

In many engineering problems, we are concerned with a random variable that follows the binomial distribution. For example, consider a production process that manufactures items that are classified as either acceptable or defective. It is usually reasonable to model the occurrence of defectives with the binomial distribution, where the binomial parameter p represents the proportion of defective items produced. Consequently, many engineering decision problems involve hypothesis testing about p.

We will consider testing

$$H_0\colon p = p_0 \qquad H_1\colon p \neq p_0 \tag{9-31}$$

An approximate test based on the normal approximation to the binomial will be given. As noted above, this approximate procedure will be valid as long as p is not extremely close to zero or one, and if the sample size is relatively large. Let X be the number of observations in a random sample of size n that belongs to the class associated with p. Then, if the null hypothesis H_0: $p = p_0$ is true, we have $X \sim N[np_0, np_0(1 - p_0)]$, approximately. To test H_0: $p = p_0$, calculate the test statistic.

Test Statistic

$$Z_0 = \frac{X - np_0}{\sqrt{np_0(1 - p_0)}} \tag{9-32}$$

and reject H_0: $p = p_0$ if

$$z_0 > z_{\alpha/2} \quad \text{or} \quad z_0 < -z_{\alpha/2}$$

Note that the standard normal distribution is the **reference distribution** for this test statistic. Critical regions for the one-sided alternative hypotheses are constructed in the usual manner.

Approximate Tests on a Binomial Proportion

Null hypothesis: H_0: $p = p_0$

Test statistic: $Z_0 = \dfrac{X - np_0}{\sqrt{np_0(1 - p_0)}}$

Alternative hypothesis	Rejection criteria
H_1: $p \neq p_0$	$z_0 > z_{\alpha/2,n-1}$ or $z_0 < -z_{\alpha/2,n-1}$
H_1: $p > p_0$	$z_0 > z_{\alpha,n-1}$
H_1: $p < p_0$	$z_0 < -z_{\alpha,n-1}$

EXAMPLE 9-10
Automobile Engine Controller

A semiconductor manufacturer produces controllers used in automobile engine applications. The customer requires that the process fallout or fraction defective at a critical manufacturing step not exceed 0.05 and that the manufacturer demonstrate process capability at this level of quality using $\alpha = 0.05$. The semiconductor manufacturer takes a random sample of 200 devices and finds that four of them are defective. Can the manufacturer demonstrate process capability for the customer?

We may solve this problem using the eight-step hypothesis-testing procedure as follows:

1. The parameter of interest is the process fraction defective p.
2. H_0: $p = 0.05$
3. H_1: $p < 0.05$
   This formulation of the problem will allow the manufacturer to make a strong claim about process capability if the null hypothesis H_0: $p = 0.05$ is rejected.
4. $\alpha = 0.05$
5. The test statistic is (from Equation 9-32)

$$z_0 = \frac{x - np_0}{\sqrt{np_0(1 - p_0)}}$$

   where $x = 4$, $n = 200$, and $p_0 = 0.05$.
6. Reject H_0: $p = 0.05$ if $z_0 < -z_{0.05} = -1.645$
7. Computations: The test statistic is

$$z_0 = \frac{4 - 200(0.05)}{\sqrt{200(0.05)(0.95)}} = -1.95$$

8. Conclusions: Since $z_0 = -1.95 < -z_{0.05} = -1.645$, we reject H_0 and conclude that the process fraction defective p is less than 0.05. The P-value for this value of the test statistic z_0 is $P = 0.0256$, which is less than $\alpha = 0.05$. We conclude that the process is capable.

Another form of the test statistic Z_0 in Equation 9-32 is occasionally encountered. Note that if X is the number of observations in a random sample of size n that belongs to a class of interest, then $\hat{P} = X/n$ is the sample proportion that belongs to that class. Now divide both numerator and denominator of Z_0 in Equation 9-32 by n, giving

$$Z_0 = \frac{X/n - p_0}{\sqrt{p_0(1 - p_0)/n}}$$

or

$$Z_0 = \frac{\hat{P} - p_0}{\sqrt{p_0(1 - p_0)/n}} \qquad (9\text{-}33)$$

This presents the test statistic in terms of the sample proportion instead of the number of items X in the sample that belongs to the class of interest.

Statistical software packages usually provide the one sample Z-test for a proportion. The Minitab output for Example 9-10 follows.

Minitab Computations

Test and CI for One Proportion

Test of $p = 0.05$ vs $p < 0.05$

Sample	X	N	Sample p	95.0% Upper Bound	Z-Value	P-Value
1	4	200	0.020000	0.036283	−1.95	0.026

* NOTE * The normal approximation may be inaccurate for small samples.

Notice that both the test statistic (and accompanying P-value) and the 95% one-sided upper confidence bound are displayed. The 95% upper confidence bound is 0.036283, which is less than 0.05. This is consistent with rejection of the null hypothesis $H_0: p = 0.05$.

9-5.2 Type II Error and Choice of Sample Size

It is possible to obtain closed-form equations for the approximate β-error for the tests in Section 9-5.1. Suppose that p is the true value of the population proportion. The approximate β-error for the two-sided alternative $H_1: p \neq p_0$ is

$$\beta = \Phi\left(\frac{p_0 - p + z_{\alpha/2}\sqrt{p_0(1-p_0)/n}}{\sqrt{p(1-p)/n}}\right) - \Phi\left(\frac{p_0 - p - z_{\alpha/2}\sqrt{p_0(1-p_0)/n}}{\sqrt{p(1-p)/n}}\right) \quad (9\text{-}34)$$

If the alternative is $H_1: p < p_0$,

$$\beta = 1 - \Phi\left(\frac{p_0 - p - z_{\alpha}\sqrt{p_0(1-p_0)/n}}{\sqrt{p(1-p)/n}}\right) \quad (9\text{-}35)$$

whereas if the alternative is $H_1: p > p_0$,

$$\beta = \Phi\left(\frac{p_0 - p + z_{\alpha}\sqrt{p_0(1-p_0)/n}}{\sqrt{p(1-p)/n}}\right) \quad (9\text{-}36)$$

These equations can be solved to find the approximate sample size n that gives a test of level α that has a specified β risk. The sample size equations are

Approximate Sample Size for a Two-Sided Test on a Binomial Proportion

$$n = \left[\frac{z_{\alpha/2}\sqrt{p_0(1-p_0)} + z_{\beta}\sqrt{p(1-p)}}{p - p_0}\right]^2 \quad (9\text{-}37)$$

for the two-sided alternative and

Approximate Sample Size for a One-Sided Test on a Binomial Proportion

$$n = \left[\frac{z_{\alpha}\sqrt{p_0(1-p_0)} + z_{\beta}\sqrt{p(1-p)}}{p - p_0}\right]^2 \quad (9\text{-}38)$$

for a one-sided alternative.

EXAMPLE 9-11 Automobile Engine Controller Type II Error

Consider the semiconductor manufacturer from Example 9-10. Suppose that its process fallout is really $p = 0.03$. What is the β-error for a test of process capability that uses $n = 200$ and $\alpha = 0.05$?

The β-error can be computed using Equation 9-35 as follows:

$$\beta = 1 - \Phi\left[\frac{0.05 - 0.03 - (1.645)\sqrt{0.05(0.95)/200}}{\sqrt{0.03(1-0.03)/200}}\right] = 1 - \Phi(-0.44) = 0.67$$

Thus, the probability is about 0.7 that the semiconductor manufacturer will fail to conclude that the process is capable if the true process fraction defective is $p = 0.03$ (3%). That is, the power of the test against this particular alternative is only about 0.3. This appears to be a large β-error (or small power), but the difference between $p = 0.05$ and $p = 0.03$ is fairly small, and the sample size $n = 200$ is not particularly large.

Suppose that the semiconductor manufacturer was willing to accept a β-error as large as 0.10 if the true value of the process fraction defective was $p = 0.03$. If the manufacturer continues to use $\alpha = 0.05$, what sample size would be required?

The required sample size can be computed from Equation 9-38 as follows:

$$n = \left[\frac{1.645\sqrt{0.05(0.95)} + 1.28\sqrt{0.03(0.97)}}{0.03 - 0.05}\right]^2$$
$$\simeq 832$$

where we have used $p = 0.03$ in Equation 9-38. Note that $n = 832$ is a very large sample size. However, we are trying to detect a fairly small deviation from the null value $p_0 = 0.05$.

Minitab will also perform power and sample size calculations for the one-sample Z-test on a proportion. Output from Minitab for the engine controllers tested in Example 9-10 follows.

Power and Sample Size

Test for One Proportion
Testing proportion = 0.05 (versus < 0.05)
Alpha = 0.05

Alternative Proportion	Sample Size	Power
3.00E-02	200	0.3287

Power and Sample Size

Test for One Proportion
Testing proportion = 0.05 (versus < 0.05)
Alpha = 0.05

Alternative Proportion	Sample Size	Target Power	Actual Power
3.00E-02	833	0.9000	0.9001

Power and Sample Size

Test for One Proportion
Testing proportion = 0.05 (versus < 0.05)
Alpha = 0.05

Alternative Proportion	Sample Size	Target Power	Actual Power
3.00E-02	561	0.7500	0.7503

The first part of the output shows the power calculation based on the situation described in Example 9-11, where the true proportion is really 0.03. The power calculation from Minitab agrees with the results from Equation 9-35 in Example 9-11. The second part of the output computes the sample size necessary for a power of 0.9 ($\beta = 0.1$) if $p = 0.03$. Again, the results agree closely with those obtained from Equation 9-38. The final portion of the display shows the sample size that would be required if $p = 0.03$ and the power requirement is relaxed to 0.75. Notice that the sample size of $n = 561$ is still quite large because the difference between $p = 0.05$ and $p = 0.03$ is fairly small.

EXERCISES FOR SECTION 9-5

9-76. Suppose that 1000 customers are surveyed and 850 are satisfied or very satisfied with a corporation's products and services.

(a) Test the hypothesis $H_0: p = 0.9$ against $H_1: p \neq 0.9$ at $\alpha = 0.05$. Find the P-value.

(b) Explain how the question in part (a) could be answered by constructing a 95% two-sided confidence interval for p.

9-77. Suppose that 500 parts are tested in manufacturing and 10 are rejected.

(a) Test the hypothesis $H_0: p = 0.03$ against $H_1: p < 0.03$ at $\alpha = 0.05$. Find the P-value.

(b) Explain how the question in part (a) could be answered by constructing a 95% one-sided confidence interval for p.

9-78. A random sample of 300 circuits generated 13 defectives.

(a) Use the data to test $H_0: p = 0.05$ versus $H_1: p \neq 0.05$. Use $\alpha = 0.05$. Find the P-value for the test.

(b) Explain how the question in part (a) could be answered with a confidence interval.

9-79. An article in *British Medical Journal* (1986, Vol. 292, pp. 879–882) "Comparison of treatment of renal calculi by operative surgery, percutaneous nephrolithotomy, and extracorporeal shock wave lithotripsy," found that percutaneous nephrolithotomy (PN) had a success rate in removing kidney stones of 289 out of 350 patients. The traditional method was 78% effective.

(a) Is there evidence that the success rate for PN is greater than the historical success rate? Find the P-value.

(b) Explain how the question in part (a) could be answered with a confidence interval.

9-80. A manufacturer of interocular lenses is qualifying a new grinding machine and will qualify the machine if there is evidence that percentage of polished lenses that contain surface defects does not exceed 2%. A random sample of 250 lenses contains six defective lenses.

(a) Formulate and test an appropriate set of hypotheses to determine if the machine can be qualified. Use $\alpha = 0.05$. Find the P-value.

(b) Explain how the question in part (a) could be answered with a confidence interval.

9-81. A researcher claims that at least 10% of all football helmets have manufacturing flaws that could potentially cause injury to the wearer. A sample of 200 helmets revealed that 16 helmets contained such defects.

(a) Does this finding support the researcher's claim? Use $\alpha = 0.01$. Find the P-value.

(b) Explain how the question in part (a) could be answered with a confidence interval.

9-82. An article in *Fortune* (September 21, 1992) claimed that nearly one-half of all engineers continue academic studies beyond the B.S. degree, ultimately receiving either an M.S. or a Ph.D. degree. Data from an article in *Engineering Horizons* (Spring 1990) indicated that 117 of 484 new engineering graduates were planning graduate study.

(a) Are the data from *Engineering Horizons* consistent with the claim reported by *Fortune*? Use $\alpha = 0.05$ in reaching your conclusions. Find the P-value for this test.

(b) Discuss how you could have answered the question in part (a) by constructing a two-sided confidence interval on p.

9-83. The advertised claim for batteries for cell phones is set at 48 operating hours, with proper charging procedures. A study of 5000 batteries is carried out and 15 stop operating prior to 48 hours. Do these experimental results support the claim that less than 0.2 percent of the company's batteries will fail during the advertised time period, with proper charging procedures? Use a hypothesis-testing procedure with $\alpha = 0.01$.

9-84. A random sample of 500 registered voters in Phoenix is asked if they favor the use of oxygenated fuels year-round to reduce air pollution. If more than 315 voters respond positively, we will conclude that at least 60% of the voters favor the use of these fuels.

(a) Find the probability of type I error if exactly 60% of the voters favor the use of these fuels.

(b) What is the type II error probability β if 75% of the voters favor this action?

9-85. In a random sample of 85 automobile engine crankshaft bearings, 10 have a surface finish roughness that exceeds the specifications. Does this data present strong evidence that the proportion of crankshaft bearings exhibiting excess surface roughness exceeds 0.10?

(a) State and test the appropriate hypotheses using $\alpha = 0.05$.

(b) If it is really the situation that $p = 0.15$, how likely is it that the test procedure in part (a) will not reject the null hypothesis?

(c) If $p = 0.15$, how large would the sample size have to be for us to have a probability of correctly rejecting the null hypothesis of 0.9?

9-6 SUMMARY TABLE OF INFERENCE PROCEDURES FOR A SINGLE SAMPLE

The table in the end papers of this book (inside back cover) presents a summary of all the single-sample inference procedures from Chapters 8 and 9. The table contains the null hypothesis statement, the test statistic, the various alternative hypotheses and the criteria for rejecting H_0, and the formulas for constructing the $100(1 - \alpha)\%$ two-sided confidence interval. It would also be helpful to refer to the roadmap table on page 281 that provides guidance to match the problem type to the information inside the back cover.

9-7 TESTING FOR GOODNESS OF FIT

The hypothesis-testing procedures that we have discussed in previous sections are designed for problems in which the population or probability distribution is known and the hypotheses involve the parameters of the distribution. Another kind of hypothesis is often encountered: we do not know the underlying distribution of the population, and we wish to test the hypothesis that a particular distribution will be satisfactory as a population model. For example, we might wish to test the hypothesis that the population is normal.

We have previously discussed a very useful graphical technique for this problem called **probability plotting** and illustrated how it was applied in the case of a normal distribution. In this section, we describe a formal **goodness-of-fit test** procedure based on the chi-square distribution.

The test procedure requires a random sample of size n from the population whose probability distribution is unknown. These n observations are arranged in a frequency histogram, having k bins or class intervals. Let O_i be the observed frequency in the ith class interval. From the hypothesized probability distribution, we compute the expected frequency in the ith class interval, denoted E_i. The test statistic is

Goodness of Fit Test Statistic

$$X_0^2 = \sum_{i=1}^{k} \frac{(O_i - E_i)^2}{E_i} \tag{9-39}$$

It can be shown that, if the population follows the hypothesized distribution, X_0^2 has, approximately, a chi-square distribution with $k - p - 1$ degrees of freedom, where p represents the number of parameters of the hypothesized distribution estimated by sample statistics. This approximation improves as n increases. We would reject the hypothesis that the distribution of the population is the hypothesized distribution if the calculated value of the test statistic $\chi_0^2 > \chi^2_{\alpha,k-p-1}$.

One point to be noted in the application of this test procedure concerns the magnitude of the expected frequencies. If these expected frequencies are too small, the test statistic X_0^2 will not reflect the departure of observed from expected, but only the small magnitude of the expected frequencies. There is no general agreement regarding the minimum value of expected frequencies, but values of 3, 4, and 5 are widely used as minimal. Some writers suggest that an expected frequency could be as small as 1 or 2, so long as most of them exceed 5. Should an expected frequency be too small, it can be combined with the expected frequency in an adjacent class interval. The corresponding observed frequencies would then also be combined, and k would be reduced by 1. Class intervals are not required to be of equal width.

We now give two examples of the test procedure.

EXAMPLE 9-12
Printed Circuit Board Defects

A Poisson Distribution

The number of defects in printed circuit boards is hypothesized to follow a Poisson distribution. A random sample of $n = 60$ printed boards has been collected, and the following number of defects observed.

Number of Defects	Observed Frequency
0	32
1	15
2	9
3	4

The mean of the assumed Poisson distribution in this example is unknown and must be estimated from the sample data. The estimate of the mean number of defects per board is the sample average, that is, $(32 \cdot 0 + 15 \cdot 1 + 9 \cdot 2 + 4 \cdot 3)/60 = 0.75$. From the Poisson distribution with parameter 0.75, we may compute p_i, the theoretical, hypothesized probability associated with the ith class interval. Since each class interval corresponds to a particular number of defects, we may find the p_i as follows:

$$p_1 = P(X = 0) = \frac{e^{-0.75}(0.75)^0}{0!} = 0.472$$

$$p_2 = P(X = 1) = \frac{e^{-0.75}(0.75)^1}{1!} = 0.354$$

$$p_3 = P(X = 2) = \frac{e^{-0.75}(0.75)^2}{2!} = 0.133$$

$$p_4 = P(X \geq 3) = 1 - (p_1 + p_2 + p_3) = 0.041$$

The expected frequencies are computed by multiplying the sample size $n = 60$ times the probabilities p_i. That is, $E_i = np_i$. The expected frequencies follow:

Number of Defects	Probability	Expected Frequency
0	0.472	28.32
1	0.354	21.24
2	0.133	7.98
3 (or more)	0.041	2.46

Since the expected frequency in the last cell is less than 3, we combine the last two cells:

Number of Defects	Observed Frequency	Expected Frequency
0	32	28.32
1	15	21.24
2 (or more)	13	10.44

The chi-square test statistic in Equation 9-39 will have $k - p - 1 = 3 - 1 - 1 = 1$ degree of freedom, because the mean of the Poisson distribution was estimated from the data.

The eight-step hypothesis-testing procedure may now be applied, using $\alpha = 0.05$, as follows:

1. The variable of interest is the form of the distribution of defects in printed circuit boards.
2. H_0: The form of the distribution of defects is Poisson.
3. H_1: The form of the distribution of defects is not Poisson.
4. $\alpha = 0.05$
5. The test statistic is

$$\chi_0^2 = \sum_{i=1}^{k} \frac{(o_i - E_i)^2}{E_i}$$

6. Reject H_0 if $\chi_0^2 > \chi_{0.05,1}^2 = 3.84$.
7. Computations:

$$\chi_0^2 = \frac{(32 - 28.32)^2}{28.32} + \frac{(15 - 21.24)^2}{21.24} + \frac{(13 - 10.44)^2}{10.44} = 2.94$$

8. Conclusions: Since $\chi_0^2 = 2.94 < \chi_{0.05,1}^2 = 3.84$, we are unable to reject the null hypothesis that the distribution of defects in printed circuit boards is Poisson. The P-value for the test is $P = 0.0864$. (This value was computed using an HP-48 calculator.)

EXAMPLE 9-13
Power Supply Distribution

Continuous Distribution

A manufacturing engineer is testing a power supply used in a notebook computer and, using $\alpha = 0.05$, wishes to determine whether output voltage is adequately described by a normal distribution. Sample estimates of the mean and standard deviation of $\bar{x} = 5.04$ V and $s = 0.08$ V are obtained from a random sample of $n = 100$ units.

A common practice in constructing the class intervals for the frequency distribution used in the chi-square goodness-of-fit test is to choose the cell boundaries so that the expected frequencies $E_i = np_i$ are equal for all cells. To use this method, we want to choose the cell boundaries $a_0, a_1, \ldots, a_k$ for the k cells so that all the probabilities

$$p_i = P(a_{i-1} \le X \le a_i) = \int_{a_{i-1}}^{a_i} f(x)\,dx$$

are equal. Suppose we decide to use $k = 8$ cells. For the standard normal distribution, the intervals that divide the scale into eight equally likely segments are [0, 0.32), [0.32, 0.675) [0.675, 1.15), [1.15, ∞) and their four "mirror image" intervals on the other side of zero. For each interval $p_i = 1/8 = 0.125$, so the expected cell frequencies are $E_i = np_i = 100(0.125) = 12.5$. The complete table of observed and expected frequencies is as follows:

Class Interval	Observed Frequency o_i	Expected Frequency E_i
$x < 4.948$	12	12.5
$4.948 \le x < 4.986$	14	12.5
$4.986 \le x < 5.014$	12	12.5
$5.014 \le x < 5.040$	13	12.5
$5.040 \le x < 5.066$	12	12.5
$5.066 \le x < 5.094$	11	12.5
$5.094 \le x < 5.132$	12	12.5
$5.132 \le x$	14	12.5
Totals	100	100

The boundary of the first class interval is $\bar{x} - 1.15s = 4.948$. The second class interval is $[\bar{x} - 1.15s, \bar{x} - 0.675s)$ and so forth. We may apply the eight-step hypothesis-testing procedure to this problem.

1. The variable of interest is the form of the distribution of power supply voltage.
2. H_0: The form of the distribution is normal.
3. H_1: The form of the distribution is nonnormal.
4. $\alpha = 0.05$
5. The test statistic is

$$\chi_0^2 = \sum_{i=1}^{k} \frac{(o_i - E_i)^2}{E_i}$$

6. Since two parameters in the normal distribution have been estimated, the chi-square statistic above will have $k - p - 1 = 8 - 2 - 1 = 5$ degrees of freedom. Therefore, we will reject H_0 if $\chi_0^2 > \chi_{0.05,5}^2 = 11.07$.
7. Computations:

$$\begin{aligned}\chi_0^2 &= \sum_{i=1}^{8} \frac{(o_i - E_i)^2}{E_i} \\ &= \frac{(12 - 12.5)^2}{12.5} + \frac{(14 - 12.5)^2}{12.5} + \cdots + \frac{(14 - 12.5)^2}{12.5} \\ &= 0.64\end{aligned}$$

8. Conclusions: Since $\chi_0^2 = 0.64 < \chi_{0.05,5}^2 = 11.07$, we are unable to reject H_0, and there is no strong evidence to indicate that output voltage is not normally distributed. The P-value for the chi-square statistic $\chi_0^2 = 0.64$ is $P = 0.9861$.

EXERCISES FOR SECTION 9-7

9-86. Consider the following frequency table of observations on the random variable X.

Values	0	1	2	3	4
Observed Frequency	24	30	31	11	4

(a) Based on these 100 observations, is a Poisson distribution with a mean of 1.2 an appropriate model? Perform a goodness-of-fit procedure with $\alpha = 0.05$.

(b) Calculate the P-value for this test.

9-87. Let X denote the number of flaws observed on a large coil of galvanized steel. Seventy-five coils are inspected and the following data were observed for the values of X:

Values	1	2	3	4	5	6	7	8
Observed Frequency	1	11	8	13	11	12	10	9

(a) Does the assumption of the Poisson distribution seem appropriate as a probability model for this data? Use $\alpha = 0.01$.

(b) Calculate the P-value for this test.

9-88. The number of calls arriving at a switchboard from noon to 1 PM during the business days Monday through Friday is monitored for six weeks (i.e., 30 days). Let X be defined as the number of calls during that one-hour period. The relative frequency of calls was recorded and reported as

Value	5	6	8	9	10
Relative Frequency	0.067	0.067	0.100	0.133	0.200
Value	11	12	13	14	15
Relative Frequency	0.133	0.133	0.067	0.033	0.067

(a) Does the assumption of a Poisson distribution seem appropriate as a probability model for this data? Use $\alpha = 0.05$.

(b) Calculate the P-value for this test.

9-89. Consider the following frequency table of observations on the random variable X:

Values	0	1	2	3	4
Frequency	4	21	10	13	2

(a) Based on these 50 observations, is a binomial distribution with $n = 6$ and $p = 0.25$ an appropriate model? Perform a goodness-of-fit procedure with $\alpha = 0.05$.
(b) Calculate the P-value for this test.

9-90. Define X as the number of underfilled bottles from a filling operation in a carton of 24 bottles. Seventy-five cartons are inspected and the following observations on X are recorded:

Values	0	1	2	3
Frequency	39	23	12	1

(a) Based on these 75 observations, is a binomial distribution an appropriate model? Perform a goodness-of-fit procedure with $\alpha = 0.05$.
(b) Calculate the P-value for this test.

9-91. The number of cars passing eastbound through the intersection of Mill and University Avenues has been tabulated by a group of civil engineering students. They have obtained the data in the adjacent table:
(a) Does the assumption of a Poisson distribution seem appropriate as a probability model for this process? Use $\alpha = 0.05$.
(b) Calculate the P-value for this test.

Vehicles per Minute	Observed Frequency	Vehicles per Minute	Observed Frequency
40	14	53	102
41	24	54	96
42	57	55	90
43	111	56	81
44	194	57	73
45	256	58	64
46	296	59	61
47	378	60	59
48	250	61	50
49	185	62	42
50	171	63	29
51	150	64	18
52	110	65	15

9-92. Reconsider Exercise 6-63. The data was the number of earthquakes per year of magnitude 7.0 and greater since 1900.

1900	13	1927	20	1954	17	1981	14
1901	14	1928	22	1955	19	1982	10
1902	8	1929	19	1956	15	1983	15
1903	10	1930	13	1957	34	1984	8
1904	16	1931	26	1958	10	1985	15
1905	26	1932	13	1959	15	1986	6
1906	32	1933	14	1960	22	1987	11
1907	27	1934	22	1961	18	1988	8
1908	18	1935	24	1962	15	1989	7
1909	32	1936	21	1963	20	1990	18
1910	36	1937	22	1964	15	1991	16
1911	24	1938	26	1965	22	1992	13
1912	22	1939	21	1966	19	1993	12
1913	23	1940	23	1967	16	1994	13
1914	22	1941	24	1968	30	1995	20
1915	18	1942	27	1969	27	1996	15
1916	25	1943	41	1970	29	1997	16
1917	21	1944	31	1971	23	1998	12
1918	21	1945	27	1972	20	1999	18
1919	14	1946	35	1973	16	2000	15
1920	8	1947	26	1974	21	2001	16
1921	11	1948	28	1975	21	2002	13
1922	14	1949	36	1976	25	2003	15
1923	23	1950	39	1977	16	2004	15
1924	18	1951	21	1978	18		
1925	17	1952	17	1979	15		
1926	19	1953	22	1980	18		

(a) Use computer software to summarize this data into a frequency distribution. Test the hypothesis that the number of earthquakes of magnitude 7.0 or greater each year follows a Poisson distribution at $\alpha = 0.05$.
(b) Calculate the P-value for the test.

9-8 CONTINGENCY TABLE TESTS

Many times, the n elements of a sample from a population may be classified according to two different criteria. It is then of interest to know whether the two methods of classification are statistically independent; for example, we may consider the population of graduating engineers, and we may wish to determine whether starting salary is independent of academic disciplines. Assume that the first method of classification has r levels and that the second method has c levels. We will let O_{ij} be the observed frequency for level i of the first classification

Table 9-2 An $r \times c$ Contingency Table

		Columns			
		1	2	...	c
Rows	1	O_{11}	O_{12}	...	O_{1c}
	2	O_{21}	O_{22}	...	O_{2c}
	⋮	⋮	⋮	⋮	⋮
	r	O_{r1}	O_{r2}	...	O_{rc}

method and level j on the second classification method. The data would, in general, appear as shown in Table 9-2. Such a table is usually called an **$r \times c$ contingency table.**

We are interested in testing the hypothesis that the row-and-column methods of classification are independent. If we reject this hypothesis, we conclude there is some interaction between the two criteria of classification. The exact test procedures are difficult to obtain, but an approximate test statistic is valid for large n. Let p_{ij} be the probability that a randomly selected element falls in the ijth cell, given that the two classifications are independent. Then $p_{ij} = u_i v_j$, where u_i is the probability that a randomly selected element falls in row class i and v_j is the probability that a randomly selected element falls in column class j. Now, assuming independence, the estimators of u_i and v_j are

$$\hat{u}_i = \frac{1}{n}\sum_{j=1}^{c} O_{ij}$$

$$\hat{v}_j = \frac{1}{n}\sum_{i=1}^{r} O_{ij} \tag{9-40}$$

Therefore, the expected frequency of each cell is

$$E_{ij} = n\hat{u}_i\hat{v}_j = \frac{1}{n}\sum_{j=1}^{c} O_{ij}\sum_{i=1}^{r} O_{ij} \tag{9-41}$$

Then, for large n, the statistic

$$\chi_0^2 = \sum_{i=1}^{r}\sum_{j=1}^{c}\frac{(O_{ij} - E_{ij})^2}{E_{ij}} \tag{9-42}$$

has an approximate chi-square distribution with $(r - 1)(c - 1)$ degrees of freedom if the null hypothesis is true. Therefore, we would reject the hypothesis of independence if the observed value of the test statistic χ_0^2 exceeded $\chi^2_{\alpha,(r-1)(c-1)}$.

EXAMPLE 9-14
Pension Plan Preference

A company has to choose among three pension plans. Management wishes to know whether the preference for plans is independent of job classification and wants to use $\alpha = 0.05$. The opinions of a random sample of 500 employees are shown in Table 9-3.

To find the expected frequencies, we must first compute $\hat{u}_1 = (340/500) = 0.68$, $\hat{u}_2 = (160/500) = 0.32$, $\hat{v}_1 = (200/500) = 0.40$, $\hat{v}_2 = (200/500) = 0.40$, and $\hat{v}_3 = (100/500) = 0.20$. The expected frequencies may now be computed from Equation 9-41. For example, the expected number of salaried workers favoring pension plan 1 is

$$E_{11} = n\hat{u}_1\hat{v}_1 = 500(0.68)(0.40) = 136$$

Table 9-3 Observed Data for Example 9-14

	Pension Plan			
Job Classification	1	2	3	Totals
Salaried workers	160	140	40	340
Hourly workers	40	60	60	160
Totals	200	200	100	500

Table 9-4 Expected Frequencies for Example 9-14

	Pension Plan			
Job Classification	1	2	3	Totals
Salaried workers	136	136	68	340
Hourly workers	64	64	32	160
Totals	200	200	100	500

The expected frequencies are shown in Table 9-4.

The eight-step hypothesis-testing procedure may now be applied to this problem.

1. The variable of interest is employee preference among pension plans.
2. H_0: Preference is independent of salaried versus hourly job classification.
3. H_1: Preference is not independent of salaried versus hourly job classification.
4. $\alpha = 0.05$
5. The test statistic is

$$\chi_0^2 = \sum_{i=1}^{r} \sum_{j=1}^{c} \frac{(o_{ij} - E_{ij})^2}{E_{ij}}$$

6. Since $r = 2$ and $c = 3$, the degrees of freedom for chi-square are $(r - 1)(c - 1) = (1)(2) = 2$, and we would reject H_0 if $\chi_0^2 > \chi_{0.05,2}^2 = 5.99$.
7. Computations:

$$\chi_0^2 = \sum_{i=1}^{2} \sum_{j=1}^{3} \frac{(o_{ij} - E_{ij})^2}{E_{ij}}$$

$$= \frac{(160 - 136)^2}{136} + \frac{(140 - 136)^2}{136} + \frac{(40 - 68)^2}{68} + \frac{(40 - 64)^2}{64} + \frac{(60 - 64)^2}{64} + \frac{(60 - 32)^2}{32} = 49.63$$

8. Conclusions: Since $\chi_0^2 = 49.63 > \chi_{0.05,2}^2 = 5.99$, we reject the hypothesis of independence and conclude that the preference for pension plans is not independent of job classification. The P-value for $\chi_0^2 = 49.63$ is $P = 1.671 \times 10^{-11}$. (This value was computed from computer software.) Further analysis would be necessary to explore the nature of the association between these factors. It might be helpful to examine the table of observed minus expected frequencies.

Using the two-way contingency table to test independence between two variables of classification in a sample from a single population of interest is only one application of contingency table methods. Another common situation occurs when there are r populations of interest and each population is divided into the same c categories. A sample is then taken from the ith population, and the counts are entered in the appropriate columns of the ith row. In this situation we want to investigate whether or not the proportions in the c categories are the same for all populations. The null hypothesis in this problem states that the populations are **homogeneous** with respect to the categories. For example, when there are only two categories, such as success and failure, defective and nondefective, and so on, the test for homogeneity is really a test of the equality of r binomial parameters. Calculation of expected frequencies, determination of degrees of freedom, and computation of the chi-square statistic for the test for homogeneity are identical to the test for independence.

EXERCISES FOR SECTION 9-8

9-93. A company operates four machines three shifts each day. From production records, the following data on the number of breakdowns are collected:

	Machines			
Shift	*A*	*B*	*C*	*D*
1	41	20	12	16
2	31	11	9	14
3	15	17	16	10

Test the hypothesis (using $\alpha = 0.05$) that breakdowns are independent of the shift. Find the P-value for this test.

9-94. Patients in a hospital are classified as surgical or medical. A record is kept of the number of times patients require nursing service during the night and whether or not these patients are on Medicare. The data are presented here:

	Patient Category	
Medicare	Surgical	Medical
Yes	46	52
No	36	43

Test the hypothesis (using $\alpha = 0.01$) that calls by surgical-medical patients are independent of whether the patients are receiving Medicare. Find the P-value for this test.

9-95. Grades in a statistics course and an operations research course taken simultaneously were as follows for a group of students.

	Operation Research Grade			
Statistics Grade	*A*	*B*	*C*	Other
A	25	6	17	13
B	17	16	15	6
C	18	4	18	10
Other	10	8	11	20

Are the grades in statistics and operations research related? Use $\alpha = 0.01$ in reaching your conclusion. What is the P-value for this test?

9-96. An experiment with artillery shells yields the following data on the characteristics of lateral deflections and ranges. Would you conclude that deflection and range are independent? Use $\alpha = 0.05$. What is the P-value for this test?

	Lateral Deflection		
Range (yards)	Left	Normal	Right
0–1,999	6	14	8
2,000–5,999	9	11	4
6,000–11,999	8	17	6

9-97. A study is being made of the failures of an electronic component. There are four types of failures possible and two mounting positions for the device. The following data have been taken:

	Failure Type			
Mounting Position	*A*	*B*	*C*	*D*
1	22	46	18	9
2	4	17	6	12

Would you conclude that the type of failure is independent of the mounting position? Use $\alpha = 0.01$. Find the P-value for this test.

9-98. A random sample of students is asked their opinions on a proposed core curriculum change. The results are as follows.

	Opinion	
Class	Favoring	Opposing
Freshman	120	80
Sophomore	70	130
Junior	60	70
Senior	40	60

Test the hypothesis that opinion on the change is independent of class standing. Use $\alpha = 0.05$. What is the P-value for this test?

9-99. An article in *British Medical Journal* (1986, Vol. 292, pp. 879–882) "Comparison of treatment of renal calculi by operative surgery, percutaneous nephrolithotomy, and extracorporeal shock wave lithotripsy," found that percutaneous nephrolithotomy (PN) had a success rate in removing kidney stones of 289 out of 350 (83%) patients. However, when the stone diameter was considered the results looked different. For stones of <2 cm, 87% (234/270) of cases were successful. For stones of ≥ 2 cm, a success rate of 69% (55/80) was observed for PN.

(a) Are the successes and size of stones independent? Use $\alpha = 0.05$.
(b) Find the P-value for this test.

Supplemental Exercises

9-100. Suppose we wish to test the hypothesis H_0: $\mu = 85$ versus the alternative H_1: $\mu > 85$ where $\sigma = 16$. Suppose that the true mean is $\mu = 86$ and that in the practical context of the problem this is not a departure from $\mu_0 = 85$ that has practical significance.
(a) For a test with $\alpha = 0.01$, compute β for the sample sizes $n = 25$, 100, 400, and 2500 assuming that $\mu = 86$.
(b) Suppose the sample average is $\bar{x} = 86$. Find the P-value for the test statistic for the different sample sizes specified in part (a). Would the data be statistically significant at $\alpha = 0.01$?
(c) Comment on the use of a large sample size in this problem.

9-101. A manufacturer of semiconductor devices takes a random sample of size n of chips and tests them, classifying each chip as defective or nondefective. Let $X_i = 0$ if the chip is nondefective and $X_i = 1$ if the chip is defective. The sample fraction defective is

$$\hat{p} = \frac{X_1 + X_2 + \cdots + X_n}{n}$$

What are the sampling distribution, the sample mean, and sample variance estimates of $\hat{p}$ when
(a) The sample size is $n = 50$?
(b) The sample size is $n = 80$?
(c) The sample size is $n = 100$?
(d) Compare your answers to parts (a)–(c) and comment on the effect of sample size on the variance of the sampling distribution.

9-102. Consider the situation of Exercise 9-101. After collecting a sample, we are interested in testing H_0: $p = 0.10$ versus H_1: $p \neq 0.10$ with $\alpha = 0.05$. For each of the following situations, compute the p-value for this test:
(a) $n = 50$, $\hat{p} = 0.095$
(b) $n = 100$, $\hat{p} = 0.095$
(c) $n = 500$, $\hat{p} = 0.095$
(d) $n = 1000$, $\hat{p} = 0.095$
(e) Comment on the effect of sample size on the observed P-value of the test.

9-103. An inspector of flow metering devices used to administer fluid intravenously will perform a hypothesis test to determine whether the mean flow rate is different from the flow rate setting of 200 milliliters per hour. Based on prior information the standard deviation of the flow rate is assumed to be known and equal to 12 milliliters per hour. For each of the following sample sizes, and a fixed $\alpha = 0.05$, find the probability of a type II error if the true mean is 205 milliliters per hour.
(a) $n = 20$
(b) $n = 50$
(c) $n = 100$
(d) Does the probability of a type II error increase or decrease as the sample size increases? Explain your answer.

9-104. Suppose that in Exercise 9-103, the experimenter had believed that $\sigma = 14$. For each of the following sample sizes, and a fixed $\alpha = 0.05$, find the probability of a type II error if the true mean is 205 milliliters per hour.
(a) $n = 20$
(b) $n = 50$
(c) $n = 100$
(d) Comparing your answers to those in Exercise 9-103, does the probability of a type II error increase or decrease with the increase in standard deviation? Explain your answer.

9-105. The marketers of shampoo products know that customers like their product to have a lot of foam. A manufacturer of shampoo claims that the foam height of his product exceeds 200 millimeters. It is known from prior experience that the standard deviation of foam height is 8 millimeters. For each of the following sample sizes, and a fixed $\alpha = 0.05$, find the power of the test if the true mean is 204 millimeters.
(a) $n = 20$
(b) $n = 50$
(c) $n = 100$
(d) Does the power of the test increase or decrease as the sample size increases? Explain your answer.

9-106. Suppose we are testing H_0: $p = 0.5$ versus H_0: $p \neq 0.5$. Suppose that p is the true value of the population proportion.
(a) Using $\alpha = 0.05$, find the power of the test for $n = 100$, 150, and 300 assuming that $p = 0.6$. Comment on the effect of sample size on the power of the test.
(b) Using $\alpha = 0.01$, find the power of the test for $n = 100$, 150, and 300 assuming that $p = 0.6$. Compare your answers to those from part (a) and comment on the effect of α on the power of the test for different sample sizes.
(c) Using $\alpha = 0.05$, find the power of the test for $n = 100$, assuming $p = 0.08$. Compare your answer to part (a) and comment on the effect of the true value of p on the power of the test for the same sample size and α level.
(d) Using $\alpha = 0.01$, what sample size is required if $p = 0.6$ and we want $\beta = 0.05$? What sample is required if $p = 0.8$ and we want $\beta = 0.05$? Compare the two sample sizes and comment on the effect of the true value of p on sample size required when β is held approximately constant.

9-107. The cooling system in a nuclear submarine consists of an assembly of welded pipes through which a coolant is circulated. Specifications require that weld strength must meet or exceed 150 psi.
(a) Suppose that the design engineers decide to test the hypothesis H_0: $\mu = 150$ versus H_1: $\mu > 150$. Explain why this choice of alternative hypothesis is better than H_1: $\mu < 150$.

(b) A random sample of 20 welds results in $\bar{x} = 153.7$ psi and $s = 11.3$ psi. What conclusions can you draw about the hypothesis in part (a)? State any necessary assumptions about the underlying distribution of the data.

9-108. The mean pull-off force of an adhesive used in manufacturing a connector for an automotive engine application should be at least 75 pounds. This adhesive will be used unless there is strong evidence that the pull-off force does not meet this requirement. A test of an appropriate hypothesis is to be conducted with sample size $n = 10$ and $\alpha = 0.05$. Assume that the pull-off force is normally distributed, and σ is not known.

(a) If the true standard deviation is $\sigma = 1$, what is the risk that the adhesive will be judged acceptable when the true mean pull-off force is only 73 pounds? Only 72 pounds?
(b) What sample size is required to give a 90% chance of detecting that the true mean is only 72 pounds when $\sigma = 1$?
(c) Rework parts (a) and (b) assuming that $\sigma = 2$. How much impact does increasing the value of σ have on the answers you obtain?

9-109. A manufacturer of precision measuring instruments claims that the standard deviation in the use of the instruments is at most 0.00002 millimeter. An analyst, who is unaware of the claim, uses the instrument eight times and obtains a sample standard deviation of 0.00001 millimeter.

(a) Confirm using a test procedure and an α level of 0.01 that there is insufficient evidence to support the claim that the standard deviation of the instruments is at most 0.00002. State any necessary assumptions about the underlying distribution of the data.
(b) Explain why the sample standard deviation, $s = 0.00001$, is less than 0.00002, yet the statistical test procedure results do not support the claim.

9-110. A biotechnology company produces a therapeutic drug whose concentration has a standard deviation of 4 grams per liter. A new method of producing this drug has been proposed, although some additional cost is involved. Management will authorize a change in production technique only if the standard deviation of the concentration in the new process is less than 4 grams per liter. The researchers chose $n = 10$ and obtained the following data in grams per liter. Perform the necessary analysis to determine whether a change in production technique should be implemented.

16.628	16.630
16.622	16.631
16.627	16.624
16.623	16.622
16.618	16.626

9-111. Consider the 40 observations collected on the number of nonconforming coil springs in production batches of size 50 given in Exercise 6-85.

(a) Based on the description of the random variable and these 40 observations, is a binomial distribution an appropriate model? Perform a goodness-of-fit procedure with $\alpha = 0.05$.
(b) Calculate the P-value for this test.

9-112. Consider the 20 observations collected on the number of errors in a string of 1000 bits of a communication channel given in Exercise 6-86.

(a) Based on the description of the random variable and these 20 observations, is a binomial distribution an appropriate model? Perform a goodness-of-fit procedure with $\alpha = 0.05$.
(b) Calculate the P-value for this test.

9-113. Consider the spot weld shear strength data in Exercise 6-23. Does the normal distribution seem to be a reasonable model for these data? Perform an appropriate goodness-of-fit test to answer this question.

9-114. Consider the water quality data in Exercise 6-24.

(a) Do these data support the claim that mean concentration of suspended solids does not exceed 50 parts per million? Use $\alpha = 0.05$.
(b) What is the P-value for the test in part (a)?
(c) Does the normal distribution seem to be a reasonable model for these data? Perform an appropriate goodness-of-fit test to answer this question.

9-115. Consider the golf ball overall distance data in Exercise 6-25.

(a) Do these data support the claim that the mean overall distance for this brand of ball does not exceed 270 yards? Use $\alpha = 0.05$.
(b) What is the P-value for the test in part (a)?
(c) Do these data appear to be well modeled by a normal distribution? Use a formal goodness-of-fit test in answering this question.

9-116. Consider the baseball coefficient of restitution data in Exercise 8-86. If the mean coefficient of restitution exceeds 0.635, the population of balls from which the sample has been taken will be too "lively" and considered unacceptable for play.

(a) Formulate an appropriate hypothesis testing procedure to answer this question.
(b) Test these hypotheses and draw conclusions, using $\alpha = 0.01$.
(c) Find the P-value for this test.
(d) In Exercise 8-86(b), you found a 99% confidence interval on the mean coefficient of restitution. Does this interval, or a one-sided CI, provide additional useful information to the decision maker? Explain why or why not.

9-117. Consider the dissolved oxygen data in Exercise 8-88. Water quality engineers are interested in knowing whether these data support a claim that mean dissolved oxygen concentration is 2.5 milligrams per liter.

(a) Formulate an appropriate hypothesis testing procedure to investigate this claim.
(b) Test these hypotheses and draw conclusions, using $\alpha = 0.05$.
(c) Find the P-value for this test.

(d) In Exercise 8-88(b) you found a 95% CI on the mean dissolved oxygen concentration. Does this interval provide useful additional information beyond that of the hypothesis testing results? Explain your answer.

9-118. An article in *Food Testing and Analysis*, "Improving Reproducibility of Refractometry Measurements of Fruit Juices" (1999, Vol. 4, No. 4, pp. 13–17) measured the sugar concentration (Brix) in clear apple juice. All readings were taken at 20°C:

11.48	11.45	11.48	11.47	11.48
11.50	11.42	11.49	11.45	11.44
11.45	11.47	11.46	11.47	11.43
11.50	11.49	11.45	11.46	11.47

(a) Test the hypothesis H_0: $\mu = 11.5$versus H_1: $\mu \neq 11.5$ using $\alpha = 0.05$. Find the P-value.
(b) Compute the power of the test if the true mean is 11.4.
(c) What sample size would be required to detect a true mean sugar concentration of 11.45 if we wanted the power of test to be at least 0.9?
(d) Explain how the question in part (a) could be answered by constructing a two-sided confidence interval on the mean sugar concentration.
(e) Is there evidence to support the assumption that the sugar concentration is normally distributed?

9-119. An article in *Growth: A Journal Devoted to Problems of Normal and Abnormal Growth*, "Comparison of Measured and Estimated Fat-Free Weight, Fat, Potassium and Nitrogen of Growing Guinea Pigs" (1982, Vol. 46, No. 4, pp. 306–321) measured the body weight (grams) for guinea pigs at their birth.

421.0	452.6	456.1	494.6	373.8
90.5	110.7	96.4	81.7	102.4
241.0	296.0	317.0	290.9	256.5
447.8	687.6	705.7	879.0	88.8
296.0	273.0	268.0	227.5	279.3
258.5	296.0			

(a) Test the hypothesis that mean body weight is 300 grams. Use $\alpha = 0.05$.
(b) What is the smallest level of significance at which you would be willing to reject the null hypothesis?
(c) Explain how you could answer the question in part (a) with a two-sided confidence interval on mean body weight.

9-120. An article in *Food Chemistry*, "A Study of Factors Affecting Extraction of Peanut (*Arachis Hypgaea* L.) Solids with Water" (1991, Vol. 42, No. 2, pp. 153–165) found the percent protein extracted from peanut milk as follows:

78.3	77.1	71.3	84.5	87.8	75.7	64.8	72.5
78.2	91.2	86.2	80.9	82.1	89.3	89.4	81.6

(a) Can you support a claim that mean percent protein extracted exceeds 80 percent? Use $\alpha = 0.05$.
(b) Is there evidence that percent protein extracted is normally distributed?
(c) What is the P-value of the test statistic computed in part (a)?

9-121. An article in *Biological Trace Element Research*, "Interaction of Dietary Calcium, Manganese, and Manganese Source (Mn Oxide or Mn Methionine Complex) or Chick Performance and Manganese Utilization" (1991, Vol. 29, No. 3, pp. 217–228) showed the following results of tissue assay for liver manganese (ppm) in chicks fed high-Ca diets.

6.02	6.08	7.11	5.73	5.32	7.10
5.29	5.84	6.03	5.99	4.53	6.81

(a) Test the hypothesis H_0: $\sigma^2 = 0.6$ versus H_1: $\sigma^2 \neq 0.6$ using $\alpha = 0.01$.
(b) What is the P-value for this test?
(c) Discuss how part (a) could be answered by constructing a 99% two-sided confidence interval for σ.

9-122. An article in *Experimental Brain Research*, "Synapses in the Granule Cell Layer of the Rat Dentate Gyrus: Serial-Sectionin Study" (1996, Vol. 112, No. 2, pp. 237–243) showed the ratio between the numbers of symmetrical and total synapses on somata and azon initial segments of reconstructed granule cells in the dentate gyrus of a 12-week-old rat:

0.65	0.90	0.78	0.94	0.40	0.94
0.91	0.86	0.53	0.84	0.42	0.50
0.50	0.68	1.00	0.57	1.00	1.00
0.84	0.9	0.91	0.92	0.96	
0.96	0.56	0.67	0.96	0.52	
0.89	0.60	0.54			

(a) Use the data to test H_0: $\sigma^2 = 0.02$ versus H_1: $\sigma^2 \neq 0.02$ using $\alpha = 0.05$.
(b) Find the P-value for the test.

9-123. An article in the *Journal of Electronic Material*, "Progress in CdZnTe Substrate Producibility and Critical Drive of IRFPA Yield Originating with CdZnTe Substrates" (1998, Vol. 27, No. 6, pp. 564–572) improved the quality of CdZnTe substrates used to produce the HgCdTe infrared focal plane arrays (IRFPAs), also defined as sensor chip assemblies (SCAs). The cut-on wave length (μm) on 11 wafers was measured and shown below:

6.06 6.16 6.57 6.67 6.98 6.17 6.17 6.93 6.73 6.87 6.76

(a) Is there evidence that the mean of cut-on wave length is not 6.50 μm?
(b) What is the P-value for this test?
(c) What sample size would be required to detect a true mean cut-on wave length of 6.25 μm with probability 95%?

(d) What is the type II error probability if the true mean cut-on wave length is 6.95 μm?

9-124. Consider the fatty acid measurements for the diet margarine described in Exercise 8-32.

(a) For the sample size $n = 6$, using a two-sided alternative hypothesis and $\alpha = 0.01$, test H_0: $\sigma^2 = 1.0$.
(b) Suppose instead of $n = 6$, the sample size was $n = 51$. Repeat the analysis performed in part (a) using $n = 51$.
(c) Compare your answers and comment on how sample size affects your conclusions drawn in parts (a) and (b).

9-125. Consider the television picture tube brightness experiment described in Exercise 8-31.

(a) For the sample size $n = 10$, do the data support the claim that the standard deviation of current is less than 20 microamps?
(b) Suppose instead of $n = 10$, the sample size was 51. Repeat the analysis performed in part (a) using $n = 51$.
(c) Compare your answers and comment on how sample size affects your conclusions drawn in parts (a) and (b).

MIND-EXPANDING EXERCISES

9-126. Suppose that we wish to test H_0: $\mu = \mu_0$ versus H_1: $\mu \neq \mu_0$, where the population is normal with known σ. Let $0 < \epsilon < \alpha$, and define the critical region so that we will reject H_0 if $z_0 > z_\epsilon$ or if $z_0 < -z_{\alpha-\epsilon}$, where z_0 is the value of the usual test statistic for these hypotheses.

(a) Show that the probability of type I error for this test is α.
(b) Suppose that the true mean is $\mu_1 = \mu_0 + \delta$. Derive an expression for β for the above test.

9-127. Derive an expression for β for the test on the variance of a normal distribution. Assume that the two-sided alternative is specified.

9-128. When $X_1, X_2, \ldots, X_n$ are independent Poisson random variables, each with parameter λ, and n is large, the sample mean $\overline{X}$ has an approximate normal distribution with mean λ and variance λ/n. Therefore,

$$Z = \frac{\overline{X} - \lambda}{\sqrt{\lambda/n}}$$

has approximately a standard normal distribution. Thus we can test H_0: $\lambda = \lambda_0$ by replacing λ in Z by λ_0. When X_i are Poisson variables, this test is preferable to the large-sample test of Section 9-2.3, which would use $S/\sqrt{n}$ in the denominator, because it is designed just for the Poisson distribution. Suppose that the number of open circuits on a semiconductor wafer has a Poisson distribution. Test data for 500 wafers indicate a total of 1038 opens. Using $\alpha = 0.05$, does this suggest that the mean number of open circuits per wafer exceeds 2.0?

9-129. When $X_1, X_2, \ldots, X_n$ is a random sample from a normal distribution and n is large, the sample standard deviation has approximately a normal distribution with mean σ and variance $\sigma^2/(2n)$. Therefore, a large-sample test for H_0: $\sigma = \sigma_0$ can be based on the statistic

$$Z = \frac{S - \sigma_0}{\sqrt{\sigma_0^2/(2n)}}$$

(a) Use this result to test H_0: $\sigma = 10$ versus H_1: $\sigma < 10$ for the golf ball overall distance data in Exercise 6-25.
(b) Find an approximately unbiased estimator of the 95 percentile $\theta = \mu + 1.645\sigma$. From the fact that $\overline{X}$ and S are independent random variables, find the standard error of the estimator of θ. How would you estimate the standard error?
(c) Consider the golf ball overall distance data in Exercise 6-25. We wish to investigate a claim that the 95 percentile of overall distance does not exceed 285 yards. Construct a test statistic that can be used for testing the appropriate hypotheses. Apply this procedure to the data from Exercise 6-25. What are your conclusions?

9-130. Let $X_1, X_2, \ldots, X_n$ be a sample from an exponential distribution with parameter λ. It can be shown that $2\lambda \sum_{i=1}^{n} X_i$ has a chi-square distribution with $2n$ degrees of freedom. Use this fact to devise a test statistic and critical region for H_0: $\lambda = \lambda_0$ versus the three usual alternatives.

IMPORTANT TERMS AND CONCEPTS

α and β
Connection between hypothesis tests and confidence intervals
Critical region for a test statistic
Hypothesis test
Inference
Null distribution
Null hypothesis
One- and two-sided alternative hypotheses
Operating characteristic (OC) curves
Power of a test
P-value
Reference distribution for a test statistic
Sample size determination for hypothesis tests
Significance level of a test
Statistical hypotheses
Statistical versus practical significance
Test for goodness of fit
Test for homogeneity
Test for independence
Test statistic
Type I and type II errors

10 Statistical Inference: Two Samples

CHAPTER OUTLINE

LEARNING OBJECTIVES

After careful study of this chapter, you should be able to do the following:

1. Structure comparative experiments involving two samples as hypothesis tests
2. Test hypotheses and construct confidence intervals on the difference in means of two normal distributions
3. Test hypotheses and construct confidence intervals on the ratio of the variances or standard deviations of two normal distributions
4. Test hypotheses and construct confidence intervals on the difference in two population proportions
5. Use the *P*-value approach for making decisions in hypotheses tests
6. Compute power, type II error probability, and make sample size decisions for two-sample tests on means, variances, and proportions
7. Explain and use the relationship between confidence intervals and hypothesis tests

10-1 INTRODUCTION

The previous two chapters presented hypothesis tests and confidence intervals for a single population parameter (the mean μ, the variance σ^2, or a proportion p). This chapter extends those results to the case of two independent populations.

The general situation is shown in Fig. 10-1. Population 1 has mean μ_1 and variance σ_1^2, while population 2 has mean μ_2 and variance σ_2^2. Inferences will be based on two random samples of sizes n_1 and n_2, respectively. That is, $X_{11}, X_{12}, p, X_{1n_1}$ is a random sample of n_1 observations from population 1, and $X_{21}, X_{22}, p, X_{2n_2}$ is a random sample of n_2 observations from population 2. Most of the practical applications of the procedures in this chapter arise in the context of **simple comparative experiments** in which the objective is to study the difference in the parameters of the two populations.

Engineers and scientists are often interested in comparing two different conditions to determine whether either condition produces a significant effect on the response that is observed. These conditions are sometimes called **treatments.** Example 10-1 illustrates such an experiment; the two different treatments are two paint formulations, and the response is the drying time. The purpose of the study is to determine whether the new formulation results in a significant effect—reducing drying time. In this situation, the product developer (the experimenter) randomly assigned 10 test specimens to one formulation and 10 test specimens to the other formulation. Then the paints were applied to the test specimens in random order until all 20 specimens were painted. This is an example of a **completely randomized experiment.**

When statistical significance is observed in a randomized experiment, the experimenter can be confident in the conclusion that it was the difference in treatments that resulted in the difference in response. That is, we can be confident that a **cause-and-effect** relationship has been found.

Sometimes the objects to be used in the comparison are not assigned at random to the treatments. For example, the September 1992 issue of *Circulation* (a medical journal

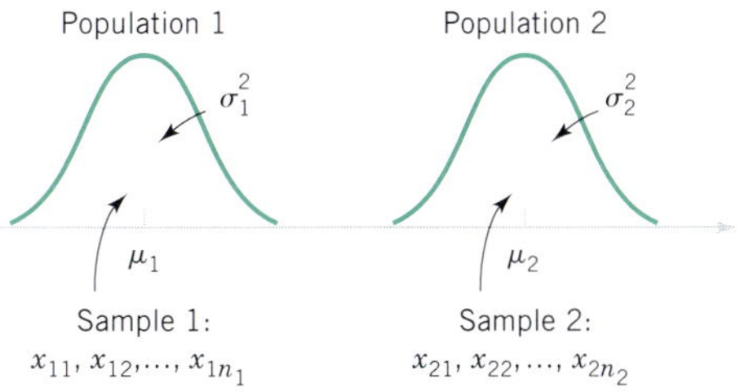

Figure 10-1 Two independent populations.

published by the American Heart Association) reports a study linking high iron levels in the body with increased risk of heart attack. The study, done in Finland, tracked 1931 men for five years and showed a statistically significant effect of increasing iron levels on the incidence of heart attacks. In this study, the comparison was not performed by randomly selecting a sample of men and then assigning some to a "low iron level" treatment and the others to a "high iron level" treatment. The researchers just tracked the subjects over time. Recall from Chapter 1 that this type of study is called an **observational study.**

It is difficult to identify causality in observational studies, because the observed statistically significant difference in response between the two groups may be due to some other underlying factor (or group of factors) that was not equalized by randomization and not due to the treatments. For example, the difference in heart attack risk could be attributable to the difference in iron levels, or to other underlying factors that form a reasonable explanation for the observed results—such as cholesterol levels or hypertension.

10-2 INFERENCE ON THE DIFFERENCE IN MEANS OF TWO NORMAL DISTRIBUTIONS, VARIANCES KNOWN

In this section we consider statistical inferences on the difference in means $\mu_1 - \mu_2$ of two normal distributions, where the variances σ_1^2 and σ_2^2 are known. The assumptions for this section are summarized as follows.

Assumptions for Two-Sample Inference

1. $X_{11}, X_{12}, \ldots, X_{1n_1}$ is a random sample from population 1.
2. $X_{21}, X_{22}, \ldots, X_{2n_2}$ is a random sample from population 2.
3. The two populations represented by X_1 and X_2 are independent.
4. Both populations are normal.

A logical point estimator of $\mu_1 - \mu_2$ is the difference in sample means $\overline{X}_1 - \overline{X}_2$. Based on the properties of expected values

$$E(\overline{X}_1 - \overline{X}_2) = E(\overline{X}_1) - E(\overline{X}_2) = \mu_1 - \mu_2$$

and the variance of $\overline{X}_1 - \overline{X}_2$ is

$$V(\overline{X}_1 - \overline{X}_2) = V(\overline{X}_1) + V(\overline{X}_2) = \frac{\sigma_1^2}{n_1} + \frac{\sigma_2^2}{n_2}$$

Based on the assumptions and the preceding results, we may state the following.

The quantity

$$Z = \frac{\overline{X}_1 - \overline{X}_2 - (\mu_1 - \mu_2)}{\sqrt{\dfrac{\sigma_1^2}{n_1} + \dfrac{\sigma_2^2}{n_2}}} \tag{10-1}$$

has a $N(0, 1)$ distribution.

This result will be used to form tests of hypotheses and confidence intervals on $\mu_1 - \mu_2$. Essentially, we may think of $\mu_1 - \mu_2$ as a parameter θ, and its estimator is $\hat{\Theta} = \overline{X}_1 - \overline{X}_2$ with variance $\sigma_{\hat{\Theta}}^2 = \sigma_1^2/n_1 + \sigma_2^2/n_2$. If θ_0 is the null hypothesis value specified for θ, the test statistic will be $(\hat{\Theta} - \theta_0)/\sigma_{\hat{\Theta}}$. Notice how similar this is to the test statistic for a single mean used in Equation 9-8 of Chapter 9.

10-2.1 Hypothesis Tests on the Difference in Means, Variances Known

We now consider hypothesis testing on the difference in the means $\mu_1 - \mu_2$ of two normal populations. Suppose that we are interested in testing that the difference in means $\mu_1 - \mu_2$ is equal to a specified value Δ_0. Thus, the null hypothesis will be stated as H_0: $\mu_1 - \mu_2 = \Delta_0$. Obviously, in many cases, we will specify $\Delta_0 = 0$ so that we are testing the equality of two means (i.e., H_0: $\mu_1 = \mu_2$). The appropriate test statistic would be found by replacing $\mu_1 - \mu_2$ in Equation 10-1 by Δ_0, and this test statistic would have a standard normal distribution under H_0. That is, the standard normal distribution is the **reference distribution** for the test statistic. Suppose that the alternative hypothesis is H_1: $\mu_1 - \mu_2 \neq \Delta_0$. Now, a sample value of $\bar{x}_1 - \bar{x}_2$ that is considerably different from Δ_0 is evidence that H_1 is true. Because Z_0 has the $N(0, 1)$ distribution when H_0 is true, we would take $-z_{\alpha/2}$ and $z_{\alpha/2}$ as the boundaries of the critical region just as we did in the single-sample hypothesis-testing problem of Section 9-2.1. This would give a test with level of significance α. Critical regions for the one-sided alternatives would be located similarly. Formally, we summarize these results below.

Tests on the Difference in Means, Variances Known

Null hypothesis: H_0: $\mu_1 - \mu_2 = \Delta_0$

Test statistic:
$$Z_0 = \frac{\overline{X}_1 - \overline{X}_2 - \Delta_0}{\sqrt{\dfrac{\sigma_1^2}{n_1} + \dfrac{\sigma_2^2}{n_2}}} \tag{10-2}$$

Alternative Hypotheses	Rejection Criterion
H_1: $\mu_1 - \mu_2 \neq \Delta_0$	$z_0 > z_{\alpha/2}$ or $z_0 < -z_{\alpha/2}$
H_1: $\mu_1 - \mu_2 > \Delta_0$	$z_0 > z_\alpha$
H_1: $\mu_1 - \mu_2 < \Delta_0$	$z_0 < -z_\alpha$

EXAMPLE 10-1
Paint Drying Time

A product developer is interested in reducing the drying time of a primer paint. Two formulations of the paint are tested; formulation 1 is the standard chemistry, and formulation 2 has a new drying ingredient that should reduce the drying time. From experience, it is known that the standard deviation of drying time is 8 minutes, and this inherent variability should be unaffected by the addition of the new ingredient. Ten specimens are painted with formulation 1, and another 10 specimens are painted with formulation 2; the 20 specimens are painted in random order. The two sample average drying times are $\bar{x}_1 = 121$ minutes and $\bar{x}_2 = 112$ minutes, respectively. What conclusions can the product developer draw about the effectiveness of the new ingredient, using $\alpha = 0.05$?

We apply the eight-step procedure to this problem as follows:

1. The quantity of interest is the difference in mean drying times, $\mu_1 - \mu_2$, and $\Delta_0 = 0$.
2. H_0: $\mu_1 - \mu_2 = 0$, or H_0: $\mu_1 = \mu_2$.
3. H_1: $\mu_1 > \mu_2$. We want to reject H_0 if the new ingredient reduces mean drying time.

4. $\alpha = 0.05$

5. The test statistic is

$$z_0 = \frac{\bar{x}_1 - \bar{x}_2 - 0}{\sqrt{\dfrac{\sigma_1^2}{n_1} + \dfrac{\sigma_2^2}{n_2}}}$$

where $\sigma_1^2 = \sigma_2^2 = (8)^2 = 64$ and $n_1 = n_2 = 10$.

6. Reject H_0: $\mu_1 = \mu_2$ if $z_0 > 1.645 = z_{0.05}$.

7. Computations: Since $\bar{x}_1 = 121$ minutes and $\bar{x}_2 = 112$ minutes, the test statistic is

$$z_0 = \frac{121 - 112}{\sqrt{\dfrac{(8)^2}{10} + \dfrac{(8)^2}{10}}} = 2.52$$

8. Conclusion: Since $z_0 = 2.52 > 1.645$, we reject H_0: $\mu_1 = \mu_2$ at the $\alpha = 0.05$ level and conclude that adding the new ingredient to the paint significantly reduces the drying time. Alternatively, we can find the P-value for this test as

$$P\text{-value} = 1 - \Phi(2.52) = 0.0059$$

Therefore, H_0: $\mu_1 = \mu_2$ would be rejected at any significance level $\alpha \geq 0.0059$.

When the population variances are unknown, the sample variances s_1^2 and s_2^2 can be substituted into the test statistic Equation 10-2 to produce a **large-sample test** for the difference in means. This procedure will also work well when the populations are not necessarily normally distributed. However, both n_1 and n_2 should exceed 40 for this large-sample test to be valid.

10-2.2 Type II Error and Choice of Sample Size

Use of Operating Characteristic Curves

The operating characteristic curves (OC) in Appendix Charts VII*a*, VII*b*, VII*c*, and VII*d* may be used to evaluate the type II error probability for the hypotheses in the display (10-2). These curves are also useful in determining sample size. Curves are provided for $\alpha = 0.05$ and $\alpha = 0.01$. For the two-sided alternative hypothesis, the abscissa scale of the operating characteristic curve in charts VII*a* and VII*b* is d, where

$$d = \frac{|\mu_1 - \mu_2 - \Delta_0|}{\sqrt{\sigma_1^2 + \sigma_2^2}} = \frac{|\Delta - \Delta_0|}{\sqrt{\sigma_1^2 + \sigma_2^2}} \tag{10-3}$$

and one must choose equal sample sizes, say, $n = n_1 = n_2$. The one-sided alternative hypotheses require the use of Charts VII*c* and VII*d*. For the one-sided alternatives H_1: $\mu_1 - \mu_2 > \Delta_0$ or H_1: $\mu_1 - \mu_2 < \Delta_0$, the abscissa scale is also given by

$$d = \frac{|\mu_1 - \mu_2 - \Delta_0|}{\sqrt{\sigma_1^2 + \sigma_2^2}} = \frac{|\Delta - \Delta_0|}{\sqrt{\sigma_1^2 + \sigma_2^2}}$$

It is not unusual to encounter problems where the costs of collecting data differ substantially between the two populations, or where one population variance is much greater than the other.

In those cases, we often use unequal sample sizes. If $n_1 \neq n_2$, the operating characteristic curves may be entered with an *equivalent* value of n computed from

$$n = \frac{\sigma_1^2 + \sigma_2^2}{\sigma_1^2/n_1 + \sigma_2^2/n_2} \tag{10-4}$$

If $n_1 \neq n_2$, and their values are fixed in advance, Equation 10-4 is used directly to calculate n, and the operating characteristic curves are entered with a specified d to obtain β. If we are given d and it is necessary to determine n_1 and n_2 to obtain a specified β, say, β^*, we guess at trial values of n_1 and n_2, calculate n in Equation 10-4, and enter the curves with the specified value of d to find β. If $\beta = \beta^*$, the trial values of n_1 and n_2 are satisfactory. If $\beta \neq \beta^*$, adjustments to n_1 and n_2 are made and the process is repeated.

EXAMPLE 10-2
Paint Drying Time, Sample Size from OC Curves

Consider the paint drying time experiment from Example 10-1. If the true difference in mean drying times is as much as 10 minutes, find the sample sizes required to detect this difference with probability at least 0.90.

The appropriate value of the abscissa parameter is (since $\Delta_0 = 0$, and $\Delta = 10$)

$$d = \frac{|\mu_1 - \mu_2|}{\sqrt{\sigma_1^2 + \sigma_2^2}} = \frac{10}{\sqrt{8^2 + 8^2}} = 0.88$$

and since the detection probability or power of the test must be at least 0.9, with $\alpha = 0.05$, we find from Appendix Chart VII*c* that $n = n_1 = n_2 \simeq 11$.

Sample Size Formulas

It is also possible to obtain formulas for calculating the sample sizes directly. Suppose that the null hypothesis H_0: $\mu_1 - \mu_2 = \Delta_0$ is false and that the true difference in means is $\mu_1 - \mu_2 = \Delta$, where $\Delta > \Delta_0$. One may find formulas for the sample size required to obtain a specific value of the type II error probability β for a given difference in means Δ and level of significance α.

For example, we first write the expression for the β-error for the two-sided alternative, which is

$$\beta = \Phi\left(z_{\alpha/2} - \frac{\Delta - \Delta_0}{\sqrt{\dfrac{\sigma_1^2}{n_1} + \dfrac{\sigma_2^2}{n_2}}} \right) - \Phi\left(-z_{\alpha/2} - \frac{\Delta - \Delta_0}{\sqrt{\dfrac{\sigma_1^2}{n_1} + \dfrac{\sigma_2^2}{n_2}}} \right)$$

The derivation for sample size closely follows the single-sample case in Section 9-2.2.

Sample Size for a Two-Sided Test on the Difference in Means with $n_1 = n_2$, Variances Known

For the two-sided alternative hypothesis with significance level α, the sample size $n_1 = n_2 = n$ required to detect a true difference in means of Δ with power at least $1 - \beta$ is

$$n \simeq \frac{(z_{\alpha/2} + z_\beta)^2(\sigma_1^2 + \sigma_2^2)}{(\Delta - \Delta_0)^2} \tag{10-5}$$

This approximation is valid when $\Phi(-z_{\alpha/2} - (\Delta - \Delta_0)\sqrt{n}/\sqrt{\sigma_1^2 + \sigma_2^2})$ is small compared to β.

Sample Size for a One-Sided Test on the Difference in Means with $n_1 = n_2$, Variances Known

For a one-sided alternative hypothesis with significance level α, the sample size $n_1 = n_2 = n$ required to detect a true difference in means of $\Delta(\neq\Delta_0)$ with power at least $1 - \beta$ is

$$n = \frac{(z_\alpha + z_\beta)^2(\sigma_1^2 + \sigma_2^2)}{(\Delta - \Delta_0)^2} \tag{10-6}$$

where Δ is the true difference in means of interest. Then by following a procedure similar to that used to obtain Equation 9-17, the expression for β can be obtained for the case where $n = n_1 = n_2$.

EXAMPLE 10-3 Paint Drying Time Sample Size

To illustrate the use of these sample size equations, consider the situation described in Example 10-1, and suppose that if the true difference in drying times is as much as 10 minutes, we want to detect this with probability at least 0.90. Under the null hypothesis, $\Delta_0 = 0$. We have a one-sided alternative hypothesis with $\Delta = 10$, $\alpha = 0.05$ (so $z_\alpha = z_{0.05} = 1.645$), and since the power is 0.9, $\beta = 0.10$ (so $z_\beta = z_{0.10} = 1.28$). Therefore we may find the required sample size from Equation 10-6 as follows:

$$n = \frac{(z_\alpha + z_\beta)^2(\sigma_1^2 + \sigma_2^2)}{(\Delta - \Delta_0)^2} = \frac{(1.645 + 1.28)^2[(8)^2 + (8)^2]}{(10 - 0)^2} = 11$$

This is exactly the same as the result obtained from using the OC curves.

10-2.3 Confidence Interval on the Difference in Means, Variances Known

The $100(1 - \alpha)\%$ confidence interval on the difference in two means $\mu_1 - \mu_2$ when the variances are known can be found directly from results given previously in this section. Recall that $X_{11}, X_{12}, \ldots, X_{1n_1}$ is a random sample of n_1 observations from the first population and $X_{21}, X_{22}, \ldots, X_{2n_2}$ is a random sample of n_2 observations from the second population. The difference in sample means $\overline{X}_1 - \overline{X}_2$ is a point estimator of $\mu_1 - \mu_2$, and

$$Z = \frac{\overline{X}_1 - \overline{X}_2 - (\mu_1 - \mu_2)}{\sqrt{\dfrac{\sigma_1^2}{n_1} + \dfrac{\sigma_2^2}{n_2}}}$$

has a standard normal distribution if the two populations are normal or is approximately standard normal if the conditions of the central limit theorem apply, respectively. This implies that $P(-z_{\alpha/2} \leq Z \leq z_{\alpha/2}) = 1 - \alpha$, or

$$P\left[-z_{\alpha/2} \leq \frac{\overline{X}_1 - \overline{X}_2 - (\mu_1 - \mu_2)}{\sqrt{\dfrac{\sigma_1^2}{n_1} + \dfrac{\sigma_2^2}{n_2}}} \leq z_{\alpha/2}\right] = 1 - \alpha$$

This can be rearranged as

$$P\left(\overline{X}_1 - \overline{X}_2 - z_{\alpha/2}\sqrt{\frac{\sigma_1^2}{n_1} + \frac{\sigma_2^2}{n_2}} \le \mu_1 - \mu_2 \le \overline{X}_1 - \overline{X}_2 + z_{\alpha/2}\sqrt{\frac{\sigma_1^2}{n_1} + \frac{\sigma_2^2}{n_2}}\right) = 1 - \alpha$$

Therefore, the $100(1 - \alpha)\%$ confidence interval for $\mu_1 - \mu_2$ is defined as follows.

Confidence Interval on the Difference in Means, Variances Known

If $\bar{x}_1$ and $\bar{x}_2$ are the means of independent random samples of sizes n_1 and n_2 from two independent normal populations with known variances σ_1^2 and σ_2^2, respectively, a **$100(1 - \alpha)\%$ confidence interval for $\mu_1 - \mu_2$** is

$$\bar{x}_1 - \bar{x}_2 - z_{\alpha/2}\sqrt{\frac{\sigma_1^2}{n_1} + \frac{\sigma_2^2}{n_2}} \le \mu_1 - \mu_2 \le \bar{x}_1 - \bar{x}_2 + z_{\alpha/2}\sqrt{\frac{\sigma_1^2}{n_1} + \frac{\sigma_2^2}{n_2}} \quad (10\text{-}7)$$

where $z_{\alpha/2}$ is the upper $\alpha/2$ percentage point of the standard normal distribution.

The confidence level $1 - \alpha$ is exact when the populations are normal. For nonnormal populations, the confidence level is approximately valid for large sample sizes.

EXAMPLE 10-4
Aluminum Tensile Strength

Tensile strength tests were performed on two different grades of aluminum spars used in manufacturing the wing of a commercial transport aircraft. From past experience with the spar manufacturing process and the testing procedure, the standard deviations of tensile strengths are assumed to be known. The data obtained are as follows: $n_1 = 10$, $\bar{x}_1 = 87.6$, $\sigma_1 = 1$, $n_2 = 12$, $\bar{x}_2 = 74.5$, and $\sigma_2 = 1.5$. If μ_1 and μ_2 denote the true mean tensile strengths for the two grades of spars, we may find a 90% confidence interval on the difference in mean strength $\mu_1 - \mu_2$ as follows:

$$\bar{x}_1 - \bar{x}_2 - z_{\alpha/2}\sqrt{\frac{\sigma_1^2}{n_1} + \frac{\sigma_2^2}{n_2}} \le \mu_1 - \mu_2 \le \bar{x}_1 - \bar{x}_2 + z_{\alpha/2}\sqrt{\frac{\sigma_1^2}{n_1} + \frac{\sigma_2^2}{n_2}}$$

$$87.6 - 74.5 - 1.645\sqrt{\frac{(1)^2}{10} + \frac{(1.5)^2}{12}} \le \mu_1 - \mu_2 \le 87.6 - 74.5 + 1.645\sqrt{\frac{(1^2)}{10} + \frac{(1.5)^2}{12}}$$

Therefore, the 90% confidence interval on the difference in mean tensile strength (in kilograms per square millimeter) is

$$12.22 \le \mu_1 - \mu_2 \le 13.98 \text{ (in kilograms per square millimeter)}$$

Notice that the confidence interval does not include zero, implying that the mean strength of aluminum grade 1 (μ_1) exceeds the mean strength of aluminum grade 2 (μ_2). In fact, we can state that we are 90% confident that the mean tensile strength of aluminum grade 1 exceeds that of aluminum grade 2 by between 12.22 and 13.98 kilograms per square millimeter.

Choice of Sample Size

If the standard deviations σ_1 and σ_2 are known (at least approximately) and the two sample sizes n_1 and n_2 are equal ($n_1 = n_2 = n$, say), we can determine the sample size required so that the error in estimating $\mu_1 - \mu_2$ by $\bar{x}_1 - \bar{x}_2$ will be less than E at $100(1 - \alpha)\%$ confidence. The required sample size from each population is

Sample Size for a Confidence Interval on the Difference in Means, Variances Known

$$n = \left(\frac{z_{\alpha/2}}{E}\right)^2 (\sigma_1^2 + \sigma_2^2) \tag{10-8}$$

Remember to round up if n is not an integer. This will ensure that the level of confidence does not drop below $100(1 - \alpha)\%$.

One-Sided Confidence Bounds

One-sided confidence bounds on $\mu_1 - \mu_2$ may also be obtained. A $100(1 - \alpha)\%$ upper-confidence bound on $\mu_1 - \mu_2$ is

One-Sided Upper Confidence Bound

$$\mu_1 - \mu_2 \le \bar{x}_1 - \bar{x}_2 + z_\alpha \sqrt{\frac{\sigma_1^2}{n_1} + \frac{\sigma_2^2}{n_2}} \tag{10-9}$$

and a $100(1 - \alpha)\%$ lower-confidence bound is

One-Sided Lower Confidence Bound

$$\bar{x}_1 - \bar{x}_2 - z_\alpha \sqrt{\frac{\sigma_1^2}{n_1} + \frac{\sigma_2^2}{n_2}} \le \mu_1 - \mu_2 \tag{10-10}$$

EXERCISES FOR SECTION 10-2

10-1. Consider the hypothesis test $H_0: \mu_1 = \mu_2$ against $H_1: \mu_1 \neq \mu_2$ with known variances $\sigma_1 = 10$ and $\sigma_2 = 5$. Suppose that sample sizes $n_1 = 10$ and $n_2 = 15$ and that $\bar{x}_1 = 4.7$ and $\bar{x}_2 = 7.8$. Use $\alpha = 0.05$.

(a) Test the hypothesis and find the P-value.
(b) Explain how the test could be conducted with a confidence interval.
(c) What is the power of the test in part (a) for a true difference in means of 3?
(d) Assuming equal sample sizes, what sample size should be used to obtain $\beta = 0.05$ if the true difference in means is 3? Assume that $\alpha = 0.05$.

10-2. Consider the hypothesis test $H_0: \mu_1 = \mu_2$ against $H_1: \mu_1 < \mu_2$ with known variances $\sigma_1 = 10$ and $\sigma_2 = 5$. Suppose that sample sizes $n_1 = 10$ and $n_2 = 15$ and that $\bar{x}_1 = 14.2$ and $\bar{x}_2 = 19.7$. Use $\alpha = 0.05$.

(a) Test the hypothesis and find the P-value.
(b) Explain how the test could be conducted with a confidence interval.
(c) What is the power of the test in part (a) if μ_1 is 4 units less than μ_2?
(d) Assuming equal sample sizes, what sample size should be used to obtain $\beta = 0.05$ if μ_1 is 4 units less than μ_2? Assume that $\alpha = 0.05$.

10-3. Consider the hypothesis test $H_0: \mu_1 = \mu_2$ against $H_1: \mu_1 > \mu_2$ with known variances $\sigma_1 = 10$ and $\sigma_2 = 5$. Suppose that sample sizes $n_1 = 10$ and $n_2 = 15$ and that $\bar{x}_1 = 24.5$ and $\bar{x}_2 = 21.3$. Use $\alpha = 0.01$.

(a) Test the hypothesis and find the P-value.
(b) Explain how the test could be conducted with a confidence interval.
(c) What is the power of the test in part (a) if μ_1 is 2 units greater than μ_2?

(d) Assuming equal sample sizes, what sample size should be used to obtain $\beta = 0.05$ if μ_1 is 2 units greater than μ_2? Assume that $\alpha = 0.05$.

10-4. Two machines are used for filling plastic bottles with a net volume of 16.0 ounces. The fill volume can be assumed normal, with standard deviation $\sigma_1 = 0.020$ and $\sigma_2 = 0.025$ ounces. A member of the quality engineering staff suspects that both machines fill to the same mean net volume, whether or not this volume is 16.0 ounces. A random sample of 10 bottles is taken from the output of each machine.

Machine 1		Machine 2	
16.03	16.01	16.02	16.03
16.04	15.96	15.97	16.04
16.05	15.98	15.96	16.02
16.05	16.02	16.01	16.01
16.02	15.99	15.99	16.00

(a) Do you think the engineer is correct? Use $\alpha = 0.05$. What is the P-value for this test?
(b) Calculate a 95% confidence interval on the difference in means. Provide a practical interpretation of this interval.
(c) What is the power of the test in part (a) for a true difference in means of 0.04?
(d) Assuming equal sample sizes, what sample size should be used to assure that $\beta = 0.05$ if the true difference in means is 0.04? Assume that $\alpha = 0.05$.

10-5. Two types of plastic are suitable for use by an electronics component manufacturer. The breaking strength of this plastic is important. It is known that $\sigma_1 = \sigma_2 = 1.0$ psi. From a random sample of size $n_1 = 10$ and $n_2 = 12$, we obtain $\bar{x}_1 = 162.5$ and $\bar{x}_2 = 155.0$. The company will not adopt plastic 1 unless its mean breaking strength exceeds that of plastic 2 by at least 10 psi.
(a) Based on the sample information, should it use plastic 1? Use $\alpha = 0.05$ in reaching a decision. Find the P-value.
(b) Calculate a 95% confidence interval on the difference in means. Suppose that the true difference in means is really 12 psi.
(c) Find the power of the test assuming that $\alpha = 0.05$.
(d) If it is really important to detect a difference of 12 psi, are the sample sizes employed in part (a) adequate, in your opinion?

10-6. The burning rates of two different solid-fuel propellants used in aircrew escape systems are being studied. It is known that both propellants have approximately the same standard deviation of burning rate; that is $\sigma_1 = \sigma_2 = 3$ centimeters per second. Two random samples of $n_1 = 20$ and $n_2 = 20$ specimens are tested; the sample mean burning rates are $\bar{x}_1 = 18$ centimeters per second and $\bar{x}_2 = 24$ centimeters per second.
(a) Test the hypothesis that both propellants have the same mean burning rate. Use $\alpha = 0.05$. What is the P-value?
(b) Construct a 95% confidence interval on the difference in means $\mu_1 - \mu_2$. What is the practical meaning of this interval?
(c) What is the β-error of the test in part (a) if the true difference in mean burning rate is 2.5 centimeters per second?
(d) Assuming equal sample sizes, what sample size is needed to obtain power of 0.9 at a true difference in means of 14 cm/s?

10-7. Two different formulations of an oxygenated motor fuel are being tested to study their road octane numbers. The variance of road octane number for formulation 1 is $\sigma_1^2 = 1.5$, and for formulation 2 it is $\sigma_2^2 = 1.2$. Two random samples of size $n_1 = 15$ and $n_2 = 20$ are tested, and the mean road octane numbers observed are $\bar{x}_1 = 89.6$ and $\bar{x}_2 = 92.5$. Assume normality.
(a) If formulation 2 produces a higher road octane number than formulation 1, the manufacturer would like to detect it. Formulate and test an appropriate hypothesis, using $\alpha = 0.05$. What is the P-value?
(b) Explain how the question in part (a) could be answered with a 95% confidence interval on the difference in mean road octane number.
(c) What sample size would be required in each population if we wanted to be 95% confident that the error in estimating the difference in mean road octane number is less than 1?

10-8. A polymer is manufactured in a batch chemical process. Viscosity measurements are normally made on each batch, and long experience with the process has indicated that the variability in the process is fairly stable with $\sigma = 20$. Fifteen batch viscosity measurements are given as follows:

724, 718, 776, 760, 745, 759, 795, 756, 742, 740, 761, 749, 739, 747, 742

A process change is made which involves switching the type of catalyst used in the process. Following the process change, eight batch viscosity measurements are taken:

735, 775, 729, 755, 783, 760, 738, 780

Assume that process variability is unaffected by the catalyst change. If the difference in mean batch viscosity is 10 or less, the manufacturer would like to detect it with a high probability.
(a) Formulate and test an appropriate hypothesis using $\alpha = 0.10$. What are your conclusions? Find the P-value.
(b) Find a 90% confidence interval on the difference in mean batch viscosity resulting from the process change.
(c) Compare the results of parts (a) and (b) and discuss your findings.

10-9. The concentration of active ingredient in a liquid laundry detergent is thought to be affected by the type of catalyst used in the process. The standard deviation of active concentration is known to be 3 grams per liter, regardless of the catalyst type. Ten observations on concentration are taken with each catalyst, and the data follow:

Catalyst 1: 57.9, 66.2, 65.4, 65.4, 65.2, 62.6, 67.6, 63.7, 67.2, 71.0

Catalyst 2: 66.4, 71.7, 70.3, 69.3, 64.8, 69.6, 68.6, 69.4, 65.3, 68.8

(a) Find a 95% confidence interval on the difference in mean active concentrations for the two catalysts. Find the *P*-value.

(b) Is there any evidence to indicate that the mean active concentrations depend on the choice of catalyst? Base your answer on the results of part (a).

(c) Suppose that the true mean difference in active concentration is 5 grams per liter. What is the power of the test to detect this difference if $\alpha = 0.05$?

(d) If this difference of 5 grams per liter is really important, do you consider the sample sizes used by the experimenter to be adequate? Does the assumption of normality seem reasonable for both samples?

(e) Does the assumption of normality of the differences seem reasonable?

10-3 INFERENCE ON THE DIFFERENCE IN MEANS OF TWO NORMAL DISTRIBUTIONS, VARIANCES UNKNOWN

We now extend the results of the previous section to the difference in means of the two distributions in Fig. 10-1 when the variances of both distributions σ_1^2 and σ_2^2 are unknown. If the sample sizes n_1 and n_2 exceed 40, the normal distribution procedures in Section 10-2 could be used. However, when small samples are taken, we will assume that the populations are normally distributed and base our hypotheses tests and confidence intervals on the t distribution. This nicely parallels the case of inference on the mean of a single sample with unknown variance.

10-3.1 Hypotheses Tests on the Difference in Means, Variances Unknown

We now consider tests of hypotheses on the difference in means $\mu_1 - \mu_2$ of two normal distributions where the variances σ_1^2 and σ_2^2 are unknown. A t-statistic will be used to test these hypotheses. As noted above and in Section 9-3, the normality assumption is required to develop the test procedure, but moderate departures from normality do not adversely affect the procedure. Two different situations must be treated. In the first case, we assume that the variances of the two normal distributions are unknown but equal; that is, $\sigma_1^2 = \sigma_2^2 = \sigma^2$. In the second, we assume that σ_1^2 and σ_2^2 are unknown and not necessarily equal.

Case 1: $\sigma_1^2 = \sigma_2^2 = \sigma^2$

Suppose we have two independent normal populations with unknown means μ_1 and μ_2, and unknown but equal variances, $\sigma_1^2 = \sigma_2^2 = \sigma^2$. We wish to test

$$\begin{aligned} H_0&: \mu_1 - \mu_2 = \Delta_0 \\ H_1&: \mu_1 - \mu_2 \neq \Delta_0 \end{aligned} \tag{10-11}$$

Let $X_{11}, X_{12}, \ldots, X_{1n_1}$ be a random sample of n_1 observations from the first population and $X_{21}, X_{22}, \ldots, X_{2n_2}$ be a random sample of n_2 observations from the second population. Let $\overline{X}_1$, $\overline{X}_2$, S_1^2, and S_2^2 be the sample means and sample variances, respectively. Now the expected value of the difference in sample means $\overline{X}_1 - \overline{X}_2$ is $E(\overline{X}_1 - \overline{X}_2) = \mu_1 - \mu_2$, so $\overline{X}_1 - \overline{X}_2$ is an unbiased estimator of the difference in means. The variance of $\overline{X}_1 - \overline{X}_2$ is

$$V(\overline{X}_1 - \overline{X}_2) = \frac{\sigma^2}{n_1} + \frac{\sigma^2}{n_2} = \sigma^2\left(\frac{1}{n_1} + \frac{1}{n_2}\right)$$

It seems reasonable to combine the two sample variances S_1^2 and S_2^2 to form an estimator of σ^2. The **pooled estimator** of σ^2 is defined as follows.

Pooled Estimator of Variance

The **pooled estimator** of σ^2, denoted by S_p^2, is defined by

$$S_p^2 = \frac{(n_1 - 1)S_1^2 + (n_2 - 1)S_2^2}{n_1 + n_2 - 2} \tag{10-12}$$

It is easy to see that the pooled estimator S_p^2 can be written as

$$S_p^2 = \frac{n_1 - 1}{n_1 + n_2 - 2} S_1^2 + \frac{n_2 - 1}{n_1 + n_2 - 2} S_2^2 = wS_1^2 + (1 - w)S_2^2$$

where $0 < w \leq 1$. Thus S_p^2 is a **weighted average** of the two sample variances S_1^2 and S_2^2, where the weights w and $1 - w$ depend on the two sample sizes n_1 and n_2. Obviously, if $n_1 = n_2 = n$, $w = 0.5$ and S_p^2 is just the arithmetic average of S_1^2 and S_2^2. If $n_1 = 10$ and $n_2 = 20$ (say), $w = 0.32$ and $1 - w = 0.68$. The first sample contributes $n_1 - 1$ degrees of freedom to S_p^2 and the second sample contributes $n_2 - 1$ degrees of freedom. Therefore, S_p^2 has $n_1 + n_2 - 2$ degrees of freedom.

Now we know that

$$Z = \frac{\overline{X}_1 - \overline{X}_2 - (\mu_1 - \mu_2)}{\sigma\sqrt{\dfrac{1}{n_1} + \dfrac{1}{n_2}}}$$

has a $N(0, 1)$ distribution. Replacing σ by S_p gives the following.

Given the assumptions of this section, the quantity

$$T = \frac{\overline{X}_1 - \overline{X}_2 - (\mu_1 - \mu_2)}{S_p\sqrt{\dfrac{1}{n_1} + \dfrac{1}{n_2}}} \tag{10-13}$$

has a t distribution with $n_1 + n_2 - 2$ degrees of freedom.

The use of this information to test the hypotheses in Equation 10-11 is now straightforward: simply replace $\mu_1 - \mu_2$ by Δ_0, and the resulting **test statistic** has a t distribution with $n_1 + n_2 - 2$ degrees of freedom under H_0: $\mu_1 - \mu_2 = \Delta_0$. Therefore, the reference distribution for the test statistic is the t distribution with $n_1 + n_2 - 2$ degrees of freedom. The location of the critical region for both two- and one-sided alternatives parallels those in the one-sample case. Because a pooled estimate of variance is used, the procedure is often called the **pooled *t*-test.**

Case 1: Tests on the Difference in Means, Variances Unknown and Equal*

Null hypothesis: $H_0: \mu_1 - \mu_2 = \Delta_0$

Test statistic:
$$T_0 = \frac{\overline{X}_1 - \overline{X}_2 - \Delta_0}{S_p\sqrt{\dfrac{1}{n_1} + \dfrac{1}{n_2}}} \tag{10-14}$$

Alternative Hypothesis	Rejection Criterion
$H_1: \mu_1 - \mu_2 \neq \Delta_0$	$t_0 > t_{\alpha/2,n_1+n_2-2}$ or $t_0 < -t_{\alpha/2,n_1+n_2-2}$
$H_1: \mu_1 - \mu_2 > \Delta_0$	$t_0 > t_{\alpha,n_1+n_2-2}$
$H_1: \mu_1 - \mu_2 < \Delta_0$	$t_0 < -t_{\alpha,n_1+n_2-2}$

EXAMPLE 10-5
Yield from a Catalyst

Two catalysts are being analyzed to determine how they affect the mean yield of a chemical process. Specifically, catalyst 1 is currently in use, but catalyst 2 is acceptable. Since catalyst 2 is cheaper, it should be adopted, providing it does not change the process yield. A test is run in the pilot plant and results in the data shown in Table 10-1. Is there any difference between the mean yields? Use $\alpha = 0.05$, and assume equal variances.

The solution using the eight-step hypothesis-testing procedure is as follows:

1. The parameters of interest are μ_1 and μ_2, the mean process yield using catalysts 1 and 2, respectively, and we want to know if $\mu_1 - \mu_2 = 0$.
2. $H_0: \mu_1 - \mu_2 = 0$, or $H_0: \mu_1 = \mu_2$
3. $H_1: \mu_1 \neq \mu_2$
4. $\alpha = 0.05$
5. The test statistic is

$$t_0 = \frac{\overline{x}_1 - \overline{x}_2 - 0}{s_p\sqrt{\dfrac{1}{n_1} + \dfrac{1}{n_2}}}$$

Table 10-1 Catalyst Yield Data, Example 10-5

Observation Number	Catalyst 1	Catalyst 2
1	91.50	89.19
2	94.18	90.95
3	92.18	90.46
4	95.39	93.21
5	91.79	97.19
6	89.07	97.04
7	94.72	91.07
8	89.21	92.75
	$\overline{x}_1 = 92.255$	$\overline{x}_2 = 92.733$
	$s_1 = 2.39$	$s_2 = 2.98$

*While we have given the development of this procedure for the case where the sample sizes could be different, there is an advantage to using equal sample sizes $n_1 = n_2 = n$. When the sample sizes are the same from both populations, the t-test is more robust to the assumption of equal variances.

6. Reject H_0 if $t_0 > t_{0.025,14} = 2.145$ or if $t_0 < -t_{0.025,14} = -2.145$.

7. Computations: From Table 10-1 we have $\bar{x}_1 = 92.255$, $s_1 = 2.39$, $n_1 = 8$, $\bar{x}_2 = 92.733$, $s_2 = 2.98$, and $n_2 = 8$. Therefore

$$s_p^2 = \frac{(n_1 - 1)s_1^2 + (n_2 - 1)s_2^2}{n_1 + n_2 - 2} = \frac{(7)(2.39)^2 + 7(2.98)^2}{8 + 8 - 2} = 7.30$$

$$s_p = \sqrt{7.30} = 2.70$$

and

$$t_0 = \frac{\bar{x}_1 - \bar{x}_2}{2.70\sqrt{\frac{1}{n_1} + \frac{1}{n_2}}} = \frac{92.255 - 92.733}{2.70\sqrt{\frac{1}{8} + \frac{1}{8}}} = -0.35$$

8. Conclusions: Since $-2.145 < t_0 = -0.35 < 2.145$, the null hypothesis cannot be rejected. That is, at the 0.05 level of significance, we do not have strong evidence to conclude that catalyst 2 results in a mean yield that differs from the mean yield when catalyst 1 is used.

A *P*-value could also be used for decision making in this example. From Appendix Table V we find that $t_{0.40,14} = 0.258$ and $t_{0.25,14} = 0.692$. Therefore, since $0.258 < 0.35 < 0.692$, we conclude that lower and upper bounds on the *P*-value are $0.50 < P < 0.80$. Therefore, since the *P*-value exceeds $\alpha = 0.05$, the null hypothesis cannot be rejected.

The Minitab two-sample *t*-test and confidence interval procedure for Example 10-5 follows:

Minitab Computations

Two-Sample T-Test and CI: Cat 1, Cat 2

Two-sample T for Cat 1 vs Cat 2

	N	Mean	StDev	SE Mean
Cat 1	8	92.26	2.39	0.84
Cat 2	8	92.73	2.99	1.1

Difference = mu Cat 1 − mu Cat 2
Estimate for difference: −0.48
95% CI for difference: (−3.37, 2.42)
T-Test of difference = 0 (vs not =): T-Value = −0.35 P-Value = 0.730 DF = 14
Both use Pooled StDev = 2.70

Notice that the numerical results are essentially the same as the manual computations in Example 10-5. The *P*-value is reported as $P = 0.73$. The two-sided CI on $\mu_1 - \mu_2$ is also reported. We will give the computing formula for the CI in Section 10-3.3. Figure 10-2 shows the normal probability plot of the two samples of yield data and comparative box plots. The normal probability plots indicate that there is no problem with the normality assumption. Furthermore, both straight lines have similar slopes, providing some verification of the assumption of equal variances. The comparative box plots indicate that there is no obvious difference in the two catalysts, although catalyst 2 has slightly greater sample variability.

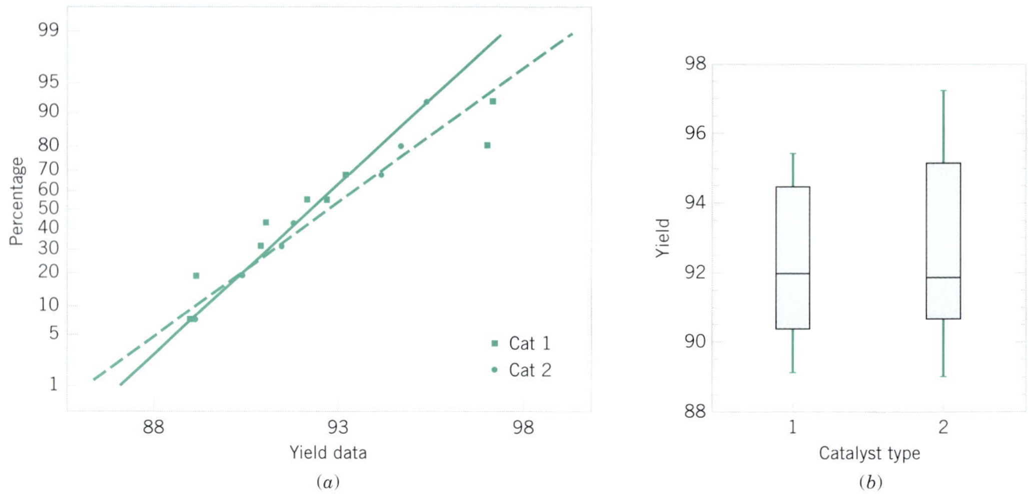

Figure 10-2 Normal probability plot and comparative box plot for the catalyst yield data in Example 10-5. (a) Normal probability plot, (b) Box plots.

Case 2: $\sigma_1^2 \neq \sigma_2^2$

In some situations, we cannot reasonably assume that the unknown variances σ_1^2 and σ_2^2 are equal. There is not an exact t-statistic available for testing H_0: $\mu_1 - \mu_2 = \Delta_0$ in this case. However, an approximate result can be applied.

Case 2: Test Statistic for the Difference in Means, Variances Unknown and Not Assumed Equal

If H_0: $\mu_1 - \mu_2 = \Delta_0$ is true, the statistic

$$T_0^* = \frac{\overline{X}_1 - \overline{X}_2 - \Delta_0}{\sqrt{\dfrac{S_1^2}{n_1} + \dfrac{S_2^2}{n_2}}} \tag{10-15}$$

is distributed approximately as t with degrees of freedom given by

$$v = \frac{\left(\dfrac{s_1^2}{n_1} + \dfrac{s_2^2}{n_2}\right)^2}{\dfrac{(s_1^2/n_1)^2}{n_1 - 1} + \dfrac{(s_2^2/n_2)^2}{n_2 - 1}} \tag{10-16}$$

If v is not an integer, round down to the nearest integer.

Therefore, if $\sigma_1^2 \neq \sigma_2^2$, the hypotheses on differences in the means of two normal distributions are tested as in the equal variances case, except that T_0^* is used as the test statistic and $n_1 + n_2 - 2$ is replaced by v in determining the degrees of freedom for the test.

EXAMPLE 10-6
Arsenic in Water

Arsenic concentration in public drinking water supplies is a potential health risk. An article in the *Arizona Republic* (Sunday, May 27, 2001) reported drinking water arsenic concentrations in parts per billion (ppb) for 10 metropolitan Phoenix communities and 10 communities in rural Arizona. The data follow:

Metro Phoenix $(\bar{x}_1 = 12.5, s_1 = 7.63)$	**Rural Arizona** $(\bar{x}_2 = 27.5, s_2 = 15.3)$
Phoenix, 3	Rimrock, 48
Chandler, 7	Goodyear, 44
Gilbert, 25	New River, 40
Glendale, 10	Apache Junction, 38
Mesa, 15	Buckeye, 33
Paradise Valley, 6	Nogales, 21
Peoria, 12	Black Canyon City, 20
Scottsdale, 25	Sedona, 12
Tempe, 15	Payson, 1
Sun City, 7	Casa Grande, 18

We wish to determine it there is any difference in mean arsenic concentrations between metropolitan Phoenix communities and communities in rural Arizona. Figure 10-3 shows a normal probability plot for the two samples of arsenic concentration. The assumption of normality appears quite reasonable, but since the slopes of the two straight lines are very different, it is unlikely that the population variances are the same.

Applying the eight-step procedure gives the following:

1. The parameters of interest are the mean arsenic concentrations for the two geographic regions, say, μ_1 and μ_2, and we are interested in determining whether $\mu_1 - \mu_2 = 0$.
2. H_0: $\mu_1 - \mu_2 = 0$, or H_0: $\mu_1 = \mu_2$
3. H_1: $\mu_1 \neq \mu_2$
4. $\alpha = 0.05$ (say)
5. The test statistic is

$$t_0^* = \frac{\bar{x}_1 - \bar{x}_2 - 0}{\sqrt{\dfrac{s_1^2}{n_1} + \dfrac{s_2^2}{n_2}}}$$

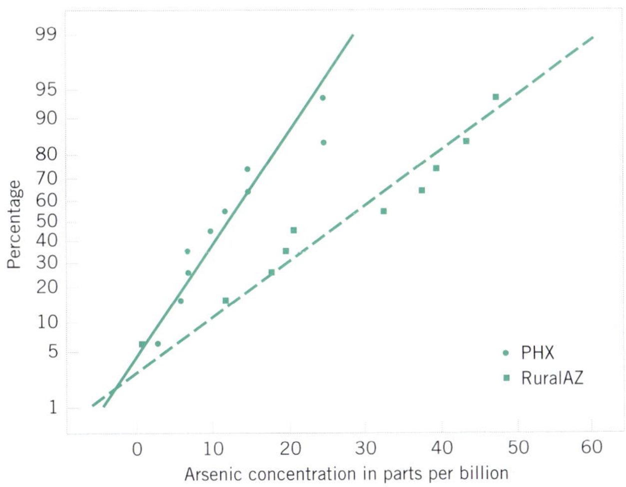

Figure 10-3 Normal probability plot of the arsenic concentration data from Example 10-6.

6. The degrees of freedom on t_0^* are found from Equation 10-16 as

$$v = \frac{\left(\dfrac{s_1^2}{n_1} + \dfrac{s_2^2}{n_2}\right)^2}{\dfrac{(s_1^2/n_1)^2}{n_1 - 1} + \dfrac{(s_2^2/n_2)^2}{n_2 - 1}} = \frac{\left[\dfrac{(7.63)^2}{10} + \dfrac{(15.3)^2}{10}\right]^2}{\dfrac{[(7.63)^2/10]^2}{9} + \dfrac{[(15.3)^2/10]^2}{9}} = 13.2 \simeq 13$$

Therefore, using $\alpha = 0.05$, we would reject H_0: $\mu_1 = \mu_2$ if $t_0^* > t_{0.025,13} = 2.160$ or if $t_0^* < -t_{0.025,13} = -2.160$

7. Computations: Using the sample data we find

$$t_0^* = \frac{\bar{x}_1 - \bar{x}_2}{\sqrt{\dfrac{s_1^2}{n_1} + \dfrac{s_2^2}{n_2}}} = \frac{12.5 - 27.5}{\sqrt{\dfrac{(7.63)^2}{10} + \dfrac{(15.3)^2}{10}}} = -2.77$$

8. Conclusions: Because $t_0^* = -2.77 < t_{0.025,13} = -2.160$, we reject the null hypothesis. Therefore, there is evidence to conclude that mean arsenic concentration in the drinking water in rural Arizona is different from the mean arsenic concentration in metropolitan Phoenix drinking water. Furthermore, the mean arsenic concentration is higher in rural Arizona communities. The P-value for this test is approximately $P = 0.016$.

The Minitab output for this example follows:

Minitab Computations

Two-Sample T-Test and CI: PHX, RuralAZ

Two-sample T for PHX vs RuralAZ

	N	Mean	StDev	SE Mean
PHX	10	12.50	7.63	2.4
RuralAZ	10	27.5	15.3	4.9

Difference = mu PHX − mu RuralAZ
Estimate for difference: −15.00
95% CI for difference: (−26.71, −3.29)
T-Test of difference = 0 (vs not =): T-Value = −2.77 P-Value = 0.016 DF = 13

The numerical results from Minitab exactly match the calculations from Example 10-6. Note that a two-sided 95% CI on $\mu_1 - \mu_2$ is also reported. We will discuss its computation in Section 10-3.3; however, note that the interval does not include zero. Indeed, the upper 95% of confidence limit is −3.29 ppb, well below zero, and the mean observed difference is $\bar{x}_1 - \bar{x}_2 = 12.5 - 27.5 = -15$ ppb.

10-3.2 Type II Error and Choice of Sample Size

The operating characteristic curves in Appendix Charts VII*e*, VII*f*, VII*g*, and VII*h* are used to evaluate the type II error for the case where $\sigma_1^2 = \sigma_2^2 = \sigma^2$. Unfortunately, when $\sigma_1^2 \neq \sigma_2^2$, the distribution of T_0^* is unknown if the null hypothesis is false, and no operating characteristic curves are available for this case.

For the two-sided alternative H_1: $\mu_1 - \mu_2 = \Delta \neq \Delta_0$, when $\sigma_1^2 = \sigma_2^2 = \sigma^2$ and $n_1 = n_2 = n$, Charts VII*e* and VII*f* are used with

$$d = \frac{|\Delta - \Delta_0|}{2\sigma} \tag{10-17}$$

where Δ is the true difference in means that is of interest. To use these curves, they must be entered with the sample size $n^* = 2n - 1$. For the one-sided alternative hypothesis, we use Charts VII*g* and VII*h* and define d and Δ as in Equation 10-17. It is noted that the parameter d is a function of σ, which is unknown. As in the single-sample t-test, we may have to rely on a prior estimate of σ or use a subjective estimate. Alternatively, we could define the differences in the mean that we wish to detect relative to σ.

EXAMPLE 10-7
Yield from Catalyst Sample Size

Consider the catalyst experiment in Example 10-5. Suppose that, if catalyst 2 produces a mean yield that differs from the mean yield of catalyst 1 by 4.0%, we would like to reject the null hypothesis with probability at least 0.85. What sample size is required?

Using $s_p = 2.70$ as a rough estimate of the common standard deviation σ, we have $d = |\Delta|/2\sigma = |4.0|/[(2)(2.70)] = 0.74$. From Appendix Chart VII*e* with $d = 0.74$ and $\beta = 0.15$, we find $n^* = 20$, approximately. Therefore, since $n^* = 2n - 1$,

$$n = \frac{n^* + 1}{2} = \frac{20 + 1}{2} = 10.5 \simeq 11(\text{say})$$

and we would use sample sizes of $n_1 = n_2 = n = 11$.

Minitab will also perform power and sample size calculations for the two-sample t-test (equal variances). The output from Example 10-7 is as follows:

Minitab Computations

Power and Sample Size

2-Sample t Test
Testing mean 1 = mean 2 (versus not =)
Calculating power for mean 1 = mean 2 + difference
Alpha = 0.05 Sigma = 2.7

Difference	Sample Size	Target Power	Actual Power
4	10	0.8500	0.8793

The results agree fairly closely with the results obtained from the O.C. curve.

10-3.3 Confidence Interval on the Difference in Means, Variances Unknown

Case 1: $\sigma_1^2 = \sigma_2^2 = \sigma^2$

To develop the confidence interval for the difference in means $\mu_1 - \mu_2$ when both variances are equal, note that the distribution of the statistic

$$T = \frac{\overline{X}_1 - \overline{X}_2 - (\mu_1 - \mu_2)}{S_p\sqrt{\dfrac{1}{n_1} + \dfrac{1}{n_2}}} \tag{10-18}$$

is the t distribution with $n_1 + n_2 - 2$ degrees of freedom. Therefore $P(-t_{\alpha/2,n_1+n_2-2} \leq T \leq t_{\alpha/2,n_1+n_2-2}) = 1 - \alpha$. Now substituting Equation 10-18 for T and manipulating the quantities inside the probability statement will lead to the $100(1 - \alpha)\%$ confidence interval on $\mu_1 - \mu_2$.

Case 1: Confidence Interval on the Difference in Means, Variances Unknowns and Equal

If $\bar{x}_1, \bar{x}_2, s_1^2$ and s_2^2 are the sample means and variances of two random samples of sizes n_1 and n_2, respectively, from two independent normal populations with unknown but equal variances, then a **100(1 − α)% confidence interval on the difference in means $\mu_1 - \mu_2$** is

$$\bar{x}_1 - \bar{x}_2 - t_{\alpha/2,n_1+n_2-2}\, s_p \sqrt{\frac{1}{n_1} + \frac{1}{n_2}}$$

$$\leq \mu_1 - \mu_2 \leq \bar{x}_1 - \bar{x}_2 + t_{\alpha/2,n_1+n_2-2}\, s_p \sqrt{\frac{1}{n_1} + \frac{1}{n_2}} \qquad (10\text{-}19)$$

where $s_p = \sqrt{[(n_1 - 1)s_1^2 + (n_2 - 1)s_2^2]/(n_1 + n_2 - 2)}$ is the pooled estimate of the common population standard deviation, and $t_{\alpha/2,n_1+n_2-2}$ is the upper $\alpha/2$ percentage point of the t distribution with $n_1 + n_2 - 2$ degrees of freedom.

EXAMPLE 10-8
Cement Hydration

An article in the journal *Hazardous Waste and Hazardous Materials* (Vol. 6, 1989) reported the results of an analysis of the weight of calcium in standard cement and cement doped with lead. Reduced levels of calcium would indicate that the hydration mechanism in the cement is blocked and would allow water to attack various locations in the cement structure. Ten samples of standard cement had an average weight percent calcium of $\bar{x}_1 = 90.0$, with a sample standard deviation of $s_1 = 5.0$, while 15 samples of the lead-doped cement had an average weight percent calcium of $\bar{x}_2 = 87.0$, with a sample standard deviation of $s_2 = 4.0$.

We will assume that weight percent calcium is normally distributed and find a 95% confidence interval on the difference in means, $\mu_1 - \mu_2$, for the two types of cement. Furthermore, we will assume that both normal populations have the same standard deviation.

The pooled estimate of the common standard deviation is found using Equation 10-12 as follows:

$$\begin{aligned} s_p^2 &= \frac{(n_1 - 1)s_1^2 + (n_2 - 1)s_2^2}{n_1 + n_2 - 2} \\ &= \frac{9(5.0)^2 + 14(4.0)^2}{10 + 15 - 2} \\ &= 19.52 \end{aligned}$$

Therefore, the pooled standard deviation estimate is $s_p = \sqrt{19.52} = 4.4$. The 95% confidence interval is found using Equation 10-19:

$$\bar{x}_1 - \bar{x}_2 - t_{0.025,23}\, s_p \sqrt{\frac{1}{n_1} + \frac{1}{n_2}} \leq \mu_1 - \mu_2 \leq \bar{x}_1 - \bar{x}_2 + t_{0.025,23}\, s_p \sqrt{\frac{1}{n_1} + \frac{1}{n_2}}$$

or upon substituting the sample values and using $t_{0.025,23} = 2.069$,

$$90.0 - 87.0 - 2.069(4.4)\sqrt{\frac{1}{10} + \frac{1}{15}} \le \mu_1 - \mu_2$$

$$\le 90.0 - 87.0 + 2.069(4.4)\sqrt{\frac{1}{10} + \frac{1}{15}}$$

which reduces to

$$-0.72 \le \mu_1 - \mu_2 \le 6.72$$

Notice that the 95% confidence interval includes zero; therefore, at this level of confidence we cannot conclude that there is a difference in the means. Put another way, there is no evidence that doping the cement with lead affected the mean weight percent of calcium; therefore, we cannot claim that the presence of lead affects this aspect of the hydration mechanism at the 95% level of confidence.

Case 2: $\sigma_1^2 \ne \sigma_2^2$

In many situations it is not reasonable to assume that $\sigma_1^2 = \sigma_2^2$. When this assumption is unwarranted, we may still find a $100(1 - \alpha)\%$ confidence interval on $\mu_1 - \mu_2$ using the fact that $T^* = [\bar{X}_1 - \bar{X}_2 - (\mu_1 - \mu_2)]/\sqrt{S_1^2/n_1 + S_2^2/n_2}$ is distributed approximately as t with degrees of freedom v given by Equation 10-16. The CI expression follows.

Case 2: Approximate Confidence Interval on the Difference in Means, Variances Unknown Are Not Assumed Equal

If $\bar{x}_1$, $\bar{x}_2$, s_1^2, and s_2^2 are the means and variances of two random samples of sizes n_1 and n_2, respectively, from two independent normal populations with unknown and unequal variances, an approximate $100(1 - \alpha)\%$ confidence interval on the difference in means $\mu_1 - \mu_2$ is

$$\bar{x}_1 - \bar{x}_2 - t_{\alpha/2,v}\sqrt{\frac{s_1^2}{n_1} + \frac{s_2^2}{n_2}} \le \mu_1 - \mu_2 \le \bar{x}_1 - \bar{x}_2 + t_{\alpha/2,v}\sqrt{\frac{s_1^2}{n_1} + \frac{s_2^2}{n_2}} \quad (10\text{-}20)$$

where v is given by Equation 10-16 and $t_{\alpha/2,v}$ is the upper $\alpha/2$ percentage point of the t distribution with v degrees of freedom.

EXERCISES FOR SECTION 10-3

10-10. Consider the hypothesis test $H_0: \mu_1 = \mu_2$ against $H_1: \mu_1 \ne \mu_2$. Suppose that sample sizes are $n_1 = 15$ and $n_2 = 15$, that $\bar{x}_1 = 4.7$ and $\bar{x}_2 = 7.8$ and that $s_1^2 = 4$ and $s_2^2 = 6.25$. Assume that $\sigma_1^2 = \sigma_2^2$ and that the data are drawn from normal distributions. Use $\alpha = 0.05$.

(a) Test the hypothesis and find the P-value.
(b) Explain how the test could be conducted with a confidence interval.
(c) What is the power of the test in part (a) for a true difference in means of 3?
(d) Assuming equal sample sizes, what sample size should be used to obtain $\beta = 0.05$ if the true difference in means is -2? Assume that $\alpha = 0.05$.

10-11. Consider the hypothesis test $H_0: \mu_1 = \mu_2$ against $H_1: \mu_1 < \mu_2$. Suppose that sample sizes $n_1 = 15$ and $n_2 = 15$, that $\bar{x}_1 = 6.2$ and $\bar{x}_2 = 7.8$ and that $s_1^2 = 4$ and $s_2^2 = 6.25$. Assume that $\sigma_1^2 = \sigma_2^2$ and that the data are drawn from normal distributions. Use $\alpha = 0.05$.

(a) Test the hypothesis and find the P-value.
(b) Explain how the test could be conducted with a confidence interval.
(c) What is the power of the test in part (a) if μ_1 is 3 units less than μ_2?
(d) Assuming equal sample sizes, what sample size should be used to obtain $\beta = 0.05$ if μ_1 is 2.5 units less than μ_2? Assume that $\alpha = 0.05$.

10-12. Consider the hypothesis test $H_0: \mu_1 = \mu_2$ against $H_1: \mu_1 > \mu_2$. Suppose that sample sizes $n_1 = 10$ and $n_2 = 10$, that $\bar{x}_1 = 7.8$ and $\bar{x}_2 = 5.6$ and that $s_1^2 = 4$ and $s_2^2 = 9$. Assume that $\sigma_1^2 = \sigma_2^2$ and that the data are drawn from normal distributions. Use $\alpha = 0.05$.

(a) Test the hypothesis and find the *P*-value.

(b) Explain how the test could be conducted with a confidence interval.

(c) What is the power of the test in part (a) if μ_1 is 3 units greater than μ_2?

(d) Assuming equal sample sizes, what sample size should be used to obtain $\beta = 0.05$ if μ_1 is 3 units greater than μ_2? Assume that $\alpha = 0.05$.

10-13. The diameter of steel rods manufactured on two different extrusion machines is being investigated. Two random samples of sizes $n_1 = 15$ and $n_2 = 17$ are selected, and the sample means and sample variances are $\bar{x}_1 = 8.73$, $s_1^2 = 0.35$, $\bar{x}_2 = 8.68$, and $s_2^2 = 0.40$, respectively. Assume that $\sigma_1^2 = \sigma_2^2$ and that the data are drawn from a normal distribution.

(a) Is there evidence to support the claim that the two machines produce rods with different mean diameters? Use $\alpha = 0.05$ in arriving at this conclusion. Find the *P*-value.

(b) Construct a 95% confidence interval for the difference in mean rod diameter. Interpret this interval.

10-14. An article in *Fire Technology* investigated two different foam expanding agents that can be used in the nozzles of fire-fighting spray equipment. A random sample of five observations with an aqueous film-forming foam (AFFF) had a sample mean of 4.7 and a standard deviation of 0.6. A random sample of five observations with alcohol-type concentrates (ATC) had a sample mean of 6.9 and a standard deviation 0.8.

(a) Can you draw any conclusions about differences in mean foam expansion? Assume that both populations are well represented by normal distributions with the same standard deviations.

(b) Find a 95% confidence interval on the difference in mean foam expansion of these two agents.

10-15. Two catalysts may be used in a batch chemical process. Twelve batches were prepared using catalyst 1, resulting in an average yield of 86 and a sample standard deviation of 3. Fifteen batches were prepared using catalyst 2, and they resulted in an average yield of 89 with a standard deviation of 2. Assume that yield measurements are approximately normally distributed with the same standard deviation.

(a) Is there evidence to support a claim that catalyst 2 produces a higher mean yield than catalyst 1? Use $\alpha = 0.01$.

(b) Find a 95% confidence interval on the difference in mean yields that can be used to test the claim in part (a).

10-16. The deflection temperature under load for two different types of plastic pipe is being investigated. Two random samples of 15 pipe specimens are tested, and the deflection temperatures observed are as follows (in °F):

Type 1: 206, 188, 205, 187, 194, 193, 207, 185, 189, 213, 192, 210, 194, 178, 205.

Type 2: 177, 197, 206, 201, 180, 176, 185, 200, 197, 192, 198, 188, 189, 203, 192.

(a) Construct box plots and normal probability plots for the two samples. Do these plots provide support of the assumptions of normality and equal variances? Write a practical interpretation for these plots.

(b) Do the data support the claim that the deflection temperature under load for type 1 pipe exceeds that of type 2? In reaching your conclusions, use $\alpha = 0.05$. Calculate a *P*-value.

(c) If the mean deflection temperature for type 2 pipe exceeds that of type 1 by as much as 5°F, it is important to detect this difference with probability at least 0.90. Is the choice of $n_1 = n_2 = 15$ adequate? Use $\alpha = 0.05$.

10-17. In semiconductor manufacturing, wet chemical etching is often used to remove silicon from the backs of wafers prior to metallization. The etch rate is an important characteristic in this process and known to follow a normal distribution. Two different etching solutions have been compared, using two random samples of 10 wafers for each solution. The observed etch rates are as follows (in mils per minute):

Solution 1		Solution 2	
9.9	10.6	10.2	10.0
9.4	10.3	10.6	10.2
9.3	10.0	10.7	10.7
9.6	10.3	10.4	10.4
10.2	10.1	10.5	10.3

(a) Construct normal probability plots for the two samples. Do these plots provide support for the assumptions of normality and equal variances? Write a practical interpretation for these plots.

(b) Do the data support the claim that the mean etch rate is the same for both solutions? In reaching your conclusions, use $\alpha = 0.05$ and assume that both population variances are equal. Calculate a *P*-value.

(c) Find a 95% confidence interval on the difference in mean etch rates.

10-18. Two suppliers manufacture a plastic gear used in a laser printer. The impact strength of these gears measured in foot-pounds is an important characteristic. A random sample of 10 gears from supplier 1 results in $\bar{x}_1 = 290$ and $s_1 = 12$, while another random sample of 16 gears from the second supplier results in $\bar{x}_2 = 321$ and $s_2 = 22$.

(a) Is there evidence to support the claim that supplier 2 provides gears with higher mean impact strength? Use $\alpha = 0.05$, and assume that both populations are normally distributed but the variances are not equal. What is the *P*-value for this test?

(b) Do the data support the claim that the mean impact strength of gears from supplier 2 is at least 25 foot-pounds higher than that of supplier 1? Make the same assumptions as in part (a).

(c) Construct a confidence interval estimate for the difference in mean impact strength, and explain how this interval could be used to answer the question posed regarding supplier-to-supplier differences.

10-19. The melting points of two alloys used in formulating solder were investigated by melting 21 samples of each material. The sample mean and standard deviation for alloy 1 was $\bar{x}_1 = 420°F$ and $s_1 = 4°F$, while for alloy 2 they were $\bar{x}_2 = 426°F$, and $s_2 = 3°F$.

(a) Do the sample data support the claim that both alloys have the same melting point? Use $\alpha = 0.05$ and assume that both populations are normally distributed and have the same standard deviation. Find the P-value for the test.

(b) Suppose that the true mean difference in melting points is 3°F. How large a sample would be required to detect this difference using an $\alpha = 0.05$ level test with probability at least 0.9? Use $\sigma_1 = \sigma_2 = 4$ as an initial estimate of the common standard deviation.

10-20. A photoconductor film is manufactured at a nominal thickness of 25 mils. The product engineer wishes to increase the mean speed of the film, and believes that this can be achieved by reducing the thickness of the film to 20 mils. Eight samples of each film thickness are manufactured in a pilot production process, and the film speed (in microjoules per square inch) is measured. For the 25-mil film the sample data result is $\bar{x}_1 = 1.15$ and $s_1 = 0.11$, while for the 20-mil film, the data yield $\bar{x}_2 = 1.06$ and $s_2 = 0.09$. Note that an increase in film speed would lower the value of the observation in microjoules per square inch.

(a) Do the data support the claim that reducing the film thickness increases the mean speed of the film? Use $\alpha = 0.10$ and assume that the two population variances are equal and the underlying population of film speed is normally distributed. What is the P-value for this test?

(b) Find a 95% confidence interval on the difference in the two means that can be used to test the claim in part (a).

10-21. Two companies manufacture a rubber material intended for use in an automotive application. The part will be subjected to abrasive wear in the field application, so we decide to compare the material produced by each company in a test. Twenty-five samples of material from each company are tested in an abrasion test, and the amount of wear after 1000 cycles is observed. For company 1, the sample mean and standard deviation of wear are $\bar{x}_1 = 20$ milligrams/1000 cycles and $s_1 = 2$ milligrams/1000 cycles, while for company 2 we obtain $\bar{x}_2 = 15$ milligrams/1000 cycles and $s_2 = 8$ milligrams/1000 cycles.

(a) Do the data support the claim that the two companies produce material with different mean wear? Use $\alpha = 0.05$, and assume each population is normally distributed but that their variances are not equal. What is the P-value for this test?

(b) Do the data support a claim that the material from company 1 has higher mean wear than the material from company 2? Use the same assumptions as in part (a).

(c) Construct confidence intervals that will address the questions in parts (a) and (b) above.

10-22. The thickness of a plastic film (in mils) on a substrate material is thought to be influenced by the temperature at which the coating is applied. A completely randomized experiment is carried out. Eleven substrates are coated at 125°F, resulting in a sample mean coating thickness of $\bar{x}_1 = 103.5$ and a sample standard deviation of $s_1 = 10.2$. Another 13 substrates are coated at 150°F, for which $\bar{x}_2 = 99.7$ and $s_2 = 20.1$ are observed. It was originally suspected that raising the process temperature would reduce mean coating thickness.

(a) Do the data support this claim? Use $\alpha = 0.01$ and assume that the two population standard deviations are not equal. Calculate an approximate P-value for this test.

(b) How could you have answered the question posed regarding the effect of temperature on coating thickness by using a confidence interval? Explain your answer.

10-23. An article in *Electronic Components and Technology Conference* (2001, Vol. 52, pp. 1167–1171) compared single versus dual spindle saw processes for copper metallized wafers. A total of 15 devices of each type were measured for the width of the backside chipouts, $\bar{x}_{single} = 66.385$, $s_{single} = 7.895$ and $\bar{x}_{double} = 45.278$, $s_{double} = 8.612$.

(a) Do the sample data support the claim that both processes have the same chip outputs? Use $\alpha = 0.05$ and assume that both populations are normally distributed and have the same variance. Find the P-value for the test.

(b) Construct a 95% two-sided confidence interval on the mean difference in spindle saw process. Compare this interval to the results in part (a).

(c) If the β-error of the test when the true difference in chip outputs is 15 should not exceed 0.1, what sample sizes must be used? Use $\alpha = 0.05$.

10-24. An article in *IEEE International Symposium on Electromagnetic Compatibility* (2002, Vol. 2, pp. 667–670) quantified the absorption of electromagnetic energy and the resulting thermal effect from cellular phones. The experimental results were obtained from in vivo experiments conducted on rats. The arterial blood pressure values (mmHg) for the control group (8 rats) during the experiment are $\bar{x}_1 = 90$, $s_1 = 5$ and for the test group (9 rats) are $\bar{x}_2 = 115$, $s_2 = 10$.

(a) Is there evidence to support the claim that the test group has higher mean blood pressure? Use $\alpha = 0.05$, and assume that both populations are normally distributed but the variances are not equal. What is the P-value for this test?

(b) Calculate a confidence interval to answer the question in part (a).

(c) Do the data support the claim that the mean blood pressure from the test group is at least 15 mmHg higher than the control group? Make the same assumptions as in part (a).
(d) Explain how the question in part (c) could be answered with a confidence interval.

10-25. An article in *Radio Engineering and Electronic Physics* (1984, Vol. 29, No. 3, pp. 63–66) investigated the behavior of a stochastic generator in the presence of external noise. The number of periods was measured in a sample of 100 trains for each of two different levels of noise voltage, 100 and 150 mV. For 100 mV, the mean number of periods in a train was 7.9 with $s = 2.6$. For 150 mV, the mean was 6.9 with $s = 2.4$.
(a) It was originally suspected that raising noise voltage would reduce mean number of periods. Do the data support this claim? Use $\alpha = 0.01$ and assume that each population is normally distributed and the two population variances are equal. What is the P-value for this test?
(b) Calculate a confidence interval to answer the question in part (a).

10-26. An article in *Technometrics* (1999, Vol. 41, pp. 202–211) studied the capability of a gauge by measuring the weights of two sheets of paper. The data are shown below.

Paper	Observations				
1	3.481	3.448	3.485	3.475	3.472
	3.477	3.472	3.464	3.472	3.470
	3.470	3.470	3.477	3.473	3.474
2	3.258	3.254	3.256	3.249	3.241
	3.254	3.247	3.257	3.239	3.250
	3.258	3.239	3.245	3.240	3.254

(a) Check the assumption that the data from each sheet are from normal distributions.
(b) Test the hypothesis that the mean weight of the two sheets are equal against the alternative that they are not (and assume equal variances). Use $\alpha = 0.05$ and assume equal variances. Find the P-value.
(c) Repeat the previous test with $\alpha = 0.10$.
(d) Compare your answers for parts (b) and (c) and explain why they are the same or different.
(e) Explain how the questions in parts (b) and (c) could be answered with confidence intervals.

10-27. The overall distance traveled by a golf ball is tested by hitting the ball with Iron Byron, a mechanical golfer with a swing that is said to emulate the legendary champion, Byron Nelson. Ten randomly selected balls of two different brands are tested and the overall distance measured. The data follow:

Brand 1: 275, 286, 287, 271, 283, 271, 279, 275, 263, 267
Brand 2: 258, 244, 260, 265, 273, 281, 271, 270, 263, 268

(a) Is there evidence that overall distance is approximately normally distributed? Is an assumption of equal variances justified?
(b) Test the hypothesis that both brands of ball have equal mean overall distance. Use $\alpha = 0.05$. What is the P-value?
(c) Construct a 95% two-sided CI on the mean difference in overall distance between the two brands of golf balls.
(d) What is the power of the statistical test in part (b) to detect a true difference in mean overall distance of 5 yards?
(e) What sample size would be required to detect a true difference in mean overall distance of 3 yards with power of approximately 0.75?

10-28. The "spring-like effect" in a golf club could be determined by measuring the coefficient of restitution (the ratio of the outbound velocity to the inbound velocity of a golf ball fired at the clubhead). Twelve randomly selected drivers produced by two clubmakers are tested and the coefficient of restitution measured. The data follow:

Club 1: 0.8406, 0.8104, 0.8234, 0.8198, 0.8235, 0.8562, 0.8123, 0.7976, 0.8184, 0.8265, 0.7773, 0.7871
Club 2: 0.8305, 0.7905, 0.8352, 0.8380, 0.8145, 0.8465, 0.8244, 0.8014, 0.8309, 0.8405, 0.8256, 0.8476

(a) Is there evidence that coefficient of restitution is approximately normally distributed? Is an assumption of equal variances justified?
(b) Test the hypothesis that both brands of clubs have equal mean coefficient of restitution. Use $\alpha = 0.05$. What is the P-value of the test?
(c) Construct a 95% two-sided CI on the mean difference in coefficient of restitution between the two brands of golf clubs.
(d) What is the power of the statistical test in part (b) to detect a true difference in mean coefficient of restitution of 0.2?
(e) What sample size would be required to detect a true difference in mean coefficient of restitution of 0.1 with power of approximately 0.8?

10-4 PAIRED *t*-TEST

A special case of the two-sample t-tests of Section 10-3 occurs when the observations on the two populations of interest are collected in **pairs.** Each pair of observations, say (X_{1j}, X_{2j}), is taken under homogeneous conditions, but these conditions may change from one pair to another. For example, suppose that we are interested in comparing two different types of tips for a hardness-testing machine. This machine presses the tip into a metal specimen with a known force. By measuring the depth of the depression caused by the tip, the hardness of the specimen can be determined. If several specimens were selected at random, half tested with tip 1, half tested with tip 2, and the pooled or independent t-test in Section 10-3 was applied, the results of the test could be erroneous. The metal specimens could have been cut from bar stock that was produced in different heats, or they might not be homogeneous in some other way that might affect hardness. Then the observed difference between mean hardness readings for the two tip types also includes hardness differences between specimens.

A more powerful experimental procedure is to collect the data in pairs—that is, to make two hardness readings on each specimen, one with each tip. The test procedure would then consist of analyzing the *differences* between hardness readings on each specimen. If there is no difference between tips, the mean of the differences should be zero. This test procedure is called the **paired *t*-test.**

Let $(X_{11}, X_{21}), (X_{12}, X_{22}), \ldots, (X_{1n}, X_{2n})$ be a set of n paired observations where we assume that the mean and variance of the population represented by X_1 are μ_1 and σ_1^2, and the mean and variance of the population represented by X_2 are μ_2 and σ_2^2. Define the differences between each pair of observations as $D_j = X_{1j} - X_{2j}$, $j = 1, 2, \ldots, n$. The D_j's are assumed to be normally distributed with mean

$$\mu_D = E(X_1 - X_2) = E(X_1) - E(X_2) = \mu_1 - \mu_2$$

and variance σ_D^2, so testing hypotheses about the difference between μ_1 and μ_2 can be accomplished by performing a one-sample t-test on μ_D. Specifically, testing H_0: $\mu_1 - \mu_2 = \Delta_0$ against H_1: $\mu_1 - \mu_2 \neq \Delta_0$ is equivalent to testing

$$H_0\colon \mu_D = \Delta_0$$
$$H_1\colon \mu_D \neq \Delta_0 \qquad (10\text{-}21)$$

The test statistic is given below.

Paired *t*-Test

Null hypothesis: H_0: $\mu_D = \Delta_0$

Test statistic: $$T_0 = \frac{\overline{D} - \Delta_0}{S_D/\sqrt{n}} \qquad (10\text{-}22)$$

Alternative Hypothesis	Rejection Region
H_1: $\mu_D \neq \Delta_0$	$t_0 > t_{\alpha/2,n-1}$ or $t_0 < -t_{\alpha/2,n-1}$
H_1: $\mu_D > \Delta_0$	$t_0 > t_{\alpha,n-1}$
H_1: $\mu_D < \Delta_0$	$t_0 < -t_{\alpha,n-1}$

In Equation 10-22, $\overline{D}$ is the sample average of the n differences $D_1, D_2, \ldots, D_n$, and S_D is the sample standard deviation of these differences.

Table 10-2 Strength Predictions for Nine Steel Plate Girders (Predicted Load/Observed Load)

Girder	Karlsruhe Method	Lehigh Method	Difference d_j
S1/1	1.186	1.061	0.125
S2/1	1.151	0.992	0.159
S3/1	1.322	1.063	0.259
S4/1	1.339	1.062	0.277
S5/1	1.200	1.065	0.135
S2/1	1.402	1.178	0.224
S2/2	1.365	1.037	0.328
S2/3	1.537	1.086	0.451
S2/4	1.559	1.052	0.507

EXAMPLE 10-9
Strength for Steel Girders

An article in the *Journal of Strain Analysis* (1983, Vol. 18, No. 2) compares several methods for predicting the shear strength for steel plate girders. Data for two of these methods, the Karlsruhe and Lehigh procedures, when applied to nine specific girders, are shown in Table 10-2. We wish to determine whether there is any difference (on the average) between the two methods.

The eight-step procedure is applied as follows:

1. The parameter of interest is the difference in mean shear strength between the two methods, say, $\mu_D = \mu_1 - \mu_2 = 0$.
2. H_0: $\mu_D = 0$
3. H_1: $\mu_D \neq 0$
4. $\alpha = 0.05$
5. The test statistic is

$$t_0 = \frac{\bar{d}}{s_D/\sqrt{n}}$$

6. Reject H_0 if $t_0 > t_{0.025,8} = 2.306$ or if $t_0 < -t_{0.025,8} = -2.306$.
7. Computations: The sample average and standard deviation of the differences d_j are $\bar{d} = 0.2739$ and $s_D = 0.1351$, so the test statistic is

$$t_0 = \frac{\bar{d}}{s_D/\sqrt{n}} = \frac{0.2739}{0.1351/\sqrt{9}} = 6.08$$

8. Conclusions: Since $t_0 = 6.08 > 2.306$, we conclude that the strength prediction methods yield different results. Specifically, the data indicate that the Karlsruhe method produces, on the average, higher strength predictions than does the Lehigh method. The P-value for $t_0 = 6.08$ is $P = 0.0003$, so the test statistic is well into the critical region.

Paired Versus Unpaired Comparisons3

In performing a comparative experiment, the investigator can sometimes choose between the paired experiment and the two-sample (or unpaired) experiment. If n measurements are to be made on each population, the two-sample t-statistic is

$$T_0 = \frac{\bar{X}_1 - \bar{X}_2 - \Delta_0}{S_p\sqrt{\dfrac{1}{n} + \dfrac{1}{n}}}$$

which would be compared to t_{2n-2}, and of course, the paired t-statistic is

$$T_0 = \frac{\overline{D} - \Delta_0}{S_D/\sqrt{n}}$$

which is compared to t_{n-1}. Notice that since

$$\overline{D} = \sum_{j=1}^{n} \frac{D_j}{n} = \sum_{j=1}^{n} \frac{(X_{1j} - X_{2j})}{n} = \sum_{j=1}^{n} \frac{X_{1j}}{n} - \sum_{j=1}^{n} \frac{X_{2j}}{n} = \overline{X}_1 - \overline{X}_2$$

the numerators of both statistics are identical. However, the denominator of the two-sample t-test is based on the assumption that X_1 and X_2 are *independent*. In many paired experiments, a strong positive correlation ρ exists between X_1 and X_2. Then it can be shown that

$$\begin{aligned} V(\overline{D}) &= V(\overline{X}_1 - \overline{X}_2 - \Delta_0) \\ &= V(\overline{X}_1) + V(\overline{X}_2) - 2\,\text{cov}(\overline{X}_1, \overline{X}_2) \\ &= \frac{2\sigma^2(1 - \rho)}{n} \end{aligned}$$

assuming that both populations X_1 and X_2 have identical variances σ^2. Furthermore, S_D^2/n estimates the variance of $\overline{D}$. Whenever there is positive correlation within the pairs, the denominator for the paired t-test will be smaller than the denominator of the two-sample t-test. This can cause the two-sample t-test to considerably understate the significance of the data if it is incorrectly applied to paired samples.

Although pairing will often lead to a smaller value of the variance of $\overline{X}_1 - \overline{X}_2$, it does have a disadvantage—namely, the paired t-test leads to a loss of $n - 1$ degrees of freedom in comparison to the two-sample t-test. Generally, we know that increasing the degrees of freedom of a test increases the power against any fixed alternative values of the parameter.

So how do we decide to conduct the experiment? Should we pair the observations or not? Although there is no general answer to this question, we can give some guidelines based on the above discussion.

1. If the experimental units are relatively homogeneous (small σ) and the correlation within pairs is small, the gain in precision attributable to pairing will be offset by the loss of degrees of freedom, so an independent-sample experiment should be used.
2. If the experimental units are relatively heterogeneous (large σ) and there is large positive correlation within pairs, the paired experiment should be used. Typically, this case occurs when the experimental units are the *same* for both treatments; as in Example 10-9, the same girders were used to test the two methods.

Implementing the rules still requires judgment, because σ and ρ are never known precisely. Furthermore, if the number of degrees of freedom is large (say, 40 or 50), the loss of $n - 1$ of them for pairing may not be serious. However, if the number of degrees of freedom is small (say, 10 or 20), losing half of them is potentially serious if not compensated for by increased precision from pairing.

Confidence Interval for μ_D

To construct the confidence interval for $\mu_D = \mu_1 - \mu_2$, note that

$$T = \frac{\overline{D} - \mu_D}{S_D/\sqrt{n}}$$

follows a t distribution with $n - 1$ degrees of freedom. Then, since $P(-t_{\alpha/2,n-1} \leq T \leq t_{\alpha/2,n-1}) = 1 - \alpha$, we can substitute for T in the above expression and perform the necessary steps to isolate $\mu_D = \mu_1 - \mu_2$ between the inequalities. This leads to the following $100(1 - \alpha)\%$ confidence interval on $\mu_1 - \mu_2$.

Confidence Interval for μ_D from Paired Samples

If $\bar{d}$ and s_D are the sample mean and standard deviation of the difference of n random pairs of normally distributed measurements, a **$100(1 - \alpha)\%$ confidence interval on the difference in means $\mu_D = \mu_1 - \mu_2$** is

$$\bar{d} - t_{\alpha/2,n-1} s_D/\sqrt{n} \leq \mu_D \leq \bar{d} + t_{\alpha/2,n-1} s_D/\sqrt{n} \tag{10-23}$$

where $t_{\alpha/2,n-1}$ is the upper $\alpha/2\%$ point of the t-distribution with $n - 1$ degrees of freedom.

This confidence interval is also valid for the case where $\sigma_1^2 \neq \sigma_2^2$, because s_D^2 estimates $\sigma_D^2 = V(X_1 - X_2)$. Also, for large samples (say, $n \geq 30$ pairs), the explicit assumption of normality is unnecessary because of the central limit theorem.

EXAMPLE 10-10
Parallel Park Cars

The journal *Human Factors* (1962, pp. 375–380) reports a study in which $n = 14$ subjects were asked to parallel park two cars having very different wheel bases and turning radii. The time in seconds for each subject was recorded and is given in Table 10-3. From the column of observed differences we calculate $\bar{d} = 1.21$ and $s_D = 12.68$. The 90% confidence interval for $\mu_D = \mu_1 - \mu_2$ is found from Equation 10-23 as follows:

$$\begin{aligned}
\bar{d} - t_{0.05,13} s_D/\sqrt{n} &\leq \mu_D \leq \bar{d} + t_{0.05,13} s_D/\sqrt{n} \\
1.21 - 1.771(12.68)/\sqrt{14} &\leq \mu_D \leq 1.21 + 1.771(12.68)/\sqrt{14} \\
-4.79 &\leq \mu_D \leq 7.21
\end{aligned}$$

Table 10-3 Time in Seconds to Parallel Park Two Automobiles

	Automobile		Difference
Subject	$1(x_{1j})$	$2(x_{2j})$	(d_j)
1	37.0	17.8	19.2
2	25.8	20.2	5.6
3	16.2	16.8	−0.6
4	24.2	41.4	−17.2
5	22.0	21.4	0.6
6	33.4	38.4	−5.0
7	23.8	16.8	7.0
8	58.2	32.2	26.0
9	33.6	27.8	5.8
10	24.4	23.2	1.2
11	23.4	29.6	−6.2
12	21.2	20.6	0.6
13	36.2	32.2	4.0
14	29.8	53.8	−24.0

Notice that the confidence interval on μ_D includes zero. This implies that, at the 90% level of confidence, the data do not support the claim that the two cars have different mean parking times μ_1 and μ_2. That is, the value $\mu_D = \mu_1 - \mu_2 = 0$ is not inconsistent with the observed data.

EXERCISES FOR SECTION 10-4

10-29. Consider the shear strength experiment described in Example 10-9.

(a) Construct a 95% confidence interval on the difference in mean shear strength for the two methods. Is the result you obtained consistent with the findings in Example 10-9? Explain why.

(b) Do each of the individual shear strengths have to be normally distributed for the paired t-test to be appropriate, or is it only the difference in shear strengths that must be normal? Use a normal probability plot to investigate the normality assumption.

10-30. Consider the parking data in Example 10-10.

(a) Use the paired t-test to investigate the claim that the two types of cars have different levels of difficulty to parallel park. Use $\alpha = 0.10$.

(b) Compare your results with the confidence interval constructed in Example 10-10 and comment on why they are the same or different.

(c) Investigate the assumption that the differences in parking times are normally distributed.

10-31. The manager of a fleet of automobiles is testing two brands of radial tires and assigns one tire of each brand at random to the two rear wheels of eight cars and runs the cars until the tires wear out. The data (in kilometers) follow. Find a 99% confidence interval on the difference in mean life. Which brand would you prefer, based on this calculation?

Car	Brand 1	Brand 2
1	36,925	34,318
2	45,300	42,280
3	36,240	35,500
4	32,100	31,950
5	37,210	38,015
6	48,360	47,800
7	38,200	37,810
8	33,500	33,215

10-32. A computer scientist is investigating the usefulness of two different design languages in improving programming tasks. Twelve expert programmers, familiar with both languages, are asked to code a standard function in both languages, and the time (in minutes) is recorded. The data follow:

(a) Is the assumption that the difference in coding time is normally distributed reasonable?

	Time	
Programmer	Design Language 1	Design Language 2
1	17	18
2	16	14
3	21	19
4	14	11
5	18	23
6	24	21
7	16	10
8	14	13
9	21	19
10	23	24
11	13	15
12	18	20

(b) Find a 95% confidence interval on the difference in mean coding times. Is there any indication that one design language is preferable?

10-33. Fifteen adult males between the ages of 35 and 50 participated in a study to evaluate the effect of diet and exercise on blood cholesterol levels. The total cholesterol was measured in each subject initially and then three months after

	Blood Cholesterol Level	
Subject	Before	After
1	265	229
2	240	231
3	258	227
4	295	240
5	251	238
6	245	241
7	287	234
8	314	256
9	260	247
10	279	239
11	283	246
12	240	218
13	238	219
14	225	226
15	247	233

participating in an aerobic exercise program and switching to a low-fat diet. The data are shown in the accompanying table

(a) Do the data support the claim that low-fat diet and aerobic exercise are of value in producing a mean reduction in blood cholesterol levels? Use $\alpha = 0.05$. Find the P-value.

(b) Calculate a one-sided confidence limit that can be used to answer the question in part (a).

10-34. An article in the *Journal of Aircraft* (Vol. 23, 1986, pp. 859–864) described a new equivalent plate analysis method formulation that is capable of modeling aircraft structures such as cranked wing boxes and that produces results similar to the more computationally intensive finite element analysis method. Natural vibration frequencies for the cranked wing box structure are calculated using both methods, and results for the first seven natural frequencies follow:

Freq.	Finite Element Cycle/s	Equivalent Plate, Cycle/s
1	14.58	14.76
2	48.52	49.10
3	97.22	99.99
4	113.99	117.53
5	174.73	181.22
6	212.72	220.14
7	277.38	294.80

(a) Do the data suggest that the two methods provide the same mean value for natural vibration frequency? Use $\alpha = 0.05$. Find the P-value.

(b) Find a 95% confidence interval on the mean difference between the two methods.

10-35. Ten individuals have participated in a diet-modification program to stimulate weight loss. Their weight both before and after participation in the program is shown in the following list.

Subject	Before	After
1	195	187
2	213	195
3	247	221
4	201	190
5	187	175
6	210	197
7	215	199
8	246	221
9	294	278
10	310	285

(a) Is there evidence to support the claim that this particular diet-modification program is effective in producing a mean weight reduction? Use $\alpha = 0.05$.

(b) Is there evidence to support the claim that this particular diet-modification program will result in a mean weight loss of at least 10 pounds? Use $\alpha = 0.05$.

(c) Suppose that, if the diet-modification program results in mean weight loss of at least 10 pounds, it is important to detect this with probability of at least 0.90. Was the use of 10 subjects an adequate sample size? If not, how many subjects should have been used?

10-36. Two different analytical tests can be used to determine the impurity level in steel alloys. Eight specimens are tested using both procedures, and the results are shown in the following tabulation.

Specimen	Test 1	Test 2
1	1.2	1.4
2	1.3	1.7
3	1.5	1.5
4	1.4	1.3
5	1.7	2.0
6	1.8	2.1
7	1.4	1.7
8	1.3	1.6

(a) Is there sufficient evidence to conclude that tests differ in the mean impurity level, using $\alpha = 0.01$?

(b) Is there evidence to support the claim that Test 1 generates a mean difference 0.1 units lower than Test 2? Use $\alpha = 0.05$.

(c) If the mean from Test 1 is 0.1 less than the mean from Test 2, it is important to detect this with probability at least 0.90. Was the use of 8 alloys an adequate sample size? If not, how many alloys should have been used?

10-37. An article in *Neurology* (1998, Vol. 50, pp. 1246–1252) discussed that monozygotic twins share numerous physical, psychological, and pathological traits. The investigators measured an intelligence score of 10 pairs of twins and the data are as follows:

Pair	Birth Order: 1	Birth Order: 2
1	6.08	5.73
2	6.22	5.80
3	7.99	8.42
4	7.44	6.84
5	6.48	6.43
6	7.99	8.76
7	6.32	6.32
8	7.60	7.62
9	6.03	6.59
10	7.52	7.67

(a) Is the assumption that the difference in score is normally distributed reasonable? Show results to support your answer.
(b) Find a 95% confidence interval on the difference in mean score. Is there any evidence that mean score depends on birth order?
(c) It is important to detect a mean difference in score of one point, with a probability of at least 0.90. Was the use of 10 pairs an adequate sample size? If not, how many pairs should have been used?

10-38. In *Biometrics* (1990, Vol. 46, pp. 673–87), the authors analyzed the circumference of five orange trees (labeled as A–E) measured on seven occasions (x_i).

Tree	x_1	x_2	x_3	x_4	x_5	x_6	x_7
A	30	58	87	115	120	142	145
B	33	69	111	156	172	203	203
C	30	51	75	108	115	139	140
D	32	62	112	167	179	209	214
E	30	49	81	125	142	174	177

(a) Compare the mean increase in circumference in periods 1 to 2 to the mean increase in periods 2 to 3. The increase is the difference in circumference in the two periods. Are these means significantly different at $\alpha = 0.10$?
(b) Is there evidence that the mean increase in period 1 to period 2 is greater than the mean increase in period 6 to period 7 at $\alpha = 0.05$?
(c) Are the assumptions of the test in part (a) violated because the same data (period 2 circumference) is used to calculate both mean increases?

10-5 INFERENCE ON THE VARIANCES OF TWO NORMAL DISTRIBUTIONS

We now introduce tests and confidence intervals for the two population variances shown in Fig. 10-1. We will assume that both populations are normal. Both the hypothesis-testing and confidence interval procedures are relatively sensitive to the normality assumption.

10-5.1 *F* Distribution

Suppose that two independent normal populations are of interest, where the population means and variances, say, μ_1, σ_1^2, μ_2, and σ_2^2, are unknown. We wish to test hypotheses about the equality of the two variances, say, H_0: $\sigma_1^2 = \sigma_2^2$. Assume that two random samples of size n_1 from population 1 and of size n_2 from population 2 are available, and let S_1^2 and S_2^2 be the sample variances. We wish to test the hypotheses

$$H_0: \sigma_1^2 = \sigma_2^2$$
$$H_1: \sigma_1^2 \neq \sigma_2^2 \tag{10-24}$$

The development of a test procedure for these hypotheses requires a new probability distribution, the F distribution. The random variable F is defined to be the ratio of two independent chi-square random variables, each divided by its number of degrees of freedom. That is,

$$F = \frac{W/u}{Y/v} \tag{10-25}$$

where W and Y are independent chi-square random variables with u and v degrees of freedom, respectively. We now formally state the sampling distribution of F.

F Distribution

Let W and Y be independent chi-square random variables with u and v degrees of freedom, respectively. Then the ratio

$$F = \frac{W/u}{Y/v} \tag{10-26}$$

has the probability density function

$$f(x) = \frac{\Gamma\left(\frac{u+v}{2}\right)\left(\frac{u}{v}\right)^{u/2} x^{(u/2)-1}}{\Gamma\left(\frac{u}{2}\right)\Gamma\left(\frac{v}{2}\right)\left[\left(\frac{u}{v}\right)x + 1\right]^{(u+v)/2}}, \qquad 0 < x < \infty \tag{10-27}$$

and is said to follow the F distribution with u degrees of freedom in the numerator and v degrees of freedom in the denominator. It is usually abbreviated as $F_{u,v}$.

The mean and variance of the F distribution are $\mu = v/(v-2)$ for $v > 2$, and

$$\sigma^2 = \frac{2v^2(u+v-2)}{u(v-2)^2(v-4)}, \qquad v > 4$$

Two F distributions are shown in Fig. 10-4. The F random variable is nonnegative, and the distribution is skewed to the right. The F distribution looks very similar to the chi-square distribution; however, the two parameters u and v provide extra flexibility regarding shape.

The percentage points of the F distribution are given in Table VI of the Appendix. Let $f_{\alpha,u,v}$ be the percentage point of the F distribution, with numerator degrees of freedom u and denominator degrees of freedom v such that the probability that the random variable F exceeds this value is

$$P(F > f_{\alpha,u,v}) = \int_{f_{\alpha,u,v}}^{\infty} f(x)\,dx = \alpha$$

This is illustrated in Fig. 10-5. For example, if $u = 5$ and $v = 10$, we find from Table V of the Appendix that

$$P(F > f_{0.05,5,10}) = P(F_{5,10} > 3.33) = 0.05$$

That is, the upper 5 percentage point of $F_{5,10}$ is $f_{0.05,5,10} = 3.33$.

Table V contains only upper-tail percentage points (for selected values of $f_{\alpha,u,v}$ for $\alpha \leq 0.25$) of the F distribution. The lower-tail percentage points $f_{1-\alpha,u,v}$ can be found as follows.

$$f_{1-\alpha,u,v} = \frac{1}{f_{\alpha,v,u}} \tag{10-28}$$

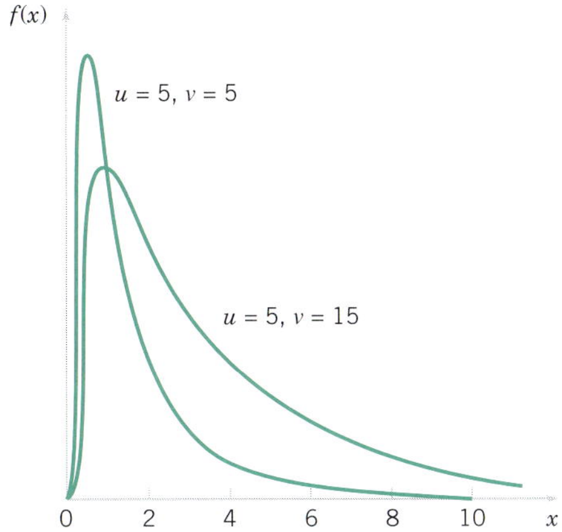

Figure 10-4 Probability density functions of two F distributions.

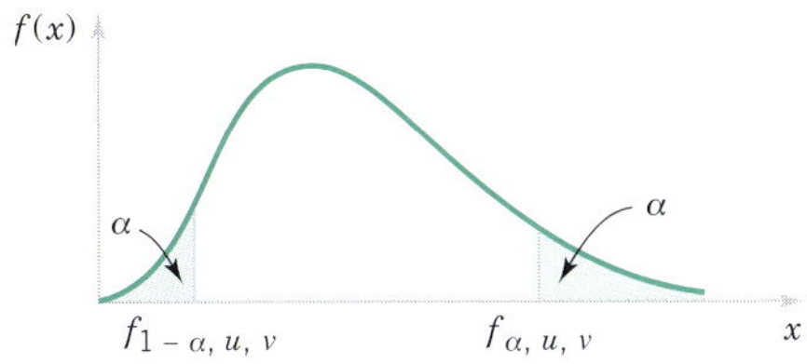

Figure 10-5 Upper and lower percentage points of the F distribution.

For example, to find the lower-tail percentage point $f_{0.95,5,10}$, note that

$$f_{0.95,5,10} = \frac{1}{f_{0.05,10,5}} = \frac{1}{4.74} = 0.211$$

10-5.2 Hypothesis Tests on the Ratio of Two Variances

A hypothesis-testing procedure for the equality of two variances is based on the following result.

Distribution of the Ratio of Sample Variances from Two Normal Distributions

Let $X_{11}, X_{12}, \ldots, X_{1n_1}$ be a random sample from a normal population with mean μ_1 and variance σ_1^2, and let $X_{21}, X_{22}, \ldots, X_{2n_2}$ be a random sample from a second normal population with mean μ_2 and variance σ_2^2. Assume that both normal populations are independent. Let S_1^2 and S_2^2 be the sample variances. Then the ratio

$$F = \frac{S_1^2/\sigma_1^2}{S_2^2/\sigma_2^2}$$

has an F distribution with $n_1 - 1$ numerator degrees of freedom and $n_2 - 1$ denominator degrees of freedom.

This result is based on the fact that $(n_1 - 1)S_1^2/\sigma_1^2$ is a chi-square random variable with $n_1 - 1$ degrees of freedom, that $(n_2 - 1)S_2^2/\sigma_2^2$ is a chi-square random variable with $n_2 - 1$ degrees of freedom, and that the two normal populations are independent. Clearly under the null hypothesis H_0: $\sigma_1^2 = \sigma_2^2$ the ratio $F_0 = S_1^2/S_2^2$ has an F_{n_1-1,n_2-1} distribution. This is the basis of the following test procedure.

Tests on the Ratio of Variances from Two Normal Distributions

Null hypothesis: $H_0: \sigma_1^2 = \sigma_2^2$

Test statistic:
$$F_0 = \frac{S_1^2}{S_2^2} \tag{10-29}$$

Alternative Hypotheses	Rejection Criterion
$H_1: \sigma_1^2 \neq \sigma_2^2$	$f_0 > f_{\alpha/2, n_1-1, n_2-1}$ or $f_0 < f_{1-\alpha/2, n_1-1, n_2-1}$
$H_1: \sigma_1^2 > \sigma_2^2$	$f_0 > f_{\alpha, n_1-1, n_2-1}$
$H_1: \sigma_1^2 < \sigma_2^2$	$f_0 < f_{1-\alpha, n_1-1, n_2-1}$

EXAMPLE 10-11
Semiconductor Etch Variability

Oxide layers on semiconductor wafers are etched in a mixture of gases to achieve the proper thickness. The variability in the thickness of these oxide layers is a critical characteristic of the wafer, and low variability is desirable for subsequent processing steps. Two different mixtures of gases are being studied to determine whether one is superior in reducing the variability of the oxide thickness. Twenty wafers are etched in each gas. The sample standard deviations of oxide thickness are $s_1 = 1.96$ angstroms and $s_2 = 2.13$ angstroms, respectively. Is there any evidence to indicate that either gas is preferable? Use $\alpha = 0.05$.

The eight-step hypothesis-testing procedure may be applied to this problem as follows:

1. The parameters of interest are the variances of oxide thickness σ_1^2 and σ_2^2. We will assume that oxide thickness is a normal random variable for both gas mixtures.
2. $H_0: \sigma_1^2 = \sigma_2^2$
3. $H_1: \sigma_1^2 \neq \sigma_2^2$
4. $\alpha = 0.05$
5. The test statistic is given by Equation 10-29:
$$f_0 = \frac{s_1^2}{s_2^2}$$
6. Since $n_1 = n_2 = 20$, we will reject $H_0: \sigma_1^2 = \sigma_2^2$ if $f_0 > f_{0.025,19,19} = 2.53$ or if $f_0 < f_{0.975,19,19} = 1/f_{0.025,19,19} = 1/2.53 = 0.40$.
7. Computations: Since $s_1^2 = (1.96)^2 = 3.84$ and $s_2^2 = (2.13)^2 = 4.54$, the test statistic is
$$f_0 = \frac{s_1^2}{s_2^2} = \frac{3.84}{4.54} = 0.85$$
8. Conclusions: Since $f_{0.975,19,19} = 0.40 < f_0 = 0.85 < f_{0.025,19,19} = 2.53$, we cannot reject the null hypothesis $H_0: \sigma_1^2 = \sigma_2^2$ at the 0.05 level of significance. Therefore, there is no strong evidence to indicate that either gas results in a smaller variance of oxide thickness.

We may also find a P-value for the F-statistic in Example 10-11. Since $f_{0.50,19,19} = 1.00$, the computed value of the test statistic $f_0 = s_1^2/s_2^2 = 3.84/4.54 = 0.85$ is nearer the lower tail of the F distribution than the upper tail. The probability that an F-random variable with 19 numerator and denominator degrees of freedom is less than 0.85 is 0.3634. Since it is arbitrary which population is identified as "one," we could have computed the test statistic as $f_0 = 4.54/3.84 = 1.18$. The probability that an F-random variable with 19 numerator and denominator degrees of freedom exceeds 1.18 is 0.3610. Therefore, the P-value for the test statistic $f_0 = 0.85$ is the sum of these two probabilities, or $P = 0.3634 + 0.3610 = 0.7244$. Since the P-value exceeds 0.05, the null hypothesis $H_0: \sigma_1^2 = \sigma_2^2$ cannot be rejected. (The probabilities given above were computed using a hand-held calculator.)

10-5.3 Type II Error and Choice of Sample Size

Appendix Charts VII*o*, VII*p*, VII*q*, and VII*r* provide operating characteristic curves for the *F*-test given in Section 10-5.1 for $\alpha = 0.05$ and $\alpha = 0.01$, assuming that $n_1 = n_2 = n$. Charts VII*o* and VII*p* are used with the two-sided alternate hypothesis. They plot β against the abscissa parameter

$$\lambda = \frac{\sigma_1}{\sigma_2} \tag{10-30}$$

for various $n_1 = n_2 = n$. Charts VII*q* and VII*r* are used for the one-sided alternative hypotheses.

EXAMPLE 10-12
Semiconductor Etch Variability Sample Size

For the semiconductor wafer oxide etching problem in Example 10-11, suppose that one gas resulted in a standard deviation of oxide thickness that is half the standard deviation of oxide thickness of the other gas. If we wish to detect such a situation with probability at least 0.80, is the sample size $n_1 = n_2 = 20$ adequate?

Note that if one standard deviation is half the other,

$$\lambda = \frac{\sigma_1}{\sigma_2} = 2$$

By referring to Appendix Chart VII*o* with $n_1 = n_2 = n = 20$ and $\lambda = 2$, we find that $\beta \simeq 0.20$. Therefore, if $\beta = 0.20$, the power of the test (which is the probability that the difference in standard deviations will be detected by the test) is 0.80, and we conclude that the sample sizes $n_1 = n_2 = 20$ are adequate.

10-5.4 Confidence Interval on the Ratio of Two Variances

To find the confidence interval on σ_1^2/σ_2^2, recall that the sampling distribution of

$$F = \frac{S_2^2/\sigma_2^2}{S_1^2/\sigma_1^2}$$

is an F with $n_2 - 1$ and $n_1 - 1$ degrees of freedom. Therefore, $P(f_{1-\alpha/2, n_2-1, n_1-1} \leq F \leq f_{\alpha/2, n_2-1, n_1-1}) = 1 - \alpha$. Substitution for F and manipulation of the inequalities will lead to the $100(1 - \alpha)\%$ confidence interval for σ_1^2/σ_2^2.

Confidence Interval on the Ratio of Variances from Two Normal Distributions

If s_1^2 and s_2^2 are the sample variances of random samples of sizes n_1 and n_2, respectively, from two independent normal populations with unknown variances σ_1^2 and σ_2^2, then a **100(1 − α)% confidence interval on the ratio σ_1^2/σ_2^2** is

$$\frac{s_1^2}{s_2^2} f_{1-\alpha/2, n_2-1, n_1-1} \leq \frac{\sigma_1^2}{\sigma_2^2} \leq \frac{s_1^2}{s_2^2} f_{\alpha/2, n_2-1, n_1-1} \tag{10-31}$$

where $f_{\alpha/2, n_2-1, n_1-1}$ and $f_{1-\alpha/2, n_2-1, n_1-1}$ are the upper and lower $\alpha/2$ percentage points of the F distribution with $n_2 - 1$ numerator and $n_1 - 1$ denominator degrees of freedom, respectively. A confidence interval on the ratio of the standard deviations can be obtained by taking square roots in Equation 10-31.

EXAMPLE 10-13 Surface Finish for Titanium Alloy

A company manufactures impellers for use in jet-turbine engines. One of the operations involves grinding a particular surface finish on a titanium alloy component. Two different grinding processes can be used, and both processes can produce parts at identical mean surface roughness. The manufacturing engineer would like to select the process having the least variability in surface roughness. A random sample of $n_1 = 11$ parts from the first process results in a sample standard deviation $s_1 = 5.1$ microinches, and a random sample of $n_2 = 16$ parts from the second process results in a sample standard deviation of $s_2 = 4.7$ microinches. We will find a 90% confidence interval on the ratio of the two standard deviations, σ_1/σ_2.

Assuming that the two processes are independent and that surface roughness is normally distributed, we can use Equation 10-31 as follows:

$$\frac{s_1^2}{s_2^2} f_{0.95,15,10} \leq \frac{\sigma_1^2}{\sigma_2^2} \leq \frac{s_1^2}{s_2^2} f_{0.05,15,10}$$

$$\frac{(5.1)^2}{(4.7)^2} 0.39 \leq \frac{\sigma_1^2}{\sigma_2^2} \leq \frac{(5.1)^2}{(4.7)^2} 2.85$$

or upon completing the implied calculations and taking square roots,

$$0.678 \leq \frac{\sigma_1}{\sigma_2} \leq 1.832$$

Notice that we have used Equation 10-28 to find $f_{0.95,15,10} = 1/f_{0.05,10,15} = 1/2.54 = 0.39$. Since this confidence interval includes unity, we cannot claim that the standard deviations of surface roughness for the two processes are different at the 90% level of confidence.

EXERCISES FOR SECTION 10-5

10-39. For an F distribution, find the following:

(a) $f_{0.25,5,10}$ (b) $f_{0.10,24,9}$
(c) $f_{0.05,8,15}$ (d) $f_{0.75,5,10}$
(e) $f_{0.90,24,9}$ (f) $f_{0.95,8,15}$

10-40. For an F distribution, find the following:

(a) $f_{0.25,7,15}$ (b) $f_{0.10,10,12}$
(c) $f_{0.01,20,10}$ (d) $f_{0.75,7,15}$
(e) $f_{0.90,10,12}$ (f) $f_{0.99,20,10}$

10-41. Consider the hypothesis test $H_0\colon \sigma_1^2 = \sigma_2^2$ against $H_1\colon \sigma_1^2 < \sigma_2^2$. Suppose that the sample sizes are $n_1 = 5$ and $n_2 = 10$ and that $s_1^2 = 23.2$ and $s_2^2 = 28.8$. Use $\alpha = 0.05$. Test the hypothesis and explain how the test could be conducted with a confidence interval on σ_1/σ_2.

10-42. Consider the hypothesis test $H_0\colon \sigma_1^2 = \sigma_2^2$ against $H_0\colon \sigma_1^2 > \sigma_2^2$. Suppose that the sample sizes are $n_1 = 20$ and $n_2 = 8$ and that $s_1^2 = 4.5$ and $s_2^2 = 2.3$. Use $\alpha = 0.01$. Test the hypothesis and explain how the test could be conducted with a confidence interval on σ_1/σ_2.

10-43. Consider the hypothesis test $H_0\colon \sigma_1^2 = \sigma_2^2$ against $H_1\colon \sigma_1^2 \neq \sigma_2^2$. Suppose that the sample sizes are $n_1 = 15$ and $n_2 = 15$ and the sample variances are $s_1^2 = 2.3$ and $s_2^2 = 1.9$. Use $\alpha = 0.05$.

(a) Test the hypothesis and explain how the test could be conducted with a confidence interval on σ_1/σ_2.

(b) What is the power of the test in part (a) if σ_1 is twice as large as σ_2?

(c) Assuming equal sample sizes, what sample size should be used to obtain $\beta = 0.05$ if the σ_2 is half of σ_1?

10-44. Two chemical companies can supply a raw material. The concentration of a particular element in this material is important. The mean concentration for both suppliers is the same, but we suspect that the variability in concentration may differ between the two companies. The standard deviation of concentration in a random sample of $n_1 = 10$ batches produced by company 1 is $s_1 = 4.7$ grams per liter, while for company 2, a random sample of $n_2 = 16$ batches yields $s_2 = 5.8$ grams per liter. Is there sufficient evidence to conclude that the two population variances differ? Use $\alpha = 0.05$.

10-45. A study was performed to determine whether men and women differ in their repeatability in assembling components on printed circuit boards. Random samples of 25 men and 21 women were selected, and each subject assembled the units. The two sample standard deviations of assembly time were $s_{\text{men}} = 0.98$ minutes and $s_{\text{women}} = 1.02$ minutes.

(a) Is there evidence to support the claim that men and women differ in repeatability for this assembly task? Use $\alpha = 0.02$ and state any necessary assumptions about the underlying distribution of the data.

(b) Find a 98% confidence interval on the ratio of the two variances. Provide an interpretation of the interval.

10-46. Consider the foam data in Exercise 10-14. Construct the following:

(a) A 90% two-sided confidence interval on σ_1^2/σ_2^2.
(b) A 95% two-sided confidence interval on σ_1^2/σ_2^2. Comment on the comparison of the width of this interval with the width of the interval in part (a).
(c) A 90% lower-confidence bound on σ_1/σ_2.

10-47. Consider the diameter data in Exercise 10-13. Construct the following:

(a) A 90% two-sided confidence interval on σ_1/σ_2.
(b) A 95% two-sided confidence interval on σ_1/σ_2. Comment on the comparison of the width of this interval with the width of the interval in part (a).
(c) A 90% lower-confidence bound on σ_1/σ_2.

10-48. Consider the gear impact strength data in Exercise 10-18. Is there sufficient evidence to conclude that the variance of impact strength is different for the two suppliers? Use $\alpha = 0.05$.

10-49. Consider the melting point data in Exercise 10-19. Do the sample data support a claim that both alloys have the same variance of melting point? Use $\alpha = 0.05$ in reaching your conclusion.

10-50. Exercise 10-22 presented measurements of plastic coating thickness at two different application temperatures. Test H_0: $\sigma_1^2 = \sigma_2^2$ against H_1: $\sigma_1^2 \neq \sigma_2^2$ using $\alpha = 0.01$.

10-51. Reconsider the overall distance data for golf balls in Exercise 10-27. Is there evidence to support the claim that the standard deviation of overall distance is the same for both brands of balls (use $\alpha = 0.05$)? Explain how this question can be answered with a 95% confidence interval on σ_1/σ_2.

10-52. Reconsider the coefficient of restitution data in Exercise 10-28. Do the data suggest that the standard deviation is the same for both brands of drivers (use $\alpha = 0.05$)? Explain how to answer this question with a confidence interval on σ_1/σ_2.

10-53. Consider the weight of paper data from Technometrics In Exercise 10-26. Is there evidence that the variance of the weight measurement differs between the sheets of paper? Use $\alpha = 0.05$. Explain how this test can be conducted with a confidence interval.

10-54. Consider the film speed data in Exercise 10-20.

(a) Test H_0: $\sigma_1^2 = \sigma_2^2$ versus H_1: $\sigma_1^2 \neq \sigma_2^2$ using $\alpha = 0.02$.
(b) Suppose that one population standard deviation is 50% larger than the other. Is the sample size $n_1 = n_2 = 8$ adequate to detect this difference with high probability? Use $\alpha = 0.01$ in answering this question.

10-55. Consider the etch rate data in Exercise 10-17.

(a) Test the hypothesis H_0: $\sigma_1^2 = \sigma_2^2$ against H_1: $\sigma_1^2 \neq \sigma_2^2$ using $\alpha = 0.05$, and draw conclusions.
(b) Suppose that if one population variance is twice as large as the other, we want to detect this with probability at least 0.90 (using $\alpha = 0.05$). Are the sample sizes $n_1 = n_2 = 10$ adequate?

10-6 INFERENCE ON TWO POPULATION PROPORTIONS

We now consider the case where there are two binomial parameters of interest, say, p_1 and p_2, and we wish to draw inferences about these proportions. We will present large-sample hypothesis testing and confidence interval procedures based on the normal approximation to the binomial.

10-6.1 Large-Sample Test on the Difference in Population Proportions

Suppose that two independent random samples of sizes n_1 and n_2 are taken from two populations, and let X_1 and X_2 represent the number of observations that belong to the class of interest in samples 1 and 2, respectively. Furthermore, suppose that the normal approximation to the binomial is applied to each population, so the estimators of the population proportions $\hat{P}_1 = X_1/n_1$ and $\hat{P}_2 = X_2/n_2$ have approximate normal distributions. We are interested in testing the hypotheses

$$H_0: p_1 = p_2$$
$$H_1: p_1 \neq p_2$$

The statistic

Test Statistic for the Difference of Two Population Proportions

$$Z = \frac{\hat{P}_1 - \hat{P}_2 - (p_1 - p_2)}{\sqrt{\dfrac{p_1(1-p_1)}{n_1} + \dfrac{p_2(1-p_2)}{n_2}}} \tag{10-32}$$

is distributed approximately as standard normal and is the basis of a test for $H_0: p_1 = p_2$. Specifically, if the null hypothesis $H_0: p_1 = p_2$ is true, using the fact that $p_1 = p_2 = p$, the random variable

$$Z = \frac{\hat{P}_1 - \hat{P}_2}{\sqrt{p(1-p)\left(\dfrac{1}{n_1} + \dfrac{1}{n_2}\right)}}$$

is distributed approximately $N(0, 1)$. An estimator of the common parameter p is

$$\hat{P} = \frac{X_1 + X_2}{n_1 + n_2}$$

The **test statistic** for $H_0: p_1 = p_2$ is then

$$Z_0 = \frac{\hat{P}_1 - \hat{P}_2}{\sqrt{\hat{P}(1-\hat{P})\left(\dfrac{1}{n_1} + \dfrac{1}{n_2}\right)}}$$

This leads to the test procedures described below.

Approximate Tests on the Difference of Two Population Proportions

Null hypothesis: $H_0: p_1 = p_2$

Test statistic:

$$Z_0 = \frac{\hat{P}_1 - \hat{P}_2}{\sqrt{\hat{P}(1-\hat{P})\left(\dfrac{1}{n_1} + \dfrac{1}{n_2}\right)}} \tag{10-33}$$

Alternative Hypotheses	Rejection Criterion
$H_1: p_1 \neq p_2$	$z_0 > z_{\alpha/2}$ or $z_0 < -z_{\alpha/2}$
$H_1: p_1 > p_2$	$z_0 > z_\alpha$
$H_1: p_1 < p_2$	$z_0 < -z_\alpha$

EXAMPLE 10-14
St. John's Wort

Extracts of St. John's Wort are widely used to treat depression. An article in the April 18, 2001 issue of the *Journal of the American Medical Association* ("Effectiveness of St. John's Wort on Major Depression: A Randomized Controlled Trial") compared the efficacy of a standard extract of St. John's Wort with a placebo in 200 outpatients diagnosed with major depression. Patients were randomly assigned to two groups; one group received the St. John's Wort, and the other received the placebo. After eight weeks, 19 of the placebo-treated patients showed improvement, whereas 27 of those treated with St. John's Wort improved. Is there any reason to believe that St. John's Wort is effective in treating major depression? Use $\alpha = 0.05$.

The eight-step hypothesis testing procedure leads to the following results:

1. The parameters of interest are p_1 and p_2, the proportion of patients who improve following treatment with St. John's Wort (p_1) or the placebo (p_2).
2. $H_0: p_1 = p_2$
3. $H_1: p_1 \neq p_2$
4. $\alpha = 0.05$
5. The test statistic is

$$z_0 = \frac{\hat{p}_1 - \hat{p}_2}{\sqrt{\hat{p}(1 - \hat{p})\left(\frac{1}{n_1} + \frac{1}{n_2}\right)}}$$

where $\hat{p}_1 = 27/100 = 0.27, \hat{p}_2 = 19/100 = 0.19, n_1 = n_2 = 100,$ and

$$\hat{p} = \frac{x_1 + x_2}{n_1 + n_2} = \frac{19 + 27}{100 + 100} = 0.23$$

6. Reject $H_0: p_1 = p_2$ if $z_0 > z_{0.025} = 1.96$ or if $z_0 < -z_{0.025} = -1.96$.
7. Computations: The value of the test statistic is

$$z_0 = \frac{0.27 - 0.19}{\sqrt{0.23(0.77)\left(\frac{1}{100} + \frac{1}{100}\right)}} = 1.34$$

8. Conclusions: Since $z_0 = 1.34$ does not exceed $z_{0.025}$, we cannot reject the null hypothesis. Note that the P-value is $P \simeq 0.179$. There is insufficient evidence to support the claim that St. John's Wort is effective in treating major depression.

The following box shows the Minitab two-sample hypothesis test and CI procedure for proportions. Notice that the 95% CI on $p_1 - p_2$ includes zero. The equation for constructing the CI will be given in Section 10-6.3.

Minitab Computations

Test and CI for Two Proportions

Sample	X	N	Sample p
1	27	100	0.270000
2	19	100	0.190000

Estimate for p(1) − p(2): 0.08
95% CI for p(1) − p(2): (−0.0361186, 0.196119)
Test for p(1) − p(2) = 0 (vs not = 0): Z = 1.35 P-Value = 0.177

10-6.2 Type II Error and Choice of Sample Size

The computation of the β-error for the large-sample test of H_0: $p_1 = p_2$ is somewhat more involved than in the single-sample case. The problem is that the denominator of the test statistic Z_0 is an estimate of the standard deviation of $\hat{P}_1 - \hat{P}_2$ under the assumption that $p_1 = p_2 = p$. When H_0: $p_1 = p_2$ is false, the standard deviation of $\hat{P}_1 - \hat{P}_2$ is

$$\sigma_{\hat{P}_1 - \hat{P}_2} = \sqrt{\frac{p_1(1-p_1)}{n_1} + \frac{p_2(1-p_2)}{n_2}} \tag{10-34}$$

Approximate Type II Error for a Two-Sided Test on the Difference of Two Population Proportions

If the alternative hypothesis is two sided, the β-error is

$$\beta = \Phi\left[\frac{z_{\alpha/2}\sqrt{\bar{p}\bar{q}(1/n_1 + 1/n_2)} - (p_1 - p_2)}{\sigma_{\hat{P}_1 - \hat{P}_2}}\right] - \Phi\left[\frac{-z_{\alpha/2}\sqrt{\bar{p}\bar{q}(1/n_1 + 1/n_2)} - (p_1 - p_2)}{\sigma_{\hat{P}_1 - \hat{P}_2}}\right] \tag{10-35}$$

where

$$\bar{p} = \frac{n_1 p_1 + n_2 p_2}{n_1 + n_2} \quad \text{and} \quad \bar{q} = \frac{n_1(1-p_1) + n_2(1-p_2)}{n_1 + n_2}$$

and $\sigma_{\hat{P}_1 - \hat{P}_2}$ is given by Equation 10-34.

Approximate Type II Error for a One-Sided Test on the Difference of Two Population Proportions

If the alternative hypothesis is H_1: $p_1 > p_2$,

$$\beta = \Phi\left[\frac{z_{\alpha}\sqrt{\bar{p}\bar{q}(1/n_1 + 1/n_2)} - (p_1 - p_2)}{\sigma_{\hat{P}_1 - \hat{P}_2}}\right] \tag{10-36}$$

and if the alternative hypothesis is H_1: $p_1 < p_2$,

$$\beta = 1 - \Phi\left[\frac{-z_{\alpha}\sqrt{\bar{p}\bar{q}(1/n_1 + 1/n_2)} - (p_1 - p_2)}{\sigma_{\hat{P}_1 - \hat{P}_2}}\right] \tag{10-37}$$

For a specified pair of values p_1 and p_2, we can find the sample sizes $n_1 = n_2 = n$ required to give the test of size α that has specified type II error β.

Approximate Sample Size for a Two-Sided Test on the Difference in Population Proportions

For the two-sided alternative, the common sample size is

$$n = \frac{\left[z_{\alpha/2}\sqrt{(p_1 + p_2)(q_1 + q_2)/2} + z_\beta\sqrt{p_1q_1 + p_2q_2}\right]^2}{(p_1 - p_2)^2} \tag{10-38}$$

where $q_1 = 1 - p_1$ and $q_2 = 1 - p_2$.

For a one-sided alternative, replace $z_{\alpha/2}$ in Equation 10-38 by z_α.

10-6.3 Confidence Interval on the Difference in Population Proportions

The confidence interval for $p_1 - p_2$ can be found directly, since we know that

$$Z = \frac{\hat{P}_1 - \hat{P}_2 - (p_1 - p_2)}{\sqrt{\dfrac{p_1(1-p_1)}{n_1} + \dfrac{p_2(1-p_2)}{n_2}}}$$

is a standard normal random variable. Thus $P(-z_{\alpha/2} \le Z \le z_{\alpha/2}) \simeq 1 - \alpha$, so we can substitute for Z in this last expression and use an approach similar to the one employed previously to find an approximate $100(1 - \alpha)\%$ two-sided confidence interval for $p_1 - p_2$.

Approximate Confidence Interval on the Difference in Population Proportions

If $\hat{p}_1$ and $\hat{p}_2$ are the sample proportions of observations in two independent random samples of sizes n_1 and n_2 that belong to a class of interest, an **approximate two-sided $100(1 - \alpha)\%$ confidence interval on the difference in the true proportions $p_1 - p_2$** is

$$\hat{p}_1 - \hat{p}_2 - z_{\alpha/2}\sqrt{\frac{\hat{p}_1(1-\hat{p}_1)}{n_1} + \frac{\hat{p}_2(1-\hat{p}_2)}{n_2}} \le p_1 - p_2 \le \hat{p}_1 - \hat{p}_2 + z_{\alpha/2}\sqrt{\frac{\hat{p}_1(1-\hat{p}_1)}{n_1} + \frac{\hat{p}_2(1-\hat{p}_2)}{n_2}} \tag{10-39}$$

where $z_{\alpha/2}$ is the upper $\alpha/2$ percentage point of the standard normal distribution.

EXAMPLE 10-15
Defective Bearings

Consider the process manufacturing crankshaft bearings described in Example 8-7. Suppose that a modification is made in the surface finishing process and that, subsequently, a second random sample of 85 bearings is obtained. The number of defective bearings in this second sample is 8. Therefore, since $n_1 = 85$, $\hat{p}_1 = 0.12$, $n_2 = 85$, and $\hat{p}_2 = 8/85 = 0.09$, we can obtain an approximate 95% confidence interval on the difference in the proportion of defective bearings produced under the two processes from Equation 10-39 as follows:

$$\hat{p}_1 - \hat{p}_2 - z_{0.025}\sqrt{\frac{\hat{p}_1(1-\hat{p}_1)}{n_1} + \frac{\hat{p}_2(1-\hat{p}_2)}{n_2}} \le p_1 - p_2 \le \hat{p}_1 - \hat{p}_2 + z_{0.025}\sqrt{\frac{\hat{p}_1(1-\hat{p}_1)}{n_1} + \frac{\hat{p}_2(1-\hat{p}_2)}{n_2}}$$

or

$$0.12 - 0.09 - 1.96\sqrt{\frac{0.12(0.88)}{85} + \frac{0.09(0.91)}{85}} \le p_1 - p_2 \le 0.12 - 0.09 + 1.96\sqrt{\frac{0.12(0.88)}{85} + \frac{0.09(0.91)}{85}}$$

This simplifies to

$$-0.06 \le p_1 - p_2 \le 0.12$$

This confidence interval includes zero, so, based on the sample data, it seems unlikely that the changes made in the surface finish process have reduced the proportion of defective crankshaft bearings being produced.

EXERCISES FOR SECTION 10-6

10-56. An article in *Knee Surgery, Sports Traumatology, Arthroscopy* (2005, Vol. 13, pp. 273–279), considered arthroscopic meniscal repair with an absorbable screw. Results showed that for tears greater than 25 millimeters, 14 of 18 (78%) repairs were successful while for shorter tears, 22 of 30 (73%) repairs were successful.

(a) Is there evidence that the success rate is greater for longer tears? Use $\alpha = 0.05$. What is the P-value?

(b) Calculate a one-sided 95% confidence bound on the difference in proportions that can be used to answer the question in part (a).

10-57. In the 2004 presidential election, exit polls from the critical state of Ohio provided the following results: for respondents with college degrees, 53% voted for Bush and 46% voted for Kerry. There were 2020 respondents.

(a) Is there a significant difference in these proportions? Use $\alpha = 0.05$. What is the P-value?

(b) Calculate a 95% confidence interval for the difference in the two proportions and comment on the use of this interval to answer the question in part (a).

10-58. Two different types of injection-molding machines are used to form plastic parts. A part is considered defective if it has excessive shrinkage or is discolored. Two random samples, each of size 300, are selected, and 15 defective parts are found in the sample from machine 1 while 8 defective parts are found in the sample from machine 2.

(a) Is it reasonable to conclude that both machines produce the same fraction of defective parts, using $\alpha = 0.05$? Find the P-value for this test.

(b) Construct a 95% confidence interval on the difference in the two fractions defective.

(c) Suppose that $p_1 = 0.05$ and $p_2 = 0.01$. With the sample sizes given here, what is the power of the test for this two-sided alternate?

(d) Supoose that $p_1 = 0.05$ and $p_2 = 0.01$. Determine the sample size needed to detect this difference with a probability of at least 0.9.

(e) Suppose that $p_1 = 0.05$ and $p_2 = 0.02$. With the sample sizes given here, what is the power of the test for this two-sided alternate?

(f) Suppose that $p_1 = 0.05$ and $p_2 = 0.02$. Determine the sample size needed to detect this difference with a probability of at least 0.9.

10-59. Two different types of polishing solutions are being evaluated for possible use in a tumble-polish operation for manufacturing interocular lenses used in the human eye following cataract surgery. Three hundred lenses were tumble polished using the first polishing solution, and of this number 253 had no polishing-induced defects. Another 300 lenses were tumble-polished using the second polishing solution, and 196 lenses were satisfactory upon completion.

(a) Is there any reason to believe that the two polishing solutions differ? Use $\alpha = 0.01$. What is the P-value for this test?

(b) Discuss how this question could be answered with a confidence interval on $p_1 - p_2$.

10-60. A random sample of 500 adult residents of Maricopa County found that 385 were in favor of increasing the highway speed limit to 75 mph, while another sample of 400 adult residents of Pima County found that 267 were in favor of the increased speed limit.

(a) Do these data indicate that there is a difference in the support for increasing the speed limit between the residents of the two counties? Use $\alpha = 0.05$. What is the P-value for this test?

(b) Construct a 95% confidence interval on the difference in the two proportions. Provide a practical interpretation of this interval.

10-7 SUMMARY TABLE AND ROADMAPS FOR INFERENCE PROCEDURES FOR TWO SAMPLES

The table in the end papers of the book summarizes all of the two-sample inference procedures given in this chapter. The table contains the null hypothesis statements, the test statistics, the criteria for rejection of the various alternative hypotheses, and the formulas for constructing the $100(1 - \alpha)\%$ confidence intervals.

The roadmap to select the appropriate confidence interval formula or hypothesis test method for one-sample problems was presented in Table 8-1. In Table 10-4, we extend the road

Table 10-4 Roadmap to Construct Confidence Intervals and Hypothesis Tests, Two-Sample Case

Function of the Parameters to be Bounded by the Confidence Interval or Tested with a Hypothesis	Symbol	Other Parameters?	Confidence Interval Section	Hypothesis Test Section	Comments
Difference in means from two normal distributions	$\mu_1 - \mu_2$	Standard deviations σ_1 and σ_2 known	10-2.3	10-2.1	
Difference in means from two arbitrary distributions with large sample sizes	$\mu_1 - \mu_2$	Sample sizes large enough that σ_1 and σ_2 are essentially known	10-2.3	10-2.1	Large sample size is often taken to be n_1 and $n_2 \geq 40$
Difference in means from two normal distributions	$\mu_1 - \mu_2$	Standard deviations σ_1 and σ_2 are unknown, and assumed equal	10-3.3	10-3.1	Case 1: $\sigma_1 = \sigma_2$
Differene in means from two normal distributions	$\mu_1 - \mu_2$	Standard deviations σ_1 and σ_2 are unknown, and NOT assumed equal	10-3.3	10-3.1	Case 2: $\sigma_1 \neq \sigma_2$
Difference in means from two normal distributions in a paired analysis	$\mu_D = \mu_1 - \mu_2$	Standard deviation of differences are unknown	10-4	10-4	Paired analysis calculates differences and uses a one-sample method for inference on the mean difference
Ratio of variances of two normal distributions	σ_1^2/σ_2^2	Means μ_1 and μ_2 unknown and estimated	10-5.4	10-5.2	
Difference in Two Population Proportions	$p_1 - p_2$	None	10-6.3	10-6.1	Normal approximation to the binomial distribution used for the tests and confidence intervals

map to two-sample problems. The primary comments stated previously also apply here (except we usually apply conclusions to a function of the parameters from each sample, such as the difference in means):

1. Determine the function of the parameters (and the distribution of the data) that is to be bounded by the confidence interval or tested by the hypothesis.
2. Check if other parameters are known or need to be estimated (and if any assumptions are made).

Supplemental Exercises

10-61. An article in the *Journal of Materials Engineering* (1989, Vol. 11, No. 4, pp. 275–282) reported the results of an experiment to determine failure mechanisms for plasma-sprayed thermal barrier coatings. The failure stress for one particular coating (NiCrAlZr) under two different test conditions is as follows:

Failure stress ($\times 10^6$ Pa) after nine 1-hour cycles: 19.8, 18.5, 17.6, 16.7, 16.7, 14.8, 15.4, 14.1, 13.6

Failure stress ($\times 10^6$ Pa) after six 1-hour cycles: 14.9, 12.7, 11.9, 11.4, 10.1, 7.9

(a) What assumptions are needed to construct confidence intervals for the difference in mean failure stress under the two different test conditions? Use normal probability plots of the data to check these assumptions.
(b) Find a 99% confidence interval on the difference in mean failure stress under the two different test conditions.
(c) Using the confidence interval constructed in part (b), does the evidence support the claim that the first test conditions yield higher results, on the average, than the second? Explain your answer.
(d) Construct a 95% confidence interval on the ratio of the variances, σ_1^2/σ_2^2, of failure stress under the two different test conditions.
(e) Use your answer in part (b) to determine whether there is a significant difference in variances of the two different test conditions. Explain your answer.

10-62. A procurement specialist has purchased 25 resistors from vendor 1 and 35 resistors from vendor 2. Each resistor's resistance is measured with the following results:

Vendor 1

96.8	100.0	100.3	98.5	98.3	98.2
99.6	99.4	99.9	101.1	103.7	97.7
99.7	101.1	97.7	98.6	101.9	101.0
99.4	99.8	99.1	99.6	101.2	98.2
98.6					

Vendor 2

106.8	106.8	104.7	104.7	108.0	102.2
103.2	103.7	106.8	105.1	104.0	106.2
102.6	100.3	104.0	107.0	104.3	105.8
104.0	106.3	102.2	102.8	104.2	103.4
104.6	103.5	106.3	109.2	107.2	105.4
106.4	106.8	104.1	107.1	107.7	

(a) What distributional assumption is needed to test the claim that the variance of resistance of product from vendor 1 is not significantly different from the variance of resistance of product from vendor 2? Perform a graphical procedure to check this assumption.
(b) Perform an appropriate statistical hypothesis-testing procedure to determine whether the procurement specialist can claim that the variance of resistance of product from vendor 1 is significantly different from the variance of resistance of product from vendor 2.

10-63. A liquid dietary product implies in its advertising that use of the product for one month results in an average weight loss of at least 3 pounds. Eight subjects use the product for one month, and the resulting weight loss data are reported below. Use hypothesis-testing procedures to answer the following questions.

Subject	Initial Weight (lb)	Final Weight (lb)
1	165	161
2	201	195
3	195	192
4	198	193
5	155	150
6	143	141
7	150	146
8	187	183

(a) Do the data support the claim of the producer of the dietary product with the probability of a type I error set to 0.05?

(b) Do the data support the claim of the producer of the dietary product with the probability of a type I error set to 0.01?

(c) In an effort to improve sales, the producer is considering changing its claim from "at least 3 pounds" to "at least 5 pounds." Repeat parts (a) and (b) to test this new claim.

10-64. The breaking strength of yarn supplied by two manufacturers is being investigated. We know from experience with the manufacturers' processes that $\sigma_1 = 5$ psi and $\sigma_2 = 4$ psi. A random sample of 20 test specimens from each manufacturer results in $\bar{x}_1 = 88$ psi and $\bar{x}_2 = 91$ psi, respectively.

(a) Using a 90% confidence interval on the difference in mean breaking strength, comment on whether or not there is evidence to support the claim that manufacturer 2 produces yarn with higher mean breaking strength.

(b) Using a 98% confidence interval on the difference in mean breaking strength, comment on whether or not there is evidence to support the claim that manufacturer 2 produces yarn with higher mean breaking strength.

(c) Comment on why the results from parts (a) and (b) are different or the same. Which would you choose to make your decision and why?

10-65. The Salk polio vaccine experiment in 1954 focused on the effectiveness of the vaccine in combating paralytic polio. Because it was felt that without a control group of children there would be no sound basis for evaluating the efficacy of the Salk vaccine, the vaccine was administered to one group, and a placebo (visually identical to the vaccine but known to have no effect) was administered to a second group. For ethical reasons, and because it was suspected that knowledge of vaccine administration would affect subsequent diagnoses, the experiment was conducted in a double-blind fashion. That is, neither the subjects nor the administrators knew who received the vaccine and who received the placebo. The actual data for this experiment are as follows:

Placebo group: $n = 201{,}299$: 110 cases of polio observed

Vaccine group: $n = 200{,}745$: 33 cases of polio observed

(a) Use a hypothesis-testing procedure to determine if the proportion of children in the two groups who contracted paralytic polio is statistically different. Use a probability of a type I error equal to 0.05.

(b) Repeat part (a) using a probability of a type I error equal to 0.01.

(c) Compare your conclusions from parts (a) and (b) and explain why they are the same or different.

10-66. Consider Supplemental Exercise 10-64. Suppose that prior to collecting the data, you decide that you want the error in estimating $\mu_1 - \mu_2$ by $\bar{x}_1 - \bar{x}_2$ to be less than 1.5 psi. Specify the sample size for the following percentage confidence:

(a) 90%

(b) 98%

(c) Comment on the effect of increasing the percentage confidence on the sample size needed.

(d) Repeat parts (a)–(c) with an error of less than 0.75 psi instead of 1.5 psi.

(e) Comment on the effect of decreasing the error on the sample size needed.

10-67. A random sample of 1500 residential telephones in Phoenix in 1990 found that 387 of the numbers were unlisted. A random sample in the same year of 1200 telephones in Scottsdale found that 310 were unlisted.

(a) Find a 95% confidence interval on the difference in the two proportions and use this confidence interval to determine if there is a statistically significant difference in proportions of unlisted numbers between the two cities.

(b) Find a 90% confidence interval on the difference in the two proportions and use this confidence interval to determine if there is a statistically significant difference in proportions of unlisted numbers between the two cities.

(c) Suppose that all the numbers in the problem description were doubled. That is, 774 residents out of 3000 sampled in Phoenix and 620 residents out of 2400 in Scottsdale had unlisted phone numbers. Repeat parts (a) and (b) and comment on the effect of increasing the sample size without changing the proportions on your results.

10-68. In a random sample of 200 Phoenix residents who drive a domestic car, 165 reported wearing their seat belt regularly, while another sample of 250 Phoenix residents who drive a foreign car revealed 198 who regularly wore their seat belt.

(a) Perform a hypothesis-testing procedure to determine if there is a statistically significant difference in seat belt usage between domestic and foreign car drivers. Set your probability of a type I error to 0.05.

(b) Perform a hypothesis-testing procedure to determine if there is a statistically significant difference in seat belt usage between domestic and foreign car drivers. Set your probability of a type I error to 0.1.

(c) Compare your answers for parts (a) and (b) and explain why they are the same or different.

(d) Suppose that all the numbers in the problem description were doubled. That is, in a random sample of 400 Phoenix residents who drive a domestic car, 330 reported wearing their seat belt regularly, while another sample of 500 Phoenix residents who drive a foreign car revealed 396 who regularly wore their seat belt. Repeat parts (a) and (b) and comment on the effect of increasing the sample size without changing the proportions on your results.

10-69. Consider the previous exercise, which summarized data collected from drivers about their seat belt usage.

(a) Do you think there is a reason not to believe these data? Explain your answer.

(b) Is it reasonable to use the hypothesis-testing results from the previous problem to draw an inference about the difference in proportion of seat belt usage

(i) of the spouses of these drivers of domestic and foreign cars? Explain your answer.

(ii) of the children of these drivers of domestic and foreign cars? Explain your answer.
(iii) of all drivers of domestic and foreign cars? Explain your answer.
(iv) of all drivers of domestic and foreign trucks? Explain your answer.

10-70. A manufacturer of a new pain relief tablet would like to demonstrate that its product works twice as fast as the competitor's product. Specifically, the manufacturer would like to test

$$H_0\colon \mu_1 = 2\mu_2$$
$$H_1\colon \mu_1 > 2\mu_2$$

where μ_1 is the mean absorption time of the competitive product and μ_2 is the mean absorption time of the new product. Assuming that the variances σ_1^2 and σ_2^2 are known, develop a procedure for testing this hypothesis.

10-71. Two machines are used to fill plastic bottles with dishwashing detergent. The standard deviations of fill volume are known to be $\sigma_1 = 0.10$ fluid ounces and $\sigma_2 = 0.15$ fluid ounces for the two machines, respectively. Two random samples of $n_1 = 12$ bottles from machine 1 and $n_2 = 10$ bottles from machine 2 are selected, and the sample mean fill volumes are $\bar{x}_1 = 30.87$ fluid ounces and $\bar{x}_2 = 30.68$ fluid ounces. Assume normality.

(a) Construct a 90% two-sided confidence interval on the mean difference in fill volume. Interpret this interval.
(b) Construct a 95% two-sided confidence interval on the mean difference in fill volume. Compare and comment on the width of this interval to the width of the interval in part (a).
(c) Construct a 95% upper-confidence interval on the mean difference in fill volume. Interpret this interval.
(d) Test the hypothesis that both machines fill to the same mean volume. Use $\alpha = 0.05$. What is the P-value?
(e) If the β-error of the test when the true difference in fill volume is 0.2 fluid ounces should not exceed 0.1, what sample sizes must be used? Use $\alpha = 0.05$.

10-72. Suppose that we are testing $H_0\colon \mu_1 = \mu_2$ versus $H_1\colon \mu_1 \neq \mu_2$, and we plan to use equal sample sizes from the two populations. Both populations are assumed to be normal with unknown but equal variances. If we use $\alpha = 0.05$ and if the true mean $\mu_1 = \mu_2 + \sigma$, what sample size must be used for the power of this test to be at least 0.90?

10-73. Consider the situation described in Exercise 10-59.

(a) Redefine the parameters of interest to be the proportion of lenses that are unsatisfactory following tumble polishing with polishing fluids 1 or 2. Test the hypothesis that the two polishing solutions give different results using $\alpha = 0.01$.
(b) Compare your answer in part (a) with that for Exercise 10-59. Explain why they are the same or different.
(c) We wish to use $\alpha = 0.01$. Suppose that if $p_1 = 0.9$ and $p_2 = 0.6$, we wish to detect this with a high probability, say, at least 0.9. What sample sizes are required to meet this objective?

10-74. Consider the fire-fighting foam expanding agents investigated in Exercise 10-14, in which five observations of each agent were recorded. Suppose that, if agent 1 produces a mean expansion that differs from the mean expansion of agent 1 by 1.5, we would like to reject the null hypothesis with probability at least 0.95.

(a) What sample size is required?
(b) Do you think that the original sample size in Exercise 10-14 was appropriate to detect this difference? Explain your answer.

10-75. A fuel-economy study was conducted for two German automobiles, Mercedes and Volkswagen. One vehicle of each brand was selected, and the mileage performance was observed for 10 tanks of fuel in each car. The data are as follows (in miles per gallon):

Mercedes		Volkswagen	
24.7	24.9	41.7	42.8
24.8	24.6	42.3	42.4
24.9	23.9	41.6	39.9
24.7	24.9	39.5	40.8
24.5	24.8	41.9	29.6

(a) Construct a normal probability plot of each of the data sets. Based on these plots, is it reasonable to assume that they are each drawn from a normal population?
(b) Suppose that it was determined that the lowest observation of the Mercedes data was erroneously recorded and should be 24.6. Furthermore, the lowest observation of the Volkswagen data was also mistaken and should be 39.6. Again construct normal probability plots of each of the data sets with the corrected values. Based on these new plots, is it reasonable to assume that they are each drawn from a normal population?
(c) Compare your answers from parts (a) and (b) and comment on the effect of these mistaken observations on the normality assumption.
(d) Using the corrected data from part (b) and a 95% confidence interval, is there evidence to support the claim that the variability in mileage performance is greater for a Volkswagen than for a Mercedes?
(e) Rework part (d) of this problem using an appropriate hypothesis-testing procedure. Did you get the same answer as you did originally? Why?

10-76. An experiment was conducted to compare the filling capability of packaging equipment at two different wineries. Ten bottles of pinot noir from Ridgecrest Vineyards were randomly selected and measured, along with 10 bottles of pinot noir from Valley View Vineyards. The data are as follows (fill volume is in milliliters):

Ridgecrest				Valley View			
755	751	752	753	756	754	757	756
753	753	753	754	755	756	756	755
752	751			755	756		

(a) What assumptions are necessary to perform a hypothesis-testing procedure for equality of means of these data? Check these assumptions.
(b) Perform the appropriate hypothesis-testing procedure to determine whether the data support the claim that both wineries will fill bottles to the same mean volume.
(c) Suppose that the true difference in mean fill volume is as much as 2 fluid ounces; did the sample sizes of 10 from each vineyard provide good detection capability when $\alpha = 0.05$? Explain your answer.

10-77. A Rockwell hardness-testing machine presses a tip into a test coupon and uses the depth of the resulting depression to indicate hardness. Two different tips are being compared to determine whether they provide the same Rockwell C-scale hardness readings. Nine coupons are tested, with both tips being tested on each coupon. The data are shown in the accompanying table.

Coupon	Tip 1	Tip 2	Coupon	Tip 1	Tip 2
1	47	46	6	41	41
2	42	40	7	45	46
3	43	45	8	45	46
4	40	41	9	49	48
5	42	43			

(a) State any assumptions necessary to test the claim that both tips produce the same Rockwell C-scale hardness readings. Check those assumptions for which you have the information.
(b) Apply an appropriate statistical method to determine if the data support the claim that the difference in Rockwell C-scale hardness readings of the two tips is significantly different from zero
(c) Suppose that if the two tips differ in mean hardness readings by as much as 1.0, we want the power of the test to be at least 0.9. For an $\alpha = 0.01$, how many coupons should have been used in the test?

10-78. Two different gauges can be used to measure the depth of bath material in a Hall cell used in smelting aluminum. Each gauge is used once in 15 cells by the same operator.

Cell	Gauge 1	Gauge 2	Cell	Gauge 1	Gauge 2
1	46 in.	47 in.	9	52	51
2	50	53	10	47	45
3	47	45	11	49	51
4	53	50	12	45	45
5	49	51	13	47	49
6	48	48	14	46	43
7	53	54	15	50	51
8	56	53			

(a) State any assumptions necessary to test the claim that both gauges produce the same mean bath depth readings. Check those assumptions for which you have the information.
(b) Apply an appropriate statistical procedure to determine if the data support the claim that the two gauges produce different mean bath depth readings.
(c) Suppose that if the two gauges differ in mean bath depth readings by as much as 1.65 inch, we want the power of the test to be at least 0.8. For $\alpha = 0.01$, how many cells should have been used?

10-79. An article in the *Journal of the Environmental Engineering Division* ("Distribution of Toxic Substances in Rivers," 1982, Vol. 108, pp. 639–649) investigated the concentration of several hydrophobic organic substances in the Wolf River in Tennessee. Measurements on hexachlorobenzene (HCB) in nanograms per liter were taken at different depths downstream of an abandoned dump site. Data for two depths follow:

Surface: 3.74, 4.61, 4.00, 4.67, 4.87, 5.12, 4.52, 5.29, 5.74, 5.48
Bottom: 5.44, 6.88, 5.37, 5.44, 5.03, 6.48, 3.89, 5.85, 6.85, 7.16

(a) What assumptions are required to test the claim that mean HCB concentration is the same at both depths? Check those assumptions for which you have the information.
(b) Apply an appropriate procedure to determine if the data support the claim in part a.
(c) Suppose that the true difference in mean concentrations is 2.0 nanograms per liter. For $\alpha = 0.05$, what is the power of a statistical test for H_0: $\mu_1 = \mu_2$ versus H_1: $\mu_1 \neq \mu_2$?
(d) What sample size would be required to detect a difference of 1.0 nanograms per liter at $\alpha = 0.05$ if the power must be at least 0.9?

MIND-EXPANDING EXERCISES

10-80. Three different pesticides can be used to control infestation of grapes. It is suspected that pesticide 3 is more effective than the other two. In a particular vineyard, three different plantings of pinot noir grapes are selected for study. The following results on yield are obtained:

Pesticide	$\bar{x}_i$ (Bushels/Plant)	s_i	n_i (Number of Plants)
1	4.6	0.7	100
2	5.2	0.6	120
3	6.1	0.8	130

If μ_i is the true mean yield after treatment with the ith pesticide, we are interested in the quantity

$$\mu = \frac{1}{2}(\mu_1 + \mu_2) - \mu_3$$

which measures the difference in mean yields between pesticides 1 and 2 and pesticide 3. If the sample sizes n_i are large, the estimator (say, $\hat{\mu}$) obtained by replacing each individual μ_i by $\bar{X}_i$ is approximately normal.

(a) Find an approximate $100(1 - \alpha)\%$ large-sample confidence interval for μ.

(b) Do these data support the claim that pesticide 3 is more effective than the other two? Use $\alpha = 0.05$ in determining your answer.

10-81. Suppose that we wish to test H_0: $\mu_1 = \mu_2$ versus H_1: $\mu_1 \neq \mu_2$, where σ_1^2 and σ_2^2 are known. The total sample size N is to be determined, and the allocation of observations to the two populations such that $n_1 + n_2 = N$ is to be made on the basis of cost. If the cost of sampling for populations 1 and 2 are C_1 and C_2, respectively, find the minimum cost sample sizes that provide a specified variance for the difference in sample means.

10-82. Suppose that we wish to test the hypothesis H_0: $\mu_1 = \mu_2$ versus H_1: $\mu_1 \neq \mu_2$, where both variances σ_1^2 and σ_2^2 are known. A total of $n_1 + n_2 = N$ observations can be taken. How should these observations be allocated to the two populations to maximize the probability that H_0 will be rejected if H_1 is true and $\mu_1 - \mu_2 = \Delta \neq 0$?

10-83. Suppose that we wish to test H_0: $\mu = \mu_0$ versus H_1: $\mu \neq \mu_0$, where the population is normal with known σ. Let $0 < \epsilon < \alpha$, and define the critical region so that we will reject H_0 if $z_0 > z_\epsilon$ or if $z_0 < -z_{\alpha-\epsilon}$, where z_0 is the value of the usual test statistic for these hypotheses.

(a) Show that the probability of type I error for this test is α.

(b) Suppose that the true mean is $\mu_1 = \mu_0 + \Delta$. Derive an expression for β for the above test.

10-84. Construct a data set for which the paired t-test statistic is very large, indicating that when this analysis is used the two population means are different, but t_0 for the two-sample t-test is very small so that the incorrect analysis would indicate that there is no significant difference between the means.

10-85. In some situations involving proportions, we are interested in the ratio $\theta = p_1/p_2$ rather than the difference $p_1 - p_2$. Let $\hat{\theta} = \hat{p}_1/\hat{p}_2$. We can show that $\ln(\hat{\theta})$ has an approximate normal distribution with the mean (n/θ) and variance $[(n_1 - x_1)/(n_1x_1) + (n_2 - x_2)/(n_2x_2)]^{1/2}$.

(a) Use the information above to derive a large-sample confidence interval for $\ln \theta$.

(b) Show how to find a large-sample CI for θ.

(c) Use the data from the St. John's Wort study in Example 10-14, and find a 95% CI on $\theta = p_1/p_2$. Provide a practical interpretation for this CI.

10-86. Derive an expression for β for the test of the equality of the variances of two normal distributions. Assume that the two-sided alternative is specified.

IMPORTANT TERMS AND CONCEPTS

Comparative experiments
Critical region for a test statistic
Identifying cause and effect
Null and alternative hypotheses
One-sided and two-sided alternative hypotheses
Operating characteristic curves
Paired t-test
Pooled t-test
P-value
Reference distribution for a test statistic
Sample size determination for hypothesis tests and confidence intervals
Statistical hypotheses
Test statistic

11 Analysis of Simple Linear Regression and Correlation

CHAPTER OUTLINE

LEARNING OBJECTIVES

After careful study of this chapter, you should be able to do the following:

1. Use simple linear regression for building empirical models to engineering and scientific data
2. Understand how the method of least squares is used to estimate the parameters in a linear regression model
3. Analyze residuals to determine if the regression model is an adequate fit to the data or to see if any underlying assumptions are violated
4. Test statistical hypotheses and construct confidence intervals on regression model parameters
5. Use the regression model to make a prediction of a future observation and construct an appropriate prediction interval on the future observation
6. Apply the correlation model
7. Use simple transformations to achieve a linear regression model

11-1 EMPIRICAL MODELS

Many problems in engineering and science involve exploring the relationships between two or more variables. **Regression analysis** is a statistical technique that is very useful for these types of problems. For example, in a chemical process, suppose that the yield of the product is related to the process-operating temperature. Regression analysis can be used to build a model to predict yield at a given temperature level. This model can also be used for process optimization, such as finding the level of temperature that maximizes yield, or for process control purposes.

As an illustration, consider the data in Table 11-1. In this table y is the purity of oxygen produced in a chemical distillation process, and x is the percentage of hydrocarbons that are present in the main condenser of the distillation unit. Figure 11-1 presents a **scatter diagram** of the data in Table 11-1. This is just a graph on which each (x_i, y_i) pair is represented as a point plotted in a two-dimensional coordinate system. This scatter diagram was produced by Minitab, and we selected an option that shows dot diagrams of the x and y variables along the top and right margins of the graph, respectively, making it easy to see the distributions of the individual variables (box plots or histograms could also be selected). Inspection of this scatter diagram indicates that, although no simple curve will pass exactly through all the points, there is a strong indication that the points lie scattered randomly around a straight line. Therefore, it is probably reasonable to assume that the mean of the random variable Y is related to x by the following straight-line relationship:

$$E(Y|x) = \mu_{Y|x} = \beta_0 + \beta_1 x$$

Table 11-1 Oxygen and Hydrocarbon Levels

Observation Number	Hydrocarbon Level x (%)	Purity y (%)
1	0.99	90.01
2	1.02	89.05
3	1.15	91.43
4	1.29	93.74
5	1.46	96.73
6	1.36	94.45
7	0.87	87.59
8	1.23	91.77
9	1.55	99.42
10	1.40	93.65
11	1.19	93.54
12	1.15	92.52
13	0.98	90.56
14	1.01	89.54
15	1.11	89.85
16	1.20	90.39
17	1.26	93.25
18	1.32	93.41
19	1.43	94.98
20	0.95	87.33

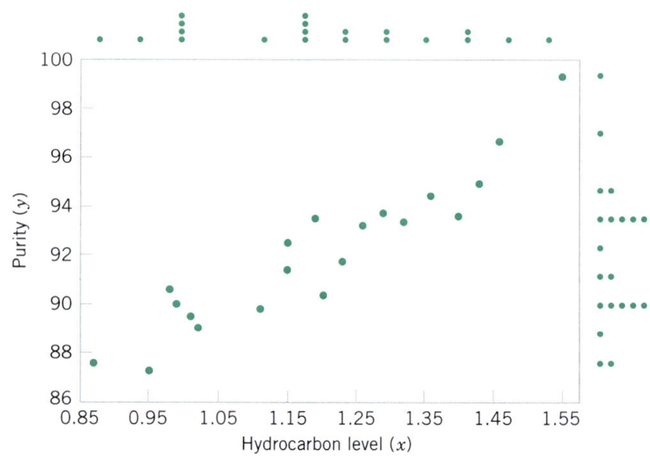

Figure 11-1 Scatter diagram of oxygen purity versus hydrocarbon level from Table 11-1.

where the slope and intercept of the line are called **regression coefficients.** While the mean of Y is a linear function of x, the actual observed value y does not fall exactly on a straight line. The appropriate way to generalize this to a probabilistic linear model is to assume that the expected value of Y is a linear function of x, but that for a fixed value of x the actual value of Y is determined by the mean value function (the linear model) plus a random error term, say,

$$Y = \beta_0 + \beta_1 x + \epsilon \tag{11-1}$$

where ϵ is the random error term. We will call this model the **simple linear regression model,** because it has only one independent variable or **regressor.** Sometimes a model like this will arise from a theoretical relationship. At other times, we will have no theoretical knowledge of the relationship between x and y, and the choice of the model is based on inspection of a scatter diagram, such as we did with the oxygen purity data. We then think of the regression model as an **empirical model.**

To gain more insight into this model, suppose that we can fix the value of x and observe the value of the random variable Y. Now if x is fixed, the random component ϵ on the right-hand side of the model in Equation 11-1 determines the properties of Y. Suppose that the mean and variance of ϵ are 0 and σ^2, respectively. Then

$$E(Y|x) = E(\beta_0 + \beta_1 x + \epsilon) = \beta_0 + \beta_1 x + E(\epsilon) = \beta_0 + \beta_1 x$$

Notice that this is the same relationship that we initially wrote down empirically from inspection of the scatter diagram in Fig. 11-1. The variance of Y given x is

$$V(Y|x) = V(\beta_0 + \beta_1 x + \epsilon) = V(\beta_0 + \beta_1 x) + V(\epsilon) = 0 + \sigma^2 = \sigma^2$$

Thus, the true regression model $\mu_{Y|x} = \beta_0 + \beta_1 x$ is a line of mean values; that is, the height of the regression line at any value of x is just the expected value of Y for that x. The slope, β_1, can be interpreted as the change in the mean of Y for a unit change in x. Furthermore, the variability of Y at a particular value of x is determined by the error variance σ^2. This implies that there is a distribution of Y-values at each x and that the variance of this distribution is the same at each x.

For example, suppose that the true regression model relating oxygen purity to hydrocarbon level is $\mu_{Y|x} = 75 + 15x$, and suppose that the variance is $\sigma^2 = 2$. Figure 11-2 illustrates this situation. Notice that we have used a normal distribution to describe the random variation in ϵ. Since Y is the sum of a constant $\beta_0 + \beta_1 x$ (the mean) and a normally distributed

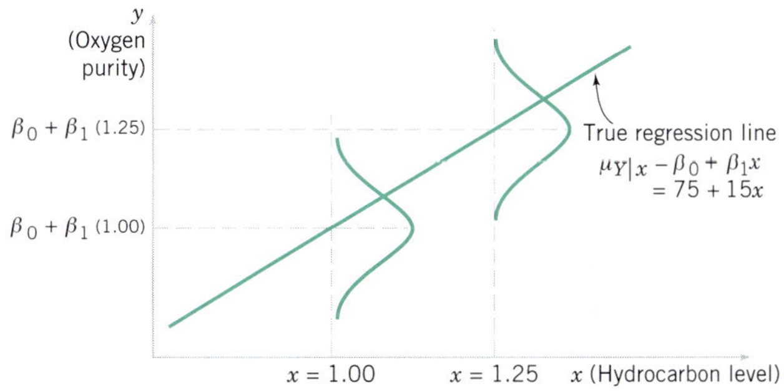

Figure 11-2 The distribution of Y for a given value of x for the oxygen purity-hydrocarbon data.

random variable, Y is a normally distributed random variable. The variance σ^2 determines the variability in the observations Y on oxygen purity. Thus, when σ^2 is small, the observed values of Y will fall close to the line, and when σ^2 is large, the observed values of Y may deviate considerably from the line. Because σ^2 is constant, the variability in Y at any value of x is the same.

The regression model describes the relationship between oxygen purity Y and hydrocarbon level x. Thus, for any value of hydrocarbon level, oxygen purity has a normal distribution with mean $75 + 15x$ and variance 2. For example, if $x = 1.25$, Y has mean value $\mu_{Y|x} = 75 + 15(1.25) = 93.75$ and variance 2.

In most real-world problems, the values of the intercept and slope (β_0, β_1) and the error variance σ^2 will not be known, and they must be estimated from sample data. Then this fitted regression equation or model is typically used in prediction of future observations of Y, or for estimating the mean response at a particular level of x. To illustrate, a chemical engineer might be interested in estimating the mean purity of oxygen produced when the hydrocarbon level is $x = 1.25\%$. This chapter discusses such procedures and applications for the simple linear regression model. Chapter 12 will discuss multiple linear regression models that involve more than one regressor.

Historical Note

Sir Francis Galton first used the term **regression analysis** in a study of the heights of fathers (x) and sons (y). Galton fit a least squares line and used it to predict the son's height from the father's height. He found that if a father's height was above average, the son's height would also be above average, but not by as much as the father's height was. A similar effect was observed for below average heights. That is, the son's height "regressed" toward the average. Consequently, Galton referred to the least squares line as a **regression line.**

Abuses of Regression

Regression is widely used and frequently misused; several common abuses of regression are briefly mentioned here. Care should be taken in selecting variables with which to construct regression equations and in determining the form of the model. It is possible to develop statistically significant relationships among variables that are completely unrelated in a **causal** sense. For example, we might attempt to relate the shear strength of spot welds with the number of empty parking spaces in the visitor parking lot. A straight line may even appear to provide a good fit to the data, but the relationship is an unreasonable one on which to rely. You can't increase the weld strength by blocking off parking spaces. A strong observed association between variables does not necessarily imply that a causal relationship exists between those variables. This type of effect is encountered fairly often in retrospective data analysis, and even in observational studies. **Designed experiments** are the only way to determine cause-and-effect relationships.

Regression relationships are valid only for values of the regressor variable within the range of the original data. The linear relationship that we have tentatively assumed may be valid over the original range of x, but it may be unlikely to remain so as we extrapolate—that is, if we use values of x beyond that range. In other words, as we move beyond the range of values of x for which data were collected, we become less certain about the validity of the assumed model. Regression models are not necessarily valid for extrapolation purposes.

Now this does not mean *don't ever extrapolate*. There are many problem situations in science and engineering where extrapolation of a regression model is the only way to even approach the problem. However, there is a strong warning to **be careful.** A modest extrapolation may be perfectly all right in many cases, but a large extrapolation will almost never produce acceptable results.

11-2 SIMPLE LINEAR REGRESSION

The case of **simple linear regression** considers a single **regressor variable** or **predictor variable** x and a dependent or **response variable** Y. Suppose that the true relationship between Y and x is a straight line and that the observation Y at each level of x is a random variable. As noted previously, the expected value of Y for each value of x is

$$E(Y|x) = \beta_0 + \beta_1 x$$

where the intercept β_0 and the slope β_1 are unknown regression coefficients. We assume that each observation, Y, can be described by the model

$$Y = \beta_0 + \beta_1 x + \epsilon \tag{11-2}$$

where ϵ is a random error with mean zero and (unknown) variance σ^2. The random errors corresponding to different observations are also assumed to be uncorrelated random variables.

Suppose that we have n pairs of observations $(x_1, y_1), (x_2, y_2), \ldots, (x_n, y_n)$. Figure 11-3 shows a typical scatter plot of observed data and a candidate for the estimated regression line. The estimates of β_0 and β_1 should result in a line that is (in some sense) a "best fit" to the data. The German scientist Karl Gauss (1777–1855) proposed estimating the parameters β_0 and β_1 in Equation 11-2 to minimize the sum of the squares of the vertical deviations in Fig. 11-3.

We call this criterion for estimating the regression coefficients the method of **least squares.** Using Equation 11-2, we may express the n observations in the sample as

$$y_i = \beta_0 + \beta_1 x_i + \epsilon_i, \qquad i = 1, 2, \ldots, n \tag{11-3}$$

and the sum of the squares of the deviations of the observations from the true regression line is

$$L = \sum_{i=1}^{n} \epsilon_i^2 = \sum_{i=1}^{n} (y_i - \beta_0 - \beta_1 x_i)^2 \tag{11-4}$$

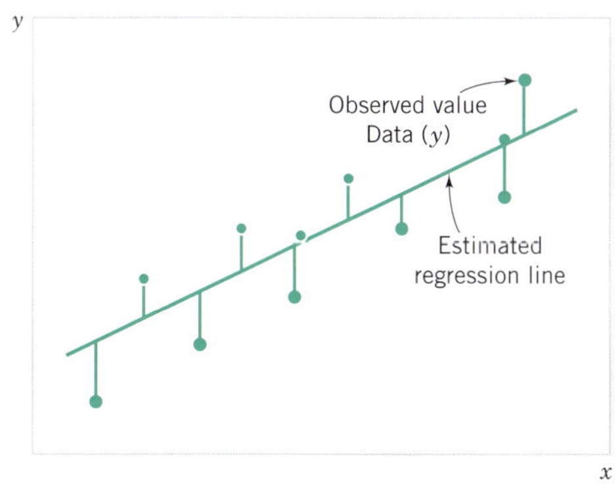

Figure 11-3 Deviations of the data from the estimated regression model.

The least squares estimators of β_0 and β_1, say, $\hat{\beta}_0$ and $\hat{\beta}_1$, must satisfy

$$\left.\frac{\partial L}{\partial \beta_0}\right|_{\hat{\beta}_0,\hat{\beta}_1} = -2\sum_{i=1}^{n}(y_i - \hat{\beta}_0 - \hat{\beta}_1 x_i) = 0$$

$$\left.\frac{\partial L}{\partial \beta_1}\right|_{\hat{\beta}_0,\hat{\beta}_1} = -2\sum_{i=1}^{n}(y_i - \hat{\beta}_0 - \hat{\beta}_1 x_i)x_i = 0 \qquad (11\text{-}5)$$

Simplifying these two equations yields

$$n\hat{\beta}_0 + \hat{\beta}_1 \sum_{i=1}^{n} x_i = \sum_{i=1}^{n} y_i$$

$$\hat{\beta}_0 \sum_{i=1}^{n} x_i + \hat{\beta}_1 \sum_{i=1}^{n} x_i^2 = \sum_{i=1}^{n} y_i x_i \qquad (11\text{-}6)$$

Equations 11-6 are called the **least squares normal equations.** The solution to the normal equations results in the least squares estimators $\hat{\beta}_0$ and $\hat{\beta}_1$.

Least Squares Estimates

The **least squares estimates** of the intercept and slope in the simple linear regression model are

$$\hat{\beta}_0 = \bar{y} - \hat{\beta}_1 \bar{x} \qquad (11\text{-}7)$$

$$\hat{\beta}_1 = \frac{\displaystyle\sum_{i=1}^{n} y_i x_i - \frac{\left(\displaystyle\sum_{i=1}^{n} y_i\right)\left(\displaystyle\sum_{i=1}^{n} x_i\right)}{n}}{\displaystyle\sum_{i=1}^{n} x_i^2 - \frac{\left(\displaystyle\sum_{i=1}^{n} x_i\right)^2}{n}} \qquad (11\text{-}8)$$

where $\bar{y} = (1/n)\sum_{i=1}^{n} y_i$ and $\bar{x} = (1/n)\sum_{i=1}^{n} x_i$.

The **fitted** or **estimated regression line** is therefore

$$\hat{y} = \hat{\beta}_0 + \hat{\beta}_1 x \qquad (11\text{-}9)$$

Note that each pair of observations satisfies the relationship

$$y_i = \hat{\beta}_0 + \hat{\beta}_1 x_i + e_i, \qquad i = 1, 2, \ldots, n$$

where $e_i = y_i - \hat{y}_i$ is called the **residual.** The residual describes the error in the fit of the model to the ith observation y_i. Later in this chapter we will use the residuals to provide information about the adequacy of the fitted model.

Notationally, it is occasionally convenient to give special symbols to the numerator and denominator of Equation 11-8. Given data $(x_1, y_1), (x_2, y_2), \ldots, (x_n, y_n)$, let

$$S_{xx} = \sum_{i=1}^{n} (x_i - \bar{x})^2 = \sum_{i=1}^{n} x_i^2 - \frac{\left(\sum_{i=1}^{n} x_i\right)^2}{n} \tag{11-10}$$

and

$$S_{xy} = \sum_{i=1}^{n} (y_i - \bar{y})(x_i - \bar{x}) = \sum_{i=1}^{n} x_i y_i - \frac{\left(\sum_{i=1}^{n} x_i\right)\left(\sum_{i=1}^{n} y_i\right)}{n} \tag{11-11}$$

EXAMPLE 11-1
Oxygen Purity

We will fit a simple linear regression model to the oxygen purity data in Table 11-1. The following quantities may be computed:

$$n = 20 \quad \sum_{i=1}^{20} x_i = 23.92 \quad \sum_{i=1}^{20} y_i = 1{,}843.21 \quad \bar{x} = 1.1960 \quad \bar{y} = 92.1605$$

$$\sum_{i=1}^{20} y_i^2 = 170{,}044.5321 \quad \sum_{i=1}^{20} x_i^2 = 29.2892 \quad \sum_{i=1}^{20} x_i y_i = 2{,}214.6566$$

$$S_{xx} = \sum_{i=1}^{20} x_i^2 - \frac{\left(\sum_{i=1}^{20} x_i\right)^2}{20} = 29.2892 - \frac{(23.92)^2}{20} = 0.68088$$

and

$$S_{xy} = \sum_{i=1}^{20} x_i y_i - \frac{\left(\sum_{i=1}^{20} x_i\right)\left(\sum_{i=1}^{20} y_i\right)}{20} = 2{,}214.6566 - \frac{(23.92)(1{,}843.21)}{20} = 10.17744$$

Therefore, the least squares estimates of the slope and intercept are

$$\hat{\beta}_1 = \frac{S_{xy}}{S_{xx}} = \frac{10.17744}{0.68088} = 14.94748$$

and

$$\hat{\beta}_0 = \bar{y} - \hat{\beta}_1 \bar{x} = 92.1605 - (14.94748)1.196 = 74.28331$$

The fitted simple linear regression model (with the coefficients reported to three decimal places) is

$$\hat{y} = 74.283 + 14.947x$$

This model is plotted in Fig. 11-4, along with the sample data.

Computer software programs are widely used in regression modeling. These programs typically carry more decimal places in the calculations. Table 11-2 shows a portion of the output from Minitab for this problem. The estimates $\hat{\beta}_0$ and $\hat{\beta}_1$ are highlighted. In subsequent sections we will provide explanations for the information provided in this computer output.

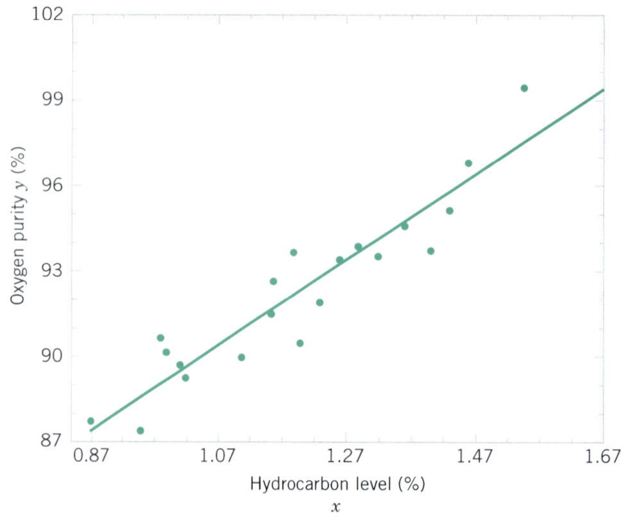

Figure 11-4 Scatter plot of oxygen purity y versus hydrocarbon level x and regression model $\hat{y} = 74.283 + 14.947x$.

Using the regression model of Example 11-1, we would predict oxygen purity of $\hat{y} = 89.23\%$ when the hydrocarbon level is $x = 1.00\%$. The purity 89.23% may be interpreted as an estimate of the true population mean purity when $x = 1.00\%$, or as an estimate of a new observation when $x = 1.00\%$. These estimates are, of course, subject to error; that is, it is unlikely that a future observation on purity would be exactly 89.23% when the hydrocarbon level is 1.00%. In subsequent sections we will see how to use confidence intervals and prediction intervals to describe the error in estimation from a regression model.

Table 11-2 Minitab Output for the Oxygen Purity Data in Example 11-1

Regression Analysis

The regression equation is

Purity = 74.3 + 14.9 HC Level

Predictor	Coef	SE Coef	T	P
Constant	74.283 ← $\hat{\beta}_0$	1.593	46.62	0.000
HC Level	14.947 ← $\hat{\beta}_1$	1.317	11.35	0.000

S = 1.087 R-Sq = 87.7% R-Sq (adj) = 87.1%

Analysis of Variance

Source	DF	SS	MS	F	P
Regression	1	152.13	152.13	128.86	0.000
Residual Error	18	21.25 ← SS_E	1.18 ← $\hat{\sigma}^2$		
Total	19	173.38			

Predicted Values for New Observations

New Obs	Fit	SE Fit	95.0% CI	95.0% PI
1	89.231	0.354	(88.486, 89.975)	(86.830, 91.632)

Values of Predictors for New Observations

New Obs	HC Level
1	1.00

Estimating σ^2

There is actually another unknown parameter in our regression model, σ^2 (the variance of the error term ϵ). The residuals $e_i = y_i - \hat{y}_i$ are used to obtain an estimate of σ^2. The sum of squares of the residuals, often called the **error sum of squares,** is

$$SS_E = \sum_{i=1}^{n} e_i^2 = \sum_{i=1}^{n} (y_i - \hat{y}_i)^2 \tag{11-12}$$

We can show that the expected value of the error sum of squares is $E(SS_E) = (n - 2)\sigma^2$. Therefore an **unbiased estimator** of σ^2 is

Estimator of Variance

$$\hat{\sigma}^2 = \frac{SS_E}{n - 2} \tag{11-13}$$

Computing SS_E using Equation 11-12 would be fairly tedious. A more convenient computing formula can be obtained by substituting $\hat{y}_i = \hat{\beta}_0 + \hat{\beta}_1 x_i$ into Equation 11-12 and simplifying. The resulting computing formula is

$$SS_E = SS_T - \hat{\beta}_1 S_{xy} \tag{11-14}$$

where $SS_T = \sum_{i=1}^{n}(y_i - \bar{y})^2 = \sum_{i=1}^{n} y_i^2 - n\bar{y}^2$ is the total sum of squares of the response variable y. Formulas such as this are presented in Section 11-4. The error sum of squares and the estimate of σ^2 for the oxygen purity data, $\hat{\sigma}^2 = 1.18$, are highlighted in the Minitab output in Table 11-2.

EXERCISES FOR SECTION 11-2

11-1. An article in *Concrete Research* ("Near Surface Characteristics of Concrete: Intrinsic Permeability," Vol. 41, 1989), presented data on compressive strength x and intrinsic permeability y of various concrete mixes and cures. Summary quantities are $n = 14$, $\sum y_i = 572$, $\sum y_i^2 = 23{,}530$, $\sum x_i = 43$, $\sum x_i^2 = 157.42$, and $\sum x_i y_i = 1697.80$. Assume that the two variables are related according to the simple linear regression model.

(a) Calculate the least squares estimates of the slope and intercept. Estimate σ^2. Graph the regression line.

(b) Use the equation of the fitted line to predict what permeability would be observed when the compressive strength is $x = 4.3$.

(c) Give a point estimate of the mean permeability when compressive strength is $x = 3.7$.

(d) Suppose that the observed value of permeability at $x = 3.7$ is $y = 46.1$. Calculate the value of the corresponding residual.

11-2. Regression methods were used to analyze the data from a study investigating the relationship between roadway surface temperature (x) and pavement deflection (y). Summary quantities were $n = 20$, $\sum y_i = 12.75$, $\sum y_i^2 = 8.86$, $\sum x_i = 1478$, $\sum x_i^2 = 143{,}215.8$, and $\sum x_i y_i = 1083.67$.

(a) Calculate the least squares estimates of the slope and intercept. Graph the regression line. Estimate σ^2.

(b) Use the equation of the fitted line to predict what pavement deflection would be observed when the surface temperature is 85°F.

(c) What is the mean pavement deflection when the surface temperature is 90°F?

(d) What change in mean pavement deflection would be expected for a 1°F change in surface temperature?

11-3. The following table presents data on the ratings of quarterbacks for the 2004 National Football League season (source: *The Sports Network*). It is suspected that the rating (y)

Player	Yards per Attempt	Rating Points
D. Culpepper, MIN	8.61	110.9
D. McNabb, PHI	8.26	104.7
B. Griese, TAM	7.83	97.5
M. Bulger, STL	8.17	93.7
B. Favre, GBP	7.57	92.4
J. Delhomme, CAR	7.29	87.3
K. Warner, NYG	7.42	86.5
M. Hasselbeck, SEA	7.14	83.1
A. Brooks, NOS	7.03	79.5
T. Rattay, SFX	6.67	78.1
M. Vick, ATL	7.21	78.1
J. Harrington, DET	6.23	77.5
V. Testaverde, DAL	7.14	76.4
P. Ramsey, WAS	6.12	74.8
J. McCown, ARI	6.15	74.1
P. Manning, IND	9.17	121.1
D. Brees, SDC	7.9	104.8
B. Roethlisberger, PIT	8.88	98.1
T. Green, KAN	8.26	95.2
T. Brady, NEP	7.79	92.6
C. Pennington, NYJ	7.22	91
B. Volek, TEN	6.96	87.1
J. Plummer, DEN	7.85	84.5
D. Carr, HOU	7.58	83.5
B. Leftwich, JAC	6.67	82.2
C. Palmer, CIN	6.71	77.3
J. Garcia, CLE	6.87	76.7
D. Bledsoe, BUF	6.52	76.6
K. Collins, OAK	6.81	74.8
K. Boller, BAL	5.52	70.9

is related to the average number of yards gained per pass attempt (x).

(a) Calculate the least squares estimates of the slope and intercept. What is the estimate of σ^2? Graph the regression model.

(b) Find an estimate of the mean rating if a quarterback averages 7.5 yards per attempt.

(c) What change in the mean rating is associated with a decrease of one yard per attempt?

(d) To increase the mean rating by 10 points, how much increase in the average yards per attempt must be generated?

(e) Given that $x = 7.21$ yards (M. Vick), find the fitted value of y and the corresponding residual.

11-4. An article in *Technometrics* by S. C. Narula and J. F. Wellington ("Prediction, Linear Regression, and a Minimum Sum of Relative Errors," Vol. 19, 1977) presents data on the selling price and annual taxes for 24 houses. The data are shown in the following table.

Sale Price/1000	Taxes (Local, School), County)/1000	Sale Price/1000	Taxes (Local, School), County)/1000
25.9	4.9176	30.0	5.0500
29.5	5.0208	36.9	8.2464
27.9	4.5429	41.9	6.6969
25.9	4.5573	40.5	7.7841
29.9	5.0597	43.9	9.0384
29.9	3.8910	37.5	5.9894
30.9	5.8980	37.9	7.5422
28.9	5.6039	44.5	8.7951
35.9	5.8282	37.9	6.0831
31.5	5.3003	38.9	8.3607
31.0	6.2712	36.9	8.1400
30.9	5.9592	45.8	9.1416

(a) Assuming that a simple linear regression model is appropriate, obtain the least squares fit relating selling price to taxes paid. What is the estimate of σ^2?

(b) Find the mean selling price given that the taxes paid are $x = 7.50$.

(c) Calculate the fitted value of y corresponding to $x = 5.8980$. Find the corresponding residual.

(d) Calculate the fitted $\hat{y}_i$ for each value of x_i used to fit the model. Then construct a graph of $\hat{y}_i$ versus the corresponding observed value y_i and comment on what this plot would look like if the relationship between y and x was a deterministic (no random error) straight line. Does the plot actually obtained indicate that taxes paid is an effective regressor variable in predicting selling price?

11-5. The number of pounds of steam used per month by a chemical plant is thought to be related to the average ambient temperature (in° F) for that month. The past year's usage and temperature are shown in the following table:

Month	Temp.	Usage/1000	Month	Temp.	Usage/1000
Jan.	21	185.79	July	68	621.55
Feb.	24	214.47	Aug.	74	675.06
Mar.	32	288.03	Sept.	62	562.03
Apr.	47	424.84	Oct.	50	452.93
May	50	454.58	Nov.	41	369.95
June	59	539.03	Dec.	30	273.98

(a) Assuming that a simple linear regression model is appropriate, fit the regression model relating steam usage (y) to the average temperature (x). What is the estimate of σ^2? Graph the regression line.
(b) What is the estimate of expected steam usage when the average temperature is 55°F?
(c) What change in mean steam usage is expected when the monthly average temperature changes by 1°F?
(d) Suppose the monthly average temperature is 47°F. Calculate the fitted value of y and the corresponding residual.

11-6. The following table presents the highway gasoline mileage performance and engine displacement for DamlerChrysler vehicles for model year 2005 (source: U.S. Environmental Protection Agency).
(a) Fit a simple linear model relating highway miles per gallon (y) to engine displacement (x) in cubic inches using least squares.
(b) Find an estimate of the mean highway gasoline mileage performance for a car with 150 cubic inches engine displacement.
(c) Obtain the fitted value of y and the corresponding residual for a car, the Neon, with an engine displacement of 122 cubic inches.

Carline	Engine Displacement (in^3)	MPG (highway)
300C/SRT-8	215	30.8
CARAVAN 2WD	201	32.5
CROSSFIRE ROADSTER	196	35.4
DAKOTA PICKUP 2WD	226	28.1
DAKOTA PICKUP 4WD	226	24.4
DURANGO 2WD	348	24.1
GRAND CHEROKEE 2WD	226	28.5
GRAND CHEROKEE 4WD	348	24.2
LIBERTY/CHEROKEE 2WD	148	32.8
LIBERTY/CHEROKEE 4WD	226	28
NEON/SRT-4/SX 2.0	122	41.3
PACIFICA 2WD	215	30.0
PACIFICA AWD	215	28.2
PT CRUISER	148	34.1
RAM 1500 PICKUP 2WD	500	18.7
RAM 1500 PICKUP 4WD	348	20.3
SEBRING 4-DR	165	35.1
STRATUS 4-DR	148	37.9
TOWN & COUNTRY 2WD	148	33.8
VIPER CONVERTIBLE	500	25.9
WRANGLER/TJ 4WD	148	26.4

11-7. An article in the *Tappi Journal* (March, 1986) presented data on green liquor Na_2S concentration (in grams per liter) and paper machine production (in tons per day). The data (read from a graph) are shown as follows:

y	40	42	49	46	44	48
x	825	830	890	895	890	910

y	46	43	53	52	54	57	58
x	915	960	990	1010	1012	1030	1050

(a) Fit a simple linear regression model with y = green liquor Na_2S concentration and x = production. Find an estimate of σ^2. Draw a scatter diagram of the data and the resulting least squares fitted model.
(b) Find the fitted value of y corresponding to $x = 910$ and the associated residual.
(c) Find the mean green liquor Na_2S concentration when the production rate is 950 tons per day.

11-8. An article in the *Journal of Sound and Vibration* (Vol. 151, 1991, pp. 383–394) described a study investigating the relationship between noise exposure and hypertension. The following data are representative of those reported in the article.

y	1	0	1	2	5	1	4	6	2	3
x	60	63	65	70	70	70	80	90	80	80

y	5	4	6	8	4	5	7	9	7	6
x	85	89	90	90	90	90	94	100	100	100

(a) Draw a scatter diagram of y (blood pressure rise in millimeters of mercury) versus x (sound pressure level in decibels). Does a simple linear regression model seem reasonable in this situation?
(b) Fit the simple linear regression model using least squares. Find an estimate of σ^2.
(c) Find the predicted mean rise in blood pressure level associated with a sound pressure level of 85 decibels.

11-9. An article in *Wear* (Vol. 152, 1992, pp. 171–181) presents data on the fretting wear of mild steel and oil viscosity. Representative data follow, with x = oil viscosity and y = wear volume (10^{-4} cubic millimeters).

y	240	181	193	155	172
x	1.6	9.4	15.5	20.0	22.0

y	110	113	75	94
x	35.5	43.0	40.5	33.0

(a) Construct a scatter plot of the data. Does a simple linear regression model appear to be plausible?
(b) Fit the simple linear regression model using least squares. Find an estimate of σ^2.
(c) Predict fretting wear when viscosity $x = 30$.
(d) Obtain the fitted value of y when $x = 22.0$ and calculate the corresponding residual.

11-10. An article in the *Journal of Environmental Engineering* (Vol. 115, No. 3, 1989, pp. 608–619) reported the results of a study on the occurrence of sodium and chloride in surface streams in central Rhode Island. The following data are chloride concentration y (in milligrams per liter) and roadway area in the watershed x (in percentage).

y	4.4	6.6	9.7	10.6	10.8	10.9
x	0.19	0.15	0.57	0.70	0.67	0.63

y	11.8	12.1	14.3	14.7	15.0	17.3
x	0.47	0.70	0.60	0.78	0.81	0.78

y	19.2	23.1	27.4	27.7	31.8	39.5
x	0.69	1.30	1.05	1.06	1.74	1.62

(a) Draw a scatter diagram of the data. Does a simple linear regression model seem appropriate here?
(b) Fit the simple linear regression model using the method of least squares. Find an estimate of σ^2.
(c) Estimate the mean chloride concentration for a watershed that has 1% roadway area.
(d) Find the fitted value corresponding to $x = 0.47$ and the associated residual.

11-11. A rocket motor is manufactured by bonding together two types of propellants, an igniter and a sustainer. The shear strength of the bond y is thought to be a linear function of the age of the propellant x when the motor is cast. Twenty observations are shown in the following table.
(a) Draw a scatter diagram of the data. Does the straight-line regression model seem to be plausible?
(b) Find the least squares estimates of the slope and intercept in the simple linear regression model. Find an estimate of σ^2.
(c) Estimate the mean shear strength of a motor made from propellant that is 20 weeks old.
(d) Obtain the fitted values $\hat{y}_i$ that correspond to each observed value y_i. Plot $\hat{y}_i$ versus y_i, and comment on what this plot would look like if the linear relationship between shear strength and age were perfectly deterministic (no error). Does this plot indicate that age is a reasonable choice of regressor variable in this model?

Observation Number	Strength y (psi)	Age x (weeks)
1	2158.70	15.50
2	1678.15	23.75
3	2316.00	8.00
4	2061.30	17.00
5	2207.50	5.00
6	1708.30	19.00
7	1784.70	24.00
8	2575.00	2.50
9	2357.90	7.50
10	2277.70	11.00
11	2165.20	13.00
12	2399.55	3.75
13	1779.80	25.00
14	2336.75	9.75
15	1765.30	22.00
16	2053.50	18.00
17	2414.40	6.00
18	2200.50	12.50
19	2654.20	2.00
20	1753.70	21.50

11-12. An article in the *Journal of the American Ceramic Society*, "Rapid Hot-Pressing of Ultrafine PSZ Powders" (1991, Vol. 74, pp. 1547–1553) considered the microstructure of the ultrafine powder of partially stabilized zirconia as a function of temperature. The data are shown below:

x = Temperature (°C): 1100 1200 1300 1100 1500 1200 1300

y = Porosity (%): 30.8 19.2 6.0 13.5 11.4 7.7 3.6

(a) Fit the simple linear regression model using the method of least squares. Find an estimate of σ^2.
(b) Estimate the mean porosity for a temperature of 1400°C.
(c) Find the fitted value corresponding to $y = 11.4$ and the associated residual.
(d) Draw a scatter diagram of the data. Does a simple linear regression model seem appropriate here? Explain.

11-13. An article in the *Journal of the Environmental Engineering Division*, "Least Squares Estimates of BOD Parameters" (1980, Vol. 106, pp. 1197–1202) took a sample from the Holston River below Kingport, TN, during August 1977. The biochemical oxygen demand (BOD) test is conducted over a period of time in days. The resulting data is shown below:

Time (days): 1 2 4 6 8 10 12 14 16 18 20

BOD (mg/liter): 0.6 0.7 1.5 1.9 2.1 2.6 2.9 3.7 3.5 3.7 3.8

(a) Assuming that a simple linear regression model is appropriate, fit the regression model relating BOD (y) to the time (x). What is the estimate of σ^2?
(b) What is the estimate of expected BOD level when the time is 15 days?
(c) What change in mean BOD is expected when the time changes by 3 days?
(d) Suppose the time used is 6 days. Calculate the fitted value of y and the corresponding residual.
(e) Calculate the fitted $\hat{y}_i$ for each value of x_i used to fit the model. Then construct a graph of $\hat{y}_i$ versus the corresponding observed values y_i and comment on what this plot would look like if the relationship between y and x was a deterministic (no random error) straight line. Does the plot actually obtained indicate that time is an effective regressor variable in predicting BOD?

11-14. An article in *Wood Science and Technology*, "Creep in Chipboard, Part 3: Initial Assessment of the Influence of Moisture Content and Level of Stressing on Rate of Creep and Time to Failure" (1981, Vol. 15, pp. 125–144) studied the deflection (mm) of particleboard from stress levels of relative humidity. Assume that the two variables are related according to the simple linear regression model. The data are shown below:

x = Stress level (%):	54	54	61	61	68
y = Deflection (mm):	16.473	18.693	14.305	15.121	13.505
x = Stress level (%):	68	75	75	75	
y = Deflection (mm):	11.640	11.168	12.534	11.224	

(a) Calculate the least square estimates of the slope and intercept. What is the estimate of σ^2? Graph the regression model and the data.
(b) Find the estimate of the mean deflection if the stress level can be limited to 65%.
(c) Estimate the change in the mean deflection associated with a 5% increment in stress level.
(d) To decrease the mean deflection by one millimeter, how much increase in stress level must be generated?
(e) Given that the stress level is 68%, find the fitted value of deflection and the corresponding residual.

11-15. In an article in *Statistics and Computing*, "An iterative Monte Carlo method for nonconjugate Bayesian analysis" (1991, pp. 119–128) Carlin and Gelfand investigated the age (x) and length (y) of 27 captured dugongs (sea cows).

x = 1.0, 1.5, 1.5, 1.5, 2.5, 4.0, 5.0, 5.0, 7.0, 8.0, 8.5, 9.0, 9.5, 9.5, 10.0, 12.0, 12.0, 13.0, 13.0, 14.5, 15.5, 15.5, 16.5, 17.0, 22.5, 29.0, 31.5

y = 1.80, 1.85, 1.87, 1.77, 2.02, 2.27, 2.15, 2.26, 2.47, 2.19, 2.26, 2.40, 2.39, 2.41, 2.50, 2.32, 2.32, 2.43, 2.47, 2.56, 2.65, 2.47, 2.64, 2.56, 2.70, 2.72, 2.57

(a) Find the least squares estimates of the slope and the intercept in the simple linear regression model. Find an estimate of σ^2.
(b) Estimate the mean length of dugongs at age 11.
(c) Obtain the fitted values $\hat{y}_i$ that correspond to each observed value y_i. Plot $\hat{y}_i$ versus y_i, and comment on what this plot would look like if the linear relationship between length and age were perfectly deterministic (no error). Does this plot indicate that age is a reasonable choice of regressor variable in this model?

11-16. Consider the regression model developed in Exercise 11-2.
(a) Suppose that temperature is measured in °C rather than °F. Write the new regression model.
(b) What change in expected pavement deflection is associated with a 1°C change in surface temperature?

11-17. Consider the regression model developed in Exercise 11-6. Suppose that engine displacement is measured in cubic centimeters instead of cubic inches.
(a) Write the new regression model.
(b) What change in gasoline mileage is associated with a 1 cm^3 change is engine displacement?

11-18. Show that in a simple linear regression model the point $(\bar{x}, \bar{y})$ lies exactly on the least squares regression line.

11-19. Consider the simple linear regression model $Y = \beta_0 + \beta_1 x + \epsilon$. Suppose that the analyst wants to use $z = x - \bar{x}$ as the regressor variable.
(a) Using the data in Exercise 11-11, construct one scatter plot of the (x_i, y_i) points and then another of the $(z_i = x_i - \bar{x}, y_i)$ points. Use the two plots to intuitively explain how the two models, $Y = \beta_0 + \beta_1 x + \epsilon$ and $Y = \beta_0^* + \beta_1^* z + \epsilon$, are related.
(b) Find the least squares estimates of β_0^* and β_1^* in the model $Y = \beta_0^* + \beta_1^* z + \epsilon$. How do they relate to the least squares estimates $\hat{\beta}_0$ and $\hat{\beta}_1$?

11-20. Suppose we wish to fit a regression model for which the true regression line passes through the point (0, 0). The appropriate model is $Y = \beta x + \epsilon$. Assume that we have n pairs of data $(x_1, y_1), (x_2, y_2), \ldots, (x_n, y_n)$.
(a) Find the least squares estimate of β.
(b) Fit the model $Y = \beta x + \epsilon$ to the chloride concentration-roadway area data in Exercise 11-10. Plot the fitted model on a scatter diagram of the data and comment on the appropriateness of the model.

11-3 PROPERTIES OF THE LEAST SQUARES ESTIMATORS

The statistical properties of the least squares estimators $\hat{\beta}_0$ and $\hat{\beta}_1$ may be easily described. Recall that we have assumed that the error term ϵ in the model $Y = \beta_0 + \beta_1 x + \epsilon$ is a random variable with mean zero and variance σ^2. Since the values of x are fixed, Y is a random variable with mean $\mu_{Y|x} = \beta_0 + \beta_1 x$ and variance σ^2. Therefore, the values of $\hat{\beta}_0$ and $\hat{\beta}_1$ depend on the observed y's; thus, the least squares estimators of the regression coefficients may be viewed as random variables. We will investigate the bias and variance properties of the least squares estimators $\hat{\beta}_0$ and $\hat{\beta}_1$.

Consider first $\hat{\beta}_1$. Because $\hat{\beta}_1$ is a linear combination of the observations Y_i, we can use properties of expectation to show that the expected value of $\hat{\beta}_1$ is

$$E(\hat{\beta}_1) = \beta_1 \tag{11-15}$$

Thus, $\hat{\beta}_1$ is an **unbiased estimator** of the true slope β_1.

Now consider the variance of $\hat{\beta}_1$. Since we have assumed that $V(\epsilon_i) = \sigma^2$, it follows that $V(Y_i) = \sigma^2$. Because $\hat{\beta}_1$ is a linear combination of the observations Y_i, the results in Section 5-5 can be applied to show that

$$V(\hat{\beta}_1) = \frac{\sigma^2}{S_{xx}} \tag{11-16}$$

For the intercept, we can show in a similar manner that

$$E(\hat{\beta}_0) = \beta_0 \quad \text{and} \quad V(\hat{\beta}_0) = \sigma^2\left[\frac{1}{n} + \frac{\bar{x}^2}{S_{xx}}\right] \tag{11-17}$$

Thus, $\hat{\beta}_0$ is an unbiased estimator of the intercept β_0. The covariance of the random variables $\hat{\beta}_0$ and $\hat{\beta}_1$ is not zero. It can be shown (see Exercise 11-92) that $\text{cov}(\hat{\beta}_0, \hat{\beta}_1) = -\sigma^2\bar{x}/S_{xx}$.

The estimate of σ^2 could be used in Equations 11-16 and 11-17 to provide estimates of the variance of the slope and the intercept. We call the square roots of the resulting variance estimators the **estimated standard errors** of the slope and intercept, respectively.

Estimated Standard Errors

In simple linear regression the **estimated standard error of the slope** and the **estimated standard error of the intercept** are

$$se(\hat{\beta}_1) = \sqrt{\frac{\hat{\sigma}^2}{S_{xx}}} \quad \text{and} \quad se(\hat{\beta}_0) = \sqrt{\hat{\sigma}^2\left[\frac{1}{n} + \frac{\bar{x}^2}{S_{xx}}\right]}$$

respectively, where $\hat{\sigma}^2$ is computed from Equation 11-13.

The Minitab computer output in Table 11-2 reports the estimated standard errors of the slope and intercept under the column heading "*SE* coeff."

11-4 HYPOTHESIS TESTS IN SIMPLE LINEAR REGRESSION

An important part of assessing the adequacy of a linear regression model is testing statistical hypotheses about the model parameters and constructing certain confidence intervals. Hypothesis testing in simple linear regression is discussed in this section, and Section 11-5 presents methods for constructing confidence intervals. To test hypotheses about the slope and intercept of the regression model, we must make the additional assumption that the error component in the model, ϵ, is normally distributed. Thus, the complete assumptions are that the errors are normally and independently distributed with mean zero and variance σ^2, abbreviated NID$(0, \sigma^2)$.

11-4.1 Use of *t*-Tests

Suppose we wish to test the hypothesis that the slope equals a constant, say, $\beta_{1,0}$. The appropriate hypotheses are

$$\begin{aligned} H_0&: \beta_1 = \beta_{1,0} \\ H_1&: \beta_1 \neq \beta_{1,0} \end{aligned} \tag{11-18}$$

where we have assumed a two-sided alternative. Since the errors ϵ_i are NID$(0, \sigma^2)$, it follows directly that the observations Y_i are NID$(\beta_0 + \beta_1 x_i, \sigma^2)$. Now $\hat{\beta}_1$ is a linear combination of independent normal random variables, and consequently, $\hat{\beta}_1$ is $N(\beta_1, \sigma^2/S_{xx})$, using the bias and variance properties of the slope discussed in Section 11-3. In addition, $(n-2)\hat{\sigma}^2/\sigma^2$ has a chi-square distribution with $n-2$ degrees of freedom, and $\hat{\beta}_1$ is independent of $\hat{\sigma}^2$. As a result of those properties, the statistic

$$T_0 = \frac{\hat{\beta}_1 - \beta_{1,0}}{\sqrt{\hat{\sigma}^2/S_{xx}}} \tag{11-19}$$

follows the t distribution with $n-2$ degrees of freedom under H_0: $\beta_1 = \beta_{1,0}$. We would reject H_0: $\beta_1 = \beta_{1,0}$ if

$$|t_0| > t_{\alpha/2,n-2} \tag{11-20}$$

where t_0 is computed from Equation 11-19. The denominator of Equation 11-19 is the standard error of the slope, so we could write the test statistic as

$$T_0 = \frac{\hat{\beta}_1 - \beta_{1,0}}{se(\hat{\beta}_1)}$$

A similar procedure can be used to test hypotheses about the intercept. To test

$$\begin{aligned} H_0&: \beta_0 = \beta_{0,0} \\ H_1&: \beta_0 \neq \beta_{0,0} \end{aligned} \tag{11-21}$$

we would use the statistic

$$T_0 = \frac{\hat{\beta}_0 - \beta_{0,0}}{\sqrt{\hat{\sigma}^2\left[\dfrac{1}{n} + \dfrac{\bar{x}^2}{S_{xx}}\right]}} = \frac{\hat{\beta}_0 - \beta_{0,0}}{se(\hat{\beta}_0)} \tag{11-22}$$

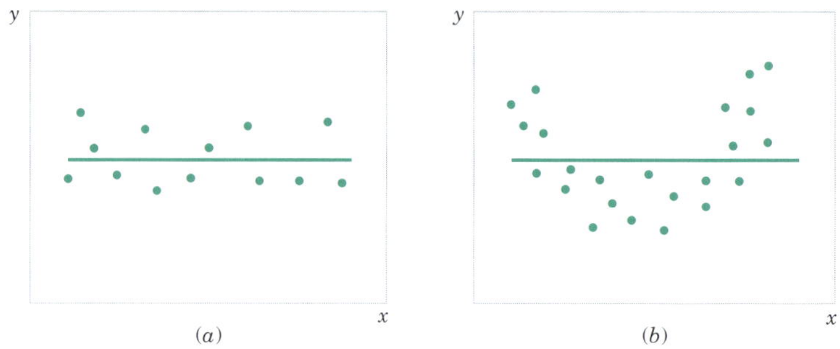

Figure 11-5 The hypothesis H_0: $\beta_1 = 0$ is not rejected.

and reject the null hypothesis if the computed value of this test statistic, t_0, is such that $|t_0| > t_{\alpha/2, n-2}$. Note that the denominator of the test statistic in Equation 11-22 is just the standard error of the intercept.

A very important special case of the hypotheses of Equation 11-18 is

$$H_0\text{: } \beta_1 = 0$$
$$H_1\text{: } \beta_1 \neq 0 \qquad (11\text{-}23)$$

These hypotheses relate to the **significance of regression.** Failure to reject H_0: $\beta_1 = 0$ is equivalent to concluding that there is no linear relationship between x and Y. This situation is illustrated in Fig. 11-5. Note that this may imply either that x is of little value in explaining the variation in Y and that the best estimator of Y for any x is $\hat{y} = \overline{Y}$ (Fig. 11-5*a*) or that the true relationship between x and Y is not linear (Fig. 11-5*b*). Alternatively, if H_0: $\beta_1 = 0$ is rejected, this implies that x is of value in explaining the variability in Y (see Fig. 11-6). Rejecting H_0: $\beta_1 = 0$ could mean either that the straight-line model is adequate (Fig. 11-6*a*) or that, although there is a linear effect of x, better results could be obtained with the addition of higher order polynomial terms in x (Fig. 11-6*b*).

EXAMPLE 11-2
Oxygen Purity Tests of Coefficients

We will test for significance of regression using the model for the oxygen purity data from Example 11-1. The hypotheses are

$$H_0\text{: } \beta_1 = 0$$
$$H_1\text{: } \beta_1 \neq 0$$

and we will use $\alpha = 0.01$. From Example 11-1 and Table 11-2 we have

$$\hat{\beta}_1 = 14.947 \quad n = 20, \quad S_{xx} = 0.68088, \quad \hat{\sigma}^2 = 1.18$$

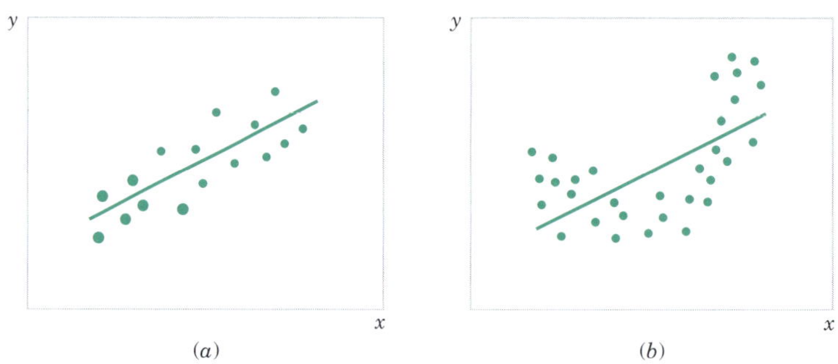

Figure 11-6 The hypothesis H_0: $\beta_1 = 0$ is rejected.

so the t-statistic in Equation 10-20 becomes

$$t_0 = \frac{\hat{\beta}_1}{\sqrt{\hat{\sigma}^2/S_{xx}}} = \frac{\hat{\beta}_1}{se(\hat{\beta}_1)} = \frac{14.947}{\sqrt{1.18/0.68088}} = 11.35$$

Since the reference value of t is $t_{0.005,18} = 2.88$, the value of the test statistic is very far into the critical region, implying that H_0: $\beta_1 = 0$ should be rejected. The P-value for this test is $P \simeq 1.23 \times 10^{-9}$. This was obtained manually with a calculator.

Table 11-2 presents the Minitab output for this problem. Notice that the t-statistic value for the slope is computed as 11.35 and that the reported P-value is $P = 0.000$. Minitab also reports the t-statistic for testing the hypothesis H_0: $\beta_0 = 0$. This statistic is computed from Equation 11-22, with $\beta_{0,0} = 0$, as $t_0 = 46.62$. Clearly, then, the hypothesis that the intercept is zero is rejected.

11-4.2 Analysis of Variance Approach to Test Significance of Regression

A method called the **analysis of variance** can be used to test for significance of regression. The procedure partitions the total variability in the response variable into meaningful components as the basis for the test. The **analysis of variance identity** is as follows:

Analysis of Variance Identity

$$\sum_{i=1}^{n} (y_i - \bar{y})^2 = \sum_{i=1}^{n} (\hat{y}_i - \bar{y})^2 + \sum_{i=1}^{n} (y_i - \hat{y}_i)^2 \tag{11-24}$$

The two components on the right-hand-side of Equation 11-24 measure, respectively, the amount of variability in y_i accounted for by the regression line and the residual variation left unexplained by the regression line. We usually call $SS_E = \sum_{i=1}^{n} (y_i - \hat{y}_i)^2$ the **error sum of squares** and $SS_R = \sum_{i=1}^{n} (\hat{y}_i - \bar{y})^2$ the **regression sum of squares.** Symbolically, Equation 11-24 may be written as

$$SS_T = SS_R + SS_E \tag{11-25}$$

where $SS_T = \sum_{i=1}^{n} (y_i - \bar{y})^2$ is the **total corrected sum of squares** of y. In Section 11-2 we noted that $SS_E = SS_T - \hat{\beta}_1 S_{xy}$ (see Equation 11-14), so since $SS_T = \hat{\beta}_1 S_{xy} + SS_E$, we note that the regression sum of squares in Equation 11-25 is $SS_R = \hat{\beta}_1 S_{xy}$. The total sum of squares SS_T has $n - 1$ degrees of freedom, and SS_R and SS_E have 1 and $n - 2$ degrees of freedom, respectively.

We may show that $E[SS_E/(n - 2)] = \sigma^2$, $E(SS_R) = \sigma^2 + \beta_1^2 S_{xx}$ and that SS_E/σ^2 and SS_R/σ^2 are independent chi-square random variables with $n - 2$ and 1 degrees of freedom, respectively. Thus, if the null hypothesis H_0: $\beta_1 = 0$ is true, the statistic

Test for Significance of Regression

$$F_0 = \frac{SS_R/1}{SS_E/(n - 2)} = \frac{MS_R}{MS_E} \tag{11-26}$$

Table 11-3 Analysis of Variance for Testing Significance of Regression

Source of Variation	Sum of Squares	Degrees of Freedom	Mean Square	F_0
Regression	$SS_R = \hat{\beta}_1 S_{xy}$	1	MS_R	MS_R/MS_E
Error	$SS_E = SS_T - \hat{\beta}_1 S_{xy}$	$n - 2$	MS_E	
Total	SS_T	$n - 1$		

Note that $MS_E = \hat{\sigma}^2$.

follows the $F_{1,n-2}$ distribution, and we would reject H_0 if $f_0 > f_{\alpha,1,n-2}$. The quantities $MS_R = SS_R/1$ and $MS_E = SS_E/(n - 2)$ are called **mean squares.** In general, a mean square is always computed by dividing a sum of squares by its number of degrees of freedom. The test procedure is usually arranged in an **analysis of variance table,** such as Table 11-3.

EXAMPLE 11-3
Oxygen Purity ANOVA

We will use the analysis of variance approach to test for significance of regression using the oxygen purity data model from Example 11-1. Recall that $SS_T = 173.38$, $\hat{\beta}_1 = 14.947$, $S_{xy} = 10.17744$, and $n = 20$. The regression sum of squares is

$$SS_R = \hat{\beta}_1 S_{xy} = (14.947)10.17744 = 152.13$$

and the error sum of squares is

$$SS_E = SS_T - SS_R = 173.38 - 152.13 = 21.25$$

The analysis of variance for testing H_0: $\beta_1 = 0$ is summarized in the Minitab output in Table 11-2. The test statistic is $f_0 = MS_R/MS_E = 152.13/1.18 = 128.86$, for which we find that the P-value is $P \simeq 1.23 \times 10^{-9}$, so we conclude that β_1 is not zero.

There are frequently minor differences in terminology among computer packages. For example, sometimes the regression sum of squares is called the "model" sum of squares, and the error sum of squares is called the "residual" sum of squares.

Note that the analysis of variance procedure for testing for significance of regression is equivalent to the t-test in Section 11-4.1. That is, either procedure will lead to the same conclusions. This is easy to demonstrate by starting with the t-test statistic in Equation 11-19 with $\beta_{1,0} = 0$, say

$$T_0 = \frac{\hat{\beta}_1}{\sqrt{\hat{\sigma}^2/S_{xx}}} \tag{11-27}$$

Squaring both sides of Equation 11-27 and using the fact that $\hat{\sigma}^2 = MS_E$ results in

$$T_0^2 = \frac{\hat{\beta}_1^2 S_{xx}}{MS_E} = \frac{\hat{\beta}_1 S_{xy}}{MS_E} = \frac{MS_R}{MS_E} \tag{11-28}$$

Note that T_0^2 in Equation 11-28 is identical to F_0 in Equation 11-26. It is true, in general, that the square of a t random variable with v degrees of freedom is an F random variable, with one and v degrees of freedom in the numerator and denominator, respectively. Thus, the test using T_0 is equivalent to the test based on F_0. Note, however, that the t-test is somewhat more flexible in that it would allow testing against a one-sided alternative hypothesis, while the F-test is restricted to a two-sided alternative.

EXERCISES FOR SECTION 11-4

11-21. Consider the data from Exercise 11-1 on x = compressive strength and y = intrinsic permeability of concrete.

(a) Test for significance of regression using $\alpha = 0.05$. Find the P-value for this test. Can you conclude that the model specifies a useful linear relationship between these two variables?

(b) Estimate σ^2 and the standard deviation of $\hat{\beta}_1$.

(c) What is the standard error of the intercept in this model?

11-22. Consider the data from Exercise 11-2 on x = roadway surface temperature and y = pavement deflection.

(a) Test for significance of regression using $\alpha = 0.05$. Find the P-value for this test. What conclusions can you draw?

(b) Estimate the standard errors of the slope and intercept.

11-23. Consider the National Football League data in Exercise 11-3.

(a) Test for significance of regression using $\alpha = 0.01$. Find the P-value for this test. What conclusions can you draw?

(b) Estimate the standard errors of the slope and intercept.

(c) Test H_0: $\beta_1 = 10$ versus H_1: $\beta_1 \neq 10$ with $\alpha = 0.01$. Would you agree with the statement that this is a test of the hypothesis that a one-yard increase in the average yards per attempt results in a mean increase of 10 rating points?

11-24. Consider the data from Exercise 11-4 on y = sales price and x = taxes paid.

(a) Test H_0: $\beta_1 = 0$ using the t-test; use $\alpha = 0.05$.

(b) Test H_0: $\beta_1 = 0$ using the analysis of variance with $\alpha = 0.05$. Discuss the relationship of this test to the test from part (a).

(c) Estimate the standard errors of the slope and intercept.

(d) Test the hypothesis that $\beta_0 = 0$.

11-25. Consider the data from Exercise 11-5 on y = steam usage and x = average temperature.

(a) Test for significance of regression using $\alpha = 0.01$. What is the P-value for this test? State the conclusions that result from this test.

(b) Estimate the standard errors of the slope and intercept.

(c) Test the hypothesis H_0: $\beta_1 = 10$ versus H_1: $\beta_1 \neq 10$ using $\alpha = 0.01$. Find the P-value for this test.

(d) Test H_0: $\beta_0 = 0$ versus H_1: $\beta_0 \neq 0$ using $\alpha = 0.01$. Find the P-value for this test and draw conclusions.

11-26. Consider the data from Exercise 11-6 on y = highway gasoline mileage and x = engine displacement.

(a) Test for significance of regression using $\alpha = 0.01$. Find the P-value for this test. What conclusions can you reach?

(b) Estimate the standard errors of the slope and intercept.

(c) Test H_0: $\beta_1 = -0.05$ versus H_1: $\beta_1 < -0.05$ using $\alpha = 0.01$ and draw conclusions. What is the P-value for this test?

(d) Test the hypothesis H_0: $\beta_0 = 0$ versus H_1: $\beta_0 \neq 0$ using $\alpha = 0.01$. What is the P-value for this test?

11-27. Consider the data from Exercise 11-7 on y = green liquor Na_2S concentration and x = production in a paper mill.

(a) Test for significance of regression using $\alpha = 0.05$. Find the P-value for this test.

(b) Estimate the standard errors of the slope and intercept.

(c) Test H_0: $\beta_0 = 0$ versus H_1: $\beta_0 \neq 0$ using $\alpha = 0.05$. What is the P-value for this test?

11-28. Consider the data from Exercise 11-8 on y = blood pressure rise and x = sound pressure level.

(a) Test for significance of regression using $\alpha = 0.05$. What is the P-value for this test?

(b) Estimate the standard errors of the slope and intercept.

(c) Test H_0: $\beta_0 = 0$ versus H_1: $\beta_0 \neq 0$ using $\alpha = 0.05$. Find the P-value for this test.

11-29. Consider the data from Exercise 11-11, on y = shear strength of a propellant and x = propellant age.

(a) Test for significance of regression with $\alpha = 0.01$. Find the P-value for this test.

(b) Estimate the standard errors of $\hat{\beta}_0$ and $\hat{\beta}_1$.

(c) Test H_0: $\beta_1 = -30$ versus H_1: $\beta_1 \neq -30$ using $\alpha = 0.01$. What is the P-value for this test?

(d) Test H_0: $\beta_0 = 0$ versus H_1: $\beta_0 \neq 0$ using $\alpha = 0.01$. What is the P-value for this test?

(e) Test H_0: $\beta_0 = 2500$ versus H_1: $\beta_0 > 2500$ using $\alpha = 0.01$. What is the P-value for this test?

11-30. Consider the data from Exercise 11-10 on y = chloride concentration in surface streams and x = roadway area.

(a) Test the hypothesis H_0: $\beta_1 = 0$ versus H_1: $\beta_1 \neq 0$ using the analysis of variance procedure with $\alpha = 0.01$.

(b) Find the P-value for the test in part (a).

(c) Estimate the standard errors of $\hat{\beta}_1$ and $\hat{\beta}_0$.

(d) Test H_0: $\beta_0 = 0$ versus H_1: $\beta_0 \neq 0$ using $\alpha = 0.01$. What conclusions can you draw? Does it seem that the model might be a better fit to the data if the intercept were removed?

11-31. Consider the data in Exercise 11-13 on y = oxygen demand and x = time.

(a) Test for significance of regression using $\alpha = 0.01$. Find the P-value for this test. What conclusions can you draw?

(b) Estimate the standard errors of the slope and intercept.

(c) Test the hypothesis that $\beta_0 = 0$.

11-32. Consider the data in Exercise 11-14 on y = deflection and x = stress level.

(a) Test for significance of regression using $\alpha = 0.01$. What is the P-value for this test? State the conclusions that result from this test.

(b) Does this model appear to be adequate?

(c) Estimate the standard errors of the slope and intercept.

11-33. An article in *The Journal of Clinical Endocrinology and Metabolism*, "Simultaneous and Continuous 24-Hour Plasma and Cerebrospinal Fluid Leptin Measurements: Dissociation of Concentrations in Central and Peripheral Compartments" (2004, Vol. 89, pp. 258–265) studied the demographics of simultaneous and continuous 24-hour plasma and cerebrospinal fluid leptin measurements. The data follow:

y = BMI (kg/m^2):	19.92	20.59	29.02	20.78	25.97	
	20.39	23.29	17.27	35.24		
x = Age (yr):	45.5	34.6	40.6	32.9	28.2	30.1
	52.1	33.3	47.0			

(a) Test for significance of regression using $\alpha = 0.05$. Find the P-value for this test. Can you conclude that the model specifies a useful linear relationship between these two variables?
(b) Estimate σ^2 and the standard deviation of $\hat{\beta}_1$.
(c) What is the standard error of the intercept in this model?

11-34. Suppose that each value of x_i is multiplied by a positive constant a, and each value of y_i is multiplied by another positive constant b. Show that the t-statistic for testing H_0: $\beta_1 = 0$ versus H_1: $\beta_1 \neq 0$ is unchanged in value.

11-35. The type II error probability for the t-test for H_0: $\beta_1 = \beta_{1,0}$ can be computed in a similar manner to the t-tests of Chapter 9. If the true value of β_1 is β_1', the value $d = |\beta_{1,0} - \beta_1'|/(\sigma\sqrt{(n-1)/S_{xx}}$ is calculated and used as the horizontal scale factor on the operating characteristic curves for the t-test (Appendix Charts VIIe through VIIh) and the type II error probability is read from the vertical scale using the curve for $n - 2$ degrees of freedom. Apply this procedure to the football data of Exercise 11-3, using $\sigma = 5.5$ and $\beta_1' = 12.5$, where the hypotheses are H_0: $\beta_1 = 10$ versus H_1: $\beta_1 \neq 10$.

11-36. Consider the no-intercept model $Y = \beta x + \epsilon$ with the ϵ's NID(0, σ^2). The estimate of σ^2 is $s^2 = \sum_{i=1}^{n}(y_i - \hat{\beta}x_i)^2/(n-1)$ and $V(\hat{\beta}) = \sigma^2/\sum_{i=1}^{n}x_i^2$.
(a) Devise a test statistic for H_0: $\beta = 0$ versus H_1: $\beta \neq 0$.
(b) Apply the test in (a) to the model from Exercise 11-20.

11-5 CONFIDENCE INTERVALS

11-5.1 Confidence Intervals on the Slope and Intercept

In addition to point estimates of the slope and intercept, it is possible to obtain **confidence interval** estimates of these parameters. The width of these confidence intervals is a measure of the overall quality of the regression line. If the error terms, ϵ_i, in the regression model are normally and independently distributed,

$$(\hat{\beta}_1 - \beta_1)/\sqrt{\hat{\sigma}^2/S_{xx}} \quad \text{and} \quad (\hat{\beta}_0 - \beta_0)/\sqrt{\hat{\sigma}^2\left[\frac{1}{n} + \frac{\bar{x}^2}{S_{xx}}\right]}$$

are both distributed as t random variables with $n - 2$ degrees of freedom. This leads to the following definition of $100(1 - \alpha)\%$ confidence intervals on the slope and intercept.

Confidence Intervals on Parameters

Under the assumption that the observations are normally and independently distributed, a $100(1 - \alpha)\%$ **confidence interval on the slope** β_1 in simple linear regression is

$$\hat{\beta}_1 - t_{\alpha/2,n-2}\sqrt{\frac{\hat{\sigma}^2}{S_{xx}}} \leq \beta_1 \leq \hat{\beta}_1 + t_{\alpha/2,n-2}\sqrt{\frac{\hat{\sigma}^2}{S_{xx}}} \tag{11-29}$$

Similarly, a $100(1 - \alpha)\%$ **confidence interval on the intercept** β_0 is

$$\hat{\beta}_0 - t_{\alpha/2,n-2}\sqrt{\hat{\sigma}^2\left[\frac{1}{n} + \frac{\bar{x}^2}{S_{xx}}\right]} \leq \beta_0 \leq \hat{\beta}_0 + t_{\alpha/2,n-2}\sqrt{\hat{\sigma}^2\left[\frac{1}{n} + \frac{\bar{x}^2}{S_{xx}}\right]} \tag{11-30}$$

EXAMPLE 11-4
Oxygen Purity Confidence Interval on the Slope

We will find a 95% confidence interval on the slope of the regression line using the data in Example 11-1. Recall that $\hat{\beta}_1 = 14.947$, $S_{xx} = 0.68088$, and $\hat{\sigma}^2 = 1.18$ (see Table 11-2). Then, from Equation 11-29 we find

$$\hat{\beta}_1 - t_{0.025,18}\sqrt{\frac{\hat{\sigma}^2}{S_{xx}}} \leq \beta_1 \leq \hat{\beta}_1 + t_{0.025,18}\sqrt{\frac{\hat{\sigma}^2}{S_{xx}}}$$

or

$$14.947 - 2.101\sqrt{\frac{1.18}{0.68088}} \leq \beta_1 \leq 14.947 + 2.101\sqrt{\frac{1.18}{0.68088}}$$

This simplifies to

$$12.181 \leq \beta_1 \leq 17.713$$

11-5.2 Confidence Interval on the Mean Response

A confidence interval may be constructed on the mean response at a specified value of x, say, x_0. This is a confidence interval about $E(Y|x_0) = \mu_{Y|x_0}$ and is often called a confidence interval about the regression line. Since $E(Y|x_0) = \mu_{Y|x_0} = \beta_0 + \beta_1 x_0$, we may obtain a point estimate of the mean of Y at $x = x_0(\mu_{Y|x_0})$ from the fitted model as

$$\hat{\mu}_{Y|x_0} = \hat{\beta}_0 + \hat{\beta}_1 x_0$$

Now $\hat{\mu}_{Y|x_0}$ is an unbiased point estimator of $\mu_{Y|x_0}$, since $\hat{\beta}_0$ and $\hat{\beta}_1$ are unbiased estimators of β_0 and β_1. The variance of $\hat{\mu}_{Y|x_0}$ is

$$V(\hat{\mu}_{Y|x_0}) = \sigma^2\left[\frac{1}{n} + \frac{(x_0 - \bar{x})^2}{S_{xx}}\right]$$

This last result follows from the fact that $\hat{\mu}_{Y|x_0} = \bar{y} + \hat{\beta}_1(x_0 - \bar{x})$ and cov $(\bar{Y}, \hat{\beta}_1) = 0$. The zero covariance result is left as a Mind-Expanding exercise. Also, $\hat{\mu}_{Y|x_0}$ is normally distributed, because $\hat{\beta}_1$ and $\hat{\beta}_0$ are normally distributed, and if we $\hat{\sigma}^2$ use as an estimate of σ^2, it is easy to show that

$$\frac{\hat{\mu}_{Y|x_0} - \mu_{Y|x_0}}{\sqrt{\hat{\sigma}^2\left[\frac{1}{n} + \frac{(x_0 - \bar{x})^2}{S_{xx}}\right]}}$$

has a t distribution with $n - 2$ degrees of freedom. This leads to the following confidence interval definition.

Confidence Interval on the Mean Response

A $100(1-\alpha)\%$ **confidence interval about the mean response** at the value of $x = x_0$, say $\mu_{Y|x_0}$, is given by

$$\hat{\mu}_{Y|x_0} - t_{\alpha/2,n-2}\sqrt{\hat{\sigma}^2\left[\frac{1}{n}+\frac{(x_0-\bar{x})^2}{S_{xx}}\right]} \le \mu_{Y|x_0} \le \hat{\mu}_{Y|x_0} + t_{\alpha/2,n-2}\sqrt{\hat{\sigma}^2\left[\frac{1}{n}+\frac{(x_0-\bar{x})^2}{S_{xx}}\right]} \quad (11\text{-}31)$$

where $\hat{\mu}_{Y|x_0} = \hat{\beta}_0 + \hat{\beta}_1 x_0$ is computed from the fitted regression model.

Note that the width of the confidence interval for $\mu_{Y|x_0}$ is a function of the value specified for x_0. The interval width is a minimum for $x_0 = \bar{x}$ and widens as $|x_0 - \bar{x}|$ increases.

EXAMPLE 11-5 Oxygen Purity Confidence Interval on the Mean Response

We will construct a 95% confidence interval about the mean response for the data in Example 11-1. The fitted model is $\hat{\mu}_{Y|x_0} = 74.283 + 14.947x_0$, and the 95% confidence interval on $\mu_{Y|x_0}$ is found from Equation 11-31 as

$$\hat{\mu}_{Y|x_0} \pm 2.101\sqrt{1.18\left[\frac{1}{20}+\frac{(x_0-1.1960)^2}{0.68088}\right]}$$

Suppose that we are interested in predicting mean oxygen purity when $x_0 = 1.00\%$. Then

$$\hat{\mu}_{Y|x_{1.00}} = 74.283 + 14.947(1.00) = 89.23$$

and the 95% confidence interval is

$$\left\{89.23 \pm 2.101\sqrt{1.18\left[\frac{1}{20}+\frac{(1.00-1.1960)^2}{0.68088}\right]}\right\}$$

or

$$89.23 \pm 0.75$$

Therefore, the 95% confidence interval on $\mu_{Y|1.00}$ is

$$88.48 \le \mu_{Y|1.00} \le 89.98$$

Minitab will also perform these calculations. Refer to Table 11-2. The predicted value of y at $x = 1.00$ is shown along with the 95% CI on the mean of y at this level of x.

By repeating these calculations for several different values for x_0 we can obtain confidence limits for each corresponding value of $\mu_{Y|x_0}$. Figure 11-7 displays the scatter diagram with the fitted model and the corresponding 95% confidence limits plotted as the upper and lower lines. The 95% confidence level applies only to the interval obtained at one value of x and not to the entire set of x-levels. Notice that the width of the confidence interval on $\mu_{Y|x_0}$ increases as $|x_0 - \bar{x}|$ increases.

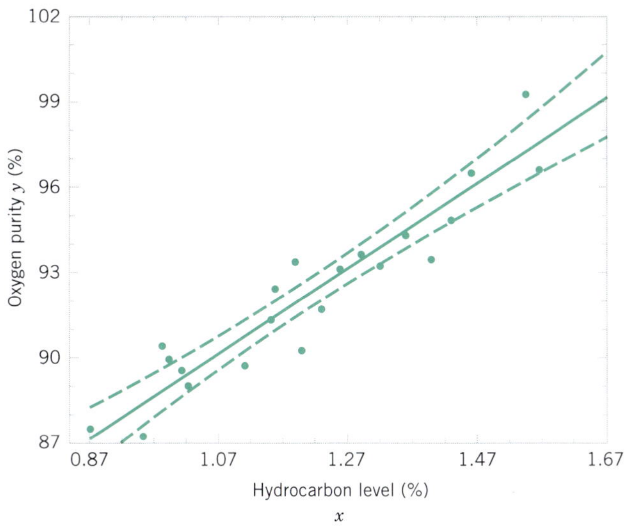

Figure 11-7 Scatter diagram of oxygen purity data from Example 11-1 with fitted regression line and 95 percent confidence limits on $\mu_{Y|x_0}$.

11-6 PREDICTION OF NEW OBSERVATIONS

An important application of a regression model is predicting new or future observations Y corresponding to a specified level of the regressor variable x. If x_0 is the value of the regressor variable of interest,

$$\hat{Y}_0 = \hat{\beta}_0 + \hat{\beta}_1 x_0 \tag{11-32}$$

is the point estimator of the new or future value of the response Y_0.

Now consider obtaining an interval estimate for this future observation Y_0. This new observation is independent of the observations used to develop the regression model. Therefore, the confidence interval for $\mu_{Y|x_0}$ in Equation 11-31 is inappropriate, since it is based only on the data used to fit the regression model. The confidence interval about $\mu_{Y|x_0}$ refers to the true mean response at $x = x_0$ (that is, a population parameter), not to future observations.

Let Y_0 be the future observation at $x = x_0$, and let $\hat{Y}_0$ given by Equation 11-32 be the estimator of Y_0. Note that the error in prediction

$$e_{\hat{p}} = Y_0 - \hat{Y}_0$$

is a normally distributed random variable with mean zero and variance

$$V(e_{\hat{p}}) = V(Y_0 - \hat{Y}_0) = \sigma^2\left[1 + \frac{1}{n} + \frac{(x_0 - \bar{x})^2}{S_{xx}}\right]$$

because Y_0 is independent of $\hat{Y}_0$. If we use $\hat{\sigma}^2$ to estimate σ^2, we can show that

$$\frac{Y_0 - \hat{Y}_0}{\sqrt{\hat{\sigma}^2\left[1 + \frac{1}{n} + \frac{(x_0 - \bar{x})^2}{S_{xx}}\right]}}$$

has a t distribution with $n - 2$ degrees of freedom. From this we can develop the following **prediction interval** definition.

Prediction Interval

A $100(1 - \alpha)$ % **prediction interval on a future observation** Y_0 at the value x_0 is given by

$$\hat{y}_0 - t_{\alpha/2,n-2}\sqrt{\hat{\sigma}^2\left[1 + \frac{1}{n} + \frac{(x_0 - \bar{x})^2}{S_{xx}}\right]}$$
$$\leq Y_0 \leq \hat{y}_0 + t_{\alpha/2,n-2}\sqrt{\hat{\sigma}^2\left[1 + \frac{1}{n} + \frac{(x_0 - \bar{x})^2}{S_{xx}}\right]} \tag{11-33}$$

The value $\hat{y}_0$ is computed from the regression model $\hat{y}_0 = \hat{\beta}_0 + \hat{\beta}_1 x_0$.

Notice that the prediction interval is of minimum width at $x_0 = \bar{x}$ and widens as $|x_0 - \bar{x}|$ increases. By comparing Equation 11-33 with Equation 11-31, we observe that the prediction interval at the point x_0 is always wider than the confidence interval at x_0. This results because the prediction interval depends on both the error from the fitted model and the error associated with future observations.

EXAMPLE 11-6 Oxygen Purity Prediction Interval

To illustrate the construction of a prediction interval, suppose we use the data in Example 11-1 and find a 95% prediction interval on the next observation of oxygen purity at $x_0 = 1.00\%$. Using Equation 11-33 and recalling from Example 11-5 that $\hat{y}_0 = 89.23$, we find that the prediction interval is

$$89.23 - 2.101\sqrt{1.18\left[1 + \frac{1}{20} + \frac{(1.00 - 1.1960)^2}{0.68088}\right]}$$
$$\leq Y_0 \leq 89.23 + 2.101\sqrt{1.18\left[1 + \frac{1}{20} + \frac{(1.00 - 1.1960)^2}{0.68088}\right]}$$

which simplifies to

$$86.83 \leq y_0 \leq 91.63$$

Minitab will also calculate prediction intervals. Refer to the output in Table 11-2. The 95% PI on the future observation at $x_0 = 1.00$ is shown in the display.

By repeating the foregoing calculations at different levels of x_0, we may obtain the 95% prediction intervals shown graphically as the lower and upper lines about the fitted regression model in Fig. 11-8. Notice that this graph also shows the 95% confidence limits on $\mu_{Y|x_0}$ calculated in Example 11-5. It illustrates that the prediction limits are always wider than the confidence limits.

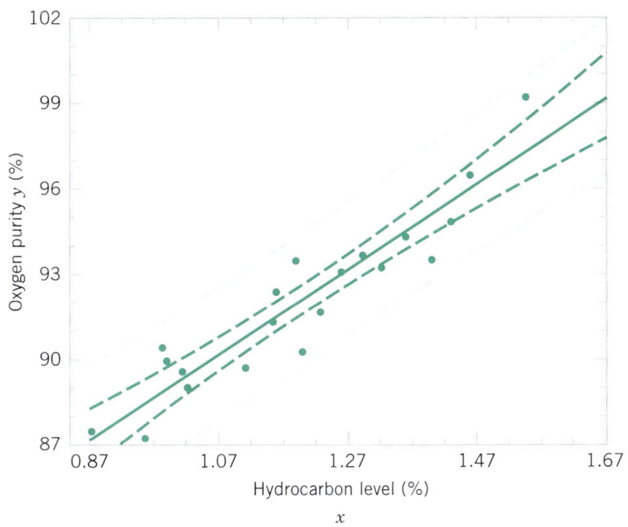

Figure 11-8 Scatter diagram of oxygen purity data from Example 11-1 with fitted regression line, 95% prediction limits (outer lines) and 95% confidence limits on $\mu_{Y|x_0}$.

EXERCISES FOR SECTIONS 11-5 AND 11-4

11-37. Refer to the data in Exercise 11-1 on y = intrinsic permeability of concrete and x = compressive strength. Find a 95% confidence interval on each of the following:
(a) Slope (b) Intercept
(c) Mean permeability when $x = 2.5$
(d) Find a 95% prediction interval on permeability when $x = 2.5$. Explain why this interval is wider than the interval in part (c).

11-38. Exercise 11-2 presented data on roadway surface temperature x and pavement deflection y. Find a 99% confidence interval on each of the following:
(a) Slope (b) Intercept
(c) Mean deflection when temperature $x = 85°F$
(d) Find a 99% prediction interval on pavement deflection when the temperature is 90°F.

11-39. Refer to the NFL quarterback ratings data in Exercise 11-3. Find a 95% confidence interval on each of the following:
(a) Slope
(b) Intercept
(c) Mean rating when the average yards per attempt is 8.0
(d) Find a 95% prediction interval on the rating when the average yards per attempt is 8.0.

11-40. Refer to the data on y = house selling price and x = taxes paid in Exercise 11-4. Find a 95% confidence interval on each of the following:
(a) β_1 (b) β_0
(c) Mean selling price when the taxes paid are $x = 7.50$
(d) Compute the 95% prediction interval for selling price when the taxes paid are $x = 7.50$.

11-41. Exercise 11-5 presented data on y = steam usage and x = monthly average temperature.
(a) Find a 99% confidence interval for β_1.
(b) Find a 99% confidence interval for β_0.
(c) Find a 95% confidence interval on mean steam usage when the average temperature is 55°F.
(d) Find a 95% prediction interval on steam usage when temperature is 55°F. Explain why this interval is wider than the interval in part (c).

11-42. Exercise 11-6 presented gasoline mileage performance for 20 cars, along with information about the engine displacement. Find a 95% confidence interval on each of the following:
(a) Slope (b) Intercept
(c) Mean highway gasoline mileage when the engine displacement is $x = 150$ in^3
(d) Construct a 95% prediction interval on highway gasoline mileage when the engine displacement is $x = 150$ in^3.

11-43. Consider the data in Exercise 11-7 on y = green liquor Na_2S concentration and x = production in a paper mill. Find a 99% confidence interval on each of the following:
(a) β_1 (b) β_0
(c) Mean Na_2S concentration when production $x = 910$ tons/day
(d) Find a 99% prediction interval on Na_2S concentration when $x = 910$ tons/day.

11-44. Exercise 11-8 presented data on y = blood pressure rise and x = sound pressure level. Find a 95% confidence interval on each of the following:
(a) β_1 (b) β_0
(c) Mean blood pressure rise when the sound pressure level is 85 decibels
(d) Find a 95% prediction interval on blood pressure rise when the sound pressure level is 85 decibels.

11-45. Refer to the data in Exercise 11-9 on y = wear volume of mild steel and x = oil viscosity. Find a 95% confidence interval on each of the following:
(a) Intercept (b) Slope
(c) Mean wear when oil viscosity $x = 30$

11-46. Exercise 11-10 presented data on chloride concentration y and roadway area x on watersheds in central Rhode Island. Find a 99% confidence interval on each of the following:
(a) β_1 (b) β_0
(c) Mean chloride concentration when roadway area $x = 1.0\%$
(d) Find a 99% prediction interval on chloride concentration when roadway area $x = 1.0\%$.

11-47. Refer to the data in Exercise 11-11 on rocket motor shear strength y and propellant age x. Find a 95% confidence interval on each of the following:
(a) Slope β_1 (b) Intercept β_0
(c) Mean shear strength when age $x = 20$ weeks
(d) Find a 95% prediction interval on shear strength when age $x = 20$ weeks.

11-48. Refer to the data in Exercise 11-12 on the microstructure of zirconia. Find a 95% confidence interval on each of the following:
(a) Slope (b) Intercept
(c) Mean length when $x = 1500$
(d) Find a 95% prediction interval on length when $x = 1500$. Explain why this interval is wider than the interval in part (c).

11-49. Refer to the data in Exercise 11-13 on oxygen demand. Find a 99% confidence interval on each of the following:
(a) β_1
(b) β_0
(c) Find a 95% confidence interval on mean BOD when the time is 8 days.

11-7 ADEQUACY OF THE REGRESSION MODEL

Fitting a regression model requires several **assumptions.** Estimation of the model parameters requires the assumption that the errors are uncorrelated random variables with mean zero and constant variance. Tests of hypotheses and interval estimation require that the errors be normally distributed. In addition, we assume that the order of the model is correct; that is, if we fit a simple linear regression model, we are assuming that the phenomenon actually behaves in a linear or first-order manner.

The analyst should always consider the validity of these assumptions to be doubtful and conduct analyses to examine the adequacy of the model that has been tentatively entertained. In this section we discuss methods useful in this respect.

11-7.1 Residual Analysis

The **residuals** from a regression model are $e_i = y_i - \hat{y}_i$, $i = 1, 2, \ldots, n$, where y_i is an actual observation and $\hat{y}_i$ is the corresponding fitted value from the regression model. Analysis of the residuals is frequently helpful in checking the assumption that the errors are approximately normally distributed with constant variance, and in determining whether additional terms in the model would be useful.

As an approximate check of normality, the experimenter can construct a frequency histogram of the residuals or a **normal probability plot of residuals.** Many computer programs will produce a normal probability plot of residuals, and since the sample sizes in regression are often too small for a histogram to be meaningful, the normal probability plotting method is preferred. It requires judgment to assess the abnormality of such plots. (Refer to the discussion of the "fat pencil" method in Section 6-6).

We may also **standardize** the residuals by computing $d_i = e_i/\sqrt{\hat{\sigma}^2}$, $i = 1, 2, \ldots, n$. If the errors are normally distributed, approximately 95% of the standardized residuals should fall in the interval $(-2, +2)$. Residuals that are far outside this interval may indicate the presence of an **outlier,** that is, an observation that is not typical of the rest of the data. Various rules have been proposed for discarding outliers. However, outliers sometimes provide

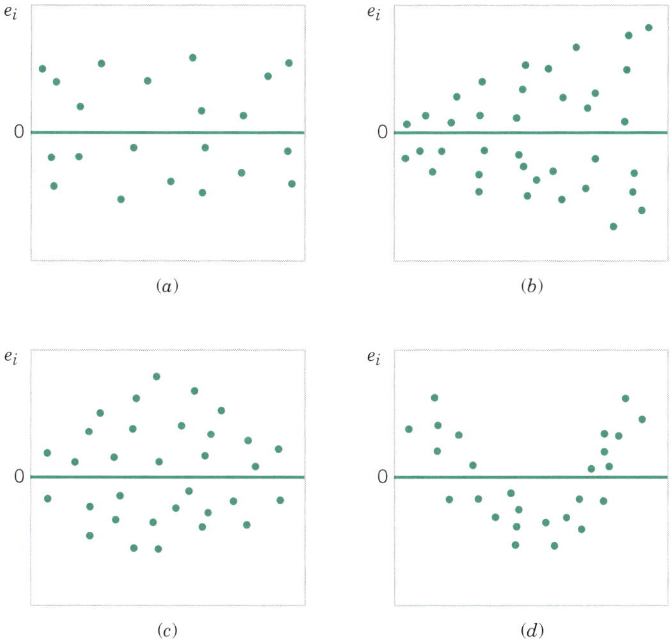

Figure 11-9 Patterns for residual plots. (a) satisfactory, (b) funnel, (c) double bow, (d) nonlinear. [Adapted from Montgomery, Peck, and Vining (2001).]

important information about unusual circumstances of interest to experimenters and should not be automatically discarded. For further discussion of outliers, see Montgomery, Peck and Vining (2001).

It is frequently helpful to plot the residuals (1) in time sequence (if known), (2), against the $\hat{y}_i$, and (3) against the independent variable x. These graphs will usually look like one of the four general patterns shown in Fig. 11-9. Pattern (*a*) in Fig. 11-9 represents the ideal situation, while patterns (*b*), (*c*), and (*d*) represent anomalies. If the residuals appear as in (*b*), the variance of the observations may be increasing with time or with the magnitude of y_i or x_i. Data transformation on the response y is often used to eliminate this problem. Widely used variance-stabilizing transformations include the use of $\sqrt{y}$, $\ln y$, or $1/y$ as the response. See Montgomery, Peck, and Vining (2001) for more details regarding methods for selecting an appropriate transformation. Plots of residuals against $\hat{y}_i$ and x_i that look like (*c*) also indicate inequality of variance. Residual plots that look like (*d*) indicate model inadequacy; that is, higher order terms should be added to the model, a transformation on the x-variable or the y-variable (or both) should be considered, or other regressors should be considered.

EXAMPLE 11-7
Oxygen Purity Residuals

The regression model for the oxygen purity data in Example 11-1 is $\hat{y} = 74.283 + 14.947x$. Table 11-4 presents the observed and predicted values of y at each value of x from this data set, along with the corresponding residual. These values were computed using Minitab and show the number of decimal places typical of computer output. A normal probability plot of the residuals is shown in Fig. 11-10. Since the residuals fall approximately along a straight line in the figure, we conclude that there is no severe departure from normality. The residuals are also plotted against the predicted value $\hat{y}_i$ in Fig. 11-11 and against the hydrocarbon levels x_i in Fig. 11-12. These plots do not indicate any serious model inadequacies.

Table 11-4 Oxygen Purity Data from Example 11-1, Predicted Values, and Residuals

	Hydrocarbon Level, x	Oxygen Purity, y	Predicted Value, $\hat{y}$	Residual $e = y - \hat{y}$		Hydrocarbon Level, x	Oxygen Purity, y	Predicted Value, $\hat{y}$	Residual $e = y - \hat{y}$
1	0.99	90.01	89.081	0.929	11	1.19	93.54	92.071	1.469
2	1.02	89.05	89.530	−0.480	12	1.15	92.52	91.473	1.047
3	1.15	91.43	91.473	−0.043	13	0.98	90.56	88.932	1.628
4	1.29	93.74	93.566	0.174	14	1.01	89.54	89.380	0.160
5	1.46	96.73	96.107	0.623	15	1.11	89.85	90.875	−1.025
6	1.36	94.45	94.612	−0.162	16	1.20	90.39	92.220	−1.830
7	0.87	87.59	87.288	0.302	17	1.26	93.25	93.117	0.133
8	1.23	91.77	92.669	−0.899	18	1.32	93.41	94.014	−0.604
9	1.55	99.42	97.452	1.968	19	1.43	94.98	95.658	−0.678
10	1.40	93.65	95.210	−1.560	20	0.95	87.33	88.483	−1.153

11-7.2 Coefficient of Determination (R^2)

A widely used measure for a regression model is a ratio of sum of squares.

R^2

The **coefficient of determination is**

$$R^2 = \frac{SS_R}{SS_T} = 1 - \frac{SS_E}{SS_T} \tag{11-34}$$

The coefficient is often used to judge the adequacy of a regression model. Subsequently, we will see that in the case where X and Y are jointly distributed random variables, R^2 is the square of the correlation coefficient between X and Y. From the analysis of variance identity in Equations 11-24 and 11-25, $0 \leq R^2 \leq 1$. We often refer loosely to R^2 as the amount of variability in the data explained or accounted for by the regression model. For the oxygen purity regression model, we have $R^2 = SS_R/SS_T = 152.13/173.38 = 0.877$; that is, the model accounts for 87.7% of the variability in the data.

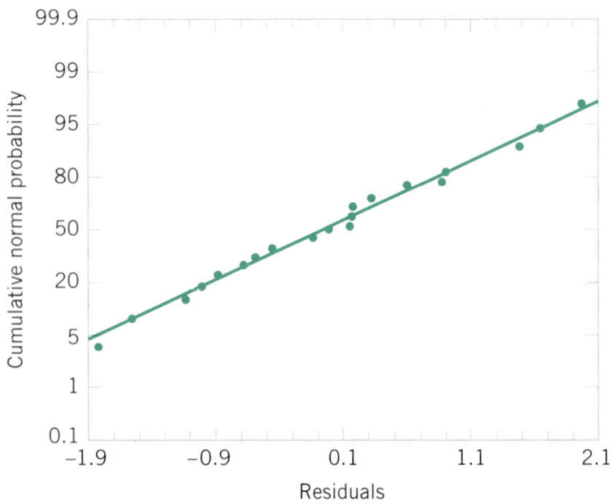

Figure 11-10 Normal probability plot of residuals, Example 11-7.

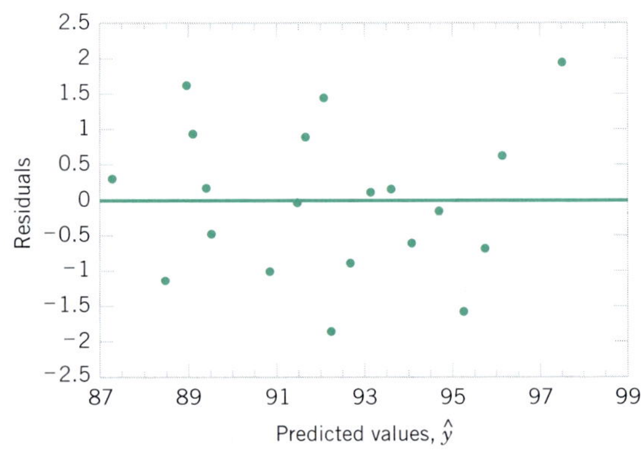

Figure 11-11 Plot of residuals versus predicted oxygen purity $\hat{y}$, Example 11-7.

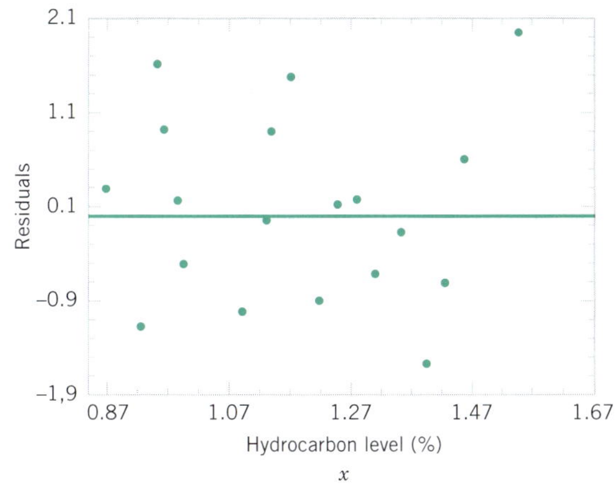

Figure 11-12 Plot of residuals versus hydrocarbon level x, Example 11-8.

The statistic R^2 should be used with caution, because it is always possible to make R^2 unity by simply adding enough terms to the model. For example, we can obtain a "perfect" fit to n data points with a polynomial of degree $n - 1$. In addition, R^2 will always increase if we add a variable to the model, but this does not necessarily imply that the new model is superior to the old one. Unless the error sum of squares in the new model is reduced by an amount equal to the original error mean square, the new model will have a larger error mean square than the old one, because of the loss of one error degree of freedom. Thus, the new model will actually be worse than the old one.

There are several misconceptions about R^2. In general, R^2 does not measure the magnitude of the slope of the regression line. A large value of R^2 does not imply a steep slope. Furthermore, R^2 does not measure the appropriateness of the model, since it can be artificially inflated by adding higher order polynomial terms in x to the model. Even if y and x are related in a nonlinear fashion, R^2 will often be large. For example, R^2 for the regression equation in Fig. 11-6(b) will be relatively large, even though the linear approximation is poor. Finally, even though R^2 is large, this does not necessarily imply that the regression model will provide accurate predictions of future observations.

EXERCISES FOR SECTION 11-7

11-50. Refer to the compressive strength data in Exercise 11-1. Use the summary statistics provided to calculate R^2 and provide a practical interpretation of this quantity.

11-51. Refer to the NFL quarterback ratings data in Exercise 11-3.

(a) Calculate R^2 for this model and provide a practical interpretation of this quantity.

(b) Prepare a normal probability plot of the residuals from the least squares model. Does the normality assumption seem to be satisfied?

(c) Plot the residuals versus the fitted values and against x. Interpret these graphs.

11-52. Refer to the data in Exercise 11-4 on house selling price y and taxes paid x.

(a) Find the residuals for the least squares model.

(b) Prepare a normal probability plot of the residuals and interpret this display.

(c) Plot the residuals versus $\hat{y}$ and versus x. Does the assumption of constant variance seem to be satisfied?

(d) What proportion of total variability is explained by the regression model?

11-53. Refer to the data in Exercise 11-5 on y = steam usage and x = average monthly temperature.

(a) What proportion of total variability is accounted for by the simple linear regression model?

(b) Prepare a normal probability plot of the residuals and interpret this graph.

(c) Plot residuals versus $\hat{y}$ and x. Do the regression assumptions appear to be satisfied?

11-54. Refer to the gasoline mileage data in Exercise 11-6.

(a) What proportion of total variability in highway gasoline mileage performance is accounted for by engine displacement?

(b) Plot the residuals versus $\hat{y}$ and x, and comment on the graphs.

(c) Prepare a normal probability plot of the residuals. Does the normality assumption appear to be satisfied?

11-55. Exercise 11-9 presents data on wear volume y and oil viscosity x.

(a) Calculate R^2 for this model. Provide an interpretation of this quantity.
(b) Plot the residuals from this model versus $\hat{y}$ and versus x. Interpret these plots.
(c) Prepare a normal probability plot of the residuals. Does the normality assumption appear to be satisfied?

11-56. Refer to Exercise 11-8, which presented data on blood pressure rise y and sound pressure level x.

(a) What proportion of total variability in blood pressure rise is accounted for by sound pressure level?
(b) Prepare a normal probability plot of the residuals from this least squares model. Interpret this plot.
(c) Plot residuals versus $\hat{y}$ and versus x. Comment on these plots.

11-57. Refer to Exercise 11-10, which presented data on chloride concentration y and roadway area x.

(a) What proportion of the total variability in chloride concentration is accounted for by the regression model?
(b) Plot the residuals versus $\hat{y}$ and versus x. Interpret these plots.
(c) Prepare a normal probability plot of the residuals. Does the normality assumption appear to be satisfied?

11-58. An article in the *Journal of the American Statistical Association*, "Markov Chain Monte Carlo Methods for Computing Bayes Factors: A Comparative Review" (2001, Vol. 96, pp. 1122–1132) analyzed the tabulated data on compressive strength parallel to the grain versus resin-adjusted density for specimens of radiata pine.

Compressive Strength	Density	Compressive Strength	Density
3040	29.2	3840	30.7
2470	24.7	3800	32.7
3610	32.3	4600	32.6
3480	31.3	1900	22.1
3810	31.5	2530	25.3
2330	24.5	2920	30.8
1800	19.9	4990	38.9
3110	27.3	1670	22.1
3160	27.1	3310	29.2
2310	24.0	3450	30.1
4360	33.8	3600	31.4
1880	21.5	2850	26.7
3670	32.2	1590	22.1
1740	22.5	3770	30.3
2250	27.5	3850	32.0
2650	25.6	2480	23.2
4970	34.5	3570	30.3
2620	26.2	2620	29.9
2900	26.7	1890	20.8
1670	21.1	3030	33.2
2540	24.1	3030	28.2

(a) Fit a regression model relating compressive strength to density.
(b) Test for significance of regression with $\alpha = 0.05$.
(c) Estimate σ^2 for this model.
(d) Calculate R^2 for this model. Provide an interpretation of this quantity.
(e) Prepare a normal probability plot of the residuals and interpret this display.
(f) Plot the residuals versus $\hat{y}$ and versus x. Does the assumption of constant variance seem to be satisfied?

11-59. Consider the rocket propellant data in Exercise 11-11.

(a) Calculate R^2 for this model. Provide an interpretation of this quantity.
(b) Plot the residuals on a normal probability scale. Do any points seem unusual on this plot?
(c) Delete the two points identified in part (b) from the sample and fit the simple linear regression model to the remaining 18 points. Calculate the value of R^2 for the new model. Is it larger or smaller than the value of R^2 computed in part (a)? Why?
(d) Did the value of $\hat{\sigma}^2$ change dramatically when the two points identified above were deleted and the model fit to the remaining points? Why?

11-60. Consider the data in Exercise 11-7 on y = green liquor Na_2S concentration and x = paper machine production. Suppose that a 14th sample point is added to the original data, where $y_{14} = 59$ and $x_{14} = 855$.

(a) Prepare a scatter diagram of y versus x. Fit the simple linear regression model to all 14 observations.
(b) Test for significance of regression with $\alpha = 0.05$.
(c) Estimate σ^2 for this model.
(d) Compare the estimate of σ^2 obtained in part (c) above with the estimate of σ^2 obtained from the original 13 points. Which estimate is larger and why?
(e) Compute the residuals for this model. Does the value of e_{14} appear unusual?
(f) Prepare and interpret a normal probability plot of the residuals.
(g) Plot the residuals versus $\hat{y}$ and versus x. Comment on these graphs.

11-61. Consider the rocket propellant data in Exercise 11-11. Calculate the standardized residuals for these data. Does this provide any helpful information about the magnitude of the residuals?

11-62. Studentized Residuals. Show that the variance of the ith residual is

$$V(e_i) = \sigma^2\left[1 - \left(\frac{1}{n} + \frac{(x_i - \bar{x})^2}{S_{xx}}\right)\right]$$

Hint:

$$\text{cov}(Y_i, \hat{Y}_i) = \sigma^2\left[\frac{1}{n} + \frac{(x_i - \bar{x})^2}{S_{xx}}\right].$$

The ith studentized residual is defined as

$$r_i = \frac{e_i}{\sqrt{\hat{\sigma}^2\left[1 - \left(\frac{1}{n} + \frac{(x_i - \bar{x})^2}{S_{xx}}\right)\right]}}$$

(a) Explain why r_i has unit standard deviation.
(b) Do the standardized residuals have unit standard deviation?
(c) Discuss the behavior of the studentized residual when the sample value x_i is very close to the middle of the range of x.
(d) Discuss the behavior of the studentized residual when the sample value x_i is very near one end of the range of x.

11-63. Show that an equivalent way to define the test for significance of regression in simple linear regression is to base the test on R^2 as follows: to test H_0: $\beta_1 = 0$ versus H_1: $\beta_1 \neq 0$, calculate

$$F_0 = \frac{R^2(n - 2)}{1 - R^2}$$

and to reject H_0: $\beta_1 = 0$ if the computed value $f_0 > f_{\alpha,1,n-2}$. Suppose that a simple linear regression model has been fit to $n = 25$ observations and $R^2 = 0.90$.

(a) Test for significance of regression at $\alpha = 0.05$.
(b) What is the smallest value of R^2 that would lead to the conclusion of a significant regression if $\alpha = 0.05$?

11-8 CORRELATION

Our development of regression analysis has assumed that x is a mathematical variable, measured with negligible error, and that Y is a random variable. Many applications of regression analysis involve situations in which both X and Y are random variables. In these situations, it is usually assumed that the observations (X_i, Y_i), $i = 1, 2, \ldots, n$ are jointly distributed random variables obtained from the distribution $f(x, y)$.

For example, suppose we wish to develop a regression model relating the shear strength of spot welds to the weld diameter. In this example, weld diameter cannot be controlled. We would randomly select n spot welds and observe a diameter (X_i) and a shear strength (Y_i) for each. Therefore (X_i, Y_i) are jointly distributed random variables.

We assume that the joint distribution of X_i and Y_i is the bivariate normal distribution presented in Chapter 5, and μ_Y and σ_Y^2 are the mean and variance of Y, μ_X and σ_X^2 are the mean and variance of X, and ρ is the **correlation coefficient** between Y and X. Recall that the correlation coefficient is defined as

$$\rho = \frac{\sigma_{XY}}{\sigma_X \sigma_Y} \tag{11-35}$$

where σ_{XY} is the covariance between Y and X.

The conditional distribution of Y for a given value of $X = x$ is

$$f_{Y|x}(y) = \frac{1}{\sqrt{2\pi}\sigma_{Y|x}} \exp\left[-\frac{1}{2}\left(\frac{y - \beta_0 - \beta_1 x}{\sigma_{Y|x}}\right)^2\right] \tag{11-36}$$

where

$$\beta_0 = \mu_Y - \mu_X \rho \frac{\sigma_Y}{\sigma_X} \tag{11-37}$$

$$\beta_1 = \frac{\sigma_Y}{\sigma_X}\rho \tag{11-38}$$

and the variance of the conditional distribution of Y given $X = x$ is

$$\sigma_{Y|x}^2 = \sigma_Y^2(1 - \rho^2) \tag{11-39}$$

That is, the conditional distribution of Y given $X = x$ is normal with mean

$$E(Y|x) = \beta_0 + \beta_1 x \tag{11-40}$$

and variance $\sigma^2_{Y|x}$. Thus, the mean of the conditional distribution of Y given $X = x$ is a simple linear regression model. Furthermore, there is a relationship between the correlation coefficient ρ and the slope β_1. From Equation 11-38 we see that if $\rho = 0$, then $\beta_1 = 0$, which implies that there is no regression of Y on X. That is, knowledge of X does not assist us in predicting Y.

The method of maximum likelihood may be used to estimate the parameters β_0 and β_1. It can be shown that the maximum likelihood estimators of those parameters are

$$\hat{\beta}_0 = \overline{Y} - \hat{\beta}_1 \overline{X} \tag{11-41}$$

and

$$\hat{\beta}_1 = \frac{\sum_{i=1}^{n} Y_i(X_i - \overline{X})}{\sum_{i=1}^{n} (X_i - \overline{X})^2} = \frac{S_{XY}}{S_{XX}} \tag{11-42}$$

We note that the estimators of the intercept and slope in Equations 11-41 and 11-42 are identical to those given by the method of least squares in the case where X was assumed to be a mathematical variable. That is, the regression model with Y and X jointly normally distributed is equivalent to the model with X considered as a mathematical variable. This follows because the random variables Y given $X = x$ are independently and normally distributed with mean $\beta_0 + \beta_1 x$ and constant variance $\sigma^2_{Y|x}$. These results will also hold for any joint distribution of Y and X such that the conditional distribution of Y given X is normal.

It is possible to draw inferences about the correlation coefficient ρ in this model. The estimator of ρ is the **sample correlation coefficient**

$$R = \frac{\sum_{i=1}^{n} Y_i(X_i - \overline{X})}{\left[\sum_{i=1}^{n} (X_i - \overline{X})^2 \sum_{i=1}^{n} (Y_i - \overline{Y})^2\right]^{1/2}} = \frac{S_{XY}}{(S_{XX} SS_T)^{1/2}} \tag{11-43}$$

Note that

$$\hat{\beta}_1 = \left(\frac{SS_T}{S_{XX}}\right)^{1/2} R \tag{11-44}$$

so the slope $\hat{\beta}_1$ is just the sample correlation coefficient R multiplied by a scale factor that is the square root of the "spread" of the Y values divided by the "spread" of the X values.Thus, $\hat{\beta}_1$ and R are closely related, although they provide somewhat different information. The sample correlation coefficient R measures the linear association between Y and X, while $\hat{\beta}_1$ measures the predicted change in the mean of Y for a unit change in X. In the case of a mathematical variable x, R has no meaning because the magnitude of R depends on the choice of spacing of x. We may also write, from Equation 11-44,

$$R^2 = \hat{\beta}_1^2 \frac{S_{XX}}{SS_T} = \frac{\hat{\beta}_1 S_{XY}}{SS_T} = \frac{SS_R}{SS_T}$$

which is just the coefficient of determination. That is, the coefficient of determination R^2 is just the square of the correlation coefficient between Y and X.

It is often useful to test the hypotheses

$$H_0\colon \rho = 0$$
$$H_1\colon \rho \neq 0 \tag{11-45}$$

The appropriate test statistic for these hypotheses is

Test Statistic for Zero Correlation

$$T_0 = \frac{R\sqrt{n-2}}{\sqrt{1-R^2}} \tag{11-46}$$

which has the t distribution with $n - 2$ degrees of freedom if $H_0\colon \rho = 0$ is true. Therefore, we would reject the null hypothesis if $|t_0| > t_{\alpha/2,n-2}$. This test is equivalent to the test of the hypothesis $H_0\colon \beta_1 = 0$ given in Section 11-5.1. This equivalence follows directly from Equation 11-46.

The test procedure for the hypothesis

$$H_0\colon \rho = \rho_0$$
$$H_1\colon \rho \neq \rho_0 \tag{11-47}$$

where $\rho_0 \neq 0$ is somewhat more complicated. For moderately large samples (say, $n \geq 25$) the statistic

$$Z = \text{arctanh } R = \frac{1}{2}\ln\frac{1+R}{1-R} \tag{11-48}$$

is approximately normally distributed with mean and variance

$$\mu_Z = \text{arctanh } \rho = \frac{1}{2}\ln\frac{1+\rho}{1-\rho} \quad \text{and} \quad \sigma_Z^2 = \frac{1}{n-3}$$

respectively. Therefore, to test the hypothesis $H_0\colon \rho = \rho_0$, we may use the test statistic

$$Z_0 = (\text{arctanh } R - \text{arctanh } \rho_0)(n-3)^{1/2} \tag{11-49}$$

and reject $H_0\colon \rho = \rho_0$ if the value of the test statistic in Equation 11-49 is such that $|z_0| > z_{\alpha/2}$.

It is also possible to construct an approximate $100(1 - \alpha)\%$ confidence interval for ρ, using the transformation in Equation 11-48. The approximate $100(1 - \alpha)\%$ confidence interval is

Confidence Interval for a Correlation Coefficient

$$\tanh\left(\text{arctanh } r - \frac{z_{\alpha/2}}{\sqrt{n-3}}\right) \leq \rho \leq \tanh\left(\text{arctanh } r + \frac{z_{\alpha/2}}{\sqrt{n-3}}\right) \tag{11-50}$$

where $\tanh u = (e^u - e^{-u})/(e^u + e^{-u})$.

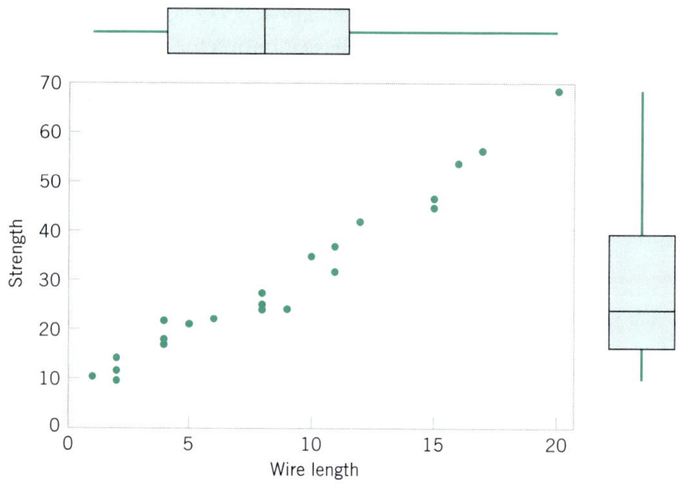

Figure 11-13 Scatter plot of wire bond strength versus wire length, Example 11-8.

EXAMPLE 11-8
Wire Bond Pull Strength

In Chapter 1 (Section 1-3) an application of regression analysis is described in which an engineer at a semiconductor assembly plant is investigating the relationship between pull strength of a wire bond and two factors: wire length and die height. In this example, we will consider only one of the factors, the wire length. A random sample of 25 units is selected and tested, and the wire bond pull strength and wire length are observed for each unit. The data are shown in Table 1-2. We assume that pull strength and wire length are jointly normally distributed.

Figure 11-13 shows a scatter diagram of wire bond strength versus wire length. We have used the Minitab option of displaying box plots of each individual variable on the scatter diagram. There is evidence of a linear relationship between the two variables.

The Minitab output for fitting a simple linear regression model to the data is shown below. Now $S_{xx} = 698.56$ and $S_{xy} = 2027.7132$, and the sample correlation coefficient is

$$r = \frac{S_{xy}}{[S_{xx}SS_T]^{1/2}} = \frac{2027.7132}{[(698.560)(6105.9)]^{1/2}} = 0.9818$$

Regression Analysis: Strength versus Length

The regression equation is
Strength = 5.11 + 2.90 Length

Predictor	Coef	SE Coef	T	P
Constant	5.115	1.146	4.46	0.000
Length	2.9027	0.1170	24.80	0.000

S = 3.093 R-Sq = 96.4% R-Sq(adj) = 96.2%
PRESS = 272.144 R-Sq(pred) = 95.54%

Analysis of Variance

Source	DF	SS	MS	F	P
Regression	1	5885.9	5885.9	615.08	0.000
Residual Error	23	220.1	9.6		
Total	24	6105.9			

Note that $r^2 = (0.9818)^2 = 0.9640$ (which is reported in the Minitab output), or that approximately 96.40% of the variability in pull strength is explained by the linear relationship to wire length.

Now suppose that we wish to test the hypothesis

$$H_0\colon \rho = 0$$
$$H_1\colon \rho \neq 0$$

with $\alpha = 0.05$. We can compute the t-statistic of Equation 11-46 as

$$t_0 = \frac{r\sqrt{n-2}}{\sqrt{1-r^2}} = \frac{0.9818\sqrt{23}}{\sqrt{1-0.9640}} = 24.8$$

This statistic is also reported in the Minitab output as a test of $H_0\colon \beta_1 = 0$. Because $t_{0.025,23} = 2.069$, we reject H_0 and conclude that the correlation coefficient $\rho \neq 0$.

Finally, we may construct an approximate 95% confidence interval on ρ from Equation 11-50. Since arctanh r = arctanh 0.9818 = 2.3452, Equation 11-50 becomes

$$\tanh\left(2.3452 - \frac{1.96}{\sqrt{22}}\right) \leq \rho \leq \tanh\left(2.3452 + \frac{1.96}{\sqrt{22}}\right)$$

which reduces to

$$0.9585 \leq \rho \leq 0.9921$$

EXERCISES FOR SECTION 11–8

11-64. Suppose data is obtained from 20 pairs of (x, y) and the sample correlation coefficient is 0.8.

(a) Test the hypothesis that $H_0\colon\rho = 0$ against $H_1\colon\rho \neq 0$ with $\alpha = 0.05$. Calculate the P-value.

(b) Test the hypothesis that $H_1\colon\rho = 0.5$ against $H_1\colon\rho \neq 0.5$ with $\alpha = 0.05$. Calculate the P-value.

(c) Construct a 95% two-sided confidence interval for the correlation coefficient. Explain how the questions in parts (a) and (b) could be answered with a confidence interval.

11-65. Suppose data is obtained from 20 pairs of (x, y) and the sample correlation coefficient is 0.75.

(a) Test the hypothesis that $H_0\colon\rho = 0$ against $H_1\colon\rho > 0$ with $\alpha = 0.05$. Calculate the P-value.

(b) Test the hypothesis that $H_1\colon\rho = 0.5$ against $H_1\colon\rho > 0.5$ with $\alpha = 0.05$. Calculate the P-value.

(c) Construct a 95% one-sided confidence interval for the correlation coefficient. Explain how the questions in parts (a) and (b) could be answered with a confidence interval.

11-66. A random sample of $n = 25$ observations was made on the time to failure of an electronic component and the temperature in the application environment in which the component was used.

(a) Given that $r = 0.83$, test the hypothesis that $\rho = 0$, using $\alpha = 0.05$. What is the P-value for this test?

(b) Find a 95% confidence interval on ρ.

(c) Test the hypothesis $H_0\colon \rho = 0.8$ versus $H_1\colon \rho \neq 0.8$, using $\alpha = 0.05$. Find the P-value for this test.

11-67. A random sample of 50 observations was made on the diameter of spot welds and the corresponding weld shear strength.

(a) Given that $r = 0.62$, test the hypothesis that $\rho = 0$, using $\alpha = 0.01$. What is the P-value for this test?

(b) Find a 99% confidence interval for ρ.

(c) Based on the confidence interval in part (b), can you conclude that $\rho = 0.5$ at the 0.01 level of significance?

11-68. The following data gave X = the water content of snow on April 1 and Y = the yield from April to July (in inches) on the Snake River watershed in Wyoming for 1919 to 1935. (The data were taken from an article in *Research Notes,* Vol. 61, 1950, Pacific Northwest Forest Range Experiment Station, Oregon)

x	y	x	y
23.1	10.5	37.9	22.8
32.8	16.7	30.5	14.1
31.8	18.2	25.1	12.9
32.0	17.0	12.4	8.8
30.4	16.3	35.1	17.4
24.0	10.5	31.5	14.9
39.5	23.1	21.1	10.5
24.2	12.4	27.6	16.1
52.5	24.9		

(a) Estimate the correlation between Y and X.
(b) Test the hypothesis that $\rho = 0$, using $\alpha = 0.05$.
(c) Fit a simple linear regression model and test for significance of regression using $\alpha = 0.05$. What conclusions can you draw? How is the test for significance of regression related to the test on ρ in part (b)?
(d) Analyze the residuals and comment on model adequacy.

11-69. The final test and exam averages for 20 randomly selected students taking a course in engineering statistics and a course in operations research follow. Assume that the final averages are jointly normally distributed.
(a) Find the regression line relating the statistics final average to the OR final average. Graph the data.
(b) Test for significance of regression using $\alpha = 0.05$.
(c) Estimate the correlation coefficient.
(d) Test the hypothesis that $\rho = 0$, using $\alpha = 0.05$.
(e) Test the hypothesis that $\rho = 0.5$, using $\alpha = 0.05$.
(f) Construct a 95% confidence interval for the correlation coefficient.

Statistics	OR	Statistics	OR	Statistics	OR
86	80	86	81	83	81
75	81	71	76	75	70
69	75	65	72	71	73
75	81	84	85	76	72
90	92	71	72	84	80
94	95	62	65	97	98
83	80	90	93		

11-70. The weight and systolic blood pressure of 26 randomly selected males in the age group 25 to 30 are shown in the following table. Assume that weight and blood pressure are jointly normally distributed.

Subject	Weight	Systolic BP	Subject	Weight	Systolic BP
1	165	130	14	172	153
2	167	133	15	159	128
3	180	150	16	168	132
4	155	128	17	174	149
5	212	151	18	183	158
6	175	146	19	215	150
7	190	150	20	195	163
8	210	140	21	180	156
9	200	148	22	143	124
10	149	125	23	240	170
11	158	133	24	235	165
12	169	135	25	192	160
13	170	150	26	187	159

(a) Find a regression line relating systolic blood pressure to weight.
(b) Test for significance of regression using $\alpha = 0.05$.
(c) Estimate the correlation coefficient.
(d) Test the hypothesis that $\rho = 0$, using $\alpha = 0.05$.
(e) Test the hypothesis that $\rho = 0.6$, using $\alpha = 0.05$.
(f) Construct a 95% confidence interval for the correlation coefficient.

11-71. In an article in *IEEE Transactions on Instrumentation and Measurement* (2001, Vol. 50, pp. 986–990) researchers studied the effects of reducing current draw in a magnetic core by electronic means. They measured the current in a magnetic winding with and without the electronics in a paired experiment and the data for the case without electronics is provided in the following table.

Supply Voltage	Current Without Electronics (mA)
0.66	7.32
1.32	12.22
1.98	16.34
2.64	23.66
3.3	28.06
3.96	33.39
4.62	34.12
3.28	39.21
5.94	44.21
6.6	47.48

(a) Graph the data and fit a regression line to predict current without electronics to supply voltage. Is there a significant regression at $\alpha = 0.05$? What is the P-value?
(b) Estimate the correlation coefficient.
(c) Test the hypothesis that $\rho = 0$ against the alternative $\rho \neq 0$ with $\alpha = 0.05$. What is the P-value?
(d) Compute a 95% confidence interval for the correlation coefficient.

11-72. The monthly absolute estimate of global (land and ocean combined) temperature indexes (degrees C) in 2000 and 2001 are (source: (http://www.ncdc.noaa.gov/oa/climate/)

2000: 12.28, 12.63, 13.22, 14.21, 15.13, 15.82, 16.05, 16.02, 15.29, 14.29, 13.16, 12.47

2001: 12.44, 12.55, 13.35, 14.22, 15.28, 15.99, 16.23, 16.17, 15.44, 14.52, 13.52, 12.61

(a) Graph the data and fit a regression line to predict 2001 temperatures from those in 2000. Is there a significant regression at $\alpha = 0.05$? What is the P-value?
(b) Estimate the correlation coefficient.
(c) Test the hypothesis that $\rho = 0.9$ against the alternative $\rho \neq 0.9$ with $\alpha = 0.05$. What is the P-value?
(d) Compute a 95% confidence interval for the correlation coefficient.

11-73 Refer to the NFL quarterback ratings data in Exercise 11-3.

(a) Estimate the correlation coefficient between the ratings and the average yards per attempt.

(b) Test the hypothesis $H_0: \rho = 0$ versus $H_1: \rho \neq 0$ using $\alpha = 0.05$. What is the P-value for this test?

(c) Construct a 95% confidence interval for ρ.

(d) Test the hypothesis $H_0: \rho = 0.7$ versus $H_1: \rho \neq 0.7$ using $\alpha = 0.05$. Find the P-value for this test.

11-74. Consider the following (x, y) data. Calculate the correlation coefficient. Graph the data and comment on the relationship between x and y. Explain why the correlation coefficient does not detect the relationship between x and y.

x	y	x	y
−4	0	0	−4
−3	−2.65	1	3.87
−3	2.65	1	−3.87
−2	−3.46	2	3.46
−2	3.46	2	−3.46
−1	−3.87	3	2.65
−1	3.87	3	−2.65
0	4	4	0

11-9 TRANSFORMATIONS

We occasionally find that the straight-line regression model $Y = \beta_0 + \beta_1 x + \epsilon$ is inappropriate because the true regression function is nonlinear. Sometimes nonlinearity is visually determined from the scatter diagram, and sometimes, because of prior experience or underlying theory, we know in advance that the model is nonlinear. Occasionally, a scatter diagram will exhibit an apparent nonlinear relationship between Y and x. In some of these situations, a nonlinear function can be expressed as a straight line by using a suitable transformation. Such nonlinear models are called **intrinsically linear.**

As an example of a nonlinear model that is intrinsically linear, consider the exponential function

$$Y = \beta_0 e^{\beta_1 x} \epsilon$$

This function is intrinsically linear, since it can be transformed to a straight line by a logarithmic transformation

$$\ln Y = \ln \beta_0 + \beta_1 x + \ln \epsilon$$

This transformation requires that the transformed error terms $\ln \epsilon$ are normally and independently distributed with mean 0 and variance σ^2.

Another intrinsically linear function is

$$Y = \beta_0 + \beta_1\left(\frac{1}{x}\right) + \epsilon$$

By using the reciprocal transformation $z = 1/x$, the model is linearized to

$$Y = \beta_0 + \beta_1 z + \epsilon$$

Sometimes several transformations can be employed jointly to linearize a function. For example, consider the function

$$Y = \frac{1}{\exp(\beta_0 + \beta_1 x + \epsilon)}$$

letting $Y^* = 1/Y$, we have the linearized form

$$\ln Y^* = \beta_0 + \beta_1 x + \epsilon$$

For examples of fitting these models, refer to Montgomery, Peck, and Vining (2001) or Myers (1990).

Transformations can be very useful in many situations where the true relationship between the response Y and the regressor x is not well approximated by a straight line. The utility of a transformation is illustrated in the following example.

EXAMPLE 11-9
Windmill Power

A research engineer is investigating the use of a windmill to generate electricity and has collected data on the DC output from this windmill and the corresponding wind velocity. The data are plotted in Figure 11-14 and listed in Table 11-5.

Inspection of the scatter diagram indicates that the relationship between DC output Y and wind velocity (x) may be nonlinear. However, we initially fit a straight-line model to the data. The regression model is

$$\hat{y} = 0.1309 + 0.2411x$$

The summary statistics for this model are $R^2 = 0.8745$, $MS_E = \hat{\sigma}^2 = 0.0557$ and $F_0 = 160.26$ (the P value is <0.0001).

A plot of the residuals versus $\hat{y}_i$ is shown in Figure 11-15. This residual plot indicates model inadequacy and implies that the linear relationship has not captured all of the information in the wind speed

Table 11-5 Observed Values y_i and Regressor Variable x_i for Example 11-9

Observation Number, i	Wind Velocity (mph), x_i	DC Output, y_i
1	5.00	1.582
2	6.00	1.822
3	3.40	1.057
4	2.70	0.500
5	10.00	2.236
6	9.70	2.386
7	9.55	2.294
8	3.05	0.558
9	8.15	2.166
10	6.20	1.866
11	2.90	0.653
12	6.35	1.930
13	4.60	1.562
14	5.80	1.737
15	7.40	2.088
16	3.60	1.137
17	7.85	2.179
18	8.80	2.112
19	7.00	1.800
20	5.45	1.501
21	9.10	2.303
22	10.20	2.310
23	4.10	1.194
24	3.95	1.144
25	2.45	0.123

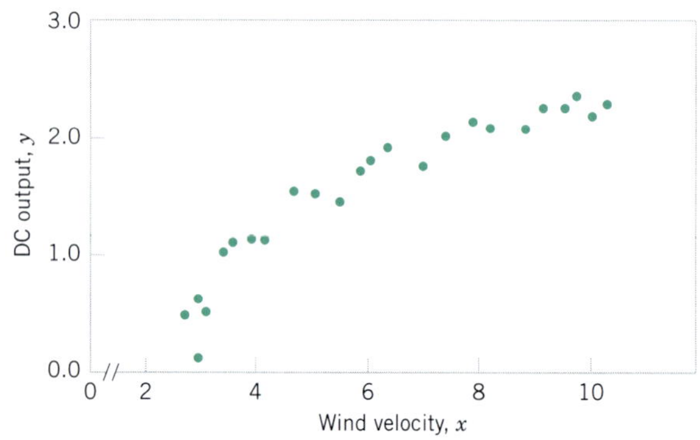

Figure 11-14 Plot of DC output y versus wind velocity x for the windmill data.

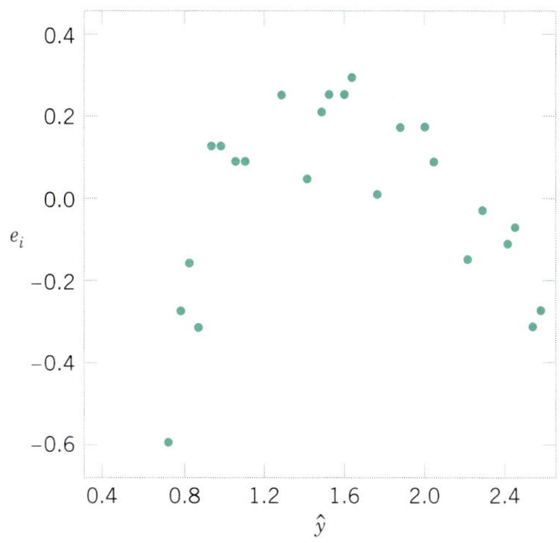

Figure 11-15 Plot of residuals e_i versus fitted values $\hat{y}_i$ for the windmill data.

variable. Note that the curvature that was apparent in the scatter diagram of Figure 11-14 is greatly amplified in the residual plots. Clearly some other model form must be considered.

We might initially consider using a quadratic model such as

$$y = \beta_0 + \beta_1 x + \beta_2 x^2 + \epsilon$$

to account for the apparent curvature. However, the scatter diagram Figure 11-14 suggests that as wind speed increases, DC output approaches an upper limit of approximately 2.5. This is also consistent with the theory of windmill operation. Since the quadratic model will eventually bend downward as wind speed increases, it would not be appropriate for these data. A more reasonable model for the windmill data that incorporates an upper asymptote would be

$$y = \beta_0 + \beta_1\left(\frac{1}{x}\right) + \epsilon$$

Figure 11-16 is a scatter diagram with the transformed variable $x' = 1/x$. This plot appears linear, indicating that the reciprocal transformation is appropriate. The fitted regression model is

$$\hat{y} = 2.9789 - 6.9345x'$$

The summary statistics for this model are $R^2 = 0.9800$, $MS_E = \hat{\sigma}^2 = 0.0089$, and $F_0 = 1128.43$ (the P value is <0.0001).

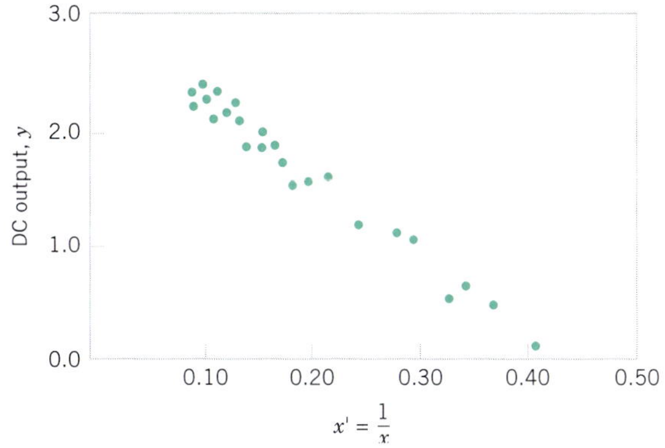

Figure 11-16 Plot of DC output versus $x' = 1/x$ for the windmill data.

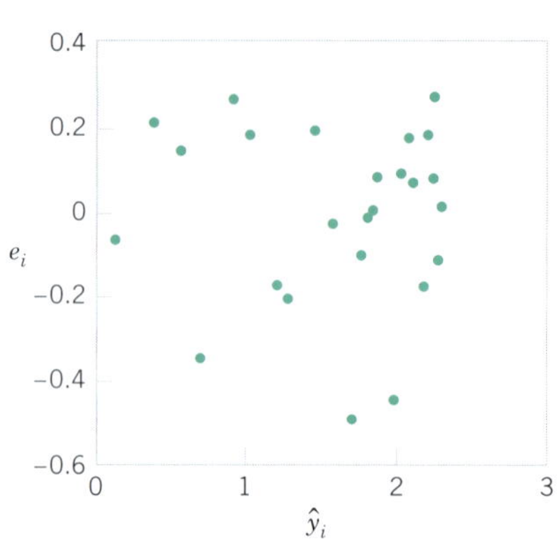

Figure 11-17 Plot of residuals versus fitted values $\hat{y}_i$ for the transformed model for the windmill data.

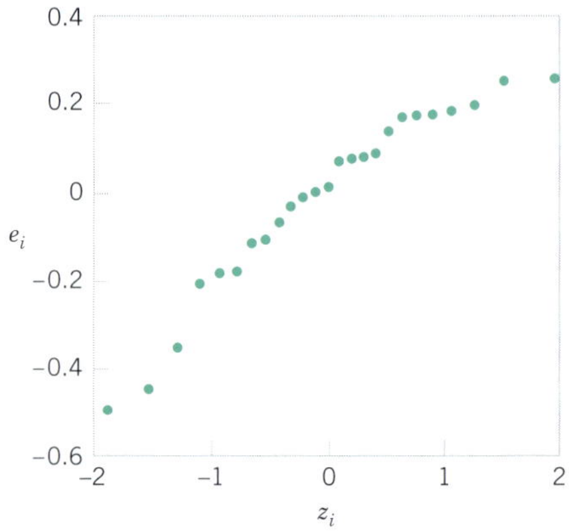

Figure 11-18 Normal probability plot of the residuals for the transformed model for the windmill data.

A plot of the residuals from the transformed model versus $\hat{y}$ is shown in Figure 11-17. This plot does not reveal any serious problem with inequality of variance. The normal probability plot, shown in Figure 11-18, gives a mild indication that the errors come from a distribution with heavier tails than the normal (notice the slight upward and downward curve at the extremes). This normal probability plot has the z-score value plotted on the horizontal axis. Since there is no strong signal of model inadequacy, we conclude that the transformed model is satisfactory.

11-9.1 Logistic Regression available at www.wiley.com/college/montgomery

EXERCISES FOR SECTION 11–9

11-75. Determine if the following models are intrinsically linear. If yes, determine the appropriate transformation to generate the linear model.

(a) $Y = \beta_0 x^{\beta_1}\epsilon$ (b) $Y = \dfrac{3 + 5x}{x} + \epsilon$

(c) $Y = \beta_0 \beta_1^x \epsilon$ (d) $Y = \dfrac{x}{\beta_0 x + \beta_1 + x\epsilon}$

11-76. The vapor pressure of water at various temperatures follows:

Observation Number, i	Temperature (K)	Vapor pressure (mm Hg)
1	273	4.6
2	283	9.2
3	293	17.5
4	303	31.8
5	313	55.3
6	323	92.5
7	333	149.4
8	343	233.7
9	353	355.1
10	363	525.8
11	373	760.0

(a) Draw a scatter diagram of these data. What type of relationship seems appropriate in relating y to x?

(b) Fit a simple linear regression model to these data.

(c) Test for significance of regression using $\alpha = 0.05$. What conclusions can you draw?

(d) Plot the residuals from the simple linear regression model versus $\hat{y}_i$. What do you conclude about model adequacy?

(e) The Clausis-Clapeyron relationship states that $\ln(P_v) \propto -\frac{1}{T}$, where P_v is the vapor pressure of water. Repeat parts (a)–(d). using an appropriate transformation.

11-77. An electric utility is interested in developing a model relating peak hour demand (y in kilowatts) to total monthly energy usage during the month (x, in kilowatt hours). Data for 50 residential customers are shown in the following table.

Customer	x	y	Customer	x	y
1	679	0.79	26	1434	0.31
2	292	0.44	27	837	4.20
3	1012	0.56	28	1748	4.88
4	493	0.79	29	1381	3.48
5	582	2.70	30	1428	7.58
6	1156	3.64	31	1255	2.63

continued

Customer	x	y	Customer	x	y
7	997	4.73	32	1777	4.99
8	2189	9.50	33	370	0.59
9	1097	5.34	34	2316	8.19
10	2078	6.85	35	1130	4.79
11	1818	5.84	36	463	0.51
12	1700	5.21	37	770	1.74
13	747	3.25	38	724	4.10
14	2030	4.43	39	808	3.94
15	1643	3.16	40	790	0.96
16	414	0.50	41	783	3.29
17	354	0.17	42	406	0.44
18	1276	1.88	43	1242	3.24
19	745	0.77	44	658	2.14
20	795	3.70	45	1746	5.71
21	540	0.56	46	895	4.12
22	874	1.56	47	1114	1.90
23	1543	5.28	48	413	0.51
24	1029	0.64	49	1787	8.33
25	710	4.00	50	3560	14.94

(a) Draw a scatter diagram of y versus x.
(b) Fit the simple linear regression model.
(c) Test for significance of regression using $\alpha = 0.05$.
(d) Plot the residuals versus $\hat{y}_i$ and comment on the underlying regression assumptions. Specifically, does it seem that the equality of variance assumption is satisfied?
(e) Find a simple linear regression model using $\sqrt{y}$ as the response. Does this transformation on y stabilize the inequality of variance problem noted in part (d) above?

Supplemental Exercises

11-78. Show that, for the simple linear regression model, the following statements are true:

(a) $\sum_{i=1}^{n}(y_i - \hat{y}_i) = 0$ (b) $\sum_{i=1}^{n}(y_i - \hat{y}_i)x_i = 0$

(c) $\frac{1}{n}\sum_{i=1}^{n}\hat{y}_i = \bar{y}$

11-79. An article in the *IEEE Transactions on Instrumentation and Measurement* ("Direct, Fast, and Accurate Measurement of V_T and K of MOS Transistor Using V_T-Sift Circuit," 1991, Vol. 40, pp. 951–955) described the use of a simple linear regression model to express drain current y (in milliamperes) as a function of ground-to-source voltage x (in volts). The data are as follows:

y	x	y	x
0.734	1.1	1.50	1.6
0.886	1.2	1.66	1.7
1.04	1.3	1.81	1.8
1.19	1.4	1.97	1.9
1.35	1.5	2.12	2.0

(a) Draw a scatter diagram of these data. Does a straight-line relationship seem plausible?
(b) Fit a simple linear regression model to these data.
(c) Test for significance of regression using $\alpha = 0.05$. What is the P-value for this test?
(d) Find a 95% confidence interval estimate on the slope.
(e) Test the hypothesis H_0: $\beta_0 = 0$ versus H_1: $\beta_0 \neq 0$ using $\alpha = 0.05$. What conclusions can you draw?

11-80. The strength of paper used in the manufacture of cardboard boxes (y) is related to the percentage of hardwood concentration in the original pulp (x). Under controlled conditions, a pilot plant manufactures 16 samples, each from a different batch of pulp, and measures the tensile strength. The data are shown in the table that follows:

y	101.4	117.4	117.1	106.2
x	1.0	1.5	1.5	1.5
y	131.9	146.9	146.8	133.9
x	2.0	2.0	2.2	2.4
y	111.0	123.0	125.1	145.2
x	2.5	2.5	2.8	2.8
y	134.3	144.5	143.7	146.9
x	3.0	3.0	3.2	3.3

(a) Fit a simple linear regression model to the data.
(b) Test for significance of regression using $\alpha = 0.05$.
(c) Construct a 90% confidence interval on the slope β_1.
(d) Construct a 90% confidence interval on the intercept β_0.
(e) Construct a 95% confidence interval on the mean strength at $x = 2.5$.
(f) Analyze the residuals and comment on model adequacy.

11-81. Consider the following data. Suppose that the relationship between Y and x is hypothesized to be $Y = (\beta_0 + \beta_1 x + \epsilon)^{-1}$. Fit an appropriate model to the data. Does the assumed model form seem reasonable?

x	10	15	18	12	9	8	11	6
y	0.1	0.13	0.09	0.15	0.20	0.21	0.18	0.24

11-82. The following data, adapted from Montgomery, Peck, and Vining (2001), present the number of certified mental defectives per 10,000 of estimated population in the United Kingdom (y) and the number of radio receiver licenses issued (x) by the BBC (in millions) for the years 1924 through 1937. Fit a regression model relating y and x. Comment on the model. Specifically, does the existence of a strong correlation imply a cause-and-effect relationship?

Year	y	x	Year	y	x
1924	8	1.350	1931	16	4.620
1925	8	1.960	1932	18	5.497
1926	9	2.270	1933	19	6.260
1927	10	2.483	1934	20	7.012
1928	11	2.730	1935	21	7.618
1929	11	3.091	1936	22	8.131
1930	12	3.674	1937	23	8.593

11-83. Consider the weight and blood pressure data in Exercise 11-70. Fit a no-intercept model to the data, and compare it to the model obtained in Exercise 11-70. Which model is superior?

11-84. An article in *Air and Waste* ("Update on Ozone Trends in California's South Coast Air Basin," Vol. 43, 1993) studied the ozone levels on the South Coast air basin of California for the years 1976–1991. The author believes that the number of days that the ozone level exceeds 0.20 parts per million depends on the seasonal meteorological index (the seasonal average 850 millibar temperature). The data follow:

Year	Days	Index	Year	Days	Index
1976	91	16.7	1984	81	18.0
1977	105	17.1	1985	65	17.2
1978	106	18.2	1986	61	16.9
1979	108	18.1	1987	48	17.1
1980	88	17.2	1988	61	18.2
1981	91	18.2	1989	43	17.3
1982	58	16.0	1990	33	17.5
1983	82	17.2	1991	36	16.6

(a) Construct a scatter diagram of the data.
(b) Fit a simple linear regression model to the data. Test for significance of regression.
(c) Find a 95% CI on the slope β_1.
(d) Analyze the residuals and comment on model adequacy.

11-85. An article in the *Journal of Applied Polymer Science* (Vol. 56, pp. 471–476, 1995) studied the effect of the mole ratio of sebacic acid on the intrinsic viscosity of copolyesters. The data follow:

Mole ratio x	1.0	0.9	0.8	0.7	0.6	0.5	0.4	0.3
Viscosity y	0.45	0.20	0.34	0.58	0.70	0.57	0.55	0.44

(a) Construct a scatter diagram of the data.
(b) Fit a simple linear repression model.
(c) Test for significance of regression. Calculate R^2 for the model.
(d) Analyze the residuals and comment on model adequacy.

11-86. Two different methods can be used for measuring the temperature of the solution in a Hall cell used in aluminum smelting, a thermocouple implanted in the cell and an indirect measurement produced from an IR device. The indirect method is preferable because the thermocouples are eventually destroyed by the solution. Consider the following 10 measurements:

Thermocouple	921	935	916	920	940
IR	918	934	924	921	945
Thermocouple	936	925	940	933	927
IR	930	919	943	932	935

(a) Construct a scatter diagram for these data, letting x = thermocouple measurement and y = IR measurement.
(b) Fit a simple linear regression model.
(c) Test for significance a regression and calculate R^2. What conclusions can you draw?
(d) Is there evidence to support a claim that both devices produce equivalent temperature measurements? Formulate and test an appropriate hypothesis to support this claim.
(e) Analyze the residuals and comment on model adequacy.

11-87. The grams of solids removed from a material (y) is thought to be related to the drying time. Ten observations obtained from an experimental study follow:

y	4.3	1.5	1.8	4.9	4.2	4.8	5.8	6.2	7.0	7.9
x	2.5	3.0	3.5	4.0	4.5	5.0	5.5	6.0	6.5	7.0

(a) Construct a scatter diagram for these data.
(b) Fit a simple linear regression model.
(c) Test for significance of regression.
(d) Based on these data, what is your estimate of the mean grams of solids removed at 4.25 hours? Find a 95% confidence interval on the mean.
(e) Analyze the residuals and comment on model adequacy.

11-88. Cesium atoms cooled by laser light could be used to build inexpensive atomic clocks. In a study in *IEEE Transactions on Instrumentation and Measurement* (2001, Vol. 50, pp. 1224–1228) the number of atoms cooled by lasers of various powers were counted.

Power (mW)	Number of Atoms (×10E9)
11	0
12	0.02
18	0.08
21	0.13
22	0.15
24	0.18
28	0.31
32	0.4
37	0.49
39	0.57
41	0.64
46	0.71
48	0.79
50	0.82
51	0.83

(a) Graph the data and fit a regression line to predict the number of atoms from laser power. Comment on the adequacy of a linear model.
(b) Is there a significant regression at $\alpha = 0.05$? What is the *P*-value?
(c) Estimate the correlation coefficient.
(d) Test the hypothesis that $\rho = 0$ against the alternative $\rho \neq 0$ with $\alpha = 0.05$. What is the *P*-value?
(e) Compute a 95% confidence interval for the slope coefficient.

11-89. The following data related diamond carats to purchase prices. It appeared in Singapore's *Business Times*, edition of February 18, 2000.

Carat	Price	Carat	Price
0.3	1302	0.33	1327
0.3	1510	0.33	1098
0.3	1510	0.34	1693
0.3	1260	0.34	1551
0.31	1641	0.34	1410
0.31	1555	0.34	1269
0.31	1427	0.34	1316
0.31	1427	0.34	1222
0.31	1126	0.35	1738

Carat	Price	Carat	Price
0.31	1126	0.35	1593
0.32	1468	0.35	1447
0.32	1202	0.35	1255
0.36	1635	0.45	1572
0.36	1485	0.46	2942
0.37	1420	0.48	2532
0.37	1420	0.5	3501
0.4	1911	0.5	3501
0.4	1525	0.5	3501
0.41	1956	0.5	3293
0.43	1747	0.5	3016

(a) Graph the data. What is the relation between carat and price? Is there an outlier?
(b) What would you say to the person who purchased the diamond that was an outlier?
(c) Fit two regression models, one with all the data and the other with unusual data omitted. Estimate the slope coefficient with a 95% confidence interval in both cases. Comment on any difference.

11-90. The following table shows the population and the average count of wood storks sighted per sample period for South Carolina from 1991–2004. Fit a regression line with population as the response and the count of wood storks as the predictor. Such an analysis might be used to evaluate the relationship between storks and babies. Is regression significant at $\alpha = 0.05$? What do you conclude about the role of regression analysis to establish a cause-and-effect relationship?

Year	Population	Stork Count
1991	3559470	0.342
1992	3600576	0.291
1993	3634507	0.291
1994	3666456	0.291
1995	3699943	0.291
1996	3738974	0.509
1997	3790066	0.294
1998	3839578	0.799
1999	3885736	0.542
2000	4012012	0.495
2001	4061209	0.859
2002	4105848	0.364
2003	4148744	0.501
2004	4198068	0.656

MIND-EXPANDING EXERCISES

11-91. Suppose that we have n pairs of observations (x_i, y_i) such that the sample correlation coefficient r is unity (approximately). Now let $z_i = y_i^2$ and consider the sample correlation coefficient for the n-pairs of data (x_i, z_i). Will this sample correlation coefficient be approximately unity? Explain why or why not.

11-92. Consider the simple linear regression model $Y = \beta_0 + \beta_1 x + \epsilon$, with $E(\epsilon) = 0$, $V(\epsilon) = \sigma^2$, and the errors ϵ uncorrelated.

(a) Show that $\text{cov}(\hat{\beta}_0, \hat{\beta}_1) = -\bar{x}\sigma^2/S_{xx}$.

(b) Show that $\text{cov}(\bar{Y}, \hat{\beta}_1) = 0$.

11-93. Consider the simple linear regression model $Y = \beta_0 + \beta_1 x + \epsilon$, with $E(\epsilon) = 0$, $V(\epsilon) = \sigma^2$, and the errors ϵ uncorrelated.

(a) Show that $E(\hat{\sigma}^2) = E(MS_E) = \sigma^2$.

(b) Show that $E(MS_R) = \sigma^2 + \beta_1^2 S_{xx}$.

11-94. Suppose that we have assumed the straight-line regression model

$$Y = \beta_0 + \beta_1 x_1 + \epsilon$$

but the response is affected by a second variable x_2 such that the true regression function is

$$E(Y) = \beta_0 + \beta_1 x_1 + \beta_2 x_2$$

Is the estimator of the slope in the simple linear regression model unbiased?

11-95. Suppose that we are fitting a line and we wish to make the variance of the regression coefficient $\hat{\beta}_1$ as small as possible. Where should the observations x_i, $i = 1, 2, \ldots, n$, be taken so as to minimize $V(\hat{\beta}_1)$? Discuss the practical implications of this allocation of the x_i.

11-96. **Weighted Least Squares.** Suppose that we are fitting the line $Y = \beta_0 + \beta_1 x + \epsilon$, but the variance of Y depends on the level of x; that is,

$$V(Y_i | x_i) = \sigma_i^2 = \frac{\sigma^2}{w_i} \qquad i = 1, 2, \ldots, n$$

where the w_i are constants, often called *weights*. Show that for an objective function in which each squared residual is multiplied by the reciprocal of the variance of the corresponding observation, the resulting **weighted** least squares normal equations are

$$\hat{\beta}_0 \sum_{i=1}^{n} w_i + \hat{\beta}_1 \sum_{i=1}^{n} w_i x_i = \sum_{i=1}^{n} w_i y_i$$

$$\hat{\beta}_0 \sum_{i=1}^{n} w_i x_i + \hat{\beta}_1 \sum_{i=1}^{n} w_i x_i^2 = \sum_{i=1}^{n} w_i x_i y_i$$

Find the solution to these normal equations. The solutions are weighted least squares estimators of β_0 and β_1.

11-97. Consider a situation where both Y and X are random variables. Let s_x and s_y be the sample standard deviations of the observed x's and y's, respectively. Show that an alternative expression for the fitted simple linear regression model $\hat{y} = \hat{\beta}_0 + \hat{\beta}_1 x$ is

$$\hat{y} = \bar{y} + r\frac{s_y}{s_x}(x - \bar{x})$$

11-98. Suppose that we are interested in fitting a simple linear regression model $Y = \beta_0 + \beta_1 x + \epsilon$, where the intercept, β_0, is known.

(a) Find the least squares estimator of β_1.

(b) What is the variance of the estimator of the slope in part (a)?

(c) Find an expression for a $100(1 - \alpha)\%$ confidence interval for the slope β_1. Is this interval longer than the corresponding interval for the case where both the intercept and slope are unknown? Justify your answer.

IMPORTANT TERMS AND CONCEPTS

Analysis of variance test in regression
Confidence interval on mean response
Correlation coefficient
Empirical model
Confidence intervals on model parameters
Intrinsically linear model
Least squares estimation of regression model parameters
Model adequacy checking
Regression analysis
Prediction interval on a future observation
Residual plots
Residuals
Scatter diagram
Significance of regression
Simple linear regression model standard errors
Statistical tests on model parameters
Transformations

12 Analysis of Multiple Linear Regression

CHAPTER OUTLINE

LEARNING OBJECTIVES

After careful study of this chapter, you should be able to do the following:

1. Use multiple regression techniques to build empirical models to engineering and scientific data
2. Understand how the method of least squares extends to fitting multiple regression models
3. Assess regression model adequacy
4. Test hypotheses and construct confidence intervals on the regression coefficients

5. Use the regression model to estimate the mean response and to make predictions and to construct confidence intervals and prediction intervals
6. Build regression models with polynomial terms
7. Use indicator variables to model categorical regressors
8. Use stepwise regression and other model building techniques to select the appropriate set of variables for a regression model

12-1 MULTIPLE LINEAR REGRESSION MODEL

12-1.1 Introduction

Many applications of regression analysis involve situations in which there are more than one regressor variable. A regression model that contains more than one regressor variable is called a **multiple regression model.**

As an example, suppose that the effective life of a cutting tool depends on the cutting speed and the tool angle. A multiple regression model that might describe this relationship is

$$Y = \beta_0 + \beta_1 x_1 + \beta_2 x_2 + \epsilon \tag{12-1}$$

where Y represents the tool life, x_1 represents the cutting speed, x_2 represents the tool angle, and ϵ is a random error term. This is a multiple linear regression model with two regressors. The term *linear* is used because Equation 12-1 is a linear function of the unknown parameters β_0, β_1, and β_2.

The regression model in Equation 12-1 describes a plane in the three-dimensional space of Y, x_1, and x_2. Figure 12-1(a) shows this plane for the regression model

$$E(Y) = 50 + 10x_1 + 7x_2$$

where we have assumed that the expected value of the error term is zero; that is $E(\epsilon) = 0$. The parameter β_0 is the **intercept** of the plane. We sometimes call β_1 and β_2 **partial regression coefficients,** because β_1 measures the expected change in Y per unit change in x_1 when x_2 is held constant, and β_2 measures the expected change in Y per unit change in x_2 when x_1 is held constant. Figure 12-1(b) shows a **contour plot** of the regression model—that is, lines

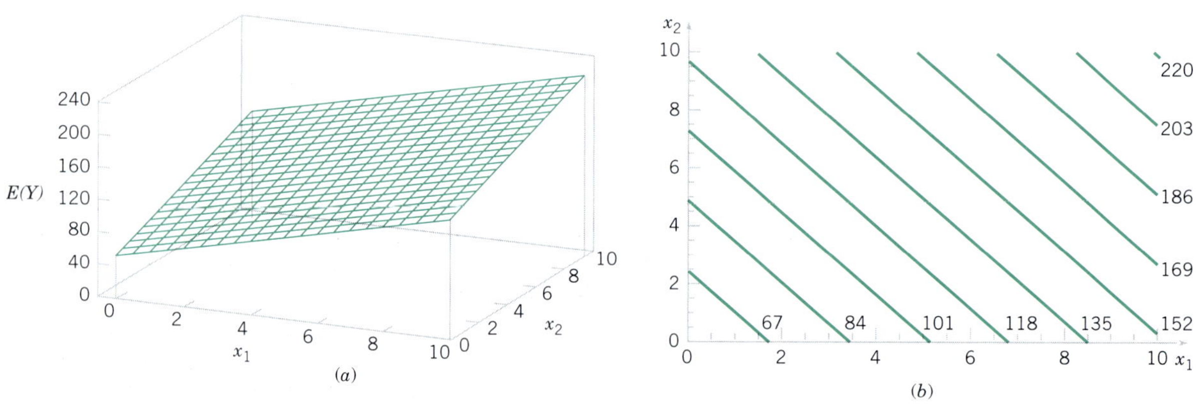

Figure 12-1 (a) The regression plane for the model $E(Y) = 50 + 10x_1 + 7x_2$. (b) The contour plot.

of constant $E(Y)$ as a function of x_1 and x_2. Notice that the contour lines in this plot are straight lines.

In general, the **dependent variable** or **response** Y may be related to k **independent** or **regressor variables.** The model

$$Y = \beta_0 + \beta_1 x_1 + \beta_2 x_2 + \cdots + \beta_k x_k + \epsilon \tag{12-2}$$

is called a multiple linear regression model with k regressor variables. The parameters β_j, $j = 0, 1, \ldots, k$, are called the regression coefficients. This model describes a hyperplane in the k-dimensional space of the regressor variables $\{x_j\}$. The parameter β_j represents the expected change in response Y per unit change in x_j when all the remaining regressors x_i $(i \neq j)$ are held constant.

Multiple linear regression models are often used as approximating functions. That is, the true functional relationship between Y and $x_1, x_2, \ldots, x_k$ is unknown, but over certain ranges of the independent variables the linear regression model is an adequate approximation.

Models that are more complex in structure than Equation 12-2 may often still be analyzed by multiple linear regression techniques. For example, consider the cubic polynomial model in one regressor variable.

$$Y = \beta_0 + \beta_1 x + \beta_2 x^2 + \beta_3 x^3 + \epsilon \tag{12-3}$$

If we let $x_1 = x, x_2 = x^2, x_3 = x^3$, Equation 12-3 can be written as

$$Y = \beta_0 + \beta_1 x_1 + \beta_2 x_2 + \beta_3 x_3 + \epsilon \tag{12-4}$$

which is a multiple linear regression model with three regressor variables.

Models that include **interaction** effects may also be analyzed by multiple linear regression methods. An interaction between two variables can be represented by a cross-product term in the model, such as

$$Y = \beta_0 + \beta_1 x_1 + \beta_2 x_2 + \beta_{12} x_1 x_2 + \epsilon \tag{12-5}$$

If we let $x_3 = x_1 x_2$ and $\beta_3 = \beta_{12}$, Equation 12-5 can be written as

$$Y = \beta_0 + \beta_1 x_1 + \beta_2 x_2 + \beta_3 x_3 + \epsilon$$

which is a linear regression model.

Figure 12-2(a) and (b) shows the three-dimensional plot of the regression model

$$Y = 50 + 10x_1 + 7x_2 + 5x_1 x_2$$

and the corresponding two-dimensional contour plot. Notice that, although this model is a linear regression model, the shape of the surface that is generated by the model is not linear. In general, **any regression model that is linear in parameters** (the β's) **is a linear regression model, regardless of the shape of the surface that it generates.**

Figure 12-2 provides a nice graphical interpretation of an interaction. Generally, interaction implies that the effect produced by changing one variable (x_1, say) depends on the level of the other variable (x_2). For example, Fig. 12-2 shows that changing x_1 from 2 to 8 produces a much smaller change in $E(Y)$ when $x_2 = 2$ than when $x_2 = 10$. Interaction effects occur

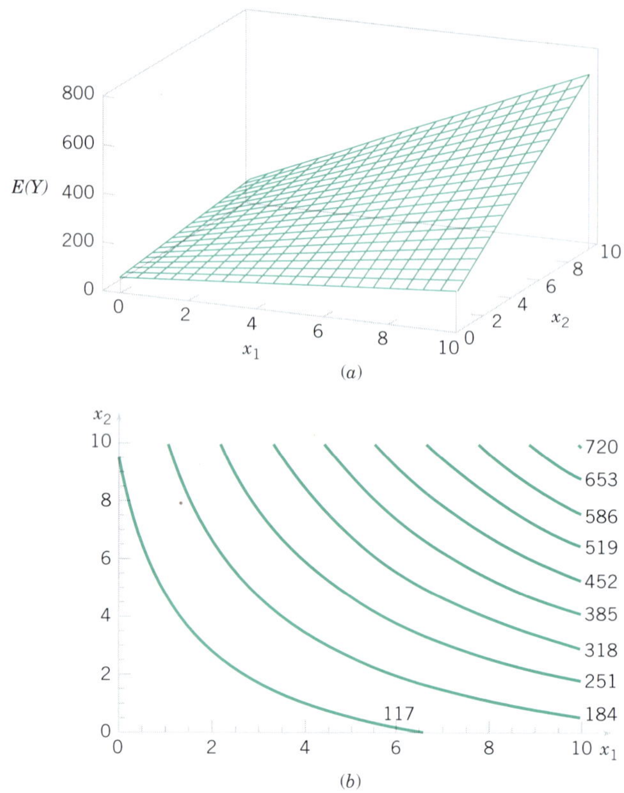

Figure 12-2 (a) Three-dimensional plot of the regression model $E(Y) = 50 + 10x_1 + 7x_2 + 5x_1x_2$. (b) The contour plot.

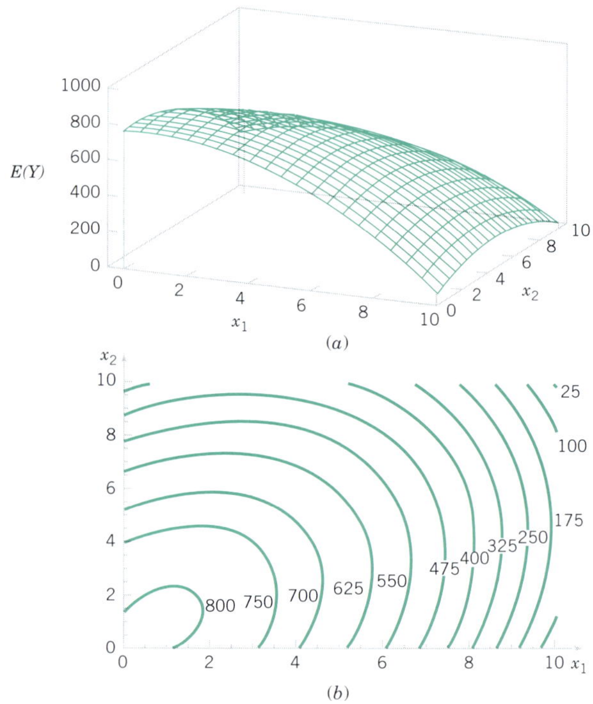

Figure 12-3 (a) Three-dimensional plot of the regression model $E(Y) = 800 + 10x_1 + 7x_2 - 8.5x_1^2 - 5x_2^2 + 4x_1x_2$. (b) The contour plot.

frequently in the study and analysis of real-world systems, and regression methods are one of the techniques that we can use to describe them.

As a final example, consider the second-order model with interaction

$$Y = \beta_0 + \beta_1 x_1 + \beta_2 x_2 + \beta_{11} x_1^2 + \beta_{22} x_2^2 + \beta_{12} x_1 x_2 + \epsilon \tag{12-6}$$

If we let $x_3 = x_1^2$, $x_4 = x_2^2$, $x_5 = x_1x_2$, $\beta_3 = \beta_{11}$, $\beta_4 = \beta_{22}$, and $\beta_5 = \beta_{12}$, Equation 12-6 can be written as a multiple linear regression model as follows:

$$Y = \beta_0 + \beta_1 x_1 + \beta_2 x_2 + \beta_3 x_3 + \beta_4 x_4 + \beta_5 x_5 + \epsilon$$

Figure 12-3(a) and (b) show the three-dimensional plot and the corresponding contour plot for

$$E(Y) = 800 + 10x_1 + 7x_2 - 8.5x_1^2 - 5x_2^2 + 4x_1x_2$$

These plots indicate that the expected change in Y when x_1 is changed by one unit (say) is a function of *both* x_1 and x_2. The quadratic and interaction terms in this model produce a mound-shaped function. Depending on the values of the regression coefficients, the second-order model with interaction is capable of assuming a wide variety of shapes; thus, it is a very flexible regression model.

12-1.2 Least Squares Estimation of the Parameters

The **method of least squares** may be used to estimate the regression coefficients in the multiple regression model, Equation 12-2. Suppose that $n > k$ observations are available, and let x_{ij} denote the ith observation or level of variable x_j. The observations are

$$(x_{i1}, x_{i2}, \ldots, x_{ik}, y_i), \qquad i = 1, 2, \ldots, n \quad \text{and} \quad n > k$$

It is customary to present the data for multiple regression in a table such as Table 12-1.

Each observation $(x_{i1}, x_{i2}, \ldots, x_{ik}, y_i)$, satisfies the model in Equation 12-2, or

$$\begin{aligned} y_i &= \beta_0 + \beta_1 x_{i1} + \beta_2 x_{i2} + \cdots + \beta_k x_{ik} + \epsilon_i \\ &= \beta_0 + \sum_{j=1}^{k} \beta_j x_{ij} + \epsilon_i \qquad i = 1, 2, \ldots, n \end{aligned} \tag{12-7}$$

The least squares function is

$$L = \sum_{i=1}^{n} \epsilon_i^2 = \sum_{i=1}^{n} \left(y_i - \beta_0 - \sum_{j=1}^{k} \beta_j x_{ij} \right)^2 \tag{12-8}$$

We want to minimize L with respect to $\beta_0, \beta_1, \ldots, \beta_k$. The **least squares estimates** of $\beta_0, \beta_1, \ldots, \beta_k$ must satisfy

$$\left.\frac{\partial L}{\partial \beta_0}\right|_{\hat{\beta}_0, \hat{\beta}_1, \ldots, \hat{\beta}_k} = -2 \sum_{i=1}^{n} \left(y_i - \hat{\beta}_0 - \sum_{j=1}^{k} \hat{\beta}_j x_{ij} \right) = 0 \tag{12-9a}$$

and

$$\left.\frac{\partial L}{\partial \beta_j}\right|_{\hat{\beta}_0, \hat{\beta}_1, \ldots, \hat{\beta}_k} = -2 \sum_{i=1}^{n} \left(y_i - \hat{\beta}_0 - \sum_{j=1}^{k} \hat{\beta}_j x_{ij} \right) x_{ij} = 0 \qquad j = 1, 2, \ldots, k \tag{12-9b}$$

Simplifying Equation 12-9, we obtain the **least squares normal equations**

$$\begin{array}{cccccc} n\hat{\beta}_0 + \hat{\beta}_1 \sum_{i=1}^{n} x_{i1} & + \hat{\beta}_2 \sum_{i=1}^{n} x_{i2} & + \cdots + \hat{\beta}_k \sum_{i=1}^{n} x_{ik} & = \sum_{i=1}^{n} y_i \\ \hat{\beta}_0 \sum_{i=1}^{n} x_{i1} + \hat{\beta}_1 \sum_{i=1}^{n} x_{i1}^2 & + \hat{\beta}_2 \sum_{i=1}^{n} x_{i1} x_{i2} & + \cdots + \hat{\beta}_k \sum_{i=1}^{n} x_{i1} x_{ik} & = \sum_{i=1}^{n} x_{i1} y_i \\ \vdots & \vdots & \vdots & \vdots \\ \hat{\beta}_0 \sum_{i=1}^{n} x_{ik} + \hat{\beta}_1 \sum_{i=1}^{n} x_{ik} x_{i1} & + \hat{\beta}_2 \sum_{i=1}^{n} x_{ik} x_{i2} & + \cdots + \hat{\beta}_k \sum_{i=1}^{n} x_{ik}^2 & = \sum_{i=1}^{n} x_{ik} y_i \end{array} \tag{12-10}$$

Table 12-1 Data for Multiple Linear Regression

y	x_1	x_2	...	x_k
y_1	x_{11}	x_{12}	...	x_{1k}
y_2	x_{21}	x_{22}	...	x_{2k}
$\vdots$	$\vdots$	$\vdots$		$\vdots$
y_n	x_{n1}	x_{n2}	...	x_{nk}

Note that there are $p = k + 1$ normal equations, one for each of the unknown regression coefficients. The solution to the normal equations will be the **least squares estimators** of the regression coefficients, $\hat{\beta}_0, \hat{\beta}_1, \ldots, \hat{\beta}_k$. The normal equations can be solved by any method appropriate for solving a system of linear equations.

EXAMPLE 12-1
Wire Bond Strength

In Chapter 1, we used data on pull strength of a wire bond in a semiconductor manufacturing process, wire length, and die height to illustrate building an empirical model. We will use the same data, repeated for convenience in Table 12-2, and show the details of estimating the model parameters. A three-dimensional scatter plot of the data is presented in Fig. 1-15. Figure 12-4 shows a matrix of two-dimensional scatter plots of the data. These displays can be helpful in visualizing the relationships among variables in a multivariable data set. For example, the plot indicates that there is a strong linear relationship between strength and wire length.

Specifically, we will fit the multiple linear regression model

$$Y = \beta_0 + \beta_1 x_1 + \beta_2 x_2 + \epsilon$$

where Y = pull strength, x_1 = wire length, and x_2 = die height. From the data in Table 12-2 we calculate

$$n = 25, \quad \sum_{i=1}^{25} y_i = 725.82$$

$$\sum_{i=1}^{25} x_{i1} = 206, \quad \sum_{i=1}^{25} x_{i2} = 8{,}294$$

$$\sum_{i=1}^{25} x_{i1}^2 = 2{,}396, \quad \sum_{i=1}^{25} x_{i2}^2 = 3{,}531{,}848$$

$$\sum_{i=1}^{25} x_{i1}x_{i2} = 77{,}177, \quad \sum_{i=1}^{25} x_{i1}y_i = 8{,}008.47, \quad \sum_{i=1}^{25} x_{i2}y_i = 274{,}816.71$$

Table 12-2 Wire Bond Data for Example 11-1

Observation Number	Pull Strength y	Wire Length x_1	Die Height x_2	Observation Number	Pull Strength y	Wire Length x_1	Die Height x_2
1	9.95	2	50	14	11.66	2	360
2	24.45	8	110	15	21.65	4	205
3	31.75	11	120	16	17.89	4	400
4	35.00	10	550	17	69.00	20	600
5	25.02	8	295	18	10.30	1	585
6	16.86	4	200	19	34.93	10	540
7	14.38	2	375	20	46.59	15	250
8	9.60	2	52	21	44.88	15	290
9	24.35	9	100	22	54.12	16	510
10	27.50	8	300	23	56.63	17	590
11	17.08	4	412	24	22.13	6	100
12	37.00	11	400	25	21.15	5	400
13	41.95	12	500				

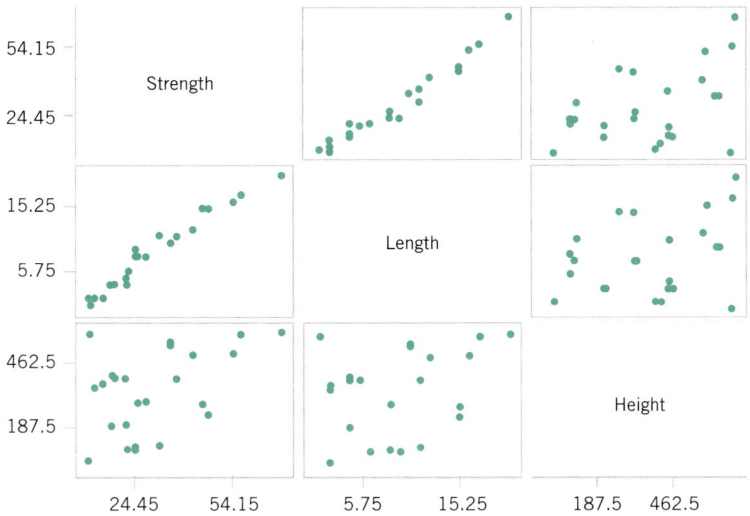

Figure 12-4 Matrix of scatter plots (from Minitab) for the wire bond pull strength data in Table 12-2.

For the model $Y = \beta_0 + \beta_1 x_1 + \beta_2 x_2 + \epsilon$, the normal equations 12-10 are

$$
\begin{aligned}
n\hat{\beta}_0 + \hat{\beta}_1 \sum_{i=1}^{n} x_{i1} + \hat{\beta}_2 \sum_{i=1}^{n} x_{i2} &= \sum_{i=1}^{n} y_i \\
\hat{\beta}_0 \sum_{i=1}^{n} x_{i1} + \hat{\beta}_1 \sum_{i=1}^{n} x_{i1}^2 + \hat{\beta}_2 \sum_{i=1}^{n} x_{i1}x_{i2} &= \sum_{i=1}^{n} x_{i1}y_i \\
\hat{\beta}_0 \sum_{i=1}^{n} x_{i2} + \hat{\beta}_1 \sum_{i=1}^{n} x_{i1}x_{i2} + \hat{\beta}_2 \sum_{i=1}^{n} x_{i2}^2 &= \sum_{i=1}^{n} x_{i2}y_i
\end{aligned}
$$

Inserting the computed summations into the normal equations, we obtain

$$
\begin{aligned}
25\hat{\beta}_0 + 206\hat{\beta}_1 + 8294\hat{\beta}_2 &= 725.82 \\
206\hat{\beta}_0 + 2396\hat{\beta}_1 + 77{,}177\hat{\beta}_2 &= 8{,}008.47 \\
8294\hat{\beta}_0 + 77{,}177\hat{\beta}_1 + 3{,}531{,}848\hat{\beta}_2 &= 274{,}816.71
\end{aligned}
$$

The solution to this set of equations is

$$\hat{\beta}_0 = 2.26379, \quad \hat{\beta}_1 = 2.74427, \quad \hat{\beta}_2 = 0.01253$$

Therefore, the fitted regression equation is

$$\hat{y} = 2.26379 + 2.74427x_1 + 0.01253x_2$$

This equation can be used to predict pull strength for pairs of values of the regressor variables wire length (x_1) and die height (x_2). This is essentially the same regression model given in Section 1-3. Figure 1-16 shows a three-dimensional plot of the plane of predicted values $\hat{y}$ generated from this equation.

12-1.3 Matrix Approach to Multiple Linear Regression

In fitting a multiple regression model, it is much more convenient to express the mathematical operations using **matrix notation.** Suppose that there are k regressor variables and n observations, $(x_{i1}, x_{i2}, \ldots, x_{ik}, y_i)$, $i = 1, 2, \ldots, n$ and that the model relating the regressors to the response is

$$y_i = \beta_0 + \beta_1 x_{i1} + \beta_2 x_{i2} + \cdots + \beta_k x_{ik} + \epsilon_i \qquad i = 1, 2, \ldots, n$$

This model is a system of n equations that can be expressed in matrix notation as

$$\mathbf{y} = \mathbf{X}\boldsymbol{\beta} + \boldsymbol{\epsilon} \tag{12-11}$$

where

$$\mathbf{y} = \begin{bmatrix} y_1 \\ y_2 \\ \vdots \\ y_n \end{bmatrix} \qquad \mathbf{X} = \begin{bmatrix} 1 & x_{11} & x_{12} & \cdots & x_{1k} \\ 1 & x_{21} & x_{22} & \cdots & x_{2k} \\ \vdots & \vdots & \vdots & & \vdots \\ 1 & x_{n1} & x_{n2} & \cdots & x_{nk} \end{bmatrix} \qquad \boldsymbol{\beta} = \begin{bmatrix} \beta_0 \\ \beta_1 \\ \vdots \\ \beta_k \end{bmatrix} \quad \text{and} \quad \boldsymbol{\epsilon} = \begin{bmatrix} \epsilon_1 \\ \epsilon_2 \\ \vdots \\ \epsilon_n \end{bmatrix}$$

In general, $\mathbf{y}$ is an $(n \times 1)$ vector of the observations, $\mathbf{X}$ is an $(n \times p)$ matrix of the levels of the independent variables, $\boldsymbol{\beta}$ is a $(p \times 1)$ vector of the regression coefficients, and $\boldsymbol{\epsilon}$ is a $(n \times 1)$ vector of random errors.

We wish to find the vector of least squares estimators, $\hat{\boldsymbol{\beta}}$, that minimizes

$$L = \sum_{i=1}^{n} \epsilon_i^2 = \boldsymbol{\epsilon}'\boldsymbol{\epsilon} = (\mathbf{y} - \mathbf{X}\boldsymbol{\beta})'(\mathbf{y} - \mathbf{X}\boldsymbol{\beta})$$

The least squares estimator $\hat{\boldsymbol{\beta}}$ is the solution for $\boldsymbol{\beta}$ in the equations

$$\frac{\partial L}{\partial \boldsymbol{\beta}} = \mathbf{0}$$

We will not give the details of taking the derivatives above; however, the resulting equations that must be solved are

Normal Equations

$$\mathbf{X}'\mathbf{X}\hat{\boldsymbol{\beta}} = \mathbf{X}'\mathbf{y} \tag{12-12}$$

Equations 12-12 are the least squares normal equations in matrix form. They are identical to the scalar form of the normal equations given earlier in Equations 12-10. To solve the normal equations, multiply both sides of Equations 12-12 by the inverse of $\mathbf{X}'\mathbf{X}$. Therefore, the least squares estimate of $\boldsymbol{\beta}$ is

Least Square Estimate of β

$$\hat{\boldsymbol{\beta}} = (\mathbf{X}'\mathbf{X})^{-1}\mathbf{X}'\mathbf{y} \tag{12-13}$$

Note that there are $p = k + 1$ normal equations in $p = k + 1$ unknowns (the values of $\hat{\beta}_0, \hat{\beta}_1, \ldots, \hat{\beta}_k$). Furthermore, the matrix $\mathbf{X}'\mathbf{X}$ is always nonsingular, as was assumed above, so the methods described in textbooks on determinants and matrices for inverting these matrices can be used to find $(\mathbf{X}'\mathbf{X})^{-1}$. In practice, multiple regression calculations are almost always performed using a computer.

It is easy to see that the matrix form of the normal equations is identical to the scalar form. Writing out Equation 12-12 in detail, we obtain

$$\begin{bmatrix} n & \sum_{i=1}^{n} x_{i1} & \sum_{i=1}^{n} x_{i2} & \cdots & \sum_{i=1}^{n} x_{ik} \\ \sum_{i=1}^{n} x_{i1} & \sum_{i=1}^{n} x_{i1}^2 & \sum_{i=1}^{n} x_{i1}x_{i2} & \cdots & \sum_{i=1}^{n} x_{i1}x_{ik} \\ \vdots & \vdots & \vdots & & \vdots \\ \sum_{i=1}^{n} x_{ik} & \sum_{i=1}^{n} x_{ik}x_{i1} & \sum_{i=1}^{n} x_{ik}x_{i2} & \cdots & \sum_{i=1}^{n} x_{ik}^2 \end{bmatrix} \begin{bmatrix} \hat{\beta}_0 \\ \hat{\beta}_1 \\ \vdots \\ \hat{\beta}_k \end{bmatrix} = \begin{bmatrix} \sum_{i=1}^{n} y_i \\ \sum_{i=1}^{n} x_{i1}y_i \\ \vdots \\ \sum_{i=1}^{n} x_{ik}y_i \end{bmatrix}$$

If the indicated matrix multiplication is performed, the scalar form of the normal equations (that is, Equation 12-10) will result. In this form it is easy to see that $\mathbf{X}'\mathbf{X}$ is a $(p \times p)$ symmetric matrix and $\mathbf{X}'\mathbf{y}$ is a $(p \times 1)$ column vector. Note the special structure of the $\mathbf{X}'\mathbf{X}$ matrix. The diagonal elements of $\mathbf{X}'\mathbf{X}$ are the sums of squares of the elements in the columns of $\mathbf{X}$, and the off-diagonal elements are the sums of cross-products of the elements in the columns of $\mathbf{X}$. Furthermore, note that the elements of $\mathbf{X}'\mathbf{y}$ are the sums of cross-products of the columns of $\mathbf{X}$ and the observations $\{y_i\}$.

The fitted regression model is

$$\hat{y}_i = \hat{\beta}_0 + \sum_{j-1}^{k} \hat{\beta}_j x_{ij} \qquad i = 1, 2, \ldots, n \tag{12-14}$$

In matrix notation, the fitted model is

$$\hat{\mathbf{y}} = \mathbf{X}\hat{\boldsymbol{\beta}}$$

The difference between the observation y_i and the fitted value $\hat{y}_i$ is a **residual,** say, $e_i = y_i - \hat{y}_i$. The $(n \times 1)$ vector of residuals is denoted by

$$\mathbf{e} = \mathbf{y} - \hat{\mathbf{y}} \tag{12-15}$$

EXAMPLE 12-2
Wire Bond Strength with Matrix Notation

In Example 12-1, we illustrated fitting the multiple regression model

$$y = \beta_0 + \beta_1 x_1 + \beta_2 x_2 + \epsilon$$

where y is the observed pull strength for a wire bond, x_1 is the wire length, and x_2 is the die height. The 25 observations are in Table 12-2. We will now use the matrix approach to fit the regression model above to these data. The $\mathbf{X}$ matrix and $\mathbf{y}$ vector for this model are

$$\mathbf{X} = \begin{bmatrix} 1 & 2 & 50 \\ 1 & 8 & 110 \\ 1 & 11 & 120 \\ 1 & 10 & 550 \\ 1 & 8 & 295 \\ 1 & 4 & 200 \\ 1 & 2 & 375 \\ 1 & 2 & 52 \\ 1 & 9 & 100 \\ 1 & 8 & 300 \\ 1 & 4 & 412 \\ 1 & 11 & 400 \\ 1 & 12 & 500 \\ 1 & 2 & 360 \\ 1 & 4 & 205 \\ 1 & 4 & 400 \\ 1 & 20 & 600 \\ 1 & 1 & 585 \\ 1 & 10 & 540 \\ 1 & 15 & 250 \\ 1 & 15 & 290 \\ 1 & 16 & 510 \\ 1 & 17 & 590 \\ 1 & 6 & 100 \\ 1 & 5 & 400 \end{bmatrix} \quad \mathbf{y} = \begin{bmatrix} 9.95 \\ 24.45 \\ 31.75 \\ 35.00 \\ 25.02 \\ 16.86 \\ 14.38 \\ 9.60 \\ 24.35 \\ 27.50 \\ 17.08 \\ 37.00 \\ 41.95 \\ 11.66 \\ 21.65 \\ 17.89 \\ 69.00 \\ 10.30 \\ 34.93 \\ 46.59 \\ 44.88 \\ 54.12 \\ 56.63 \\ 22.13 \\ 21.15 \end{bmatrix}$$

The $\mathbf{X}'\mathbf{X}$ matrix is

$$\mathbf{X}'\mathbf{X} = \begin{bmatrix} 1 & 1 & \cdots & 1 \\ 2 & 8 & \cdots & 5 \\ 50 & 110 & \cdots & 400 \end{bmatrix} \begin{bmatrix} 1 & 2 & 50 \\ 1 & 8 & 110 \\ \vdots & \vdots & \vdots \\ 1 & 5 & 400 \end{bmatrix} = \begin{bmatrix} 25 & 206 & 8{,}294 \\ 206 & 2{,}396 & 77{,}177 \\ 8{,}294 & 77{,}177 & 3{,}531{,}848 \end{bmatrix}$$

and the $\mathbf{X}'\mathbf{y}$ vector is

$$\mathbf{X}'\mathbf{y} = \begin{bmatrix} 1 & 1 & \cdots & 1 \\ 2 & 8 & \cdots & 5 \\ 50 & 110 & \cdots & 400 \end{bmatrix} \begin{bmatrix} 9.95 \\ 24.45 \\ \vdots \\ 21.15 \end{bmatrix} = \begin{bmatrix} 725.82 \\ 8{,}008.47 \\ 274{,}816.71 \end{bmatrix}$$

The least squares estimates are found from Equation 12-13 as

$$\hat{\boldsymbol{\beta}} = (\mathbf{X}'\mathbf{X})^{-1}\mathbf{X}'\mathbf{y}$$

or

$$\begin{bmatrix} \hat{\beta}_0 \\ \hat{\beta}_1 \\ \hat{\beta}_2 \end{bmatrix} = \begin{bmatrix} 25 & 206 & 8,294 \\ 206 & 2,396 & 77,177 \\ 8,294 & 77,177 & 3,531,848 \end{bmatrix}^{-1} \begin{bmatrix} 725.82 \\ 8,008.37 \\ 274,811.31 \end{bmatrix}$$

$$= \begin{bmatrix} 0.214653 & -0.007491 & -0.000340 \\ -0.007491 & 0.001671 & -0.000019 \\ -0.000340 & -0.000019 & +0.0000015 \end{bmatrix} \begin{bmatrix} 725.82 \\ 8,008.47 \\ 274,811.31 \end{bmatrix} = \begin{bmatrix} 2.26379143 \\ 2.74426964 \\ 0.01252781 \end{bmatrix}$$

Therefore, the fitted regression model with the regression coefficients rounded to five decimal places is

$$\hat{y} = 2.26379 + 2.74427x_1 + 0.01253x_2$$

This is identical to the results obtained in Example 12-1.

This regression model can be used to predict values of pull strength for various values of wire length (x_1) and die height (x_2). We can also obtain the **fitted values** $\hat{y}_i$ by substituting each observation (x_{i1}, x_{i2}), $i = 1, 2, \ldots, n$, into the equation. For example, the first observation has $x_{11} = 2$ and $x_{12} = 50$, and the fitted value is

$$\begin{aligned} \hat{y}_1 &= 2.26379 + 2.74427x_{11} + 0.01253x_{12} \\ &= 2.26379 + 2.74427(2) + 0.01253(50) \\ &= 8.38 \end{aligned}$$

The corresponding observed value is $y_1 = 9.95$. The *residual* corresponding to the first observation is

$$\begin{aligned} e_1 &= y_1 - \hat{y}_1 \\ &= 9.95 - 8.38 \\ &= 1.57 \end{aligned}$$

Table 12-3 displays all 25 fitted values $\hat{y}_i$ and the corresponding residuals. The fitted values and residuals are calculated to the same accuracy as the original data.

Table 12-3 Observations, Fitted Values, and Residuals for Example 12-2

Observation Number	y_i	$\hat{y}_i$	$e_i = y_i - \hat{y}_i$	Observation Number	y_i	$\hat{y}_i$	$e_i = y_i - \hat{y}_i$
1	9.95	8.38	1.57	14	11.66	12.26	−0.60
2	24.45	25.60	−1.15	15	21.65	15.81	5.84
3	31.75	33.95	−2.20	16	17.89	18.25	−0.36
4	35.00	36.60	−1.60	17	69.00	64.67	4.33
5	25.02	27.91	−2.89	18	10.30	12.34	−2.04
6	16.86	15.75	1.11	19	34.93	36.47	−1.54
7	14.38	12.45	1.93	20	46.59	46.56	0.03
8	9.60	8.40	1.20	21	44.88	47.06	−2.18
9	24.35	28.21	−3.86	22	54.12	52.56	1.56
10	27.50	27.98	−0.48	23	56.63	56.31	0.32
11	17.08	18.40	−1.32	24	22.13	19.98	2.15
12	37.00	37.46	−0.46	25	21.15	21.00	0.15
13	41.95	41.46	0.49				

Table 12-4 Minitab Multiple Regression Output for the Wire Bond Pull Strength Data

Regression Analysis: Strength versus Length, Height

The regression equation is
Strength = 2.26 + 2.74 Length + 0.0125 Height

Predictor	Coef	SE Coef	T	P	VIF
Constant	$\hat{\beta}_0 \rightarrow$ 2.264	1.060	2.14	0.044	
Length	$\hat{\beta}_1 \rightarrow$ 2.74427	0.09352	29.34	0.000	1.2
Height	$\hat{\beta}_2 \rightarrow$ 0.012528	0.002798	4.48	0.000	1.2

S = 2.288   R-Sq = 98.1%   R-Sq (adj) = 97.9%
PRESS = 156.163   R-Sq (pred) = 97.44%

Analysis of Variance

Source	DF	SS	MS	F	P
Regression	2	5990.8	2995.4	572.17	0.000
Residual Error	22	115.2	5.2 $\leftarrow \hat{\sigma}^2$		
Total	24	6105.9			

Source	DF	Seq SS
Length	1	5885.9
Height	1	104.9

Predicted Values for New Observations

New Obs	Fit	SE Fit	95.0% CI	95.0% PI
1	27.663	0.482	(26.663, 28.663)	(22.814, 32.512)

Values of Predictors for New Observations

New Obs	Length	Height
1	8.00	275

Computers are almost always used in fitting multiple regression models. Table 12-4 presents some annotated output from Minitab for the least squares regression model for wire bond pull strength data. The upper part of the table contains the numerical estimates of the regression coefficients. The computer also calculates several other quantities that reflect important information about the regression model. In subsequent sections, we will define and explain the quantities in this output.

Estimating σ^2

Just as in simple linear regression, it is important to estimate σ^2, the variance of the error term ϵ, in a multiple regression model. Recall that in simple linear regression the estimate of σ^2 was obtained by dividing the sum of the squared residuals by $n - 2$. Now there are two parameters in the simple linear regression model, so in multiple linear regression with p parameters a logical estimator for σ^2 is

Estimator of Variance

$$\hat{\sigma}^2 = \frac{\sum_{i=1}^{n} e_i^2}{n - p} = \frac{SS_E}{n - p} \tag{12-16}$$

This is an **unbiased estimator** of σ^2. Just as in simple linear regression, the estimate of σ^2 is usually obtained from the **analysis of variance** for the regression model. The numerator of Equation 12-16 is called the **error** or **residual sum of squares**, and the denominator $n - p$ is called the **error** or **residual degrees of freedom.**

We can find a computing formula for SS_E as follows:

$$SS_E = \sum_{i=1}^{n} (y_i - \hat{y}_i)^2 = \sum_{i=1}^{n} e_i^2 = \mathbf{e}'\mathbf{e}$$

Substituting $\mathbf{e} = \mathbf{y} - \hat{\mathbf{y}} = \mathbf{y} - \mathbf{X}\hat{\boldsymbol{\beta}}$ into the above, we obtain

$$\begin{aligned} SS_E &= \mathbf{y}'\mathbf{y} - \hat{\boldsymbol{\beta}}'\mathbf{X}'\mathbf{y} \\ &= 27{,}178.5316 - 27{,}063.3581 = 115.174 \end{aligned} \tag{12-17}$$

Table 12-4 shows that the estimate of σ^2 for the wire bond pull strength regression model is $\hat{\sigma}^2 = 115.2/22 = 5.2364$. The Minitab output rounds the estimate to $\hat{\sigma}^2 = 5.2$.

12-1.4 Properties of the Least Squares Estimators

The statistical properties of the least squares estimators $\hat{\beta}_0, \hat{\beta}_1, \ldots, \hat{\beta}_k$ may be easily found, under certain assumptions on the error terms $\epsilon_1, \epsilon_2, \ldots, \epsilon_n$, in the regression model. Paralleling the assumptions made in Chapter 11, we assume that the errors ϵ_i are statistically independent with mean zero and variance σ^2. Under these assumptions, the least squares estimators $\hat{\beta}_0, \hat{\beta}_1, \ldots, \hat{\beta}_k$ are **unbiased estimators** of the regression coefficients $\beta_0, \beta_1, \ldots, \beta_k$. This property may be shown as follows:

$$\begin{aligned} E(\hat{\boldsymbol{\beta}}) &= E[(\mathbf{X}'\mathbf{X})^{-1}\mathbf{X}'\mathbf{Y}] \\ &= E[(\mathbf{X}'\mathbf{X})^{-1}\mathbf{X}'(\mathbf{X}\boldsymbol{\beta} + \boldsymbol{\epsilon})] \\ &= E[(\mathbf{X}'\mathbf{X})^{-1}\mathbf{X}'\mathbf{X}\boldsymbol{\beta} + (\mathbf{X}'\mathbf{X})^{-1}\mathbf{X}'\boldsymbol{\epsilon}] \\ &= \boldsymbol{\beta} \end{aligned}$$

since $E(\boldsymbol{\epsilon}) = \mathbf{0}$ and $(\mathbf{X}'\mathbf{X})^{-1}\mathbf{X}'\mathbf{X} = \mathbf{I}$, the identity matrix. Thus, $\hat{\boldsymbol{\beta}}$ is an unbiased estimator of $\boldsymbol{\beta}$.

The variances of the $\hat{\boldsymbol{\beta}}$'s are expressed in terms of the elements of the inverse of the $\mathbf{X}'\mathbf{X}$ matrix. The inverse of $\mathbf{X}'\mathbf{X}$ times the constant σ^2 represents the **covariance matrix** of the regression coefficients $\hat{\boldsymbol{\beta}}$. The diagonal elements of $\sigma^2(\mathbf{X}'\mathbf{X})^{-1}$ are the variances of $\hat{\beta}_0, \hat{\beta}_1, \ldots, \hat{\beta}_k$, and the off-diagonal elements of this matrix are the covariances. For example, if we have $k = 2$ regressors, such as in the pull-strength problem,

$$\mathbf{C} = (\mathbf{X}'\mathbf{X})^{-1} = \begin{bmatrix} C_{00} & C_{01} & C_{02} \\ C_{10} & C_{11} & C_{12} \\ C_{20} & C_{21} & C_{22} \end{bmatrix}$$

which is symmetric ($C_{10} = C_{01}$, $C_{20} = C_{02}$, and $C_{21} = C_{12}$) because $(\mathbf{X}'\mathbf{X})^{-1}$ is symmetric, and we have

$$\begin{aligned} V(\hat{\beta}_j) &= \sigma^2 C_{jj}, \qquad j = 0, 1, 2 \\ \text{cov}(\hat{\beta}_i, \hat{\beta}_j) &= \sigma^2 C_{ij}, \qquad i \neq j \end{aligned}$$

In general, the covariance matrix of $\hat{\boldsymbol{\beta}}$ is a $(p \times p)$ symmetric matrix whose jjth element is the variance of $\hat{\beta}_j$ and whose i, jth element is the covariance between $\hat{\beta}_i$ and $\hat{\beta}_j$, that is,

$$\text{cov}(\hat{\boldsymbol{\beta}}) = \sigma^2(\mathbf{X}'\mathbf{X})^{-1} = \sigma^2\mathbf{C}$$

The estimates of the variances of these regression coefficients are obtained by replacing σ^2 with an estimate. When σ^2 is replaced by its estimate $\hat{\sigma}^2$, the square root of the estimated variance of the jth regression coefficient is called the **estimated standard error** of $\hat{\beta}_j$ or $se(\hat{\beta}_j) = \sqrt{\hat{\sigma}^2 C_{jj}}$. These standard errors are a useful measure of the **precision of estimation** for the regression coefficients; small standard errors imply good precision.

Multiple regression computer programs usually display these standard errors. For example, the Minitab output in Table 12-4 reports $se(\hat{\beta}_0) = 1.060$, $se(\hat{\beta}_1) = 0.09352$, and $se(\hat{\beta}_2) = 0.002798$. The slope estimate is about twice the magnitude of its standard error, and $\hat{\beta}_1$ and $\hat{\beta}_2$ are considerably larger than $se(\hat{\beta}_1)$ and $se(\hat{\beta}_2)$. This implies reasonable precision of estimation, although the parameters β_1 and β_2 are much more precisely estimated than the intercept (this is not unusual in multiple regression).

EXERCISES FOR SECTION 12-1

12-1. A study was performed to investigate the shear strength of soil (y) as it related to depth in feet (x_1) and % moisture content (x_2). Ten observations were collected, and the following summary quantities obtained: $n = 10$, $\sum x_{i1} = 223$, $\sum x_{i2} = 553$, $\sum y_i = 1{,}916$, $\sum x_{i1}^2 = 5{,}200.9$, $\sum x_{i2}^2 = 31{,}729$, $\sum x_{i1}x_{i2} = 12{,}352$, $\sum x_{i1}y_i = 43{,}550.8$, $\sum x_{i2}y_i = 104{,}736.8$, and $\sum y_i^2 = 371{,}595.6$.

(a) Set up the least squares normal equations for the model $Y = \beta_0 + \beta_1 x_1 + \beta_2 x_2 + \epsilon$.

(b) Estimate the parameters in the model in part (a).

(c) What is the predicted strength when $x_1 = 18$ feet and $x_2 = 43\%$?

12-2. A regression model is to be developed for predicting the ability of soil to absorb chemical contaminants. Ten observations have been taken on a soil absorption index (y) and two regressors: x_1 = amount of extractable iron ore and x_2 = amount of bauxite. We wish to fit the model $Y = \beta_0 + \beta_1 x_1 + \beta_2 x_2 + \epsilon$. Some necessary quantities are:

$$(\mathbf{X}'\mathbf{X})^{-1} = \begin{bmatrix} 1.17991 & -7.30982\text{ E-3} & 7.3006\text{ E-4} \\ -7.30982\text{ E-3} & 7.9799\text{ E-5} & -1.23713\text{ E-4} \\ 7.3006\text{ E-4} & -1.23713\text{ E-4} & 4.6576\text{ E-4} \end{bmatrix},$$

$$\mathbf{X}'\mathbf{y} = \begin{bmatrix} 220 \\ 36{,}768 \\ 9{,}965 \end{bmatrix}$$

(a) Estimate the regression coefficients in the model specified above.

(b) What is the predicted value of the absorption index y when $x_1 = 200$ and $x_2 = 50$?

12-3. A chemical engineer is investigating how the amount of conversion of a product from a raw material (y) depends on reaction temperature (x_1) and the reaction time (x_2). He has developed the following regression models:

1. $\hat{y} = 100 + 2x_1 + 4x_2$
2. $\hat{y} = 95 + 1.5x_1 + 3x_2 + 2x_1x_2$

Both models have been built over the range $0.5 \le x_2 \le 10$.

(a) What is the predicted value of conversion when $x_2 = 2$? Repeat this calculation for $x_2 = 8$. Draw a graph of the predicted values for both conversion models. Comment on the effect of the interaction term in model 2.

(b) Find the expected change in the mean conversion for a unit change in temperature x_1 for model 1 when $x_2 = 5$. Does this quantity depend on the specific value of reaction time selected? Why?

(c) Find the expected change in the mean conversion for a unit change in temperature x_1 for model 2 when $x_2 = 5$. Repeat this calculation for $x_2 = 2$ and $x_2 = 8$. Does the result depend on the value selected for x_2? Why?

12-4. The electric power consumed each month by a chemical plant is thought to be related to the average ambient temperature (x_1), the number of days in the month (x_2), the average product purity (x_3), and the tons of product produced (x_4). The past year's historical data are available and are presented in the following table:

y	x_1	x_2	x_3	x_4
240	25	24	91	100
236	31	21	90	95
270	45	24	88	110
274	60	25	87	88
301	65	25	91	94
316	72	26	94	99
300	80	25	87	97
296	84	25	86	96
267	75	24	88	110
276	60	25	91	105
288	50	25	90	100
261	38	23	89	98

Table 12-5 DaimlerChrysler Fuel Economy and Emissions

mfr	carline	car/truck	cid	rhp	trns	drv	od	etw	cmp	axle	n/v	a/c	hc	co	co2	mpg
20	300C/SRT-8	C	215	253	L5	4	2	4500	9.9	3.07	30.9	Y	0.011	0.09	288	30.8
20	CARAVAN 2WD	T	201	180	L4	F	2	4500	9.3	2.49	32.3	Y	0.014	0.11	274	32.5
20	CROSSFIRE ROADSTER	C	196	168	L5	R	2	3375	10	3.27	37.1	Y	0.001	0.02	250	35.4
20	DAKOTA PICKUP 2WD	T	226	210	L4	R	2	4500	9.2	3.55	29.6	Y	0.012	0.04	316	28.1
20	DAKOTA PICKUP 4WD	T	226	210	L4	4	2	5000	9.2	3.55	29.6	Y	0.011	0.05	365	24.4
20	DURANGO 2WD	T	348	345	L5	R	2	5250	8.6	3.55	27.2	Y	0.023	0.15	367	24.1
20	GRAND CHEROKEE 2WD	T	226	210	L4	R	2	4500	9.2	3.07	30.4	Y	0.006	0.09	312	28.5
20	GRAND CHEROKEE 4WD	T	348	230	L5	4	2	5000	9	3.07	24.7	Y	0.008	0.11	369	24.2
20	LIBERTY/CHEROKEE 2WD	T	148	150	M6	R	2	4000	9.5	4.1	41	Y	0.004	0.41	270	32.8
20	LIBERTY/CHEROKEE 4WD	T	226	210	L4	4	2	4250	9.2	3.73	31.2	Y	0.003	0.04	317	28
20	NEON/SRT-4/SX 2.0	C	122	132	L4	F	2	3000	9.8	2.69	39.2	Y	0.003	0.16	214	41.3
20	PACIFICA 2WD	T	215	249	L4	F	2	4750	9.9	2.95	35.3	Y	0.022	0.01	295	30
20	PACIFICA AWD	T	215	249	L4	4	2	5000	9.9	2.95	35.3	Y	0.024	0.05	314	28.2
20	PT CRUISER	T	148	220	L4	F	2	3625	9.5	2.69	37.3	Y	0.002	0.03	260	34.1
20	RAM 1500 PICKUP 2WD	T	500	500	M6	R	2	5250	9.6	4.1	22.3	Y	0.01	0.1	474	18.7
20	RAM 1500 PICKUP 4WD	T	348	345	L5	4	2	6000	8.6	3.92	29	Y	0	0	0	20.3
20	SEBRING 4-DR	C	165	200	L4	F	2	3625	9.7	2.69	36.8	Y	0.011	0.12	252	35.1
20	STRATUS 4-DR	C	148	167	L4	F	2	3500	9.5	2.69	36.8	Y	0.002	0.06	233	37.9
20	TOWN & COUNTRY 2WD	T	148	150	L4	F	2	4250	9.4	2.69	34.9	Y	0	0.09	262	33.8
20	VIPER CONVERTIBLE	C	500	501	M6	R	2	3750	9.6	3.07	19.4	Y	0.007	0.05	342	25.9
20	WRANGLER/TJ 4WD	T	148	150	M6	4	2	3625	9.5	3.73	40.1	Y	0.004	0.43	337	26.4

mfr: mfr code
carline-car line name (test vehicle model name)
car/truck-'C' for passenger vehicle and 'T' for truck
cid-cubic inch displacement of test vehicle
rhp-rated horsepower
trns-transmission code
drv-drive system code
od-overdrive code
etw-equivalent test weight
cmp-compression ratio
axle-axle ratio
n/v-n/v ratio (engine speed versus vehicle speed at 50 mph)
a/c-indicates air conditioning simulation
hc-HC(hydrocarbon emissions) Test level composite results
co-CO(carbon monoxide emissions) Test level composite results
co2-CO2(carbon dioxide emissions) Test level composite results
mpg-mpg(fuel economy, miles per gallon)

(a) Fit a multiple linear regression model to these data.
(b) Estimate σ^2.
(c) Compute the standard errors of the regression coefficients.
(d) Predict power consumption for a month in which $x_1 = 75°\text{F}$, $x_2 = 24$ days, $x_3 = 90\%$, and $x_4 = 98$ tons.

12-5. Table 12.5 provides the highway gasoline mileage test results for 2005 model year vehicles from DaimlerChrysler. The full table of data (available on the book's Web site) contains the same data for 2005 models from over 250 vehicles from many manufacturers (source: Environmental Protection Agency Web site www.epa.gov/otaq/cert/mpg/testcars/database).

(a) Fit a multiple linear regression model to these data to estimate gasoline mileage that uses the following regressors: *cid, rhp, etw, cmp, axle, n/v.*
(b) Estimate σ^2 and the standard errors of the regression coefficients.
(c) Predict the gasoline mileage for the first vehicle in the table.

12-6. The pull strength of a wire bond is an important characteristic. The following table gives information on pull strength (y), die height (x_1), post height (x_2), loop height (x_3), wire length (x_4), bond width on the die (x_5), and bond width on the post (x_6).

(a) Fit a multiple linear regression model using x_2, x_3, x_4, and x_5 as the regressors.
(b) Estimate σ^2.

y	x_1	x_2	x_3	x_4	x_5	x_6
8.0	5.2	19.6	29.6	94.9	2.1	2.3
8.3	5.2	19.8	32.4	89.7	2.1	1.8
8.5	5.8	19.6	31.0	96.2	2.0	2.0
8.8	6.4	19.4	32.4	95.6	2.2	2.1
9.0	5.8	18.6	28.6	86.5	2.0	1.8
9.3	5.2	18.8	30.6	84.5	2.1	2.1
9.3	5.6	20.4	32.4	88.8	2.2	1.9
9.5	6.0	19.0	32.6	85.7	2.1	1.9
9.8	5.2	20.8	32.2	93.6	2.3	2.1
10.0	5.8	19.9	31.8	86.0	2.1	1.8
10.3	6.4	18.0	32.6	87.1	2.0	1.6
10.5	6.0	20.6	33.4	93.1	2.1	2.1
10.8	6.2	20.2	31.8	83.4	2.2	2.1
11.0	6.2	20.2	32.4	94.5	2.1	1.9
11.3	6.2	19.2	31.4	83.4	1.9	1.8
11.5	5.6	17.0	33.2	85.2	2.1	2.1
11.8	6.0	19.8	35.4	84.1	2.0	1.8
12.3	5.8	18.8	34.0	86.9	2.1	1.8
12.5	5.6	18.6	34.2	83.0	1.9	2.0

(c) Find the $se(\hat{\beta}_j)$. How precisely are the regression coefficients estimated, in your opinion?

(d) Use the model from part (a) to predict pull strength when $x_2 = 20$, $x_3 = 30$, $x_4 = 90$, and $x_5 = 2.0$.

12-7. An engineer at a semiconductor company wants to model the relationship between the device HFE (y) and three parameters: Emitter-RS (x_1), Base-RS (x_2), and Emitter-to-Base RS (x_3). The data are shown in the following table.

x_1 Emitter-RS	x_2 Base-RS	x_3 E-B-RS	y HFE-1M-5V
14.620	226.00	7.000	128.40
15.630	220.00	3.375	52.62
14.620	217.40	6.375	113.90
15.000	220.00	6.000	98.01
14.500	226.50	7.625	139.90
15.250	224.10	6.000	102.60
16.120	220.50	3.375	48.14
15.130	223.50	6.125	109.60
15.500	217.60	5.000	82.68
15.130	228.50	6.625	112.60
15.500	230.20	5.750	97.52
16.120	226.50	3.750	59.06
15.130	226.60	6.125	111.80
15.630	225.60	5.375	89.09
15.380	229.70	5.875	101.00
14.380	234.00	8.875	171.90
15.500	230.00	4.000	66.80
14.250	224.30	8.000	157.10
14.500	240.50	10.870	208.40
14.620	223.70	7.375	133.40

(a) Fit a multiple linear regression model to the data.

(b) Estimate σ^2.

(c) Find the standard errors $se(\hat{\beta}_j)$.

(d) Predict HFE when $x_1 = 14.5$, $x_2 = 220$, and $x_3 = 5.0$.

12-8. Heat treating is often used to carburize metal parts, such as gears. The thickness of the carburized layer is considered a crucial feature of the gear and contributes to the overall reliability of the part. Because of the critical nature of this feature, two different lab tests are performed on each furnace load. One test is run on a sample pin that accompanies each load. The other test is a destructive test, where an actual part is cross-sectioned. This test involves running a carbon analysis on the surface of both the gear pitch (top of the gear tooth) and the gear root (between the gear teeth). Table 12-6 shows the results of the pitch carbon analysis test for 32 parts.

The regressors are furnace temperature (TEMP), carbon concentration and duration of the carburizing cycle (SOAKPCT, SOAKTIME), and carbon concentration and duration of the diffuse cycle (DIFFPCT, DIFFTIME).

(a) Fit a linear regression model relating the results of the pitch carbon analysis test (PITCH) to the five regressor variables.

(b) Estimate σ^2.

(c) Find the standard errors $se(\hat{\beta}_j)$.

(d) Use the model in part (a) to predict PITCH when TEMP = 1650, SOAKTIME = 1.00, SOAKPCT = 1.10, DIFFTIME = 1.00, and DIFFPCT = 0.80.

12-9. An article in *Electronic Packaging and Production* (2002, Vol. 42) considered the effect of X-ray inspection of integrated circuits. The rads (radiation dose) were studied as a function of current (in milliamps) and exposure time (in minutes).

Rads	mAmps	Exposure Time
7.4	10	0.25
14.8	10	0.5
29.6	10	1
59.2	10	2
88.8	10	3
296	10	10
444	10	15
592	10	20
11.1	15	0.25
22.2	15	0.5
44.4	15	1

continued

Rads	mAmps	Exposure Time
88.8	15	2
133.2	15	3
444	15	10
666	15	15
888	15	20
14.8	20	0.25
29.6	20	0.5
59.2	20	1
118.4	20	2
177.6	20	3
592	20	10
888	20	15
1184	20	20
22.2	30	0.25
44.4	30	0.5
88.8	30	1
177.6	30	2
266.4	30	3
888	30	10
1332	30	15
1776	30	20
29.6	40	0.25
59.2	40	0.5
118.4	40	1
236.8	40	2
355.2	40	3
1184	40	10
1776	40	15
2368	40	20

(a) Fit a multiple linear regression model to these data with rads as the response.
(b) Estimate σ^2 and the standard errors of the regression coefficients.
(c) Use the model to predict rads when the current is 15 milliamps and the exposure time is 5 seconds.

12-10. An article in *Cancer Epidemiology, Biomarkers and Prevention* (1996, Vol. 5, pp. 849–852) conducted a pilot study to assess the use of toenail arsenic concentrations as an indicator of ingestion of arsenic-containing water. Twenty-one participants were interviewed regarding use of their private (unregulated) wells for drinking and cooking, and each provided a sample of water and toenail clippings. The table below showed the data of age (years), sex of person (1 = male, 2 = female), proportion of times household well used for drinking ($1 \leq 1/4$, $2 = 1/4$, $3 = 1/2$, $4 = 3/4$, $5 \geq 3/4$), proportion of times household well used for cooking ($1 \leq 1/4$, $2 = 1/4$, $3 = 1/2$, $4 = 3/4$, $5 \geq 3/4$), arsenic in water (ppm), and arsenic in toenails (ppm) respectively.

Age	Sex	Drink Use	Cook Use	Arsenic Water	Arsenic Nails
44	2	5	5	0.00087	0.119
45	2	4	5	0.00021	0.118
44	1	5	5	0	0.099
66	2	3	5	0.00115	0.118
37	1	2	5	0	0.277
45	2	5	5	0	0.358
47	1	5	5	0.00013	0.08
38	2	4	5	0.00069	0.158
41	2	3	2	0.00039	0.31
49	2	4	5	0	0.105
72	2	5	5	0	0.073
45	2	1	5	0.046	0.832
53	1	5	5	0.0194	0.517
86	2	5	5	0.137	2.252
8	2	5	5	0.0214	0.851
32	2	5	5	0.0175	0.269
44	1	5	5	0.0764	0.433
63	2	5	5	0	0.141
42	1	5	5	0.0165	0.275
62	1	5	5	0.00012	0.135
36	1	5	5	0.0041	0.175

(a) Fit a multiple linear regression model using arsenic concentration in nails as the response and age, drink use, cook use, and arsenic in the water as the regressors.
(b) Estimate σ^2 and the standard errors of the regression coefficients.
(c) Use the model to predict the arsenic in nails when the age is 30, the drink use is category 5, the cook use is category 5, and arsenic in the water is 0.135 ppm.

12-11. In an article in *IEEE Transactions on Instrumentation and Measurement* (2001, Vol. 50, pp. 2033–2040) powdered mixtures of coal and limestone were analyzed for permittivity. The errors in the density measurement was the response.

Density	Dielectric Constant	Loss Factor
0.749	2.05	0.016
0.798	2.15	0.02
0.849	2.25	0.022
0.877	2.3	0.023
0.929	2.4	0.026
0.963	2.47	0.028
0.997	2.54	0.031
1.046	2.64	0.034
1.133	2.85	0.039
1.17	2.94	0.042
1.215	3.05	0.045

Table 12-6

TEMP	SOAKTIME	SOAKPCT	DIFFTIME	DIFFPCT	PITCH
1650	0.58	1.10	0.25	0.90	0.013
1650	0.66	1.10	0.33	0.90	0.016
1650	0.66	1.10	0.33	0.90	0.015
1650	0.66	1.10	0.33	0.95	0.016
1600	0.66	1.15	0.33	1.00	0.015
1600	0.66	1.15	0.33	1.00	0.016
1650	1.00	1.10	0.50	0.80	0.014
1650	1.17	1.10	0.58	0.80	0.021
1650	1.17	1.10	0.58	0.80	0.018
1650	1.17	1.10	0.58	0.80	0.019
1650	1.17	1.10	0.58	0.90	0.021
1650	1.17	1.10	0.58	0.90	0.019
1650	1.17	1.15	0.58	0.90	0.021
1650	1.20	1.15	1.10	0.80	0.025
1650	2.00	1.15	1.00	0.80	0.025
1650	2.00	1.10	1.10	0.80	0.026
1650	2.20	1.10	1.10	0.80	0.024
1650	2.20	1.10	1.10	0.80	0.025
1650	2.20	1.15	1.10	0.80	0.024
1650	2.20	1.10	1.10	0.90	0.025
1650	2.20	1.10	1.10	0.90	0.027
1650	2.20	1.10	1.50	0.90	0.026
1650	3.00	1.15	1.50	0.80	0.029
1650	3.00	1.10	1.50	0.70	0.030
1650	3.00	1.10	1.50	0.75	0.028
1650	3.00	1.15	1.66	0.85	0.032
1650	3.33	1.10	1.50	0.80	0.033
1700	4.00	1.10	1.50	0.70	0.039
1650	4.00	1.10	1.50	0.70	0.040
1650	4.00	1.15	1.50	0.85	0.035
1700	12.50	1.00	1.50	0.70	0.056
1700	18.50	1.00	1.50	0.70	0.068

(a) Fit a multiple linear regression model to these data with the density as the response.
(b) Estimate σ^2 and the standard errors of the regression coefficients.
(c) Use the model to predict the density when the dielectric constant is 2.5 and the loss factor is 0.03.

12-12. An article in *Biotechnology Progress* (2001, Vol. 17, pp. 366–368) reported on an experiment to investigate and optimize nisin extraction in aqueous two-phase systems (ATPS). The nisin recovery was the dependent variable (y). The two regressor variables were concentration (%) of PEG 4000 (denoted as x_1) and concentration (%) of Na_2SO_4 (denoted as x_2).
(a) Fit a multiple linear regression model to these data.
(b) Estimate σ^2 and the standard errors of the regression coefficients.

x_1	x_2	y
13	11	62.8739
15	11	76.1328
13	13	87.4667
15	13	102.3236
14	12	76.1872
14	12	77.5287
14	12	76.7824
14	12	77.4381
14	12	78.7417

(c) Use the model to predict the nisin recovery when $x_1 = 14.5$ and $x_2 = 12.5$.

12-13. An article in *Optical Engineering,* "Operating curve extraction of a correlator's filter" (2004, Vol. 43, pp. 2775–2779) reported on use of an optical correlator to perform an experiment by varying brightness and contrast. The resulting modulation is characterized by the useful range of gray levels. The data are shown below:

Brightness (%):	54	61	65	100	100	100	50	57	54
Contrast (%):	56	80	70	50	65	80	25	35	26
Useful range (ng):	96	50	50	112	96	80	155	144	255

(a) Fit a multiple linear regression model to these data.
(b) Estimate σ^2.
(c) Compute the standard errors of the regression coefficients.
(d) Predict the useful range when brightness = 80 and contrast = 75.

12-14. An article in *Technometrics* (1974, Vol. 16, pp. 523–531) considered the following stack-loss data from a plant oxidizing ammonia to nitric acid. Twenty-one daily responses of stack loss y (the amount of ammonia escaping) were measured with air flow x_1, temperature x_2, and acid concentration x_3.

Stack loss y = 42, 37, 37, 28, 18, 18, 19, 20, 15, 14, 14, 13, 11, 12, 8, 7, 8, 8, 9, 15, 15
x_1 = 80, 80, 75, 62, 62, 62, 62, 62, 58, 58, 58, 58, 58, 58, 50, 50, 50, 50, 50, 56, 70
x_2 = 27, 27, 25, 24, 22, 23, 24, 24, 23, 18, 18, 17, 18, 19, 18, 18, 19, 19, 20, 20, 20
x_3 = 89, 88, 90, 87, 87, 87, 93, 93, 87, 80, 89, 88, 82, 93, 89, 86, 72, 79, 80, 82, 91

(a) Fit a linear regression model relating the results of the stack loss to the three regressor varilables.
(b) Estimate σ^2.
(c) Find the standard error $se(\hat{\beta}_j)$.
(d) Use the model in part (a) to predict stack loss when $x_1 = 60$, $x_2 = 26$ and $x_3 = 85$.

12-15. Table 12-7 presents quarterback ratings for the 2004 National Football League season (source: *The Sports Network*).
(a) Fit a multiple regression model to relate the quarterback rating to the percentage of completions, the percentage of TDs, and the percentage of interceptions.
(b) Estimate σ^2.
(c) What are the standard errors of the regression coefficients?
(d) Use the model to predict the rating when the percentage of completions is 60%, the percentage of TDs is 4%, and the percentage of interceptions is 3%.

12-16. Table 12-8 presents statistics for the National Hockey League teams from the 2003–2004 season (source: *The Sports Network*). Fit a multiple linear regression model that relates *Wins* to the variables *GF* through *FG.* Because teams play 82 games $W = 82 - L - T - OTL$, but such a model does not help build a better team. Estimate σ^2 and find the standard errors of the regression coefficients for your model.

12-17. A study was performed on wear of a bearing y and its relationship to x_1 = oil viscosity and x_2 = load. The following data were obtained.

y	x_1	x_2
293	1.6	851
230	15.5	816
172	22.0	1058
91	43.0	1201
113	33.0	1357
125	40.0	1115

(a) Fit a multiple linear regression model to these data.
(b) Estimate σ^2 and the standard errors of the regression coefficients.
(c) Use the model to predict wear when $x_1 = 25$ and $x_2 = 1000$.
(d) Fit a multiple linear regression model with an interaction term to these data.
(e) Estimate σ^2 and $se(\hat{\beta}_j)$ for this new model. How did these quantities change? Does this tell you anything about the value of adding the interaction term to the model?
(f) Use the model in (d) to predict when $x_1 = 25$ and $x_2 = 1000$. Compare this prediction with the predicted value from part (c) above.

12-18. Consider the linear regression model

$$Y_i = \beta_0' + \beta_1(x_{i1} - \bar{x}_1) + \beta_2(x_{i2} - \bar{x}_2) + \epsilon_i$$

where $\bar{x}_1 = \sum x_{i1}/n$ and $\bar{x}_2 = \sum x_{i2}/n$.
(a) Write out the least squares normal equations for this model.
(b) Verify that the least squares estimate of the intercept in this model is $\hat{\beta}_0' = \sum y_i/n = \bar{y}$.
(c) Suppose that we use $y_i - \bar{y}$ as the response variable in the model above. What effect will this have on the least squares estimate of the intercept?

Table 12-7 Quarterback Ratings for the 2004 National Football League Season

Player	Att	Comp	Pct Comp	Yds	Yds per Att	TD	Pct TD	Long	Int	Pct Int	Rating Pts
D. Culpepper, MIN	548	379	69.2	4717	8.61	39	7.1	82	11	2	110.9
D. McNabb, PHI	469	300	64	3875	8.26	31	6.6	80	8	1.7	104.7
B. Griese, TAM	336	233	69.3	2632	7.83	20	6	68	12	3.6	97.5
M. Bulger, STL	485	321	66.2	3964	8.17	21	4.3	56	14	2.9	93.7
B. Favre, GBP	540	346	64.1	4088	7.57	30	5.6	79	17	3.1	92.4
J. Delhomme, CAR	533	310	58.2	3886	7.29	29	5.4	63	15	2.8	87.3
K. Warner, NYG	277	174	62.8	2054	7.42	6	2.2	62	4	1.4	86.5
M. Hasselbeck, SEA	474	279	58.9	3382	7.14	22	4.6	60	15	3.2	83.1
A. Brookes, NOS	542	309	57	3810	7.03	21	3.9	57	16	3	79.5
T. Rattary, SFX	325	198	60.9	2169	6.67	10	3.1	65	10	3.1	78.1
M. Vick, ATL	321	181	56.4	2313	7.21	14	4.4	62	12	3.7	78.1
J. Harrington, DET	489	274	56	3047	6.23	19	3.9	62	12	2.5	77.5
V. Testaverde, DAL	495	297	60	3532	7.14	17	3.4	53	20	4	76.4
P. Ramsey, WAS	272	169	62.1	1665	6.12	10	3.7	51	11	4	74.8
J. McCown, ARI	408	233	57.1	2511	6.15	11	2.7	48	10	2.5	74.1
P. Manning, IND	497	336	67.6	4557	9.17	49	9.9	80	10	2	121.1
D. Brees, SDC	400	262	65.5	3159	7.9	27	6.8	79	7	1.8	104.8
B. Roethlisberger, PIT	295	196	66.4	2621	8.88	17	5.8	58	11	3.7	98.1
T. Green, KAN	556	369	66.4	4591	8.26	27	4.9	70	17	3.1	95.2
T. Brady, NEP	474	288	60.8	3692	7.79	28	5.9	50	14	3	92.6
C. Pennington, NYJ	370	242	65.4	2673	7.22	16	4.3	48	9	2.4	91
B. Volek, TEN	357	218	61.1	2486	6.96	18	5	48	10	2.8	87.1
J. Plummer, DEN	521	303	58.2	4089	7.85	27	5.2	85	20	3.8	84.5
D. Carr, HOU	466	285	61.2	3531	7.58	16	3.4	69	14	3	83.5
B. Leftwich, JAC	441	267	60.5	2941	6.67	15	3.4	65	10	2.3	82.2
C. Palmer, CIN	432	263	60.9	2897	6.71	18	4.2	76	18	4.2	77.3
J. Garcia, CLF	252	144	57.1	1731	6.87	10	4	99	9	3.6	76.7
D. Bledsoe, BUF	450	256	56.9	2932	6.52	20	4.4	69	16	3.6	76.6
K. Collins, OAK	513	289	56.3	3495	6.81	21	4.1	63	20	3.9	74.8
K. Boller, BAL	464	258	55.6	2559	5.52	13	2.8	57	11	2.4	70.9

Att	Attempts (number of pass attempts)
Comp	Completed passes
Pct Comp	Percentage of completed passes
Yds	Yards gained passing
Yds per Att	Yards gained per pass attempt
TD	Number of touchdown passes
Pct TD	Percentage of attempts that are touchdowns
Long	Longest pass completion
Int	Number of interceptions
Pct Int	Percentage of attempts that are interceptions
Rating Pts	Rating points

Table 12-8 Team Statistics for the 2003–2004 National Hockey League Season

Team	W	L	T	OTL	PTS	GF	GA	ADV	PPGF	PCTG	PEN	BMI	AVG	SHT	PPGA	PKPCT	SHGF	SHGA	FG
Anaheim	29	35	10	8	76	184	213	310	56	18.1	1131	14	13.8	353	54	84.7	5	4	38
Atlanta	33	37	8	4	78	214	243	409	60	14.7	1505	32	18.4	391	58	85.2	11	7	31
Boston	41	19	15	7	104	209	188	300	48	16	1208	22	14.7	335	55	83.6	7	13	34
Buffalo	37	34	7	4	85	220	221	367	58	15.8	1289	28	15.7	348	53	84.8	7	9	46
Calgary	42	30	7	3	94	200	176	357	54	15.1	1428	42	17.4	346	53	84.7	11	2	41
Carolina	28	34	14	6	76	172	209	384	41	10.7	1102	16	13.4	359	66	81.6	9	9	30
Chicago	20	43	11	8	59	188	259	349	55	15.8	1318	18	16.1	381	63	83.5	5	14	33
Colorado	40	22	13	7	100	236	198	363	74	20.4	1293	28	15.8	357	58	83.8	7	14	47
Columbus	25	45	8	4	62	177	238	401	58	14.5	1209	22	14.7	337	50	85.2	5	14	37
Dallas	41	26	13	2	97	194	175	328	52	15.9	1143	20	13.9	311	44	85.9	5	11	41
Detroit	48	21	11	2	109	255	189	314	63	20.1	966	22	11.8	317	42	86.8	15	5	47
Edmonton	36	29	12	5	89	221	208	338	44	13	1220	22	14.9	349	66	81.1	13	8	35
Florida	28	35	15	4	75	188	221	310	52	16.8	1192	22	14.5	374	51	86.4	7	12	36
Los Angeles	28	29	16	9	81	205	217	356	50	14	1163	6	14.2	336	65	80.7	9	4	44
Minnesota	30	29	20	3	83	188	183	338	46	13.6	1035	44	12.6	283	44	84.5	7	7	46
Montreal	41	30	7	4	93	208	192	319	55	17.2	1039	16	12.7	314	55	82.5	5	6	51
Nashville	38	29	11	4	91	216	217	426	72	16.9	1360	28	16.6	266	64	85.3	8	5	41
New Jersey	43	25	12	2	100	213	164	312	51	16.3	894	14	10.9	351	39	81.8	14	8	48
NY Islanders	38	29	11	4	91	237	210	323	54	16.7	1168	16	14.2	359	52	85.5	11	5	39
NY Rangers	27	40	7	8	69	206	250	322	46	14.3	1459	30	17.8	388	80	79.4	6	8	40
Ottawa	43	23	10	6	102	262	189	371	80	21.6	1270	20	15.5	347	57	83.6	6	8	45
Philadelphia	40	21	15	6	101	229	186	314	65	20.7	1357	22	16.5	348	58	83.3	7	3	44
Phoenix	22	36	18	6	68	188	245	358	51	14.2	1300	24	15.9	398	71	82.2	9	8	36
Pittsburgh	23	47	8	4	58	190	303	360	65	18.1	1270	32	15.5	369	84	77.2	7	15	31
San Jose	43	21	12	6	104	219	183	337	56	16.6	1091	18	13.3	319	47	85.3	9	9	42
St. Louis	39	30	11	2	91	191	198	341	61	17.9	1274	28	15.5	369	57	84.6	7	11	45
Tampa Bay	46	22	8	6	106	245	192	339	55	16.2	985	34	12	278	42	84.9	15	4	50
Toronto	45	24	10	3	103	242	204	373	75	20.1	1452	20	17.7	386	64	83.4	8	8	43
Vancouver	43	24	10	5	101	235	194	374	56	15	1274	28	15.5	357	50	86	5	8	48
Washington	23	46	10	3	59	186	253	334	64	19.2	1282	12	15.6	401	75	81.3	4	5	36

W	Wins
L	Losses during regular time
T	Ties
OTL	Overtime losses
PTS	Points. Two points for winning a game, one point for a tie or losing in overtime, zero points for losing in regular time.
GF	Goals for
GA	Goals against
ADV	Total advantages. Power play opportunities.
PPGF	Power-play goals for. Goals scored while on power play.
PCTG	Power play percentage. Power-play goals divided by total advantages.
PEN	Total penalty minutes including bench minutes
BMI	Total bench minor minutes
AVG	Average penalty minutes per game
SHT	Total times short-handed. Measures opponent opportunities.
PPGA	Power-play goals against
PKPCT	Penalty killing percentage. Measures a team's ability to prevent goals while its opponent is on a power play. Opponent opportunities minus power play goals divided by opponent's opportunities.
SHGF	Short-handed goals for
SHGA	Short-handed goals against
FG	Games scored first

12-2 HYPOTHESIS TESTS IN MULTIPLE LINEAR REGRESSION

In multiple linear regression problems, certain tests of hypotheses about the model parameters are useful in measuring model adequacy. In this section, we describe several important hypothesis-testing procedures. As in the simple linear regression case, hypothesis testing requires that the error terms ϵ_i in the regression model are normally and independently distributed with mean zero and variance σ^2.

12-2.1 Test for Significance of Regression

The test for significance of regression is a test to determine whether a linear relationship exists between the response variable y and a subset of the regressor variables $x_1, x_2, \ldots, x_k$. The appropriate hypotheses are

Hypotheses for ANOVA Test

$$H_0: \beta_1 = \beta_2 = \cdots = \beta_k = 0$$
$$H_1: \beta_j \neq 0 \quad \text{for at least one } j \tag{12-17}$$

Rejection of H_0: $\beta_1 = \beta_2 = \cdots = \beta_k = 0$ implies that at least one of the regressor variables $x_1, x_2, \ldots, x_k$ contributes significantly to the model.

The test for significance of regression is a generalization of the procedure used in simple linear regression. The total sum of squares SS_T is partitioned into a sum of squares due to regression and a sum of squares due to error, say,

$$SS_T = SS_R + SS_E$$

Now if H_0: $\beta_1 = \beta_2 = \cdots = \beta_k = 0$ is true, SS_R/σ^2 is a chi-square random variable with k degrees of freedom. Note that the number of degrees of freedom for this chi-square random variable is equal to the number of regressor variables in the model. We can also show that the SS_E/σ^2 is a chi-square random variable with $n - p$ degrees of freedom, and that SS_E and SS_R are independent. The test statistic for H_0: $\beta_1 = \beta_2 = \cdots = \beta_k = 0$ is

Test Statistic for ANOVA

$$F_0 = \frac{SS_R/k}{SS_E/(n-p)} = \frac{MS_R}{MS_E} \tag{12-18}$$

We should reject H_0 if the computed value of the test statistic in Equation 12-18, f_0, is greater than $f_{\alpha,k,n-p}$. The procedure is usually summarized in an analysis of variance table such as Table 12-9.

A computational formula for SS_R may be found easily. Now since $SS_T = \sum_{i=1}^n y_i^2 - (\sum_{i=1}^n y_i)^2/n = \mathbf{y}'\mathbf{y} - (\sum_{i=1}^n y_i)^2/n$, we may rewrite Equation 12-19 as

$$SS_E = \mathbf{y}'\mathbf{y} - \frac{\left(\sum_{i=1}^{n} y_i\right)^2}{n} - \left[\hat{\boldsymbol{\beta}}'\mathbf{X}'\mathbf{y} - \frac{\left(\sum_{i=1}^{n} y_i\right)^2}{n}\right]$$

Table 12-9 Analysis of Variance for Testing Significance of Regression in Multiple Regression

Source of Variation	Sum of Squares	Degrees of Freedom	Mean Square	F_0
Regression	SS_R	k	MS_R	MS_R/MS_E
Error or residual	SS_E	$n - p$	MS_E	
Total	SS_T	$n - 1$		

or

$$SS_E = SS_T - SS_R$$

Therefore, the regression sum of squares is

$$SS_R = \hat{\boldsymbol{\beta}}'\mathbf{X}'\mathbf{y} - \frac{\left(\sum_{i=1}^{n} y_i\right)^2}{n} \tag{12-20}$$

EXAMPLE 12-3
Wire Bond Strength ANOVA

We will test for significance of regression (with $\alpha = 0.05$) using the wire bond pull strength data from Example 12-1. The total sum of squares is

$$SS_T = \mathbf{y}'\mathbf{y} - \frac{\left(\sum_{i=1}^{n} y_i\right)^2}{n} = 27{,}178.5316 - \frac{(725.82)^2}{25} = 6105.9447$$

The regression sum of squares is computed from Equation 12-20 as follows:

$$SS_R = \hat{\boldsymbol{\beta}}'\mathbf{X}'\mathbf{y} - \frac{\left(\sum_{i=1}^{n} y_i\right)^2}{n} = 27{,}063.3581 - \frac{(725.82)^2}{25} = 5990.7712$$

and by subtraction

$$SS_E = SS_T - SS_R = \mathbf{y}'\mathbf{y} - \hat{\boldsymbol{\beta}}'\mathbf{X}'\mathbf{y} = 115.1716$$

The analysis of variance is shown in Table 12-10. To test H_0: $\beta_1 = \beta_2 = 0$, we calculate the statistic

$$f_0 = \frac{MS_R}{MS_E} = \frac{2995.3856}{5.2352} = 572.17$$

Since $f_0 > f_{0.05,2,22} = 3.44$ (or since the P-value is considerably smaller than $\alpha = 0.05$), we reject the null hypothesis and conclude that pull strength is linearly related to either wire length or die height, or both. However, we note that this does not necessarily imply that the relationship found is an appropriate model for predicting pull strength as a function of wire length and die height. Further tests of model adequacy are required before we can be comfortable using this model in practice.

Most multiple regression computer programs provide the test for significance of regression in their output display. The middle portion of Table 12-4 is the Minitab output for this example. Compare Tables 12-4 and 12-10 and note their equivalence apart from rounding. The P-value is rounded to zero in the computer output.

Table 12-10 Test for Significance of Regression for Example 12-3

Source of Variation	Sum of Squares	Degrees of Freedom	Mean Square	f_0	P-value
Regression	5990.7712	2	2995.3856	572.17	1.08E-19
Error or residual	115.1735	22	5.2352		
Total	6105.9447	24			

R^2 and Adjusted R^2

We may also use the **coefficient of multiple determination** R^2 as a global statistic to assess the fit of the model. Computationally,

$$R^2 = \frac{SS_R}{SS_T} = 1 - \frac{SS_E}{SS_T} \tag{12-21}$$

For the wire bond pull strength data, we find that $R^2 = SS_R/SS_T = 5990.7712/6105.9447 = 0.9811$. Thus the model accounts for about 98% of the variability in the pull strength response (refer to the Minitab output in Table 12-4). The R^2 statistic is somewhat problematic as a measure of the quality of the fit for a multiple regression model because it always increases when a variable is added to a model.

To illustrate, consider the model fit to the wire bond pull strength data in Example 11-8. This was a simple linear regression model with x_1 = wire length as the regressor. The value of R^2 for this model is $R^2 = 0.9640$. Therefore, adding x_2 = die height to the model increases R^2 by $0.9811 - 0.9640 = 0.0171$, a very small amount. Since R^2 always increases when a regressor is added, it can be difficult to judge whether the increase is telling us anything useful about the new regressor. It is particularly hard to interpret a small increase, such as observed in the pull strength data.

Many regression users prefer to use an **adjusted** R^2 statistic:

Adjusted R^2

$$R^2_{\text{adj}} = 1 - \frac{SS_E/(n-p)}{SS_T/(n-1)} \tag{12-22}$$

Because $SS_E/(n-p)$ is the error or residual mean square and $SS_T/(n-1)$ is a constant, R^2_{adj} will only increase when a variable is added to the model if the new variable reduces the error mean square. Note that for the multiple regression model for the pull strength data $R^2_{\text{adj}} = 0.979$ (see the Minitab output in Table 12-4), whereas in Example 11-8 the adjusted R^2 for the one-variable model is $R^2_{\text{adj}} = 0.962$. Therefore, we would conclude that adding x_2 = die height to the model does result in a meaningful reduction in unexplained variability in the response.

The **adjusted** R^2 statistic essentially penalizes the analyst for adding terms to the model. It is an easy way to guard against **overfitting,** that is, including regressors that are not really useful. Consequently, it is very useful in comparing and evaluating competing regression models. We will use R^2_{adj} for this when we discuss **variable selection** in regression in Section 12-6.3.

12-2.2 Tests on Individual Regression Coefficients and Subsets of Coefficients

We are frequently interested in testing hypotheses on the individual regression coefficients. Such tests would be useful in determining the potential value of each of the regressor variables in the regression model. For example, the model might be more effective with the inclusion of additional variables or perhaps with the deletion of one or more of the regressors presently in the model.

The hypothesis to test if an individual regression coefficient, say β_j equals a value β_{j0} is

$$H_0: \beta_j = \beta_{j0}$$
$$H_1: \beta_j \neq \beta_{j0} \tag{12-23}$$

The test statistic for this hypothesis is

$$T_0 = \frac{\hat{\beta}_j - \beta_{j0}}{\sqrt{\sigma^2 C_{jj}}} = \frac{\hat{\beta}_j - \beta_{j0}}{se(\beta_{j0})} \tag{12-24}$$

where C_{jj} is the diagonal element of $(\mathbf{X}'\mathbf{X})^{-1}$ corresponding to $\hat{\beta}_j$. Notice that the denominator of Equation 12-24 is the standard error of the regression coefficient $\hat{\beta}_j$. The null hypothesis H_0: $\beta_j = \beta_{j0}$ is rejected if $|t_0| > t_{\alpha/2,n-p}$. This is called a **partial** or **marginal test** because the regression coefficient $\hat{\beta}_j$ depends on all the other regressor variables $x_i (i \neq j)$ that are in the model. More will be said about this in the following example.

An important special case of the previous hypothesis occurs for $\beta_{j0} = 0$. If H_0: $\beta_j = 0$ is not rejected, this indicates that the regressor x_j can be deleted from the model. Adding a variable to a regression model always causes the sum of squares for regression to increase and the error sum of squares to decrease (this is why R^2 always increases when a variable is added). We must decide whether the increase in the regression sum of squares is large enough to justify using the additional variable in the model. Furthermore, adding an unimportant variable to the model can actually increase the error mean square, indicating that adding such a variable has actually made the model a poorer fit to the data (this is why R^2_{adj} is a better measure of global model fit then the ordinary R^2).

EXAMPLE 12-4
Wire Bond Strength Coefficient Test

Consider the wire bond pull strength data, and suppose that we want to test the hypothesis that the regression coefficient for x_2 (die height) is zero. The hypotheses are

$$H_0: \beta_2 = 0$$
$$H_1: \beta_2 \neq 0$$

The main diagonal element of the $(\mathbf{X}'\mathbf{X})^{-1}$ matrix corresponding to $\hat{\beta}_2$ is $C_{22} = 0.0000015$, so the t-statistic in Equation 12-24 is

$$t_0 = \frac{\hat{\beta}_2}{\sqrt{\hat{\sigma}^2 C_{22}}} = \frac{0.01253}{\sqrt{(5.2352)(0.0000015)}} = 4.477$$

Note that we have used the estimate of σ^2 reported to four decimal places in Table 12-10. Since $t_{0.025,22} = 2.074$, we reject H_0: $\beta_2 = 0$ and conclude that the variable x_2 (die height) contributes significantly to the model. We could also have used a P-value to draw conclusions. The P-value for $t_0 = 4.477$ is

$P = 0.0002$, so with $\alpha = 0.05$ we would reject the null hypothesis. Note that this test measures the marginal or partial contribution of x_2 given that x_1 is in the model. That is, the t-test measures the contribution of adding the variable x_2 = die height to a model that already contains x_1 = wire length. Table 12-4 shows the value of the t-test computed by Minitab. The Minitab t-test statistic is reported to two decimal places. Note that the computer produces a t-test for each regression coefficient in the model. These t-tests indicate that both regressors contribute to the model.

EXAMPLE 12-5
Wire Bond Strength One-Sided Coefficient Test

There is an interest in the effect of die height on strength. This can be evaluated by the magnitude of the coefficient for die height. To conclude that the coefficient for die height exceeds 0.01 the hypotheses become

$$H_0{:}\beta_2 = 0.01 \qquad H_1{:}\beta_2 > 0.01$$

For such a test computer software can complete much of the hard work. We only need to assemble the pieces. From the Minitab output in Table 12-4, $\hat{\beta}_2 = 0.012528$ and the standard error of $\hat{\beta}_2 = 0.002798$. Therefore the t-statistic is

$$t_0 = \frac{0.012528 - 0.01}{0.002798} = 0.9035$$

with 22 degrees of freedom (error degrees of freedom). From Table IV in Appendix A, $t_{0.25,22} = 0.686$ and $t_{0.1,22} = 1.321$. Therefore, the P-value can be bounded as $0.1 < P\text{-value} < 0.25$. One cannot conclude that the coefficient exceeds 0.01 at common levels of significance.

There is another way to test the contribution of an individual regressor variable to the model. This approach determines the increase in the regression sum of squares obtained by adding a variable x_j (say) to the model, given that other variables $x_i (i \neq j)$ are already included in the regression equation.

The procedure used to do this is called the **general regression significance test**, or the **extra sum of squares method.** This procedure can also be used to investigate the contribution of a *subset* of the regressor variables to the model. Consider the regression model with k regressor variables

$$\mathbf{y} = \mathbf{X}\boldsymbol{\beta} + \boldsymbol{\epsilon} \tag{12-25}$$

where $\mathbf{y}$ is $(n \times 1)$, $\mathbf{X}$ is $(n \times p)$, $\boldsymbol{\beta}$ is $(p \times 1)$, $\boldsymbol{\epsilon}$ is $(n \times 1)$, and $p = k + 1$. We would like to determine if the subset of regressor variables $x_1, x_2, \ldots, x_r$ $(r < k)$ as a whole contributes significantly to the regression model. Let the vector of regression coefficients be partitioned as follows:

$$\boldsymbol{\beta} = \begin{bmatrix} \boldsymbol{\beta}_1 \\ \boldsymbol{\beta}_2 \end{bmatrix} \tag{12-26}$$

where $\boldsymbol{\beta}_1$ is $(r \times 1)$ and $\boldsymbol{\beta}_2$ is $[(p - r) \times 1]$. We wish to test the hypotheses

Hypotheses for General Regression Test

$$H_0{:}\ \boldsymbol{\beta}_1 = \mathbf{0}$$
$$H_1{:}\ \boldsymbol{\beta}_1 \neq \mathbf{0} \tag{12-27}$$

where $\mathbf{0}$ denotes a vector of zeroes. The model may be written as

$$\mathbf{y} = \mathbf{X}\boldsymbol{\beta} + \boldsymbol{\epsilon} = \mathbf{X}_1\boldsymbol{\beta}_1 + \mathbf{X}_2\boldsymbol{\beta}_2 + \boldsymbol{\epsilon} \tag{12-28}$$

where $\mathbf{X}_1$ represents the columns of $\mathbf{X}$ associated with $\boldsymbol{\beta}_1$ and $\mathbf{X}_2$ represents the columns of $\mathbf{X}$ associated with $\boldsymbol{\beta}_2$.

For the **full model** (including both $\boldsymbol{\beta}_1$ and $\boldsymbol{\beta}_2$), we know that $\hat{\boldsymbol{\beta}} = (\mathbf{X}'\mathbf{X})^{-1}\mathbf{X}'\mathbf{y}$. In addition, the regression sum of squares for all variables including the intercept is

$$SS_R(\boldsymbol{\beta}) = \hat{\boldsymbol{\beta}}'\mathbf{X}'\mathbf{y} \qquad (p = k + 1 \text{ degrees of freedom})$$

and

$$MS_E = \frac{\mathbf{y}'\mathbf{y} - \hat{\boldsymbol{\beta}}\mathbf{X}'\mathbf{y}}{n - p}$$

$SS_R(\boldsymbol{\beta})$ is called the regression sum of squares due to $\boldsymbol{\beta}$. To find the contribution of the terms in $\boldsymbol{\beta}_1$ to the regression, fit the model assuming the null hypothesis H_0: $\boldsymbol{\beta}_1 = \mathbf{0}$ to be true. The **reduced model** is found from Equation 12-28 as

$$\mathbf{y} = \mathbf{X}_2\boldsymbol{\beta}_2 + \boldsymbol{\epsilon} \tag{12-29}$$

The least squares estimate of $\boldsymbol{\beta}_2$ is $\hat{\boldsymbol{\beta}}_2 = (\mathbf{X}_2'\mathbf{X}_2)^{-1}\mathbf{X}_2'\mathbf{y}$, and

$$SS_R(\boldsymbol{\beta}_2) = \hat{\boldsymbol{\beta}}_2'\mathbf{X}_2'\mathbf{y} \qquad (p - r \text{ degrees of freedom}) \tag{12-30}$$

The regression sum of squares due to $\boldsymbol{\beta}_1$ given that $\boldsymbol{\beta}_2$ is already in the model is

$$SS_R(\boldsymbol{\beta}_1|\boldsymbol{\beta}_2) = SS_R(\boldsymbol{\beta}) - SS_R(\boldsymbol{\beta}_2) \tag{12-31}$$

This sum of squares has r degrees of freedom. It is sometimes called the extra sum of squares due to $\boldsymbol{\beta}_1$. Note that $SS_R(\boldsymbol{\beta}_1|\boldsymbol{\beta}_2)$ is the increase in the regression sum of squares due to including the variables $x_1, x_2, \ldots, x_r$ in the model. Now $SS_R(\boldsymbol{\beta}_1|\boldsymbol{\beta}_2)$ is independent of MS_E, and the null hypothesis $\boldsymbol{\beta}_1 = \mathbf{0}$ may be tested by the statistic

F Statistic for General Regression Test

$$F_0 = \frac{SS_R(\boldsymbol{\beta}_1|\boldsymbol{\beta}_2)/r}{MS_E} \tag{12-32}$$

If the computed value of the test statistic $f_0 > f_{\alpha,r,n-p}$, we reject H_0, concluding that at least one of the parameters in $\boldsymbol{\beta}_1$ is not zero and, consequently, at least one of the variables $x_1, x_2, \ldots, x_r$ in $\mathbf{X}_1$ contributes significantly to the regression model. Some authors call the test in Equation 12-32 a **partial *F*-test.**

The partial F-test is very useful. We can use it to measure the contribution of each individual regressor x_j as if it were the last variable added to the model by computing

$$SS_R(\beta_j|\beta_0, \beta_1, \ldots, \beta_{j-1}, \beta_{j+1}, \ldots, \beta_k), \qquad j = 1, 2, \ldots, k$$

This is the increase in the regression sum of squares due to adding x_j to a model that already includes $x_1, \ldots, x_{j-1}, x_{j+1}, \ldots, x_k$. The partial F-test is a more general procedure in that we can measure the effect of sets of variables. In Section 12-6.3 we show how the partial F-test plays a major role in *model building*—that is, in searching for the best set of regressor variables to use in the model.

EXAMPLE 12-6
Wire Bond Strength General Regression Test

Consider the wire bond pull-strength data in Example 12-1. We will investigate the contribution of two new variables, x_3 and x_4, to the model using the partial F-test approach. The new variables are explained at the end of this example. That is, we wish to test

$$H_0: \beta_3 = \beta_4 = 0 \qquad H_1: \beta_3 \neq 0 \text{ or } \beta_4 \neq 0$$

To test this hypothesis, we need the extra sum of squares due to β_3 and β_4 or

$$\begin{aligned} SS_R(\beta_4, \beta_3 | \beta_2, \beta_1, \beta_0) &= SS_R(\beta_4, \beta_3, \beta_2, \beta_1, \beta_0) - SS_R(\beta_2, \beta_1, \beta_0) \\ &= SS_R(\beta_4, \beta_3, \beta_2, \beta_1 | \beta_0) - SS_R(\beta_2, \beta_1 | \beta_0) \end{aligned}$$

In Example 12-3 we calculated

$$SS_R(\beta_2, \beta_1 | \beta_0) = \boldsymbol{\beta}'\mathbf{X}'\mathbf{y} - \frac{\sum_{i=1}^{n} y_i}{n} = 5990.7712 \text{ (two degrees of freedom)}$$

Also, Table 12-4 shows the Minitab output for the model with only x_1 and x_2 as predictors. In the analysis of variance table, we can see that $SS_R = 5990.8$ and this agrees with our calculation. In practice, the computer output would be used to obtain this sum of squares.

If we fit the model $Y = \beta_0 + \beta_1 x_1 + \beta_2 x_2 + \beta_3 x_3 + \beta_4 x_4$, we can use the same matrix formula. Alternatively, we can look at SS_R from computer output for this model. The analysis of variance table for this model is shown in Table 12-11 and we see that

$$SS_R(\beta_4, \beta_3, \beta_2, \beta_1 | \beta_0) = 6024.0 \text{ (four degrees of freedom)}$$

Therefore,

$$SS_R(\beta_4, \beta_3 | \beta_2, \beta_1, \beta_0) = 6024.0 - 5990.8 = 33.2 \text{ (two degrees of freedom)}$$

This is the increase in the regression sum of squares due to adding x_3 and x_4 to a model already containing x_1 and x_2. To test H_0, calculate the test statistic

$$f_0 = \frac{SS_R(\beta_4, \beta_3 | \beta_2, \beta_1, \beta_0)/2}{MS_E} = \frac{33.2/2}{4.1} = 4.05$$

Note that MS_E from the full model using x_1, x_2, x_3 and x_4 is used in the denominator of the test statistic. Because $f_{0.05,2,20} = 3.49$, we reject H_0 and conclude that at least one of the new variables contributes significantly to the model. Further analysis and tests will be needed to refine the model and determine if one or both of x_3 and x_4 are important.

The mystery of the new variables can now be explained. These are quadratic powers of the original predictors wire length and wire height. That is, $x_3 = x_1^2$ and $x_4 = x_2^2$. A test for quadratic terms is a common use of partial F-tests. With this information and the original data for x_1 and x_2, you can use computer software to reproduce these calculations. Multiple regression allows models to be extended in such a simple manner that the real meaning of x_3 and x_4 did not even enter into the test procedure. Polynomial models such as this are discussed further in Section 12-6.

If a partial F-test is applied to a single variable, it is equivalent to a t-test. To see this, consider the Minitab regression output for the wire bond pull strength in Table 12-4. Just below the analysis of variance summary in this table, the quantity labeled "'SeqSS'' shows the sum

Table 12-11 Regression Analysis: y versus x1, x2, x3, x4

The regression equation is y = 5.00 + 1.90 x1 + 0.0151 x2 + 0.0460 x3 − 0.000008 x4

Predictor	Coef	SE Coef	T	P
Constant	4.996	1.655	3.02	0.007
x1	1.9049	0.3126	6.09	0.000
x2	0.01513	0.01051	1.44	0.165
x3	0.04595	0.01666	2.76	0.012
x4	−0.00000766	0.00001641	−0.47	0.646

S = 2.02474 R−Sq = 98.7% R−Sq (adj) = 98.4%

Analysis of Variance

Source	DF	SS	MS	F	P
Regression	4	6024.0	1506.0	367.35	0.000
Residual Error	20	82.0	4.1		
Total	24	6105.9			

Source	DF	Seq SS
x1	1	5885.9
x2	1	104.9
x3	1	32.3
x4	1	0.9

of squares obtained by fitting x_1 alone (5885.9) and the sum of squares obtained by fitting x_2 after x_1 (104.9). In out notation, these are referred to as $SS_R(\beta_1|\beta_0)$ and $SS_R(\beta_2, \beta_1|\beta_0)$, respectively. Therefore, to test $H_0: \beta_2 = 0$ $H_1: \beta_2 \neq 0$ the partial F-test is

$$f_0 = \frac{SS_R(\beta_2|\beta_1, \beta_0)/1}{MS_E} = \frac{104.92}{5.24} = 20.2$$

where MS_E is the mean square for residual in the computer output in Table 12.4. This statistic should be compared to an F-distribution with 1 and 22 degrees of freedom in the numerator and denominator, respectively. From Table 12-4, the t-test for the same hypothesis is $t_0 = 4.48$. Note that $t_0^2 = 4.48^2 = 20.07 = f_0$, except for round-off error. Furthermore, the square of a t-random variable with ν degrees of freedom is an F-random variable with one and ν degrees of freedom. Consequently, the t-test provides an equivalent method to test a single variable for contribution to a model. Because the t-test is typically provided by computer output, it is the preferred method to test a single variable.

EXERCISES FOR SECTION 12-2

12-19. Consider the regression model fit to the soil shear strength data in Exercise 12-1.

(a) Test for significance of regression using $\alpha = 0.05$. What is the P-value for this test?

(b) Construct the t-test on each regression coefficient. What are your conclusions, using $\alpha = 0.05$? Calculate P-values.

12-20. Consider the absorption index data in Exercise 12-2. The total sum of squares for y is $SS_T = 742.00$.

(a) Test for significance of regression using $\alpha = 0.01$. What is the P-value for this test?

(b) Test the hypothesis H_0: $\beta_1 = 0$ versus H_1: $\beta_1 \neq 0$ using $\alpha = 0.01$. What is the P-value for this test?

(c) What conclusion can you draw about the usefulness of x_1 as a regressor in this model?

12-21. A regression model $Y = \beta_0 + \beta_1 x_1 + \beta_2 x_2 + \beta_3 x_3 + \epsilon$ has been fit to a sample of $n = 25$ observations. The calculated t-ratios $\hat{\beta}_j / se(\hat{\beta}_j), j = 1, 2, 3$ are as follows: for β_1, $t_0 = 4.82$, for β_2, $t_0 = 8.21$ and for β_3, $t_0 = 0.98$.

(a) Find P-values for each of the t-statistics.

(b) Using $\alpha = 0.05$, what conclusions can you draw about the regressor x_3? Does it seem likely that this regressor contributes significantly to the model?

12-22. Consider the electric power consumption data in Exercise 12-4.

(a) Test for significance of regression using $\alpha = 0.05$. What is the P-value for this test?

(b) Use the t-test to assess the contribution of each regressor to the model. Using $\alpha = 0.05$, what conclusions can you draw?

12-23. Consider the gasoline mileage data in Exercise 12-5.

(a) Test for significance of regression using $\alpha = 0.05$. What conclusions can you draw?

(b) Find the t-test statistic for each regressor. Using $\alpha = 0.05$, what conclusions can you draw? Does each regressor contribute to the model?

12-24. Consider the wire bond pull strength data in Exercise 12-6.

(a) Test for significance of regression using $\alpha = 0.05$. Find the P-value for this test. What conclusions can you draw?

(b) Calculate the t-test statistic for each regression coefficient. Using $\alpha = 0.05$, what conclusions can you draw? Do all variables contribute to the model?

12-25. Reconsider the semiconductor data in Exercise 12-7.

(a) Test for significance of regression using $\alpha = 0.05$. What conclusions can you draw?

(b) Calculate the t-test statistic and P-value for each regression coefficient. Using $\alpha = 0.05$, what conclusions can you draw?

12-26. Consider the regression model fit to the arsenic data in Exercise 12-10. Use arsenic in nails as the response and age, drink use, and cook use as the regressors.

(a) Test for significance of regression using $\alpha = 0.05$. What is the P-value for this test?

(b) Construct a t-test on each regression coefficient. What conclusions can you draw about the variables in this model? Use $\alpha = 0.05$.

12-27. Consider the regression model fit to the X-ray inspection data in Exercise 12-9. Use rads as the response.

(a) Test for significance of regression using $\alpha = 0.05$. What is the P-value for this test?

(b) Construct a t-test on each regression coefficient. What conclusions can you draw about the variables in this model? Use $\alpha = 0.05$.

12-28. Consider the regression model fit to the nisin extraction data in Exercise 12-12. Use nisin extraction as the response.

(a) Test for significance of regression using $\alpha = 0.05$. What is the P-value for this test?

(b) Construct a t-test on each regression coefficient. What conclusions can you draw about the variables in this model? Use $\alpha = 0.05$.

(c) Comment on the effect of a small sample size to the tests in the previous parts.

12-29. Consider the regression model fit to the grey range modulation data in Exercise 12-13. Use the useful range as the response.

(a) Test for significance of regression using $\alpha = 0.05$. What is the P-value for this test?

(b) Construct a t-test on each regression coefficient. What conclusions can you draw about the variables in this model? Use $\alpha = 0.05$.

12-30. Consider the regression model fit to the stack loss data in Exercise 12-14. Use stack loss as the response.

(a) Test for significance of regression using $\alpha = 0.05$. What is the P-value for this test?

(b) Construct a t-test on each regression coefficient. What conclusions can you draw about the variables in this model? Use $\alpha = 0.05$.

12-31. Consider the NFL data in Exercise 12-15.

(a) Test for significance of regression using $\alpha = 0.05$. What is the P-value for this test?

(b) Conduct the t-test for each regression coefficient. Using $\alpha = 0.05$, what conclusions can you draw about the variables in this model?

(c) Find the amount by which the regressor x_2 (TD percentage) increases the regression sum of squares and conduct an F-test for H_0: $\beta_2 = 0$ versus H_1: $\beta_2 \neq 0$ using $\alpha = 0.05$. What is the P-value for this test? what conclusions can you draw?

12-32. Exercise 12-8 presents data on heat treating gears.

(a) Test the regression model for significance of regression. Using $\alpha = 0.05$, find the P-value for the test and draw conclusions.

(b) Evaluate the contribution of each regressor to the model using the t-test with $\alpha = 0.05$.

(c) Fit a new model to the response PITCH using new regressors x_1 = SOAKTIME × SOAKPCT and x_2 = DIFFTIME × DIFFPCT.

(d) Test the model in part (c) for significance of regression using $\alpha = 0.05$. Also calculate the t-test for each regressor and draw conclusions.

(e) Estimate σ^2 for the model from part (c) and compare this to the estimate of σ^2 for the model in part (a). Which estimate is smaller? Does this offer any insight regarding which model might be preferable?

12-33. Consider the bearing wear data in Exercise 12-17.

(a) For the model with no interaction, test for significance of regression using $\alpha = 0.05$. What is the P-value for this test? What are your conclusions?

(b) For the model with no interaction, compute the t-statistics for each regression coefficient. Using $\alpha = 0.05$, what conclusions can you draw?
(c) For the model with no interaction, use the extra sum of squares method to investigate the usefulness of adding x_2 = load to a model that already contains x_1 = oil viscosity. Use $\alpha = 0.05$.
(d) Refit the model with an interaction term. Test for significance of regression using $\alpha = 0.05$.
(e) Use the extra sum of squares method to determine whether the interaction term contributes significantly to the model. Use $\alpha = 0.05$.
(f) Estimate σ^2 for the interaction model. Compare this to the estimate of σ^2 from the model in part (a).

12-34. Data on National Hockey League team performance was presented in Exercise 12-16.
(a) Test the model from this exercise for significance of regression using $\alpha = 0.05$. What conclusions can you draw?
(b) Use the t-test to evaluate the contribution of each regressor to the model. Does it seem that all regressors are necessary? Use $\alpha = 0.05$.
(c) Fit a regression model relating the number of games won to the number of goals for and the number of power play goals for. Does this seem to be a logical choice of regressors, considering your answer to part (b)? Test this new model for significance of regression and evaluate the contribution of each regressor to the model using the t-test. Use $\alpha = 0.05$.

12-3 CONFIDENCE INTERVALS IN MULTIPLE LINEAR REGRESSION

12-3.1 Confidence Intervals on Individual Regression Coefficients

In multiple regression models, it is often useful to construct confidence interval estimates for the regression coefficients $\{\beta_j\}$. The development of a procedure for obtaining these confidence intervals requires that the errors $\{\epsilon_i\}$ are normally and independently distributed with mean zero and variance σ^2. This is the same assumption required in hypothesis testing. Therefore, the observations $\{Y_i\}$ are normally and independently distributed with mean $\beta_0 + \sum_{j=1}^{k} \beta_j x_{ij}$ and variance σ^2. Since the least squares estimator $\hat{\boldsymbol{\beta}}$ is a linear combination of the observations, it follows that $\hat{\boldsymbol{\beta}}$ is normally distributed with mean vector $\boldsymbol{\beta}$ and covariance matrix $\sigma^2(\mathbf{X}'\mathbf{X})^{-1}$. Then each of the statistics

$$T = \frac{\hat{\beta}_j - \beta_j}{\sqrt{\hat{\sigma}^2 C_{jj}}} \qquad j = 0, 1, \ldots, k \tag{12-33}$$

has a t distribution with $n - p$ degrees of freedom, where C_{jj} is the jjth element of the $(\mathbf{X}'\mathbf{X})^{-1}$ matrix, and $\hat{\sigma}^2$ is the estimate of the error variance, obtained from Equation 12-16. This leads to the following $100(1 - \alpha)\%$ confidence interval for the regression coefficient $\beta_j, j = 0, 1, \ldots, k$.

Confidence Interval on a Regression Coefficient

A $100(1 - \alpha)\%$ **confidence interval on the regression coefficient** β_j, $j = 0, 1, \ldots, k$ in the multiple linear regression model is given by

$$\hat{\beta}_j - t_{\alpha/2,n-p}\sqrt{\hat{\sigma}^2 C_{jj}} \le \beta_j \le \hat{\beta}_j + t_{\alpha/2,n-p}\sqrt{\hat{\sigma}^2 C_{jj}} \tag{12-34}$$

Because $\sqrt{\hat{\sigma}^2 C_{jj}}$ is the standard error of the regression coefficient $\hat{\beta}_j$, we would also write the CI formula as $\hat{\beta}_j - t_{\alpha/2,n-p}\, se(\hat{\beta}_j) \le \beta_j \le \hat{\beta}_j + t_{\alpha/2,n-p}\, se(\hat{\beta}_j)$.

EXAMPLE 12-7
Wire Bond Strength Confidence Interval

We will construct a 95% confidence interval on the parameter β_1 in the wire bond pull strength problem. The point estimate of β_1 is $\hat{\beta}_1 = 2.74427$ and the diagonal element of $(\mathbf{X}'\mathbf{X})^{-1}$ corresponding to β_1 is $C_{11} = 0.001671$. The estimate of σ^2 is $\hat{\sigma}^2 = 5.2352$, and $t_{0.025,22} = 2.074$. Therefore, the 95% CI on β_1 is computed from Equation 12-34 as

$$2.74427 - (2.074)\sqrt{(5.2352)(.001671)} \leq \beta_1 \leq 2.74427 + (2.074)\sqrt{(5.2352)(.001671)}$$

which reduces to

$$2.55029 \leq \beta_1 \leq 2.93825$$

Also, computer software such as Minitab can be used to help calculate this confidence interval. From the regression output in Table 10-4, $\hat{\beta}_1 = 2.74427$ and the standard error of $\hat{\beta}_1 = 0.0935$. This standard error is the multiplier of the t-table constant in the confidence interval. That is, $0.0935 = \sqrt{(5.2352)(0.001671)}$. Consequently, all the numbers are available from the computer output to construct the interval and this is the typical method used in practice.

12-3.2 Confidence Interval on the Mean Response

We may also obtain a confidence interval on the mean response at a particular point, say, $x_{01}, x_{02}, \ldots, x_{0k}$. To estimate the mean response at this point, define the vector

$$\mathbf{x}_0 = \begin{bmatrix} 1 \\ x_{01} \\ x_{02} \\ \vdots \\ x_{0k} \end{bmatrix}$$

The mean response at this point is $E(Y|\mathbf{x}_0) = \mu_{Y|\mathbf{x}_0} = \mathbf{x}_0'\boldsymbol{\beta}$, which is estimated by

$$\hat{\mu}_{Y|\mathbf{x}_0} = \mathbf{x}_0'\hat{\boldsymbol{\beta}} \tag{12-35}$$

This estimator is unbiased, since $E(\mathbf{x}_0'\hat{\boldsymbol{\beta}}) = \mathbf{x}_0'\boldsymbol{\beta} = E(Y|\mathbf{x}_0) = \mu_{Y|\mathbf{x}_0}$ and the variance of $\hat{\mu}_{Y|\mathbf{x}_0}$ is

$$V(\hat{\mu}_{Y|\mathbf{x}_0}) = \sigma^2\mathbf{x}_0'(\mathbf{X}'\mathbf{X})^{-1}\mathbf{x}_0 \tag{12-36}$$

A $100(1-\alpha)$ % CI on $\mu_{Y|\mathbf{x}_0}$ can be constructed from the statistic

$$\frac{\hat{\mu}_{Y|\mathbf{x}_0} - \mu_{Y|\mathbf{x}_0}}{\sqrt{\hat{\sigma}^2\mathbf{x}_0'(\mathbf{X}'\mathbf{X})^{-1}\mathbf{x}_0}} \tag{12-37}$$

Confidence Interval on the Mean Response

For the multiple linear regression model, a $100(1-\alpha)$% **confidence interval on the mean response** at the point $x_{01}, x_{02}, \ldots, x_{0k}$ is

$$\hat{\mu}_{Y|\mathbf{x}_0} - t_{\alpha/2,n-p}\sqrt{\hat{\sigma}^2\mathbf{x}_0'(\mathbf{X}'\mathbf{X})^{-1}\mathbf{x}_0} \leq \mu_{Y|\mathbf{x}_0} \leq \hat{\mu}_{Y|\mathbf{x}_0} + t_{\alpha/2,n-p}\sqrt{\hat{\sigma}^2\mathbf{x}_0'(\mathbf{X}'\mathbf{X})^{-1}\mathbf{x}_0} \tag{12-38}$$

Equation 12-38 is a CI about the regression plane (or hyperplane). It is the multiple regression generalization of Equation 11-31.

EXAMPLE 12-8 Wire Bond Strength Confidence Interval on the Mean Response

The engineer in Example 12-1 would like to construct a 95% CI on the mean pull strength for a wire bond with wire length $x_1 = 8$ and die height $x_2 = 275$. Therefore,

$$\mathbf{x}_0 = \begin{bmatrix} 1 \\ 8 \\ 275 \end{bmatrix}$$

The estimated mean response at this point is found from Equation 12-35 as

$$\hat{\mu}_{Y|\mathbf{x}_0} = \mathbf{x}_0'\hat{\boldsymbol{\beta}} = [1 \quad 8 \quad 275] \begin{bmatrix} 2.26379 \\ 2.74427 \\ 0.01253 \end{bmatrix} = 27.66$$

The variance of $\hat{\mu}_{Y|\mathbf{x}_0}$ is estimated by

$$\hat{\sigma}^2\mathbf{x}_0'(\mathbf{X}'\mathbf{X})^{-1}\mathbf{x}_0 = 5.2352\,[1\ 8\ 275] \times \begin{bmatrix} .214653 & -.007491 & -.000340 \\ -.007491 & .001671 & -.000019 \\ -.000340 & -.000019 & .0000015 \end{bmatrix} \begin{bmatrix} 1 \\ 8 \\ 275 \end{bmatrix}$$

$$= 5.2352\,(0.0444) = 0.23244$$

Therefore, a 95% CI on the mean pull strength at this point is found from Equation 12-38 as

$$27.66 - 2.074\sqrt{0.23244} \le \mu_{Y|\mathbf{x}_0} \le 27.66 + 2.074\sqrt{0.23244}$$

which reduces to

$$26.66 \le \mu_{Y|\mathbf{x}_0} \le 28.66$$

Some computer software packages will provide estimates of the mean for a point of interest $\mathbf{x}_0$ and the associated CI. Table 12-4 shows the Minitab output for Example 12-8. Both the estimate of the mean and the 95% CI are provided.

12-4 PREDICTION OF NEW OBSERVATIONS

A regression model can be used to predict new or **future observations** on the response variable Y corresponding to particular values of the independent variables, say, $x_{01}, x_{02}, \ldots, x_{0k}$. If $\mathbf{x}_0' = [1, x_{01}, x_{02}, \ldots, x_{0k}]$, a point estimate of the future observation Y_0 at the point x_{01}, $x_{02}, \ldots, x_{0k}$ is

$$\hat{y}_0 = \mathbf{x}_0'\hat{\boldsymbol{\beta}} \qquad (12\text{-}39)$$

Prediction Interval

A 100(1 − α)% **prediction interval for this future observation** is

$$\hat{y}_0 - t_{\alpha/2,n-p}\sqrt{\hat{\sigma}^2(1 + \mathbf{x}_0'(\mathbf{X}'\mathbf{X})^{-1}\mathbf{x}_0)}$$
$$\leq Y_0 \leq \hat{y}_0 + t_{\alpha/2,n-p}\sqrt{\hat{\sigma}^2(1 + \mathbf{x}_0'(\mathbf{X}'\mathbf{X})^{-1}\mathbf{x}_0)} \quad (12\text{-}40)$$

This prediction interval is a generalization of the prediction interval given in Equation 11-33 for a future observation in simple linear regression. If you compare the prediction interval Equation 12-40 with the expression for the confidence interval on the mean, Equation 12-38, you will observe that the prediction interval is always wider than the confidence interval. The confidence interval expresses the error in estimating the mean of a distribution, while the prediction interval expresses the error in predicting a future observation from the distribution at the point $\mathbf{x}_0$. This must include the error in estimating the mean at that point, as well as the inherent variability in the random variable Y at the same value $\mathbf{x} = \mathbf{x}_0$.

Also, one might want to predict the mean of several values of Y, say m, all at the same value $\mathbf{x} = \mathbf{x}_0$. Because the variance of a sample mean is σ^2/m, Equation 12-40 is modified as follows. Replace the constant 1 under the square root with $1/m$ to reflect the lower variability in the mean of m observations. This results in a narrower interval.

In predicting new observations and in estimating the mean response at a given point $x_{01}, x_{02}, \ldots, x_{0k}$, we must be careful about **extrapolating** beyond the region containing the original observations. It is very possible that a model that fits well in the region of the original data will no longer fit well outside of that region. In multiple regression it is often easy to inadvertently extrapolate, since the levels of the variables $(x_{i1}, x_{i2}, \ldots, x_{ik})$, $i = 1, 2, \ldots, n$, jointly define the region containing the data. As an example, consider Fig. 12-5, which illustrates the region containing the observations for a two-variable regression model. Note that the point (x_{01}, x_{02}) lies within the ranges of both regressor variables x_1 and x_2, but it is outside the region that is actually spanned by the original observations. This is sometimes called a **hidden extrapolation.** Either predicting the value of a new observation or estimating the mean response at this point is an extrapolation of the original regression model.

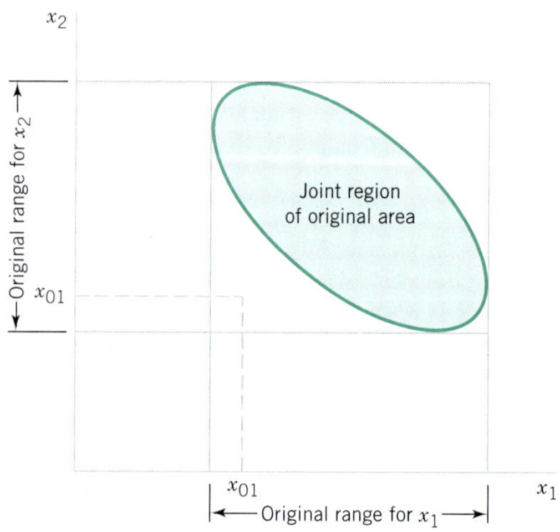

Figure 12-5 An example of extrapolation in multiple regression.

EXAMPLE 12-9
Wire Bond Strength Confidence Interval

Suppose that the engineer in Example 12-1 wishes to construct a 95% prediction interval on the wire bond pull strength when the wire length is $x_1 = 8$ and the die height is $x_2 = 275$. Note that $\mathbf{x}_0' = [1 \quad 8 \quad 275]$, and the point estimate of the pull strength is $\hat{y}_0 = \mathbf{x}_0'\hat{\beta} = 27.66$. Also, in Example 12-8 we calculated $\mathbf{x}_0'(\mathbf{X}'\mathbf{X})^{-1}\mathbf{x}_0 = 0.04444$. Therefore, from Equation 12-40 we have

$$27.66 - 2.074\sqrt{5.2352(1 + 0.0444)} \leq Y_0 \leq 27.66 + 2.074\sqrt{5.2352(1 + 0.0444)}$$

and the 95% prediction interval is

$$22.81 \leq Y_0 \leq 32.51$$

Notice that the prediction interval is wider than the confidence interval on the mean response at the same point, calculated in Example 12-8. The Minitab output in Table 12-4 also displays this prediction interval.

EXERCISES FOR SECTIONS 12-3 AND 12-4

12-35. Consider the regression model fit to the shear strength of soil in Exercise 12-1.
(a) Calculate 95% confidence intervals on each regression coefficient.
(b) Calculate a 95% confidence interval on mean strength when $x_1 = 18$ feet and $x_2 = 43\%$.
(c) Calculate 95% prediction interval on strength for the same values of the regressors used in the previous part.

12-36. Consider the soil absorption data in Exercise 12-2.
(a) Find 95% confidence intervals on the regression coefficients.
(b) Find a 95% confidence interval on mean soil absorption index when $x_1 = 200$ and $x_2 = 50$.
(c) Find a 95% prediction interval on the soil absorption index when $x_1 = 200$ and $x_2 = 50$.

12-37. Consider the semiconductor data in Exercise 12-7.
(a) Find 99% confidence intervals on the regression coefficients.
(b) Find a 99% prediction interval on HFE when $x_1 = 14.5$, $x_2 = 220$, and $x_3 = 5.0$.
(c) Find a 99% confidence interval on mean HFE when $x_1 = 14.5$, $x_2 = 220$, and $x_3 = 5.0$.

12-38. Consider the electric power consumption data in Exercise 12-4.
(a) Find 95% confidence intervals on β_1, β_2, β_3, and β_4.
(b) Find a 95% confidence interval on the mean of Y when $x_1 = 75$, $x_2 = 24$, $x_3 = 90$, and $x_4 = 98$.
(c) Find a 95% prediction interval on the power consumption when $x_1 = 75$, $x_2 = 24$, $x_3 = 90$, and $x_4 = 98$.

12-39. Consider the bearing wear data in Exercise 12-17.
(a) Find 99% confidence intervals on β_1 and β_2.
(b) Recompute the confidence intervals in part (a) after the interaction term x_1x_2 is added to the model. Compare the lengths of these confidence intervals with those computed in part (a). Do the lengths of these intervals provide any information about the contribution of the interaction term in the model?

12-40. Consider the wire bond pull strength data in Exercise 12-6.
(a) Find 95% confidence interval on the regression coefficients.
(b) Find a 95% confidence interval on mean pull strength when $x_2 = 20$, $x_3 = 30$, $x_4 = 90$ and $x_5 = 2.0$.
(c) Find a 95% prediction interval on pull strength when $x_2 = 20$, $x_3 = 30$, $x_4 = 90$, and $x_5 = 2.0$.

12-41. Consider the regression model fit to the X-ray inspection data in Exercise 12-9. Use rads as the response.
(a) Calculate 95% confidence intervals on each regression coefficient.
(b) Calculate a 99% confidence interval on mean rads at 15 milliamps and 1 second on exposure time.
(c) Calculate a 99% prediction interval on rads for the same values of the regressors used in the previous part.

12-42. Consider the regression model fit to the arsenic data in Exercise 12-10. Use arsenic in nails as the response and age, drink use, and cook use as the regressors.
(a) Calculate 99% confidence intervals on each regression coefficient.
(b) Calculate a 99% confidence interval on mean arsenic concentration in nails when age = 30, drink use = 4, and cook use = 4.
(c) Calculate a prediction interval on arsenic concentration in nails for the same values of the regressors used in the previous part.

12-43. Consider the regression model fit to the coal and limestone mixture data in Exercise 12-11. Use density as the response.
(a) Calculate 90% confidence intervals on each regression coefficient.
(b) Calculate a 90% confidence interval on mean density when the dielectric constant = 2.3 and the loss factor = 0.025.
(c) Calculate a prediction interval on density for the same values of the regressors used in the previous part.

12-44. Consider the regression model fit to the nisin extraction data in Exercise 12-12.

(a) Calculate 95% confidence intervals on each regression coefficient.

(b) Calculate a 95% confidence interval on mean nisin extraction when $x_1 = 15.5$ and $x_2 = 16$.

(c) Calculate a prediction interval on nisin extraction for the same values of the regressors used in the previous part.

(d) Comment on the effect of a small sample size to the widths of these intervals.

12-45. Consider the regression model fit to the grey range modulation data in Exercise 12-13. Use the useful range as the response.

(a) Calculate 99% confidence intervals on each regression coefficient.

(b) Calculate a 99% confidence interval on mean useful range when brightness = 70 and contrast = 80.

(c) Calculate a prediction interval on useful range for the same values of the regressors used in the previous part.

(d) Calculate a 99% confidence interval and a 99% a prediction interval on useful range when brightness = 50 and contrast = 25. Compare the widths of these intervals to those calculated in parts (b) and (c). Explain any differences in widths.

12-46. Consider the stack loss data in Exercise 12-14.

(a) Calculate 95% confidence intervals on each regression coefficient.

(b) Calculate a 95% confidence interval on mean stack loss when $x_1 = 80$, $x_2 = 25$ and $x_3 = 90$.

(c) Calculate a prediction interval on stack loss for the same values of the regressors used in the previous part.

(d) Calculate a 95% confidence interval and a 95% prediction interval on stack loss when $x_1 = 80$, $x_2 = 19$ and $x_3 = 93$. Compare the widths of these intervals to those calculated in parts (b) and (c). Explain any differences in widths.

12-47. Consider the NFL data in Exercise 12-15.

(a) Find 95% confidence interval on the regression coefficients.

(b) What is the estimated standard error of $\hat{\mu}_{Y|\mathbf{x}_0}$ when the percentage of completions is 60%, the percentage of TDs is 4%, and the percentage of interceptions is 3%.

(c) Find a 95% confidence interval on the mean rating when the percentage of completions is 60%, the percentage of TDs is 4%, and the percentage of interceptions is 3%.

12-48. Consider the heat treating data from Exercise 12-8.

(a) Find 95% confidence intervals on the regression coefficients.

(b) Find a 95% confidence interval on mean PITCH when TEMP = 1650, SOAKTIME = 1.00, SOAKPCT = 1.10, DIFFTIME = 1.00, and DIFFPCT = 0.80.

(c) Fit a model to PITCH using regressors x_1 = SOAKTIME × SOAKPCT and x_2 = DIFFTIME × DIFFPCT. Using the model with regressors x_1 and x_2, find a 95% confidence interval on mean PITCH when SOAKTIME = 1.00, SOAKPCT = 1.10, DIFFTIME = 1.00, and DIFFPCT = 0.80.

(d) Compare the length of this confidence interval with the length of the confidence interval on mean PITCH at the same point from part (b), where an additive model in SOAKTIME, SOAKPCT, DIFFTIME, and DIFFPCT was used. Which confidence interval is shorter? Does this tell you anything about which model is preferable?

12-49. Consider the gasoline mileage data in Exercise 12-5.

(a) Find 99% confidence intervals on the regression coefficients.

(b) Find a 99% confidence interval on the mean of Y for the regressor values in the first row of data.

(c) Fit a new regression model to these data using *cid*, *etw*, and *axle* as the regressors. Find 99% confidence intervals on the regression coefficients in this new model.

(d) Compare the lengths of the confidence intervals from part (c) with those found in part (a). Which intervals are longer? Does this offer any insight about which model is preferable?

12-50. Consider the NHL data in Exercise 12-16.

(a) Find a 95% confidence interval on the regression coefficient for the variable *GF*.

(b) Fit a simple linear regression model relating the response variable *W* to the regressor *GF*.

(c) Find a 95% confidence interval on the slope for the simple linear regression model from part (b).

(d) Compare the lengths of the two confidence intervals computed in parts (a) and (c). Which interval is shorter? Does this tell you anything about which model is preferable?

12-5 MODEL ADEQUACY CHECKING

12-5.1 Residual Analysis

The **residuals** from the multiple regression model, defined by $e_i = y_i - \hat{y}_i$, play an important role in judging model adequacy just as they do in simple linear regression. As noted in Section 11-7.1, several residual plots are often useful; these are illustrated in Example 12-10.

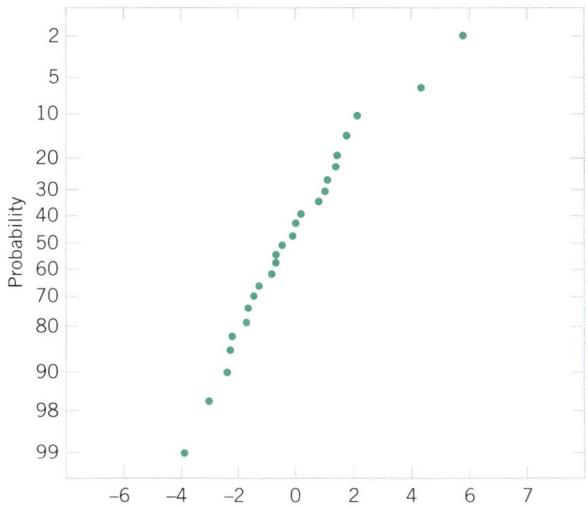

Figure 12-6 Normal probability plot of residuals.

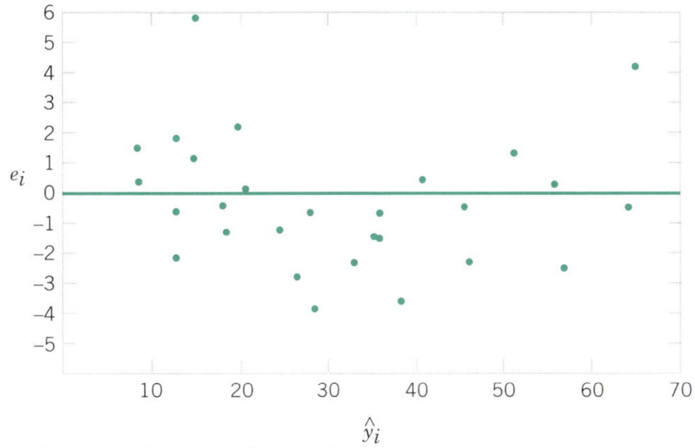

Figure 12-7 Plot of residuals against $\hat{y}$.

It is also helpful to plot the residuals against variables not presently in the model that are possible candidates for inclusion. Patterns in these plots may indicate that the model may be improved by adding the candidate variable.

EXAMPLE 12-10
Wire Bond Strength Residuals

The residuals for the model from Example 12-1 are shown in Table 12-3. A normal probability plot of these residuals is shown in Fig. 12-6. No severe deviations from normality are obviously apparent, although the two largest residuals ($e_{15} = 5.84$ and $e_{17} = 4.33$) do not fall extremely close to a straight line drawn through the remaining residuals.

The **standardized residuals**

Standardized Residual

$$d_i = \frac{e_i}{\sqrt{MS_E}} = \frac{e_i}{\sqrt{\hat{\sigma}^2}} \tag{12-41}$$

are often more useful than the ordinary residuals when assessing residual magnitude. The standardized residuals corresponding to e_{15} and e_{17} are $d_{15} = 5.84/\sqrt{5.2352} = 2.55$ and $d_{17} = 4.33/\sqrt{5.2352} = 1.89$, and they do not seem unusually large. Inspection of the data does not reveal any error in collecting observations 15 and 17, nor does it produce any other reason to discard or modify these two points.

The residuals are plotted against $\hat{y}$ in Fig. 12-7, and against x_1 and x_2 in Figs. 12-8 and 12-9, respectively.* The two largest residuals, e_{15} and e_{17}, are apparent. Figure 12-8 gives some indication that the model underpredicts the pull strength for assemblies with short wire length ($x_1 \leq 6$) and long wire length ($x_1 \geq 15$) and overpredicts the strength for assemblies with intermediate wire length ($7 \leq x_1 \leq 14$). The same impression is obtained from Fig. 12-7. Either the relationship between strength and wire length is not linear (requiring that a term involving x_1^2, say, be added to the model), or other regressor variables not presently in the model affected the response.

*There are other methods, described in Montgomery, Peck, and Vining (2001) and Myers (1990), that plot a modified version of the residual, called a **partial residual,** against each regressor. These partial residual plots are useful in displaying the relationship between the response y and each individual regressor.

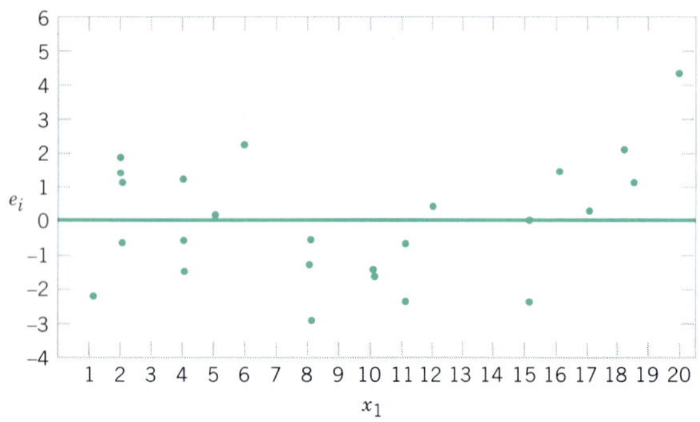

Figure 12-8 Plot of residuals against x_1.

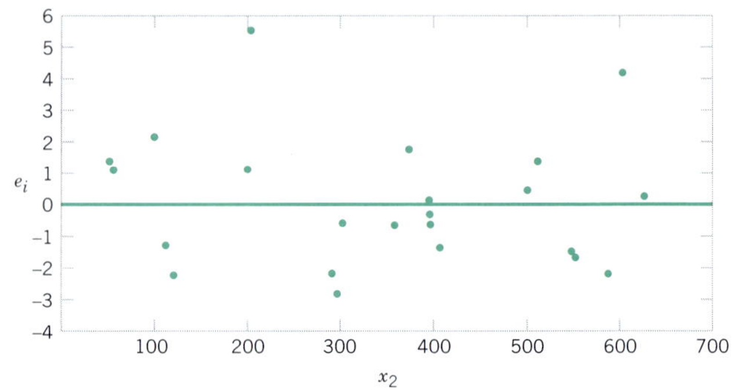

Figure 12-9 Plot of residuals against x_2.

In Example 12-10 we used the standardized residuals $d_i = e_i/\sqrt{\hat{\sigma}^2}$ as a measure of residual magnitude. Some analysts prefer to plot standardized residuals instead of ordinary residuals, because the standardized residuals are scaled so that their standard deviation is approximately unity. Consequently, large residuals (that may indicate possible outliers or unusual observations) will be more obvious from inspection of the residual plots.

Many regression computer programs compute other types of scaled residuals. One of the most popular are the **studentized residuals**

Studentized Residual

$$r_i = \frac{e_i}{\sqrt{\hat{\sigma}^2(1 - h_{ii})}} \qquad i = 1, 2, \ldots, n \tag{12-42}$$

where h_{ii} is the ith diagonal element of the matrix

$$\mathbf{H} = \mathbf{X}(\mathbf{X}'\mathbf{X})^{-1}\mathbf{X}'$$

The $\mathbf{H}$ matrix is sometimes called the **"hat" matrix,** since

$$\hat{\mathbf{y}} = \mathbf{X}\hat{\boldsymbol{\beta}} = \mathbf{X}(\mathbf{X}'\mathbf{X})^{-1}\mathbf{X}'\mathbf{y} = \mathbf{H}\mathbf{y}$$

Thus $\mathbf{H}$ transforms the observed values of $\mathbf{y}$ into a vector of fitted values $\hat{\mathbf{y}}$.

Since each row of the matrix $\mathbf{X}$ corresponds to a vector, say $\mathbf{x}_i' = [1, x_{i1}, x_{i2}, \ldots, x_{ik}]$, another way to write the diagonal elements of the hat matrix is

Diagonal Elements of Hat Matrix

$$h_{ii} = \mathbf{x}_i'(\mathbf{X}'\mathbf{X})^{-1}\mathbf{x}_i \tag{12-43}$$

Note that apart from σ^2, h_{ii} is the variance of the fitted value $\hat{y}_i$. The quantities h_{ii} were used in the computation of the confidence interval on the mean response in Section 12-3.2.

Under the usual assumptions that the model errors are independently distributed with mean zero and variance σ^2, we can show that the variance of the ith residual e_i is

$$V(e_i) = \sigma^2(1 - h_{ii}), \qquad i = 1, 2, \ldots, n$$

Furthermore, the h_{ii} elements must fall in the interval $0 < h_{ii} \leq 1$. This implies that the standardized residuals understate the true residual magnitude; thus, the studentized residuals would be a better statistic to examine in evaluating potential **outliers.**

To illustrate, consider the two observations identified in Example 12-10 as having residuals that might be unusually large, observations 15 and 17. The standardized residuals are

$$d_{15} = \frac{e_{15}}{\sqrt{\hat{\sigma}^2}} = \frac{5.84}{\sqrt{5.2352}} = 2.55 \quad \text{and} \quad d_{17} = \frac{e_{17}}{\sqrt{MS_E}} = \frac{4.33}{\sqrt{5.2352}} = 1.89$$

Now $h_{15,15} = 0.0737$ and $h_{17,17} = 0.2593$, so the studentized residuals are

$$r_{15} = \frac{e_{15}}{\sqrt{\hat{\sigma}^2(1 - h_{15,15})}} = \frac{5.84}{\sqrt{5.2352(1 - 0.0737)}} = 2.65$$

and

$$r_{17} = \frac{e_{17}}{\sqrt{\hat{\sigma}^2(1 - h_{17,17})}} = \frac{4.33}{\sqrt{5.2352(1 - 0.2593)}} = 2.20$$

Notice that the studentized residuals are larger than the corresponding standardized residuals. However, the studentized residuals are still not so large as to cause us serious concern about possible outliers.

12-5.2 Influential Observations

When using multiple regression, we occasionally find that some subset of the observations is unusually influential. Sometimes these influential observations are relatively far away from the vicinity where the rest of the data were collected. A hypothetical situation for two variables is depicted in Fig. 12-10, where one observation in x-space is remote from the rest of the data. The disposition of points in the x-space is important in determining the properties of the model. For example, point (x_{i1}, x_{i2}) in Fig. 12-10 may be very influential in determining R^2, the estimates of the regression coefficients, and the magnitude of the error mean square.

We would like to examine the influential points to determine whether they control many model properties. If these influential points are "bad" points, or erroneous in any way, they

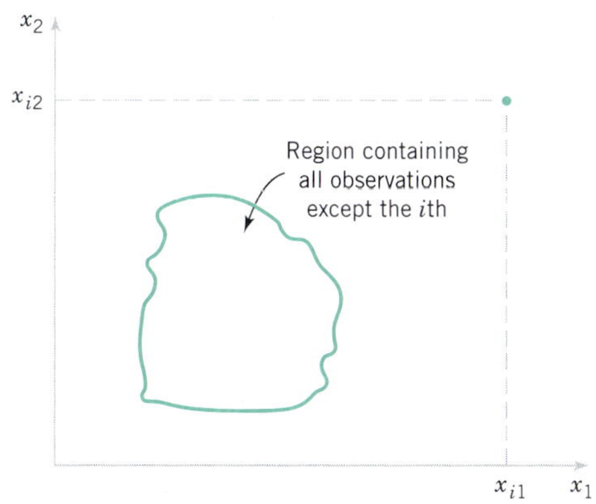

Figure 12-10 A point that is remote in x-space.

should be eliminated. On the other hand, there may be nothing wrong with these points, but at least we would like to determine whether or not they produce results consistent with the rest of the data. In any event, even if an influential point is a valid one, if it controls important model properties, we would like to know this, since it could have an impact on the use of the model.

Montgomery, Peck, and Vining (2001) and Myers (1990) describe several methods for detecting influential observations. An excellent diagnostic is the **distance measure** developed by Dennis R. Cook. This is a measure of the squared distance between the usual least squares estimate of $\boldsymbol{\beta}$ based on all n observations and the estimate obtained when the ith point is removed, say, $\hat{\boldsymbol{\beta}}_{(i)}$. The **Cook's distance measure** is

Cook's Distance

$$D_i = \frac{(\hat{\boldsymbol{\beta}}_{(i)} - \hat{\boldsymbol{\beta}})'\mathbf{X}'\mathbf{X}(\hat{\boldsymbol{\beta}}_{(i)} - \hat{\boldsymbol{\beta}})}{p\hat{\sigma}^2} \qquad i = 1, 2, \ldots, n$$

Clearly, if the ith point is influential, its removal will result in $\hat{\boldsymbol{\beta}}_{(i)}$ changing considerably from the value $\hat{\boldsymbol{\beta}}$. Thus, a large value of D_i implies that the ith point is influential. The statistic D_i is actually computed using

Cook's Distance Formula

$$D_i = \frac{r_i^2}{p}\frac{h_{ii}}{(1 - h_{ii})} \qquad i = 1, 2, \ldots, n \qquad \text{(12-44)}$$

From Equation 12-44 we see that D_i consists of the squared studentized residual, which reflects how well the model fits the ith observation y_i [recall that $r_i = e_i/\sqrt{\hat{\sigma}^2(1 - h_{ii})}$] and a component that measures how far that point is from the rest of the data [$h_{ii}/(1 - h_{ii})$ is a measure of the distance of the ith point from the centroid of the remaining $n - 1$ points]. A value of $D_i > 1$ would indicate that the point is influential. Either component of D_i (or both) may contribute to a large value.

EXAMPLE 12-11 Wire Bond Strength Cook's Distances

Table 12-12 lists the values of the hat matrix diagonals h_{ii} and Cook's distance measure D_i for the wire bond pull strength data in Example 12-1. To illustrate the calculations, consider the first observation:

$$\begin{aligned} D_1 &= \frac{r_1^2}{p} \cdot \frac{h_{11}}{(1 - h_{11})} \\ &= \frac{[e_1/\sqrt{MS_E(1 - h_{11})}]^2}{p} \cdot \frac{h_{11}}{(1 - h_{11})} \\ &= \frac{[1.57/\sqrt{5.2352(1 - 0.1573)}]^2}{3} \cdot \frac{0.1573}{(1 - 0.1573)} \\ &= 0.035 \end{aligned}$$

The Cook distance measure D_i does not identify any potentially influential observations in the data, for no value of D_i exceeds unity.

Table 12-12 Influence Diagnostics for the Wire Bond Pull Strength Data 2

Observations i	h_{ii}	Cook's Distance Measure D_i	Observations i	h_{ii}	Cook's Distance Measure D_i
1	0.1573	0.035	14	0.1129	0.003
2	0.1116	0.012	15	0.0737	0.187
3	0.1419	0.060	16	0.0879	0.001
4	0.1019	0.021	17	0.2593	0.565
5	0.0418	0.024	18	0.2929	0.155
6	0.0749	0.007	19	0.0962	0.018
7	0.1181	0.036	20	0.1473	0.000
8	0.1561	0.020	21	0.1296	0.052
9	0.1280	0.160	22	0.1358	0.028
10	0.0413	0.001	23	0.1824	0.002
11	0.0925	0.013	24	0.1091	0.040
12	0.0526	0.001	25	0.0729	0.000
13	0.0820	0.001			

EXERCISES FOR SECTION 12-5

12-51. Consider the gasoline mileage data in Exercise 12-5.

(a) What proportion of total variability is explained by this model?
(b) Construct a normal probability plot of the residuals and comment on the normality assumption.
(c) Plot residuals versus $\hat{y}$ and versus each regressor. Discuss these residual plots.
(d) Calculate Cook's distance for the observations in this data set. Are any observations influential?

12-52. Consider the electric power consumption data in Exercise 12-4.

(a) Calculate R^2 for this model. Interpret this quantity.
(b) Plot the residuals versus $\hat{y}$ and versus each regressor. Interpret this plot.
(c) Construct a normal probability plot of the residuals and comment on the normality assumption.

12-53. Consider the regression model for the NFL data in Exercise 12-15.

(a) What proportion of total variability is explained by this model?
(b) Construct a normal probability plot of the residuals. What conclusion can you draw from this plot?
(c) Plot the residuals versus $\hat{y}$ and versus each regressor, and comment on model adequacy.
(d) Are there any influential points in these data?

12-54. Consider the regression model for the heat treating data in Exercise 12-8.

(a) Calculate the percent of variability explained by this model.
(b) Construct a normal probability plot for the residuals. Comment on the normality assumption.
(c) Plot the residuals versus $\hat{y}$ and interpret the display.
(d) Calculate Cook's distance for each observation and provide an interpretation of this statistic.

12-55. Consider the regression model fit to the X-ray inspection data in Exercise 12-9. Use rads as the response.

(a) What proportion of total variability is explained by this model?
(b) Construct a normal probability plot of the residuals. What conclusion can you draw from this plot?
(c) Plot the residuals versus $\hat{y}$ and versus each regressor, and comment on model adequacy.
(d) Calculate Cook's distance for the observations in this data set. Are there any influential points in these data?

12-56. Consider the regression model fit to the arsenic data in Exercise 12-10. Use arsenic in nails as the response and age, drink use, and cook use as the regressors.

(a) What proportion of total variability is explained by this model?
(b) Construct a normal probability plot of the residuals. What conclusion can you draw from this plot?
(c) Plot the residuals versus $\hat{y}$ and versus each regressor, and comment on model adequacy.
(d) Calculate Cook's distance for the observations in this data set. Are there any influential points in these data?

12-57. Consider the regression model fit to the coal and limestone mixture data in Exercise 12-11. Use density as the response.

(a) What proportion of total variability is explained by this model?
(b) Construct a normal probability plot of the residuals. What conclusion can you draw from this plot?
(c) Plot the residuals versus $\hat{y}$ and versus each regressor, and comment on model adequacy.
(d) Calculate Cook's distance for the observations in this data set. Are there any influential points in these data?

12-58. Consider the regression model fit to the nisin extraction data in Exercise 12-12.
(a) What proportion of total variability is explained by this model?
(b) Construct a normal probability plot of the residuals. What conclusion can you draw from this plot?
(c) Plot the residuals versus $\hat{y}$ and versus each regressor, and comment on model adequacy.
(d) Calculate Cook's distance for the observations in this data set. Are there any influential points in these data?

12-59. Consider the regression model fit to the grey range modulation data in Exercise 12-13. Use the useful range as the response.
(a) What proportion of total variability is explained by this model?
(b) Construct a normal probability plot of the residuals. What conclusion can you draw from this plot?
(c) Plot the residuals versus $\hat{y}$ and versus each regressor, and comment on model adequacy.
(d) Calculate Cook's distance for the observations in this data set. Are there any influential points in these data?

12-60. Consider the stack loss data in Exercise 12-14.
(a) What proportion of total variability is explained by this model?
(b) Construct a normal probability plot of the residuals. What conclusion can you draw from this plot?
(c) Plot the residuals versus $\hat{y}$ and versus each regressor, and comment on model adequacy.
(d) Calculate Cook's distance for the observations in this data set. Are there any influential points in these data?

12-61. Consider the bearing wear data in Exercise 12-17.
(a) Find the value of R^2 when the model uses the regressors x_1 and x_2.
(b) What happens to the value of R^2 when an interaction term x_1x_2 is added to the model? Does this necessarily imply that adding the interaction term is a good idea?

12-62. Fit a model to the response PITCH in the heat treating data of Exercise 12-8 using new regressors $x_1 =$ SOAKTIME $\times$ SOAKPCT and $x_2 =$ DIFFTIME $\times$ DIFFPCT.
(a) Calculate the R^2 for this model and compare it to the value of R^2 from the original model in Exercise 12-10. Does this provide some information about which model is preferable?
(b) Plot the residuals from this model versus $\hat{y}$ and on a normal probability scale. Comment on model adequacy.
(c) Find the values of Cook's distance measure. Are any observations unusually influential?

12-63. Consider the semiconductor HFE data in Exercise 12-7.
(a) Plot the residuals from this model versus $\hat{y}$. Comment on the information in this plot.
(b) What is the value of R^2 for this model?
(c) Refit the model using log HFE as the response variable.
(d) Plot the residuals versus predicted log HFE for the model in part (c). Does this give any information about which model is preferable?
(e) Plot the residuals from the model in part (d) versus the regressor x_3. Comment on this plot.
(f) Refit the model to log HFE using x_1, x_2, and $1/x_3$, as the regressors. Comment on the effect of this change in the model.

12-64. Consider the regression model for the NHL data from Exercise 12-16.
(a) Fit a model using *GF* as the only regressor.
(b) How much variability is explained by this model?
(c) Plot the residuals versus $\hat{y}$ and comment on model adequacy.
(d) Plot the residuals from part (a) versus *PPG*, the points scored while in power play. Does this indicate that the model would be better if this variable were included?

12-65. The diagonal elements of the hat matrix are often used to denote **leverage**—that is, a point that is unusual in its location in the x-space and that may be influential. Generally, the ith point is called a **leverage point** if its hat diagonal h_{ii} exceeds $2p/n$, which is twice the average size of all the hat diagonals. Recall that $p = k + 1$.
(a) Table 12-12 contains the hat diagonal for the wire bond pull strength data used in Example 12-1. Find the average size of these elements.
(b) Based on the criterion above, are there any observations that are leverage points in the data set?

12-6 ASPECTS OF MULTIPLE REGRESSION MODELING

In this section we briefly discuss several other aspects of building multiple regression models. For more extensive presentations of these topics and additional examples refer to Montgomery, Peck, and Vining (2001) and Myers (1990).

12-6.1 Polynomial Regression Models

The linear model $\mathbf{Y} = \mathbf{X}\boldsymbol{\beta} + \boldsymbol{\epsilon}$ is a general model that can be used to fit any relationship that is **linear in the unknown parameters** $\boldsymbol{\beta}$. This includes the important class of **polynomial regression models.** For example, the second-degree polynomial in one variable

$$Y = \beta_0 + \beta_1 x + \beta_{11} x^2 + \epsilon \tag{12-45}$$

and the second-degree polynomial in two variables

$$Y = \beta_0 + \beta_1 x_1 + \beta_2 x_2 + \beta_{11} x_1^2 + \beta_{22} x_2^2 + \beta_{12} x_1 x_2 + \epsilon \tag{12-46}$$

are linear regression models.

Polynomial regression models are widely used when the response is curvilinear, because the general principles of multiple regression can be applied. The following example illustrates some of the types of analyses that can be performed.

EXAMPLE 12-12
Airplane Sidewall Panels

Sidewall panels for the interior of an airplane are formed in a 1500-ton press. The unit manufacturing cost varies with the production lot size. The data shown below give the average cost per unit (in hundreds of dollars) for this product (y) and the production lot size (x). The scatter diagram, shown in Fig. 12-11, indicates that a second-order polynomial may be appropriate.

y	1.81	1.70	1.65	1.55	1.48	1.40	1.30	1.26	1.24	1.21	1.20	1.18
x	20	25	30	35	40	50	60	65	70	75	80	90

We will fit the model

$$Y = \beta_0 + \beta_1 x + \beta_{11} x^2 + \epsilon$$

The **y** vector, **X** matrix, and **β** vector are as follows:

$$\mathbf{y} = \begin{bmatrix} 1.81 \\ 1.70 \\ 1.65 \\ 1.55 \\ 1.48 \\ 1.40 \\ 1.30 \\ 1.26 \\ 1.24 \\ 1.21 \\ 1.20 \\ 1.18 \end{bmatrix} \qquad \mathbf{X} = \begin{bmatrix} 1 & 20 & 400 \\ 1 & 25 & 625 \\ 1 & 30 & 900 \\ 1 & 35 & 1225 \\ 1 & 40 & 1600 \\ 1 & 50 & 2500 \\ 1 & 60 & 3600 \\ 1 & 65 & 4225 \\ 1 & 70 & 4900 \\ 1 & 75 & 5625 \\ 1 & 80 & 6400 \\ 1 & 90 & 8100 \end{bmatrix} \qquad \boldsymbol{\beta} = \begin{bmatrix} \beta_0 \\ \beta_1 \\ \beta_{11} \end{bmatrix}$$

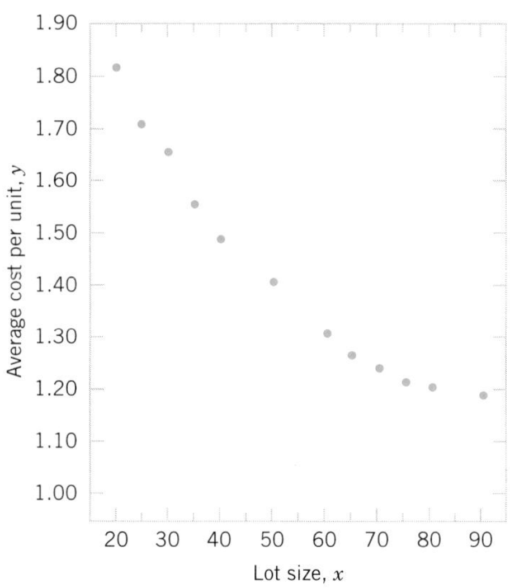

Figure 12-11 Data for Example 12-11.

Table 12-13 Test for Significance of Regression for the Second-Order Model in Example 12-12

Source of Variation	Sum of Squares	Degrees of Freedom	Mean Square	f_0	P-value
Regression	0.52516	2	0.26258	1762.28	2.12E-12
Error	0.00134	9	0.00015		
Total	0.5265	11			

Solving the normal equations $\mathbf{X}'\mathbf{X}\hat{\boldsymbol{\beta}} = \mathbf{X}'\mathbf{y}$ gives the fitted model

$$\hat{y} = 2.19826629 - 0.02252236x + 0.00012507x^2$$

The test for significance of regression is shown in Table 12-13. Since $f_0 = 1762.3$ is significant at 1%, we conclude that at least one of the parameters β_1 and β_{11} is not zero. Furthermore, the standard tests for model adequacy do not reveal any unusual behavior, and we would conclude that this is a reasonable model for the sidewall panel cost data.

In fitting polynomials, we generally like to use the **lowest-degree model** consistent with the data. In this example, it would seem logical to investigate the possibility of dropping the quadratic term from the model. That is, we would like to test

$$H_0: \beta_{11} = 0$$
$$H_1: \beta_{11} \neq 0$$

The general regression significance test can be used to test this hypothesis. We need to determine the "extra sum of squares" due to β_{11}, or

$$SS_R(\beta_{11}|\beta_1,\beta_0) = SS_R(\beta_1,\beta_{11}|\beta_0) - SS_R(\beta_1|\beta_0)$$

The sum of squares $SS_R(\beta_1,\beta_{11}|\beta_0) = 0.52516$ from Table 12-13. To find $SS_R(\beta_1|\beta_0)$, we fit a simple linear regression model to the original data, yielding

$$\hat{y} = 1.90036313 - 0.00910056x$$

It can be easily verified that the regression sum of squares for this model is

$$SS_R(\beta_1|\beta_0) = 0.4942$$

Therefore, the extra sum of the squares due to β_{11}, given that β_1 and β_0 are in the model, is

$$\begin{aligned} SS_R(\beta_{11}|\beta_1,\beta_0) &= SS_R(\beta_1,\beta_{11}|\beta_0) - SS_R(\beta_1|\beta_0) \\ &= 0.5252 - 0.4942 \\ &= 0.031 \end{aligned}$$

The analysis of variance, with the test of H_0: $\beta_{11} = 0$ incorporated into the procedure, is displayed in Table 12-14. Note that the quadratic term contributes significantly to the model.

12-6.2 Categorical Regressors and Indicator Variables

The regression models presented in previous sections have been based on **quantitative** variables, that is, variables that are measured on a numerical scale. For example, variables such as tempera-

Table 12-14 Analysis of Variance for Example 12-12, Showing the Test for H_0: $\beta_{11} = 0$

Source of Variation	Sum of Squares	Degrees of Freedom	Mean Square	f_0	P-value
Regression	$SS_R(\beta_1,\beta_{11}\|\beta_0) = 0.52516$	2	0.26258	1767.40	2.09E-12
Linear	$SS_R(\beta_1\|\beta_0) = 0.49416$	1	0.49416	2236.12	7.13E-13
Quadratic	$SS_R(\beta_{11}\|\beta_0,\beta_1) = 0.03100$	1	0.03100	208.67	1.56E-7
Error	0.00133	9	0.00015		
Total	0.5265	11			

ture, pressure, distance, and voltage are quantitative variables. Occasionally, we need to incorporate **categorical,** or **qualitative,** variables in a regression model. For example, suppose that one of the variables in a regression model is the operator who is associated with each observation y_i. Assume that only two operators are involved. We may wish to assign different levels to the two operators to account for the possibility that each operator may have a different effect on the response.

The usual method of accounting for the different levels of a qualitative variable is to use **indicator variables.** For example, to introduce the effect of two different operators into a regression model, we could define an indicator variable as follows:

$$x = \begin{cases} 0 \text{ if the observation is from operator 1} \\ 1 \text{ if the observation is from operator 2} \end{cases}$$

In general, a qualitative variable with r-levels can be modeled by $r - 1$ indicator variables, which are assigned the value of either zero or one. Thus, if there are *three* operators, the different levels will be accounted for by the *two* indicator variables defined as follows:

x_1	x_2	
0	0	if the observation is from operator 1
1	0	if the observation is from operator 2
0	1	if the observation is from operator 3

Indicator variables are also referred to as **dummy** variables. The following example [from Montgomery, Peck, and Vining (2001)] illustrates some of the uses of indicator variables; for other applications, see Montgomery, Peck, and Vining (2001).

EXAMPLE 12-13
Surface Finish

A mechanical engineer is investigating the surface finish of metal parts produced on a lathe and its relationship to the speed (in revolutions per minute) of the lathe. The data are shown in Table 12-15. Note that the data have been collected using two different types of cutting tools. Since the type of cutting tool likely affects the surface finish, we will fit the model

$$Y = \beta_0 + \beta_1 x_1 + \beta_2 x_2 + \epsilon$$

where Y is the surface finish, x_1 is the lathe speed in revolutions per minute, and x_2 is an indicator variable denoting the type of cutting tool used; that is,

$$x_2 = \begin{cases} 0, \text{ for tool type 302} \\ 1, \text{ for tool type 416} \end{cases}$$

The parameters in this model may be easily interpreted. If $x_2 = 0$, the model becomes

$$Y = \beta_0 + \beta_1 x_1 + \epsilon$$

Table 12-15 Surface Finish Data for Example 12-13

Observation Number, i	Surface Finish y_i	RPM	Type of Cutting Tool	Observation Number, i	Surface Finish y_i	RPM	Type of Cutting Tool
1	45.44	225	302	11	33.50	224	416
2	42.03	200	302	12	31.23	212	416
3	50.10	250	302	13	37.52	248	416
4	48.75	245	302	14	37.13	260	416
5	47.92	235	302	15	34.70	243	416
6	47.79	237	302	16	33.92	238	416
7	52.26	265	302	17	32.13	224	416
8	50.52	259	302	18	35.47	251	416
9	45.58	221	302	19	33.49	232	416
10	44.78	218	302	20	32.29	216	416

which is a straight-line model with slope β_1 and intercept β_0. However, if $x_2 = 1$, the model becomes

$$Y = \beta_0 + \beta_1 x_1 + \beta_2(1) + \epsilon = (\beta_0 + \beta_2) + \beta_1 x_1 + \epsilon$$

which is a straight-line model with slope β_1 and intercept $\beta_0 + \beta_2$. Thus, the model $Y = \beta_0 + \beta_1 x + \beta_2 x_2 + \epsilon$ implies that surface finish is linearly related to lathe speed and that the slope β_1 does not depend on the type of cutting tool used. However, the type of cutting tool does affect the intercept, and β_2 indicates the change in the intercept associated with a change in tool type from 302 to 416.

The **X** matrix and **y** vector for this problem are as follows:

$$\mathbf{X} = \begin{bmatrix} 1 & 225 & 0 \\ 1 & 200 & 0 \\ 1 & 250 & 0 \\ 1 & 245 & 0 \\ 1 & 235 & 0 \\ 1 & 237 & 0 \\ 1 & 265 & 0 \\ 1 & 259 & 0 \\ 1 & 221 & 0 \\ 1 & 218 & 0 \\ 1 & 224 & 1 \\ 1 & 212 & 1 \\ 1 & 248 & 1 \\ 1 & 260 & 1 \\ 1 & 243 & 1 \\ 1 & 238 & 1 \\ 1 & 224 & 1 \\ 1 & 251 & 1 \\ 1 & 232 & 1 \\ 1 & 216 & 1 \end{bmatrix} \quad \mathbf{y} = \begin{bmatrix} 45.44 \\ 42.03 \\ 50.10 \\ 48.75 \\ 47.92 \\ 47.79 \\ 52.26 \\ 50.52 \\ 45.58 \\ 44.78 \\ 33.50 \\ 31.23 \\ 37.52 \\ 37.13 \\ 34.70 \\ 33.92 \\ 32.13 \\ 35.47 \\ 33.49 \\ 32.29 \end{bmatrix}$$

The fitted model is

$$\hat{y} = 14.27620 + 0.14115x_1 - 13.28020x_2$$

Table 12-16 Analysis of Variance for Example 12-13

Source of Variation	Sum of Squares	Degrees of Freedom	Mean Square	f_0	P-value
Regression	1012.0595	2	506.0297	1103.69	1.02E-18
$SS_R(\beta_1\|\beta_0)$	130.6091	1	130.6091	284.87	4.70E-12
$SS_R(\beta_2\|\beta_1,\beta_0)$	881.4504	1	881.4504	1922.52	6.24E-19
Error	7.7943	17	0.4585		
Total	1019.8538	19			

The analysis of variance for this model is shown in Table 12-16. Note that the hypothesis H_0: $\beta_1 = \beta_2 = 0$ (significance of regression) would be rejected at any reasonable level of significance because the P-value is very small. This table also contains the sums of squares

$$\begin{aligned} SS_R &= SS_R(\beta_1,\beta_2|\beta_0) \\ &= SS_R(\beta_1|\beta_0) + SS_R(\beta_2|\beta_1,\beta_0) \end{aligned}$$

so a test of the hypothesis H_0: $\beta_2 = 0$ can be made. Since this hypothesis is also rejected, we conclude that tool type has an effect on surface finish.

It is also possible to use indicator variables to investigate whether tool type affects both the slope and intercept. Let the model be

$$Y = \beta_0 + \beta_1 x_1 + \beta_2 x_2 + \beta_3 x_1 x_2 + \epsilon$$

where x_2 is the indicator variable. Now if tool type 302 is used, $x_2 = 0$, and the model is

$$Y = \beta_0 + \beta_1 x_1 + \epsilon$$

If tool type 416 is used, $x_2 = 1$, and the model becomes

$$\begin{aligned} Y &= \beta_0 + \beta_1 x_1 + \beta_2 + \beta_3 x_1 + \epsilon \\ &= (\beta_0 + \beta_2) + (\beta_1 + \beta_3)x_1 + \epsilon \end{aligned}$$

Note that β_2 is the change in the intercept and that β_3 is the change in slope produced by a change in tool type.

Another method of analyzing these data is to fit separate regression models to the data for each tool type. However, the indicator variable approach has several advantages. First, only one regression model must be fit. Second, by pooling the data on both tool types, more degrees of freedom for error are obtained. Third, tests of both hypotheses on the parameters β_2 and β_3 are just special cases of the extra sum of squares method.

12-6.3 Selection of Variables and Model Building

An important problem in many applications of regression analysis involves selecting the set of regressor variables to be used in the model. Sometimes previous experience or underlying theoretical considerations can help the analyst specify the set of regressor variables to use in a particular situation. Usually, however, the problem consists of selecting an appropriate set of regressors from a set that quite likely includes all the important variables, but we are sure that not all these candidate regressors are necessary to adequately model the response Y.

In such a situation, we are interested in **variable selection;** that is, screening the candidate variables to obtain a regression model that contains the "best" subset of regressor variables. We would like the final model to contain enough regressor variables so that in the intended use of the model (prediction, for example) it will perform satisfactorily. On the other hand, to keep model maintenance costs to a minimum and to make the model easy to use, we would like the model to use as few regressor variables as possible. The compromise between these conflicting objectives is often called finding the "best" regression equation. However, in most problems, no single regression model is "best" in terms of the various evaluation criteria that have been proposed. A great deal of judgment and experience with the system being modeled is usually necessary to select an appropriate set of regressor variables for a regression equation.

No single algorithm will always produce a good solution to the variable selection problem. Most of the currently available procedures are search techniques, and to perform satisfactorily, they require interaction with judgment by the analyst. We now briefly discuss some of the more popular variable selection techniques. We assume that there are K candidate regressors, $x_1, x_2, \ldots, x_K$, and a single response variable y. All models will include an intercept term β_0, so the model with *all* variables included would have $K + 1$ terms. Furthermore, the functional form of each candidate variable (for example, $x_1 = 1/x, x_2 = \ln x$, etc.) is assumed to be correct.

All Possible Regressions

This approach requires that the analyst fit all the regression equations involving one candidate variable, all regression equations involving two candidate variables, and so on. Then these equations are evaluated according to some suitable criteria to select the "best" regression model. If there are K candidate regressors, there are 2^K total equations to be examined. For example, if $K = 4$, there are $2^4 = 16$ possible regression equations; while if $K = 10$, there are $2^{10} = 1024$ possible regression equations. Hence, the number of equations to be examined increases rapidly as the number of candidate variables increases. However, there are some very efficient computing algorithms for all possible regressions available and they are widely implemented in statistical software, so it is a very practical procedure unless the number of candidate regressors is fairly large. Look for a menu choice such as "Best Subsets" regression.

Several criteria may be used for evaluating and comparing the different regression models obtained. A commonly used criterion is based on the value of R^2 or the value of the adjusted R^2, R^2_{adj}. Basically, the analyst continues to increase the number of variables in the model until the increase in R^2 or the adjusted R^2_{adj} is small. Often, we will find that the R^2_{adj} will stabilize and actually begin to decrease as the number of variables in the model increases. Usually, the model that maximizes R^2_{adj} is considered to be a good candidate for the best regression equation. Because we can write $R^2_{\text{adj}} = 1 - \{MS_E/[SS_T/(n - 1)]\}$ and $SS_T/(n - 1)$ is a constant, the model that maximizes the R^2_{adj} value also minimizes the mean square error, so this is a very attractive criterion.

Another criterion used to evaluate regression models is the C_p **statistic,** which is a measure of the total mean square error for the regression model. We define the total standardized mean square error for the regression model as

$$\begin{aligned}\Gamma_p &= \frac{1}{\sigma^2}\sum_{i=1}^{n} E[\hat{Y}_i - E(Y_i)]^2 \\ &= \frac{1}{\sigma^2}\left\{\sum_{i=1}^{n}[E(Y_i) - E(\hat{Y}_i)]^2 + \sum_{i=1}^{n} V(\hat{Y}_i)\right\} \\ &= \frac{1}{\sigma^2}[(\text{bias})^2 + \text{variance}]\end{aligned}$$

We use the mean square error from the *full* $K + 1$ term model as an estimate of σ^2; that is, $\hat{\sigma}^2 = MS_E(K + 1)$. Then an estimator of Γ_p is [see Montgomery, Peck, and Vining (2001) or Myers (1990) for the details]:

C_p Statistic

$$C_p = \frac{SS_E(p)}{\hat{\sigma}^2} - n + 2p \tag{12-47}$$

If the p-term model has negligible bias, it can be shown that

$$E(C_p | \text{zero bias}) = p$$

Therefore, the values of C_p for each regression model under consideration should be evaluated relative to p. The regression equations that have negligible bias will have values of C_p that are close to p, while those with significant bias will have values of C_p that are significantly greater than p. We then choose as the "best" regression equation either a model with *minimum* C_p or a model with a slightly larger C_p, that does not contain as much bias (i.e., $C_p \cong p$).

The **PRESS** statistic can also be used to evaluate competing regression models. PRESS is an acronym for **Prediction Error Sum of Squares,** and it is defined as the sum of the squares of the differences between each observation y_i and the corresponding predicted value based on a model fit to the *remaining* $n - 1$ points, say $\hat{y}_{(i)}$. So PRESS provides a measure of how well the model is likely to perform when predicting *new* data, or data that was not used to fit the regression model. The computing formula for PRESS is

Prediction Error Sum of Squares (PRESS)

$$\text{PRESS} = \sum_{i=1}^{n} (y_i - \hat{y}_{(i)})^2 = \sum_{i=1}^{n} \left(\frac{e_i}{1 - h_{ii}} \right)^2$$

where $e_i = y_i - \hat{y}_i$ is the usual residual. Thus PRESS is easy to calculate from the standard least squares regression results. Models that have small values of PRESS are preferred.

EXAMPLE 12-14
Wine Quality

Table 12-17 presents data on taste-testing 38 brands of pinot noir wine (the data were first reported in an article by Kwan, Kowalski, and Skogenboe in an article in the *Journal of Agricultural and Food Chemistry,* Vol. 27, 1979, and it also appears as one of the default data sets in Minitab). The response variable is y = quality, and we wish to find the "best" regression equation that relates quality to the other five parameters.

Figure 12-12 is the matrix of scatter plots for the wine quality data, as constructed by Minitab. We notice that there are some indications of possible linear relationships between quality and the regressors, but there is no obvious visual impression of which regressors would be appropriate. Table 12-18 lists the all possible regressions output from Minitab. In this analysis, we asked Minitab to present the best three equations for each subset size. Note that Minitab reports the values of R^2, R^2_{adj}, C_p, and $S = \sqrt{MS_E}$ for each model. From Table 12-18 we see that the three-variable equation with x_2 = aroma, x_4 = flavor, and x_5 = oakiness produces the minimum C_p equation, whereas the four-variable

Table 12-17 Wine Quality Data

	x_1 Clarity	x_2 Aroma	x_3 Body	x_4 Flavor	x_5 Oakiness	y Quality
1	1.0	3.3	2.8	3.1	4.1	9.8
2	1.0	4.4	4.9	3.5	3.9	12.6
3	1.0	3.9	5.3	4.8	4.7	11.9
4	1.0	3.9	2.6	3.1	3.6	11.1
5	1.0	5.6	5.1	5.5	5.1	13.3
6	1.0	4.6	4.7	5.0	4.1	12.8
7	1.0	4.8	4.8	4.8	3.3	12.8
8	1.0	5.3	4.5	4.3	5.2	12.0
9	1.0	4.3	4.3	3.9	2.9	13.6
10	1.0	4.3	3.9	4.7	3.9	13.9
11	1.0	5.1	4.3	4.5	3.6	14.4
12	0.5	3.3	5.4	4.3	3.6	12.3
13	0.8	5.9	5.7	7.0	4.1	16.1
14	0.7	7.7	6.6	6.7	3.7	16.1
15	1.0	7.1	4.4	5.8	4.1	15.5
16	0.9	5.5	5.6	5.6	4.4	15.5
17	1.0	6.3	5.4	4.8	4.6	13.8
18	1.0	5.0	5.5	5.5	4.1	13.8
19	1.0	4.6	4.1	4.3	3.1	11.3
20	0.9	3.4	5.0	3.4	3.4	7.9
21	0.9	6.4	5.4	6.6	4.8	15.1
22	1.0	5.5	5.3	5.3	3.8	13.5
23	0.7	4.7	4.1	5.0	3.7	10.8
24	0.7	4.1	4.0	4.1	4.0	9.5
25	1.0	6.0	5.4	5.7	4.7	12.7
26	1.0	4.3	4.6	4.7	4.9	11.6
27	1.0	3.9	4.0	5.1	5.1	11.7
28	1.0	5.1	4.9	5.0	5.1	11.9
29	1.0	3.9	4.4	5.0	4.4	10.8
30	1.0	4.5	3.7	2.9	3.9	8.5
31	1.0	5.2	4.3	5.0	6.0	10.7
32	0.8	4.2	3.8	3.0	4.7	9.1
33	1.0	3.3	3.5	4.3	4.5	12.1
34	1.0	6.8	5.0	6.0	5.2	14.9
35	0.8	5.0	5.7	5.5	4.8	13.5
36	0.8	3.5	4.7	4.2	3.3	12.2
37	0.8	4.3	5.5	3.5	5.8	10.3
38	0.8	5.2	4.8	5.7	3.5	13.2

model, which adds x_1 = clarity to the previous three regressors, results in maximum R^2_{adj} (or minimum MS_E). The three-variable model is

$$\hat{y} = 6.47 + 0.580x_2 + 1.20x_4 - 0.602x_5$$

and the four-variable model is

$$\hat{y} = 4.99 + 1.79x_1 + 0.530x_2 + 1.26x_4 - 0.659x_5$$

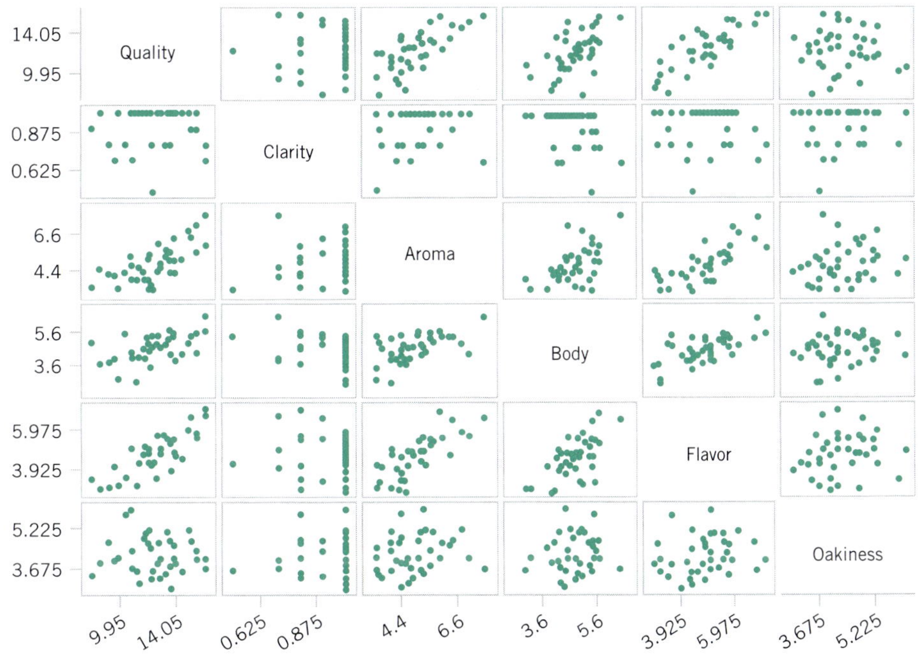

Figure 12-12 A Matrix of Scatter Plots from Minitab for the Wine Quality Data.

Table 12-18 Minitab All Possible Regressions Output for the Wine Quality Data

Best Subsets Regression: Quality versus Clarity, Aroma, . . .

Response is Quality

Vars	R-Sq	R-Sq (adj)	C–p	S	Clarity	Aroma	Body	Flavor	Oakiness
1	62.4	61.4	9.0	1.2712				X	
1	50.0	48.6	23.2	1.4658		X			
1	30.1	28.2	46.0	1.7335			X		
2	66.1	64.2	6.8	1.2242				X	X
2	65.9	63.9	7.1	1.2288		X		X	
2	63.3	61.2	10.0	1.2733	X			X	
3	70.4	67.8	3.9	1.1613		X		X	X
3	68.0	65.2	6.6	1.2068	X			X	X
3	66.5	63.5	8.4	1.2357			X	X	X
4	71.5	68.0	4.7	1.1568	X	X		X	X
4	70.5	66.9	5.8	1.1769		X	X	X	X
4	69.3	65.6	7.1	1.1996	X		X	X	X
5	72.1	67.7	6.0	1.1625	X	X	X	X	X

These models should now be evaluated further using residuals plots and the other techniques discussed earlier in the chapter, to see if either model is satisfactory with respect to the underlying assumptions and to determine if one of them is preferable. It turns out that the residual plots do not reveal any major problems with either model. The value of PRESS for the three-variable model is 56.0524 and for the four-variable model it is 60.3327. Since PRESS is smaller in the model with three regressors, and since it is the model with the smallest number of predictors, it would likely be the preferred choice.

Stepwise Regression

Stepwise regression is probably the most widely used variable selection technique. The procedure iteratively constructs a sequence of regression models by adding or removing variables at each step. The criterion for adding or removing a variable at any step is usually expressed in terms of a partial F-test. Let f_{in} be the value of the F-random variable for adding a variable to the model, and let f_{out} be the value of the F-random variable for removing a variable from the model. We must have $f_{\text{in}} \geq f_{\text{out}}$, and usually $f_{\text{in}} = f_{\text{out}}$.

Stepwise regression begins by forming a one-variable model using the regressor variable that has the highest correlation with the response variable Y. This will also be the regressor producing the largest F-statistic. For example, suppose that at this step, x_1 is selected. At the second step, the remaining $K - 1$ candidate variables are examined, and the variable for which the partial F-statistic

$$F_j = \frac{SS_R(\beta_j | \beta_1, \beta_0)}{MS_E(x_j, x_1)} \tag{12-48}$$

is a maximum is added to the equation, provided that $f_j > f_{\text{in}}$. In equation 12-48, $MS_E(x_j, x_1)$ denotes the mean square for error for the model containing both x_1 and x_j. Suppose that this procedure indicates that x_2 should be added to the model. Now the stepwise regression algorithm determines whether the variable x_1 added at the first step should be removed. This is done by calculating the F-statistic

$$F_1 = \frac{SS_R(\beta_1 | \beta_2, \beta_0)}{MS_E(x_1, x_2)} \tag{12-49}$$

If the calculated value $f_1 < f_{\text{out}}$, the variable x_1 is removed; otherwise it is retained, and we would attempt to add a regressor to the model containing both x_1 and x_2.

In general, at each step the set of remaining candidate regressors is examined, and the regressor with the largest partial F-statistic is entered, provided that the observed value of f exceeds f_{in}. Then the partial F-statistic for each regressor in the model is calculated, and the regressor with the smallest observed value of F is deleted if the observed $f < f_{\text{out}}$. The procedure continues until no other regressors can be added to or removed from the model.

Stepwise regression is almost always performed using a computer program. The analyst exercises control over the procedure by the choice of f_{in} and f_{out}. Some stepwise regression computer programs require that numerical values be specified for f_{in} and f_{out}. Since the number of degrees of freedom on MS_E depends on the number of variables in the model, which changes from step to step, a fixed value of f_{in} and f_{out} causes the type I and type II error rates to vary. Some computer programs allow the analyst to specify the type I error levels for f_{in} and f_{out}. However, the "advertised" significance level is not the true level, because the variable selected is the one that maximizes (or minimizes) the partial F-statistic at that stage. Sometimes it is useful to experiment with different values of f_{in} and f_{out} (or different advertised

Table 12-19 Minitab Stepwise Regression Output for the Wine Quality Data

Stepwise Regression: Quality versus Clarity, Aroma, . . .

Alpha-to-Enter: 0.15 Alpha-to-Remove: 0.15

Response is Quality on 5 predictors, with N = 38

Step	1	2	3
Constant	4.941	6.912	6.467
Flavor	1.57	1.64	1.20
T-Value	7.73	8.25	4.36
P-Value	0.000	0.000	0.000
Oakiness		−0.54	−0.60
T-Value		−1.95	−2.28
P-Value		0.059	0.029
Aroma			0.58
T-Value			2.21
P-Value			0.034
S	1.27	1.22	1.16
R-Sq	62.42	66.11	70.38
R-Sq(adj)	61.37	64.17	67.76
C–p	9.0	6.8	3.9

type I error rates) in several different runs to see if this substantially affects the choice of the final model.

EXAMPLE 12-15
Wine Quality Stepwise Regression

Table 12-19 gives the Minitab stepwise regression output for the wine quality data. Minitab uses fixed values of α for entering and removing variables. The default level is $\alpha = 0.15$ for both decisions. The output in Table 12-19 uses the default value. Notice that the variables were entered in the order Flavor (step 1), Oakiness (step 2), and Aroma (step 3) and that no variables were removed. No other variable could be entered, so the algorithm terminated. This is the three-variable model found by all possible regressions that results in a minimum value of C_p.

Forward Selection

The **forward selection** procedure is a variation of stepwise regression and is based on the principle that regressors should be added to the model one at a time until there are no remaining candidate regressors that produce a significant increase in the regression sum of squares. That is, variables are added one at a time as long as their partial F-value exceeds f_{in}. Forward selection is a simplification of stepwise regression that omits the partial F-test for deleting variables from the model that have been added at previous steps. This is a potential weakness of forward selection; that is, the procedure does not explore the effect that adding a regressor at the current step has on regressor variables added at earlier steps. Notice that if we were to apply forward selection to the wine quality data, we would obtain exactly the same results as we did with stepwise regression in Example 12-15, since stepwise regression terminated without deleting a variable.

Table 12-20 Minitab Backward Elimination Output for the Wine Quality Data

Stepwise Regression: Quality versus Clarity, Aroma, . . .

Backward elimination. Alpha-to-Remove: 0.1

Response is Quality on 5 predictors, with N = 38

Step	1	2	3
Constant	3.997	4.986	6.467
Clarity	2.3	1.8	
T-Value	1.35	1.12	
P-Value	0.187	0.269	
Aroma	0.48	0.53	0.58
T-Value	1.77	2.00	2.21
P-Value	0.086	0.054	0.034
Body	0.27		
T-Value	0.82		
P-Value	0.418		
Flavor	1.17	1.26	1.20
T-Value	3.84	4.52	4.36
P-Value	0.001	0.000	0.000
Oakiness	−0.68	−0.66	−0.60
T-Value	−2.52	−2.46	−2.28
P-Value	0.017	0.019	0.029
S	1.16	1.16	1.16
R-Sq	72.06	71.47	70.38
R-Sq(adj)	67.69	68.01	67.76
C–p	6.0	4.7	3.9

Backward Elimination

The **backward elimination** algorithm begins with all K candidate regressors in the model. Then the regressor with the smallest partial F-statistic is deleted if this F-statistic is insignificant, that is, if $f < f_{out}$. Next, the model with $K - 1$ regressors is fit, and the next regressor for potential elimination is found. The algorithm terminates when no further regressor can be deleted.

Table 12-20 shows the Minitab output for backward elimination applied to the wine quality data. The α value for removing a variable is $\alpha = 0.10$. Notice that this procedure removes Body at step 1 and then Clarity at step 2, terminating with the three-variable model found previously.

Some Comments on Final Model Selection

We have illustrated several different approaches to the selection of variables in multiple linear regression. The final model obtained from any model-building procedure should be subjected to the usual adequacy checks, such as residual analysis, lack-of-fit testing, and examination of the effects of influential points. The analyst may also consider augmenting the original set of candidate variables with cross-products, polynomial terms, or other transformations of the original variables that might improve the model. A major criticism of variable selection methods such as stepwise regression is that the analyst may conclude there is one "best" regression equation. Generally, this is not the case, because several equally good regression models can often be used. One way to avoid this problem is to use several different model-building techniques and see if different models result. For example, we have found the same model for the wine quality data using stepwise regression, forward selection, and backward elimination. The same model was also one of the two best found from all possible regressions. The results from variable selection methods frequently do not agree, so this is a good indication that the three-variable model is the best regression equation.

If the number of candidate regressors is not too large, the all-possible regressions method is recommended. We usually recommend using the minimum MS_E and C_p evaluation criteria in conjunction with this procedure. The all-possible regressions approach can find the "best" regression equation with respect to these criteria, while stepwise-type methods offer no such assurance. Furthermore, the all-possible regressions procedure is not distorted by dependencies among the regressors, as stepwise-type methods are.

12-6.4 Multicollinearity

In multiple regression problems, we expect to find dependencies between the response variable Y and the regressors x_j. In most regression problems, however, we find that there are also dependencies among the regressor variables x_j. In situations where these dependencies are strong, we say that **multicollinearity** exists. Multicollinearity can have serious effects on the estimates of the regression coefficients and on the general applicability of the estimated model.

The effects of multicollinearity may be easily demonstrated. The diagonal elements of the matrix $\mathbf{C} = (\mathbf{X}'\mathbf{X})^{-1}$ can be written as

$$C_{jj} = \frac{1}{(1 - R_j^2)} \qquad j = 1, 2, \ldots, k$$

where R_j^2 is the coefficient of multiple determination resulting from regressing x_j on the other $k - 1$ regressor variables. Clearly, the stronger the linear dependency of x_j on the remaining regressor variables, and hence the stronger the multicollinearity, the larger the value of R_j^2 will be. Recall that $V(\hat{\beta}_j) = \sigma^2 C_{jj}$. Therefore, we say that the variance of $\hat{\beta}_j$ is "inflated" by the quantity $(1 - R_j^2)^{-1}$. Consequently, we define the **variance inflation factor** for β_j as

Variance Inflation Factor (VIF)

$$VIF(\beta_j) = \frac{1}{(1 - R_j^2)} \qquad j = 1, 2, \ldots, k \qquad (12\text{-}50)$$

These factors are an important measure of the extent to which multicollinearity is present.

Although the estimates of the regression coefficients are very imprecise when multicollinearity is present, the fitted model equation may still be useful. For example, suppose we wish to predict new observations on the response. If these predictions are interpolations in the original region of the x-space where the multicollinearity is in effect, satisfactory predictions will often be obtained, because while individual β_j may be poorly estimated, the function $\sum_{j=1}^{k} \beta_j x_{ij}$ may be estimated quite well. On the other hand, if the prediction of new observations requires extrapolation beyond the original region of the x-space where the data were collected, generally we would expect to obtain poor results. Extrapolation usually requires good estimates of the individual model parameters.

Multicollinearity arises for several reasons. It will occur when the analyst collects data such that a linear constraint holds approximately among the columns of the **X** matrix. For example, if four regressor variables are the components of a mixture, such a constraint will always exist because the sum of the components is always constant. Usually, these constraints do not hold exactly, and the analyst might not know that they exist.

The presence of multicollinearity can be detected in several ways. Two of the more easily understood of these will be discussed briefly.

1. The **variance inflation factors,** defined in equation 12-50, are very useful measures of multicollinearity. The larger the variance inflation factor, the more severe the multicollinearity. Some authors have suggested that if any variance inflation factor exceeds 10, multicollinearity is a problem. Other authors consider this value too liberal and suggest that the variance inflation factors should not exceed 4 or 5. Minitab will calculate the variance inflation factors. Table 12-4 presents the Minitab multiple regression output for the wire bond pull strength data. Since both VIF_1 and VIF_2 are small, there is no problem with multicollinearity.
2. If the F-test for significance of regression is significant, but tests on the individual regression coefficients are not significant, multicollinearity may be present.

Several remedial measures have been proposed for solving the problem of multicollinearity. Augmenting the data with new observations specifically designed to break up the approximate linear dependencies that currently exist is often suggested. However, this is sometimes impossible because of economic reasons or because of the physical constraints that relate the x_j. Another possibility is to delete certain variables from the model, but this approach has the disadvantage of discarding the information contained in the deleted variables.

Since multicollinearity primarily affects the stability of the regression coefficients, it would seem that estimating these parameters by some method that is less sensitive to multicollinearity than ordinary least squares would be helpful. Several methods have been suggested. One alternative to ordinary least squares, **ridge regression,** can be useful in combating multicollinearity. For more details on ridge regression, there are more extensive presentations in Montgomery, Peck, and Vining (2001) and Myers (1990).

EXERCISES FOR SECTION 12-6

12-66. An article entitled "A Method for Improving the Accuracy of Polynomial Regression Analysis" in the *Journal of Quality Technology* (1971, pp. 149–155) reported the following data on y = ultimate shear strength of a rubber compound (psi) and x = cure temperature (°F).

y	770	800	840	810
x	280	284	292	295
y	735	640	590	560
x	298	305	308	315

(a) Fit a second-order polynomial to these data.
(b) Test for significance of regression using $\alpha = 0.05$.
(c) Test the hypothesis that $\beta_{11} = 0$ using $\alpha = 0.05$.
(d) Compute the residuals from part (a) and use them to evaluate model adequacy.

12-67. Consider the following data, which result from an experiment to determine the effect of x = test time in hours at a particular temperature on y = change in oil viscosity:
(a) Fit a second-order polynomial to the data.

y	−1.42	−1.39	−1.55	−1.89	−2.43
x	.25	.50	.75	1.00	1.25
y	−3.15	−4.05	−5.15	−6.43	−7.89
x	1.50	1.75	2.00	2.25	2.50

(b) Test for significance of regression using $\alpha = 0.05$.
(c) Test the hypothesis that $\beta_{11} = 0$ using $\alpha = 0.05$.
(d) Compute the residuals from part (a) and use them to evaluate model adequacy.

12-68. The following data were collected during an experiment to determine the change in thrust efficiency

(y, in percent) as the divergence angle of a rocket nozzle (x) changes:

y	24.60	24.71	23.90	39.50	39.60	57.12
x	4.0	4.0	4.0	5.0	5.0	6.0
y	67.11	67.24	67.15	77.87	80.11	84.67
x	6.5	6.5	6.75	7.0	7.1	7.3

(a) Fit a second-order model to the data.
(b) Test for significance of regression and lack of fit using $\alpha = 0.05$.
(c) Test the hypothesis that $\beta_{11} = 0$, using $\alpha = 0.05$.
(d) Plot the residuals and comment on model adequacy.
(e) Fit a cubic model, and test for the significance of the cubic term using $\alpha = 0.05$.

12-69. An article in the *Journal of Pharmaceuticals Sciences* (Vol. 80, 1991, pp. 971–977) presents data on the observed mole fraction solubility of a solute at a constant temperature and the dispersion, dipolar, and hydrogen bonding Hansen partial solubility parameters. The data are as shown in the following table, where y is the negative logarithm of the mole fraction solubility, x_1 is the dispersion partial solubility, x_2 is the dipolar partial solubility, and x_3 is the hydrogen bonding partial solubility.

Observation Number	y	x_1	x_2	x_3
1	0.22200	7.3	0.0	0.0
2	0.39500	8.7	0.0	0.3
3	0.42200	8.8	0.7	1.0
4	0.43700	8.1	4.0	0.2
5	0.42800	9.0	0.5	1.0
6	0.46700	8.7	1.5	2.8
7	0.44400	9.3	2.1	1.0
8	0.37800	7.6	5.1	3.4
9	0.49400	10.0	0.0	0.3
10	0.45600	8.4	3.7	4.1
11	0.45200	9.3	3.6	2.0
12	0.11200	7.7	2.8	7.1
13	0.43200	9.8	4.2	2.0
14	0.10100	7.3	2.5	6.8
15	0.23200	8.5	2.0	6.6
16	0.30600	9.5	2.5	5.0
17	0.09230	7.4	2.8	7.8
18	0.11600	7.8	2.8	7.7
19	0.07640	7.7	3.0	8.0
20	0.43900	10.3	1.7	4.2
21	0.09440	7.8	3.3	8.5
22	0.11700	7.1	3.9	6.6
23	0.07260	7.7	4.3	9.5
24	0.04120	7.4	6.0	10.9
25	0.25100	7.3	2.0	5.2
26	0.00002	7.6	7.8	20.7

(a) Fit the model $Y = \beta_0 + \beta_1 x_1 + \beta_2 x_2 + \beta_3 x_3 + \beta_{12}x_1x_2 + \beta_{13}x_1x_3 + \beta_{23}x_2x_3 + \beta_{11}x_1^2 + \beta_{22}x_2^2 + \beta_{33}x_3^2 + \epsilon$.
(b) Test for significance of regression using $\alpha = 0.05$.
(c) Plot the residuals and comment on model adequacy.
(d) Use the extra sum of squares method to test the contribution of the second-order terms using $\alpha = 0.05$.

12-70. Consider the arsenic concentration data in Exercise 12-10.
(a) Discuss how you would model the information about the person's sex.
(b) Fit a regression model to the arsenic in nails using age, drink use, cook use, and the person's sex as the regressors.
(c) Is there evidence that the person's sex affects arsenic in the nails? Why?

12-71. Consider the gasoline mileage data in Exercise 12-5.
(a) Discuss how you would model the information about the type of transmission in the car.
(b) Fit a regression model to the gasoline mileage using *cid, etw,* and the type of transmission in the car as the regressors.
(c) Is there evidence that the type of transmission (L4, L5, or M6) affects gasoline mileage performance?

12-72. Consider the surface finish data in Example 12-13. Test the hypothesis that two different regression models (with different slopes and intercepts) are required to adequately model the data. Use indicator variables in answering this question.

12-73. Consider the X-ray inspection data in Exercise 12-9. Use rads as the response. Build regression models for the data using the following techniques:
(a) All possible regressions.
(b) Stepwise regression.
(c) Forward selection.
(d) Backward elimination.
(e) Comment on the models obtained. Which model would you prefer? Why?

12-74. Consider the electric power data in Exercise 12-4. Build regression models for the data using the following techniques:
(a) All possible regressions. Find the minimum C_p and minimum MS_E equations.
(b) Stepwise regression.
(c) Forward selection.
(d) Backward elimination.
(e) Comment on the models obtained. Which model would you prefer?

12-75. Consider the regression model fit to the coal and limestone mixture data in Exercise 12-11. Use density as the response. Build regression models for the data using the following techniques:
(a) All possible regressions.
(b) Stepwise regression.

(c) Forward selection.
(d) Backward elimination.
(e) Comment on the models obtained. Which model would you prefer? Why?

12-76. Consider the wire bond pull strength data in Exercise 12-6. Build regression models for the data using the following methods:
(a) All possible regressions. Find the minimum C_p and minimum MS_E equations.
(b) Stepwise regression.
(c) Forward selection.
(d) Backward elimination.
(e) Comment on the models obtained. Which model would you prefer?

12-77. Consider the grey range modulation data in Exercise 12-13. Use the useful range as the response. Build regression models for the data using the following techniques:
(a) All possible regressions.
(b) Stepwise regression.
(c) Forward selection.
(d) Backward elimination.
(e) Comment on the models obtained. Which model would you prefer? Why?

12-78. Consider the nisin extraction data in Exercise 12-12. Build regression models for the data using the following techniques:
(a) All possible regressions.
(b) Stepwise regression.
(c) Forward selection.
(d) Backward elimination.
(e) Comment on the models obtained. Which model would you prefer? Why?

12.79. Consider the stack loss data in Exercise 12-14. Build regression models for the data using the following techniques:
(a) All possible regressions.
(b) Stepwise regression.
(c) Forward selection.
(d) Backward elimination.
(e) Comment on the models obtained. Which model would you prefer? Why?
(f) Remove any influential data points and repeat the model building in the previous parts? Does your conclusion in part (e) change?

12-80. Consider the NHL data in Exercise 12-16. Build regression models for these data with regressors *GF* through *FG* using the following methods:
(a) All possible regressions. Find the minimum C_p and minimum MS_E equations.
(b) Stepwise regression.
(c) Forward selection.
(d) Backward elimination.
(e) Which model would you prefer?

12-81. Use the football data in Exercise 12-15 to build regression models using the following techniques:
(a) All possible regressions. Find the equations that minimize MS_E and that minimize C_p.
(b) Stepwise regression.
(c) Forward selection.
(d) Backward elimination.
(e) Comment on the various models obtained. Which model seems "best," and why?

12-82. Consider the arsenic data in Exercise 12-10. Use arsenic in nails as the response and age, drink use, and cook use as the regressors. Build regression models for the data using the following techniques:
(a) All possible regressions.
(b) Stepwise regression.
(c) Forward selection.
(d) Backward elimination.
(e) Comment on the models obtained. Which model would you prefer? Why?
(f) Now construct an indicator variable and add the person's sex to the list of regressors. Repeat the model building in the previous parts. Does your conclusion in part (e) change?

12-83. Consider the gas mileage data in Exercise 12-5. Build regression models for the data from the numerical regressors using the following techniques:
(a) All possible regressions.
(b) Stepwise regression.
(c) Forward selection.
(d) Backward elimination.
(e) Comment on the models obtained. Which model would you prefer? Why?
(f) Now construct indicator variable for *trns* and *drv* and add these to the list of regressors. Repeat the model building in the previous parts. Does your conclusion in part (e) change?

12-84. When fitting polynomial regression models, we often subtract $\bar{x}$ from each x value to produce a "centered" regressor $x' = x - \bar{x}$. This reduces the effects of dependencies among the model terms and often leads to more accurate estimates of the regression coefficients. Using the data from Exercise 12-66, fit the model $Y = \beta_0^* + \beta_1^* x' + \beta_{11}^*(x')^2 + \epsilon$.
(a) Use the results to estimate the coefficients in the uncentered model $Y = \beta_0 + \beta_1 x + \beta_{11} x^2 + \epsilon$. Predict y when $x = 285°\text{F}$. Suppose that we use a standardized variable $x' = (x - \bar{x})/s_x$, where s_x is the standard deviation of x, in constructing a polynomial regression model. Fit the model $Y = \beta_0^* + \beta_1^* x' + \beta_{11}^*(x')^2 + \epsilon$.
(b) What value of y do you predict when $x = 285°\text{F}$?
(c) Estimate the regression coefficients in the unstandardized model $Y = \beta_0 + \beta_1 x + \beta_{11} x^2 + \epsilon$.
(d) What can you say about the relationship between SS_E and R^2 for the standardized and unstandardized models?
(e) Suppose that $y' = (y - \bar{y})/s_y$ is used in the model along with x'. Fit the model and comment on the relationship

between SS_E and R^2 in the standardized model and the unstandardized model.

12-85. Consider the data in Exercise 12-69. Use all the terms in the full quadratic model as the candidate regressors.

(a) Use forward selection to identify a model.
(b) Use backward elimination to identify a model.
(c) Compare the two models obtained in parts (a) and (b). Which model would you prefer and why?

12-86. We have used a sample of 30 observations to fit a regression model. The full model has nine regressors, the variance estimate is $\hat{\sigma}^2 = MS_E = 100$, and $R^2 = 0.92$.

(a) Calculate the F-statistic for testing significance of regression. Using $\alpha = 0.05$, what would you conclude?
(b) Suppose that we fit another model using only four of the original regressors and that the error sum of squares for this new model is 2200. Find the estimate of σ^2 for this new reduced model. Would you conclude that the reduced model is superior to the old one? Why?
(c) Find the value of C_p for the reduced model in part (b). Would you conclude that the reduced model is better than the old model?

12-87. A sample of 25 observations is used to fit a regression model in seven variables. The estimate of σ^2 for this full model is $MS_E = 10$.

(a) A forward selection algorithm has put three of the original seven regressors in the model. The error sum of squares for the three-variable model is $SS_E = 300$. Based on C_p, would you conclude that the three-variable model has any remaining bias?
(b) After looking at the forward selection model in part (a), suppose you could add one more regressor to the model. This regressor will reduce the error sum of squares to 275. Will the addition of this variable improve the model? Why?

Supplemental Exercises

12-88. The data shown in Table 12-21 represent the thrust of a jet-turbine engine (y) and six candidate regressors: x_1 = primary speed of rotation, x_2 = secondary speed of rotation, x_3 = fuel flow rate, x_4 = pressure, x_5 = exhaust temperature, and x_6 = ambient temperature at time of test.

(a) Fit a multiple linear regression model using x_3 = fuel flow rate, x_4 = pressure, and x_5 = exhaust temperature as the regressors.
(b) Test for significance of regression using $\alpha = 0.01$. Find the P-value for this test. What are your conclusions?
(c) Find the t-test statistic for each regressor. Using $\alpha = 0.01$, explain carefully the conclusion you can draw from these statistics.
(d) Find R^2 and the adjusted statistic for this model.
(e) Construct a normal probability plot of the residuals and interpret this graph.
(f) Plot the residuals versus $\hat{y}$. Are there any indications of inequality of variance or nonlinearity?
(g) Plot the residuals versus x_3. Is there any indication of nonlinearity?
(h) Predict the thrust for an engine for which $x_3 = 28900$, $x_4 = 170$, and $x_5 = 1589$.

12-89. Consider the engine thrust data in Exercise 12-88. Refit the model using $y^* = \ln y$ as the response variable and $x_3^* = \ln x_3$ as the regressor (along with x_4 and x_5).

(a) Test for significance of regression using $\alpha = 0.01$. Find the P-value for this test and state your conclusions.
(b) Use the t-statistic to test H_0: $\beta_j = 0$ versus H_1: $\beta_j \neq 0$ for each variable in the model. If $\alpha = 0.01$, what conclusions can you draw?
(c) Plot the residuals versus $\hat{y}^*$ and versus x_3^*. Comment on these plots. How do they compare with their counterparts obtained in Exercise 12-88 parts (f) and (g)?

12-90. Transient points of an electronic inverter are influenced by many factors. Table 12-22 gives data on the transient point (y, in volts) of PMOS-NMOS inverters and five candidate regressors: x_1 = width of the NMOS device, x_2 = length of the NMOS device, x_3 = width of the PMOS device, x_4 = length of the PMOS device, and x_5 = temperature (°C).

Table 12-22 Transient Point of an Electronic Inverter

Observation Number	x_1	x_2	x_3	x_4	x_5	y
1	3	3	3	3	0	0.787
2	8	30	8	8	0	0.293
3	3	6	6	6	0	1.710
4	4	4	4	12	0	0.203
5	8	7	6	5	0	0.806
6	10	20	5	5	0	4.713
7	8	6	3	3	25	0.607
8	6	24	4	4	25	9.107
9	4	10	12	4	25	9.210
10	16	12	8	4	25	1.365
11	3	10	8	8	25	4.554
12	8	3	3	3	25	0.293
13	3	6	3	3	50	2.252
14	3	8	8	3	50	9.167
15	4	8	4	8	50	0.694
16	5	2	2	2	50	0.379
17	2	2	2	3	50	0.485
18	10	15	3	3	50	3.345
19	15	6	2	3	50	0.208
20	15	6	2	3	75	0.201
21	10	4	3	3	75	0.329
22	3	8	2	2	75	4.966
23	6	6	6	4	75	1.362
24	2	3	8	6	75	1.515
25	3	3	8	8	75	0.751

Table 12-21 Thrust of a Jet-Turbine Engine

Observation Number	y	x_1	x_2	x_3	x_4	x_5	x_6
1	4540	2140	20640	30250	205	1732	99
2	4315	2016	20280	30010	195	1697	100
3	4095	1905	19860	29780	184	1662	97
4	3650	1675	18980	29330	164	1598	97
5	3200	1474	18100	28960	144	1541	97
6	4833	2239	20740	30083	216	1709	87
7	4617	2120	20305	29831	206	1669	87
8	4340	1990	19961	29604	196	1640	87
9	3820	1702	18916	29088	171	1572	85
10	3368	1487	18012	28675	149	1522	85
11	4445	2107	20520	30120	195	1740	101
12	4188	1973	20130	29920	190	1711	100
13	3981	1864	19780	29720	180	1682	100
14	3622	1674	19020	29370	161	1630	100
15	3125	1440	18030	28940	139	1572	101
16	4560	2165	20680	30160	208	1704	98
17	4340	2048	20340	29960	199	1679	96
18	4115	1916	19860	29710	187	1642	94
19	3630	1658	18950	29250	164	1576	94
20	3210	1489	18700	28890	145	1528	94
21	4330	2062	20500	30190	193	1748	101
22	4119	1929	20050	29960	183	1713	100
23	3891	1815	19680	29770	173	1684	100
24	3467	1595	18890	29360	153	1624	99
25	3045	1400	17870	28960	134	1569	100
26	4411	2047	20540	30160	193	1746	99
27	4203	1935	20160	29940	184	1714	99
28	3968	1807	19750	29760	173	1679	99
29	3531	1591	18890	29350	153	1621	99
30	3074	1388	17870	28910	133	1561	99
31	4350	2071	20460	30180	198	1729	102
32	4128	1944	20010	29940	186	1692	101
33	3940	1831	19640	29750	178	1667	101
34	3480	1612	18710	29360	156	1609	101
35	3064	1410	17780	28900	136	1552	101
36	4402	2066	20520	30170	197	1758	100
37	4180	1954	20150	29950	188	1729	99
38	3973	1835	19750	29740	178	1690	99
39	3530	1616	18850	29320	156	1616	99
40	3080	1407	17910	28910	137	1569	100

(a) Fit a multiple linear regression model that uses all regressors to these data. Test for significance of regression using $\alpha = 0.01$. Find the P-value for this test and use it to draw your conclusions.
(b) Test the contribution of each variable to the model using the t-test with $\alpha = 0.05$. What are your conclusions?
(c) Delete x_5 from the model. Test the new model for significance of regression. Also test the relative contribution of each regressor to the new model with the t-test. Using $\alpha = 0.05$, what are your conclusions?
(d) Notice that the MS_E for the model in part (c) is smaller than the MS_E for the full model in part (a). Explain why this has occurred.
(e) Calculate the studentized residuals. Do any of these seem unusually large?
(f) Suppose that you learn that the second observation was incorrectly recorded. Delete this observation and refit the model using x_1, x_2, x_3, and x_4 as the regressors. Notice that the R^2 for this model is considerably higher than the R^2 for either of the models fitted previously. Explain why the R^2 for this model has increased.
(g) Test the model from part (f) for significance of regression using $\alpha = 0.05$. Also investigate the contribution of each regressor to the model using the t-test with $\alpha = 0.05$. What conclusions can you draw?
(h) Plot the residuals from the model in part (f) versus $\hat{y}$ and versus each of the regressors x_1, x_2, x_3, and x_4. Comment on the plots.

12-91. Consider the inverter data in Exercise 12-90. Delete observation 2 from the original data. Define new variables as follows: $y^* = \ln y$, $x_1^* = 1/\sqrt{x_1}$, $x_2^* = \sqrt{x_2}$, $x_3^* = 1/\sqrt{x_3}$, and $x_4^* = \sqrt{x_4}$.
(a) Fit a regression model using these transformed regressors (do not use x_5).
(b) Test the model for significance of regression using $\alpha = 0.05$. Use the t-test to investigate the contribution of each variable to the model ($\alpha = 0.05$). What are your conclusions?
(c) Plot the residuals versus $\hat{y}^*$ and versus each of the transformed regressors. Comment on the plots.

12-92. Following are data on y = green liquor (g/l) and x = paper machine speed (feet per minute) from a Kraft paper machine. (The data were read from a graph in an article in the *Tappi Journal,* March 1986.)

y	16.0	15.8	15.6	15.5	14.8
x	1700	1720	1730	1740	1750
y	14.0	13.5	13.0	12.0	11.0
x	1760	1770	1780	1790	1795

(a) Fit the model $Y = \beta_0 + \beta_1 x + \beta_2 x^2 + \epsilon$ using least squares.
(b) Test for significance of regression using $\alpha = 0.05$. What are your conclusions?
(c) Test the contribution of the quadratic term to the model, over the contribution of the linear term, using an F-statistic. If $\alpha = 0.05$, what conclusion can you draw?
(d) Plot the residuals from the model in part (a) versus $\hat{y}$. Does the plot reveal any inadequacies?
(e) Construct a normal probability plot of the residuals. Comment on the normality assumption.

12-93. Consider the jet engine thrust data in Exercise 12-88 and 12-89. Define the response and regressors as in Exercise 12-89.
(a) Use all possible regressions to select the best regression equation, where the model with the minimum value of MS_E is to be selected as "best."
(b) Repeat part (a) using the C_P criterion to identify the best equation.
(c) Use stepwise regression to select a subset regression model.
(d) Compare the models obtained in parts (a), (b), and (c) above.
(e) Consider the three-variable regression model. Calculate the variance inflation factors for this model. Would you conclude that multicollinearity is a problem in this model?

12-94. Consider the electronic inverter data in Exercise 12-90 and 12-91. Define the response and regressors variables as in Exercise 12-91, and delete the second observation in the sample.
(a) Use all possible regressions to find the equation that minimizes C_p.
(b) Use all possible regressions to find the equation that minimizes MS_E.
(c) Use stepwise regression to select a subset regression model.
(d) Compare the models you have obtained.

12-95. A multiple regression model was used to relate y = viscosity of a chemical product to x_1 = temperature and x_2 = reaction time. The data set consisted of $n = 15$ observations.
(a) The estimated regression coefficients were $\hat{\beta}_0 = 300.00$, $\hat{\beta}_1 = 0.85$, and $\hat{\beta}_2 = 10.40$. Calculate an estimate of mean viscosity when $x_1 = 100°F$ and $x_2 = 2$ hours.
(b) The sums of squares were $SS_T = 1230.50$ and $SS_E = 120.30$. Test for significance of regression using $\alpha = 0.05$. What conclusion can you draw?
(c) What proportion of total variability in viscosity is accounted for by the variables in this model?
(d) Suppose that another regressor, x_3 = stirring rate, is added to the model. The new value of the error sum of squares is $SS_E = 117.20$. Has adding the new variable resulted in a smaller value of MS_E? Discuss the significance of this result.
(e) Calculate an F-statistic to assess the contribution of x_3 to the model. Using $\alpha = 0.05$, what conclusions do you reach?

12-96. Tables 12-23 and 12-24 present statistics for the Major League Baseball 2005 season (source: *The Sports Network*).
(a) Consider the batting data. Use model-building methods to predict *Wins* from the other variables. Check that the assumptions for your model are valid.

Table 12-23 Major League Baseball 2005 Season

	American League Batting													
Team	W	AVG	R	H	2B	3B	HR	RBI	BB	SO	SB	GIDP	LOB	OBP
Chicago	99	0.262	741	1450	253	23	200	713	435	1002	137	122	1032	0.322
Boston	95	0.281	910	1579	339	21	199	863	653	1044	45	135	1249	0.357
LA Angels	95	0.27	761	1520	278	30	147	726	447	848	161	125	1086	0.325
New York	95	0.276	886	1552	259	16	229	847	637	989	84	125	1264	0.355
Cleveland	93	0.271	790	1522	337	30	207	760	503	1093	62	128	1148	0.334
Oakland	88	0.262	772	1476	310	20	155	739	537	819	31	148	1170	0.33
Minnesota	83	0.259	688	1441	269	32	134	644	485	978	102	155	1109	0.323
Toronto	80	0.265	775	1480	307	39	136	735	486	955	72	126	1118	0.331
Texas	79	0.267	865	1528	311	29	260	834	495	1112	67	123	1104	0.329
Baltimore	74	0.269	729	1492	296	27	189	700	447	902	83	145	1103	0.327
Detroit	71	0.272	723	1521	283	45	168	678	384	1038	66	137	1077	0.321
Seattle	69	0.256	699	1408	289	34	130	657	466	986	102	115	1076	0.317
Tampa Bay	67	0.274	750	1519	289	40	157	717	412	990	151	133	1065	0.329
Kansas City	56	0.263	701	1445	289	34	126	653	424	1008	53	139	1062	0.32
	National League Batting													
Team	W	AVG	R	H	2B	3B	HR	RBI	BB	SO	SB	GIDP	LOB	OBP
St. Louis	100	0.27	805	1494	287	26	170	757	534	947	83	127	1152	0.339
Atlanta	90	0.265	769	1453	308	37	184	733	534	1084	92	146	1114	0.333
Houston	89	0.256	693	1400	281	32	161	654	481	1037	115	116	1136	0.322
Philadelphia	88	0.27	807	1494	282	35	167	760	639	1083	116	107	1251	0.348
Florida	83	0.272	717	1499	306	32	128	678	512	918	96	144	1181	0.339
New York	83	0.258	722	1421	279	32	175	683	486	1075	153	103	1122	0.322
San Diego	82	0.257	684	1416	269	39	130	655	600	977	99	122	1220	0.333
Milwaukee	81	0.259	726	1413	327	19	175	689	531	1162	79	137	1120	0.331
Washington	81	0.252	639	1367	311	32	117	615	491	1090	45	130	1137	0.322
Chicago	79	0.27	703	1506	323	23	194	674	419	920	65	131	1133	0.324
Arizona	77	0.256	696	1419	291	27	191	670	606	1094	67	132	1247	0.332
San Francisco	75	0.261	649	1427	299	26	128	617	431	901	71	147	1093	0.319
Cincinnati	73	0.261	820	1453	335	15	222	784	611	1303	72	116	1176	0.339
Los Angeles	71	0.253	685	1374	284	21	149	653	541	1094	58	139	1135	0.326
Colorado	67	0.267	740	1477	280	34	150	704	509	1103	65	125	1197	0.333
Pittsburgh	67	0.259	680	1445	292	38	139	656	471	1092	73	130	1193	0.322

Batting

W	Wins
AVG	Batting average
R	Runs
H	Hits
2B	Doubles
3B	Triples
HR	Home runs
RBI	Runs batted in
BB	Walks
SO	Strikeouts
SB	Stolen bases
GID	Grounded into double play
LOB	Left on base
OBP	On-base percentage

Pitching

ERA	Earned run average
SV	Saves
H	Hits
R	Runs
ER	Earned runs
HR	Home runs
BB	Walks
SO	Strikeouts
AVG	Opponent batting average

Table 12-24 Major League Baseball 2005 Season

Team	American League Pitching									
	W	ERA	SV	H	R	ER	HR	BB	SO	AVG
Chicago	99	3.61	54	1392	645	592	167	459	1040	0.249
Boston	95	4.74	38	1550	805	752	164	440	959	0.276
LA Angels	95	3.68	54	1419	643	598	158	443	1126	0.254
New York	95	4.52	46	1495	789	718	164	463	985	0.269
Cleveland	93	3.61	51	1363	642	582	157	413	1050	0.247
Oakland	88	3.69	38	1315	658	594	154	504	1075	0.241
Minnesota	83	3.71	44	1458	662	604	169	348	965	0.261
Toronto	80	4.06	35	1475	705	653	185	444	958	0.264
Texas	79	4.96	46	1589	858	794	159	522	932	0.279
Baltimore	74	4.56	38	1458	800	724	180	580	1052	0263
Detroit	71	4.51	37	1504	787	719	193	461	907	0.272
Seattle	69	4.49	39	1483	751	712	179	496	892	0.268
Tampa Bay	67	5.39	43	1570	936	851	194	615	949	0.28
Kansas City	56	5.49	25	1640	935	862	178	580	924	0.291

Team	National League Pitching									
	W	ERA	SV	H	R	ER	HR	BB	SO	AVG
St. Louis	100	3.49	48	1399	634	560	153	443	974	0.257
Atlanta	90	3.98	38	1487	674	639	145	520	929	0.268
Houston	89	3.51	45	1336	609	563	155	440	1164	0.246
Philadelphia	88	4.21	40	1379	726	672	189	487	1159	0.253
Florida	83	4.16	42	1459	732	666	116	563	1125	0.266
New York	83	3.76	38	1390	648	599	135	491	1012	0.255
San Diego	82	4.13	45	1452	726	668	146	503	1133	0.259
Milwaukee	81	3.97	46	1382	697	635	169	569	1173	0.251
Washington	81	3.87	51	1456	673	627	140	539	997	0.262
Chicago	79	4.19	39	1357	714	671	186	576	1256	0.25
Arizona	77	4.84	45	1580	856	783	193	537	1038	0.278
San Francisco	75	4.33	46	1456	745	695	151	592	972	0.263
Cincinnati	73	5.15	31	1657	889	820	219	492	955	0.29
Los Angeles	71	4.38	40	1434	755	695	182	471	1004	0.263
Colorado	67	5.13	37	1600	862	808	175	604	981	0.287
Pittsburgh	67	4.42	35	1456	769	706	162	612	958	0.267

Batting

W	Wins
AVG	Batting average
R	Runs
H	Hits
2B	Doubles
3B	Triples
HR	Home runs
RBI	Runs batted in
BB	Walks
SO	Strikeouts
SB	Stolen bases
GID	Grounded into double play
LOB	Left on base
OBP	On-base percentage

Pitching

ERA	Earned run average
SV	Saves
H	Hits
R	Runs
ER	Earned runs
HR	Home runs
BB	Walks
SO	Strikeouts
AVG	Opponent batting average

(b) Repeat part (a) for the pitching data.
(c) Use both the batting and pitching data to build a model to predict *Wins*. What variables are most important? Check that the assumptions for your model are valid.

12-97. An article in the *Journal of the American Ceramics Society* (1992, Vol. 75, pp. 112–116) describes a process for immobilizing chemical or nuclear wastes in soil by dissolving the contaminated soil into a glass block. The authors mix CaO and Na_2O with soil and model viscosity and electrical conductivity. The electrical conductivity model involves six regressors, and the sample consists of $n = 14$ observations.

(a) For the six-regressor model, suppose that $SS_T = 0.50$ and $R^2 = 0.94$. Find SS_E and SS_R, and use this information to test for significance of regression with $\alpha = 0.05$. What are your conclusions?
(b) Suppose that one of the original regressors is deleted from the model, resulting in $R^2 = 0.92$. What can you conclude about the contribution of the variable that was removed? Answer this question by calculating an F-statistic.
(c) Does deletion of the regressor variable in part (b) result in a smaller value of MS_E for the five-variable model, in comparison to the original six-variable model? Comment on the significance of your answer.

MIND-EXPANDING EXERCISES

12-98. Consider a multiple regression model with k regressors. Show that the test statistic for significance of regression can be written as

$$F_0 = \frac{R^2/k}{(1 - R^2)/(n - k - 1)}$$

Suppose that $n = 20$, $k = 4$, and $R^2 = 0.90$. If $\alpha = 0.05$, what conclusion would you draw about the relationship between y and the four regressors?

12-99. A regression model is used to relate a response y to $k = 4$ regressors with $n = 20$. What is the smallest value of R^2 that will result in a significant regression if $\alpha = 0.05$? Use the results of the previous exercise. Are you surprised by how small the value of R^2 is?

12-100. Show that we can express the residuals from a multiple regression model as $\mathbf{e} = (\mathbf{I} - \mathbf{H})\mathbf{y}$, where $\mathbf{H} = \mathbf{X}(\mathbf{X}'\mathbf{X})^{-1}\mathbf{X}'$.

12-101. Show that the variance of the ith residual e_i in a multiple regression model is $\sigma^2(1 - h_{ii})$ and that the covariance between e_i and e_j is $-\sigma^2 h_{ij}$, where the h's are the elements of $\mathbf{H} = \mathbf{X}(\mathbf{X}\,\mathbf{X})^{-1}\mathbf{X}'$.

12-102. Consider the multiple linear regression model $\mathbf{y} = \mathbf{X}\boldsymbol{\beta} + \boldsymbol{\epsilon}$. If $\hat{\boldsymbol{\beta}}$ denotes the least squares estimator of $\boldsymbol{\beta}$, show that $\hat{\boldsymbol{\beta}} = \boldsymbol{\beta} + \mathbf{R}\boldsymbol{\epsilon}$, where $\mathbf{R} = (\mathbf{X}'\mathbf{X})^{-1}\mathbf{X}'$.

12-103. **Constrained Least Squares.** Suppose we wish to find the least squares estimator of $\boldsymbol{\beta}$ in the model $\mathbf{y} = \mathbf{X}\boldsymbol{\beta} + \boldsymbol{\epsilon}$ subject to a set of equality constraints, say, $\mathbf{T}\boldsymbol{\beta} = \mathbf{c}$.

(a) Show that the estimator is

$$\hat{\boldsymbol{\beta}}_c = \hat{\boldsymbol{\beta}} + (\mathbf{X}'\mathbf{X})^{-1} \times \mathbf{T}'[\mathbf{T}(\mathbf{X}'\mathbf{X})^{-1}\mathbf{T}']^{-1}(\mathbf{c} - \mathbf{T}\hat{\boldsymbol{\beta}})$$

where $\hat{\boldsymbol{\beta}} = (\mathbf{X}'\mathbf{X})^{-1}\mathbf{X}'\mathbf{y}$.
(b) Discuss situations where this model might be appropriate.

12-104. **Piecewise Linear Regression.** Suppose that y is piecewise linearly related to x. That is, different linear relationships are appropriate over the intervals $-\infty < x \le x^*$ and $x^* < x < \infty$.

(a) Show how indicator variables can be used to fit such a piecewise linear regression model, assuming that the point x^* is known.
(b) Suppose that at the point x^* a discontinuity occurs in the regression function. Show how indicator variables can be used to incorporate the discontinuity into the model.
(c) Suppose that the point x^* is not known with certainty and must be estimated. Suggest an approach that could be used to fit the piecewise linear regression model.

IMPORTANT TERMS AND CONCEPTS

All possible regressions
Analysis of variance test in multiple regression
Categorical variables
Confidence interval on the mean response
Cp statistic
Extra sum of squares method
Hidden extrapolation
Indicator variables
Inference (test and intervals) on individual model parameters
Influential observations
Model parameters and their interpretation in multiple regression
Multicollinearity
Multiple Regression
Outliers
Polynomial regression model
Prediction interval on a future observation
PRESS statistic
Residual analysis and model adequacy checking
Significance of regression
Stepwise regression and related methods
Variance Inflation Factor (VIF)

13 Designing and Analyzing Single-Factor Experiments

CHAPTER OUTLINE

LEARNING OBJECTIVES

After careful study of this chapter, you should be able to do the following:

1. Design and conduct engineering experiments involving a single factor with an arbitrary number of levels
2. Understand how the analysis of variance is used to analyze the data from these experiments
3. Assess model adequacy with residual plots
4. Use multiple comparison procedures to identify specific differences between means
5. Make decisions about sample size in single-factor experiments
6. Understand the difference between fixed and random factors
7. Estimate variance components in an experiment involving random factors
8. Understand the blocking principle and how it is used to isolate the effect of nuisance factors
9. Design and conduct experiments involving the randomized complete block design

13-1 DESIGNING ENGINEERING EXPERIMENTS

Experiments are a natural part of the engineering and scientific decision-making process. Suppose, for example, that a civil engineer is investigating the effects of different curing methods on the mean compressive strength of concrete. The experiment would consist of making up several test specimens of concrete using each of the proposed curing methods and then testing the compressive strength of each specimen. The data from this experiment could be used to determine which curing method should be used to provide maximum mean compressive strength.

If there are only two curing methods of interest, this experiment could be designed and analyzed using the statistical hypothesis methods for two samples introduced in Chapter 10. That is, the experimenter has a single **factor** of interest—curing methods—and there are only two **levels** of the factor. If the experimenter is interested in determining which curing method produces the maximum compressive strength, the number of specimens to test can be determined from the operating characteristic curves in Appendix Chart VII, and the *t*-test can be used to decide if the two means differ.

Many single-factor experiments require that more than two levels of the factor be considered. For example, the civil engineer may want to investigate five different curing methods. In this chapter we show how the **analysis of variance** (frequently abbreviated ANOVA) can be used for comparing means when there are more than two levels of a single factor. We will also discuss **randomization** of the experimental runs and the important role this concept plays in the overall experimentation strategy. In the next chapter, we will show how to design and analyze experiments with several factors.

Statistically based experimental design techniques are particularly useful in the engineering world for improving the performance of a manufacturing process. They also have extensive applications in the development of new processes. Most processes can be described in terms of several **controllable variables,** such as temperature, pressure, and feed rate. By using designed experiments, engineers can determine which subset of the process variables has the greatest influence on process performance. The results of such an experiment can lead to

- Improved process yield
- Reduced variability in the process and closer conformance to nominal or target requirements
- Reduced design and development time
- Reduced cost of operation

Experimental design methods are also useful in **engineering design** activities, where new products are developed and existing ones are improved. Some typical applications of statistically designed experiments in engineering design include

- Evaluation and comparison of basic design configurations
- Evaluation of different materials
- Selection of design parameters so that the product will work well under a wide variety of field conditions (or so that the design will be robust)
- Determination of key product design parameters that affect product performance

The use of experimental design in the engineering design process can result in products that are easier to manufacture, products that have better field performance and reliability than their competitors, and products that can be designed, developed, and produced in less time.

Designed experiments are usually employed **sequentially.** That is, the first experiment with a complex system (perhaps a manufacturing process) that has many controllable variables is often a **screening experiment** designed to determine which variables are most important. Subsequent experiments are used to refine this information and determine which adjustments to these critical variables are required to improve the process. Finally, the objective of the experimenter is optimization, that is, to determine which levels of the critical variables result in the best process performance.

Every experiment involves a sequence of activities:

1. **Conjecture**—the original hypothesis that motivates the experiment.
2. **Experiment**—the test performed to investigate the conjecture.
3. **Analysis**—the statistical analysis of the data from the experiment.
4. **Conclusion**—what has been learned about the original conjecture from the experiment. Often the experiment will lead to a revised conjecture, and a new experiment, and so forth.

The statistical methods introduced in this chapter and Chapter 14 are essential to good experimentation. **All experiments are designed experiments;** unfortunately, some of them are poorly designed, and as a result, valuable resources are used ineffectively. Statistically designed experiments permit efficiency and economy in the experimental process, and the use of statistical methods in examining the data results in **scientific objectivity** when drawing conclusions.

13-2 COMPLETELY RANDOMIZED SINGLE-FACTOR EXPERIMENT

13-2.1 Example: Tensile Strength

A manufacturer of paper used for making grocery bags is interested in improving the tensile strength of the product. Product engineering thinks that tensile strength is a function of the hardwood concentration in the pulp and that the range of hardwood concentrations of practical interest is between 5 and 20%. A team of engineers responsible for the study decides to investigate four levels of hardwood concentration: 5%, 10%, 15%, and 20%. They decide to make up six test specimens at each concentration level, using a pilot plant. All 24 specimens are tested on a laboratory tensile tester, in random order. The data from this experiment are shown in Table 13-1.

This is an example of a completely randomized single-factor experiment with four levels of the factor. The levels of the factor are sometimes called **treatments,** and each treatment has six observations or **replicates.** The role of **randomization** in this experiment is extremely

Table 13-1 Tensile Strength of Paper (psi)

Hardwood Concentration (%)	Observations 1	2	3	4	5	6	Totals	Averages
5	7	8	15	11	9	10	60	10.00
10	12	17	13	18	19	15	94	15.67
15	14	18	19	17	16	18	102	17.00
20	19	25	22	23	18	20	127	21.17
							383	15.96

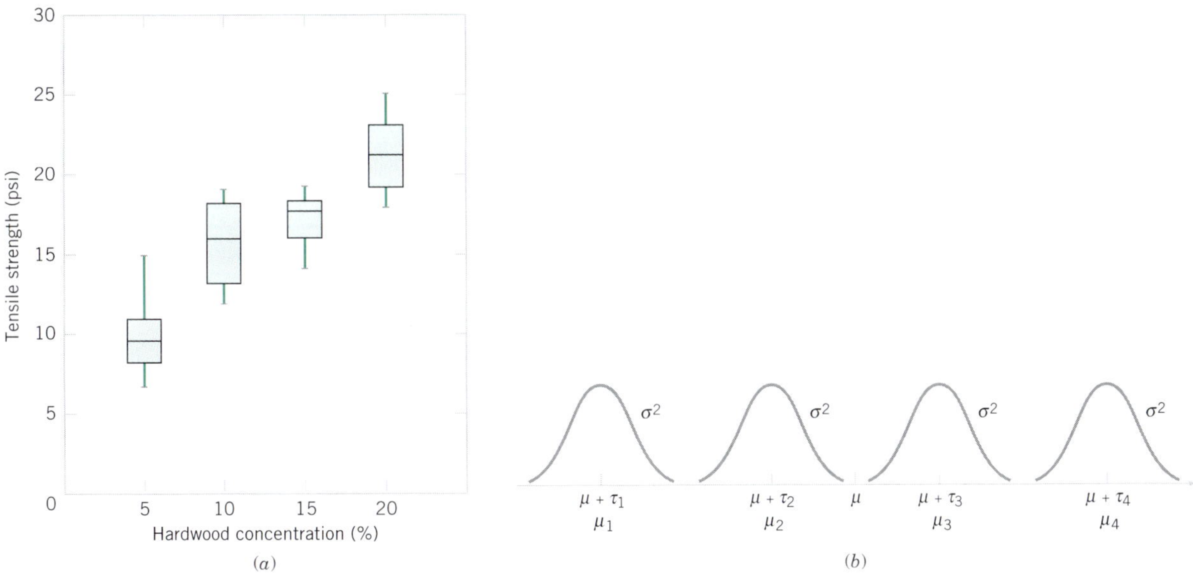

Figure 13-1 (a) Box plots of hardwood concentration data. (b) Display of the model in Equation 13-1 for the completely randomized single-factor experiment.

important. By randomizing the order of the 24 runs, the effect of any nuisance variable that may influence the observed tensile strength is approximately balanced out. For example, suppose that there is a warm-up effect on the tensile testing machine; that is, the longer the machine is on, the greater the observed tensile strength. If all 24 runs are made in order of increasing hardwood concentration (that is, all six 5% concentration specimens are tested first, followed by all six 10% concentration specimens, etc.), any observed differences in tensile strength could also be due to the warm-up effect. The role of randomization to identify causality was discussed in Section 10-1.

It is important to graphically analyze the data from a designed experiment. Figure 13-1(a) presents box plots of tensile strength at the four hardwood concentration levels. This figure indicates that changing the hardwood concentration has an effect on tensile strength; specifically, higher hardwood concentrations produce higher observed tensile strength. Furthermore, the distribution of tensile strength at a particular hardwood level is reasonably symmetric, and the variability in tensile strength does not change dramatically as the hardwood concentration changes.

Graphical interpretation of the data is always useful. Box plots show the variability of the observations *within* a treatment (factor level) and the variability *between* treatments. We now discuss how the data from a single-factor randomized experiment can be analyzed statistically.

13-2.2 Analysis of Variance

Suppose we have a different levels of a single factor that we wish to compare. Sometimes, each factor level is called a **treatment,** a very general term that can be traced to the early applications of experimental design methodology in the agricultural sciences. The **response** for each of the a treatments is a random variable. The observed data would appear as shown

Table 13-2 Typical Data for a Single-Factor Experiment

Treatment	Observations				Totals	Averages
1	y_{11}	y_{12}	$\cdots$	y_{1n}	$y_{1\cdot}$	$\bar{y}_{1\cdot}$
2	y_{21}	y_{22}	$\cdots$	y_{2n}	$y_{2\cdot}$	$\bar{y}_{2\cdot}$
$\vdots$	$\vdots$	$\vdots$	$\vdots\vdots\vdots$	$\vdots$	$\vdots$	$\vdots$
a	y_{a1}	y_{a2}	$\cdots$	y_{an}	$y_{a\cdot}$	$\bar{y}_{a\cdot}$
					$y_{\cdot\cdot}$	$\bar{y}_{\cdot\cdot}$

in Table 13-2. An entry in Table 13-2, say y_{ij}, represents the jth observation taken under treatment i. We initially consider the case in which there are an equal number of observations, n, on each treatment.

We may describe the observations in Table 13-2 by the **linear statistical model**

$$Y_{ij} = \mu + \tau_i + \epsilon_{ij} \begin{cases} i = 1, 2, \ldots, a \\ j = 1, 2, \ldots, n \end{cases} \tag{13-1}$$

where Y_{ij} is a random variable denoting the (ij)th observation, μ is a parameter common to all treatments called the **overall mean,** τ_i is a parameter associated with the ith treatment called the ith **treatment effect,** and ϵ_{ij} is a random error component. Notice that the model could have been written as

$$Y_{ij} = \mu_i + \epsilon_{ij} \begin{cases} i = 1, 2, \ldots, a \\ j = 1, 2, \ldots, n \end{cases}$$

where $\mu_i = \mu + \tau_i$ is the mean of the ith treatment. In this form of the model, we see that each treatment defines a population that has mean μ_i, consisting of the overall mean μ plus an effect τ_i that is due to that particular treatment. We will assume that the errors ϵ_{ij} are normally and independently distributed with mean zero and variance σ^2. Therefore, each treatment can be thought of as a normal population with mean μ_i and variance σ^2. See Fig. 13-1(b).

Equation 13-1 is the underlying model for a single-factor experiment. Furthermore, since we require that the observations are taken in random order and that the environment (often called the experimental units) in which the treatments are used is as uniform as possible, this experimental design is called a **completely randomized design.**

The a factor levels in the experiment could have been chosen in two different ways. First, the experimenter could have specifically chosen the a treatments. In this situation, we wish to test hypotheses about the treatment means, and conclusions cannot be extended to similar treatments that were not considered. In addition, we may wish to estimate the treatment effects. This is called the **fixed-effects model.** Alternatively, the a treatments could be a random sample from a larger population of treatments. In this situation, we would like to be able to extend the conclusions (which are based on the sample of treatments) to all treatments in the population, whether or not they were explicitly considered in the experiment. Here the treatment effects τ_i are random variables, and knowledge about the particular ones investigated is relatively unimportant. Instead, we test hypotheses about the variability of the τ_i and try to estimate this variability. This is called the **random effects,** or **components of variance,** model.

In this section we develop the **analysis of variance** for the fixed-effects model. The analysis of variance is not new to us; it was used previously in the presentation of regression analysis. However, in this section we show how it can be used to test for equality of treatment effects. In the fixed-effects model, the treatment effects τ_i are usually defined as deviations from the overall mean μ, so that

$$\sum_{i=1}^{a} \tau_i = 0 \tag{13-2}$$

Let $y_{i\cdot}$ represent the total of the observations under the ith treatment and $\bar{y}_{i\cdot}$ represent the average of the observations under the ith treatment. Similarly, let $y_{\cdot\cdot}$ represent the grand total of all observations and $\bar{y}_{\cdot\cdot}$ represent the grand mean of all observations. Expressed mathematically,

$$y_{i\cdot} = \sum_{j=1}^{n} y_{ij} \qquad \bar{y}_{i\cdot} = y_{i\cdot}/n \qquad i = 1, 2, \ldots, a$$

$$y_{\cdot\cdot} = \sum_{i=1}^{a} \sum_{j=1}^{n} y_{ij} \qquad \bar{y}_{\cdot\cdot} = y_{\cdot\cdot}/N \tag{13-3}$$

where $N = an$ is the total number of observations. Thus, the "dot" subscript notation implies summation over the subscript that it replaces.

We are interested in testing the equality of the a treatment means $\mu_1, \mu_2, \ldots, \mu_a$. Using Equation 13-2, we find that this is equivalent to testing the hypotheses

$$\begin{aligned} &H_0\colon \tau_1 = \tau_2 = \cdots = \tau_a = 0 \\ &H_1\colon \tau_i \neq 0 \quad \text{for at least one } i \end{aligned} \tag{13-4}$$

Thus, if the null hypothesis is true, each observation consists of the overall mean μ plus a realization of the random error component ϵ_{ij}. This is equivalent to saying that all N observations are taken from a normal distribution with mean μ and variance σ^2. Therefore, if the null hypothesis is true, changing the levels of the factor has no effect on the mean response.

The ANOVA partitions the total variability in the sample data into two component parts. Then, the test of the hypothesis in Equation 13-4 is based on a comparison of two independent estimates of the population variance. The total variability in the data is described by the **total sum of squares**

$$SS_T = \sum_{i-1}^{a} \sum_{j-1}^{n} (y_{ij} - \bar{y}_{\cdot\cdot})^2$$

The partition of the total sum of squares is given in the following definition.

ANOVA Sum of Squares Identity: Single Factor

The **sum of squares identity** is

$$\sum_{i=1}^{a} \sum_{j=1}^{n} (y_{ij} - \bar{y}_{\cdot\cdot})^2 = n \sum_{i=1}^{a} (\bar{y}_{i\cdot} - \bar{y}_{\cdot\cdot})^2 + \sum_{i=1}^{a} \sum_{j=1}^{n} (y_{ij} - \bar{y}_{i\cdot})^2 \tag{13-5}$$

or symbolically

$$SS_T = SS_{\text{Treatments}} + SS_E \tag{13-6}$$

The identity in Equation 13-5 shows that the total variability in the data, measured by the total corrected sum of squares SS_T, can be partitioned into a sum of squares of differences between treatment means and the grand mean denoted $SS_{\text{Treatments}}$ and a sum of squares of differences of observations within a treatment from the treatment mean denoted SS_E. Differences between observed treatment means and the grand mean measure the differences between treatments, while differences of observations within a treatment from the treatment mean can be due only to random error.

We can gain considerable insight into how the analysis of variance works by examining the expected values of $SS_{\text{Treatments}}$ and SS_E. This will lead us to an appropriate statistic for testing the hypothesis of no differences among treatment means (or all $\tau_i = 0$).

Expected Values of Sums of Squares: Single Factor

The expected value of the treatment sum of squares is

$$E(SS_{\text{Treatments}}) = (a - 1)\sigma^2 + n \sum_{i=1}^{a} \tau_i^2$$

and the expected value of the error sum of squares is

$$E(SS_E) = a(n - 1)\sigma^2$$

There is also a partition of the number of degrees of freedom that corresponds to the sum of squares identity in Equation 13-5. That is, there are $an = N$ observations; thus, SS_T has $an - 1$ degrees of freedom. There are a levels of the factor, so $SS_{\text{Treatments}}$ has $a - 1$ degrees of freedom. Finally, within any treatment there are n replicates providing $n - 1$ degrees of freedom with which to estimate the experimental error. Since there are a treatments, we have $a(n - 1)$ degrees of freedom for error. Therefore, the degrees of freedom partition is

$$an - 1 = a - 1 + a(n - 1)$$

The ratio

$$MS_{\text{Treatments}} = SS_{\text{Treatments}}/(a - 1)$$

is called the **mean square for treatments.** Now if the null hypothesis H_0: $\tau_1 = \tau_2 = \cdots = \tau_a = 0$ is true, $MS_{\text{Treatments}}$ is an unbiased estimator of σ^2 because $\sum_{i=1}^{a} \tau_i = 0$. However, if H_1 is true, $MS_{\text{Treatments}}$ estimates σ^2 plus a positive term that incorporates variation due to the systematic difference in treatment means.

Note that the **error mean square**

$$MS_E = SS_E/[a(n - 1)]$$

is an unbiased estimator of σ^2 regardless of whether or not H_0 is true. We can also show that $MS_{\text{Treatments}}$ and MS_E are independent. Consequently, we can show that if the null hypothesis H_0 is true, the ratio

ANOVA *F*-Test

$$F_0 = \frac{SS_{\text{Treatments}}/(a - 1)}{SS_E/[a(n - 1)]} = \frac{MS_{\text{Treatments}}}{MS_E} \tag{13-7}$$

has an F-distribution with $a - 1$ and $a(n - 1)$ degrees of freedom. Furthermore, from the expected mean squares, we know that MS_E is an unbiased estimator of σ^2. Also, under the null hypothesis, $MS_{\text{Treatments}}$ is an unbiased estimator of σ^2. However, if the null hypothesis is false, the expected value of $MS_{\text{Treatments}}$ is greater than σ^2. Therefore, under the alternative hypothesis, the expected value of the numerator of the test statistic (Equation 13-7) is greater than the expected value of the denominator. Consequently, we should reject H_0 if the statistic is large. This implies an upper-tail, one-tail critical region. Therefore, we would reject H_0 if $f_0 > f_{\alpha, a-1, a(n-1)}$ where f_0 is the computed value of F_0 from Equation 13-7.

Efficient computational formulas for the sums of squares may be obtained by expanding and simplifying the definitions of $SS_{\text{Treatments}}$ and SS_T. This yields the following results.

Computing Formulas for ANOVA: Single Factor with Equal Sample Sizes

The sums of squares computing formulas for the ANOVA with equal sample sizes in each treatment are

$$SS_T = \sum_{i=1}^{a} \sum_{j=1}^{n} y_{ij}^2 - \frac{y_{..}^2}{N} \qquad (13\text{-}8)$$

and

$$SS_{\text{Treatments}} = \sum_{i=1}^{a} \frac{y_{i.}^2}{n} - \frac{y_{..}^2}{N} \qquad (13\text{-}9)$$

The error sum of squares is obtained by subtraction as

$$SS_E = SS_T - SS_{\text{Treatments}} \qquad (13\text{-}10)$$

The computations for this test procedure are usually summarized in tabular form as shown in Table 13-3. This is called an **analysis of variance** (or **ANOVA**) **table.**

EXAMPLE 13-1
Tensile Strength ANOVA

Consider the paper tensile strength experiment described in Section 13-2.1. We can use the analysis of variance to test the hypothesis that different hardwood concentrations do not affect the mean tensile strength of the paper.

The hypotheses are

$$H_0\colon \tau_1 = \tau_2 = \tau_3 = \tau_4 = 0$$

$$H_1\colon \tau_i \neq 0 \text{ for at least one } i.$$

Table 13-3 The Analysis of Variance for a Single-Factor Experiment, Fixed-Effects Model

Source of Variation	Sum of Squares	Degrees of Freedom	Mean Square	F_0
Treatments	$SS_{\text{Treatments}}$	$a - 1$	$MS_{\text{Treatments}}$	$\dfrac{MS_{\text{Treatments}}}{MS_E}$
Error	SS_E	$a(n - 1)$	MS_E	
Total	SS_T	$an - 1$		

Table 13-4 ANOVA for the Tensile Strength Data

Source of Variation	Sum of Squares	Degrees of Freedom	Mean Square	f_0	P-value
Hardwood concentration	382.79	3	127.60	19.60	3.59 E-6
Error	130.17	20	6.51		
Total	512.96	23			

We will use $\alpha = 0.01$. The sums of squares for the analysis of variance are computed from Equations 13-8, 13-9, and 13-10 as follows:

$$SS_T = \sum_{i=1}^{4} \sum_{j=1}^{6} y_{ij}^2 - \frac{y_{..}^2}{N}$$

$$= (7)^2 + (8)^2 + \cdots + (20)^2 - \frac{(383)^2}{24} = 512.96$$

$$SS_{\text{Treatments}} = \sum_{i=1}^{4} \frac{y_{i.}^2}{n} - \frac{y_{..}^2}{N}$$

$$= \frac{(60)^2 + (94)^2 + (102)^2 + (127)^2}{6} - \frac{(383)^2}{24} = 382.79$$

$$SS_E = SS_T - SS_{\text{Treatments}}$$

$$= 512.96 - 382.79 = 130.17$$

The ANOVA is summarized in Table 13-4. Since $f_{0.01,3,20} = 4.94$, we reject H_0 and conclude that hardwood concentration in the pulp significantly affects the mean strength of the paper. We can also find a P-value for this test statistic as follows:

$$P = P(F_{3,20} > 19.60) \simeq 3.59 \times 10^{-6}$$

Since $P \simeq 3.59 \times 10^{-6}$ is considerably smaller than $\alpha = 0.01$, we have strong evidence to conclude that H_0 is not true.

Minitab Output

Many software packages have the capability to analyze data from designed experiments using the analysis of variance. Table 13-5 presents the output from the Minitab one-way analysis of variance routine for the paper tensile strength experiment in Example 13-1. The results agree closely with the manual calculations reported previously in Table 13-4.

The Minitab output also presents 95% confidence intervals on each individual treatment mean. The mean of the ith treatment is defined as

$$\mu_i = \mu + \tau_i \qquad i = 1, 2, \ldots, a$$

A point estimator of μ_i is $\hat{\mu}_i = \overline{Y}_{i.}$. Now, if we assume that the errors are normally distributed, each treatment average is normally distributed with mean μ_i and variance σ^2/n. Thus, if σ^2 were known, we could use the normal distribution to construct a CI. Using MS_E as an estimator of σ^2

Table 13-5 Minitab Analysis of Variance Output for Example 13-1

```
One-Way ANOVA: Strength versus CONC
Analysis of Variance for Strength

Source    DF       SS        MS        F        P
Conc       3    382.79    127.60    19.61    0.000
Error     20    130.17      6.51
Total     23    512.96                Individual 95% CIs For Mean
                                      Based on Pooled StDev
Level      N     Mean      StDev      --+------+------+------+--
  5        6    10.000     2.828     (—*—)
 10        6    15.667     2.805             (—*—)
 15        6    17.000     1.789                 (—*—)
 20        6    21.167     2.639                     (—*—)
                                      --+--------+--------+--------+-
Pooled StDev =   2.551                 10.0     15.0     20.0     25.0

Fisher's pairwise comparisons

    Family error rate = 0.192
Individual error rate = 0.0500

Critical value = 2.086

Intervals for (column level mean) − (row level mean)
                  5           10           15
   10        −8.739
             −2.594

   15       −10.072       −4.406
             −3.928        1.739

   20       −14.239       −8.572       −7.239
             −8.094       −2.428       −1.094
```

(The square root of MS_E is the "Pooled StDev" referred to in the Minitab output), we would base the CI on the t distribution, since

$$T = \frac{\overline{Y}_{i\cdot} - \mu_i}{\sqrt{MS_E/n}}$$

has a t-distribution with $a(n - 1)$ degrees of freedom. This leads to the following definition of the confidence interval.

Confidence Interval on a Treatment Mean

A $100(1 - \alpha)$ percent confidence interval on the mean of the ith treatment μ_i is

$$\bar{y}_{i\cdot} - t_{\alpha/2,a(n-1)}\sqrt{\frac{MS_E}{n}} \leq \mu_i \leq \bar{y}_{i\cdot} + t_{\alpha/2,a(n-1)}\sqrt{\frac{MS_E}{n}} \tag{13-11}$$

Equation 13-11 is used to calculate the 95% CIs shown graphically in the Minitab output of Table 13-5. For example, at 20% hardwood the point estimate of the mean is $\bar{y}_{4\cdot} = 21.167$, $MS_E = 6.51$, and $t_{0.025,20} = 2.086$, so the 95% CI is

$$[\bar{y}_{4\cdot} \pm t_{0.025,20}\sqrt{MS_E/n}]$$
$$[21.167 \pm (2.086)\sqrt{6.51/6}]$$

or

$$19.00 \text{ psi} \leq \mu_4 \leq 23.34 \text{ psi}$$

It can also be interesting to find confidence intervals on the difference in two treatment means, say, $\mu_i - \mu_j$. The point estimator of $\mu_i - \mu_j$ is $\bar{Y}_{i\cdot} - \bar{Y}_{j\cdot}$, and the variance of this estimator is

$$V(\bar{Y}_{i\cdot} - \bar{Y}_{j\cdot}) = \frac{\sigma^2}{n} + \frac{\sigma^2}{n} = \frac{2\sigma^2}{n}$$

Now if we use MS_E to estimate σ^2,

$$T = \frac{\bar{Y}_{i\cdot} - \bar{Y}_{j\cdot} - (\mu_i - \mu_j)}{\sqrt{2MS_E/n}}$$

has a t-distribution with $a(n - 1)$ degrees of freedom. Therefore, a CI on $\mu_i - \mu_j$ may be based on the t-distribution.

Confidence Interval on a Difference in Treatment Means

A $100(1 - \alpha)$ percent confidence interval on the difference in two treatment means $\mu_i - \mu_j$ is

$$\bar{y}_{i\cdot} - \bar{y}_{j\cdot} - t_{\alpha/2,a(n-1)}\sqrt{\frac{2MS_E}{n}} \leq \mu_i - \mu_j \leq \bar{y}_{i\cdot} - \bar{y}_{j\cdot} + t_{\alpha/2,a(n-1)}\sqrt{\frac{2MS_E}{n}} \tag{13-12}$$

A 95% CI on the difference in means $\mu_3 - \mu_2$ is computed from Equation 13-12 as follows:

$$[\bar{y}_{3\cdot} - \bar{y}_{2\cdot} \pm t_{0.025,20}\sqrt{2MS_E/n}]$$
$$[17.00 - 15.67 \pm (2.086)\sqrt{2(6.51)/6}]$$

or

$$-1.74 \leq \mu_3 - \mu_2 \leq 4.40$$

Since the CI includes zero, we would conclude that there is no difference in mean tensile strength at these two particular hardwood levels.

The bottom portion of the computer output in Table 13-5 provides additional information concerning which specific means are different. We will discuss this in more detail in Section 13-2.3.

An Unbalanced Experiment

In some single-factor experiments, the number of observations taken under each treatment may be different. We then say that the design is **unbalanced.** In this situation, slight modifications must be made in the sums of squares formulas. Let n_i observations be taken under treatment i ($i = 1, 2, \ldots, a$), and let the total number of observations $N = \sum_{i=1}^{a} n_i$. The computational formulas for SS_T and $SS_{\text{Treatments}}$ are as shown in the following definition.

Computing Formulas for ANOVA: Single Factor with Unequal Sample Sizes

The sums of squares computing formulas for the ANOVA with unequal sample sizes n_i in each treatment are

$$SS_T = \sum_{i=1}^{a} \sum_{j=1}^{n_i} y_{ij}^2 - \frac{y_{..}^2}{N} \tag{13-13}$$

$$SS_{\text{Treatments}} = \sum_{i=1}^{a} \frac{y_{i\cdot}^2}{n_i} - \frac{y_{..}^2}{N} \tag{13-14}$$

and

$$SS_E = SS_T - SS_{\text{Treatments}} \tag{13-15}$$

Choosing a balanced design has two important advantages. First, the ANOVA is relatively insensitive to small departures from the assumption of equality of variances if the sample sizes are equal. This is not the case for unequal sample sizes. Second, the power of the test is maximized if the samples are of equal size.

13-2.3 Multiple Comparisons Following the ANOVA

When the null hypothesis H_0: $\tau_1 = \tau_2 = \cdots = \tau_a = 0$ is rejected in the ANOVA, we know that some of the treatment or factor level means are different. However, the ANOVA doesn't identify which means are different. Methods for investigating this issue are called **multiple comparisons methods.** Many of these procedures are available. Here we describe a very simple one, Fisher's **least significant difference** (LSD) method and a graphical method. Montgomery (2005) presents these and other methods and provides a comparative discussion.

The Fisher LSD method compares all pairs of means with the null hypotheses H_0: $\mu_i = \mu_j$ (for all $i \neq j$) using the t-statistic

$$t_0 = \frac{\bar{y}_{i\cdot} - \bar{y}_{j\cdot}}{\sqrt{\dfrac{2MS_E}{n}}}$$

Assuming a two-sided alternative hypothesis, the pair of means μ_i and μ_j would be declared significantly different if

$$|\bar{y}_{i\cdot} - \bar{y}_{j\cdot}| > \text{LSD}$$

where LSD, the **least significant difference,** is

Least Significant Difference for Multiple Comparisons

$$\text{LSD} = t_{\alpha/2, a(n-1)} \sqrt{\frac{2MS_E}{n}} \tag{13-16}$$

If the sample sizes are different in each treatment, the LSD is defined as

$$\text{LSD} = t_{\alpha/2,N-a}\sqrt{MS_E\left(\frac{1}{n_i} + \frac{1}{n_j}\right)}$$

EXAMPLE 13-2

We will apply the Fisher LSD method to the hardwood concentration experiment. There are $a = 4$ means, $n = 6$, $MS_E = 6.51$, and $t_{0.025,20} = 2.086$. The treatment means are

$$\begin{aligned}\bar{y}_{1\cdot} &= 10.00 \text{ psi}\\ \bar{y}_{2\cdot} &= 15.67 \text{ psi}\\ \bar{y}_{3\cdot} &= 17.00 \text{ psi}\\ \bar{y}_{4\cdot} &= 21.17 \text{ psi}\end{aligned}$$

The value of LSD is LSD $= t_{0.025,20}\sqrt{2MS_E/n} = 2.086\sqrt{2(6.51)/6} = 3.07$. Therefore, any pair of treatment averages that differs by more than 3.07 implies that the corresponding pair of treatment means are different.

The comparisons among the observed treatment averages are as follows:

$$\begin{aligned}4 \text{ vs. } 1 &= 21.17 - 10.00 = 11.17 > 3.07\\ 4 \text{ vs. } 2 &= 21.17 - 15.67 = 5.50 > 3.07\\ 4 \text{ vs. } 3 &= 21.17 - 17.00 = 4.17 > 3.07\\ 3 \text{ vs. } 1 &= 17.00 - 10.00 = 7.00 > 3.07\\ 3 \text{ vs. } 2 &= 17.00 - 15.67 = 1.33 < 3.07\\ 2 \text{ vs. } 1 &= 15.67 - 10.00 = 5.67 > 3.07\end{aligned}$$

From this analysis, we see that there are significant differences between all pairs of means except 2 and 3. This implies that 10 and 15% hardwood concentration produce approximately the same tensile strength and that all other concentration levels tested produce different tensile strengths. It is often helpful to draw a graph of the treatment means, such as in Fig. 13-2, with the means that are *not* different underlined. This graph clearly reveals the results of the experiment and shows that 20% hardwood produces the maximum tensile strength.

The Minitab output in Table 13-5 shows the Fisher LSD method under the heading "Fisher's pairwise comparisons." The critical value reported is actually the value of $t_{0.025,20} = 2.086$. Minitab implements Fisher's LSD method by computing **confidence intervals** on all pairs of treatment means using Equation 13-12. The lower and upper 95% confidence limits are shown at the bottom of the table. Notice that the only pair of means for which the confidence interval includes zero is for μ_{10} and μ_{15}. This implies that μ_{10} and μ_{15} are not significantly different, the same result found in Example 13-2.

Table 13-5 also provides a "family error rate," equal to 0.192 in this example. When all possible pairs of means are tested, the probability of at least one type I error can be much greater than for a single test. We can interpret the family error rate as follows. The probability is $1 - 0.192 = 0.808$ that there are no type I errors in the six comparisons. The family error

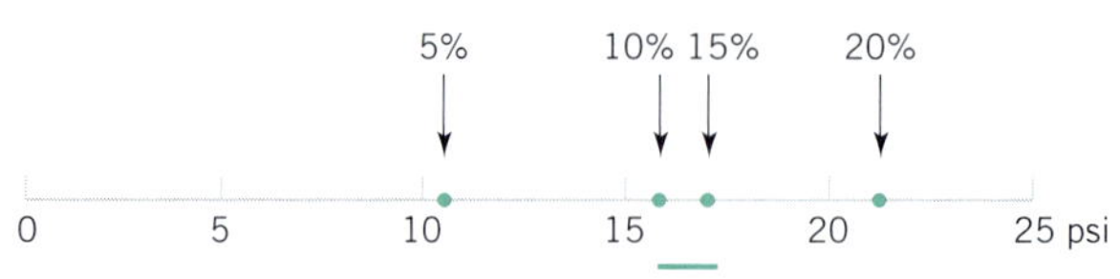

Figure 13-2 Results of Fisher's LSD method in Example 13-2.

rate in Table 13-5 is based on the distribution of the range of the sample means. See Montgomery (2001) for details. Alternatively, Minitab permits you to specify a family error rate and will then calculate an individual error rate for each comparison.

Graphical Comparison of Means

It is easy to compare treatment means graphically, following the analysis of variance. Suppose that the factor has a levels and that $\bar{y}_{1\cdot}, \bar{y}_{2\cdot}, \ldots, \bar{y}_{a\cdot}$ are the observed averages for these factor levels. Each treatment average has standard deviation $\sigma/\sqrt{n}$, where σ is the standard deviation of an individual observation. If all treatment means are equal, the observed means $\bar{y}_{i\cdot}$ would behave as if they were a set of observations drawn at random from a normal distribution with mean μ and standard deviation $\sigma/\sqrt{n}$.

Visualize this normal distribution capable of being slid along an axis below which the treatment means $\bar{y}_{1\cdot}, \bar{y}_{2\cdot}, \ldots, \bar{y}_{a\cdot}$ are plotted. If all treatment means are equal, there should be some position for this distribution that makes it obvious that the $\bar{y}_{i\cdot}$ values were drawn from the same distribution. If this is not the case, the $\bar{y}_{i\cdot}$ values that do not appear to have been drawn from this distribution are associated with treatments that produce different mean responses.

The only flaw in this logic is that σ is unknown. However, we can use $\sqrt{MS_E}$ from the analysis of variance to estimate σ. This implies that a t-distribution should be used instead of the normal in making the plot, but since the t looks so much like the normal, sketching a normal curve that is approximately $6\sqrt{MS_E/n}$ units wide will usually work very well.

Figure 13-3 shows this arrangement for the hardwood concentration experiment in Example 13-1. The standard deviation of this normal distribution is

$$\sqrt{MS_E/n} = \sqrt{6.51/6} = 1.04$$

If we visualize sliding this distribution along the horizontal axis, we note that there is no location for the distribution that would suggest that all four observations (the plotted means) are typical, randomly selected values from that distribution. This, of course, should be expected, because the analysis of variance has indicated that the means differ, and the display in Fig. 13-3 is just a graphical representation of the analysis of variance results. The figure does indicate that treatment 4 (20% hardwood) produces paper with higher mean tensile strength than do the other treatments, and treatment 1 (5% hardwood) results in lower mean tensile strength than do the other treatments. The means of treatments 2 and 3 (10 and 15% hardwood, respectively) do not differ.

This simple procedure is a rough but very effective multiple comparison technique. It works well in many situations.

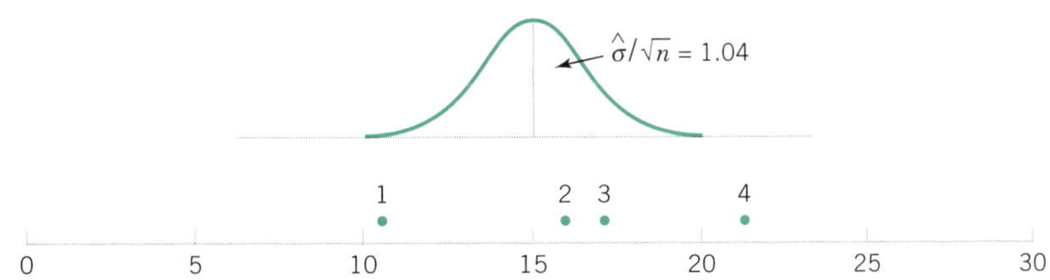

Figure 13-3 Tensile strength averages from the hardwood concentration experiment in relation to a normal distribution with standard deviation $\sqrt{MS_E/n} = \sqrt{6.51/6} = 1.04$.

Table 13-6 Residuals for the Tensile Strength Experiment

Hardwood Concentration (%)	Residuals					
5	−3.00	−2.00	5.00	1.00	−1.00	0.00
10	−3.67	1.33	−2.67	2.33	3.33	−0.67
15	−3.00	1.00	2.00	0.00	−1.00	1.00
20	−2.17	3.83	0.83	1.83	−3.17	−1.17

13-2.4 Residual Analysis and Model Checking

The analysis of variance assumes that the observations are normally and independently distributed with the same variance for each treatment or factor level. These assumptions should be checked by examining the residuals. A **residual** is the difference between an observation y_{ij} and its estimated (or fitted) value from the statistical model being studied, denoted as $\hat{y}_{ij}$. For the completely randomized design $\hat{y}_{ij} = \bar{y}_{i\cdot}$ and each residual is $e_{ij} = y_{ij} - \bar{y}_{i\cdot}$, that is, the difference between an observation and the corresponding observed treatment mean. The residuals for the paper tensile strength experiment are shown in Table 13-6. Using $\bar{y}_{i\cdot}$ to calculate each residual essentially removes the effect of hardwood concentration from the data; consequently, the residuals contain information about unexplained variability.

The normality assumption can be checked by constructing a **normal probability plot** of the residuals. To check the assumption of equal variances at each factor level, plot the residuals against the factor levels and compare the spread in the residuals. It is also useful to plot the residuals against $\bar{y}_{i\cdot}$ (sometimes called the fitted value); the variability in the residuals should not depend in any way on the value of $\bar{y}_{i\cdot}$. Most statistical software packages will construct these plots on request. When a pattern appears in these plots, it usually suggests the need for a transformation, that is, analyzing the data in a different metric. For example, if the variability in the residuals increases with $\bar{y}_{i\cdot}$, a transformation such as $\log y$ or $\sqrt{y}$ should be considered. In some problems, the dependency of residual scatter on the observed mean $\bar{y}_{i\cdot}$ is very important information. It may be desirable to select the factor level that results in maximum response; however, this level may also cause more variation in response from run to run.

The independence assumption can be checked by plotting the residuals against the time or run order in which the experiment was performed. A pattern in this plot, such as sequences of positive and negative residuals, may indicate that the observations are not independent. This suggests that time or run order is important or that variables that change over time are important and have not been included in the experimental design.

A normal probability plot of the residuals from the paper tensile strength experiment is shown in Fig. 13-4. Figures 13-5 and 13-6 present the residuals plotted against the factor levels and the fitted value $\bar{y}_{i\cdot}$ respectively. These plots do not reveal any model inadequacy or unusual problem with the assumptions.

13-2.5 Determining Sample Size

In any experimental design problem, the choice of the sample size or number of replicates to use is important. **Operating characteristic curves** can be used to provide guidance in making this selection. Recall that an operating characteristic curve is a plot of the probability of a type II error (β) for various sample sizes against values of the parameters under test. The

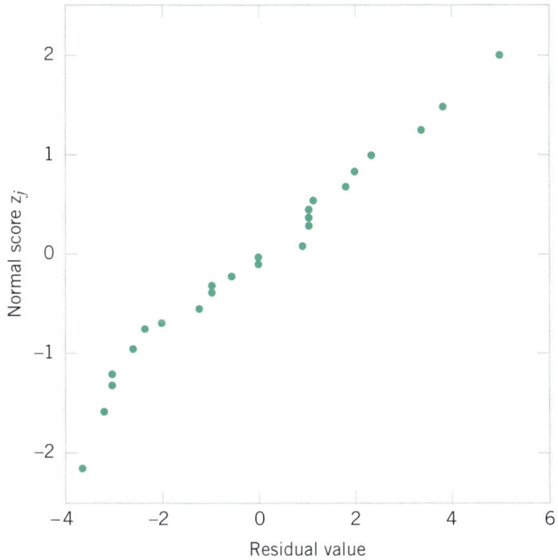

Figure 13-4 Normal probability plot of residuals from the hardwood concentration experiment.

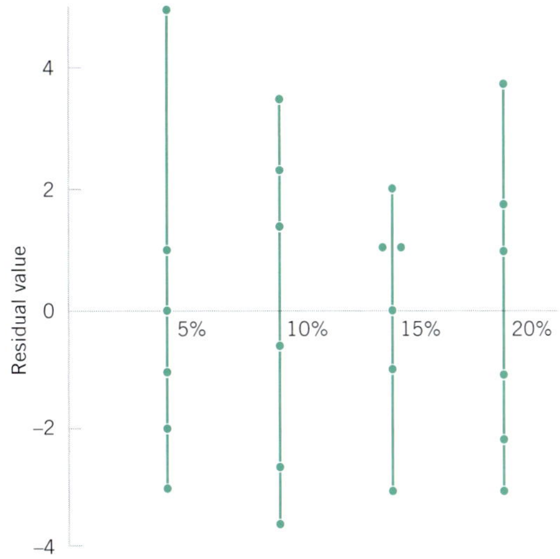

Figure 13-5 Plot of residuals versus factor levels (hardwood concentration).

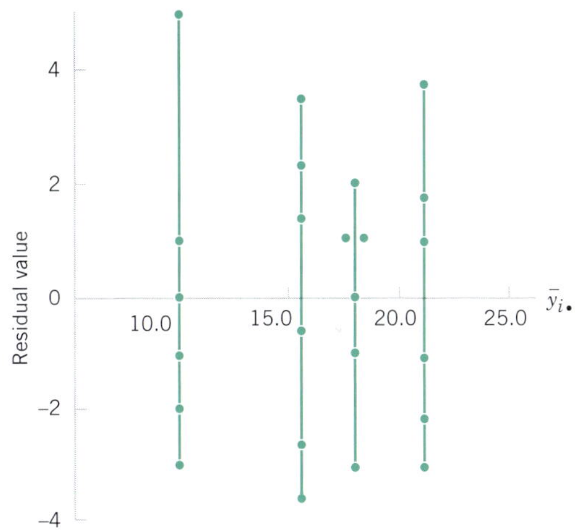

Figure 13-6 Plot of residuals versus $\bar{y}_{i.}$

operating characteristic curves can be used to determine how many replicates are required to achieve adequate sensitivity.

The power of the ANOVA test is

$$\begin{aligned} 1 - \beta &= P\{\text{Reject } H_0 \mid H_0 \text{ is false}\} \\ &= P\{F_0 > f_{\alpha,a-1,a(n-1)} \mid H_0 \text{ is false}\} \end{aligned} \tag{13-17}$$

To evaluate this probability statement, we need to know the distribution of the test statistic F_0 if the null hypothesis is false. Because ANOVA compares several means, the null hypothesis can be false in different ways. For example, possibly $\tau_1 > 0$, $\tau_2 = 0$, $\tau_3 < 0$, and so forth. It can be shown that the power for ANOVA in Equation 13-17 depends on the τ_i's only through the function

$$\Phi^2 = \frac{n \sum_{i=1}^{a} \tau_i^2}{a\sigma^2}$$

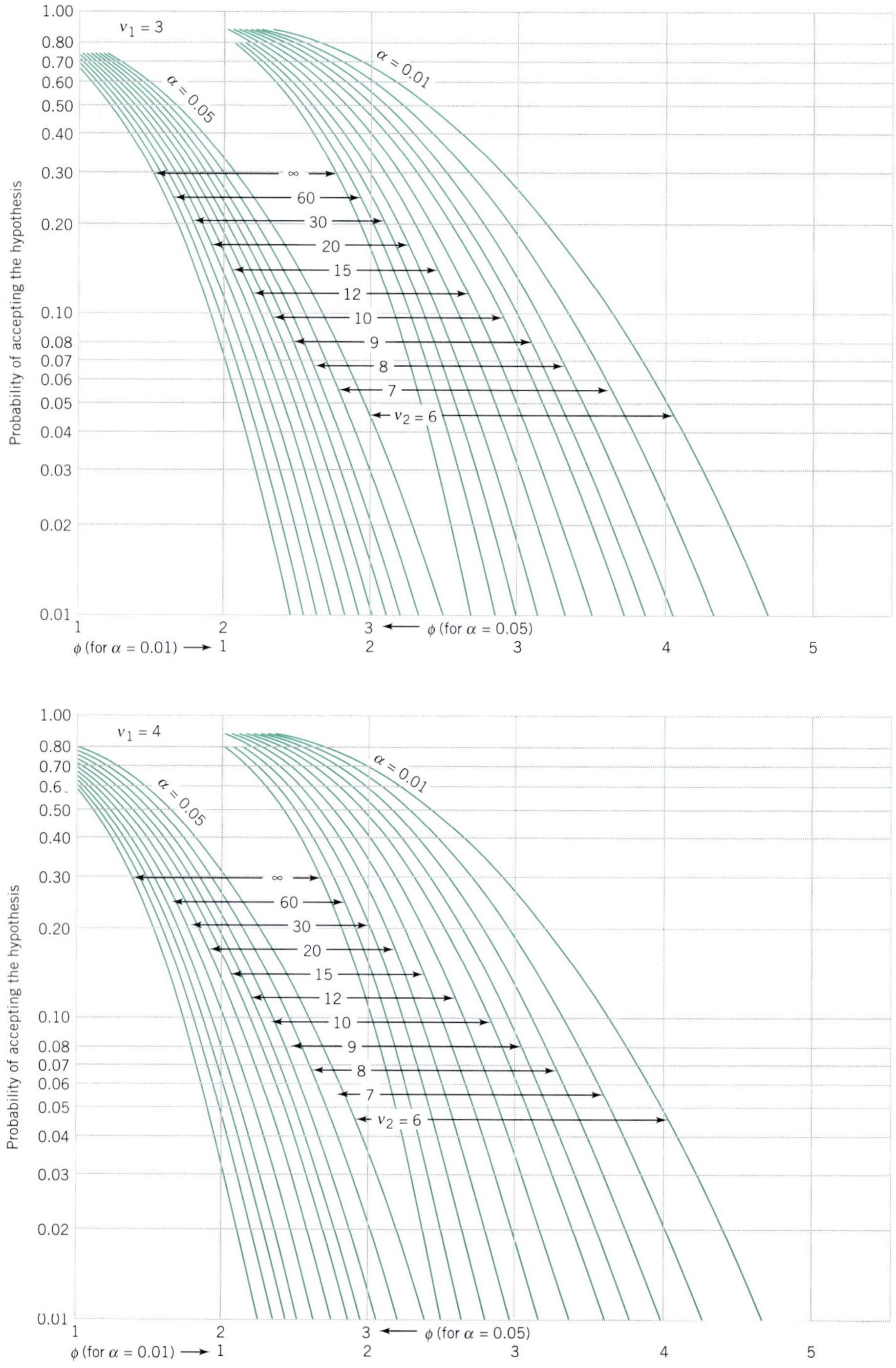

Figure 13-7 Two Operating Characteristic curves for the fixed-effects model analysis of variance. Top curves for four treatments and bottom curves for five treatments.

Therefore, alternative hypotheses for the τ_i's can be used to calculate Φ^2 and this in turn can be used to calculate the power. Specifically, it can be shown that if H_0 is false, the statistic $F_0 = MS_{\text{Treatments}}/MS_E$ has a **noncentral F distribution,** with $a - 1$ and $n(a - 1)$ degrees of freedom and a noncentrality parameter that depends on Φ^2. Instead of tables for the noncentral F distribution, operating characteristic curves are used to evaluate β defined in Equation 13-17. These curves plot β against Φ.

Curves are available for $\alpha = 0.05$ and $\alpha = 0.01$ and for several values of the number of degrees of freedom for numerator (denoted v_1) and denominator (denoted v_2). Figure 13-7 gives representative O.C. curves, one for $a = 4$ ($v_1 = 3$) and one for $a = 5$ ($v_1 = 4$) treatments. Notice that for each value of a there are curves for $\alpha = 0.05$ and $\alpha = 0.01$.

In using the operating curves, we must define the difference in means that we wish to detect in terms of $\sum_{i=1}^{a} \tau_i^2$. Also, the error variance σ^2 is usually unknown. In such cases, we must choose ratios of $\sum_{i=1}^{a} \tau_i^2/\sigma^2$ that we wish to detect. Alternatively, if an estimate of σ^2 is available, one may replace σ^2 with this estimate. For example, if we were interested in the sensitivity of an experiment that has already been performed, we might use MS_E as the estimate of σ^2.

EXAMPLE 13-3

Suppose that five means are being compared in a completely randomized experiment with $\alpha = 0.01$. The experimenter would like to know how many replicates to run if it is important to reject H_0 with probability at least 0.90 if $\sum_{i=1}^{5} \tau_i^2/\sigma^2 = 5.0$. The parameter Φ^2 is, in this case,

$$\Phi^2 = \frac{n \sum_{i=1}^{a} \tau_i^2}{a\sigma^2} = \frac{n}{5}(5) = n$$

and for the operating characteristic curve with $v_1 = a - 1 = 5 - 1 = 4$, and $v_2 = a(n - 1) = 5(n - 1)$ error degrees of freedom refer to the lower curve in Figure 13-7. As a first guess, try $n = 4$ replicates. This yields $\Phi^2 = 4$, $\Phi = 2$, and $v_2 = 5(3) = 15$ error degrees of freedom. Consequently, from Figure 13-7, we find that $\beta \simeq 0.38$. Therefore, the power of the test is approximately $1 - \beta = 1 - 0.38 = 0.62$, which is less than the required 0.90, and so we conclude that $n = 4$ replicates is not sufficient. Proceeding in a similar manner, we can construct the following table:

n	Φ^2	Φ	$a(n-1)$	β	Power $= (1 - \beta)$
4	4	2.00	15	0.38	0.62
5	5	2.24	20	0.18	0.82
6	6	2.45	25	0.06	0.94

Thus, at least $n = 6$ replicates must be run in order to obtain a test with the required power.

EXERCISES FOR SECTION 13-2

13-1. In *Design and Analysis of Experiments,* 6th edition (John Wiley & Sons, 2005) D. C. Montgomery describes an experiment in which the tensile strength of a synthetic fiber is of interest to the manufacturer. It is suspected that strength is related to the percentage of cotton in the fiber. Five levels of cotton percentage are used, and five replicates are run in random order, resulting in the data below.

Cotton Percentage	Observations				
	1	2	3	4	5
15	7	7	15	11	9
20	12	17	12	18	18
25	14	18	18	19	19
30	19	25	22	19	23
35	7	10	11	15	11

(a) Does cotton percentage affect breaking strength? Draw comparative box plots and perform an analysis of variance. Use $\alpha = 0.05$.
(b) Plot average tensile strength against cotton percentage and interpret the results.
(c) Analyze the residuals and comment on model adequacy.

13-2. In "Orthogonal Design for Process Optimization and Its Application to Plasma Etching" (*Solid State Technology,* May 1987), G. Z. Yin and D. W. Jillie describe an experiment to determine the effect of C_2F_6 flow rate on the uniformity of the etch on a silicon wafer used in integrated circuit manufacturing. Three flow rates are used in the experiment, and the resulting uniformity (in percent) for six replicates is shown below.

C_2F_6 Flow (SCCM)	**Observations**					
	1	2	3	4	5	6
125	2.7	4.6	2.6	3.0	3.2	3.8
160	4.9	4.6	5.0	4.2	3.6	4.2
200	4.6	3.4	2.9	3.5	4.1	5.1

(a) Does C_2F_6 flow rate affect etch uniformity? Construct box plots to compare the factor levels and perform the analysis of variance. Use $\alpha = 0.05$.
(b) Do the residuals indicate any problems with the underlying assumptions?

13-3. The compressive strength of concrete is being studied, and four different mixing techniques are being investigated. The following data have been collected.

Mixing Technique	Compressive Strength (psi)			
1	3129	3000	2865	2890
2	3200	3300	2975	3150
3	2800	2900	2985	3050
4	2600	2700	2600	2765

(a) Test the hypothesis that mixing techniques affect the strength of the concrete. Use $\alpha = 0.05$.
(b) Find the *P*-value for the *F*-statistic computed in part (a).
(c) Analyze the residuals from this experiment.

13-4. The response time in milliseconds was determined for three different types of circuits in an electronic calculator. The results are recorded here.

Circuit Type	Response				
1	19	22	20	18	25
2	20	21	33	27	40
3	16	15	18	26	17

(a) Using $\alpha = 0.01$, test the hypothesis that the three circuit types have the same response time.
(b) Analyze the residuals from this experiment.
(c) Find a 95% confidence interval on the response time for circuit three.

13-5. An electronics engineer is interested in the effect on tube conductivity of five different types of coating for cathode ray tubes in a telecommunications system display device. The following conductivity data are obtained.

Coating Type	Conductivity			
1	143	141	150	146
2	152	149	137	143
3	134	133	132	127
4	129	127	132	129
5	147	148	144	142

(a) Is there any difference in conductivity due to coating type? Use $\alpha = 0.01$.
(b) Analyze the residuals from this experiment.
(c) Construct a 95% interval estimate of the coating type 1 mean. Construct a 99% interval estimate of the mean difference between coating types 1 and 4.

13-6. An article in *Environment International* (Vol. 18, No. 4, 1992) describes an experiment in which the amount of radon released in showers was investigated. Radon-enriched water was used in the experiment, and six different orifice diameters were tested in shower heads. The data from the experiment are shown in the following table.

Orifice Diameter	Radon Released (%)			
0.37	80	83	83	85
0.51	75	75	79	79
0.71	74	73	76	77
1.02	67	72	74	74
1.40	62	62	67	69
1.99	60	61	64	66

(a) Does the size of the orifice affect the mean percentage of radon released? Use $\alpha = 0.05$.
(b) Find the *P*-value for the *F*-statistic in part (a).
(c) Analyze the residuals from this experiment.
(d) Find a 95% confidence interval on the mean percent of radon released when the orifice diameter is 1.40.

13-7. An article in the *ACI Materials Journal* (Vol. 84, 1987, pp. 213–216) describes several experiments investigating the rodding of concrete to remove entrapped air. A 3-inch $\times$ 6-inch cylinder was used, and the number of times this rod was used is the design variable. The resulting compressive strength of the concrete specimen is the response. The data are shown in the following table.

Rodding Level	Compressive Strength		
10	1530	1530	1440
15	1610	1650	1500
20	1560	1730	1530
25	1500	1490	1510

(a) Is there any difference in compressive strength due to the rodding level?
(b) Find the P-value for the F-statistic in part (a).
(c) Analyze the residuals from this experiment. What conclusions can you draw about the underlying model assumptions?

13-8. An article in the *Materials Research Bulletin* (Vol. 26, No. 11, 1991) investigated four different methods of preparing the superconducting compound $PbMo_6S_8$. The authors contend that the presence of oxygen during the preparation process affects the material's superconducting transition temperature T_c. Preparation methods 1 and 2 use techniques that are designed to eliminate the presence of oxygen, while methods 3 and 4 allow oxygen to be present. Five observations on T_c (in °K) were made for each method, and the results are as follows:

Preparation Method	Transition Temperature T_c(°K)				
1	14.8	14.8	14.7	14.8	14.9
2	14.6	15.0	14.9	14.8	14.7
3	12.7	11.6	12.4	12.7	12.1
4	14.2	14.4	14.4	12.2	11.7

(a) Is there evidence to support the claim that the presence of oxygen during preparation affects the mean transition temperature? Use $\alpha = 0.05$.
(b) What is the P-value for the F-test in part (a)?
(c) Analyze the residuals from this experiment.
(d) Find a 95% confidence interval on mean T_c when method 1 is used to prepare the material.

13-9. A paper in the *Journal of the Association of Asphalt Paving Technologists* (Vol. 59, 1990) describes an experiment to determine the effect of air voids on percentage retained strength of asphalt. For purposes of the experiment, air voids are controlled at three levels; low (2–4%), medium (4–6%), and high (6–8%). The data are shown in the following table.

Air Voids	Retained Strength (%)							
Low	106	90	103	90	79	88	92	95
Medium	80	69	94	91	70	83	87	83
High	78	80	62	69	76	85	69	85

(a) Do the different levels of air voids significantly affect mean retained strength? Use $\alpha = 0.01$.
(b) Find the P-value for the F-statistic in part (a).
(c) Analyze the residuals from this experiment.
(d) Find a 95% confidence interval on mean retained strength where there is a high level of air voids.
(e) Find a 95% confidence interval on the difference in mean retained strength at the low and high levels of air voids.

13-10. An article in *Quality Engineering*, "Estimating sources of variation: a case study from polyurethane product research" (1999–2000, Vol. 12, pp. 89–96) studied the effects of additives on final polymer properties. In this case, polyurethane additives were referred to as cross-linkers. The average domain spacing was the measurement of the polymer property. The data are as follows:

Cross-Linker Level	Domain Spacing (nm)					
−1	8.2	8	8.2	7.9	8.1	8
−0.75	8.3	8.4	8.3	8.2	8.3	8.1
−0.5	8.9	8.7	8.9	8.4	8.3	8.5
0	8.5	8.7	8.7	8.7	8.8	8.8
0.5	8.8	9.1	9.0	8.7	8.9	8.5
1	8.6	8.5	8.6	8.7	8.8	8.8

(a) Is there a difference in the cross-linker level? Draw comparative boxplots and perform an analysis of variance. Use $\alpha = 0.05$.
(b) Find the P-value of the test. Estimate the variability due to random error.
(c) Plot average domain spacing against cross-linker level and interpret the results.
(d) Analyze the residuals from this experiment and comment on model adequacy.

13-11. In the book *Analysis of Longitudinal Data, 2nd ed.*, (2002, *Oxford University Press*), by Diggle, Heagerty, Liang, and Zeger, the authors analyzed the effects of three diets on the protein content of cow's milk. The data shown here were collected after one week and include 25 cows on the barley diet, and 27 cows each on the other two diets:

(a) Does diet affect the protein content of cow's milk? Draw comparative boxplots and perform an analysis of variance. Use $\alpha = 0.05$.

Diet	Protein Content of Cow's Milk													
Barley	3.63	3.24	3.98	3.66	4.34	4.36	4.17	4.4	3.4	3.75	4.2	4.02	4.02	3.9
Barley+lupins	3.38	3.8	4.17	4.59	4.07	4.32	3.56	3.67	4.15	3.51	4.2	4.12	3.52	4.08
Lupins	3.69	4.2	3.31	3.13	3.73	4.32	3.04	3.84	3.98	4.18	4.2	4.1	3.25	3.34
Diet (continued)														
Barley	3.81	3.62	3.66	4.44	4.23	3.82	3.53	4.47	3.93	3.27	3.3			
Barley+lupins	4.02	3.18	4.11	3.27	3.27	3.97	3.31	4.12	3.92	3.78	4	4.37	3.79	
Lupins	3.5	4.13	3.21	3.9	3.5	4.1	2.69	4.3	4.06	3.88	4	3.67	4.27	

(b) Find the P-value of the test. Estimate the variability due to random error.

(c) Plot average protein content against diets and interpret the results.

(d) Analyze the residuals and comment on model adequacy.

13-12. An article in *Journal of Food Science* (2001, Vol. 66, No. 3, pp. 472–477) studied potato spoilage based on different conditions of acidified oxine (AO), which is a mixture of chlorite and chlorine dioxide. The data are shown below:

AO Solution (ppm)	% Spoilage		
50	100	50	60
100	60	30	30
200	60	50	29
400	25	30	15

(a) Do the AO solutions differ in the spoilage percentage? Use $\alpha = 0.05$.

(b) Find the P-value of the test. Estimate the variability due to random error.

(c) Plot average spoilage against AO solution and interpret the results. Which AO solution would you recommend for use in practice?

(d) Analyze the residuals from this experiment.

13-13. An experiment was run to determine whether four specific firing temperatures affect the density of a certain type of brick. The experiment led to the following data.

Temperature (°F)	Density						
100	21.8	21.9	21.7	21.6	21.7	21.5	21.8
125	21.7	21.4	21.5	21.5	—	—	—
150	21.9	21.8	21.8	21.6	21.5	—	—
175	21.9	21.7	21.8	21.7	21.6	21.8	—

(a) Does the firing temperature affect the density of the bricks? Use $\alpha = 0.05$.

(b) Find the P-value for the F-statistic computed in part (a).

(c) Analyze the residuals from the experiment.

13-14. (a) Use Fisher's LSD method with $\alpha = 0.05$ to analyze the means of the three flow rates in Exercise 13-2.

(b) Use the graphical method to compare means described in this section and compare your conclusions to those from Fisher's LSD method.

13-15. (a) Use Fisher's LSD method with $\alpha = 0.05$ to analyze the means of the five different levels of cotton content in Exercise 13-1.

(b) Use the graphical method to compare means described in this section and compare your conclusions to those from Fisher's LSD method.

13-16. (a) Use Fisher's LSD method with $\alpha = 0.01$ to analyze the mean response times for the three circuits described in Exercise 13-4.

(b) Use the graphical method to compare means described in this section and compare your conclusions to those from Fisher's LSD method.

13-17. (a) Use Fisher's LSD method with $\alpha = 0.05$ to analyze the mean compressive strength of the four mixing techniques in Exercise 13-3.

(b) Use the graphical method to compare means described in this section and compare your conclusions to those from Fisher's LSD method.

13-18. (a) Use Fisher's LSD method with $\alpha = 0.05$ to analyze the mean amounts of radon released in the experiment described in Exercise 13-6.

(b) Use the graphical method to compare means described in this section and compare your conclusions to those from Fisher's LSD method.

13-19. (a) Use Fisher's LSD method with $\alpha = 0.01$ to analyze the five means for the coating types described in Exercise 13-5.

(b) Use the graphical method to compare means described in this section and compare your conclusions to those from Fisher's LSD method.

13-20. (a) Apply Fisher's LSD method with $\alpha = 0.05$ to the superconducting material experiment described in Exercise 13-8. Which preparation methods differ?

(b) Use the graphical method to compare means described in this section and compare your conclusions to those from Fisher's LSD method.

13-21. (a) Apply Fisher's LSD method to the air void experiment described in Exercise 13-9. Using $\alpha = 0.05$, which treatment means are different?

(b) Use the graphical method to compare means described in this section and compare your conclusions to those from Fisher's LSD method.

13-22. (a) Apply Fisher's LSD method to the domain spacing data in Exercise 13-10. Which cross-linker levels differ? Use $\alpha = 0.05$.

(b) Use the graphical method to compare means described in this section and compare your conclusions to those from Fisher's LSD method.

13-23. (a) Apply Fisher's LSD method to the data on protein content of milk in Exercise 13-11. Which diets differ? Use $\alpha = 0.01$.

(b) Use the graphical method to compare means described in this section and compare your conclusions to those from Fisher's LSD method.

13-24. Suppose that four normal populations have common variance $\sigma^2 = 25$ and means $\mu_1 = 50$, $\mu_2 = 60$, $\mu_3 = 50$, and $\mu_4 = 60$. How many observations should be taken on each population so that the probability of rejecting the hypothesis of equality of means is at least 0.90? Use $\alpha = 0.05$.

13-25. Suppose that five normal populations have common variance $\sigma^2 = 100$ and means $\mu_1 = 175$, $\mu_2 = 190$, $\mu_3 = 160$, $\mu_4 = 200$, and $\mu_5 = 215$. How many observations per population must be taken so that the probability of rejecting the hypothesis of equality of means is at least 0.95? Use $\alpha = 0.01$.

13-3 THE RANDOM-EFFECTS MODEL

13-3.1 Fixed Versus Random Factors

In many situations, the factor of interest has a large number of possible levels. The analyst is interested in drawing conclusions about the entire population of factor levels. If the experimenter randomly selects a of these levels from the population of factor levels, we say that the factor is a **random factor.** Because the levels of the factor actually used in the experiment were chosen randomly, the conclusions reached will be valid for the entire population of factor levels. We will assume that the population of factor levels is either of infinite size or is large enough to be considered infinite. Notice that this is a very different situation than we encountered in the fixed-effects case, where the conclusions apply only for the factor levels used in the experiment.

13-3.2 ANOVA and Variance Components

The linear statistical model is

$$Y_{ij} = \mu + \tau_i + \epsilon_{ij} \begin{cases} i = 1, 2, \ldots, a \\ j = 1, 2, \ldots, n \end{cases} \tag{13-18}$$

where the treatment effects τ_i and the errors ϵ_{ij} are independent random variables. Note that the model is identical in structure to the fixed-effects case, but the parameters have a different interpretation. If the variance of the treatment effects τ_i is σ_τ^2, by independence the variance of the response is

$$V(Y_{ij}) = \sigma_\tau^2 + \sigma^2 \tag{13-19}$$

The variancces σ_τ^2 and σ^2 are called **variance components,** and the model, Equation 13-19, is called the **components of variance model** or the **random-effects model.** To test hypotheses in this model, we assume that the errors ϵ_{ij} are normally and independently distributed with mean 0 and variance σ^2 and that the treatment effects τ_i are normally and independently distributed with mean zero and variance σ_τ^2.*

*The assumption that the $\{\tau_i\}$ are independent random variables implies that the usual assumption of $\sum_{i=1}^{a} \tau_i = 0$ from the fixed-effects model does not apply to the random-effects model.

For the random-effects model, testing the hypothesis that the individual treatment effects are zero is meaningless. It is more appropriate to test hypotheses about σ_τ^2. Specifically,

$$H_0\colon \sigma_\tau^2 = 0$$
$$H_1\colon \sigma_\tau^2 > 0$$

If $\sigma_\tau^2 = 0$, all treatments are identical; but if $\sigma_\tau^2 > 0$, there is variability between treatments.

The ANOVA decomposition of total variability is still valid; that is,

$$SS_T = SS_{\text{Treatments}} + SS_E \tag{13-20}$$

However, the expected values of the mean squares for treatments and error are somewhat different than in the fixed-effects case.

Expected Values of Mean Squares: Random Effects

In the random-effects model for a single-factor, completely randomized experiment, the expected mean square for treatments is

$$E(MS_{\text{Treatments}}) = E\left(\frac{SS_{\text{Treatments}}}{a-1}\right)$$
$$= \sigma^2 + n\sigma_\tau^2 \tag{13-21}$$

and the expected mean square for error is

$$E(MS_E) = E\left[\frac{SS_E}{a(n-1)}\right]$$
$$= \sigma^2 \tag{13-22}$$

From examining the expected mean squares, it is clear that both MS_E and $MS_{\text{Treatments}}$ estimate σ^2 when $H_0\colon \sigma_\tau^2 = 0$ is true. Furthermore, MS_E and $MS_{\text{Treatments}}$ are independent. Consequently, the ratio

$$F_0 = \frac{MS_{\text{Treatments}}}{MS_E} \tag{13-23}$$

is an F random variable with $a - 1$ and $a(n - 1)$ degrees of freedom when H_0 is true. The null hypothesis would be rejected at the α-level of significance if the computed value of the test statistic $f_0 > f_{\alpha,a-1,a(n-1)}$.

The computational procedure and construction of the ANOVA table for the random-effects model are identical to the fixed-effects case. The conclusions, however, are quite different because they apply to the entire population of treatments.

Usually, we also want to estimate the variance components (σ^2 and σ_τ^2) in the model. The procedure that we will use to estimate σ^2 and σ_τ^2 is called the **analysis of variance method** because it uses the information in the analysis of variance table. It does not require the normality assumption on the observations. The procedure consists of equating the expected mean squares to their observed values in the ANOVA table and solving for the variance components. When equating observed and expected mean squares in the one-way classification random-effects model, we obtain

$$MS_{\text{Treatments}} = \sigma^2 + n\sigma_\tau^2 \quad \text{and} \quad MS_E = \sigma^2$$

Therefore, the estimators of the variance components are

Variance Components Estimates

$$\hat{\sigma}^2 = MS_E \tag{13-24}$$

and

$$\hat{\sigma}_\tau^2 = \frac{MS_{\text{Treatments}} - MS_E}{n} \tag{13-25}$$

Sometimes the analysis of variance method produces a negative estimate of a variance component. Since variance components are by definition nonnegative, a negative estimate of a variance component is disturbing. One course of action is to accept the estimate and use it as evidence that the true value of the variance component is zero, assuming that sampling variation led to the negative estimate. While this approach has intuitive appeal, it will disturb the statistical properties of other estimates. Another alternative is to reestimate the negative variance component with a method that always yields nonnegative estimates. Still another possibility is to consider the negative estimate as evidence that the assumed linear model is incorrect, requiring that a study of the model and its assumptions be made to find a more appropriate model.

EXAMPLE 13-4
Textile Manufacturing

In *Design and Analysis of Experiments*, 6th edition (John Wiley, 2005), D. C. Montgomery describes a single-factor experiment involving the random-effects model in which a textile manufacturing company weaves a fabric on a large number of looms. The company is interested in loom-to-loom variability in tensile strength. To investigate this variability, a manufacturing engineer selects four looms at random and makes four strength determinations on fabric samples chosen at random from each loom. The data are shown in Table 13-7 and the ANOVA is summarized in Table 13-8.

From the analysis of variance, we conclude that the looms in the plant differ significantly in their ability to produce fabric of uniform strength. The variance components are estimated by $\hat{\sigma}^2 = 1.90$ and

$$\hat{\sigma}_\tau^2 = \frac{29.73 - 1.90}{4} = 6.96$$

Therefore, the variance of strength in the manufacturing process is estimated by

$$\widehat{V(Y_{ij})} = \hat{\sigma}_\tau^2 + \hat{\sigma}^2 = 6.96 + 1.90 = 8.86$$

Most of this variability is attributable to differences between looms.

Table 13-7 Strength Data for Example 13-4

	Observations					
Loom	1	2	3	4	Total	Average
1	98	97	99	96	390	97.5
2	91	90	93	92	366	91.5
3	96	95	97	95	383	95.8
4	95	96	99	98	388	97.0
					1527	95.45

Table 13-8 Analysis of Variance for the Strength Data

Source of Variation	Sum of Squares	Degrees of Freedom	Mean Square	f_0	*P*-value
Looms	89.19	3	29.73	15.68	1.88 E-4
Error	22.75	12	1.90		
Total	111.94	15			

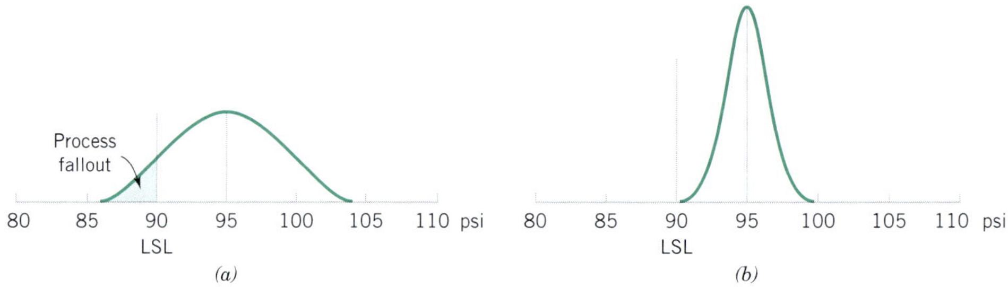

Figure 13-8 The distribution of fabric strength. (a) Current process, (b) improved process.

This example illustrates an important application of the analysis of variance—the isolation of different sources of variability in a manufacturing process. Problems of excessive variability in critical functional parameters or properties frequently arise in quality-improvement programs. For example, in the previous fabric strength example, the process mean is estimated by $\bar{y} = 95.45$ psi, and the process standard deviation is estimated by $\hat{\sigma}_y = \sqrt{\hat{V}(Y_{ij})} = \sqrt{8.86} = 2.98$ psi. If strength is approximately normally distributed, the distribution of strength in the outgoing product would look like the normal distribution shown in Fig. 13-8(a). If the lower specification limit (LSL) on strength is at 90 psi, a substantial proportion of the process output is **fallout**—that is, scrap or defective material that must be sold as second quality, and so on. This fallout is directly related to the excess variability resulting from differences between looms. Variability in loom performance could be caused by faulty setup, poor maintenance, inadequate supervision, poorly trained operators, and so forth. The engineer or manager responsible for quality improvement must identify and remove these sources of variability from the process. If this can be done, strength variability will be greatly reduced, perhaps as low as $\hat{\sigma}_Y = \sqrt{\hat{\sigma}^2} = \sqrt{1.90} = 1.38$ psi, as shown in Fig. 13-8(b). In this improved process, reducing the variability in strength has greatly reduced the fallout, resulting in lower cost, higher quality, a more satisfied customer, and enhanced competitive position for the company.

EXERCISES FOR SECTION 13-3

13-26. An article in the *Journal of the Electrochemical Society* (Vol. 139, No. 2, 1992, pp. 524–532) describes an experiment to investigate the low-pressure vapor deposition of polysilicon. The experiment was carried out in a large-capacity reactor at Sematech in Austin, Texas. The reactor has several wafer positions, and four of these positions are selected at random. The response variable is film thickness uniformity. Three replicates of the experiment were run, and the data are as follows:

Wafer Position	Uniformity		
1	2.76	5.67	4.49
2	1.43	1.70	2.19
3	2.34	1.97	1.47
4	0.94	1.36	1.65

(a) Is there a difference in the wafer positions? Use $\alpha = 0.05$.
(b) Estimate the variability due to wafer positions.
(c) Estimate the random error component.
(d) Analyze the residuals from this experiment and comment on model adequacy.

13-27. A textile mill has a large number of looms. Each loom is supposed to provide the same output of cloth per minute. To investigate this assumption, five looms are chosen at random, and their output is measured at different times. The following data are obtained:

Loom	Output (lb/min)				
1	4.0	4.1	4.2	4.0	4.1
2	3.9	3.8	3.9	4.0	4.0
3	4.1	4.2	4.1	4.0	3.9
4	3.6	3.8	4.0	3.9	3.7
5	3.8	3.6	3.9	3.8	4.0

(a) Are the looms similar in output? Use $\alpha = 0.05$.
(b) Estimate the variability between looms.

(c) Estimate the experimental error variance.
(d) Analyze the residuals from this experiment and check for model adequacy.

13-28. In the book *Bayesian Inference in Statistical Analysis* (1973, John Wiley and Sons.) by Box and Tiao, the total product yield was determined for five samples randomly selected from each of six randomly chosen batches of raw material.

Batch	Yield (in grams)				
1	1545	1440	1440	1520	1580
2	1540	1555	1490	1560	1495
3	1595	1550	1605	1510	1560
4	1445	1440	1595	1465	1545
5	1595	1630	1515	1635	1625
6	1520	1455	1450	1480	1445

(a) Do the different batches of raw material significantly affect mean yield? Use $\alpha = 0.01$.
(b) Estimate the variability between batches.
(c) Estimate the variability between samples within batches.
(d) Analyze the residuals from this experiment and check for model adequacy.

13-29. An article in the *Journal of Quality Technology* (Vol. 13, No. 2, 1981, pp. 111–114) describes an experiment that investigates the effects of four bleaching chemicals on pulp brightness. These four chemicals were selected at random from a large population of potential bleaching agents. The data are as follows:

Chemical	Pulp Brightness				
1	77.199	74.466	92.746	76.208	82.876
2	80.522	79.306	81.914	80.346	73.385
3	79.417	78.017	91.596	80.802	80.626
4	78.001	78.358	77.544	77.364	77.386

(a) Is there a difference in the chemical types? Use $\alpha = 0.05$.
(b) Estimate the variability due to chemical types.
(c) Estimate the variability due to random error.
(d) Analyze the residuals from this experiment and comment on model adequacy.

13-30. Consider the vapor-deposition experiment described in Exercise 13-26.
(a) Estimate the total variability in the uniformity response.
(b) How much of the total variability in the uniformity response is due to the difference between positions in the reactor?
(c) To what level could the variability in the uniformity response be reduced, if the position-to-position variability in the reactor could be eliminated? Do you believe this is a significant reduction?

13-31. Reconsider Exercise 13-11 in which the effect of different diets on the protein content of cow's milk was investigated. Suppose that the three diets reported were selected at random from a large number of diets. To simplify, delete the last two observations in the diets with $n = 27$ (to make equal sample sizes).
(a) How does this change the interpretation of the experiment?
(b) What is an appropriate statistical model for this experiment?
(c) Estimate the parameters of this model.

13-4 RANDOMIZED COMPLETE BLOCK DESIGN

13-4.1 Design and Statistical Analysis

In many experimental design problems, it is necessary to design the experiment so that the variability arising from a nuisance factor can be controlled. For example, consider the situation of Example 10-9, where two different methods were used to predict the shear strength of steel plate girders. Because each girder has different strength (potentially), and this variability in strength was not of direct interest, we designed the experiment by using the two test methods on each girder and then comparing the average difference in strength readings on each girder to zero using the paired t-test. The paired t-test is a procedure for comparing two treatment means when all experimental runs cannot be made under homogeneous conditions. Alternatively, we can view the paired t-test as a method for reducing the background noise in the experiment by blocking out a **nuisance factor** effect. The block is the nuisance factor, and in this case, the nuisance factor is the actual **experimental unit**—the steel girder specimens used in the experiment.

The randomized block design is an extension of the paired t-test to situations where the factor of interest has more than two levels; that is, more than two treatments must be compared. For example, suppose that three methods could be used to evaluate the strength readings on steel plate girders. We may think of these as three treatments, say t_1, t_2, and t_3. If we use four girders as the experimental units, a **randomized complete block design**

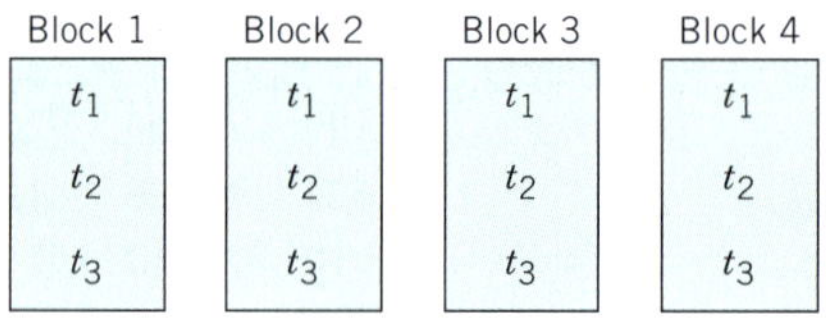

Figure 13-9 A randomized complete block design.

Table 13-9 A Randomized Complete Block Design

Treatments (Method)	Block (Girder) 1	2	3	4
1	y_{11}	y_{12}	y_{13}	y_{14}
2	y_{21}	y_{22}	y_{23}	y_{24}
3	y_{31}	y_{32}	y_{33}	y_{34}

would appear as shown in Fig. 13-9. The design is called a randomized complete block design because each block is large enough to hold all the treatments and because the actual assignment of each of the three treatments within each block is done randomly. Once the experiment has been conducted, the data are recorded in a table, such as is shown in Table 13-9. The observations in this table, say y_{ij}, represent the response obtained when method i is used on girder j.

The general procedure for a randomized complete block design consists of selecting b blocks and running a complete replicate of the experiment in each block. The data that result from running a randomized complete block design for investigating a single factor with a levels and b blocks are shown in Table 13-10. There will be a observations (one per factor level) in each block, and the order in which these observations are run is randomly assigned within the block.

We will now describe the statistical analysis for a randomized complete block design. Suppose that a single factor with a levels is of interest and that the experiment is run in b blocks. The observations may be represented by the **linear statistical model**

$$Y_{ij} = \mu + \tau_i + \beta_j + \epsilon_{ij} \begin{cases} i = 1, 2, \ldots, a \\ j = 1, 2, \ldots, b \end{cases} \tag{13-26}$$

where μ is an overall mean, τ_i is the effect of the ith treatment, β_j is the effect of the jth block, and ϵ_{ij} is the random error term, which is assumed to be normally and independently distributed with mean zero and variance σ^2. Furthermore, the treatment and block effects are defined as deviations from the overall mean, so $\sum_{i=1}^{a} \tau_i = 0$ and $\sum_{j=1}^{b} \beta_j = 0$. This was the same type of definition used for completely randomized experiments in Section 13-2. We also assume that treatments and blocks do not interact. That is, the effect of treatment i is the same regardless of which block (or blocks) it is tested in. We are interested in testing the equality of the treatment effects. That is

$$H_0: \tau_1 = \tau_2 = \cdots = \tau_a = 0$$
$$H_1: \tau_i \neq 0 \text{ at least one } i$$

Table 13-10 A Randomized Complete Block Design with a Treatments and b Blocks

Treatments	Blocks 1	2	…	b	Totals	Averages
1	y_{11}	y_{12}	…	y_{1b}	$y_{1\cdot}$	$\bar{y}_{1\cdot}$
2	y_{21}	y_{22}	…	y_{2b}	$y_{2\cdot}$	$\bar{y}_{2\cdot}$
⋮	⋮	⋮		⋮	⋮	⋮
a	y_{a1}	y_{a2}	…	y_{ab}	$y_{a\cdot}$	$\bar{y}_{a\cdot}$
Totals	$y_{\cdot 1}$	$y_{\cdot 2}$	…	$y_{\cdot b}$	$y_{\cdot\cdot}$	
Averages	$\bar{y}_{\cdot 1}$	$\bar{y}_{\cdot 2}$	…	$\bar{y}_{\cdot b}$		$\bar{y}_{\cdot\cdot}$

The analysis of variance can be extended to the randomized complete block design. The procedure uses a sum of squares identity that partitions the total sum of squares into three components.

ANOVA Sums of Squares Identity: Randomized Block

The **sum of squares identity for the randomized complete block design** is

$$\sum_{i=1}^{a}\sum_{j=1}^{b}(y_{ij}-\bar{y}_{..})^2 = b\sum_{i=1}^{a}(\bar{y}_{i\cdot}-\bar{y}_{..})^2 + a\sum_{j=1}^{b}(\bar{y}_{\cdot j}-\bar{y}_{..})^2 + \sum_{i=1}^{a}\sum_{j=1}^{b}(y_{ij}-\bar{y}_{\cdot j}-\bar{y}_{i\cdot}+\bar{y}_{..})^2 \tag{13-27}$$

or symbolically

$$SS_T = SS_{\text{Treatments}} + SS_{\text{Blocks}} + SS_E$$

Furthermore, the degrees of freedom corresponding to these sums of squares are

$$ab - 1 = (a-1) + (b-1) + (a-1)(b-1)$$

For the randomized block design, the relevant mean squares are

$$MS_{\text{Treatments}} = \frac{SS_{\text{Treatments}}}{a-1}$$

$$MS_{\text{Blocks}} = \frac{SS_{\text{Blocks}}}{b-1}$$

$$MS_E = \frac{SS_E}{(a-1)(b-1)}$$

The expected values of these mean squares can be shown to be as follows:

Expected Mean Squares: Randomized Block

$$E(MS_{\text{Treatments}}) = \sigma^2 + \frac{b\sum_{i=1}^{a}\tau_i^2}{a-1}$$

$$E(MS_{\text{Blocks}}) = \sigma^2 + \frac{a\sum_{j=1}^{b}\beta_j^2}{b-1}$$

$$E(MS_E) = \sigma^2$$

Therefore, if the null hypothesis H_0 is true so that all treatment effects $\tau_i = 0$, $MS_{\text{Treatments}}$ is an unbiased estimator of σ^2, while if H_0 is false, $MS_{\text{Treatments}}$ overestimates σ^2. The mean square for error is always an unbiased estimate of σ^2. To test the null hypothesis that the treatment effects are all zero, we use the ratio

$$F_0 = \frac{MS_{\text{Treatments}}}{MS_E} \tag{13-28}$$

which has an F-distribution with $a - 1$ and $(a - 1)(b - 1)$ degrees of freedom if the null hypothesis is true. We would reject the null hypothesis at the α-level of significance if the computed value of the test statistic in Equation 13-28 is $f_0 > f_{\alpha,a-1,(a-1)(b-1)}$.

In practice, we compute SS_T, $SS_{\text{Treatments}}$ and SS_{Blocks} and then obtain the error sum of squares SS_E by subtraction. The appropriate computing formulas are as follows.

Computing Formulas for ANOVA: Randomized Block

The computing formulas for the sums of squares in the analysis of variance for a randomized complete block design are

$$SS_T = \sum_{i=1}^{a}\sum_{j=1}^{b} y_{ij}^2 - \frac{y_{..}^2}{ab} \tag{13-29}$$

$$SS_{\text{Treatments}} = \frac{1}{b}\sum_{i=1}^{a} y_{i.}^2 - \frac{y_{..}^2}{ab} \tag{13-30}$$

$$SS_{\text{Blocks}} = \frac{1}{a}\sum_{j=1}^{b} y_{.j}^2 - \frac{y_{..}^2}{ab} \tag{13-31}$$

and

$$SS_E = SS_T - SS_{\text{Treatments}} - SS_{\text{Blocks}} \tag{13-32}$$

The computations are usually arranged in an ANOVA table, such as is shown in Table 13-11. Generally, a computer software package will be used to perform the analysis of variance for the randomized complete block design.

EXAMPLE 13-5 Fabric Strength

An experiment was performed to determine the effect of four different chemicals on the strength of a fabric. These chemicals are used as part of the permanent press finishing process. Five fabric samples were selected, and a randomized complete block design was run by testing each chemical type once in random order on each fabric sample. The data are shown in Table 13-12. We will test for differences in means using an ANOVA with $\alpha = 0.01$.

The sums of squares for the analysis of variance are computed as follows:

$$SS_T = \sum_{i=1}^{4}\sum_{j=1}^{5} y_{ij}^2 - \frac{y_{..}^2}{ab}$$

$$= (1.3)^2 + (1.6)^2 + \cdots + (3.4)^2 - \frac{(39.2)^2}{20} = 25.69$$

Table 13-11 ANOVA for a Randomized Complete Block Design

Source of Variation	Sum of Squares	Degrees of Freedom	Mean Square	F_0
Treatments	$SS_{\text{Treatments}}$	$a - 1$	$\dfrac{SS_{\text{Treatments}}}{a-1}$	$\dfrac{MS_{\text{Treatments}}}{MS_E}$
Blocks	SS_{Blocks}	$b - 1$	$\dfrac{SS_{\text{Blocks}}}{b-1}$	
Error	SS_E (by subtraction)	$(a - 1)(b - 1)$	$\dfrac{SS_E}{(a-1)(b-1)}$	
Total	SS_T	$ab - 1$		

Table 13-12 Fabric Strength Data—Randomized Complete Block Design

	Fabric Sample					Treatment Totals	Treatment Averages
Chemical Type	1	2	3	4	5	$y_{i\cdot}$	$\bar{y}_{i\cdot}$
1	1.3	1.6	0.5	1.2	1.1	5.7	1.14
2	2.2	2.4	0.4	2.0	1.8	8.8	1.76
3	1.8	1.7	0.6	1.5	1.3	6.9	1.38
4	3.9	4.4	2.0	4.1	3.4	17.8	3.56
Block totals $y_{\cdot j}$	9.2	10.1	3.5	8.8	7.6	39.2($y_{\cdot\cdot}$)	
Block averages $\bar{y}_{\cdot j}$	2.30	2.53	0.88	2.20	1.90		1.96($\bar{y}_{\cdot\cdot}$)

$$SS_{\text{Treatments}} = \sum_{i=1}^{4} \frac{y_{i\cdot}^2}{b} - \frac{y_{\cdot\cdot}^2}{ab}$$

$$= \frac{(5.7)^2 + (8.8)^2 + (6.9)^2 + (17.8)^2}{5} - \frac{(39.2)^2}{20} = 18.04$$

$$SS_{\text{Blocks}} = \sum_{j=1}^{5} \frac{y_{\cdot j}^2}{a} - \frac{y_{\cdot\cdot}^2}{ab}$$

$$= \frac{(9.2)^2 + (10.1)^2 + (3.5)^2 + (8.8)^2 + (7.6)^2}{4} - \frac{(39.2)^2}{20} = 6.69$$

$$SS_E = SS_T - SS_{\text{Blocks}} - SS_{\text{Treatments}}$$
$$= 25.69 - 6.69 - 18.04 = 0.96$$

The ANOVA is summarized in Table 13-13. Since $f_0 = 75.13 > f_{0.01,3,12} = 5.95$ (the P-value is 4.79×10^{-8}), we conclude that there is a significant difference in the chemical types so far as their effect on strength is concerned.

When Is Blocking Necessary?

Suppose an experiment is conducted as a randomized block design, and blocking was not really necessary. There are ab observations and $(a - 1)(b - 1)$ degrees of freedom for error. If the experiment had been run as a completely randomized single-factor design with b replicates, we would have had $a(b - 1)$ degrees of freedom for error. Therefore, blocking has cost $a(b - 1) - (a - 1)(b - 1) = b - 1$ degrees of freedom for error. Thus, since the loss in error degrees of freedom is usually small, if there is a reasonable chance that block effects may be important, the experimenter should use the randomized block design.

Table 13-13 Analysis of Variance for the Randomized Complete Block Experiment

Source of Variation	Sum of Squares	Degrees of Freedom	Mean Square	f_0	P-value
Chemical types (treatments)	18.04	3	6.01	75.13	4.79 E-8
Fabric samples (blocks)	6.69	4	1.67		
Error	0.96	12	0.08		
Total	25.69	19			

Table 13-14 Minitab Analysis of Variance for the Randomized Complete Block Design in Example 13-5

Analysis of Variance (Balanced Designs)

Factor	Type	Levels	Values				
Chemical	fixed	4	1	2	3	4	
Fabric S	fixed	5	1	2	3	4	5

Analysis of Variance for strength

Source	DF	SS	MS	F	P
Chemical	3	18.0440	6.0147	75.89	0.000
Fabric S	4	6.6930	1.6733	21.11	0.000
Error	12	0.9510	0.0792		
Total	19	25.6880			

F-test with denominator: Error
Denominator MS = 0.079250 with 12 degrees of freedom

Numerator	DF	MS	F	P
Chemical	3	6.015	75.89	0.000
Fabric S	4	1.673	21.11	0.000

For example, consider the experiment described in Example 13-5 as a single-factor experiment with no blocking. We would then have 16 degrees of freedom for error. In the randomized block design, there are 12 degrees of freedom for error. Therefore, blocking has cost only 4 degrees of freedom, which is a very small loss considering the possible gain in information that would be achieved if block effects are really important. The block effect in Example 13-5 is large, and if we had not blocked, SS_{Blocks} would have been included in the error sum of squares for the completely randomized analysis. This would have resulted in a much larger MS_E, making it more difficult to detect treatment differences. As a general rule, when in doubt as to the importance of block effects, the experimenter should block and gamble that the block effect does exist. If the experimenter is wrong, the slight loss in the degrees of freedom for error will have a negligible effect, unless the number of degrees of freedom is very small.

Computer Solution

Table 13-14 presents the computer output from Minitab for the randomized complete block design in Example 13-5. We used the analysis of variance menu for balanced designs to solve this problem. The results agree closely with the hand calculations from Table 13-13.

Notice that Minitab computes an F-statistic for the blocks (the fabric samples). The validity of this ratio as a test statistic for the null hypothesis of no block effects is doubtful because the blocks represent a restriction on randomization; that is, we have only randomized within the blocks. If the blocks are not chosen at random, or if they are not run in random order, the F-ratio for blocks may not provide reliable information about block effects. For more discussion see Montgomery (2005, Chapter 4).

13-4.2 Multiple Comparisons

When the ANOVA indicates that a difference exists between the treatment means, we may need to perform some follow-up tests to isolate the specific differences. Any multiple comparison method, such as Fisher's LSD method, could be used for this purpose.

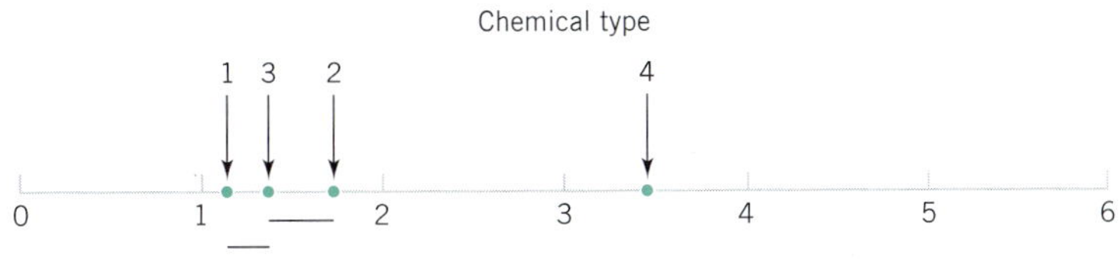

Figure 13-10 Results of Fisher's LSD method.

We will illustrate Fisher's LSD method. The four chemical type averages from Example 13-5 are:

$$\bar{y}_{1\cdot} = 1.14 \qquad \bar{y}_{2\cdot} = 1.76 \qquad \bar{y}_{3\cdot} = 1.38 \qquad \bar{y}_{4\cdot} = 3.56$$

Each treatment average uses $b = 5$ observations (one from each block). We will use $\alpha = 0.05$, so $t_{0.025,12} = 2.179$. Therefore the value of the LSD is

$$\text{LSD} = t_{0.025,12}\sqrt{\frac{2MS_E}{b}} = 2.179\sqrt{\frac{2(0.08)}{5}} = 0.39$$

Any pair of treatment averages that differ by 0.39 or more indicates that this pair of treatment means is significantly different. The comparisons are shown below:

$$\begin{aligned}
4 \text{ vs. } 1 &= \bar{y}_{4\cdot} - \bar{y}_{1\cdot} = 3.56 - 1.14 = 2.42 > 0.39\\
4 \text{ vs. } 3 &= \bar{y}_{4\cdot} - \bar{y}_{3\cdot} = 3.56 - 1.38 = 2.18 > 0.39\\
4 \text{ vs. } 2 &= \bar{y}_{4\cdot} - \bar{y}_{2\cdot} = 3.56 - 1.76 = 1.80 > 0.39\\
2 \text{ vs. } 1 &= \bar{y}_{2\cdot} - \bar{y}_{1\cdot} = 1.76 - 1.14 = 0.62 > 0.39\\
2 \text{ vs. } 3 &= \bar{y}_{2\cdot} - \bar{y}_{3\cdot} = 1.76 - 1.38 = 0.38 < 0.39\\
3 \text{ vs. } 1 &= \bar{y}_{3\cdot} - \bar{y}_{1\cdot} = 1.38 - 1.14 = 0.24 < 0.39
\end{aligned}$$

Figure 13-10 presents the results graphically. The underlined pairs of means are not different. The LSD procedure indicates that chemical type 4 results in significantly different strengths than the other three types do. Chemical types 2 and 3 do not differ, and types 1 and 3 do not differ. There may be a small difference in strength between types 1 and 2.

13-4.3 Residual Analysis and Model Checking

In any designed experiment, it is always important to examine the residuals and to check for violation of basic assumptions that could invalidate the results. As usual, the residuals for the randomized complete block design are just the difference between the observed and estimated (or fitted) values from the statistical model, say,

$$e_{ij} = y_{ij} - \hat{y}_{ij} \tag{13-33}$$

and the fitted values are

$$\hat{y}_{ij} = \bar{y}_{i\cdot} + \bar{y}_{\cdot j} - \bar{y}_{\cdot\cdot}$$

The fitted value represents the estimate of the mean response when the ith treatment is run in the jth block. The residuals from the chemical type experiment are shown in Table 13-15.

Figures 13-11, 13-12, 13-13, and 13-14 present the important residual plots for the experiment. These residual plots are usually constructed by computer software packages. There

Table 13-15 Residuals from the Randomized Complete Block Design

Chemical Type	Fabric Sample				
	1	2	3	4	5
1	−0.18	−0.10	0.44	−0.18	0.02
2	0.10	0.08	−0.28	0.00	0.10
3	0.08	−0.24	0.30	−0.12	−0.02
4	0.00	0.28	−0.48	0.30	−0.10

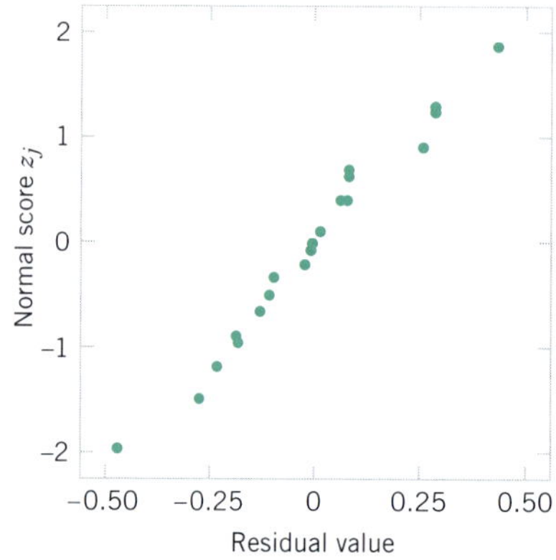

Figure 13-11 Normal probability plot of residuals from the randomized complete block design.

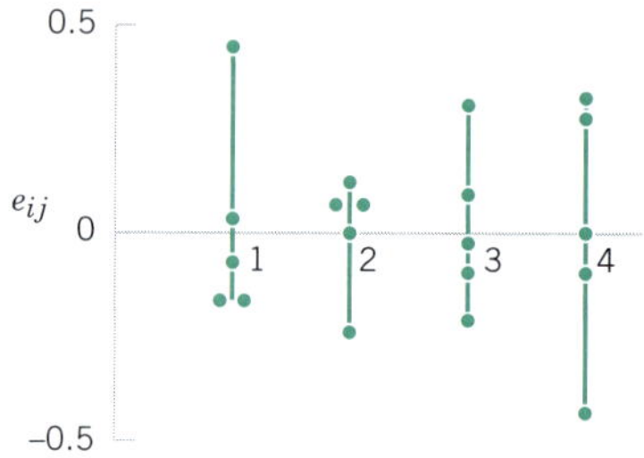

Figure 13-12 Residuals by treatment.

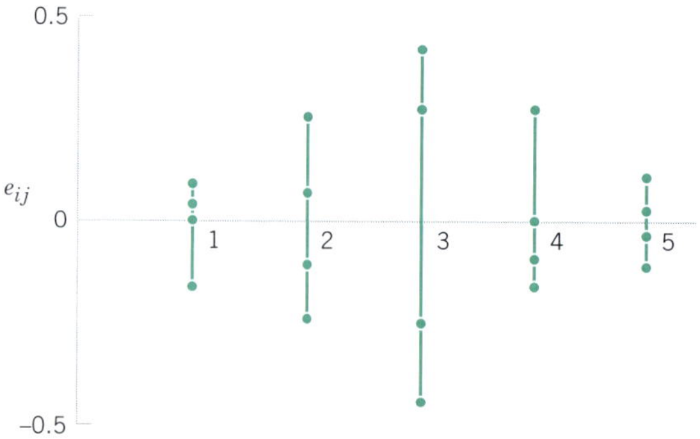

Figure 13-13 Residuals by block.

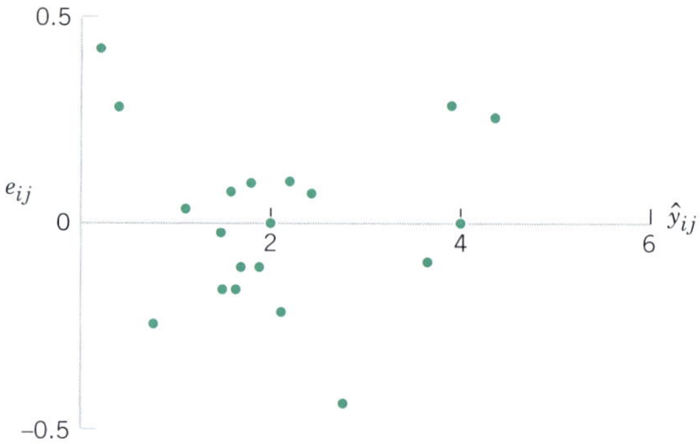

Figure 13-14 Residuals versus $\hat{y}_{ij}$.

is some indication that fabric sample (block) 3 has greater variability in strength when treated with the four chemicals than the other samples. Chemical type 4, which provides the greatest strength, also has somewhat more variability in strength. Followup experiments may be necessary to confirm these findings, if they are potentially important.

EXERCISES FOR SECTION 13-4

13-32. An article in *Quality Engineering* "Designed experiment to stabilize blood glucose levels," (1999–2000, Vol. 12, pp. 83–87) described an experiment to minimize variations in blood glucose levels. The treatment was the exercise time on a Nordic Track cross-country skier (10 or 20 min). The experiment was blocked for time of day. The data are as follows:

Exercise (min)	Time of Day	Average Blood Glucose
10	pm	71.5
10	am	103
20	am	83.5
20	pm	126
10	am	125.5
10	pm	129.5
20	pm	95
20	am	93

(a) Is there an effect of exercise time on the average blood glucose? Use $\alpha = 0.05$.
(b) Find the P-value for the test in part (a).
(c) Analyze the residuals from this experiment.

13-33. In "The Effect of Nozzle Design on the Stability and Performance of Turbulent Water Jets" (*Fire Safety Journal,* Vol. 4, August 1981), C. Theobald describes an experiment in which a shape measurement was determined for several different nozzle types at different levels of jet efflux velocity. Interest in this experiment focuses primarily on nozzle type, and velocity is a nuisance factor. The data are as follows:

Nozzle Type	**Jet Efflux Velocity (m/s)**					
	11.73	14.37	16.59	20.43	23.46	28.74
1	0.78	0.80	0.81	0.75	0.77	0.78
2	0.85	0.85	0.92	0.86	0.81	0.83
3	0.93	0.92	0.95	0.89	0.89	0.83
4	1.14	0.97	0.98	0.88	0.86	0.83
5	0.97	0.86	0.78	0.76	0.76	0.75

(a) Does nozzle type affect shape measurement? Compare the nozzles with box plots and the analysis of variance.
(b) Use Fisher's LSD method to determine specific differences between the nozzles. Does a graph of the average (or standard deviation) of the shape measurements versus nozzle type assist with the conclusions?
(c) Analyze the residuals from this experiment.

13-34. In *Design and Analysis of Experiments,* 6th edition (John Wiley & Sons, 2005), D. C. Montgomery describes an experiment that determined the effect of four different types of tips in a hardness tester on the observed hardness of a metal alloy. Four specimens of the alloy were obtained, and each tip was tested once on each specimen, producing the following data:

Type of Tip	**Specimen**			
	1	2	3	4
1	9.3	9.4	9.6	10.0
2	9.4	9.3	9.8	9.9
3	9.2	9.4	9.5	9.7
4	9.7	9.6	10.0	10.2

(a) Is there any difference in hardness measurements between the tips?
(b) Use Fisher's LSD method to investigate specific differences between the tips.
(c) Analyze the residuals from this experiment.

13-35. An article in the *American Industrial Hygiene Association Journal* (Vol. 37, 1976, pp. 418–422) described a field test for detecting the presence of arsenic in urine samples. The test has been proposed for use among forestry workers because of the increasing use of organic arsenics in that industry. The experiment compared the test as performed by both a trainee and an experienced trainer to an analysis at a remote laboratory. Four subjects were selected for testing and are considered as blocks. The response variable is arsenic content (in ppm) in the subject's urine. The data are as follows:

Test	**Subject**			
	1	2	3	4
Trainee	0.05	0.05	0.04	0.15
Trainer	0.05	0.05	0.04	0.17
Lab	0.04	0.04	0.03	0.10

(a) Is there any difference in the arsenic test procedure?
(b) Analyze the residuals from this experiment.

13-36. An article in the *Food Technology Journal* (Vol. 10, 1956, pp. 39–42) described a study on the protopectin content of tomatoes during storage. Four storage times were selected, and samples from nine lots of tomatoes were analyzed. The protopectin content (expressed as hydrochloric acid soluble fraction mg/kg) is in the following table.

Table 13-16 Protopectin Content of Tomatoes in Storage

Storage Time	**Lot**								
	1	2	3	4	5	6	7	8	9
0 days	1694.0	989.0	917.3	346.1	1260.0	965.6	1123.0	1106.0	1116.0
7 days	1802.0	1074.0	278.8	1375.0	544.0	672.2	818.0	406.8	461.6
14 days	1568.0	646.2	1820.0	1150.0	983.7	395.3	422.3	420.0	409.5
21 days	415.5	845.4	377.6	279.4	447.8	272.1	394.1	356.4	351.2

(a) The researchers in this study hypothesized that mean protopectin content would be different at different storage times. Can you confirm this hypothesis with a statistical test using $\alpha = 0.05$?
(b) Find the P-value for the test in part (a).
(c) Which specific storage times are different? Would you agree with the statement that protopectin content decreases as storage time increases?
(d) Analyze the residuals from this experiment.

13-37. An experiment was conducted to investigate leaking current in a SOS MOSFETS device. The purpose of the experiment was to investigate how leakage current varies as the channel length changes. Four channel lengths were selected. For each channel length, five different widths were also used, and width is to be considered a nuisance factor. The data are as follows:

Channel Length	Width 1	2	3	4	5
1	0.7	0.8	0.8	0.9	1.0
2	0.8	0.8	0.9	0.9	1.0
3	0.9	1.0	1.7	2.0	4.0
4	1.0	1.5	2.0	3.0	20.0

(a) Test the hypothesis that mean leakage voltage does not depend on the channel length, using $\alpha = 0.05$.
(b) Analyze the residuals from this experiment. Comment on the residual plots.
(c) The observed leakage voltage for channel length 4 and width 5 was erroneously recorded. The correct observation is 4.0. Analyze the corrected data from this experiment. Is there evidence to conclude that mean leakage voltage increases with channel length?

Supplemental Exercises

13-38. An article in *Lubrication Engineering* (December 1990) described the results of an experiment designed to investigate the effects of carbon material properties on the progression of blisters on carbon face seals. The carbon face seals are used extensively in equipment such as air turbine starters. Five different carbon materials were tested, and the surface roughness was measured. The data are as follows:

Carbon Material Type	Surface Roughness					
EC10	0.50	0.55	0.55	0.36		
EC10A	0.31	0.07	0.25	0.18	0.56	0.20
EC4	0.20	0.28	0.12			
EC1	0.10	0.16				

(a) Does carbon material type have an effect on mean surface roughness? Use $\alpha = 0.05$.
(b) Find the residuals for this experiment. Does a normal probability plot of the residuals indicate any problem with the normality assumption?
(c) Plot the residuals versus $\hat{y}_{ij}$. Comment on the plot.
(d) Find a 95% confidence interval on the difference between the mean surface roughness between the EC10 and the EC1 carbon grades.
(e) Apply the Fisher LSD method to this experiment. Summarize your conclusions regarding the effect of material type on surface roughness.

13-39. An article in the *IEEE Transactions on Components, Hybrids, and Manufacturing Technology* (Vol. 15, No. 2, 1992, pp. 146–153) described an experiment in which the contact resistance of a brake-only relay was studied for three different materials (all were silver-based alloys). The data are as follows:

Alloy	Contact Resistance				
1	95	97	99	98	99
	99	99	94	95	98
2	104	102	102	105	99
	102	111	103	100	103
3	119	130	132	136	141
	172	145	150	144	135

(a) Does the type of alloy affect mean contact resistance? Use $\alpha = 0.01$.
(b) Use Fisher's LSD method to determine which means differ.
(c) Find a 99% confidence interval on the mean contact resistance for alloy 3.
(d) Analyze the residuals for this experiment.

13-40. An article in the *Journal of Quality Technology* (Vol. 14, No. 2, 1982, pp. 80–89) described an experiment in which three different methods of preparing fish are evaluated on the basis of sensory criteria and a quality score is assigned. Assume that these methods have been randomly selected from a large population of preparation methods. The data are in the following table:

Method	Score			
1	24.4	23.2	25.0	19.7
	22.2	24.4	23.8	18.0
2	22.1	19.5	17.3	19.7
	22.3	23.2	21.4	22.6
3	23.3	22.8	22.4	23.7
	20.4	23.5	20.8	24.1

(a) Is there any difference in preparation methods? Use $\alpha = 0.05$.
(b) Calculate the P-value for the F-statistic in part (a).
(c) Analyze the residuals from this experiment and comment on model adequacy.
(d) Estimate the components of variance.

13-41. An article in the *Journal of Agricultural Engineering Research* (Vol. 52, 1992, pp. 53–76) described an experiment to investigate the effect of drying temperature of wheat grain on the baking quality of bread. Three temperature levels were used, and the response variable measured was the volume of the loaf of bread produced. The data are as follows:

Temperature (°C)	Volume (CC)				
70.0	1245	1235	1285	1245	1235
75.0	1235	1240	1200	1220	1210
80.0	1225	1200	1170	1155	1095

(a) Does drying temperature affect mean bread volume? Use $\alpha = 0.01$.
(b) Find the P-value for this test.
(c) Use the Fisher LSD method to determine which means are different.
(d) Analyze the residuals from this experiment and comment on model adequacy.

13-42. An article in *Agricultural Engineering* (December 1964, pp. 672–673) described an experiment in which the daily weight gain of swine is evaluated at different levels of housing temperature. The mean weight of each group of swine at the start of the experiment is considered to be a nuisance factor. The data from this experiment are as follows:

Mean Weight (lbs)	Housing Air Temperatures (degrees F)					
	50	60	70	80	90	100
100	1.37	1.58	2.00	1.97	1.40	0.39
150	1.47	1.75	2.16	1.82	1.14	−0.19
200	1.19	1.91	2.22	1.67	0.88	−0.77

(a) Does housing air temperature affect mean weight gain? Use $\alpha = 0.05$.
(b) Use Fisher's LSD method to determine which temperature levels are different.
(c) Analyze the residuals from this experiment and comment on model adequacy.

13-43. An article in *Communications of the ACM* (Vol. 30, No. 5, 1987) studied different algorithms for estimating software development costs. Six algorithms were applied to eight software development projects and the percent error in estimating the development cost was observed. The data are in Table 13-17.
(a) Do the algorithms differ in their mean cost estimation accuracy? Use $\alpha = 0.05$.
(b) Analyze the residuals from this experiment.
(c) Which algorithm would you recommend for use in practice?

13-44. An article in *Nature Genetics* (2003, Vol. 34(1), pp. 85–90) "Treatment-specific changes in gene expression discriminate in vivo drug response in human leukemia cells" studied gene expression as a function of different treatments for leukemia. Three treatment groups are: mercaptopurine (MP) only; low-dose methotrexate (LDMTX) and MP; and high-dose methotrexate (HDMTX) and MP. Each group contained ten subjects. The responses from a specific gene are shown in Table 13-18.

Table 13-17 Software Development Costs

Algorithm	Project 1	2	3	4	5	6	7	8
1(SLIM)	1244	21	82	2221	905	839	527	122
2(COCOMO-A)	281	129	396	1306	336	910	473	199
3(COCOMO-R)	220	84	458	543	300	794	488	142
4(COCOMO-C)	225	83	425	552	291	826	509	153
5(FUNCTION POINTS)	19	11	−34	121	15	103	87	−17
6(ESTIMALS)	−20	35	−53	170	104	199	142	41

Table 13-18 Treatment-Specific Changes in Gene Expression

Treatments	Observations									
MP ONLY	334.5	31.6	701	41.2	61.2	69.6	67.5	66.6	120.7	881.9
MP+HDMTX	919.4	404.2	1024.8	54.1	62.8	671.6	882.1	354.2	321.9	91.1
MP+LDMTX	108.4	26.1	240.8	191.1	69.7	242.8	62.7	396.9	23.6	290.4

(a) Check the normality of the data. Can we assume these samples are from normal populations?
(b) Take the logarithm of the raw data and check the normality of the transformed data. Is there evidence to support the claim that the treatment means differ for the transformed data? Use $\alpha = 0.1$.
(c) Analyze the residuals from the transformed data and comment on model adequacy.

13-45. Consider an ANOVA situation with $a = 5$ treatments. Let $\sigma^2 = 9$ and $\alpha = 0.05$, and suppose that $n = 4$.
(a) Find the power of the ANOVA F-test when $\mu_1 = \mu_2 = \mu_3 = 1$, $\mu_4 = 3$, and $\mu_5 = 2$.
(b) What sample size is required if we want the power of the F-test in this situation to be at least 0.90?

13-46. Consider an ANOVA situation with $a = 4$ means $\mu_1 = 1$, $\mu_2 = 5$, $\mu_3 = 8$, and $\mu_4 = 4$. Suppose that $\sigma^2 = 4$, $n = 4$, and $\alpha = 0.05$.
(a) Find the power of the ANOVA F-test.
(b) How large would the sample size have to be if we want the power of the F-test for detecting this difference in means to be at least 0.90?

MIND-EXPANDING EXERCISES

13-47. Show that in the fixed-effects model analysis of variance $E(MS_E) = \sigma^2$. How would your development change if the random-effects model had been specified?

13-48. Consider testing the equality of the means of two normal populations where the variances are unknown but are assumed equal. The appropriate test procedure is the two-sample t-test. Show that the two-sample t-test is equivalent to the single-factor analysis of variance F-test.

13-49. Consider the ANOVA with $a = 2$ treatments. Show that the MS_E in this analysis is equal to the pooled variance estimate used in the two-sample t-test.

13-50. Show that the variance of the linear combination

$$\sum_{i=1}^{a} c_i Y_{i\cdot} \text{ is } \sigma^2 \sum_{i=1}^{a} n_i c_i^2.$$

13-51. In a fixed-effects model, suppose that there are n observations for each of four treatments. Let Q_1^2, Q_2^2, and Q_3^2 be single-degree-of-freedom sums of squares for orthogonal contrasts. A contrast is a linear combination of the treatment means with coefficients that sum to zero. The coefficient vectors of orthogonal contrasts are orthogonal vectors. Prove that $SS_{\text{Treatments}} = Q_1^2 + Q_2^2 + Q_3^2$.

13-52. Consider the single-factor completely randomized design with a treatments and n replicates. Show that if the difference between any two treatment means is as large as D, the minimum value that the OC curve parameter Φ^2 can take is

$$\Phi^2 = \frac{nD^2}{2a\sigma^2}$$

13-53. Consider the single-factor completely randomized design. Show that a $100(1 - \alpha)$ percent confidence interval for σ^2 is

$$\frac{(N-a)MS_E}{\chi^2_{\alpha/2,N-a}} \le \sigma^2 \le \frac{(N-a)MS_E}{\chi^2_{1-\alpha/2,N-a}}$$

where N is the total number of observations in the experimental design.

13-54. Consider the random-effect model for the single-factor completely randomized design. Show that a $100(1 - \alpha)\%$ confidence interval on the ratio of variance components σ_τ^2/σ^2 is given by

$$L \le \frac{\sigma_\tau^2}{\sigma^2} \le U$$

where

$$L = \frac{1}{n}\left[\frac{MS_{\text{Treatments}}}{MS_E} \times \left(\frac{1}{f_{\alpha/2,a-1,N-a}}\right) - 1\right]$$

and

$$U = \frac{1}{n}\left[\frac{MS_{\text{Treatments}}}{MS_E} \times \left(\frac{1}{f_{1-\alpha/2,a-1,N-a}}\right) - 1\right]$$

13-55. Consider a random-effects model for the single-factor completely randomized design. (a) Show

MIND-EXPANDING EXERCISES

that a $100(1 - \alpha)\%$ confidence interval on the ratio $\sigma_\tau^2/(\sigma^2 + \sigma_\tau^2)$ is

$$\frac{L}{1 + L} \le \frac{\sigma_\tau^2}{\sigma^2 + \sigma_\tau^2} \le \frac{U}{1 + U}$$

where L and U are as defined in Exercise 13-54.

(b) Use the results of part (a) to find a $100(1 - \alpha)\%$ confidence interval for $\sigma^2/(\sigma^2 + \sigma_\tau^2)$.

13-56. Consider the fixed-effect model of the completely randomized single-factor design. The model parameters are restricted by the constraint $\sum_{i=1}^{a} \tau_i = 0$. (Actually, other restrictions could be used, but this one is simple and results in intuitively pleasing estimates for the model parameters.) For the case of unequal sample size $n_1, n_2, \ldots, n_a$, the restriction is $\sum_{i=1}^{a} n_i\tau_i = 0$. Use this to show that

$$E(MS_{\text{Treatments}}) = \sigma^2 + \frac{\sum_{i=1}^{a} n_i\tau_i^2}{a - 1}$$

Does this suggest that the null hypothesis in this model is $H_0\colon n_1\tau_1 = n_2\tau_2 = \cdots = n_a\tau_a = 0$?

13-57. **Sample Size Determination.** In the single-factor completely randomized design, the accuracy of a $100(1 - \alpha)\%$ confidence interval on the difference in any two treatment means is $t_{\alpha/2,a(n-1)}\sqrt{2MS_E/n}$.

(a) Show that if A is the desired accuracy of the interval, the sample size required is

$$n = \frac{2F_{\alpha/2,1,a(n-1)}MS_E}{A^2}$$

(b) Suppose that in comparing $a = 5$ means we have a preliminary estimate of σ^2 of 4. If we want the 95% confidence interval on the difference in means to have an accuracy of 2, how many replicates should we use?

IMPORTANT TERMS AND CONCEPTS

Analysis of variance (ANOVA)
Blocking
Completely randomized experiment
Expected mean squares
Fisher's least significant difference (LSD) method
Fixed factor
Graphical comparison of means
Levels of a factor
Mean square
Multiple comparisons
Nuisance factors
Random factor
Randomization
Randomized complete block design
Residual analysis and model adequacy checking
Sample size and replication in an experiment
Treatment effect
Variance component

14 Designing Experiments with Several Factors

CHAPTER OUTLINE

LEARNING OBJECTIVES

After careful study of this chapter, you should be able to do the following:

1. Design and conduct engineering experiments involving several factors using the factorial design approach
2. Know how to analyze and interpret main effects and interactions
3. Understand how the ANOVA is used to analyze the data from these experiments
4. Assess model adequacy with residual plots
5. Know how to use the two-level series of factorial designs
6. Understand how two-level factorial designs can be run in blocks
7. Design and conduct two-level fractional factorial designs

8. Test for curvature in two-level factorial designs by using center points
9. Use response surface methodology for process optimization experiments

14-1 INTRODUCTION

An **experiment** is just a **test** or series of tests. Experiments are performed in all engineering and scientific disciplines and are an important part of the way we learn about how systems and processes work. The validity of the conclusions that are drawn from an experiment depends to a large extent on how the experiment was conducted. Therefore, the **design** of the experiment plays a major role in the eventual solution of the problem that initially motivated the experiment.

In this chapter we focus on experiments that include two or more factors that the experimenter thinks may be important. A **factorial experiment** is a powerful technique for this type of problem. Generally, in a factorial experimental design, experimental trials (or runs) are performed at all combinations of factor levels. For example, if a chemical engineer is interested in investigating the effects of reaction time and reaction temperature on the yield of a process, and if two levels of time (1 and 1.5 hours) and two levels of temperature (125 and 150°F) are considered important, a factorial experiment would consist of making experimental runs at each of the four possible combinations of these levels of reaction time and reaction temperature.

Experimental design is an extremely important tool for engineers and scientists who are interested in improving the performance of a manufacturing process. It also has extensive application in the development of new processes and in new product design. We now give some examples.

Process Characterization Experiment

In an article in *IEEE Transactions* on "Electronics Packaging Manufacturing" (2001, Vol. 24(4), pp. 249–254), the authors discussed the change to lead-free solder in surface mount technology (SMT). SMT is a process to assemble electronic components to a printed circuit board. Solder paste is printed through a stencil onto the printed circuit board. The stencil-printing machine has squeegees; the paste rolls in front of the squeegee and fills the apertures in the stencil. The squeegee shears off the paste in the apertures as it moves over the stencil. Once the print stroke is completed, the board is separated mechanically from the stencil. Electronic components are placed on the deposits and the board is heated so that the paste reflows to form the solder joints.

The current SMT soldering process is based on tin–lead solders, and it has been well developed and refined over the years to operate at a competitive cost. The process will have several (perhaps many) variables, and all of them are not equally important. The initial list of candidate variables to be included in an experiment is constructed by combining the knowledge and information about the process from all team members. For example, engineers would conduct a **brainstorming** session and invite manufacturing personnel with SMT experience to participate. SMT has several variables that can be controlled. These include

> (1) squeegee speed, (2) squeegee pressure, (3) squeegee angle, (3) metal or polyurethane squeegee, (4) squeegee vibration, (5) delay time before the squeegee lifts from the stencil, (6) stencil separation speed, (7) print gap, (8) solder paste alloy, (9) paste pretreatment (10) paste particle size, (12) flux type, (11) reflow temperature, (12) reflow time, and so forth.

In addition to these controllable factors, there are several other factors that cannot be easily controlled during routine manufacturing, including

> (1) thickness of the printed circuit board, (2) types of components used on the board and aperture width and length, (3) layout of the components on the board, (4) paste density variation, (5) environmental factors, (5) squeegee wear, (6) cleanliness, and so forth.

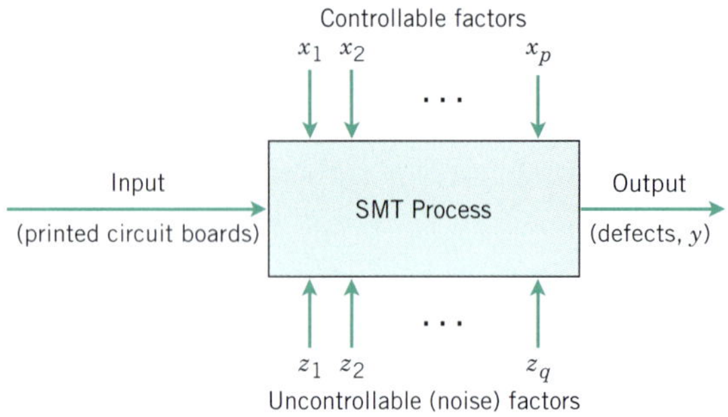

Figure 14-1 The flow solder experiment.

Sometimes we call the uncontrollable factors *noise* factors. A schematic representation of the process is shown in Fig. 14-1. In this situation, the engineer wants to **characterize** the SMT process; that is, to determine the factors (both controllable and uncontrollable) that affect the occurrence of defects on the printed circuit boards. To determine these factors, an experiment can be designed to estimate the magnitude and direction of the factor effects. Sometimes we call such an experiment a **screening experiment.** The information from this characterization study, or screening experiment, can help determine the critical process variables, as well as the direction of adjustment for these factors in order to reduce the number of defects, and assist in determining which process variables should be carefully controlled during manufacturing to prevent high defect levels and erratic process performance.

Optimization Experiment

In a characterization experiment, we are interested in determining which factors affect the response. A logical next step is to determine the region in the important factors that leads to an **optimum response.** For example, if the response is cost, we will look for a region of minimum cost.

As an illustration, suppose that the yield of a chemical process is influenced by the operating temperature and the reaction time. We are currently operating the process at 155°F and 1.7 hours of reaction time, and the current process yield is around 75%. Figure 14-2 shows a view of the time–temperature space from above. In this graph we have connected points of constant yield with lines. These lines are yield **contours,** and we have shown the contours at 60, 70, 80, 90, and 95% yield. To locate the optimum, we might begin with a factorial experiment such as we describe below, with the two factors, time and temperature, run at two levels each at 10°F and 0.5 hours above and below the current operating conditions. This two-factor factorial design is shown in Fig. 14-2. The average responses observed at the four points in the experiment (145°F, 1.2 hours; 145°F, 2.2 hours; 165°F, 1.2 hours; and 165°F, 2.2 hours) indicate that we should move in the general direction of increased temperature and lower reaction time to increase yield. A few additional runs could be performed in this direction to locate the region of maximum yield.

A Product Design Example

We can also use experimental design in the development of new products. For example, suppose that a group of engineers are designing a door hinge for an automobile. The product characteristic is the check effort, or the holding ability, of the latch that prevents the door from swinging closed when the vehicle is parked on a hill. The check mechanism consists of a leaf spring and a roller. When the door is opened, the roller travels through an arc causing the leaf spring to be compressed. To close the door, the spring must be forced aside, and this creates the check effort. The engineering team thinks that check effort is a function of the following factors:

> (1) roller travel distance, (2) spring height from pivot to base, (3) horizontal distance from pivot to spring, (4) free height of the reinforcement spring, (5) free height of the main spring.

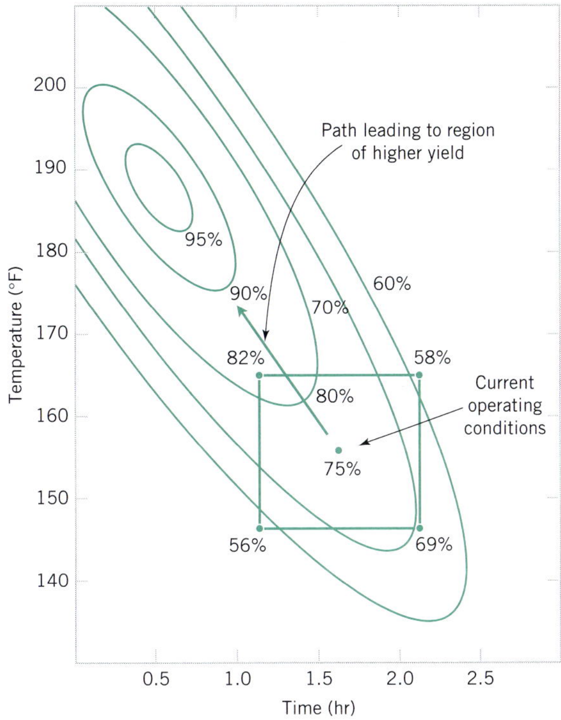

Figure 14-2 Contour plot of yield as a function of reaction time and reaction temperature, illustrating an optimization experiment.

The engineers can build a prototype hinge mechanism in which all these factors can be varied over certain ranges. Once appropriate levels for these five factors have been identified, an experiment can be designed consisting of various combinations of the factor levels, and the prototype can be tested at these combinations. This will produce information concerning which factors are most influential on the latch check effort, and through analysis of this information, the latch design can be improved.

Most of the statistical concepts introduced in Chapter 13 for single-factor experiments can be extended to the factorial experiments of this chapter. The **analysis of variance (ANOVA),** in particular, will continue to be used as one of the primary tools for statistical data analysis. We will also introduce several graphical methods that are useful in analyzing the data from designed experiments.

14-2 FACTORIAL EXPERIMENTS

When several factors are of interest in an experiment, a factorial experiment should be used. As noted previously, in these experiments factors are varied together.

Factorial Experiment

> By a **factorial experiment** we mean that in each complete trial or replicate of the experiment all possible combinations of the levels of the factors are investigated.

Thus, if there are two factors A and B with a levels of factor A and b levels of factor B, each replicate contains all ab treatment combinations.

The effect of a factor is defined as the change in response produced by a change in the level of the factor. It is called a **main effect** because it refers to the primary factors in the study. For example, consider the data in Table 14-1. This is a factorial experiment with two factors, A and B, each at two levels (A_{low}, A_{high}, and B_{low}, B_{high}). The main effect of factor A is the difference between the average response at the high level of A and the average response at the low level of A, or

$$A = \frac{30 + 40}{2} - \frac{10 + 20}{2} = 20$$

That is, changing factor A from the low level to the high level causes an average response increase of 20 units. Similarly, the main effect of B is

$$B = \frac{20 + 40}{2} - \frac{10 + 30}{2} = 10$$

In some experiments, the difference in response between the levels of one factor is not the same at all levels of the other factors. When this occurs, there is an **interaction** between the factors. For example, consider the data in Table 14-2. At the low level of factor B, the A effect is

$$A = 30 - 10 = 20$$

and at the high level of factor B, the A effect is

$$A = 0 - 20 = -20$$

Since the effect of A depends on the level chosen for factor B, there is interaction between A and B.

When an interaction is large, the corresponding main effects have very little practical meaning. For example, by using the data in Table 14-2, we find the main effect of A as

$$A = \frac{30 + 0}{2} - \frac{10 + 20}{2} = 0$$

and we would be tempted to conclude that there is no factor A effect. However, when we examined the effects of A at *different levels of factor B*, we saw that this was not the case. The effect of factor A depends on the levels of factor B. Thus, knowledge of the AB interaction is more useful than knowledge of the main effect. A significant interaction can mask the significance of main effects. Consequently, when interaction is present, the main effects of the factors involved in the interaction may not have much meaning.

Table 14-1 A Factorial Experiment with Two Factors

Factor A	**Factor B** B_{low}	B_{high}
A_{low}	10	20
A_{high}	30	40

Table 14-2 A Factorial Experiment with Interaction

Factor A	**Factor B** B_{low}	B_{high}
A_{low}	10	20
A_{high}	30	0

It is easy to estimate the interaction effect in factorial experiments such as those illustrated in Tables 14-1 and 14-2. In this type of experiment, when both factors have two levels, the AB interaction effect is the difference in the diagonal averages. This represents one-half the difference between the A effects at the two levels of B. For example, in Table 14-1, we find the AB interaction effect to be

$$AB = \frac{20 + 30}{2} - \frac{10 + 40}{2} = 0$$

Thus, there is no interaction between A and B. In Table 14-2, the AB interaction effect is

$$AB = \frac{20 + 30}{2} - \frac{10 + 0}{2} = 20$$

As we noted before, the interaction effect in these data is very large.

The concept of interaction can be illustrated graphically in several ways. Figure 14-3 plots the data in Table 14-1 against the levels of A for both levels of B. Note that the B_{low} and B_{high} lines are approximately parallel, indicating that factors A and B do not interact significantly. Figure 14-4 presents a similar plot for the data in Table 14-2. In this graph, the B_{low} and B_{high} lines are not parallel, indicating the interaction between factors A and B. Such graphical displays are called **two-factor interaction plots.** They are often useful in presenting the results of experiments, and many computer software programs used for analyzing data from designed experiments will construct these graphs automatically.

Figures 14-5 and 14-6 present another graphical illustration of the data from Tables 14-1 and 14-2. In Fig. 14-3 we have shown a **three-dimensional surface plot** of the data from Table 14-1. These data contain no interaction, and the surface plot is a plane lying above the A-B space. The slope of the plane in the A and B directions is proportional to the main effects of factors A and B, respectively. Figure 14-6 is a surface plot of the data from Table 14-2. Notice that the effect of the interaction in these data is to "twist" the plane, so that there is curvature in the response function. **Factorial experiments are the only way to discover interactions between variables.**

An alternative to the factorial design that is (unfortunately) used in practice is to change the factors ***one at a time*** rather than to vary them simultaneously. To illustrate this one-factor-at-a-time procedure, suppose that an engineer is interested in finding the values of temperature and pressure that maximize yield in a chemical process. Suppose that we fix temperature at 155°F (the current operating level) and perform five runs at different levels of time, say,

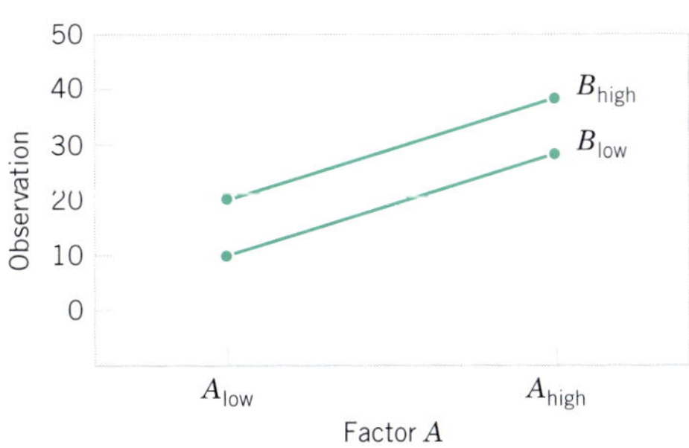

Figure 14-3 Factorial experiment, no interaction.

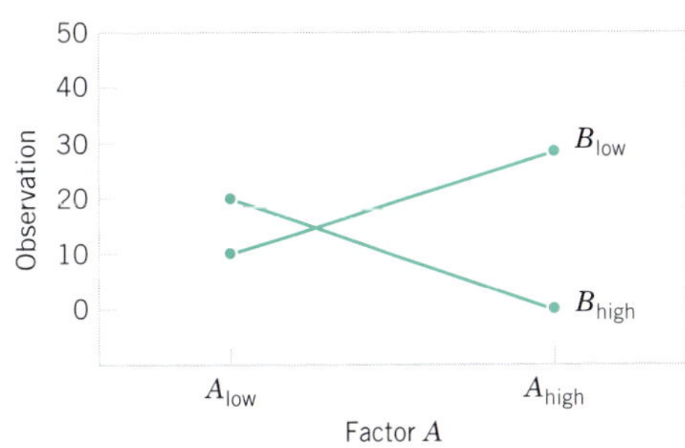

Figure 14-4 Factorial experiment, with interaction.

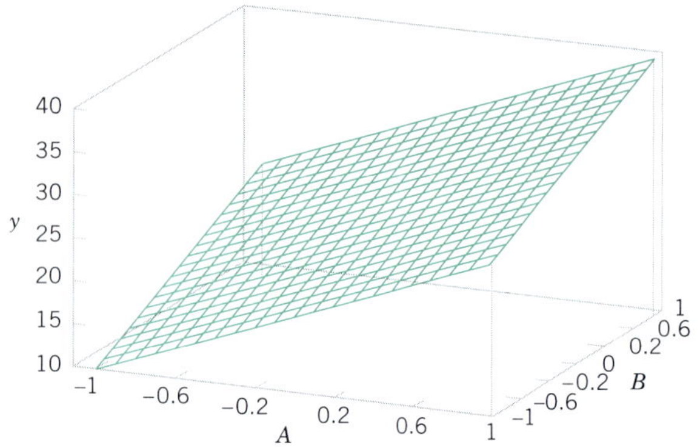

Figure 14-5 Three-dimensional surface plot of the data from Table 14-1, showing main effects of the two factors A and B.

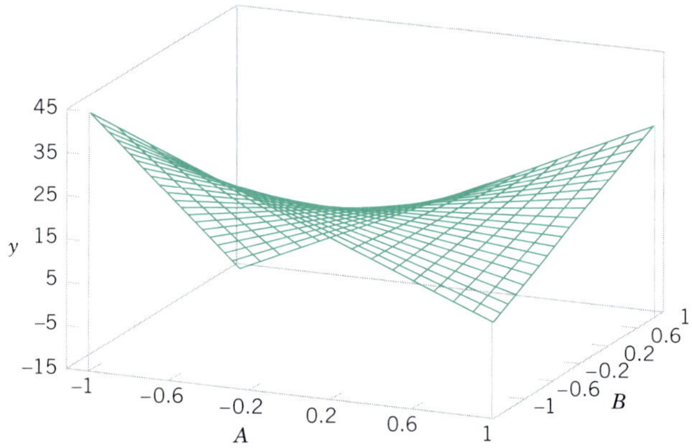

Figure 14-6 Three-dimensional surface plot of the data from Table 14-2 showing the effect of the A and B interaction.

0.5, 1.0, 1.5, 2.0, and 2.5 hours. The results of this series of runs are shown in Fig. 14-7. This figure indicates that maximum yield is achieved at about 1.7 hours of reaction time. To optimize temperature, the engineer then fixes time at 1.7 hours (the apparent optimum) and performs five runs at different temperatures, say, 140, 150, 160, 170, and 180°F. The results of this set of runs are plotted in Fig. 14-8. Maximum yield occurs at about 155°F. Therefore, we would conclude that running the process at 155°F and 1.7 hours is the best set of operating conditions, resulting in yields of around 75%.

Figure 14-9 displays the contour plot of actual process yield as a function of temperature and time with the one-factor-at-a-time experiments superimposed on the contours. Clearly, this one-factor-at-a-time approach has failed dramatically here, as the true optimum is at least 20 yield points higher and occurs at much lower reaction times and higher temperatures. The failure to discover the importance of the shorter reaction times is particularly important because this could have significant impact on production volume or capacity, production planning, manufacturing cost, and total productivity.

The one-factor-at-a-time approach has failed here because it cannot detect the interaction between temperature and time. Factorial experiments are the only way to detect interactions. Furthermore, the one-factor-at-a-time method is inefficient. It will require more

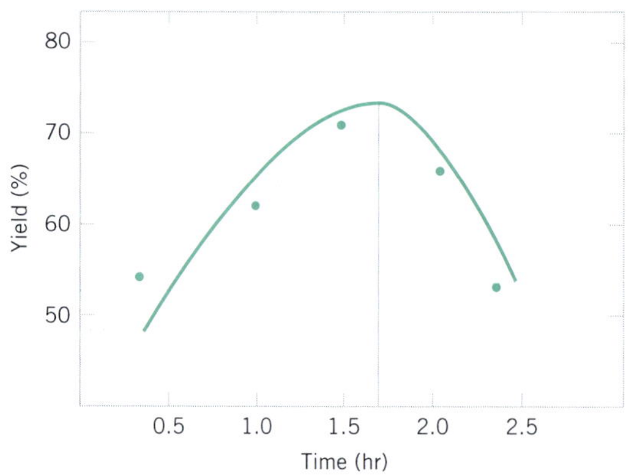

Figure 14-7 Yield versus reaction time with temperature constant at 155 °F.

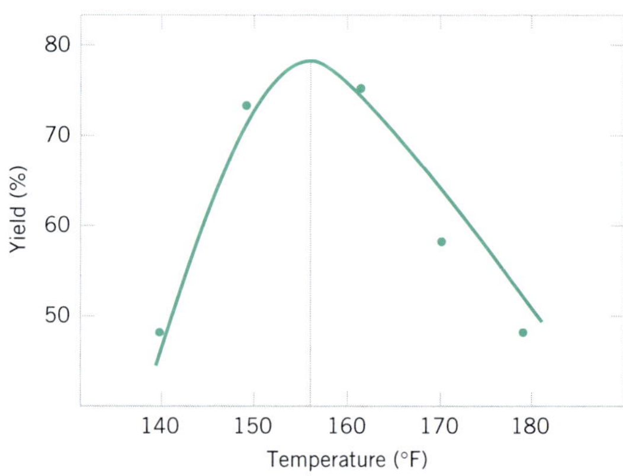

Figure 14-8 Yield versus temperature with reaction time constant at 1.7 hours.

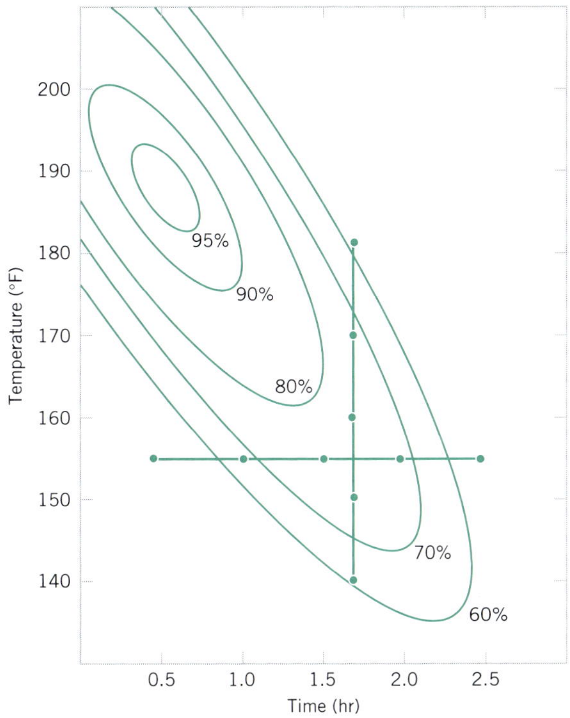

Figure 14-9 Optimization experiment using the one-factor-at-a-time method.

experimentation than a factorial, and as we have just seen, there is no assurance that it will produce the correct results.

14-3 TWO-FACTOR FACTORIAL EXPERIMENTS

The simplest type of factorial experiment involves only two factors, say A, and B. There are a levels of factor A and b levels of factor B. This two-factor factorial is shown in Table 14-3. The experiment has n **replicates,** and each replicate contains all ab treatment combinations.

Table 14-3 Data Arrangement for a Two-Factor Factorial Design

		Factor B					
		1	2	$\cdots$	b	Totals	Averages
Factor A	1	$y_{111}, y_{112}, \ldots, y_{11n}$	$y_{121}, y_{122}, \ldots, y_{12n}$		$y_{1b1}, y_{1b2}, \ldots, y_{1bn}$	$y_{1\cdot\cdot}$	$\bar{y}_{1\cdot\cdot}$
	2	$y_{211}, y_{212}, \ldots, y_{21n}$	$y_{221}, y_{222}, \ldots, y_{22n}$		$y_{2b1}, y_{2b2}, \ldots, y_{2bn}$	$y_{2\cdot\cdot}$	$\bar{y}_{2\cdot\cdot}$
	$\vdots$						
	a	$y_{a11}, y_{a12}, \ldots, y_{a1n}$	$y_{a21}, y_{a22}, \ldots, y_{a2n}$		$y_{ab1}, y_{ab2}, \ldots, y_{abn}$	$y_{a\cdot\cdot}$	$\bar{y}_{a\cdot\cdot}$
Totals		$y_{\cdot 1\cdot}$	$y_{\cdot 2\cdot}$		$y_{\cdot b\cdot}$	$y_{\cdots}$	
Averages		$\bar{y}_{\cdot 1\cdot}$	$\bar{y}_{\cdot 2\cdot}$		$\bar{y}_{\cdot b\cdot}$		$\bar{y}_{\cdots}$

The observation in the ijth cell for the kth replicate is denoted by y_{ijk}. In performing the experiment, the abn observations would be run in **random order.** Thus, like the single-factor experiment studied in Chapter 13, the two-factor factorial is a *completely randomized design*.

The observations may be described by the linear statistical model

$$Y_{ijk} = \mu + \tau_i + \beta_j + (\tau\beta)_{ij} + \epsilon_{ijk} \begin{cases} i = 1, 2, \ldots, a \\ j = 1, 2, \ldots, b \\ k = 1, 2, \ldots, n \end{cases} \tag{14-1}$$

where μ is the overall mean effect, τ_i is the effect of the ith level of factor A, β_j is the effect of the jth level of factor B, $(\tau\beta)_{ij}$ is the effect of the interaction between A and B, and ϵ_{ijk} is a random error component having a normal distribution with mean zero and variance σ^2. We are interested in testing the hypotheses of no main effect for factor A, no main effect for B, and no AB interaction effect. As with the single-factor experiments of Chapter 13, the analysis of variance (ANOVA) will be used to test these hypotheses. Since there are two factors in the experiment, the test procedure is sometimes called the two-way analysis of variance.

14-3.1 Statistical Analysis of the Fixed-Effects Model

Suppose that A and B are **fixed factors.** That is, the a levels of factor A and the b levels of factor B are specifically chosen by the experimenter, and inferences are confined to these levels only. In this model, it is customary to define the effects τ_i, β_j, and $(\tau\beta)_{ij}$ as deviations from the mean, so that $\sum_{i=1}^{a}\tau_i = 0$, $\sum_{j=1}^{b}\beta_j = 0$, $\sum_{i=1}^{a}(\tau\beta)_{ij} = 0$, and $\sum_{j=1}^{b}(\tau\beta)_{ij} = 0$.

The **analysis of variance** can be used to test hypotheses about the main factor effects of A and B and the AB interaction. To present the ANOVA, we will need some symbols, some of which are illustrated in Table 14-3. Let $y_{i\cdot\cdot}$ denote the total of the observations taken at the ith level of factor A; $y_{\cdot j\cdot}$ denote the total of the observations taken at the jth level of factor B; $y_{ij\cdot}$ denote the total of the observations in the ijth cell of Table 14-3; and $y_{\cdots}$ denote the grand total of all the observations. Define $\bar{y}_{i\cdot\cdot}$, $\bar{y}_{\cdot j\cdot}$, $\bar{y}_{ij\cdot}$, and $\bar{y}_{\cdots}$ as the corresponding row, column, cell, and grand averages. That is,

Notation for Totals and Means

$$y_{i\cdot\cdot} = \sum_{j=1}^{b}\sum_{k=1}^{n} y_{ijk} \qquad \bar{y}_{i\cdot\cdot} = \frac{y_{i\cdot\cdot}}{bn} \qquad i = 1, 2, \ldots, a$$

$$y_{\cdot j\cdot} = \sum_{i=1}^{a}\sum_{k=1}^{n} y_{ijk} \qquad \bar{y}_{\cdot j\cdot} = \frac{y_{\cdot j\cdot}}{an} \qquad j = 1, 2, \ldots, b$$

$$y_{ij\cdot} = \sum_{k=1}^{n} y_{ijk} \qquad \bar{y}_{ij\cdot} = \frac{y_{ij\cdot}}{n} \qquad \begin{matrix} i = 1, 2, \ldots, a \\ j = 1, 2, \ldots, b \end{matrix}$$

$$y_{\cdots} = \sum_{i=1}^{a}\sum_{j=1}^{b}\sum_{k=1}^{n} y_{ijk} \qquad \bar{y}_{\cdots} = \frac{y_{\cdots}}{abn}$$

The hypotheses that we will test are as follows:

1. H_0: $\tau_1 = \tau_2 = \cdots = \tau_a = 0$ (no main effect of factor A)
H_1: at least one $\tau_i \neq 0$

2. H_0: $\beta_1 = \beta_2 = \cdots = \beta_b = 0$ (no main effect of factor B) (14-2)
H_1: at least one $\beta_j \neq 0$

3. H_0: $(\tau\beta)_{11} = (\tau\beta)_{12} = \cdots = (\tau\beta)_{ab} = 0$ (no interaction)
H_1: at least one $(\tau\beta)_{ij} \neq 0$

As before, the ANOVA tests these hypotheses by decomposing the total variability in the data into component parts and then comparing the various elements in this decomposition. Total variability is measured by the total sum of squares of the observations

$$SS_T = \sum_{i=1}^{a} \sum_{j=1}^{b} \sum_{k=1}^{n} (y_{ijk} - \bar{y}_{...})^2$$

and the sum of squares decomposition is defined below.

ANOVA Sum of Squares Identity: Two Factors

The **sum of squares identity for a two-factor ANOVA** is

$$\sum_{i=1}^{a} \sum_{j=1}^{b} \sum_{k=1}^{n} (y_{ijk} - \bar{y}_{...})^2 = bn \sum_{i=1}^{a} (\bar{y}_{i..} - \bar{y}_{...})^2$$
$$+ an \sum_{j=1}^{b} (\bar{y}_{.j.} - \bar{y}_{...})^2$$
$$+ n \sum_{i=1}^{a} \sum_{j=1}^{b} (\bar{y}_{ij.} - \bar{y}_{i..} - \bar{y}_{.j.} + \bar{y}_{...})^2$$
$$+ \sum_{i=1}^{a} \sum_{j=1}^{b} \sum_{k=1}^{n} (y_{ijk} - \bar{y}_{ij.})^2 \tag{14-3}$$

or symbolically,

$$SS_T = SS_A + SS_B + SS_{AB} + SS_E \tag{14-4}$$

Equations 14-3 and 14-4 state that the total sum of squares SS_T is partitioned into a sum of squares for the row factor A (SS_A), a sum of squares for the column factor B (SS_B), a sum of squares for the interaction between A and B (SS_{AB}), and an error sum of squares (SS_E). There are $abn - 1$ total degrees of freedom. The main effects A and B have $a - 1$ and $b - 1$ degrees of freedom, while the interaction effect AB has $(a - 1)(b - 1)$ degrees of freedom. Within each of the ab cells in Table 14-3, there are $n - 1$ degrees of freedom between the n replicates, and observations in the same cell can differ only because of random error. Therefore, there are $ab(n - 1)$ degrees of freedom for error. Therefore, the degrees of freedom are partitioned according to

$$abn - 1 = (a - 1) + (b - 1) + (a - 1)(b - 1) + ab(n - 1)$$

If we divide each of the sum of squares on the right-hand side of Equation 14-4 by the corresponding number of degrees of freedom, we obtain the **mean squares** for A, B, the interaction, and error:

$$MS_A = \frac{SS_A}{a-1} \qquad MS_B = \frac{SS_B}{b-1} \qquad MS_{AB} = \frac{SS_{AB}}{(a-1)(b-1)} \qquad MS_E = \frac{SS_E}{ab(n-1)}$$

Assuming that factors A and B are fixed factors, it is not difficult to show that the **expected values** of these mean squares are

Expected Values of Mean Squares: Two Factors

$$E(MS_A) = E\left(\frac{SS_A}{a-1}\right) = \sigma^2 + \frac{bn\sum_{i=1}^{a}\tau_i^2}{a-1} \qquad E(MS_B) = E\left(\frac{SS_B}{b-1}\right) = \sigma^2 + \frac{an\sum_{j=1}^{b}\beta_j^2}{b-1}$$

$$E(MS_{AB}) = E\left(\frac{SS_{AB}}{(a-1)(b-1)}\right) = \sigma^2 + \frac{n\sum_{i=1}^{a}\sum_{j=1}^{b}(\tau\beta)_{ij}^2}{(a-1)(b-1)}$$

$$E(MS_E) = E\left(\frac{SS_E}{ab(n-1)}\right) = \sigma^2$$

From examining these expected mean squares, it is clear that if the null hypotheses about main effects H_0: $\tau_i = 0$, H_0: $\beta_j = 0$, and the interaction hypothesis H_0: $(\tau\beta)_{ij} = 0$ are all true, all four mean squares are unbiased estimates of σ^2.

To test that the row factor effects are all equal to zero (H_0: $\tau_i = 0$), we would use the ratio

F Test for Factor A

$$F_0 = \frac{MS_A}{MS_E}$$

which has an F-distribution with $a-1$ and $ab(n-1)$ degrees of freedom if H_0: $\tau_i = 0$ is true. This null hypothesis is rejected at the α level of significance if $f_0 > f_{\alpha,a-1,ab(n-1)}$. Similarly, to test the hypothesis that all the column factor effects are equal to zero (H_0: $\beta_j = 0$), we would use the ratio

F Test for Factor B

$$F_0 = \frac{MS_B}{MS_E}$$

which has an F-distribution with $b-1$ and $ab(n-1)$ degrees of freedom if H_0: $\beta_j = 0$ is true. This null hypothesis is rejected at the α level of significance if $f_0 > f_{\alpha,b-1,ab(n-1)}$. Finally, to test the hypothesis H_0: $(\tau\beta)_{ij} = 0$, which is the hypothesis that all interaction effects are zero, we use the ratio

F Test for AB Interaction

$$F_0 = \frac{MS_{AB}}{MS_E}$$

which has an F-distribution with $(a-1)(b-1)$ and $ab(n-1)$ degrees of freedom if the null hypothesis H_0: $(\tau\beta)_{ij} = 0$. This hypothesis is rejected at the α level of significance if $f_0 > f_{\alpha,(a-1)(b-1),ab(n-1)}$.

It is usually best to conduct the test for interaction first and then to evaluate the main effects. If interaction is not significant, interpretation of the tests on the main effects is straightforward. However, as noted in Section 14-3, when interaction is significant, the main effects of the factors involved in the interaction may not have much practical interpretative value. Knowledge of the interaction is usually more important than knowledge about the main effects.

Computational formulas for the sums of squares are easily obtained.

Computing Formulas for ANOVA: Two Factors

Computing formulas for the sums of squares in a two-factor analysis of variance.

$$SS_T = \sum_{i=1}^{a}\sum_{j=1}^{b}\sum_{k=1}^{n} y_{ijk}^2 - \frac{y_{\cdots}^2}{abn} \tag{14-5}$$

$$SS_A = \sum_{i=1}^{a} \frac{y_{i\cdot\cdot}^2}{bn} - \frac{y_{\cdots}^2}{abn} \tag{14-6}$$

$$SS_B = \sum_{j=1}^{b} \frac{y_{\cdot j\cdot}^2}{an} - \frac{y_{\cdots}^2}{abn} \tag{14-7}$$

$$SS_{AB} = \sum_{i=1}^{a}\sum_{j=1}^{b} \frac{y_{ij\cdot}^2}{n} - \frac{y_{\cdots}^2}{abn} - SS_A - SS_B \tag{14-8}$$

$$SS_E = SS_T - SS_{AB} - SS_A - SS_B \tag{14-9}$$

The computations are usually displayed in an ANOVA table, such as Table 14-4.

EXAMPLE 14-1 Aircraft Primer Paint

Aircraft primer paints are applied to aluminum surfaces by two methods: dipping and spraying. The purpose of the primer is to improve paint adhesion, and some parts can be primed using either application method. The process engineering group responsible for this operation is interested in learning whether three different primers differ in their adhesion properties. A factorial experiment was performed to

Table 14-4 ANOVA Table for a Two-Factor Factorial, Fixed-Effects Model

Source of Variation	Sum of Squares	Degrees of Freedom	Mean Square	F_0
A treatments	SS_A	$a-1$	$MS_A = \dfrac{SS_A}{a-1}$	$\dfrac{MS_A}{MS_E}$
B treatments	SS_B	$b-1$	$MS_B = \dfrac{SS_B}{b-1}$	$\dfrac{MS_B}{MS_E}$
Interaction	SS_{AB}	$(a-1)(b-1)$	$MS_{AB} = \dfrac{SS_{AB}}{(a-1)(b-1)}$	$\dfrac{MS_{AB}}{MS_E}$
Error	SS_E	$ab(n-1)$	$MS_E = \dfrac{SS_E}{ab(n-1)}$	
Total	SS_T	$abn-1$		

Table 14-5 Adhesion Force Data for Example 14-1

Primer Type	Dipping		Spraying		$y_{i\cdot\cdot}$
1	4.0, 4.5, 4.3	(12.8)	5.4, 4.9, 5.6	(15.9)	28.7
2	5.6, 4.9, 5.4	(15.9)	5.8, 6.1, 6.3	(18.2)	34.1
3	3.8, 3.7, 4.0	(11.5)	5.5, 5.0, 5.0	(15.5)	27.0
$y_{\cdot j\cdot}$	40.2		49.6		$89.8 = y_{\cdots}$

investigate the effect of paint primer type and application method on paint adhesion. For each combination of primer type and application method, three specimens were painted, then a finish paint was applied, and the adhesion force was measured. The data from the experiment are shown in Table 14-5. The circled numbers in the cells are the cell totals $y_{ij\cdot}$. The sums of squares required to perform the ANOVA are computed as follows:

$$SS_T = \sum_{i=1}^{a}\sum_{j=1}^{b}\sum_{k=1}^{n} y_{ijk}^2 - \frac{y_{\cdots}^2}{abn}$$

$$= (4.0)^2 + (4.5)^2 + \cdots + (5.0)^2 - \frac{(89.8)^2}{18} = 10.72$$

$$SS_{\text{types}} = \sum_{i=1}^{a} \frac{y_{i\cdot\cdot}^2}{bn} - \frac{y_{\cdots}^2}{abn}$$

$$= \frac{(28.7)^2 + (34.1)^2 + (27.0)^2}{6} - \frac{(89.8)^2}{18} = 4.58$$

$$SS_{\text{methods}} = \sum_{j=1}^{b} \frac{y_{\cdot j\cdot}^2}{an} - \frac{y_{\cdots}^2}{abn}$$

$$= \frac{(40.2)^2 + (49.6)^2}{9} - \frac{(89.8)^2}{18} = 4.91$$

$$SS_{\text{interaction}} = \sum_{i=1}^{a}\sum_{j=1}^{b} \frac{y_{ij\cdot}^2}{n} - \frac{y_{\cdots}^2}{abn} - SS_{\text{types}} - SS_{\text{methods}}$$

$$= \frac{(12.8)^2 + (15.9)^2 + (11.5)^2 + (15.9)^2 + (18.2)^2 + (15.5)^2}{3}$$

$$- \frac{(89.8)^2}{18} - 4.58 - 4.91 = 0.24$$

and

$$SS_E = SS_T - SS_{\text{types}} - SS_{\text{methods}} - SS_{\text{interaction}}$$
$$= 10.72 - 4.58 - 4.91 - 0.24 = 0.99$$

The ANOVA is summarized in Table 14-6. The experimenter has decided to use $\alpha = 0.05$. Since $f_{0.05,2,12} = 3.89$ and $f_{0.05,1,12} = 4.75$, we conclude that the main effects of primer type and application method affect adhesion force. Furthermore, since $1.5 < f_{0.05,2,12}$, there is no indication of interaction between these factors. The last column of Table 14-6 shows the P-value for each F-ratio. Notice that the P-values

Table 14-6 ANOVA for Example 14-1

Source of Variation	Sum of Squares	Degrees of Freedom	Mean Square	f_0	P-Value
Primer types	4.58	2	2.29	28.63	2.7 × E-5
Application methods	4.91	1	4.91	61.38	4.7 × E-6
Interaction	0.24	2	0.12	1.50	0.2621
Error	0.99	12	0.08		
Total	10.72	17			

for the two test statistics for the main effects are considerably less than 0.05, while the P-value for the test statistic for the interaction is greater than 0.05.

A graph of the cell adhesion force averages $\{\bar{y}_{ij\cdot}\}$ versus levels of primer type for each application method is shown in Fig. 14-10. The no-interaction conclusion is obvious in this graph, because the two lines are nearly parallel. Furthermore, since a large response indicates greater adhesion force, we conclude that spraying is the best application method and that primer type 2 is most effective.

Tests on Individual Means

When both factors are fixed, comparisons between the individual means of either factor may be made using any multiple comparison technique such as Fisher's LSD method (described in Chapter 13). When there is no interaction, these comparisons may be made using either the row averages $\bar{y}_{i\cdot\cdot}$ or the column averages $\bar{y}_{\cdot j\cdot}$. However, when interaction is significant, comparisons between the means of one factor (say, A) may be obscured by the AB interaction. In this case, we could apply a procedure such as Fisher's LSD method to the means of factor A, with factor B set at a particular level.

Minitab Output

Table 14-7 shows some of the output from the Minitab analysis of variance procedure for the aircraft primer paint experiment in Example 14-1. The upper portion of the table gives factor name and level information, and the lower portion of the table presents the analysis of variance for the adhesion force response. The results are identical to the manual calculations displayed in Table 14-6 apart from rounding.

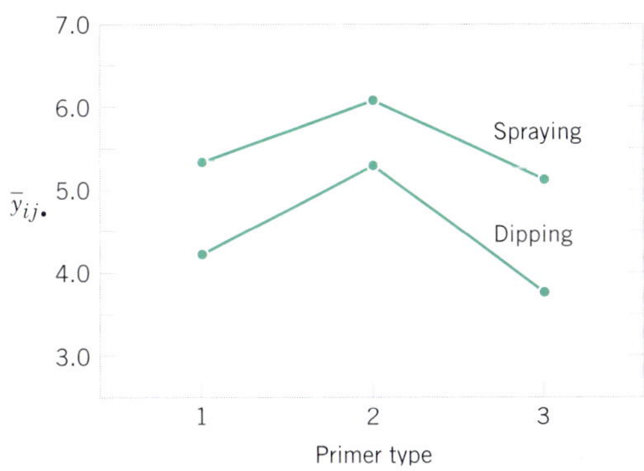

Figure 14-10 Graph of average adhesion force versus primer types for both application methods.

Table 14-7 Analysis of Variance From Minitab for Example 14-1

ANOVA (Balanced Designs)

Factor	Type	Levels	Values		
Primer	fixed	3	1	2	3
Method	fixed	2	Dip	Spray	

Analysis of Variance for Adhesion

Source	DF	SS	MS	F	P
Primer	2	4.5811	2.2906	27.86	0.000
Method	1	4.9089	4.9089	59.70	0.000
Primer *Method	2	0.2411	0.1206	1.47	0.269
Error	12	0.9867	0.0822		
Total	17	10.7178			

14-3.2 Model Adequacy Checking

Just as in the single-factor experiments discussed in Chapter 13, the **residuals** from a factorial experiment play an important role in assessing **model adequacy.** The residuals from a two-factor factorial are

$$e_{ijk} = y_{ijk} - \bar{y}_{ij\cdot}$$

That is, the residuals are just the difference between the observations and the corresponding cell averages.

Table 14-8 presents the residuals for the aircraft primer paint data in Example 14-1. The normal probability plot of these residuals is shown in Fig. 14-11. This plot has tails that do not fall exactly along a straight line passing through the center of the plot, indicating some potential problems with the normality assumption, but the deviation from normality does not appear severe. Figures 14-12 and, 14-13 plot the residuals versus the levels of primer types and application methods, respectively. There is some indication that primer type 3 results in slightly lower variability in adhesion force than the other two primers. The graph of residuals versus fitted values in Fig. 14-14 does not reveal any unusual or diagnostic pattern.

14-3.3 One Observation per Cell

In some cases involving a two-factor factorial experiment, we may have only one replicate—that is, only one observation per cell. In this situation, there are exactly as many parameters in the analysis of variance model as observations, and the error degrees of freedom are zero. Thus, we cannot test hypotheses about the main effects and interactions unless some additional

Table 14-8 Residuals for the Aircraft Primer Experiment in Example 14-1

	Application Method	
Primer Type	Dipping	Spraying
1	−0.27, 0.23, 0.03	0.10, −0.40, 0.30
2	0.30, −0.40, 0.10	−0.27, 0.03, 0.23
3	−0.03, −0.13, 0.17	0.33, −0.17, −0.17

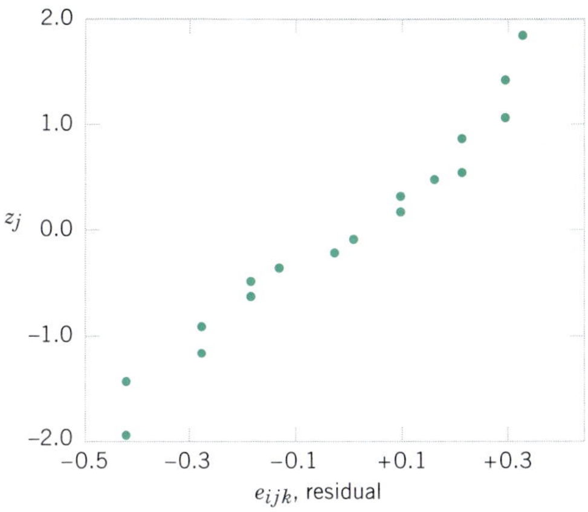

Figure 14-11 Normal probability plot of the residuals from Example 14-1.

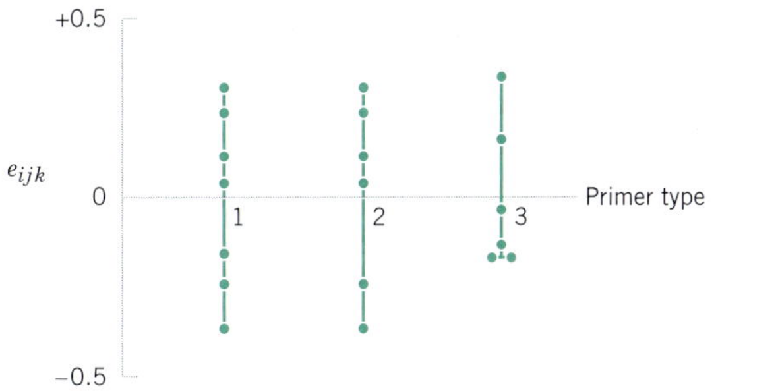

Figure 14-12 Plot of residuals versus primer type.

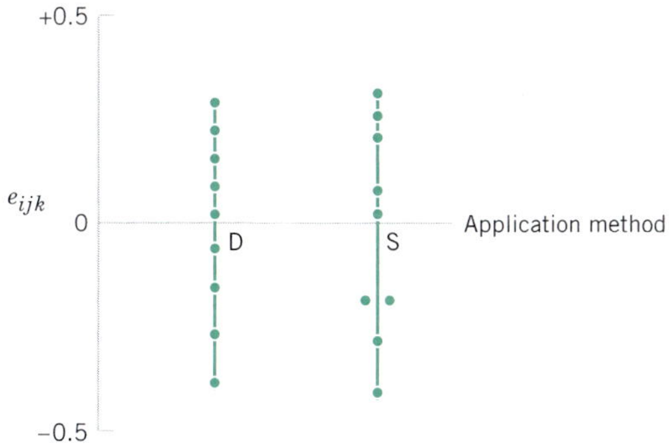

Figure 14-13 Plot of residuals versus application method.

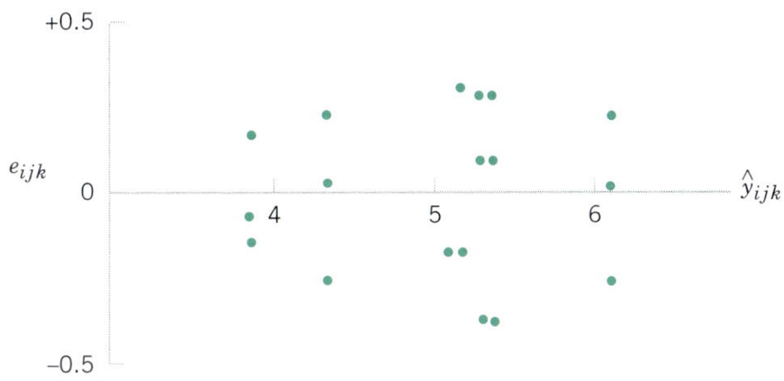

Figure 14-14 Plot of residuals versus predicted values $\hat{y}_{ijk}$.

assumptions are made. One possible assumption is to assume the interaction effect is negligible and use the interaction mean square as an error mean square. Thus, the analysis is equivalent to the analysis used in the randomized block design. This no-interaction assumption can be dangerous, and the experimenter should carefully examine the data and the residuals for indications as to whether or not interaction is present. For more details, see Montgomery (2005).

EXERCISES FOR SECTION 14-3

14-1. An article in *Industrial Quality Control* (1956, pp. 5–8) describes an experiment to investigate the effect of two factors (glass type and phosphor type) on the brightness of a television tube. The response variable measured is the current (in microamps) necessary to obtain a specified brightness level. The data are shown in the following table:

(a) State the hypotheses of interest in this experiment.

(b) Test the above hypotheses and draw conclusions using the analysis of variance with $\alpha = 0.05$.

(c) Analyze the residuals from this experiment.

Glass Type	Phosphor Type 1	2	3
1	280	300	290
	290	310	285
	285	295	290
2	230	260	220
	235	240	225
	240	235	230

14-2. An engineer suspects that the surface finish of metal parts is influenced by the type of paint used and the drying time. He selected three drying times—20, 25, and 30 minutes—and used two types of paint. Three parts are tested with each combination of paint type and drying time. The data are as follows:

	Drying Time (min)		
Paint	20	25	30
1	74	73	78
	64	61	85
	50	44	92
2	92	98	66
	86	73	45
	68	88	85

(a) State the hypotheses of interest in this experiment.
(b) Test the above hypotheses and draw conclusions using the analysis of variance with $\alpha = 0.05$.
(c) Analyze the residuals from this experiment.

14-3. In his book (*Design and Analysis of Experiments*, 6th edition, 2005 John Wiley & Sons), D. C. Montgomery presents the results of an experiment involving a storage battery used in the launching mechanism of a shoulder-fired ground-to-air missile. Three material types can be used to make the battery plates. The objective is to design a battery that is relatively unaffected by the ambient temperature. The output response from the battery is effective life in hours. Three temperature levels are selected, and a factorial experiment with four replicates is run. The data are as follows:

	Temperature (°F)					
Material	Low		Medium		High	
1	130	155	34	40	20	70
	74	180	80	75	82	58
2	150	188	136	122	25	70
	159	126	106	115	58	45
3	138	110	174	120	96	104
	168	160	150	139	82	60

(a) Test the appropriate hypotheses and draw conclusions using the analysis of variance with $\alpha = 0.05$.
(b) Graphically analyze the interaction.
(c) Analyze the residuals from this experiment.

14-4. An experiment was conducted to determine whether either firing temperature or furnace position affects the baked density of a carbon anode. The data are as follows:

	Temperature (°C)		
Position	800	825	850
1	570	1063	565
	565	1080	510
	583	1043	590
2	528	988	526
	547	1026	538
	521	1004	532

(a) State the hypotheses of interest.
(b) Test the above hypotheses using the analysis of variance with $\alpha = 0.05$. What are your conclusions?
(c) Analyze the residuals from this experiment.
(d) Using Fisher's LSD method, investigate the differences between the mean baked anode density at the three different levels of temperature. Use $\alpha = 0.05$.

14-5. An article in *Technometrics*, "Exact analysis of means with unequal variances" (2002, Vol. 44, pp. 152–160) described the technique of the analysis of means (ANOM) and presented the results of an experiment on insulation. Four insulation types were tested at three different temperatures. The data are as follows:

	Temperature (F)					
Insulation	1		2		3	
	6.6	4	4.5	2.2	2.3	0.9
	2.7	6.2	5.5	2.7	5.6	4.9
1	6	5	4.8	5.8	2.2	3.4
	3	3.2	3	1.5	1.3	3.3
	2.1	4.1	2.5	2.6	0.5	1.1
2	5.9	2.5	0.4	3.5	1.7	0.1
	5.7	4.4	8.9	7.7	2.6	9.9
	3.2	3.2	7	7.3	11.5	10.5
3	5.3	9.7	8	2.2	3.4	6.7
	7	8.9	12	9.7	8.3	8
	7.3	9	8.5	10.8	10.4	9.7
4	8.6	11.3	7.9	7.3	10.6	7.4

(a) Write down a model for this experiment.
(b) Test the appropriate hypotheses and draw conclusions using the analysis of variance with $\alpha = 0.05$.
(c) Graphically analyze the interaction.
(d) Analyze the residuals from the experiment.
(e) Use Fisher's LSD method to investigate the differences between mean effects of insulation type. Use $\alpha = 0.05$.

14-6. Johnson and Leone (*Statistics and Experimental Design in Engineering and the Physical Sciences*, John Wiley, 1977) describe an experiment conducted to investigate warping of copper plates. The two factors studied were temperature and the copper content of the plates. The response variable is the amount of warping. The data are as follows:

Temperature (°C)	Copper Content (%)			
	40	60	80	100
50	17, 20	16, 21	24, 22	28, 27
75	12, 9	18, 13	17, 12	27, 31
100	16, 12	18, 21	25, 23	30, 23
125	21, 17	23, 21	23, 22	29, 31

(a) Is there any indication that either factor affects the amount of warping? Is there any interaction between the factors? Use $\alpha = 0.05$.
(b) Analyze the residuals from this experiment.
(c) Plot the average warping at each level of copper content and compare the levels using Fisher's LSD method. Describe the differences in the effects of the different levels of copper content on warping. If low warping is desirable, what level of copper content would you specify?
(d) Suppose that temperature cannot be easily controlled in the environment in which the copper plates are to be used. Does this change your answer for part (c)?

14-7. An article in the *IEEE Transactions on Electron Devices* (November 1986, p. 1754) describes a study on the effects of two variables—polysilicon doping and anneal conditions (time and temperature)—on the base current of a bipolar transistor. The data from this experiment follows below.
(a) Is there any evidence to support the claim that either polysilicon doping level or anneal conditions affect base current? Do these variables interact? Use $\alpha = 0.05$.
(b) Graphically analyze the interaction.
(c) Analyze the residuals from this experiment.
(d) Use Fisher's LSD method to isolate the effects of anneal conditions on base current, with $\alpha = 0.05$.

14-8. An article in the *Journal of Testing and Evaluation* (1988, Vol. 16, pp. 508–515) investigated the effects of cyclic loading frequency and environment conditions on fatigue crack growth at a constant 22 MPa stress for a particular material. The data from the experiment follow. The response variable is fatigue crack growth rate.

Frequency	Environment: Air	H_2O	Salt H_2O
10	2.29	2.06	1.90
	2.47	2.05	1.93
	2.48	2.23	1.75
	2.12	2.03	2.06
1	2.65	3.20	3.10
	2.68	3.18	3.24
	2.06	3.96	3.98
	2.38	3.64	3.24
0.1	2.24	11.00	9.96
	2.71	11.00	10.01
	2.81	9.06	9.36
	2.08	11.30	10.40

(a) Is there indication that either factor affects crack growth rate? Is there any indication of interaction? Use $\alpha = 0.05$.
(b) Analyze the residuals from this experiment.
(c) Repeat the analysis in part (a) using $ln(y)$ as the response. Analyze the residuals from this new response variable and comment on the results.

14-9. Consider a two-factor factorial experiment. Develop a formula for finding a $100(1 - \alpha)\%$ confidence interval on the difference between any two means for either a row or column factor. Apply this formula to find a 95% CI on the difference in mean warping at the levels of copper content 60 and 80% in Exercise 14-6.

Polysilicon doping	Anneal (temperature/time): 900 / 60	900 / 180	950 / 60	1000 / 15	1000 / 30
1×10^{20}	4.40	8.30	10.15	10.29	11.01
	4.60	8.90	10.20	10.30	10.58
2×10^{20}	3.20	7.81	9.38	10.19	10.81
	3.50	7.75	10.02	10.10	10.60

14-4 GENERAL FACTORIAL EXPERIMENTS

Many experiments involve more than two factors. In this section we introduce the case where there are a levels of factor A, b levels of factor B, c levels of factor C, and so on, arranged in a factorial experiment. In general, there will be $abc \ldots n$ total observations, if there are n replicates of the complete experiment.

Table 14-9 Analysis of Variance Table for the Three-Factor Fixed Effects Model

Source of Variation	Sum of Squares	Degrees of Freedom	Mean Square	Expected Mean Squares	F_0
A	SS_A	$a-1$	MS_A	$\sigma^2 + \dfrac{bcn\Sigma\tau_i^2}{a-1}$	$\dfrac{MS_A}{MS_E}$
B	SS_B	$b-1$	MS_B	$\sigma^2 + \dfrac{acn\Sigma\beta_j^2}{b-1}$	$\dfrac{MS_B}{MS_E}$
C	SS_C	$c-1$	MS_C	$\sigma^2 + \dfrac{abn\Sigma\gamma_k^2}{c-1}$	$\dfrac{MS_C}{MS_E}$
AB	SS_{AB}	$(a-1)(b-1)$	MS_{AB}	$\sigma^2 + \dfrac{cn\Sigma\Sigma(\tau\beta)_{ij}^2}{(a-1)(b-1)}$	$\dfrac{MS_{AB}}{MS_E}$
AC	SS_{AC}	$(a-1)(c-1)$	MS_{AC}	$\sigma^2 + \dfrac{bn\Sigma\Sigma(\tau\gamma)_{ik}^2}{(a-1)(c-1)}$	$\dfrac{MS_{AC}}{MS_E}$
BC	SS_{BC}	$(b-1)(c-1)$	MS_{BC}	$\sigma^2 + \dfrac{an\Sigma\Sigma(\beta\gamma)_{jk}^2}{(b-1)(c-1)}$	$\dfrac{MS_{BC}}{MS_E}$
ABC	SS_{ABC}	$(a-1)(b-1)(c-1)$	MS_{ABC}	$\sigma^2 + \dfrac{n\Sigma\Sigma\Sigma(\tau\beta\gamma)_{ijk}^2}{(a-1)(b-1)(c-1)}$	$\dfrac{MS_{ABC}}{MS_E}$
Error	SS_E	$abc(n-1)$	MS_E	σ^2	
Total	SS_T	$abcn-1$			

For example, consider the **three-factor-factorial experiment,** with underlying model

$$Y_{ijkl} = \mu + \tau_i + \beta_j + \gamma_k + (\tau\beta)_{ij} + (\tau\gamma)_{ik} + (\beta\gamma)_{jk} + (\tau\beta\gamma)_{ijk} + \epsilon_{ijkl} \quad \begin{cases} i = 1, 2, \ldots, a \\ j = 1, 2, \ldots, b \\ k = 1, 2, \ldots, c \\ l = 1, 2, \ldots, n \end{cases} \tag{14-10}$$

Notice that the model contains three main effects, three two-factor interactions, a three-factor interaction, and an error term. Assuming that A, B, and C are fixed factors, the analysis of variance is shown in Table 14-9. Note that there must be at least two replicates ($n \geq 2$) to compute an error sum of squares. The F-test on main effects and interactions follows directly from the expected mean squares. These ratios follow F distributions under the respective null hypotheses.

EXAMPLE 14-2
Surface Roughness

A mechanical engineer is studying the surface roughness of a part produced in a metal-cutting operation. Three factors, feed rate (A), depth of cut (B), and tool angle (C), are of interest. All three factors have been assigned two levels, and two replicates of a factorial design are run. The coded data are shown in Table 14-10.

The ANOVA is summarized in Table 14-11. Since manual ANOVA computations are tedious for three-factor experiments, we have used Minitab for the solution of this problem. The F-ratios for all three

Table 14-10 Coded Surface Roughness Data for Example 14-2

Feed Rate (A)	Depth of Cut (B) 0.025 inch, Tool Angle (C) 15°	0.025 inch, 25°	0.040 inch, Tool Angle (C) 15°	0.040 inch, 25°	$y_{i\cdots}$
20 inches per minute	9 7	11 10	9 11	10 8	75
30 inches per minute	10 12	10 13	12 15	16 14	102

main effects and the interactions are formed by dividing the mean square for the effect of interest by the error mean square. Since the experimenter has selected $\alpha = 0.05$, the critical value for each of these F-ratios is $f_{0.05,1,8} = 5.32$. Alternately, we could use the P-value approach. The P-values for all the test statistics are shown in the last column of Table 14-11. Inspection of these P-values is revealing. There is a strong main effect of feed rate, since the F-ratio is well into the critical region. However, there is some indication of an effect due to the depth of cut, since $P = 0.0710$ is not much greater than $\alpha = 0.05$. The next largest effect is the AB or feed rate $\times$ depth of cut interaction. Most likely, both feed rate and depth of cut are important process variables.

Obviously, factorial experiments with three or more factors can require many runs, particularly if some of the factors have several (more than two) levels. This point of view leads us to the class of factorial designs considered in Section 14-5 with all factors at two levels. These designs are easy to set up and analyze, and they may be used as the basis of many other useful experimental designs.

Table 14-11 Minitab ANOVA for Example 14-2

ANOVA (Balanced Designs)

Factor	Type	Levels	Values	
Feed	fixed	2	20	30
Depth	fixed	2	0.025	0.040
Angle	fixed	2	15	25

Analysis of Variance for Roughness

Source	DF	SS	MS	F	P
Feed	1	45.563	45.563	18.69	0.003
Depth	1	10.563	10.563	4.33	0.071
Angle	1	3.063	3.063	1.26	0.295
Feed*Depth	1	7.563	7.563	3.10	0.116
Feed*Angle	1	0.062	0.062	0.03	0.877
Depth*Angle	1	1.563	1.563	0.64	0.446
Feed*Depth*Angle	1	5.062	5.062	2.08	0.188
Error	8	19.500	2.437		
Total	15	92.938			

EXERCISES FOR SECTION 14-4

14-10. The quality control department of a fabric finishing plant is studying the effects of several factors on dyeing for a blended cotton/synthetic cloth used to manufacture shirts. Three operators, three cycle times, and two temperatures were selected, and three small specimens of cloth were dyed under each set of conditions. The finished cloth was compared to a standard, and a numerical score was assigned. The results are shown in the following table.

(a) State and test the appropriate hypotheses using the analysis of variance with $\alpha = 0.05$.

(b) The residuals may be obtained from $e_{ijkl} = y_{ijkl} - \bar{y}_{ijk\cdot}$. Graphically analyze the residuals from this experiment.

	Temperature					
	300°			350°		
	Operator			Operator		
Cycle Time	1	2	3	1	2	3
	23	27	31	24	38	34
40	24	28	32	23	36	36
	25	26	28	28	35	39
	36	34	33	37	34	34
50	35	38	34	39	38	36
	36	39	35	35	36	31
	28	35	26	26	36	28
60	24	35	27	29	37	26
	27	34	25	25	34	34

14-11. The percentage of hardwood concentration in raw pulp, the freeness, and the cooking time of the pulp are being investigated for their effects on the strength of paper. The data from a three-factor factorial experiment are shown in the following table.

(a) Analyze the data using the analysis of variance assuming that all factors are fixed. Use $\alpha = 0.05$.

(b) Find P-values for the F-ratios in part (a).

(c) The residuals are found by $e_{ijkl} = y_{ijkl} - \bar{y}_{ijk\cdot}$. Graphically analyze the residuals from this experiment.

Percentage of Hardwood Concentration	Cooking Time 1.5 hours			Cooking Time 2.0 hours		
	Freeness			Freeness		
	350	500	650	350	500	650
10	96.6	97.7	99.4	98.4	99.6	100.6
	96.0	96.0	99.8	98.6	100.4	100.9
15	98.5	96.0	98.4	97.5	98.7	99.6
	97.2	96.9	97.6	98.1	96.0	99.0
20	97.5	95.6	97.4	97.6	97.0	98.5
	96.6	96.2	98.1	98.4	97.8	99.8

14-5 2^k FACTORIAL DESIGNS

Factorial designs are frequently used in experiments involving several factors where it is necessary to study the joint effect of the factors on a response. However, several special cases of the general factorial design are important because they are widely employed in research work and because they form the basis of other designs of considerable practical value.

The most important of these special cases is that of k factors, each at only two levels. These levels may be quantitative, such as two values of temperature, pressure, or time; or they may be qualitative, such as two machines, two operators, the "high" and "low" levels of a factor, or perhaps the presence and absence of a factor. A complete replicate of such a design requires $2 \times 2 \times \cdots \times 2 = 2^k$ observations and is called a **2^k factorial design.**

The 2^k design is particularly useful in the early stages of experimental work, when many factors are likely to be investigated. It provides the smallest number of runs for which k factors can be studied in a complete factorial design. Because there are only two levels for each factor, we must assume that the response is approximately linear over the range of the factor levels chosen.

14-5.1 2^2 Design

The simplest type of 2^k design is the 2^2—that is, two factors A and B, each at two levels. We usually think of these levels as the low and high levels of the factor. The 2^2 design is shown in Fig. 14-15. Note that the design can be represented geometrically as a square with the $2^2 = 4$ runs, or treatment combinations, forming the corners of the square. In the 2^2 design it is customary to denote the low and high levels of the factors A and B by the signs $-$ and $+$, respectively. This is sometimes called the **geometric notation** for the design.

A special notation is used to label the treatment combinations. In general, a treatment combination is represented by a series of lowercase letters. If a letter is present, the corresponding factor is run at the high level in that treatment combination; if it is absent, the factor is run at its low level. For example, treatment combination a indicates that factor A is at the high level and factor B is at the low level. The treatment combination with both factors at the low level is represented by (1). This notation is used throughout the 2^k design series. For example, the treatment combination in a 2^4 with A and C at the high level and B and D at the low level is denoted by ac.

The effects of interest in the 2^2 design are the main effects A and B and the two-factor interaction AB. Let the letters (1), a, b, and ab also represent the totals of all n observations taken at these design points. It is easy to estimate the effects of these factors. To estimate the main effect of A, we would average the observations on the right side of the square in Fig. 14-15 where A is at the high level, and subtract from this the average of the observations on the left side of the square, where A is at the low level, or

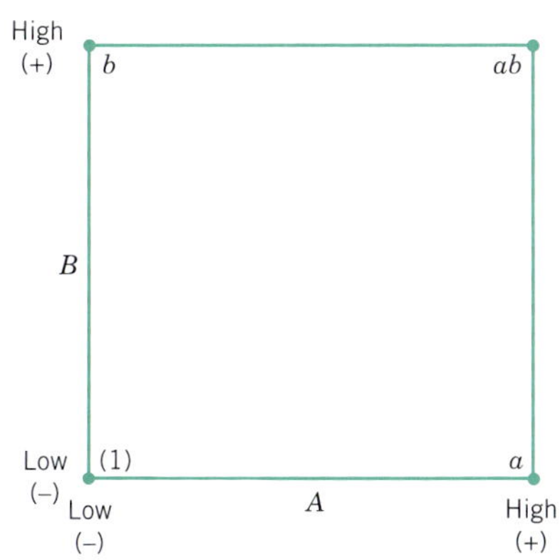

Treatment	A	B
(1)	−	−
a	+	−
b	−	+
ab	+	+

Figure 14-15 The 2^2 factorial design.

Main Effect of Factor A: 2^2 Design

$$A = \bar{y}_{A+} - \bar{y}_{A-} = \frac{a + ab}{2n} - \frac{b + (1)}{2n} = \frac{1}{2n}[a + ab - b - (1)] \tag{14-11}$$

Similarly, the main effect of B is found by averaging the observations on the top of the square, where B is at the high level, and subtracting the average of the observations on the bottom of the square, where B is at the low level:

Main Effect of Factor B: 2^2 Design

$$B = \bar{y}_{B+} - \bar{y}_{B-} = \frac{b + ab}{2n} - \frac{a + (1)}{2n} = \frac{1}{2n}[b + ab - a - (1)] \tag{14-12}$$

Finally, the AB interaction is estimated by taking the difference in the diagonal averages in Fig. 14-15, or

Interaction Effect AB: 2^2 Design

$$AB = \frac{ab + (1)}{2n} - \frac{a + b}{2n} = \frac{1}{2n}[ab + (1) - a - b] \tag{14-13}$$

The quantities in brackets in Equations 14-11, 14-12, and 14-13 are called **contrasts.** For example, the A contrast is

$$\text{Contrast}_A = a + ab - b - (1)$$

In these equations, the contrast coefficients are always either $+1$ or -1. A table of plus and minus signs, such as Table 14-12, can be used to determine the sign on each treatment

Table 14-12 Signs for Effects in the 2^2 Design

Treatment Combination	Factorial Effect			
	I	A	B	AB
(1)	+	−	−	+
a	+	+	−	−
b	+	−	+	−
ab	+	+	+	+

combination for a particular contrast. The column headings for Table 14-12 are the main effects A and B, the AB interaction, and I, which represents the total. The row headings are the treatment combinations. Note that the signs in the AB column are the product of signs from columns A and B. To generate a contrast from this table, multiply the signs in the appropriate column of Table 14-12 by the treatment combinations listed in the rows and add. For example, $\text{contrast}_{AB} = [(1)] + [-a] + [-b] + [ab] = ab + (1) - a - b$.

Contrasts are used in calculating both the effect estimates and the sums of squares for A, B, and the AB interaction. The sums of squares formulas are

ANOVA Sums of Squares: 2^2 Design

$$SS_A = \frac{[a + ab - b - (1)]^2}{4n}$$
$$SS_B = \frac{[b + ab - a - (1)]^2}{4n} \qquad (14\text{-}14)$$
$$SS_{AB} = \frac{[ab + (1) - a - b]^2}{4n}$$

The analysis of variance is completed by computing the total sum of squares SS_T (with $4n - 1$ degrees of freedom) as usual, and obtaining the error sum of squares SS_E [with $4(n - 1)$ degrees of freedom] by subtraction.

EXAMPLE 14-3 Wafer Layer

An article in the *AT&T Technical Journal* (Vol. 65, March/April 1986, pp. 39–50) describes the application of two-level factorial designs to integrated circuit manufacturing. A basic processing step in this industry is to grow an epitaxial layer on polished silicon wafers. The wafers are mounted on a susceptor and positioned inside a bell jar. Chemical vapors are introduced through nozzles near the top of the jar. The susceptor is rotated, and heat is applied. These conditions are maintained until the epitaxial layer is thick enough.

Table 14-13 presents the results of a 2^2 factorial design with $n = 4$ replicates using the factors A = deposition time and B = arsenic flow rate. The two levels of deposition time are $-$ = short and $+$ = long, and the two levels of arsenic flow rate are $-$ = 55% and $+$ = 59%. The response variable is epitaxial layer thickness (μm). We may find the estimates of the effects using Equations 14-11, 14-12, and 14-13 as follows:

$$A = \frac{1}{2n}[a + ab - b - (1)]$$
$$= \frac{1}{2(4)}[59.299 + 59.156 - 55.686 - 56.081] = 0.836$$

$$B = \frac{1}{2n}[b + ab - a - (1)]$$
$$= \frac{1}{2(4)}[55.686 + 59.156 - 59.299 - 56.081] = -0.067$$

$$AB = \frac{1}{2n}[ab + (1) - a - b]$$
$$AB = \frac{1}{2(4)}[59.156 + 56.081 - 59.299 - 55.686] = 0.032$$

Table 14-13 The 2^2 Design for the Epitaxial Process Experiment

Treatment Combination	Design Factors A	B	AB	Thickness (μm)				Thickness (μm) Total	Average
(1)	−	−	+	14.037	14.165	13.972	13.907	56.081	14.020
a	+	−	−	14.821	14.757	14.843	14.878	59.299	14.825
b	−	+	−	13.880	13.860	14.032	13.914	55.686	13.922
ab	+	+	+	14.888	14.921	14.415	14.932	59.156	14.789

The numerical estimates of the effects indicate that the effect of deposition time is large and has a positive direction (increasing deposition time increases thickness), since changing deposition time from low to high changes the mean epitaxial layer thickness by 0.836 μm. The effects of arsenic flow rate (B) and the AB interaction appear small.

The importance of these effects may be confirmed with the analysis of variance. The sums of squares for A, B, and AB are computed as follows:

$$SS_A = \frac{[a + ab - b - (1)]^2}{16} = \frac{[6.688]^2}{16} = 2.7956$$

$$SS_B = \frac{[b + ab - a - (1)]^2}{16} = \frac{[-0.538]^2}{16} = 0.0181$$

$$SS_{AB} = \frac{[ab + (1) - a - b]^2}{16} = \frac{[0.252]^2}{16} = 0.0040$$

$$SS_T = 14.037^2 + \cdots + 14.932^2 - \frac{(56.081 + \cdots + 59.156)^2}{16}$$
$$= 3.0672$$

The analysis of variance is summarized in Table 14-14 and confirms our conclusions obtained by examining the magnitude and direction of the effects. Deposition time is the only factor that significantly affects epitaxial layer thickness, and from the direction of the effect estimates we know that longer deposition times lead to thicker epitaxial layers.

Residual Analysis

It is easy to obtain the residuals from a 2^k design by fitting a **regression model** to the data. For the epitaxial process experiment, the regression model is

$$Y = \beta_0 + \beta_1 x_1 + \epsilon$$

Table 14-14 Analysis of Variance for the Epitaxial Process Experiment

Source of Variation	Sum of Squares	Degrees of Freedom	Mean Square	f_0	*P*-Value
A (deposition time)	2.7956	1	2.7956	134.40	7.07 E-8
B (arsenic flow)	0.0181	1	0.0181	0.87	0.38
AB	0.0040	1	0.0040	0.19	0.67
Error	0.2495	12	0.0208		
Total	3.0672	15			

since the only active variable is deposition time, which is represented by a coded variable x_1. The low and high levels of deposition time are assigned values $x_1 = -1$ and $x_1 = +1$, respectively. The least squares fitted model is

$$\hat{y} = 14.389 + \left(\frac{0.836}{2}\right) x_1$$

where the intercept $\hat{\beta}_0$ is the grand average of all 16 observations ($\bar{y}$) and the slope $\hat{\beta}_1$is one-half the effect estimate for deposition time. (The regression coefficient is one-half the effect estimate because regression coefficients measure the effect of a unit change in x_1 on the mean of Y, and the effect estimate is based on a two-unit change from -1 to $+1$.)

This model can be used to obtain the predicted values at the four points that form the corners of the square in the design. For example, consider the point with low deposition time ($x_1 = -1$) and low arsenic flow rate. The predicted value is

$$\hat{y} = 14.389 + \left(\frac{0.836}{2}\right)(-1) = 13.971 \ \mu\text{m}$$

and the residuals for the four runs at that design point are

$$\begin{aligned} e_1 &= 14.037 - 13.971 = 0.066 \\ e_2 &= 14.165 - 13.971 = 0.194 \\ e_3 &= 13.972 - 13.971 = 0.001 \\ e_4 &= 13.907 - 13.971 = -0.064 \end{aligned}$$

The remaining predicted values and residuals at the other three design points are calculated in a similar manner.

A normal probability plot of these residuals is shown in Fig. 14-16. This plot indicates that one residual $e_{15} = -0.392$ is an **outlier**. Examining the four runs with high deposition time and high arsenic flow rate reveals that observation $y_{15} = 14.415$ is considerably smaller than the other three observations at that treatment combination. This adds some additional evidence to the tentative conclusion that observation 15 is an outlier. Another possibility is that some process variables affect the *variability* in epitaxial layer thickness. If we could discover which variables produce this effect, we could perhaps adjust these variables to levels that would minimize the variability in epitaxial layer thickness. This could have important implications in subsequent manufacturing stages. Figures 14-17 and 14-18 are plots of residuals versus deposition time and arsenic flow rate, respectively. Apart from that unusually large residual associated with y_{15}, there is no strong evidence that either deposition time or arsenic flow rate influences the variability in epitaxial layer thickness.

Figure 14-19 shows the standard deviation of epitaxial layer thickness at all four runs in the 2^2 design. These standard deviations were calculated using the data in Table 14-13. Notice that the standard deviation of the four observations with A and B at the high level is considerably larger than the standard deviations at any of the other three design points. Most of this difference is attributable to the unusually low thickness measurement associated with y_{15}. The standard deviation of the four observations with A and B at the low level is also somewhat larger than the standard deviations at the remaining two runs. This could indicate that other process variables not included in this experiment may affect the variability in epitaxial layer thickness. Another experiment to study this possibility, involving other process variables, could be designed and conducted. (The original paper in the *AT&T Technical Journal* shows that two additional factors, not considered in this example, affect process variability.)

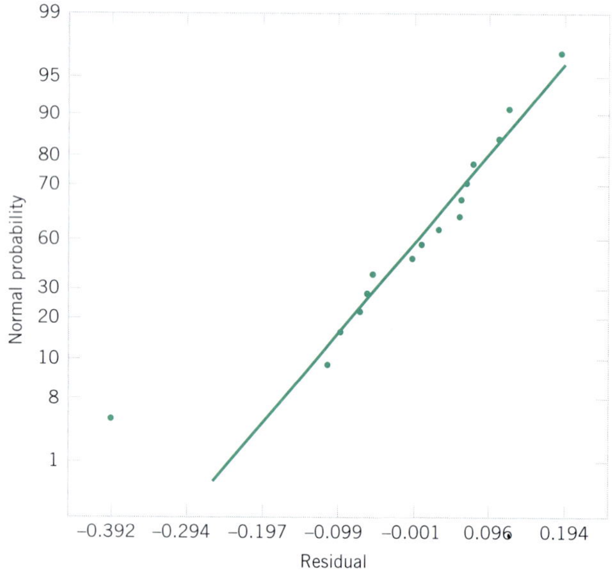

Figure 14-16 Normal probability plot of residuals for the epitaxial process experiment.

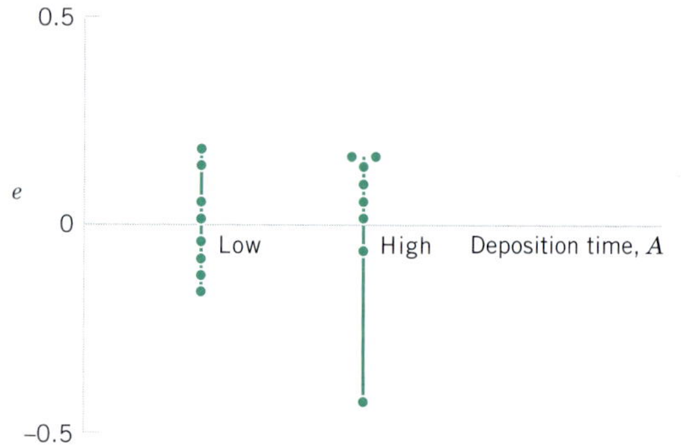

Figure 14-17 Plot of residuals versus deposition time.

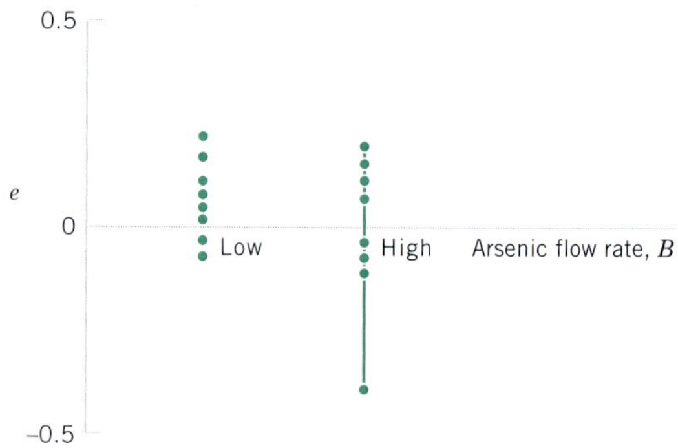

Figure 14-18 Plot of residuals versus arsenic flow rate.

14-5.2 2^k Design for $k \geq 3$ Factors

The methods presented in the previous section for factorial designs with $k = 2$ factors each at two levels can be easily extended to more than two factors. For example, consider $k = 3$ factors, each at two levels. This design is a 2^3 factorial design, and it has eight runs or treatment combinations. Geometrically, the design is a cube as shown in Fig. 14-20(a), with the eight runs forming the corners of the cube. Figure 14-20(b) lists the eight runs in a table, with each row representing one of the runs and the − and + settings indicating the low and high levels

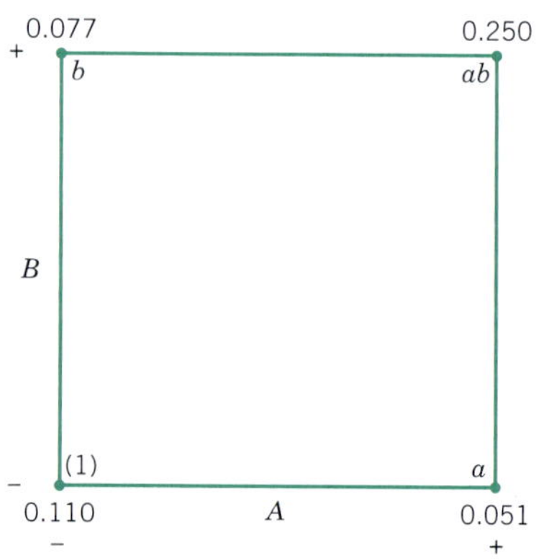

Figure 14-19 The standard deviation of epitaxial layer thickness at the four runs in the 2^2 design.

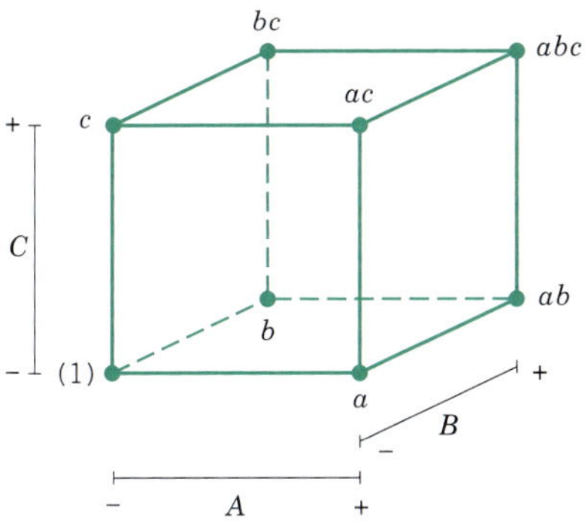

(a) Geometric view

Run	A	B	C
1	–	–	–
2	+	–	–
3	–	+	–
4	+	+	–
5	–	–	+
6	+	–	+
7	–	+	+
8	+	+	+

(b) The 2^3 design matrix

Figure 14-20 The 2^3 design.

for each of the three factors. This table is sometimes called the **design matrix.** This design allows three main effects to be estimated (A, B, and C) along with three two-factor interactions (AB, AC, and BC) and a three-factor interaction (ABC).

The main effects can easily be estimated. Remember that the lowercase letters (1), a, b, ab, c, ac, bc, and abc represent the total of all n replicates at each of the eight runs in the design. As seen in Fig. 14-21(a), the main effect of A can be estimated by averaging the four treatment combinations on the right-hand side of the cube, where A is at the high level, and by

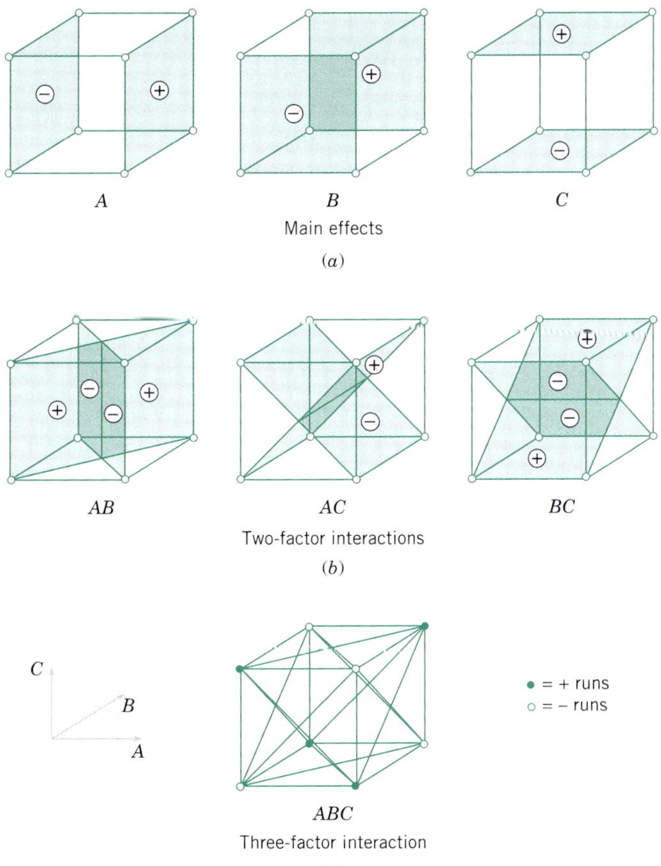

Figure 14-21 Geometric presentation of contrasts corresponding to the main effects and interaction in the 2^3 design. (a) Main effects. (b) Two-factor interactions. (c) Three-factor interaction.

subtracting from this quantity the average of the four treatment combinations on the left-hand side of the cube where A is at the low level. This gives

$$A = \bar{y}_{A+} - \bar{y}_{A-}$$
$$= \frac{a + ab + ac + abc}{4n} - \frac{(1) + b + c + bc}{4n}$$

This equation can be rearranged as

Main Effect of Factor A: 2^3 Design

$$A = \bar{y}_{A+} - \bar{y}_{A-}$$
$$= \frac{1}{4n}[a + ab + ac + abc - (1) - b - c - bc] \quad (14\text{-}15)$$

In a similar manner, the effect of B is the difference in averages between the four treatment combinations in the back face of the cube (Fig. 14-19a), and the four in the front. This yields

Main Effect of Factor B: 2^3 Design

$$B = \bar{y}_{B+} - \bar{y}_{B-}$$
$$= \frac{1}{4n}[b + ab + bc + abc - (1) - a - c - ac] \quad (14\text{-}16)$$

The effect of C is the difference in average response between the four treatment combinations in the top face of the cube in Figure 14-19(a) and the four in the bottom, that is,

Main Effect of Factor C: 2^3 Design

$$C = \bar{y}_{C+} - \bar{y}_{C-}$$
$$= \frac{1}{4n}[c + ac + bc + abc - (1) - a - b - ab] \quad (14\text{-}17)$$

The two-factor interaction effects may be computed easily. A measure of the AB interaction is the difference between the average A effects at the two levels of B. By convention, one-half of this difference is called the AB interaction. Symbolically,

B	Average A Effect
High (+)	$\dfrac{[(abc - bc) + (ab - b)]}{2n}$
Low (−)	$\dfrac{\{(ac - c) + [a - (1)]\}}{2n}$
Difference	$\dfrac{[abc - bc + ab - b - ac + c - a + (1)]}{2n}$

Since the AB interaction is one-half of this difference,

Two-Factor Interaction Effect: 2^3 Design

$$AB = \frac{1}{4n}[abc - bc + ab - b - ac + c - a + (1)] \qquad (14\text{-}18)$$

We could write Equation 14-18 as follows:

$$AB = \frac{abc + ab + c + (1)}{4n} - \frac{bc + b + ac + a}{4n}$$

In this form, the AB interaction is easily seen to be the difference in averages between runs on two diagonal planes in the cube in Fig. 14-19(b). Using similar logic and referring to Fig. 14-19(b), we find that the AC and BC interactions are

Two-Factor Interaction Effect: 2^3 Design

$$AC = \frac{1}{4n}[(1) - a + b - ab - c + ac - bc + abc] \qquad (14\text{-}19)$$

$$BC = \frac{1}{4n}[(1) + a - b - ab - c - ac + bc + abc] \qquad (14\text{-}20)$$

The ABC interaction is defined as the average difference between the AB interaction for the two different levels of C. Thus,

$$ABC = \frac{1}{4n}\{[abc - bc] - [ac - c] - [ab - b] + [a - (1)]\}$$

or

Three-Factor Interaction Effect: 2^3 Design

$$ABC = \frac{1}{4n}[abc - bc - ac + c - ab + b + a - (1)] \qquad (14\text{-}21)$$

As before, we can think of the ABC interaction as the difference in two averages. If the runs in the two averages are isolated, they define the vertices of the two tetrahedra that comprise the cube in Fig. 14-21(c).

In Equations 14-15 through 14-21, the quantities in brackets are **contrasts** in the treatment combinations. A table of plus and minus signs can be developed from the contrasts and is shown in Table 14-15. Signs for the main effects are determined directly from the test matrix in Figure 14-20(b). Once the signs for the main effect columns have been established, the signs for the remaining columns can be obtained by multiplying the appropriate

Table 14-15 Algebraic Signs for Calculating Effects in the 2^3 Design

Treatment Combination	Factorial Effect							
	I	A	B	AB	C	AC	BC	ABC
(1)	+	−	−	+	−	+	+	−
a	+	+	−	−	−	−	+	+
b	+	−	+	−	−	+	−	+
ab	+	+	+	+	−	−	−	−
c	+	−	−	+	+	−	−	+
ac	+	+	−	−	+	+	−	−
bc	+	−	+	−	+	−	+	−
abc	+	+	+	+	+	+	+	+

main effect row by row. For example, the signs in the AB column are the products of the A and B column signs in each row. The contrast for any effect can easily be obtained from this table.

Table 14-15 has several interesting properties:

1. Except for the identity column I, each column has an equal number of plus and minus signs.
2. The sum of products of signs in any two columns is zero; that is, the columns in the table are **orthogonal.**
3. Multiplying any column by column I leaves the column unchanged; that is, I is an **identity element.**
4. The product of any two columns yields a column in the table, for example $A \times B = AB$, and $AB \times ABC = A^2B^2C = C$, since any column multiplied by itself is the identity column.

The estimate of any main effect or interaction in a 2^k design is determined by multiplying the treatment combinations in the first column of the table by the signs in the corresponding main effect or interaction column, by adding the result to produce a contrast, and then by dividing the contrast by one-half the total number of runs in the experiment. For any 2^k design with n replicates, the effect estimates are computed from

Relationship Between a Contrast and an Effect

$$\text{Effect} = \frac{\text{Contrast}}{n2^{k-1}} \tag{14-22}$$

and the sum of squares for any effect is

Sum of Squares for an Effect

$$SS = \frac{(\text{Contrast})^2}{n2^k} \tag{14-23}$$

EXAMPLE 14-4
Surface Roughness

Consider the surface roughness experiment originally described in Example 14-2. This is a 2^3 factorial design in the factors feed rate (A), depth of cut (B), and tool angle (C), with $n = 2$ replicates. Table 14-16 presents the observed surface roughness data.

The main effects may be estimated using Equations 14-15 through 14-21. The effect of A, for example, is

$$\begin{aligned} A &= \frac{1}{4n}[a + ab + ac + abc - (1) - b - c - bc] \\ &= \frac{1}{4(2)}[22 + 27 + 23 + 30 - 16 - 20 - 21 - 18] \\ &= \frac{1}{8}[27] = 3.375 \end{aligned}$$

and the sum of squares for A is found using Equation 14-23:

$$SS_A = \frac{(\text{Contrast}_A)^2}{n2^k} = \frac{(27)^2}{2(8)} = 45.5625$$

It is easy to verify that the other effects are

$$\begin{aligned} B &= 1.625 \\ C &= 0.875 \\ AB &= 1.375 \\ AC &= 0.125 \\ BC &= -0.625 \\ ABC &= 1.125 \end{aligned}$$

Examining the magnitude of the effects clearly shows that feed rate (factor A) is dominant, followed by depth of cut (B) and the AB interaction, although the interaction effect is relatively small. The analysis of variance, summarized in Table 14-17, confirms our interpretation of the effect estimates.

Minitab will analyze 2^k factorial designs. The output from the Minitab DOE (Design of Experiments) module for this experiment is shown in Table 14-18. The upper portion of the table displays the effect estimates and regression coefficients for each factorial effect. However, a t-statistic is reported for each effect instead of the F-statistic used in Table 14-17. Now the square of a t random variable with

Table 14-16 Surface Roughness Data for Example 14-4

Treatment Combinations	Design Factors			Surface Roughness	Totals
	A	B	C		
(1)	−1	−1	−1	9, 7	16
a	1	−1	−1	10, 12	22
b	−1	1	−1	9, 11	20
ab	1	1	−1	12, 15	27
c	−1	−1	1	11, 10	21
ac	1	−1	1	10, 13	23
bc	−1	1	1	10, 8	18
abc	1	1	1	16, 14	30

Table 14-17 Analysis of Variance for the Surface Finish Experiment

Source of Variation	Sum of Squares	Degrees of Freedom	Mean Square	f_0	P-Value
A	45.5625	1	45.5625	18.69	0.0025
B	10.5625	1	10.5625	4.33	0.0709
C	3.0625	1	3.0625	1.26	0.2948
AB	7.5625	1	7.5625	3.10	0.1162
AC	0.0625	1	0.0625	0.03	0.8784
BC	1.5625	1	1.5625	0.64	0.4548
ABC	5.0625	1	5.0625	2.08	0.1875
Error	19.5000	8	2.4375		
Total	92.9375	15			

d degrees of freedom is an F random variable with 1 numerator and d denominator degrees of freedom. Thus the square of the t-statistic reported by Minitab will be equal (apart from rounding errors) to the F-statistic in Table 14-17. To illustrate, for the main effect of feed Minitab reports $t = 4.32$ (with eight degrees of freedom), and $t^2 = (4.32)^2 = 18.66$, which is approximately equal to the F-ratio for feed reported in Table 14-17 ($F = 18.69$). This F-ratio has one numerator and eight denominator degrees of freedom.

The lower panel of the Minitab output in Table 14-18 is an analysis of variance summary focusing on the types of terms in the model. A regression model approach is used in the presentation. You might find it helpful to review Section 12-2.2, particularly the material on the partial F-test. The row entitled "main effects" under source refers to the three main effects feed, depth, and angle, each having a single degree of freedom, giving the total 3 in the column headed "DF." The column headed "Seq SS" (an

Table 14-18 Minitab Analysis for Example 14-4

Estimated Effects and Coefficients for Roughness

Term	Effect	Coef	StDev Coef	T	P
Constant		11.0625	0.3903	28.34	0.000
Feed	3.3750	1.6875	0.3903	4.32	0.003
Depth	1.6250	0.8125	0.3903	2.08	0.071
Angle	0.8750	0.4375	0.3903	1.12	0.295
Feed*Depth	1.3750	0.6875	0.3903	1.76	0.116
Feed*Angle	0.1250	0.0625	0.3903	0.16	0.877
Depth*Angle	−0.6250	−0.3125	0.3903	−0.80	0.446
Feed*Depth*Angle	1.1250	0.5625	0.3903	1.44	0.188

Analysis of Variance for Roughness

Source	DF	Seq SS	Adj SS	Adj MS	F	P
Main Effects	3	59.188	59.188	19.729	8.09	0.008
2-Way Interactions	3	9.187	9.187	3.062	1.26	0.352
3-Way Interactions	1	5.062	5.062	5.062	2.08	0.188
Residual Error	8	19.500	19.500	2.437		
Pure Error	8	19.500	19.500	2.437		
Total	15	92.938				

abbreviation for sequential sum of squares) reports how much the model sum of squares increases when each group of terms is added to a model that contains the terms listed *above* the groups. The first number in the "Seq SS" column presents the model sum of squares for fitting a model having only the three main effects. The row labeled "2-Way Interactions" refers to AB, AC, and BC, and the sequential sum of squares reported here is the increase in the model sum of squares if the interaction terms are added to a model containing only the main effects. Similarly, the sequential sum of squares for the three-way interaction is the increase in the model sum of squares that results from adding the term ABC to a model containing all other effects. The column headed "Adj SS" (an abbreviation for adjusted sum of squares) reports how much the model sum of squares increases when each group of terms is added to a model that contains *all* the other terms. Now since any 2^k design with an equal number of replicates in each cell is an orthogonal design, the adjusted sum of squares will equal the sequential sum of squares. Therefore, the F-tests for each row in the Minitab analysis of variance table are testing the significance of each group of terms (main effects, two-factor interactions, and three-factor interactions) as if they were the last terms to be included in the model. Clearly, only the main effect terms are significant. The t-tests on the individual factor effects indicate that feed rate and depth of cut have large main effects, and there may be some mild interaction between these two factors. Therefore, the Minitab output is in agreement with the results given previously.

Residual Analysis

We may obtain the residuals from a 2^k design by using the method demonstrated earlier for the 2^2 design. As an example, consider the surface roughness experiment. The three largest effects are A, B, and the AB interaction. The regression model used to obtain the predicted values is

$$Y = \beta_0 + \beta_1 x_1 + \beta_2 x_2 + \beta_{12} x_1 x_2 + \epsilon$$

where x_1 represents factor A, x_2 represents factor B, and $x_1 x_2$ represents the AB interaction. The regression coefficients β_1, β_2, and β_{12} are estimated by one-half the corresponding effect estimates, and β_0 is the grand average. Thus

$$\begin{aligned}\hat{y} &= 11.0625 + \left(\frac{3.375}{2}\right)x_1 + \left(\frac{1.625}{2}\right)x_2 + \left(\frac{1.375}{2}\right)x_1 x_2 \\ &= 11.0625 + 1.6875x_1 + 0.8125x_2 + 0.6875x_1 x_2\end{aligned}$$

Note that the regression coefficients are presented by Minitab in the upper panel of Table 14-18. The predicted values would be obtained by substituting the low and high levels of A and B into this equation. To illustrate this, at the treatment combination where A, B, and C are all at the low level, the predicted value is

$$\hat{y} = 11.0625 + 1.6875(-1) + 0.8125(-1) + 0.6875(-1)(-1) = 9.25$$

Since the observed values at this run are 9 and 7, the residuals are $9 - 9.25 = -0.25$ and $7 - 9.25 = -2.25$. Residuals for the other 14 runs are obtained similarly.

A normal probability plot of the residuals is shown in Fig. 14-22. Since the residuals lie approximately along a straight line, we do not suspect any problem with normality in the data. There are no indications of severe outliers. It would also be helpful to plot the residuals versus the predicted values and against each of the factors A, B, and C.

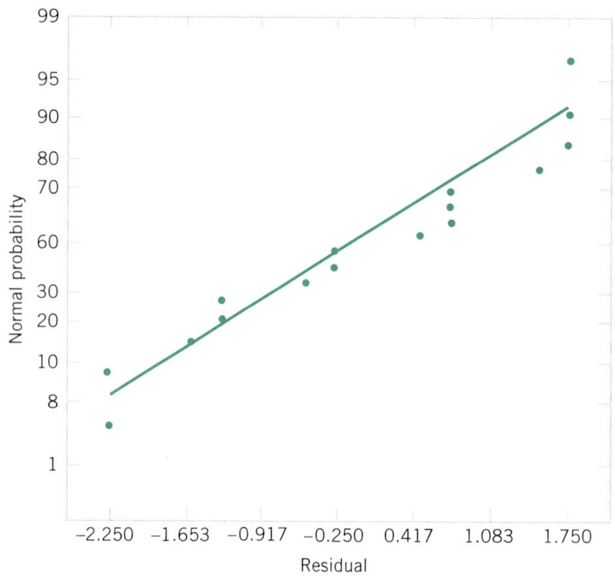

Figure 14-22 Normal probability plot of residuals from the surface roughness experiment.

Projection of 2^k Designs

Any 2^k design will collapse or project into another 2^k design in fewer variables if one or more of the original factors are dropped. Sometimes this can provide additional insight into the remaining factors. For example, consider the surface roughness experiment. Since factor C and all its interactions are negligible, we could eliminate factor C from the design. The result is to collapse the cube in Fig. 14-20 into a square in the $A - B$ plane; therefore, each of the four runs in the new design has four replicates. In general, if we delete h factors so that $r = k - h$ factors remain, the original 2^k design with n replicates will project into a 2^r design with $n2^h$ replicates.

14-5.3 Single Replicate of the 2^k Design

As the number of factors in a factorial experiment grows, the number of effects that can be estimated also grows. For example, a 2^4 experiment has 4 main effects, 6 two-factor interactions, 4 three-factor interactions, and 1 four-factor interaction, while a 2^6 experiment has 6 main effects, 15 two-factor interactions, 20 three-factor interactions, 15 four-factor interactions, 6 five-factor interactions, and 1 six-factor interaction. In most situations the **sparsity of effects principle** applies; that is, the system is usually dominated by the main effects and low-order interactions. The three-factor and higher order interactions are usually negligible. Therefore, when the number of factors is moderately large, say, $k \geq 4$ or 5, a common practice is to run only a single replicate of the 2^k design and then pool or combine the higher order interactions as an estimate of error. Sometimes a single replicate of a 2^k design is called an **unreplicated** 2^k factorial design.

When analyzing data from unreplicated factorial designs, occasionally real high-order interactions occur. The use of an error mean square obtained by pooling high-order interactions is inappropriate in these cases. A simple method of analysis can be used to overcome this problem. Construct a plot of the estimates of the effects on a normal probability scale. The effects that are negligible are normally distributed, with mean zero and variance σ^2 and will tend to fall along a straight line on this plot, whereas significant effects will have nonzero means and will not lie along the straight line. We will illustrate this method in the next example.

EXAMPLE 14-5
Plasma Etch

An article in *Solid State Technology* ("Orthogonal Design for Process Optimization and Its Application in Plasma Etching," May 1987, pp. 127–132) describes the application of factorial designs in developing a nitride etch process on a single-wafer plasma etcher. The process uses C_2F_6 as the reactant gas. It is possible to vary the gas flow, the power applied to the cathode, the pressure in the reactor chamber, and the spacing between the anode and the cathode (gap). Several response variables would usually be of interest in this process, but in this example we will concentrate on etch rate for silicon nitride.

We will use a single replicate of a 2^4 design to investigate this process. Since it is unlikely that the three- and four-factor interactions are significant, we will tentatively plan to combine them as an estimate of error. The factor levels used in the design are shown below:

	Design Factor			
Level	Gap (cm)	Pressure (mTorr)	C_2F_6 Flow (SCCM)	Power (w)
Low (−)	0.80	450	125	275
High (+)	1.20	550	200	325

Table 14-19 presents the data from the 16 runs of the 2^4 design. Table 14-20 is the table of plus and minus signs for the 2^4 design. The signs in the columns of this table can be used to estimate the factor effects. For example, the estimate of factor A is

$$\begin{aligned} A &= \frac{1}{8}[a + ab + ac + abc + ad + abd + acd + abcd - (1) - b \\ &\quad - c - bc - d - bd - cd - bcd] \\ &= \frac{1}{8}[669 + 650 + 642 + 635 + 749 + 868 + 860 + 729 \\ &\quad - 550 - 604 - 633 - 601 - 1037 - 1052 - 1075 - 1063] \\ &= -101.625 \end{aligned}$$

Table 14-19 The 2^4 Design for the Plasma Etch Experiment

A (Gap)	B (Pressure)	C (C_2F_6 Flow)	D (Power)	Etch Rate (Å/min)
−1	−1	−1	−1	550
1	−1	−1	−1	669
−1	1	−1	−1	604
1	1	−1	−1	650
−1	−1	1	−1	633
1	−1	1	−1	642
−1	1	1	−1	601
1	1	1	−1	635
−1	−1	−1	1	1037
1	−1	−1	1	749
−1	1	−1	1	1052
1	1	−1	1	868
−1	−1	1	1	1075
1	−1	1	1	860
−1	1	1	1	1063
1	1	1	1	729

Table 14-20 Contrast Constants for the 2^4 Design

	A	*B*	*AB*	*C*	*AC*	*BC*	*ABC*	*D*	*AD*	*BD*	*ABD*	*CD*	*ACD*	*BCD*	*ABCD*
(1)	−	−	+	−	+	+	−	−	+	+	−	+	−	−	+
a	+	−	−	−	−	+	+	−	−	+	+	+	+	−	−
b	−	+	−	−	+	−	+	−	+	−	+	+	−	+	−
ab	+	+	+	−	−	−	−	−	−	−	−	+	+	+	+
c	−	−	+	+	−	−	+	−	+	+	−	−	+	+	−
ac	+	−	−	+	+	−	−	−	−	+	+	−	−	+	+
bc	−	+	−	+	−	+	−	−	+	−	+	−	+	−	+
abc	+	+	+	+	+	+	+	−	−	−	−	−	−	−	−
d	−	−	+	−	+	+	−	+	−	−	+	−	+	+	−
ad	+	−	−	−	−	+	+	+	+	−	−	−	−	+	+
bd	−	+	−	−	+	−	+	+	−	+	−	−	+	−	+
abd	+	+	+	−	−	−	−	+	+	+	+	−	−	−	−
cd	−	−	+	+	−	−	+	+	−	−	+	+	−	−	+
acd	+	−	−	+	+	−	−	+	+	−	−	+	+	−	−
bcd	−	+	−	+	−	+	−	+	−	+	−	+	−	+	−
abcd	+	+	+	+	+	+	+	+	+	+	+	+	+	+	+

Thus, the effect of increasing the gap between the anode and the cathode from 0.80 to 1.20 centimeters is to decrease the etch rate by 101.625 angstroms per minute.

It is easy to verify (using Minitab, for example) that the complete set of effect estimates is

$$
\begin{aligned}
A &= -101.625 & AD &= -153.625\\
B &= -1.625 & BD &= -0.625\\
AB &= -7.875 & ABD &= 4.125\\
C &= 7.375 & CD &= -2.125\\
AC &= -24.875 & ACD &= 5.625\\
BC &= -43.875 & BCD &= -25.375\\
ABC &= -15.625 & ABCD &= -40.125\\
D &= 306.125 & &
\end{aligned}
$$

The normal probability plot of these effects from the plasma etch experiment is shown in Fig. 14-23. Clearly, the main effects of *A* and *D* and the *AD* interaction are significant, because they fall far from the line passing through the other points. The analysis of variance summarized in Table 14-21 confirms these findings. Notice that in the analysis of variance we have pooled the three- and four-factor interactions to form the error mean square. If the normal probability plot had indicated that any of these interactions were important, they would not have been included in the error term.

Since $A = -101.625$, the effect of increasing the gap between the cathode and anode is to decrease the etch rate. However, $D = 306.125$; thus, applying higher power levels will increase the etch rate. Figure 14-24 is a plot of the *AD* interaction. This plot indicates that the effect of changing the gap width at low power settings is small, but that increasing the gap at high power settings dramatically reduces the etch rate. High etch rates are obtained at high power settings and narrow gap widths.

The residuals from the experiment can be obtained from the regression model

$$\hat{y} = 776.0625 - \left(\frac{101.625}{2}\right)x_1 + \left(\frac{306.125}{2}\right)x_4 - \left(\frac{153.625}{2}\right)x_1x_4$$

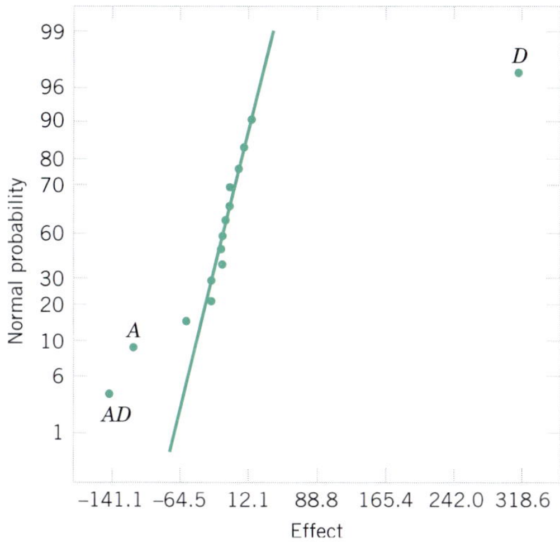

Figure 14-23 Normal probability plot of effects from the plasma etch experiment.

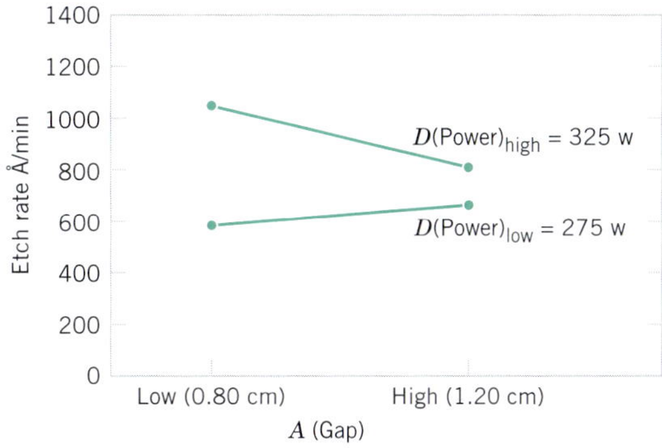

Figure 14-24 *AD* (Gap-Power) interaction from the plasma etch experiment.

For example, when both A and D are at the low level, the predicted value is

$$\hat{y} = 776.0625 - \left(\frac{101.625}{2}\right)(-1) + \left(\frac{306.125}{2}\right)(-1) - \left(\frac{153.625}{2}\right)(-1)(-1)$$
$$= 597$$

and the four residuals at this treatment combination are

$$e_1 = 550 - 597 = -47 \qquad e_2 = 604 - 597 = 7$$
$$e_3 = 633 - 597 = \quad 36 \qquad e_4 = 601 - 597 = 4$$

The residuals at the other three treatment combinations (A high, D low), (A low, D high), and (A high, D high) are obtained similarly. A normal probability plot of the residuals is shown in Fig. 14-25. The plot is satisfactory.

Table 14-21 Analysis of Variance for the Plasma Etch Experiment

Source of Variation	Sum of Squares	Degrees of Freedom	Mean Square	f_0	P-Value
A	41,310.563	1	41,310.563	20.28	0.0064
B	10.563	1	10.563	<1	—
C	217.563	1	217.563	<1	—
D	374,850.063	1	374,850.063	183.99	0.0000
AB	248.063	1	248.063	<1	—
AC	2,475.063	1	2,475.063	1.21	0.3206
AD	94,402.563	1	94,402.563	46.34	0.0010
BC	7,700.063	1	7,700.063	3.78	0.1095
BD	1.563	1	1.563	<1	—
CD	18.063	1	18.063	<1	—
Error	10,186.813	5	2,037.363		
Total	531,420.938	15			

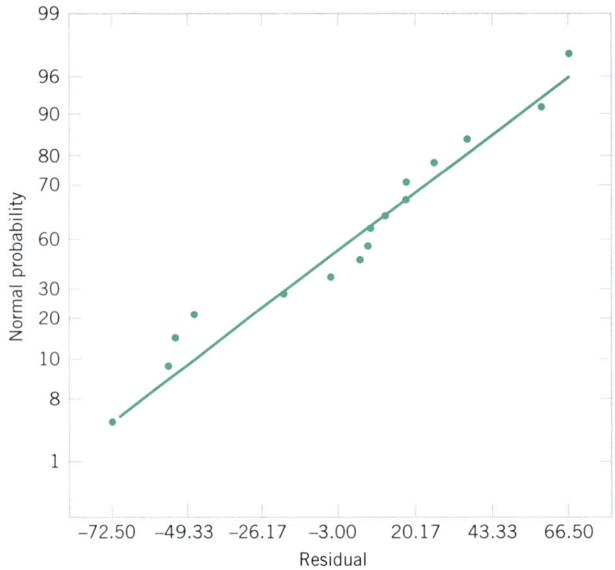

Figure 14-25 Normal probability plot of residuals from the plasma etch experiment.

14-5.4 Addition of Center Points to a 2^k Design

A potential concern in the use of two-level factorial designs is the assumption of linearity in the factor effects. Of course, perfect linearity is unnecessary, and the 2^k system will work quite well even when the linearity assumption holds only approximately. However, there is a method of replicating certain points in the 2^k factorial that will provide protection against curvature as well as allow an independent estimate of error to be obtained. The method consists of adding **center points** to the 2^k design. These consist of n_C replicates run at the point $x_i = 0$ $(i = 1, 2, \ldots, k)$. One important reason for adding the replicate runs at the design center is that center points do not affect the usual effects estimates in a 2^k design. We assume that the k factors are quantitative.

To illustrate the approach, consider a 2^2 design with one observation at each of the factorial points $(-, -)$, $(+, -)$, $(-, +)$, and $(+, +)$ and n_C observations at the center points $(0, 0)$. Figure 14-26 illustrates the situation. Let $\overline{y}_F$ be the average of the four runs at the four factorial points, and let $\overline{y}_C$ be the average of the n_C run at the center point. If the difference

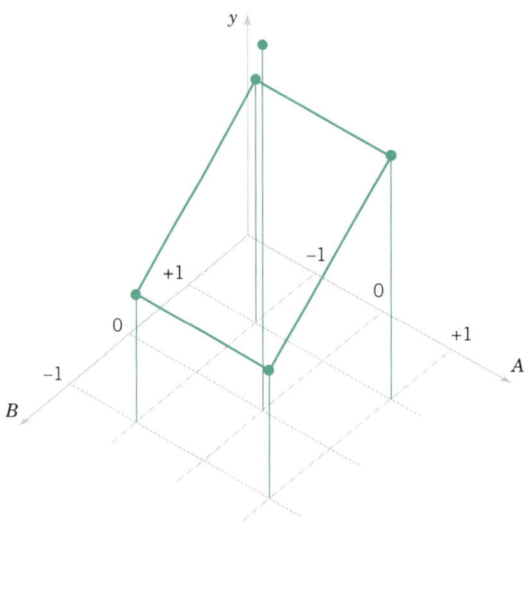

Figure 14-26 A 2^2 design with center points.

$\bar{y}_F - \bar{y}_C$ is small, the center points lie on or near the plane passing through the factorial points, and there is no curvature. On the other hand, if $\bar{y}_F - \bar{y}_C$ is large, curvature is present. A single-degree-of-freedom sum of squares for curvature is given by

Curvature Sum of Squares

$$SS_{\text{Curvature}} = \frac{n_F n_C(\bar{y}_F - \bar{y}_C)^2}{n_F + n_C} = \left(\frac{\bar{y}_F - \bar{y}_C}{\sqrt{\dfrac{1}{n_F} + \dfrac{1}{n_C}}}\right)^2 \tag{14-24}$$

where, in general, n_F is the number of factorial design points. This quantity may be compared to the error mean square to test for curvature. Notice that when Equation 14-24 is divided by $\hat{\sigma}^2 = MS_E$, the result is similar to the square of the t statistic used to compare two means.

More specifically, when points are added to the center of the 2^k design, the model we may entertain is

$$Y = \beta_0 + \sum_{j=1}^{k} \beta_j x_j + \sum_{i<j}\sum \beta_{ij} x_i x_j + \sum_{j=1}^{k} \beta_{jj} x_j^2 + \epsilon$$

where the β_{jj} are pure quadratic effects. The test for curvature actually tests the hypotheses

$$H_0\text{: } \sum_{j=1}^{k} \beta_{jj} = 0$$

$$H_1\text{: } \sum_{j=1}^{k} \beta_{jj} \neq 0$$

Furthermore, if the factorial points in the design are unreplicated, we may use the n_C center points to construct an estimate of error with $n_C - 1$ degrees of freedom.

EXAMPLE 14-6
Process Yield

A chemical engineer is studying the percentage of conversion or yield of a process. There are two variables of interest, reaction time and reaction temperature. Because she is uncertain about the assumption of linearity over the region of exploration, the engineer decides to conduct a 2^2 design (with a single replicate of each factorial run) augmented with five center points. The design and the yield data are shown in Fig. 14-27.

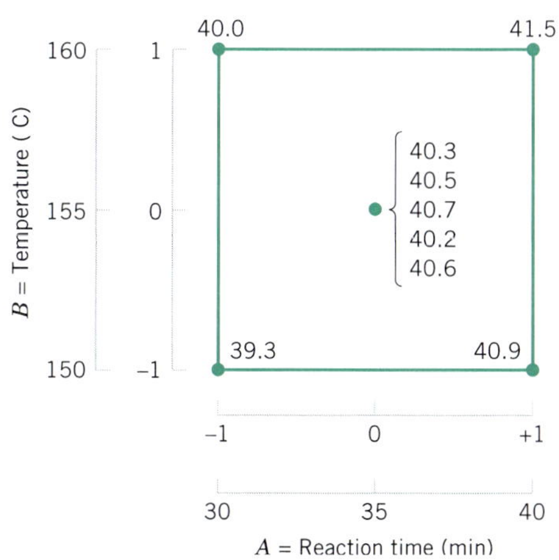

Figure 14-27 The 2^2 design with five center points for Example 14-6.

Table 14-22 Analysis of Variance for Example 14-6

Source of Variation	Sum of Squares	Degrees of Freedom	Mean Square	f_0	P-Value
A (Time)	2.4025	1	2.4025	55.87	0.0017
B (Temperature)	0.4225	1	0.4225	9.83	0.0350
AB	0.0025	1	0.0025	0.06	0.8237
Curvature	0.0027	1	0.0027	0.06	0.8163
Error	0.1720	4	0.0430		
Total	3.0022	8			

Table 14-22 summarizes the analysis of variance for this experiment. The mean square error is calculated from the center points as follows:

$$MS_E = \frac{SS_E}{n_C - 1} = \frac{\sum_{\text{Center points}} (y_i - \bar{y}_C)^2}{n_C - 1}$$

$$= \frac{\sum_{i=1}^{5} (y_i - 40.46)^2}{4} = \frac{0.1720}{4} = 0.0430$$

The average of the points in the factorial portion of the design is $\bar{y}_F = 40.425$, and the average of the points at the center is $\bar{y}_C = 40.46$. The difference $\bar{y}_F - \bar{y}_C = 40.425 - 40.46 = -0.035$ appears to be small. The curvature sum of squares in the analysis of variance table is computed from Equation 14-24 as follows:

$$SS_{\text{Curvature}} = \frac{n_F n_C(\bar{y}_F - \bar{y}_C)^2}{n_F + n_C}$$

$$= \frac{(4)(5)(-0.035)^2}{4 + 5} = 0.0027$$

The analysis of variance indicates that both factors exhibit significant main effects, that there is no interaction, and that there is no evidence of curvature in the response over the region of exploration. That is, the null hypothesis H_0: $\sum_{j=1}^{k} \beta_{jj} = 0$ cannot be rejected.

EXERCISES FOR SECTION 14-5

14-12. An engineer is interested in the effect of cutting speed (A), metal hardness (B), and cutting angle (C) on the life of a cutting tool. Two levels of each factor are chosen, and two replicates of a 2^3 factorial design are run. The tool life data (in hours) are shown in the table at right.

(a) Analyze the data from this experiment.

(b) Find an appropriate regression model that explains tool life in terms of the variables used in the experiment.

(c) Analyze the residuals from this experiment.

Treatment Combination	Replicate	
	I	II
(1)	221	311
a	325	435
b	354	348
ab	552	472
c	440	453
ac	406	377
bc	605	500
abc	392	419

14-13. Four factors are thought to influence the taste of a soft-drink beverage: type of sweetener (A), ratio of syrup to water (B), carbonation level (C), and temperature (D). Each factor can be run at two levels, producing a 2^4 design. At each run in the design, samples of the beverage are given to a test panel consisting of 20 people. Each tester assigns the beverage a point score from 1 to 10. Total score is the response variable, and the objective is to find a formulation that maximizes total score. Two replicates of this design are run, and the results are shown in the table. Analyze the data and draw conclusions. Use $\alpha = 0.05$ in the statistical tests.

Treatment Combination	Replicate I	Replicate II
(1)	159	163
a	168	175
b	158	163
ab	166	168
c	175	178
ac	179	183
bc	173	168
abc	179	182
d	164	159
ad	187	189
bd	163	159
abd	185	191
cd	168	174
acd	197	199
bcd	170	174
abcd	194	198

14-14. The data shown here represent a single replicate of a 2^5 design that is used in an experiment to study the compressive strength of concrete. The factors are mix (A), time (B), laboratory (C), temperature (D), and drying time (E).

(1)	= 700	*e*	= 800
a	= 900	*ae*	= 1200
b	= 3400	*be*	= 3500
ab	= 5500	*abe*	= 6200
c	= 600	*ce*	= 600
ac	= 1000	*ace*	= 1200
bc	= 3000	*bce*	= 3006
abc	= 5300	*abce*	= 5500
d	= 1000	*de*	= 1900
ad	= 1100	*ade*	= 1500
bd	= 3000	*bde*	= 4000
abd	= 6100	*abde*	= 6500
cd	= 800	*cde*	= 1500
acd	= 1100	*acde*	= 2000
bcd	= 3300	*bcde*	= 3400
abcd	= 6000	*abcde*	= 6800

(a) Estimate the factor effects.
(b) Which effects appear important? Use a normal probability plot.
(c) If it is desirable to maximize the strength, in which direction would you adjust the process variables?
(d) Analyze the residuals from this experiment.

14-15. An article in the *IEEE Transactions on Semiconductor Manufacturing* (Vol. 5, 1992, pp. 214–222) describes an experiment to investigate the surface charge on a silicon wafer. The factors thought to influence induced surface charge are cleaning method (spin rinse dry or SRD and spin dry or SD) and the position on the wafer where the charge was measured. The surface charge ($\times 10^{11}$ q/cm^3) response data are as shown.

Cleaning Method		Test Position L	Test Position R
	SD	1.66	1.84
		1.90	1.84
		1.92	1.62
	SRD	−4.21	−7.58
		−1.35	−2.20
		−2.08	−5.36

(a) Estimate the factor effects.
(b) Which factors appear important? Use $\alpha = 0.05$.
(c) Analyze the residuals from this experiment.

14-16. An article in *Oikos: A Journal of Ecology*, "Regulation of root vole population dynamics by food supply and predation: a two-factor experiment" (2005, Vol. 109, pp. 387–395) investigated how food supply interacts with predation in the regulation of root vole (Microtus oeconomus Pallas) population dynamics. A replicated two-factor field experiment manipulating both food supply and predation condition for root voles was conducted. Four treatments were applied: −P, +F (no-predator, food-supplemented); +P, +F (predator-access, food-supplemented); −P, −F (no-predator, non-supplemented); +P, −F (predator-access, food-supplemented). The population density of root voles (voles ha^{-1}) for each treatment combination in each is shown below.

Food Supply (F)	Predation (P)	Replicates		
+1	−1	88.589	114.059	200.979
+1	+1	56.949	97.079	78.759
−1	−1	65.439	89.089	172.339
−1	+1	40.799	47.959	74.439

(a) What is an appropriate statistical model for this experiment?
(b) Analyze the data and draw conclusions.

(c) Analyze the residuals from this experiment. Are there any problems with model adequacy?

14-17. An experiment was run in a semiconductor fabrication plant in an effort to increase yield. Five factors, each at two levels, were studied. The factors (and levels) were A = aperture setting (small, large), B = exposure time (20% below nominal, 20% above nominal), C = development time (30 and 45 seconds), D = mask dimension (small, large), and E = etch time (14.5 and 15.5 minutes). The following unreplicated 2^5 design was run:

(1)	= 7	*e*	= 8
a	= 9	*ae*	= 12
b	= 34	*be*	= 35
ab	= 55	*abe*	= 52
c	= 16	*ce*	= 15
ac	= 20	*ace*	= 22
bc	= 40	*bce*	= 45
abc	= 60	*abce*	= 65
d	= 8	*de*	= 6
ad	= 10	*ade*	= 10
bd	= 32	*bde*	= 30
abd	= 50	*abde*	= 53
cd	= 18	*cde*	= 15
acd	= 21	*acde*	= 20
bcd	= 44	*bcde*	= 41
abcd	= 61	*abcde*	= 63

(a) Construct a normal probability plot of the effect estimates. Which effects appear to be large?
(b) Conduct an analysis of variance to confirm your findings for part (a).
(c) Construct a normal probability plot of the residuals. Is the plot satisfactory?
(d) Plot the residuals versus the predicted yields and versus each of the five factors. Comment on the plots.
(e) Interpret any significant interactions.
(f) What are your recommendations regarding process operating conditions?
(g) Project the 2^5 design in this problem into a 2^r for $r < 5$ design in the important factors. Sketch the design and show the average and range of yields at each run. Does this sketch aid in data interpretation?

14-18. An experiment described by M. G. Natrella in the National Bureau of Standards *Handbook of Experimental Statistics* (No. 91, 1963) involves flame testing fabrics after applying fire-retardant treatments. The four factors considered are type of fabric (A), type of fire-retardant treatment (B), laundering condition (C—the low level is no laundering, the high level is after one laundering), and method of conducting the flame test (D). All factors are run at two levels, and the response variable is the inches of fabric burned on a standard size test sample. The data are:

(1)	= 42	*d*	= 40
a	= 31	*ad*	= 30
b	= 45	*bd*	= 50
ab	= 29	*abd*	= 25
c	= 39	*cd*	= 40
ac	= 28	*acd*	= 25
bc	= 46	*bcd*	= 50
abc	= 32	*abcd*	= 23

(a) Estimate the effects and prepare a normal plot of the effects.
(b) Construct an analysis of variance table based on the model tentatively identified in part (a).
(c) Construct a normal probability plot of the residuals and comment on the results.

14-19. Consider the data from Exercise 14-12. I suppose that the data from the second replicate was not available. Analyze the data from replicate I only and comment on your findings.

14-20. A 2^4 factorial design was run in a chemical process. The design factors are A = time, B = concentration, C = pressure, and D = temperature. The response variable is yield. The data follows:

Run	A	B	C	D	Yield (pounds)	**Factor Levels**	−	+
1	−	−	−	−	12	A (hours)	2	3
2	+	−	−	−	18	B (%)	14	18
3	−	+	−	−	13	C (psi)	60	80
4	+	+	−	−	16	D (°C)	200	250
5	−	−	+	−	17			
6	+	−	+	−	15			
7	−	+	+	−	20			
8	+	+	+	−	15			
9	−	−	−	+	10			
10	+	−	−	+	25			
11	−	+	−	+	13			
12	+	+	−	+	24			
13	−	−	+	+	19			
14	+	−	+	+	21			
15	−	+	+	+	17			
16	+	+	+	+	23			

(a) Estimate the factor effects. Based on a normal probability plot of the effect estimates, identify a model for the data from this experiment.
(b) Conduct an ANOVA based on the model identified in part (a). What are your conclusions?
(c) Analyze the residuals and comment on model adequacy.
(d) Find a regression model to predict yield in terms of the actual factor levels.

(e) Can this design be projected into a 2^3 design with two replicates? If so, sketch the design and show the average and range of the two yield values at each cube corner. Discuss the practical value of this plot.

14-21. An experiment has run a single replicate of a 2^4 design and calculated the following factor effects:

$A = 80.25$	$AB = 53.25$	$ABC = -2.95$
$B = -65.50$	$AC = 11.00$	$ABD = -8.00$
$C = -9.25$	$AD = 9.75$	$ACD = 10.25$
$D = -20.50$	$BC = 18.36$	$BCD = -7.95$
	$BD = 15.10$	$ABCD = -6.25$
	$CD = -1.25$	

(a) Construct a normal probability plot of the effects.
(b) Identify a tentative model, based on the plot of effects in part (a).
(c) Estimate the regression coefficients in this model, assuming that $\bar{y} = 400$.

14-22. A two-level factorial experiment in four factors was conducted by Chrysler and described in the article "Sheet Molded Compound Process Improvement" by P. I. Hsieh and D. E. Goodwin (*Fourth Symposium on Taguchi Methods*, American Supplier Institute, Dearborn, MI, 1986, pp. 13–21). The purpose was to reduce the number of defects in the finish of sheet-molded grill opening panels. A portion of the experimental design, and the resulting number of defects, y_i observed on each run is shown in the table following. This is a single replicate of the 2^4 design.
(a) Estimate the factor effects and use a normal probability plot to tentatively identify the important factors.
(b) Fit an appropriate model using the factors identified in part (a) above.
(c) Plot the residuals from this model versus the predicted number of defects. Also, prepare a normal probability plot of the residuals. Comment on the adequacy of these plots.
(d) The table also shows the square root of the number of defects. Repeat part (a) and (c) of the analysis using the square root of the number of defects as the response. Does this change the conclusions?

The Gril Defects Experiment

Run	A	B	C	D	y	$\sqrt{y}$
1	−	−	−	−	56	7.48
2	+	−	−	−	17	4.12
3	−	+	−	−	2	1.41
4	+	+	−	−	4	2.00
5	−	−	+	−	3	1.73
6	+	−	+	−	4	2.00
7	−	+	+	−	50	7.07
8	+	+	+	−	2	1.41
9	−	−	−	+	1	1.00
10	+	−	−	+	0	0.00
11	−	+	−	+	3	1.73
12	+	+	−	+	12	3.46
13	−	−	+	+	3	1.73
14	+	−	+	+	4	2.00
15	−	+	+	+	0	0.00
16	+	+	+	+	0	0.00

14-23. Consider a 2^2 factorial experiment with four center points. The data are $(1) = 21$, $a = 125$, $b = 154$, $ab = 352$, and the responses at the center point are 92, 130, 98, 152. Compute an ANOVA with the sum of squares for curvature and conduct an F-test for curvature. Use $\alpha = 0.05$.

14-24. Consider the experiment in Exercise 14-14. Suppose that a center point with five replicates is added to the factorial runs and the responses are 2800, 5600, 4500, 5400, 3600. Compute an ANOVA with the sum of squares for curvature and conduct an F-test for curvature. Use $\alpha = 0.05$.

14-25. Consider the experiment in Exercise 14-17. Suppose that a center point with five replicates is added to the factorial runs and the responses are 45, 40, 41, 47, and 43.
(a) Estimate the experimental error using the center points. Compare this to the estimate obtained originally in Exercise 14-17 by pooling apparently nonsignificant effects.
(b) Test for curvature with $\alpha = 0.05$.

14-6 BLOCKING AND CONFOUNDING IN THE 2^k DESIGN

It is often impossible to run all the observations in a 2^k factorial design under homogeneous conditions. Blocking is the design technique that is appropriate for this general situation. However, in many situations the block size is smaller than the number of runs in the complete replicate. In these cases, **confounding** is a useful procedure for running the 2^k design in 2^p blocks where the number of runs in a block is less than the number of treatment combinations in one complete replicate. The technique causes certain interaction effects to be indistinguishable from blocks or **confounded with blocks.** We will illustrate confounding in the 2^k factorial design in 2^p blocks, where $p < k$.

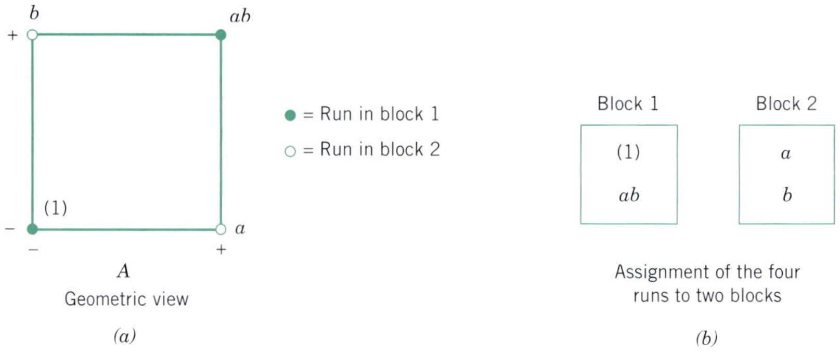

Figure 14-28 A 2^2 design in two blocks. (a) Geometric view. (b) Assignment of the four runs to two blocks.

Consider a 2^2 design. Suppose that each of the $2^2 = 4$ treatment combinations requires four hours of laboratory analysis. Thus, two days are required to perform the experiment. If days are considered as blocks, we must assign two of the four treatment combinations to each day.

This design is shown in Fig. 14-28. Notice that block 1 contains the treatment combinations (1) and ab and that block 2 contains a and b. The contrasts for estimating the main effects of factors A and B are

$$\text{Contrast}_A = ab + a - b - (1)$$
$$\text{Contrast}_B = ab + b - a - (1)$$

Note that these contrasts are unaffected by blocking since in each contrast there is one plus and one minus treatment combination from each block. That is, any difference between block 1 and block 2 that increases the readings in one block by an additive constant cancels out. The contrast for the AB interaction is

$$\text{Contrast}_{AB} = ab + (1) - a - b$$

Since the two treatment combinations with the plus signs, ab and (1), are in block 1 and the two with the minus signs, a and b, are in block 2, the block effect and the AB interaction are identical. That is, the AB interaction is confounded with blocks.

The reason for this is apparent from the table of plus and minus signs for the 2^2 design shown in Table 14-12. From the table we see that all treatment combinations that have a plus on AB are assigned to block 1, whereas all treatment combinations that have a minus sign on AB are assigned to block 2.

This scheme can be used to confound any 2^k design in two blocks. As a second example, consider a 2^3 design, run in two blocks. From the table of plus and minus signs, shown in Table 14-15, we assign the treatment combinations that are minus in the ABC column to block 1 and those that are plus in the ABC column to block 2. The resulting design is shown in Fig. 14-29.

There is a more general method of constructing the blocks. The method employs a **defining contrast,** say

$$L = \alpha_1 x_1 + \alpha_2 x_2 + \dots + \alpha_k x_k \tag{14-25}$$

where x_i is the level of the ith factor appearing in a treatment combination and α_i is the exponent appearing on the ith factor in the effect that is to be confounded with blocks. For the 2^k system, we have either $\alpha_i = 0$ or 1, and either $x_i = 0$ (low level) or $x_i = 1$ (high level). Treatment combinations that produce the same value of L (modulus 2) will be placed in the

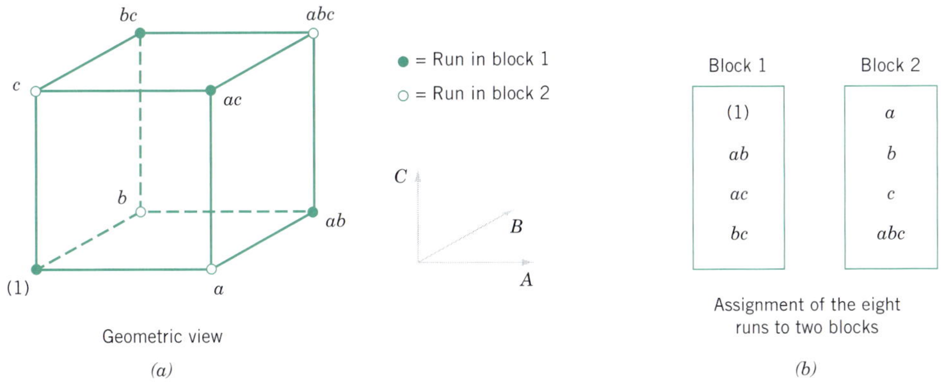

Figure 14-29 The 2^3 design in two blocks with ABC confounded. (a) Geometric View. (b) Assignment of the eight runs to two blocks.

same block. Since the only possible values of L (mod 2) are 0 and 1, this will assign the 2^k treatment combinations to exactly two blocks.

As an example, consider the 2^3 design with ABC confounded with blocks. Here x_1 corresponds to A, x_2 to B, x_3 to C, and $\alpha_1 = \alpha_2 = \alpha_3 = 1$. Thus, the defining contrast that would be used to confound ABC with blocks is

$$L = x_1 + x_2 + x_3$$

To assign the treatment combinations to the two blocks, we substitute the treatment combinations into the defining contrast as follows:

$$\begin{aligned}
(1)&: L = 1(0) + 1(0) + 1(0) = 0 = 0 \text{ (mod 2)} \\
a&: \quad L = 1(1) + 1(0) + 1(0) = 1 = 1 \text{ (mod 2)} \\
b&: \quad L = 1(0) + 1(1) + 1(0) = 1 = 1 \text{ (mod 2)} \\
ab&: L = 1(1) + 1(1) + 1(0) = 2 = 0 \text{ (mod 2)} \\
c&: \quad L = 1(0) + 1(0) + 1(1) = 1 = 1 \text{ (mod 2)} \\
ac&: L = 1(1) + 1(0) + 1(1) = 2 = 0 \text{ (mod 2)} \\
bc&: L = 1(0) + 1(1) + 1(1) = 2 = 0 \text{ (mod 2)} \\
abc&: L = 1(1) + 1(1) + 1(1) = 3 = 1 \text{ (mod 2)}
\end{aligned}$$

Thus (1), ab, ac, and bc are run in block 1, and a, b, c, and abc are run in block 2. This same design is shown in Fig. 14-29.

A shortcut method is useful in constructing these designs. The block containing the treatment combination (1) is called the **principal block.** Any element [except (1)] in the principal block may be generated by multiplying two other elements in the principal block modulus 2 on the exponents. For example, consider the principal block of the 2^3 design with ABC confounded, shown in Fig. 14-29. Note that

$$ab \cdot ac = a^2bc = bc$$

$$ab \cdot bc = ab^2c = ac$$

$$ac \cdot bc = abc^2 = ab$$

Treatment combinations in the other block (or blocks) may be generated by multiplying one element in the new block by each element in the principal block modulus 2 on the exponents. For the 2^3 with ABC confounded, since the principal block is (1), ab, ac, and bc, we know that the treatment combination b is in the other block. Thus, elements of this second block are

$$\begin{aligned} b \cdot (1) &= b \\ b \cdot ab = ab^2 &= a \\ b \cdot ac &= abc \\ b \cdot bc = b^2c &= c \end{aligned}$$

EXAMPLE 14-7
Missile Miss Distance

An experiment is performed to investigate the effect of four factors on the terminal miss distance of a shoulder-fired ground-to-air-missile. The four factors are target type (A), seeker type (B), target altitude (C), and target range (D). Each factor may be conveniently run at two levels, and the optical tracking system will allow terminal miss distance to be measured to the nearest foot. Two different operators or gunners are used in the flight test and, since there may be differences between operators, the test engineers decided to conduct the 2^4 design in two blocks with $ABCD$ confounded. Thus, the defining contrast is

$$L = x_1 + x_2 + x_3 + x_4$$

The experimental design and the resulting data are shown in Fig. 14-30. The effect estimates obtained from Minitab are shown in Table 14-23. A normal probability plot of the effects in Fig. 14-31 reveals that A (target type), D (target range), AD, and AC have large effects. A confirming analysis of variance, pooling the three-factor interactions as error, is shown in Table 14-24. Since the AC and AD interactions are significant, it is logical to conclude that A (target type), C (target altitude), and D (target range) all have important effects on the miss distance and that there are interactions between target type and altitude and target type and range. Notice that the $ABCD$ effect is treated as blocks in this analysis.

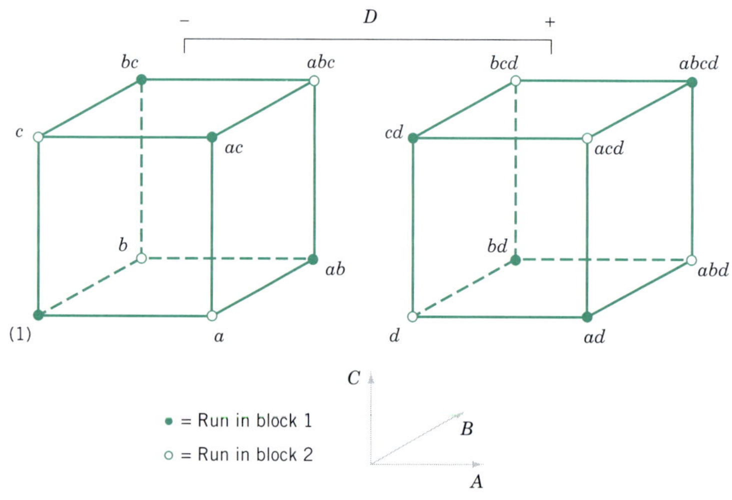

Geometric view

(a)

Block 1	Block 2
(1) = 3	a = 7
ab = 7	b = 5
ac = 6	c = 6
bc = 8	d = 4
ad = 10	abc = 6
bd = 4	bcd = 7
cd = 8	acd = 9
$abcd$ = 9	abd = 12

Assignment of the sixteen runs to two blocks

(b)

Figure 14-30 The 2^4 design in two blocks for Example 14-7. (a) Geometric view. (b) Assignment of the 16 runs to two blocks.

Table 14-23 Minitab Effect Estimates for Example 14-7

Estimated Effects and Coefficients for Distance		
Term	Effect	Coef
Constant		6.938
Block		0.063
A	2.625	1.312
B	0.625	0.313
C	0.875	0.438
D	1.875	0.938
AB	−0.125	−0.063
AC	−2.375	−1.187
AD	1.625	0.813
BC	−0.375	−0.188
BD	−0.375	−0.187
CD	−0.125	−0.062
ABC	−0.125	−0.063
ABD	0.875	0.438
ACD	−0.375	−0.187
BCD	−0.375	−0.187

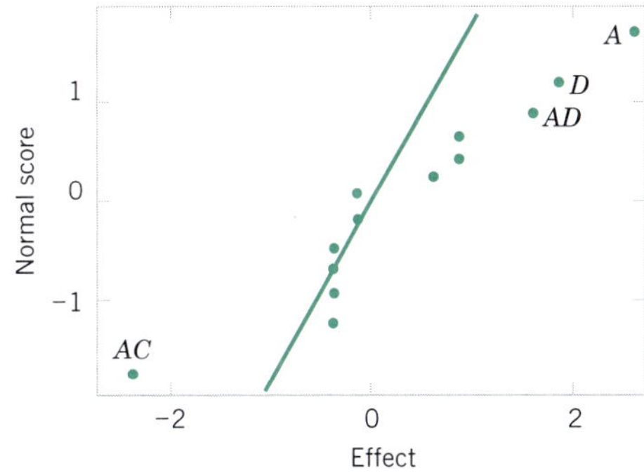

Figure 14-31 Normal probability plot of the effects from Minitab, Example 14-6.

It is possible to confound the 2^k design in four blocks of 2^{k-2} observations each. To construct the design, two effects are chosen to confound with blocks, and their defining contrasts are obtained. A third effect, the **generalized interaction** of the two effects initially chosen, is also confounded with blocks. The generalized interaction of two effects is found by multiplying their respective letters and reducing the exponents modulus 2.

For example, consider the 2^4 design in four blocks. If AC and BD are confounded with blocks, their generalized interaction is $(AC)(BD) = ABCD$. The design is constructed by using

Table 14-24 Analysis of Variance for Example 14-7

Source of Variation	Sum of Squares	Degrees of Freedom	Mean Square	f_0	P-Value
Blocks ($ABCD$)	0.0625	1	0.0625	0.06	—
A	27.5625	1	27.5625	25.94	0.0070
B	1.5625	1	1.5625	1.47	0.2920
C	3.0625	1	3.0625	2.88	0.1648
D	14.0625	1	14.0625	13.24	0.0220
AB	0.0625	1	0.0625	0.06	—
AC	22.5625	1	22.5625	21.24	0.0100
AD	10.5625	1	10.5625	9.94	0.0344
BC	0.5625	1	0.5625	0.53	—
BD	0.5625	1	0.5625	0.53	—
CD	0.0625	1	0.0625	0.06	—
Error ($ABC + ABD + ACD + BCD$)	4.2500	4	1.0625		
Total	84.9375	15			

the defining contrasts for AC and BD:

$$L_1 = x_1 + x_3$$
$$L_2 = x_2 + x_4$$

It is easy to verify that the four blocks are

Block 1 $L_1 = 0, L_2 = 0$	Block 2 $L_1 = 1, L_2 = 0$	Block 3 $L_1 = 0, L_2 = 1$	Block 4 $L_1 = 1, L_2 = 1$
(1)	a	b	ab
ac	c	abc	bc
bd	abd	d	ad
$abcd$	bcd	acd	cd

This general procedure can be extended to confounding the 2^k design in 2^p blocks, where $p < k$. Start by selecting p effects to be confounded, such that no effect chosen is a generalized interaction of the others. Then the blocks can be constructed from the p defining contrasts $L_1, L_2, \ldots, L_p$ that are associated with these effects. In addition to the p effects chosen to be confounded, exactly $2^p - p - 1$ additional effects are confounded with blocks; these are the generalized interactions of the original p effects chosen. Care should be taken so as not to confound effects of potential interest.

For more information on confounding in the 2^k factorial design, refer to Montgomery (2005, Chapter 7). This book contains guidelines for selecting factors to confound with blocks so that main effects and low-order interactions are not confounded. In particular, the book contains a table of suggested confounding schemes for designs with up to seven factors and a range of block sizes, some of which are as small as two runs.

EXERCISES FOR SECTION 14-6

14-26. Consider the data from the first replicate of Exercise 14-12.

(a) Suppose that these observations could not all be run under the same conditions. Set up a design to run these observations in two blocks of four observations each, with ABC confounded.

(b) Analyze the data.

14-27. Consider the data from the first replicate of Exercise 14-13.

(a) Construct a design with two blocks of eight observations each, with $ABCD$ confounded.

(b) Analyze the data.

14-28. Consider the data from Exercise 14-18.

(a) Construct the design that would have been used to run this experiment in two blocks of eight runs each.

(b) Analyze the data and draw conclusions.

14-29. Construct a 2^5 design in two blocks. Select the $ABCDE$ interaction to be confounded with blocks.

14-30. Consider the data from the first replicate of Exercise 14-13, assuming that four blocks are required. Confound ABD and ABC (and consequently CD) with blocks.

(a) Construct a design with four blocks of four observations each.

(b) Analyze the data.

14-31. Construct a 2^5 design in four blocks. Select the appropriate effects to confound so that the highest possible interactions are confounded with blocks.

14-32. Consider the 2^6 factorial design. Set up a design to be run in four blocks of 16 runs each. Show that a design that confounds three of the four-factor interactions with blocks is the best possible blocking arrangement.

14-33. An article in *Quality Engineering* (1999–2000, Vol. 12, pp. 83–87) "Designed experiment to stabilize blood glucose levels," reported on an experiment to minimize variations in blood glucose levels. The factors were: volume of juice intake before exercise (4 or 8 oz), amount of exercise on a Nordic Track cross-country skier (10 or 20 min), and delay between the time of juice intake (0 or 20 min) and the beginning of the exercise period. The experiment was blocked for time of day. The data follow.

(a) What effects are confounded with blocks? Comment on any concerns with the confounding in this design.

(b) Analyze the data and draw conclusions.

Run	Juice (oz)	Exercise (min)	Delay (min)	Time of Day	Average Blood Glucose
1	4	10	0	pm	71.5
2	8	10	0	am	103
3	4	20	0	am	83.5
4	8	20	0	pm	126
5	4	10	20	am	125.5
6	8	10	20	pm	129.5
7	4	20	20	pm	95
8	8	20	20	am	93

14-34. An article in *Industrial and Engineering Chemistry* ("Factorial Experiments in Pilot Plant Studies," 1951, pp. 1300–1306) reports on an experiment to investigate the effect of temperature (A), gas throughput (B), and concentration (C) on the strength of product solution in a recirculation unit. Two blocks were used with ABC confounded, and the experiment was replicated twice. The data follow.

(a) Analyze the data from this experiment.

Replicate 1

Block 1	Block 2
$(1) = 99$	$a = 18$
$ab = 52$	$b = 51$
$ac = 42$	$c = 108$
$bc = 95$	$abc = 35$

Replicate 2

Block 3	Block 4
$(1) = 46$	$a = 18$
$ab = 47$	$b = 62$
$ac = 22$	$c = 104$
$bc = 67$	$abc = 36$

(b) Analyze the residuals and comment on model adequacy.

(c) Comment on the efficiency of this design. Note that we have replicated the experiment twice, yet we have no information on the ABC interaction.

(d) Suggest a better design, specifically, one that would provide some information on *all* interactions.

14-7 FRACTIONAL REPLICATION OF THE 2^k DESIGN

As the number of factors in a 2^k factorial design increases, the number of runs required increases rapidly. For example, a 2^5 requires 32 runs. In this design, only 5 degrees of freedom correspond to main effects, and 10 degrees of freedom correspond to two-factor interactions. Sixteen of the 31 degrees of freedom are used to estimate high-order interactions—that is, three-factor and higher order interactions. Often there is little interest in these high-order interactions, particularly when we first begin to study a process or system. If we can assume that certain high-order interactions are negligible, a **fractional factorial design** involving fewer than the complete set of 2^k runs can be used to obtain information on the main effects and low-order interactions. In this section, we will introduce fractional replications of the 2^k design.

A major use of fractional factorials is in **screening experiments.** These are experiments in which many factors are considered with the purpose of identifying those factors (if any) that have large effects. Screening experiments are usually performed in the early stages of a project when it is likely that many of the factors initially considered have little or no effect on the response. The factors that are identified as important are then investigated more thoroughly in subsequent experiments.

14-7.1 One-Half Fraction of the 2^k Design

A one-half fraction of the 2^k design contains 2^{k-1} runs and is often called a 2^{k-1} fractional factorial design. As an example, consider the 2^{3-1} design—that is, a one-half fraction of the 2^3. This design has only four runs, in contrast to the full factorial that would require eight runs. The table of plus and minus signs for the 2^3 design is shown in Table 14-25. Suppose we select the four treatment combinations a, b, c, and abc, as our one-half fraction. These treatment combinations are shown in the top half of Table 14-25 and in Fig. 14-32(a).

Notice that the 2^{3-1} design is formed by selecting only those treatment combinations that yield a plus on the ABC effect. Thus, ABC is called the **generator** of this particular fraction.

Table 14-25 Plus and Minus Signs for the 2^3 Factorial Design

Treatment Combination	Factorial Effect							
	I	A	B	C	AB	AC	BC	ABC
a	+	+	−	−	−	−	+	+
b	+	−	+	−	−	+	−	+
c	+	−	−	+	+	−	−	+
abc	+	+	+	+	+	+	+	+
ab	+	+	+	−	+	−	−	−
ac	+	+	−	+	−	+	−	−
bc	+	−	+	+	−	−	+	−
(1)	+	−	−	−	+	+	+	−

Furthermore, the identity element I is also plus for the four runs, so we call

$$I = ABC$$

the **defining relation** for the design.

The treatment combinations in the 2^{3-1} design yields three degrees of freedom associated with the main effects. From the upper half of Table 14-25, we obtain the estimates of the main effects as linear combinations of the observations, say,

$$A = \tfrac{1}{2}[a - b - c + abc]$$
$$B = \tfrac{1}{2}[-a + b - c + abc]$$
$$C = \tfrac{1}{2}[-a - b + c + abc]$$

It is also easy to verify that the estimates of the two-factor interactions should be the following linear combinations of the observations:

$$BC = \tfrac{1}{2}[a - b - c + abc]$$
$$AC = \tfrac{1}{2}[-a + b - c + abc]$$
$$AB = \tfrac{1}{2}[-a - b + c + abc]$$

Thus, the linear combination of observations in column A, ℓ_A, estimates both the main effect of A and the BC interaction. That is, the linear combination ℓ_A estimates the sum of these two effects $A + BC$. Similarly, ℓ_B estimates $B + AC$, and ℓ_C estimates $C + AB$. Two or more

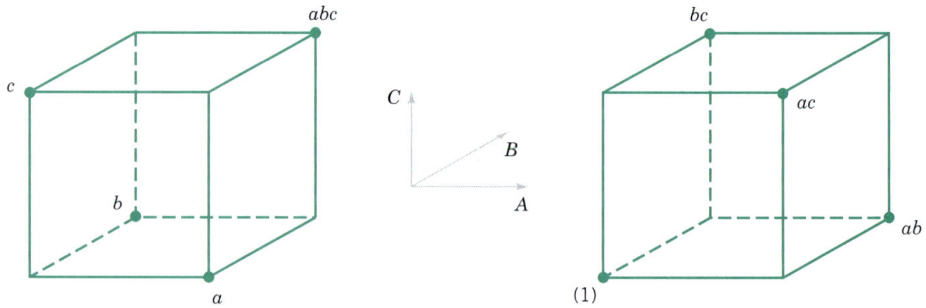

The principal fraction, $I = +ABC$

(a)

The alternate fraction, $I = -ABC$

(b)

Figure 14-32 The one-half fractions of the 2^3 design. (a) The principal fraction, $I = +ABC$. (b) The alternate fraction, $I = -ABC$.

effects that have this property are called **aliases.** In our 2^{3-1} design, A and BC are aliases, B and AC are aliases, and C and AB are aliases. Aliasing is the direct result of fractional replication. In many practical situations, it will be possible to select the fraction so that the main effects and low-order interactions that are of interest will be aliased only with high-order interactions (which are probably negligible).

The alias structure for this design is found by using the defining relation $I = ABC$. Multiplying any effect by the defining relation yields the aliases for that effect. In our example, the alias of A is

$$A = A \cdot ABC = A^2BC = BC$$

since $A \cdot I = A$ and $A^2 = I$. The aliases of B and C are

$$B = B \cdot ABC = AB^2C = AC$$

and

$$C = C \cdot ABC = ABC^2 = AB$$

Now suppose that we had chosen the other one-half fraction, that is, the treatment combinations in Table 14-25 associated with minus on ABC. These four runs are shown in the lower half of Table 14-25 and in Fig. 14-32(b). The defining relation for this design is $I = -ABC$. The aliases are $A = -BC$, $B = -AC$, and $C = -AB$. Thus, estimates of A, B, and C that result from this fraction really estimate $A - BC$, $B - AC$, and $C - AB$. In practice, it usually does not matter which one-half fraction we select. The fraction with the plus sign in the defining relation is usually called the **principal fraction,** and the other fraction is usually called the **alternate fraction.**

Note that if we had chosen AB as the generator for the fractional factorial,

$$A = A \cdot AB = B$$

and the two main effects of A and B would be aliased. This typically loses important information.

Sometimes we use **sequences** of fractional factorial designs to estimate effects. For example, suppose we had run the principal fraction of the 2^{3-1} design with generator ABC. From this design we have the following effect estimates:

$$\ell_A = A + BC$$
$$\ell_B = B + AC$$
$$\ell_C = C + AB$$

Suppose that we are willing to assume at this point that the two-factor interactions are negligible. If they are, the 2^{3-1} design has produced estimates of the three main effects A, B, and C. However, if after running the principal fraction we are uncertain about the interactions, it is possible to estimate them by running the *alternate* fraction. The alternate fraction produces the following effect estimates:

$$\ell_A' = A - BC$$
$$\ell_B' = B - AC$$
$$\ell_C' = C - AB$$

We may now obtain de-aliased estimates of the main effects and two-factor interactions by adding and subtracting the linear combinations of effects estimated in the two individual fractions. For example, suppose we want to de-alias A from the two-factor interaction BC. Since $\ell_A = A + BC$ and $\ell_A' = A - BC$, we can combine these effect estimates as follows:

$$\frac{1}{2}(\ell_A + \ell_A') = \frac{1}{2}(A + BC + A - BC) = A$$

and

$$\frac{1}{2}(\ell_A - \ell_A') = \frac{1}{2}(A + BC - A + BC) = BC$$

For all three pairs of effect estimates, we would obtain the following results:

Effect, i	**from $\frac{1}{2}(l_i + l_i')$**	**from $\frac{1}{2}(l_i - l_i')$**
$i = A$	$\frac{1}{2}(A + BC + A - BC) = A$	$\frac{1}{2}[A + BC - (A - BC)] = BC$
$i = B$	$\frac{1}{2}(B + AC + B - AC) = B$	$\frac{1}{2}[B + AC - (B - AC)] = AC$
$i = C$	$\frac{1}{2}(C + AB + C - AB) = C$	$\frac{1}{2}[C + AB - (C - AB)] = AB$

Thus, by combining a sequence of two fractional factorial designs, we can isolate both the main effects and the two-factor interactions. This property makes the fractional factorial design highly useful in experimental problems since we can run sequences of small, efficient experiments, combine information across *several* experiments, and take advantage of learning about the process we are experimenting with as we go along. This is an illustration of the concept of sequential experimentation.

A 2^{k-1} design may be constructed by writing down the treatment combinations for a full factorial with $k - 1$ factors, called the **basic design,** and then adding the kth factor by identifying its plus and minus levels with the plus and minus signs of the highest order interaction. Therefore, a 2^{3-1} fractional factorial is constructed by writing down the basic design as a full 2^2 factorial and then equating factor C with the $\pm AB$ interaction. Thus, to construct the principal fraction, we would use $C = +AB$ as follows:

Basic Design		**Fractional Design**		
Full 2^2		**$2^{3-1}, I = +ABC$**		
A	B	A	B	$C = AB$
−	−	−	−	+
+	−	+	−	−
−	+	−	+	−
+	+	+	+	+

To obtain the alternate fraction we would equate the last column to $C = -AB$.

EXAMPLE 14-8
Plasma Etch

To illustrate the use of a one-half fraction, consider the plasma etch experiment described in Example 14-5. Suppose that we decide to use a 2^{4-1} design with $I = ABCD$ to investigate the four factors gap (A), pressure (B), C_2F_6 flow rate (C), and power setting (D). This design would be constructed by writing down as the basic design a 2^3 in the factors A, B, and C and then setting the levels of the fourth factor $D = ABC$. The design and the resulting etch rates are shown in Table 14-26. The design is shown graphically in Fig. 14-33.

Table 14-26 The 2^{4-1} Design with Defining Relation $I = ABCD$

A	B	C	$D = ABC$	Treatment Combination	Etch Rate
−	−	−	−	(1)	550
+	−	−	+	*ad*	749
−	+	−	+	*bd*	1052
+	+	−	−	*ab*	650
−	−	+	+	*cd*	1075
+	−	+	−	*ac*	642
−	+	+	−	*bc*	601
+	+	+	+	*abcd*	729

In this design, the main effects are aliased with the three-factor interactions; note that the alias of A is

$$A \cdot I = A \cdot ABCD \qquad \text{or} \qquad A = A^2BCD = BCD$$

and similarly $B = ACD$, $C = ABD$, and $D = ABC$.

The two-factor interactions are aliased with each other. For example, the alias of AB is CD:

$$AB \cdot I = AB \cdot ABCD \qquad \text{or} \qquad AB = A^2B^2CD = CD$$

The other aliases are $AC = BD$ and $AD = BC$.

The estimates of the main effects and their aliases are found using the four columns of signs in Table 14-26. For example, from column A we obtain the estimated effect

$$\begin{aligned}\ell_A = A + BCD &= \tfrac{1}{4}(-550 + 749 - 1052 + 650 - 1075 + 642 - 601 + 729)\\ &= -127.00\end{aligned}$$

The other columns produce

$$\ell_B = B + ACD = 4.00 \quad \ell_C = C + ABD = 11.50 \quad \text{and} \quad \ell_D = D + ABC = 290.50$$

Clearly, ℓ_A and ℓ_D are large, and if we believe that the three-factor interactions are negligible, the main effects A (gap) and D (power setting) significantly affect etch rate.

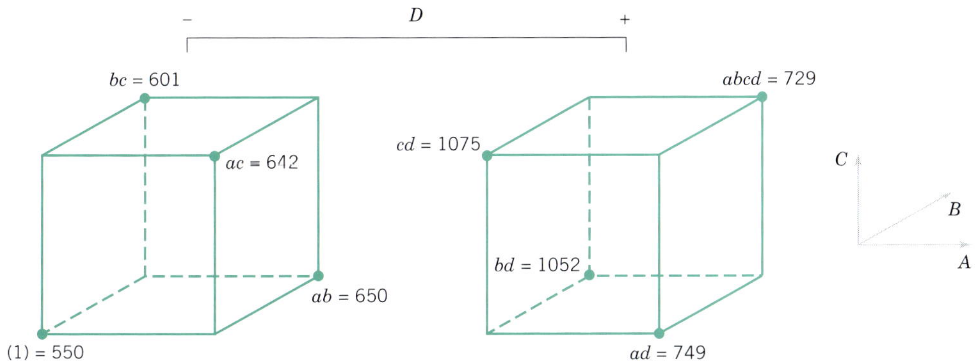

Figure 14-33 The 2^{4-1} design for the experiment of Example 14-8.

The interactions are estimated by forming the AB, AC, and AD columns and adding them to the table. For example, the signs in the AB column are $+, -, -, +, +, -, -, +$, and this column produces the estimate

$$\ell_{AB} = AB + CD = \tfrac{1}{4}(550 - 749 - 1052 + 650 + 1075 - 642 - 601 + 729) = -10$$

From the AC and AD columns we find

$$\ell_{AC} = AC + BD = -25.50 \qquad \text{and} \qquad \ell_{AD} = AD + BC = -197.50$$

The ℓ_{AD} estimate is large; the most straightforward interpretation of the results is that since A and D are large, this is the AD interaction. Thus, the results obtained from the 2^{4-1} design agree with the full factorial results in Example 14-5.

Computer Solution

Fractional factorial designs are usually analyzed with a software package. Table 14-26 shows the effect estimates obtained from Minitab for Example 14-8. They are in agreement with the hand calculation reported earlier.

Normal Probability Plot of Effects

The normal probability plot is very useful in assessing the significance of effects from a fractional factorial design, particularly when many effects are to be estimated. We strongly recommend examining this plot. Figure 14-34 presents the normal probability plot of the effects from Example 14-8. This plot was obtained from Minitab. Notice that the A, D, and AD interaction effects stand out clearly in this graph.

Residual Analysis

The residuals can be obtained from a fractional factorial by the regression model method shown previously. Note that the Minitab output for Example 14-8 in Table 14-27 shows the regression coefficients. The residuals should be graphically analyzed as we have discussed before, both to assess the validity of the underlying model assumptions and to gain additional insight into the experimental situation.

Projection of the 2^{k-1} Design

If one or more factors from a one-half fraction of a 2^k can be dropped, the design will project into a full factorial design. For example, Fig. 14-35 presents a 2^{3-1} design. Notice that this

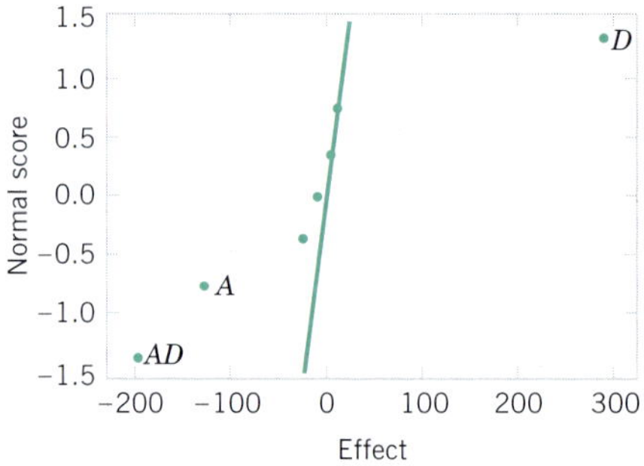

Figure 14-34 Normal probability plot of the effects from Minitab, Example 14-8.

Table 14-27 Effect Estimates from Minitab, Example 14-8

Fractional Factorial Fit

Estimated Effects and Coefficients for Etch Rt

Term	Effect	Coef
Constant		756.00
Gap	−127.00	−63.50
Pressure	4.00	2.00
F	11.50	5.75
Power	290.50	145.25
Gap*Pressure	−10.00	−5.00
Gap*F	−25.50	−12.75
Gap*Power	−197.50	−98.75

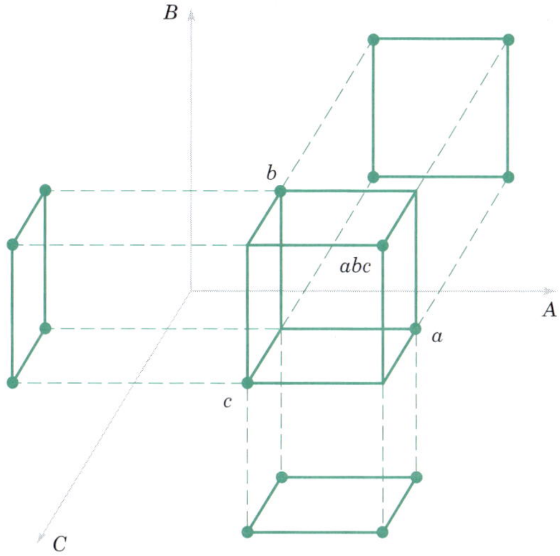

Figure 14-35 Projection of a 2^{3-1} design into three 2^2 designs.

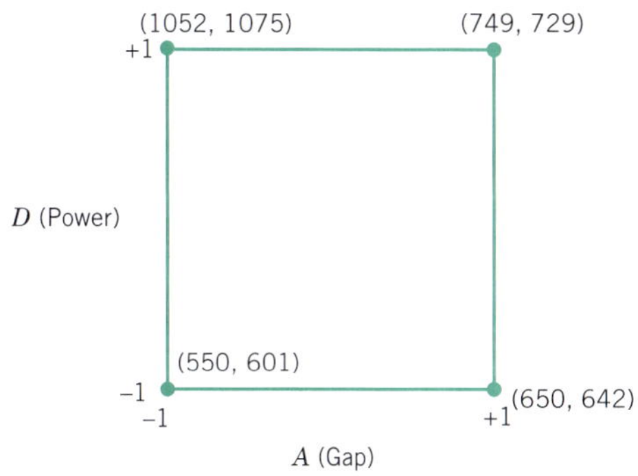

Figure 14-36 The 2^2 design obtained by dropping factors B and C from the plasma etch experiment in Example 14-7.

design will project into a full factorial in any two of the three original factors. Thus, if we think that at most two of the three factors are important, the 2^{3-1} design is an excellent design for identifying the significant factors. This **projection property** is highly useful in factor screening, because it allows negligible factors to be eliminated, resulting in a stronger experiment in the active factors that remain.

In the 2^{4-1} design used in the plasma etch experiment in Example 14-8, we found that two of the four factors (B and C) could be dropped. If we eliminate these two factors, the remaining columns in Table 14-24 form a 2^2 design in the factors A and D, with two replicates. This design is shown in Fig. 14-36. The main effects of A and D and the strong two-factor AD interaction are clearly evident from this graph.

Design Resolution

The concept of design resolution is a useful way to catalog fractional factorial designs according to the alias patterns they produce. Designs of resolution III, IV, and V are particularly important. The definitions of these terms and an example of each follow.

1. **Resolution III Designs.** These are designs in which no main effects are aliased with any other main effect, but main effects are aliased with two-factor interactions and some two-factor interactions may be aliased with each other. The 2^{3-1} design with $I = ABC$ is a resolution III design. We usually employ a Roman numeral subscript to indicate design resolution; thus, this one-half fraction is a 2_{III}^{3-1} design.
2. **Resolution IV Designs.** These are designs in which no main effect is aliased with any other main effect or two-factor interactions, but two-factor interactions are aliased with each other. The 2^{4-1} design with $I = ABCD$ used in Example 14-8 is a resolution IV design (2_{IV}^{4-1}).
3. **Resolution V Designs.** These are designs in which no main effect or two-factor interaction is aliased with any other main effect or two-factor interaction, but two-factor interactions are aliased with three-factor interactions. The 2^{5-1} design with $I = ABCDE$ is a resolution V design (2_{V}^{5-1}).

Resolution III and IV designs are particularly useful in factor screening experiments. A resolution IV design provides good information about main effects and will provide some information about all two-factor interactions.

14-7.2 Smaller Fractions: The 2^{k-p} Fractional Factorial

Although the 2^{k-1} design is valuable in reducing the number of runs required for an experiment, we frequently find that smaller fractions will provide almost as much useful information at even greater economy. In general, a 2^k design may be run in a $1/2^p$ fraction called a 2^{k-p} fractional factorial design. Thus, a $1/4$ fraction is called a 2^{k-2} design, a $1/8$ fraction is called a 2^{k-3} design, a $1/16$ fraction a 2^{k-4} design, and so on.

To illustrate the $1/4$ fraction, consider an experiment with six factors and suppose that the engineer is primarily interested in main effects but would also like to get some information about the two-factor interactions. A 2^{6-1} design would require 32 runs and would have 31 degrees of freedom for estimating effects. Since there are only six main effects and 15 two-factor interactions, the one-half fraction is inefficient—it requires too many runs. Suppose we consider a $1/4$ fraction, or a 2^{6-2} design. This design contains 16 runs and, with 15 degrees of freedom, will allow all six main effects to be estimated, with some capability for examining the two-factor interactions.

To generate this design, we would write down a 2^4 design in the factors A, B, C, and D as the basic design and then add two columns for E and F. To find the new columns we could select the two **design generators** $I = ABCE$ and $I = BCDF$. Thus, column E would be found from $E = ABC$, and column F would be $F = BCD$. That is, columns $ABCE$ and $BCDF$ are equal to the identity column. However, we know that the product of any two columns in the table of plus and minus signs for a 2^k design is just another column in the table; therefore, the product of $ABCE$ and $BCDF$ or $ABCE(BCDF) = AB^2C^2DEF = ADEF$ is also an identity column. Consequently, the **complete defining relation** for the 2^{6-2} design is

$$I = ABCE = BCDF = ADEF$$

We refer to each term in a defining relation (such as $ABCE$ above) as a **word.** To find the alias of any effect, simply multiply the effect by each word in the foregoing defining relation. For example, the alias of A is

$$A = BCE = ABCDF = DEF$$

The complete alias relationships for this design are shown in Table 14-28. In general, the resolution of a 2^{k-p} design is equal to the number of letters in the shortest word in the complete

Table 14-28 Alias Structure for the 2^{6-2}_{IV} Design with $I = ABCE = BCDF = ADEF$

$A = BCE = DEF = ABCDF$	$AB = CE = ACDF = BDEF$
$B = ACE = CDF = ABDEF$	$AC = BE = ABDF = CDEF$
$C = ABE = BDF = ACDEF$	$AD = EF = BCDE = ABCF$
$D = BCF = AEF = ABCDE$	$AE = BC = DF = ABCDEF$
$E = ABC = ADF = BCDEF$	$AF = DE = BCEF = ABCD$
$F = BCD = ADE = ABCEF$	$BD = CF = ACDE = ABEF$
$ABD = CDE = ACF = BEF$	$BF = CD = ACEF = ABDE$
$ACD = BDE = ABF = CEF$	

Table 14-29 A 2_{IV}^{6-2} Design for the Injection-Molding Experiment

Run	A	B	C	D	$E = ABC$	$F = BCD$	Observed Shrinkage (×10)
1	−	−	−	−	−	−	6
2	+	−	−	−	+	−	10
3	−	+	−	−	+	+	32
4	+	+	−	−	−	+	60
5	−	−	+	−	+	+	4
6	+	−	+	−	−	+	15
7	−	+	+	−	−	−	26
8	+	+	+	−	+	−	60
9	−	−	−	+	−	+	8
10	+	−	−	+	+	+	12
11	−	+	−	+	+	−	34
12	+	+	−	+	−	−	60
13	−	−	+	+	+	−	16
14	+	−	+	+	−	−	5
15	−	+	+	+	−	+	37
16	+	+	+	+	+	+	52

defining relation. Therefore, this is a resolution IV design; main effects are aliased with three-factor and higher interactions, and two-factor interactions are aliased with each other. This design would provide good information on the main effects and would give some idea about the strength of the two-factor interactions. The construction and analysis of the design are illustrated in Example 14-9.

EXAMPLE 14-9
Injection Molding

Parts manufactured in an injection-molding process are showing excessive shrinkage, which is causing problems in assembly operations upstream from the injection-molding area. In an effort to reduce the shrinkage, a quality-improvement team has decided to use a designed experiment to study the injection-molding process. The team investigates six factors—mold temperature (A), screw speed (B), holding time (C), cycle time (D), gate size (E), and holding pressure (F)—each at two levels, with the objective of learning how each factor affects shrinkage and obtaining preliminary information about how the factors interact.

The team decides to use a 16-run two-level fractional factorial design for these six factors. The design is constructed by writing down a 2^4 as the basic design in the factors A, B, C, and D and then setting $E = ABC$ and $F = BCD$ as discussed above. Table 14-29 shows the design, along with the observed shrinkage (×10) for the test part produced at each of the 16 runs in the design.

A normal probability plot of the effect estimates from this experiment is shown in Fig. 14-37. The only large effects are A (mold temperature), B (screw speed), and the AB interaction. In light of the alias relationship in Table 14-28, it seems reasonable to tentatively adopt these conclusions. The plot of the AB interaction in Fig. 14-38 shows that the process is insensitive to temperature if the screw speed is at the low level but sensitive to temperature if the screw speed is at the high level. With the screw speed at a low level, the process should produce an average shrinkage of around 10% regardless of the temperature level chosen.

Based on this initial analysis, the team decides to set both the mold temperature and the screw speed at the low level. This set of conditions should reduce the mean shrinkage of parts to around 10%. However, the variability in shrinkage from part to part is still a potential problem. In effect, the mean shrinkage can be adequately reduced by the above modifications; however, the part-to-part variability in shrinkage over a production run could still cause problems in assembly. One way to address this issue is to see if any of the process factors affect the variability in parts shrinkage.

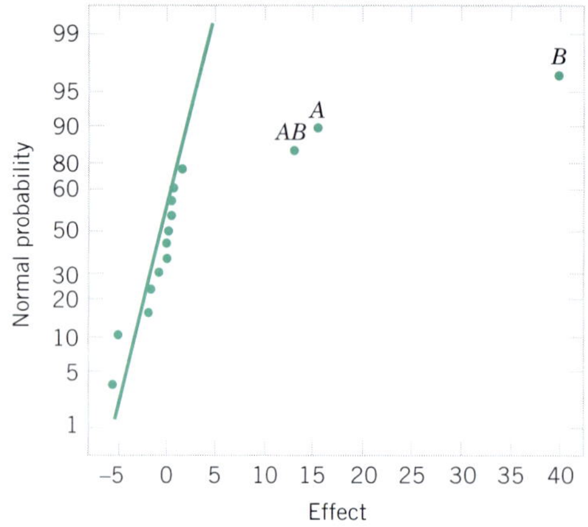

Figure 14-37 Normal probability plot of effects for Example 14-9.

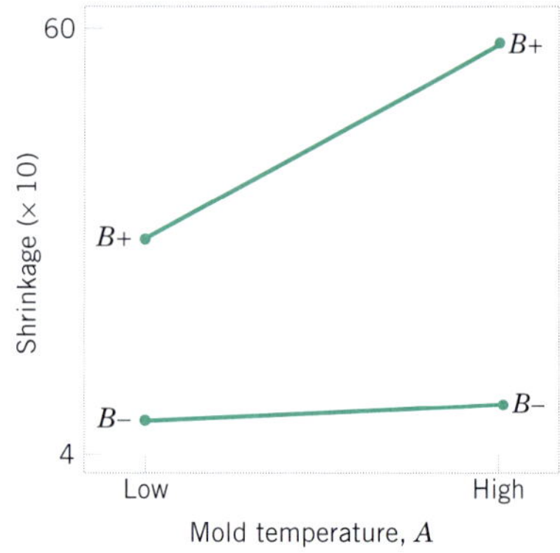

Figure 14-38 Plot of *AB* (mold temperature–screw speed) interaction for Example 14-9.

Figure 14-39 presents the normal probability plot of the residuals. This plot appears satisfactory. The plots of residuals versus each factor were then constructed. One of these plots, that for residuals versus factor *C* (holding time), is shown in Fig. 14-40. The plot reveals much less scatter in the residuals at the low holding time than at the high holding time. These residuals were obtained in the usual way from a model for predicted shrinkage

$$\begin{aligned}\hat{y} &= \hat{\beta}_0 + \hat{\beta}_1 x_1 + \hat{\beta}_2 x_2 + \hat{\beta}_{12} x_1 x_2 \\ &= 27.3125 + 6.9375 x_1 + 17.8125 x_2 + 5.9375 x_1 x_2\end{aligned}$$

where x_1, x_2, and x_1x_2 are coded variables that correspond to the factors *A* and *B* and the *AB* interaction. The residuals are then

$$e = y - \hat{y}$$

The regression model used to produce the residuals essentially removes the location effects of *A*, *B*, and *AB* from the data; the residuals therefore contain information about unexplained variability.

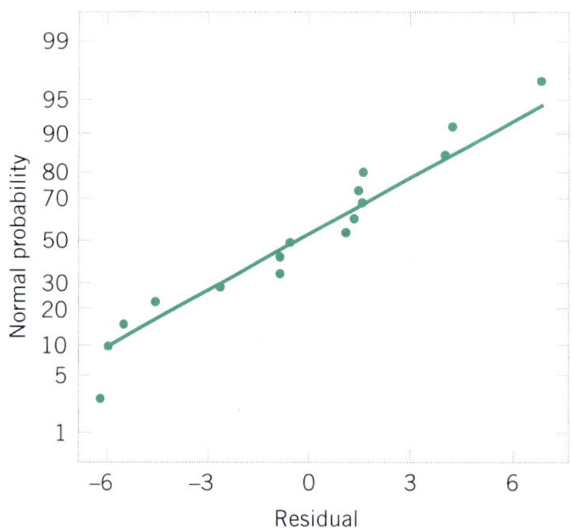

Figure 14-39 Normal probability plot of residuals for Example 14-9.

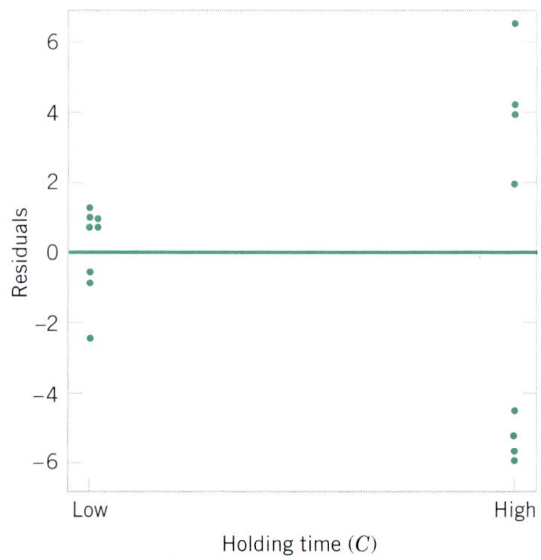

Figure 14-40 Residuals versus holding time (*C*) for Example 14-9.

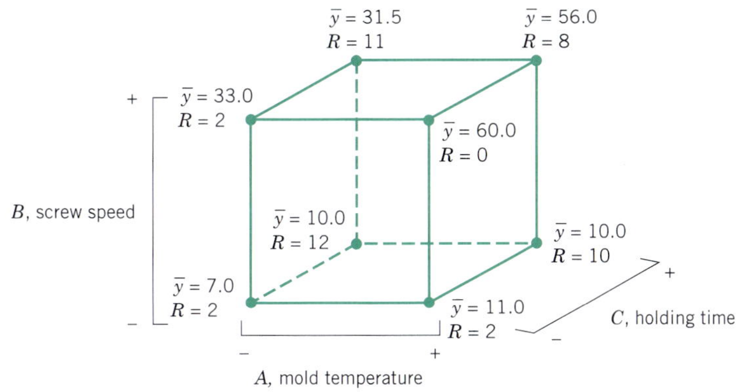

Figure 14-41 Average shrinkage and range of shrinkage in factors A, B, and C for Example 14-9.

Figure 14-40 indicates that there is a pattern in the variability and that the variability in the shrinkage of parts may be smaller when the holding time is at the low level.

Figure 14-41 shows the data from this experiment projected onto a cube in the factors A, B, and C. The average observed shrinkage and the range of observed shrinkage are shown at each corner of the cube. From inspection of this figure, we see that running the process with the screw speed (B) at the low level is the key to reducing average parts shrinkage. If B is low, virtually any combination of temperature (A) and holding time (C) will result in low values of average parts shrinkage. However, from examining the ranges of the shrinkage values at each corner of the cube, it is immediately clear that setting the holding time (C) at the low level is the most appropriate choice if we wish to keep the part-to-part variability in shrinkage low during a production run.

The concepts used in constructing the 2^{6-2} fractional factorial design in Example 14-9 can be extended to the construction of any 2^{k-p} fractional factorial design. In general, a 2^k fractional factorial design containing 2^{k-p} runs is called a $1/2^p$ fraction of the 2^k design or, more simply, a 2^{k-p} fractional factorial design. These designs require the selection of p independent generators. The defining relation for the design consists of the p generators initially chosen and their $2^p - p - 1$ generalized interactions.

The alias structure may be found by multiplying each effect column by the defining relation. Care should be exercised in choosing the generators so that effects of potential interest are not aliased with each other. Each effect has $2^p - 1$ aliases. For moderately large values of k, we usually assume higher order interactions (say, third- or fourth-order or higher) to be negligible, and this greatly simplifies the alias structure.

It is important to select the p generators for the 2^{k-p} fractional factorial design in such a way that we obtain the best possible alias relationships. A reasonable criterion is to select the generators so that the resulting 2^{k-p} design has the highest possible design resolution. Montgomery (2005) presents a table of recommended generators for 2^{k-p} fractional factorial designs for $k \leq 15$ factors and up to as many as $n \leq 128$ runs. A portion of his table is reproduced here as Table 14-30. In this table, the generators are shown with either + or − choices; selection of all generators as + will give a principal fraction, while if any generators are − choices, the design will be one of the alternate fractions for the same family. The suggested generators in this table will result in a design of the highest possible resolution. Montgomery (2005) also gives a table of alias relationships for these designs.

EXAMPLE 14-10
Aliases with Seven Factors

To illustrate the use of Table 14-30, suppose that we have seven factors and that we are interested in estimating the seven main effects and obtaining some insight regarding the two-factor interactions. We are willing to assume that three-factor and higher interactions are negligible. This information suggests that a resolution IV design would be appropriate.

Table 14-30 Selected 2^{k-p} Fractional Factorial Designs

Number of Factors k	Fraction	Number of Runs	Design Generators	Number of Factors k	Fraction	Number of Runs	Design Generators
3	2^{3-1}_{III}	4	$C = \pm AB$	10			$H = \pm ABCG$
4	2^{4-1}_{IV}	8	$D = \pm ABC$				$J = \pm ACDE$
5	2^{5-1}_{V}	16	$E = \pm ABCD$		2^{10-3}_{V}	128	$K = \pm ACDF$
	2^{5-2}_{III}	8	$D = \pm AB$				$G = \pm BCDF$
			$E = \pm AC$				$H = \pm ACDF$
6	2^{6-1}_{VI}	32	$F = \pm ABCDE$				$J = \pm ABDE$
	2^{6-2}_{IV}	16	$E = \pm ABC$		2^{10-4}_{IV}	64	$K = \pm ABCE$
			$F = \pm BCD$				$F = \pm ABCD$
	2^{6-3}_{III}	8	$D = \pm AB$				$G = \pm ABCE$
			$E = \pm AC$				$H = \pm ABDE$
			$F = \pm BC$				$J = \pm ACDE$
7	2^{7-1}_{VII}	64	$G = \pm ABCDEF$		2^{10-5}_{IV}	32	$K = \pm BCDE$
	2^{7-2}_{IV}	32	$E = \pm ABC$				$E = \pm ABC$
			$G = \pm ABDE$				$F = \pm BCD$
	2^{7-3}_{IV}	16	$E = \pm ABC$				$G = \pm ACD$
			$F = \pm BCD$				$H = \pm ABD$
			$G = \pm ACD$				$J = \pm ABCD$
	2^{7-4}_{III}	8	$D = \pm AB$		2^{10-6}_{III}	16	$K = \pm AB$
			$E = \pm AC$	11			$G = \pm CDE$
			$F = \pm BC$				$H = \pm ABCD$
			$G = \pm ABC$				$J = \pm ABF$
8	2^{8-2}_{V}	64	$G = \pm ABCD$				$K = \pm BDEF$
			$H = \pm ABEF$		2^{11-5}_{IV}	64	$L = \pm ADEF$
	2^{8-3}_{IV}	32	$F = \pm ABC$				$F = \pm ABC$
			$G = \pm ABD$				$G = \pm BCD$
			$H = \pm BCDE$				$H = \pm CDE$
	2^{8-4}_{IV}	16	$E = \pm BCD$				$J = \pm ACD$
			$F = \pm ACD$				$K = \pm ADE$
			$G = \pm ABC$		2^{11-6}_{IV}	32	$L = \pm BDE$
			$H = \pm ABD$				$E = \pm ABC$
9	2^{9-2}_{VI}	128	$H = \pm ACDFG$				$F = \pm BCD$
			$J = \pm BCEFG$				$G = \pm ACD$
	2^{9-3}_{IV}	64	$G = \pm ABCD$				$H = \pm ABD$
			$H = \pm ACEF$				$J = \pm ABCD$
			$J = \pm CDEF$				$K = \pm AB$
	2^{9-4}_{IV}	32	$F = \pm BCDE$		2^{11-7}_{III}	16	$L = \pm AC$
			$G = \pm ACDE$				
			$H = \pm ABDE$				
			$J = \pm ABCE$				
	2^{9-5}_{III}	16	$E = \pm ABC$				
			$F = \pm BCD$				
			$G = \pm ACD$				
			$H = \pm ABD$				
			$J = \pm ABCD$				

Source: Montgomery (2005)

Table 14-31 Generators, Defining Relation, and Aliases for the 2_{IV}^{7-3} Fractional Factorial Design

Generators and Defining Relation	
$E = ABC, \quad F = BCD, \quad G = ACD$	
$I = ABCE = BCDF = ADEF = ACDG = BDEG = ABFG = CEFG$	
Aliases	
$A = BCE = DEF = CDG = BFG$	$AB = CE = FG$
$B = ACE = CDF = DEG = AFG$	$AC = BE = DG$
$C = ABE = BDF = ADG = EFG$	$AD = EF = CG$
$D = BCF = AEF = ACG = BEG$	$AE = BC = DF$
$E = ABC = ADF = BDG = CFG$	$AF = DE = BG$
$F = BCD = ADE = ABG = CEG$	$AG = CD = BF$
$G = ACD = BDE = ABF = CEF$	$BD = CF = EG$
$ABD = CDE = ACF = BEF = BCG = AEG = DFG$	

Table 14-30 shows that two resolution IV fractions are available: the 2_{IV}^{7-2} with 32 runs and the 2_{IV}^{7-3} with 16 runs. The aliases involving main effects and two- and three-factor interactions for the 16-run design are presented in Table 14-31. Notice that all seven main effects are aliased with three-factor interactions. All the two-factor interactions are aliased in groups of three. Therefore, this design will satisfy our objectives; that is, it will allow the estimation of the main effects, and it will give some insight regarding two-factor interactions. It is not necessary to run the 2_{IV}^{7-2} design, which would require 32 runs. The construction of the 2_{IV}^{7-3} design is shown in Table 14-32. Notice that it was constructed by starting with the 16-run 2^4 design in A, B, C, and D as the basic design and then adding the three columns $E = ABC$, $F = BCD$, and $G = ACD$ as suggested in Table 14-30. Thus, the generators for this design are $I = ABCE$, $I = BCDF$, and $I = ACDG$. The complete defining relation is $I = ABCE = BCDF = ADEF = ACDG = BDEG = CEFG = ABFG$. This defining relation was used to produce the aliases in Table 14-31. For example, the alias relationship of A is

$$A = BCE = ABCDF = DEF = CDG = ABDEG = ACEFG = BFG$$

which, if we ignore interactions higher than three factors, agrees with Table 14-31.

Table 14-32 A 2_{IV}^{7-3} Fractional Factorial Design

	Basic Design						
Run	A	B	C	D	$E = ABC$	$F = BCD$	$G = ACD$
1	−	−	−	−	−	−	−
2	+	−	−	−	+	−	+
3	−	+	−	−	+	+	−
4	+	+	−	−	−	+	+
5	−	−	+	−	+	+	+
6	+	−	+	−	−	+	−
7	−	+	+	−	−	−	+
8	+	+	+	−	+	−	−
9	−	−	−	+	−	+	+
10	+	−	−	+	+	+	−
11	−	+	−	+	+	−	+
12	+	+	−	+	−	−	−
13	−	−	+	+	+	−	−
14	+	−	+	+	−	−	+
15	−	+	+	+	−	+	−
16	+	+	+	+	+	+	+

Table 14-33 A 2_{III}^{7-4} Fractional Factorial Design

A	B	C	$D = AB$	$E = AC$	$F = BC$	$G = ABC$
−	−	−	+	+	+	−
+	−	−	−	−	+	+
−	+	−	−	+	−	+
+	+	−	+	−	−	−
−	−	+	+	−	−	+
+	−	+	−	+	−	−
−	+	+	−	−	+	−
+	+	+	+	+	+	+

For seven factors, we can reduce the number of runs even further. The 2^{7-4} design is an eight-run experiment accommodating seven variables. This is a 1/16th fraction and is obtained by first writing down a 2^3 design as the basic design in the factors A, B, and C, and then forming the four new columns from $I = ABD$, $I = ACE$, $I = BCF$, and $I = ABCG$, as suggested in Table 14-30. The design is shown in Table 14-33.

The complete defining relation is found by multiplying the generators together two, three, and finally four at a time, producing

$$I = ABD = ACE = BCF = ABCG = BCDE = ACDF = CDG = ABEF$$
$$= BEG = AFG = DEF = ADEG = CEFG = BDFG = ABCDEFG$$

The alias of any main effect is found by multiplying that effect through each term in the defining relation. For example, the alias of A is

$$A = BD = CE = ABCF = BCG = ABCDE = CDF = ACDG$$
$$= BEF = ABEG = FG = ADEF = DEG = ACEFG = ABDFG = BCDEFG$$

This design is of resolution III, since the main effect is aliased with two-factor interactions. If we assume that all three-factor and higher interactions are negligible, the aliases of the seven main effects are

$$\ell_A = A + BD + CE + FG$$
$$\ell_B = B + AD + CF + EG$$
$$\ell_C = C + AE + BF + DG$$
$$\ell_D = D + AB + CG + EF$$
$$\ell_E = E + AC + BG + DF$$
$$\ell_F = F + BC + AG + DE$$
$$\ell_G = G + CD + BE + AF$$

This 2_{III}^{7-4} design is called a **saturated fractional factorial,** because all the available degrees of freedom are used to estimate main effects. It is possible to combine sequences of these resolution III fractional factorials to separate the main effects from the two-factor interactions. The procedure is illustrated in Montgomery (2005) and in Box, Hunter, and Hunter (1978).

EXERCISES FOR SECTION 14-7

14-35. Consider the problem in Exercise 14-17. Suppose that only half of the 32 runs could be made.

(a) Choose the half that you think should be run.
(b) Write out the alias relationships for your design.
(c) Estimate the factor effects.
(d) Plot the effect estimates on normal probability paper and interpret the results.
(e) Set up an analysis of variance for the factors identified as potentially interesting from the normal probability plot in part (d).
(f) Analyze the residuals from the model.
(g) Provide a practical interpretation of the results.

14-36. Suppose that in Exercise 14-20 it was possible to run only a $\frac{1}{2}$ fraction of the 2^4 design. Construct the design and use only the data from the eight runs you have generated to perform the analysis.

14-37. An article by L. B. Hare ("In the Soup: A Case Study to Identify Contributors to Filling Variability," *Journal of Quality Technology*, Vol. 20, pp. 36–43) describes a factorial experiment used to study filling variability of dry soup mix packages. The factors are A = number of mixing ports through which the vegetable oil was added (1, 2), B = temperature surrounding the mixer (cooled, ambient), C = mixing time (60, 80 sec), D = batch weight (1500, 2000 lb), and E = number of days of delay between mixing and packaging (1, 7). Between 125 and 150 packages of soup were sampled over an eight-hour period for each run in the design, and the standard deviation of package weight was used as the response variable. The design and resulting data follow.

Std Order	A Mixer Ports	B Temp	C Time	D Batch Weight	E Delay	y Std Dev
1	−	−	−	−	−	1.13
2	+	−	−	−	+	1.25
3	−	+	−	−	+	0.97
4	+	+	−	−	−	1.7
5	−	−	+	−	+	1.47
6	+	−	+	−	−	1.28
7	−	+	+	−	−	1.18
8	+	+	+	−	+	0.98
9	−	−	−	+	+	0.78
10	+	−	−	+	−	1.36
11	−	+	−	+	−	1.85
12	+	+	−	+	+	0.62
13	−	−	+	+	−	1.09
14	+	−	+	+	+	1.1
15	−	+	+	+	+	0.76
16	+	+	+	+	−	2.1

(a) What is the generator for this design?
(b) What is the resolution of this design?
(c) Estimate the factor effects. Which effects are large?
(d) Does a residual analysis indicate any problems with the underlying assumptions?
(e) Draw conclusions about this filling process.

14-38. Montgomery (2005) describes a 2^{4-1} fractional factorial design used to study four factors in a chemical process. The factors are A = temperature, B = pressure, C = concentration, and D = stirring rate, and the response is filtration rate. The design and the data are as follows:

Run	A	B	C	$D = ABC$	Treatment Combination	Filtration Rate
1	−	−	−	−	(1)	45
2	+	−	−	+	*ad*	100
3	−	+	−	+	*bd*	45
4	+	+	−	−	*ab*	65
5	−	−	+	+	*cd*	75
6	+	−	+	−	*ac*	60
7	−	+	+	−	*bc*	80
8	+	+	+	+	*abcd*	96

(a) Write down the alias relationships.
(b) Estimate the factor effects. Which factor effects appear large?
(c) Project this design into a full factorial in the three apparently important factors and provide a practical interpretation of the results.

14-39. R. D. Snee ("Experimenting with a Large Number of Variables," in *Experiments in Industry: Design, Analysis and Interpretation of Results*, by R. D. Snee, L. B. Hare, and J. B. Trout, eds., ASQC, 1985) describes an experiment in which a 2^{5-1} design with $I = ABCDE$ was used to investigate the effects of five factors on the color of a chemical product. The factors are A = solvent/reactant, B = catalyst/reactant, C = temperature, D = reactant purity, and E = reactant pH. The results obtained are as follows:

$$e = -0.63 \quad d = 6.79$$
$$a = 2.51 \quad ade = 6.47$$
$$b = -2.68 \quad bde = 3.45$$
$$abe = 1.66 \quad abd = 5.68$$
$$c = 2.06 \quad cde = 5.22$$
$$ace = 1.22 \quad acd = 4.38$$
$$bce = -2.09 \quad bcd = 4.30$$
$$abc = 1.93 \quad abcde = 4.05$$

(a) Prepare a normal probability plot of the effects. Which factors are active?

(b) Calculate the residuals. Construct a normal probability plot of the residuals and plot the residuals versus the fitted values. Comment on the plots.
(c) If any factors are negligible, collapse the 2^{5-1} design into a full factorial in the active factors. Comment on the resulting design, and interpret the results.

14-40. An article in *Quality Engineering,* "A comparison of multi-response optimization: sensitivity to parameter selection" (1999, Vol. 11, pp. 405–415) conducted a half replicate of a 2^5 factorial design to optimize the retort process of beef stew MREs, a military ration. The design factors are x_1 = Sauce Viscosity, x_2 = Residual Gas, x_3 = Solid/Liquid Ratio, x_4 = Net Weight, x_5 = Rotation Speed. The response variable is the heating rate index, a measure of heat penetration, and there are two replicates.

Run	x_1	x_2	x_3	x_4	x_5	Heating Rate Index I	Heating Rate Index II
1	−1	−1	−1	−1	1	8.46	9.61
2	1	−1	−1	−1	−1	15.68	14.68
3	−1	1	−1	−1	−1	14.94	13.09
4	1	1	−1	−1	1	12.52	12.71
5	−1	−1	1	−1	−1	17	16.36
6	1	−1	1	−1	1	11.44	11.83
7	−1	1	1	−1	1	10.45	9.22
8	1	1	1	−1	−1	19.73	16.94
9	−1	−1	−1	1	−1	17.37	16.36
10	1	−1	−1	1	1	14.98	11.93
11	−1	1	−1	1	1	8.4	8.16
12	1	1	−1	1	−1	19.08	15.40
13	−1	−1	1	1	1	13.07	10.55
14	1	−1	1	1	−1	18.57	20.53
15	−1	1	1	1	−1	20.59	21.19
16	1	1	1	1	1	14.03	11.31

(a) Estimate the factor effects. Based on a normal probability plot of the effect estimates, identify a model for the data from this experiment.
(b) Conduct an ANOVA based on the model identified in part (a). What are your conclusions?
(c) Analyze the residuals and comment on model adequacy.
(d) Find a regression model to predict yield in terms of the coded factor levels.
(e) This experiment was replicated, so an ANOVA could have been conducted without using a normal plot of the effects to tentatively identify a model. What model would be appropriate? Use the ANOVA to analyze this model and compare the results with those obtained from the normal probability plot approach.

14-41. An article in *Industrial and Engineering Chemistry* ("More on Planning Experiments to Increase Research Efficiency," 1970, pp. 60–65) uses a 2^{5-2} design to investigate the effect on process yield of A = condensation temperature, B = amount of material 1, C = solvent volume, D = condensation time, and E = amount of material 2. The results obtained are as follows:

$$e = 23.2 \quad cd = 23.8$$
$$ab = 15.5 \quad ace = 23.4$$
$$ad = 16.9 \quad bde = 16.8$$
$$bc = 16.2 \quad abcde = 18.1$$

(a) Verify that the design generators used were $I = ACE$ and $I = BDE$.
(b) Write down the complete defining relation and the aliases from the design.
(c) Estimate the main effects.
(d) Prepare an analysis of variance table. Verify that the AB and AD interactions are available to use as error.
(e) Plot the residuals versus the fitted values. Also construct a normal probability plot of the residuals. Comment on the results.

14-42. Suppose that in Exercise 14-14 only a ¼ fraction of the 2^5 design could be run. Construct the design and analyze the data that are obtained by selecting only the response for the eight runs in your design.

14-43. For each of the following designs write down the aliases, assuming that only main effects and two factor interactions are of interest.
(a) 2^{6-3}_{III} (b) 2^{8-4}_{IV}

14-44. Consider the 2^{6-2} design in Table 14-29.
(a) Suppose that after analyzing the original data, we find that factors C and E can be dropped. What type of 2^k design is left in the remaining variables?
(b) Suppose that after the original data analysis, we find that factors D and F can be dropped. What type of 2^k design is left in the remaining variables? Compare the results with part (a). Can you explain why the answers are different?

14-8 RESPONSE SURFACE METHODS AND DESIGNS

Response surface methodology, or RSM, is a collection of mathematical and statistical techniques that are useful for modeling and analysis in applications where a response of interest is influenced by several variables and the objective is to **optimize this response.** For example, suppose that a chemical engineer wishes to find the levels of temperature (x_1) and feed

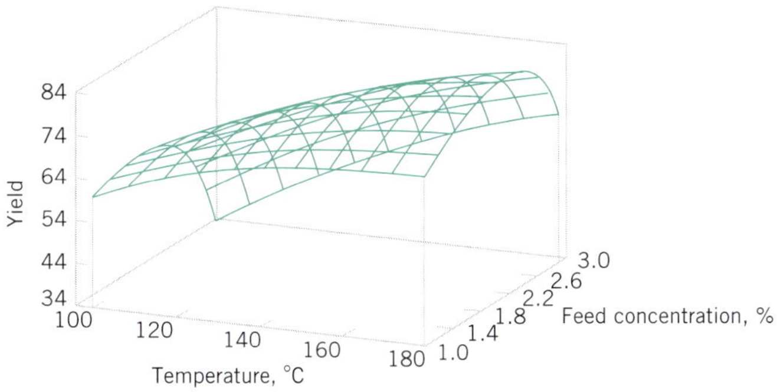

Figure 14-42 A three-dimensional response surface showing the expected yield as a function of temperature and feed concentration.

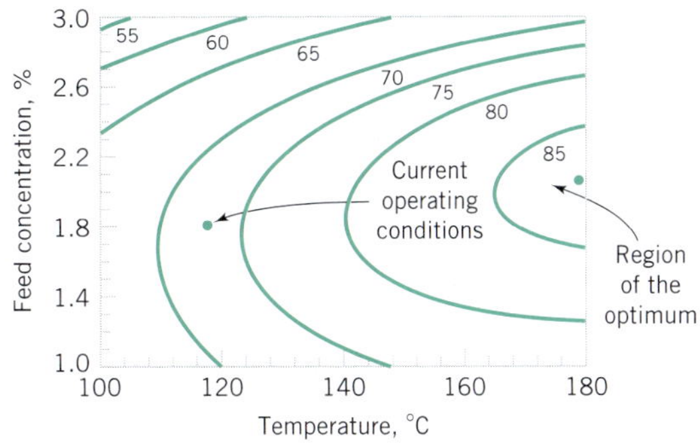

Figure 14-43 A contour plot of the yield response surface in Figure 14-42.

concentration (x_2) that maximize the yield (y) of a process. The process yield is a function of the levels of temperature and feed concentration, say,

$$Y = f(x_1, x_2) + \epsilon$$

where ϵ represents the noise or error observed in the response Y. If we denote the expected response by $E(Y) = f(x_1, x_2) = \eta$, then the surface represented by

$$\eta = f(x_1, x_2)$$

is called a **response surface.**

We may represent the response surface graphically as shown in Fig. 14-42, where η is plotted versus the levels of x_1 and x_2. Notice that the response is represented as a surface plot in a three-dimensional space. To help visualize the shape of a response surface, we often plot the contours of the response surface as shown in Fig. 14-43. In the contour plot, lines of constant response are drawn in the x_1, x_2 plane. Each contour corresponds to a particular height of the response surface. The contour plot is helpful in studying the levels of x_1 and x_2 that result in changes in the shape or height of the response surface.

In most RSM problems, the form of the relationship between the response and the independent variables is unknown. Thus, the first step in RSM is to find a suitable approximation for the true relationship between Y and the independent variables. Usually, a low-order polynomial in some region of the independent variables is employed. If the response is well modeled by a linear function of the independent variables, the approximating function is the **first-order model**

$$Y = \beta_0 + \beta_1 x_1 + \beta_2 x_2 + \cdots + \beta_k x_k + \epsilon \tag{14-26}$$

If there is curvature in the system, then a polynomial of higher degree must be used, such as the **second-order model**

$$Y = \beta_0 + \sum_{i=1}^{k} \beta_i x_i + \sum_{i=1}^{k} \beta_{ii} x_i^2 + \sum_{i<j}\sum \beta_{ij} x_i x_j + \epsilon \tag{14-27}$$

Many RSM problems use one or both of these approximating polynomials. Of course, it is unlikely that a polynomial model will be a reasonable approximation of the true functional relationship over the entire space of the independent variables, but for a relatively small region they usually work quite well.

The method of least squares, discussed in Chapters 11 and 12, is used to estimate the parameters in the approximating polynomials. The response surface analysis is then done in terms of the fitted surface. If the fitted surface is an adequate approximation of the true response function, analysis of the fitted surface will be approximately equivalent to analysis of the actual system.

RSM is a **sequential** procedure. Often, when we are at a point on the response surface that is remote from the optimum, such as the current operating conditions in Fig. 14-43, there is little curvature in the system and the first-order model will be appropriate. Our objective here is to lead the experimenter rapidly and efficiently to the general vicinity of the optimum. Once the region of the optimum has been found, a more elaborate model such as the second-order model may be employed, and an analysis may be performed to locate the optimum. From Fig. 14-43, we see that the analysis of a response surface can be thought of as "climbing a hill," where the top of the hill represents the point of maximum response. If the true optimum is a point of minimum response, we may think of "descending into a valley."

The eventual objective of RSM is to determine the optimum operating conditions for the system or to determine a region of the factor space in which operating specifications are satisfied. Also, note that the word "optimum" in RSM is used in a special sense. The "hill climbing" procedures of RSM guarantee convergence to a local optimum only.

Method of Steepest Ascent

Frequently, the initial estimate of the optimum operating conditions for the system will be far from the actual optimum. In such circumstances, the objective of the experimenter is to move rapidly to the general vicinity of the optimum. We wish to use a simple and economically efficient experimental procedure. When we are remote from the optimum, we usually assume that a first-order model is an adequate approximation to the true surface in a small region of the x's.

The method of **steepest ascent** is a procedure for moving sequentially along the path of steepest ascent, that is, in the direction of the maximum increase in the response. Of course, if **minimization** is desired, we are talking about the method of **steepest descent.** The fitted first-order model is

$$\hat{y} = \hat{\beta}_0 + \sum_{i=1}^{k} \hat{\beta}_i x_i \tag{14-28}$$

and the first-order response surface, that is, the contours of $\hat{y}$, is a series of parallel lines such as that shown in Fig. 14-44. The direction of steepest ascent is the direction in which

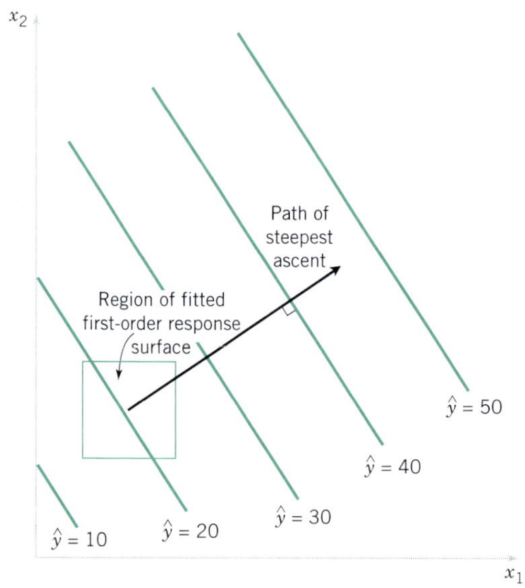

Figure 14-44 First-order response surface and path of steepest ascent.

$\hat{y}$ increases most rapidly. This direction is normal to the fitted response surface contours. We usually take as the path of **steepest ascent** the line through the center of the region of interest and normal to the fitted surface contours. Thus, the steps along the path are proportional to the regression coefficients $\{\hat{\beta}_i\}$. The experimenter determines the actual step size based on process knowledge or other practical considerations.

Experiments are conducted along the path of steepest ascent until no further increase in response is observed. Then a new first-order model may be fit, a new direction of steepest ascent determined, and further experiments conducted in that direction until the experimenter feels that the process is near the optimum.

EXAMPLE 14-11
Process Yield Steepest Ascent

In Example 14-6 we described an experiment on a chemical process in which two factors, reaction time (x_1) and reaction temperature (x_2), affect the percent conversion or yield (Y). Figure 14-27 shows the 2^2 design plus five center points used in this study. The engineer found that both factors were important, there was no interaction, and there was no curvature in the response surface. Therefore, the first-order model

$$Y = \beta_0 + \beta_1 x_1 + \beta_2 x_2 + \epsilon$$

should be appropriate. Now the effect estimate of time is 1.55 hours and the effect estimate of temperature is 0.65°F, and since the regression coefficients $\hat{\beta}_1$ and $\hat{\beta}_2$ are one-half of the corresponding effect estimates, the fitted first-order model is

$$\hat{y} = 40.44 + 0.775x_1 + 0.325x_2$$

Figure 14-45(a) and (b) show the contour plot and three-dimensional surface plot of this model. Figure 14-45 also shows the relationship between the **coded variables** x_1 and x_2 (that defined the high and low levels of the factors) and the original variables, time (in minutes) and temperature (in °F).

From examining these plots (or the fitted model), we see that to move away from the design center—the point ($x_1 = 0, x_2 = 0$)—along the path of steepest ascent, we would move 0.775 unit in the x_1 direction for every 0.325 unit in the x_2 direction. Thus, the path of steepest ascent passes through the point ($x_1 = 0, x_2 = 0$) and has a slope 0.325/0.775. The engineer decides to use 5 minutes of reaction

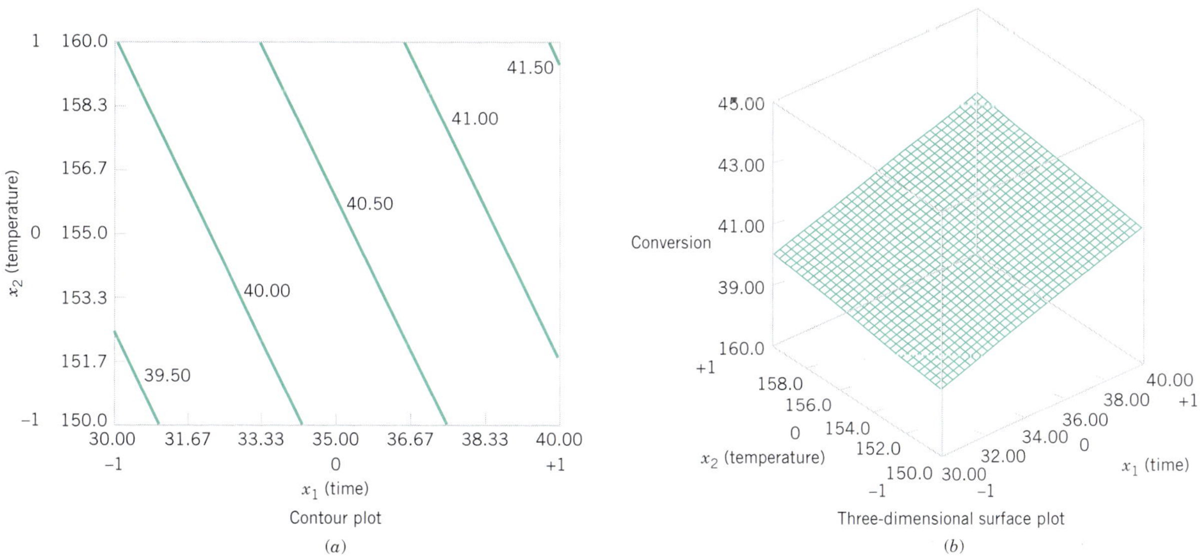

Figure 14-45 Response surface plots for the first-order model in Example S14-3.

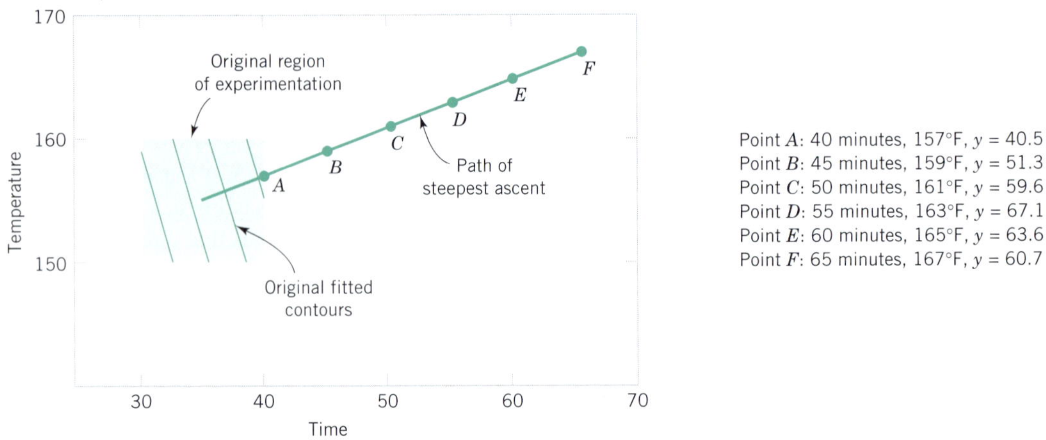

Figure 14-46 Steepest ascent experiment for Example 14-11.

time as the basic step size. Now, 5 minutes of reaction time is equivalent to a step in the *coded* variable x_1 of $\Delta x_1 = 1$. Therefore, the steps along the path of steepest ascent are $\Delta x_1 = 1.0000$ and $\Delta x_2 = (0.325/0.775)\Delta x_1 = 0.42$. A change of $\Delta x_2 = 0.42$ in the coded variable x_2 is equivalent to about 2°F in the original variable temperature. Therefore, the engineer will move along the path of steepest ascent by increasing reaction time by 5 minutes and temperature by 2°F. An actual observation on yield will be determined at each point.

Figure 14-46 shows several points along this path of steepest ascent and the yields actually observed from the process at those points. At points *A–D* the observed yield increases steadily, but beyond point *D*, the yield decreases. Therefore, steepest ascent would terminate in the vicinity of 55 minutes of reaction time and 163°F with an observed percent conversion of 67%.

Analysis of a Second-Order Response Surface

When the experimenter is relatively close to the optimum, a second-order model is usually required to approximate the response because of curvature in the true response surface. The fitted second-order model is

$$\hat{y} = \hat{\beta}_0 + \sum_{i=1}^{k} \hat{\beta}_i x_i + \sum_{i=1}^{k} \hat{\beta}_{ii} x_i^2 + \sum_{i<j}\sum \hat{\beta}_{ij} x_i x_j$$

where $\hat{\beta}$ denotes the least squares estimate of β. In this section we show how to use this fitted model to find the optimum set of operating conditions for the x's and to characterize the nature of the response surface.

EXAMPLE 14-12 Process Yield Central Composite Design

Continuation of Example 14-11

Consider the chemical process from Example 14-11, where the method of steepest ascent terminated at a reaction time of 55 minutes and a temperature of 163°F. The experimenter decides to fit a sccond-order model in this region. Table 14-34 and Fig. 14-47 show the experimental design, which consists of a 2^2 design centered at 55 minutes and 165°F, five center points, and four runs along the coordinate axes called axial runs. This type of design is called a **central composite design,** and it is a very popular design for fitting second-order response surfaces.

Two response variables were measured during this phase of the experiment: percentage conversion (yield) and viscosity. The least-squares quadratic model for the yield response is

$$\hat{y}_1 = 69.1 + 1.633x_1 + 1.083x_2 - 0.969x_1^2 - 1.219x_2^2 + 0.225x_1x_2$$

Table 14-34 Central Composite Design for Example 14-12

Observation Number	Time (minutes)	Temperature (°F)	Coded Variables x_1	x_2	Conversion (percent) Response 1	Viscosity (mPa-sec) Response 2
1	50	160	−1	−1	65.3	35
2	60	160	1	−1	68.2	39
3	50	170	−1	1	66	36
4	60	170	1	1	69.8	43
5	48	165	−1.414	0	64.5	30
6	62	165	1.414	0	69	44
7	55	158	0	−1.414	64	31
8	55	172	0	1.414	68.5	45
9	55	165	0	0	68.9	37
10	55	165	0	0	69.7	34
11	55	165	0	0	68.5	35
12	55	165	0	0	69.4	36
13	55	165	0	0	69	37

The analysis of variance for this model is shown in Table 14-35.

Figure 14-48 shows the response surface contour plot and the three-dimensional surface plot for this model. From examination of these plots, the maximum yield is about 70%, obtained at approximately 60 minutes of reaction time and 167°F.

The viscosity response is adequately described by the first-order model

$$\hat{y}_2 = 37.08 + 3.85x_1 + 3.10x_2$$

Table 14-36 summarizes the analysis of variance for this model. The response surface is shown graphically in Fig. 14-49. Notice that viscosity increases as both time and temperature increase.

As in most response surface problems, the experimenter in this example had conflicting objectives regarding the two responses. The objective was to maximize yield, but the acceptable range for viscosity was $38 \leq y_2 \leq 42$. When there are only a few independent variables, an easy way to solve this problem

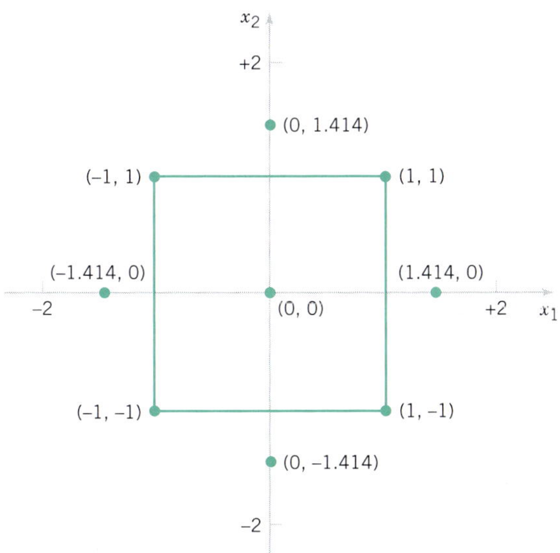

Figure 14-47 Central composite design for Example 14-12.

Table 14-35 Analysis of Variance for the Quadratic Model, Yield Response

Source of Variation	Sum of Squares	Degrees of Freedom	Mean Square	f_0	*P*-Value
Model	45.89	5	9.178	14.93	0.0013
Residual	4.30	7	0.615		
Total	50.19	12			

Independent Variable	Coefficient Estimate	Standard Error	*t* for H_0 Coefficient = 0	*P*-Value
Intercept	69.100	0.351	197.1	
x_1	1.633	0.277	5.891	0.0006
x_2	1.083	0.277	3.907	0.0058
x_1^2	−0.969	0.297	−3.259	0.0139
x_2^2	−1.219	0.297	−4.100	0.0046
x_1x_2	0.225	0.392	0.5740	0.5839

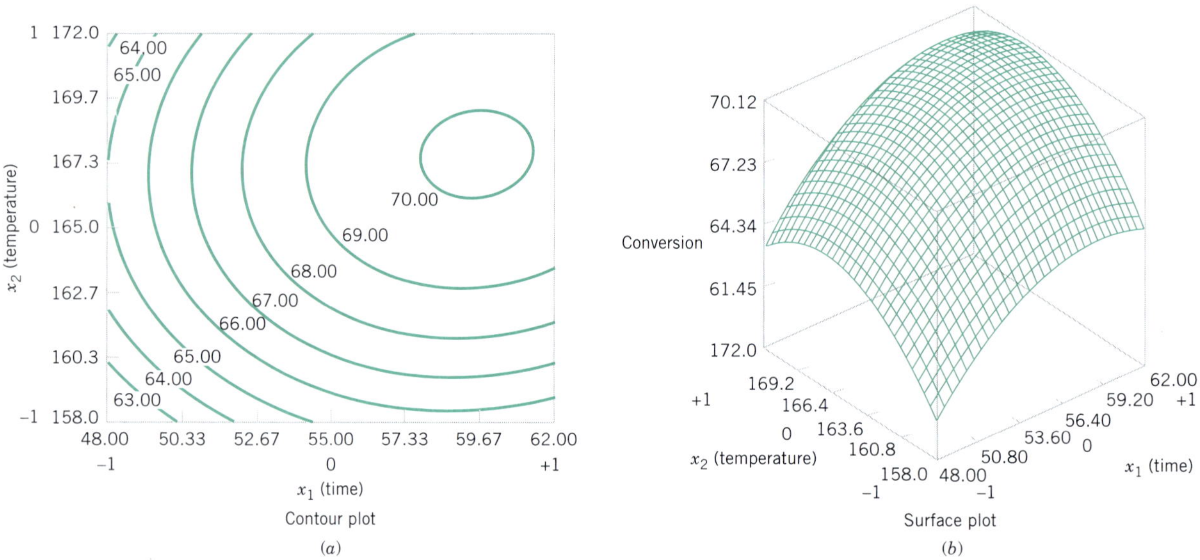

Figure 14-48 Response surface plots for the yield response, Example 14-12.

Table 14-36 Analysis of Variance for the First-Order Model, Viscosity Response

Source of Variation	Sum of Squares	Degrees of Freedom	Mean Square	f_0	*P*-Value
Model	195.4	2	97.72	15.89	0.0008
Residual	61.5	10	6.15		
Total	256.9	12			

Independent Variable	Coefficient Estimate	Degrees of Freedom	Standard Error	*t* for H_0 Coefficient = 0	*P*-Value
Intercept	37.08	1	0.69	53.91	
x_1	3.85	1	0.88	4.391	0.0014
x_2	3.10	1	0.88	3.536	0.0054

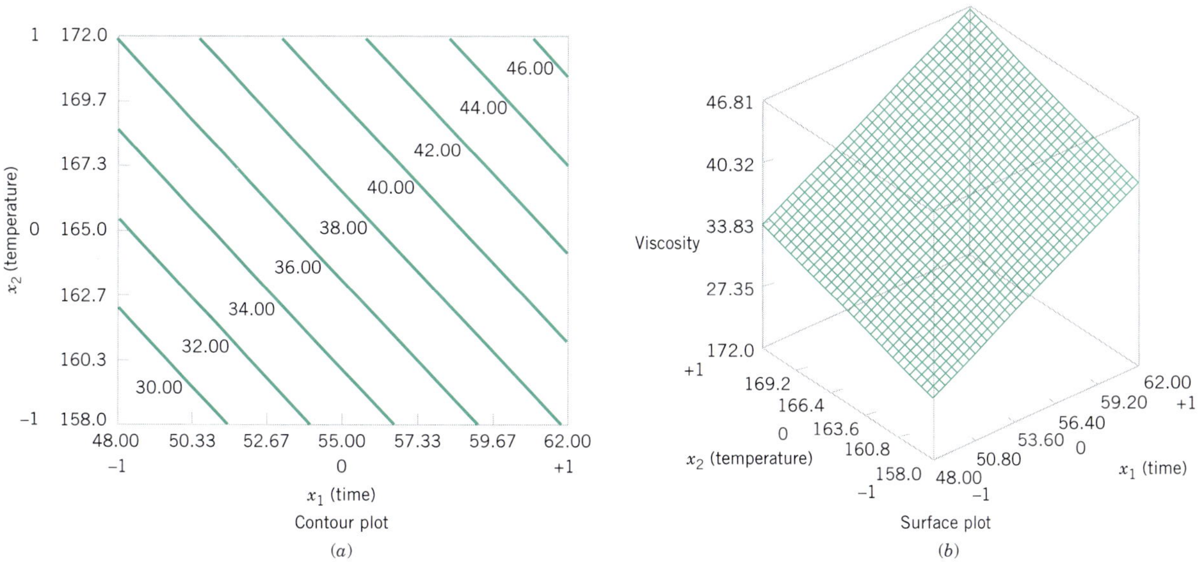

Figure 14-49 Response surface plots for the viscosity response, Example 14-12.

is to overlay the response surfaces to find the optimum. Figure 14-50 shows the overlay plot of both responses, with the contours $y_1 = 69\%$ conversion, $y_2 = 38$, and $y_2 = 42$ highlighted. The shaded areas on this plot identify unfeasible combinations of time and temperature. This graph shows that several combinations of time and temperature will be satisfactory.

Example 14-12 illustrates the use of a **central composite design** (CCD) for fitting a second-order response surface model. These designs are widely used in practice because they are relatively efficient with respect to the number of runs required. In general, a CCD in k factors requires 2^k factorial runs, $2k$ axial runs, and at least one center point (three to five center points are typically used). Designs for $k = 2$ and $k = 3$ factors are shown in Fig. 14-51.

The central composite design may be made **rotatable** by proper choice of the axial spacing α in Fig. 14-51. If the design is rotatable, the standard deviation of predicted response

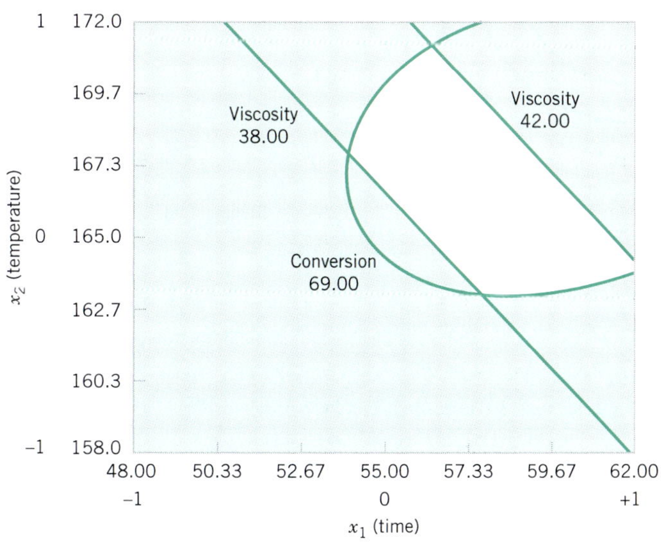

Figure 14-50 Overlay of yield and viscosity response surfaces, Example 14-12.

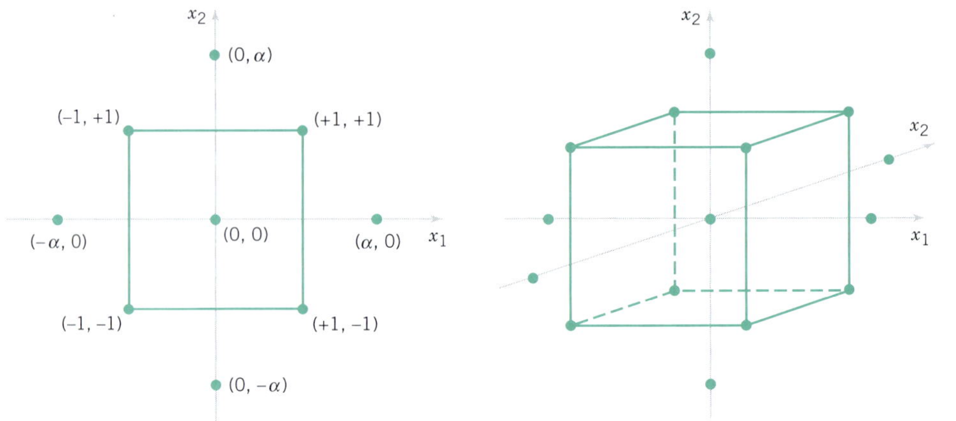

Figure 14-51 Central composite designs for $k = 2$ and $k = 3$.

$\hat{y}$ is constant at all points that are the same distance from the center of the design. For rotatability, choose $\alpha = (F)^{1/4}$, where F is the number of points in the factorial part of the design (usually $F = 2^k$). For the case of $k = 2$ factors, $\alpha = (2^2)^{1/4} = 1.414$, as was used in the design in Example 14-12. Figure 14-52 presents a contour plot and a surface plot of the standard deviation of prediction for the quadratic model used for the yield response. Notice that the contours are concentric circles, implying that yield is predicted with equal precision for all points that are the same distance from the center of the design. Also, as one would expect, the precision decreases with increasing distance from the design center.

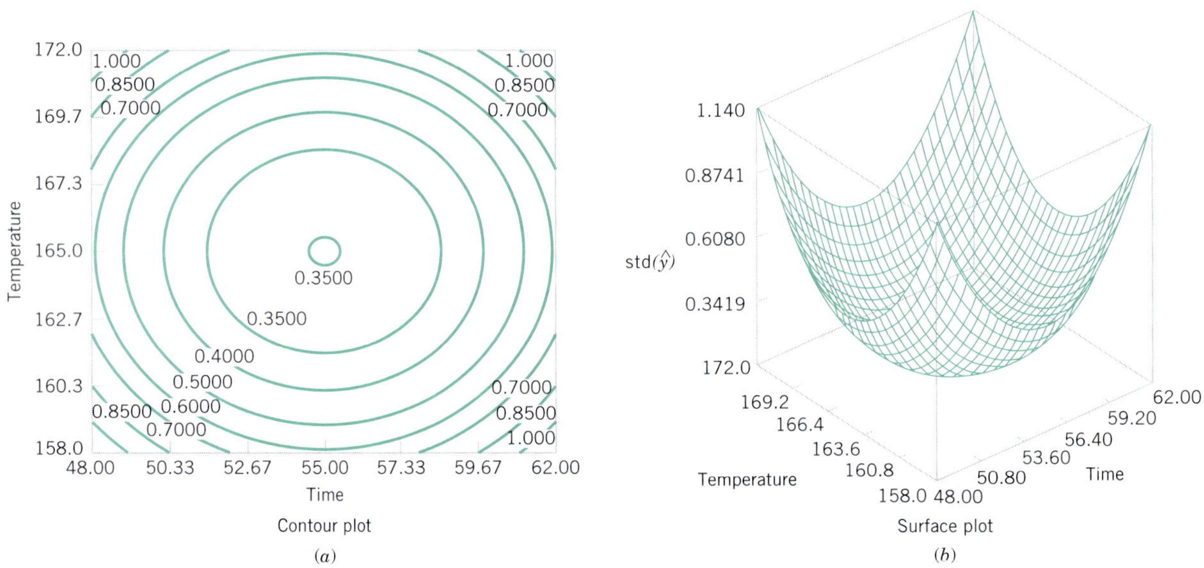

Figure 14-52 Plots of constant $\sqrt{V(\hat{y})}$ for a rotatable central composite design.

EXERCISES FOR SECTION 14-8

14-45. An article in *Rubber Age* (Vol. 89, 1961, pp. 453–458) describes an experiment on the manufacture of a product in which two factors were varied. The factors are reaction time (hr) and temperature (°C). These factors are coded as x_1 = (time − 12)/8 and x_2 = (temperature − 250)/30. The following data were observed where y is the yield (in percent):

Run Number	x_1	x_2	y
1	−1	0	83.8
2	1	0	81.7
3	0	0	82.4
4	0	0	82.9
5	0	−1	84.7
6	0	1	75.9
7	0	0	81.2
8	−1.414	−1.414	81.3
9	−1.414	1.414	83.1
10	1.414	−1.414	85.3
11	1.414	1.414	72.7
12	0	0	82.0

(a) Plot the points at which the experimental runs were made.
(b) Fit a second-order model to the data. Is the second-order model adequate?
(c) Plot the yield response surface. What recommendations would you make about the operating conditions for this process?

14-46. An article in *Quality Engineering*, "Mean and variance modeling with qualitative responses: a case study" (1998–1999, Vol. 11, pp. 141–148) studied how three active ingredients of a particular food affect the overall taste of the product. The measure of the overall taste is the overall mean liking score (MLS). The three ingredients are identified by the variables x_1, x_2, and x_3. The data are shown in the following table.

Run	x_1	x_2	x_3	MLS
1	1	1	−1	6.3261
2	1	1	1	6.2444
3	0	0	0	6.5909
4	0	−1	0	6.3409
5	1	−1	1	5.907
6	1	−1	−1	6.488
7	0	0	−1	5.9773
8	0	1	0	6.8605
9	−1	−1	1	6.0455
10	0	0	1	6.3478
11	1	0	0	6.7609
12	−1	−1	−1	5.7727
13	−1	1	−1	6.1805
14	−1	1	1	6.4894
15	−1	0	0	6.8182

(a) Fit a second-order response surface model to the data.
(b) Construct contour plots and response surface plots for MLS. What are your conclusions?
(c) Analyze the residuals from this experiment. Does your analysis indicate any potential problems?
(d) This design has only a single center point. Is this a good design in your opinion?

14-47. Consider the first-order model

$$\hat{y} = 50 + 1.5x_1 - 0.8x_2$$

where $-1 \le x_i \le 1$. Find the direction of steepest ascent.

14-48. A manufacturer of cutting tools has developed two empirical equations for tool life (y_1) and tool cost (y_2). Both models are functions of tool hardness (x_1) and manufacturing time (x_2). The equations are

$$\hat{y}_1 = 10 + 5x_1 + 2x_2$$
$$\hat{y}_2 = 23 + 3x_1 + 4x_2$$

and both equations are valid over the range $-1.5 \le x_i \le 1.5$. Suppose that tool life must exceed 12 hours and cost must be below \$27.50.
(a) Is there a feasible set of operating conditions?
(b) Where would you run this process?

14-49. An article in *Tappi* (1960, Vol. 43, pp. 38–44) describes an experiment that investigated the ash value of paper pulp (a measure of inorganic impurities). Two variables, temperature T in degrees Celsius and time t in hours, were studied, and some of the results are shown in the following table. The coded predictor variables shown are

$$x_1 = \frac{(T - 775)}{115}, \qquad x_2 = \frac{(t - 3)}{1.5}$$

and the response y is (dry ash value in %) $\times 10^3$.

x_1	x_2	y	x_1	x_2	y
−1	1	211	0	−1.5	168
1	−1	92	0	1.5	179
−1	1	216	0	0	122
1	1	99	0	0	175
−1.5	0	222	0	0	157
1.5	0	48	0	0	146

(a) What type of design has been used in this study? Is the design rotatable?
(b) Fit a quadratic model to the data. Is this model satisfactory?
(c) If it is important to minimize the ash value, where would you run the process?

14-50. In their book *Empirical Model Building and Response Surfaces* (John Wiley, 1987), G. E. P. Box and N. R. Draper describe an experiment with three factors. The data shown in the following table are a variation of the original experiment on page 247 of their book. Suppose that these data were collected in a semiconductor manufacturing process.

(a) The response y_1 is the average of three readings on resistivity for a single wafer. Fit a quadratic model to this response.
(b) The response y_2 is the standard deviation of the three resistivity measurements. Fit a linear model to this response.
(c) Where would you recommend that we set x_1, x_2, and x_3 if the objective is to hold mean resistivity at 500 and minimize the standard deviation?

x_1	x_2	x_3	y_1	y_2	x_1	x_2	x_3	y_1	y_2
−1	−1	−1	24.00	12.49	1	0	0	501.67	92.50
0	−1	−1	120.33	8.39	−1	1	0	264.00	63.50
1	−1	−1	213.67	42.83	0	1	0	427.00	88.61
−1	0	−1	86.00	3.46	1	1	0	730.67	21.08
0	0	−1	136.63	80.41	−1	−1	1	220.67	133.82
1	0	−1	340.67	16.17	0	−1	1	239.67	23.46
−1	1	−1	112.33	27.57	1	−1	1	422.00	18.52
0	1	−1	256.33	4.62	−1	0	1	199.00	29.44
1	1	−1	271.67	23.63	0	0	1	485.33	44.67
−1	−1	0	81.00	0.00	1	0	1	673.67	158.21
0	−1	0	101.67	17.67	−1	1	1	176.67	55.51
1	−1	0	357.00	32.91	0	1	1	501.00	138.94
−1	0	0	171.33	15.01	1	1	1	1010.00	142.45
0	0	0	372.00	0.00					

Supplemental Exercises

14-51. An article in *Process Engineering* (1992, No. 71, pp. 46–47) presents a two-factor factorial experiment used to investigate the effect of pH and catalyst concentration on product viscosity (cSt). The data are as follows:

	Catalyst Concentration	
pH	2.5	2.7
5.6	192, 199, 189, 198	178, 186, 179, 188
5.9	185, 193, 185, 192	197, 196, 204, 204

(a) Test for main effects and interactions using $\alpha = 0.05$. What are your conclusions?
(b) Graph the interaction and discuss the information provided by this plot.
(c) Analyze the residuals from this experiment.

14-52. Heat treating of metal parts is a widely used manufacturing process. An article in the *Journal of Metals* (Vol. 41, 1989) describes an experiment to investigate flatness distortion from heat treating for three types of gears and two heat-treating times. The data are as follows:

	Time (minutes)	
Gear Type	90	120
20-tooth	0.0265	0.0560
	0.0340	0.0650
24-tooth	0.0430	0.0720
	0.0510	0.0880
28-tooth	0.0405	0.0620
	0.0575	0.0825

(a) Is there any evidence that flatness distortion is different for the different gear types? Is there any indication that heat treating time affects the flatness distortion? Do these factors interact? Use $\alpha = 0.05$.
(b) Construct graphs of the factor effects that aid in drawing conclusions from this experiment.
(c) Analyze the residuals from this experiment. Comment on the validity of the underlying assumptions.

14-53. An article in the *Textile Research Institute Journal* (1984, Vol. 54, pp. 171–179) reported the results of an experiment that studied the effects of treating fabric with selected inorganic salts on the flammability of the material. Two application levels of each salt were used, and a vertical burn test was used on each sample. (This finds the temperature at which each sample ignites.) The burn test data follow.

Level	Salt					
	Untreated	$MgCl_2$	NaCl	$CaCO_3$	$CaCl_2$	Na_2CO_3
1	812	752	739	733	725	751
	827	728	731	728	727	761
	876	764	726	720	719	755
2	945	794	741	786	756	910
	881	760	744	771	781	854
	919	757	727	779	814	848

(a) Test for differences between salts, application levels, and interactions. Use $\alpha = 0.01$.
(b) Draw a graph of the interaction between salt and application level. What conclusions can you draw from this graph?
(c) Analyze the residuals from this experiment.

14-54. An article in the *IEEE Transactions on Components, Hybrids, and Manufacturing Technology* (1992, Vol. 15) describes an experiment for investigating a method for aligning optical chips onto circuit boards. The method involves placing solder bumps onto the bottom of the chip. The experiment used three solder bump sizes and three alignment methods. The response variable is alignment accuracy (in micrometers). The data are as follows:

Solder Bump Size (diameter in μm)	Alignment Method		
	1	2	3
75	4.60	1.55	1.05
	4.53	1.45	1.00
130	2.33	1.72	0.82
	2.44	1.76	0.95
260	4.95	2.73	2.36
	4.55	2.60	2.46

(a) Is there any indication that either solder bump size or alignment method affects the alignment accuracy? Is there any evidence of interaction between these factors? Use $\alpha = 0.05$.
(b) What recommendations would you make about this process?
(c) Analyze the residuals from this experiment. Comment on model adequacy.

14-55. An article in *Solid State Technology* (Vol. 29, 1984, pp. 281–284) describes the use of factorial experiments in photolithography, an important step in the process of manufacturing integrated circuits. The variables in this experiment (all at two levels) are prebake temperature (A), prebake time (B), and exposure energy (C), and the response variable is delta line width, the difference between the line on the mask and the printed line on the device. The data are as follows: $(1) = -2.30$, $a = -9.87$, $b = -18.20$, $ab = -30.20$, $c = -23.80$, $ac = -4.30$, $bc = -3.80$, and $abc = -14.70$.
(a) Estimate the factor effects.
(b) Use a normal probability plot of the effect estimates to identity factors that may be important.
(c) What model would you recommend for predicting the delta line width response, based on the results of this experiment?
(d) Analyze the residuals from this experiment, and comment on model adequacy.

14-56. An article in the *Journal of Coatings Technology* (Vol. 60, 1988, pp. 27–32) describes a 2^4 factorial design used for studying a silver automobile basecoat. The response variable is distinctness of image (DOI). The variables used in the experiment are

A = Percentage of polyester by weight of polyester/melamine (low value = 50%, high value = 70%)
B = Percentage of cellulose acetate butyrate carboxylate (low value = 15%, high value = 30%)
C = Percentage of aluminum stearate (low value = 1%, high value = 3%)
D = Percentage of acid catalyst (low value = 0.25%, high value = 0.50%)

The responses are $(1) = 63.8$, $a = 77.6$, $b = 68.8$, $ab = 76.5$, $c = 72.5$, $ac = 77.2$, $bc = 77.7$, $abc = 84.5$, $d = 60.6$, $ad = 64.9$, $bd = 72.7$, $abd = 73.3$, $cd = 68.0$, $acd = 76.3$, $bcd = 76.0$, and $abcd = 75.9$.
(a) Estimate the factor effects.
(b) From a normal probability plot of the effects, identify a tentative model for the data from this experiment.
(c) Using the apparently negligible factors as an estimate of error, test for significance of the factors identified in part (b). Use $\alpha = 0.05$.
(d) What model would you use to describe the process, based on this experiment? Interpret the model.
(e) Analyze the residuals from the model in part (d) and comment on your findings.

14-57. An article in the *Journal of Manufacturing Systems* (Vol. 10, 1991, pp. 32–40) describes an experiment to investigate the effect of four factors P = waterjet pressure, F = abrasive flow rate, G = abrasive grain size, and V = jet traverse speed on the surface roughness of a waterjet cutter. A 2^4 design follows.
(a) Estimate the factor effects.
(b) Form a tentative model by examining a normal probability plot of the effects.

(c) Is the model in part (b) a reasonable description of the process? Is lack of fit significant? Use $\alpha = 0.05$.
(d) Interpret the results of this experiment.
(e) Analyze the residuals from this experiment.

Run	Factors V (in/min)	F (lb/min)	P (kpsi)	G (Mesh No.)	Surface Roughness (μm)
1	6	2.0	38	80	104
2	2	2.0	38	80	98
3	6	2.0	30	80	103
4	2	2.0	30	80	96
5	6	1.0	38	80	137
6	2	1.0	38	80	112
7	6	1.0	30	80	143
8	2	1.0	30	80	129
9	6	2.0	38	170	88
10	2	2.0	38	170	70
11	6	2.0	30	170	110
12	2	2.0	30	170	110
13	6	1.0	38	170	102
14	2	1.0	38	170	76
15	6	1.0	30	170	98
16	2	1.0	30	170	68

14-58. Construct a 2_{IV}^{4-1} design for the problem in Exercise 14-56. Select the data for the eight runs that would have been required for this design. Analyze these runs and compare your conclusions to those obtained in Exercise 14-56 for the full factorial.

14-59. Construct a 2_{IV}^{4-1} design for the problem in Exercise 14-57. Select the data for the eight runs that would have been required for this design. Analyze these data and compare your conclusions to those obtained in Exercise 14-57 for the full factorial.

14-60. Construct a 2_{IV}^{8-4} design in 16 runs. What are the alias relationships in this design?

14-61. Construct a 2_{III}^{5-2} design in eight runs. What are the alias relationships in this design?

14-62. An article in the *Journal of Quality Technology* (Vol. 17, 1985, pp. 198–206) describes the use of a replicated fractional factorial to investigate the effect of five factors on the free height of leaf springs used in an automotive application. The factors are A = furnace temperature, B = heating time, C = transfer time, D = hold down time, and E = quench oil temperature. The data are shown in the following table.

(a) What is the generator for this fraction? Write out the alias structure.
(b) Analyze the data. What factors influence mean free height?
(c) Calculate the range of free height for each run. Is there any indication that any of these factors affect variability in free height?
(d) Analyze the residuals from this experiment and comment on your findings.

A	B	C	D	E	Free Height		
−	−	−	−	−	7.78	7.78	7.81
+	−	−	+	−	8.15	8.18	7.88
−	+	−	+	−	7.50	7.56	7.50
+	+	−	−	−	7.59	7.56	7.75
−	−	+	+	−	7.54	8.00	7.88
+	−	+	−	−	7.69	8.09	8.06
−	+	+	−	−	7.56	7.52	7.44
+	+	+	+	−	7.56	7.81	7.69
−	−	−	−	+	7.50	7.56	7.50
+	−	−	+	+	7.88	7.88	7.44
−	+	−	+	+	7.50	7.56	7.50
+	+	−	−	+	7.63	7.75	7.56
−	−	+	+	+	7.32	7.44	7.44
+	−	+	−	+	7.56	7.69	7.62
−	+	+	−	+	7.18	7.18	7.25
+	+	+	+	+	7.81	7.50	7.59

14-63. An article in *Rubber Chemistry and Technology* (Vol. 47, 1974, pp. 825–836) describes an experiment that studies the Mooney viscosity of rubber to several variables, including silica filler (parts per hundred) and oil filler (parts per hundred). Data typical of that reported in this experiment follow, where

$$x_1 = \frac{\text{silica} - 60}{15}, \qquad x_2 = \frac{\text{oil} - 21}{15}$$

Coded levels		
x_1	x_2	y
−1	−1	13.71
1	−1	14.15
−1	1	12.87
1	1	13.53
−1	−1	13.90
1	−1	14.88
−1	1	12.25
−1	1	13.35

(a) What type of experimental design has been used?
(b) Analyze the data and draw appropriate conclusions.

14-64. An article in *Tropical Science,* "Proximate Composition of the Seeds of *Acacia Nilotica* var *Adansonti* (Bagaruwa) and Extraction of Its Protein" (1992, Vol. 32, No. 3, pp. 263–268) reported on research extracting and concentrating the proteins of the bagaruwa seeds in livestock feeding in Nigeria to eliminate the toxic substances from the seeds. The following are the effects of extraction time and flour to solvent ratio on protein extractability of the bagaruwa seeds in distilled water:

Flour: Solvent Ratio (w/v) (%)	Percentage of Protein Extracted at Time (min)			
	30	60	90	120
3	30.5	45.7	30.5	31.0
	36.9	44.3	29.5	22.1
7	32.9	42.4	28.2	23.5
	37.5	40.9	27.3	34.1
11	29.0	39.5	29.0	29.0
	32.7	43.6	30.5	28.4

All values are means of three determinations.

(a) Test the appropriate hypotheses and draw conclusions using the analysis of variance with $\alpha = 0.5$.
(b) Graphically analyze the interaction.
(c) Analyze the residuals from this experiment.

14-65. An article in *Plant Disease,* "Effect of Nitrogen and Potassium Fertilizer Rates on Severity of Xanthomonas Blight of *Syngonium Podophyllum*" (1989, Vol. 73, No. 12, pp. 972–975) showed the effect of the variable nitrogen and potassium rates on the growth of "White Butterfly" and the mean percentage of disease. Data representative of that collected in this experiment is provided in the following table.

Nitrogen (mg/pot/wk)	Potassium (mg/pot/wk)					
	30		90		120	
50	61.0	61.3	45.5	42.5	59.5	58.2
150	54.5	55.9	53.5	51.9	34.0	35.9
250	42.7	40.4	36.5	37.4	32.5	33.8

(a) State the appropriate hypotheses.
(b) Use the analysis of variance to test these hypotheses with $\alpha = 0.05$.
(c) Graphically analyze the residuals from this experiment.
(d) Estimate the appropriate variance components.

14-66. An article in *Biotechnology Progress* (2001, Vol. 17, pp. 366–368) reported on an experiment to investigate and optimize the operating conditions of the nisin extraction in aqueous two-phase systems (ATPS). A 2^2 full factorial design with center points was used to verify the most significant factors affecting the nisin recovery. The factor x_1 was the concentration (% w/w) of PEG 4000 and x_2 was the concentration (% w/w) of Na_2SO_4. The range and levels of the variables investigated in this study are presented below. Nisin extraction is a ratio representing the concentration of nisin and this was the response y.

Trial	x_1	x_2	y
1	13	11	62.874
2	15	11	76.133
3	13	13	87.467
4	15	13	102.324
5	14	12	76.187
6	14	12	77.523
7	14	12	76.782
8	14	12	77.438
9	14	12	78.742

(a) Compute an ANOVA table for the effects and test for curvature with $\alpha = 0.05$. Is curvature important in this region of the factors?
(b) Calculate residuals from the linear model and test for adequacy of your model.

(c) In a new region of factor space a central composite design (CCD) was used to perform second order optimization. The results are shown in the following table. Fit a second order model to this data and make conclusions.

	Coded		Uncoded		
Trial	x_1	x_2	x_1	x_2	y
1	−1	−1	15	14	102.015
2	1	−1	16	14	106.868
3	−1	1	15	16	108.13
4	1	1	16	16	110.176
5	−1.414	0	14.793	15	105.236
6	1.414	0	16.207	15	110.289
7	0	−1.414	15.5	13.586	103.999
8	0	1.414	15.5	16.414	110.171
9	0	0	15.5	15	108.044
10	0	0	15.5	15	109.098
11	0	0	15.5	15	107.824
12	0	0	15.5	15	108.978
13	0	0	15.5	15	109.169

MIND-EXPANDING EXERCISES

14-67. Consider an unreplicated 2^k factorial, and suppose that one of the treatment combinations is missing. One logical approach to this problem is to estimate the missing value with a number that makes the highest order interaction estimate zero. Apply this technique to the data in Example 14-5, assuming that *ab* is missing. Compare the results of the analysis of these data with the results in Example 14-5.

14-68. What blocking scheme would you recommend if it were necessary to run a 2^4 design in four blocks of four runs each?

14-69. Consider a 2^2 design in two blocks with *AB* confounded with blocks. Prove algebraically that $SS_{AB} = SS_{\text{Blocks}}$.

14-70. Consider a 2^3 design. Suppose that the largest number of runs that can be made in one block is four, but we can afford to perform a total of 32 observations.

(a) Suggest a blocking scheme that will provide some information on all interactions.

(b) Show an outline (source of variability, degrees of freedom only) for the analysis of variance for this design.

14-71. Construct a 2^{5-1} design. Suppose that it is necessary to run this design in two blocks of eight runs each. Show how this can be done by confounding a two-factor interaction (and its aliased three-factor interaction) with blocks.

14-72. Construct a 2^{7-2}_{IV} design. Show how this design may be confounded in four blocks of eight runs each. Are any two-factor interactions confounded with blocks?

14-73. Construct a 2^{7-3}_{IV} design. Show how this design can be confounded in two blocks of eight runs each without losing information on any of the two-factor interactions.

14-74. Set up a 2^{7-4}_{III} design using $D = AB$, $E = AC$, $F = BC$, and $G = ABC$ as the design generators. Ignore all interaction above the two factors.

(a) Verify that each main effect is aliased with three two-factor interactions.

(b) Suppose that a second 2^{7-4}_{III} design with generators $D = -AB$, $E = -AC$, $F = -BC$, and $G = ABC$ is run. What are the aliases of the main effects in this design?

(c) What factors may be estimated if the two sets of factor effect estimates above are combined?

14-75. Consider the square root of the sum of squares for curvature and divide by the square root of mean squared error. Explain why the statistic that results has a *t* distribution and why it can be used to conduct a *t* test for curvature that is equivalent to the *F* test in the ANOVA.

IMPORTANT TERMS AND CONCEPTS

Analysis of variance (ANOVA)
Blocking and nuisance factors
Center points
Central composite design
Confounding
Contrast
Defining relation
Design matrix
Factorial Experiment
Fractional factorial design
Generator
Interaction
Main effect
Normal probability plot of factor effects
Optimization experiment
Orthogonal design
Regression model
Residual analysis
Resolution
Response surface
Screening experiment
Steepest ascent (or descent)
2^k factorial design
Two-level factorial design

15 Introduction to Nonparametric Statistics

CHAPTER OUTLINE

LEARNING OBJECTIVES

After careful study of this chapter, you should be able to do the following:

1. Determine situations where nonparametric procedures are better alternatives to the *t*-test and ANOVA
2. Use one- and two-sample nonparametric tests
3. Use nonparametric alternatives to the single-factor ANOVA
4. Understand how nonparametric tests compare to the *t*-test in terms of relative efficiency

15-1 INTRODUCTION

Most of the hypothesis-testing and confidence interval procedures discussed in previous chapters are based on the assumption that we are working with random samples from normal populations. Traditionally, we have called these procedures **parametric methods** because they are based on a particular parametric family of distributions—in this case, the normal. Alternately, sometimes we say that these procedures are not *distribution-free* because they depend on the assumption of normality. Fortunately, most of these procedures are relatively insensitive to slight departures from normality. In general, the t- and F-tests and the t-confidence intervals will have actual levels of significance or confidence levels that differ from the nominal or advertised levels chosen by the experimenter, although the difference between the actual and advertised levels is usually fairly small when the underlying population is not too different from the normal.

In this chapter we describe procedures called **nonparametric** and **distribution-free methods,** and we usually make no assumptions about the distribution of the underlying population other than that it is continuous. These procedures have actual level of significance α or confidence level $100(1 - \alpha)\%$ for many different types of distributions. These procedures have considerable appeal. One of their advantages is that the data need not be quantitative but can be categorical (such as yes or no, defective or nondefective) or rank data. Another advantage is that nonparametric procedures are usually very quick and easy to perform.

The procedures described in this chapter are alternatives to the parametric t- and F-procedures described earlier. Consequently, it is important to compare the performance of both parametric and nonparametric methods under the assumptions of both normal and nonnormal populations. In general, nonparametric procedures do not utilize all the information provided by the sample. As a result, a nonparametric procedure will be less efficient than the corresponding parametric procedure when the underlying population is normal. This loss of efficiency is reflected by a requirement of a larger sample size for the nonparametric procedure than would be required by the parametric procedure in order to achieve the same power. On the other hand, this loss of efficiency is usually not large, and often the difference in sample size is very small. When the underlying distributions are not close to normal, nonparametric methods have much to offer. They often provide considerable improvement over the normal-theory parametric methods.

Generally, if both parametric and nonparametric methods are applicable to a particular problem, we should use the more efficient parametric procedure. However, the assumptions for the parametric method may be difficult or impossible to justify. For example, the data may be in the form of **ranks.** These situations frequently occur in practice. For instance, a panel of judges may be used to evaluate 10 different formulations of a soft-drink beverage for overall quality, with the "best" formulation assigned rank 1, the "next-best" formulation assigned rank 2, and so forth. It is unlikely that rank data satisfy the normality assumption. Many nonparametric methods involve the analysis of ranks and consequently are ideally suited to this type of problem.

15-2 SIGN TEST

15-2.1 Description of the Test

The **sign test** is used to test hypotheses about the **median** $\tilde{\mu}$ of a continuous distribution. The median of a distribution is a value of the random variable X such that the probability is 0.5 that an observed value of X is less than or equal to the median, and the probability is 0.5 that an observed value of X is greater than or equal to the median. That is, $P(X \leq \tilde{\mu}) = P(X \geq \tilde{\mu}) = 0.5$.

Since the normal distribution is symmetric, the mean of a normal distribution equals the median. Therefore, the sign test can be used to test hypotheses about the mean of a normal distribution. This is the same problem for which we used the t-test in Chapter 9. We will discuss the relative merits of the two procedures in Section 15-2.4. Note that, although the t-test was designed for samples from a normal distribution, the sign test is appropriate for samples from any continuous distribution. Thus, the sign test is a nonparametric procedure.

Suppose that the hypotheses are

$$\begin{aligned} H_0&: \tilde{\mu} = \tilde{\mu}_0 \\ H_1&: \tilde{\mu} < \tilde{\mu}_0 \end{aligned} \tag{15-1}$$

The test procedure is easy to describe. Suppose that $X_1, X_2, \ldots, X_n$ is a random sample from the population of interest. Form the differences

$$X_i - \tilde{\mu}_0, \qquad i = 1, 2, \ldots, n \tag{15-2}$$

Now if the null hypothesis H_0: $\tilde{\mu} = \tilde{\mu}_0$ is true, any difference $X_i - \tilde{\mu}_0$ is equally likely to be positive or negative. An appropriate test statistic is the number of these differences that are positive, say R^+. Therefore, to test the null hypothesis we are really testing that the number of plus signs is a value of a binomial random variable that has the parameter $p = 1/2$. A P-value for the observed number of plus signs r^+ can be calculated directly from the binomial distribution. For instance, in testing the hypotheses in Equation 15-1, we will reject H_0 in favor of H_1 only if the proportion of plus signs is sufficiently less than 1/2 (or equivalently, whenever the observed number of plus signs r^+ is too small). Thus, if the computed P-value

$$P = P\left(R^+ \leq r^+ \text{ when } p = \frac{1}{2}\right)$$

is less than or equal to some preselected significance level α, we will reject H_0 and conclude H_1 is true.

To test the other one-sided hypothesis

$$\begin{aligned} H_0&: \tilde{\mu} = \tilde{\mu}_0 \\ H_1&: \tilde{\mu} > \tilde{\mu}_0 \end{aligned} \tag{15-3}$$

we will reject H_0 in favor of H_1 only if the observed number of plus signs, say r^+, is large or, equivalently, whenever the observed fraction of plus signs is significantly greater than 1/2. Thus, if the computed P-value

$$P = P\left(R^+ \geq r^+ \text{ when } p = \frac{1}{2}\right)$$

is less than α, we will reject H_0 and conclude that H_1 is true.

The two-sided alternative may also be tested. If the hypotheses are

$$\begin{aligned} H_0&: \tilde{\mu} = \tilde{\mu}_0 \\ H_1&: \tilde{\mu} \neq \tilde{\mu}_0 \end{aligned} \tag{15-4}$$

we should reject H_0: $\tilde{\mu} = \tilde{\mu}_0$ if the proportion of plus signs is significantly different (either less than or greater than) from 1/2. This is equivalent to the observed number of plus signs r^+ being either sufficiently large or sufficiently small. Thus, if $r^+ < n/2$ the P-value is

$$P = 2P\left(R^+ \leq r^+ \text{ when } p = \frac{1}{2}\right)$$

and if $r^+ > n/2$ the P-value is

$$P = 2P\left(R^+ \geq r^+ \text{ when } p = \frac{1}{2}\right)$$

If the P-value is less than some preselected level α, we will reject H_0 and conclude that H_1 is true.

EXAMPLE 15-1
Propellant Shear Strength Sign Test

Montgomery, Peck, and Vining (2001) report on a study in which a rocket motor is formed by binding an igniter propellant and a sustainer propellant together inside a metal housing. The shear strength of the bond between the two propellant types is an important characteristic. The results of testing 20 randomly selected motors are shown in Table 15-1. We would like to test the hypothesis that the median shear strength is 2000 psi, using $\alpha = 0.05$.

This problem can be solved using the eight-step hypothesis-testing procedure introduced in Chapter 9:

1. The parameter of interest is the median of the distribution of propellant shear strength.
2. H_0: $\tilde{\mu} = 2000$ psi
3. H_1: $\tilde{\mu} \neq 2000$ psi
4. $\alpha = 0.05$

Table 15-1 Propellant Shear Strength Data

Observation i	Shear Strength x_i	Differences $x_i - 2000$	Sign
1	2158.70	+158.70	+
2	1678.15	−321.85	−
3	2316.00	+316.00	+
4	2061.30	+61.30	+
5	2207.50	+207.50	+
6	1708.30	−291.70	−
7	1784.70	−215.30	−
8	2575.10	+575.10	+
9	2357.90	+357.90	+
10	2256.70	+256.70	+
11	2165.20	+165.20	+
12	2399.55	+399.55	+
13	1779.80	−220.20	−
14	2336.75	+336.75	+
15	1765.30	−234.70	−
16	2053.50	+53.50	+
17	2414.40	+414.40	+
18	2200.50	+200.50	+
19	2654.20	+654.20	+
20	1753.70	−246.30	−

5. The test statistic is the observed number of plus differences in Table 15-1, or $r^+ = 14$.
6. We will reject H_0 if the P-value corresponding to $r^+ = 14$ is less than or equal to $\alpha = 0.05$.
7. Computations: Since $r^+ = 14$ is greater than $n/2 = 20/2 = 10$, we calculate the P-value from

$$P = 2P\left(R^+ \geq 14 \text{ when } p = \frac{1}{2}\right)$$
$$= 2\sum_{r=14}^{20}\binom{20}{r}(0.5)^r(0.5)^{20-r}$$
$$= 0.1153$$

8. Conclusions: Since $P = 0.1153$ is not less than $\alpha = 0.05$, we cannot reject the null hypothesis that the median shear strength is 2000 psi. Another way to say this is that the observed number of plus signs $r^+ = 14$ was not large or small enough to indicate that median shear strength is different from 2000 psi at the $\alpha = 0.05$ level of significance.

It is also possible to construct a table of critical values for the sign test. This table is shown as Appendix Table VIII. The use of this table for the two-sided alternative hypothesis in Equation 15-4 is simple. As before, let R^+ denote the number of the differences $(X_i - \tilde{\mu}_0)$ that are positive and let R^- denote the number of these differences that are negative. Let $R = \min(R^+, R^-)$. Appendix Table VIII presents critical values r_α^* for the sign test that ensure that P (type I error) $= P$ (reject H_0 when H_0 is true) $= \alpha$ for $\alpha = 0.01$, $\alpha = 0.05$ and $\alpha = 0.10$. If the observed value of the test statistic $r \leq r_\alpha^*$, the null hypothesis H_0: $\tilde{\mu} = \tilde{\mu}_0$ should be rejected.

To illustrate how this table is used, refer to the data in Table 15-1 that was used in Example 15-1. Now $r^+ = 14$ and $r^- = 6$; therefore, $r = \min(14, 6) = 6$. From Appendix Table VIII with $n = 20$ and $\alpha = 0.05$, we find that $r_{0.05}^* = 5$. Since $r = 6$ is not less than or equal to the critical value $r_{0.05}^* = 5$, we cannot reject the null hypothesis that the median shear strength is 2000 psi.

We can also use Appendix Table VIII for the sign test when a one-sided alternative hypothesis is appropriate. If the alternative is H_1: $\tilde{\mu} > \tilde{\mu}_0$, reject H_0: $\tilde{\mu} = \tilde{\mu}_0$ if $r^- \leq r_\alpha^*$; if the alternative is H_1: $\tilde{\mu} > \tilde{\mu}_0$, reject H_0: $\tilde{\mu} = \tilde{\mu}_0$ if $r^+ \leq r_\alpha^*$. The level of significance of a one-sided test is one-half the value for a two-sided test. Appendix Table VIII shows the one-sided significance levels in the column headings immediately below the two-sided levels.

Finally, note that when a test statistic has a discrete distribution such as R does in the sign test, it may be impossible to choose a critical value r_α^* that has a level of significance exactly equal to α. The approach used in Appendix Table VIII is to choose r_α^* to yield an α that is as close to the advertised significance level α as possible.

Ties in the Sign Test

Since the underlying population is assumed to be continuous, there is a zero probability that we will find a "tie"—that is, a value of X_i exactly equal to $\tilde{\mu}_0$. However, this may sometimes happen in practice because of the way the data are collected. When ties occur, they should be set aside and the sign test applied to the remaining data.

The Normal Approximation

When $p = 0.5$, the binomial distribution is well approximated by a normal distribution when n is at least 10. Thus, since the mean of the binomial is np and the variance is $np(1 - p)$, the distribution of R^+ is approximately normal with mean $0.5n$ and variance $0.25n$ whenever n is moderately large. Therefore, in these cases the null hypothesis H_0: $\tilde{\mu} = \tilde{\mu}_0$ can be tested using the statistic

Normal Approximation for Sign Test Statistic

$$Z_0 = \frac{R^+ - 0.5n}{0.5\sqrt{n}} \quad (15\text{-}5)$$

The two-sided alternative would be rejected if the observed value of the test statistic $|z_0| > z_{\alpha/2}$, and the critical regions of the one-sided alternative would be chosen to reflect the sense of the alternative. (If the alternative is H_1: $\tilde{\mu} > \tilde{\mu}_0$, reject H_0 if $z_0 > z_\alpha$, for example.)

EXAMPLE 15-2
Propellant Shear Strength Normal Approximation

We will illustrate the normal approximation procedure by applying it to the problem in Example 15-1. Recall that the data for this example are in Table 15-1. The eight-step procedure follows:

1. The parameter of interest is the median of the distribution of propellant shear strength.
2. H_0: $\tilde{\mu} = 2000$ psi
3. H_1: $\tilde{\mu} \neq 2000$ psi
4. $\alpha = 0.05$
5. The test statistic is

$$z_0 = \frac{r^+ - 0.5n}{0.5\sqrt{n}}$$

6. Since $\alpha = 0.05$, we will reject H_0 in favor of H_1 if $|z_0| > z_{0.025} = 1.96$.
7. Computations: Since $r^+ = 14$, the test statistic is

$$z_0 = \frac{14 - 0.5(20)}{0.5\sqrt{20}} = 1.789$$

8. Conclusions: Since $z_0 = 1.789$ is not greater than $z_{0.025} = 1.96$, we cannot reject the null hypothesis. Thus, our conclusions are identical to those in Example 15-1.

15-2.2 Sign Test for Paired Samples

The sign test can also be applied to paired observations drawn from continuous populations. Let (X_{1j}, X_{2j}), $j = 1, 2, \ldots, n$ be a collection of paired observations from two continuous populations, and let

$$D_j = X_{1j} - X_{2j} \qquad j = 1, 2, \ldots, n$$

be the paired differences. We wish to test the hypothesis that the two populations have a common median, that is, that $\tilde{\mu}_1 = \tilde{\mu}_2$. This is equivalent to testing that the median of the differences $\tilde{\mu}_D = 0$. This can be done by applying the sign test to the n observed differences d_j, as illustrated in the following example.

Table 15-2 Performance of Flow Metering Devices

	Metering Device			
Car	1	2	Difference, d_j	Sign
1	17.6	16.8	0.8	+
2	19.4	20.0	−0.6	−
3	19.5	18.2	1.3	+
4	17.1	16.4	0.7	+
5	15.3	16.0	−0.7	−
6	15.9	15.4	0.5	+
7	16.3	16.5	−0.2	−
8	18.4	18.0	0.4	+
9	17.3	16.4	0.9	+
10	19.1	20.1	−1.0	−
11	17.8	16.7	1.1	+
12	18.2	17.9	0.3	+

EXAMPLE 15-3
Flow Metering Paired Sign Test

An automotive engineer is investigating two different types of metering devices for an electronic fuel injection system to determine whether they differ in their fuel mileage performance. The system is installed on 12 different cars, and a test is run with each metering device on each car. The observed fuel mileage performance data, corresponding differences, and their signs are shown in Table 15-2. We will use the sign test to determine whether the median fuel mileage performance is the same for both devices using $\alpha = 0.05$. The eight-step-procedure follows:

1. The parameters of interest are the median fuel mileage performance for the two metering devices.
2. H_0: $\tilde{\mu}_1 = \tilde{\mu}_2$, or, equivalently, H_0: $\tilde{\mu}_D = 0$
3. H_1: $\tilde{\mu}_1 \neq \tilde{\mu}_2$, or, equivalently, H_1: $\tilde{\mu}_D \neq 0$
4. $\alpha = 0.05$
5. We will use Appendix Table VIII to conduct the test, so the test statistic is $r = \min(r^+, r^-)$.
6. Since $\alpha = 0.05$ and $n = 12$, Appendix Table VIII gives the critical values as $r^*_{0.05} = 2$. We will reject H_0 in favor of H_1 if $r \leq 2$.
7. Computations: Table 15-2 shows the differences and their signs, and we note that $r^+ = 8$, $r^- = 4$, and so $r = \min(8, 4) = 4$.
8. Conclusions: Since $r = 4$ is not less than or equal to the critical value $r^*_{0.05} = 2$, we cannot reject the null hypothesis that the two devices provide the same median fuel mileage performance.

15-2.3 Type II Error for the Sign Test

The sign test will control the probability of type I error at an advertised level α for testing the null hypothesis H_0: $\tilde{\mu} = \tilde{\mu}$ for any continuous distribution. As with any hypothesis-testing procedure, it is important to investigate the probability of a type II error, β. The test should be able to effectively detect departures from the null hypothesis, and a good measure of this effectiveness is the value of β for departures that are important. A small value of β implies an effective test procedure.

In determining β, it is important to realize not only that a particular value of $\tilde{\mu}$, say $\tilde{\mu}_0 + \Delta$, must be used but also that the **form** of the underlying distribution will affect the calculations. To

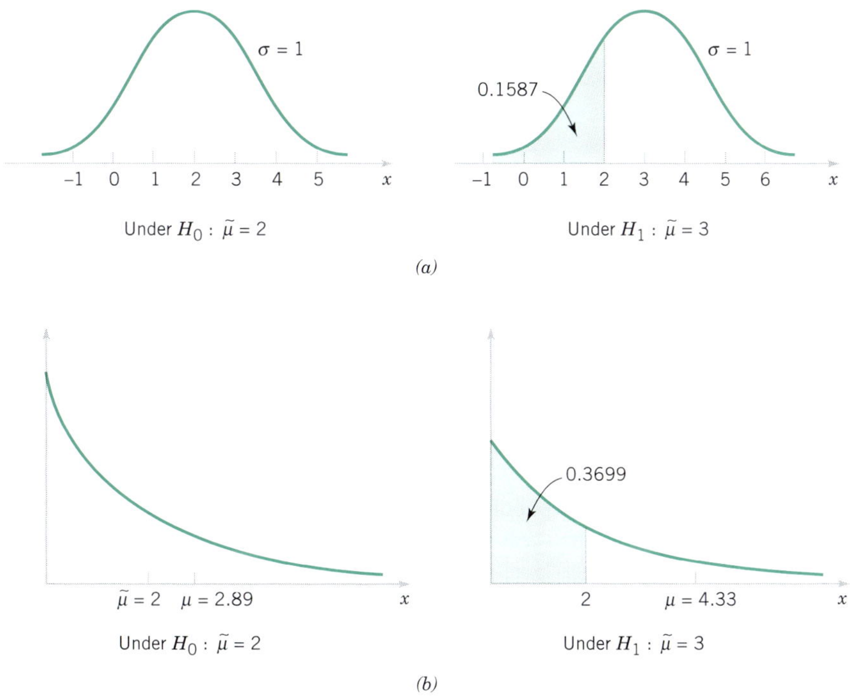

Figure 15-1 Calculation of β for the sign test. (a) Normal distributions. (b) Exponential distributions.

illustrate, suppose that the underlying distribution is normal with $\sigma = 1$ and we are testing the hypothesis H_0: $\tilde{\mu} = 2$ versus H_1: $\tilde{\mu} > 2$. (Since $\tilde{\mu} = \mu$ in the normal distribution, this is equivalent to testing that the mean equals 2.) Suppose that it is important to detect a departure from $\tilde{\mu} = 2$ to $\tilde{\mu} = 3$. The situation is illustrated graphically in Fig. 15-1(a). When the alternative hypothesis is true (H_1: $\tilde{\mu} = 3$), the probability that the random variable X is less than or equal to the value 2 is

$$p = P(X \leq 2) = P(Z \leq -1) = \Phi(-1) = 0.1587$$

Suppose we have taken a random sample of size 12. At the $\alpha = 0.05$ level, Appendix Table VIII indicates that we would reject H_0: $\tilde{\mu} = 2$ if $r^- \leq r^*_{0.05} = 2$. Therefore, β is the probability that we do not reject H_0: $\tilde{\mu} = 2$ when in fact $\tilde{\mu} = 3$, or

$$\beta = 1 - \sum_{x=0}^{2} \binom{12}{x} (0.1587)^x (0.8413)^{12-x} = 0.2944$$

If the distribution of X had been exponential rather than normal, the situation would be as shown in Fig. 15-1(b), and the probability that the random variable X is less than or equal to the value $x = 2$ when $\tilde{\mu} = 3$ (note that when the median of an exponential distribution is 3, the mean is 4.33) is

$$p = P(X \leq 2) = \int_0^2 \frac{1}{4.33} e^{-\frac{1}{4.33}x} \, dx = 0.3699$$

In this case,

$$\beta = 1 - \sum_{x=0}^{2} \binom{12}{x} (0.3699)^x (0.6301)^{12-x} = 0.8794$$

Thus, β for the sign test depends not only on the alternative value of $\tilde{\mu}$ but also on the area to the right of the value specified in the null hypothesis under the population probability distribution. This area is highly dependent on the shape of that particular probability distribution. In this example, β is large so the ability of the test to detect this departure from the null hypothesis with the current sample size is poor.

15-2.4 Comparison to the *t*-Test

If the underlying population is normal, either the sign test or the *t*-test could be used to test H_0: $\tilde{\mu} = \tilde{\mu}_0$. The *t*-test is known to have the smallest value of β possible among all tests that have significance level α for the one-sided alternative and for tests with symmetric critical regions for the two-sided alternative, so it is superior to the sign test in the normal distribution case. When the population distribution is symmetric and nonnormal (but with finite mean $\mu = \tilde{\mu}$), the *t*-test will have a smaller β (or a higher power) than the sign test, unless the distribution has very heavy tails compared with the normal. Thus, the sign test is usually considered a test procedure for the median rather than as a serious competitor for the *t*-test. The **Wilcoxon signed-rank test** discussed in the next section is preferable to the sign test and compares well with the *t*-test for symmetric distributions.

EXERCISES FOR SECTION 15-2

15-1. Ten samples were taken from a plating bath used in an electronics manufacturing process, and the bath pH was determined. The sample pH values are 7.91, 7.85, 6.82, 8.01, 7.46, 6.95, 7.05, 7.35, 7.25, 7.42. Manufacturing engineering believes that pH has a median value of 7.0.

(a) Do the sample data indicate that this statement is correct? Use the sign test with $\alpha = 0.05$ to investigate this hypothesis. Find the *P*-value for this test.

(b) Use the normal approximation for the sign test to test H_0: $\tilde{\mu} = 7.0$ versus H_1: $\tilde{\mu} \neq 7.0$. What is the *P*-value for this test?

15-2. The titanium content in an aircraft-grade alloy is an important determinant of strength. A sample of 20 test coupons reveals the following titanium content (in percent):

8.32, 8.05, 8.93, 8.65, 8.25, 8.46, 8.52, 8.35, 8.36, 8.41, 8.42, 8.30, 8.71, 8.75, 8.60, 8.83, 8.50, 8.38, 8.29, 8.46

The median titanium content should be 8.5%.

(a) Use the sign test with $\alpha = 0.05$ to investigate this hypothesis. Find the *P*-value for this test.

(b) Use the normal approximation for the sign test to test H_0: $\tilde{\mu} = 8.5$ versus H_1: $\tilde{\mu} \neq 8.5$, with $\alpha = 0.05$. What is the *P*-value for this test?

15-3. The impurity level (in ppm) is routinely measured in an intermediate chemical product. The following data were observed in a recent test:

2.4, 2.5, 1.7, 1.6, 1.9, 2.6, 1.3, 1.9, 2.0, 2.5, 2.6, 2.3, 2.0, 1.8, 1.3, 1.7, 2.0, 1.9, 2.3, 1.9, 2.4, 1.6

Can you claim that the median impurity level is less than 2.5 ppm?

(a) State and test the appropriate hypothesis using the sign test with $\alpha = 0.05$. What is the *P*-value for this test?

(b) Use the normal approximation for the sign test to test H_0: $\tilde{\mu} = 2.5$ versus H_1: $\tilde{\mu} < 2.5$. What is the *P*-value for this test?

15-4. Consider the margarine fat content data in Exercise 8-32. Use the sign test to test H_0: $\tilde{\mu} = 17.0$ versus H_1: $\tilde{\mu} \neq 17.0$, with $\alpha = 0.05$.

(a) Find the *P*-value for the test statistic and use this quantity to make your decision.

(b) Use the normal approximation to test the same hypothesis that you formulated in part (a). What is the *P*-value for this test?

15-5. Consider the compressive strength data in Exercise 8-33.

(a) Use the sign test to investigate the claim that the median strength is at least 2250 psi. Use $\alpha = 0.05$.
(b) Use the normal approximation to test the same hypothesis that you formulated in part (a). What is the P-value for this test?

15-6. The diameter of a ball bearing was measured by 12 inspectors, each using two different kinds of calipers. The results were as follows:

Inspector	Caliper 1	Caliper 2
1	0.265	0.264
2	0.265	0.265
3	0.266	0.264
4	0.267	0.266
5	0.267	0.267
6	0.265	0.268
7	0.267	0.264
8	0.267	0.265
9	0.265	0.265
10	0.268	0.267
11	0.268	0.268
12	0.265	0.269

(a) Is there a significant difference between the medians of the population of measurements represented by the two samples? Use $\alpha = 0.05$.
(b) Use the normal approximation for the sign test. With $\alpha = 0.05$, what conclusions can you draw?

15-7. Two different types of tips can be used in a Rockwell hardness tester. Eight coupons from test ingots of a nickel-based alloy are selected, and each coupon is tested twice, once with each tip. The Rockwell C-scale hardness readings are shown in the following table.
(a) Use the sign test with $\alpha = 0.05$ to determine whether or not the two tips produce equivalent hardness readings.
(b) Find the P-value for this statistic.

Coupon	Tip 1	Tip 2
1	63	60
2	52	51
3	58	56
4	60	59
5	55	58
6	57	54
7	53	52
8	59	61

15-8. Two different formulations of primer paint can be used on aluminum panels. The drying time of these two formulations is an important consideration in the manufacturing process. Twenty panels are selected; half of each panel is painted with primer 1, and the other half is painted with primer 2. The drying times are observed and reported in the following table.

	Drying Times (in hr)	
Panel	Formulation 1	Formulation 2
1	1.6	1.8
2	1.3	1.5
3	1.5	1.5
4	1.6	1.7
5	1.7	1.6
6	1.9	2.0
7	1.8	2.1
8	1.6	1.7
9	1.4	1.6
10	1.8	1.9
11	1.9	2.0
12	1.8	1.9
13	1.7	1.5
14	1.5	1.7
15	1.6	1.6
16	1.4	1.2
17	1.3	1.6
18	1.6	1.8
19	1.5	1.6
20	1.8	2.0

(a) Is there evidence that the median drying times of the two formulations are different? Use the sign test with $\alpha = 0.01$.
(b) Use the normal approximation to the sign test. What conclusions can you draw?
(c) Find the P-value for the statistic in part (a). Compare it to the P-value for the normal approximation test statistic computed in part (b).

15-9. The distribution time between arrivals in a telecommunication system is exponential, and the system manager wishes to test the hypothesis that H_0: $\tilde{\mu} = 3.5$ minutes versus H_1: $\tilde{\mu} > 3.5$ minutes.
(a) What is the value of the mean of the exponential distribution under H_0: $\tilde{\mu} = 3.5$?
(b) Suppose that we have taken a sample of $n = 10$ observations and we observe $r^- = 3$. Would the sign test reject H_0 at $\alpha = 0.05$?
(c) What is the type II error probability of this test if $\tilde{\mu} = 4.5$?

15-10. Consider the blood cholesterol data in Exercise 10-33.

(a) Use the sign test to determine whether there is any difference between the medians of the two groups of measurements, with $\alpha = 0.05$. What practical conclusion would you draw from this study?

(b) Use the normal approximation to the sign test. With $\alpha = 0.05$, what conclusions can you draw?

15-11. Suppose that we take a sample of $n = 10$ measurements from a normal distribution with $\sigma = 1$. We wish to test H_0: $\mu = 0$ against H_1: $\mu > 0$. The normal test statistic is $Z_0 = \overline{X}/(\sigma/\sqrt{n})$, and we decide to use a critical region of 1.96 (that is, reject H_0 if $z_0 \geq 1.96$).

(a) What is α for this test?

(b) What is β for this test, if $\mu = 1$?

(c) If a sign test is used, specify the critical region that gives an α value consistent with α for the normal test.

(d) What is the β value for the sign test, if $\mu = 1$? Compare this with the result obtained in part (b).

15-3 WILCOXON SIGNED-RANK TEST

The sign test makes use only of the plus and minus signs of the differences between the observations and the median $\tilde{\mu}_0$ (or the plus and minus signs of the differences between the observations in the paired case). It does not take into account the size or magnitude of these differences. Frank Wilcoxon devised a test procedure that uses both direction (sign) and magnitude. This procedure, now called the **Wilcoxon signed-rank test,** is discussed and illustrated in this section.

The Wilcoxon signed-rank test applies to the case of **symmetric continuous distributions.** Under these assumptions, the mean equals the median, and we can use this procedure to test the null hypothesis $\mu = \mu_0$. We now show how to do this.

15-3.1 Description of the Test

We are interested in testing H_0: $\mu = \mu_0$ against the usual alternatives. Assume that $X_1, X_2, \ldots, X_n$ is a random sample from a continuous and symmetric distribution with mean (and median) μ. Compute the differences $X_i - \mu_0$, $i = 1, 2, \ldots, n$. Rank the absolute differences $|X_i - \mu_0|$, $i = 1, 2, \ldots, n$ in ascending order, and then give the ranks the signs of their corresponding differences. Let W^+ be the sum of the positive ranks and W^- be the absolute value of the sum of the negative ranks, and let $W = \min(W^+, W^-)$. Appendix Table IX contains critical values of W, say w^*_α. If the alternative hypothesis is H_1: $\mu \neq \mu_0$, then if the observed value of the statistic $w \leq w^*_\alpha$, the null hypothesis H_0: $\mu = \mu_0$ is rejected. Appendix Table IX provides significance levels of $\alpha = 0.10$, $\alpha = 0.05$, $\alpha = 0.02$, $\alpha = 0.01$ for the two-sided test.

For one-sided tests, if the alternative is H_1: $\mu > \mu_0$, reject H_0: $\mu = \mu_0$ if $w^- \leq w^*_\alpha$; and if the alternative is H_1: $\mu < \mu_0$, reject H_0: $\mu = \mu_0$ if $w^+ \leq w^*_\alpha$. The significance levels for one-sided tests provided in Appendix Table IX are $\alpha = 0.05, 0.025, 0.01$, and 0.005.

EXAMPLE 15-4 Propellant Shear Strength Wilcoxon Signed-Rank Test

We will illustrate the Wilcoxon signed-rank test by applying it to the propellant shear strength data from Table 15-1. Assume that the underlying distribution is a continuous symmetric distribution. The eight-step procedure is applied as follows:

1. The parameter of interest is the mean (or median) of the distribution of propellant shear strength.
2. H_0: $\mu = 2000$ psi
3. H_1: $\mu \neq 2000$ psi
4. $\alpha = 0.05$

5. The test statistic is

$$w = \min(w^+, w^-)$$

6. We will reject H_0 if $w \le w^*_{0.05} = 52$ from Appendix Table IX.

7. Computations: The signed ranks from Table 15-1 are shown in the following table:

Observation	Difference $x_i - 2000$	Signed Rank
16	+53.50	+1
4	+61.30	+2
1	+158.70	+3
11	+165.20	+4
18	+200.50	+5
5	+207.50	+6
7	−215.30	−7
13	−220.20	−8
15	−234.70	−9
20	−246.30	−10
10	+256.70	+11
6	−291.70	−12
3	+316.00	+13
2	−321.85	−14
14	+336.75	+15
9	+357.90	+16
12	+399.55	+17
17	+414.40	+18
8	+575.10	+19
19	+654.20	+20

The sum of the positive ranks is $w^+ = (1 + 2 + 3 + 4 + 5 + 6 + 11 + 13 + 15 + 16 + 17 + 18 + 19 + 20) = 150$, and the sum of the absolute values of the negative ranks is $w^- = (7 + 8 + 9 + 10 + 12 + 14) = 60$. Therefore,

$$w = \min(150, 60) = 60$$

8. Conclusions: Since $w = 60$ is not less than or equal to the critical value $w_{0.05} = 52$, we cannot reject the null hypothesis that the mean (or median, since the population is assumed to be symmetric) shear strength is 2000 psi.

Ties in the Wilcoxon Signed-Rank Test

Because the underlying population is continuous, ties are theoretically impossible, although they will sometimes occur in practice. If several observations have the same absolute magnitude, they are assigned the average of the ranks that they would receive if they differed slightly from one another.

15-3.2 Large-Sample Approximation

If the sample size is moderately large, say $n > 20$, it can be shown that W^+ (or W^-) has approximately a normal distribution with mean

$$\mu_{W^+} = \frac{n(n+1)}{4}$$

and variance

$$\sigma^2_{W^+} = \frac{n(n+1)(2n+1)}{24}$$

Therefore, a test of H_0: $\mu = \mu_0$ can be based on the statistic

Normal Approximation for Wilcoxon Signed-Rank Statistic

$$Z_0 = \frac{W^+ - n(n+1)/4}{\sqrt{n(n+1)(2n+1)/24}} \tag{15-6}$$

An appropriate critical region for either the two-sided or one-sided alternative hypotheses can be chosen from a table of the standard normal distribution.

15-3.3 Paired Observations

The Wilcoxon signed-rank test can be applied to paired data. Let (X_{1j}, X_{2j}), $j = 1, 2, \ldots, n$ be a collection of paired observations from two continuous distributions that differ only with respect to their means. (It is not necessary that the distributions of X_1 and X_2 be symmetric.) This assures that the distribution of the differences $D_j = X_{1j} - X_{2j}$ is continuous and symmetric. Thus, the null hypothesis is H_0: $\mu_1 = \mu_2$, which is equivalent to H_0: $\mu_D = 0$. We initially consider the two-sided alternative H_1: $\mu_1 \neq \mu_2$ (or H_1: $\mu_D \neq 0$).

To use the Wilcoxon signed-rank test, the differences are first ranked in ascending order of their absolute values, and then the ranks are given the signs of the differences. Ties are assigned average ranks. Let W^+ be the sum of the positive ranks and W^- be the absolute value of the sum of the negative ranks, and $W = \min(W^+, W^-)$. If the observed value $w \leq w^*_\alpha$, the null hypothesis H_0: $\mu_1 = \mu_2$ (or H_0: $\mu_D = 0$) is rejected where w^*_α is chosen from Appendix Table VIII.

For one-sided tests, if the alternative is H_1: $\mu_1 > \mu_2$ (or H_1: $\mu_D > 0$), reject H_0 if $w^- \leq w^*_\alpha$; and if H_1: $\mu_1 < \mu_2$ (or H_1: $\mu_D < 0$), reject H_0 if $w^+ \leq w^*_\alpha$. Be sure to use the one-sided test significance levels shown in Appendix Table VIII.

EXAMPLE 15-5 Flow Metering Paired Wilcoxon Signed-Rank Test

We will apply the Wilcoxon signed-rank test to the fuel-metering device test data used previously in Example 15-3. The eight-step hypothesis-testing procedure can be applied as follows:

1. The parameters of interest are the mean fuel mileage performance for the two metering devices.
2. H_0: $\mu_1 = \mu_2$ or, equivalently, H_0: $\mu_D = 0$
3. H_1: $\mu_1 \neq \mu_2$ or, equivalently, H_1: $\mu_D \neq 0$

4. $\alpha = 0.05$
5. The test statistic is

$$w = \min(w^+, w^-)$$

where w^+ and w^- are the sums of the positive and negative ranks of the differences in Table 15-2.
6. Since $\alpha = 0.05$ and $n = 12$, Appendix Table IX gives the critical value as $w^*_{0.05} = 13$. We will reject H_0: $\mu_D = 0$ if $w \leq 13$.
7. Computations: Using the data in Table 15-2, we compute the following signed ranks:

Car	Difference	Signed Rank
7	−0.2	−1
12	0.3	2
8	0.4	3
6	0.5	4
2	−0.6	−5
4	0.7	6.5
5	−0.7	−6.5
1	0.8	8
9	0.9	9
10	−1.0	−10
11	1.1	11
3	1.3	12

Note that $w^+ = 55.5$ and $w^- = 22.5$. Therefore,

$$w = \min(55.5, 22.5) = 22.5$$

8. Conclusions: Since $w = 22.5$ is not less than or equal to $w^*_{0.05} = 13$, we cannot reject the null hypothesis that the two metering devices produce the same mileage performance.

15-3.4 Comparison to the *t*-Test

When the underlying population is normal, either the *t*-test or the Wilcoxon signed-rank test can be used to test hypotheses about μ. As mentioned earlier, the *t*-test is the best test in such situations in the sense that it produces a minimum value of β for all tests with significance level α. However, since it is not always clear that the normal distribution is appropriate, and since in many situations it is inappropriate, it is of interest to compare the two procedures for both normal and nonnormal populations.

Unfortunately, such a comparison is not easy. The problem is that β for the Wilcoxon signed-rank test is very difficult to obtain, and β for the *t*-test is difficult to obtain for nonnormal distributions. Because type II error comparisons are difficult, other measures of comparison have been developed. One widely used measure is **asymptotic relative efficiency** (ARE). The ARE of one test relative to another is the limiting ratio of the sample sizes necessary to obtain identical error probabilities for the two procedures. For example, if the ARE of one test relative to the competitor is 0.5, when sample sizes are large, the first test will require twice as large a sample as the second one to obtain similar error performance. While this does not tell us anything for small sample sizes, we can say the following:

1. For normal populations, the ARE of the Wilcoxon signed-rank test relative to the t-test is approximately 0.95.
2. For nonnormal populations, the ARE is at least 0.86, and in many cases it will exceed unity.

Although these are large-sample results, we generally conclude that the Wilcoxon signed-rank test will never be much worse than the t-test and that in many cases where the population is nonnormal it may be superior. Thus, the Wilcoxon signed-rank test is a useful alternate to the t-test.

EXERCISES FOR SECTION 15-3

15-12. Consider the data in Exercise 15-2.

(a) Suppose that the distribution of titanium content is symmetric and continuous. Use the Wilcoxon signed-rank test with $\alpha = 0.05$ to test the hypothesis H_0: $\mu = 8.5$ versus H_1: $\mu \neq 8.5$.

(b) Use the large-sample approximation for the Wilcoxon signed-rank test to test the hypothesis H_0: $\mu = 8.5$ versus H_1: $\mu \neq 8.5$. Use $\alpha = 0.05$. Assume that the distribution of titanium content is continuous and symmetric.

15-13. Consider the data in Exercise 15-1 and assume that the distribution of pH is symmetric and continuous. Use the Wilcoxon signed-rank test with $\alpha = 0.05$ to test the hypothesis H_0: $\mu = 7$ against H_1: $\mu \neq 7$.

15-14. Consider the caliper measurement data in Exercise 15-6. Apply the Wilcoxon signed-rank test. Use $\alpha = 0.05$ and assume that the two distributions of measurements are continuous.

15-15. Consider the data in Exercise 15-3. Use the Wilcoxon signed-rank test to test the hypothesis H_0: $\mu = 2.5$ ppm versus H_1: $\mu < 2.5$ ppm with $\alpha = 0.05$. Assume that the distribution of impurity level is continuous and symmetric.

15-16. Consider the paint drying time data in Exercise 15-8. Assume that both populations are continuous, and use the Wilcoxon signed-rank test to test that the difference in mean drying times between the two formulations is zero. Use $\alpha = 0.01$.

15-17. Consider the Rockwell hardness test data in Exercise 15-7. Assume that both distributions are continuous and use the Wilcoxon signed-rank test to test that the mean difference in hardness readings between the two tips is zero. Use $\alpha = 0.05$.

15-4 WILCOXON RANK-SUM TEST

Suppose that we have two independent continuous populations X_1 and X_2 with means μ_1 and μ_2. Assume that the distributions of X_1 and X_2 have the same shape and spread and differ only (possibly) in their locations. The Wilcoxon rank-sum test can be used to test the hypothesis H_0: $\mu_1 = \mu_2$. This procedure is sometimes called the Mann-Whitney test, although the Mann-Whitney test statistic is usually expressed in a different form.

15-4.1 Description of the Test

Let $X_{11}, X_{12}, \ldots, X_{1n_1}$ and $X_{21}, X_{22}, \ldots, X_{2n_2}$ be two independent random samples of sizes $n_1 \leq n_2$ from the continuous populations X_1 and X_2 described earlier. We wish to test the hypotheses

$$H_0: \mu_1 = \mu_2$$
$$H_1: \mu_1 \neq \mu_2$$

The test procedure is as follows. Arrange all $n_1 + n_2$ observations in ascending order of magnitude and assign ranks to them. If two or more observations are tied (identical), use the mean of the ranks that would have been assigned if the observations differed.

Let W_1 be the sum of the ranks in the smaller sample (1), and define W_2 to be the sum of the ranks in the other sample. Then,

$$W_2 = \frac{(n_1 + n_2)(n_1 + n_2 + 1)}{2} - W_1 \tag{15-7}$$

Now if the sample means do not differ, we will expect the sum of the ranks to be nearly equal for both samples after adjusting for the difference in sample size. Consequently, if the sums of the ranks differ greatly, we will conclude that the means are not equal.

Appendix Table X contains the critical value of the rank sums for $\alpha = 0.05$ and $\alpha = 0.01$ assuming the two-sided alternative above. Refer to Appendix Table X with the appropriate sample sizes n_1 and n_2, and the critical value w_α can be obtained. The null H_0: $\mu_1 = \mu_2$ is rejected in favor of H_1: $\mu_1 \neq \mu_2$ if either of the observed values w_1 or w_2 is less than or equal to the tabulated critical value w_α.

The procedure can also be used for one-sided alternatives. If the alternative is H_1: $\mu_1 < \mu_2$, reject H_0 if $w_1 \leq w_\alpha$; for H_1: $\mu_1 > \mu_2$, reject H_0 if $w_2 \leq w_\alpha$. For these one-sided tests, the tabulated critical values w_α correspond to levels of significance of $\alpha = 0.025$ and $\alpha = 0.005$.

EXAMPLE 15-6
Axial Stress

The mean axial stress in tensile members used in an aircraft structure is being studied. Two alloys are being investigated. Alloy 1 is a traditional material, and alloy 2 is a new aluminum-lithium alloy that is much lighter than the standard material. Ten specimens of each alloy type are tested, and the axial stress is measured. The sample data are assembled in Table 15-3. Using $\alpha = 0.05$, we wish to test the hypothesis that the means of the two stress distributions are identical.

We will apply the eight-step hypothesis-testing procedure to this problem:

1. The parameters of interest are the means of the two distributions of axial stress.
2. H_0: $\mu_1 = \mu_2$
3. H_1: $\mu_1 \neq \mu_2$
4. $\alpha = 0.05$
5. We will use the Wilcoxon rank-sum test statistic in Equation 15-7,

$$w_2 = \frac{(n_1 + n_2)(n_1 + n_2 + 1)}{2} - w_1$$

6. Since $\alpha = 0.05$ and $n_1 = n_2 = 10$, Appendix Table X gives the critical value as $w_{0.05} = 78$. If either w_1 or w_2 is less than or equal to $w_{0.05} = 78$, we will reject H_0: $\mu_1 = \mu_2$.

Table 15-3 Axial Stress for Two Aluminum-Lithium Alloys

Alloy 1		Alloy 2	
3238 psi	3254 psi	3261 psi	3248 psi
3195	3229	3187	3215
3246	3225	3209	3226
3190	3217	3212	3240
3204	3241	3258	3234

7. Computations: The data from Table 15-3 are analyzed in ascending order and ranked as follows:

Alloy Number	Axial Stress	Rank
2	3187 psi	1
1	3190	2
1	3195	3
1	3204	4
2	3209	5
2	3212	6
2	3215	7
1	3217	8
1	3225	9
2	3226	10
1	3229	11
2	3234	12
1	3238	13
2	3240	14
1	3241	15
1	3246	16
2	3248	17
1	3254	18
2	3258	19
2	3261	20

The sum of the ranks for alloy 1 is

$$w_1 = 2 + 3 + 4 + 8 + 9 + 11 + 13 + 15 + 16 + 18 = 99$$

and for alloy 2

$$w_2 = \frac{(n_1 + n_2)(n_1 + n_2 + 1)}{2} - w_1 = \frac{(10 + 10)(10 + 10 + 1)}{2} - 99 = 111$$

8. Conclusions: Since neither w_1 nor w_2 is less than or equal to $w_{0.05} = 78$, we cannot reject the null hypothesis that both alloys exhibit the same mean axial stress.

15-4.2 Large-Sample Approximation

When both n_1 and n_2 are moderately large, say, greater than 8, the distribution of w_1 can be well approximated by the normal distribution with mean

$$\mu_{W_1} = \frac{n_1(n_1 + n_2 + 1)}{2}$$

and variance

$$\sigma^2_{W_1} = \frac{n_1 n_2(n_1 + n_2 + 1)}{12}$$

Therefore, for n_1 and $n_2 > 8$, we could use

Normal Approximation for Wilcoxon Rank-Sum Test Statistic

$$Z_0 = \frac{W_1 - \mu_{W_1}}{\sigma_{W_1}} \tag{15-8}$$

as a statistic, and the appropriate critical region is $|z_0| > z_{\alpha/2}$, $z_0 > z_\alpha$, or $z_0 < -z_\alpha$, depending on whether the test is a two-tailed, upper-tail, or lower-tail test.

15-4.3 Comparison to the *t*-Test

In Section 15-3.4 we discussed the comparison of the *t*-test with the Wilcoxon signed-rank test. The results for the two-sample problem are identical to the one-sample case. That is, when the normality assumption is correct, the Wilcoxon rank-sum test is approximately 95% as efficient as the *t*-test in large samples. On the other hand, regardless of the form of the distributions, the Wilcoxon rank-sum test will always be at least 86% as efficient. The efficiency of the Wilcoxon test relative to the *t*-test is usually high if the underlying distribution has heavier tails than the normal, because the behavior of the *t*-test is very dependent on the sample mean, which is quite unstable in heavy-tailed distributions.

EXERCISES FOR SECTION 15-4

15-18. An electrical engineer must design a circuit to deliver the maximum amount of current to a display tube to achieve sufficient image brightness. Within her allowable design constraints, she has developed two candidate circuits and tests prototypes of each. The resulting data (in microamperes) are as follows:

Circuit 1:	251, 255, 258, 257, 250, 251, 254, 250, 248
Circuit 2:	250, 253, 249, 256, 259, 252, 260, 251

(a) Use the Wilcoxon rank-sum test to test H_0: $\mu_1 = \mu_2$ against the alternative H_1: $\mu_1 > \mu_2$. Use $\alpha = 0.025$.

(b) Use the normal approximation for the Wilcoxon rank-sum test. Assume that $\alpha = 0.05$. Find the approximate *P*-value for this test statistic.

15-19. One of the authors travels regularly to Seattle, Washington. He uses either Delta or Alaska. Flight delays are sometimes unavoidable, but he would be willing to give most of his business to the airline with the best on-time arrival record. The number of minutes that his flight arrived late for the last six trips on each airline follows. Is there evidence that either airline has superior on-time arrival performance? Use $\alpha = 0.01$ and the Wilcoxon rank-sum test.

Delta:	13, 10, 1, −4, 0, 9 (minutes late)
Alaska:	15, 8, 3, −1, −2, 4 (minutes late)

15-20. The manufacturer of a hot tub is interested in testing two different heating elements for his product. The element that produces the maximum heat gain after 15 minutes would be preferable. He obtains 10 samples of each heating unit and tests each one. The heat gain after 15 minutes (in °F) follows.

(a) Is there any reason to suspect that one unit is superior to the other? Use $\alpha = 0.05$ and the Wilcoxon rank-sum test.

Unit 1:	25, 27, 29, 31, 30, 26, 24, 32, 33, 38
Unit 2:	31, 33, 32, 35, 34, 29, 38, 35, 37, 30

(b) Use the normal approximation for the Wilcoxon rank-sum test. Assume that $\alpha = 0.05$. What is the approximate *P*-value for this test statistic?

15-21. Consider the chemical etch rate data in Exercise 10-17.

(a) Use the Wilcoxon rank-sum test to investigate the claim that the mean etch rate is the same for both solutions. If $\alpha = 0.05$, what are your conclusions?

(b) Use the normal approximation for the Wilcoxon rank-sum test. Assume that $\alpha = 0.05$. Find the approximate *P*-value for this test.

15-22. Consider the pipe deflection data in Exercise 10-16.

(a) Use the Wilcoxon rank-sum test for the pipe deflection temperature experiment. If $\alpha = 0.05$, what are your conclusions?

(b) Use the normal approximation for the Wilcoxon rank-sum test. Assume that $\alpha = 0.05$. Find the approximate *P*-value for this test.

15-23. Consider the distance traveled by a golf ball in Exercise 10-27.

(a) Use the Wilcoxon rank-sum test to investigate if the means differ. Use $\alpha = 0.05$.

(b) Use the normal approximation for the Wilcoxon rank-sum test with $\alpha = 0.05$. Find the approximate *P*-value for this test.

15-5 NONPARAMETRIC METHODS IN THE ANALYSIS OF VARIANCE

15-5.1 Kruskal-Wallis Test

The single-factor analysis of variance model developed in Chapter 13 for comparing a population means is

$$Y_{ij} = \mu + \tau_i + \epsilon_{ij} \begin{cases} i = 1, 2, \ldots, a \\ j = 1, 2, \ldots, n_i \end{cases} \tag{15-9}$$

In this model, the error terms ϵ_{ij} are assumed to be normally and independently distributed with mean zero and variance σ^2. The assumption of normality led directly to the F-test described in Chapter 13. The Kruskal-Wallis test is a nonparametric alternative to the F-test; it requires only that the ϵ_{ij} have the same continuous distribution for all factor levels $i = 1, 2, \ldots, a$.

Suppose that $N = \sum_{i=1}^{a} n_i$ is the total number of observations. Rank all N observations from smallest to largest, and assign the smallest observation rank 1, the next smallest rank 2, . . . , and the largest observation rank N. If the null hypothesis

$$H_0\colon \mu_1 = \mu_2 = \cdots = \mu_a$$

is true, the N observations come from the same distribution, and all possible assignments of the N ranks to the a samples are equally likely, we would expect the ranks $1, 2, \ldots, N$ to be mixed throughout the a samples. If, however, the null hypothesis H_0 is false, some samples will consist of observations having predominantly small ranks, while other samples will consist of observations having predominantly large ranks. Let R_{ij} be the rank of observation Y_{ij}, and let $R_{i\cdot}$ and $\bar{R}_{i\cdot}$ denote the total and average of the n_i ranks in the ith treatment. When the null hypothesis is true,

$$E(R_{ij}) = \frac{N+1}{2}$$

and

$$E(\bar{R}_{i\cdot}) = \frac{1}{n_i} \sum_{j=1}^{n_i} E(R_{ij}) = \frac{N+1}{2}$$

The Kruskal-Wallis test statistic measures the degree to which the actual observed average ranks $\bar{R}_{i\cdot}$ differ from their expected value $(N+1)/2$. If this difference is large, the null hypothesis H_0 is rejected. The test statistic is

Kruskal-Wallis Test Statistic

$$H = \frac{12}{N(N+1)} \sum_{i=1}^{a} n_i \left(\bar{R}_{i\cdot} - \frac{N+1}{2} \right)^2 \tag{15-10}$$

An alternative computing formula that is occasionally more convenient is

Computing Formula for Kruskal-Wallis Test Statistic

$$H = \frac{12}{N(N+1)} \sum_{i=1}^{a} \frac{R_{i\cdot}^2}{n_i} - 3(N+1) \tag{15-11}$$

We would usually prefer Equation 15-11 to Equation 15-10 because it involves the rank totals rather than the averages.

The null hypothesis H_0 should be rejected if the sample data generate a large value for H. The null distribution for H has been obtained by using the fact that under H_0 each possible assignment of ranks to the a treatments is equally likely. Thus, we could enumerate all possible assignments and count the number of times each value of H occurs. This has led to tables of the critical values of H, although most tables are restricted to small sample sizes n_i. In practice, we usually employ the following large-sample approximation. Whenever H_0 is true and either

$$a = 3 \quad \text{and} \quad n_i \geq 6 \quad \text{for } i = 1, 2, 3$$

$$a > 3 \quad \text{and} \quad n_i \geq 5 \quad \text{for } i = 1, 2, \ldots, a$$

H has approximately a chi-square distribution with $a - 1$ degrees of freedom. Since large values of H imply that H_0 is false, we will reject H_0 if the observed value

$$h \geq \chi^2_{\alpha, a-1}$$

The test has approximate significance level α.

Ties in the Kruskal-Wallis Test

When observations are tied, assign an average rank to each of the tied observations. When there are ties, we should replace the test statistic in Equation 15-11 by

$$H = \frac{1}{S^2}\left[\sum_{i=1}^{a} \frac{R_{i\cdot}^2}{n_i} - \frac{N(N+1)^2}{4}\right] \tag{15-12}$$

where n_i is the number of observations in the ith treatment, N is the total number of observations, and

$$S^2 = \frac{1}{N-1}\left[\sum_{i=1}^{a} \sum_{j=1}^{n_i} R_{ij}^2 - \frac{N(N+1)^2}{4}\right] \tag{15-13}$$

Note that S^2 is just the variance of the ranks. When the number of ties is moderate, there will be little difference between Equations 15-11 and 15-12 and the simpler form (Equation 15-11) may be used.

EXAMPLE 15-7
Fabric Tensile Strength Kruskal-Wallis Test

Montgomery (2001) presented data from an experiment in which five different levels of cotton content in a synthetic fiber were tested to determine whether cotton content has any effect on fiber tensile strength. The sample data and ranks from this experiment are shown in Table 15-4. We will apply the Kruskal-Wallis test to these data, using $\alpha = 0.01$.

Table 15-4 Data and Ranks for the Fabric Tensile Strength Experiment

Percentage of Cotton							$r_{i\cdot}$
15	y_{1j}	7	7	15	11	9	
ranks	r_{1j}	2.0	2.0	12.5	7.0	4.0	27.5
20	y_{2j}	12	17	12	18	18	
ranks	r_{2j}	9.5	14.0	9.5	16.5	16.5	66.0
25	y_{3j}	14	18	18	19	19	
ranks	r_{3j}	11.0	16.5	16.5	20.5	20.5	85.0
30	y_{4j}	19	25	22	19	23	
ranks	r_{4j}	20.5	25.0	23.0	20.5	24.0	113.0
35	y_{5j}	7	10	11	15	11	
ranks	r_{5j}	2.0	5.0	7.0	12.5	7.0	33.5

Since there is a fairly large number of ties, we use Equation 15-12 as the test statistic. From Equation 15-13 we find

$$s^2 = \frac{1}{N-1}\left[\sum_{i=1}^{a}\sum_{j=1}^{n_i} r_{ij}^2 - \frac{N(N+1)^2}{4}\right]$$

$$= \frac{1}{24}\left[5510 - \frac{25(26)^2}{4}\right]$$

$$= 53.54$$

and the test statistic is

$$h = \frac{1}{s^2}\left[\sum_{i=1}^{a}\frac{r_{i\cdot}^2}{n_i} - \frac{N(N+1)^2}{4}\right]$$

$$= \frac{1}{53.54}\left[5245.7 - \frac{25(26)^2}{4}\right]$$

$$= 19.06$$

Since $h > \chi^2_{0.01,4} = 13.28$, we would reject the null hypothesis and conclude that treatments differ. This same conclusion is given by the usual analysis of variance F-test.

15-5.2 Rank Transformation

The procedure used in the previous section whereby the observations are replaced by their ranks is called the **rank transformation.** It is a very powerful and widely useful technique. If we were to apply the ordinary F-test to the ranks rather than to the original data, we would obtain

$$F_0 = \frac{H/(a-1)}{(N-1-H)/(N-a)}$$

as the test statistic. Note that as the Kruskal-Wallis statistic H increases or decreases, F_0 also increases or decreases. Now, since the distribution of F_0 is approximated by the F-distribution, the Kruskal-Wallis test is approximately equivalent to applying the usual analysis of variance to the ranks.

The rank transformation has wide applicability in experimental design problems for which no nonparametric alternative to the analysis of variance exists. If the data are ranked and the ordinary F-test is applied, an approximate procedure results, but one that has good statistical properties. When we are concerned about the normality assumption or the effect of outliers or "wild" values, we recommend that the usual analysis of variance be performed on both the original data and the ranks. When both procedures give similar results, the analysis of variance assumptions are probably satisfied reasonably well, and the standard analysis is satisfactory. When the two procedures differ, the rank transformation should be preferred since it is less likely to be distorted by nonnormality and unusual observations. In such cases, the experimenter may want to investigate the use of transformations for nonnormality and examine the data and the experimental procedure to determine whether outliers are present and why they have occurred.

EXERCISES FOR SECTION 15-5

15-24. An article in the *Quality Control Handbook*, 3rd edition (McGraw-Hill, 1962) presents the results of an experiment performed to investigate the effect of three different conditioning methods on the breaking strength of cement briquettes. The data are shown in the following table.

Conditioning Method	Breaking Strength (lb/in.^{2})				
1	553	550	568	541	537
2	553	599	579	545	540
3	492	530	528	510	571

(a) Using $\alpha = 0.05$, is there any indication that conditioning method affects breaking strength?
(b) Find the approximate P-value for the test statistic.

15-25. Montgomery (2001) presented the results of an experiment to compare four different mixing techniques on the tensile strength of portland cement. The results are shown in the following table.

Mixing Technique	Tensile Strength (lb/in.^{2})			
1	3129	3000	2865	2890
2	3200	3000	2975	3150
3	2800	2900	2985	3050
4	2600	2700	2600	2765

(a) Is there any indication that mixing technique affects the strength? Use $\alpha = 0.05$.
(b) Find the approximate P-value for the test statistic.

15-26. In *Statistics for Research* (John Wiley & Sons, 1983), S. Dowdy and S. Wearden presented the results of an experiment to measure the performance of hand-held chain saws. The experimenters measured the kickback angle through which the saw is deflected when it begins to cut a 3-inch stock synthetic board. Shown in the following table are deflection angles for five saws chosen at random from each of four different manufacturers.

Manufacturer	Kickback Angle				
A	42	17	24	39	43
B	28	50	44	32	61
C	57	45	48	41	54
D	29	40	22	34	30

(a) Is there any evidence that the manufacturers' products differ with respect to kickback angle? Use $\alpha = 0.01$.
(b) Find the approximate P-value for the test statistic.

15-27. Consider the conductivity data in Exercise 13-5.
(a) Use the Kruskal-Wallis procedure with $\alpha = 0.05$ to test for differences in means.
(b) Find the approximate P-value for the test statistic.

15-28. Consider the etch rate data in Exercise 13-2.
(a) Use the Kruskal-Wallis procedure with $\alpha = 0.05$ to test for differences between mean uniformity at the three different gas flow rates.
(b) Find the approximate P-value for the test statistic.

15-6 RUNS TEST

Another nonparametric test is simple, and widely applicable to a number of different hypotheses. We first present it for a two-sample test of equal distributions and then comment on additional uses. The Wilcoxon rank-sum test was used to compare the means between two samples. But suppose we would like to more generally check whether two samples were obtained from different distributions. The **runs test,** also known as the **Wald-Wolfowitz test,** can be used in this type of problem.

Let $X_{11}, X_{12}, \ldots, X_{1n_1}$ and $X_{21}, X_{22}, \ldots, X_{2n_2}$ be two independent random samples of sizes n_1 and n_2 from the continuous populations X_1 and X_2 with distribution functions $F_1(x)$ and $F_2(x)$. We wish to test the hypotheses that the distributions are the same versus the alternative that they differ in some manner. They might differ because the mean of X_1 is greater than the mean of X_2 or because there is greater variability in X_1, or for any other characteristic, although the test that will be used is more sensitive to some differences than others. Because a probability distribution function uniquely describes a distribution we can state the hypothesis of equal distributions as

$$H_0\colon F_1(x) = F_2(x)$$
$$H_1\colon F_1(x) \neq F_2(x)$$

The test procedure begins with a ranking of the data in a manner similar to the Wilcoxon rank-sum test. Arrange all $n_1 + n_2$ observations in ascending order of magnitude. However, instead of assigning ranks to the data, simply record whether the observation is from sample 1 or 2. This results in an ordered sequence of the numbers 1 and 2. For example, this might produce

1,1,2,1,1,1,2,2,1,1,1,2,2,1,2,1,1,2,2,2,1

Let the random variable R denote the number of runs in the sequence. A run is a consecutive sequence of the same sample label. In the simple sequence shown, there is a run of two 1s followed by a run of one 2 followed by a run of three 1s, and so forth. The total number of runs is $r = 11$. The value of R is not affected by ties within the same sample. If two or more observations from different samples are tied (identical), randomly assign one sample or the other.

If there are two many or too few runs, we conclude that the distributions are not equal. Under the null hypothesis we expect the data from the samples to mix randomly so that the distribution of R can be calculated. For example, suppose that the distribution of X_1 is normal with mean 0 and variance 1, and suppose that the distribution of X_2 is also normal with mean 0, but with variance 4. Then we expect there to be more observed values from sample 2 than from sample 1 at the beginning and end of the sequence. This would result in fewer total runs and a smaller value observed for R than if the distributions were equal. The distribution of R has been determined and some computer software packages provide the exact distribution. Such a package can calculate the probability under the null hypothesis of $P(R > r)$ or $P(R < r)$ for any constant r and this can be used to determine critical values or P-values. Specifically, with some work it can be shown that

$$P(R = 2k) = \frac{2\binom{n_1-1}{k-1}\binom{n_2-1}{k-1}}{\binom{n_1+n_2}{n_1}}$$

and

$$P(R = 2k + 1) = \frac{2\binom{n_1-1}{k-1}\binom{n_2-1}{k} + \binom{n_1-1}{k}\binom{n_2-1}{k-1}}{\binom{n_1+n_2}{n_1}}$$

A large-sample approximation is often used when n_1 or n_2 is greater than 20. Let $N = n_1 + n_2$. The approximation uses the fact that the mean and variance of R are

Approximate Distribution of *R*

$$\mu_R = \frac{2n_1n_2 + N}{N} \quad \text{and} \quad \sigma_R^2 = \frac{2n_1n_2(2n_1n_2 - N)}{N^2(N-1)}$$

and the distribution of R is approximately normally distributed. A z-statistic to test H_0 is calculated as

$$z = \frac{r - \mu_R}{\sigma_R}$$

and a critical region or P-value is calculated from this statistic.

Another use of the runs test is to check for randomness in a sequence of observations. Let $X_1, X_2, \ldots, X_n$ denote measurements from a process over time. Select a constant value that is used to define runs. For example, if the constant is the process mean μ, the number of runs above and below the mean are calculated. Let n_1 and n_2 denote the number of observations below and above the mean, respectively. Under the assumption that the measurements are independent, the distribution of the number of runs is the same distribution that was used for the two-sample problem.

EXAMPLE 15-8 Runs for Earthquakes

Consider the earthquake data in Exercise 6-63. The number of earthquakes (of magnitude 7.0 or greater) per year is provided for 105 years from 1900–2004. The runs test can be used to evaluate the randomness in the number of earthquakes per year.

Year	1900	1901	1902	1903	1904	1905	1906	1907	1908	1909	…
Number	13	14	8	10	16	26	32	27	18	32	…
Coded Result	0	0	0	0	0	1	1	1	0	1	…

It can be checked that the median number of earthquakes per year is 18. Then each year can be evaluated as either below or above the median, and this is coded as either 0 or 1, respectively, in the partial data shown.

Ties from years with exactly 18 earthquakes are handled as follows. Years with exactly 18 earthquakes are alternately coded as 0 (below) or 1 (above) the median. From the full sequence of 105 years the runs statistic is computed to be 35. There are 49 years coded 0 and 56 years coded 1. The mean and variance of the run statistic are calculated to be

$$\mu_R = \frac{2 \times 49 \times 56 + 105}{105} = 53.27 \quad \text{and} \quad \sigma_R^2 = \frac{2 \times 49 \times 56(2 \times 49 \times 56 - 105)}{105^2(105 - 1)} = 25.76$$

Therefore,

$$z = \frac{35 - 53.27}{\sqrt{25.76}} = -3.60$$

and the normal approximation for R rejects the randomness hypothesis at most reasonable significance levels. The number of earthquakes per year do not appear to be random. There are fewer runs than the randomness hypothesis supports.

EXERCISES FOR SECTION 15-6

15-29. Consider the flight arrival data in Exercise 15-19.
(a) Use the exact distribution of the runs statistic to calculate the P-value for the hypotheses in the flight arrival exercise.
(b) Use the Wilcoxon ranked-sum test for this problem with $\alpha = 0.05$. What hypotheses are being tested in this problem?

15-30. Consider the circuit data in Exercise 15-18.
(a) Use the exact distribution of the runs statistic to calculate the P-value for the hypotheses in the circuit exercise.
(b) Use the Wilcoxon ranked-sum test for this problem with $\alpha = 0.05$. What conclusions can you draw? Does the hypothesis you are testing now differ from the one tested in part (a)?

15-31. Consider the axial stress data in Example 15-6. Use the normal approximation to the runs statistic to test for a difference in distributions at $\alpha = 0.05$.

15-32. Consider the propellant shear strength data in Table 15-1. Assume that the observations are in time order. Use the normal approximation to the runs statistic to test the data for randomness. Generate runs about the value 2000 psi.

Supplemental Exercises

15-33. The surface finish of 10 metal parts produced in a grinding process is as follows: (in microinches): 10.32, 9.68, 9.92, 10.10, 10.20, 9.87, 10.14, 9.74, 9.80, 10.26.
(a) Do the data support the claim that the median value of surface finish is 10 microinches? Use the sign test with $\alpha = 0.05$. What is the P-value for this test?
(b) Use the normal approximation for the sign test. Find the P-value for this test. What are your conclusions if $\alpha = 0.05$?

15-34. Fluoride emissions (in ppm) from a chemical plant are monitored routinely. The following are 15 observations based on air samples taken randomly during one month of production: 7, 3, 4, 2, 5, 6, 9, 8, 7, 3, 4, 4, 3, 2, 6.
(a) Can you claim that the median fluoride impurity level is less than 6 ppm? State and test the appropriate hypotheses using the sign test with $\alpha = 0.05$. What is the P-value for this test?
(b) Use the normal approximation for the sign test. What is the P-value for this test?

15-35. Use the Wilcoxon signed-rank test with $\alpha = 0.05$ for the diet-modification experiment described in Exercise 10-35. State carefully the conclusions that you can draw from this experiment.

15-36. Consider the data in Exercise 10-36. Use the sign test with $\alpha = 0.05$ to determine whether there is a difference in median impurity readings between the two analytical tests.

15-37. Consider the fuel economy data in Exercise 10-75.
(a) Use the Wilcoxon rank-sum test with $\alpha = 0.01$. What conclusions can you draw about the difference in mean mileage performance for the two vehicles in this study?
(b) Use the large-sample approximation for the Wilcoxon rank-sum test. What conclusions can you draw about the difference in means if $\alpha = 0.01$? Find the P-value for this test.

15-38. Consider the fill-capability experiment described in Exercise 10-76.
(a) Use the Wilcoxon rank-sum test with $\alpha = 0.025$. What conclusions can you draw about the capability of the two fillers?
(b) Use the large-sample approximation for the Wilcoxon rank-sum test with $\alpha = 0.025$. Find the P-value for this test. What conclusions can you draw?

15-39. Consider the contact resistance experiment in Exercise 13-39. Use the Kruskal-Wallis test to test for differences in mean contact resistance among the three alloys. If $\alpha = 0.01$, what are your conclusions? Find the P-value for this test.

15-40. Consider the experiment described in Exercise 13-36. Use the Kruskal-Wallis test for this experiment with $\alpha = 0.05$. What conclusions would you draw? Find the P-value for this test.

15-41. Consider the bread quality experiment in Exercise 13-41. Use the Kruskal-Wallis test with $\alpha = 0.01$ to analyze the data from this experiment. Find the P-value for this test. What conclusions can you draw?

15-42. Consider the gene expression data in Exercise 13-44. Use the Kruskal-Wallis test for the transformed data with $\alpha = 0.1$. Find the P-value for this test.

MIND-EXPANDING EXERCISES

15-43. For the large-sample approximation to the Wilcoxon signed-rank test, derive the mean and standard deviation of the test statistic used in the procedure.

15-44. **Testing for Trends.** A turbocharger wheel is manufactured using an investment-casting process. The shaft fits into the wheel opening, and this wheel opening is a critical dimension. As wheel wax patterns are formed, the hard tool producing the wax patterns wears. This may cause growth in the wheel-opening dimension. Ten wheel-opening measurements, in time order of production, are 4.00 (millimeters), 4.02, 4.03, 4.01, 4.00, 4.03, 4.04, 4.02, 4.03, 4.03.

(a) Suppose that p is the probability that observation X_{i+5} exceeds observation X_i. If there is no upward or downward trend, X_{i+5} is no more or less likely to exceed X_i or lie below X_i. What is the value of p?

(b) Let V be the number of values of i for which $X_{i+5} > X_i$. If there is no upward or downward trend in the measurements, what is the probability distribution of V?

(c) Use the data above and the results of parts (a) and (b) to test H_0: there is no trend, versus H_1: there is upward trend. Use $\alpha = 0.05$.

Note that this test is a modification of the sign test. It was developed by Cox and Stuart.

15-45. Consider the Wilcoxon signed-rank test, and suppose that $n = 5$. Assume that H_0: $\mu = \mu_0$ is true.

(a) How many different sequences of signed ranks are possible? Enumerate these sequences.

(b) How many different values of W^+ are there? Find the probability associated with each value of W^+.

(c) Suppose that we define the critical region of the test to be to reject H_0 if $w^+ > w^*_\alpha$ and $w^*_\alpha = 13$. What is the approximate α level of this test?

(d) Does this exercise show how the critical values for the Wilcoxon signed-rank test were developed? Explain.

IMPORTANT TERMS AND CONCEPTS

Asymptotic relative efficiency
Kruskal-Wallis test
Nonparametric or distribution free procedures
Normal approximation to nonparametric tests
Paired observations
Ranks
Rank transformation in ANOVA
Runs test
Sign test
Tests on the mean of a symmetric continuous distribution
Tests on the median of a continuous distribution
Wilcoxon rank-sum test
Wilcoxon signed-rank test

16 Introduction to Statistical Process Control

CHAPTER OUTLINE

LEARNING OBJECTIVES

After careful study of this chapter, you should be able to do the following:

1. Understand the role of statistical tools in quality improvement
2. Understand the different types of variability, rational subgroups, and how a control chart is used to detect assignable causes
3. Understand the general form of a Shewhart control chart and how to apply zone rules (such as the Western Electric rules) and pattern analysis to detect assignable causes
4. Construct and interpret control charts for variables such as $\overline{X}$, R, S, and individuals charts

5. Construct and interpret control charts for attributes such as *P* and *U* charts
6. Calculate and interpret process capability ratios
7. Calculate the ARL performance for a Shewhart control chart
8. Construct and interpret a cumulative sum and exponentially weighted moving average control chart
9. Use other statistical process control problem-solving tools

16-1 QUALITY IMPROVEMENT AND STATISTICS

The quality of products and services has become a major decision factor in most businesses today. Regardless of whether the consumer is an individual, a corporation, a military defense program, or a retail store, when the consumer is making purchase decisions, he or she is likely to consider quality of equal importance to cost and schedule. Consequently, **quality improvement** has become a major concern to many U.S. corporations. This chapter is about **statistical quality control,** a collection of tools that are essential in quality-improvement activities.

Quality means **fitness for use.** For example, you or I may purchase automobiles that we expect to be free of manufacturing defects and that should provide reliable and economical transportation, a retailer buys finished goods with the expectation that they are properly packaged and arranged for easy storage and display, or a manufacturer buys raw material and expects to process it with no rework or scrap. In other words, all consumers expect that the products and services they buy will meet their requirements. Those requirements define fitness for use.

Quality or fitness for use is determined through the interaction of **quality of design** and **quality of conformance.** By quality of design we mean the different grades or levels of performance, reliability, serviceability, and function that are the result of deliberate engineering and management decisions. By quality of conformance, we mean the systematic **reduction of variability** and **elimination of defects** until every unit produced is identical and defect-free.

Some confusion exists in our society about quality improvement; some people still think that it means gold-plating a product or spending more money to develop a product or process. This thinking is wrong. Quality improvement means the systematic **elimination of waste.** Examples of waste include scrap and rework in manufacturing, inspection and testing, errors on documents (such as engineering drawings, checks, purchase orders, and plans), customer complaint hotlines, warranty costs, and the time required to do things over again that could have been done right the first time. A successful quality-improvement effort can eliminate much of this waste and lead to lower costs, higher productivity, increased customer satisfaction, increased business reputation, higher market share, and ultimately higher profits for the company.

Statistical methods play a vital role in quality improvement. Some applications are outlined below:

1. In product design and development, statistical methods, including designed experiments, can be used to compare different materials, components, or ingredients, and to help determine both system and component tolerances. This application can significantly lower development costs and reduce development time.
2. Statistical methods can be used to determine the capability of a manufacturing process. Statistical process control can be used to systematically improve a process by reducing variability.
3. Experimental design methods can be used to investigate improvements in the process. These improvements can lead to higher yields and lower manufacturing costs.

4. Life testing provides reliability and other performance data about the product. This can lead to new and improved designs and products that have longer useful lives and lower operating and maintenance costs.

Some of these applications have been illustrated in earlier chapters of this book. It is essential that engineers, scientists, and managers have an in-depth understanding of these statistical tools in any industry or business that wants to be a high-quality, low-cost producer. In this chapter we provide an introduction to the basic methods of statistical quality control that, along with experimental design, form the basis of a successful quality-improvement effort.

16-2 STATISTICAL QUALITY CONTROL

The field of statistical quality control can be broadly defined as those statistical and engineering methods that are used in measuring, monitoring, controlling, and improving quality. Statistical quality control is a field that dates back to the 1920s. Dr. Walter A. Shewhart of the Bell Telephone Laboratories was one of the early pioneers of the field. In 1924 he wrote a memorandum showing a modern control chart, one of the basic tools of statistical process control. Harold F. Dodge and Harry G. Romig, two other Bell System employees, provided much of the leadership in the development of statistically based sampling and inspection methods. The work of these three men forms much of the basis of the modern field of statistical quality control. World War II saw the widespread introduction of these methods to U.S. industry. Dr. W. Edwards Deming and Dr. Joseph M. Juran have been instrumental in spreading statistical quality-control methods since World War II.

The Japanese have been particularly successful in deploying statistical quality-control methods and have used statistical methods to gain significant advantage over their competitors. In the 1970s American industry suffered extensively from Japanese (and other foreign) competition; that has led, in turn, to renewed interest in statistical quality-control methods in the United States. Much of this interest focuses on *statistical process control* and *experimental design*. Many U.S. companies have begun extensive programs to implement these methods in their manufacturing, engineering, and other business organizations.

16-3 STATISTICAL PROCESS CONTROL

It is impractical to inspect quality into a product; the product must be built right the first time. The manufacturing process must therefore be stable or repeatable and capable of operating with little variability around the target or nominal dimension. Online statistical process control is a powerful tool for achieving process stability and improving capability through the reduction of variability.

It is customary to think of **statistical process control (SPC)** as a set of problem-solving tools that may be applied to any process. The major tools of SPC* are

1. Histogram
2. Pareto chart
3. Cause-and-effect diagram

* Some prefer to include the experimental design methods discussed previously as part of the SPC toolkit. We did not do so, because we think of SPC as an online approach to quality improvement using techniques founded on passive observation of the process, while design of experiments is an active approach in which deliberate changes are made to the process variables. As such, designed experiments are often referred to as offline quality control.

4. Defect-concentration diagram
5. Control chart
6. Scatter diagram
7. Check sheet

Although these tools are an important part of SPC, they comprise only the technical aspect of the subject. An equally important element of SPC is attitude—a desire of all individuals in the organization for continuous improvement in quality and productivity through the systematic reduction of variability. The control chart is the most powerful of the SPC tools.

16-4 INTRODUCTION TO CONTROL CHARTS

16-4.1 Basic Principles

In any production process, regardless of how well-designed or carefully maintained it is, a certain amount of inherent or natural variability will always exist. This natural variability or "background noise" is the cumulative effect of many small, essentially unavoidable causes. When the background noise in a process is relatively small, we usually consider it an acceptable level of process performance. In the framework of statistical quality control, this natural variability is often called a "stable system of chance causes." A process that is operating with only **chance causes** of variation present is said to be in statistical control. In other words, the chance causes are an inherent part of the process.

Other kinds of variability may occasionally be present in the output of a process. This variability in key quality characteristics usually arises from three sources: improperly adjusted machines, operator errors, or defective raw materials. Such variability is generally large when compared to the background noise, and it usually represents an unacceptable level of process performance. We refer to these sources of variability that are not part of the chance cause pattern as **assignable causes.** A process that is operating in the presence of assignable causes is said to be out of control.*

Production processes will often operate in the in-control state, producing acceptable product for relatively long periods of time. Occasionally, however, assignable causes will occur, seemingly at random, resulting in a "shift" to an out-of-control state where a large proportion of the process output does not conform to requirements. A major objective of statistical process control is to quickly detect the occurrence of assignable causes or process shifts so that investigation of the process and corrective action may be undertaken before many nonconforming units are manufactured. The control chart is an online process-monitoring technique widely used for this purpose.

Recall the following from Chapter 1. Figure 1-11 illustrates that adjustments to common causes of variation increase the variation of a process whereas Fig. 1-12 illustrates that actions should be taken in response to assignable causes of variation. Control charts may also be used to estimate the parameters of a production process and, through this information, to determine the capability of a process to meet specifications. The control chart can also provide information that is useful in improving the process. Finally, remember that the eventual goal of statistical process control is the *elimination of variability in the process*. Although it may not be possible to eliminate variability completely, the control chart helps reduce it as much as possible.

---

* The terminology *chance* and *assignable* causes was developed by Dr. Walter A. Shewhart. Today, some writers use *common* cause instead of *chance* cause and *special* cause instead of *assignable* cause.

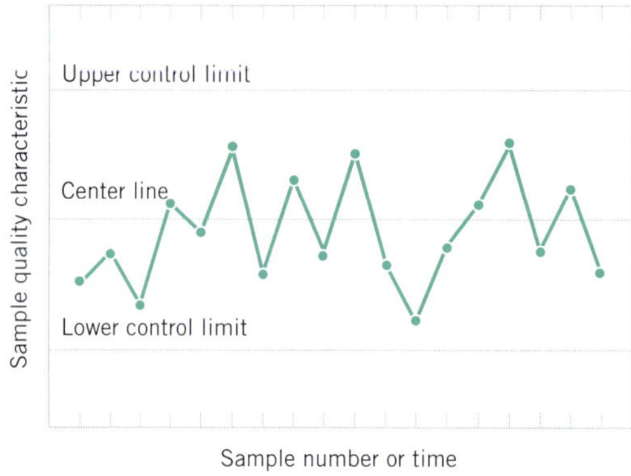

Figure 16-1 A typical control chart.

A typical control chart is shown in Fig. 16-1, which is a graphical display of a quality characteristic that has been measured or computed from a sample versus the sample number or time. Often, the samples are selected at periodic intervals such as every hour. The chart contains a center line (CL) that represents the average value of the quality characteristic corresponding to the in-control state. (That is, only chance causes are present.) Two other horizontal lines, called the upper control limit (UCL) and the lower control limit (LCL), are also shown on the chart. These control limits are chosen so that if the process is in control, nearly all of the sample points will fall between them. In general, as long as the points plot within the control limits, the process is assumed to be in control, and no action is necessary. However, a point that plots outside of the control limits is interpreted as evidence that the process is out of control, and investigation and corrective action are required to find and eliminate the assignable cause or causes responsible for this behavior. The sample points on the control chart are usually connected with straight-line segments so that it is easier to visualize how the sequence of points has evolved over time.

Even if all the points plot inside the control limits, if they behave in a systematic or nonrandom manner, this is an indication that the process is out of control. For example, if 18 of the last 20 points plotted above the center line but below the upper control limit and only two of these points plotted below the center line but above the lower control limit, we would be very suspicious that something was wrong. If the process is in control, all the plotted points should have an essentially random pattern. Methods designed to find sequences or nonrandom patterns can be applied to control charts as an aid in detecting out-of-control conditions. A particular nonrandom pattern usually appears on a control chart for a reason, and if that reason can be found and eliminated, process performance can be improved.

There is a close connection between control charts and hypothesis testing. Essentially, the control chart is a test of the hypothesis that the process is in a state of statistical control. A point plotting within the control limits is equivalent to failing to reject the hypothesis of statistical control, and a point plotting outside the control limits is equivalent to rejecting the hypothesis of statistical control.

We give a general *model* for a control chart. Let W be a sample statistic that measures some quality characteristic of interest, and suppose that the mean of W is μ_W and the standard deviation of W is σ_W.* Then the center line, the upper control limit, and the lower control limit become

*Note that "sigma" refers to the standard deviation of the statistic plotted on the chart (i.e., σ_W), not the standard deviation of the quality characteristic.

Control Chart Model

$$
\begin{aligned}
UCL &= \mu_W + k\sigma_W \\
CL &= \mu_W \\
LCL &= \mu_W - k\sigma_W
\end{aligned}
\qquad (16\text{-}1)
$$

where k is the "distance" of the control limits from the center line, expressed in standard deviation units. A common choice is $k = 3$. This general theory of control charts was first proposed by Dr. Walter A. Shewhart, and control charts developed according to these principles are often called **Shewhart control charts.**

The control chart is a device for describing exactly what is meant by statistical control; as such, it may be used in a variety of ways. In many applications, it is used for online process monitoring. That is, sample data are collected and used to construct the control chart, and if the sample values of $\bar{x}$ (say) fall within the control limits and do not exhibit any systematic pattern, we say the process is in control at the level indicated by the chart. Note that we may be interested here in determining *both* whether the past data came from a process that was in control and whether future samples from this process indicate statistical control.

The most important use of a control chart is to *improve* the process. We have found that, generally

1. Most processes do not operate in a state of statistical control.
2. Consequently, the routine and attentive use of control charts will identify assignable causes. If these causes can be eliminated from the process, variability will be reduced and the process will be improved.

This process-improvement activity using the control chart is illustrated in Fig. 16-2. Notice that:

3. The control chart will only *detect* assignable causes. Management, operator, and engineering *action* will usually be necessary to eliminate the assignable cause. An action plan for responding to control chart signals is vital.

In identifying and eliminating assignable causes, it is important to find the underlying **root cause** of the problem and to attack it. A cosmetic solution will not result in any real, long-term

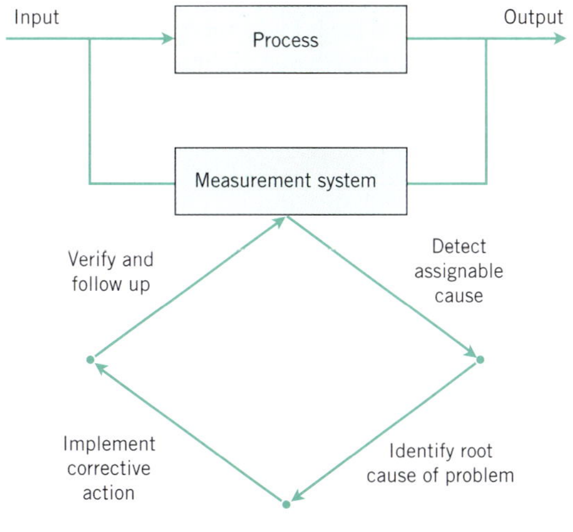

Figure 16-2 Process improvement using the control chart.

process improvement. Developing an effective system for corrective action is an essential component of an effective SPC implementation.

We may also use the control chart as an *estimating device*. That is, from a control chart that exhibits statistical control, we may estimate certain process parameters, such as the mean, standard deviation, and fraction nonconforming or fallout. These estimates may then be used to determine the *capability* of the process to produce acceptable products. Such **process capability studies** have considerable impact on many management decision problems that occur over the product cycle, including make-or-buy decisions, plant and process improvements that reduce process variability, and contractual agreements with customers or suppliers regarding product quality.

Control charts may be classified into two general types. Many quality characteristics can be measured and expressed as numbers on some continuous scale of measurement. In such cases, it is convenient to describe the quality characteristic with a measure of central tendency and a measure of variability. Control charts for central tendency and variability are collectively called **variables control charts.** The $\overline{X}$ chart is the most widely used chart for monitoring central tendency, whereas charts based on either the sample range or the sample standard deviation are used to control process variability. Many quality characteristics are not measured on a continuous scale or even a quantitative scale. In these cases, we may judge each unit of product as either conforming or nonconforming on the basis of whether or not it possesses certain attributes, or we may count the number of nonconformities (defects) appearing on a unit of product. Control charts for such quality characteristics are called **attributes control charts.**

Control charts have had a long history of use in industry. There are at least five reasons for their popularity:

1. **Control charts are a proven technique for improving productivity.** A successful control chart program will reduce scrap and rework, which are the primary productivity killers in *any* operation. If you reduce scrap and rework, productivity increases, cost decreases, and production capacity (measured in the number of *good* parts per hour) increases.
2. **Control charts are effective in defect prevention.** The control chart helps keep the process in control, which is consistent with the "do it right the first time" philosophy. It is never cheaper to sort out the "good" units from the "bad" later on than it is to build them correctly initially. If you do not have effective process control, you are paying someone to make a nonconforming product.
3. **Control charts prevent unnecessary process adjustments.** A control chart can distinguish between background noise and abnormal variation; no other device, including a human operator, is as effective in making this distinction. If process operators adjust the process based on periodic tests unrelated to a control chart program, they will often overreact to the background noise and make unneeded adjustments. These unnecessary adjustments can result in a deterioration of process performance. In other words, the control chart is consistent with the "if it isn't broken, don't fix it" philosophy.
4. **Control charts provide diagnostic information.** Frequently, the pattern of points on the control chart will contain information that is of diagnostic value to an experienced operator or engineer. This information allows the operator to implement a change in the process that will improve its performance.
5. **Control charts provide information about process capability.** The control chart provides information about the value of important process parameters and their stability over time. This allows an estimate of process capability to be made. This information is of tremendous use to product and process designers.

Control charts are among the most effective management control tools, and they are as important as cost controls and material controls. Modern computer technology has made it easy to implement control charts in any type of process, because data collection and analysis can be performed on a microcomputer or a local area network terminal in real time, online at the work center.

16-4.2 Design of a Control Chart

To illustrate these ideas, we give a simplified example of a control chart. In manufacturing automobile engine piston rings, the inside diameter of the rings is a critical quality characteristic. The process mean inside ring diameter is 74 millimeters, and it is known that the standard deviation of ring diameter is 0.01 millimeters. A control chart for average ring diameter is shown in Fig. 16-3. Every hour a random sample of five rings is taken, the average ring diameter of the sample (say $\bar{x}$) is computed, and $\bar{x}$ is plotted on the chart. Because this control chart utilizes the sample mean $\bar{X}$ to monitor the process mean, it is usually called an $\bar{X}$ control chart. Note that all the points fall within the control limits, so the chart indicates that the process is in statistical control.

Consider how the control limits were determined. The process average is 74 millimeters, and the process standard deviation is $\sigma = 0.01$ millimeters. Now if samples of size $n = 5$ are taken, the standard deviation of the sample average $\bar{X}$ is

$$\sigma_{\bar{X}} = \frac{\sigma}{\sqrt{n}} = \frac{0.01}{\sqrt{5}} = 0.0045$$

Therefore, if the process is in control with a mean diameter of 74 millimeters, by using the central limit theorem to assume that $\bar{X}$ is approximately normally distributed, we would expect approximately $100(1 - \alpha)\%$ of the sample mean diameters $\bar{X}$ to fall between $74 + z_{\alpha/2}(0.0045)$ and $74 - z_{\alpha/2}(0.0045)$. As discussed above, we customarily choose the constant $z_{\alpha/2}$ to be 3, so the upper and lower control limits become

$$UCL = 74 + 3(0.0045) = 74.0135$$

and

$$LCL = 74 - 3(0.0045) = 73.9865$$

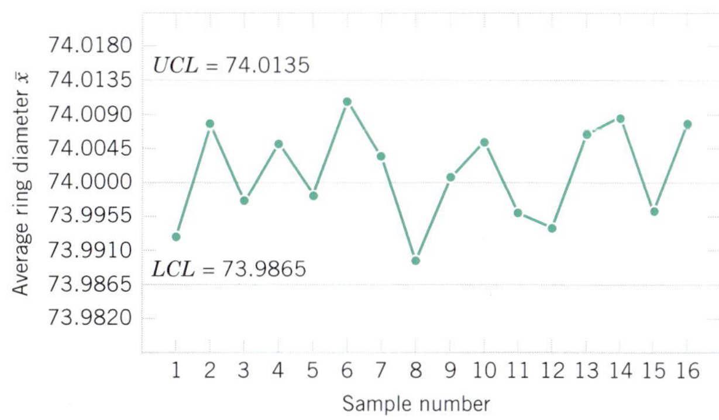

Figure 16-3 $\bar{X}$ control chart for piston ring diameter.

as shown on the control chart. These are the 3-sigma control limits referred to above. Note that the use of 3-sigma limits implies that $\alpha = 0.0027$; that is, the probability that the point plots outside the control limits when the process is in control is 0.0027. The width of the control limits is inversely related to the sample size n for a given multiple of sigma. Choosing the control limits is equivalent to setting up the critical region for testing the hypothesis

$$H_0: \mu = 74$$
$$H_1: \mu \neq 74$$

where $\sigma = 0.01$ is known. Essentially, the control chart tests this hypothesis repeatedly at different points in time.

In designing a control chart, we must specify both the sample size to use and the frequency of sampling. In general, larger samples will make it easier to detect small shifts in the process. When choosing the sample size, we must keep in mind the size of the shift that we are trying to detect. If we are interested in detecting a relatively large process shift, we use smaller sample sizes than those that would be employed if the shift of interest were relatively small.

We must also determine the frequency of sampling. The most desirable situation from the point of view of detecting shifts would be to take large samples very frequently; however, this is usually not economically feasible. The general problem is one of *allocating sampling effort*. That is, either we take small samples at short intervals or larger samples at longer intervals. Current industry practice tends to favor smaller, more frequent samples, particularly in high-volume manufacturing processes or where a great many types of assignable causes can occur. Furthermore, as automatic sensing and measurement technology develops, it is becoming possible to greatly increase frequencies. Ultimately, every unit can be tested as it is manufactured. This capability will not eliminate the need for control charts because the test system will not prevent defects. The increased data will increase the effectiveness of process control and improve quality.

When preliminary samples are used to construct limits for control charts, these limits are customarily treated as trial values. Therefore, the sample statistics should be plotted on the appropriate charts, and any points that exceed the control limits should be investigated. If assignable causes for these points are discovered, they should be eliminated and new limits for the control charts determined. In this way, the process may be eventually brought into statistical control and its inherent capabilities assessed. Other changes in process centering and dispersion may then be contemplated.

16-4.3 Rational Subgroups

A fundamental idea in the use of control charts is to collect sample data according to what Shewhart called the **rational subgroup** concept. Generally, this means that subgroups or samples should be selected so that to the extent possible, the variability of the observations within a subgroup should include all the chance or natural variability and exclude the assignable variability. Then, the control limits will represent bounds for all the chance variability and not the assignable variability. Consequently, assignable causes will tend to generate points that are outside of the control limits, while chance variability will tend to generate points that are within the control limits.

When control charts are applied to production processes, the time order of production is a logical basis for rational subgrouping. Even though time order is preserved, it is still possible to form subgroups erroneously. If some of the observations in the subgroup are taken at the end of one 8-hour shift and the remaining observations are taken at the start of the next 8-hour shift, any differences between shifts might not be detected. Time order is frequently a good basis for forming subgroups because it allows us to detect assignable causes that occur over time.

Two general approaches to constructing rational subgroups are used. In the first approach, each subgroup consists of units that were produced at the same time (or as closely together as possible). This approach is used when the primary purpose of the control chart is to detect process shifts. It minimizes variability due to assignable causes *within* a sample, and it maximizes variability *between* samples if assignable causes are present. It also provides better estimates of the standard deviation of the process in the case of variables control charts. This approach to rational subgrouping essentially gives a "snapshot" of the process at each point in time where a sample is collected.

In the second approach, each sample consists of units of product that are representative of *all* units that have been produced since the last sample was taken. Essentially, each subgroup is a *random sample* of *all* process output over the sampling interval. This method of rational subgrouping is often used when the control chart is employed to make decisions about the acceptance of all units of product that have been produced since the last sample. In fact, if the process shifts to an out-of-control state and then back in control again *between* samples, it is sometimes argued that the first method of rational subgrouping defined above will be ineffective against these types of shifts, and so the second method must be used.

When the rational subgroup is a random sample of all units produced over the sampling interval, considerable care must be taken in interpreting the control charts. If the process mean drifts between several levels during the interval between samples, the range of observations within the sample may consequently be relatively large. It is the within-sample variability that determines the width of the control limits on an $\overline{X}$ chart, so this practice will result in wider limits on the $\overline{X}$ chart. This makes it harder to detect shifts in the mean. In fact, we can often make *any* process appear to be in statistical control just by stretching out the interval between observations in the sample. It is also possible for shifts in the process average to cause points on a control chart for the range or standard deviation to plot out of control, even though no shift in process variability has taken place.

There are other bases for forming rational subgroups. For example, suppose a process consists of several machines that pool their output into a common stream. If we sample from this common stream of output, it will be very difficult to detect whether or not some of the machines are out of control. A logical approach to rational subgrouping here is to apply control chart techniques to the output for each individual machine. Sometimes this concept needs to be applied to different heads on the same machine, different workstations, different operators, and so forth.

The rational subgroup concept is very important. The proper selection of samples requires careful consideration of the process, with the objective of obtaining as much useful information as possible from the control chart analysis.

16-4.4 Analysis of Patterns on Control Charts

A control chart may indicate an out-of-control condition either when one or more points fall beyond the control limits, or when the plotted points exhibit some nonrandom pattern of behavior. For example, consider the $\overline{X}$ chart shown in Fig. 16-4. Although all 25 points fall within the control limits, the points do not indicate statistical control because their pattern is very nonrandom in appearance. Specifically, we note that 19 of the 25 points plot below the center line, while only 6 of them plot above. If the points are truly random, we should expect a more even distribution of them above and below the center line. We also observe that following the fourth point, five points in a row increase in magnitude. This arrangement of points is called a **run.** Since the observations are increasing, we could call it a run up; similarly, a sequence of decreasing points

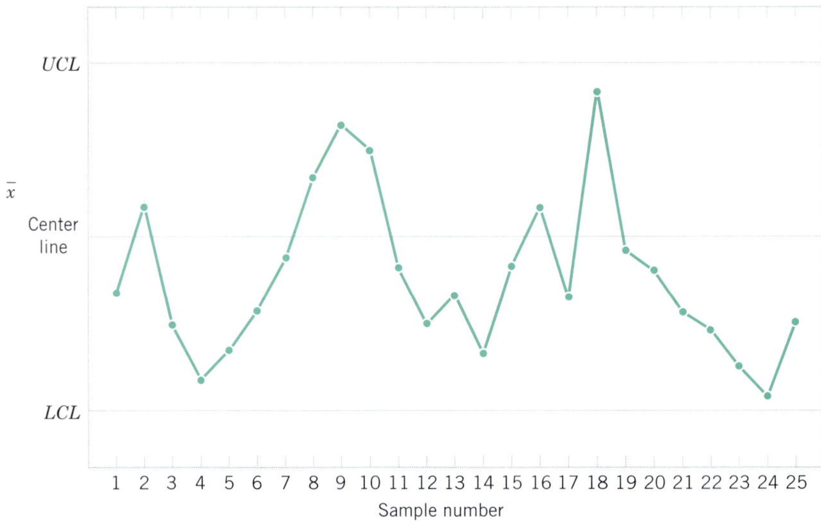

Figure 16-4 An $\overline{X}$ control chart.

is called a run down. This control chart has an unusually long run up (beginning with the fourth point) and an unusually long run down (beginning with the eighteenth point).

In general, we define a run as a sequence of observations of the same type. In addition to runs up and runs down, we could define the types of observations as those above and below the center line, respectively, so two points in a row above the center line would be a run of length 2.

A run of length 8 or more points has a very low probability of occurrence in a random sample of points. Consequently, any type of run of length 8 or more is often taken as a signal of an out-of-control condition. For example, eight consecutive points on one side of the center line will indicate that the process is out of control.

Although runs are an important measure of nonrandom behavior on a control chart, other types of patterns may also indicate an out-of-control condition. For example, consider the $\overline{X}$ chart in Fig. 16-5. Note that the plotted sample averages exhibit a cyclic behavior, yet they all fall within the control limits. Such a pattern may indicate a problem with the process, such as operator fatigue, raw material deliveries, and heat or stress buildup. The yield may be improved by eliminating or reducing the sources of variability causing this cyclic behavior (see Fig. 16-6). In Fig. 16-6, *LSL* and *USL* denote the lower and upper specification limits of the process. These limits represent bounds within which acceptable product must fall and they are often based on customer requirements.

The problem is one of **pattern recognition,** that is, recognizing systematic or nonrandom patterns on the control chart and identifying the reason for this behavior. The ability to interpret

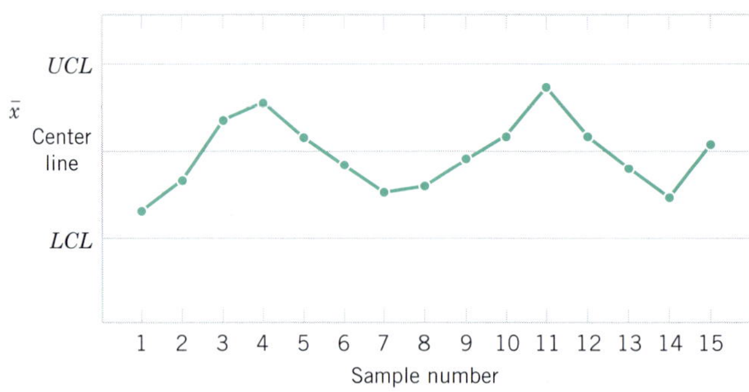

Figure 16-5 An $\overline{X}$ chart with a cyclic pattern.

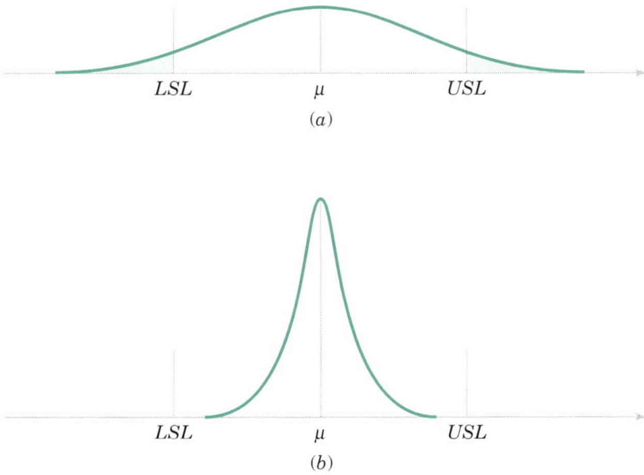

Figure 16-6 (a) Variability with the cyclic pattern. (b) Variability with the cyclic pattern eliminated.

a particular pattern in terms of assignable causes requires experience and knowledge of the process. That is, we must not only know the statistical principles of control charts, but we must also have a good understanding of the process.

The Western Electric Handbook (1956) suggests a set of decision rules for detecting nonrandom patterns on control charts. Specifically, the **Western Electric rules** would conclude that the process is out of control if either

1. One point plots outside 3-sigma control limits.
2. Two out of three consecutive points plot beyond a 2-sigma limit.
3. Four out of five consecutive points plot at a distance of 1-sigma or beyond from the center line.
4. Eight consecutive points plot on one side of the center line.

We have found these rules very effective in practice for enhancing the sensitivity of control charts. Rules 2 and 3 apply to one side of the center line at a time. That is, a point above the *upper* 2-sigma limit followed immediately by a point below the *lower* 2-sigma limit would not signal an out-of-control alarm.

Figure 16-7 shows an $\overline{X}$ control chart for the piston ring process with the 1-sigma, 2-sigma, and 3-sigma limits used in the Western Electric procedure. Notice that these inner

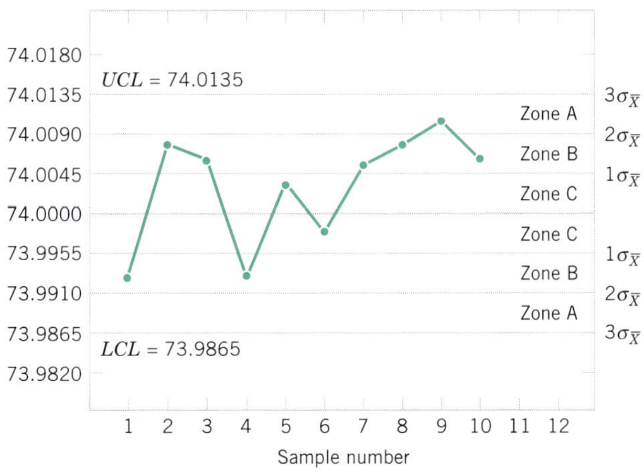

Figure 16-7 The Western Electric zone rules.

limits (sometimes called **warning limits**) partition the control chart into three zones A, B, and C on each side of the center line. Consequently, the Western Electric rules are sometimes called the **run rules** for control charts. Notice that the last four points fall in zone B or beyond. Thus, since four of five consecutive points exceed the 1-sigma limit, the Western Electric procedure will conclude that the pattern is nonrandom and the process is out of control.

16-5 $\overline{X}$ AND R OR S CONTROL CHARTS

When dealing with a quality characteristic that can be expressed as a measurement, it is customary to monitor both the mean value of the quality characteristic and its variability. Control over the average quality is exercised by the control chart for averages, usually called the $\overline{X}$ chart. Process variability can be controlled by either a range chart (R chart) or a standard deviation chart (S chart), depending on how the population standard deviation is estimated.

Suppose that the process mean and standard deviation μ and σ are known and that we can assume that the quality characteristic has a normal distribution. Consider the $\overline{X}$ chart. As discussed previously, we can use μ as the center line for the control chart, and we can place the upper and lower 3-sigma limits at

$$\begin{aligned} UCL &= \mu + 3\sigma/\sqrt{n} \\ LCL &= \mu - 3\sigma/\sqrt{n} \\ CL &= \mu \end{aligned} \qquad (16\text{-}2)$$

When the parameters μ and σ are unknown, we usually estimate them on the basis of preliminary samples, taken when the process is thought to be in control. We recommend the use of at least 20 to 25 preliminary samples. Suppose m preliminary samples are available, each of size n. Typically, n will be 4, 5, or 6; these relatively small sample sizes are widely used and often arise from the construction of rational subgroups. Let the sample mean for the ith sample be $\overline{X}_i$. Then we estimate the mean of the population, μ, by the **grand mean**

$$\hat{\mu} = \overline{\overline{X}} = \frac{1}{m}\sum_{i=1}^{m}\overline{X}_i \qquad (16\text{-}3)$$

Thus, we may take $\overline{\overline{X}}$ as the center line on the $\overline{X}$ control chart.

We may estimate σ from either the standard deviation or the range of the observations within each sample. The sample size is relatively small, so there is little loss in efficiency in estimating σ from the sample ranges.

The relationship between the range R of a sample from a normal population with known parameters and the standard deviation of that population is needed. Since R is a random variable, the quantity $W = R/\sigma$, called the relative range, is also a random variable. The parameters of the distribution of W have been determined for any sample size n. The mean of the distribution of W is called d_2, and a table of d_2 for various n is given in Appendix Table XI. The standard deviation of W is called d_3. Because $R = \sigma W$

$$\mu_R = d_2\sigma \qquad \sigma_R = d_3\sigma \qquad (16\text{-}4)$$

Let R_i be the range of the ith sample, and let

$$\overline{R} = \frac{1}{m}\sum_{i=1}^{m} R_i \tag{16-5}$$

be the average range. Then $\overline{R}$ is an estimator of μ_R and from Equation 16-4 an unbiased estimator of σ is

Estimator of σ from R Chart

$$\hat{\sigma} = \frac{\overline{R}}{d_2} \tag{16-6}$$

Therefore, we may use as our upper and lower control limits for the $\overline{X}$ chart

$$UCL = \overline{\overline{X}} + \frac{3}{d_2\sqrt{n}}\overline{R} \qquad LCL = \overline{\overline{X}} - \frac{3}{d_2\sqrt{n}}\overline{R} \tag{16-7}$$

Define the constant

$$A_2 = \frac{3}{d_2\sqrt{n}} \tag{16-8}$$

Now, once we have computed the sample values $\overline{\overline{x}}$ and $\overline{r}$, the $\overline{X}$ control chart may be defined as follows:

$\overline{X}$ Control Chart (from $\overline{R}$)

The center line and upper and lower control limits for an $\overline{X}$ control chart are

$$UCL = \overline{\overline{x}} + A_2\overline{r} \qquad CL = \overline{\overline{x}} \qquad LCL = \overline{\overline{x}} - A_2\overline{r} \tag{16-9}$$

where the constant A_2 is tabulated for various sample sizes in Appendix Table XI.

The parameters of the R chart may also be easily determined. The center line will obviously be $\overline{R}$. To determine the control limits, we need an estimate of σ_R, the standard deviation of R. Once again, assuming the process is in control, the distribution of the relative range, W, will be useful. We may estimate σ_R from Equation 16-4 as

$$\hat{\sigma}_R = d_3\hat{\sigma} = d_3\frac{\overline{R}}{d_2} \tag{16-10}$$

and we would use as the upper and lower control limits on the R chart

$$UCL = \overline{R} + \frac{3d_3}{d_2}\overline{R} = \left(1 + \frac{3d_3}{d_2}\right)\overline{R}$$

$$LCL = \overline{R} - \frac{3d_3}{d_2}\overline{R} = \left(1 - \frac{3d_3}{d_2}\right)\overline{R} \tag{16-11}$$

Setting $D_3 = 1 - 3d_3/d_2$ and $D_4 = 1 + 3d_3/d_2$ leads to the following definition.

R Chart

The center line and upper and lower control limits for an R chart are

$$UCL = D_4\bar{r} \qquad CL = \bar{r} \qquad LCL = D_3\bar{r} \tag{16-12}$$

where $\bar{r}$ is the sample average range, and the constants D_3 and D_4 are tabulated for various sample sizes in Appendix Table XI.

The LCL for an R chart can be a negative number. In that case, it is customary to set LCL to zero. Because the points plotted on an R chart are nonnegative, no points can fall below an LCL of zero. Also, we often study the R chart first because if the process variability is not constant over time the control limits calculated for the $\overline{X}$ chart can be misleading.

Rather than base control charts on ranges, a more modern approach is to calculate the standard deviation of each subgroup and plot these standard deviations to monitor the process standard deviation σ. This is called an S chart. When an S chart is used, it is common to use these standard deviations to develop control limits for the $\overline{X}$ chart. Typically, the sample size used for subgroups is small (fewer than 10) and in that case there is usually little difference in the $\overline{X}$ chart generated from ranges or standard deviations. However, because computer software is often used to implement control charts, S charts are quite common. Details to construct these charts follow.

In Section 7-3, it was stated that S is a biased estimator of σ. That is, $E(S) = c_4\sigma$ where c_4 is a constant that is near, but not equal to, 1. Furthermore, a calculation similar to the one used for $E(S)$ can derive the standard deviation of the statistic S with the result $\sigma\sqrt{1 - c_4^2}$. Therefore, the center line and three-sigma control limits for S are

$$LCL = c_4\sigma - 3\sigma\sqrt{1 - c_4^2} \qquad CL = c_4\sigma \qquad UCL = c_4\sigma + 3\sigma\sqrt{1 - c_4^2} \tag{16-13}$$

Assume that there are m preliminary samples available, each of size n, and let S_i denote the standard deviation of the ith sample. Define

$$\overline{S} = \frac{1}{m}\sum_{i=1}^{m} S_i \tag{16-14}$$

Because $E(\overline{S}) = c_4\sigma$, an unbiased estimator of σ is $\overline{S}/c_4$ That is,

Estimator of σ from S Chart

$$\hat{\sigma} = \overline{S}/c_4 \tag{16-15}$$

A control chart for standard deviations follows.

S Chart

$$UCL = \bar{s} + 3\frac{\bar{s}}{c_4}\sqrt{1 - c_4^2} \qquad CL = \bar{s} \qquad LCL = \bar{s} - 3\frac{\bar{s}}{c_4}\sqrt{1 - c_4^2} \tag{16-16}$$

The LCL for an S chart can be a negative number, in that case, it is customary to set LCL to zero. When an S chart is used, the estimate for σ in Equation 16-15 is commonly used to calculate the control limits for an $\overline{X}$ chart. This produces the following control limits for an $\overline{X}$ chart.

$\overline{X}$ Control Chart (from $\overline{S}$)

$$UCL = \overline{\overline{x}} + 3\frac{\overline{s}}{c_4\sqrt{n}} \qquad CL = \overline{\overline{x}} \qquad LCL = \overline{s} - 3\frac{\overline{s}}{c_4\sqrt{n}} \qquad (16\text{-}17)$$

EXAMPLE 16-1
Vane Opening

A component part for a jet aircraft engine is manufactured by an investment casting process. The vane opening on this casting is an important functional parameter of the part. We will illustrate the use of $\overline{X}$ and R control charts to assess the statistical stability of this process. Table 16-1 presents 20 samples of five parts each. The values given in the table have been coded by using the last three digits of the dimension; that is, 31.6 should be 0.50316 inch.

The quantities $\overline{\overline{x}} = 33.3$ and $\overline{r} = 5.8$ are shown at the foot of Table 16-1. The value of A_2 for samples of size 5 is $A_2 = 0.577$. Then the trial control limits for the $\overline{X}$ chart are

$$\overline{\overline{x}} \pm A_2\overline{r} = 33.32 \pm (0.577)(5.8) = 33.32 \pm 3.35$$

or

$$UCL = 36.67 \qquad LCL = 29.97$$

Table 16-1 Vane-Opening Measurements

Sample Number	x_1	x_2	x_3	x_4	x_5	$\overline{x}$	r	s
1	33	29	31	32	33	31.6	4	1.67332
2	33	31	35	37	31	33.4	6	2.60768
3	35	37	33	34	36	35.0	4	1.58114
4	30	31	33	34	33	32.2	4	1.64317
5	33	34	35	33	34	33.8	2	0.83666
6	38	37	39	40	38	38.4	3	1.14018
7	30	31	32	34	31	31.6	4	1.51658
8	29	39	38	39	39	36.8	10	4.38178
9	28	33	35	36	43	35.0	15	5.43139
10	38	33	32	35	32	34.0	6	2.54951
11	28	30	28	32	31	29.8	4	1.78885
12	31	35	35	35	34	34.0	4	1.73205
13	27	32	34	35	37	33.0	10	3.80789
14	33	33	35	37	36	34.8	4	1.78885
15	35	37	32	35	39	35.6	7	2.60768
16	33	33	27	31	30	30.8	6	2.48998
17	35	34	34	30	32	33.0	5	2.00000
18	32	33	30	30	33	31.6	3	1.51658
19	25	27	34	27	28	28.2	9	3.42053
20	35	35	36	33	30	33.8	6	2.38747
						$\overline{\overline{x}} = 33.32$	$\overline{r} = 5.8$	$\overline{s} = 2.345$

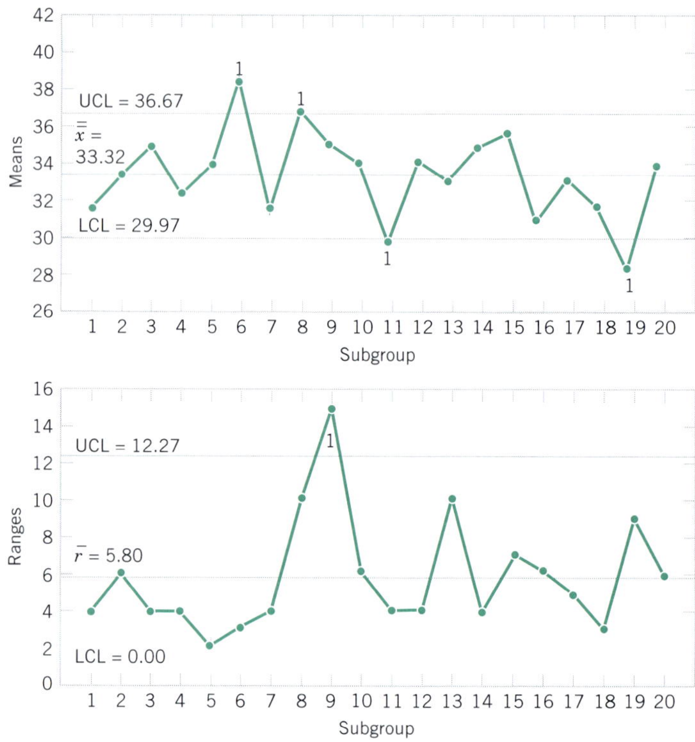

Figure 16-8 The $\overline{X}$ and R control charts for vane opening.

For the R chart, the trial control limits are

$$UCL = D_4\bar{r} = (2.115)(5.8) = 12.27$$
$$LCL = D_3\bar{r} = (0)(5.8) = 0$$

The $\overline{X}$ and R control charts with these trial control limits are shown in Fig. 16-8. Notice that samples 6, 8, 11, and 19 are out of control on the $\overline{X}$ chart and that sample 9 is out of control on the R chart. (These points are labeled with a "1" because they violate the first Western Electric rule.)

For the S chart, the value of $c_4 = 0.94$. Therefore,

$$\frac{3\bar{s}}{c_4}\sqrt{1 - c_4^2} = \frac{3(2.345)}{0.94}\sqrt{1 - 0.94^2} = 2.553$$

and the trial control limits are

$$UCL = 2.345 + 2.553 = 4.898$$
$$LCL = 2.345 - 2.553 = -0.208$$

The LCL is set to zero. If $\bar{s}$ is used to determine the control limits for the $\overline{X}$ chart,

$$\bar{\bar{x}} \pm \frac{3\bar{s}}{c_4\sqrt{n}} = 33.32 \pm \frac{3(2.345)}{0.94\sqrt{5}} = 33.32 \pm 3.35$$

and this result is nearly the same as from $\bar{r}$. The S chart is shown in Fig. 16-9. Because the control limits for the $\overline{X}$ chart calculated from $\bar{s}$ are nearly the same as from $\bar{r}$, the chart is not shown.

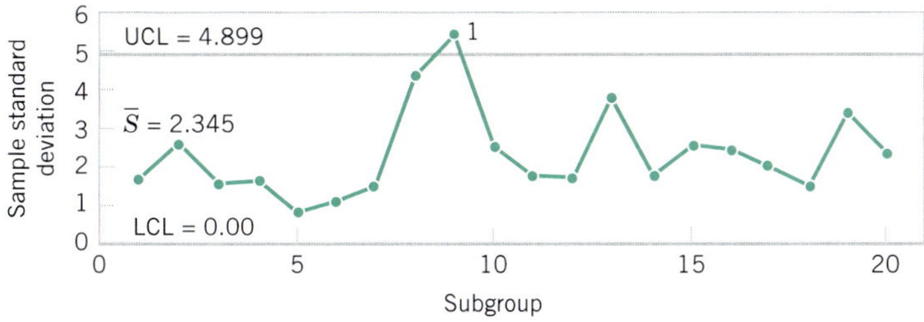

Figure 16-9. The S control chart for vane opening.

Suppose that all of these assignable causes can be traced to a defective tool in the wax-molding area. We should discard these five samples and recompute the limits for the $\overline{X}$ and R charts. These new revised limits are, for the $\overline{X}$ chart,

$$UCL = \bar{\bar{x}} + A_2\bar{r} = 33.21 + (0.577)(5.0) = 36.10$$

$$LCL = \bar{\bar{x}} - A_2\bar{r} = 33.21 - (0.577)(5.0) = 30.33$$

and for the R chart,

$$UCL = D_4\bar{r} = (2.115)(5.0) = 10.57$$

$$LCL = D_3\bar{r} = (0)(5.0) = 0$$

The revised control charts are shown in Fig. 16-10. Notice that we have treated the first 20 preliminary samples as **estimation data** with which to establish control limits. These limits can now be used to judge the statistical control of future production. As each new sample becomes available, the values of $\bar{x}$ and r should be computed and plotted on the control charts. It may be desirable to revise the limits periodically, even if the process remains stable. The limits should always be revised when process improvements are made.

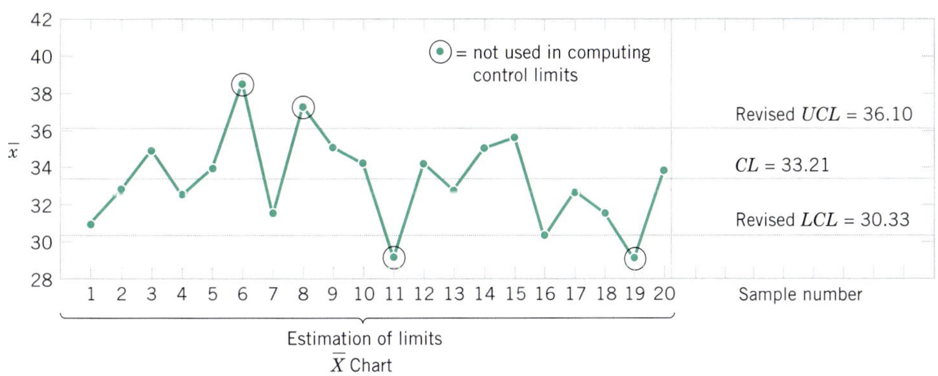

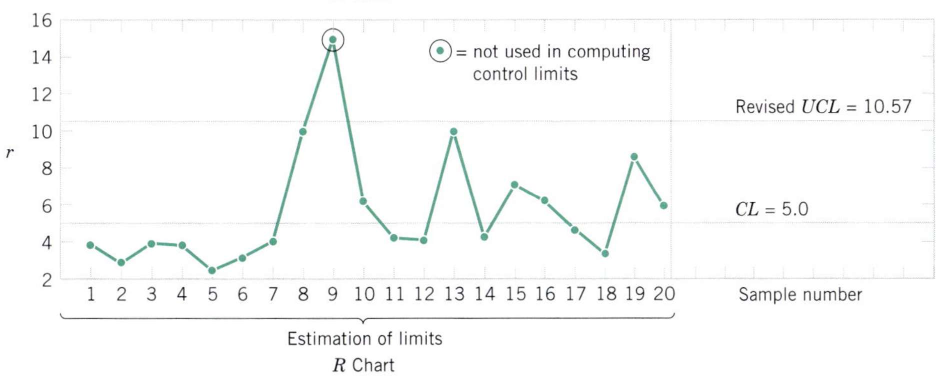

Figure 16-10 The $\overline{X}$ and R control charts for vane opening, revised limits.

Table 16-2 Summary Report from Minitab for the Vane-Opening Data

Test Results for Xbar Chart
TEST 1. One point more than 3.00 sigmas from center line.
Test Failed at points: 6 8 11 19

Test Results for R Chart
TEST 1. One point more than 3.00 sigmas from center line.
Test Failed at points: 9

Computer Construction of $\overline{X}$ and R Control Charts

Many computer programs construct $\overline{X}$ and R control charts. Figures 16-8 and 16-10 show charts similar to those produced by Minitab for the vane-opening data. This program will allow the user to select any multiple of sigma as the width of the control limits and use the Western Electric rules to detect out-of-control points. The program will also prepare a summary report as in Table 16-2 and exclude subgroups from the calculation of the control limits.

EXERCISES FOR SECTION 16-5

16-1. Control charts for $\overline{X}$ and R are to be set up for an important quality characteristic. The sample size is $n = 5$, and $\overline{x}$ and r are computed for each of 35 preliminary samples. The summary data are

$$\sum_{i=1}^{35} \overline{x}_i = 7805 \qquad \sum_{i=1}^{35} r_i = 1200$$

(a) Find trial control limits for $\overline{X}$ and R charts.
(b) Assuming that the process is in control, estimate the process mean and standard deviation.

16-2. Twenty-five samples of size 5 are drawn from a process at one-hour intervals, and the following data are obtained:

$$\sum_{i=1}^{25} \overline{x}_i = 362.75 \qquad \sum_{i=1}^{25} r_i = 8.60 \qquad \sum_{i=1}^{25} s_i = 3.64$$

(a) Find trial control limits for $\overline{X}$ and R charts.
(b) Repeat part (a) for $\overline{X}$ and S charts.

16-3. Control charts are to be constructed for samples of size $n = 4$, and $\overline{x}$ and s are computed for each of 20 preliminary samples as follows:

$$\sum_{i=1}^{20} \overline{x}_i = 4460 \qquad \sum_{i=1}^{20} s_i = 271.6$$

(a) Determine trial control limits for $\overline{X}$ and S charts.
(b) Assuming the process is in control, estimate the process mean and standard deviation.

16-4. Samples of size $n = 6$ are collected from a process every hour. After 20 samples have been collected, we calculate $\overline{\overline{x}} = 20.0$ and $\overline{r}/d_2 = 1.4$.
(a) Find trial control limits for $\overline{X}$ and R charts.
(b) If $\overline{s}/c_4 = 1.5$, determine trial control limits for $\overline{X}$ and S charts.

16-5. An extrusion die is used to produce aluminum rods. The diameter of the rods is a critical quality characteristic. The following table shows $\overline{x}$ and r values for 20 samples of five rods each. Specifications on the rods are 0.5035 ± 0.0010 inch. The values given are the last three digits of the measurement; that is, 34.2 is read as 0.50342.

Sample	$\overline{x}$	r
1	34.2	3
2	31.6	4
3	31.8	4
4	33.4	5
5	35.0	4
6	32.1	2
7	32.6	7
8	33.8	9
9	34.8	10
10	38.6	4
11	35.4	8
12	34.0	6
13	36.0	4
14	37.2	7
15	35.2	3
16	33.4	10
17	35.0	4
18	34.4	7
19	33.9	8
20	34.0	4

(a) Using all the data, find trial control limits for $\overline{X}$ and R charts, construct the chart, and plot the data.
(b) Use the trial control limits from part (a) to identify out-of-control points. If necessary, revise your control limits, assuming that any samples that plot outside the control limits can be eliminated. Estimate σ.

16-6. The copper content of a plating bath is measured three times per day, and the results are reported in ppm. The $\bar{x}$ and r values for 25 days are shown in the following table:

Day	$\bar{x}$	r	Day	$\bar{x}$	r
1	5.45	1.21	14	7.01	1.45
2	5.39	0.95	15	5.83	1.37
3	6.85	1.43	16	6.35	1.04
4	6.74	1.29	17	6.05	0.83
5	5.83	1.35	18	7.11	1.35
6	7.22	0.88	19	7.32	1.09
7	6.39	0.92	20	5.90	1.22
8	6.50	1.13	21	5.50	0.98
9	7.15	1.25	22	6.32	1.21
10	5.92	1.05	23	6.55	0.76
11	6.45	0.98	24	5.90	1.20
12	5.38	1.36	25	5.95	1.19
13	6.03	0.83			

(a) Using all the data, find trial control limits for $\overline{X}$ and R charts, construct the chart, and plot the data. Is the process in statistical control?
(b) If necessary, revise the control limits computed in part (a), assuming that any samples that plot outside the control limits can be eliminated.

16-7. The pull strength of a wire-bonded lead for an integrated circuit is monitored. The following table provides data for 20 samples each of size three.

Sample Number	x_1	x_2	x_3
1	15.4	15.6	15.3
2	15.4	17.1	15.2
3	16.1	16.1	13.5
4	13.5	12.5	10.2
5	18.3	16.1	17.0
6	19.2	17.2	19.4
7	14.1	12.4	11.7
8	15.6	13.3	13.6
9	13.9	14.9	15.5
10	18.7	21.2	20.1
11	15.3	13.1	13.7
12	16.6	18.0	18.0
13	17.0	15.2	18.1
14	16.3	16.5	17.7
15	8.4	7.7	8.4
16	11.1	13.8	11.9
17	16.5	17.1	18.5
18	18.0	14.1	15.9
19	17.8	17.3	12.0
20	11.5	10.8	11.2

(a) Use all the data to determine trial control limits for $\overline{X}$ and R charts, construct the control limits, and plot the data.
(b) Use the control limits from part (a) to identify out-of-control points. If necessary, revise your control limits assuming that any samples that plot outside of the control limits can be eliminated.
(c) Repeat parts (a) and (b) for $\overline{X}$ and S charts.

16-8. The following data was considered in *Quality Engineering* (1999–2000, Vol. 12(1)) "An SPC case study on stabilizing syringe lengths." The syringe length is measured during a pharmaceutical manufacturing process. The following table provides data (in inches) for 20 samples each of size five.

Sample	x_1	x_2	x_3	x_4	x_5
1	4.960	4.946	4.950	4.956	4.958
2	4.958	4.927	4.935	4.940	4.950
3	4.971	4.929	4.965	4.952	4.938
4	4.940	4.982	4.970	4.953	4.960
5	4.964	4.950	4.953	4.962	4.956
6	4.969	4.951	4.955	4.966	4.954
7	4.960	4.944	4.957	4.948	4.951
8	4.969	4.949	4.963	4.952	4.962
9	4.984	4.928	4.960	4.943	4.955
10	4.970	4.934	4.961	4.940	4.965
11	4.975	4.959	4.962	4.971	4.968
12	4.945	4.977	4.950	4.969	4.954
13	4.976	4.964	4.970	4.968	4.972
14	4.970	4.954	4.964	4.959	4.968
15	4.982	4.962	4.968	4.975	4.963
16	4.961	4.943	4.950	4.949	4.957
17	4.980	4.970	4.975	4.978	4.977
18	4.975	4.968	4.971	4.969	4.972
19	4.977	4.966	4.969	4.973	4.970
20	4.975	4.967	4.969	4.972	4.972

(a) Using all the data, find trial control limits for $\overline{X}$ and R charts, construct the chart, and plot the data. Is this process in statistical control?
(b) Use the trial control limits from part (a) to identify out-of-control points. If necessary, revise your control limits assuming that any samples that plot outside of the control limits can be eliminated.
(c) Repeat parts (a) and (b) for $\overline{X}$ and S charts.

16-9. The thickness of a metal part is an important quality parameter. Data on thickness (in inches) are given in the following table, for 25 samples of five parts each.

Sample Number	x_1	x_2	x_3	x_4	x_5
1	0.0629	0.0636	0.0640	0.0635	0.0640
2	0.0630	0.0631	0.0622	0.0625	0.0627
3	0.0628	0.0631	0.0633	0.0633	0.0630
4	0.0634	0.0630	0.0631	0.0632	0.0633
5	0.0619	0.0628	0.0630	0.0619	0.0625
6	0.0613	0.0629	0.0634	0.0625	0.0628
7	0.0630	0.0639	0.0625	0.0629	0.0627
8	0.0628	0.0627	0.0622	0.0625	0.0627
9	0.0623	0.0626	0.0633	0.0630	0.0624
10	0.0631	0.0631	0.0633	0.0631	0.0630
11	0.0635	0.0630	0.0638	0.0635	0.0633
12	0.0623	0.0630	0.0630	0.0627	0.0629
13	0.0635	0.0631	0.0630	0.0630	0.0630
14	0.0645	0.0640	0.0631	0.0640	0.0642
15	0.0619	0.0644	0.0632	0.0622	0.0635
16	0.0631	0.0627	0.0630	0.0628	0.0629
17	0.0616	0.0623	0.0631	0.0620	0.0625
18	0.0630	0.0630	0.0626	0.0629	0.0628
19	0.0636	0.0631	0.0629	0.0635	0.0634
20	0.0640	0.0635	0.0629	0.0635	0.0634
21	0.0628	0.0625	0.0616	0.0620	0.0623
22	0.0615	0.0625	0.0619	0.0619	0.0622
23	0.0630	0.0632	0.0630	0.0631	0.0630
24	0.0635	0.0629	0.0635	0.0631	0.0633
25	0.0623	0.0629	0.0630	0.0626	0.0628

(a) Using all the data, find trial control limits for $\overline{X}$ and R charts, construct the chart, and plot the data. Is the process in statistical control?

(b) Use the trial control limits from part (a) to identify out-of-control points. If necessary, revise your control limits assuming that any samples that plot outside of the control limits can be eliminated.

(c) Repeat parts (a) and (b) for $\overline{X}$ and S charts.

16-10. Apply the Western Electric Rules to the following $\bar{x}$ control chart. The warning limits are shown as dotted lines. Describe any rule violations.

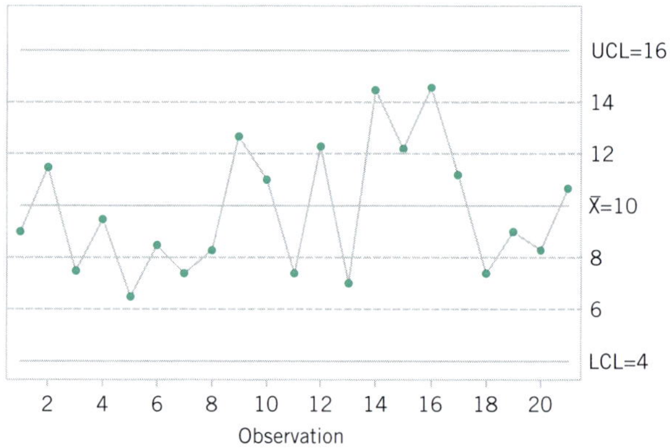

16-11. Consider the data in Exercise 16-7. Calculate the sample standard deviation of all 60 measurements and compare this result to the estimate of σ obtained from your revised $\overline{X}$ and R charts. Explain any differences.

16-12. Consider the data in Exercise 16-8. Calculate the sample standard deviation of all 100 measurements and compare this result to the estimate of σ obtained from your revised $\overline{X}$ and R charts. Explain any differences.

16-6 CONTROL CHARTS FOR INDIVIDUAL MEASUREMENTS

In many situations, the sample size used for process control is $n = 1$; that is, the sample consists of an individual unit. Some examples of these situations are as follows:

1. Automated inspection and measurement technology is used, and every unit manufactured is analyzed.
2. The production rate is very slow, and it is inconvenient to allow sample sizes of $n > 1$ to accumulate before being analyzed.
3. Repeat measurements on the process differ only because of laboratory or analysis error, as in many chemical processes.

4. In process plants, such as papermaking, measurements on some parameters such as coating thickness *across* the roll will differ very little and produce a standard deviation that is much too small if the objective is to control coating thickness *along* the roll.

In such situations, the **individuals control chart** (also called an ***X* chart**) is useful. The control chart for individuals uses the **moving range** of two successive observations to estimate the process variability. The moving range is defined as $MR_i = |X_i - X_{i-1}|$.

An estimate of σ is

$$\hat{\sigma} = \frac{\overline{MR}}{d_2} = \frac{\overline{MR}}{1.128} \tag{16-18}$$

because $d_2 = 1.128$ when two consecutive observations are used to calculate a moving range. It is also possible to establish a control chart on the moving range using D_3 and D_4 for $n = 2$. The parameters for these charts are defined as follows.

Individuals Control Chart

The center line and upper and lower control limits for a control chart for individuals are

$$\begin{aligned} UCL &= \bar{x} + 3\frac{\overline{mr}}{d_2} = \bar{x} + 3\frac{\overline{mr}}{1.128} \\ CL &= \bar{x} \\ LCL &= \bar{x} - 3\frac{\overline{mr}}{d_2} = \bar{x} - 3\frac{\overline{mr}}{1.128} \end{aligned} \tag{16-19}$$

and for a control chart for moving ranges

$$\begin{aligned} UCL &= D_4\overline{mr} = 3.267\overline{mr} \\ CL &= \overline{mr} \\ LCL &= D_3\overline{mr} = 0 \end{aligned}$$

Note that LCL for this moving range chart is always zero. The procedure is illustrated in the following example.

EXAMPLE 16-2 Chemical Process Concentration

Table 16-3 shows 20 observations on concentration for the output of a chemical process. The observations are taken at one-hour intervals. If several observations are taken at the same time, the observed concentration reading will differ only because of measurement error. Since the measurement error is small, only one observation is taken each hour.

To set up the control chart for individuals, note that the sample average of the 20 concentration readings is $\bar{x} = 99.1$ and that the average of the moving ranges of two observations shown in the last column of Table 16-3 is $\overline{mr} = 2.59$. To set up the moving-range chart, we note that $D_3 = 0$ and $D_4 = 3.267$ for $n = 2$. Therefore, the moving-range chart has center line $\overline{mr} = 2.59$, $LCL = 0$, and $UCL = D_4\overline{mr} = (3.267)(2.59) = 8.46$. The control chart is shown as the lower control chart in Fig. 16-11. This control chart was constructed by Minitab. Because no points exceed the upper control limit, we may now set up the control chart for individual concentration measurements. If a moving range of $n = 2$

Table 16-3 Chemical Process Concentration Measurements

Observation	Concentration x	Moving Range mr
1	102.0	
2	94.8	7.2
3	98.3	3.5
4	98.4	0.1
5	102.0	3.6
6	98.5	3.5
7	99.0	0.5
8	97.7	1.3
9	100.0	2.3
10	98.1	1.9
11	101.3	3.2
12	98.7	2.6
13	101.1	2.4
14	98.4	2.7
15	97.0	1.4
16	96.7	0.3
17	100.3	3.6
18	101.4	1.1
19	97.2	4.2
20	101.0	3.8
	$\bar{x} = 99.1$	$\overline{mr} = 2.59$

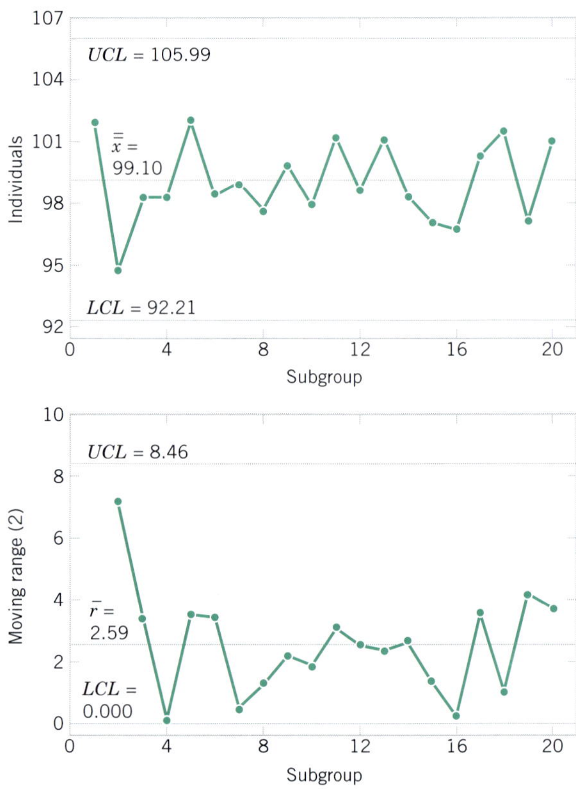

Figure 16-11 Control charts for individuals and the moving range (from Minitab) for the chemical process concentration data.

observations is used, $d_2 = 1.128$. For the data in Table 16-3 we have

$$UCL = \bar{x} + 3\frac{\overline{mr}}{d_2} = 99.1 + 3\frac{2.59}{1.128} = 105.99$$

$$CL = \bar{x} = 99.1$$

$$LCL = \bar{x} - 3\frac{\overline{mr}}{d_2} = 99.1 - 3\frac{2.59}{1.128} = 92.21$$

The control chart for individual concentration measurements is shown as the upper control chart in Fig. 16-11. There is no indication of an out-of-control condition.

The chart for individuals can be interpreted much like an ordinary $\overline{X}$ control chart. A shift in the process average will result in either a point (or points) outside the control limits, or a pattern consisting of a run on one side of the center line.

Some care should be exercised in interpreting patterns on the moving-range chart. The moving ranges are correlated, and this correlation may often induce a pattern of runs or cycles on the chart. The individual measurements are assumed to be uncorrelated, however, and any apparent pattern on the individuals' control chart should be carefully investigated.

The control chart for individuals is very insensitive to small shifts in the process mean. For example, if the size of the shift in the mean is one standard deviation, the average number of points to detect this shift is 43.9. This result is shown later in the chapter. While the performance of the control chart for individuals is much better for large shifts, in many situations the shift of interest is not large and more rapid shift detection is desirable. In these cases, we recommend the *cumulative sum control chart* (discussed in Section 16-10) or an *exponentially weighted moving-average chart* (Montgomery, 2005).

Some individuals have suggested that limits narrower than 3-sigma be used on the chart for individuals to enhance its ability to detect small process shifts. This is a dangerous suggestion, for narrower limits will dramatically increase false alarms such that the charts may be ignored and become useless. If you are interested in detecting small shifts, use the cumulative sum or exponentially weighted moving-average control chart referred to on the previous page.

EXERCISES FOR SECTION 16-6

16-13. Twenty successive hardness measurements are made on a metal alloy, and the data are shown in the following table.

Observation	Hardness	Observation	Hardness
1	51	11	51
2	52	12	57
3	54	13	58
4	55	14	50
5	55	15	53
6	51	16	52
7	52	17	54
8	50	18	50
9	51	19	56
10	56	20	53

(a) Using all the data, compute trial control limits for individual observations and moving-range charts. Construct the chart and plot the data. Determine whether the process is in statistical control. If not, assume assignable causes can be found to eliminate these samples and revise the control limits.

(b) Estimate the process mean and standard deviation for the in-control process.

16-14. In a semiconductor manufacturing process CVD metal thickness was measured on 30 wafers obtained over approximately two weeks. Data are shown in the following table.

(a) Using all the data, compute trial control limits for individual observations and moving-range charts. Construct the chart and plot the data. Determine whether the process is in statistical control. If not, assume assignable causes can be found to eliminate these samples and revise the control limits.

(b) Estimate the process mean and standard deviation for the in-control process.

Wafer	x	Wafer	x
1	16.8	16	15.4
2	14.9	17	14.3
3	18.3	18	16.1
4	16.5	19	15.8
5	17.1	20	15.9
6	17.4	21	15.2
7	15.9	22	16.7
8	14.4	23	15.2
9	15.0	24	14.7
10	15.7	25	17.9
11	17.1	26	14.8
12	15.9	27	17.0
13	16.4	28	16.2
14	15.8	29	15.6
15	15.4	30	16.3

16-15. The diameter of holes is measured in consecutive order by an automatic sensor. The results of measuring 25 holes are in the following table.

Sample	Diameter	Sample	Diameter
1	9.94	14	9.99
2	9.93	15	10.12
3	10.09	16	9.81
4	9.98	17	9.73
5	10.11	18	10.14
6	9.99	19	9.96
7	10.11	20	10.06
8	9.84	21	10.11
9	9.82	22	9.95
10	10.38	23	9.92
11	9.99	24	10.09
12	10.41	25	9.85
13	10.36		

(a) Using all the data, compute trial control limits for individual observations and moving-range charts. Construct the control chart and plot the data. Determine whether the process is in statistical control. If not, assume assignable causes can be found to eliminate these samples and revise the control limits.
(b) Estimate the process mean and standard deviation for the in-control process.

16-16. The viscosity of a chemical intermediate is measured every hour. Twenty samples each of size $n = 1$, are in the following table.

Sample	Viscosity	Sample	Viscosity
1	495	11	493
2	491	12	507
3	501	13	503
4	501	14	475
5	512	15	497
6	540	16	499
7	492	17	468
8	504	18	486
9	542	19	511
10	508	20	487

(a) Using all the data, compute trial control limits for individual observations and moving-range charts. Determine whether the process is in statistical control. If not, assume assignable causes can be found to eliminate these samples and revise the control limits.
(b) Estimate the process mean and standard deviation for the in-control process.

16-17. The following table of data was analyzed in *Quality Engineering* (1991–1992, Vol. 4(1)). The average particle size of raw material was obtained from 25 successive samples.

Observation	Size	Observation	Size
1	96.1	14	100.5
2	94.4	15	103.1
3	116.2	16	93.1
4	98.8	17	93.7
5	95.0	18	72.4
6	120.3	19	87.4
7	104.8	20	96.1
8	88.4	21	97.1
9	106.8	22	95.7
10	96.8	23	94.2
11	100.9	24	102.4
12	117.7	25	131.9
13	115.6		

(a) Using all the data, compute trial control limits for individual observations and moving-range charts. Construct the chart and plot the data. Determine whether the process is in statistical control. If not, assume assignable causes can be found to eliminate these samples and revise the control limits.
(b) Estimate the process mean and standard deviation for the in-control process.

16-7 PROCESS CAPABILITY

It is usually necessary to obtain some information about the process capability, that is, the performance of the process when it is operating in control. Two graphical tools, the **tolerance chart** (or tier chart) and the **histogram,** are helpful in assessing process capability. The tolerance chart for all 20 samples from the vane-manufacturing process is shown in Fig. 16-12. The specifications on vane opening are 0.5030 ± 0.0010 in. In terms of the coded data, the upper specification limit is $USL = 40$ and the lower specification limit is $LSL = 20$, and these limits are shown on the chart in Fig. 16-12. Each measurement is plotted on the tolerance chart. Measurements from the same subgroup are connected with lines. The tolerance chart is useful in revealing patterns over time in the individual measurements, or it may show that a particular value of $\bar{x}$ or r was produced by one or two unusual observations in the sample. For example, note the two unusual observations in sample 9 and the single unusual observation in sample 8. Note also that it is appropriate to plot the specification limits on the tolerance chart, since it is a chart of individual measurements. **It is never appropriate to plot specification limits on a control chart or to use the specifications in determining the control limits.** Specification limits and control limits are unrelated. Finally, note from Fig. 16-12 that the process is running off-center from the nominal dimension of 30 (or 0.5030 inch).

The histogram for the vane-opening measurements is shown in Fig. 16-13. The observations from samples 6, 8, 9, 11, and 19 (corresponding to out of-control points on either the $\overline{X}$ or R chart) have been deleted from this histogram. The general impression from examining this histogram is that the process is capable of meeting the specification but that it is running off-center.

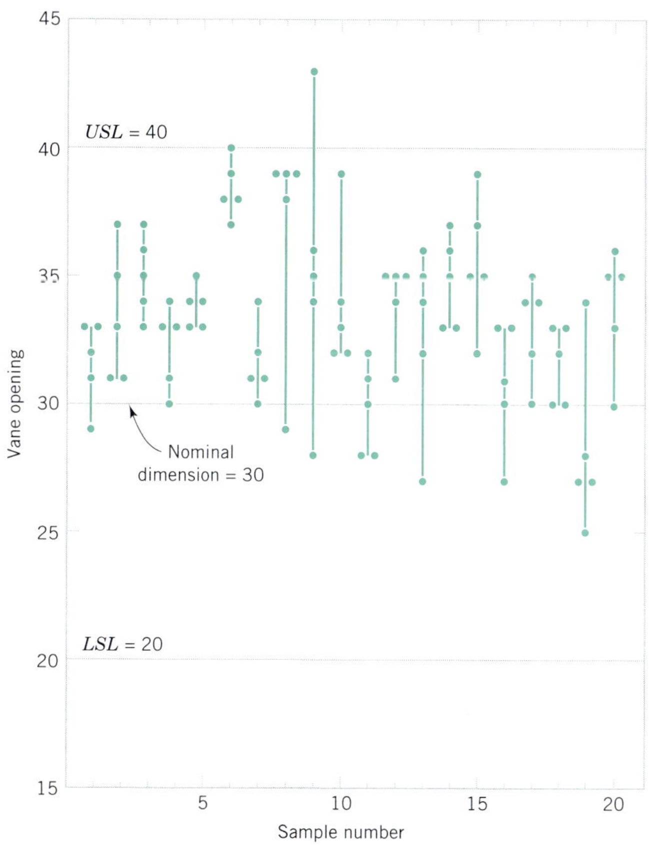

Figure 16-12
Tolerance diagram of vane openings.

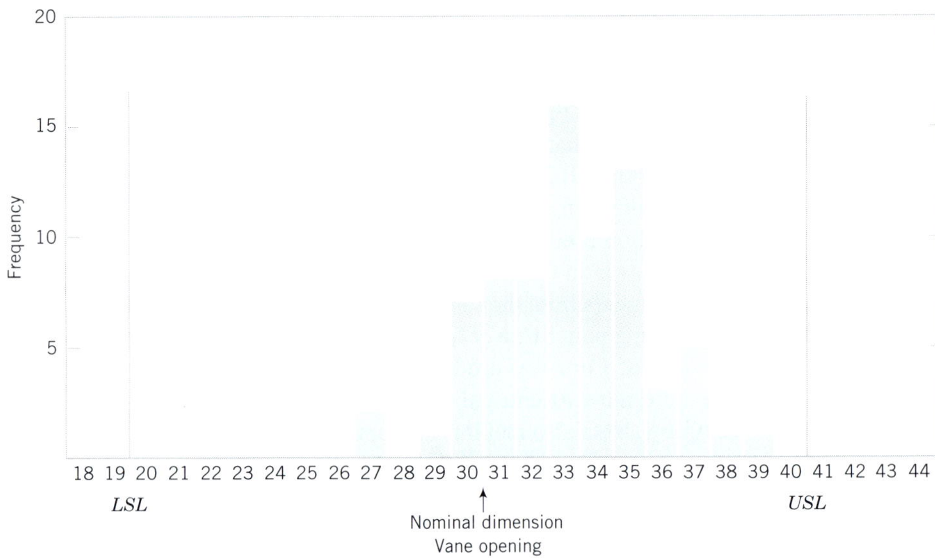

Figure 16-13 Histogram for vane opening.

Another way to express process capability is in terms of an index that is defined as follows.

Process Capability Ratio

The **process capability ratio** (PCR) is

$$PCR = \frac{USL - LSL}{6\sigma} \tag{16-20}$$

The numerator of PCR is the width of the specifications. The limits 3σ on either side of the process mean are sometimes called **natural tolerance limits,** for these represent limits that an in-control process should meet with most of the units produced. Consequently, 6σ is often referred to as the width of the process. For the vane opening, where our sample size is 5, we could estimate σ as

$$\hat{\sigma} = \frac{\bar{r}}{d_2} = \frac{5.0}{2.326} = 2.15$$

Therefore, the PCR is estimated to be

$$PCR = \frac{USL - LSL}{6\hat{\sigma}} = \frac{40 - 20}{6(2.15)} = 1.55$$

The PCR has a natural interpretation: $(1/PCR)100\%$ is just the percentage of the specifications' width used by the process. Thus, the vane-opening process uses approximately $(1/1.55)100\% = 64.5\%$ of the specifications' width.

Figure 16-14(a) shows a process for which the PCR exceeds unity. Since the process natural tolerance limits lie inside the specifications, very few defective or nonconforming

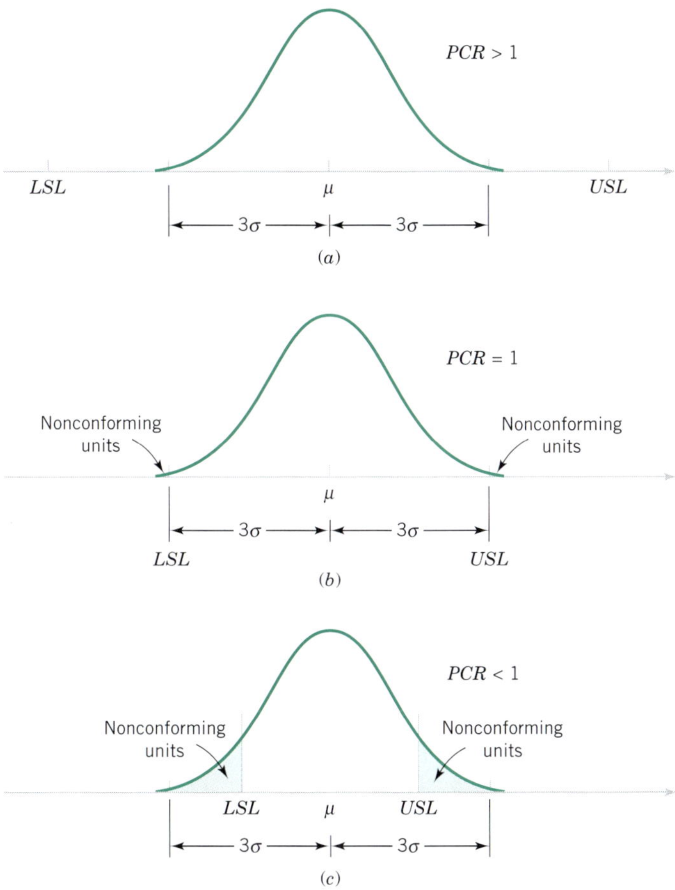

Figure 16-14 Process fallout and the process capability ratio (PCR).

units will be produced. If $PCR = 1$, as shown in Fig. 16-14(b), more nonconforming units result. In fact, for a normally distributed process, if $PCR = 1$, the fraction nonconforming is 0.27%, or 2700 parts per million. Finally, when the PCR is less than unity, as in Fig. 16-14(c), the process is very yield-sensitive and a large number of nonconforming units will be produced.

The definition of the PCR given in Equation 16-20 implicitly assumes that the process is centered at the nominal dimension. If the process is running off-center, its **actual capability** will be less than indicated by the PCR. It is convenient to think of PCR as a measure of **potential capability,** that is, capability with a centered process. If the process is not centered, a measure of actual capability is often used. This ratio, called PCR_k, is defined below.

PCR_k

$$PCR_k = \min\left[\frac{USL - \mu}{3\sigma}, \frac{\mu - LSL}{3\sigma}\right] \qquad (16\text{-}21)$$

In effect, PCR_k is a one-sided process capability ratio that is calculated relative to the specification limit nearest to the process mean. For the vane-opening process, we find that the estimate of the process capability ratio PCR_k (after deleting the samples corresponding to out-of-control points) is

$$\widehat{PCR}_k = \min\left[\frac{USL - \bar{\bar{x}}}{3\hat{\sigma}}, \frac{\bar{\bar{x}} - LSL}{3\hat{\sigma}}\right]$$

$$= \min\left[\frac{40 - 33.21}{3(2.15)} = 1.06, \frac{33.21 - 20}{3(2.15)} = 2.04\right] = 1.05$$

Note that if $PCR = PCR_k$, the process is centered at the nominal dimension. Since $\widehat{PCR}_k = 1.05$ for the vane-opening process and $\widehat{PCR} = 1.55$, the process is obviously running off-center, as was first noted in Figs. 16-10 and 16-13. This off-center operation was ultimately traced to an oversized wax tool. Changing the tooling resulted in a substantial improvement in the process (Montgomery, 2005).

The fractions of nonconforming output (or fallout) below the lower specification limit and above the upper specification limit are often of interest. Suppose that the output from a normally distributed process in statistical control is denoted as X. The fractions are determined from

$$P(X < LSL) = P(Z < (LSL - \mu)/\sigma) \qquad P(X > USL) = P(Z > (USL - \mu)/\sigma)$$

EXAMPLE 16-3
Electrical Current

For an electronic manufacturing process a current has specifications of 100 ± 10 milliamperes. The process mean μ and standard deviation σ are 107.0 and 1.5, respectively. The process mean is nearer to the *USL*. Consequently,

$$PCR = (110 - 90)/(6 \cdot 1.5) = 2.22 \quad \text{and} \quad PCR_k = (110 - 107)/(3 \cdot 1.5) = 0.67$$

The small PCR_k indicates that the process is likely to produce currents outside of the specification limits. From the normal distribution in Appendix Table II

$$P(X < LSL) = P(Z < (90 - 107)/1.5) = P(Z < -11.33) = 0$$

$$P(X > USL) = P(Z > (110 - 107)/1.5) = P(Z > 2) = 0.023$$

For this example, the relatively large probability of exceeding the *USL* is a warning of potential problems with this criterion even if none of the measured observations in a preliminary sample exceed this limit. We emphasize that the fraction-nonconforming calculation assumes that the observations are normally distributed and the process is in control. Departures from normality can seriously affect the results. The calculation should be interpreted as an approximate guideline for process performance. To make matters worse, μ and σ need to be estimated from the data available and a small sample size can result in poor estimates that further degrade the calculation.

Montgomery (2005) provides guidelines on appropriate values of the *PCR* and a table relating fallout for a normally distributed process in statistical control to the value of *PCR*. Many U.S. companies use $PCR = 1.33$ as a minimum acceptable target and $PCR = 1.66$ as a minimum target for strength, safety, or critical characteristics. Some companies require that internal processes and those at suppliers achieve a $PCR_k = 2.0$. Figure 16-15 illustrates a process with $PCR = PCR_k = 2.0$. Assuming a normal distribution, the calculated fallout for this process is 0.0018 parts per million. A process with $PCR_k = 2.0$ is referred to as a **six-sigma process** because the distance from the process mean to the nearest specification is six standard deviations. The reason that such a large process capability is often required is that it is difficult to maintain a process mean at the center of the specifications for long periods of

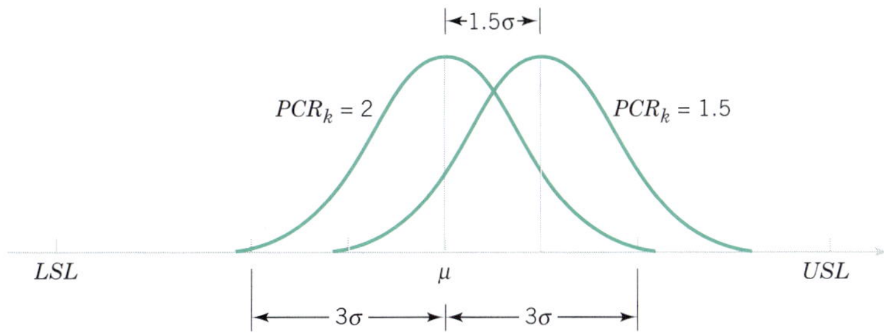

Figure 16-15 Mean of a six-sigma process shifts by 1.5 standard deviations.

time. A common model that is used to justify the importance of a six-sigma process is illustrated by referring to Fig. 16-15. If the process mean shifts off-center by 1.5 standard deviations, the PCR_k decreases to $4.5\sigma/3\sigma = 1.5$. Assuming a normally distributed process, the fallout of the shifted process is **3.4 parts per million.** Consequently, the mean of a 6-sigma process can shift 1.5 standard deviations from the center of the specifications and still maintain a fallout of 3.4 parts per million.

In addition, some U.S. companies, particularly the automobile industry, have adopted the terminology $C_p = PCR$ and $C_{pk} = PCR_k$. Because C_p has another meaning in statistics (in multiple regression) we prefer the traditional notation PCR and PCR_k.

We repeat that process capability calculations are meaningful only for stable processes; that is, processes that are in control. A process capability ratio indicates whether or not the natural or chance variability in a process is acceptable relative to the specifications.

EXERCISES FOR SECTION 16-7

16-18. Suppose that a quality characteristic is normally distributed with specifications at 100 ± 20. The process standard deviation is 6.

(a) Suppose that the process mean is 100. What are the natural tolerance limits? What is the fraction defective? Calculate PCR and PCR_k and interpret these ratios.

(b) Suppose that the process mean is 106. What are the natural tolerance limits? What is the fraction defective? Calculate PCR and PCR_k and interpret these ratios.

16-19. Suppose that a quality characteristic is normally distributed with specifications from 20 to 32 units.

(a) What value is needed for σ to achieve a PCR of 1.5?

(b) What value for the process mean minimizes the fraction defective? Does this choice for the mean depend on the value of σ?

16-20. Suppose that a quality characteristic is normally distributed with specifications from 10 to 30 units. The process standard deviation is 2 units.

(a) Calculate the natural tolerance limits, fraction defective, PCR, and PCR_k when the process mean is 20.

(b) Suppose the process mean shifts higher by 1.5 standard deviations. Recalculate the quantities in part (b).

(c) Compare the results in parts (a) and (b) and comment on any differences.

16-21. A normally distributed process uses 66.7% of the specification band. It is centered at the nominal dimension, located halfway between the upper and lower specification limits.

(a) Estimate PCR and PCR_k. Interpret these ratios.

(b) What fallout level (fraction defective) is produced?

16-22. A normally distributed process uses 85% of the specification band. It is centered at the nominal dimension, located halfway between the upper and lower specification limits.

(a) Estimate PCR and PCR_k. Interpret these ratios.

(b) What fallout level (fraction defective) is produced?

16-23. Reconsider Exercise 16-1. Suppose that the quality characteristic is normally distributed with specification at 220 ± 40. What is the fallout level? Estimate PCR and PCR_k and interpret these ratios.

16-24. Reconsider Exercise 16-2, where the specification limits are 14.50 ± 0.50.

(a) What conclusions can you draw about the ability of the process to operate within these limits? Estimate the percentage of defective items that will be produced.

(b) Estimate PCR and PCR_k. Interpret these ratios.

16-25. Reconsider Exercise 16-3. Suppose that the variable is normally distributed with specifications at 220 ± 50. What is the proportion out of specifications? Estimate and interpret PCR and PCR_k.

16-26. Reconsider Exercise 16-4(a). Assuming that both charts exhibit statistical control and that the process specifications are at 20 ± 5, estimate PCR and PCR_k and interpret these ratios.

16-27. Reconsider Exercise 16-5. Use the revised control limits and process estimates.

(a) Estimate PCR and PCR_k. Interpret these ratios.
(b) What percentage of defectives is being produced by this process?

16-28. Reconsider Exercise 16-6. Given that the specifications are at 6.0 ± 1.0, estimate PCR and PCR_k and interpret these ratios.

16-29. Reconsider Exercise 16-7. Using the process estimates, what is the fallout level if the specifications are 10 ± 5 mm? Estimate PCR and interpret this ratio.

16-30. Reconsider 16-16. The viscosity specifications are at 500 ± 25. Calculate estimates of the process capability ratios PCR and PCR_k for this process and provide an interpretation.

16-8 ATTRIBUTE CONTROL CHARTS

16-8.1 *P* Chart (Control Chart for Proportions)

Often it is desirable to classify a product as either defective or nondefective on the basis of comparison with a standard. This classification is usually done to achieve economy and simplicity in the inspection operation. For example, the diameter of a ball bearing may be checked by determining whether it will pass through a gauge consisting of circular holes cut in a template. This kind of measurement would be much simpler than directly measuring the diameter with a device such as a micrometer. Control charts for attributes are used in these situations. Attribute control charts often require a considerably larger sample size than do their variable measurements counterparts. In this section, we will discuss the **fraction-defective control chart,** or ***P* chart.** Sometimes the P chart is called the control chart for fraction nonconforming.

Suppose D is the number of defective units in a random sample of size n. We assume that D is a binomial random variable with unknown parameter p. The fraction defective

$$P = \frac{D}{n}$$

of each sample is plotted on the chart. Furthermore, the variance of the statistic P is

$$\sigma_P^2 = \frac{p(1-p)}{n}$$

Therefore, a P chart for fraction defective could be constructed using p as the center line and control limits at

$$UCL = p + 3\sqrt{\frac{p(1-p)}{n}} \qquad LCL = p - 3\sqrt{\frac{p(1-p)}{n}} \tag{16-22}$$

However, the true process fraction defective is almost always unknown and must be estimated using the data from preliminary samples.

Suppose that m preliminary samples each of size n are available, and let D_i be the number of defectives in the ith sample. Then $P_i = D_i/n$ is the sample fraction defective in the ith sample. The average fraction defective is

$$\overline{P} = \frac{1}{m}\sum_{i=1}^{m} P_i = \frac{1}{mn}\sum_{i=1}^{m} D_i \tag{16-23}$$

Now $\overline{P}$ may be used as an estimator of p in the center line and control limit formulas.

P Chart

The center line and upper and lower control limits for the P chart are

$$UCL = \bar{p} + 3\sqrt{\frac{\bar{p}(1-\bar{p})}{n}} \quad CL = \bar{p} \quad LCL = \bar{p} - 3\sqrt{\frac{\bar{p}(1-\bar{p})}{n}} \tag{16-24}$$

where $\bar{p}$ is the observed value of the average fraction defective.

These control limits are based on the normal approximation to the binomial distribution. When p is small, the normal approximation may not always be adequate. In such cases, we may use control limits obtained directly from a table of binomial probabilities. If $\bar{p}$ is small, the lower control limit obtained from the normal approximation may be a negative number. If this should occur, it is customary to consider zero as the lower control limit.

EXAMPLE 16-4
Ceramic Substrate

Suppose we wish to construct a fraction-defective control chart for a ceramic substrate production line. We have 20 preliminary samples, each of size 100; the number of defectives in each sample is shown in Table 16-4. Assume that the samples are numbered in the sequence of production. Note that $\bar{p} = (800/2000) = 0.40$; therefore, the trial parameters for the control chart are

$$UCL = 0.40 + 3\sqrt{\frac{(0.40)(0.60)}{100}} = 0.55 \qquad CL = 0.40$$

$$LCL = 0.40 - 3\sqrt{\frac{(0.40)(0.60)}{100}} = 0.25$$

The control chart is shown in Fig. 16-16. All samples are in control. If they were not, we would search for assignable causes of variation and revise the limits accordingly. This chart can be used for controlling future production.

Although this process exhibits statistical control, its defective rate ($\bar{p} = 0.40$) is very poor. We should take appropriate steps to investigate the process to determine why such a large number of defective units is being produced. Defective units should be analyzed to determine the specific types of defects present. Once the defect types are known, process changes should be investigated to determine their impact on defect levels. Designed experiments may be useful in this regard.

Table 16-4 Number of Defectives in Samples of 100 Ceramic Substrates

Sample	No. of Defectives	Sample	No. of Defectives
1	44	11	36
2	48	12	52
3	32	13	35
4	50	14	41
5	29	15	42
6	31	16	30
7	46	17	46
8	52	18	38
9	44	19	26
10	48	20	30

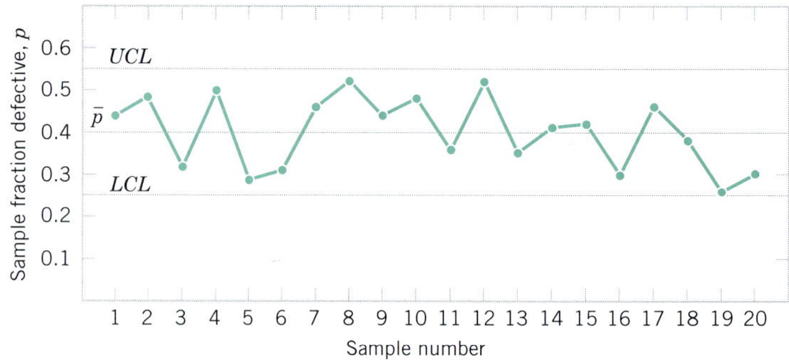

Figure 16-16 *P* chart for a ceramic substrate.

Computer software also produces an ***NP* chart.** This is just a control chart of $nP = D$, the number of defectives in a sample. The points, center line, and control limits for this chart are just multiples (times n) of the corresponding elements of a P chart. The use of an NP chart avoids the fractions in a P chart.

16-8.2 *U* Chart (Control Chart for Defects per Unit)

It is sometimes necessary to monitor the number of defects in a unit of product rather than the fraction defective. Suppose that in the production of cloth it is necessary to control the number of defects per yard or that in assembling an aircraft wing the number of missing rivets must be controlled. In these situations we may use the control chart for defects per unit, or the ***U* chart.** Many defects-per-unit situations can be modeled by the Poisson distribution.

If each sample consists of n units and there are C total defects in the sample,

$$U = \frac{C}{n}$$

is the average number of defects per unit. A U chart may be constructed for such data.

If the number of defects in a unit is a Poisson random variable with parameter λ, the mean and variance of this distribution are both λ. Each point on the chart is U, the average number of defects per unit from a sample of n units. Therefore, the mean of U is λ and the variance of U is λ/n. The control limits for the U chart with known λ are:

$$UCL = \lambda + 3\sqrt{\frac{\lambda}{n}}$$

$$LCL = \lambda - 3\sqrt{\frac{\lambda}{n}} \tag{16-25}$$

If there are m preliminary samples, and the number of defects per unit in these samples are $U_1, U_2, \ldots, U_m$, the estimator of the average number of defects per unit is

$$\overline{U} = \frac{1}{m}\sum_{i=1}^{m} U_i \tag{16-26}$$

Now $\overline{U}$ may be used as an estimator of λ in the centerline and control limit formulas.

U Chart

The center line and upper and lower control limits on the U chart are

$$UCL = \bar{u} + 3\sqrt{\frac{\bar{u}}{n}} \qquad CL = \bar{u} \qquad LCL = \bar{u} - 3\sqrt{\frac{\bar{u}}{n}} \tag{16-27}$$

where $\bar{u}$ is the average number of defects per unit.

These control limits are based on the normal approximation to the Poisson distribution. When λ is small, the normal approximation may not always be adequate. In such cases, we may use control limits obtained directly from a table of Poisson probabilities. If $\bar{u}$ is small, the lower control limit obtained from the normal approximation may be a negative number. If this should occur, it is customary to use zero as the lower control limit.

EXAMPLE 16-5
Printed Circuit Boards

Printed circuit boards are assembled by a combination of manual assembly and automation. Surface Mount Technology (SMT) is used to make the mechanical and electrical connections of the components to the board. Every hour five boards are selected and inspected for process-control purposes. The number of defects in each sample of five boards is noted. Results for 20 samples are shown in Table 16-5.

The center line for the U chart is

$$\bar{u} = \frac{1}{20}\sum_{i=1}^{20} u_i = \frac{32.0}{20} = 1.6$$

and the upper and lower control limits are

$$UCL = \bar{u} + 3\sqrt{\frac{\bar{u}}{n}} = 1.6 + 3\sqrt{\frac{1.6}{5}} = 3.3$$

$$LCL = \bar{u} - 3\sqrt{\frac{\bar{u}}{n}} = 1.6 - 3\sqrt{\frac{1.6}{5}} < 0$$

The control chart is plotted in Fig. 16-17. Because LCL is negative, it is set to 0. From the control chart in Fig. 16-17, we see that the process is in control. However, eight defects per group of five circuit boards are too many (about $8/5 = 1.6$ defects/board), and the process needs improvement. An investigation needs to be made of the specific types of defects found on the printed circuit boards. This will usually suggest potential avenues for process improvement.

Table 16-5 Number of Defects in Samples of Five Printed Circuit Boards

Sample	Number of Defects	Defects per Unit u_i	Sample	Number of Defects	Defects per Unit u_i
1	6	1.2	11	9	1.8
2	4	0.8	12	15	3.0
3	8	1.6	13	8	1.6
4	10	2.0	14	10	2.0
5	9	1.8	15	8	1.6
6	12	2.4	16	2	0.4
7	16	3.2	17	7	1.4
8	2	0.4	18	1	0.2
9	3	0.6	19	7	1.4
10	10	2.0	20	13	2.6

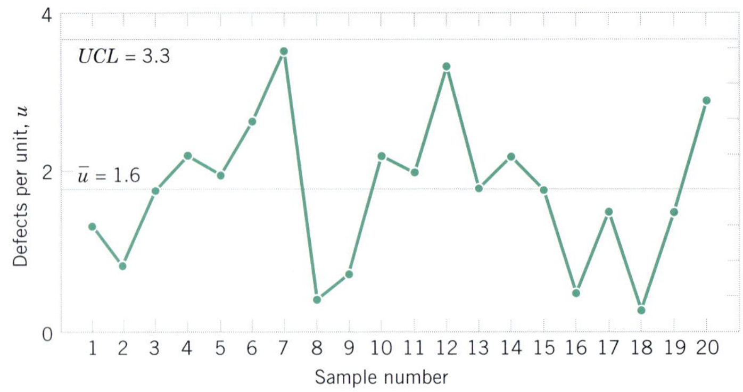

Figure 16-17 *U* chart of defects per unit on printed circuit boards.

Computer software also produces a ***C* chart.** This is just a control chart of *C*, the total of defects in a sample. The points, center line, and control limits for this chart are just multiples (times *n*) of the corresponding elements of a *U* chart. The use of a *C* chart avoids the fractions that can occur in a *U* chart.

EXERCISES FOR SECTION 16-8

16-31. An early example of SPC was described in *Industrial Quality Control* (1944, Vol. 1(1), pp. 4–9) "The introduction of quality control at Colonial Radio Corporation." The following are the fractions defective of shaft and washer assemblies during the month of April in samples of $n = 1500$ each:

Sample	Fraction Defective	Sample	Fraction Defective
1	0.11	11	0.03
2	0.06	12	0.03
3	0.1	13	0.04
4	0.11	14	0.07
5	0.14	15	0.04
6	0.11	16	0.04
7	0.14	17	0.04
8	0.03	18	0.03
9	0.02	19	0.06
10	0.03	20	0.06

(a) Set up a *P* chart for this process. Is this process in statistical control?

(b) Suppose that instead of $n = 1500$, $n = 100$. Use the data given to set up a *P* chart for this process. Revise the control limits if necessary.

(c) Compare your control limits for the *P* charts in parts (a) and (b). Explain why they differ. Also, explain why your assessment about statistical control differs for the two sizes of *n*.

16-32. Suppose the following fraction defective has been found in successive samples of size 100 (read down):

0.09	0.03	0.12
0.10	0.05	0.14
0.13	0.13	0.06
0.08	0.10	0.05
0.14	0.14	0.14
0.09	0.07	0.11
0.10	0.06	0.09
0.15	0.09	0.13
0.13	0.08	0.12
0.06	0.11	0.09

(a) Using all the data, compute trial control limits for a fraction-defective control chart, construct the chart, and plot the data.

(b) Determine whether the process is in statistical control. If not, assume assignable causes can be found and out-of-control points eliminated. Revise the control limits.

16-33. The following are the numbers of defective solder joints found during successive samples of 500 solder joints:

Day	No. of Defectives	Day	No. of Defectives
1	106	12	37
2	116	13	25
3	164	14	88
4	89	15	101
5	99	16	64
6	40	17	51
7	112	18	74
8	36	19	71
9	69	20	43
10	74	21	80
11	42		

(a) Using all the data, compute trial control limits for a fraction-defective control chart, construct the chart, and plot the data.
(b) Determine whether the process is in statistical control. If not, assume assignable causes can be found and out-of-control points eliminated. Revise the control limits.

16-34. The following represent the number of defects per 1000 feet in rubber-covered wire: 1, 1, 3, 7, 8, 10, 5, 13, 0, 19, 24, 6, 9, 11, 15, 8, 3, 6, 7, 4, 9, 20, 11, 7, 18, 10, 6, 4, 0, 9, 7, 3, 1, 8, 12. Do the data come from a controlled process?

16-35. The following represent the number of solder defects observed on 24 samples of five printed circuit boards: 7, 6, 8, 10, 24, 6, 5, 4, 8, 11, 15, 8, 4, 16, 11, 12, 8, 6, 5, 9, 7, 14, 8, 21.
(a) Using all the data, compute trial control limits for a U control chart, construct the chart, and plot the data.
(b) Can we conclude that the process is in control using a U chart? If not, assume assignable causes can be found, list points and revise the control limits.

16-36. Consider the data on the number of earthquakes of magnitude 7.0 or greater by year in Exercise 9-92.
(a) Construct a U chart for this data with a sample size of $n = 1$?
(b) Do the data appear to be generated by an in-control process? Explain.

16-9 CONTROL CHART PERFORMANCE

Specifying the control limits is one of the critical decisions that must be made in designing a control chart. By moving the control limits further from the center line, we decrease the risk of a type I error—that is, the risk of a point falling beyond the control limits, indicating an out-of-control condition when no assignable cause is present. However, widening the control limits will also increase the risk of a type II error—that is, the risk of a point falling between the control limits when the process is really out of control. If we move the control limits closer to the center line, the opposite effect is obtained: The risk of type I error is increased, while the risk of type II error is decreased.

The control limits on a Shewhart control chart are customarily located a distance of plus or minus three standard deviations of the variable plotted on the chart from the center line. That is, the constant k in equation 16-1 should be set equal to 3. These limits are called **3-sigma control limits.**

A way to evaluate decisions regarding sample size and sampling frequency is through the **average run length (ARL)** of the control chart. Essentially, the ARL is the average number of points that must be plotted before a point indicates an out-of-control condition. For any Shewhart control chart, the ARL can be calculated from the mean of a geometric random variable (Montgomery 2005). Suppose that p is the probability that any point exceeds the control limits. Then

Average Run Length

$$\text{ARL} = \frac{1}{p} \tag{16-28}$$

Thus, for an $\overline{X}$ chart with 3-sigma limits, $p = 0.0027$ is the probability that a single point falls outside the limits when the process is in control, so

$$\text{ARL} = \frac{1}{p} = \frac{1}{0.0027} \cong 370$$

is the average run length of the $\overline{X}$ chart when the process is in control. That is, even if the process remains in control, an out-of-control signal will be generated every 370 points, on the average.

Consider the piston ring process discussed in Section 16-4.2, and suppose we are sampling every hour. Thus, we will have a **false alarm** about every 370 hours on the average. Suppose we are using a sample size of $n = 5$ and that when the process goes out of control the mean shifts to 74.0135 millimeters. Then, the probability that $\overline{X}$ falls between the control limits of Fig. 16-3 is equal to

$$\begin{aligned} P[73.9865 &\leq \overline{X} \leq 74.0135 \text{ when } \mu = 74.0135] \\ &= P\left[\frac{73.9865 - 74.0135}{0.0045} \leq Z \leq \frac{74.0135 - 74.0135}{0.0045}\right] \\ &= P[-6 \leq Z \leq 0] = 0.5 \end{aligned}$$

Therefore, p in Equation 16-28 is 0.50, and the out-of-control ARL is

$$\text{ARL} = \frac{1}{p} = \frac{1}{0.5} = 2$$

That is, the control chart will require two samples to detect the process shift, on the average, so two hours will elapse between the shift and its detection (*again on the average*). Suppose this approach is unacceptable, because production of piston rings with a mean diameter of 74.0135 millimeters results in excessive scrap costs and delays final engine assembly. How can we reduce the time needed to detect the out-of-control condition? One method is to sample more frequently. For example, if we sample every half hour, only one hour will elapse (on the average) between the shift and its detection. The second possibility is to increase the sample size. For example, if we use $n = 10$, the control limits in Fig. 16-3 narrow to 73.9905 and 74.0095. The probability of $\overline{X}$ falling between the control limits when the process mean is 74.0135 millimeters is approximately 0.1, so $p = 0.9$, and the out-of-control ARL is

$$\text{ARL} = \frac{1}{p} = \frac{1}{0.9} = 1.11$$

Thus, the larger sample size would allow the shift to be detected about twice as quickly as the old one. If it became important to detect the shift in the first hour after it occurred, two control chart designs would work:

Design 1	Design 2
Sample size: $n = 5$	Sample size: $n = 10$
Sampling frequency: every half hour	Sampling frequency: every hour

Table 16-6 provides average run lengths for an $\overline{X}$ chart with 3-sigma control limits. The average run lengths are calculated for shifts in the process mean from 0 to 3.0σ and for sample

Table 16-6 Average Run Length (ARL) for an $\overline{X}$ Chart with 3-Sigma Control Limits

Magnitude of Process Shift	ARL $n = 1$	ARL $n = 4$
0	370.4	370.4
0.5σ	155.2	43.9
1.0σ	43.9	6.3
1.5σ	15.0	2.0
2.0σ	6.3	1.2
3.0σ	2.0	1.0

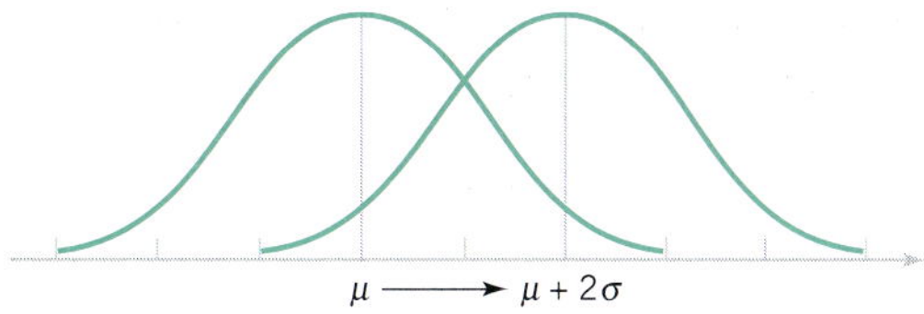

Figure 16-18 Process mean shift of 2σ.

sizes of $n = 1$ and $n = 4$ by using $1/p$, where p is the probability that a point plots outside of the control limits. Figure 16-18 illustrates a shift in the process mean of 2σ.

EXERCISES FOR SECTION 16-9

16-37. An X chart uses samples of size 1. The center line is at 100 and the upper and lower 3-sigma limits are at 112 and 88, respectively.
(a) What is the process σ?
(b) Suppose the process mean shifts to 96. Find the probability that this shift will be detected on the next sample.
(c) Find the ARL to detect the shift in part (b).

16-38. An $\overline{X}$ chart uses samples of size 4. The center line is at 100, and the upper and lower 3-sigma control limits are at 106 and 94, respectively.
(a) What is the process σ?
(b) Suppose the process mean shifts to 96. Find the probability that this shift will be detected on the next sample.
(c) Find the ARL to detect the shift in part (b).

16-39. Consider the $\overline{X}$ control chart in Fig. 16-3. Suppose that the mean shifts to 74.010 millimeters.
(a) What is the probability that this shift will be detected on the next sample?
(b) What is the ARL after the shift?

16-40. Consider an $\overline{X}$ control chart with $\bar{r} = 0.344$, $UCL = 14.708$, $LCL = 14.312$, and $n = 5$. Suppose that the mean shifts to 14.6.
(a) What is the probability that this shift will be detected on the next sample?
(b) What is the ARL after the shift?

16-41. Consider an $\overline{X}$ control chart with $\bar{r} = 34.286$, $UCL = 242.780$, $LCL = 203.220$, and $n = 5$. Suppose that the mean shifts to 210.
(a) What is the probability that this shift will be detected on the next sample?
(b) What is the ARL after the shift?

16-42. Consider an $\overline{X}$ control chart with $\hat{\sigma} = 1.40$, $UCL = 21.71$, $LCL = 18.29$, and $n = 6$. Suppose that the mean shifts to 17.
(a) What is the probability that this shift will be detected on the next sample?
(b) What is the ARL after the shift?

16-43. Consider an $\overline{X}$ control chart with $\hat{\sigma} = 2.466$, $UCL = 37.404$, $LCL = 30.780$, and $n = 5$. Suppose that the mean shifts to 36.
(a) What is the probability that this shift will be detected on the next sample?
(b) What is the ARL after the shift?

16-44. Consider an $\overline{X}$ control chart with $\bar{r} = 2.25$, $UCL = 17.40$, $LCL = 12.79$, and $n = 3$. Suppose that the mean shifts to 13.
(a) What is the probability that this shift will be detected on the next sample?
(b) What is the ARL after the shift?

16-45. Consider the revised $\overline{X}$ control chart in Exercise 16-9 with $\bar{r} = 0.000924$, $UCL = 0.0635$, $LCL = 0.0624$, and $n = 5$. Suppose that the mean shifts to 0.0625.
(a) What is the probability that this shift will be detected on the next sample?
(b) What is the ARL after the shift?

16-46. Consider the revised $\overline{X}$ control chart in Exercise 16-6 with $\hat{\sigma} = 0.669$, $UCL = 7.443$, $LCL = 5.125$, and $n = 3$. Suppose that the mean shifts to 5.5.
(a) What is the probability that this shift will be detected on the next sample?
(b) What is the ARL after the shift?

16-10 TIME-WEIGHTED CHARTS

16-10.1 Cumulative Sum Control Chart

In Sections 16-5 and 16-6 we have presented basic types of **Shewhart control charts.** A major disadvantage of any Shewhart control chart is that the chart is relatively insensitive to small shifts in the process, say, on the order of about 1.5σ or less. One reason for this relatively poor performance in detecting small process shifts is that the Shewhart chart makes use of only the information in the last plotted point, and it ignores the information in the sequence of points. This problem can be addressed, to some extent by adding criteria such as the **Western Electric rules** to a Shewhart chart, but the use of these rules reduces the simplicity and ease of interpretation of the chart. These rules would also cause the in-control average run length of a Shewhart chart to drop below 370. This increase in the false alarm rate can have serious practical consequences.

A very effective alternative to the Shewhart control chart is the **cumulative sum control chart** (or **CUSUM**). This chart has much better performance (in terms of ARL) for detecting small shifts than the Shewhart chart, but it does not cause the in-control ARL to drop significantly. This section will illustrate the use of the CUSUM for sample averages and individual measurements.

The CUSUM chart plots the cumulative sums of the deviations of the sample values from a target value. For example, suppose that samples of size $n \geq 1$ are collected, and $\overline{X}_j$ is the average of the jth sample. Then if μ_0 is the target for the process mean, the cumulative sum control chart is formed by plotting the quantity

$$S_i = \sum_{j=1}^{i} (\overline{X}_j - \mu_0) \tag{16-29}$$

against the sample number i. Now, S_i is called the cumulative sum up to and including the ith sample. Because they combine information from *several* samples, cumulative sum charts are more effective than Shewhart charts for detecting small process shifts. Furthermore, they are particularly effective with samples of $n = 1$. This makes the cumulative sum control chart a good candidate for use in the chemical and process industries where rational subgroups are frequently of size 1, as well as in discrete parts manufacturing with automatic measurement of each part and online control using a microcomputer directly at the work center.

If the process remains in control at the target value μ_0, the cumulative sum defined in equation 16-29 should fluctuate around zero. However, if the mean shifts upward to some value $\mu_1 > \mu_0$, say, an upward or positive drift will develop in the cumulative sum S_i. Conversely, if the mean shifts downward to some $\mu_1 < \mu_0$, a downward or negative drift in S_i will develop. Therefore, if a trend develops in the plotted points either upward or downward, we should consider this as evidence that the process mean has shifted, and a search for the assignable cause should be performed.

This theory can easily be demonstrated by applying the CUSUM to the chemical process concentration data in Table 16-3. Since the concentration readings are individual measurements, we would take $\overline{X}_j = X_j$ in computing the CUSUM. Suppose that the target value for the concentration is $\mu_0 = 99$. Then the CUSUM is

$$\begin{aligned} S_i &= \sum_{j=1}^{i} (X_j - 99) \\ &= (X_i - 99) + \sum_{j=1}^{i-1} (X_j - 99) \\ &= (X_i - 99) + S_{i-1} \end{aligned}$$

Table 16-7 shows the computation of this CUSUM, where the starting value of the CUSUM, S_0, is taken to be zero. Figure 16-19 plots the CUSUM from the last column of Table 16-7. Notice that the CUSUM fluctuates around the value of 0.

Table 16-7 CUSUM Computations for the Chemical Process Concentration Data in Table 16-3

Observation, i	x_i	$x_i - 99$	$s_i = (x_i - 99) + s_{i-1}$
1	102.0	3.0	3.0
2	94.8	−4.2	−1.2
3	98.3	−0.7	−1.9
4	98.4	−0.6	−2.5
5	102.0	3.0	0.5
6	98.5	−0.5	0.0
7	99.0	0.0	0.0
8	97.7	−1.3	−1.3
9	100.0	1.0	−0.3
10	98.1	−0.9	−1.2
11	101.3	2.3	1.1
12	98.7	−0.3	0.8
13	101.1	2.1	2.9
14	98.4	−0.6	2.3
15	97.0	−2.0	0.3
16	96.7	−2.3	−2.0
17	100.3	1.3	−0.7
18	101.4	2.4	1.7
19	97.2	−1.8	−0.1
20	101.0	2.0	1.9

The graph in Fig. 16-19 is not a control chart because it lacks control limits. There are two general approaches to devising control limits for CUSUMS. The older of these two methods is the V-mask procedure. A typical V mask is shown in Fig. 16-20(a). It is a V-shaped notch in a plane that can be placed at different locations on the CUSUM chart. The decision procedure consists of placing the V mask on the cumulative sum control chart with

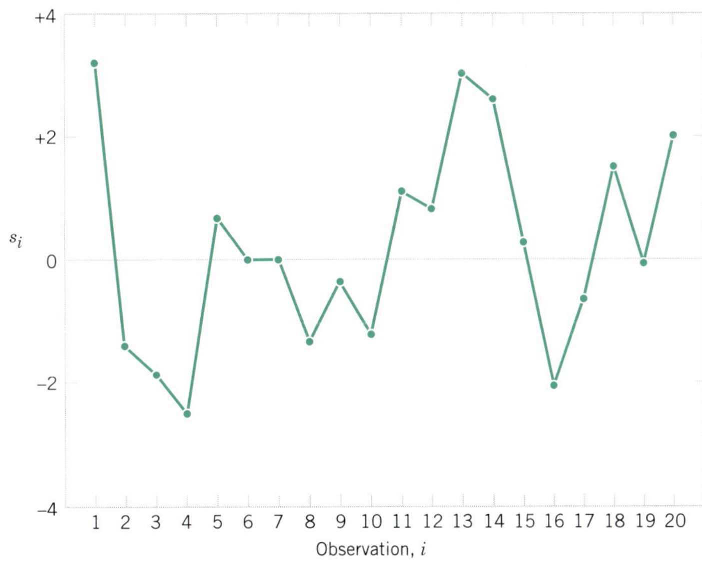

Figure 16-19 Plot of the cumulative sum for the concentration data, Table 16-7.

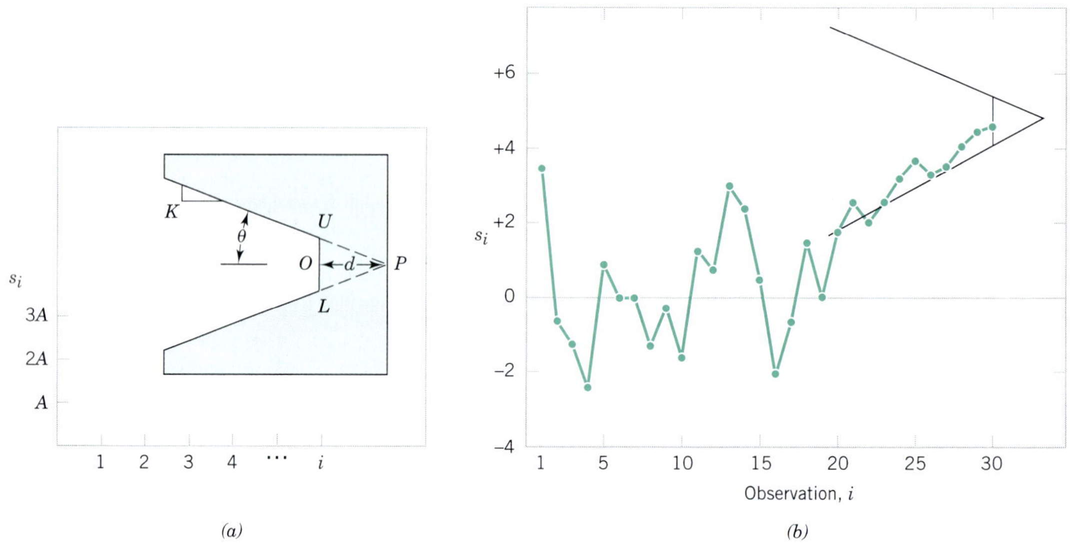

Figure 16-20 The cumulative sum control chart. (a) The V-mask and scaling. (b) The cumulative sum control chart in operation.

the point O on the last value of s_i and the line OP parallel to the horizontal axis. If all the previous cumulative sums, $s_1, s_2, \ldots, s_{i-1}$, lie within the two arms of the V mask, the process is in control. However, if any s_i lies outside the arms of the mask, the process is considered to be out of control. In actual use, the V mask would be applied to each new point on the CUSUM chart as soon as it was plotted. In the example shown in Fig. 16-20(b), an upward shift in the mean is indicated, since at least one of the points that have occurred earlier than sample 22 now lies below the lower arm of the mask, when the V mask is centered on the thirtieth observation. If the point lies above the upper arm, a downward shift in the mean is indicated. Thus, the V mask forms a visual frame of reference similar to the control limits on an ordinary Shewhart control chart. For the technical details of designing the V mask, see Montgomery (2005).

While some computer programs plot CUSUMS with the V-mask control scheme, we feel that the other approach to CUSUM control, the **tabular CUSUM,** is superior. The tabular procedure is particularly attractive when the CUSUM is implemented on a computer.

Let $S_H(i)$ be an upper one-sided CUSUM for period i and $S_L(i)$ be a lower one-sided CUSUM for period i. These quantities are calculated from

CUSUM Control Chart

$$s_H(i) = \max[0, \bar{x}_i - (\mu_0 + K) + s_H(i-1)] \tag{16-30}$$

and

$$s_L(i) = \max[0, (\mu_0 - K) - \bar{x}_i + s_L(i-1)] \tag{16-31}$$

where the starting values $s_H(0) = s_L(0) = 0$.

In Equations 16-30 and 16-31 K is called the **reference value,** which is usually chosen about halfway between the target μ_0 and the value of the mean corresponding to the out-of-control state, $\mu_1 = \mu_0 + \Delta$. That is, K is about one-half the magnitude of the shift we are interested in, or

$$K = \frac{\Delta}{2}$$

Notice that $S_H(i)$ and $S_L(i)$ accumulate deviations from the target value that are greater than K, with both quantities reset to zero upon becoming negative. If either $S_H(i)$ or $S_L(i)$ exceeds a constant H, the process is out of control. This constant H is usually called the **decision interval.**

EXAMPLE 16-6
Chemical Process Concentration CUSUM

A Tabular CUSUM

We will illustrate the tabular CUSUM by applying it to the chemical process concentration data in Table 16-7. The process target is $\mu_0 = 99$, and we will use $K = 1$ as the reference value and $H = 10$ as the decision interval. The reasons for these choices will be explained later.

Table 16-8 shows the tabular CUSUM scheme for the chemical process concentration data. To illustrate the calculations, note that

$$\begin{aligned} s_H(i) &= \max[0, x_i - (\mu_0 + K) + s_H(i-1)] = \max[0, x_i - (99 + 1) + s_H(i-1)] \\ &= \max[0, x_i - 100 + s_H(i-1)] \end{aligned}$$

$$\begin{aligned} s_L(i) &= \max[0, (\mu_0 - K) - x_i + s_L(i-1)] = \max[0, (99 - 1) - x_i + s_L(i-1)] \\ &= \max[0, 98 - x_i + s_L(i-1)] \end{aligned}$$

Therefore, for observation 1 the CUSUMS are

$$s_H(1) = \max[0, x_1 - 100 + s_H(0)] = \max[0, 102.0 - 100 + 0] = 2.0$$

Table 16-8 The Tabular CUSUM for the Chemical Process Concentration Data

Observation	Upper CUSUM				Lower CUSUM		
i	x_i	$x_i - 100$	$s_H(i)$	n_H	$98 - x_i$	$s_L(i)$	n_L
1	102.0	2.0	2.0	1	−4.0	0.0	0
2	94.8	−5.2	0.0	0	3.2	3.2	1
3	98.3	−1.7	0.0	0	−0.3	2.9	2
4	98.4	−1.6	0.0	0	−0.4	2.5	3
5	102.0	2.0	2.0	1	−4.0	0.0	0
6	98.5	−1.5	0.5	2	−0.5	0.0	0
7	99.0	−1.0	0.0	0	−1.0	0.0	0
8	97.7	−2.3	0.0	0	0.3	0.3	1
9	100.0	0.0	0.0	0	−2.0	0.0	0
10	98.1	−1.9	0.0	0	−0.1	0.0	0
11	101.3	1.3	1.3	1	−3.3	0.0	0
12	98.7	−1.3	0.0	0	−0.7	0.0	0
13	101.1	1.1	1.1	1	−3.1	0.0	0
14	98.4	−1.6	0.0	0	−0.4	0.0	0
15	97.0	−3.0	0.0	0	1.0	1.0	1
16	96.7	−3.3	0.0	0	1.3	2.3	2
17	100.3	0.3	0.3	1	−2.3	0.0	0
18	101.4	1.4	1.7	2	−3.4	0.0	0
19	97.2	−2.8	0.0	0	0.8	0.8	1
20	101.0	1.0	1.0	0	−3.0	0.0	0

and

$$s_L(1) = \max[0, 98 - x_1 + s_L(0)] = \max[0, 98 - 102.0 + 0] = 0$$

as shown in Table 16-8. The quantities n_H and n_L in Table 16-8 indicate the number of periods that the CUSUM $s_H(i)$ or $s_L(i)$ have been nonzero. Notice that the CUSUMS in this example never exceed the decision interval $H = 10$. We would therefore conclude that the process is in control.

When the tabular CUSUM indicates that the process is out of control, we should search for the assignable cause, take any corrective actions indicated, and restart the CUSUMS at zero. It may be helpful to have an estimate of the new process mean following the shift. This can be computed from

$$\hat{\mu} = \begin{cases} \mu_0 + K + \dfrac{s_H(i)}{n_H}, & \text{if } s_H(i) > H \\ \mu_0 - K - \dfrac{s_L(i)}{n_L}, & \text{if } s_L(i) > H \end{cases} \tag{16-32}$$

It is also useful to present a graphical display of the tabular CUSUMS, which are sometimes called CUSUM status charts. They are constructed by plotting $s_H(i)$ and $s_L(i)$ versus the sample number. Figure 16-21 shows the CUSUM status chart for the data in Example 16-6. Each vertical bar represents the value of $s_H(i)$ and $s_L(i)$ in period i. With the decision interval plotted on the chart, the CUSUM status chart resembles a Shewhart control chart. We have also plotted the sample statistics x_i for each period on the CUSUM status chart as the solid dots. This frequently helps the user of the control chart to visualize the actual process performance that has led to a particular value of the CUSUM.

The tabular CUSUM is designed by choosing values for the reference value K and the decision interval H. We recommend that these parameters be selected to provide good average

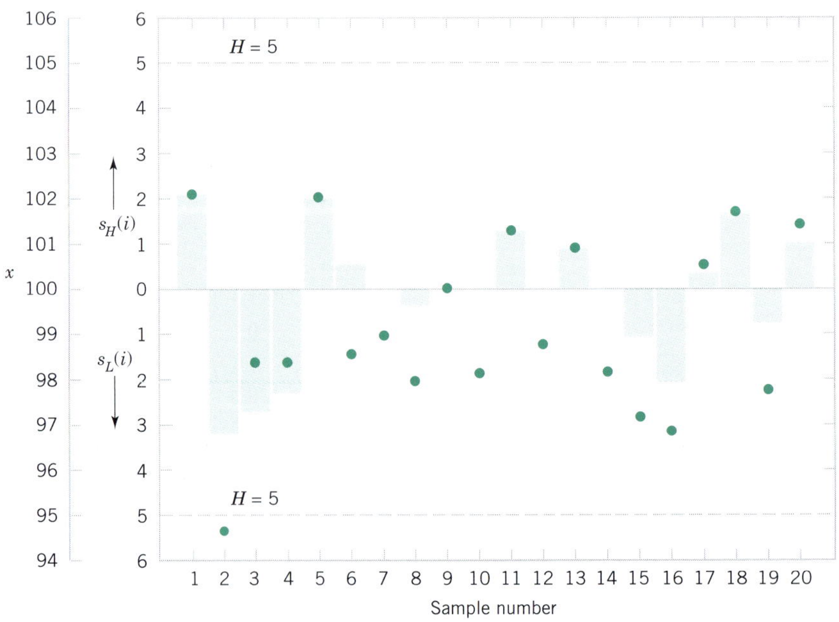

Figure 16-21 The CUSUM status chart for Example 16-6.

Table 16-9 Average Run Lengths for a CUSUM Control Chart with $k = 1/2$

Shift in Mean (multiple of $\sigma_{\overline{X}}$)	$h = 4$	$h = 5$
0	168	465
0.25	74.2	139
0.50	26.6	38.0
0.75	13.3	17.0
1.00	8.38	10.4
1.50	4.75	5.75
2.00	3.34	4.01
2.50	2.62	3.11
3.00	2.19	2.57
4.00	1.71	2.01

run-length values. There have been many analytical studies of CUSUM ARL performance. Based on these studies, we may give some general recommendations for selecting H and K. Define $H = h\sigma_{\overline{X}}$ and $K = k\sigma_{\overline{X}}$, where $\sigma_{\overline{X}}$ is the standard deviation of the sample variable used in forming the CUSUM (if $n = 1$, $\sigma_{\overline{X}} = \sigma_X$). Using $h = 4$ or $h = 5$ and $k = 1/2$ will generally provide a CUSUM that has good ARL properties against a shift of about $1\sigma_{\overline{X}}$ (or $1\sigma_X$) in the process mean. If much larger or smaller shifts are of interest, set $k = \delta/2$, where δ is the size of the shift in standard deviation units.

To illustrate how well the recommendations of $h = 4$ or $h = 5$ with $k = 1/2$ work, consider these average run lengths in Table 16-9. Notice that a shift of $1\sigma_{\overline{X}}$ would be detected in either 8.38 samples (with $k = 1/2$ and $h = 4$) or 10.4 samples (with $k = 1/2$ and $h = 5$). By comparison, Table 16-6 shows that an $\overline{X}$ chart would require approximately 43.9 samples, on the average, to detect this shift.

These design rules were used for the CUSUM in Example 16-6. We assumed that the process standard deviation $\sigma = 2$. (This is a reasonable value; see Example 16-2.) Then with $k = 1/2$ and $h = 5$, we would use

$$K = k\sigma = \tfrac{1}{2}(2) = 1 \qquad \text{and} \qquad H = h\sigma = 5(2) = 10$$

in the tabular CUSUM procedure.

Finally, we should note that supplemental procedures such as the Western Electric rules cannot be safely applied to the CUSUM, because successive values of $S_H(i)$ and $S_L(i)$ are not independent. In fact, the CUSUM can be thought of as a weighted average, where the weights are stochastic or random. In effect, all the CUSUM values are highly correlated, thereby causing the Western Electric rules to give too many false alarms.

16-10.2 Exponentially Weighted Moving Average Control Chart

Data collected in time order is often averaged over several time periods. For example, economic data is often presented as an average over the last four quarters. That is, at time t the average of the last four measurements can be written as

$$\bar{x}_t(4) = \frac{1}{4}x_t + \frac{1}{4}x_{t-1} + \frac{1}{4}x_{t-2} + \frac{1}{4}x_{t-3}$$

This average places weight of 1/4 on each of the most recent observations, and zero weight on older observations. It is called a **moving average** and in this case a *window* of size 4 is used. An average of the recent data is used to smooth the noise in the data to generate a better estimate of the process mean than only the most recent observation. However, in a dynamic environment where the process mean may change, the number of observations used to construct the average is kept to a modest size so that the estimate can adjust to any change in the process mean. Therefore, the window size is a compromise between a better statistical estimate from an average and a response to a mean change. If a window of size 10 were used in a moving average, the statistic $\bar{x}_t(10)$ would have lower variability, but it would not adjust as well to a change in mean.

For statistical process control rather than use a fixed window size, it is useful to place the greatest weight on the most recent observation or subgroup average, and then gradually decrease the weight on older observations. One average of this type can be constructed by a multiplicative decrease in the weights. Let $\lambda \leq 1$ denote a constant and μ_0 denote the process target or historical mean. Suppose that samples of size $n \geq 1$ are callected and $\bar{x}_t$ is the average of the sample at time t. The **exponentially weighted moving average (EWMA)** is

$$z_t = \lambda\bar{x}_t + \lambda(1-\lambda)\bar{x}_{t-1} + \lambda(1-\lambda)^2\bar{x}_{t-2} + \cdots + \lambda(1-\lambda)^{t-1}\bar{x}_1 + (1-\lambda)^t\mu_0$$
$$= \sum_{k=0}^{t} \lambda(1-\lambda)^k\bar{x}_{t-k} + (1-\lambda)^t\mu_0$$

Each older observation has its weight decreased by the factor $(1 - \lambda)$. The weight on the starting value μ_0 is selected so that the weights sum to one. It is also sometimes called a geometric average.

The value of λ determines the compromise between noise reduction and response to a mean change. For example, the series of weights when $\lambda = 0.8$ and $\lambda = 0.2$ are

0.8, 0.16, 0.032, 0.0064, 0.00128, . . .

0.2, 0.16, 0.128, 0.1024, 0.0819, . . .

When $\lambda = 0.8$ the weights decrease rapidly. Most of the weight is placed on the most recent observation, with modest contributions to the EWMA from older measurements. In this case, the EWMA does not average noise much, but it responds to a mean change. However, when $\lambda = 0.2$ the weights decrease much more slowly and the EWMA has substantial contributions from the more recent observations. In this case, the EWMA averages noise more, but it responds more slowly to a change in the mean. Fig. 16-22 displays a series of observations with

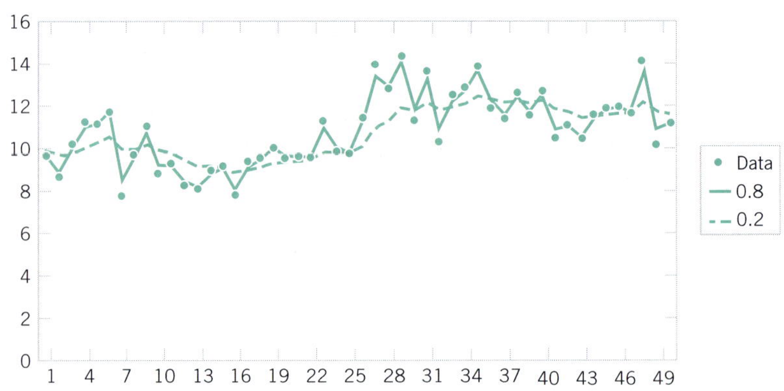

Figure 16-22 EWMAs with $\lambda = 0.8$ and $\lambda = 0.2$ show a compromise between a smooth curve and a response to a shift.

a mean shift in the middle on the series. Notice that the EWMA with $\lambda = 0.2$ smooths the data more, but that the EWMA with $\lambda = 0.8$ adjusts the estimate to the mean shift more quickly.

It appears that it is difficult to calculate an EWMA because at every time t a new weighted average of all previous data is required. However, there is an easy method to calculate an EWMA z_t based on a simple recursive equation. Let $z_0 = \mu_0$. Then it can be shown that

EWMA Update Equation

$$z_t = \lambda \bar{x}_t + (1 - \lambda) z_{t-1} \tag{16-33}$$

Consequently, only a brief computation is needed at each time t.

To develop a control chart from an EWMA, control limits are needed for Z_t. The control limits are defined in a straightforward manner. They are placed at three standard deviations around the mean of the plotted statistic Z_t. This follows the general approach for a control chart in Equation 16-1. An EWMA control chart may be applied to individual measurements as an extension to an X chart or to subgroup averages. Formulas here are developed for the more general case with an average from a subgroup of size n. For individual measurements $n = 1$.

Because Z_t is a linear function of the independent observations $X_1, X_2, \ldots, X_t$ (and μ_0) the results from Chapter 5 can be used to show that

$$E(Z_t) = \mu_0 \qquad \text{and} \qquad V(Z_t) = \frac{\sigma^2}{n}\frac{\lambda}{2 - \lambda}\left[1 - (1 - \lambda)^{2t}\right]$$

where n is the subgroup size. Therefore an EWMA control chart uses estimates of μ_0 and σ in the following formulas

EWMA Control Chart

$$\begin{aligned} LCL &= \mu_0 - 3\frac{\sigma}{\sqrt{n}}\sqrt{\frac{\lambda}{2 - \lambda}[1 - (1 - \lambda)^{2t}]} \\ CL &= \mu_0 \\ UCL &= \mu_0 + 3\frac{\sigma}{\sqrt{n}}\sqrt{\frac{\lambda}{2 - \lambda}[1 - (1 - \lambda)^{2t}]} \end{aligned} \tag{16-34}$$

Note that the control limits are not of equal width about the centerline. The control limits are calculated from the variance of Z_t and that changes with time. However, for large t the variance of Z_t converges to

$$\lim_{t \to \infty} V(Z_t) = \frac{\sigma^2}{n}\left(\frac{\lambda}{2 - \lambda}\right)$$

so that the control limits tend to be parallel lines about the centerline as t increases.

The parameters μ_0 and σ are estimated by the same statistics used in $\bar{X}$ or X charts. That is, for subgroups

$$\hat{\mu}_0 = \bar{\bar{X}} \qquad \text{and} \qquad \hat{\sigma} = \bar{R}/d_2 \qquad \text{or} \qquad \hat{\sigma} = \bar{S}/c_4$$

and for $n = 1$

$$\hat{\mu}_0 = \bar{X} \qquad \text{and} \qquad \hat{\sigma} = \overline{MR}/1.128$$

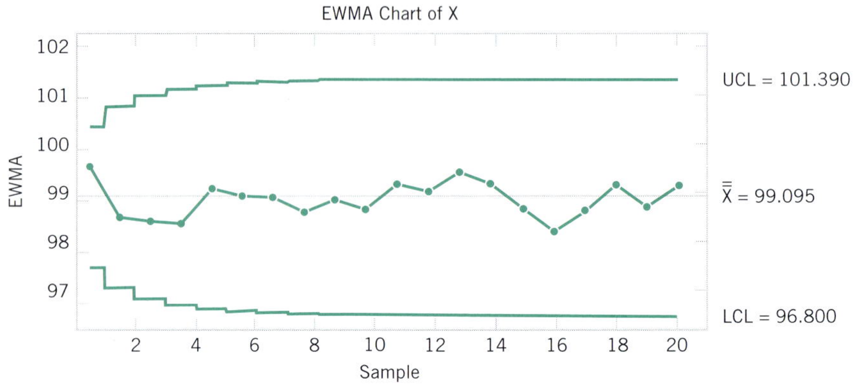

Figure 16-23 EWMA Control Chart for the Chemical Process Concentration Data from Minitab.

EXAMPLE 16-7
Chemical Process Concentration EWMA

Consider the concentration data shown in Table 16-3. Construct an EWMA control chart with $\lambda = 0.2$ with $n = 1$. It was determined that $\bar{x} = 99.1$ and $\overline{mr} = 2.59$. Therefore $\hat{\mu}_0 = 99.1$ and $\hat{\sigma} = 2.59/1.128 = 2.30$. The control limits for z_1 are

$$LCL = 99.1 - 3(2.30)\sqrt{\frac{0.2}{2 - 0.2}[1 - (1 - 0.2)^2]} = 98.19$$

$$LCL = 99.1 + 3(2.30)\sqrt{\frac{0.2}{2 - 0.2}[1 - (1 - 0.2)^2]} = 100.01$$

The first few values of z_t along with the corresponding control limits are

t	1	2	3	4	5
x_t	102.0	94.8	98.3	98.4	102.0
z_t	99.68	98.70	98.62	98.58	99.26
LCL	97.72	97.33	97.12	97.00	96.93
UCL	100.48	100.87	101.08	101.20	101.27

The chart generated by Minitab is shown in Figure 16-23. Notice that the control limits widen as time increases but quickly stabilize. Each point is within its set of corresponding control limits so there are no signals from the chart.

Similar to a CUSUM chart, the points plotted on an EWMA control chart are not independent. Therefore, run rules should not be applied to an EWMA control chart. Information in the history of the data that is considered by run rules is to a large extent incorporated into the EWMA that is calculated at each time t.

The value of λ is usually chosen from the range $0.1 < \lambda < 0.5$. A common choice is $\lambda = 0.2$. Smaller values for λ provide more sensitivity for small shifts and larger values better tune the chart for larger shifts. This performance can be seen in the average run lengths in Table 16-10. Here, $\lambda = 0.1$ and 0.5 are compared. The multiplier of the standard deviation, denoted L in the table, is adjusted so that the average run length (ARL) equals 500 for both choices for λ. That is, the control limits are placed at $E(Z_t) \pm L\sqrt{V(Z_t)}$ and L is chosen so the ARL without a mean shift is 500 in both cases.

The EWMA ARLs in the table indicate that the smaller value for λ is preferred when the magnitude of the shift is small. Also, the EWMA performance is in general much better than results for a Shewhart control chart (in Table 16-6) and the results are comparable to a CUSUM control chart (in Table 16-9). However, these are average results. At the time of an

Table 16-10 Average Run Lengths for an EWMA Control Chart

Shift in Mean (multiple of $\sigma_{\overline{X}}$)	$\lambda = 0.5$ $L = 3.07$	$\lambda = 0.1$ $L = 2.81$
0	500	500
0.25	255	106
0.5	88.8	31.3
0.75	35.9	15.9
1	17.5	10.3
1.5	6.53	6.09
2	3.63	4.36
3	1.93	2.87

increase in the process mean, z_t might be negative and there would be some performance penalty to first increase z_t to near zero and then further increase it to a signal above the *UCL*. Such a penalty provides an advantage to CUSUM control charts that is not accounted for in these ARL tables. A more refined analysis can be used to quantify this penalty, but the conclusion is that the EWMA penalty is moderate to small in most applications.

EXERCISES FOR SECTION 16-10

16-47. The following data was considered in *Quality Engineering* (1992, Vol. 4(4)) "Parabolic control limits for the exponentially weighted moving average control charts in Quality Engineering." In a chemical plant, the data for one of the quality characteristics (viscosity) was obtained for each 12-hour batch at the completion of the batch. The results of 15 consecutive measurements are shown in the table below.

Batch	Viscosity	Batch	Viscosity
1	13.3	11	14.3
2	14.5	12	16.1
3	15.3	13	13.1
4	15.3	14	15.5
5	14.3	15	12.6
6	14.8		
7	15.2		
8	14.9		
9	14.6		
10	14.1		

(a) Set up a CUSUM control chart for this process. Assume the desired process target 14.1. Does the process appear to be in control?

(b) Suppose that the next five observations are 14.6, 15.3, 15.7, 16.1, and 16.8. Apply the CUSUM in part (a) to these new observations. Is there any evidence that the process has shifted out of control?

16-48. The purity of a chemical product is measured every two hours. The results of 20 consecutive measurements are as follows:

Sample	Purity	Sample	Purity
1	89.11	11	88.55
2	90.59	12	90.43
3	91.03	13	91.04
4	89.46	14	88.17
5	89.78	15	91.23
6	90.05	16	90.92
7	90.63	17	88.86
8	90.75	18	90.87
9	89.65	19	90.73
10	90.15	20	89.78

(a) Set up a CUSUM control chart for this process. Use $\sigma = 0.8$ in setting up the procedure, and assume that the desired process target is 90. Does the process appear to be in control?

(b) Suppose that the next five observations are 90.75, 90.00, 91.15, 90.95, and 90.86. Apply the CUSUM in part (a) to these new observations. Is there any evidence that the process has shifted out of control?

16-49. The diameter of holes is measured in consecutive order by an automatic sensor. The results of measuring 25 holes follow.

Sample	Diameter	Sample	Diameter
1	9.94	14	9.99
2	9.93	15	10.12
3	10.09	16	9.81
4	9.98	17	9.73
5	10.11	18	10.14
6	9.99	19	9.96
7	10.11	20	10.06
8	9.84	21	10.11
9	9.82	22	9.95
10	10.38	23	9.92
11	9.99	24	10.09
12	10.41	25	9.85
13	10.36		

(a) Estimate the process standard deviation.
(b) Set up a CUSUM control procedure, assuming that the target diameter is 10.0 millimeters. Does the process appear to be operating in a state of statistical control at the desired target level?

16-50. The concentration of a chemical product is measured by taking four samples from each batch of material. The average concentration of these measurements is shown for the last 20 batches in the following table:

Batch	Concentration	Batch	Concentration
1	104.5	11	95.4
2	99.9	12	94.5
3	106.7	13	104.5
4	105.2	14	99.7
5	94.8	15	97.7
6	94.6	16	97.0
7	104.4	17	95.8
8	99.4	18	97.4
9	100.3	19	99.0
10	100.3	20	102.6

(a) Suppose that the process standard deviation is $\sigma = 8$ and that the target value of concentration for this process is 100. Design a CUSUM scheme for the process. Does the process appear to be in control at the target?
(b) How many batches would you expect to be produced with off-target concentration before it would be detected by the CUSUM control chart if the concentration shifted to 104? Use Table 16-9.

16-51. Consider a CUSUM with $h = 5$ and $k = 1/2$. Samples are taken every two hours from the process. The target value for the process is $\mu_0 = 50$ and $\sigma = 2$. Use Table 16-9.

(a) If the sample size is $n = 1$, how many samples would be required to detect a shift in the process mean to $\mu = 51$ on average?
(b) If the sample size is increased to $n = 4$, how does this affect the average run length to detect the shift to $\mu = 51$ that you determined in part (a)?

16-52. Consider the purity data in Exercise 16-48. Use $\sigma = 0.8$ and assume that the desired process target is 90.
(a) Construct an EWMA control chart with $\lambda = 0.2$. Does the process appear to be in control?
(b) Construct an EWMA control chart with $\lambda = 0.5$. Compare your results to part (a).
(c) Suppose that the next five observations are 90.75, 90.00, 91.15, 90.95, and 90.86. Apply the EWMAs in part (a) and (b) to these new observations. Is there any evidence that the process has shifted out of control?

16-53. Consider the diameter data in Exercise 16-49. Assume that the desired process target is 10.0 millimeters.
(a) Estimate the process standard deviation.
(b) Construct an EWMA control chart with $\lambda = 0.2$. Does the process appear to be in control?
(c) Construct an EWMA control chart with $\lambda = 0.5$. Compare your results to part (a).

16-54. Consider the concentration data in Exercise 16-50. Use $\sigma = 8$ and assume that the desired process target is 100.
(a) Construct an EWMA control chart with $\lambda = 0.2$. Does the process appear to be in control?
(b) Construct an EWMA control chart with $\lambda = 0.5$. Compare your results to part (a).
(c) If the concentration shifted to 104, would you prefer the chart in part (a) or (b)? Explain.

16-55. Consider an EMWA control chart. The target value for the process is $\mu_0 = 50$ and $\sigma = 2$. Use Table 16-10.
(a) If the sample size is $n = 1$, would you prefer an EWMA chart with $\lambda = 0.1$ and $L = 2.81$ or $\lambda = 0.5$ and $L = 3.07$ to detect a shift in the process mean to $\mu = 52$ on average? Why?
(b) If the sample size is increased to $n = 4$, which chart in part (a) do you prefer? Why?
(c) If an EWMA chart with $\lambda = 0.1$ and $L = 2.81$ is used, what sample size is needed to detect a shift to $\mu = 52$ in approximately 3 samples on average?

16-56. A process has a target of $\mu_0 = 100$ and a standard deviation of $\sigma = 4$. Samples of size $n = 1$ are taken every two hours. Use Table 16-9.
(a) Suppose the process mean shifts to $\mu = 102$. How many hours of production will occur before the process shift is detected by a CUSUM with $h = 5$ and $k = 1/2$?
(b) It is important to detect the shift defined in part (a) more quickly. A proposal is made to reduce the sampling frequency to 0.5 hour. How will this affect the CUSUM control procedure? How much more quickly will the shift be detected?

(c) Suppose that the 0.5 hour sampling interval in part (b) is adopted. How often will false alarms occur with this new sampling interval? How often did they occur with the old interval of two hours?

(d) A proposal is made to increase the sample size to $n = 4$ and retain the two-hour sampling interval. How does this suggestion compare in terms of average detection time to the suggestion of decreasing the sampling interval to 0.5 hour?

16-11 OTHER SPC PROBLEM-SOLVING TOOLS

While the control chart is a very powerful tool for investigating the causes of variation in a process, it is most effective when used with other SPC problem-solving tools. In this section we illustrate some of these tools, using the printed circuit board defect data in Example 16-5.

Figure 16-17 shows a *U* chart for the number of defects in samples of five printed circuit boards. The chart exhibits statistical control, but the number of defects must be reduced. The average number of defects per board is $8/5 = 1.6$, and this level of defects would require extensive rework.

The first step in solving this problem is to construct a **Pareto diagram** of the individual defect types. The Pareto diagram, shown in Fig. 16-24, indicates that insufficient solder and solder balls are the most frequently occurring defects, accounting for $(109/160)\ 100 = 68\%$ of the observed defects. Furthermore, the first five defect categories on the Pareto chart are all solder-related defects. This points to the flow solder process as a potential opportunity for improvement.

To improve the surface mount process, a team consisting of the operator, the shop supervisor, the manufacturing engineer responsible for the process, and a quality engineer meets to study potential causes of solder defects. They conduct a brainstorming session and produce the cause-and-effect diagram shown in Fig. 16-25. The cause-and-effect diagram is widely used to display the various potential causes of defects in products and their interrelationships. They are useful in summarizing knowledge about the process.

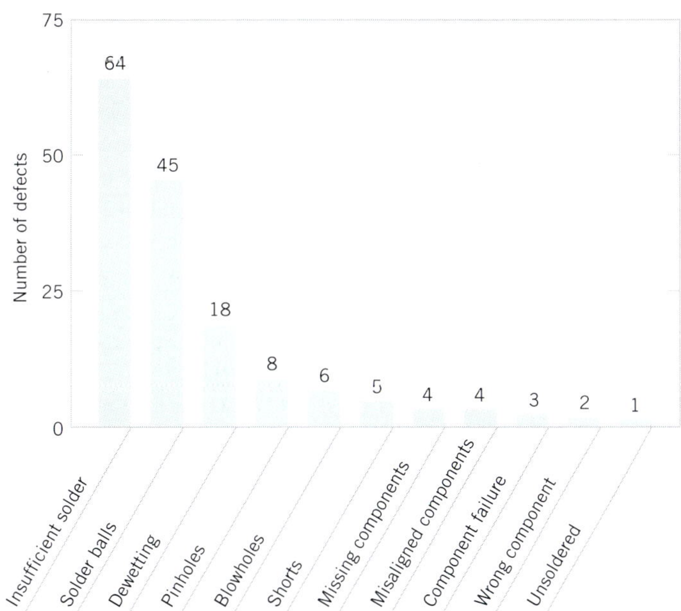

Figure 16-24 Pareto diagram for printed circuit board defects.

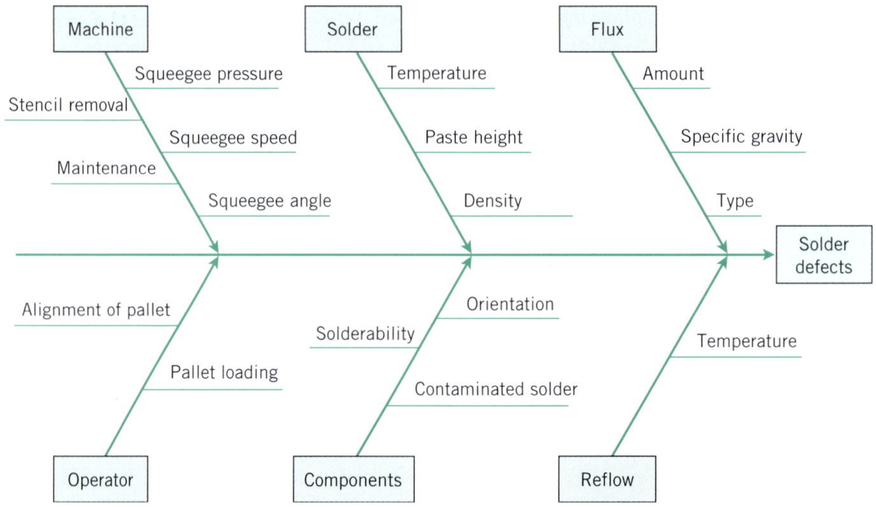

Figure 16-25 Cause-and-effect diagram for the printed circuit board flow solder process.

As a result of the brainstorming session, the team tentatively identifies the following variables as potentially influential in creating solder defects:

1. Flux specific gravity
2. Reflow temperature
3. Squeegee speed
4. Squeegee angle
5. Paste height
6. Reflow temperature
7. Board loading method

A statistically **designed experiment** could be used to investigate the effect of these seven variables on solder defects.

In addition, the team constructed a **defect concentration diagram** for the product. A defect concentration diagram is just a sketch or drawing of the product, with the most frequently occurring defects shown on the part. This diagram is used to determine whether defects occur in the same location on the part. The defect concentration diagram for the printed circuit board is shown in Fig. 16-26. This diagram indicates that most of the insufficient solder defects are near the front edge of the board. Further investigation showed that one of the pallets used to carry the boards was bent, causing the front edge of the board to make poor contact with the squeegee.

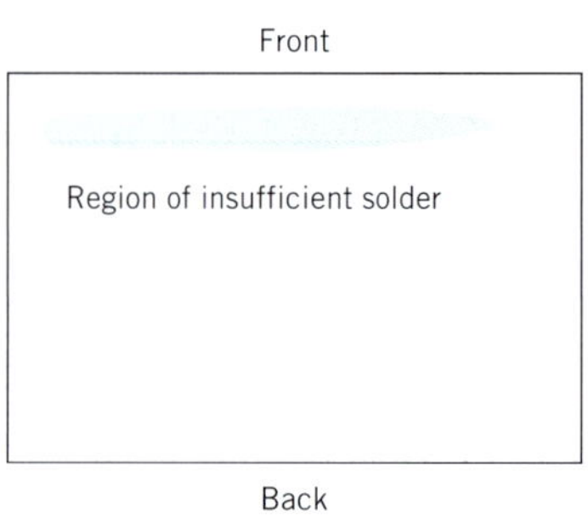

Figure 16-26 Defect concentration diagram for a printed circuit board.

When the defective pallet was replaced, a designed experiment was used to investigate the seven variables discussed earlier. The results of this experiment indicated that several of these factors were influential and could be adjusted to reduce solder defects. After the results of the experiment were implemented, the percentage of solder joints requiring rework was reduced from 1% to under 100 parts per million (0.01%).

16-12 IMPLEMENTING SPC

The methods of statistical process control can provide significant payback to those companies that can successfully implement them. While SPC seems to be a collection of statistically based problem-solving tools, there is more to the successful use of SPC than simply learning and using these tools. Management involvement and commitment to the quality-improvement process is the most vital component of SPC's potential success. Management is a role model, and others in the organization will look to management for guidance and as an example. A team approach is also important, for it is usually difficult for one person alone to introduce process improvements. Many of the "magnificent seven" problem-solving tools are helpful in building an improvement team, including cause-and-effect diagrams, Pareto charts, and defect concentration diagrams. The basic SPC problem-solving tools must become widely known and widely used throughout the organization. Continuous training in SPC and quality improvement is necessary to achieve this widespread knowledge of the tools.

The objective of an SPC-based quality-improvement program is continuous improvement on a weekly, quarterly, and annual basis. SPC is not a one-time program to be applied when the business is in trouble and later abandoned. Quality improvement must become part of the culture of the organization.

The control chart is an important tool for process improvement. Processes do not naturally operate in an in-control state, and the use of control charts is an important step that must be taken early in an SPC program to eliminate assignable causes, reduce process variability, and stabilize process performance. To improve quality and productivity, we must begin to manage with facts and data, and not just rely on judgment. Control charts are an important part of this change in management approach.

In implementing a company-wide SPC program, we have found that the following elements are usually present in all successful efforts:

1. Management leadership
2. A team approach
3. Education of employees at all levels
4. Emphasis on continuous improvement
5. A mechanism for recognizing success

We cannot overemphasize the importance of management leadership and the team approach. Successful quality improvement is a "top-down" management-driven activity. It is also important to measure progress and success and to spread knowledge of this success throughout the organization. When successful improvements are communicated throughout the company, this can provide motivation and incentive to improve other processes and to make continuous improvement a normal part of the way of doing business.

The philosophy of W. Edwards **Deming** provides an important framework for implementing quality and productivity improvement. Deming's philosophy is summarized in his 14 points for management. The adherence to these management principles has been an important factor in Japan's industrial success and continues to be the catalyst in that nation's quality- and productivity-improvement efforts. This philosophy has also now spread rapidly in the West. **Deming's 14 points** are as follows.

1. **Create a constancy of purpose focused on the improvement of products and services.** Constantly try to improve product design and performance. Investment in research, development, and innovation will have a long-term payback to the organization.

2. **Adopt a new philosophy of rejecting poor workmanship, defective products, or bad service.** It costs as much to produce a defective unit as it does to produce a good one (and sometimes more). The cost of dealing with scrap, rework, and other losses created by defectives is an enormous drain on company resources.

3. **Do not rely on mass inspection to "control" quality.** All inspection can do is sort out defectives, and at this point it is too late because we have already paid to produce these defectives. Inspection occurs too late in the process, it is expensive, and it is often ineffective. Quality results from the prevention of defectives through process improvement, not inspection.

4. **Do not award business to suppliers on the basis of price alone, but also consider quality.** Price is a meaningful measure of a supplier's product only if it is considered in relation to a measure of quality. In other words, the total cost of the item must be considered, not just the purchase price. When quality is considered, the lowest bidder is frequently not the low-cost supplier. Preference should be given to suppliers who use modern methods of quality improvement in their business and who can demonstrate process control and capability.

5. **Focus on continuous improvement.** Constantly try to improve the production and service system. Involve the workforce in these activities and make use of statistical methods, particularly the SPC problem-solving tools discussed in the previous section.

6. **Practice modern training methods and invest in training for all employees.** Everyone should be trained in the technical aspects of their job, as well as in modern quality- and productivity-improvement methods. The training should encourage all employees to practice these methods every day.

7. **Practice modern supervision methods.** Supervision should not consist merely of passive surveillance of workers, but should be focused on helping the employees improve the system in which they work. The first goal of supervision should be to improve the work system and the product.

8. **Drive out fear.** Many workers are afraid to ask questions, report problems, or point out conditions that are barriers to quality and effective production. In many organizations the economic loss associated with fear is large; only management can eliminate fear.

9. **Break down the barriers between functional areas of the business.** Teamwork among different organizational units is essential for effective quality and productivity improvement to take place.

10. **Eliminate targets, slogans, and numerical goals for the workforce.** A target such as "zero defects" is useless without a plan as to how to achieve this objective. In fact, these slogans and "programs" are usually counterproductive. Work to improve the system and provide information on that.

11. **Eliminate numerical quotas and work standards.** These standards have historically been set without regard to quality. Work standards are often symptoms of management's inability to understand the work process and to provide an effective management system focused on improving this process.

12. **Remove the barriers that discourage employees from doing their jobs.** Management must listen to employee suggestions, comments, and complaints. The person who is doing the job is the one who knows the most about it, and usually has valuable ideas about how to make the process work more effectively. The workforce is an important participant in the business, and not just an opponent in collective bargaining.

13. **Institute an ongoing program of training and education for all employees.** Education in simple, powerful statistical techniques should be mandatory for all employees. Use of the basic SPC problem-solving tools, particularly the control chart, should become widespread in the business. As these charts become widespread, and as employees understand their uses, they will be more likely to look for the causes of poor quality and to identify process improvements. Education is a way of making everyone partners in the quality-improvement process.

14. **Create a structure in top management that will vigorously advocate the first 13 points.**

As we read Deming's 14 points, we notice two things. First, there is a strong emphasis on change. Second, the role of management in guiding this change process is of dominating importance. But what should be changed, and how should this change process be started? For example, if we want to improve the yield of a semiconductor manufacturing process, what should we do? It is in this area that statistical methods most frequently come into play. To improve the semiconductor process, we must determine which controllable factors in the process influence the number of defective units produced. To answer this question, we must collect data on the process and see how the system reacts to changes in the process variables. Statistical methods, including the SPC and experimental design techniques in this book, can contribute to this knowledge.

SUPPLEMENTAL EXERCISES

16-57. The diameter of fuse pins used in an aircraft engine application is an important quality characteristic. Twenty-five samples of three pins each are shown as follows:

Sample Number	Diameter		
1	64.030	64.002	64.019
2	63.995	63.992	64.001
3	63.988	64.024	64.021
4	64.002	63.996	63.993
5	63.992	64.007	64.015
6	64.009	63.994	63.997
7	63.995	64.006	63.994
8	63.985	64.003	63.993
9	64.008	63.995	64.009
10	63.998	74.000	63.990
11	63.994	63.998	63.994
12	64.004	64.000	64.007
13	63.983	64.002	63.998
14	64.006	63.967	63.994
15	64.012	64.014	63.998
16	64.000	63.984	64.005
17	63.994	64.012	63.986
18	64.006	64.010	64.018
19	63.984	64.002	64.003
20	64.000	64.010	64.013
21	63.988	64.001	64.009
22	64.004	63.999	63.990
23	64.010	63.989	63.990
24	64.015	64.008	63.993
25	63.982	63.984	63.995

(a) Set up $\overline{X}$ and R charts for this process. If necessary, revise limits so that no observations are out-of-control.
(b) Estimate the process mean and standard deviation.
(c) Suppose the process specifications are at 64 ± 0.02. Calculate an estimate of PCR. Does the process meet a minimum capability level of $PCR \geq 1.33$?
(d) Calculate an estimate of PCR_k. Use this ratio to draw conclusions about process capability.
(e) To make this process a six-sigma process, the variance σ^2 would have to be decreased such that $PCR_k = 2.0$. What should this new variance value be?
(f) Suppose the mean shifts to 64.01. What is the probability that this shift will be detected on the next sample? What is the ARL after the shift?

16-58. Rework Exercise 16-57 with $\overline{X}$ and S charts.

16-59. Plastic bottles for liquid laundry detergent are formed by blow molding. Twenty samples of $n = 100$ bottles are inspected in time order of production, and the fraction defective in each sample is reported. The data are as follows:

Sample	Fraction Defective
1	0.12
2	0.15
3	0.18
4	0.10
5	0.12
6	0.11
7	0.05
8	0.09
9	0.13
10	0.13
11	0.10
12	0.07
13	0.12
14	0.08
15	0.09
16	0.15
17	0.10
18	0.06
19	0.12
20	0.13

(a) Set up a P chart for this process. Is the process in statistical control?
(b) Suppose that instead of $n = 100$, $n = 200$. Use the data given to set up a P chart for this process. Revise the control limits if necessary.
(c) Compare your control limits for the P charts in parts (a) and (b). Explain why they differ. Also, explain why your assessment about statistical control differs for the two sizes of n.

16-60. Cover cases for a personal computer are manufactured by injection molding. Samples of five cases are taken from the process periodically, and the number of defects is noted. Twenty-five samples follow:

Sample	No. of Defects	Sample	No. of Defects
1	3	14	8
2	2	15	0
3	0	16	2
4	1	17	4
5	4	18	3
6	3	19	5
7	2	20	0
8	4	21	2
9	1	22	1
10	0	23	9
11	2	24	3
12	3	25	2
13	2		

(a) Using all the data, find trial control limits for a U chart for the process.
(b) Use the trial control limits from part (a) to identify out-of-control points. If necessary, revise your control limits.
(c) Suppose that instead of samples of 5 cases, the sample size was 10. Repeat parts (a) and (b). Explain how this change alters your answers to parts (a) and (b).

16-61. An article in *Quality Engineering* (1992, Vol. 4(4)) "Is the process capable? Tables and graphs in assessing C_{pm}" considered manufacturing data. Specifications for the outer diameter of the hubs were 60.3265 ± 0.001 mm. A random sample with size $n = 20$ was taken and the data are shown in the following table.

Sample	x	Sample	x
1	60.3262	11	60.3262
2	60.3262	12	60.3262
3	60.3262	13	60.3269
4	60.3266	14	60.3261
5	60.3263	15	60.3265
6	60.3260	16	60.3266
7	60.3262	17	60.3265
8	60.3267	18	60.3268
9	60.3263	19	60.3262
10	60.3269	20	60.3266

(a) Construct a control chart for individual measurements. Revise the control limits if necessary.
(b) Compare your chart in part (a) to one that uses only the last (least significant) digit of each diameter as the measurement. Explain your conclusion.
(c) Estimate μ and σ from the moving range of the revised chart and use this value to estimate PCR and PCR_k and interpret these ratios.

16-62. The following data from the U.S. Department of Energy Web site (http://www.eia.doe.gov/emeu/mer/dataunits.html) reported the total U.S. *renewable* energy consumption by year (trillion BTU) from 1973 to 2004.

Year	Total Renewable Energy Consumption (Trillion BTU)	Year	Total Renewable Energy Consumption (Trillion Btu)
1973	4433.121	1989	6294.209
1974	4769.395	1990	6132.572
1975	4723.494	1991	6158.087
1976	4767.792	1992	5907.147
1977	4249.002	1993	6155.959
1978	5038.938	1994	6064.779
1979	5166.379	1995	6669.261
1980	5494.42	1996	7136.799
1981	5470.574	1997	7075.152
1982	5985.352	1998	6560.632
1983	6487.898	1999	6598.63
1984	6430.646	2000	6158.232
1985	6032.728	2001	5328.335
1986	6131.542	2002	5835.339
1987	5686.932	2003	6081.722
1988	5488.649	2004	6116.287

(a) Using all the data, find trial control limits for a control chart for individual measurements, construct the chart, and plot the data.
(b) Do the data appear to be generated from an in-control process? Comment on any patterns on the chart.

16-63. The following dataset was considered in *Quality Engineering* (1994–1995, Vol. 7(2)) "Analytic examination of variance components." A quality characteristic for cement mortar briquettes was monitored. Samples of size $n = 6$ were taken from the process and 25 samples from the process are shown in the following table.
(a) Using all the data, find trial control limits for $\overline{X}$ and S charts. Is the process in control?

Batch	$\bar{x}$	s
1	572.00	73.25
2	583.83	79.30
3	720.50	86.44
4	368.67	98.62
5	374.00	92.36
6	580.33	93.50
7	388.33	110.23
8	559.33	74.79
9	562.00	76.53
10	729.00	49.80
11	469.00	40.52
12	566.67	113.82
13	578.33	58.03
14	485.67	103.33
15	746.33	107.88
16	436.33	98.69
17	556.83	99.25
18	390.33	117.35
19	562.33	75.69
20	675.00	90.10
21	416.50	89.27
22	568.33	61.36
23	762.67	105.94
24	786.17	65.05
25	530.67	99.42

(b) Suppose that the specifications are at 580 ± 250. What statements can you make about process capability? Compute estimates of the appropriate process capability ratios.
(c) To make this process a "6-sigma process," the variance σ^2 would have to be decreased such that $PCR_k = 2.0$. What should this new variance value be?
(d) Suppose the mean shifts to 600. What is the probability that this shift will be detected on the next sample? What is the ARL after the shift?

16-64. Suppose that an $\overline{X}$ control chart with 2-sigma limits is used to control a process. Find the probability that a false out-of-control signal will be produced on the next sample. Compare this with the corresponding probability for the chart with 3-sigma limits and discuss. Comment on when you would prefer to use 2-sigma limits instead of 3-sigma limits.

16-65. Consider the diameter data in Exercise 16-57.
(a) Construct an EWMA control chart with $\lambda = 0.2$ and $L = 3$. Comment on process control.
(b) Construct an EWMA control chart with $\lambda = 0.5$ and $L = 3$ and compare your conclusion to part (a).

16-66. Consider the renewable energy data in Exercise 16-62.

(a) Construct an EWMA control chart with $\lambda = 0.2$ and $L = 3$. Do the data appear to be generated from an in-control process?

(b) Construct an EWMA control chart with $\lambda = 0.5$ and $L = 3$ and compare your conclusion to part (a).

16-67. Consider the hub data in Exercise 16-61.

(a) Construct an EWMA control chart with $\lambda = 0.2$ and $L = 3$. Comment on process control.

(b) Construct an EWMA control chart with $\lambda = 0.5$ and $L = 3$ and compare your conclusion to part (a).

16-68. Consider the data in Exercise 16-14. Set up a CUSUM scheme for this process assuming that $\mu = 16$ is the process target. Explain how you determined your estimate of σ and the CUSUM parameters K and H.

16-69. Consider the hardness measurement data in Exercise 16-13. Set up a CUSUM scheme for this process using $\mu = 50$ and $\sigma = 2$, so that $K = 1$ and $H = 10$. Is the process in control?

16-70. Reconsider the data in Exercise 16-16. Construct a CUSUM control chart for this process using $\mu_0 = 500$ as the process target. Explain how you determined your estimate of σ and the CUSUM parameters H and K.

16-71. The following data was considered in *Quality Progress* (1990, May, pp. 66–68) "Digidot Plots for Process Surveillance." Measurements of center thickness (in mils) from 25 contact lenses sampled from the production process at regular intervals are shown in the following table.

Sample	x	Sample	x
1	0.3978	14	0.3999
2	0.4019	15	0.4062
3	0.4031	16	0.4048
4	0.4044	17	0.4071
5	0.3984	18	0.4015
6	0.3972	19	0.3991
7	0.3981	20	0.4021
8	0.3947	21	0.4009
9	0.4012	22	0.3988
10	0.4043	23	0.3994
11	0.4051	24	0.4016
12	0.4016	25	0.4010
13	0.3994		

(a) Construct a CUSUM scheme for this process with the target $\mu_0 = 0.4$. Explain how you determined your estimate of σ and the CUSUM parameters H and K. Is the process in control?

(b) Construct an EWMA control chart with $\lambda = 0.5$ and $L = 3$ and compare your conclusions to part (a).

16-72. Suppose that a process is in control and an $\overline{X}$ chart is used with a sample size of 4 to monitor the process. Suddenly there is a mean shift of 1.5σ.

(a) If 3-sigma control limits are in use on the $\overline{X}$ chart, what is the probability that this shift will remain undetected for three consecutive samples?

(b) If 2-sigma control limits are in use on the $\overline{X}$ chart, what is the probability that this shift will remain undetected for three consecutive samples?

(c) Compare your answers to parts (a) and (b) and explain why they differ. Also, which limits you would recommend using and why?

16-73. Consider the control chart for individuals with 3-sigma limits.

(a) Suppose that a shift in the process mean of magnitude σ occurs. Verify that the ARL for detecting the shift is ARL = 43.9.

(b) Find the ARL for detecting a shift of magnitude 2σ in the process mean.

(c) Find the ARL for detecting a shift of magnitude 3σ in the process mean.

(d) Compare your answers to parts (a), (b), and (c) and explain why the ARL for detection is decreasing as the magnitude of the shift increases.

16-74. Consider a control chart for individuals, applied to a continuous 24-hour chemical process with observations taken every hour.

(a) If the chart has 3-sigma limits, verify that the in-control ARL is ARL = 370. How many false alarms would occur each 30-day month, on the average, with this chart?

(b) Suppose that the chart has 2-sigma limits. Does this reduce the ARL for detecting a shift in the mean of magnitude σ? (Recall that the ARL for detecting this shift with 3-sigma limits is 43.9.)

(c) Find the in-control ARL if 2-sigma limits are used on the chart. How many false alarms would occur each month with this chart? Is this in-control ARL performance satisfactory? Explain your answer.

16-75. The depth of a keyway is an important part quality characteristic. Samples of size $n = 5$ are taken every four hours from the process and 20 samples are summarized in the following table.

(a) Using all the data, find trial control limits for $\overline{X}$ and R charts. Is the process in control?

(b) Use the trial control limits from part (a) to identify out-of-control points. If necessary, revise your control limits. Then, estimate the process standard deviation.

(c) Suppose that the specifications are at 140 ± 2. Using the results from part (b), what statements can you make about process capability? Compute estimates of the appropriate process capability ratios.

(d) To make this process a "6-sigma process," the variance σ^2 would have to be decreased such that $PCR_k = 2.0$. What should this new variance value be?

Sample	$\overline{X}$	r
1	139.7	1.1
2	139.8	1.4
3	140.0	1.3
4	140.1	1.6
5	139.8	0.9
6	139.9	1.0
7	139.7	1.4
8	140.2	1.2
9	139.3	1.1
10	140.7	1.0
11	138.4	0.8
12	138.5	0.9
13	137.9	1.2
14	138.5	1.1
15	140.8	1.0
16	140.5	1.3
17	139.4	1.4
18	139.9	1.0
19	137.5	1.5
20	139.2	1.3

(e) Suppose the mean shifts to 139.7. What is the probability that this shift will be detected on the next sample? What is the ARL after the shift?

16-76. Consider a control chart for individuals with 3-sigma limits. What is the probability that there will not be a signal in three samples? In 6 samples? In 10 samples?

16-77. Suppose a process has a $PCR = 2$, but the mean is exactly three standard deviations above the upper specification limit. What is the probability of making a product outside the specification limits?

16-78. A process is controlled by a P chart using samples of size 100. The center line on the chart is 0.05.

(a) What is the probability that the control chart detects a shift to 0.08 on the first sample following the shift?
(b) What is the probability that the control chart does not detect a shift to 0.08 on the first sample following the shift but does detect it on the second sample?
(c) Suppose that instead of a shift in the mean to 0.08, the mean shifts to 0.10. Repeat parts (a) and (b).
(d) Compare your answers for a shift to 0.08 and for a shift to 0.10. Explain why they differ. Also, explain why a shift to 0.10 is easier to detect.

16-79. Suppose the average number of defects in a unit is known to be 8. If the mean number of defects in a unit shifts to 16, what is the probability that it will be detected by a U chart on the first sample following the shift

(a) if the sample size is $n = 4$?
(b) if the sample size is $n = 10$?

Use a normal approximation for U.

16-80. Suppose the average number of defects in a unit is known to be 10. If the mean number of defects in a unit shifts to 14, what is the probability that it will be detected by a U chart on the first sample following the shift

(a) if the sample size is $n = 1$?
(b) if the sample size is $n = 4$?

Use a normal approximation for U.

16-81. An EWMA chart with $\lambda = 0.5$ and $L = 3.07$ is to be used to monitor a process. Suppose that the process mean is $\mu_0 = 10$ and $\sigma = 2$.

(a) Assume that $n = 1$. What is the ARL without any shift in the process mean? What is the ARL to detect a shift to $\mu = 12$.
(b) Assume that $n = 4$. Repeat part (a) and comment on your conclusions.

16-82. The following table provides the costs for gasoline by month in the U.S. over recent years and the percentage of the cost due to refining, distribution and marketing, taxes, and crude oil. The table is from the U.S. Department of Energy Web site (http://tonto.eia.doe.gov/oog/info/gdu/gaspump.html). There is some concern that the refining or distribution and marketing percentages of the retail price have shown patterns over time.

(a) Construct separate control charts for the refining percentage of the retail price and the distribution and marketing percentage of the retail price. Use control charts for individual measurements. Comment on any signs of assignable causes on these charts.
(b) Construct a control chart for the crude oil percentage of the retail price. Use a control chart for individual measurements. Comment on any signs of assignable causes on this chart.
(c) Another way to study the data is to calculate refining, distribution and marketing, and tax as costs directly. The costs of these categories might not depend strongly on the crude oil cost. Use the percentages provided in the table to calculate the cost each month associated with refining and distribution and marketing. Construct separate control charts for the refining and the distribution and marketing costs each month. Use control charts for individual measurements. Comment on any signs of assignable causes on these charts and comment on any differences between these charts and the ones constructed in part (a).

What We Pay for in a Gallon of Regular Gasoline

Mo/Year	Retail Price (Cents per gallon)	Refining (percentage)	Distribution and Marketing (percentage)	Taxes (percentage)	Crude Oil (percentage)
Jan-00	128.9	7.8	13.0	32.1	47.1
Feb-00	137.7	17.9	7.5	30.1	44.6
Mar-00	151.7	15.4	12.8	27.3	44.6
Apr-00	146.5	10.1	20.2	28.3	41.4
May-00	148.5	20.2	9.2	27.9	42.7
Jun-00	163.3	22.2	8.8	25.8	43.1
Jul-00	155.1	13.2	15.8	27.2	43.8
Aug-00	146.5	15.8	7.5	28.8	47.8
Sep-00	155.0	15.4	9.0	27.2	48.3
Oct-00	153.2	13.7	10.1	27.5	48.6
Nov-00	151.7	10.4	11.8	27.8	50.0
Dec-00	144.3	8.0	17.9	29.2	44.8
Jan-01	144.7	17.8	10.4	29.2	42.7
Feb-01	145.0	17.3	11.0	29.1	42.6
Mar-01	140.9	18.8	9.7	30.0	41.5
Apr-01	155.2	31.6	4.6	27.1	36.7
May-01	170.2	26.4	14.0	24.7	35.0
Jun-01	161.6	13.2	24.1	26.0	36.7
Jul-01	142.1	10.0	20.0	30.0	40.0
Aug-01	142.1	20.0	9.0	30.0	41.0
Sep-01	152.2	18.0	17.0	28.0	37.0
Oct-01	131.5	10.0	20.8	31.9	37.2
Nov-01	117.1	10.0	18.0	36.0	36.0
Dec-01	108.6	11.7	12.7	38.7	36.9
Jan-02	110.7	13.0	11.8	37.9	37.2
Feb-02	111.4	12.1	11.2	37.7	39.1
Mar-02	124.9	19.4	6.1	33.6	40.9
Apr-02	139.7	15.5	13.0	30.1	41.4
May-02	139.2	11.9	14.2	30.2	43.7
Jun-02	138.2	15.0	13.0	30.4	41.6
Jul-02	139.7	15.0	12.6	30.1	42.3
Aug-02	139.6	11.4	13.4	30.0	45.0
Sep-02	140.0	10.8	12.6	30.0	46.7
Oct-02	144.5	13.9	11.7	29.1	45.3
Nov-02	141.9	11.1	18.0	29.6	41.3
Dec-02	138.6	11.7	12.3	30.3	45.7
Jan-03	145.8	11.5	10.3	28.8	49.4
Feb-03	161.3	15.0	9.5	26.0	49.5
Mar-03	169.3	14.8	14.8	24.8	45.5
Apr-03	158.9	13.2	19.8	26.4	40.5
May-03	149.7	15.3	16.3	28.1	40.4
Jun-03	149.3	15.1	12.3	28.1	44.5
Jul-03	151.3	15.3	11.9	27.8	44.9
Aug-03	162.0	22.5	8.2	25.9	43.3
Sept-03	167.9	13.9	22.7	25.0	38.3
Oct-03	156.4	14.9	16.1	26.9	42.2

What We Pay for in a Gallon of Regular Gasoline

Mo/Year	Retail Price (Cents per gallon)	Refining (percentage)	Distribution and Marketing (percentage)	Taxes (percentage)	Crude Oil (percentage)
Nov-03	151.2	11.7	15.3	27.8	45.2
Dec-03	147.9	11.5	12.6	28.4	47.5
Jan-04	157.2	15.9	9.9	26.7	47.5
Feb-04	164.8	19.1	9.2	25.5	46.2
Mar-04	173.6	19.0	11.3	24.2	45.5
Apr-04	179.8	22.0	9.9	23.4	44.6
May-04	198.3	30.6	7.8	21.2	40.4
Jun-04	196.9	21.3	16.7	21.3	40.7
Jul-04	191.1	20.9	11.3	21.9	45.8
Aug-04	187.8	13.9	12.2	22.4	51.5
Sep-04	187.0	14.8	9.1	22.5	53.6
Oct-04	200.0	13.0	9.3	21.0	56.7
Nov-04	197.9	10.7	14.6	21.2	53.6
Dec-04	184.1	8.9	18.1	23.9	49.1
Jan-05	183.1	17.7	7.3	24.0	50.9
Feb-05	191.0	16.1	9.3	23.0	51.6
Mar-05	207.9	19.3	6.2	21.2	53.4
Apr-05	224.3	20.9	9.6	19.6	49.8
May-05	216.1	17.9	12.8	20.4	49.0
Jun-05	215.6	18.5	6.9	20.4	54.2
Jul-05	229.0	17.9	8.0	19.2	54.9

MIND-EXPANDING EXERCISES

16-83. Suppose a process is in control, and 3-sigma control limits are in use on the $\overline{X}$ chart. Let the mean shift by 1.5σ. What is the probability that this shift will remain undetected for three consecutive samples? What would its probability be if 2-sigma control limits were used? The sample size is 4.

16-84. Consider an $\overline{X}$ control chart with k-sigma control limits. Develop a general expression for the probability that a point will plot outside the control limits when the process mean has shifted by δ units from the center line.

16-85. Suppose that an $\overline{X}$ chart is used to control a normally distributed process and that samples of size n are taken every n hours and plotted on the chart, which has k-sigma limits.

(a) Find a general expression for the expected number of samples and time that will be taken until a false action signal is generated.

(b) Suppose that the process mean shifts to an out-of-control state, say $\mu_1 = \mu_0 + \delta\sigma$. Find an expression for the expected number of samples that will be taken until a false action is generated.

(c) Evaluate the in-control ARL for $k = 3$. How, does this change if $k = 2$? What do you think about the use of 2-sigma limits in practice?

(d) Evaluate the out-of-control ARL for a shift of 1 sigma, given that $n = 5$.

16-86. Suppose a P chart with center line at $\overline{p}$ with k-sigma control limits is used to control a process. There is a critical fraction defective p_c that must be detected with probability 0.50 on the first sample following the shift to this state. Derive a general formula for the sample size that should be used on this chart.

16-87. Suppose that a P chart with center line at $\overline{p}$ and k-sigma control limits is used to control a process. What

MIND-EXPANDING EXERCISES

is the smallest sample size that can be used on this control chart to ensure that the lower control limit is positive?

16-88. A process is controlled by a P chart using samples of size 100. The center line on the chart is 0.05. What is the probability that the control chart detects a shift to 0.08 on the first sample following the shift? What is the probability that the shift is detected by at least the third sample following the shift?

16-89. Consider a process where specifications on a quality characteristic are 100 ± 15. We know that the standard deviation of this normally distributed quality characteristic is 5. Where should we center the process to minimize the fraction defective produced? Now suppose the mean shifts to 105 and we are using a sample size of 4 on an $\overline{X}$ chart. What is the probability that such a shift will be detected on the first sample following the shift? What is the average number of samples until an out-of-control point occurs? Compare this result to the average number of observations until a defective occurs (assuming normality).

16-90. ***NP* Control Chart.** An alternative to the control chart for fraction defective is a control chart based on the number of defectives, or the NP control chart. The chart has centerline at $n\overline{p}$, and the control limits are

$$UCL = n\overline{p} + 3\sqrt{n\overline{p}(1 - \overline{p})}$$
$$LCL = n\overline{p} - 3\sqrt{n\overline{p}(1 - \overline{p})}$$

and the number of defectives for each sample is plotted on the chart.

(a) Verify that the control limits given above are correct.
(b) Apply this control chart to the data in Example 16-4.
(c) Will this chart always provide results that are equivalent to the usual P chart?

16-91. ***C* Control Chart.** An alternative to the U chart is a chart based on the number of defects. The chart has center line at $n\overline{u}$, and the control limits are

$$UCL = n\overline{u} + 3\sqrt{n\overline{u}}$$
$$LCL = n\overline{u} - 3\sqrt{n\overline{u}}$$

(a) Apply this chart to the data in Example 16-5.
(b) Will this chart always provide results equivalent to the U chart?

16-92. **Standardized Control Chart.** Consider the P chart with the usual 3-sigma control limits. Suppose that we define a new variable:

$$Z_i = \frac{\hat{P}_i - \overline{P}}{\sqrt{\dfrac{\overline{P}(1 - \overline{P})}{n}}}$$

as the quantity to plot on a control chart. It is proposed that this new chart will have a center line at 0 with the upper and lower control limits at ± 3. Verify that this standardized control chart will be equivalent to the original P chart.

16-93. **Unequal Sample Sizes.** One application of the standardized control chart introduced in Exercise 16-92 is to allow unequal sample sizes on the control chart. Provide details concerning how this procedure would be implemented and illustrate using the following data:

Sample, i	1	2	3	4	5	6	7	8	9	10
n_i	20	25	20	25	50	30	25	25	25	20
p_i	0.2	0.16	0.25	0.08	0.3	0.1	0.12	0.16	0.12	0.15

IMPORTANT TERMS AND CONCEPTS

ARL
Assignable causes
Attributes control charts
Average run length
C chart
Cause-and-effect diagram
Center line
Chance causes
Control chart
Control limits
Cumulative sum control chart
Defect concentration diagram
Defects-per-unit chart
Deming's 14 points
Exponentially weighted moving average control chart (EWMA)
False alarm

Fraction-defective control chart
Implementing SPC
Individuals control chart (*X* chart)
Moving range
NP chart
P chart
Pareto diagram
PCR
PCR_k
Problem-solving tools
Process capability
Process capability ratio
Quality control
R chart
Rational subgroup
Run rule
S chart
Shewhart control chart
Six-sigma process
Specification limits
Statistical process control (SPC)
Statistical quality control
U chart
V mask
Variables control charts
Warning limits
Western Electric rules
$\overline{X}$ chart

APPENDICES

Appendix A Statistical Tables and Charts

Table I Summary of Common Probability Distributions

Name	Probability Distribution	Mean	Variance	Section in Book
Discrete				
Uniform	$\frac{1}{n}, a \le b$	$\frac{(b+a)}{2}$	$\frac{(b-a+1)^2-1}{12}$	3-5
Binomial	$\binom{n}{x} p^x(1-p)^{n-x}$, $x = 0, 1, \ldots, n, 0 \le p \le 1$	np	$np(1-p)$	3-6
Geometric	$(1-p)^{x-1}p$, $x = 1, 2, \ldots, 0 \le p \le 1$	$1/p$	$(1-p)/p^2$	3-7.1
Negative binomial	$\binom{x-1}{r-1}(1-p)^{x-r}p^r$ $x = r, r+1, r+2, \ldots, 0 \le p \le 1$	r/p	$r(1-p)/p^2$	3-7.2
Hypergeometric	$\frac{\binom{K}{x}\binom{N-K}{n-x}}{\binom{N}{n}}$ $x = \max(0, n-N+K), 1, \ldots$ $\min(K, n), K \le N, n \le N$	np, where $p = \frac{K}{N}$	$np(1-p)\left(\frac{N-n}{N-1}\right)$	3-8
Poisson	$\frac{e^{-\lambda}\lambda^x}{x!}, x = 0, 1, 2, \ldots, 0 < \lambda$	λ	λ	3-9
Continuous				
Uniform	$\frac{1}{b-a}, a \le x \le b$	$\frac{(b+a)}{2}$	$\frac{(b-a)^2}{12}$	4-5
Normal	$\frac{1}{\sigma\sqrt{2\pi}} e^{-\frac{1}{2}\left(\frac{x-\mu}{\sigma}\right)^2}$ $-\infty < x < \infty, -\infty < \mu < \infty, 0 < \sigma$	μ	σ^2	4-6
Exponential	$\lambda e^{-\lambda x}, 0 \le x, 0 < \lambda$	$1/\lambda$	$1/\lambda^2$	4-8
Erlang	$\frac{\lambda^r x^{r-1} e^{-\lambda x}}{(r-1)!}, 0 < x, r = 1, 2, \ldots$	r/λ	r/λ^2	4-9.1
Gamma	$\frac{\lambda x^{r-1} e^{-\lambda x}}{\Gamma(r)}, 0 < x, 0 < r, 0 < \lambda$	r/λ	r/λ^2	4-9.2
Weibull	$\frac{\beta}{\delta}\left(\frac{x}{\delta}\right)^{\beta-1} e^{-(x/\delta)^\beta}$, $0 < x, 0 < \beta, 0 < \delta$	$\delta\Gamma\left(1+\frac{1}{\beta}\right)$	$\delta^2\Gamma\left(1+\frac{2}{\beta}\right) - \delta^2\left[\Gamma\left(1+\frac{1}{\beta}\right)\right]^2$	4-10
Lognormal	$\frac{1}{x\omega\sqrt{2\pi}} \exp\left(\frac{-[\ln(x)-\theta]^2}{2\omega^2}\right)$	$e^{\theta+\omega^2/2}$	$e^{2\theta+\omega^2}(e^{\omega^2}-1)$	4-11

Table II Cumulative Binomial Probabilities $P(X <= x)$

		P										
n	x	0.1	0.2	0.3	0.4	0.5	0.6	0.7	0.8	0.9	0.95	0.99
1	0	0.9000	0.8000	0.7000	0.6000	0.5000	0.4000	0.3000	0.2000	0.1000	0.0500	0.0100
2	0	0.8100	0.6400	0.4900	0.3600	0.2500	0.1600	0.0900	0.0400	0.0100	0.0025	0.0001
	1	0.9900	0.9600	0.9100	0.8400	0.7500	0.6400	0.5100	0.3600	0.1900	0.0975	0.0199
3	0	0.7290	0.5120	0.3430	0.2160	0.1250	0.0640	0.0270	0.0080	0.0010	0.0001	0.0000
	1	0.9720	0.8960	0.7840	0.6480	0.5000	0.3520	0.2160	0.1040	0.0280	0.0073	0.0003
	2	0.9990	0.9920	0.9730	0.9360	0.8750	0.7840	0.6570	0.4880	0.2710	0.1426	0.0297
4	0	0.6561	0.4096	0.2401	0.1296	0.0625	0.0256	0.0081	0.0016	0.0001	0.0000	0.0000
	1	0.9477	0.8192	0.6517	0.4752	0.3125	0.1792	0.0837	0.0272	0.0037	0.0005	0.0000
	2	0.9963	0.9728	0.9163	0.8208	0.6875	0.5248	0.3483	0.1808	0.0523	0.0140	0.0006
	3	0.9999	0.9984	0.9919	0.9744	0.9375	0.8704	0.7599	0.5904	0.3439	0.1855	0.0394
5	0	0.5905	0.3277	0.1681	0.0778	0.0313	0.0102	0.0024	0.0003	0.0000	0.0000	0.0000
	1	0.9185	0.7373	0.5282	0.3370	0.1875	0.0870	0.0308	0.0067	0.0005	0.0000	0.0000
	2	0.9914	0.9421	0.8369	0.6826	0.5000	0.3174	0.1631	0.0579	0.0086	0.0012	0.0000
	3	0.9995	0.9933	0.9692	0.9130	0.8125	0.6630	0.4718	0.2627	0.0815	0.0226	0.0010
	4	1.0000	0.9997	0.9976	0.9898	0.6988	0.9222	0.8319	0.6723	0.4095	0.2262	0.0490
6	0	0.5314	0.2621	0.1176	0.0467	0.0156	0.0041	0.0007	0.0001	0.0000	0.0000	0.0000
	1	0.8857	0.6554	0.4202	0.2333	0.1094	0.0410	0.0109	0.0016	0.0001	0.0000	0.0000
	2	0.9842	0.9011	0.7443	0.5443	0.3438	0.1792	0.0705	0.0170	0.0013	0.0001	0.0000
	3	0.9987	0.9830	0.9295	0.8208	0.6563	0.4557	0.2557	0.0989	0.0159	0.0022	0.0000
	4	0.9999	0.9984	0.9891	0.9590	0.9806	0.7667	0.5798	0.3446	0.1143	0.0328	0.0015
	5	1.0000	0.9999	0.9993	0.9959	0.9844	0.9533	0.8824	0.7379	0.4686	0.2649	0.0585
7	0	0.4783	0.2097	0.0824	0.0280	0.0078	0.0016	0.0002	0.0000	0.0000	0.0000	0.0000
	1	0.8503	0.5767	0.3294	0.1586	0.0625	0.0188	0.0038	0.0004	0.0000	0.0000	0.0000
	2	0.9743	0.8520	0.6471	0.4199	0.2266	0.0963	0.0288	0.0047	0.0002	0.0000	0.0000
	3	0.9973	0.9667	0.8740	0.7102	0.5000	0.2898	0.1260	0.0333	0.0027	0.0002	0.0000
	4	0.9998	0.9953	0.9712	0.9037	0.7734	0.5801	0.3529	0.1480	0.0257	0.0038	0.0000
	5	1.0000	0.9996	0.9962	0.9812	0.9375	0.8414	0.6706	0.4233	0.1497	0.0444	0.0020
	6	1.0000	1.0000	0.9998	0.9984	0.9922	0.9720	0.9176	0.7903	0.5217	0.3017	0.0679
8	0	0.4305	0.1678	0.0576	0.0168	0.0039	0.0007	0.0001	0.0000	0.0000	0.0000	0.0000
	1	0.8131	0.5033	0.2553	0.1064	0.0352	0.0085	0.0013	0.0001	0.0000	0.0000	0.0000
	2	0.9619	0.7969	0.5518	0.3154	0.1445	0.0498	0.0113	0.0012	0.0000	0.0000	0.0000
	3	0.9950	0.9437	0.8059	0.5941	0.3633	0.1737	0.0580	0.0104	0.0004	0.0000	0.0000
	4	0.9996	0.9896	0.9420	0.8263	0.6367	0.4059	0.1941	0.0563	0.0050	0.0004	0.0000
	5	1.0000	0.9988	0.9887	0.9502	0.8555	0.6846	0.4482	0.2031	0.0381	0.0058	0.0001
	6	1.0000	0.9999	0.9987	0.9915	0.9648	0.8936	0.7447	0.4967	0.1869	0.0572	0.0027
	7	1.0000	1.0000	0.9999	0.9993	0.9961	0.9832	0.9424	0.8322	0.5695	0.3366	0.0773
9	0	0.3874	0.1342	0.0404	0.0101	0.0020	0.0003	0.0000	0.0000	0.0000	0.0000	0.0000
	1	0.7748	0.4362	0.1960	0.0705	0.0195	0.0038	0.0004	0.0000	0.0000	0.0000	0.0000
	2	0.9470	0.7382	0.4628	0.2318	0.0889	0.0250	0.0043	0.0003	0.0000	0.0000	0.0000
	3	0.9917	0.9144	0.7297	0.4826	0.2539	0.0994	0.0253	0.0031	0.0001	0.0000	0.0000
	4	0.9991	0.9804	0.9012	0.7334	0.5000	0.2666	0.0988	0.0196	0.0009	0.0000	0.0000
	5	0.9999	0.9969	0.9747	0.9006	0.7461	0.5174	0.2703	0.0856	0.0083	0.0006	0.0000
	6	1.0000	0.9997	0.9957	0.9750	0.9102	0.7682	0.5372	0.2618	0.0530	0.0084	0.0001
	7	1.0000	1.0000	0.9996	0.9962	0.9805	0.9295	0.8040	0.5638	0.2252	0.0712	0.0034
	8	1.0000	1.0000	1.0000	0.9997	0.9980	0.9899	0.9596	0.8658	0.6126	0.3698	0.0865

Table II Cumulative Binomial Probabilities $P(X<=x)$ *(continued)*

		P										
n	x	0.1	0.2	0.3	0.4	0.5	0.6	0.7	0.8	0.9	0.95	0.99
10	0	0.3487	0.1074	0.0282	0.0060	0.0010	0.0001	0.0000	0.0000	0.0000	0.0000	0.0000
	1	0.7361	0.3758	0.1493	0.0464	0.0107	0.0017	0.0001	0.0000	0.0000	0.0000	0.0000
	2	0.9298	0.6778	0.3828	0.1673	0.0547	0.0123	0.0016	0.0001	0.0000	0.0000	0.0000
	3	0.9872	0.8791	0.6496	0.3823	0.1719	0.0548	0.0106	0.0009	0.0000	0.0000	0.0000
	4	0.9984	0.9672	0.8497	0.6331	0.3770	0.1662	0.0473	0.0064	0.0001	0.0000	0.0000
	5	0.9999	0.9936	0.9527	0.8338	0.6230	0.3669	0.1503	0.0328	0.0016	0.0001	0.0000
	6	1.0000	0.9991	0.9894	0.9452	0.8281	0.6177	0.3504	0.1209	0.0128	0.0010	0.0000
	7	1.0000	0.9999	0.9984	0.9877	0.9453	0.8327	0.6172	0.3222	0.0702	0.0115	0.0001
	8	1.0000	1.0000	0.9999	0.9983	0.9893	0.9536	0.8507	0.6242	0.2639	0.0861	0.0043
	9	1.0000	1.0000	1.0000	0.9999	0.9990	0.9940	0.9718	0.8926	0.6513	0.4013	0.0956
11	0	0.3138	0.0859	0.0198	0.0036	0.0005	0.0000	0.0000	0.0000	0.0000	0.0000	0.0000
	1	0.6974	0.3221	0.1130	0.0302	0.0059	0.0007	0.0000	0.0000	0.0000	0.0000	0.0000
	2	0.9104	0.6174	0.3127	0.1189	0.0327	0.0059	0.0006	0.0000	0.0000	0.0000	0.0000
	3	0.9815	0.8389	0.5696	0.2963	0.1133	0.0293	0.0043	0.0002	0.0000	0.0000	0.0000
	4	0.9972	0.9496	0.7897	0.5328	0.2744	0.0994	0.0216	0.0020	0.0000	0.0000	0.0000
	5	0.9997	0.9883	0.9218	0.7535	0.5000	0.2465	0.0782	0.0117	0.0003	0.0000	0.0000
	6	1.0000	0.9980	0.9784	0.9006	0.7256	0.4672	0.2103	0.0504	0.0028	0.0001	0.0000
	7	1.0000	0.9998	0.9957	0.9707	0.8867	0.7037	0.4304	0.1611	0.0185	0.0016	0.0000
	8	1.0000	1.0000	0.9994	0.9941	0.9673	0.8811	0.6873	0.3826	0.0896	0.0152	0.0002
	9	1.0000	1.0000	1.0000	0.9993	0.9941	0.9698	0.8870	0.6779	0.3026	0.1019	0.0052
	10	1.0000	1.0000	1.0000	1.0000	0.9995	0.9964	0.9802	0.9141	0.6862	0.4312	0.1047
12	0	0.2824	0.0687	0.0138	0.0022	0.0002	0.0000	0.0000	0.0000	0.0000	0.0000	0.0000
	1	0.6590	0.2749	0.0850	0.0196	0.0032	0.0003	0.0000	0.0000	0.0000	0.0000	0.0000
	2	0.8891	0.5583	0.2528	0.0834	0.0193	0.0028	0.0002	0.0000	0.0000	0.0000	0.0000
	3	0.9744	0.7946	0.4925	0.2253	0.0730	0.0153	0.0017	0.0001	0.0000	0.0000	0.0000
	4	0.9957	0.9274	0.7237	0.4382	0.1938	0.0573	0.0095	0.0006	0.0000	0.0000	0.0000
	5	0.9995	0.9806	0.8822	0.6652	0.3872	0.1582	0.0386	0.0039	0.0001	0.0000	0.0000
	6	0.9999	0.9961	0.9614	0.8418	0.6128	0.3348	0.1178	0.0194	0.0005	0.0000	0.0000
	7	1.0000	0.9994	0.9905	0.9427	0.8062	0.5618	0.2763	0.0726	0.0043	0.0002	0.0000
	8	1.0000	0.9999	0.9983	0.9847	0.9270	0.7747	0.5075	0.2054	0.0256	0.0022	0.0000
	9	1.0000	1.0000	0.9998	0.9972	0.9807	0.9166	0.7472	0.4417	0.1109	0.0196	0.0002
	10	1.0000	1.0000	1.0000	0.9997	0.9968	0.9804	0.9150	0.7251	0.3410	0.1184	0.0062
	11	1.0000	1.0000	1.0000	1.0000	0.9998	0.9978	0.9862	0.9313	0.7176	0.4596	0.1136
13	0	0.2542	0.0550	0.0097	0.0013	0.0001	0.0000	0.0000	0.0000	0.0000	0.0000	0.0000
	1	0.6213	0.2336	0.0637	0.0126	0.0017	0.0001	0.0000	0.0000	0.0000	0.0000	0.0000
	2	0.8661	0.5017	0.2025	0.0579	0.0112	0.0013	0.0001	0.0000	0.0000	0.0000	0.0000
	3	0.9658	0.7473	0.4206	0.1686	0.0461	0.0078	0.0007	0.0000	0.0000	0.0000	0.0000
	4	0.9935	0.9009	0.6543	0.3530	0.1334	0.0321	0.0040	0.0002	0.0000	0.0000	0.0000
	5	0.9991	0.9700	0.8346	0.5744	0.2905	0.0977	0.0182	0.0012	0.0000	0.0000	0.0000
	6	0.9999	0.9930	0.9376	0.7712	0.5000	0.2288	0.0624	0.0070	0.0001	0.0000	0.0000
	7	1.0000	0.9988	0.9818	0.9023	0.7095	0.4256	0.1654	0.0300	0.0009	0.0000	0.0000
	8	1.0000	0.9988	0.9960	0.9679	0.8666	0.6470	0.3457	0.0991	0.0065	0.0003	0.0000
	9	1.0000	1.0000	0.9993	0.9922	0.9539	0.8314	0.5794	0.2527	0.0342	0.0031	0.0000
	10	1.0000	1.0000	0.9999	0.9987	0.9888	0.9421	0.7975	0.4983	0.1339	0.0245	0.0003
	11	1.0000	1.0000	1.0000	0.9999	0.9983	0.9874	0.9363	0.7664	0.3787	0.1354	0.0072
	12	1.0000	1.0000	1.0000	1.0000	0.9999	0.9987	0.9903	0.9450	0.7458	0.4867	0.1225

Table II Cumulative Binomial Probabilities $P(X<=x)$ *(continued)*

		P										
n	*x*	0.1	0.2	0.3	0.4	0.5	0.6	0.7	0.8	0.9	0.95	0.99
14	0	0.2288	0.0440	0.0068	0.0008	0.0001	0.0000	0.0000	0.0000	0.0000	0.0000	0.0000
	1	0.5846	0.1979	0.0475	0.0081	0.0009	0.0001	0.0000	0.0000	0.0000	0.0000	0.0000
	2	0.8416	0.4481	0.1608	0.0398	0.0065	0.0006	0.0000	0.0000	0.0000	0.0000	0.0000
	3	0.9559	0.6982	0.3552	0.1243	0.0287	0.0039	0.0002	0.0000	0.0000	0.0000	0.0000
	4	0.9908	0.8702	0.5842	0.2793	0.0898	0.0175	0.0017	0.0000	0.0000	0.0000	0.0000
	5	0.9985	0.9561	0.7805	0.4859	0.2120	0.0583	0.0083	0.0004	0.0000	0.0000	0.0000
	6	0.9998	0.9884	0.9067	0.6925	0.3953	0.1501	0.0315	0.0024	0.0000	0.0000	0.0000
	7	1.0000	0.9976	0.9685	0.8499	0.6047	0.3075	0.0933	0.0116	0.0002	0.0000	0.0000
	8	1.0000	0.9996	0.9917	0.9417	0.7880	0.5141	0.2195	0.0439	0.0015	0.0000	0.0000
	9	1.0000	1.0000	0.9983	0.9825	0.9102	0.7207	0.4158	0.1298	0.0092	0.0004	0.0000
	10	1.0000	1.0000	0.9998	0.9961	0.9713	0.8757	0.6448	0.3018	0.0441	0.0042	0.0000
	11	1.0000	1.0000	1.0000	0.9994	0.9935	0.9602	0.8392	0.5519	0.1584	0.0301	0.0003
	12	1.0000	1.0000	1.0000	0.9999	0.9991	0.9919	0.9525	0.8021	0.4154	0.1530	0.0084
	13	1.0000	1.0000	1.0000	1.0000	0.9999	0.9992	0.9932	0.9560	0.7712	0.5123	0.1313
15	0	0.2059	0.0352	0.0047	0.0005	0.0000	0.0000	0.0000	0.0000	0.0000	0.0000	0.0000
	1	0.5490	0.1671	0.0353	0.0052	0.0005	0.0000	0.0000	0.0000	0.0000	0.0000	0.0000
	2	0.8159	0.3980	0.1268	0.0271	0.0037	0.0003	0.0000	0.0000	0.0000	0.0000	0.0000
	3	0.9444	0.6482	0.2969	0.0905	0.0176	0.0019	0.0001	0.0000	0.0000	0.0000	0.0000
	4	0.9873	0.8358	0.5155	0.2173	0.0592	0.0093	0.0007	0.0000	0.0000	0.0000	0.0000
	5	0.9978	0.9389	0.7216	0.4032	0.1509	0.0338	0.0037	0.0001	0.0000	0.0000	0.0000
	6	0.9997	0.9819	0.8689	0.6098	0.3036	0.0950	0.0152	0.0008	0.0000	0.0000	0.0000
	7	1.0000	0.9958	0.9500	0.7869	0.5000	0.2131	0.0500	0.0042	0.0000	0.0000	0.0000
	8	1.0000	0.9992	0.9848	0.9050	0.6964	0.3902	0.1311	0.0181	0.0003	0.0000	0.0000
	9	1.0000	0.9999	0.9963	0.9662	0.8491	0.5968	0.2784	0.0611	0.0022	0.0001	0.0000
	10	1.0000	1.0000	0.9993	0.9907	0.9408	0.7827	0.4845	0.1642	0.0127	0.0006	0.0000
	11	1.0000	1.0000	0.9999	0.9981	0.9824	0.9095	0.7031	0.3518	0.0556	0.0055	0.0000
	12	1.0000	1.0000	1.0000	0.9997	0.9963	0.9729	0.8732	0.6020	01841	0.0362	0.0004
	13	1.0000	1.0000	1.0000	1.0000	0.9995	0.9948	0.9647	0.8329	0.4510	0.1710	0.0096
	14	1.0000	1.0000	1.0000	1.0000	1.0000	0.9995	0.9953	0.9648	0.7941	0.5367	0.1399
20	0	0.1216	0.0115	0.0008	0.0000	0.0000	0.0000	0.0000	0.0000	0.0000	0.0000	0.0000
	1	0.3917	0.0692	0.0076	0.0005	0.0000	0.0000	0.0000	0.0000	0.0000	0.0000	0.0000
	2	0.6769	0.2061	0.0355	0.0036	0.0002	0.0000	0.0000	0.0000	0.0000	0.0000	0.0000
	3	0.8670	0.4114	0.1071	0.0160	0.0013	0.0000	0.0000	0.0000	0.0000	0.0000	0.0000
	4	0.9568	0.6296	0.2375	0.0510	0.0059	0.0003	0.0000	0.0000	0.0000	0.0000	0.0000
	5	0.9887	0.8042	0.4164	0.1256	0.0207	0.0016	0.0000	0.0000	0.0000	0.0000	0.0000
	6	0.9976	0.9133	0.6080	0.2500	0.0577	0.0065	0.0003	0.0000	0.0000	0.0000	0.0000
	7	0.9996	0.9679	0.7723	0.4159	0.1316	0.0210	0.0013	0.0000	0.0000	0.0000	0.0000
	8	0.9999	0.9900	0.8867	0.5956	0.2517	0.0565	0.0051	0.0001	0.0000	0.0000	0.0000
	9	1.0000	0.9974	0.9520	0.7553	0.4119	0.1275	0.0171	0.0006	0.0000	0.0000	0.0000
	10	1.0000	0.9994	0.9829	0.8725	0.5881	0.2447	0.0480	0.0026	0.0000	0.0000	0.0000
	11	1.0000	0.9999	0.9949	0.9435	0.7483	0.4044	0.1133	0.0100	0.0001	0.0000	0.0000
	12	1.0000	1.0000	0.9987	0.9790	0.8684	0.5841	0.2277	0.0321	0.0004	0.0000	0.0000
	13	1.0000	1.0000	0.9997	0.9935	0.9423	0.7500	0.3920	0.0867	0.0024	0.0000	0.0000
	14	1.0000	1.0000	1.0000	0.9984	0.9793	0.8744	0.5836	0.1958	0.0113	0.0003	0.0000
	15	1.0000	1.0000	1.0000	0.9997	0.9941	0.9490	0.7625	0.3704	0.0432	0.0026	0.0000
	16	1.0000	1.0000	1.0000	1.0000	0.9987	0.9840	0.8929	0.5886	0.1330	0.0159	0.0000
	17	1.0000	1.0000	1.0000	1.0000	0.9998	0.9964	0.9645	0.7939	0.3231	0.0755	0.0010
	18	1.0000	1.0000	1.0000	1.0000	1.0000	0.9995	0.9924	0.9308	0.6083	0.2642	0.0169
	19	1.0000	1.0000	1.0000	1.0000	1.0000	1.0000	0.9992	0.9885	0.8784	0.6415	0.1821

$$\Phi(z) = P(Z \leq z) = \int_{-\infty}^{z} \frac{1}{\sqrt{2\pi}} e^{-\frac{1}{2}u^2} du$$

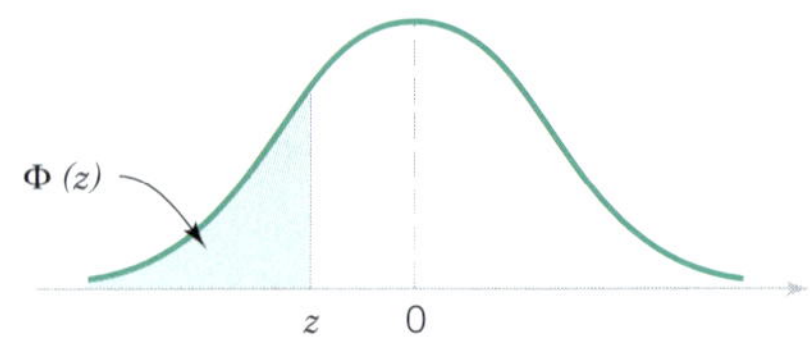

Table III Cumulative Standard Normal Distribution

z	−0.09	−0.08	−0.07	−0.06	−0.05	−0.04	−0.03	−0.02	−0.01	−0.00
−3.9	0.000033	0.000034	0.000036	0.000037	0.000039	0.000041	0.000042	0.000044	0.000046	0.000048
−3.8	0.000050	0.000052	0.000054	0.000057	0.000059	0.000062	0.000064	0.000067	0.000069	0.000072
−3.7	0.000075	0.000078	0.000082	0.000085	0.000088	0.000092	0.000096	0.000100	0.000104	0.000108
−3.6	0.000112	0.000117	0.000121	0.000126	0.000131	0.000136	0.000142	0.000147	0.000153	0.000159
−3.5	0.000165	0.000172	0.000179	0.000185	0.000193	0.000200	0.000208	0.000216	0.000224	0.000233
−3.4	0.000242	0.000251	0.000260	0.000270	0.000280	0.000291	0.000302	0.000313	0.000325	0.000337
−3.3	0.000350	0.000362	0.000376	0.000390	0.000404	0.000419	0.000434	0.000450	0.000467	0.000483
−3.2	0.000501	0.000519	0.000538	0.000557	0.000577	0.000598	0.000619	0.000641	0.000664	0.000687
−3.1	0.000711	0.000736	0.000762	0.000789	0.000816	0.000845	0.000874	0.000904	0.000935	0.000968
−3.0	0.001001	0.001035	0.001070	0.001107	0.001144	0.001183	0.001223	0.001264	0.001306	0.001350
−2.9	0.001395	0.001441	0.001489	0.001538	0.001589	0.001641	0.001695	0.001750	0.001807	0.001866
−2.8	0.001926	0.001988	0.002052	0.002118	0.002186	0.002256	0.002327	0.002401	0.002477	0.002555
−2.7	0.002635	0.002718	0.002803	0.002890	0.002980	0.003072	0.003167	0.003264	0.003364	0.003467
−2.6	0.003573	0.003681	0.003793	0.003907	0.004025	0.004145	0.004269	0.004396	0.004527	0.004661
−2.5	0.004799	0.004940	0.005085	0.005234	0.005386	0.005543	0.005703	0.005868	0.006037	0.006210
−2.4	0.006387	0.006569	0.006756	0.006947	0.007143	0.007344	0.007549	0.007760	0.007976	0.008198
−2.3	0.008424	0.008656	0.008894	0.009137	0.009387	0.009642	0.009903	0.010170	0.010444	0.010724
−2.2	0.011011	0.011304	0.011604	0.011911	0.012224	0.012545	0.012874	0.013209	0.013553	0.013903
−2.1	0.014262	0.014629	0.015003	0.015386	0.015778	0.016177	0.016586	0.017003	0.017429	0.017864
−2.0	0.018309	0.018763	0.019226	0.019699	0.020182	0.020675	0.021178	0.021692	0.022216	0.022750
−1.9	0.023295	0.023852	0.024419	0.024998	0.025588	0.026190	0.026803	0.027429	0.028067	0.028717
−1.8	0.029379	0.030054	0.030742	0.031443	0.032157	0.032884	0.033625	0.034379	0.035148	0.035930
−1.7	0.036727	0.037538	0.038364	0.039204	0.040059	0.040929	0.041815	0.042716	0.043633	0.044565
−1.6	0.045514	0.046479	0.047460	0.048457	0.049471	0.050503	0.051551	0.052616	0.053699	0.054799
−1.5	0.055917	0.057053	0.058208	0.059380	0.060571	0.061780	0.063008	0.064256	0.065522	0.066807
−1.4	0.068112	0.069437	0.070781	0.072145	0.073529	0.074934	0.076359	0.077804	0.079270	0.080757
−1.3	0.082264	0.083793	0.085343	0.086915	0.088508	0.090123	0.091759	0.093418	0.095098	0.096801
−1.2	0.098525	0.100273	0.102042	0.103835	0.105650	0.107488	0.109349	0.111233	0.113140	0.115070
−1.1	0.117023	0.119000	0.121001	0.123024	0.125072	0.127143	0.129238	0.131357	0.133500	0.135666
−1.0	0.137857	0.140071	0.142310	0.144572	0.146859	0.149170	0.151505	0.153864	0.156248	0.158655
−0.9	0.161087	0.163543	0.166023	0.168528	0.171056	0.173609	0.176185	0.178786	0.181411	0.184060
−0.8	0.186733	0.189430	0.192150	0.194894	0.197662	0.200454	0.203269	0.206108	0.208970	0.211855
−0.7	0.214764	0.217695	0.220650	0.223627	0.226627	0.229650	0.232695	0.235762	0.238852	0.241964
−0.6	0.245097	0.248252	0.251429	0.254627	0.257846	0.261086	0.264347	0.267629	0.270931	0.274253
−0.5	0.277595	0.280957	0.284339	0.287740	0.291160	0.294599	0.298056	0.301532	0.305026	0.308538
−0.4	0.312067	0.315614	0.319178	0.322758	0.326355	0.329969	0.333598	0.337243	0.340903	0.344578
−0.3	0.348268	0.351973	0.355691	0.359424	0.363169	0.366928	0.370700	0.374484	0.378281	0.382089
−0.2	0.385908	0.389739	0.393580	0.397432	0.401294	0.405165	0.409046	0.412936	0.416834	0.420740
−0.1	0.424655	0.428576	0.432505	0.436441	0.440382	0.444330	0.448283	0.452242	0.456205	0.460172
0.0	0.464144	0.468119	0.472097	0.476078	0.480061	0.484047	0.488033	0.492022	0.496011	0.500000

$$\Phi(z) = P(Z \le z) = \int_{-\infty}^{z} \frac{1}{\sqrt{2\pi}} e^{-\frac{1}{2}u^2}\, du$$

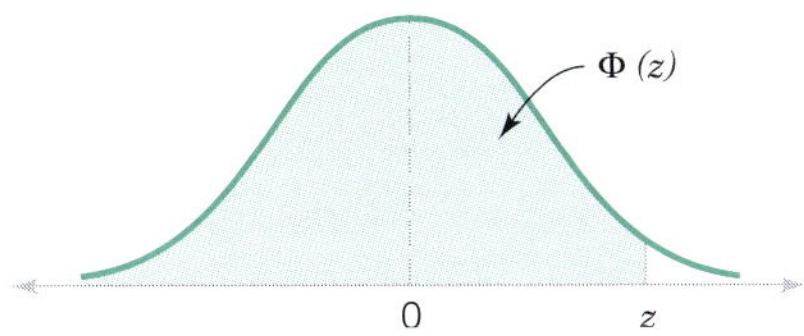

Table III Cumulative Standard Normal Distribution (*continued*)

z	0.00	0.01	0.02	0.03	0.04	0.05	0.06	0.07	0.08	0.09
0.0	0.500000	0.503989	0.507978	0.511967	0.515953	0.519939	0.532922	0.527903	0.531881	0.535856
0.1	0.539828	0.543795	0.547758	0.551717	0.555760	0.559618	0.563559	0.567495	0.571424	0.575345
0.2	0.579260	0.583166	0.587064	0.590954	0.594835	0.598706	0.602568	0.606420	0.610261	0.614092
0.3	0.617911	0.621719	0.625516	0.629300	0.633072	0.636831	0.640576	0.644309	0.648027	0.651732
0.4	0.655422	0.659097	0.662757	0.666402	0.670031	0.673645	0.677242	0.680822	0.684386	0.687933
0.5	0.691462	0.694974	0.698468	0.701944	0.705401	0.708840	0.712260	0.715661	0.719043	0.722405
0.6	0.725747	0.729069	0.732371	0.735653	0.738914	0.742154	0.745373	0.748571	0.751748	0.754903
0.7	0.758036	0.761148	0.764238	0.767305	0.770350	0.773373	0.776373	0.779350	0.782305	0.785236
0.8	0.788145	0.791030	0.793892	0.796731	0.799546	0.802338	0.805106	0.807850	0.810570	0.813267
0.9	0.815940	0.818589	0.821214	0.823815	0.826391	0.828944	0.831472	0.833977	0.836457	0.838913
1.0	0.841345	0.843752	0.846136	0.848495	0.850830	0.853141	0.855428	0.857690	0.859929	0.862143
1.1	0.864334	0.866500	0.868643	0.870762	0.872857	0.874928	0.876976	0.878999	0.881000	0.882977
1.2	0.884930	0.886860	0.888767	0.890651	0.892512	0.894350	0.896165	0.897958	0.899727	0.901475
1.3	0.903199	0.904902	0.906582	0.908241	0.909877	0.911492	0.913085	0.914657	0.916207	0.917736
1.4	0.919243	0.920730	0.922196	0.923641	0.925066	0.926471	0.927855	0.929219	0.930563	0.931888
1.5	0.933193	0.934478	0.935744	0.936992	0.938220	0.939429	0.940620	0.941792	0.942947	0.944083
1.6	0.945201	0.946301	0.947384	0.948449	0.949497	0.950529	0.951543	0.952540	0.953521	0.954486
1.7	0.955435	0.956367	0.957284	0.958185	0.959071	0.959941	0.960796	0.961636	0.962462	0.963273
1.8	0.964070	0.964852	0.965621	0.966375	0.967116	0.967843	0.968557	0.969258	0.969946	0.970621
1.9	0.971283	0.971933	0.972571	0.973197	0.973810	0.974412	0.975002	0.975581	0.976148	0.976705
2.0	0.977250	0.977784	0.978308	0.978822	0.979325	0.979818	0.980301	0.980774	0.981237	0.981691
2.1	0.982136	0.982571	0.982997	0.983414	0.983823	0.984222	0.984614	0.984997	0.985371	0.985738
2.2	0.986097	0.986447	0.986791	0.987126	0.987455	0.987776	0.988089	0.988396	0.988696	0.988989
2.3	0.989276	0.989556	0.989830	0.990097	0.990358	0.990613	0.990863	0.991106	0.991344	0.991576
2.4	0.991802	0.992024	0.992240	0.992451	0.992656	0.992857	0.993053	0.993244	0.993431	0.993613
2.5	0.993790	0.993963	0.994132	0.994297	0.994457	0.994614	0.994766	0.994915	0.995060	0.995201
2.6	0.995339	0.995473	0.995604	0.995731	0.995855	0.995975	0.996093	0.996207	0.996319	0.996427
2.7	0.996533	0.996636	0.996736	0.996833	0.996928	0.997020	0.997110	0.997197	0.997282	0.997365
2.8	0.997445	0.997523	0.997599	0.997673	0.997744	0.997814	0.997882	0.997948	0.998012	0.998074
2.9	0.998134	0.998193	0.998250	0.998305	0.998359	0.998411	0.998462	0.998511	0.998559	0.998605
3.0	0.998650	0.998694	0.998736	0.998777	0.998817	0.998856	0.998893	0.998930	0.998965	0.998999
3.1	0.999032	0.999065	0.999096	0.999126	0.999155	0.999184	0.999211	0.999238	0.999264	0.999289
3.2	0.999313	0.999336	0.999359	0.999381	0.999402	0.999423	0.999443	0.999462	0.999481	0.999499
3.3	0.999517	0.999533	0.999550	0.999566	0.999581	0.999596	0.999610	0.999624	0.999638	0.999650
3.4	0.999663	0.999675	0.999687	0.999698	0.999709	0.999720	0.999730	0.999740	0.999749	0.999758
3.5	0.999767	0.999776	0.999784	0.999792	0.999800	0.999807	0.999815	0.999821	0.999828	0.999835
3.6	0.999841	0.999847	0.999853	0.999858	0.999864	0.999869	0.999874	0.999879	0.999883	0.999888
3.7	0.999892	0.999896	0.999900	0.999904	0.999908	0.999912	0.999915	0.999918	0.999922	0.999925
3.8	0.999928	0.999931	0.999933	0.999936	0.999938	0.999941	0.999943	0.999946	0.999948	0.999950
3.9	0.999952	0.999954	0.999956	0.999958	0.999959	0.999961	0.999963	0.999964	0.999966	0.999967

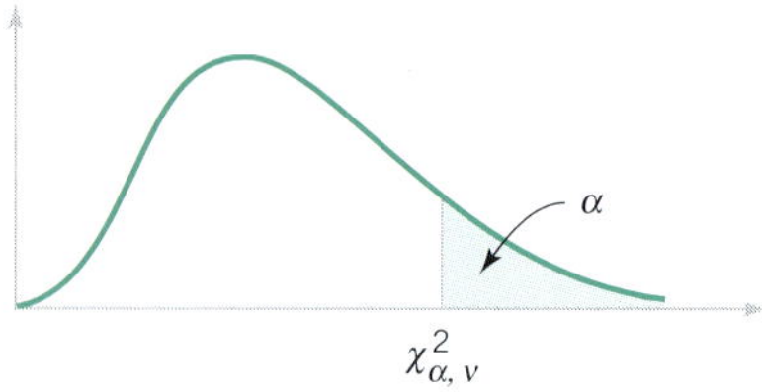

Table IV Percentage Points $\chi^2_{\alpha,\nu}$ of the Chi-Squared Distribution

ν \ α	.995	.990	.975	.950	.900	.500	.100	.050	.025	.010	.005
1	.00+	.00+	.00+	.00+	.02	.45	2.71	3.84	5.02	6.63	7.88
2	.01	.02	.05	.10	.21	1.39	4.61	5.99	7.38	9.21	10.60
3	.07	.11	.22	.35	.58	2.37	6.25	7.81	9.35	11.34	12.84
4	.21	.30	.48	.71	1.06	3.36	7.78	9.49	11.14	13.28	14.86
5	.41	.55	.83	1.15	1.61	4.35	9.24	11.07	12.83	15.09	16.75
6	.68	.87	1.24	1.64	2.20	5.35	10.65	12.59	14.45	16.81	18.55
7	.99	1.24	1.69	2.17	2.83	6.35	12.02	14.07	16.01	18.48	20.28
8	1.34	1.65	2.18	2.73	3.49	7.34	13.36	15.51	17.53	20.09	21.96
9	1.73	2.09	2.70	3.33	4.17	8.34	14.68	16.92	19.02	21.67	23.59
10	2.16	2.56	3.25	3.94	4.87	9.34	15.99	18.31	20.48	23.21	25.19
11	2.60	3.05	3.82	4.57	5.58	10.34	17.28	19.68	21.92	24.72	26.76
12	3.07	3.57	4.40	5.23	6.30	11.34	18.55	21.03	23.34	26.22	28.30
13	3.57	4.11	5.01	5.89	7.04	12.34	19.81	22.36	24.74	27.69	29.82
14	4.07	4.66	5.63	6.57	7.79	13.34	21.06	23.68	26.12	29.14	31.32
15	4.60	5.23	6.27	7.26	8.55	14.34	22.31	25.00	27.49	30.58	32.80
16	5.14	5.81	6.91	7.96	9.31	15.34	23.54	26.30	28.85	32.00	34.27
17	5.70	6.41	7.56	8.67	10.09	16.34	24.77	27.59	30.19	33.41	35.72
18	6.26	7.01	8.23	9.39	10.87	17.34	25.99	28.87	31.53	34.81	37.16
19	6.84	7.63	8.91	10.12	11.65	18.34	27.20	30.14	32.85	36.19	38.58
20	7.43	8.26	9.59	10.85	12.44	19.34	28.41	31.41	34.17	37.57	40.00
21	8.03	8.90	10.28	11.59	13.24	20.34	29.62	32.67	35.48	38.93	41.40
22	8.64	9.54	10.98	12.34	14.04	21.34	30.81	33.92	36.78	40.29	42.80
23	9.26	10.20	11.69	13.09	14.85	22.34	32.01	35.17	38.08	41.64	44.18
24	9.89	10.86	12.40	13.85	15.66	23.34	33.20	36.42	39.36	42.98	45.56
25	10.52	11.52	13.12	14.61	16.47	24.34	34.28	37.65	40.65	44.31	46.93
26	11.16	12.20	13.84	15.38	17.29	25.34	35.56	38.89	41.92	45.64	48.29
27	11.81	12.88	14.57	16.15	18.11	26.34	36.74	40.11	43.19	46.96	49.65
28	12.46	13.57	15.31	16.93	18.94	27.34	37.92	41.34	44.46	48.28	50.99
29	13.12	14.26	16.05	17.71	19.77	28.34	39.09	42.56	45.72	49.59	52.34
30	13.79	14.95	16.79	18.49	20.60	29.34	40.26	43.77	46.98	50.89	53.67
40	20.71	22.16	24.43	26.51	29.05	39.34	51.81	55.76	59.34	63.69	66.77
50	27.99	29.71	32.36	34.76	37.69	49.33	63.17	67.50	71.42	76.15	79.49
60	35.53	37.48	40.48	43.19	46.46	59.33	74.40	79.08	83.30	88.38	91.95
70	43.28	45.44	48.76	51.74	55.33	69.33	85.53	90.53	95.02	100.42	104.22
80	51.17	53.54	57.15	60.39	64.28	79.33	96.58	101.88	106.63	112.33	116.32
90	59.20	61.75	65.65	69.13	73.29	89.33	107.57	113.14	118.14	124.12	128.30
100	67.33	70.06	74.22	77.93	82.36	99.33	118.50	124.34	129.56	135.81	140.17

ν = degrees of freedom.

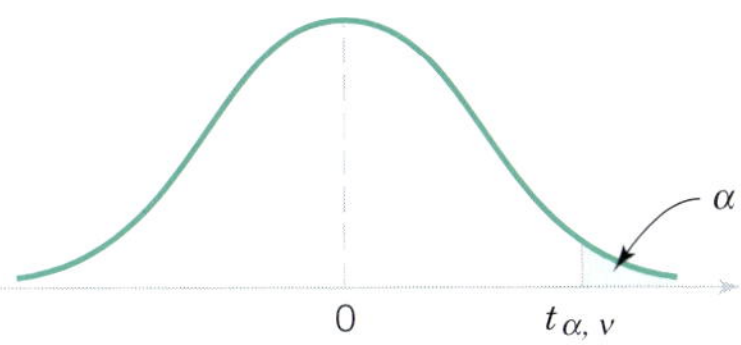

Table V Percentage Points $t_{\alpha,\nu}$ of the t-Distribution

ν \ α	.40	.25	.10	.05	.025	.01	.005	.0025	.001	.0005
1	.325	1.000	3.078	6.314	12.706	31.821	63.657	127.32	318.31	636.62
2	.289	.816	1.886	2.920	4.303	6.965	9.925	14.089	23.326	31.598
3	.277	.765	1.638	2.353	3.182	4.541	5.841	7.453	10.213	12.924
4	.271	.741	1.533	2.132	2.776	3.747	4.604	5.598	7.173	8.610
5	.267	.727	1.476	2.015	2.571	3.365	4.032	4.773	5.893	6.869
6	.265	.718	1.440	1.943	2.447	3.143	3.707	4.317	5.208	5.959
7	.263	.711	1.415	1.895	2.365	2.998	3.499	4.029	4.785	5.408
8	.262	.706	1.397	1.860	2.306	2.896	3.355	3.833	4.501	5.041
9	.261	.703	1.383	1.833	2.262	2.821	3.250	3.690	4.297	4.781
10	.260	.700	1.372	1.812	2.228	2.764	3.169	3.581	4.144	4.587
11	.260	.697	1.363	1.796	2.201	2.718	3.106	3.497	4.025	4.437
12	.259	.695	1.356	1.782	2.179	2.681	3.055	3.428	3.930	4.318
13	.259	.694	1.350	1.771	2.160	2.650	3.012	3.372	3.852	4.221
14	.258	.692	1.345	1.761	2.145	2.624	2.977	3.326	3.787	4.140
15	.258	.691	1.341	1.753	2.131	2.602	2.947	3.286	3.733	4.073
16	.258	.690	1.337	1.746	2.120	2.583	2.921	3.252	3.686	4.015
17	.257	.689	1.333	1.740	2.110	2.567	2.898	3.222	3.646	3.965
18	.257	.688	1.330	1.734	2.101	2.552	2.878	3.197	3.610	3.922
19	.257	.688	1.328	1.729	2.093	2.539	2.861	3.174	3.579	3.883
20	.257	.687	1.325	1.725	2.086	2.528	2.845	3.153	3.552	3.850
21	.257	.686	1.323	1.721	2.080	2.518	2.831	3.135	3.527	3.819
22	.256	.686	1.321	1.717	2.074	2.508	2.819	3.119	3.505	3.792
23	.256	.685	1.319	1.714	2.069	2.500	2.807	3.104	3.485	3.767
24	.256	.685	1.318	1.711	2.064	2.492	2.797	3.091	3.467	3.745
25	.256	.684	1.316	1.708	2.060	2.485	2.787	3.078	3.450	3.725
26	.256	.684	1.315	1.706	2.056	2.479	2.779	3.067	3.435	3.707
27	.256	.684	1.314	1.703	2.052	2.473	2.771	3.057	3.421	3.690
28	.256	.683	1.313	1.701	2.048	2.467	2.763	3.047	3.408	3.674
29	.256	.683	1.311	1.699	2.045	2.462	2.756	3.038	3.396	3.659
30	.256	.683	1.310	1.697	2.042	2.457	2.750	3.030	3.385	3.646
40	.255	.681	1.303	1.684	2.021	2.423	2.704	2.971	3.307	3.551
60	.254	.679	1.296	1.671	2.000	2.390	2.660	2.915	3.232	3.460
120	.254	.677	1.289	1.658	1.980	2.358	2.617	2.860	3.160	3.373
∞	.253	.674	1.282	1.645	1.960	2.326	2.576	2.807	3.090	3.291

ν = degrees of freedom.

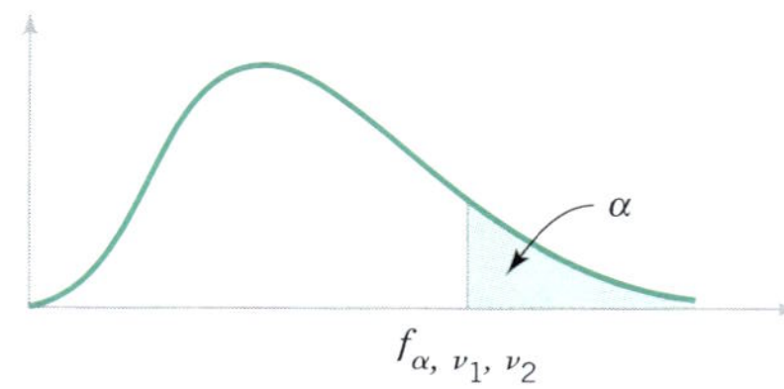

Table VI Percentage Points f_{α,ν_1,ν_2} of the F-Distribution

$f_{0.25,\nu_1,\nu_2}$

ν_2 \ ν_1	Degrees of freedom for the numerator (ν_1)																		
Degrees of freedom for the denominator (ν_2)	1	2	3	4	5	6	7	8	9	10	12	15	20	24	30	40	60	120	∞
1	5.83	7.50	8.20	8.58	8.82	8.98	9.10	9.19	9.26	9.32	9.41	9.49	9.58	9.63	9.67	9.71	9.76	9.80	9.85
2	2.57	3.00	3.15	3.23	3.28	3.31	3.34	3.35	3.37	3.38	3.39	3.41	3.43	3.43	3.44	3.45	3.46	3.47	3.48
3	2.02	2.28	2.36	2.39	2.41	2.42	2.43	2.44	2.44	2.44	2.45	2.46	2.46	2.46	2.47	2.47	2.47	2.47	2.47
4	1.81	2.00	2.05	2.06	2.07	2.08	2.08	2.08	2.08	2.08	2.08	2.08	2.08	2.08	2.08	2.08	2.08	2.08	2.08
5	1.69	1.85	1.88	1.89	1.89	1.89	1.89	1.89	1.89	1.89	1.89	1.89	1.88	1.88	1.88	1.88	1.87	1.87	1.87
6	1.62	1.76	1.78	1.79	1.79	1.78	1.78	1.78	1.77	1.77	1.77	1.76	1.76	1.75	1.75	1.75	1.74	1.74	1.74
7	1.57	1.70	1.72	1.72	1.71	1.71	1.70	1.70	1.70	1.69	1.68	1.68	1.67	1.67	1.66	1.66	1.65	1.65	1.65
8	1.54	1.66	1.67	1.66	1.66	1.65	1.64	1.64	1.63	1.63	1.62	1.62	1.61	1.60	1.60	1.59	1.59	1.58	1.58
9	1.51	1.62	1.63	1.63	1.62	1.61	1.60	1.60	1.59	1.59	1.58	1.57	1.56	1.56	1.55	1.54	1.54	1.53	1.53
10	1.49	1.60	1.60	1.59	1.59	1.58	1.57	1.56	1.56	1.55	1.54	1.53	1.52	1.52	1.51	1.51	1.50	1.49	1.48
11	1.47	1.58	1.58	1.57	1.56	1.55	1.54	1.53	1.53	1.52	1.51	1.50	1.49	1.49	1.48	1.47	1.47	1.46	1.45
12	1.46	1.56	1.56	1.55	1.54	1.53	1.52	1.51	1.51	1.50	1.49	1.48	1.47	1.46	1.45	1.45	1.44	1.43	1.42
13	1.45	1.55	1.55	1.53	1.52	1.51	1.50	1.49	1.49	1.48	1.47	1.46	1.45	1.44	1.43	1.42	1.42	1.41	1.40
14	1.44	1.53	1.53	1.52	1.51	1.50	1.49	1.48	1.47	1.46	1.45	1.44	1.43	1.42	1.41	1.41	1.40	1.39	1.38
15	1.43	1.52	1.52	1.51	1.49	1.48	1.47	1.46	1.46	1.45	1.44	1.43	1.41	1.41	1.40	1.39	1.38	1.37	1.36
16	1.42	1.51	1.51	1.50	1.48	1.47	1.46	1.45	1.44	1.44	1.43	1.41	1.40	1.39	1.38	1.37	1.36	1.35	1.34
17	1.42	1.51	1.50	1.49	1.47	1.46	1.45	1.44	1.43	1.43	1.41	1.40	1.39	1.38	1.37	1.36	1.35	1.34	1.33
18	1.41	1.50	1.49	1.48	1.46	1.45	1.44	1.43	1.42	1.42	1.40	1.39	1.38	1.37	1.36	1.35	1.34	1.33	1.32
19	1.41	1.49	1.49	1.47	1.46	1.44	1.43	1.42	1.41	1.41	1.40	1.38	1.37	1.36	1.35	1.34	1.33	1.32	1.30
20	1.40	1.49	1.48	1.47	1.45	1.44	1.43	1.42	1.41	1.40	1.39	1.37	1.36	1.35	1.34	1.33	1.32	1.31	1.29
21	1.40	1.48	1.48	1.46	1.44	1.43	1.42	1.41	1.40	1.39	1.38	1.37	1.35	1.34	1.33	1.32	1.31	1.30	1.28
22	1.40	1.48	1.47	1.45	1.44	1.42	1.41	1.40	1.39	1.39	1.37	1.36	1.34	1.33	1.32	1.31	1.30	1.29	1.28
23	1.39	1.47	1.47	1.45	1.43	1.42	1.41	1.40	1.39	1.38	1.37	1.35	1.34	1.33	1.32	1.31	1.30	1.28	1.27
24	1.39	1.47	1.46	1.44	1.43	1.41	1.40	1.39	1.38	1.38	1.36	1.35	1.33	1.32	1.31	1.30	1.29	1.28	1.26
25	1.39	1.47	1.46	1.44	1.42	1.41	1.40	1.39	1.38	1.37	1.36	1.34	1.33	1.32	1.31	1.29	1.28	1.27	1.25
26	1.38	1.46	1.45	1.44	1.42	1.41	1.39	1.38	1.37	1.37	1.35	1.34	1.32	1.31	1.30	1.29	1.28	1.26	1.25
27	1.38	1.46	1.45	1.43	1.42	1.40	1.39	1.38	1.37	1.36	1.35	1.33	1.32	1.31	1.30	1.28	1.27	1.26	1.24
28	1.38	1.46	1.45	1.43	1.41	1.40	1.39	1.38	1.37	1.36	1.34	1.33	1.31	1.30	1.29	1.28	1.27	1.25	1.24
29	1.38	1.45	1.45	1.43	1.41	1.40	1.38	1.37	1.36	1.35	1.34	1.32	1.31	1.30	1.29	1.27	1.26	1.25	1.23
30	1.38	1.45	1.44	1.42	1.41	1.39	1.38	1.37	1.36	1.35	1.34	1.32	1.30	1.29	1.28	1.27	1.26	1.24	1.23
40	1.36	1.44	1.42	1.40	1.39	1.37	1.36	1.35	1.34	1.33	1.31	1.30	1.28	1.26	1.25	1.24	1.22	1.21	1.19
60	1.35	1.42	1.41	1.38	1.37	1.35	1.33	1.32	1.31	1.30	1.29	1.27	1.25	1.24	1.22	1.21	1.19	1.17	1.15
120	1.34	1.40	1.39	1.37	1.35	1.33	1.31	1.30	1.29	1.28	1.26	1.24	1.22	1.21	1.19	1.18	1.16	1.13	1.10
∞	1.32	1.39	1.37	1.35	1.33	1.31	1.29	1.28	1.27	1.25	1.24	1.22	1.19	1.18	1.16	1.14	1.12	1.08	1.00

Table VI Percentage Points of the F-Distribution (*continued*)

$f_{0.10,v_1,v_2}$

Degrees of freedom for the numerator (v_1); Degrees of freedom for the denominator (v_2)

v_2 \ v_1	1	2	3	4	5	6	7	8	9	10	12	15	20	24	30	40	60	120	∞
1	39.86	49.50	53.59	55.83	57.24	58.20	58.91	59.44	59.86	60.19	60.71	61.22	61.74	62.00	62.26	62.53	62.79	63.06	63.33
2	8.53	9.00	9.16	9.24	9.29	9.33	9.35	9.37	9.38	9.39	9.41	9.42	9.44	9.45	9.46	9.47	9.47	9.48	9.49
3	5.54	5.46	5.39	5.34	5.31	5.28	5.27	5.25	5.24	5.23	5.22	5.20	5.18	5.18	5.17	5.16	5.15	5.14	5.13
4	4.54	4.32	4.19	4.11	4.05	4.01	3.98	3.95	3.94	3.92	3.90	3.87	3.84	3.83	3.82	3.80	3.79	3.78	3.76
5	4.06	3.78	3.62	3.52	3.45	3.40	3.37	3.34	3.32	3.30	3.27	3.24	3.21	3.19	3.17	3.16	3.14	3.12	3.10
6	3.78	3.46	3.29	3.18	3.11	3.05	3.01	2.98	2.96	2.94	2.90	2.87	2.84	2.82	2.80	2.78	2.76	2.74	2.72
7	3.59	3.26	3.07	2.96	2.88	2.83	2.78	2.75	2.72	2.70	2.67	2.63	2.59	2.58	2.56	2.54	2.51	2.49	2.47
8	3.46	3.11	2.92	2.81	2.73	2.67	2.62	2.59	2.56	2.54	2.50	2.46	2.42	2.40	2.38	2.36	2.34	2.32	2.29
9	3.36	3.01	2.81	2.69	2.61	2.55	2.51	2.47	2.44	2.42	2.38	2.34	2.30	2.28	2.25	2.23	2.21	2.18	2.16
10	3.29	2.92	2.73	2.61	2.52	2.46	2.41	2.38	2.35	2.32	2.28	2.24	2.20	2.18	2.16	2.13	2.11	2.08	2.06
11	3.23	2.86	2.66	2.54	2.45	2.39	2.34	2.30	2.27	2.25	2.21	2.17	2.12	2.10	2.08	2.05	2.03	2.00	1.97
12	3.18	2.81	2.61	2.48	2.39	2.33	2.28	2.24	2.21	2.19	2.15	2.10	2.06	2.04	2.01	1.99	1.96	1.93	1.90
13	3.14	2.76	2.56	2.43	2.35	2.28	2.23	2.20	2.16	2.14	2.10	2.05	2.01	1.98	1.96	1.93	1.90	1.88	1.85
14	3.10	2.73	2.52	2.39	2.31	2.24	2.19	2.15	2.12	2.10	2.05	2.01	1.96	1.94	1.91	1.89	1.86	1.83	1.80
15	3.07	2.70	2.49	2.36	2.27	2.21	2.16	2.12	2.09	2.06	2.02	1.97	1.92	1.90	1.87	1.85	1.82	1.79	1.76
16	3.05	2.67	2.46	2.33	2.24	2.18	2.13	2.09	2.06	2.03	1.99	1.94	1.89	1.87	1.84	1.81	1.78	1.75	1.72
17	3.03	2.64	2.44	2.31	2.22	2.15	2.10	2.06	2.03	2.00	1.96	1.91	1.86	1.84	1.81	1.78	1.75	1.72	1.69
18	3.01	2.62	2.42	2.29	2.20	2.13	2.08	2.04	2.00	1.98	1.93	1.89	1.84	1.81	1.78	1.75	1.72	1.69	1.66
19	2.99	2.61	2.40	2.27	2.18	2.11	2.06	2.02	1.98	1.96	1.91	1.86	1.81	1.79	1.76	1.73	1.70	1.67	1.63
20	2.97	2.59	2.38	2.25	2.16	2.09	2.04	2.00	1.96	1.94	1.89	1.84	1.79	1.77	1.74	1.71	1.68	1.64	1.61
21	2.96	2.57	2.36	2.23	2.14	2.08	2.02	1.98	1.95	1.92	1.87	1.83	1.78	1.75	1.72	1.69	1.66	1.62	1.59
22	2.95	2.56	2.35	2.22	2.13	2.06	2.01	1.97	1.93	1.90	1.86	1.81	1.76	1.73	1.70	1.67	1.64	1.60	1.57
23	2.94	2.55	2.34	2.21	2.11	2.05	1.99	1.95	1.92	1.89	1.84	1.80	1.74	1.72	1.69	1.66	1.62	1.59	1.55
24	2.93	2.54	2.33	2.19	2.10	2.04	1.98	1.94	1.91	1.88	1.83	1.78	1.73	1.70	1.67	1.64	1.61	1.57	1.53
25	2.92	2.53	2.32	2.18	2.09	2.02	1.97	1.93	1.89	1.87	1.82	1.77	1.72	1.69	1.66	1.63	1.59	1.56	1.52
26	2.91	2.52	2.31	2.17	2.08	2.01	1.96	1.92	1.88	1.86	1.81	1.76	1.71	1.68	1.65	1.61	1.58	1.54	1.50
27	2.90	2.51	2.30	2.17	2.07	2.00	1.95	1.91	1.87	1.85	1.80	1.75	1.70	1.67	1.64	1.60	1.57	1.53	1.49
28	2.89	2.50	2.29	2.16	2.06	2.00	1.94	1.90	1.87	1.84	1.79	1.74	1.69	1.66	1.63	1.59	1.56	1.52	1.48
29	2.89	2.50	2.28	2.15	2.06	1.99	1.93	1.89	1.86	1.83	1.78	1.73	1.68	1.65	1.62	1.58	1.55	1.51	1.47
30	2.88	2.49	2.28	2.14	2.03	1.98	1.93	1.88	1.85	1.82	1.77	1.72	1.67	1.64	1.61	1.57	1.54	1.50	1.46
40	2.84	2.44	2.23	2.09	2.00	1.93	1.87	1.83	1.79	1.76	1.71	1.66	1.61	1.57	1.54	1.51	1.47	1.42	1.38
60	2.79	2.39	2.18	2.04	1.95	1.87	1.82	1.77	1.74	1.71	1.66	1.60	1.54	1.51	1.48	1.44	1.40	1.35	1.29
120	2.75	2.35	2.13	1.99	1.90	1.82	1.77	1.72	1.68	1.65	1.60	1.55	1.48	1.45	1.41	1.37	1.32	1.26	1.19
∞	2.71	2.30	2.08	1.94	1.85	1.77	1.72	1.67	1.63	1.60	1.55	1.49	1.42	1.38	1.34	1.30	1.24	1.17	1.00

Table VI Percentage Points of the F-Distribution (*continued*)

$f_{0.05,\nu_1,\nu_2}$

Degrees of freedom for the numerator (ν_1); Degrees of freedom for the denominator (ν_2)

ν_2 \ ν_1	1	2	3	4	5	6	7	8	9	10	12	15	20	24	30	40	60	120	∞
1	161.4	199.5	215.7	224.6	230.2	234.0	236.8	238.9	240.5	241.9	243.9	245.9	248.0	249.1	250.1	251.1	252.2	253.3	254.3
2	18.51	19.00	19.16	19.25	19.30	19.33	19.35	19.37	19.38	19.40	19.41	19.43	19.45	19.45	19.46	19.47	19.48	19.49	19.50
3	10.13	9.55	9.28	9.12	9.01	8.94	8.89	8.85	8.81	8.79	8.74	8.70	8.66	8.64	8.62	8.59	8.57	8.55	8.53
4	7.71	6.94	6.59	6.39	6.26	6.16	6.09	6.04	6.00	5.96	5.91	5.86	5.80	5.77	5.75	5.72	5.69	5.66	5.63
5	6.61	5.79	5.41	5.19	5.05	4.95	4.88	4.82	4.77	4.74	4.68	4.62	4.56	4.53	4.50	4.46	4.43	4.40	4.36
6	5.99	5.14	4.76	4.53	4.39	4.28	4.21	4.15	4.10	4.06	4.00	3.94	3.87	3.84	3.81	3.77	3.74	3.70	3.67
7	5.59	4.74	4.35	4.12	3.97	3.87	3.79	3.73	3.68	3.64	3.57	3.51	3.44	3.41	3.38	3.34	3.30	3.27	3.23
8	5.32	4.46	4.07	3.84	3.69	3.58	3.50	3.44	3.39	3.35	3.28	3.22	3.15	3.12	3.08	3.04	3.01	2.97	2.93
9	5.12	4.26	3.86	3.63	3.48	3.37	3.29	3.23	3.18	3.14	3.07	3.01	2.94	2.90	2.86	2.83	2.79	2.75	2.71
10	4.96	4.10	3.71	3.48	3.33	3.22	3.14	3.07	3.02	2.98	2.91	2.85	2.77	2.74	2.70	2.66	2.62	2.58	2.54
11	4.84	3.98	3.59	3.36	3.20	3.09	3.01	2.95	2.90	2.85	2.79	2.72	2.65	2.61	2.57	2.53	2.49	2.45	2.40
12	4.75	3.89	3.49	3.26	3.11	3.00	2.91	2.85	2.80	2.75	2.69	2.62	2.54	2.51	2.47	2.43	2.38	2.34	2.30
13	4.67	3.81	3.41	3.18	3.03	2.92	2.83	2.77	2.71	2.67	2.60	2.53	2.46	2.42	2.38	2.34	2.30	2.25	2.21
14	4.60	3.74	3.34	3.11	2.96	2.85	2.76	2.70	2.65	2.60	2.53	2.46	2.39	2.35	2.31	2.27	2.22	2.18	2.13
15	4.54	3.68	3.29	3.06	2.90	2.79	2.71	2.64	2.59	2.54	2.48	2.40	2.33	2.29	2.25	2.20	2.16	2.11	2.07
16	4.49	3.63	3.24	3.01	2.85	2.74	2.66	2.59	2.54	2.49	2.42	2.35	2.28	2.24	2.19	2.15	2.11	2.06	2.01
17	4.45	3.59	3.20	2.96	2.81	2.70	2.61	2.55	2.49	2.45	2.38	2.31	2.23	2.19	2.15	2.10	2.06	2.01	1.96
18	4.41	3.55	3.16	2.93	2.77	2.66	2.58	2.51	2.46	2.41	2.34	2.27	2.19	2.15	2.11	2.06	2.02	1.97	1.92
19	4.38	3.52	3.13	2.90	2.74	2.63	2.54	2.48	2.42	2.38	2.31	2.23	2.16	2.11	2.07	2.03	1.98	1.93	1.88
20	4.35	3.49	3.10	2.87	2.71	2.60	2.51	2.45	2.39	2.35	2.28	2.20	2.12	2.08	2.04	1.99	1.95	1.90	1.84
21	4.32	3.47	3.07	2.84	2.68	2.57	2.49	2.42	2.37	2.32	2.25	2.18	2.10	2.05	2.01	1.96	1.92	1.87	1.81
22	4.30	3.44	3.05	2.82	2.66	2.55	2.46	2.40	2.34	2.30	2.23	2.15	2.07	2.03	1.98	1.94	1.89	1.84	1.78
23	4.28	3.42	3.03	2.80	2.64	2.53	2.44	2.37	2.32	2.27	2.20	2.13	2.05	2.01	1.96	1.91	1.86	1.81	1.76
24	4.26	3.40	3.01	2.78	2.62	2.51	2.42	2.36	2.30	2.25	2.18	2.11	2.03	1.98	1.94	1.89	1.84	1.79	1.73
25	4.24	3.39	2.99	2.76	2.60	2.49	2.40	2.34	2.28	2.24	2.16	2.09	2.01	1.96	1.92	1.87	1.82	1.77	1.71
26	4.23	3.37	2.98	2.74	2.59	2.47	2.39	2.32	2.27	2.22	2.15	2.07	1.99	1.95	1.90	1.85	1.80	1.75	1.69
27	4.21	3.35	2.96	2.73	2.57	2.46	2.37	2.31	2.25	2.20	2.13	2.06	1.97	1.93	1.88	1.84	1.79	1.73	1.67
28	4.20	3.34	2.95	2.71	2.56	2.45	2.36	2.29	2.24	2.19	2.12	2.04	1.96	1.91	1.87	1.82	1.77	1.71	1.65
29	4.18	3.33	2.93	2.70	2.55	2.43	2.35	2.28	2.22	2.18	2.10	2.03	1.94	1.90	1.85	1.81	1.75	1.70	1.64
30	4.17	3.32	2.92	2.69	2.53	2.42	2.33	2.27	2.21	2.16	2.09	2.01	1.93	1.89	1.84	1.79	1.74	1.68	1.62
40	4.08	3.23	2.84	2.61	2.45	2.34	2.25	2.18	2.12	2.08	2.00	1.92	1.84	1.79	1.74	1.69	1.64	1.58	1.51
60	4.00	3.15	2.76	2.53	2.37	2.25	2.17	2.10	2.04	1.99	1.92	1.84	1.75	1.70	1.65	1.59	1.53	1.47	1.39
120	3.92	3.07	2.68	2.45	2.29	2.17	2.09	2.02	1.96	1.91	1.83	1.75	1.66	1.61	1.55	1.55	1.43	1.35	1.25
∞	3.84	3.00	2.60	2.37	2.21	2.10	2.01	1.94	1.88	1.83	1.75	1.67	1.57	1.52	1.46	1.39	1.32	1.22	1.00

Table VI Percentage Points of the F-Distribution (*continued*)

$f_{0.025,v_1,v_2}$

v_2 \ v_1	1	2	3	4	5	6	7	8	9	10	12	15	20	24	30	40	60	120	∞
	Degrees of freedom for the numerator (v_1)																		
1	647.8	799.5	864.2	899.6	921.8	937.1	948.2	956.7	963.3	968.6	976.7	984.9	993.1	997.2	1001	1006	1010	1014	1018
2	38.51	39.00	39.17	39.25	39.30	39.33	39.36	39.37	39.39	39.40	39.41	39.43	39.45	39.46	39.46	39.47	39.48	39.49	39.50
3	17.44	16.04	15.44	15.10	14.88	14.73	14.62	14.54	14.47	14.42	14.34	14.25	14.17	14.12	14.08	14.04	13.99	13.95	13.90
4	12.22	10.65	9.98	9.60	9.36	9.20	9.07	8.98	8.90	8.84	8.75	8.66	8.56	8.51	8.46	8.41	8.36	8.31	8.26
5	10.01	8.43	7.76	7.39	7.15	6.98	6.85	6.76	6.68	6.62	6.52	6.43	6.33	6.28	6.23	6.18	6.12	6.07	6.02
6	8.81	7.26	6.60	6.23	5.99	5.82	5.70	5.60	5.52	5.46	5.37	5.27	5.17	5.12	5.07	5.01	4.96	4.90	4.85
7	8.07	6.54	5.89	5.52	5.29	5.12	4.99	4.90	4.82	4.76	4.67	4.57	4.47	4.42	4.36	4.31	4.25	4.20	4.14
8	7.57	6.06	5.42	5.05	4.82	4.65	4.53	4.43	4.36	4.30	4.20	4.10	4.00	3.95	3.89	3.84	3.78	3.73	3.67
9	7.21	5.71	5.08	4.72	4.48	4.32	4.20	4.10	4.03	3.96	3.87	3.77	3.67	3.61	3.56	3.51	3.45	3.39	3.33
10	6.94	5.46	4.83	4.47	4.24	4.07	3.95	3.85	3.78	3.72	3.62	3.52	3.42	3.37	3.31	3.26	3.20	3.14	3.08
11	6.72	5.26	4.63	4.28	4.04	3.88	3.76	3.66	3.59	3.53	3.43	3.33	3.23	3.17	3.12	3.06	3.00	2.94	2.88
12	6.55	5.10	4.47	4.12	3.89	3.73	3.61	3.51	3.44	3.37	3.28	3.18	3.07	3.02	2.96	2.91	2.85	2.79	2.72
13	6.41	4.97	4.35	4.00	3.77	3.60	3.48	3.39	3.31	3.25	3.15	3.05	2.95	2.89	2.84	2.78	2.72	2.66	2.60
14	6.30	4.86	4.24	3.89	3.66	3.50	3.38	3.29	3.21	3.15	3.05	2.95	2.84	2.79	2.73	2.67	2.61	2.55	2.49
15	6.20	4.77	4.15	3.80	3.58	3.41	3.29	3.20	3.12	3.06	2.96	2.86	2.76	2.70	2.64	2.59	2.52	2.46	2.40
16	6.12	4.69	4.08	3.73	3.50	3.34	3.22	3.12	3.05	2.99	2.89	2.79	2.68	2.63	2.57	2.51	2.45	2.38	2.32
17	6.04	4.62	4.01	3.66	3.44	3.28	3.16	3.06	2.98	2.92	2.82	2.72	2.62	2.56	2.50	2.44	2.38	2.32	2.25
18	5.98	4.56	3.95	3.61	3.38	3.22	3.10	3.01	2.93	2.87	2.77	2.67	2.56	2.50	2.44	2.38	2.32	2.26	2.19
19	5.92	4.51	3.90	3.56	3.33	3.17	3.05	2.96	2.88	2.82	2.72	2.62	2.51	2.45	2.39	2.33	2.27	2.20	2.13
20	5.87	4.46	3.86	3.51	3.29	3.13	3.01	2.91	2.84	2.77	2.68	2.57	2.46	2.41	2.35	2.29	2.22	2.16	2.09
21	5.83	4.42	3.82	3.48	3.25	3.09	2.97	2.87	2.80	2.73	2.64	2.53	2.42	2.37	2.31	2.25	2.18	2.11	2.04
22	5.79	4.38	3.78	3.44	3.22	3.05	2.93	2.84	2.76	2.70	2.60	2.50	2.39	2.33	2.27	2.21	2.14	2.08	2.00
23	5.75	4.35	3.75	3.41	3.18	3.02	2.90	2.81	2.73	2.67	2.57	2.47	2.36	2.30	2.24	2.18	2.11	2.04	1.97
24	5.72	4.32	3.72	3.38	3.15	2.99	2.87	2.78	2.70	2.64	2.54	2.44	2.33	2.27	2.21	2.15	2.08	2.01	1.94
25	5.69	4.29	3.69	3.35	3.13	2.97	2.85	2.75	2.68	2.61	2.51	2.41	2.30	2.24	2.18	2.12	2.05	1.98	1.91
26	5.66	4.27	3.67	3.33	3.10	2.94	2.82	2.73	2.65	2.59	2.49	2.39	2.28	2.22	2.16	2.09	2.03	1.95	1.88
27	5.63	4.24	3.65	3.31	3.08	2.92	2.80	2.71	2.63	2.57	2.47	2.36	2.25	2.19	2.13	2.07	2.00	1.93	1.85
28	5.61	4.22	3.63	3.29	3.06	2.90	2.78	2.69	2.61	2.55	2.45	2.34	2.23	2.17	2.11	2.05	1.98	1.91	1.83
29	5.59	4.20	3.61	3.27	3.04	2.88	2.76	2.67	2.59	2.53	2.43	2.32	2.21	2.15	2.09	2.03	1.96	1.89	1.81
30	5.57	4.18	3.59	3.25	3.03	2.87	2.75	2.65	2.57	2.51	2.41	2.31	2.20	2.14	2.07	2.01	1.94	1.87	1.79
40	5.42	4.05	3.46	3.13	2.90	2.74	2.62	2.53	2.45	2.39	2.29	2.18	2.07	2.01	1.94	1.88	1.80	1.72	1.64
60	5.29	3.93	3.34	3.01	2.79	2.63	2.51	2.41	2.33	2.27	2.17	2.06	1.94	1.88	1.82	1.74	1.67	1.58	1.48
120	5.15	3.80	3.23	2.89	2.67	2.52	2.39	2.30	2.22	2.16	2.05	1.94	1.82	1.76	1.69	1.61	1.53	1.43	1.31
∞	5.02	3.69	3.12	2.79	2.57	2.41	2.29	2.19	2.11	2.05	1.94	1.83	1.71	1.64	1.57	1.48	1.39	1.27	1.00

Degrees of freedom for the denominator (v_2)

Table VI Percentage Points of the F-Distribution (*continued*)

$f_{0.01,\nu_1,\nu_2}$

ν_2 \ ν_1	Degrees of freedom for the numerator (ν_1)																		
Degrees of freedom for the denominator (ν_2)	1	2	3	4	5	6	7	8	9	10	12	15	20	24	30	40	60	120	∞
1	4052	4999.5	5403	5625	5764	5859	5928	5982	6022	6056	6106	6157	6209	6235	6261	6287	6313	6339	6366
2	98.50	99.00	99.17	99.25	99.30	99.33	99.36	99.37	99.39	99.40	99.42	99.43	99.45	99.46	99.47	99.47	99.48	99.49	99.50
3	34.12	30.82	29.46	28.71	28.24	27.91	27.67	27.49	27.35	27.23	27.05	26.87	26.69	26.00	26.50	26.41	26.32	26.22	26.13
4	21.20	18.00	16.69	15.98	15.52	15.21	14.98	14.80	14.66	14.55	14.37	14.20	14.02	13.93	13.84	13.75	13.65	13.56	13.46
5	16.26	13.27	12.06	11.39	10.97	10.67	10.46	10.29	10.16	10.05	9.89	9.72	9.55	9.47	9.38	9.29	9.20	9.11	9.02
6	13.75	10.92	9.78	9.15	8.75	8.47	8.26	8.10	7.98	7.87	7.72	7.56	7.40	7.31	7.23	7.14	7.06	6.97	6.88
7	12.25	9.55	8.45	7.85	7.46	7.19	6.99	6.84	6.72	6.62	6.47	6.31	6.16	6.07	5.99	5.91	5.82	5.74	5.65
8	11.26	8.65	7.59	7.01	6.63	6.37	6.18	6.03	5.91	5.81	5.67	5.52	5.36	5.28	5.20	5.12	5.03	4.95	4.46
9	10.56	8.02	6.99	6.42	6.06	5.80	5.61	5.47	5.35	5.26	5.11	4.96	4.81	4.73	4.65	4.57	4.48	4.40	4.31
10	10.04	7.56	6.55	5.99	5.64	5.39	5.20	5.06	4.94	4.85	4.71	4.56	4.41	4.33	4.25	4.17	4.08	4.00	3.91
11	9.65	7.21	6.22	5.67	5.32	5.07	4.89	4.74	4.63	4.54	4.40	4.25	4.10	4.02	3.94	3.86	3.78	3.69	3.60
12	9.33	6.93	5.95	5.41	5.06	4.82	4.64	4.50	4.39	4.30	4.16	4.01	3.86	3.78	3.70	3.62	3.54	3.45	3.36
13	9.07	6.70	5.74	5.21	4.86	4.62	4.44	4.30	4.19	4.10	3.96	3.82	3.66	3.59	3.51	3.43	3.34	3.25	3.17
14	8.86	6.51	5.56	5.04	4.69	4.46	4.28	4.14	4.03	3.94	3.80	3.66	3.51	3.43	3.35	3.27	3.18	3.09	3.00
15	8.68	6.36	5.42	4.89	4.36	4.32	4.14	4.00	3.89	3.80	3.67	3.52	3.37	3.29	3.21	3.13	3.05	2.96	2.87
16	8.53	6.23	5.29	4.77	4.44	4.20	4.03	3.89	3.78	3.69	3.55	3.41	3.26	3.18	3.10	3.02	2.93	2.84	2.75
17	8.40	6.11	5.18	4.67	4.34	4.10	3.93	3.79	3.68	3.59	3.46	3.31	3.16	3.08	3.00	2.92	2.83	2.75	2.65
18	8.29	6.01	5.09	4.58	4.25	4.01	3.84	3.71	3.60	3.51	3.37	3.23	3.08	3.00	2.92	2.84	2.75	2.66	2.57
19	8.18	5.93	5.01	4.50	4.17	3.94	3.77	3.63	3.52	3.43	3.30	3.15	3.00	2.92	2.84	2.76	2.67	2.58	2.59
20	8.10	5.85	4.94	4.43	4.10	3.87	3.70	3.56	3.46	3.37	3.23	3.09	2.94	2.86	2.78	2.69	2.61	2.52	2.42
21	8.02	5.78	4.87	4.37	4.04	3.81	3.64	3.51	3.40	3.31	3.17	3.03	2.88	2.80	2.72	2.64	2.55	2.46	2.36
22	7.95	5.72	4.82	4.31	3.99	3.76	3.59	3.45	3.35	3.26	3.12	2.98	2.83	2.75	2.67	2.58	2.50	2.40	2.31
23	7.88	5.66	4.76	4.26	3.94	3.71	3.54	3.41	3.30	3.21	3.07	2.93	2.78	2.70	2.62	2.54	2.45	2.35	2.26
24	7.82	5.61	4.72	4.22	3.90	3.67	3.50	3.36	3.26	3.17	3.03	2.89	2.74	2.66	2.58	2.49	2.40	2.31	2.21
25	7.77	5.57	4.68	4.18	3.85	3.63	3.46	3.32	3.22	3.13	2.99	2.85	2.70	2.62	2.54	2.45	2.36	2.27	2.17
26	7.72	5.53	4.64	4.14	3.82	3.59	3.42	3.29	3.18	3.09	2.96	2.81	2.66	2.58	2.50	2.42	2.33	2.23	2.13
27	7.68	5.49	4.60	4.11	3.78	3.56	3.39	3.26	3.15	3.06	2.93	2.78	2.63	2.55	2.47	2.38	2.29	2.20	2.10
28	7.64	5.45	4.57	4.07	3.75	3.53	3.36	3.23	3.12	3.03	2.90	2.75	2.60	2.52	2.44	2.35	2.26	2.17	2.06
29	7.60	5.42	4.54	4.04	3.73	3.50	3.33	3.20	3.09	3.00	2.87	2.73	2.57	2.49	2.41	2.33	2.23	2.14	2.03
30	7.56	5.39	4.51	4.02	3.70	3.47	3.30	3.17	3.07	2.98	2.84	2.70	2.55	2.47	2.39	2.30	2.21	2.11	2.01
40	7.31	5.18	4.31	3.83	3.51	3.29	3.12	2.99	2.89	2.80	2.66	2.52	2.37	2.29	2.20	2.11	2.02	1.92	1.80
60	7.08	4.98	4.13	3.65	3.34	3.12	2.95	2.82	2.72	2.63	2.50	2.35	2.20	2.12	2.03	1.94	1.84	1.73	1.60
120	6.85	4.79	3.95	3.48	3.17	2.96	2.79	2.66	2.56	2.47	2.34	2.19	2.03	1.95	1.86	1.76	1.66	1.53	1.38
∞	6.63	4.61	3.78	3.32	3.02	2.80	2.64	2.51	2.41	2.32	2.18	2.04	1.88	1.79	1.70	1.59	1.47	1.32	1.00

Chart VII Operating Characteristic Curves

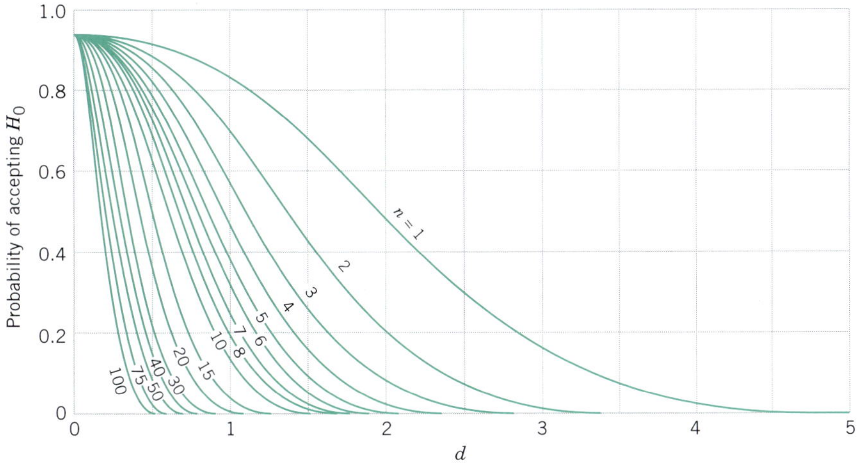

(*a*) *O.C.* curves for different values of *n* for the two-sided normal test for a level of significance $\alpha = 0.05$.

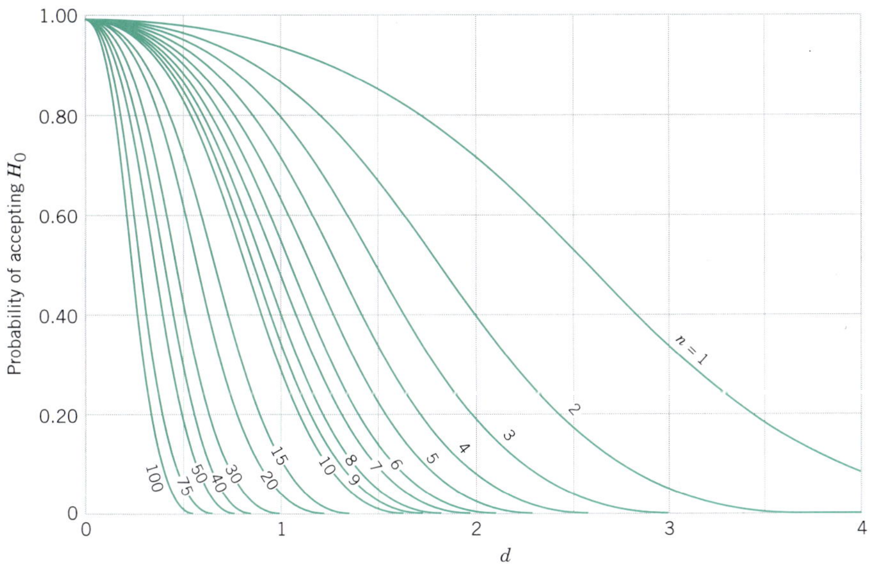

(*b*) *O.C.* curves for different values of *n* for the two-sided normal test for a level of significance $\alpha = 0.01$.

Source: Charts VI*a, e, f, k, m,* and *q* are reproduced with permission from "Operating Characteristics for the Common Statistical Tests of Significance," by C. L. Ferris, F. E. Grubbs, and C. L. Weaver, *Annals of Mathematical Statistics,* June 1946.

Charts VI*b, c, d, g, h, i, j, l, n, o, p,* and *r* are reproduced with permission from *Engineering Statistics,* 2nd Edition, by A. H. Bowker and G. J. Lieberman, Prentice-Hall, 1972.

Chart VII Operating Characteristic Curves (*continued*)

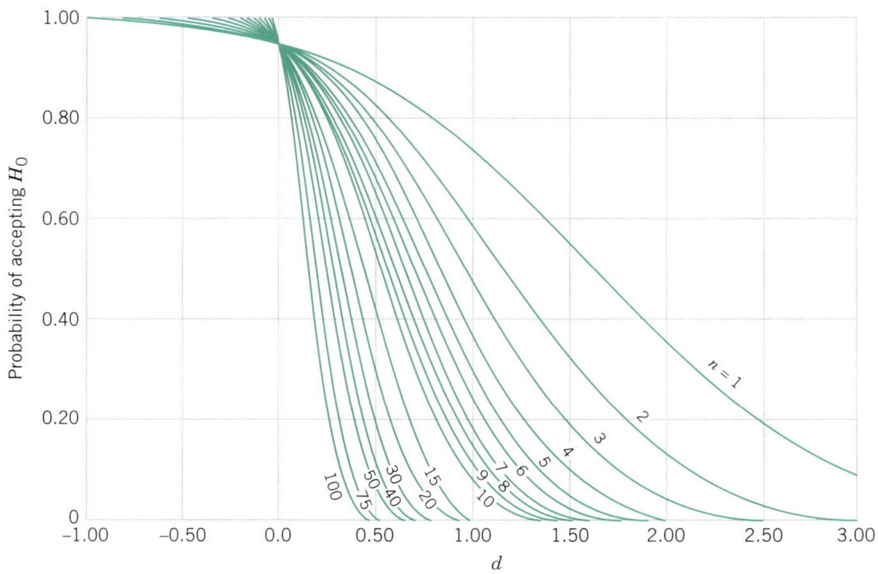

(*c*) *O.C.* curves for different values of *n* for the one-sided normal test for a level of significance $\alpha = 0.05$.

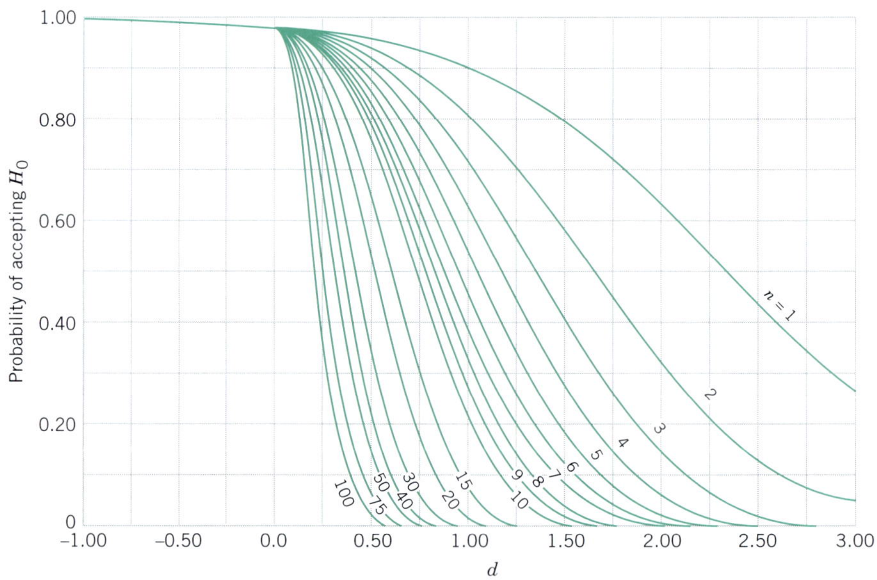

(*d*) *O.C.* curves for different values of *n* for the one-sided normal test for a level of significance $\alpha = 0.01$.

Chart VII Operating Characteristic Curves (*continued*)

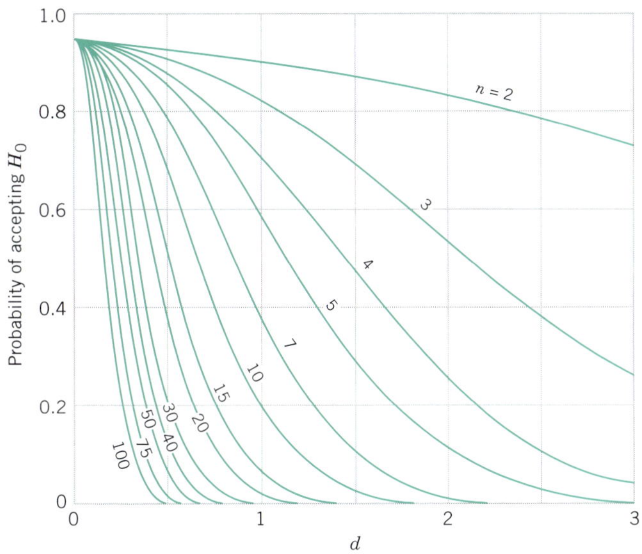

(e) $O.C.$ curves for different values of n for the two-sided t-test for a level of significance $\alpha = 0.05$.

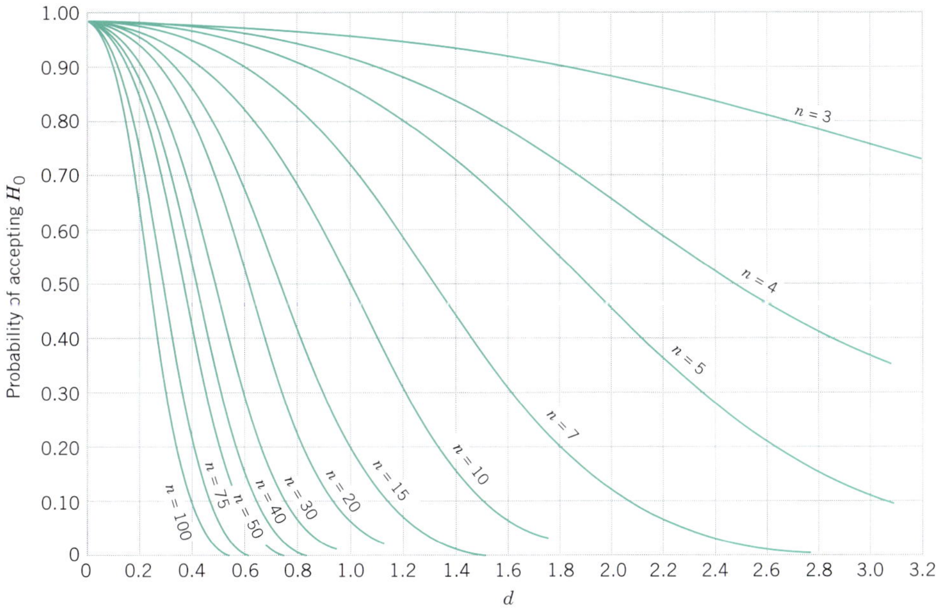

(f) $O.C.$ curves for different values of n for the two-sided t-test for a level of significance $\alpha = 0.01$.

Chart VII Operating Characteristic Curves (*continued*)

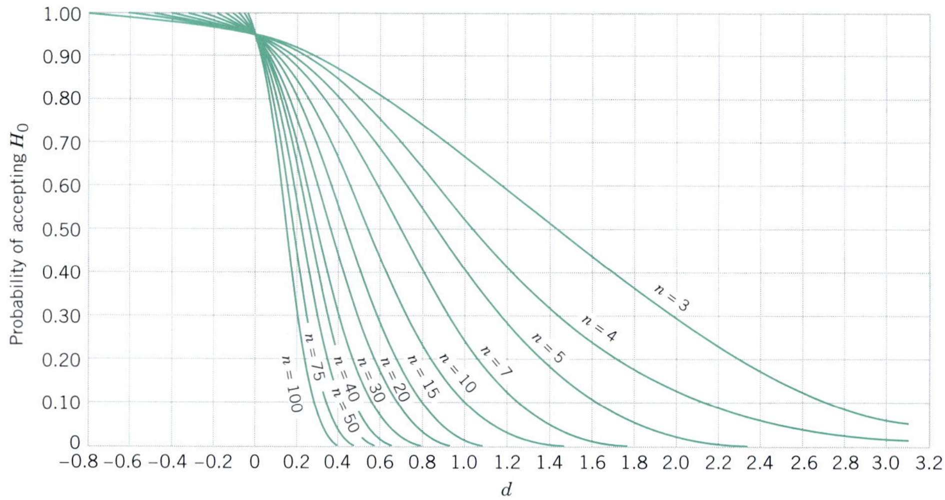

(*g*) *O.C.* curves for different values of *n* for the one-sided *t*-test for a level of significance $\alpha = 0.05$.

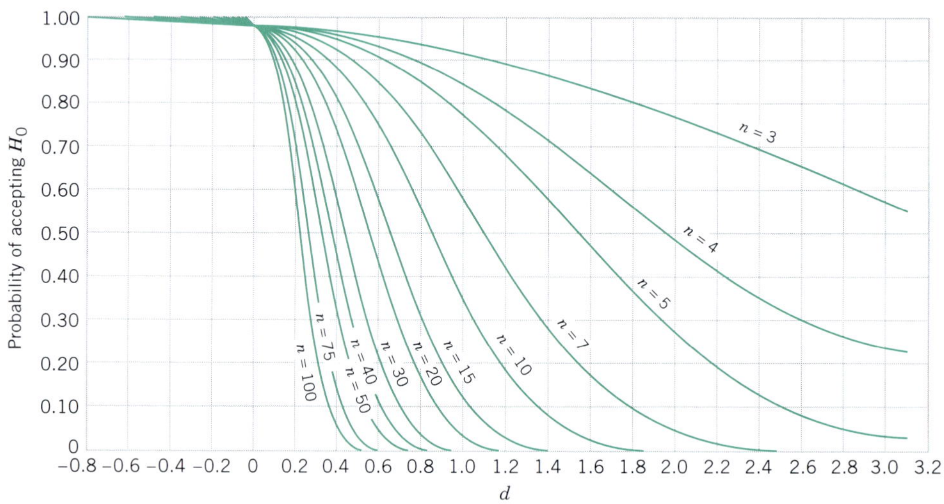

(*h*) *O.C.* curves for different values of *n* for the one-sided *t*-test for a level of significance $\alpha = 0.01$.

Chart VII Operating Characteristic Curves (*continued*)

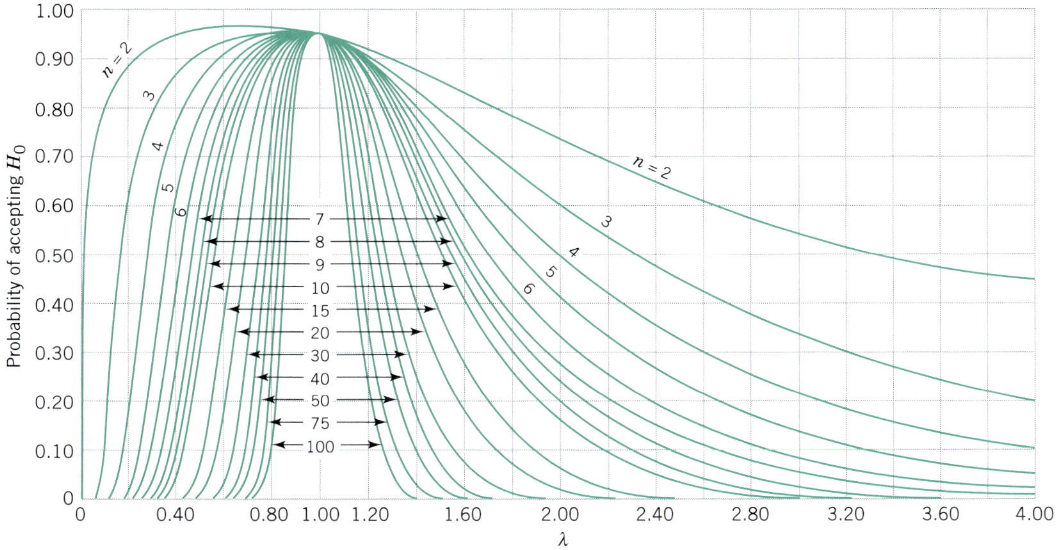

(*i*) *O.C.* curves for different values of *n* for the two-sided chi-square test for a level of significance $\alpha = 0.05$.

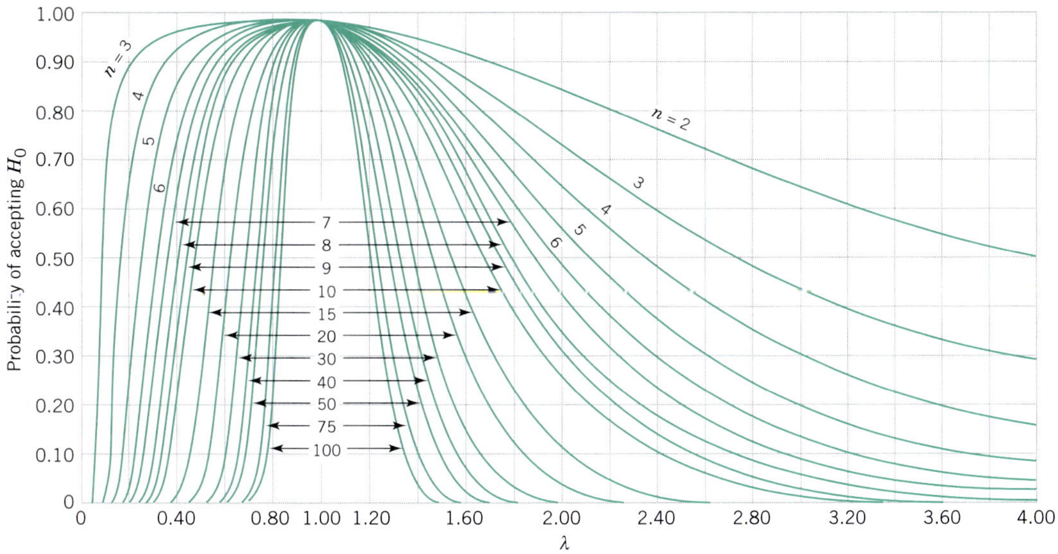

(*j*) *O.C.* curves for different values of *n* for the two-sided chi-square test for a level of significance $\alpha = 0.01$.

Chart VII Operating Characteristic Curves (*continued*)

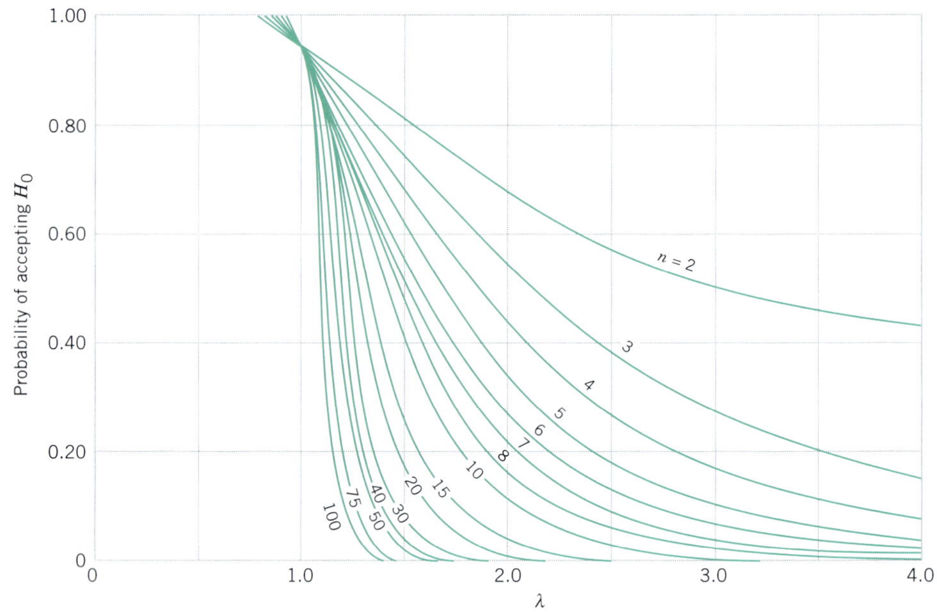

(*k*) *O.C.* curves for different values of *n* for the one-sided (upper tail) chi-square test for a level of significance $\alpha = 0.05$.

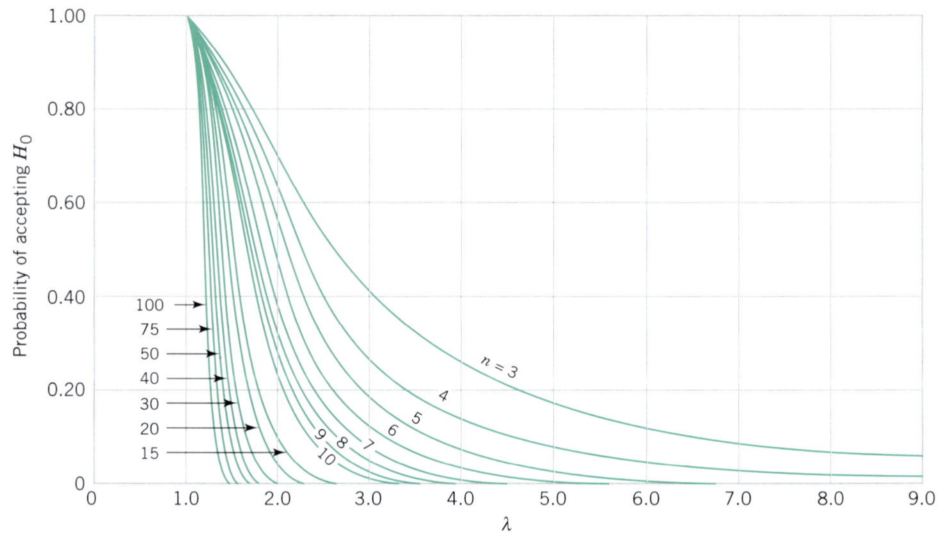

(*l*) *O.C.* curves for different values of *n* for the one-sided (upper tail) chi-square test for a level of significance $\alpha = 0.01$.

Chart VII Operating Characteristic Curves (*continued*)

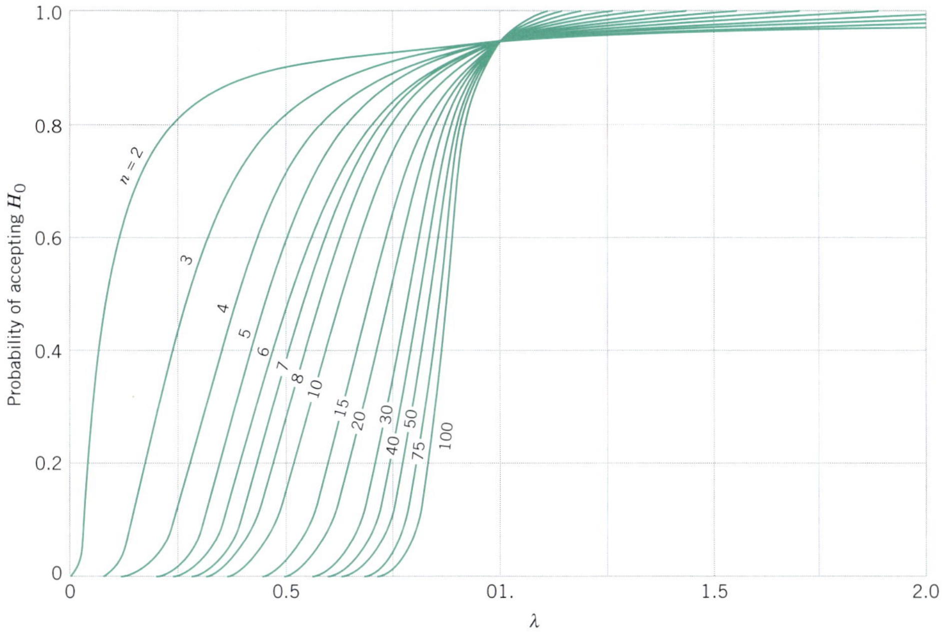

(*m*) *O.C.* curves for different values of *n* for the one-sided (lower tail) chi-square test for a level of significance $\alpha = 0.05$.

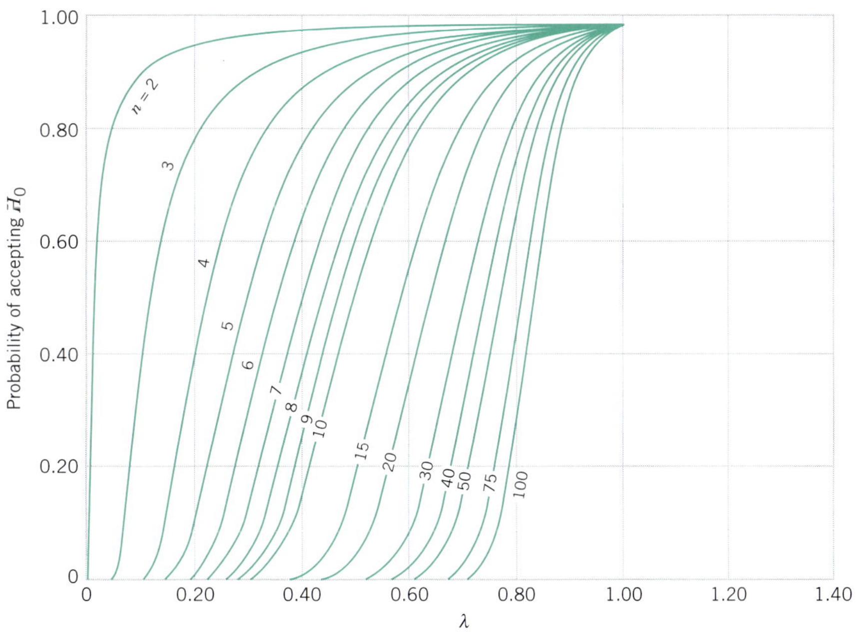

(*n*) *O.C.* curves for different values of *n* for the one-sided (lower tail) chi-square test for a level of significance $\alpha = 0.01$.

Chart VII Operating Characteristic Curves (*continued*)

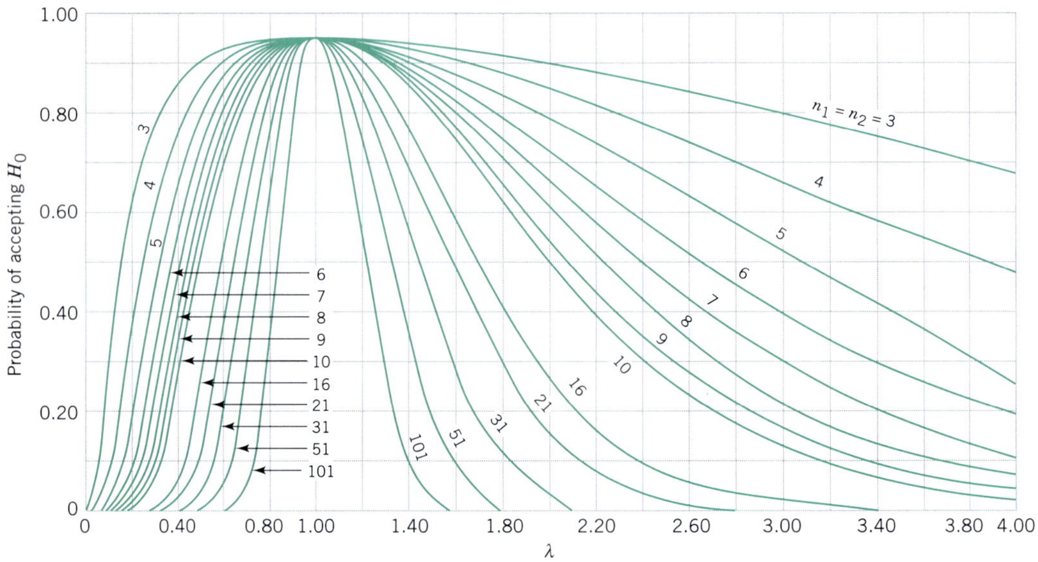

(*o*) *O.C.* curves for different values of *n* for the two-sided *F*-test for a level of significance $\alpha = 0.05$.

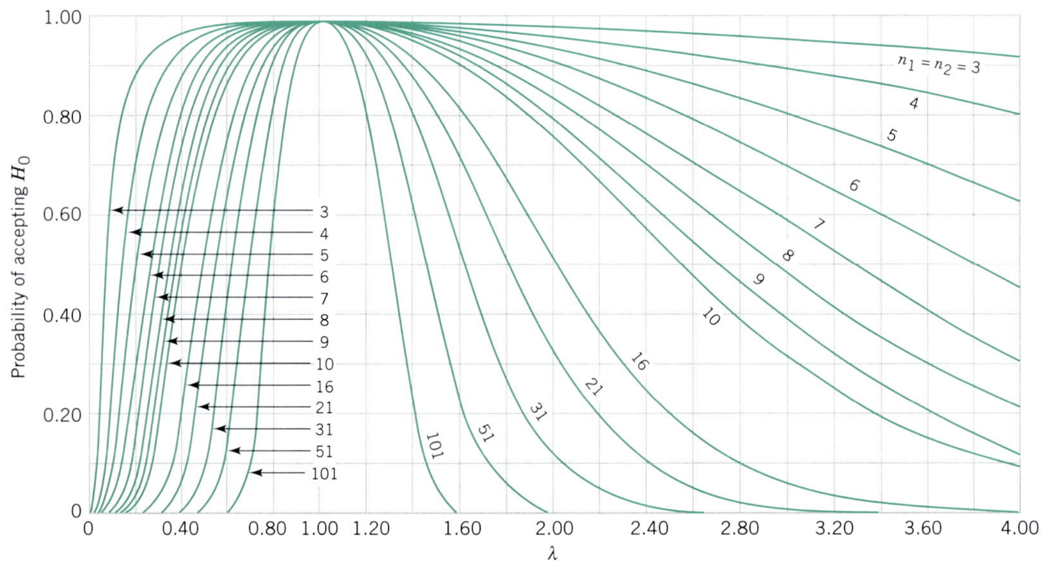

(*p*) O.C. curves for different values of *n* for the two-sided *F*-test for a level of significance $\alpha = 0.01$.

Chart VII Operating Characteristic Curves (*continued*)

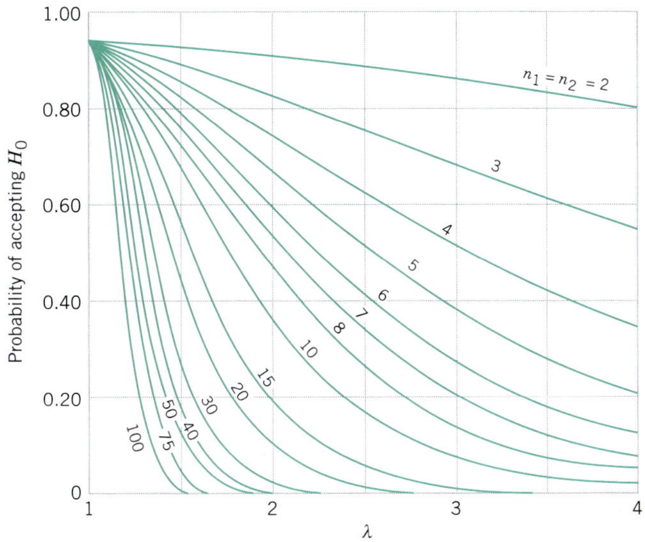

(*q*) *O.C.* curves for different values of *n* for the one-sided *F*-test for a level of significance $\alpha = 0.05$.

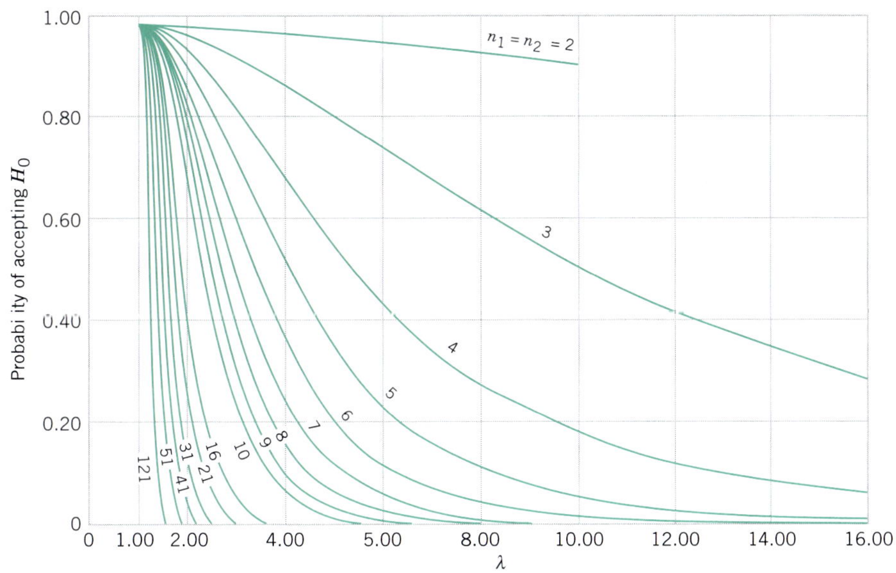

(*r*) *O.C.* curves for different values of *n* for the one-sided *F*-test for a level of significance $\alpha = 0.01$.

Table VIII Critical Values for the Sign Test

r^*_α

n \ α	0.10 0.05	0.05 0.025	0.01 0.005	Two-sided tests One-sided tests	n \ α	0.10 0.05	0.05 0.025	0.01 0.005	Two-sided tests One-sided tests
5	0				23	7	6	4	
6	0	0			24	7	6	5	
7	0	0			25	7	7	5	
8	1	0	0		26	8	7	6	
9	1	1	0		27	8	7	6	
10	1	1	0		28	9	8	6	
11	2	1	0		29	9	8	7	
12	2	2	1		30	10	9	7	
13	3	2	1		31	10	9	7	
14	3	2	1		32	10	9	8	
15	3	3	2		33	11	10	8	
16	4	3	2		34	11	10	9	
17	4	4	2		35	12	11	9	
18	5	4	3		36	12	11	9	
19	5	4	3		37	13	12	10	
20	5	5	3		38	13	12	10	
21	6	5	4		39	13	12	11	
22	6	5	4		40	14	13	11	

Table IX Critical Values for the Wilcoxon Signed-Rank Test

w^*_α

n^* \ α	0.10 0.05	0.05 0.025	0.02 0.01	0.01 0.005	Two-sided tests One-sided tests
4					
5	0				
6	2	0			
7	3	2	0		
8	5	3	1	0	
9	8	5	3	1	
10	10	8	5	3	
11	13	10	7	5	
12	17	13	9	7	
13	21	17	12	9	
14	25	21	15	12	
15	30	25	19	15	
16	35	29	23	19	
17	41	34	27	23	
18	47	40	32	27	
19	53	46	37	32	
20	60	52	43	37	
21	67	58	49	42	
22	75	65	55	48	
23	83	73	62	54	
24	91	81	69	61	
25	100	89	76	68	

* If $n > 25$, W^- (or W^-) is approximately normally distributed with mean $n(n + 1)/4$ and variance $n(n + 1)(2n + 1)/24$.

Table X Critical Values for the Wilcoxon Rank-Sum Test

$w_{0.05}$

n_2 \ n_1*	4	5	6	7	8	9	10	11	12	13	14	15
4	10											
5	11	17										
6	12	18	26									
7	13	20	27	36								
8	14	21	29	38	49							
9	15	22	31	40	51	63						
10	15	23	32	42	53	65	78					
11	16	24	34	44	55	68	81	96				
12	17	26	35	46	58	71	85	99	115			
13	18	27	37	48	60	73	88	103	119	137		
14	19	28	38	50	63	76	91	106	123	141	160	
15	20	29	40	52	65	79	94	110	127	145	164	185
16	21	31	42	54	67	82	97	114	131	150	169	
17	21	32	43	56	70	84	100	117	135	154		
18	22	33	45	58	72	87	103	121	139			
19	23	34	46	60	74	90	107	124				
20	24	35	48	62	77	93	110					
21	25	37	50	64	79	95						
22	26	38	51	66	82							
23	27	39	53	68								
24	28	40	55									
25	28	42										
26	29											
27												
28												

*For n_1 and $n_2 > 8$, W_1 is approximately normally distributed with mean $\frac{1}{2}n_1(n_1 + n_2 + 1)$ and variance $n_1 n_2(n_1 + n_2 + 1)/12$.

Table X Critical Values for the Wilcoxon Rank-Sum Test *(continued)*

$w_{0.01}$

n_2 \ n_1	4	5	6	7	8	9	10	11	12	13	14	15
5		15										
6	10	16	23									
7	10	17	24	32								
8	11	17	25	34	43							
9	11	18	26	35	45	56						
10	12	19	27	37	47	58	71					
11	12	20	28	38	49	61	74	87				
12	13	21	30	40	51	63	76	90	106			
13	14	22	31	41	53	65	79	93	109	125		
14	14	22	32	43	54	67	81	96	112	129	147	
15	15	23	33	44	56	70	84	99	115	133	151	171
16	15	24	34	46	58	72	86	102	119	137	155	
17	16	25	36	47	60	74	89	105	122	140		
18	16	26	37	49	62	76	92	108	125			
19	17	27	38	50	64	78	94	111				
20	18	28	39	52	66	81	97					
21	18	29	40	53	68	83						
22	19	29	42	55	70							
23	19	30	43	57								
24	20	31	44									
25	20	32										
26	21											
27												
28												

Table XI Factors for Constructing Variables Control Charts

	Factor for Control Limits						
	$\overline{X}$ Chart			R Chart		S Chart	
n^*	A_1	A_2	d_2	D_3	D_4	c_4	n
2	3.760	1.880	1.128	0	3.267	0.7979	2
3	2.394	1.023	1.693	0	2.575	0.8862	3
4	1.880	.729	2.059	0	2.282	0.9213	4
5	1.596	.577	2.326	0	2.115	0.9400	5
6	1.410	.483	2.534	0	2.004	0.9515	6
7	1.277	.419	2.704	.076	1.924	0.9594	7
8	1.175	.373	2.847	.136	1.864	0.9650	8
9	1.094	.337	2.970	.184	1.816	0.9693	9
10	1.028	.308	3.078	.223	1.777	0.9727	10
11	.973	.285	3.173	.256	1.744	0.9754	11
12	.925	.266	3.258	.284	1.716	0.9776	12
13	.884	.249	3.336	.308	1.692	0.9794	13
14	.848	.235	3.407	.329	1.671	0.9810	14
15	.816	.223	3.472	.348	1.652	0.9823	15
16	.788	.212	3.532	.364	1.636	0.9835	16
17	.762	.203	3.588	.379	1.621	0.9845	17
18	.738	.194	3.640	.392	1.608	0.9854	18
19	.717	.187	3.689	.404	1.596	0.9862	19
20	.697	.180	3.735	.414	1.586	0.9869	20
21	.679	.173	3.778	.425	1.575	0.9876	21
22	.662	.167	3.819	.434	1.566	0.9882	22
23	.647	.162	3.858	.443	1.557	0.9887	23
24	.632	.157	3.895	.452	1.548	0.9892	24
25	.619	.153	3.931	.459	1.541	0.9896	25

$^*n > 25$: $A_1 = 3/\sqrt{n}$ where n = number of observations in sample.

Table XII Factors for Tolerance Intervals

Values of k for Two-Sided Intervals									
	Confidence Level								
	0.90			0.95			0.99		
	Probability of Coverage								
Sample Size	0.90	0.95	0.99	0.90	0.95	0.99	0.90	0.95	0.99
2	15.978	18.800	24.167	32.019	37.674	48.430	160.193	188.491	242.300
3	5.847	6.919	8.974	8.380	9.916	12.861	18.930	22.401	29.055
4	4.166	4.943	6.440	5.369	6.370	8.299	9.398	11.150	14.527
5	3.949	4.152	5.423	4.275	5.079	6.634	6.612	7.855	10.260
6	3.131	3.723	4.870	3.712	4.414	5.775	5.337	6.345	8.301
7	2.902	3.452	4.521	3.369	4.007	5.248	4.613	5.488	7.187
8	2.743	3.264	4.278	3.136	3.732	4.891	4.147	4.936	6.468
9	2.626	3.125	4.098	2.967	3.532	4.631	3.822	4.550	5.966
10	2.535	3.018	3.959	2.839	3.379	4.433	3.582	4.265	5.594
11	2.463	2.933	3.849	2.737	3.259	4.277	3.397	4.045	5.308
12	2.404	2.863	3.758	2.655	3.162	4.150	3.250	3.870	5.079
13	2.355	2.805	3.682	2.587	3.081	4.044	3.130	3.727	4.893
14	2.314	2.756	3.618	2.529	3.012	3.955	3.029	3.608	4.737
15	2.278	2.713	3.562	2.480	2.954	3.878	2.945	3.507	4.605
16	2.246	2.676	3.514	2.437	2.903	3.812	2.872	3.421	4.492
17	2.219	2.643	3.471	2.400	2.858	3.754	2.808	3.345	4.393
18	2.194	2.614	3.433	2.366	2.819	3.702	2.753	3.279	4.307
19	2.172	2.588	3.399	2.337	2.784	3.656	2.703	3.221	4.230
20	2.152	2.564	3.368	2.310	2.752	3.615	2.659	3.168	4.161
21	2.135	2.543	3.340	2.286	2.723	3.577	2.620	3.121	4.100
22	2.118	2.524	3.315	2.264	2.697	3.543	2.584	3.078	4.044
23	2.103	2.506	3.292	2.244	2.673	3.512	2.551	3.040	3.993
24	2.089	2.489	3.270	2.225	2.651	3.483	2.522	3.004	3.947
25	2.077	2.474	3.251	2.208	2.631	3.457	2.494	2.972	3.904
30	2.025	2.413	3.170	2.140	2.529	3.350	2.385	2.841	3.733
40	1.959	2.334	3.066	2.052	2.445	3.213	2.247	2.677	3.518
50	1.916	2.284	3.001	1.996	2.379	3.126	2.162	2.576	3.385
60	1.887	2.248	2.955	1.958	2.333	3.066	2.103	2.506	3.293
70	1.865	2.222	2.920	1.929	2.299	3.021	2.060	2.454	3.225
80	1.848	2.202	2.894	1.907	2.272	2.986	2.026	2.414	3.173
90	1.834	2.185	2.872	1.889	2.251	2.958	1.999	2.382	3.130
100	1.822	2.172	2.854	1.874	2.233	2.934	1.977	2.355	3.096

Table XII Factors for Tolerance Intervals (*continued*)

	Values of k for One-Sided Intervals								
	Confidence Level								
	0.90			0.95			0.99		
Sample	Probability of Coverage								
Size	0.90	0.95	0.99	0.90	0.95	0.99	0.90	0.95	0.99
2	10.253	13.090	18.500	20.581	26.260	37.094	103.029	131.426	185.617
3	4.258	5.311	7.340	6.155	7.656	10.553	13.995	17.370	23.896
4	3.188	3.957	5.438	4.162	5.144	7.042	7.380	9.083	12.387
5	2.742	3.400	4.666	3.407	4.203	5.741	5.362	6.578	8.939
6	2.494	3.092	4.243	3.006	3.708	5.062	4.411	5.406	7.335
7	2.333	2.894	3.972	2.755	3.399	4.642	3.859	4.728	6.412
8	2.219	2.754	3.783	2.582	3.187	4.354	3.497	4.285	5.812
9	2.133	2.650	3.641	2.454	3.031	4.143	3.240	3.972	5.389
10	2.066	2.568	3.532	2.355	2.911	3.981	3.048	3.738	5.074
11	2.011	2.503	3.443	2.275	2.815	3.852	2.898	3.556	4.829
12	1.966	2.448	3.371	2.210	2.736	3.747	2.777	3.410	4.633
13	1.928	2.402	3.309	2.155	2.671	3.659	2.677	3.290	4.472
14	1.895	2.363	3.257	2.109	2.614	3.585	2.593	3.189	4.337
15	1.867	2.329	3.212	2.068	2.566	3.520	2.521	3.102	4.222
16	1.842	2.299	3.172	2.033	2.524	3.464	2.459	3.028	4.123
17	1.819	2.272	3.137	2.002	2.486	3.414	2.405	2.963	4.037
18	1.800	2.249	3.105	1.974	2.453	3.370	2.357	2.905	3.960
19	1.782	2.227	3.077	1.949	2.423	3.331	2.314	2.854	3.892
20	1.765	2.028	3.052	1.926	2.396	3.295	2.276	2.808	3.832
21	1.750	2.190	3.028	1.905	2.371	3.263	2.241	2.766	3.777
22	1.737	2.174	3.007	1.886	2.349	3.233	2.209	2.729	3.727
23	1.724	2.159	2.987	1.869	2.328	3.206	2.180	2.694	3.681
24	1.712	2.145	2.969	1.853	2.309	3.181	2.154	2.662	3.640
25	1.702	2.132	2.952	1.838	2.292	3.158	2.129	2.633	3.601
30	1.657	2.080	2.884	1.777	2.220	3.064	2.030	2.515	3.447
40	1.598	2.010	2.793	1.697	2.125	2.941	1.902	2.364	3.249
50	1.559	1.965	2.735	1.646	2.065	2.862	1.821	2.269	3.125
60	1.532	1.933	2.694	1.609	2.022	2.807	1.764	2.202	3.038
70	1.511	1.909	2.662	1.581	1.990	2.765	1.722	2.153	2.974
80	1.495	1.890	2.638	1.559	1.964	2.733	1.688	2.114	2.924
90	1.481	1.874	2.618	1.542	1.944	2.706	1.661	2.082	2.883
100	1.470	1.861	2.601	1.527	1.927	2.684	1.639	2.056	2.850

Appendix B Answers to Selected Exercises

CHAPTER 2

Section 2-1

2-1. Let a, b denote a part above, below the specification, respectively
$S = \{aaa, aab, aba, abb, baa, bab, bba, bbb\}$

2-3. Let a denote an acceptable power supply
Let f, m, c denote a supply with a functional, minor, or cosmetic error, respectively.
$S = \{a, f, m, c\}$

2-5. If only the number of tracks with errors is of interest, then $S = \{0, 1, 2, \ldots, 24\}$

2-7. S is the sample space of 100 possible two digit integers.

2-9. $S = \{0, 1, 2, \ldots, 1E09\}$ in ppb

2-11. $S = \{1.0, 1.1, 1.2, \ldots, 14.0\}$

2-13. $S = \{0, 1, 2, \ldots,\}$ in milliseconds

2-17. c = connect, b = busy, $S = \{c, bc, bbc, bbbc, bbbbc, \ldots\}$

2-21. (a) S = nonnegative integers from 0 to the largest integer that can be displayed by the scale $S = \{0, 1, 2, 3, \ldots\}$
(b) S (c) $\{12, 13, 14, 15\}$
(d) $\{0, 1, 2, \ldots, 11\}$
(e) S
(f) $\{0, 1, 2, \ldots, 7\}$
(g) $\varnothing$
(h) $\varnothing$ (i) $\{8, 9, 10, \ldots\}$

2-23. Let d denoted a distorted bit and let o denote a bit that is not distorted.

(a) $S = \left\{\begin{matrix} dddd, dodd, oddd, oodd, \\ dddo, dodo, oddo, oodo, \\ ddod, dood, odod, oood, \\ ddoo, dooo, odoo, oooo \end{matrix}\right\}$

(b) No, for example $A_1 \cap A_2 = \{dddd, dddo, ddod, ddoo\}$

(c) $A_1 = \left\{\begin{matrix} dddd, dodd, \\ dddo, dodo \\ ddod, dood \\ ddoo, dooo \end{matrix}\right\}$

(d) $A_1' = \left\{\begin{matrix} oddd, oodd, \\ oddo, oodo, \\ odod, oood, \\ odoo, oooo \end{matrix}\right\}$

(e) $A_1 \cap A_2 \cap A_3 \cap A_4 = \{dddd\}$

(f) $(A_1 \cap A_2) \cup (A_3 \cap A_4) = \{dddd, dodd, dddo, oddd, ddod, oodd, ddoo\}$

2-25. Let P denote being positive and let N denote being negative. The sample space is $\{PPP, PPN, PNP, NPP, PNN, NPN, NNP, NNN\}$.
(a) $A = \{PPP\}$
(b) $B = \{NNN\}$
(c) $A \cap B = \varnothing$
(d) $A \cup B = \{PPP, NNN\}$

2-27. (a) $A' \cap B = 10$, $B' = 10$, $A \cup B = 92$

2-29. (a) $A' = \{x \mid x \geq 72.5\}$
(b) $B' = \{x \mid x \leq 52.5\}$
(c) $A \cap B = \{x \mid 52.5 < x < 72.5\}$
(d) $A \cup B = \{x \mid x > 0\}$

2-31. Let g denote a good board, m a board with minor defects, and j a board with major defects.
(a) $S = \{gg, gm, gj, mg, mm, mj, jg, jm, jj\}$
(b) $S = \{gg, gm, gj, mg, mm, mj, jg, jm\}$

2-35. 120
2-37. 144
2-39. 14,400
2-41. (a) 416,965,528 (b) 113,588,800 (c) 130,721,752
2-43. (a) 21 (b) 2520 (c) 720
2-45. (a) 1000 (b) 160 (c) 720
2-47. (a) 0.416 (b) 0.712 (c) 0.206
2-49. (a) 0.0082 (b) 0.0082

Section 2-2

2-51. (a) 0.4 (b) 0.8 (c) 0.6 (d) 1 (e) 0.2
2-53. (a) 1/10 (b) 5/10
2-55. (a) $S = \{1, 2, 3, 4, 5, 6, 7, 8\}$ (b) 2/8 (c) 6/8
2-57. (a) 0.83 (b) 0.85
2-59. $(1/10^3)*(1/26^3) = 5.7 \times 10^{-8}$
2-61. (a) $4 + 4 \times 3 + 4 \times 3 \times 3 = 52$ (b) 36/52 (c) No
2-63. (a) 0.30 (b) 0.77 (c) 0.70 (d) 0.22 (e) 0.85 (f) 0.92

Section 2-3

2-67. (a) 0.9 (b) 0 (c) 0 (d) 0 (e) 0.1
2-69. (a) 0.70 (b) 0.95 (c) No
2-71. (a) 350/370 (b) 362/370 (c) 358/370 (d) 345/370
2-73. (a) 13/130 (b) 0.90, No

Section 2-4

2-75. (a) 86/100 (b) 79/100 (c) 70/79 (d) 70/86
2-77. (a) 0.903 (b) 0.591
2-79. (a) 12/100 (b) 12/28 (c) 34/122
2-81. (a) 0.5625 (b) 0.1918 (c) 0.3333
2-83. (a) 20/100 (b) 19/99 (c) 0.038 (d) 0.2
2-85. (a) 0.02 (b) 0.000458 (c) 0.9547

Section 2-5

2-89. (a) 0.2 (b) 0.3
2-91. 0.014
2-93. 0.028
2-95. (a) 0.2376 (b) 0.0078
2-97. (a) 0.2 (b) 0.2

Section 2-6

2-103. (a) *not* independent (b) yes
2-107. (a) 0.59 (b) 0.328 (c) 0.41
2-109. (a) 0.0048 (b) 0.0768
2-111. (a) 0.01 (b) 0.49 (c) 0.09
2-113. (a) 0.00003 (b) 0.00024 (c) 0.00107
2-115. 0.9702

Section 2-7

2-117. 0.89
2-119. (a) 0.97638 (b) 0.207552
2-121. (a) 0.615 (b) 0.618 (c) 0.052
2-123. (a) 0.9847 (b) 0.1184

Supplemental Exercises

2-125. 0.014
2-127. (a) 0.82 (b) 0.90 (c) 0.18 (d) 0.80 (e) 0.92 (f) 0.98
2-131. (a) 0.2 (b) 0.202 (c) 0.638 (d) 0.2
2-133. (a) 0.03 (b) 0.97 (c) 0.40 (d) 0.05 (e) 0.012 (f) 0.018 (g) 0.0605
2-135. (a) 0.18143 (b) 0.005976 (c) 0.86494
2-137. 0.000008
2-139. (a) 50 (b) 37 (c) 93
2-143. (a) 0.19 (b) 0.15 (c) 0.99 (d) 0.80 (e) 0.158
2-145. (a) No (b) No (c) 40/240 (d) 200/240 (e) 234/240 (f) 1
2-147. (a) 0.282 (b) 0.718
2-149. 0.996
2-151. (a) 0.0037 (b) 0.8108
2-153. (a) 0.0778 (b) 0.00108 (c) 0.947
2-155. (a) 0.9764 (b) 0.3159
2-157. (a) 0.207 (b) 0.625
2-159. (a) 0.453 (b) 0.262 (c) 0.881 (d) 0.547 (e) 0.783 (f) 0.687
2-161. 1.58×10^{-7}
2-163. (a) 0.67336 (b) 2.646×10^{-8} (c) 0.99973
2-165. (a) 36^7 (b) $70(26^6)$ (c) $100(26^5)$
2-167. (a) 0.994, 0.995 (b) 0.99, 0.985 (c) 0.998, 0.9975

Mind-Expanding Exercises

2-169. (a) $n = 3$ (b) $n = 3$
2-171. 0.306, 0.694

CHAPTER 3

Section 3-1

3-1. $\{0, 1, 2, \ldots, 1000\}$
3-3. $\{0, 1, 2, \ldots, 99999\}$
3-5. $\{1, 2, \ldots, 491\}$
3-7. $\{0, 1, 2, \ldots\}$
3-9. $\{0, 1, 2, \ldots, 15\}$
3-11. $\{0, 1, 2, \ldots, 10000\}$
3-13. $\{0, 1, 2, \ldots, 40000\}$

Section 3-2

3-15. (a) 1 (b) 7/8 (c) 3/4 (d) 1/2
3-17. (a) 9/25 (b) 4/25 (c) 12/25 (d) 1
3-19. $f(0) = 0.033, f(1) = 0.364, f(2) = 0.603$
3-21. $P(X = 0) = 0.008$, $P(X = 1) = 0.096$, $P(X = 2) = 0.384$, $P(X = 3) = 0.512$
3-23. $P(X = 50) = 0.5$, $P(X = 25) = 0.3$, $P(X = 10) = 0.2$
3-25. $P(X = 15) = 0.6$, $P(X = 5) = 0.3$, $P(X = -0.5) = 0.1$
3-27. $P(X = 0) = 0.00001$, $P(X = 1) = 0.00167$, $P(X = 2) = 0.07663$, $P(X = 3) = 0.92169$

Section 3-3

3-29. (a) 7/8 (b) 1 (c) 3/4 (d) 3/8

3-31. $$F(x) = \begin{cases} 0, & x < 0 \\ 0.008, & 0 \le x < 1 \\ 0.104, & 1 \le x < 2 \\ 0.488, & 2 \le x < 3 \\ 1, & 3 \le x \end{cases}$$
where $f_X(0) = 0.008$, $f_X(1) = 0.096$, $f_X(2) = 0.384$, $f_X(3) = 0.512$

3-33. $$F(x) = \begin{cases} 0, & x < 10 \\ 0.2, & 10 \le x < 25 \\ 0.5, & 25 \le x < 50 \\ 1, & 50 \le x \end{cases}$$
where $P(X = 50 \text{ million}) = 0.5$, $P(X = 25 \text{ million}) = 0.3$, $P(X = 10 \text{ million}) = 0.2$

3-35. (a) 1 (b) 0.5 (c) 0.5 (d) 0.5

3-37. (a) 1 (b) 0.75 (c) 0.25 (d) 0.25 (e) 0 (f) 0

Section 3-4

3-39. $\mu = 2, \sigma^2 = 2$
3-41. $\mu = 0, \sigma^2 = 1.5$
3-43. $\mu = 2.8, \sigma^2 = 1.36$
3-45. $\mu = 1.57, \sigma^2 = 0.311$
3-47. 24
3-49. $\mu = 0.0004, \sigma^2 = 0.00039996$
3-51. (a) $\mu = 18.694$, $\sigma^2 = 735.9644$, $\sigma = 27.1287$ (b) $\mu = 37.172$, $\sigma^2 = 2947.996$, $\sigma = 54.2955$

Section 3-5

3-53. $\mu = 2, \sigma^2 = 0.667$
3-55. $\mu = 300, \sigma^2 = 6666.67$
3-57. (a) $\mu = 687.5, \sigma^2 = 56.25$ (b) $\mu = 87.5, \sigma^2 = 56.25$
3-59. $E(X) = 4.5, E(Y) = 22.5$, $\sigma_Y = 14.36$

Section 3-6

3-63. (a) 0.9298 (b) 0 (c) 0.0112 (d) 0.0016
3-65. (a) 2.40×10^{-8} (b) 0.9999 (c) 9.91×10^{-18} (d) 1.138×10^{-4}
3-67. (a) 0 (b) 10
3-71. (a) 0.215 (b) 0.994 (c) 4
3-73. (a) 0.410 (b) 0.218 (c) 0.37
3-75. (a) 1 (b) 0.999997 (c) $E(X) = 12.244$, $V(X) = 2.179$
3-77. (a) Binomial, $p = 10^4/36^9$, $n = $ 1E09 (b) 0 (c) $E(X) = 4593.9$, $V(X) = 4593.9$
3-79. (a) 0.9961 (b) 0.9886

Section 3-7

3-81. (a) 0.5 (b) 0.0625 (c) 0.0039 (d) 0.75 (e) 0.25
3-83. (a) 5 (b) 5
3-85. (a) 0.0064 (b) 0.9984 (c) 0.008
3-87. (a) 0.0167 (b) 0.9039 (c) 50
3-89. (a) 0.13 (b) 0.098 (c) $7.69 \approx 8$
3-91. (a) 3.91×10^{-19} (b) 200 (c) 2.56×10^{18}
3-93. (a) 3×10^8 (b) 3×10^{16}

Section 3-8

3-97. (a) 0.4191 (b) 0 (c) 0.001236 (d) $E(X) = 0.8$, $V(X) = 0.6206$
3-101. (a) 0.1201 (b) 0.8523
3-103. (a) 0.087 (b) 0.9934 (c) 0.297 (d) 0.9998
3-105. (a) 0.7069 (b) 0.0607 (c) 0.2811.

Section 3-9

3-107. (a) 0.0183 (b) 0.2381 (c) 0.1954 (d) 0.0298
3-109. $E(X) = V(X) = 2.996$.
3-111. (a) 0.264 (b) 48
3-113. (a) 0.4566 (b) 0.047
3-115. (a) 0.2 (b) 99.89%
3-117. (a) 0.6065 (b) 0.0067 (c) $P(W = 0) = 0.0067$, $P(W = 1) = 0.0437$, $P(W \leq 1) = 0.0504$

Supplemental Exercises

3-119. $E(X) = 1/4$, $V(X) = 0.0104$
3-121. (a) $n = 50, p = 0.1$ (b) 0.112 (c) $P(X \geq 49) = 4.51 \times 10^{-48}$
3-123. (a) 0.000224 (b) 0.2256 (c) 0.4189
3-125. (a) 0.1024 (b) 0.1074
3-127. (a) 3000 (b) 1731.18
3-129. (a) $P(X = 0) = 0.0498$ (b) 0.5768 (c) $P(X \leq x) \geq 0.9, x = 5$ (d) $\sigma^2 = \lambda = 6$. Not appropriate.
3-131. (a) 0.1877 (b) 0.4148 (c) 15
3-133. (a) 0.0110 (b) 8/3
3-135. 40000
3-137. 0.1330
3-139. (a) 500 (b) 222.49
3-141. 0.1
3-143. $f_X(0) = 0.16, f_X(1) = 0.19$, $f_X(2) = 0.20, f_X(3) = 0.31$, $f_X(4) = 0.14$
3-145. $f_x(2) = 0.2, f_x(5.7) = 0.3$, $f_x(6.5) = 0.3, f_x(8.5) = 0.2$
3-147. (a) 0.0433 (b) 3.58
3-149. (a) $f_X(0) = 0.2357$, $f_X(1) = 0.3971$, $f_X(2) = 0.2647$, $f_X(3) = 0.0873$, $f_X(4) = 0.01424$, $f_X(5) = 0.00092$ (b) $f_X(0) = 0.0546$, $f_X(1) = 0.1866$, $f_X(2) = 0.2837$, $f_X(3) = 0.2528$, $f_X(4) = 0.1463$, $f_X(5) = 0.0574$, $f_X(6) = 0.0155$, $f_X(7) = 0.0028$, $f_X(8) = 0.0003$, $f_X(9) = 0.0000$, $f_X(10) = 0.0000$
3-151. 37.8 seconds

CHAPTER 4

Section 4-2

4-1. (a) 0.3679 (b) 0.2858 (c) 0 (d) 0.9817 (e) 0.0498
4-3. (a) 0.4375 (b) 0.7969 (c) 0.5625 (d) 0.7031 (e) 0.5
4-5. (a) 0.5 (b) 0.4375 (c) 0.125 (d) 0 (e) 1 (f) 0.9655
4-7. (a) 0.5 (b) 49.8
4-9. (a) 0.1

Section 4-3

4-11. (a) 0.56 (b) 0.7 (c) 0 (d) 0
4-13. $F(x) = 0$ for $x < 0$; $0.25x^2$ for $0 \leq x < 2$; 1 for $2 \leq x$
4-15. $F(x) = 0$ for $x \leq 0$; $1 - e^{-x}$ for $x > 0$
4-17. $F(x) = 0$ for $x \leq 4$; $1 - e^{-(x-4)}$ for $x > 4$
4-19. 0.2
4-21. $f(x) = 0.2$ for $0 < x < 4$; $f(x) = 0.04$ for $4 \leq x < 9$

Section 4-4

4-23. $E(X) = 2, V(X) = 4/3$
4-25. $E(X) = 0, V(X) = 0.6$
4-27. $E(X) = 2$
4-29. (a) $E(X) = 109.39, V(X) = 33.19$ (b) 54.70
4-31. (a) $E(X) = 5.1, V(X) = 0.01$ (b) 0.3679

Section 4-5

4-33. (a) $E(X) = 0$, $V(X) = 0.577$
(b) 0.90 (c) $F(x) = 0$ for $x < -1$; $0.5x + 0.5$ for $-1 \le x < 1$; 1 for $1 \le x$

4-35. (a) $F(x) = 0$ for $x < 0.95$; $10x - 9.5$ for $0.95 \le x < 1.05$; 1 for $1.05 \le x$
(b) 0.3 (c) 0.96
(d) $E(X) = 1.00$, $V(X) = 0.00083$

4-37. (a) $F(x) = 0$ for $x < 0.2050$; $100x - 20.50$ for $0.2050 \le x < 0.2150$; 1 for $0.21.50 \le x$
(b) 0.25 (c) 0.2140
(d) $E(X) = 0.2100$, $V(X) = 8.33 \times 10^{-6}$

4-39. (a) $F(X) = x/90$ for $0 \le x \le 90$
(b) $E(X) = 45$, $V(X) = 675$
(c) 0.667 (d) 0.333

Section 4-6

4-41. (a) 0.90658 (b) 0.99865
(c) 0.07353 (d) 0.98422
(e) 0.95116

4-43. (a) 0.90 (b) 0.5 (c) 1.28
(d) −1.28 (e) 1.33

4-45. (a) 0.93319 (b) 0.69146
(c) 0.9545 (d) 0.00132
(e) 0.15866

4-47. (a) 0.93319 (b) 0.89435
(c) 0.38292 (d) 0.80128
(e) 0.54674

4-49. (a) 0.99379 (b) 0.13591
(c) 5835

4-51. (a) 0.0228 (b) 0.019
(c) 152.028
(d) small (less than 5%)

4-53. (a) 0.0082 (b) 0.72109
(c) 0.564

4-55. (a) 12.309 (b) 12.1545

4-57. (a) 0.00621 (b) 0.308538
(c) 133.33

4-59. (a) 0.1587 (b) 1.3936
(c) 0.9545

4-61. (a) 0.00043 (b) 6016
(c) 1/8

4-63. (a) 0.02275 (b) 0.324
(c) 11.455

Section 4-7

4-65. (a) 0.0853 (b) 0.8293
(c) 0.0575

4-67. (a) 0.1908 (b) 0.5
(c) 0.3829

4-69. (a) 0.2743 (b) 0.8417

4-71. 0.022

4-73. 0.498

4-75. (a) 0 (b) 0.156
(c) 10,233
(d) 8.1 days/year
(e) 0.0071

Section 4-8

4-77. (a) 0.3679 (b) 0.1353
(c) 0.9502
(d) 0.95, $x = 29.96$

4-79. (a) 0.3679 (b) 0.2835
(c) 0.1170

4-81. (a) 0.1353 (b) 0.4866
(c) 0.2031 (d) 34.54

4-83. (a) 0.0498 (b) 0.8775

4-85. (a) 0.0025 (b) 0.6321
(c) 23.03 (d) same as part (c)
(e) 6.93

4-87. (a) 15.625 (b) 0.1629
(c) 3×10^{-6}

4-89. (a) 0.2212 (b) 0.2865
(c) 0.2212 (d) 0.9179
(e) 0.2337

4-91. (a) 0.3528 (b) 0.04979
(c) 46.05 (d) 6.14×10^{-6}
(e) e^{-12} (f) same

4-93. (a) 0.3679 (b) 0.1353
(c) 0.0498
(d) not depend on θ

Section 4-9

4-95. (a) 120 (b) 1.32934
(c) 11.6317

4-97. (a) Erlang $\lambda = 5$ calls/min, $r = 10$ (b) $E(X) = 2$, $V(X) = 0.4$
(c) 0.2 minute (d) 0.1755
(e) 0.2643

4-99. (a) 50000 (b) 0.6767

4-101. (a) 5×10^5 (b) $V(X) = 5 \times 10^{10}$, $\sigma = 223607$
(c) 0.0803

4-103. (a) 0.1429 (b) 0.1847

Section 4-10

4-107. $E(X) = 12000$, $V(X) = 3.61 \times 10^{10}$

4-109. 1000

4-111. (a) 803.68 hours
(b) 85319.64 (c) 0.1576

4-113. (a) 443.11 (b) 53650.5
(c) 0.2212

Section 4-11

4-117. (a) 0.0016 (b) 0.0029
(c) $E(X) = 12.1825$, $V(X) = 1202455.87$

4-119. (a) 0.03593 (b) 1.65
(c) 2.7183
12.6965

4-121. (a) $\theta = 8.4056$, $\omega^2 = 1.6094$
(b) 0.2643 (c) 881.65

Supplemental Exercises

4-125. (a) 0.99379 (b) 0.621%

4-127. (a) 0.15866 (b) 90.0
(c) 0.9973 (d) $(0.9973)^{10}$
(e) 9.973

4-129. (a) 0.0217 (b) 0.9566
(c) 229.5

4-131. 0.8488

4-133. (a) 620.4 (b) 105154.9
(c) 0.4559

4-135. (a) 0.0625 (b) 0.75 (c) 0.5
(d) $F(x) = 0$ for $x < 2$; $x/4 - x + 1$ for $2 \le x < 4$; 1 for $4 \le x$
(e) $E(X) = 10/3$, $V(X) = 0.2222$

4-137. (a) 0.3935 (b) 0.9933

4-139. (a) $\theta = 3.43$, $\omega^2 = 0.96$
(b) 0.946301

4-141. (a) 0.6915 (b) 0.683
(c) 1.86

4-143. (a) 0.0062 (b) 0.012
(c) 5.33

4-145. 0.0008 to 0.0032

4-147. (a) 0.5633 (b) 737.5

4-149. (a) 0.9906 (b) 0.8790

Mind-Expanding Exercises

4-153. (a) $k\sigma^2$ (b) $k\sigma^2 + k(\mu - m)^2$

CHAPTER 5

Section 5-1

5-1. (a) 3/8 (b) 5/8 (c) 3/8
(d) 1/8
(e) $V(X) = 0.4961$ $V(Y) = 1.8594$
(f) $f(1) = 1/4$, $f(1.5) = 3/8$, $f(2.5) = 1/4$, $f(3) = 1/8$
(g) $f(2) = 1/3$, $f(3) = 2/3$
(h) 1 (i) 2 1/3
(j) Not independent

5-3. (a) 3/8 (b) 3/8 (c) 7/8
(d) 5/8 (e) $V(X) = 0.4219$ $V(Y) = 1.6875$

(f) $f(-1) = 1/8, f(-0.5) = 1/4,$ $f(0.5) = 1/2, f(1) = 1/8$
(g) 1 (h) 1 (i) 0.5
(j) Not independent

5-5. (b) $f_X(0) = 0.970299,$ $f_X(1) = 0.029403,$ $f_X(2) = 0.000297,$ $f_X(3) = 0.000001$
(c) 0.03 (d) $f(0) = 0.920824,$ $f(1) = 0.077543, f(2) =$ 0.001632 (e) 0.080807
(g) Not independent

5-7. (b) $f(0) = 2.40 \times 10^{-9},$ $f(1) = 1.36 \times 10^{-6},$ $f(2) = 2.899 \times 10^{-4},$ $f(3) = 0.0274, f(4) = 0.972$
(c) 3.972 (d) equals $f(y)$
(e) 3.988 (f) 0.0120
(g) Independent

5-9. (b) $f_X(0) = 0.2511,$ $f_X(1) = 0.0405,$ $f_X(2) = 0.0063,$ $f_X(3) = 0.0009,$ $f_X(4) = 0.0001$
(c) 0.0562
(d) $f_{Y|3}(0) = 2/3, f_{Y|3}(1) = 1/3,$ $f_{Y|3}(2) = f_{Y|3}(3) = f_{Y|3}(4) = 0$
(e) 0.0003 (f) 0.0741
(g) Not independent

5-11. (a) $p_1 = 0.05, p_2 = 0.85,$ $p_3 = 0.10$
(d) $E(X) = 1, V(X) = 0.95$
(f) 0.07195
(g) 0.7358
(h) $E(Y) = 17$
(i) 0
(j) $P(X = 2, Y = 17) =$ $0.0540, P(X = 2 \mid Y = 17)$ $= 0.2224$
(k) $E(X \mid Y = 17) = 1$

5-13. (b) 0.1944 (c) 0.0001
(e) $E(X) = 2.4$
(f) $E(Y) = 1.2$ (g) 0.7347
(h) 0
(i) $P(X = 0 \mid Y = 2) = 0.0204,$ $P(X = 1 \mid Y = 2) = 0.2449,$ $P(X = 2 \mid Y = 2) = 0.7347$
(j) 1.7143

5-15. (c) 0.308 (d) 5.7

Section 5-2

5-17. (a) 0.4444 (b) 0.6944
(c) 0.5833 (d) 0.3733
(e) 2 (f) 0
(g) $f_X(x) = 2x/9; 0 < x < 3$
(h) $f_{Y|1.5}(y) = 2y/9; 0 < y < 3$
(i) 2 (j) 4/9
(k) $f_{X|2}(x) = 2x/9; 0 < x < 3$

5-19. (a) 1/81 (b) 5/27
(c) 0.01235 (d) 16/81
(e) 12/5 (f) 8/5
(g) $f(x) = 4x^3/81; 0 < x < 3$
(h) $f_{Y|X=1}(y) = 2y; 0 < y < 1$
(i) 1 (j) 0
(k) $f_{X|Y=2}(x) = 2x/9; 0 < x < 3$

5-21. (a) 0.9879 (b) 0.0067
(c) 0.000308 (d) 0.9939
(e) 0.04 (f) 8/15
(g) $f(x) = 5e^{-5x}; x > 0$
(h) $f_{Y|X=1}(y) = 3e^{3-3y}; 1 < y$
(i) 4/3
(j) $0.9502, f_Y(2) = 15e^{-6}/2;$ $0 < y$
(k) $f_{X|Y=2}(y) = 2e^{-2x}; 0 < x < 2$

5-23. (a) 1/30 (b) 1/12 (c) 19/9
(d) 97/45 (g) 1 (h) 0.25

5-25. (a) $P(X > 5, Y > 5) = 0.0439,$ $P(X > 10, Y > 10) = 0.0019$
(b) 0.0655

5-27. (a) 0.25 (b) 0.0625 (c) 1
(d) 1 (e) 2/3 (f) 0.25
(g) 0.0625
(h) $f_{X|YZ}(y, z) = 2x; 0 < x < 1$
(i) 0.25

5-29. (a) 0.75 (b) 3/4 (c) 0.875
(d) 0.25 (g) 1 for $x > 0$

5-31. (a) 0.032 (b) 0.0267

Section 5-3

5-33. 0.8851
5-35. $c = 1/36, \rho = -0.0435$
5-37. $\rho = -0.5$, negative
5-39. $c = 8/81, \rho = 0.4924$
5-41. $\sigma_{XY} = \rho_{XY} = 0$
5-43. $\sigma_{XY} = \rho_{XY} = 0$

Section 5-4

5-47. (a) 0.7887 (b) 0.7887
(c) 0.6220
5-49. 0.8270
5-55. (a) 18 (b) 77 (c) 0.5
(d) 0.873

Section 5-5

5-57. (a) $E(T) = 4, \sigma_T = 0.1414$
5-59. (a) 0 (b) 1
5-61. $E(X) = 1290, V(X) = 19600$
5-63. (a) 0.002 (b) $n = 6$
(c) 0.9612
5-65. (a) 0.0027 (b) No (c) 0

Section 5-6

5-67. $f_Y(y) = ¼; y = 3, 5, 7, 9$
5-69. (b) 18

Supplemental Exercises

5-75. (a) 3/8 (b) 3/4 (c) 3/4
(d) 3/8
(e) $E(X) = 7/8, V(X) = 39/64,$ $E(Y) = 7/8, V(Y) = 39/64$
(h) 2/3 (i) Not independent
(j) 0.7949

5-77. (a) 0.0560 (b) Z~Bin(20, 0.1)
(c) 2 (d) 0.863
(g) Not independent (f) 1

5-79. (a) 1/108 (b) 0.5787
(c) 3/4 (d) 0.2199
(e) 9/4 (f) 4/3

5-81. 3/4

5-83. (a) 0.085 (b) Z~Bin(10, 0.3)
(c) 3

5-85. (a) 0.499 (b) 0.5

5-87. (a) 0.057 (b) 0.057

5-91. (a) $E(T) = 1.5, V(T) = 0.078$
(b) 0.0367
(c) $E(P) = 4, V(P) = 0.568$

CHAPTER 6

Section 6-1

6-1. $\bar{x} = 74.0044, s = 0.00473$
6-3. $\bar{x} = 7068.1, s = 226.5$
6-5. $\bar{x} = 43.975, s = 12.294$
6-7. $\mu = 5.44$
6-9. $\bar{x}_1 = 287.89, s_1 = 160.154$ $\bar{x}_2 = 325.01, s_2 = 121.20$
6-11. $\bar{x} = 7.184, s = 0.02066$
6-13. (a) $\bar{x} = 65.86, s = 12.16$
(b) $\bar{x} = 66.86, s = 10.74$

Section 6-2

6-15. $\tilde{x} = 1436.5$, lower quartile: $Q_1 = 1097.8$, and upper quartile: $Q_3 = 1735.0$
6-17. $\tilde{x} = 89.250$, lower quartile: $Q_1 = 86.100$, and upper quartile: $Q_3 = 93.125$
6-19. median: $\tilde{x} = 1436.5$, mode: 1102, 1315, and 1750, mean: $\bar{x} = 1403.7$
6-21. $\bar{x} = 366.57, s = 940.02,$ and $\tilde{x} = 41.455$

6-23. 95th percentile = 5479
6-25. $\bar{x} = 260.3$, $s = 13.41$, $\tilde{x} = 260.85$, and 90th percentile $= 277.2$
6-27. $\bar{x} = 89.45$, $s = 2.8$, $\tilde{x} = 90$, and proportion $= 22/40 = 55\%$

Section 6-4

6-45. (a) $\bar{x} = 2.415$, $s^2 = 0.285$, and $s = 0.543$
6-47. (a) $\bar{x} = 952.44$, $s^2 = 9.55$, and $s = 3.09$
(b) $\tilde{x} = 953$
6-49. (a) $\tilde{x} = 67.50$, lower quartile: $Q_1 = 58.50$, and upper quartile: $Q_3 = 75.00$
(b) $\tilde{x} = 68.00$, lower quartile: $Q_1 = 60.00$, and upper quartile: $Q_3 = 75.00$

Supplemental Exercises

6-79. (a) Sample 1 Range = 4
Sample 2 Range = 4
(b) Sample 1: $s = 1.604$
Sample 2: $s = 1.852$
6-85. (b) $\bar{x} = 9.325$, $s = 4.4858$

Mind-Expanding Exercises

6-95. $s^2\ (old) = 50.61$, $s^2\ (new) = 5061.1$
6-99. $\bar{y} = 431.89$, $s_y^2 = 34.028$
6-101. $\tilde{x} = 69$
6-103. (a) $\bar{x} = 89.29$ (b) $\bar{x} = 89.19$

CHAPTER 7

Section 7-2

7-1. 0.8186
7-3. 0.4306
7-5. 0.191
7-7. $n = 12$
7-9. 0.2312
7-11. (a) 0.5885 (b) 0.1759
7-13. 0.983
7-15. $V(\hat{\theta}_1) = \sigma^2/7$ is smallest
7-21. Bias $= \sigma^2/n$
7-23. (a) 423.33 (b) 9.08 (c) 1.85 (d) 424 (e) 0.2917
7-27. (d) 0.01 (e) 0.0413

Section 7-3

7-43. (a) 5.046 (b) 5.05

Section 7-4

7-29. $\bar{x}$
7-33. unbiased
7-35. (a) $\hat{\theta} = \Sigma x_i^2/(2n)$
(b) same as part (a)
7-41. (b) $\lambda_0(m + x + 1)/(m + \lambda_0 + 1)$

Supplemental Exercises

7-49. 0.8664
7-51. 5.6569
7-53. $n = 100$
7-55. $\hat{\theta} = \bar{x}/3$

CHAPTER 8

Section 8-2

8-1. (a) 96.76% (b) 98.72% (c) 93.56%
8-3. (a) 1.29 (b) 1.65 (c) 2.33
8-5. (a) 4 (b) 7
8-7. (a) Longer (b) No (c) Yes
8-9. [87.85, 93.11]
8-11. (a) [74.0353, 74.0367]
(b) [1005, ∞)
8-13. (a) [3232.11.3267.89]
(b) [3226.4, 3273.6]
8-15. 267
8-17. 2^2
8-19. (a) [13.383, 14.157]
(b) [13.521, ∞)
(c) 1 (d) 2

Section 8-3

8-21. (a) 2.179 (b) 2.064 (c) 3.012 (d) 4.073
8-23. [58197.33, 62082.07]
8-25. [1.094, 1.106]
8-27. (−∞, 125.312]
8-29. [443.520, 528.080]
8-31. [301.06, 333.34]
8-33. (b) [2237.3, 2282.5]
(c) [2241.4, ∞)
8-35. (b) [4.051, 4.575]
(c) [4.099, ∞)
8-37. (b) [2.813, 2.991]

Section 8-4

8-39. [0.00003075, ∞)
8-41. [0.31, 0.46]
8-43. [0.626, 1.926]
8-45. [0.0061. 0.0131]

Section 8-5

8-47. (a) [0.02029, 0.06637]
(b) [−∞, 0.0627]
8-49. (a) [0.501, 0.571]
(b) [0.506, ∞)
8-51. (a) [0.225, 0.575] (b) 2305
(c) 2401
8-53. 666

Section 8-7

8-55. [52131.1, 68148.3]
8-57. [1.068, 1.13]
8-59. [292.049, 679.551]
8-61. [263.7, 370.7]
8-63. [2193.5, 2326.5]
8-65. 90% PI = [2.71, 3.09]
90% CI = [2.85, 2.95]
90% CI = [2.81, 2.99]
8-67. [49555.54, 70723.86]
8-69. [1.06, 1.14]
8-71. TI = [237.18, 734.42]
CI = [443.42, 528.08]
8-73. TI = [247.60, 386.60]
CI = [301.06, 333.34]
8-75. TI = [2.49, 3.31]
CI = [2.84, 2.96]

Supplemental Exercises

8-79. (a) 0.0997 and 0.064
(b) 0.044 and 0.014
(c) 0.0051 and < 0.001
8-81. (a) Normality
(b) [16.99, ∞)
(c) [16.99, 33.25]
(d) (−∞, 343.74]
(e) [28.23, 343.74]
(f) $16.91 < \mu < 29.09$
$15.85 < \sigma^2 < 192.97$
(g) mean: [16.88, 33.12], variance: [28.16, 342.94]
8-83. (a) [13.74, 16.92]
(b) [13.24, 17.42]
8-85. (a) Yes
(b) [197.84, 208.56]
(c) [185.41, 220.99]
(d) [171.21, 235.19]
8-87. [0.0956, ∞)
8-89. (a) Yes
(b) [1.501, 1.557]
(c) [1370, 1.688]
(d) [1.339, 1.719]
8-91. (a) [0.0004505, 0.009549]
(b) 518
(c) 26002
8-93. (a) Normality
(c) [18.478, 26.982]
(e) [19.565, 123.289]

Mind-Expanding Exercises

8-95. (b) [28.62, 101.98]

8-97. (a) 46 (b) [10.19, 10.41], $p = 0.6004$

8-99. 950 of CIs and 0.9963

CHAPTER 9

Section 9-1

9-1. (a) Yes (b) No (c) No (d) No (e) No

9-3. (a) $H_0{:}\mu = 20$nm $H_1{:}\mu < 20$nm (b) No

9-5. (a) $\alpha = 0.02275.$ (b) $\beta = 0.15866$ (c) $\beta = 0.5$

9-7. (a) $11.4175 \le \overline{X}_c \le 11.42$ (b) $11.5875 \le \overline{X}_c \le 11.59$ (c) $11.7087 \le \overline{X}_c \le 11.71$ (d) $11.7937 \le \overline{X}_c \le 11.84$

9-9. (a) P-value $= 0.0135$ (b) P-value ≤ 0.000034 (c) P-value $= 0.158655$

9-11. (a) 0.09296 (b) $\beta = 0.04648$ (c) $\beta = 0.00005$

9-13. (a) $\beta \cong 0.005543$ (b) $\beta \cong 0.082264$ (c) As n increases, β decreases

9-15. (a) $\alpha = 0.05705$ (b) $\beta = 0.5$ (c) $\beta = 0.05705.$

9-17. (a) $\mu \le 191.40$ (b) $\mu \le 185.37$ (c) $\mu \le 186.6$ (d) $\mu \le 183.2$

9-19. (a) P-value $= 0.2148$ (b) P-value $= 0.008894$ (c) P-value $= 0.785236$

9-21. (a) $\alpha = 0.0164$ (b) $1 - \beta = 0.21186$ (c) α will increase and the power will increase with increased sample size.

9-23. (a) P-value $= 0.238$ (b) P-value $= 0.0007$ (c) P-value $= 0.2585$

9-25. (a) $\alpha = 0.29372$ (b) $\beta = 0.25721$

9-27. (a) $\alpha \approx 0$ (b) $\beta = 0.99506$

Section 9-2

9-29. (a) $H_0: \mu = 10, H_1: \mu > 10$ (b) $H_0: \mu = 7, H_1: \mu \ne 7$ (c) $H_0: \mu = 5, H_1: \mu < 5$

9-31. (a) $a = z_\alpha \cong -2.33$ (b) $a = z_\alpha \cong -1.64$ (c) $a = z_\alpha \cong -1.29$

9-33. (a) P-value $\cong 0.04$ (b) P-value $\cong 0.066$ (c) P-value $\cong 0.69$

9-35. (a) P-value $\cong 0.98$ (b) P-value $\cong 0.03$ (c) P-value $\cong 0.65$

9-37. (a) $z_0 = -0.95 > -1.96$, do not reject H_0 (b) $\beta = 0.80939$ (c) $n \cong 16$

9-39. (a) $z_0 = 1.26 < 1.65$ do not reject H_0 (b) P-value $= 0.1038$ (c) $\beta \cong 0.000325$ (d) $n \cong 1$ (e) $39.85 \le \mu$

9-41. (a) $z_0 = 1.56 > -1.65$, do not reject H_0 (b) P-value $\cong 0.94$ (c) Power $= 0.97062$ (d) $n \cong 5$ (e) $\mu \le 104.53$

9-43. (a) $z_0 = 1.77 > 1.65$, reject H_0 (b) P-value $\cong 0.04$ (c) Power $= 1$ (d) $n \cong 2$ (e) $4.003 \le \mu$

9-45. (a) critical value > 2.539 (b) critical value > 1.796 (c) critical value > 1.345

9-47. (a) $0.05 \le p \le 0.1$ (b) $0.05 \le p \le 0.1$ (c) $0.5 \le p \le 0.8$

9-49. (a) $0.95 \le p \le 0.975$ (b) $0.025 \le p \le 0.05$ (c) $0.6 \le p \le 0.75$

Section 9-3

9-51. (a) $|t_0| = 3.48 > 2.064$ reject; H_0 P-value $= 0.002$ (b) Yes (c) power $\cong 1$ (d) $n = 20$ (e) $98.065 \le \mu \le 98.463$

9-53. (a) $|t_0| = 1.456 < 2.064$, do not reject H_0; $0.1 < P$-value $< 0.2.$ (b) Yes (c) power $= 0.80$ (d) $n = 100$ (e) $129.406 \le \mu \le 130.100$

9-55. (a) $|t_0| = 1.55 < 2.861$, do not reject H_0; $0.10 < P$-value < 0.20 (b) Yes, see normal probability plot (c) power $= 0.30$ (d) $n = 40$ (e) $1.9 \le \mu \le 4.62$

9-57. (a) $t_0 = 0.15 < 1.753.$, do not reject H_0; P-value $> 0.40.$ (b) $n = 4.$ Yes

9-59. $t_0 = 3.46 > 1.833$, reject H_0 $0.0025 < P$-value < 0.005

9-61. (a) $t_0 = -14.69 < 1.6604$, do not reject H_0; P-value > 0.995 (b) Yes (c) power $= 1$ (d) $n = 15$

Section 9-4

9-63. (a) critical values 6.84 and 38.58 (b) critical values 3.82 and 21.92 (c) critical values 6.57 and 23.68

9-65. (a) $\chi^2_{1-\alpha,n-1} = 7.63$ (b) $\chi^2_{1-\alpha,n-1} = 4.57$ (c) $\chi^2_{1-\alpha,n-1} = 7.79$

9-67. (a) $0.5 < P$-value < 0.9 (b) $0.5 < P$-value < 0.9 (c) $0.005 < P$-value < 0.01

9-69. (a) $\chi^2_0 = 0.23 < 26.30$ do not reject H_0; P-value > 0.995 (b) $0.07 \le \sigma$

9-71. (a) $\chi^2_0 = 109.52 > 79.49$, reject H_0; P-value < 0.01 (b) $0.31 < \sigma < 0.46$

9-73. (a) $\chi^2_0 = 12.46 > 7.26$, do not reject H_0; $0.1 < P$-value < 0.4 (b) $\sigma \le 5240$

9-75. (a) $\chi^2_0 = 11.52 < 19.02$ do not reject H_0; $0.2 < P$-value (b) $n = 30$ (c) $10.90 \le \sigma^2 \le 76.80$

Section 9-5

9-77. (a) $z_0 = -1.31 > -1.65$, do not reject H_0; P-value $= 0.095$ (b) $p \le 0.0303$

9-79. (a) $z_0 = 2.06 > 1.65$, reject H_0; P-value $= 0.0196$ (b) $0.7926 \le p$

9-81. (a) $z_0 = -0.94 < 2.33$ do not reject H_0; P-value $= 0.826$
(b) $0.035 \le p$

9-83. $z_0 = 1.58 > -2.33$ do not reject H_0

9-85. (a) $z_0 = 0.54 < 1.65$, do not reject H_0; P-value $= 0.295$
(b) $\beta = 0.639$, $n \cong 118$

Section 9-7

9-87. (a) $\chi_0^2 = 6.955 < 15.09$ do not reject H_0
(b) P-value $= 0.2237$ (from Minitab)

9-89. (a) $\chi_0^2 = 10.39 > 7.81$ reject H_0
(b) P-value $= 0.0155$

9-91. (a) $\chi_0^2 = 769.57 >>> 36.42$, reject H_0
(b) P-value $= 0$

Section 9-8

9-93. $\chi_0^2 \not> \chi_{0.05,6}^2$, do not reject H_0; P-value $= 0.070$ (from Minitab)

9-95. $\chi_0^2 > \chi_{0.01,9}^2$ reject H_0; P-value $= 0.002$

9-97. $\chi_0^2 \not> \chi_{0.01,3}^2$, do not reject H_0; P-value $= 0.013$

9-99. (a) $\chi_0^2 > \chi_{0.05,1}^2$
(b) P-value < 0.005

Supplemental Exercises

9-101. (a) 50, Normal, p, $\dfrac{p(1-p)}{50}$
(b) 80, Normal, p, $\dfrac{p(1-p)}{80}$
(c) 100, Normal, p, $\dfrac{p(1-p)}{100}$
(d) As the sample size increases, the variance of the sampling distribution decreases.

9-103. (a) $\beta = 0.564$
(b) $\beta = 0.161$
(c) $\beta = 0.0116$
(d) β, decreases as the sample size increases; larger n

9-105. (a) $\beta = 0.38974$
power $= 1 - \beta = 0.61026$
(b) $\beta = 0.00494$
power $= 1 - \beta = 0.995$
(c) $\beta = 0.00118$
power $= 1 - \beta = 0.9988$
(d) β decreases, power increases

9-107. (a) Place what we are trying to demonstrate in the alternative hypothesis.
(b) $0.05 < P$-value < 0.10

9-109. (a) $\chi_0^2 = 1.75 > 1.24$, do not reject H_0
(b) 0.00001 could have occurred as a result of sampling variation.

9-111. (a) $\chi_0^2 = 17.929 > 14.07$ reject H_0
(b) P-value $= 0.0123$

9-113. $\chi_o^2 > \chi_{0.05,5}^2$, reject normality

9-115. (a) $z_0 = -7.23 < -1.65$, reject H_0
(b) P-value $\cong 0$
(c) $\chi_o^2 = 12 > \chi_{0.05,5}^2 = 11.07$ reject normality

9-117. (b) $t_0 = 1.608 < 2.093$, do not reject H_0
(c) $0.1 < P$-value < 0.2
(d) $2.270 \le \mu \le 4.260$

9-119. (a) $t_0 = 0.667 < 2.059$, do not reject H_0
(b) $0.5 < P$-value < 0.8
(c) $246.84 \le \mu \le 404.15$

9-121. (a) $2.6 < \chi_0^2 = 10.53 < 26.76$ do not reject H_0
(b) $0.2 < P$-value
(c) $0.236 \le \sigma^2 \le 2.43$

9-123. (a) $t_0 = 0.47 < 2.228$, do not reject H_0
(b) $0.5 < P$-value < 0.8
(c) n should be at least 50
(d) $\beta \approx 0.1$

9-125. (a) $0.1 < P$-value < 0.5, do not reject H_0
(b) $0.01 < P$-value < 0.025, reject H_0

Mind-Expanding Exercises

9-129. (a) $z_0 = 4.82 > -1.65$, do not reject H_0
(c) $\hat{\theta} = 282.36$, $z_0 = -1.28 > -1.65$, do not reject H_0

CHAPTER 10

Section 10-2

10-1. (a) $-1.96 < z_0 = -0.9 < 1.96$, do not reject H_0; P-value $= 0.368$
(b) $-9.79 \le \mu_1 - \mu_2 \le 3.59$
(c) Power $= 0.14$
(d) $n_1 = n_2 = 180$

10-3. (a) $z_0 = 0.937 < 2.325$, do not reject H_0; P-value $= 0.174$
(b) $\mu_1 - \mu_2 \ge -4.74$
(c) Power $= 0.04$
(d) Use $n_1 = n_2 = 339$

10-5. (a) $z_0 = -5.84 < 1.645$, do not reject H_0; P-value $= 1$
(b) $\mu_1 - \mu_2 \ge 6.8$
(c) Power $= 0.9988$
(d) Sample size is adequate

10-7. (a) $z_0 = -7.25 < -1.645$ reject H_0; P-value $\cong 0$
(b) $-3.684 \le \mu_1 - \mu_2 \le -2.116$
(c) $n_1 = n_2 = 11$

10-9. (a) $-5.83 \le \mu_1 - \mu_2 \le -0.57$; P-value $= 0.0173$
(b) Yes
(c) Power $= 0.9616$; $n \cong 10$
(d) Normal
(e) Normal

10-11. (a) $t_0 = -1.94 < -1.701$ $0.025 < P$-value < 0.05
(b) $\mu_1 - \mu_2 \le -0.196$
(c) Power $= 0.95$
(d) $n = n_1 = n_2 = 21$

Section 10-3

10-13. (a) $-2.042 < t_0 = 0.230 < 2.042$, do not reject H_0; P-value > 0.80
(b) $-0.394 \le \mu_1 - \mu_2 \le 0.494$

10-15. (a) $t_0 = -3.11 < -2.485$, reject H_0
(b) $-5.688 \le \mu_1 - \mu_2 \le -0.3122$

10-17. (a) Assumptions verified
(b) $t_0 = -2.83 < -2.101$ reject H_0; $0.010 < P$-value < 0.020
(c) $-0.7495 \le \mu_1 - \mu_2 \le -0.1105$

10-19. (a) $t_0 = -5.498 < -2.021$ reject H_0; P-value < 0.0010
(b) $n_1 = n_2 = 38$

10-21. (a) $t_0 = 3.03 > 2.056$ reject H_0; $0.005 < P$-value < 0.010
(b) $t_0 = 3.03 > 1.706$, reject H_0

10-23. (a) $t_0 = 7.0 > 2.048$, reject H_0; P-value $\cong 0$

(b) $14.93 \leq \mu_1 - \mu_2 \leq 27.28$
(c) $n > 8$

10-25. (a) $t_0 = 2.82 > 2.326$ reject H_0; P-value $\cong 0.025$
(b) $\mu_1 - \mu_2 \geq 0.178$

10-27. (a) Normal
(b) $t_0 = 2.558 > 2.101$ reject H_0; P-value ≈ 0.02
(c) $1.86 \leq \mu_1 - \mu_2 \leq 18.94$
(d) Power = 0.05
(e) $n = 51$

Section 10-4

10-29. (a) $0.1699 \leq \mu_d \leq 0.3776$
(b) t-test is appropriate.

10-31. $-727.46 \leq \mu_d \leq 2464.21$

10-33. (a) $t_0 = 5.465 > 1.761$ reject H_0; P-value $\cong 0$
(b) $18.20 \leq \mu_d$

10-35. (a) $t_0 = 8.387 > 1.833$ reject H_0
(b) $t_0 = 3.45 > 1.833$ reject H_0
(c) Yes

10.37. (a) Normal
(b) $-0.379 \leq \mu_d \leq 0.349$
(c) $6 \leq n$

Section 10-5

10-39. (a) $f_{0.25,5,10} = 1.59$
(b) $f_{0.10,24,9} = 2.28$
(c) $f_{0.05,8,15} = 2.64$
(d) $f_{0.75,5,10} = 0.529$
(e) $f_{0.90,24,9} = 0.525$
(f) $f_{0.95,8,15} = 0.311$

10-41. $f_0 = 0.805 > 0.166$, do not reject H_0;
$0.22 \leq \frac{\sigma_1^2}{\sigma_2^2}$

10-43. (a) $f_0 = 1.21 > 0.333$, do not reject H_0;
$0.403 \leq \frac{\sigma_1^2}{\sigma_2^2} \leq 3.63$
(b) Power = 0.65
(c) $n \approx 31$

10-45. (a) $f_0 = 0.923 > 0.365$, do not reject H_0
(b) $0.3369 \leq \frac{\sigma_1^2}{\sigma_2^2} \leq 2.640$

10-47. (a) $0.6004 \leq \frac{\sigma_1}{\sigma_2} \leq 1.428$
(b) $0.5468 \leq \frac{\sigma_1}{\sigma_2} \leq 1.5710$
(c) $0.661 \leq \frac{\sigma_1}{\sigma_2}$

10-49. $0.4058 < f_0 = 1.78 < 2.46$, do not reject H_0

10-51. $0.248 < f_0 = 0.640 < 4.04$, do not reject H_0;
$0.159 \leq \frac{\sigma_1^2}{\sigma_2^2} \leq 2.579$

10-53. $0.333 < f_0 = 1.35 < 3$, do not reject H_0;
$0.45 \leq \frac{\sigma_1^2}{\sigma_2^2} \leq 4.05$

10-55. (a) $0.248 < f_0 = 3.337 < 4.03$, do not reject H_0
(b) No

Section 10-6

10-57. (a) $z_0 = 4.45 > 1.96$ reject H_0; P-value ≈ 0
(b) $0.039 \leq p_1 - p_2 \leq 0.1$

10-59. (a) $z_0 = 5.36 > 2.58$ reject H_0; P-value = 0

Supplemental Exercises

10-61. (a) normality, equality of variance, and independence of the observations.
(b) $1.40 \leq \mu_1 - \mu_2 \leq 8.36$
(c) Yes
(d) $0.1582 \leq \frac{\sigma_1^2}{\sigma_2^2} \leq 5.157$
(e) No

10-63. (a) $t_0 = 2.554 > 1.895$, reject H_0
(b) $t_0 = 2.554 < 2.998$, do not reject H_0
(c) $t_0 = -1.986 < 1.895$, do not reject H_0
(d) $t_0 = -1.986 < 2.998$, do not reject H_0

10-65. (a) $z_0 = 6.55 > 1.96$, reject H_0
(b) $z_0 = 6.55 > 2.58$, reject H_0
(c) z_0 is so large

10-67. (a) $-0.0335 \leq p_1 - p_2 \leq 0.0329$
(b) $-0.0282 \leq p_1 - p_2 \leq 0.0276$
(c) 95% CI: $-0.0238 \leq p_1 - p_2 \leq 0.0232$
90% CI: $-0.0201 \leq p_1 - p_2 \leq 0.0195$

10-69. (a) Yes
(b) Yes if similar populations

10-71. (a) $0.0987 \leq \mu_1 - \mu_2 \leq 0.2813$
(b) $0.0812 \leq \mu_1 - \mu_2 \leq 0.299$
(c) $\mu_1 - \mu_2 \leq 0.2813$
(d) $z_0 = 3.42 > 1.96$, reject H_0; P-value = 0.00062
(e) $n_1 = n_2 = 9$

10-73. (a) $z_0 = -5.36 < -2.58$, reject H_0
(b) conclusions are the same
(c) $n = 60$

10-75. (a) No
(b) data appear normal with equal variances
(c) It is more apparent the data follow normal distributions.
(d) $18.114 < \frac{\sigma_V^2}{\sigma_M^2} < 294.35$
(e) $f_0 = 72.78 > 4.03$, reject H_0

10-77. (a) Normality appears valid.
(b) $0.50 < P$-value < 0.80, do not reject H_0
(c) $n = 30$

10-79. (a) It may not be assumed that $\sigma_1^2 = \sigma_2^2$
(b) $t_0 = -2.74 < -2.131$, reject H_0
(c) Power = 0.95
(d) $n = 26$

Mind-Expanding Exercises

10-85. (c) $-1.317 \leq \theta \leq 4.157$

CHAPTER 11

Section 11-2

11-1. (a) $\hat{\beta}_0 = 48.013$, $\hat{\beta}_1 = -2.330$
(b) 37.99 (c) 39.39 (d) 6.71

11-3. (a) $\hat{\beta}_0 = -5.56$, $\hat{\beta}_1 = 12.652$, $\sigma^2 = 32.6$
(b) 89.33 (c) 12.652 (d) 0.79
(e) -7.56

11-5. (a) $\hat{\beta}_0 = -6.3355$, $\hat{\beta}_1 = 9.20836$, $\sigma^2 = 3.7746$
(b) 500.124 (c) 9.20836
(d) -1.618

11-7. (a) $\hat{\beta}_0 = -16.5093$, $\hat{\beta}_1 = 0.0693554$, $\sigma^2 = 7.3212$
(b) 1.39592 (c) 49.38

11-9. (b) $\hat{\beta}_0 = 234.071$, $\hat{\beta}_1 = -3.50856$, $\sigma^2 = 398.25$

(c) 128.814
(d) 156.833 and 15.1175

11-11. (b) $\hat{\beta}_0 = 2625.39$, $\hat{\beta}_1 = 36.962$, $\sigma^2 = 9811.2$
(c) 1886.15

11-13. (a) $\hat{\beta}_0 = 0.658$, $\hat{\beta}_1 = 0.178$, $\sigma^2 = 0.083$
(b) 3.328 (c) 0.534
(d) 1.726 and 0.174

11-15. (b) $\hat{\beta}_0 = 2.02$, $\hat{\beta}_1 = 0.0287$, $\sigma^2 = 0.0253$

11-17. (a) $\hat{y} = 39.2 - 0.0025x$
(b) $\hat{\beta}_1 = -0.0025$

11-19. (b) $\hat{\beta}_0^* = 2132.41$, $\hat{\beta}_1^* = -36.9618$

Section 11-4

11-21. (a) $f_0 = 74.63$, P-value $=$ 0.000002, reject H_0
(b) $\hat{\sigma}^2 = 1.8436$, $se(\hat{\beta}_1) = 0.2696$
(c) $se(\hat{\beta}_0) = 0.9043$

11-23. (a) $f_0 = 103.53$, P-value $=$ 0.0000, reject H_0
(b) $\hat{\sigma}^2 = 32.63$, $se(\hat{\beta}_1) = 1.243$ $se(\hat{\beta}_0) = 9.159$
(c) $t_0 = 2.134$, do not reject H_0

11-25. (a) $f_0 = 74334.4$, P-value ≈ 0, reject H_0
(b) $se(\hat{\beta}_0) = 1.66765$, $se(\hat{\beta}_1) = 0.0337744$
(c) $t_0 = -23.37$, P-value $=$ 0.000, reject H_0
(d) $t_0 = -3.8$, P-value $<$ 0.005, reject H_0

11-27. (a) $f_0 = 44.0279$, P-value $=$ 0.00004, reject H_0
(b) $se(\hat{\beta}_0) = 9.84346$, $se(\hat{\beta}_1) = 0.0104524$
(c) $t_0 = -1.67718$, P-value $=$ 0.12166, fail to reject H_0

11-29. (a) $f_0 = 155.2$, P-value $<$ 0.00001, reject H_0
(b) $se(\hat{\beta}_0) = 2.96681$, $se(\hat{\beta}_1) = 45.3468$
(c) $t_0 = -2.3466$, P-value $=$ 0.0306, fail to reject H_0
(d) $t_0 = 57.8957$, P-value $<$ 0.00001, reject H_0
(e) $t_0 = 2.7651$, P-value $=$ 0.0064, reject H_0

11-31. (a) P-Value $=$ 0.0000, reject H_0
(b) $\hat{\sigma}^2 = 0.083$ $se(\hat{\beta}_0) = 0.1657$, $se(\hat{\beta}_1) = 0.014$
(c) Reject H_0

11-33. (a) P-value $=$ 0.310, No
(b) $\hat{\sigma}^2 = 30.69$
(c) $se(\hat{\beta}_0) = 9.141$

11-35. 0.70

Sections 11-5 and 11-6

11-37. (a) $(-2.9713, -1.7423)$
(b) (46.7145, 49.3115)
(c) (41.3293, 43.0477)
(d) (39.1275, 45.2513)

11-39. (a) (10.106, 15.198)
(b) $(-24.316, 13.200)$
(c) $\hat{\mu} = 95.656$, (92.903, 98.409)
(d) (83.642, 107.67)

11-41. (a) (9.10130, 9.31543)
(b) $(-11.6219, -1.04911)$
(c) (498.72024, 501.52776)
(d) (495.57344, 504.67456)

11-43. (a) (0.03689, 0.10183)
(b) $(-47.0877, 14.0691)$
(c) (44.0897, 49.1185)
(d) (37.8298, 55.3784)

11-45. (a) (201.552, 266.590)
(b) $(-4.67015, -2.34696)$
(c) (111.8339, 145.7941)

11-47. (a) $(-43.1964, -30.7272)$
(b) (2530.09, 2720.68)
(c) (1823.7833, 1948.5247)
(d) (1668.9013, 2103.4067)

11-49. (a) (0.1325, 0.2235)
(b) (0.119, 1.196)
(c) (1.87, 2.29)

Section 11-7

11-51. (a) $R^2 = 78.7\%$

11-53. (a) $R^2 = 99.986\%$

11-55. (a) $R^2 = 87.94\%$

11-57. (a) $R^2 = 85.22\%$

11-59. (a) $R^2 = 89.6081\%$
(c) $R^2 = 95.73\%$
(d) $\hat{\sigma}^2$ old $= 9811.21$, $\hat{\sigma}^2$ new $= 4022.93$

11-63. (a) $f_0 = 207$, reject H_0

Section 11-8

11-65. (a) $t_0 = 4.81$, P-value $<$ 0.0005, reject H_0
(b) $z_0 = 1.747$, P-value $=$ 0.04, reject H_0
(c) $\rho \geq 2.26$, reject H_0

11-67. (a) $t_0 = 5.475$, P-value $=$ 0.000, reject H_0
(b) (0.3358, 0.8007)
(c) Yes

11-69. (a) $\hat{y} = -0.0280411 + 0.990987x$
(b) $f_0 = 79.838$, reject H_0
(c) 0.903
(d) $t_0 = 8.9345$, reject H_0
(e) $z_0 = 3.879$, reject H_0
(f) (0.7677, 0.9615)

11-71. (a) $\hat{y} = 5.50 + 6.73x$
(b) 0.948
(c) $t_0 = 8.425$, reject H_0
(d) (0.7898, 0.9879)

11-73. (a) $r = 0.887$
(b) $t_0 = 10.164$, reject H_0, P-value $<$ 0.005
(c) (0.774, 0.945)
(d) $z_0 = 2.808$, reject H_0, P-value $=$ 0.005

Section 11-9

11-75. (a) Yes (b) No (c) Yes
(d) Yes

11-77. (b) $\hat{y} = -0.8819 + 0.00385x$
(c) $f_0 = 122.03$, reject H_0

11-79. (b) $\hat{y} = -0.966824 + 1.54376x$
(c) $f_0 = 2555263.9$, P-value $\approx$ 0, reject H_0
(d) $[-0.97800, -0.95565]$
(e) $t_0 = -1.9934$, reject H_0

Supplemental Exercises

11-81. $y^* = 1.2232 + 0.5075x$ where $y^* = 1/y$

11-83. $\hat{y} = 0.7916x$

11-85. (b) $\hat{y} = 0.6714 + -2964x$
(c) $R^2 = 21.5\%$

11-87. (b) $\hat{y} = -0.699 + 1.66x$
(c) $f_0 = 22.75$, reject H_0, P-value $=$ 0.001
(d) (3.399, 5.114)

11-89. (c) all data: (7741.74, 10956.26), outlier removed: (8345.22, 11272.79)

CHAPTER 12

Section 12-1

12-1. (b) $\hat{y} = 171.055 + 3.714x_1 - 1.126x_2$
(c) 189.49

12-3. (b) 2

12-5. (a) $\hat{y} = 49.90 - 0.01045x_1 - 0.0012x_2 - 0.00324x_3 + 0.292x_4 - 3.855x_5 + 0.1897x_6$
(b) $\hat{\sigma}^2 = 4.965$
$se(\hat{\beta}_0) = 19.67$,
$se(\hat{\beta}_1) = 0.02338$,
$se(\hat{\beta}_2) = 0.01631$,
$se(\hat{\beta}_3) = 0.0009459$,
$se(\hat{\beta}_4) = 1.765$,
$se(\hat{\beta}_5) = 1.329$,
$se(\hat{\beta}_6) = 0.273$
(c) 29.867

12-7. (a) $\hat{y} = 47.82 - 9.604x_1 + 0.44152x_2 + 18.294x_3$
(b) $\hat{\sigma}^2 = 12.3$
(c) $se(\hat{\beta}_0) = 49.94$,
$se(\hat{\beta}_1) = 3.723$,
$se(\hat{\beta}_2) = 0.2261$, and
$se(\hat{\beta}_3) = 1.323$
(d) 91.38

12-9. (a) $\hat{y} = -440.39 + 19.147x_1 + 68.080x_2$
(b) $\hat{\sigma}^2 = 55563$ $se(\hat{\beta}_0) = 94.20$,
$se(\hat{\beta}_1) = 3.460$, and
$se(\hat{\beta}_2) = 5.241$
(c) 186.675

12-11. (a) $\hat{y} = -0.1105 + 0.4072x_1 + 2.108x_2$
(b) $\hat{\sigma}^2 = 0.00008$
$se(\hat{\beta}_0) = 0.2501$,
$se(\hat{\beta}_1) = 0.1682$, and
$se(\hat{\beta}_2) = 5.834$
(c) 0.97074

12-13. (a) $\hat{y} = 238.56 + 0.3339x_1 - 2.7167x_2$
(b) $\hat{\sigma}^2 = 1321$
(c) $se(\hat{\beta}_0) = 45.23$,
$se(\hat{\beta}_1) = 0.6763$, and
$se(\hat{\beta}_2) = 0.6887$
(d) 61.5195

12-15. (a) $\hat{y} = 3.22 + 1.224x_3 + 4.423x_7 - 4.091x_{10}$
(b) $\hat{\sigma}^2 = 3.7$
(c) $se(\hat{\beta}_0) = 6.519$,
$se(\hat{\beta}_3) = 0.1080$,
$se(\hat{\beta}_7) = 0.2799$, and
$se(\hat{\beta}_{10}) = 0.4953$
(d) 82.079

12-17. (a) $\hat{y} = 383.80 - 3.6381x_1 - 0.1119x_2$
(b) $\hat{\sigma}^2 = 153.0$,
$se(\hat{\beta}_0) = 36.22$,
$se(\hat{\beta}_1) = 0.5665$,
$se(\hat{\beta}_2) = 0.04338$
(c) 180.95
(d) $\hat{y} = 484.0 - 7.656x_1 - 0.222x_2 - 0.0041x_{12}$
(e) $\hat{\sigma}^2 = 147.0$,
$se(\hat{\beta}_0) = 101.3$,
$se(\hat{\beta}_1) = 3.846$,
$se(\hat{\beta}_2) = 0.113$,
$se(\hat{\beta}_{12}) = 0.0039$
(f) 173.1

Section 12-2

12-19. (a) $f_0 = 184.25$, P-value $= 0.000$, Reject H_0
(b) $t_0(\hat{\beta}_1) = 16.21$, P-value < 0.001, reject H_0
$t_0(\hat{\beta}_2) = -11.04$, P-value < 0.001, reject H_0

12-21. (a) P-value $= 8.16$ E−5, 3.82 E−8, and 0.3378
(b) $t_0 = 0.98$, do not reject H_0

12-23. (a) $f_0 = 19.53$, reject H_0

12-25. (a) $f_0 = 828.31$, reject H_0
(b) $t_0 = -2.58$, reject H_0
$t_0 = 1.84$, do not reject H_0
$t_0 = 13.82$, reject H_0

12-27. (a) $f_0 = 99.67$, reject H_0, P-value ≈ 0
(b) $t_0 = 5.539$, reject H_0
$t_0 = 12.99$, reject H_0

12-29. (a) $f_0 = 9.28$, reject H_0, P-value $= 0.015$
(b) $t_0 = 0.49$, do not reject H_0
$t_0 = -3.94$, reject H_0

12-31. (a) $f_0 = 379.07$, reject H_0, P-value ≈ 0
(b) $t_0 = 11.33$, reject H_0
$t_0 = 15.80$, reject H_0
$t_0 = -8.26$, reject H_0
(c) $f_0 = 249.1$, reject H_0

12-33. (a) $f_0 = 97.59$, reject H_0, P-value $= 0.002$
(b) $t_0 = -6.42$, reject H_0
$t_0 = -2.57$, do not reject H_0
(c) $f_0 = 6.629$, do not reject H_0
(d) $f_0 = 7.714$, do not reject H_0
(e) $f_0 = 1.11$, do not reject H_0
(f) 147.0

Sections 12-3 and 12-4

12-35. (a) $(49.927 < \beta_0 < 292.183)$
$(0.033 < \beta_1 < 7.393)$
$(-2.765 < \beta_2 < 0.513)$
(b) (159.58, 219.362)
(c) (127.646, 251.296)

12-37. (a) $(-20.477 < \beta_1 < 1.269)$
$(-0.245 < \beta_2 < 1.076)$
$(14.428 < \beta_3 < 22.159)$
(b) (77.582, 105.162)
(c) (82.133, 100.611)

12-39. (a) $(-6.9467 < \beta_1 < -0.3295)$
$(-0.3651 < \beta_2 < 0.1417)$
(b) $(-45.8276 < \beta_1 < 30.5156)$
$(-1.3426 < \beta_2 < 0.8984)$
$(-0.03433 < \beta_3 < 0.04251)$

12-41. (a) $(12.1363 < \beta_1 < 26.1577)$
$(57.4607 < \beta_2 < 78.6993)$
(b) (−233.4, 63.2)
(c) (−742.09, 571.89)

12-43. (a) $(0.0943 < \beta_1 < 0.7201)$
$(-8.743 < \beta_2 < 12.959)$
(b) (0.861, 0.896)
(c) (0.855, 0.903)

12-45. (a) $(-2.173 < \beta_1 < 2.841)$
$(-5.270 < \beta_2 < -0.164)$
(b) (−36.7, 125.8)
(c) (−112.8, 202.0)
(d) CI: (107.4, 267.2)
PI: (30.7, 344.0)

12-47. (a) $-10.183 < \beta_0 < 16.623$
$(1.001 < \beta_3 < 1.447)$
$(3.845 < \beta_7 < 5.001)$
$(-5.113 < \beta_{10} < -3.069)$
(b) 0.3877
(c) (81.2788, 82.8792)

12-49. (a) $-8.658 \le \beta_0 \le 108.458$
$-0.08 \le \beta_2 \le 0.059$
$-0.05 \le \beta_3 \le 0.047$
$-0.006 \le \beta_7 \le 0$
$-4.962 \le \beta_8 \le 5.546$
$-7.811 \le \beta_9 \le 0.101$
$1.102 \le \beta_{10} \le -0.523$
(b) (25.557, 33.863)
(c) $\hat{y} = 61.001 - 0.0208x_2 - 0.0035x_7 - 3.457x_9$
$50.855 \le \beta_0 \le 71.147$
$-0.036 \le \beta_2 \le -0.006$
$-0.006 \le \beta_7 \le -0.001$
$-6.485 \le \beta_9 \le -0.429$

Section 12-5

12-51. (a) 0.893

12-53. (a) 0.978

12-55. (a) 0.843

12-57. (a) 0.997

12-59. (a) 0.756

12-61. (a) 0.985 (b) 0.99

12-63. (b) 0.9937, 0.9925

12-65. (a) 0.12 (b) Yes

Section 12-6

12-67. (a) $\hat{y} = -1.633 + 1.232x - 1.495x^2$
(b) $f_0 = 1.858613$, reject H_0
(c) $t_0 = -601.64$, reject H_0

12-69. (a) $\hat{y} = -1.769 + 0.421x_1 + 0.222x_2 - 0.128x_3 - 0.02x_{12} + 0.009x_{13} + 0.003x_{23} - 0.019x_1^2 - 0.007x_2^2 + 0.001x_3^2$
(b) $f_0 = 19.628$, reject H_0
(d) $f_0 = 1.612$, do not reject H_0

12-71. (b) $\hat{y} = 56.677 - 0.1457x_1 - 0.00525x_2 - 0.138x_3 - 4.179x_4$

12-73. (a) Min $C_p: x_1, x_2$
$C_p = 3.0$, $MS_E = 55563.92$
$\hat{y} = -440.39 + 19.147x_1 + 68.080x_2$
Min MS_E is same as Min C_p
(b) Same as part (a)
(c) Same as part (a)
(d) Same as part (a)
(e) All models are the same

12-75. (a) Min C_p: x_1
$C_p = 1.1$,
$MS_E = 0.0000705$
$\hat{y} = -0.20052 + 0.467864x_1$
Min MS_E is same as Min C_p
(b) Same as part (a)
(c) Same as part (a)
(d) Same as part (a)
(e) All models are the same

12-77. (a) Min C_p: x_2
$C_p = 1.2$, $MS_E = 1178.55$
$\hat{y} = 253.06 - 2.5453x_2$
Min MS_E is same as Min C_p
(b) Same as part (a)
(c) Same as part (a)
(d) Same as part (a)
(e) All models are the same

12-79. (a) Min C_p: x_1, x_2
$C_p = 2.9$, $MS_E = 10.49$
$\hat{y} = -50.4 + 0.671x_1 + 1.30x_2$
Min MS_E is same as Min C_p
(b) Same as part (a)
(c) Same as part (a)
(d) Same as part (a)
(e) All models are the same

12-81. (a) Min $C_p = 5.4$
$\hat{y} = 3.2992 + 0.82871x_{pctcomp} + 0.00056x_{yds} + 3.828x_{yds/att} - 0.08458x_{td} + 3.715x_{pcttd} - 4.26352x_{pct\ int}$
Min $MS_E = 0.0227$
$\hat{y} = 0.9196 + 0.005207x_{alt} + 0.83061x_{pctcomp} + 4.008689x_{yds/att} - 0.05211x_{td} + 3.5494x_{pcttd} - 0.0793x_{\text{int}} - 3.9205x_{pct\ int}$
(b) $\hat{y} = 3.158 + 4.048x_{yds/att} - 4.209x_{pct\ int} + 3.358x_{pcttd} + 0.83x_{pctcomp}$
(c) Same as part (b)
(d) Same as Min MSE model in part (a)

12-83. (a) Min $C_p = 1.3$:
$\hat{y} = 61.001 - 0.02076x_{cid} - 0.00354x_{etw} - 3.457x_{axle}$
Min $MS_E = 4.0228$
$\hat{y} = 49.5 - 0.017547x_{cid} - 0.0034252x_{etw} + 1.29x_{cmp} - 3.184x_{axle} - 0.0096x_{c02}$
(b) $\hat{y} = 63.31 - 0.0178x_{cid} - 0.00375x_{etw} - 3.3x_{axle} - 0.0084x_{c02}$
(c) Same as Min MSE model in part (a)
(d) $\hat{y} = 45.18 - 0.00321x_{etw} - 4.4x_{axle} + 0.385x_{n/v}$
(e) Min C_p model is preferred
(f) Min $C_p = 4.0$,
Min $MS_E = 2.267$
$\hat{y} = 10 - 0.0038023x_{etw} + 3.936x_{cmp} + 15.216x_{co} - 0.011118x_{c02} - 7.401x_{trans} + 3.6131x_{drv1} + 2.342x_{drv2}$
Stepwise:
$\hat{y} = 39.12 - 0.0044x_{etw} + 0.271x_{n/v} - 4.5x_{trns} + 3.2x_{drv1} + 1.7x_{drv2}$
Forward selection:
$\hat{y} = 41.12 - 0.00377x_{etw} + 0.336x_{n/v} - 2.1x_{axle} - 3.4x_{trans} + 2.1x_{drv1} + 2x_{drv2}$
Backward elimination is same as Min C_p and Min MS_E

12-85. (a) $\hat{y} = -0.304 + 0.083x_1 - 0.031x_3 + 0.004x_2^2$
$C_p = 4.04$ $MS_E = 0.004$
(b) $\hat{y} = -0.256 + 0.078x_1 + 0.022x_2 - 0.042x_3 + 0.0008x_3^2$
$C_p = 4.66$ $MS_E = 0.004$
(c) Prefer the model in part (a)

Supplemental Exercises

12-89. (a) $f_0 = 1321.39$, reject H_0, P-value < 0.00001
(b) $t_0 = -1.45$, do not reject H_0
$t_0 = 19.95$, reject H_0
$t_0 = 2.53$, fail to reject H_0

12-91. (a) $\hat{y} = 4.87 + 6.12x_1^* - 6.53x_2^* - 3.56x_3^* - 1.44x_4^*$
(b) $f_0 = 21.79$, reject H_0
$t_0 = 5.76$, reject H_0
$t_0 = -5.96$, reject H_0
$t_0 = -2.90$, reject H_0
$t_0 = -4.99$, reject H_0

12-93. (a) $\hat{y}^* = 21.068 - 1.404x_3^* + 0.0055x_4 + 0.000418x_5$
$MS_E = 0.013156$ $C_p = 4.0$
(b) Same as part (a)
(c) x_4, x_5 with $C_p = 4.1$ and $MS_E = 0.0134$
(d) The part (c) model is preferable
(e) Yes

12-95. (a) $\hat{y} = 300.0 + 0.85x_1 + 10.4x_2$, $\hat{y} = 405.8$
(b) $f_0 = 55.37$, reject H_0
(c) 0.9022
(d) $MS_E = 10.65$
(e) $f_0 = 0.291$, do not reject H_0

12-97. (a) $f_0 = 18.28$, reject H_0
(b) $f_0 = 2$, do not reject H_0
(c) $MS_E(\text{reduced}) = 0.005$
$MS_E(\text{Full}) = 0.004$

Mind-Expanding Exercises

12-99. $R^2 \geq 0.449$

CHAPTER 13

Section 13-2

13-1. (a) $f_0 = 14.76$, reject H_0

13-3. (a) $f_0 = 12.73$, reject H_0
(b) P-value $\cong 0$

13-5. (a) $f_0 = 16.35$, reject H_0
(c) 95%: (140.71, 149.29)
99%: (7.36, 24.14)

13-7. (a) $f_0 = 1.86$, do not reject H_0
(b) P-value $= 0.214$

13-9. (a) $f_0 = 8.30$, reject H_0
(b) P-value = 0.002
(d) (69.17, 81.83)
(e) (8.42, 26.33)

13-11. (a) $f_0 = 0.72$, do not reject H_0
(b) P-value = 0.486, $SS_E = 0.146$

13-13. (a) $f_0 = 2.62$, do not reject H_0
(b) P-value = 0.083

13-25. $n = 3$

Section 13-3

13-27. (a) $f_0 = 5.77$, reject H_0
(b) 0.01412
(c) 0.0148

13-29. (a) $f_0 = 0.75$, do not reject H_0
(b) 0 (c) 24

13-31. (a) H_0: $\sigma_\tau^2 = 0$ H_1: $\sigma_\tau^2 \neq 0$
(c) Set equal sample sizes, $\sigma^2 = 0.164$, $\sigma_\tau^2 = 0$

Section 13-4

13-33. (a) $f_0 = 8.92$, reject H_0

13-35. (a) $f_0 = 3.00$, do not reject H_0

13-37. (a) $f_0 = 1.61$, do not reject H_0

Supplemental Exercises

13-39. (a) $f_0 = 76.09$, reject H_0
(c) (132.97, 147.83)

13-41. (a) $f_0 = 7.84$, reject H_0
(b) P-value = 0.007

13-43. (a) $f_0 = 6.23$, reject H_0
(c) Algorithm 5

13-45. (a) Power = 0.2 (b) $n = 50$

CHAPTER 14

Section 14-3

14-1. (a) 1. $H_0: \tau_1 = \tau_2 = 0$
H_1: at least one $\tau_i \neq 0$
2. $H_0: \beta_1 = \beta_2 = \beta_3 = 0$
H_1: at least one $\beta_j \neq 0$
(b) $f(G) = 273.79$, $f(P) = 8.84$, $f(GP) = 1.26$ reject H_0 for only main effects

14-3. (a) $f(M) = 7.91$, $f(T) = 28.97$, $f(MT) = 3.56$ reject H_0 for both main effects and the interaction

14-5. (a) $Y_{ijk} = \mu + \tau_i + \beta_j + (\tau\beta)_{ij} + \epsilon_{ijk} \begin{cases} i = 1, 2, 3 \\ j = 1, 2, 3, 4 \\ k = 1, 2, 3, 4, 5, 6 \end{cases}$
(b) $f(I) = 40.07$, $f(T) = 0.32$, $f(IT) = 1.70$ reject H_0 for only insulation

14-7. (a) $f(D) = 25.23$, $f(A) = 543.52$, $f(DA) = 3.54$ reject H_0 for both main effects and the interaction

14-9. $(-3.40, 7.64)$

Section 14-4

14-11. (a) $f(H) = 7.64$, $f(C) = 31.66$, $f(F) = 19.92$, $f(HC) = 2.92$, $f(HF) = 2.97$, $f(CF) = 0.96$
All these main effects are significant and the hardwood concentration-freeness interaction is significant at $\alpha = 0.05$. The P-value for the hardwood-cooking time interaction is 0.075.

Section 14-5

14-13. Significant effects:
$A = 17.00$, $C = 10.875$, $D - 8.375$, $AD - 9.125$

14-15. (a) Cleaning Method = −5.93
Test Position = −1.280
Clean*test = −1.220
(b) Cleaning Method is the only significant factor

14-17. (a) Significant effects:
$A = 11.8125$, $B = 33.9375$, $C = 9.6875$, $AB = 7.9375$

14-19. None of the effects are significant

14-21. (b) A, B and AB
(c) $\hat{y} = 400 + 40.124x_1 - 32.75x_2 + 26.625x_1x_2$

14-23. $f_0 = 5.11$, do not reject H_0

14-25. (a) For model with A, B, C, AB, $s = 2.92$; s (center pts) = 2.86
(b) $f_0 = 192.71$, curvature is significant

Section 14-6

14-27. Significant effects:
$A = 15.625$, $C = 10.625$, $D = 8.875$, $AD = 8.875$, $CD = -3.125$, $ACD = 1.875$

14-33. (a) Effect JED is confounded with blocks
(b) Marginal significant effects:
$J = 19.0$, $D = 14.75$, $JD = -18.0$, $ED = -25.5$,

14-35. (a) 2^{5-1}
(b) $A = 10.8750$, $B = 33.6250$, $C = 10.6250$, $D = -0.6250$, $E = 0.3750$, $AB + CDE = 7.1250$,

14-37. (a) $E = -ABCD$
(b) Resolution V
(c) $E = -0.4700$, $BE = -0.4050$, $DE = -0.3150$

14-39. (c) $A = 1.435$, $B = -1.465$, $D = 4.545$, $AB = 1.15$, $AD = -1.23$

14-41. (b) Design Generator: $D = BE$, $E = AC$
Defining Relation:
$I = ACE = BDE = ABCDE$
Aliases
$A = CE = BCDE = ABDE$
$B = DE = ACDE = ABCE$
$C = AE = ABDE = BCDE$
$D = BE = ABCE = ACDE$
$E = AC = BD = ABCD$
(c) $A = -1.525$, $B = -5.175$, $C = 2.275$, $D = -0.675$, $E = 2.275$

14-43. (a) Alias Structure:
$I + ABD + ACE + BCF + DEF + ABEF + ACDF + BCDE$
$A + BD + CE$
$B + AD + CF$
$C + AE + BF$
$D + AB + EF$
$E + AC + DF$
$F + BC + DE$
$AF + BE + CD$
(b) Alias Structure: $I + ABCG + ABDH + ABEF + ACDF + ACEH$

$+ ADEG + AFGH$
$+ BCDE + BCFH$
$+ BDFG + BEGH$
$+ CDGH + CEFG$
$+ DEFH$
A
B
C
D
E
F
G
H
$AB + CG + DH + EF$
$AC + BG + DF + EH$
$AD + BH + CF + EG$
$AE + BF + CH + DG$
$AF + BE + CD + GH$
$AG + BC + DE + FH$
$AH + BD + CE + FG$

14-45. (b) $\hat{y} = 82.024 - 1.115x_1 - 2.408x_2 + 0.861x_1^2 - 1.59x_2^2 - 1.801x_1x_2$

14-47. Path of steepest ascent passes through the point (0, 0) and has slope $-0.8/1.5 = -0.533$

14-49. (a) Central composite design, not rotatable
(b) $\hat{y} = 150.04 - 58.47x_1 + 3.35x_2 - 6.53x_1^2 + 10.58x_2^2 + 0.50x_1x_2$
The linear terms are significant while both the square terms and interaction terms are insignificant

Supplemental Exercises

14-51. (a) $t(P) = 2.54$, $t(PC) = 5.02$
The main effect of pH and the interaction of pH and Catalyst Concentration are significant

14-53. (a) $f(L) = 63.24$, $f(S) = 39.75$, $f(LS) = 5.29$
There is a significant difference between the application levels, the salts, and there is a significant interaction between the two factors

14-55. (a) $A = -2.74$, $B = -6.66$, $C = 3.49$, $AB = -8.71$, $AC = 7.04$, $BC = 11.46$, $ABC = -6.49$, not significant

14-57. (a) $V = -15.75$, $P = 10.75$, $G = -25.00$, $PG = -19.25$,
(b) $\hat{y} = 102.75 - 7.87x_1 + 5.37x_3 - 12.50x_4 - 9.63x_{34}$

14-59. $V = 3.25$, $F = -10.25$, $P = -10.25$, $G = -23.75$, $PG = -11.75$

14-61. Design Generators: $D = AB$
$E = AC$
Alias Structure
$I + ABD + ACE + BCDE$
$A + BD + CE + ABCDE$
$B + AD + CDE + ABCE$
$C + AE + BDE + ABCD$
$D + AB + BCE + ACDE$
$E + AC + BCD + ABDE$
$BC + DE + ABE + ACD$
$BE + CD + ABC + ADE$

14-63. (a) 2^2 factorial with two replicates
(b) Significant effects: $x_1 = 0.795$, $x_2 = -1.160$
(c) $\hat{y} = 0.1994 + 0.07688x_1$

14-65. (a) $H_0: \tau_1 = \tau_2 = \tau_3 = 0$
$H_0: \beta_1 = \beta_2 = \beta_3 = 0$
$H_0: (\tau\beta)_{11} = \ldots = (\tau\beta)_{33} = 0$
(b) $f(N) = 311.71$, $f(P) = 119.17$, $f(PN) = 92.94$
(d) $\hat{\sigma} = 1.22$

14-67. (c) $\hat{y} = 108.623 + 1.756x_1 + 2.269x_2 - 0.587x_1^2 - 0.925x_2^2 - 0.702x_1x_2$

CHAPTER 15

Section 15-2

15-1. (a) $r^+ = 8$, fail to reject H_0
(b) $z_0 = 1.90$, fail to reject H_0, P-value $= 0.0574$

15-3. (a) $r^+ = 2$, reject H_0
(b) $z_0 = -3.84$, reject H_0, P-value $= 0.000062$

15-5. (a) $r^+ = 7$, fail to reject H_0
(b) $z_0 = 0.577$, fail to reject H_0, P-value $= 0.281$

15-7. (a) $r = \min(6,2) = 2$, fail to reject H_0
(b) Fail to reject H_0, P-value $= 0.289$

15-9. (a) $E(X) = 1/\lambda = 5.05$
(b) $r^* = 3$ fail to reject H_0
(c) $\beta = 0.963$

15-11. (a) $\alpha = 0.025$
(b) $\beta = 0.115$
(c) Reject H_0 if $R^- \leq 1$
(d) $\beta = 0.1587$

Section 15-3

15-13. $w = \min(50.5, 4.5) = 4.5$, reject H_0

15-15. $w = \min(205, 5) = 5$, reject H_0

15-17. $w = \min(24.5, 11.5) = 11.5$, fail to reject H_0

Section 15-4

15-19. $w_1 = 40$ and $w_2 = 38$, fail to reject H_0

15-21. (a) $w_1 = 73$ and $w_2 = 137$, reject H_0
(b) $z_0 = 1.06$, fail to reject H_0, P-value $= 0.2891$

15-23. (a) $w_1 = 135.5$ and $w_2 = 74.5$, reject H_0
(b) $z_0 = 2.31$, reject H_0, P-value $= 0.021$

Section 15-5

15-25. (a) $h = 10.00$, reject H_0
(b) P-value $= 0.018$

15-27. (a) $h = 14.02$, reject H_0
(b) P-value $= 0.007$

15-29. (a) $z_0 = -0.6055$, fail to reject H_0
(b) $w_1 = 40$ and, $w_2 = 38$, fail to reject H_0

15-31. $z_0 = -1.67$, fail to reject H_0

15-33. (a) $r^+ = 5$, fail to reject H_0
(b) $z_0 = 0$, fail to reject H_0, P-value $= 1$

15-35. $w_1 = \min(55, 0) = 0$, reject H_0

15-37. (a) $w_1 = 55$ and $w_2 = 155$, reject H_0
(b) $z_0 = -3.78$, reject H_0, P-value $= 0.00016$

15-39. $h = 25.30$, reject H_0, P-value $\cong 0$

15-41. $h = 9.46$, reject H_0

Mind-Expanding Exercises

15-45. (a) 32 (c) 2/32

CHAPTER 16

Section 16-5

16-1. (a) $\bar{x}$ chart: $UCL = 242.78$, $CL = 223$, $LCL = 203.22$, R chart: $UCL = 72.51$, $CL = 34.286$, $LCL = 0$
(b) $\hat{\mu} = 223$
$\hat{\sigma} = 14.74$

16-3. (a) $\bar{x}$ chart: $UCL = 245.11$, $CL = 223$, $LCL = 200.89$
S chart: $UCL = 30.77$, $CL = 13.58$, $LCL = 0$
(b) $\hat{\mu} = 223$, $\hat{\sigma} = 14.74$

16-5. (a) $\bar{x}$ chart: $UCL = 37.5789$, $CL = 34.32$, $LCL = 31.0611$
R chart: $UCL = 11.9461$, $CL = 5.65$, $LCL = 0$
(b) $\bar{x}$ chart: $UCL = 37.4038$, $CL = 34.0947$, $LCL = 30.7857$
R chart: $UCL = 12.1297$, $CL = 5.73684$, $LCL = 0$
$\hat{\sigma} = 2.4664$

16-7. (a) $\bar{x}$ chart: $UCL = 17.4$, $CL = 15.09$, $LCL = 12.79$
R chart: $UCL = 5.792$, $CL = 2.25$, $LCL = 0$
(b) $\bar{x}$ chart: $UCL = 17.96$, $CL = 15.78$, $LCL = 13.62$
R chart: $UCL = 5.453$, $CL = 2.118$, $LCL = 0$
(c) $\bar{x}$ chart: $UCL = 17.42$, $CL = 15.09$, $LCL = 12.77$
S chart: $UCL = 3.051$, $CL = 1.188$, $LCL = 0$
Revised
$\bar{x}$ chart: $UCL = 17.95$, $CL = 15.78$, $LCL = 13.62$
S chart: $UCL = 2.848$, $CL = 1.109$, $LCL = 0$

16-9. (a) $\bar{x}$ chart: $UCL = 0.063505$, $CL = 0.062938$, $LCL = 0.062370$
R chart: $UCL = 0.00208$, $CL = 0.000984$, $LCL = 0$
(b) $\bar{x}$ chart: $UCL = 0.063312$, $CL = 0.062988$, $LCL = 0.062664$
R chart: $UCL = 0.001187$, $CL = 0.000562$, $LCL = 0$
(c) $\bar{x}$ chart: $UCL = 0.063461$, $CL = 0.062938$, $LCL = 0.062414$
S chart: $UCL = 0.000766$, $CL = 0.000367$, $LCL = 0$
Revised:
$\bar{x}$ chart: $UCL = 0.063320$, $CL = 0.062988$, $LCL = 0.062656$
S chart: $UCL = 0.000486$, $CL = 0.000233$, $LCL = 0$

16-11. $s = 2.956$, $\bar{r}/d_2 = 1.251$

Section 16-6

16-13. (a) Individual chart: $UCL = 60.8887$, $CL = 53.05$, $LCL = 45.2113$
MR chart: $UCL = 9.63382$, $CL = 2.94737$, $LCL = 0$
The process appears to be in control.
(b) $\hat{\mu} = 53.05$, $\hat{\sigma} = 2.613$

16-15. (a) Individual chart: $UCL = 10.5358$, $CL = 10.0272$, $LCL = 9.51856$
MR chart: $UCL = 0.625123$, $CL = 0.19125$, $LCL = 0$
The process appears to be in control.
(b) $\hat{\mu} = 10.0272$, $\hat{\sigma} = 0.16955$

16-17. (a) Initial study: Individual chart: $UCL = 130.5$, $CL = 100.78$, $LCL = 71.06$
MR chart: $UCL = 36.51$, $CL = 11.18$, $LCL = 0$
Revised: Individual chart $UCL = 127.08$, $CL = 99.48$, $LCL = 71.88$
MR chart: $UCL = 33.91$, $CL = 10.38$, $LCL = 0$
(b) $\hat{\mu} = 99.4792$, $\hat{\sigma} = 9.20059$

Section 16-7

16-19. (a) 1.3333 (b) 26

16-21. (a) $PC = PCR_K = 1.5$ (b) 0

16-23. Proportion nonconforming is 0.00779
$PCR = 0.905$
$PCR_K = 0.837$

16-25. 0.0009
$PCR = 1.13$
$PCR_K = 1.06$

16-27. (a) $PCR = 1.35$
$PCR_K = 1.217$
(b) 0.00013

16-29. 0.512
$PCR = 0.985$

Section 16-8

16-31. (a) This process is out of control $UCL = 0.0835$, $CL = 0.0645$, $LCL = 0.0455$
(b) Revised P-chart: $UCL = 0.1252$, $CL = 0.0561$, $LCL = 0$

16-33. (a) P chart: $UCL = 0.198553$, $CL = 0.150571$, $LCL = 0.10259$
(b) Revised P chart: $UCL = 0.206184$, $CL = 0.157333$, $LCL = 0.108482$

16-35. (a) The limits need to be revised. $UCL = 3.811$, $CL = 1.942$, $LCL = 0.072$, sample 5 and 24 exceed limits
(b) U chart: $UCL = 3.463$, $CL = 1.709$, $LCL = 0$

Section 16-9

16-37. (a) 4 (b) 0.0228.
(c) 43.8596

16-39. (a) 0.2177 (b) $ARL = 4.6$

16-41. (a) 0.1515 (b) $ARL = 6.6$

16-43. (a) 0.1020 (b) $ARL = 9.8$

16-45. (a) 0.2877 (b) $ARL = 3.48$

Section 16-10

16-47. (a) $h = 4$, $k = 0.5$, $UCL = 3.875$, $LCL = -3.875$
Yes, this process is in-control.
(b) Observation 20 is out of control, CUSUM = 6.08

16-49. (a) $\hat{\sigma} = 0.1736$
(b) $h = 4$, $k = 0.5$, $UCL = 0.0678$, $LCL = -0.0678$, out of control at the specified target level

16-51. (a) $ARL = 38.0$
(b) $ARL = 10.4$

16-53. (a) 0.169548
(b) The process appears to be in control. $UCL = 10.17$, $CL = 10$, $LCL = 9.83$
(c) Out of control at observation 13, $UCL = 10.29$, $CL = 10$, $LCL = 9.71$

16-55. (a) prefer $\lambda = 0.1$ and $L = 2.81$
(a) prefer $\lambda = 0.5$ and $L = 3.07$
(c) 9

Supplemental Exercises

16-57. (a) $\bar{x}$ chart: $UCL = 64.0181$, $CL = 64$, $LCL = 63.982$
R chart: $UCL = 0.0453972$, $CL = 0.01764$, $LCL = 0$
(b) $\hat{\mu} = 64$, $\hat{\sigma} = 0.0104$
(c) $PCR = 0.641$
(d) $PCR_k = 0.641$
(e) $\sigma^2 = (0.0033)^2 = 0.000011$
(f) $ARL = 5.87$

16-59. (a) The process appears to be in control.
(b) P chart: $UCL = 0.1717$, $CL = 0.1063$, $LCL = 0.04093$

16-61. (a) Individual chart: $UCL = 60.327362$, $CL = 60.32641$, $LCL = 60.325458$
MR chart: $UCL = 0.001169$, $CL = 0.000358$, $LCL = 0$
(b) Individual chart: $UCL = 0.001362$, $CL = 00.00041$, $LCL = 0.000542$
MR chart: $UCL = 0.001169$, $CL = 0.000358$, $LCL = 0$
(c) $\hat{\mu} = 60.3264$
$\hat{\sigma} = 0.0003173$
$PCR = 1.0505$
$PCR_k = 0.9455$

16-63. (a) Trial control limits
S chart: $UCL = 170.25$, $CL = 86.42$, $LCL = 2.59$
$\bar{x}$ chart: $UCL = 670.00$, $CL = 558.77$, $LCL = 447.53$
Revised S chart: $UCL = 158.93$, $CL = 80.68$, $LCL = 2.42$
$\bar{x}$ chart: $UCL = 655.79$, $CL = 551.95$, $LCL = 448.10$
(b) $PCR_k = 0.8725$
(c) $\hat{\sigma} = 36.9917$
(d) $ARL = 18.6$

16-73. (a) $ARL = 43.9$
(b) $ARL = 6.30$
(c) $ARL = 2.00$

16-75. (a) $\bar{x}$ chart: $UCL = 140.168$, $CL = 139.49$, $LCL = 138.812$
R chart: $UCL = 2.48437$, $CL = 1.175$, $LCL = 0$
(b) Revised: $\bar{x}$ chart: $UCL = 140.417$, $CL = 139.709$, $LCL = 139.001$
R chart: $UCL = 2.596$, $CL = 1.227$, $LCL = 0$
$\hat{\sigma} = 0.5276$
(c) $PCR = 1.26$
$PCR_k = 1.08$
(d) $\sigma^2 = 0.081$
(e) $ARL = 5.55$

16-77. $P(X < USL) = 0.00135$

16-79. (a) $P(\bar{U} > 12.24 \text{ when } \lambda = 16) = 0.96995$
(b) $P(U > 10.68 \text{ when } \lambda = 16) = 1$

16-81. (a) $ARL = 17.5$
(b) $ARL = 3.63$

Mind-Expanding Exercises

16-83. 0.125, 0.004

16-85. (b) $ARL = 1/p$ where $p = 1 - \Phi(k - \sqrt{n}\delta) + \Phi(-k - \sqrt{n}\delta)$
(c) $ARL = 22.0$ for $k = 2$
(d) $ARL = 4.47$

16-87. $n = \dfrac{k^2(1 - \bar{p})}{\bar{p}}$

16-91. (a) $\bar{C}$ chart
$CL = 8$
$UCL = 16.49$
$LCL = 0$
(b) Yes

Glossary

2^k factorial experiment. A full factorial experiment with k factors and all factors tested at only two levels (settings) each.

2^{k-p} factorial experiment. A fractional factorial experiment with k factors tested in a 2^{-p} fraction with all factors tested at only two levels (settings) each.

Acceptance region. In hypothesis testing, a region in the sample space of the test statistic such that if the test statistic falls within it, the null hypothesis is accepted (better terminology is that the null hypothesis cannot be rejected, since rejection is always a strong conclusion and acceptance is generally a weak conclusion).

Addition rule. A formula used to determine the probability of the union of two (or more) events from the probabilities of the events and their intersection(s).

Additivity property of χ^2. If two independent random variables X_1 and X_2 are distributed as chi-square with v_1 and v_2 degrees of freedom respectively, $Y = X_1 + X_2$ is a chi-square random variable with $u = v_1 + v_2$ degrees of freedom. This generalizes to any number of independent chi-square random variables.

Adjusted R^2. A variation of the R^2 statistic that compensates for the number of parameters in a regression model. Essentially, the adjustment is a penalty for increasing the number of parameters in the model.

Alias. In a fractional factorial experiment when certain factor effects cannot be estimated uniquely, they are said to be aliased.

All possible (subsets) regressions. A method of variable selection in regression that examines all possible subsets of the candidate regressor variables. Efficient computer algorithms have been developed for implementing all possible regressions.

Alternative hypothesis. In statistical hypothesis testing, this is a hypothesis other than the one that is being tested. The alternative hypothesis contains feasible conditions, whereas the null hypothesis specifies conditions that are under test.

Analysis of variance (ANOVA). A method of decomposing the total variability in a set of observations, as measured by the sum of the squares of these observations from their average, into component sums of squares that are associated with specific defined sources of variation.

Analytic study. A study in which a sample from a population is used to make inference to a future population. Stability needs to be assumed. *See* enumerative study.

Arithmetic mean. The arithmetic mean of a set of numbers $x_1, x_2, \ldots, x_n$ is their sum divided by the number of observations, or $(1/n)\sum_{i=1}^{n} x_i$. The arithmetic mean is usually denoted by $\bar{x}$, and is often called the average.

Assignable cause. The portion of the variability in a set of observations that can be traced to specific causes, such as operators, materials, or equipment. Also called a special cause.

Asymptotic relative efficiency (ARE). Used to compare hypothesis tests. The ARE of one test relative to another is the limiting ratio of the sample sizes necessary to obtain identical error probabilities for the two procedures.

Attribute. A qualitative characteristic of an item or unit, usually arising in quality control. For example, classifying production units as defective or nondefective results in attributes data.

Attribute control chart. Any control chart for a discrete random variable. *See* variables control charts.

Average. *See* Arithmetic Mean.

Average run length, or ARL. The average number of samples taken in a process monitoring or inspection scheme until the scheme signals that the process is operating at a level different from the level in which it began.

Axioms of probability. A set of rules that probabilities defined on a sample space must follow. *See* probability.

Backward elimination. A method of variable selection in regression that begins with all of the candidate regressor variables in the model and eliminates the insignificant regressors one at a time until only significant regressors remain.

Bayes estimator. An estimator for a parameter obtained from a Bayesian method that uses a prior distribution for the parameter along with the conditional distribution of the data given the parameter to obtain the posterior distribution of the parameter. The estimator is obtained from the posterior distribution.

Bayes' theorem. An equation for a conditional probability such as $P(A \mid B)$ in terms of the reverse conditional probability $P(B \mid A)$.

Bernoulli trials. Sequences of independent trials with only two outcomes, generally called "success" and "failure," in which the probability of success remains constant.

Bias. An effect that systematically distorts a statistical result or estimate, preventing it from representing the true quantity of interest.

Biased estimator. *See* Unbiased estimator.

Bimodal distribution. A distribution with two modes.

Binomial random variable. A discrete random variable that equals the number of successes in a fixed number of Bernoulli trials.

Bivariate distribution. The joint probability distribution of two random variables.

Bivariate normal distribution. The joint distribution of two normal random variables.

Block. In experimental design, a group of experimental units or material that is relatively homogeneous. The purpose of dividing experimental units into blocks is to produce an experimental design wherein variability within blocks is smaller than variability between blocks. This allows the factors of interest to be compared in a environment that has less variability than in an unblocked experiment.

Box plot (or box and whisker plot). A graphical display of data in which the box contains the middle 50% of the data (the interquartile range) with the median dividing it, and the whiskers extend to the smallest and largest values (or some defined lower and upper limits).

***C* chart.** An attribute control chart that plots the total number of defects per unit in a subgroup. Similar to a defects-per-unit or *U* chart.

Categorical data. Data consisting of counts or observations that can be classified into categories. The categories may be descriptive.

Causal variable. When $y = f(x)$ and y is considered to be caused by x, x is sometimes called a causal variable.

Cause-and-effect diagram. A chart used to organize the various potential causes of a problem. Also called a fishbone diagram.

Center line. A horizontal line on a control chart at the value that estimates the mean of the statistic plotted on the chart. *See* Control chart.

Central composite design (CCD). A second-order response surface design in k variables consisting of a two-level factorial, $2k$ axial runs, and one or more center points. The two-level factorial portion of a CCD can be a fractional factorial design when k is large. The CCD is the most widely used design for fitting a second-order model.

Central limit theorem. The simplest form of the central limit theorem states that the sum of n independently distributed random variables will tend to be normally distributed as n becomes large. It is a necessary and sufficient condition that none of the variances of the individual random variables are large in comparison to their sum. There are more general forms of the central theorem that allow infinite variances and correlated random variables, and there is a multivariate version of the theorem

Central tendency. The tendency of data to cluster around some value. Central tendency is usually expressed by a measure of location such as the mean, median, or mode.

Chance cause. The portion of the variability in a set of observations that is due to only random forces and which cannot be traced to specific sources, such as operators, materials, or equipment. Also called a common cause.

Chi-square (or chi-squared) random variable. A continuous random variable that results from the sum of squares of independent standard normal random variables. It is a special case of a gamma random variable.

Chi-squared test. Any test of significance based on the chi-square distribution. The most common chi-square tests are (1) testing hypotheses about the variance or standard deviation of a normal distribution and (2) testing goodness of fit of a theoretical distribution to sample data.

Coefficient of determination. *See* R^2.

Combination. A subset selected without replacement from a set used to determine the number of outcomes in events and sample spaces.

Comparative experiment. An experiment in which the treatments (experimental conditions) that are to be studied are included in the experiment. The data from the experiment are used to evaluate the treatments.

Completely randomized design (or experiment). A type of experimental design in which the treatments or design factors are assigned to the experimental units in a random manner. In designed experiments, a completely randomized design results from running all of the treatment combinations in random order.

Components of variance. The individual components of the total variance that are attributable to specific sources. This usually refers to the individual variance components arising from a random or mixed model analysis of variance.

Conditional mean. The mean of the conditional probability distribution of a random variable.

Conditional probability. The probability of an event given that the random experiment produces an outcome in another event.

Conditional probability density function. The probability density function of the conditional probability distribution of a continuous random variable.

Conditional probability distribution. The distribution of a random variable given that the random experiment produces an outcome in an event. The given event might specify values for one or more other random variables.

Conditional probability mass function. The probability mass function of the conditional probability distribution of a discrete random variable.

Conditional variance. The variance of the conditional probability distribution of a random variable.

Confidence coefficient. The probability $1 - \alpha$ associated with a confidence interval expressing the probability that the stated interval will contain the true parameter value.

Confidence interval. If it is possible to write a probability statement of the form

$$P(L \leq \theta \leq U) = 1 - \alpha$$

where L and U are functions of only the sample data and θ is a parameter, then the interval between L and U is called a confidence interval (or a $100(1 - \alpha)\%$ confidence interval). The interpretation is that a statement that the parameter θ lies in this interval will be true $100(1 - \alpha)\%$ of the times that such a statement is made.

Confidence level. Another term for the confidence coefficient.

Confounding. When a factorial experiment is run in blocks and the blocks are too small to contain a complete replicate of the experiment, one can run a fraction of the replicate in each block, but this results in losing information on some effects. These effects are linked with or confounded with the blocks. In general, when two factors are varied such that their individual effects cannot be determined separately, their effects are said to be confounded.

Consistent estimator. An estimator that converges in probability to the true value of the estimated parameter as the sample size increases.

Contingency table. A tabular arrangement expressing the assignment of members of a data set according to two or more categories or classification criteria.

Continuity correction. A correction factor used to improve the approximation to binomial probabilities from a normal distribution.

Continuous distribution. A probability distribution for a continuous random variable.

Continuous random variable. A random variable with an interval (either finite or infinite) of real numbers for its range.

Continuous uniform random variable. A continuous random variable with range of a finite interval and a constant probability density function.

Contour plot. A two-dimensional graphic used for a bivariate probability density function that displays curves for which the probability density function is constant.

Contrast. A linear function of treatment means with coefficients that total zero. A contrast is a summary of treatment means that is of interest in an experiment.

Control chart. A graphical display used to monitor a process. It usually consists of a horizontal center line corresponding to the in-control value of the parameter that is being monitored and lower and upper control limits. The control limits are determined by statistical criteria and are not arbitrary nor are they related to specification limits. If sample points fall within the control limits, the process is said to be in-control, or free from assignable causes. Points beyond the control limits indicate an out-of-control process; that is, assignable causes are likely present. This signals the need to find and remove the assignable causes.

Control limits. *See* Control chart.

Convolution. A method to derive the probability density function of the sum of two independent random variables from an integral (or sum) of probability density (or mass) functions.

Cook's distance. In regression, Cook's distance is a measure of the influence of each individual observation on the estimates of the regression model parameters. It expresses the distance that the vector of model parameter estimates with the ith observation removed lies from the vector of model parameter estimates based on all observations. Large values of Cook's distance indicate that the observation is influential.

Correction factor. A term used for the quantity $(1/n)(\sum_{i=1}^{n} x_i)^2$ that is subtracted from $\sum_{i=1}^{n} x_i^2$ to give the corrected sum of squares defined as $(1/n)\sum_{i=1}^{n}(x_i - \bar{x})^2$. The correction factor can also be written as $n\bar{x}^2$.

Correlation. In the most general usage, a measure of the interdependence among data. The concept may include more than two variables. The term is most commonly used in a narrow sense to express the relationship between quantitative variables or ranks.

Correlation coefficient. A dimensionless measure of the linear association between two variables, usually lying in the interval from -1 to $+1$, with zero indicating the absence of correlation (but not necessarily the independence of the two variables).

Correlation matrix. A square matrix that contains the correlations among a set of random variables, say $X_1, X_2, \ldots, X_k$. The main diagonal elements of the matrix are unity and the off diagonal elements r_{ij} are the correlations between X_i and X_j.

Counting techniques. Formulas used to determine the number of elements in sample spaces and events.

Covariance. A measure of association between two random variables obtained as the expected value of the product of the two random variables around their means; that is, $\text{Cov}(X, Y) = E[(X - \mu_X)(Y - \mu_Y)]$.

Covariance matrix. A square matrix that contains the variances and covariances among a set of random variables, say $X_1, X_2, \ldots, X_k$. The main diagonal elements of the matrix are the variances of the random variables and the off diagonal elements are the covariances between X_i and X_j. Also called the variance-covariance matrix. When the random variables are standardized to have unit variances, the covariance matrix becomes the correlation matrix.

Critical region. In hypothesis testing, this is the portion of the sample space of a test statistic that will lead to rejection of the null hypothesis.

Critical value(s). The value of a statistic corresponding to a stated significance level as determined from the sampling distribution. For example, if $P(Z \geq z_{0.05}) = P(Z \geq 1.96) = 0.05$, then $z_{0.05} = 1.96$ is the critical value of z at the 0.05 level of significance.

Crossed factors. Another name for factors that are arranged in a factorial experiment.

Cumulative distribution function. For a random variable X, the function of X defined as $P(X \leq x)$ that is used to specify the probability distribution.

Cumulative normal distribution function. The cumulative distribution of the standard normal distribution, often denoted as $\Phi(x)$ and tabulated in Appendix Table II.

Cumulative sum control chart (CUSUM). A control chart in which the point plotted at time t is the sum of the measured deviations from target for all statistics up to time t.

Curvilinear regression. An expression sometimes used for nonlinear regression models or polynomial regression models.

Decision interval. A parameter set in a Tabular CUSUM algorithm that is determined from a trade-off between false alarms and the detection of assignable causes.

Defect. Used in statistical quality control, a defect is a particular type of nonconformance to specifications or requirements. Sometimes defects are classified into types, such as appearance defects and functional defects.

Defect concentration diagram. A quality tool that graphically shows the location of defects on a part or in a process.

Defects-per-unit control chart. *See* U chart.

Defining relation. A subset of effects in a fractional factorial design that define the aliases in the design.

Degrees of freedom. The number of independent comparisons that can be made among the elements of a sample. The term is analogous to the number of degrees of freedom for an object in a dynamic system, which is the number of independent coordinates required to determine the motion of the object.

Deming. W. Edwards Deming (1900–1993) was a leader in the use of statistical quality control.

Deming's 14 points. A management philosophy promoted by W. Edwards Deming that emphasizes the importance of change and quality.

Density function. Another name for a probability density function.

Dependent variable. The response variable in regression or a designed experiment.

Design matrix. A matrix that provides the tests that are to be conducted in an experiment.

Designed experiment. An experiment in which the tests are planned in advance and the plans usually incorporate statistical models. *See* Experiment.

Discrete distribution. A probability distribution for a discrete random variable.

Discrete random variable. A random variable with a finite (or countably infinite) range.

Discrete uniform random variable. A discrete random variable with a finite range and constant probability mass function.

Dispersion. The amount of variability exhibited by data.

Distribution free methods(s). Any method of inference (hypothesis testing or confidence interval construction) that does not depend on the form of the underlying distribution of the observations. Sometimes called nonparametric method(s).

Distribution function. Another name for a cumulative distribution function.

Efficiency. A concept in parameter estimation that uses the variances of different estimators; essentially, an estimator is more efficient than another estimator if it has smaller variance. When estimators are biased, the concept requires modification.

Empirical model. A model to relate a response to one or more regressors or factors that is developed from data obtained from the system.

Enumerative study. A study in which a sample from a population is used to make inference to the population. *See* Analytic study.

Erlang random variable. A continuous random variable that is the sum of a fixed number of independent, exponential random variables.

β-error (or β-risk). In hypothesis testing, an error incurred by failing to reject a null hypothesis when it is actually false (also called a type II error).

α-error (or α-risk). In hypothesis testing, an error incurred by rejecting a null hypothesis when it is actually true (also called a type I error).

Error mean square. The error sum of squares divided by its number of degrees of freedom.

Error of estimation. The difference between an estimated value and the true value.

Error propagation. An analysis of how the variance of the random variable that represents that output of a system depends on the variances of the inputs. A formula exists when the output is a linear function of the inputs and the formula is simplified if the inputs are assumed to be independent.

Error sum of squares. In analysis of variance, this is the portion of total variability that is due to the random component in the data. It is usually based on replication of observations at certain treatment combinations in the experiment. It is sometimes called the residual sum of squares, although this is really a better term to use only when the sum of squares is based on the remnants of a model fitting process and not on replication.

Error variance. The variance of an error term or component in a model.

Estimate (or point estimate). The numerical value of a point estimator.

Estimator (or point estimator). A procedure for producing an estimate of a parameter of interest. An estimator is usually a function of only sample data values, and when these data values are available, it results in an estimate of the parameter of interest.

Event. A subset of a sample space.

Exhaustive. A property of a collection of events that indicates that their union equals the sample space.

Expected value. The expected value of a random variable X is its long-term average or mean value. In the continuous case, the expected value of X is $E(X) = \int_{-\infty}^{\infty} xf(x)dx$ where $f(x)$ is the density function of the random variable X.

Experiment. A series of tests in which changes are made to the system under study.

Exponential random variable. A continuous random variable that is the time between events in a Poisson process.

Extra sum of squares method. A method used in regression analysis to conduct a hypothesis test for the additional contribution of one or more variables to a model.

Factorial experiment. A type of experimental design in which every level of one factor is tested in combination with every level of another factor. In general, in a factorial experiment, all possible combinations of factor levels are tested.

False alarm. A signal from a control chart when no assignable causes are present.

***F*-distribution.** The distribution of the random variable defined as the ratio of two independent chi-square random variables each divided by their number of degrees of freedom.

Finite population correction factor. A term in the formula for the variance of a hypergeometric random variable.

First-order model. A model that contains only first-order terms. For example, the first-order response surface model in two variables is $y = \beta_0 + \beta_1 x_1 + \beta_2 x_2 + \epsilon$. A first-order model is also called a main effects model.

Fisher's least significant difference (LSD) method. A series of pair-wise hypothesis tests of treatment means in an experiment to determine which means differ.

Fixed factor (or fixed effect). In analysis of variance, a factor or effect is considered fixed if all the levels of interest for that factor are included in the experiment. Conclusions are then valid about this set of levels only, although when the factor is quantitative, it is customary to fit a model to the data for interpolating between these levels.

Forward selection. A method of variable selection in regression, where variables are inserted one at a time into the model until no other variables that contribute significantly to the model can be found.

Fraction defective control chart. *See P* chart.

Fraction defective. In statistical quality control, that portion of a number of units or the output of a process that is defective.

Fractional factorial experiment. A type of factorial experiment in which not all possible treatment combinations are run. This is usually done to reduce the size of an experiment with several factors.

Frequency distribution. An arrangement of the frequencies of observations in a sample or population according to the values that the observations take on.

***F*-test.** Any test of significance involving the F-distribution. The most common F-tests are (1) testing hypotheses about the variances or standard deviations of two independent normal distributions, (2) testing hypotheses about treatment means or variance components in the analysis of variance, and (3) testing significance of regression or tests on subsets of parameters in a regression model.

Gamma function. A function used in the probability density function of a gamma random variable that can be considered to extend factorials.

Gamma random variable. A random variable that generalizes an Erlang random variable to noninteger values of the parameter r.

Gaussian distribution. Another name for the normal distribution, based on the strong connection of Karl F. Gauss to the normal distribution; often used in physics and electrical engineering applications.

Generating function. A function that is used to determine properties of the probability distribution of a random variable. *See* Moment generating function.

Generator. Effects in a fractional factorial experiment that are used to construct the experimental tests used in the experiment. The generators also define the aliases.

Geometric mean. The geometric mean of a set of n positive data values is the nth root of the product of the data values; that is $\bar{g} = (\prod_{i=1}^{n} x_i)^{1/n}$.

Geometric random variable. A discrete random variable that is the number of Bernoulli trials until a success occurs.

Goodness of fit. In general, the agreement of a set of observed values and a set of theoretical values that depend on some hypothesis. The term is often used in fitting a theoretical distribution to a set of observations.

Harmonic mean. The harmonic mean of a set of data values is the reciprocal of the arithmetic mean of the reciprocals of the data values; that is, $\bar{h} = \left(\frac{1}{n}\sum_{i=1}^{n}\frac{1}{x_i}\right)^{-1}$.

Hat matrix. In multiple regression, the matrix $\mathbf{H} = \mathbf{X}(\mathbf{X}'\mathbf{X})^{-1}\mathbf{X}'$. This a projection matrix that maps the vector of observed response values into a vector of fitted values by $\hat{\mathbf{y}} = \mathbf{X}(\mathbf{X}'\mathbf{X})^{-1}\mathbf{X}'\mathbf{y} = \mathbf{H}\mathbf{y}$.

Hidden extrapolation. An extrapolation is a prediction in a regression analysis that is made at point $(x_1, x_2, \ldots, x_k)$ that is remote from the data used to generate the model. Hidden extrapolation occurs when it is not obvious that the point is remote. This can occur when multicollinearity is present in the data used to construct the model.

Histogram. A univariate data display that uses rectangles proportional in area to class frequencies to visually exhibit features of data such as location, variability, and shape.

Homogeneity test. In a two-way (r by c) contingency table, this tests if the proportions in the c categories are the same for all r populations.

Hypergeometric random variable. A discrete random variable that is the number of success obtained from a sample drawn without replacement from a finite populations.

Hypothesis (as in statistical hypothesis). A statement about the parameters of a probability distribution or a model, or a statement about the form of a probability distribution.

Hypothesis testing. Any procedure used to test a statistical hypothesis.

Independence. A property of a probability model and two (or more) events that allows the probability of the intersection to be calculated as the product of the probabilities.

Independence test. In a two-way (r by c) contingency table, this tests if the row and column categories are independent.

Independent random variables. Random variables for which $P(X \in A, Y \in B) = P(X \in A)P(Y \in B)$ for any sets A and B in the range of X and Y, respectively. There are several equivalent descriptions of independent random variables.

Independent variable. The predictor or regressor variables in a regression model.

Inference. Conclusion from a statistical analysis. It usually refers to the conclusion from a hypothesis test or an interval estimate.

Indicator variable(s). Variables that are assigned numerical values to identify the levels of a qualitative or categorical response. For example, a response with two categorical levels (yes and no) could be represented with an indicator variable taking on the values 0 and 1.

Individuals control chart. A Shewhart control chart in which each plotted point is an individual measurement, rather than a summary statistic. *See* control chart, Shewhart control chart.

Influential observation. An observation in a regression analysis that has a large effect on estimated parameters in the model. Influence is measured by the change in parameters when the influential observation is included and excluded in the analysis.

Interaction. In factorial experiments, two factors are said to interact if the effect of one variable is different at different levels of the other variables. In general, when variables operate independently of each other, they do not exhibit interaction.

Intercept. The constant term in a regression model.

Interquartile range. The difference between the third and first quartiles if a sample of data. The interquartile range is less sensitive to extreme data values than the usual sample range.

Interval estimation. The estimation of a parameter by a range of values between lower and upper limits, in contrast to point estimation, where the parameter is estimated by a single numerical value. A confidence interval is a typical interval estimation procedure.

Intrinsically linear model. In regression analysis, a nonlinear function that can be expressed as a linear function after a suitable transformation is called intrinsically linear.

Jacobian. A matrix of partial derivatives that is used to determine the distribution of transformed random variables.

Joint probability density function. A function used to calculate probabilities for two or more continuous random variables.

Joint probability distribution. The probability distribution for two or more random variables in a random experiment. *See* Joint probability mass function and Joint probability density function.

Joint probability mass function. A function used to calculate probabilities for two or more discrete random variables.

Kruskal-Wallis test. A nonparametric test method used for the analysis of variance as an alternative to an F-test.

Kurtosis. A measure of the degree to which a unimodal distribution is peaked.

Lack of memory property. A property of a Poisson process. The probability of a count in an interval depends only on the length of the interval (and not on the starting point of the interval). A similar property holds for a series of Bernoulli trials. The probability of a success in a specified number of trials depends only on the number of trials (and not on the starting trial).

Least significance difference test (or Fisher's LSD test). An application of the t-test to compare pairs of means following rejection of the null hypothesis in an analysis of variance. The error rate is difficult to calculate exactly because the comparisons are not all independent.

Least squares (method of). A method of parameter estimation in which the parameters of a system are estimated by minimizing the sum of the squares of the differences between the observed values and the fitted or predicted values from the system.

Least squares estimator. Any estimator obtained by the method of least squares.

Level of significance. If Z is the test statistic for a hypothesis, and the distribution of Z when the hypothesis is true are known, then we can find the probabilities $P(Z \leq z_L)$ and $P(Z \geq z_U)$. Rejection of the hypothesis is usually expressed in terms of the observed value of Z falling outside the interval from z_L to z_U. The probabilities $P(Z \leq z_L)$ and $P(Z \geq z_U)$ are usually chosen to have small values, such as 0.01, 0.025, 0.05, or 0.10, and are called levels of significance. The actual levels chosen are somewhat arbitrary and are often expressed in percentages, such as a 5% level of significance.

Levels of a factor. The settings (or conditions) used for a factor in an experiment.

Likelihood function. Suppose that the random variables $X_1, X_2, \ldots, X_n$ have a joint distribution given by $f(x_1, x_2, \ldots, x_n; \theta_1, \theta_2, \ldots, \theta_p)$ where the θs are unknown parameters. This joint distribution, considered as a function of the θs for fixed x's, is called the likelihood function.

Likelihood principle. This principle states that the information about a model given by a set of data is completely contained in the likelihood.

Likelihood ratio. Let $x_1, x_2, \ldots, x_n$ be a random sample from the population $f(x; \theta)$. The likelihood function for this sample is $L = \prod_{i=1}^{n} f(x_i; \theta)$. We wish to test the hypothesis H_0: $\theta \in \omega$, where ω is a subset of the possible values Ω for θ. Let the maximum value of L with respect to θ over the entire set of values that the parameter can take on be denoted by $L(\hat{\Omega})$, and let the maximum value of L with θ restricted to the set of values given by ω be $L(\hat{\omega})$. The null hypothesis is tested by using the likelihood ratio $\lambda = L(\hat{\omega})/L(\hat{\Omega})$, or a simple function of it. Large values of the likelihood ratio are consistent with the null hypothesis.

Likelihood ratio test. A test of a null hypothesis versus an alternative hypothesis using a test statistic derived from a likelihood ratio.

Linear function of random variables. A random variable that is defined as a linear function of several random variables.

Linear model. A model in which the observations are expressed as a linear function of the unknown parameters. For example, $y = \beta_0 + \beta_1 x + \epsilon$ and $y = \beta_0 + \beta_1 x + \beta_2 x^2 + \epsilon$ are linear models.

Location parameter. A parameter that defines a central value in a sample or a probability distribution. The mean and the median are location parameters.

Logistic regression. A regression model that is used to model a categorical response. For a binary (0, 1) response the model assumes that the logarithm of the ratio of probabilities (for zero and one) is linearly related to the regressor variables.

Lognormal random variable. A continuous random variable with probability distribution equal to that of $\exp(W)$ for a normal random variable W.

Main effect. An estimate of the effect of a factor (or variable) that independently expresses the change in response due to a change in that factor, regardless of other factors that may be present in the system.

Marginal probability density function. The probability density function of a continuous random variable obtained from the joint probability distribution of two or more random variables.

Marginal probability distribution. The probability distribution of a random variable obtained from the joint probability distribution of two or more random variables.

Marginal probability mass function. The probability mass function of a discrete random variable obtained from the joint probability distribution of two or more random variables.

Maximum likelihood estimation. A method of parameter estimation that maximizes the likelihood function of a sample.

Mean. The mean usually refers either to the expected value of a random variable or to the arithmetic average of a set of data.

Mean square. In general, a mean square is determined by dividing a sum of squares by the number of degrees of freedom associated with the sum of squares.

Mean square(d) error. The expected squared deviation of an estimator from the true value of the parameter it estimates. The mean square error can be decomposed into the variance of the estimator plus the square of the bias; that is, $MSE(\hat{\Theta}) = E(\hat{\Theta} - \theta)^2 = V(\hat{\Theta}) + [E(\hat{\Theta}) - \theta]^2$.

Mechanistic model. A model developed from theoretical knowledge or experience in contrast to a model developed from data. *See* Empirical model.

Median. The median of a set of data is that value that divides the data into two equal halves. When the number of observations is even, say $2n$, it is customary to define the median as the average of the nth and $(n + 1)$st rank-ordered values. The median can also be defined for a random variable. For example, in the case of a continuous random variable X, the median M can be defined as $\int_{-\infty}^{M} f(x)\,dx = \int_{M}^{\infty} f(x)\,dx = 1/2$.

Method of steepest ascent. A technique that allows an experimenter to move efficiently towards a set of optimal operating conditions by following the gradient direction. The method of steepest ascent is usually employed in conjunction with fitting a first-order response surface and deciding that the current region of operation is inappropriate.

Mixed model. In an analysis of variance context, a mixed model contains both random and fixed factors.

Mode. The mode of a sample is that observed value that occurs most frequently. In a probability distribution $f(x)$ with continuous first derivative, the mode is a value of x for which $df(x)/dx = 0$ and $d^2f(x)/dx^2 < 0$. There may be more than one mode of either a sample or a distribution.

Moment (or population moment). The expected value of a function of a random variable such as $E(X - c)^r$ for constants c and r. When $c = 0$, it is said that the moment is about the origin. *See* Moment generating function.

Moment estimator. A method of estimating parameters by equating sample moments to population moments. Since the population moments will be functions of the unknown parameters, this results in equations that may be solved for estimates of the parameters.

Moment generating function. A function that is used to determine properties (such as moments) of the probability distribution of a random variable. It is the expected value of $\exp(tX)$. See generating function and moment.

Moving range. The absolute value of the difference between successive observations in time-ordered data. Used to estimate chance variation in an individuals control chart.

Multicollinearity. A condition occurring in multiple regression where some of the predictor or regressor variables are nearly linearly dependent. This condition can lead to instability in the estimates of the regression model parameters.

Multinomial distribution. The joint probability distribution of the random variables that count the number of results in each of k classes in a random experiment with a series of independent trials with constant probability of each class on each trial. It generalizes a binomial distribution.

Multiplication rule. For probability, A formula used to determine the probability of the intersection of two (or more) events. For counting techniques, a formula used to determine the numbers of ways to complete an operation from the number of ways to complete successive steps.

Mutually exclusive events. A collection of events whose intersections are empty.

Natural tolerance limits. A set of symmetric limits that are three times the process standard deviation from the process mean.

Negative binomial random variable. A discrete random variable that is the number of trials until a specified number of successes occur in Bernoulli trials.

Nonlinear regression model. A regression model that is nonlinear in the parameters. It is sometimes applied to regression models that are nonlinear in the regressors or predictors, but this is an incorrect usage.

Nonparametric statistical method(s). *See* Distribution free method(s).

Normal approximation. A method to approximate probabilities for binomial and Poisson random variables.

Normal equations. The set of simultaneous linear equations arrived at in parameter estimation using the method of least squares.

Normal probability plot. A specially constructed plot for a variable x (usually on the abscissa) in which y (usually on the ordinate) is scaled so that the graph of the normal cumulative distribution is a straight line.

Normal random variable. A continuous random variable that is the most important one in statistics because it results from the central limit theorem. *See* Central limit theorem.

***NP* chart.** An attribute control chart that plots the total of defective units in a subgroup. Similar to a fraction-defective chart or P chart.

Nuisance factor. A factor that probably influences the response variable, but which is of no interest in the current study. When the levels of the nuisance factor can be controlled, blocking is the design technique that is customarily used to remove its effect.

Null distribution. In a hypothesis test, the distribution of the test statistic when the null hypothesized is assumed to be true.

Null hypothesis. This term generally relates to a particular hypothesis that is under test, as distinct from the alternative hypothesis (which defines other conditions that are feasible but not being tested). The null hypothesis determines the probability of type I error for the test procedure.

Observational study. A system is observed and data might be collected, but changes are not made to the system. *See* Experiment.

Odds ratio. The odds equals the ratio of two probabilities. In logistic regression, the logarithm of the odds is modeled as a linear function of the regressors. Given values for the regressors at a point, the odds can be calculated. The odds ratio is the odds at one point divided by the odds at another.

One-way model. In an analysis of variance context, this involves a single variable or factor with a different levels.

Operating characteristic curves (OC curves). A plot of the probability of type II error versus some measure of the extent to which the null hypothesis is false. Typically, one OC curve is used to represent each sample size of interest.

Optimization experiment. A experiment conducted to improve (or optimize) a system or process. It is assumed that the important factors are known.

Orthogonal. There are several related meanings, including the mathematical sense of perpendicular, two variables being said to be orthogonal if they are statistically independent, or in experimental design where a design is orthogonal if it admits statistically independent estimates of effects.

Orthogonal design. *See* Orthogonal.

Outcome. An element of a sample space.

Outlier(s). One or more observations in a sample that are so far from the main body of data that they give rise to the question that they may be from another population.

Overcontrol. Unnecessary adjustments made to processes that increase the deviations from target.

Overfitting. Adding more parameters to a model than is necessary.

***P* chart.** An attribute control chart that plots the proportion of defective units in a subgroup. Also called a fraction-defective control chart. Similar to an NP chart.

Parameter estimation. The process of estimating the parameters of a population or probability distribution. Parameter estimation, along with hypothesis testing, is one of the two major techniques of statistical inference.

Parameter. An unknown quantity that may vary over a set of values. Parameters occur in probability distributions and in statistical models, such as regression models.

Pareto chart. A bar chart used to rank the causes of a problem.

***PCR*.** A process capability ratio with numerator equal to the difference between the product specification limits and denominator equal to six times the process standard deviation. Said to measure the potential capability of the process because the process mean is not considered. See process capability, process capability ratio, process capability study and PCR_k. Sometimes denoted as C_p in other references.

PCR_k. A process capability ratio with numerator equal to the difference between the product target and the nearest specification limit and denominator equal to three times the process standard deviation. Said to measure the actual capability of the process because the process mean is considered. See process capability, process capability ratio, process capability study, and PCR. Sometimes denoted as C_{pk} in other references.

Percentage point. A particular value of a random variable determined from a probability (expressed as a percentage). For example, the upper 5 percentage point of the standard normal random variable is $z_{0.05} = 1.645$.

Percentile. The set of values that divide the sample into 100 equal parts.

Permutation. An ordered sequence of the elements in a set used to determine the number of outcomes in events and sample spaces.

Point estimator. *See* Estimator.

Poisson process. A random experiment with events that occur in an interval and satisfy the following assumptions. The interval can be partitioned into subintervals such that the probability of more than one event in a subinterval is zero, the probability of an event in a subinterval is proportional to the length of the subinterval, and the event in each subinterval is independent of other subintervals.

Poisson random variable. A discrete random variable that is the number of events that occur in a Poisson process.

Pooled t-test. A hypothesis to compare the means of two populations with the variances assumed to be equal.

Pooling. When several sets of data can be thought of as having been generated from the same model, it is possible to combine them, usually for purposes of estimating one or more parameters. Combining the samples for this purpose is usually called *pooling* and it is commonly used to estimate a variance.

Population standard deviation. *See* standard deviation.

Population variance. *See* variance.

Population. Any finite or infinite collection of individual units or objects.

Posterior distribution. The probability distribution for a parameter in a Bayesian analysis calculated from the prior distribution and the conditional distribution of the data given the parameter.

Power. The power of a statistical test is the probability that the test rejects the null hypothesis when the null hypothesis is indeed false. Thus the power is equal to one minus the probability of type II error.

Prediction. The process of determining the value of one or more statistical quantities at some future point in time. In a regression model, predicting the response y for some specified set of regressors or predictor variables also leads to a predicted value, although there may be no temporal element to the problem.

Prediction interval. The interval between a set of upper and lower limits associated with a predicted value designed to show on a probability basis the range of error associated with the prediction.

Predictor variable(s). The independent or regressor variables in a regression model.

PRESS statistic. In regression analysis, the predicted residual sum of squares. Delete each point and estimate the parameters of the model from the data that remains. Estimate the deleted point from this model. Restore the point and then delete the next point. Each point is estimated once and the sum of squares of these errors is calculated.

Prior distribution. The initial probability distribution assumed for a parameter in a Bayesian analysis.

Probability. A numerical measure between 0 and 1 assigned to events in a sample space. Higher numbers indicate the event is more likely to occur. See axioms of probability.

Probability density function. A function used to calculate probabilities and to specify the probability distribution of a continuous random variable.

Probability distribution. For a sample space, a description of the set of possible outcomes along with a method to determine probabilities. For a random variable, a probability distribution is a description of the range along with a method to determine probabilities.

Probability mass function. A function that provides probabilities for the values in the range of a discrete random variable.

Probability plot. A scatter plot used to judge if data can reasonably be assumed to follow a particular probability distribution. A normal probability plot is often used to evaluate the normality assumption of data or residuals.

Process capability ratio. A ratio that relates the width of the product specification limits to measures of process performance. Used to quantify the capability of the process to produce product within specifications. See process capability, process capability study, PCR and PCR_k.

Process capability study. A study that collects data to estimate process capability. See process capability, process capability ratio, PCR and PCR_k.

Process capability. The capability of a process to produce product within specification limits. See process capability ratio, process capability study, PCR, and PCR_k.

P-Value. The exact significance level of a statistical test; that is, the probability of obtaining a value of the test statistic that is at least as extreme as that observed when the null hypothesis is true.

Qualitative (data). Data derived from nonnumeric attributes, such as sex, ethnic origin or nationality, or other classification variable.

Quality control. Systems and procedures used by an organization to assure that the outputs from processes satisfy customers.

Quantiles. The set of $n - 1$ values of a variable that partition it into a number n of equal proportions. For example, $n - 1 = 3$ values partition data into four quantiles with the central value usually called the median and the lower and upper values usually called the lower and upper quartiles, respectively.

Quantitative (data). Data in the form of numerical measurements or counts.

Quartile(s). The three values of a variable that partition it into four equal parts. The central value is usually called the median and the lower and upper values are usually called the lower and upper quartiles, respectively. *Also see* Quantiles.

R^2. A quantity used in regression models to measure the proportion of total variability in the response accounted for by the model. Computationally, $R^2 = SS_{\text{Regression}}/SS_{\text{Total}}$, and large values of R^2 (near unity) are considered good. However, it is possible to have large values of R^2 and find that the model is unsatisfactory. R^2 is also called the coefficient of determination (or the coefficient of multiple determination in multiple regression).

R chart. A control chart that plots the range of the measurements in a subgroup that is used to monitor the variance of the process.

Random. Nondeterministic, occurring purely by chance, or independent of the occurrence of other events.

Random effects model. In an analysis of variance context, this refers to a model that involves only random factors.

Random error. An error (usually a term in a statistical model) that behaves as if it were drawn at random from a particular probability distribution.

Random experiment. An experiment that can result in different outcomes, even though it is repeated in the same manner each time.

Random factor. In analysis of variance, a factor whose levels are chosen at random from some population of factor levels.

Random order. A sequence or order for a set of objects that is carried out in such a way that every possible ordering is equally likely. In experimental design the runs of the experiment are typically arranged and carried out in random order.

Random sample. A sample is said to be random if it is selected in such a way so that every possible sample has the same probability of being selected.

Random variable. A function that assigns a real number to each outcome in the sample space of a random experiment.

Randomization. Randomly assign treatments to experimental units or conditions in an experiment. This is done to reduce the opportunity for a treatment to be favored or disfavored (biased) by test conditions.

Randomized complete block design. A type of experimental design in which treatment (or factor levels) are assigned to blocks in a random manner.

Range. The largest minus the smallest of a set of data values. The range is a simple measure of variability and is widely used in quality control.

Range (control) chart. A control chart used to monitor the variability (dispersion) in a process. *See* Control chart.

Rank. In the context of data, the rank of a single observation is its ordinal number when all data values are ordered according to some criterion, such as their magnitude.

Rational subgroup. A sample of data selected in a manner to include chance sources of variation and to exclude assignable sources of variation, to the extent possible.

Reference distribution. The distribution of a test statistic when the null hypothesis is true. Sometimes a reference distribution is called the null distribution of the test statistic.

Reference value. A parameter set in a Tabular CUSUM algorithm that is determined from the magnitude of the process shift that should be detected.

Regression. The statistical methods used to investigate the relationship between a dependent or response variable y and one or more independent variables x. The independent variables are usually called regressor variables or predictor variables.

Regression coefficient(s). The parameter(s) in a regression model.

Regression diagnostics. Techniques for examining a fitted regression model to investigate the adequacy of the fit and to determine if any of the underlying assumptions have been violated.

Regression line (or curve). A graphical display of a regression model, usually with the response y on the ordinate and the regressor x on the abscissa.

Regression sum of squares. The portion of the total sum of squares attributable to the model that has been fit to the data.

Regressor variable. The independent or predictor variable in a regression model.

Rejection region. In hypothesis testing, this is the region in the sample space of the test statistic that leads to rejection of the null hypothesis when the test statistic falls in this region.

Relative frequency. The relative frequency of an event is the proportion of times the event occurred in a series of trial of a random experiment.

Reliability. The probability that a specified mission will be completed. It usually refers to the probability that a lifetime of a continuous random variable exceeds a specified time limit.

Replicates. One of the independent repetitions of one or more treatment combinations in an experiment.

Replication. The independent execution of an experiment more than once.

Reproductive property of the normal distribution. A linear combination of independent, normal random variables is a normal random variable.

Residual. Generally this is the difference between the observed and the predicted value of some variable. For example, in regression a residual is the difference between the observed value of the response and the corresponding predicted value obtained from the regression model.

Residual analysis (and plots). Any technique that uses the residuals, usually to investigate the adequacy of the model that was used to generate the residuals.

Residual sum of squares. *See* Error sum of squares.

Resolution. A measure of severity of aliasing in a fractional factorial design. We commonly consider resolution III, IV, and V designs.

Response (variable). The dependent variable in a regression model or the observed output variable in a designed experiment.

Response surface. When a response y depends on a function of k quantitative variables $x_1, x_2, \ldots, x_k$, the values of the response may be viewed as a surface in $k + 1$ dimensions. This surface is called a response surface. Response surface methodology is a subset of experimental design concerned with approximating this surface with a model and using the resulting model to optimize the system or process.

Response surface designs. Experimental designs that have been developed to work well in fitting response surfaces. These are usually designs for fitting a first- or second-order model. The central composite design is a widely used second-order response surface design.

Ridge regression. A method for fitting a regression model that is intended to overcome the problems associated with using standard (or ordinary) least squares when there is a problem with multicollinearity in the data.

Rotatable design. In a rotatable design, the variance of the predicted response is the same at all points that are the same distance from the center of the design.

Run rules. A set of rules applied to the points plotted on a Shewhart control chart that are used to make the chart more sensitized to assignable causes. *See* control chart, Shewhart control chart.

Runs test. A nonparametric test to compare two distributions, or check for independence of measurements.

***S* chart.** A control chart that plots the standard deviation of the measurements in a subgroup that is used to monitor the variance of the process.

Sample. Any subset of the elements of a population.

Sample mean. The arithmetic average or mean of the observations in a sample. If the observations are $x_1, x_2, \ldots, x_n$ then the sample mean is $(1/n)\sum_{i=1}^{n} x_i$. The sample mean is usually denoted by $\overline{x}$.

Sample moment. The quantity $(1/n)\sum_{i=1}^{n} x_i^k$ is called the kth sample moment.

Sample range. *See* range.

Sample size. The number of observations in a sample.

Sample space. The set of all possible outcomes of a random experiment.

Sample standard deviation. The positive square root of the sample variance. The sample standard deviation is the most widely used measure of variability of sample data.

Sample variance. A measure of variability of sample data, defined as $s^2 = [1/(n-1)]\sum_{i=1}^{n}(x_i - \overline{x})^2$, where $\overline{x}$ is the sample mean.

Sampling distribution. The probability distribution of a statistic. For example, the sampling distribution of the sample mean $\overline{X}$ is the normal distribution.

Scatter diagram. A diagram displaying observations on two variables, x and y. Each observation is represented by a point showing its x-y coordinates. The scatter diagram can be very effective in revealing the joint variability of x and y or the nature of the relationship between them.

Screening experiment. An experiment designed and conducted for the purpose of screening out or isolating a promising set of factors for future experimentation. Many screening experiments are fractional factorials, such as two-level fractional factorial designs.

Second-order model. A model that contains second-order terms. For example, the second-order response surface model in two variables is $y = \beta_0 + \beta_1 x_1 + \beta_2 x_2 + \beta_{12} x_1 x_2 + \beta_{11} x_1^2 + \beta_{22} x_2^2 + \epsilon$. The second order terms in this model are $\beta_{12} x_1 x_2$, $\beta_{11} x_1^2$, and $\beta_{22} x_2^2$.

Shewhart control chart. A specific type of control chart developed by Walter A. Shewhart. Typically, each plotted point is a summary statistic calculated from the data in a rational subgroup. *See* control chart.

Sign test. A statistical test based on the signs of certain functions of the observations and not their magnitudes.

Signed-rank test. A statistical test based on the differences within a set of paired observations. Each difference has a sign and a rank, and the test uses the sum of the differences with regard to sign.

Significance. In hypothesis testing, an effect is said to be significant if the value of the test statistic lies in the critical region.

Significance level. *See* Level of significance.

Six-sigma process. Originally used to describe a process with the mean at least six standard deviations from the nearest specification limits. It has now been used to describe any process with a defect rate of 3.4 parts per million.

Skewness. A term for asymmetry usually employed with respect to a histogram of data or a probability distribution.

Specification limits. Numbers that define the region of measurement for acceptable product. Usually there is an upper and lower limit, but one-sided limits can also be used.

Standard deviation. The positive square root of the variance. The standard deviation is the most widely used measure of variability.

Standard error. The standard deviation of the estimator of a parameter. The standard error is also the standard deviation of the sampling distribution of the estimator of a parameter.

Standard normal random variable. A normal random variable with mean zero and variance one that has its cumulative distribution function tabulated in Appendix Table II.

Standardize. The transformation of a normal random variable that subtracts its mean and divides by its standard deviation to generate a standard normal random variable.

Standardized residual. In regression, the standardized residual is computed by dividing the ordinary residual by the square root of the residual mean square. This produces scaled residuals that have, approximately, a unit variance.

Statistic. A summary value calculated from a sample of observations. Usually, a statistic is an estimator of some population parameter.

Statistical inference. *See* Inference.

Statistical Process Control (SPC). A set of problem-solving tools based on data that are used to improve a process.

Statistical quality control. Statistical and engineering methods used to measure, monitor, control, and improve quality.

Statistics. The science of collecting, analyzing, interpreting, and drawing conclusions from data.

Steepest ascent (or descent). A strategy for a series of tests to optimize a response used along with response surface models.

Stem and leaf diagram. A method of displaying data in which the stem corresponds to a range of data values and the leaf represents the next digit. It is an alternative to the histogram but displays the individual observations rather than sorting them into bins.

Stepwise regression. A method of selecting variables for inclusion in a regression model. It operates by introducing the candidate variables one at a time (as in forward selection) and then attempting to remove variables following each forward step.

Studentized range. The range of a sample divided by the sample standard deviation.

Studentized residual. In regression, the studentized residual is calculated by dividing the ordinary residual by its exact standard deviation, producing a set of scaled residuals that have, exactly, unit standard deviation.

Sufficient statistic. An estimator is said to be a sufficient statistic for an unknown parameter if the distribution of the sample given the statistic does not depend on the unknown parameter. This means that the distribution of the estimator contains all of the useful information about the unknown parameter.

Tabular CUSUM. A numerical algorithm used to detect assignable causes on a cumulative sum control chart. *See V* mask.

Tampering. Another name for overcontrol.

t-distribution. The distribution of the random variable defined as the ratio of two independent random variables. The numerator is a standard normal random variable and the denominator is the square root of a chi-square random variable divided by its number of degrees of freedom.

Test statistic. A function of a sample of observations that provides the basis for testing a statistical hypothesis.

Time series. A set of ordered observations taken at points in time.

Tolerance interval. An interval that contains a specified proportion of a population with a stated level of confidence.

Tolerance limits. A set of limits between which some stated proportion of the values of a population must fall with specified level of confidence.

Total probability rule. Given a collection of mutually exclusive events whose union is the sample space, the probability of an event can be written as the sum of the probabilities of the intersections of the event with the members of this collection.

Treatment. In experimental design, a treatment is a specific level of a factor of interest. Thus if the factor is temperature, the treatments are the specific temperature levels used in the experiment.

Treatment effect. The mean change to the response due to the presence of the treatment.

Treatment sum of squares. In analysis of variance, this is the sum of squares that accounts for the variability in the response variable due to the different treatments that have been applied.

t-test. Any test of significance based on the t distribution. The most common t-tests are (1) testing hypotheses about the mean of a normal distribution with unknown variance, (2) testing hypotheses about the means of two normal distributions and (3) testing hypotheses about individual regression coefficients.

Two-level factorial experiment. A full or fractional factorial experiment with all factors tested at only two levels (settings) each. *See* 2^k factorial experiment.

Type I error. In hypothesis testing, an error incurred by rejecting a null hypothesis when it is actually true (also called an α-error).

Type II error. In hypothesis testing, an error incurred by failing to reject a null hypothesis when it is actually false (also called a β-error).

U chart. An attribute control chart that plots the average number of defects per unit in a subgroup. Also called a defects-per-unit control chart. Similar to a C chart.

Unbiased estimator. An estimator that has its expected value equal to the parameter that is being estimated is said to be unbiased.

Uniform random variable. Refers to either a discrete or continuous uniform random variable.

Uniqueness property of moment generating function. Refers to the fact that random variables with the same moment generating function have the same distribution.

Universe. Another name for *population.*

V mask. A geometrical figure used to detect assignable causes on a cumulative sum control chart. With appropriate values for parameters, identical conclusions can be made from a V mask and a tabular CUSUM.

Variable selection. The problem of selecting a subset of variables for a model from a candidate list that contains all or most of the useful information about the response in the data.

Variables control chart. Any control chart for a continuous random variable. See attributes control charts.

Variance. A measure of variability defined as the expected value of the square of the random variable around its mean.

Variance component. In analysis of variance models involving random effects, one of the objectives is to determine how much variability can be associated with each of the potential sources of variability defined by the experimenters. It is customary to define a variance associated with each of these sources. These variances in some sense sum to the total variance of the response, and are usually called variance components.

Variance inflation factors. Quantities used in multiple regression to assess the extent of multicollinearity (or near linear dependence) in the regressors. The variance inflation factor for the ith regressor VIF_i can be defined as $VIF_i = [1/(1 - R_i^2)]$, where R_i^2 is the coefficient of determination obtained when x_i is regressed on the other regressor variables. Thus when x_i is nearly linearly dependent on a subset of the other regressors R_i^2 will be close to unity and the value of the corresponding variance inflation factor will be large. Values of the variance inflation factors that exceed 10 are usually taken as a signal that multicollinearity is present.

Warning limits. Horizontal lines added to a control chart (in addition to the control limits) that are used to make the chart more sensitive to assignable causes.

Weibull random variable. A continuous random variable that is often used to model the time until failure of a physical system. The parameters of the distribution are flexible enough that the probability density function can assume many different shapes.

Western Electric rules. A specific set of run rules that were developed at Western Electric Corporation. See run rules.

Wilcoxon rank-sum test. A nonparametric test for the equality of means in two populations. This is sometimes called the Mann-Whitney test.

Wilcoxon signed-rank test. A distribution-free test of the equality of the location parameters of two otherwise identical distributions. It is an alternative to the two-sample t-test for nonnormal populations.

With replacement. A method to select samples in which items are replaced between successive selections.

Without replacement. A method to select samples in which items are *not* replaced between successive selections.

$\overline{X}$ chart. A control chart that plots the average of the measurements in a subgroup that is used to monitor the process mean.

Index

Applications in Examples and Exercises continued

MEDICAL

PHYSICS

SEMICONDUCTOR MANUFACTURING

SPORTS

TEXTILE MANUFACTURING

INDUSTRIAL ENGINEERING AND ECONOMICS

DEFECTIVES, FLAWS, ERRORS

LIFE AND FAILURE TIMES

LENGTH OF

THICKNESS OF

WEIGHT OF

Summary of One-Sample Hypothesis-Testing Procedures

Case	Null Hypothesis	Test Statistic	Alternative Hypothesis	Criteria for Rejection	OC Curve Parameter	OC Curve Appendix Chart VII
1.	H_0: $\mu = \mu_0$ σ^2 known	$z_0 = \dfrac{\bar{x} - \mu_0}{\sigma/\sqrt{n}}$	H_1: $\mu \neq \mu_0$ H_1: $\mu > \mu_0$ H_1: $\mu < \mu_0$	$\|z_0\| > z_{\alpha/2}$ $z_0 > z_\alpha$ $z_0 < -z_\alpha$	$d = \|\mu - \mu_0\|/\sigma$ $d = (\mu - \mu_0)/\sigma$ $d = (\mu_0 - \mu)/\sigma$	a, b c, d c, d
2.	H_0: $\mu = \mu_0$ σ^2 unknown	$t_0 = \dfrac{\bar{x} - \mu_0}{s/\sqrt{n}}$	H_1: $\mu \neq \mu_0$ H_1: $\mu > \mu_0$ H_1: $\mu < \mu_0$	$\|t_0\| > t_{\alpha/2,n-1}$ $t_0 > t_{\alpha,n-1}$ $t_0 < -t_{\alpha,n-1}$	$d = \|\mu - \mu_0\|/\sigma$ $d = (\mu - \mu_0)/\sigma$ $d = (\mu_0 - \mu)/\sigma$	e, f g, h g, h
3.	H_0: $\sigma^2 = \sigma_0^2$	$x_0^2 = \dfrac{(n-1)s^2}{\sigma_0^2}$	H_1: $\sigma^2 \neq \sigma_0^2$ H_1: $\sigma^2 > \sigma_0^2$ H_1: $\sigma^2 < \sigma_0^2$	$\chi_0^2 > \chi^2_{\alpha/2,n-1}$ or $\chi_0^2 < \chi^2_{1-\alpha/2,n-1}$ $\chi_0^2 > \chi^2_{\alpha,n-1}$ $\chi_0^2 < \chi^2_{1-\alpha,n-1}$	$\lambda = \sigma/\sigma_0$ $\lambda = \sigma/\sigma_0$ $\lambda = \sigma/\sigma_0$	i, j k, l m, n
4.	H_0: $p = p_0$	$z_0 = \dfrac{x - np_0}{\sqrt{np_0(1-p_0)}}$	H_1: $p \neq p_0$ H_1: $p > p_0$ H_1: $p < p_0$	$\|z_0\| > z_{\alpha/2}$ $z_0 > z_\alpha$ $z_0 < -z_\alpha$	— — —	— — —

Summary of One-Sample Confidence Interval Procedures

Case	Problem Type	Point Estimate	Two-sided $100(1-\alpha)$ Percent Confidence Interval
1.	Mean μ, variance σ^2 known	$\bar{x}$	$\bar{x} - z_{\alpha/2}\sigma/\sqrt{n} \leq \mu \leq \bar{x} + z_{\alpha/2}\sigma/\sqrt{n}$
2.	Mean μ of a normal distribution, variance σ^2 unknown	$\bar{x}$	$\bar{x} - t_{\alpha/2,n-1}s/\sqrt{n} \leq \mu \leq \bar{x} + t_{\alpha/2,n-1}s/\sqrt{n}$
3.	Variance σ^2 of a normal distribution	s^2	$\dfrac{(n-1)s^2}{\chi^2_{\alpha/2,n-1}} \leq \sigma^2 \leq \dfrac{(n-1)s^2}{\chi^2_{1-\alpha/2,n-1}}$
4.	Proportion or parameter of a binomial distribution p	$\hat{p}$	$\hat{p} - z_{\alpha/2}\sqrt{\dfrac{\hat{p}(1-\hat{p})}{n}} \leq p \leq \hat{p} + z_{\alpha/2}\sqrt{\dfrac{\hat{p}(1-\hat{p})}{n}}$

Summary of Two-Sample Hypothesis-Testing Procedures

Case	Null Hypothesis	Test Statistic	Alternative Hypothesis	Criteria for Rejection	OC Curve Parameter	OC Curve Appendix Chart VII
1.	$H_0: \mu_1 - \mu_2 = \Delta_0$ σ_1^2 and σ_2^2 known	$z_0 = \dfrac{\bar{x}_1 - \bar{x}_2 - \Delta_0}{\sqrt{\dfrac{\sigma_1^2}{n_1} + \dfrac{\sigma_2^2}{n_2}}}$	$H_1: \mu_1 - \mu_2 \neq \Delta_0$ $H_1: \mu_1 - \mu_2 > \Delta_0$ $H_1: \mu_1 - \mu_2 < \Delta_0$	$\|z_0\| > z_{\alpha/2}$ $z_0 > z_\alpha$ $z_0 < -z_\alpha$	$d = \dfrac{\|\mu_1 - \mu_2 - \Delta_0\|}{\sqrt{\sigma_1^2 + \sigma_2^2}}$ $d = \dfrac{\mu_1 - \mu_2 - \Delta_0}{\sqrt{\sigma_1^2 + \sigma_2^2}}$ $d = \dfrac{\mu_2 - \mu_1 - \Delta_0}{\sqrt{\sigma_1^2 + \sigma_2^2}}$	a, b c, d c, d
2.	$H_0: \mu_1 - \mu_2 = \Delta_0$ $\sigma_1^2 = \sigma_2^2$ unknown	$t_0 = \dfrac{\bar{x}_1 - \bar{x}_2 - \Delta_0}{s_p\sqrt{\dfrac{1}{n_1} - \dfrac{1}{n_2}}}$	$H_1: \mu_1 - \mu_2 \neq \Delta_0$ $H_1: \mu_1 - \mu_2 > \Delta_0$ $H_1: \mu_1 - \mu_2 < \Delta_0$	$\|t_0\| > t_{\alpha/2, n_1+n_2-2}$ $t_0 > t_{\alpha, n_1+n_2-2}$ $t_0 < -t_{\alpha, n_1+n_2-2}$	$d = \|\Delta - \Delta_0\|/2\sigma$ $d = (\Delta - \Delta_0)/2\sigma$ $d = (\Delta_0 - \Delta)/2\sigma$ where $\Delta = \mu_1 - \mu_2$	e, f g, h g, h
3.	$H_0: \mu_1 - \mu_2 = \Delta_0$ $\sigma_1^2 \neq \sigma_2^2$ unknown	$t_0 = \dfrac{\bar{x}_1 - \bar{x}_2 - \Delta_0}{\sqrt{\dfrac{s_1^2}{n_1} + \dfrac{s_2^2}{n_2}}}$ $v = \dfrac{\left(\dfrac{s_1^2}{n_1} + \dfrac{s_2^2}{n_2}\right)^2}{\dfrac{(s_1^2/n_1)^2}{n_1 - 1} + \dfrac{(s_2^2/n_2)^2}{n_2 - 1}}$	$H_1: \mu_1 - \mu_2 \neq \Delta_0$ $H_1: \mu_1 - \mu_2 > \Delta_0$ $H_1: \mu_1 - \mu_2 < \Delta_0$	$\|t_0\| > t_{\alpha/2, v}$ $t_0 > t_{\alpha, v}$ $t_0 < -t_{\alpha, v}$	— — —	— — —
4.	Paired data $H_0: \mu_D = 0$	$t_0 = \dfrac{\bar{d}}{s_d/\sqrt{n}}$	$H_1: \mu_d \neq 0$ $H_1: \mu_d > 0$ $H_1: \mu_d < 0$	$\|t_0\| > t_{\alpha/2, n-1}$ $t_0 > t_{\alpha, n-1}$ $t_0 < -t_{\alpha, n-1}$	— — —	— — —
5.	$H_0: \sigma_1^2 = \sigma_2^2$	$f_0 = s_1^2/s_2^2$	$H_1: \sigma_1^2 \neq \sigma_2^2$ $H_1: \sigma_1^2 > \sigma_2^2$	$f_0 > f_{\alpha/2, n_1-1, n_2-1}$ or $f_0 < f_{1-\alpha/2, n_1-1, n_2-1}$ $f_0 > f_{\alpha, n_1-1, n_2-1}$	$\lambda = \sigma_1/\sigma_2$ $\lambda = \sigma_1/\sigma_2$	o, p q, r
6.	$H_0: p_1 = p_2$	$z_0 = \dfrac{\hat{p}_1 - \hat{p}_2}{\sqrt{\hat{p}(1 - \hat{p})\left[\dfrac{1}{n_1} + \dfrac{1}{n_2}\right]}}$	$H_1: p_1 \neq p_2$ $H_1: p_1 > p_2$ $H_1: p_1 < p_2$	$\|z_0\| > z_{\alpha/2}$ $z_0 > z_\alpha$ $z_0 < -z_\alpha$	— — —	— — —

Summary of Two-Sample Confidence Interval Procedures

Case	Problem Type	Point Estimate	Two-Sided 100(1 − α) Percent Confidence Interval
1.	Difference in two means μ_1 and μ_2, variances σ_1^2 and σ_2^2 known	$\bar{x}_1 - \bar{x}_2$	$\bar{x}_1 - \bar{x}_2 - z_{\alpha/2}\sqrt{\frac{\sigma_1^2}{n_1} + \frac{\sigma_2^2}{n_2}} \le \mu_1 - \mu_2 \le \bar{x}_1 - \bar{x}_2 + z_{\alpha/2}\sqrt{\frac{\sigma_1^2}{n_1} + \frac{\sigma_2^2}{n_2}}$
2.	Difference in means of two normal distributions $\mu_1 - \mu_2$, variances $\sigma_1^2 = \sigma_2^2$ and unknown	$\bar{x}_1 - \bar{x}_2$	$\bar{x}_1 - \bar{x}_2 - t_{\alpha/2,n_1+n_2-2}s_p\sqrt{\frac{1}{n_1} + \frac{1}{n_2}} \le \mu_1 - \mu_2 \le \bar{x}_1 - \bar{x}_2 + t_{\alpha/2,n_1+n_2-2}s_p\sqrt{\frac{1}{n_1} + \frac{1}{n_2}}$ where $s_p = \sqrt{\frac{(n_1 - 1)s_1^2 + (n_2 - 1)s_2^2}{n_1 + n_2 - 2}}$
3.	Difference in means of two normal distributions $\mu_1 - \mu_2$, variances $\sigma_1^2 \neq \sigma_2^2$ and unknown	$\bar{x}_1 - \bar{x}_2$	$\bar{x}_1 - \bar{x}_2 - t_{\alpha/2,v}\sqrt{\frac{s_1^2}{n_1} + \frac{s_2^2}{n_2}} \le \mu_1 - \mu_2 \le \bar{x}_1 - \bar{x}_2 + t_{\alpha/2,v}\sqrt{\frac{s_1^2}{n_1} + \frac{s_2^2}{n_2}}$ where $v = \frac{(s_1^2/n_1 + s_2^2/n_2)^2}{\frac{(s_1^2/n_1)^2}{n_1-1} + \frac{(s_2^2/n_2)^2}{n_2-1}}$
4.	Difference in means of two normal distributions for paired samples $\mu_0 = \mu_1 - \mu_2$	$\bar{d}$	$\bar{d} - t_{\alpha/2,n-1}s_d/\sqrt{n} \le \mu_D \le \bar{d} + t_{\alpha/2,n-1}s_d/\sqrt{n}$
5.	Ratio of the variances σ_1^2/σ_2^2 of two normal distributions	$\frac{s_1^2}{s_2^2}$	$\frac{s_1^2}{s_2^2} f_{1-\alpha/2,n_2-1,n_1-1} \le \frac{\sigma_1^2}{\sigma_2^2} \le \frac{s_1^2}{s_2^2} f_{\alpha/2,n_2-1,n_1-1}$ where $f_{1-\alpha/2,n_2-1,n_1-1} = \frac{1}{f_{\alpha/2,n_1-1,n_2-1}}$
6.	Difference in two proportions of two binominal parameters $p_1 - p_2$	$\hat{p}_1 - \hat{p}_2$	$\hat{p}_1 - \hat{p}_2 - z_{\alpha/2}\sqrt{\frac{\hat{p}_1(1 - \hat{p}_1)}{n_1} + \frac{\hat{p}_2(1 - \hat{p}_2)}{n_2}} \le p_1 - p_2 \le \hat{p}_1 - \hat{p}_2 + z_{\alpha/2}\sqrt{\frac{\hat{p}_1(1 - \hat{p}_1)}{n_1} + \frac{\hat{p}_2(1 - \hat{p}_2)}{n_2}}$